ENCYCLOPEDIA OF
science
technology
AND ethics

EDITORS AND CONSULTANTS

ENCYCLOPEDIA OF
science
technology
AND ethics

EDITED BY
CARL MITCHAM

volume

4

s–z
appendices
index

MACMILLAN REFERENCE USA
An imprint of Thomson Gale, a part of The Thomson Corporation

THOMSON
GALE

Detroit • New York • San Francisco • San Diego • New Haven, Conn. • Waterville, Maine • London • Munich

Encyclopedia of Science, Technology, and Ethics

Carl Mitcham, Editor in Chief

LIBRARY OF CONGRESS CATALOGING-IN-PUBLICATION DATA

Encyclopedia of science, technology, and ethics / edited by Carl Mitcham.
    p. cm.
    Includes bibliographical references and index.
    ISBN 0-02-865831-0 (set, hardcover : alk. paper)—ISBN 0-02-865832-9 (v. 1) —
    ISBN 0-02-865833-7 (v. 2)—ISBN 0-02-865834-5 (v. 3)—ISBN 0-02-865901-5 (v. 4)
    1. Science—Moral and ethical aspects—Encyclopedias.
    2. Technology—Moral and ethical aspects–Encyclopedias.
    I. Mitcham, Carl. Q175.35.E53 2005
503—dc22   005006968

This title is also available as an e-book.
ISBN 0-02-865991-0
Contact your Thomson Gale representative for ordering information.

Printed in the United States of America
10 9 8 7 6 5 4 3 2 1

S

SAFETY ENGINEERING

•••

Historical Emergence
Practices

HISTORICAL EMERGENCE

The protection of people from harm increasingly has been a focus of many fields of engineering since the nineteenth century. At the dawn of the Industrial Revolution (c. 1750–1850) engineers, as the term is used today, devoted their efforts almost entirely to making devices that functioned reliably and profitably, but with little attention to safety. One notable exception is James Watt (1736–1819), the so-called inventor of the steam engine. Despite introducing numerous improvements on the Newcomen steam engine, Watt intentionally resisted building a high-pressure engine because of the dangers it posed to those working with it. In fact, when Richard Trevithick (1771–1833) began experiments with the high-pressure steam engine, which increased both efficiency and power, Watt (and his partner Matthew Boulton) petitioned Parliament to pass an act outlawing the use of such engines as a public danger.

The second generation masters of steam power for railroads and steam boats thus brought with them boiler explosions, brakeman maimings, and wrecks causing astonishing loss of life. In *Life on the Mississippi* (1883) and again in *Huckleberry Finn* (1894) Mark Twain described in vivid detail the explosion of steam ships and the resultant death and injury of passengers. Manufactories too subjected workers (and often those living nearby) to industrial accidents, toxic fumes, and loss of hearing. Although those risks were hardly unknown,

they were accepted by workers and the public as a necessary concomitant to technological progress.

However, over the course of the nineteenth century the protection of human safety became an increasingly important priority for engineers, companies, and eventually federal and state governments. Indeed, the first scientific research contract from the federal government was issued to the Franklin Institute in Philadelphia in 1830 to investigate the causes of steamboat boiler explosions and to propose solutions (Burke 1966).

As each new technology matured to the point where advances in performance were incremental, a poor safety record became a barrier to increased public acceptance and use. Workers began to organize into unions and insist that they be better protected from workplace hazards. Engineering societies, whose original charters tended to stress the promotion and facilitation of the profession's work, by the mid-twentieth century began to impose safety as a primary ethical duty of the engineer. The end of the nineteenth century also witnessed the development of safety codes and standards governing the use of natural gas and electricity, the design of building and steam boilers, and the storage and use of explosives.

In the twenty-first century nearly every engineering code of ethics stresses the safety of workers and the public. The American Nuclear Society's Code of Ethics (2003) states:

> We hold paramount the safety, health, and welfare of the public and fellow workers, work to protect the environment, and strive to comply with the principles of sustainable development in the performance of our professional duties. The first commitment in the Code of Ethics for the Insti-

tute of Electrical and Electronic Engineers mandates that members ... accept responsibility in making engineering decisions consistent with the safety, health and welfare of the public, and to disclose promptly factors that might endanger the public or the environment (Institute of Electrical and Electronic Engineers 1990).

All licensed professional engineers are bound by the Code of Ethics for Engineers promulgated by the National Society of Professional Engineers. Both Fundamental Canon No. 1 and the first Rule of Practice impose on the engineer a duty to "hold paramount the safety, health and welfare of the public" (National Society of Professional Engineers 2003).

Apart from these commitments by long-standing communities of engineers there are many engineers whose work is devoted entirely to the protection of the public and workers from the hazards of technology and natural phenomena: Fire protection engineering, automobile safety engineering, and industrial safety engineering are a few examples. Safety engineering is itself an engineering discipline; its practitioners attempt to understand the ways in which technological systems fail and discover ways to prevent such failures. The American Society of Safety Engineers, founded in 1911 and now numbering over 30,000 members, is devoted to being "the premier organization and resource for those engaged in the practice of protecting people, property and the environment, and to lead the profession globally" (American Society of Safety Engineers 2004).

The intertwining of engineering and safety probably will intensify in the future in response to constantly rising public expectations. Two prominent engineering scholars in Lancaster University's Department of Engineering have observed the large gap between the safety expectations of today and those in the early days of modern technologies:

> Safety is rapidly becoming a means by which the public and governments judge the viability of organisations involved in safety-related processes, possibly more so than environmental issues. Many large organisations could not afford a single, large-scale incident as a result of an inferior safety culture, despite buoyant economics. This is a significant dynamic departure from past public acceptability of fatal incidents (Joyce and Seward 2004).

The dedication of the engineering profession to safety as a primary goal and an ethical duty is in accordance with this change in public expectations.

WILLIAM M. SHIELDS

SEE ALSO Engineering Ethics; Safety Factors.

BIBLIOGRAPHY

Burke, John G. (1966). "Bursting Boilers and the Federal Power." Technology and Culture 7:1 (1–23). A widely reprinted article. For critical comment, see Richard N. Langlois, David J. Denault, and Samson M. Kimenyi, "Bursting Boilers and the Federal Power Redux: The Evolution of Safety on the Western Waters," University of Connecticut Department of Economics Working Papers Series (May 1994).

Joyce, Malcolm J., and Derek W. Seward. (2002). "Innovative M.Sc. in Safety Engineering—A Model for Industry-Based Courses in the 21st Century?" Engineering Education 2002: Professional Engineering Scenarios 2002/056: (2: 28/6).

INTERNET RESOURCES

American Nuclear Society. Code of Ethics. (2003). Available at www.ans.org.

American Society of Safety Engineers. (2004). Vision Statement. Available at www.asse.org.

Institute of Electrical and Electronic Engineers. Code of Ethics. (1990). Available at www.ieee.org.

National Society of Professional Engineers. Code of Ethics. (2003). Available at www.nspe.org.

PRACTICES

Safety is one of the primary goals of engineering. In most ethical codes for engineers safety is mentioned as an essential area of professional competence and responsibility.

In everyday language, the term *safety* is often used to denote *absolute safety*, that is, certainty that accidents or other harms will not occur. In engineering practice, safety is an ideal that can be approached, but never fully attained. What can be achieved is *relative safety*, meaning that it is unlikely but not impossible that harm will occur. The safety requirements in regulations and standards represent different (and mostly high) levels of relative safety. Industries with high safety ambitions, such as airway traffic, are characterized by continuous endeavors to improve the level of safety.

The ambiguity between absolute and relative safety is a common cause of misunderstandings between experts and the public. Both concepts are useful, but it is essential to distinguish between them.

In decision theory, lack of knowledge is divided into the two major categories: "risk" and "uncertainty." In decision-making under risk, the probabilities of possible outcomes are known, whereas in decision-making under uncertainty, probabilities are either unknown or known with insufficient precision. In engineering practice, both risk and uncertainty have to be taken into account. Even when engineers have a good estimate of the probability (risk) of failure, some uncertainty remains about the correctness of this estimate.

Safety has often been defined as the antonym of risk, but that is only part of the truth. In order to achieve safety in practical applications, the dangers that originate in uncertainty are equally important to eliminate or reduce as those that can be expressed in terms of risk. Many safety measures in engineering are taken to diminish the damages that would follow from possible unknown sources of failures. Such measures protect against uncertainty rather than risk.

Several methods are used by engineers to achieve safety in the design and operation of potentially dangerous technology.

Inherently safe design. The first step in safety engineering should always be to minimize the inherent dangers in the process as far as possible. Dangerous substances or reactions can be replaced by less dangerous ones. Fireproof materials can be used instead of flammable ones. In some cases, temperature or pressure can be reduced.

Safety reserves. Constructions should be strong enough to resist loads and disturbances exceeding those that are intended. In most cases, the best way to obtain sufficient safety reserves is to employ explicitly chosen safety factors.

Negative feedback. Dangerous operations should have negative feedback mechanisms that lead to a self-shutdown in critical accident situations or when the operator loses control. Two classical examples are the safety valve that lets out steam when the pressure becomes too high in a steam boiler and the "dead man's handle" that stops the train when the driver falls asleep. One of the most important safety measures in the nuclear energy industry is to ensure that a nuclear reactor closes down automatically when a meltdown approaches.

Multiple independent safety barriers. In order to avert serious dangers, a chain of barriers is needed, each of which is independent of its predecessors so that if the first fails, then the second is still intact, and so on. Typically the first barriers are measures to prevent an accident, after which follow barriers that limit the consequences of an accident, and finally rescue services as the last resort. One of the major lessons from the *Titanic* disaster (1912) is that an improvement of the early barriers is no excuse for reducing the later barriers (such as access to lifeboats).

Maintenance and inspections. Many severe accidents have resulted from insufficient maintenance of installa-

tions or pieces of equipment that were originally in excellent shape. Regular inspections by persons with sufficient competence and mandate are an efficient means to prevent this from happening.

Educated and responsible operators. Human mistakes are an important source of accidents. An efficient countermeasure is to educate workers, authorize them to temporarily stop processes they consider to be acutely dangerous, and encourage them to take initiatives to improve safety.

Incidence reporting. Experience from air traffic and nuclear energy shows that systems for reporting and analyzing safety incidents are an efficient means to prevent accidents. Systems for anonymous reporting facilitate the reporting of human mistakes.

Safety management. Safety can be achieved only in an organization whose top management gives priority to safety and aims at continuous improvement.

SVEN OVE HANSSON

SEE ALSO *Airplanes; Automobiles; Aviation Regulatory Agencies; Building Destruction and Collapse; Engineering Ethics; Fire; Regulatory Toxicology; Robot Toys; Safety Factors.*

BIBLIOGRAPHY

Marshall, Gilbert. (2000). *Safety Engineering,* 3rd edition. Des Plaines, IL: American Society of Safety Engineers. General safety principles and their application on industrial workplaces.

SAFETY FACTORS

• • •

A safety factor (also called an uncertainty factor or assessment factor) is a number by which some variable such as load or dose is multiplied or divided in order to increase safety. Safety factors are used in engineering design, toxicology, and other disciplines to avoid various types of failure.

The sources of failure that safety factors are intended to protect against can be divided into two major categories: (a) the *variability* of conditions that influence the risk of failure, such as variations in the strength of steel and in the sensitivity of humans to toxic substances, and (b) the *uncertainty* of human knowledge, including the possibility that the models used for risk assessment may be inaccurate.

Safety factors are used to obtain a safety reserve, a margin between actual conditions and those that would lead to failure. Safety reserves can also be obtained without the use of explicitly chosen safety factors.

At least since antiquity, builders have obtained safety reserves by adding extra strength to their constructions. The earliest known use of explicit safety factors in engineering dates from the 1860s. In modern engineering, safety factors are used to compensate for five types of failure:

(1) higher loads than those foreseen,

(2) worse properties of the material than foreseen,

(3) imperfect theory of the failure mechanism in question,

(4) possibly unknown failure mechanisms, and

(5) human error in design or calculations.

The first two of these can in general be classified as variabilities, whereas the last three belong to the category of (genuine) uncertainty.

In order to be an efficient guide for safe design, safety factors should be applied to all the integrity-threatening mechanisms that can occur. For instance, one safety factor may be required for resistance to plastic deformation and another for fatigue resistance. A safety factor is most commonly expressed as the ratio between a measure of the maximal load not leading to the specified type of failure and a corresponding measure of the applied load. In some cases it may be preferable to express the safety factor as the ratio between the estimated design life and the actual service life.

The use of explicit safety factors in regulatory toxicology dates from the middle of the twentieth century. In 1954 Arnold J. Lehman and O. Garth Fitzhugh, two U.S. Food and Drug Administration (FDA) toxicologists, proposed that ADIs (acceptable daily intakes) for food additives be obtained by dividing the lowest dose causing no harm in experimental animals (counted per kilogram body weight) by 100. This value of 100 is still widely used. It is now often accounted for as being the product of two subfactors: one factor of 10 for interspecies (animal to human) variability in response to the toxicity and another factor of 10 for intraspecies (human) variability in the same respect. Higher safety factors such as 1,000, 2,000, and even 5,000 can be used in the regulation of substances believed to induce severe toxic effects in humans.

The effect of a safety factor on the actual risk depends on the dose–response relationship. If the risk is proportionate to the dose (linear dose–response rela-

tionship), then the risk reduction will be proportionate to the safety factor. If the dose–response relationship is nonlinear, then the reduction in risk can be either more or less than proportionate. Because the dose–response relationship at very low doses is always unknown, the exact effect of using a safety factor cannot be known with certainty.

Natural organisms often have safety reserves that can be described in terms of safety factors. Structural safety factors have been calculated for mammalian bones, crab claws, shells of limpets, and tree stems. Natural safety reserves make the organism better able to survive unusual conditions. Hence, the extra strength of tree stems makes it possible for them to withstand storms even if they have been damaged by insects. But safety reserves also have their costs. Trees with large safety reserves are better able to resist storms, but in the competition for light reception, they may lose out to tender and high trees with smaller safety reserves.

At least two important lessons can learned from nature in this context. First, resistance to unusual loads is essential for survival. Second, a balance will nevertheless always have to be struck between the dangers of having too little reserve capacity and the costs of having an unused reserve capacity. Perfect safety cannot be obtained, but a chosen balance between safety and costs can be implemented with the help of safety factors and other regulation instruments.

SVEN OVE HANSSON

SEE ALSO Bioengineering Ethics; Engineering Ethics; Safety Engineering.

BIBLIOGRAPHY

Dourson, Michael L., and Jerry F. Stara. (1983). "Regulatory History and Experimental Support of Uncertainty (Safety) Factors." Regulatory Toxicology and Pharmacology 3(3): 224–238. Safety factors in toxicology.

Randall, F. A. (1976). "The Safety Factor of Structures in History." Professional Safety, January: 12–28. Safety factors in engineering design.

SAKHAROV, ANDREI

• • •

Theoretical physicist and the "father of the Soviet H-bomb," Andrei Sakharov (1921–1989), who was born in Moscow on May 21, became a prominent human rights

Andrei Sakharov, 1921–1989. Sakharov, one of the Soviet Union's leading theoretical physicists and regarded in scientific circles as the "father of the Soviet atomic bomb," also became Soviet Russia's most prominent political dissident in the 1970s. (© Bettmann/ Corbis.)

activist and the first Russian to win the Nobel Peace Prize.

Sakharov's father was a physics teacher and popular science author. World War II shortened his study of physics at Moscow University. After two years of work in a munitions factory, in 1945 he went on to graduate study in theoretical physics under Igor Tamm (1895–1971). In 1948 the Soviet government assigned Tamm's group, including Sakharov, to research the feasibility of a thermonuclear bomb. In a few months Sakharov suggested a new idea that was instrumental in the development of the first Soviet thermonuclear bomb (which was tested in 1953). In 1951 he pioneered the research of controlled thermonuclear fusion that led to the tokamak reactor. He was also the main developer of the full-fledged Soviet H-bomb tested in 1955: Unlike the 1953 design, the yield of the 1955 design was potentially unlimited. He was amply rewarded by 'the government, with membership of the Soviet Academy of Sciences (1953), three Hero of Socialist Labor medals (1954, 1956, and 1962), the Stalin Prize and Lenin prize, and a luxury dacha, or villa.

In 1958 Sakharov calculated the number of casualties that would result from an atmospheric test of the "cleanest" H-bomb: 6,600 victims per megaton for 8,000 years. "What moral and political conclusions must be drawn from these numbers?" he asked in an article published that year. He answered: "The cessation of tests will lead directly to the saving of the lives of hundreds of thousands of people and will have the more important indirect result of aiding in reducing international tensions and the danger of nuclear war" (1958, p. 576). Sakharov was proud of his contribution to the 1963 test ban treaty, which stopped atmospheric nuclear testing of the United States, the USSR, and the United Kingdom.

In the 1960s Sakharov returned to pure physics. His most important contribution was a 1966 explanation of the disparity of matter and antimatter in the universe, or baryon asymmetry. The major turn in Sakharov's political evolution took place in 1967 to 1968, when antiballistic missile (ABM) defense became a key issue in U.S.-Soviet relations. Sakharov wrote the Soviet leadership to argue that the moratorium proposed by the United States on ABM work would benefit the Soviet Union, because an arms race in this new technology would increase the likelihood of nuclear war. The government ignored his letter and refused to let him initiate a public discussion of ABM in the Soviet press.

An insider's view of how the upper echelons of the Soviet regime functioned led Sakharov to the conclusion that the goals of peace, progress, and human rights were inextricably linked. He made his views public in the 1968 essay "Reflections on Progress, Peaceful Coexistence, and Intellectual Freedom," published in samizdat (underground self-publishing in the Soviet Union) and in the West in the summer of 1968. The secret father of the Soviet H-bomb emerged as an open advocate of peace and human rights.

Sakharov was immediately dismissed from the military-scientific complex. He then concentrated on theoretical physics and human rights activity. The latter brought him the Nobel Peace Prize in 1975 and internal exile in 1980, after he had been stripped of all honors including the title of Hero of Socialist Labor. In 1985 the European Parliament established the annual Sakharov Prize for Freedom of Thought, given for outstanding contributions to human rights.

In December 1986 the new Soviet leader Mikhail Gorbachev (b. 1931) released Sakharov from internal exile. Upon his return he enjoyed three years of freedom, including seven months of professional politics as a member of the Soviet parliament. The latter were the last months of his life.

For many years Sakharov lived intoxicated by socialist idealism. He later said in his memoirs that he "had subconsciously ... created an illusory world to justify" himself. Totalitarian control over information enabled Soviet propaganda to brainwash even the most intelligent. Sakharov wanted to make his country strong enough to ensure peace after a horrible war. Experience brought him to a "theory of symmetry": All governments are bad and all nations face common dangers. In his dissident years he realized that the symmetry "between a normal cell and a cancerous one" could not be perfect, although he kept thinking that the theory of symmetry did contain a measure of truth.

Sakharov saw "striking parallels" between his own life and the lives of the two American physicists Robert Oppenheimer (1904–1967) and Edward Teller (1908–2003), who crossed in the "Oppenheimer Affair" (1953–1954). Sakharov did not believe that he had "known sin," in Oppenheimer's expression, by creating nuclear weapons. Nor did he try to persuade the government, as did Teller, of the need for a hydrogen bomb. Having disagreed with Teller on the prominent issues of nuclear testing and antimissile defense (e.g., the "Star Wars" program), Sakharov, nevertheless, believed that American physicists had been unfair in their attitude toward Teller following his clash with Oppenheimer. Sakharov felt that in this "tragic confrontation of two outstanding people," both deserved equal respect, because "each of them was certain he had right on his side and was morally obligated to go to the end in the name of truth" (Memoirs).

For Sakharov the statement that "the future is unpredictable" was meaningful far beyond quantum physics. It supported his personal responsibility for the future of humanity. For him knowledge was not only power but also professional and moral responsibility.

GENNADY GORELIK

SEE ALSO Oppenheimer, J. Robert; Russian Perspectives; Teller, Edward; Weapons of Mass Destruction.

BIBLIOGRAPHY

Gorelik, Gennady. (1999). "The Metamorphosis of Andrei Sakharov." Scientific American 280(3): 98–101. An explanation of Sakharov's transformation into public and political figure.

Gorelik, Gennady, with Antonina W. Bouis. (2004). The World of Andrei Sakharov: A Russian Physicist's Path to Free-dom. New York: Oxford University Press. The first authoritative study of Andrei Sakharov as scientist as well as public figure relies on previously inaccessible documents, recently declassified archives, and personal accounts by Sakharov's friends and colleagues.

Sakharov, Andrei D. (1958). "Radioaktivnyi uglerod iadernykh vzryvov i neporogovye biologicheskie effekty" [Radioactive carbon in nuclear explosions and nonthreshold biological effects]. Atomnaia energiia 4, no. 6: 576–580. In Soviet Scientists on the Danger of Nuclear Tests, ed. A. V. Lebedinsky. Moscow: Foreign Lang. Publ. House, 1960. The first of Sakharov's writing involving moral and political issues.

Sakharov, Andrei D. (1968a). "Thoughts on Progress, Peaceful Coexistence and Intellectual Freedom." New York Times Jul 22, 1968, 14–17. The first instance of Sakharov's writing manifesting his political dissent with the Soviet regime.

Sakharov, Andrei D. (1968b). Progress, Coexistence, and Intellectual Freedom. New York: Norton. Includes the essay of the previous citation.

Sakharov, Andrei. (1990). Memoirs, trans. Richard Lourie. New York: Knopf.

INTERNET RESOURCE

"Andrei Sakharov: Soviet Physics, Nuclear Weapons, and Human Rights." American Institute of Physics, Center for History of Physics. Available from http://www.aip.org/history/sakharov/.

SANGER, MARGARET
• • •

Margaret Sanger (1879–1966), born in Corning, New York on September 14, was an internationally renowned leader in the movement to secure reproductive rights for women. Founder of the first birth-control clinic in the United States and later, of the Planned Parenthood Federation of America and the International Planned Parenthood Federation, Sanger was a controversial figure with militant feminist and socialist views, working for change in areas of strong traditional values and cultural resistance.

Sanger was the sixth of eleven children born to a devout Catholic Irish-American family. To escape what she saw as a grim class heritage, she worked her way through school and chose a career in nursing. Although she married and had three children, Sanger maintained an intellectual and professional independence. She immersed herself in the radical bohemian culture of intellectuals and artists that flourished in New York City's Greenwich Village. She also joined the Women's

Margaret Sanger, 1879–1966. The pioneering work of this American crusader for scientific contraception, family planning, and population control, made her a world-renowned figure. (*The Library of Congress.*)

Committee of the New York Socialist Party and participated in labor strikes organized by the Industrial Workers of the World.

Working with poor families on the Lower East Side of New York City, Sanger increasingly focused her attention on sex education and women's health and reproductive rights. She argued that a woman's right to control her own body was the foundation of her human rights, that limiting family size would liberate working-class women from the economic burdens associated with unwanted pregnancies, and that women are as much entitled to sexual pleasure and fulfillment as men.

Sanger's ideas have remained controversial. Those who oppose family planning point to her adherence to certain popular ideas of her time as proof that the movement is fundamentally flawed. Sanger advocated birth control as a means of reducing genetically transmitted mental and physical defects, even going so far as to call for the sterilization of the mentally incompetent. But her thinking differed significantly from the reactionary eugenics that eventually became the centerpiece of the

Nazi party platform. Sanger never condoned eugenics based on race, class, or ethnicity, and in fact her writings were among the first banned and burned in Adolf Hitler's Germany.

Sanger called for the reversal of the Comstock Law and related state laws banning the dissemination of information on human sexuality and contraception. In 1914, indicted for distributing a publication that violated postal obscenity laws, she fled to England, where she was deeply influenced by the social and economic theories of Britain's radical feminist and neo-Malthusian intelligentsia. Separated from her husband and exploring her own sexual liberation, Sanger had affairs with several men including the psychologist Havelock Ellis (1859–1939) and the author and historian H. G. Wells (1866–1946). She returned to the United States in 1915 to face the charges against her, hoping to use her trial to capture media attention. But the sudden death of her five-year-old daughter generated public sympathy, and the government dropped the charges. She then embarked on a national tour and was arrested in several cities, attracting even greater publicity for herself and the birth-control movement.

Sanger founded a number of important organizations and institutions to advance the cause of reproductive rights. In 1916 she opened the first birth-control clinic in the United States in the Brownsville section of Brooklyn, New York. Nine days later, Sanger and her staff were arrested. She then opened a second clinic, the Birth Control Clinical Research Bureau, staffed by female doctors and social workers, which became important in collecting clinical data on the effectiveness of contraceptives. In 1921 Sanger founded the American Birth Control League, which later merged with the Birth Control Clinical Research Bureau to form the Birth Control Federation of American, forerunner of the Planned Parenthood Federation of America. In 1930 she founded a clinic in Harlem, and she later founded "the Negro Project," serving African Americans in the rural South. Of Sanger's work, Martin Luther King Jr. (1929–1968) said, "the struggle for equality by nonviolent direct action may not have been so resolute without the tradition established by Margaret Sanger and people like her."

After World War II, Sanger shifted her concerns to global population growth, especially in the Third World. She helped found the International Planned Parenthood Federation, serving as its president until 1959. Sanger helped find critical development funding for the birth-control pill and fostered a variety of other research efforts including the development of spermicidal jellies and spring-form diaphragms. She died only a few months after birth control became legal for married couples, a 1965 decision that reflected the influence of Sanger's long years of dedication to radical, visionary social reform.

JENNIFER CHESWORTH

SEE ALSO *Birth Control; Eugenics.*

BIBLIOGRAPHY

Chesler, Ellen. (1992). *Woman of Valor: Margaret Sanger and the Birth Control Movement in America.* New York: Simon & Schuster.

King, Martin Luther Jr. (1966). "Family Planning—A Special and Urgent Concern." King's acceptance speech upon receiving the Margaret Sanger Award from the Planned Parenthood Federation of America.

Valenza, Charles. (1985). "Was Margaret Sanger a Racist?" *Family Planning Perspectives* 17(1): 44–46.

INTERNET RESOURCE

The Margaret Sanger Papers Project. Available from http://www.nyu.edu/projects/sanger/index.html.

SARTRE, JEAN-PAUL

SEE *Existentialism.*

SCANDINAVIAN AND NORDIC PERSPECTIVES

• • •

The term "Scandinavia" traditionally includes the so-called Scandinavian countries Denmark, Norway and Sweden. Sometimes "Scandinavia" is given a broader definition that also covers the two remaining "Nordic" countries Finland and Iceland. The Scandinavian and Nordic countries are highly industrialized countries that have attempted to combine economic development with social welfare and democratic planning. Technological change has been considered in relation to competing values and interests, and ethics has played a role in this context.

The development of technology and ethics in Scandinavian and Nordic countries is characterised by some general trends that are very similar to Denmark, Norway, Sweden, Finland and Iceland. Traditionally there has been a lot of scientific and cultural exchange among these countries and therefore one finds similar theoretical trends and movements among the Nordic countries. In particular can be mentioned positivistic and instrumental positions, Marxistic postions, positions from applied ethics traditions, critical environmental positions, and positions from postmodern continental philosophy.

Historical Background

The most famous case of science and technology ethics in the Nordic countries is the criticism of the Danish physicist and Nobel Prize winner Niels Bohr (1885–1962). Bohr was paradoxically one of the physicians participating in the "Manhattan Project" during World War II that lead to the creation of the nuclear bomb. Bohr has said that it was only after that the United States dropped the bomb on Hiroshima and Nagasaki that he fully became aware of the ethical responsibility of science (Rendtorff 2003). After he realized the deadly consequences of the use of nuclear bombs Bohr became an active opponent of nuclear arms and he sent several letters to the United Nations urging avoidance spread of nuclear mass destruction weapons and prevention of a nuclear war.

Although many Nordic scientists joined Bohr in his criticism of the military use of science and technology, the spirit of science and technology during the first part of the twentieth century was in general determined by a belief in the norms of science as universal and neutral creation of knowledge for the benefit of humankind.

During the 1960s there was a general belief in technology in the Scandinavian and Nordic countries. This period was characterized by a strong belief in the progress of science and technology. The spirit of research was instrumental, pragmatic and positivistic. In the 1970s, however, many critical movements emerged. In particular, many Marxist criticisms of technology were published. Marxist critiques treated technology as an aspect of the increasing oppression of people by a capitalist society. Marxist positions were influential because they contributed to the establishment of classes on society and technology in many universities.

The well known Finnish philosopher Georg Henrik von Wright published a path-breaking critical work in technology ethics in 1986, one of the most important contributions to technology ethics in Finland and perhaps also in the rest of the Nordic countries. In his book about science and rationality the basic argument is a deep scepticism towards the possibilities of humanity to deal with technological progress and its problems. A true humanism must be based on a deep understanding of human nature and the acceptance of the natural limits on human activities and the interventions of beings in their natural and cultural environment (von Wright 1986).

In Denmark there have also been many publications on the limits of growth. The theologian Ole Jensen (1976) wrote *I Vækstens Vold* (Submitted to growth) on that subject and the philosopher Villy Sørensen and colleagues (1978) proposed a discussion aimed at overcoming the Marxist opposition to the role of technology in society and proposing a new vision of a society in harmony with technology.

In addition to Marxist positions there emerged a strong ecological movement focusing on the negative environmental consequences of science and technology in an industrial society. Discussions of environmental ethics were extensive, and in Norway the deep ecology movement represented by the philosopher Arne Næss (1976) proposed a paradigm of the relationship between humankind and nature that became influential worldwide.

During the 1980s the Danish philosopher Peter Kemp attempted to integrate the humanities and technology. Drawing on the philosophies of Hans Jonas (1903–1993), Paul Ricoeur (b. 1913) and Emmanuel Levinas (1906–1995), he argued for a symbiotic relationship between the two cultures and an ethics of technology in *The Irreplacable* (1991), which was his second doctoral habilitation at the university of Göteborg.

Bioethics

During the 1990s the focus shifted from technology ethics to bioethics and medical ethics. In Norway a debate on principles resulted from discussions about the national biotechnology legislation that was enacted at the beginning of the decade. The Norwegian parliament invented the concept of "mixed ethics," a collection of deontological, utilitarian, and cultural approaches, as the basis for biotechnology legislation. Sweden discussed these matters in the framework of the Swedish Council for medical ethics, an advisory body to the Swedish government.

In Norway technology ethics and bioethics were integrated in the so-called Ethics Research program of the Norwegian Government, which opened opportunities for many doctoral candidates to start a carrier in technology ethics. That program also involved strengthening bioethics research. The professor of medical ethics Jan Helge Solbakk (1994) was influential in developing

medical ethics in that country on the basis of the work of one of the founders of Norwegian medical ethics, Knud-Erik Tranøy (1992).

In Sweden utilitarian bioethics was defended by the consequentialist Torbjörn Tjansöe, who became a professor of philosophy in Stockholm. Tjansöe has radical views on bioethics and once was a dogmatic Marxist. A Kantian position in favor of human dignity has been defended by Matts Hannson (1991), who is the director of the Swedish ELSA program (Ethical, Legal, and Social Aspects of genetic technologies) based in Uppsala. In addition, there is an influential interdisciplinary research unit on bioethics and technology ethics at Linköping University, where the Danish professor Thomas Achen has worked on gene technology and law in Scandinavia (Achen 1997).

In Denmark discussions of bioethics emerged from debates in the Danish Council of Ethics, which was established in 1987. Two research programs that were sponsored by five Danish Research Councils in 1993 were especially important in the development of the bioethics research environment in that country.

The first program, Gran (Foundations and Applications of Bioethics) explored the foundations and applications of ethics and collaborated closely with the Danish Council of Ethics by arranging hearings about bioethics issues. Svend Andersen, a professor of theology at the University of Aarhus, who had been one of the first members of the Danish National Council of Ethics, directed this research project. The Danish philosopher and theologian Knud Ejler Løgstrup was the inspiration for Andersen's position on theoretical ethics. Andersen had also been responsible for an important report on research ethics for the ministry of research in 1994 (Andersen 1994, Rendtorff 2003). However, Andersen also collaborated with Peter Sandøe, a consequentialist who later worked on animal bioethics and in 1998 established a Center for Risk Assessment for Human and Animal Biotechnology based in the Royal Danish Vetenary School.

The second project, which was based in the Center for Ethics and Law at the University of Copenhagen, explored the relationship between biotechnology, ethics, and the law. It also collaborated with the Danish Council of Ethics in the organization of international conferences on bioethics and biolaw. Peter Kemp, a technology ethicist who in the 1980s had done work on medical ethics, became the director of the center, which published several works on bioethics and law. This project applied a phenomenological approach to the ethics of biotechnology (Rendtorff 1999). In addition, the

Center for Ethics and Law was responsible for a European research project sponsored by the BIOMED-II program of the European Commission, Basic Ethical Principles in European Bioethcis and Biolaw, that led to the publication of a two-volume research report (Rendtorff and Kemp 2000). The report investigated the ideas of autonomy, dignity, integrity, and vulnerability as guiding ideas for future European bioethics and biolaw.

In Finland there has also been much public debate about different issues of bioethics: abortion, euthanasia, genetic engineering, inequalities in health, decline of the natural environment, overpopulation, and scarcity of medical resources. Like many European countries, Finland has established a national council of ethics to advise government about ethical issues in health care, science, and technology. Academic debates about bioethics in Finland has mostly been inspired by the Anglo-American approaches in the field. The discussions are characterized by confrontations between consequentialist and deontological and right-based approaches to applied ethics (Rendtorff and Kemp 2000).

Icelandic approaches to bioethics follow the same patterns of confrontation between principles and pragmaticism. Recent discussions have been focussed on the development of an Icelandic biotechnology industry. A thought-provoking case is the fact that the Icelandic government has allowed a privately-owned enterprise to make a bio-bank with blood samples and genetic information from the 280,000 citizens of Iceland (Rendtorff 2003). The Icelandic genetic patrimony is unique because of the small genetic variation within a homogenous population; therefore there might be opportunities to discover new knowledge about genetics. The firm "decode" collaborates with international biotechnology companies; they have procured a number of patents and other rights to the genetic samples that constitute a unique opportunity to do research in genetic basis of disease and possible improvement of medicines for treatment of genetic diseases. Critical voices in the public debate have argued that this common gene pool poses serious problems of data protection, privacy, and anonymity. Moreover, it is stated that the Icelandic government has been too quick in allowing extended commercialization of genetic information and private ownership of blood samples from human bodies. However, this debate about bio-banks and uses of genetic technologies represent features that seems to be fairly common among all the Nordic countries.

Technology Ethics

Parallel to the discussions in bioethics, a scholarly literature has evolved that is concerned with the relationship of technology and society. In this literature attempts are made to understand the interrelationships between technological change and social concerns. The concept of ethics also is important in this context, but it is not always used in the strict philosophical sense of the word.

The Scandinavian and Nordic countries all have a tradition of social planning. All three countries were industrialized at a relatively late stage and at a slow pace. This has allowed for peaceful processes of industrialization with attention paid to the welfare state and social welfare. As a consequence, labor unions, among other groups, have played a crucial role in social development and various traditions of democracy and welfare planning have evolved that have a strong influence on Scandinavian societies.

This may explain why several issues in ethics, social policy, and technology have been formulated in a relatively constructive and formative rather than reactive way. In the initial stages two scholarly traditions seemed important: working life science and a critique of technology.

WORKING LIFE SCIENCE. This tradition began in the late 1960s. In 1971 the Norwegian Iron and Metal Workers' Union initiated an important project with Kresten Nygaard that dealt with planning methods for the trade unions (Fuglsang 1993). The aim of the project was to strengthen the trade unions' influence on new computer technologies. In 1975 the Swedish National Federation of Labour Unions (LO) sponsored a similar project, DEMOS, which dealt with democratic control and planning in working life. The aim of the project was to support workers' influence on the new technology. In Denmark Project DUE, which dealt with democracy, development, and data processing, was initiated. Some of these projects were inspired in part by Harry Braverman's work on the degrading and controlling aspects of work (Braverman 1976), but their aim clearly went beyond Braverman's objectives. They were not limited to studying the negative consequences of technology but instead were intended to formulate an approach to a constructive development of technology.

One of the computer scientists who took part in those discussions, Pelle Ehn, published a book explaining these aims (Ehn 1988). In that book the Scandinavian approach was seen as standing in opposition to the so-called sociotechnical approach, a functional approach in which social and technical systems were understood as being interdependent. By contrast, in Ehn's view workers should be able to participate directly in the development of computer systems.

CRITIQUE OF TECHNOLOGY. This tradition evolved from a combination of philosophical and sociological approaches. In Norway, Arne Næss developed his eco-philosophy, which was concerned, among other things, with the inability of engineers to take into consideration the wholeness of humankind and nature in which they were situated (Næss 1976). Sigmund Kvaløy (1976) developed a critique of the complexity of industrialism. The sociologist Dag Østerberg (1974) was concerned with the way in which technology could be understood as materialized social relations interacting with human activity.

In Denmark, Hans Siggard Jensen and Ole Skovs-mose published a critique of technology in which they argued for a nonteleological or deontological ethical approach to technology (Jensen and Skovmose 1986). They positioned themselves in relation to the work of the philosophers Immanuel Kant (1724–1804) and Jürgen Habermas (b. 1929). Anker Brink Lund, Robin Cheesman, and Oluf Danielsen published a book in which they criticized technocratic approaches, particularly in the area of electronic media, and pointed to possibilities for a more democratic model of technological change (Lund et al. 1981).

Tarja Cronberg (1987) has developed a distinct approach to technology that focuses on the relationship of technology and everyday life. Cronberg came to see Danish social experiments with technology as a kind of laboratory for dialogue and research inspired by phe-nomenological approaches and critical theories of communication (Habermas 1984).

In Sweden, Andrew Jamison and Aant Elzinga have tried to work out historical perspectives on science and technology policy. They also stress the impact of culture (Elzinga and Jamison 1981). Jamison (1982) has been interested in the concept of "national styles" in an attempt to determine how national culture plays a formative role in relation to science and technology; this is implicitly a deontological approach.

The two initial traditions of working life science and technology critique have been conducted in various ways in small scholarly communities. In computer science the tradition of working life science has involved differing understandings of computer design and human-computer interactions. The journal *Computer Supported Cooperative Work* has been important in this work. An influential semiethical orientation in Scandinavian computer design is "activity theory," which is present in the work of the Danish working life scientist Susanne Bødker. Technology is seen as a tool that mediates between an individual and a social object or social role in an organization. For this relationship to become meaningful, it is necessary to design and integrate computer programs in an artful way. In Finland, this tradition of activity theory has become a very important contribution to work development research through the work of Yrjö Engeström (Engeström et al. 1999) and his Centre for Activity Theory and Developmental Work Research at the University of Helsinki.

A critique of technology seems not to have developed in a systematic way in Scandinavian philosophy. Some works have been published, but they have not led to the development of distinct philosophical traditions. At the Department of Management, Politics and Philosophy in the Copenhagen Business School in Denmark some scholars have developed the notion of "ethical budgets" and values-driven management for firms, which seems to be related to technology and ethics (Ole Thyssen 1997), and other philosophical contributions in the areas of ethics, innovation, and technology have been produced.

In Finland, a tradition of engineering ethics and responsibility of scientists has developed through such organizations as the Finish nongovernmental organization Technology for Life, and the Association of Swedish-Speaking Engineers in Finland, which has created a code of ethics for its members. Attempts are here made to sustain civil courage and find ways for engineers to demonstrate loyalty to third party (the future, the nature, humankind) rather than merely to business or within professions. Engineering ethics is taught in some engineering schools and technical universities in the Scandinavian countries even though these courses are not, or at most are seldom, compulsory. At the Helsinki University of Technology, a one-year course has been created with the help of Technology for Life.

Science, Technology, and Society Studies

A small tradition of science and technology studies (STS) has developed primarily in the three Scandinavian countries (Denmark, Norway, Sweden). It has, in parallel with working life science, attempted to focus more on the development of than on the impact of technology. In Norway two STS institutions have been created that serve as examples of this work.

One is the Center for Technology and Human Values (now the Centre for Technology, Innovation and Culture), which was headed by Francis Sejersted in the period 1988–1998. Sejersted (1993) examined how a special form of capitalism has developed in Norway that is anchored in democratic, egalitarian, and local values in contrast to Chandler's (1990) notions of corporate and competitive capitalism in Germany and United States. Other researchers at this institution have shown how the transfer of technology to Norway as well as innovation processes can be seen as being intertwined with regional social structures and local values, leading to special forms of localized innovation (Wicken 1998).

A second STS institution is at the Norwegian University of Science and Technology in Trondheim, headed by Knut H. Sørensen. In his research Sørensen has been occupied with studying what he calls the domestication and cultural appropriation of technology in everyday life, which may be seen as part of a deontological, nonteleological tradition (Lie and Sørensen 1996, Sørensen 1994, Andersen and Sørensen 1992).

In Sweden several STS units have been created, such as Tema T in Linköping and Science and Technology Studies at Göteborg University. Those groups conduct research on various aspects of technology and ethics, such as the role of expertise, technology in everyday life, technology and gender, technology and identity, technology and large technological systems, and public engagements with science.

These institutions focus largely on technology *development* rather than the *consequences* of technology, and in terms of ethics they may be seen to underline mostly a deontological approach in which social values come first and technology comes second.

In Denmark and later in Norway a tradition of technology *assessment* has developed. The most important contribution in this field is probably the Danish "consensus conference," which involves laypeople in the ethical assessment of technology. The laypeople are appointed much as a jury is appointed in a court. They question experts during a three-day session. Afterward they withdraw and formulate a verdict in the form of a consensus report. This approach can be associated with a nonteleological or deontological approach to ethics and technology.

Ethics of Science

In Scandinavia debates on the ethics of science have involved research on both ethics in technology and bioethics research. However, only with the establishment of specific committees for the ethics of science has this become an integrated part of work on the ethics of technology.

In Denmark the ethics of science was prominently present in the medical research community, which had to deal with serious problems with scientific fraud. The central committee on the ethics of science was influential in resolving problems among scientists with regard to this issue.

In 1998 the Danish Committee on Scientific Fraud and Integrity in Science (Udvalgene Vedrørende Videnskabelig Redelighed) was established as a subcommittee to the national committee for medical research. This committee formulated a number of rules for the ethics of science and publication ethics. The committee was allowed to process individual complaints against scientists (Rendtorff 2003, p. 63).

In this context, an intense debate about the ethics of science emerged as a reaction to the work of the political scientist Bjørn Lomborg (2002), director of a newly established Institute for Assessment of Environmental Protection. Lomborg had argued that most of the environmental sciences had been too pessimistic with regard to their conceptions of the dangers of an environmental crisis. Lomborg's work was brought to the committee in 2002 by a number of scientists who complained that Lomborg was guilty on scientific fraud because they did not believe in his methods and research results. It was argued that Lomborg did not work with a satisfactory scientific method. Lomborg had illustrated his argument with statistical material, and many ecological scientists thought that this constituted scientific fraud because he used statistical material to illustrate arguments that, according to the ecologists, could not be defended on those grounds. Lomborg's opponents argued that Lomborg's book could not be regarded as science, but rather as a contribution to the public debate. Moreover, it was argued that Lomborg as a social scientist did not have sufficient knowledge, which led to incorrect and hasty conclusions. The Committee on Scientific Rraud and Integrity investigated the issue, based on dialogue with international experts, and in spring 2003 (Rendtorff 2003, p. 9–10) Lomborg was judged by the committee to have committed not subjective but objective scientific fraud; according to the committee, he did not understand his research subject. This led to a violent debate about environmental technology in Denmark, and after that time the ethics of science became a very widely discussed subject.

In January 2004 the Ministry for Research of the Danish liberal-conservative government intervened. They came up with a very critical assessment of the decision in the Lomborg case. However, the Ministry wanted to protect people who were charged of scientific fraud; it therefore did not accept the decision of the Committee for Scientific Fraud in the Lomborg case. So Lomborg, in the end, was not convicted of scientific fraud and the official inquiry ended in January 2002. But even though the case of Lomborg did not get a clear closing and decision about whether it really was a case of scientific fraud, it illustrates many of the basic dilemmas of the ethics of science in Scandinavian countries: problems of the definition of scientific fraud and the integration of the public in scientific debates.

LARS FUGLSANG
JACOB DAHL RENDTORFF

SEE ALSO *Bioethics; Environmental Ethics; Kierkegaard, Søren; Marxism; Science, Technology, and Society Studies; von Wright, Georg Henrik.*

BIBLIOGRAPHY

Achen, Thomas. (1997). *Den bioetiske udfordring. Et retspolitisk studie af forholdet mellem etik, politik og ret i det lovforberedende arbejde vedrørende bio- og genteknologi i Danmark, Norge og Sverige* [The bioethical challenge. A study in legal politics of the preparatory work on laws on bio- and gene technology in Denmark, Norway and Sweden]. Linköping, Sweden: Linköpings Universitet.

Andersen, Håkon With, and Knut Holten Sørensen. (1992). *Frankensteins dilemma. En bok om teknologi, miljø og verdier.* [The dilemma of Frankenstein. A book about technology, environment and values]. Oslo: Gyldendal.

Andersen, Svend. (1994). *Forskningsetik. En Udredning* [Research ethics. A report]. Copenhagen: Forskningsministeriet.

Braverman, Harry. (1976). *Labour and Monopoly Capital: The Degradation of Work in the Twentieth Century.* New York: Monthly Review Press.

Bødker, Susanne. (1987). *Through the Interface: A Human Activity Approach to User Interface Design.* Århus, Denmark: Århus University, Computer Science Department: DAIMI PB (224).

Chandler, Alfred Dupont. (1990). *Scale and Scope: The Dynamics of Industrial Capitalism.* Cambridge, MA: Belknap Press of Harvard University Press.

Cronberg, Tarja. (1987). *Det teknologiske spillerum i hverdagen. En beskrivelse af hvordan telefonen, vaskemaskinen og bilen har påvirket hverdagslivet, og en modelteoretisk analyse heraf.* Copenhagen: Nyt fra samfundsvidenskaberne.

Ehn, Pelle. (1988). *Work-Oriented Design of Computer Artifacts.* Stockholm: Arbetslivscentrum.

Elzinga, Aant, and Andrew Jamison. (1981). *Cultural Components in the Scientific Attitude to Nature: Eastern and Western Modes?* Lund, Sweden: Reprocentralen.

Engeström, Yrjö, Reijo Miettinen, and Raija-Leena Punamäki-Gitai, eds. (1999). *Perspectives on Activity Theory.* Cambridge, UK: Cambridge University Press.

Fuglsang, Lars. (1993). *Technology and New Institutions: A Comparison of Strategic Choices and Technology Studies in the United States, Denmark, and Sweden.* Copenhagen: Academic Press.

Habermas, Jürgen. (1984). *The Theory of Communicative Action.* Boston: Beacon Press.

Hansson, Mats. (1991). *Human Dignity and Animal Well-Being: A Kantian Contribution to Biomedical Ethics.* Uppsala, Sweden: Almquist & Wiksell International.

Hybel, Ulla. (1998). *Forsøgspersoner. Om den retlige beskyttelse af mennesker, der deltager I biomedicinsk forskning* [Experimental subjects. About the legal protection of persons who participate in biomedical research]. Copenhagen: Jurist og Økonomforbundets forlag.

Jamison, Andrew. (1982). *National Components of Scientific Knowledge: A Contribution to the Social Theory of Science.* Lund, Sweden: Research Policy Institute.

Jensen, Hans Siggaard, and Ole Skovmose. (1986). *Teknologikritik.* Herning, Denmark: Systime.

Jensen, Ole (1976). *I Vækstens Vold* [Submitted to growth]. Copenhagen: Fremad.

Kemp, Peter. (1991). *Det uerstattelige. En teknologi etik* [The irreplacable. A technology ethics]. Copenhagen: Spektrum. Swedish translation, *Det Oerstättliga. En teknologie-tik.* Stockholm/Skåne: Brutus Östlings Bokforlag Symposion. German translation (1992), *Das Unersetzliche. Eine Technologie-Ethik.* Berlin: Wichern. Norwegian translation (1996), *Det uerstattelige. En teknologi-etikk.* Olso: Gyldendal. French translation (1997), *L'irremplaçable.* Paris: Éditions du Cerf.

Kemp, Peter; Mette Lebech; and Jacob Rendtorff. (1997). *Den bioetiske vending.* Copenhagen: Spektrum.

Kemp, Peter; Jacob Rendtorff; and Niels Mattson Johansen. (2000). *Bioethics and Biolaw,* vols. I and II. Copenhagen: Rhodos International Publishers.

Kvaløy, Sigmund. (1976). *Økokrise, natur og menneske: En innföring i økofilosofi og økopolitikk.* Oslo: Tapir.

Lie, Merete, and Knut H. Sørensen. (1996). *Making Technology Our Own? Domesticating Technology into Everyday Life.* Oslo: Scandinavian University Press.

Lomborg, Bjørn. (2002). *The Sceptical Environmentalist.* Cambridge, UK: Cambridge University Press.

Lund, Anker Bring; Robin Cheesman; and Oluf Danielsen. (1981). *Fagre elektroniske verden.* Copenhagen: Fremad.

Næss, Arne. (1976). *Økologi, samfunn og livsstil* [Ecology, society, and lifestyle]. Olso: Scandinavian University Press.

Østerberg, Dag. (1974). *Makt og materiell.* Oslo: Pax.

Rendtorff, Jacob Dahl. (1999). *Bioetik og ret. Kroppen mellem person og ting* [Bioethics and law. The body between person and thing]. Copenhagen: Gyldendal.

Rendtorff, Jacob Dahl. (2003). *Videnskabsetik* [The ethics of science]. Roskilde: Roskilde Universitetsforlag.

Rendtorff, Jacob Dahl, and Peter Kemp. (2000). *Basic Ethical Principles in European Bioethics and Biolaw,* vols. I and II. Copenhagen and Barcelona: Centre for Ethics and Law and Institut borja di bioètica.

Sejersted, Francis. (1993). *Demokratisk kapitalisme*. Oslo: Universtetsforlaget.

Solbakk, Jan-Helge. (1994). *Medicinen som møtested og markedsplads* [Medicine as meeting place and market place]. Oslo: Forum.

Sørensen, Knut Holten, ed. (1994). *The Car and Its Environments: The Past, Present and Future of the Motorcar in Europe*. Brussels: COST Social Sciences Directorate General of Science, Research and Development.

Sørensen, Villy; Niels I. Meyer; and K. Helweg Petersen. (1978). *Oprør fra midten* [Revolt from the center]. Copenhagen: Gylden.

Tranøy, Knut-Erik. (1992). *Medisinsk etikk i vår tid* [Medical ethics in our time]. Oslo: Sigma forlag.

von Wright, Georg Henrik. (1986). *Vetenskapen Och Förnuftet* [Reason and science]. Stockholm: MånPocket.

Thyssen, Ole. (1997). *Værdiledelse* [Values-driven management]. Copenhagen: Gyldendal.

Wicken, Olav. (1998). "Regional Industrialization and Political Mobilization: Regions in Politcal Party Systems and Industrialization in Norway." *Comparative Social Research* 17: 241–271.

SCHWEITZER, ALBERT

• • •

Albert Schweitzer (1875–1965) was born in Kaysersberg, Germany (now part of France) on January 14, and became a theologian, physician, musician, and philosopher whose ethical theory argued the centrality of reverence for life. After a doctorate in philosophy from the University of Strasbourg (1899), Schweitzer received his licentiate in theology (1900), and from 1901 to 1912 held administrative posts in the Theological College of St. Thomas. In 1913, having earned an M.D. degree, he founded a hospital at Lambaréné, French Equatorial Africa (now Gabon). As a German citizen, he became a French prisoner during World War I, but returned to Lambaréné in 1924, where he spent the remainder of his life expanding, administering, and improving the hospital. Recipient of the 1952 Nobel Peace Prize, Schweitzer worked during his later years in the struggle to end the proliferation and testing of nuclear weapons. He died on September 4 and was buried at Lambaréné.

From Music to Philosophy

In 1905 Schweitzer, an accomplished organist, wrote a biography of Johann Sebastian Bach (1685–1750), and in 1906 *The Quest of the Historical Jesus* established him as a theological scholar. As a Christian, his faith guided his life as a physician at Lambaréné, where he unself-

Albert Schweitzer, 1875–1965. Schweitzer was a German religious philosopher, musicologist, and medical missionary in Africa. He was known especially for founding the Schweitzer Hospital, which provided unprecedented medical care for the natives of Lambaréné in Gabon. (*AP/NYWTS/The Library of Congress.*)

ishly treated thousands of patients, including lepers. Although successful in diverse fields, Schweitzer considered his contributions to philosophy to be his most important achievements.

Schweitzer's philosophy of culture and ethics sought to reorient material progress toward humanity as a normative ideal. In his *The Decay and the Restoration of Civilization* (1923) and *Civilization and Ethics* (1923)—brought together in *The Philosophy of Civilization* (1949)—Schweitzer interpreted World War I as the sign of a deep-rooted crisis of European culture. The Enlightenment ideals of progress and rationality had decayed and lost their ability to control the trajectory of science and technology. Philosophy and religion no longer provided intellectual and spiritual guidance. Human powers had outstripped human capacities for reason.

This asymmetry between human powers and the ability to wisely constrain and channel those powers for compassionate action underpinned Schweitzer's ethics. In *Civilization and Ethics,* he writes:

The disastrous feature of our civilization is that it is far more developed materially than spiritually. ... Through the discoveries which now place the forces of Nature at our disposal in such an unprecedented way, the relations to each other of individuals, of social groups, and of States have undergone a revolutionary change.... Advances in knowledge and power work out their effects on us almost as if they were natural occurrences.... Paradoxical as it may seem, our progress in knowledge and power makes true civilization not easier but more difficult. (pp. 86–87)

He did not conceive of his own ethical theory as completely novel, but rather as the revitalization and reformation of the ethical legacy of humanity in the twentieth century. His goal was to restore the binding character of humanity and humanitarianism as the common assets of world civilizations. Schweitzer drew not only from the Christian commandment of love but also from Asian philosophies. He held that his main principle of "devotion toward life born from reverence for life" was a plausible ethical guideline for any individual regardless of his or her culture or religion.

In contrast to the rational a priori approach of Immanuel Kant (1724–1804), Schweitzer grounded his ethics in the experience of life as an empirical hypothesis, and is in this sense closely related to Friedrich Nietzsche (1844–1900) and Arthur Schopenhauer (1788–1860). Reflecting upon life in this way, Schweitzer believed, would lead to the perspectives of reverence and responsibility. An experience of one's own "will to life," and the effort to avoid pain and seek pleasure, rationally compels an individual, under the auspices of a quasi-Kantian truthfulness, to acknowledge the same volition in others (see Meyer and Bergel 2002). This consciousness of being connected with other lives demands that people respect the moral rights of others, including plants and animals.

Schweitzer's ethics is contextual and situation-oriented and leads to a practical law that serves "concrete" humanity. He does not require an unbounded ethical responsibility beyond one's capability, but rather insists that it is most important to practice reverence for life within one's scope of action. He believed "abstraction is the demise of ethics" and that concrete humanity should always be promoted.

Ethics and Technology

Schweitzer was aware that life presented conflicting demands and that technological and scientific developments in modern civilization posed difficult challenges for practical responsibility. Yet he did not believe that this warranted the construction of dubious hierarchies and theoretical rankings of values that only solve problems in the abstract. His ethics docs not promise a methodical and self-evident solution to difficult problems. Instead, the principle of reverence for life should be used as a general guideline for the process of critical thinking.

Schweitzer's ethics serves as a compass in the complex geography of modern problems to orient practical action toward responsibility and reverence for life. In his autobiography, Out of My Life and Thought (1990), Schweitzer describes the moment when the concept of reverence for life dawned upon him as he traveled through an African jungle in September 1915. He remembers, "Late on the third day, at the very moment when, at sunset, we were making our way through a herd of hippopotamuses, there flashed upon my mind, unforeseen and unsought, the phrase 'reverence for life.' ... Now I had found my way to the principle in which affirmation of the world and ethics are joined together!" (p. 155).

Although he did not develop a special ethics for science and technology, Schweitzer's humanitarianism and reverence for life can be easily transferred to the moral problems in this field. For instance, he argued that because nuclear technology could not be controlled, it could by the same token not be responsibly used—a position that would, of course, have to be qualified by specific situations and contexts (Schweitzer 1958). In general, Schweitzer's advice for solving ethical problems, including those presented by science and technology, was to rely on and use practical reasoning, individual responsibility, and the ideal of concrete humanity.

CLAUS GÜNZLER
HANS LENK

SEE ALSO Development Ethics; Environmental Ethics; Life.

BIBLIOGRAPHY

Groos, H. (1974). Albert Schweitzer—Größe und Grenzen [Albert Schweitzer: greatness and limits]. Munich, Basel: E. Reinhardt.

Günzler, Claus. (1996). Albert Schweitzer: Einführung in sein Denken [Albert Schweitzer: introduction to his thought]. Munich: C. H. Beck.

Lenk, Hans. (2000). Albert Schweitzer: Ethik als konkrete Humanität [Albert Schweitzer: ethics as concrete humanity]. Münster, Germany: LIT.

Meyer, Marvin, and Kurt Bergel, eds. (2002). *Reverence for Life: The Ethics of Albert Schweitzer for the Twenty-First Century.* Syracuse, NY: Syracuse University Press. Includes excerpts from Schweitzer on the concept of reverence for life, contributions from others assessing the concept, and a concluding section on its import for education.

Miller, David C., and James Pouilliard, eds. (1992). *The Relevance of Albert Schweitzer at the Dawn of the Twenty-First Century.* Lanham, MD: University Press of America. Especially relevant to science, technology, and ethics, with chapters on nuclear arms reduction, medicine, human rights, and the environment.

Schweitzer, Albert. (1949). *The Philosophy of Civilization,* trans. C. T. Campion. New York: Macmillan. His major philosophical treatise, composed of *The Decay and the Restoration of Civilization* and *Civilization and Ethics,* both originally published in 1923 as *Verfall und Wiederaufbau der Kultur* and *Kultur and Ethik.*

Schweitzer, Albert. (1958). *Peace or Atomic War?* New York: Henry Holt. An essay divided into three parts—nuclear tests, atomic war, and negotiations—originally delivered as three broadcasts from Oslo, Norway.

Schweitzer, Albert. (1965). *The Teaching of Reverence for Life,* trans. Richard Winston and Clara Winston. New York: Holt, Rinehart and Winston. Begins with a basic treatment of ethics and develops Schweitzer's ethic concerning personal relations, culture, and atomic weapons.

Schweitzer, Albert. (1971–1974). *Gesammelte Werke* [Collected works], ed. R. Grabs. 5 vols. Munich: C.H. Beck.

Schweitzer, Albert. (1990). *Out of My Life and Thought: An Autobiography,* trans. Antje Bultmann Lemke. Rev. edition. New York: Henry Holt. Originally published in 1931.

Schweitzer, Albert. (1995–2005). *Werke aus dem Nachlass* [Posthumous works], ed. Richard Brüllmann, Erich Gräßer, Claus Günzler, et al. 9 vols. Munich: C.H. Beck. Includes the main philosophical work "Die Weltanschauung der Ehrfurcht vor dem Leben."

Schweitzer, Albert. (1999–2000). *Kulturphilosophie III,* ed. Claus Günzler and Johann Zürcher. Munich: C. H. Beck.

Steffahn, Harald. (1979). *Albert Schweitzer.* Reinbek, Hamburg: Rowohlt.

SCIENCE AND ENGINEERING INDICATORS

• • •

Science and Engineering Indicators is a term referring to efforts to measure the pursuit, support, and performance of science and engineering on scales that geographically extend from the local to the international. Their goal is usually to help direct policy programs in research, education, and industrial support.. The most prominent and celebrated of these is Science and Engineering Indicators (referred from here on as *Indicators*) published every two years in the United States by the National Science Board (NSB). NSB is the body that oversees the budget and policies of the National Science Foundation (NSF) and the report itself is prepared by NSF's Science and Resources Directorate.

As an NSF publication, *Indicators* was conceived after Congress, in 1968, broadened the NSF Charter to include more engineering and social sciences in the agency's support portfolio. Legislators desired a sense of the impact government support for research was having on the "health" of the national research system, and NSF, which already had an active statistics branch, broadened its ambitions to large-scale endeavors.

The first *Indicators* report was issued in 1972 as simply "Science Indicators" and ever since it has been the worldwide standard reference and model for the statistical treatment of science, engineering, and technology. *Engineering* appeared in its name in 1986 when the NSF, under Congressional pressure, sharply raised its budget for engineering research and elevated its interest in supporting partnerships between U.S. universities and industry.

No mandate, however, was established for assessing the social and economic impact of science and engineering. Editors of *Indicators* have been conscious of and curious about returns on government research investment. But they believe the report is already extensive enough and that performance indicators that assess such outcomes are, and always were, imposingly difficult areas to measure. Quantified data will probably always constitute the core of the *Indicators* endeavor.

As the research system has grown and changed over the years, *Indicators* has evolved in style, content, and presentation. The 1976 edition, reflecting a relatively simple time in the measurement of science and technology for policy, contained chapters titled "International Indicators of Science and Technology," "Resources for Research and Development," "Resources for Basic Research, Industrial R&D and Innovation," "Science and Engineering Personnel," and "Public Attitudes toward Science and Technology."

By comparison, the more voluminous and finely rendered 2002 edition mirrored the rise of new technologies, the increasing globalization of science and technology, and the wider mingling of corporate, university and government interests. Its chapters included "Elementary and Secondary Education," "Higher Education Science and Engineering," "Science and Engineering Workforce," "Funding and Alliances in U.S. and International Research and Development," "Academic Research and Development," "Industry, Technology, and the Global Marketplace," "Public Attitudes and Public Understanding of Science and Technology," and a special chapter entitled "Significance of Information Technology." By the increasing specificity of the chapter titles it was becoming clear that the *Indicators* editors were being nudged toward treating the facts and figures of science and engineering as more than self-referential measures of the enterprise.

The 2004 edition extended the publication's reach by introducing a chapter on state-by-state research and development statistics, mainly to reflect the importance states place on science and engineering for their economic development. But as to actual state-by-state outcomes, *Indicators* once more begged off entering with any sense of resoluteness an area in which statistics are, to them, impossible to gather.

The era of the Internet has improved the currency and relevance of *Indicators*. NSF has taken advantage of Internet technology by continually updating the data in its interactive online version. Thus, readers can no longer object, as they would in the past, that the publication's data were too out of date to be useful. Their objection was a valid one for scholarship: Upon the date of publication, many of *Indicators* data were often more than a year out of date.

Identifying exactly what science, engineering, and technology ought to indicate is a subject that is without a consensus but is ripe for speculation, especially in the ethical dimensions of the technical universe. Its chapters draw conclusions and projections, but the publication largely leaves it to the readers to interpret what the numbers mean. One certainty is that *Indicators* confirms that science and technology have shown huge growth both in complexity and scope since the report was first issued, raising issues related to how scientific and technological change affect, and indeed can improve on, human life.

As an information tool for ethical studies of science and technology, the best that can be said is that *Indicators* offers mountains of data for the taking—levels of funding by field of study, patent activity by universities, size of university department, and so on. But if the ethical subject is conflict of interest by scientists in universities, for example, *Indicators* will provide enough data on the extent of private funding for academic research, but offer nothing in the way of, for example, numbers of universities that require their faculties to adhere to a code of behavior in dealings with industry. If the query is numbers of litigation cases between universities and corporations over intellectual property, again, *Indicators* fails the test.

But on balance, a point can be reached where too much is asked of a report that was always meant to be statistical. *Indicators* is widely praised, universally used, and admiringly emulated. The problem for users with an interest in ethics and the social sciences is that the publication does not address societal and economic outcomes, leaving the reader with the sense that science mainly looks inward while growing in size and importance worldwide. As for technological growth, the reader has no guidance for judging its relative social benefits.

Science and engineering are such powerful forces for change that their statistical treatment will continue to evolve. Very little systematic research, however, has been done to better reflect the vast ramifications of science and technology on society and economies, raising the issue of what *Indicators* is in fact supposed to indicate. The Organization of Economic Cooperation and Development in Paris, established after World War II, began such metrics as part of the post-war reconstruction of Europe. The work of that organization continues with its periodic reports on various fields of technology, and their social and economic importance. And, of course, other countries, as mentioned, confidently persist in attempting to measure the social impact of science and technology.

By 2005 every industrial country as well as the twenty-five-member European Union (EU) had issued its own science and engineering indicators. The EU, Japan, and most of the large but less developed countries such as Brazil, India, and China tended to stress the societal dimensions as well as the purely statistical treatment of science and technology. The popularity of *Indicators* seems to support the notion that science and technology are increasingly indispensable tools of economic progress and that countries more than ever feel the need to keep pace with one another.

WIL LEPKOWSKI

SEE ALSO *Education; Social Indicators.*

BIBLIOGRAPHY

Godin, Benoît. (2002). *Are Statistics Really Useful? Myths and Politics of Science and Technology Indicators.* Project on the History and Sociology of S&T Statistics, Paper no. 20, 39 pages. Quebec City, Canada: National Institute of Scientific Research, University of Quebec.

Godin, Benoît. (2003). *The Most Cherished Indicator: Gross Domestic Expenditure on R&D (GERD)*. Project on the History and Sociology of S&T Statistics: Paper no. 22, 25 pages. Quebec City, Canada: National Institute of Scientific Research, University of Quebec.

Science and Public Policy. (1992): 19(5 and 6). Special issue.

SCIENCE EDUCATION

SEE *Activist Science Education; Education.*

SCIENCE FICTION

• • •

From its beginnings as a literary genre science fiction has displayed ambivalence toward the ethical implications of scientific discovery and technological development. As a form of literature devoted in large part to evoking the potential futures and possible worlds engendered by mechanical innovation, science fiction (SF) has emerged over the last century as the preeminent site within Euro-American popular culture where the social consequences of modern technology may be explored creatively and interrogated critically.

As Brooks Landon has argued, SF "considers the impact of science and technology on humanity" by constructing "zones of possibility" where that impact can be represented and narratively extrapolated (Landon 1997, pp. 31, 17). Landon's understanding of the genre builds on James Gunn's definition of SF as the "literature of change," a mode of writing that investigates the outcome of technological progress at a level "greater than the individual or the community; often civilization or the race itself is in danger" (Gunn 1979, p. 1). This broad focus on the promises and perils of techno-scientific transformation requires a degree of concern, however implicit, for its moral repercussions, and the best SF has not shrunk from ethical engagement.

From *Frankenstein* to *Brave New World*

If, as several critics have argued, Mary Shelley's *Frankenstein, or the Modern Prometheus* (1816) was the first true SF novel, the genre's founding text provides a paradigm of moral ambivalence toward the processes and products of scientific inquiry. Driven by an urge to unlock the secrets of nature, Victor Frankenstein is at once the genre's first heroic visionary and its first mad scientist. Indeed, these roles are inseparable: Frankenstein's bold commitment to unfettered experimentation

makes him capable of both wondrous accomplishment—the creation of an artificial person endowed with superhuman strength and intelligence—and blinkered amorality. Unable to contain or control his creation, whose prodigious powers have been turned toward destructive ends, Frankenstein comes to fear that he has unleashed "a race of devils ... upon the earth, who might make the very existence of the species of man a condition precarious and full of terror" (Shelley 1982, p. 163). *Frankenstein*, through its many cinematic incarnations, has bequeathed to contemporary popular culture an enduring myth of science as an epochal threat for humanity and a source of moral corruption.

Throughout the nineteenth century the maturing genre continued to manifest that dualistic response: on the one hand limning a world transformed by the relentless advance of modern science and industry and on the other hand depicting the corrosive effects of that transformation on traditional values and forms of life. Jules Verne's popular series of "Extraordinary Voyages," with their celebration of the wonders of technology, represented the former trend, whereas H.G. Wells's darker and more skeptical series of scientific romances, beginning with *The Time Machine* (1895), epitomized the latter response. Although Verne's *Twenty Thousand Leagues Under the Sea* (1870) contains a kind of mad scientist, Captain Nemo, he is more a misunderstood genius than a figure of Frankensteinian evil, and his futuristic submarine, the *Nautilus*, is more a marvel of invention than a lurking monster. That powerful machine may inspire fear, but this is the result of ignorance rather than intrinsic threat. By contrast, the eponymous character in Wells's *The Island of Dr. Moreau* (1896) is a power-mad fanatic whose creations, a horde of human-animal hybrids, clearly descend from Frankenstein's fiendish invention. Twisted parodies of natural forms, they point up the moral limitations of experimental science: Moreau's brilliance can mold a beast into a human semblance, but it cannot endow the result with virtue or a functioning conscience.

Emblematic though he may be of the ethical predicament of modern science, Dr. Moreau, like Victor Frankenstein, is just one man, and an isolated one, exiled on his island. In the twentieth century SF began to explore the possibility that individual overreaching might be generalized, wedding scientific novelty with industrial mass production to generate in the ironic title of Aldous Huxley's novel, a *Brave New World* (1932). Huxley's satirical vision of a future in which babies are grown in vats and emotions are managed technocratically by drugs and the mass media offers a wide-ranging

Scene from the 1954 science-fiction film "Gog." The human-vs.-robots theme is common in science fiction. *(The Kobal Collection.)*

indictment of a regimented society from which morality has been purged in favor of a coldly instrumental scientism. A triumph of scientific and social engineering, filled with technological marvels, that false utopia is ethically atrophied and spiritually void. Huxley's depiction of the dystopian implications of techno-scientific development in the capitalist west were echoed in Yvgeny Zamiatin's *We* (1924), which projected a future socialist Russia dominated by a grim totalitarianism. Though capable of tremendous feats of industrial engineering, this regime dehumanized its citizens, ruthlessly suppressing their artistic impulses, their sexual drives, and their moral aspirations.

A similar vision of simultaneous technological achievement and moral impoverishment is offered in Karel Čapek's *R.U.R.* (1920). That popular play coined the term *robots* to describe the mass-produced workers who, like Frankenstein's monster, finally rebel against their creators in an orgy of destruction. Čapek's robots, like the test-tube babies in Huxley's novel, are actually synthetic humans rather than the clanking machines their name implies. More conventional mechanical creatures figure in SF texts of the 1920s and 1930s, the most famous being the humanoid robot in Fritz Lang's *Metropolis* (1927), a sinister automaton used to manipulate and control the masses. In all its varieties the artificial person, following in the wake of *Frankenstein*, continued to provide a potent icon of moral ambivalence within the genre: Physically and intellectually superior creatures that symbolize at once the titanic capacities of modern technology and the potential perfectibility of humanity, they are ultimately soulless, wholly lacking in moral will.

An American Affirmation

Not all SF produced during that period was equally pessimistic, however. In the United States a more technophilic strain developed, associated with popular pulp magazines whose titles—*Amazing, Astounding, Won-*

der—suggest their wide-eyed enthusiasm for technological innovation. However, despite the celebratory tone of much of that material, a more cautionary note sometimes was sounded; indeed, the best pulp SF carried forward the ambivalence toward the moral implications of scientific progress that the European tradition had pioneered.

This attitude is especially visible in pulp SF depictions of artificial persons, such as Isaac Asimov's influential series of robot stories, published during the 1930s and 1940s and eventually gathered into his book *I, Robot* (1950). A large part of Asimov's purpose in the series is to overcome popular anxieties about mechanical beings as uncontrollable Frankenstein's monsters; to this end he develops an ethical code—"The Three Laws of Robotics"—that, hardwired into his robots' brains, ensures their virtuous behavior as protectors and servants of humanity. However, much of the narrative suspense of the stories lies in the various contraventions of the laws, with disobedient robots taking advantage of conflicts within the moral norms governing their operation. Clearly, if left to their own devices (i.e., if not programmed with ethical precepts), the robots would, as in Čapek's play, turn against humanity or at least refuse to accept their own servile status. Another pulp writer, Jack Williamson, pursued the logic of Asimov's Three Laws as moral safeguards to their *reductio ad absurdam* in his story "With Folded Hands" (1947), in which robots take their charge of protecting human beings from harm so seriously that they prohibit all risk taking, mandating comfort and safety through a regime of moralistic totalitarianism.

Still, within American pulp SF these moments of doubt about the ethical consequences of technological advancement were far outweighed by a resolutely affirmative vision of the overall role of science in reordering human life. John W. Campbell, Jr., who became the editor of *Astounding* in 1937 and presided over what has come to be known as SF's Golden Age in the subsequent decade, was famous for championing scientific literacy within the genre and embracing technocratic solutions to social problems. In the pages of *Astounding* and other SF pulps scientists and engineers emerged as an intellectual elite; as John Huntington has argued, a "myth of genius" (1989, p. 44) predominates, with readers encouraged to identify with superior, powerful technocrats whose expertise and pragmatic skill presumably transcend ethical doubts and hesitations. The writers most closely associated with this upbeat vision were Asimov, Robert A. Heinlein, and L. Sprague de Camp, all of whom were trained scientists.

In Heinlein's collection *The Man Who Sold the Moon* (1950) an entrepreneurial genius single-handedly pioneers space travel as a commercial venture, bypassing government control. The ethical-political complications surrounding this move into space are neatly evaded by associating moral questioning with bureaucratic inertia, a collective stagnation the confident capitalist transcends through bold individual action. De Camp's classic alternative-history novel *Lest Darkness Fall* (1941) contains a similar portrait of intrepid genius as a technologically adept time traveler from the twentieth century visits ancient Rome, deploying his expert knowledge to forestall the Dark Ages.

Such sweeping visions of techno-scientific accomplishment seemingly untroubled by ethical qualms were characteristic of much Golden Age SF, although, as Asimov's robot stories showed, a lurking anxiety about the potential perils of technological breakthrough could not be dispelled entirely.

The Return to Questions

That lingering subtext rose to the surface in American SF during the 1950s as the global repercussions of the atomic bombings that ended World War II began to be perceived fully. New SF magazines such as *Galaxy* and *Fantasy and Science Fiction* emerged as rivals to *Astounding*, and the stories they featured began to question, if not openly reject, Campbell's staunch commitment to the technocratic ideal. Although *Astounding* had published stories dealing with the coming dangers of atomic energy such as Lester Del Rey's tense novella "Nerves" (1942), which described an accident in a nuclear power plant, those tales generally had depicted enlightened engineers steadily learning to master the technology. After the horrors of Hiroshima and in the throes of a looming confrontation between rival superpowers armed with high-tech weapons, American SF began to doubt not only the moral competence of technocrats in their stewardship of the atomic age but also the very capacity of humanity to avert its self-destruction.

Still, as Paul Brians has argued, science seldom was blamed for that awful crisis: "Many science fiction writers understood that the power of the new weapon threatened civilization and perhaps human survival, but they placed the responsibility for the coming holocaust on the shoulders of politicians or military men and argued that science still provided humanity's best hope for the future" (Brians 1987, p. 29).

Nonetheless, by showing the likelihood as well as the catastrophic effects of global war, tales of nuclear holocaust strongly suggested that humans lacked the ethical resources needed to control this powerful new technology. For example, Judith Merril's novel *Shadow on the Hearth* (1950) focuses on the personal costs of atomic devastation for one typical American family, whose moral strength, although admirable, is insufficient in the face of a breakdown of civilized order. On a broader scale *A Canticle for Leibowitz* (1960) by Walter M. Miller, Jr., depicts a postholocaust culture governed by a Catholic Church unable to forestall, because of to the inherent sinfulness of human nature, a cyclical repetition of nuclear disaster.

At the same time such stories were appearing popular SF films began to deal with the nuclear menace, offering a series of alarmist portraits of the imagined effects of atomic radiation that ranged from giant mutant insects (e.g., *Them* [1954]) to *The Incredible Shrinking Man* (1957). Even the most optimistic cinematic handling of the postwar atomic threat, *The Day the Earth Stood Still* (1951), in which an alien representative of a cosmic civilization intervenes to prevent global war, suggests that human beings, if left to their own devices, are not fit to govern their planet or themselves.

During the 1960s and 1970s that downbeat attitude, in which humanity's technological reach is seen to escape its moral grasp, gained strength as a new generation of writers began to challenge the technophilia of their pulp forebears. The technocratic legacy of Campbell was interrogated skeptically, and in some cases definitively rejected, by what came to be known as SF's New Wave, a loosely affiliated cohort of authors, many writing for the British magazine *New Worlds*, who began to question if not the core values of scientific inquiry the larger social processes to which they had been conjoined in the service of state and corporate power. New Wave SF arraigned technocracy from a perspective influenced by the counterculture discourses of that period, such as student activism, second-wave feminism, anticolonial struggles, and ecological causes and in the process developed a more radical ethical-political agenda—as well as a more sophisticated aesthetic approach—than the genre had featured previously. As a result the New Wave established a crucial benchmark for modern SF's engagement with the serious moral issues surrounding science and technology.

New Wave stories with feminist, ecological, or antiwar agendas were often dire in their predictions of future developments, but their critiques of technocracy were guided by implicit ethics of gender equity, natural balance, and nonviolence. Often those different agendas were wedded, as in Ursula K. Le Guin's short novel *The Word for World Is Forest* (1976), in which the brutal military occupation of another planet directly involves the devastation of its physical environment by hypermacho men, and Thomas M. Disch's *Camp Concentration* (1968), which explores the roots of high-tech warfare in the flaws and insecurities of masculinity. The work of Alice Sheldon, most of it published under the pseudonym James Tiptree, Jr., also probes the nexus of gender hierarchy and militarist and ecological violence, seeming at times to endorse a despairing sociobiological vision in which male sexuality expresses itself through technologically augmented aggression.

The New Wave's ethical idealism thus often was tempered by pessimism, a grim assessment of the dystopian futures portended by out-of-control technology. A key New Wave theme involved the extrapolation of contemporary urban problems to hypertrophied extremes as humans find themselves immured in vast concrete prisons of their own making. Novels such as David R. Bunch's *Moderan* (1971) and Robert Silverberg's *The World Inside* (1971) present such grim portraits of claustrophobic environments that they verge on the Gothic: In these texts the universal triumph of technology predicted and celebrated in Golden Age SF has culminated in a brutal cityscape where beleaguered, stunted spirits struggle to preserve the tattered shreds of conscience and dignity. In the work of the British author J. G. Ballard the modern city emerges as a psychic disaster area. His controversial 1973 novel *Crash*, for example, depicts a denatured humanity bleakly coupling with machines, with the enveloping landscape of metal and concrete having unleashed a perverse eroticism that seeks fulfillment in violent auto wrecks. SF films of that period, such as *THX 1138* (1971), contained similarly harsh indictments of regimented megalopolises that have co-opted or paralyzed ethical judgment.

The Future of Humankind

Long-standing anxieties regarding high technology were amplified during that period by the new science of cybernetics, which claimed that no meaningful distinctions could be drawn between humans and complex machines. The emergence of so-called artificial intelligence posed a challenge to humanity's presumed supremacy, and SF took up that challenge largely by emphasizing the moral superiority of human beings over their intellectually advanced creations. Ernst Jünger's *The Glass Bees* (1957), for example, derives its satirical power from a pointed contrast between the eponymous robots, who dutifully pursue their assigned tasks, and the skeptical narrator, whose ethical questioning suggests a cognitive and spiritual autonomy denied to mere machines, however skillful or complex.

The work of the British author Arthur C. Clarke, such as his story "The Nine Billion Names of God" (1953), had long engaged the possibility that humanity might have spawned its betters in the form of powerful information machines. In 1969 Clarke collaborated with the director Stanley Kubrick to produce the popular film *2001: A Space Odyssey*, in which a sentient computer, the HAL 9000, displays at once its cognitive power and its ethical limitations, conspiring to take over an interplanetary mission, only to be foiled by human pluck and ingenuity. *2001* established a cinematic trend in which the supercomputer emerged as an instrument driven by an urge to domination, as in *Colossus: The Forbin Project* (1970).

If computers threatened to supplant human mental functions, sophisticated new forms of artificial persons seemed poised to replace humanity entirely. Philip K. Dick's novel *Do Androids Dream of Electric Sheep?* (1968) deals with this imminent danger as its policeman protagonist hunts down a group of renegade androids, synthetic duplicates that are indistinguishable on the surface from normal people. However, there is a crucial difference, and it is essentially an ethical one: Androids are incapable of genuine empathy for others. The moral quandary in the novel is that humans are seldom empathetic; moreover, the protagonist's job requires that he be efficient and ruthless—"something merciless that carried a printed list and a gun, that moved machinelike through the flat, bureaucratic job of killing" (Dick 1996, p. 158)—making him as coldly unfeeling as the androids he seeks to slay. Thus, even when a bright moral line seems to distinguish humans from machines, a technocratically regimented social system serves to obscure if not efface it.

Androids was filmed by Ridley Scott as *Blade Runner* (1982), a film that effectively captures the novel's morally ambiguous tone while pointing forward to subsequent "cyberpunk" treatments. The movie's bleak urban milieu, populated by cynical humans and idealistic machines, offers essentially the same fraught moral landscape that would be featured in novels such as William Gibson's *Neuromancer* (1984), in which artificial intelligences and other cybernetic entities seem more deeply invested with values such as freedom and autonomy than do the human characters.

Cyberpunk fictions of the 1980s and 1990s by Gibson, Bruce Sterling, Pat Cadigan, and others brought to a potent climax the trend toward ethical ambivalence that has marked SF's engagement with new technologies. Extrapolating the social futures portended by the proliferation of computers and their spin-off appliances, cyberpunk displays a humanity so morally compromised by high-tech interfaces—including powerful "wetware," machinic implants that radically alter the body and mind—that the capacity for ethical judgment has perhaps been lost. Yet even amid this spiritual collapse cyberpunk's antiheroes manage to salvage scraps of the decaying moral order, as occurs when the protagonist of *Neuromancer* refuses the quasisatanic lure of cybernetic immortality, affirming the finitude of the mortal self as an enduring ethical center, preserved somehow against the sweetest blandishments and the sternest threats of technology.

For nearly 200 years science fiction has provided windows onto futures transformed by modern science and technology. In that process it has shown both the resiliency and the limitations of ethical consciousness in confronting these potentially overwhelming changes.

ROB LATHAM

SEE ALSO *Asimov, Isaac; Brave New World; Frankenstein; Huxley, Aldous; Science, Technology, and Literature; Utopia and Dystopia; Zamyatin, Yevgeny Ivanovich.*

BIBLIOGRAPHY

Asimov, Isaac. (1950). *I, Robot*. New York: Gnome Press.

Ballard, J. G. (1973). *Crash*. London: Jonathan Cape.

Brians, Paul. (1987). *Nuclear Holocausts: Atomic War in Fiction, 1895–1984*. Kent, OH: Kent State University Press.

Bunch, David R. (1971). *Moderan*. New York: Avon.

Čapek, Karel. (1923 [1920]). *R.U.R.: A Fantastic Melodrama*. New York: Doubleday.

Clarke, Arthur C. (1953). "The Nine Billion Names of God." In *Star Science Fiction Stories*, ed. Frederik Pohl. New York: Ballantine.

De Camp, L. Sprague. (1941). *Lest Darkness Fall*. New York: Holt.

Del Rey, Lester. (1942). "Nerves." *Astounding Science Fiction* 30, no. 1 (September 1942): 54–90.

Dick, Philip K. (1996 [1968]). *Do Androids Dream of Electric Sheep?* New York: Del Rey.

Disch, Thomas M. (1968). *Camp Concentration*. London: Hart-Davis.

Gibson, William. (1984). *Neuromancer*. New York: Ace.

Gunn, James, ed. (1979). *The Road to Science Fiction #2: From Wells to Heinlein*. New York: Mentor.

Heinlein, Robert. (1950). *The Man Who Sold the Moon*. Chicago: Shasta.

Huntington, John. (1989). *Rationalizing Genius: Ideological Strategies in the Classic American Science Fiction Short Story*. New Brunswick, NJ: Rutgers University Press.

Huxley, Aldous. (1932). *Brave New World*. New York: Doubleday.

Jünger, Ernst. (1960 [1957]). *The Glass Bees*. Translated by Louise Bogan and Elizabeth Mayer. New York: Noonday.

Landon, Brooks. (1997). *Science Fiction after 1900: From the Steam Man to the Stars*. New York: Twayne.

Le Guin, Ursula K. (1976). *The Word for World Is Forest*. New York: Berkley.

Merril, Judith. (1950). *Shadow on the Hearth*. New York: Doubleday.

Miller, Walter M., Jr. (1960). *A Canticle for Leibowitz*. New York: Lippincott.

Shelley, Mary. (1982). *Frankenstein, or the Modern Prometheus*. Chicago: University of Chicago Press. Originally published in 1816 by Lackington, Allan and Company, London.

Silverberg, Robert. (1971). *The World Inside*. New York: Doubleday.

Verne, Jules. (1870). *Twenty Thousand Leagues Under the Sea*. Paris: Hetzel.

Wells, H.G. (1896). *The Island of Dr. Moreau*. London: Heineman.

Williamson, Jack. "With Folded Hands." *Astounding Science Fiction* 36, no. 2 (July 1947): 6–45.

Zamiatin, Evgeny. (1924). *We*. Translated by Gregory Zilboorg. New York: Dutton.

FILM

Blade Runner. (1982). Directed by Ridley Scott. Distributed by Columbia Tristar Pictures.

Colossus: The Forbin Project. (1970). Directed by Joseph Sargent. Distributed by Universal Pictures.

The Day the Earth Stood Still. (1951). Directed by Robert Wise. Distributed by Twentieth Century Fox.

The Incredible Shrinking Man. (1957). Directed by Jack Arnold. Distributed by Universal Pictures.

Metropolis. (1927). Directed by Fritz Lang. Distributed by Universum Film, A.G.

Them. (1954). Directed by Gordon Douglas. Distributed by Warner Brothers.

THX-1138. (1971). Directed by George Lucas. Distributed by Warner Brothers.

2001: A Space Odyssey. (1969). Directed by Stanley Kubrick. Distributed by Metro-Goldwyn-Mayer.

SCIENCE LITERACY

SEE *Public Understanding of Science*.

SCIENCE MUSEUMS

SEE *Museums of Science and Technology*.

SCIENCE: OVERVIEW

• • •

Science looms as large as any aspect of the contemporary world, with multiple moral and political engagements on its own as well as through its associations with technology. Both as a positive feature of the human world and as a phenomenon against which there are many reactions, science is a distinguishing feature of the contemporary ethical and political landscape. An overview of this landscape is facilitated by distinctions between science as a body of knowledge and as a human activity. As an activity science may be further examined as both a cognitive and a social process. Ethics is implicated in all three senses: knowledge, cognitive activity, and social process.

Body of Knowledge

In the public mind relations between science and ethics are commonly associated with the ethical and religious challenges from certain types of scientific knowledge— about the origins of life or the cosmos, about brain chemistry as the basis of mind, and more. But scientific knowledge can also be adopted to support received religious traditions and basic ethical assumptions—as when the Big Bang theory is interpreted as evidence of divine creation or quantum indeterminacy as the basis of free will.

RELIGIOUS ISSUES. Historically there have been persistent tensions between claims to revelation and knowledge acquired by natural means. During the Middle Ages Christian theology at one point sought to delimit Aristotelian natural science; specific propositions from Thomas Aquinas's effort to synthesize revelation and Aristotelian science were condemned by the bishop of Paris in 1277 (and not formally revoked until 1325). The trial of Galileo Galilei for his support of Copernican astronomy is another widely cited example. (The 1633 edict of the Inquisition was not formally revoked until 1992.) The 1925 trial of *Tennessee v. John Thomas Scopes* concerned with the teaching of Darwinian evolution in the public schools is yet another celebrated case, as is mentioned in an entry on its contemporary echo, the "Evolution–Creationism Debate."

Analyzing these and related cases scholars have distinguished a spectrum of possible interactions between science and religion, some focusing more on theological issues, others on ethics. No one has done more to parse these debates than the physicist and theologian Ian G. Barbour, winner of the 1999 Templeton Prize for Progress in Religion. According to Barbour (2000), there are at least four distinctive relations between science and religion: conflict, independence, dialogue, and integration. In a series of books published over a forty-year period, Barbour explores such relations across history, in different theological communities, and in diverse branches of science such as astronomy and cosmology, quantum physics, evolutionary biology, and genetics. At the same time, in contrast to evolutionary biologist Stephen J. Gould (1999) who argues for the independence of "non-overlapping magisterial (NOMA)" between science and religion, Barbour defends a relationship of dialogue and integration. The entry on "Christian Perspectives" makes further use of a version of this range of possibilities. Similar alternatives are also exemplified in entries on other religious traditions such as "Buddhist Perspectives" and "Jewish Perspectives."

ETHICAL ISSUES. As with religion, relations between scientific knowledge and ethics fall out into a number of different possible models: opposition (substantive ethical criticisms of science), separation (as in the fact/value dichotomy), reductionism (of ethics to science), and cooperation or partnership (in efforts to develop a scientific ethics or to use scientific knowledge to achieve ethical ends). A host of *Encyclopedia of Science, Technology, and Ethics* entries illustrate and deepen each of these models. Entries on particular branches of science, from "Astronomy" to "Psychology," tend to stress opportunities for syntheses. Entries on concepts such as "Determinism" and the "Fact/Value Dichotomy" highlight separations. Entries on "Evolutionary Ethics" and "Scientific Ethics" argue possibilities for basing ethics on science.

Increasing recognition within the scientific community of the importance of issues related to the human interpretation of scientific knowledge is reflected in the founding by the American Association for the Advancement of Science of a special Dialogue on Science, Ethics, and Religion, as described in the entry on the "American Association for the Advancement of Science." Substantive interpretations of the meaning of scientific knowledge remain an ongoing concern that has not been fully met by either scientific humanism, religious apologetics, or humanities reflection on the achievements of science—all of which are approaches represented in the present encyclopedia.

Cognitive Activity

Assessing science as a cognitive activity is the primary task of the philosophy of science and obviously overlaps with critical reflections on science as a body of knowledge. Yet in the philosophy of science the emphasis is less on the human or social meanings of scientific knowledge and more on examining the structure of such knowledge and analyzing its epistemological claims. Analyses of the structure of scientific knowledge involve three broad problem sets dealing with demarcation, confirmation, and explanation. How is scientific knowledge distinguished from pretensions to science (that is, pseudoscience) and other types of knowledge (using appeals to certainty, objectivity, reproducibility, predictive power)? What are the methods of scientific knowledge production (deduction, induction, verification, confirmation, falsification)? How do scientific explanations function (in their integration of observations, laws, and theories)?

With regard to epistemological claims, there are two major views of science: realism and instrumentalism. Realism argues that scientific propositions in some manner reflect the way the world really is, meaning they correspond to reality. By contrast, instrumentalism argues that scientific propositions are simply tools for explaining or manipulating phenomena. For the realist, the model of the atom provides a picture of what atoms actually look like. For the instrumentalist or antirealist, the differential equations used to predict the path of the Moon around Earth have no direct correspondence to the forces that actually move the Moon.

All basic philosophy of science texts cover these topic sets, as well as the debate between Thomas Kuhn and Karl Popper over the historical character of science that has been so prominent since the mid-1960s (see, e.g., the entries on "Kuhn, Thomas" and "Popper, Karl"). Increasingly there are also modest inclusions of arguments about values, especially the way gender bias may be operative in science. But in respect to values and ethics in science as a cognitive or knowledge-producing activity, it is discussions of fraud and misconduct in science, as covered by entries on "Scientific Integrity" and "Responsible Conduct of Research," that are most relevant. The most widely used introduction to these issues is the pamphlet *On Being a Scientist* (2nd edition, 1995), prepared by the U.S. National Academy of Sciences, National Academy of Engineering, and Institute of Medicine.

Social Process

Science is not only a cognitive activity but also a social process involving interactions on several levels from individual laboratories to academic disciplines and from corporations to national and international science policymaking organizations. Examination of these interactions has taken on increased importance as science has grown from a small community of practitioners to an abundant and widely dispersed "metropolis"—from small science to big technoscience. The focus of early modern philosophers, however, was on cognitive at the expense of social activities, and it was not until the 1930s that Robert Merton undertook to pursue the sociology of science.

According to Merton (as considered in the entry on "Merton, Robert"), science as a social institution rests on a normative structure that best flourishes in a democratic society because of a common ethos. Moreover, scientists ought to participate in the social order rather than pretend to a "sanguine isolationism." Indeed, World War II brought about a new era of increased participation by scientists in military and political affairs. Not only did this raise questions about their responsibility for the knowledge they produced and the products, processes, and systems such knowledge made possible, but it also posed dilemmas about the appropriate roles for scientists in political controversies. It was in the midst of such dilemmas that the "scientists' movement" (as described in Mitcham 2003) arose to help direct scientific developments toward particular ends.

Social disillusionment with science and technology in the 1960s and 1970s spurred the public understanding of science movement, which has made common cause with older traditions in the popularization of science. (See the entry on "Public Understanding of Science.") It was also related to developments in the history and philosophy of science. Against more rational reconstructionist arguments such as those of Popper, Kuhn argued that science does not progress toward reality or truth simply by the accretion of new discoveries. Rather scientific knowledge is best viewed as the product of a historically contingent group of practitioners operating from shared rules applied to a certain range of acceptable problems.

Though not his intention, Kuhn's work stimulated theories about the socially constructed nature of scientific knowledge, which in its strong form leads to relativism or antirealism, because scientific facts are deemed to be the result of network building and negotiating rather than approximating reality. But in its weak form the contextualization of science leads to the rather noncontroversial notion that knowledge is a product both of nature (a reality "out there") and human cultural and theoretical interests that condition particular trajectories of research. The move from internalist studies of science to contextual interpretations has given rise to interdisciplinary fields including science, technology, and society (STS) studies, the sociology of scientific

knowledge (SSK), and rhetoric of science, all of which challenge the Mertonian ideals as fully adequate descriptions of the real social processes in science. (For more details, see the entries on "Science, Technology, and Society Studies" and "Rhetoric of Science and Technology.")

A perennial theme of science as a social process is the extent to which planning the agenda of (especially publicly funded) scientific research to meet explicit social and economic goals is feasible or desirable. In the United Kingdom during the 1930s this debate flared between supporters of Michael Polanyi and those who backed J. D. Bernal. (The encyclopedia has entries on both men.) Polanyi argued that autonomy and self-governance by science was the best way to meet social goals, whereas Bernal held that autonomous science was inefficient and needed external guidance. The same debate occurred in the United States after World War II between Vannevar Bush and Senator Harley Kilgore regarding the appropriate relationship between science and the federal government during peacetime. (See the entry on "Bush, Vannevar," as well as that on "Science Policy.") At issue are the criteria by which to judge scientific success and whether they should be internalist (e.g., peer review) or some external measure based on societal concerns.

Pressure to increase the social and fiscal accountability of publicly funded science emerged at the end of the Cold War. Related developments included science shops in Europe and other efforts to democratize science. In the United States, examples included the Office of Technology Assessment, the Ethical, Legal, and Social Implications (ELSI) research as part of the Human Genome Project and federally funded nanotechnology research, and the "broader impacts" criterion implemented by the National Science Foundation in 1997. (Further discussion can be found in entries on "Human Genome Organization," "Science Shops," "U.S. National Science Foundation," and related entries.)

Many of these developments are reactions to the fact that scientific research, despite its numerous benefits, does not yield unmitigated goods. Health and environmental risks as well as escalating arms races are familiar unintended consequences. Additionally, scientific knowledge can complicate decision making without always improving it, and has made its own share of mistakes with regard to recommendations of public interest. But the possibility of new "subversive truths" from genomic research, uncharacterized risks from nanotechnology, and the global threat of terrorism all raise the stakes of seeking new knowledge and crafting arrangements for directing it toward common goods.

Assessment

Throughout discussions of the relationship between science and ethics one core issue that remains is the proper extent and nature of scientific autonomy. David H. Guston (2000) has identified four reasons why science is often defended as special, each of which requires a degree of autonomy for its protection. Epistemological specialness refers to the notion that science searches for objective truth. Sociological specialness is the claim that science has a unique normative order that provides for self-governance. Platonic specialness refers to its esoteric, technical nature far removed from the knowledge of common citizens. Economic specialness is the claim that investments in science are crucial for productivity.

In each case there is some truth to the claims of specialness, which require the recognition of science as a unique enterprise needing some degree of separation from other social activities to ensure its smooth functioning. But as scientists as diverse as the physicist Alvin M. Weinberg (1967) and the geologist Daniel Sarewitz (1996) have argued, none of these cases should be taken as a license for absolute autonomy. Indeed the big science of the twenty-first century is so dependent on corporate and public investments that isolation is not a real option. More fundamentally, scientific knowledge is just one good to be considered among many competing goods. The ambiguity about the right level of autonomy has led to several interpretations about the proper role of science in society within various contexts, as well as criticisms of the ways in which scientific disciplines sometimes reinforce the self-perpetuating pursuit of new knowledge in the form of what Daniel Callahan (2003) has criticized as a "research imperative."

CARL MITCHAM
ADAM BRIGGLE

SEE ALSO Ethics: Overview; Evolution-Creationism Debate; Expertise; Governance of Science; Humanization and Dehumanization; Technology: Overview; Unintended Consequences.

BIBLIOGRAPHY

Balashov, Yuri, and Alex Rosenberg, eds. (2002). *Philosophy of Science: Contemporary Readings*. London: Routledge. Twenty-nine readings divided into six parts, the last of which considers science as a historical and social process.

Barbour, Ian G. (1966). *Issues in Science and Religion*. Englewood Cliffs, NJ: Prentice Hall.

Barbour, Ian G. (1974). *Myths, Models, and Paradigms: A Comparative Study in Science and Religion*. New York: Harper and Row.

Barbour, Ian G. (2000). *When Science Meets Religion: Enemies, Strangers, or Partners?* San Francisco: HarperSanFrancisco. This is a revised and (in the best sense) popular summary of the basic arguments from his Gifford Lectures volume, *Religion in an Age of Science* (San Francisco: Harper and Row, 1990), which was previously revised and expanded as *Religion and Science: Historical and Contemporary Issues* (San Francisco: HarperSanFrancisco, 1997).

Callahan, Daniel. (2003). *What Price Better Health? Hazards of the Research Imperative*. Berkeley and Los Angeles: University of California Press; New York: Milbank Memorial Fund.

Collins, Harry, and Trevor Pinch. (1998). *The Golem: What You Should Know about Science*, 2nd edition. Cambridge, UK: Cambridge University Press. Seven case studies arguing the social construction of scientific knowledge.

Curd, Martin, and J. A. Cover, eds. (1998). *Philosophy of Science: The Central Issues*. New York: Norton. A comprehensive anthology with forty-nine selections divided into nine sections.

Gould, Stephen J. (1999). *Rocks of Ages: Science and Religion in the Fullness of Life*. New York: Ballentine.

Guston, David H. (2000). *Between Politics and Science: Assuring the Integrity and Productivity of Research*. Cambridge, UK: Cambridge University Press.

Kitcher, Philip. (2001). *Science, Truth, and Democracy*. Oxford: Oxford University Press.

Klemke, E. D.; Robert Hollinger; David Wÿss Rudge; and A. David Kline, eds. (1998). *Introductory Readings in the Philosophy of Science*, 3rd edition. Amherst, NY: Prometheus Books. Thirty-three readings divided into six parts, the last of which deals with "Science and Values."

Mitcham, Carl. (2003). "Professional Idealism among Scientists and Engineers: A Neglected Tradition in STS Studies." *Technology in Society* 25(2): 249–262.

Newton-Smith, W. H., ed. (2000). *A Companion to the Philosophy of Science*. Malden, MA: Blackwell. Eighty-one articles covering key concepts and philosophers, with one entry on "Values in Science."

Rosenberg, Alex. (2000). *The Philosophy of Science: A Contemporary Introduction*. London: Routledge. An introductory monograph.

Salmon, Merrilee H.; John Earman; Clark Glymour; et al. (1992). *Introduction to the Philosophy of Science*. Englewood Cliffs, NJ: Prentice Hall.

Sarewitz, Daniel. (1996). *Frontiers of Illusion: Science, Technology, and the Politics of Progress*. Philadelphia: Temple University Press.

U.S. National Academy of Sciences; National Academy of Engineering; and Institute of Medicine. Committee on Science, Engineering, and Public Policy. (1995). *On Being a Scientist: Responsible Conduct in Research*, 2nd edition. Washington, DC: National Academy Press.

Weinberg, Alvin M. (1967). "The Choices of Big Science." Chap. 3 in *Reflections on Big Science*. Cambridge, MA: MIT Press.

SCIENCE POLICY

• • •

Science policy involves considerations of two fundamental human activities: science and policy. People make decisions in pursuit of valued outcomes, so thinking about science policy necessarily implicates science, its close associate science-based technology, and ethics. Although science policy is a topic central to all societies, particularly developed countries that devote significant public resources to science, for two reasons the focus here is on the United States. First, the United States is responsible for the largest share of global spending on science and technology. Second, for better or worse, the budgetary leadership role of the United States in science and technology since World War II has shaped how people around the world think about science, policy, and politics.

To place United States science and technology expenditures into context, consider that according to the Organisation for Economic Co-operation and Development (OECD) in 2003 the United States provided 38 percent of the approximately $740 billion world total (public and private) investment in research and development. The next largest funders were Japan with 15 percent, China with 8 percent, and Germany with 7 percent of the world total. Measured as a fraction of national economic activities, in 2001 total (public and private) expenditures on research and development varied from more than 4 percent in Sweden to 1.93 percent for the European Union (EU) to 2.82 percent in the United States. No country invests more than 1 percent of public funds in research and development, with Sweden investing 0.90 percent, the EU 0.65 percent, and the United States 0.81 percent.

Of course science policy is more than science budgets. The institutional structures and purposes of science are also issues of science policy. If science refers to the systematic pursuit of knowledge, and policy refers to a particular type of decision making, then the phrase science policy involves all decision making related to the systematic pursuit of knowledge. Harvey Brooks (1964) characterized this relation as twofold: *Science for policy* refers to the use of knowledge to facilitate or improve decision making; *policy for science* refers to decision making about how to fund or structure the systematic pursuit of knowledge.

Brooks's characterization of science policy as including both policy for science and science for policy has shaped thinking about science policy ever since, reinforcing a perception that science and policy are separate activities subject to multiple relations. But while Brooks's distinction has proved useful, reality is more complex, because the way society views science policy itself shapes the sorts of questions that arise in science policy debates. Science for policy and policy for science are each activities that shape the other—in academic jargon they are *coproduced*. Policy for science decisions about the structure, functions, and priorities of science directly influence the kind of science that will be available in science for policy applications, and the ways science is used in policy formation will influence in turn the policies formulated for science. Policy for science and science for policy are subsets of what might be more accurately described as a *policy for science for policy* (Pielke and Betsill 1997). To the extent that thinking about science policy separates decisions about knowledge from the role of knowledge in decision making, it reinforces a practical separation of science from policy.

From such a perspective, David Guston (2000) has argued the need to develop a new language to talk about science policy, one that recognizes how science and policy are in important respects inextricably intertwined; separation is impossible. Instead, however, the artificial separation of science from policy is frequently reinforced with calls for a new *social contract* between science and society. As Guston notes, "Based on a misapprehension of the recent history of science policy and on a failed model of the interaction between politics and science, such evocations insist on a pious rededication of the polity to science, a numbing rearticulation of the rationale for the public support of research, or an obscurantist resystemization of research nomenclature" (Guston 2000 Internet site)

The present analysis of science policy in the United States, with a particular focus on federally-funded science, thus begins by examining the value structure that underlies science and its relationship to decision making, and focuses on how science and policy have come to be viewed as separate enterprises in need of connection. This will set the stage for a discussion of an ongoing revolution in science policy that challenges conventional understandings of science in society. In the early years of the twenty-first century it is unclear how this revolution will play out. But a few trends seem well established. First, the science policies that have shaped thinking and action over the past fifty years are unlikely to continue for the next fifty years. Second, decision makers and society more generally have elevated expectations about the role that science ought to play in contributing to the challenges facing the world. Third, the scientific community nevertheless struggles to manage and meet these expectations. Together these trends suggest that more than ever society needs systematic thinking about science policy—that is research on science policy itself. And such research should center on issues of ethics and values.

Axiology of Science

A value structure is part of any culture, and the culture of science is no different. Alvin Weinberg (1970) suggests four explicitly normative *axiological attitudes*—statements of value—which scientists hold about their profession. Whereas Weinberg's concern was the physical sciences, such perspectives are broadly applicable to all aspects of science:

- Pure is better than applied.
- General is better than particular.
- Search is better than codification.
- Paradigm breaking is better than spectroscopy.

For Weinberg, these attitudes are "so deeply a part of the scientist's prejudices as hardly to be recognized as implying" a theory of value (Weinberg 1970, p. 613). But these values are critical factors for understanding both thinking about and the practice of science policy in the United States. And understanding why science policy is currently undergoing dramatic change requires an understanding of how Weinberg's theory of value, if not breaking down, is currently being challenged by an alternative axiology of science.

Understanding the contemporary context of science in the United States requires a brief sojourn into the history of science. In the latter part of the 1800s, scientists began to resent "dependence on values extraneous to science," (Daniels 1967, p. 1699) in what has been called "the rise of the pure science ideal" (Daniels 1967, p. 1703). The period saw such resentment come to a head.

> The decade, in a word, witnessed the development, as a generally shared ideology, of the notion of science for science's sake. Science was no longer to be pursued as a means of solving some material problem or illustrating some Biblical text; it was to be pursued simply because the truth—which was what science was thought to be uniquely about—was lovely in itself, and because

it was praiseworthy to add what one could to the always developing cathedral of knowledge. (Daniels 1967, p. 1699)

Like many other groups during this era, the scientific community began to organize in ways that would facilitate making demands on the federal government for public resources. Science had become an interest group. Scientists who approached the federal government for support of research activities clashed with a federal government expressing the need for any such investments to be associated with practical benefits to society.

Expressing a value structure that goes back at least to Aristotle, U.S. scientists of the late-nineteenth century believed that the pursuit of knowledge associated with the pursuit of unfettered curiosity represented a higher calling than the development of tools and techniques associated with the use of knowledge. Hence, the phrase *pure research* came to refer to this higher calling with *purity* serving as a euphemism for the lack of attention to practical, real-world concerns (Daniels 1967). The first editorial published in *Science* magazine in 1883 clearly expressed a value structure:

> Research is none the less genuine, investigation none the less worthy, because the truth it discovers is utilizable for the benefit of mankind. Granting, even, that the discovery of truth for its own sake is a nobler pursuit. It may readily be conceded that the man who discovers nothing himself, but only applies to useful purposes the principle which others have discovered, stands upon a lower plane than the investigator (Editorial 1883, p. 1).

Some scientists of the period, including Thomas Henry Huxley and Louis Pasteur, resisted what they saw as a false distinction between *pure* and *applied* science (Huxley 1882, Stokes 1995). Some policy makers of the period also rejected such a distinction. For them, utility was the ultimate test of the value of science (Dupree 1957). The late 1800s saw different perspectives on the role of science and society coexisting simultaneously. But Weinberg's axiology of science emerged from the period as the value structure that would shape the further development of U.S. science policies in the first half of the twentieth century.

From Pure to Basic Research

In a well-documented transition, Weinberg's axiology of science stressed the primacy not so much of pure as of *basic* research. The term basic research was not in frequent use prior to the 1930s. But after World War II the concept became so fundamental to science policy that it is difficult to discuss the subject without invoking the corresponding axiology. The notion of basic research arose in parallel with both the growing significance of science in policy and the growing sophistication of scientists in politics. By the end of World War II and the detonation of the first nuclear weapons the acceleration of the development of science-based technology was inescapable. Throughout society science was recognized as a source of change and progress whose benefits, even if not always equally shared, were hard to dismiss.

The new context of science in society provided both opportunity and challenge. Members of the scientific community, often valuing the pursuit of pure science for itself alone, found themselves in a bind. The government valued science almost exclusively for the practical benefits that were somehow connected to research and development. Policymakers had little interest in funding science simply for the sake of knowledge production at a level desired by the scientific community, which itself had become considerably larger as a result of wartime investments. Support for pure research was unthinkable.

Congressional reticence to invest in pure science frustrated those in the scientific community who believed that, historically, advances in knowledge had been important, if not determining, factors in many practical advances. Therefore the scientific community began to develop a two-birds-with-one-stone argument to justify its desire to pursue truth and the demands of politics for practical benefits. The argument held that pure research was the basis for many practical benefits, but that those benefits (expected or realized) ought not to be the standard for evaluating scientific work. Because if practical benefits were used as the standard of scientific accountability under the U.S. system of government, then science could easily be steered away from its ideal—the pursuit of knowledge.

The scientific community took advantage of the window of opportunity presented by the demonstrable contributions of science to the war effort and successfully altered science policy perspectives. The effect was to replace the view held by most policymakers that science for knowledge's sake was of no use, and replaced it with the idea that *all research* could potentially lead to practical benefits. In the words of Vannevar Bush, the leading formulator of this postwar science policy perspective: "Statistically it is certain that important and highly useful discoveries will result from some fraction of the work undertaken [by pure scientists]; but the results of any one particular investigation cannot be predicted with accuracy" (Bush 1945, p. 81).

Central to this change in perspective was acceptance of the phrase basic research and, at least in policy and political settings, the gradual obsolescence of the term pure research. The term basic came without the pejorative notion associated with lack of purity imputed to practically focused work. More importantly, the term basic means in a dictionary-definition sense *fundamental, essential,* or *a starting point.* Research that was basic could easily be interpreted by a policymaker as being fundamental to practical benefits.

The Linear/Reservoir Model

Basic research would be connected to societal benefits through what has become frequently called the *linear model* of science. The linear model holds that basic research leads to applied research, which in turn leads to development and application (Pielke and Byerly 1998). To increase the output (that is, societal benefits) of the linear model, it is necessary to increase the input (support for science).

Bush's seminal report *Science—The Endless Frontier* (1945) "implied that in return for the privilege of receiving federal support, the researcher was obligated to produce and share knowledge freely to benefit—in mostly unspecified and long-term ways—the public good" (Office of Technology Assessment 1991, p. 4). One of the fundamental assumptions of postwar science policy is that science provides a reservoir or fund of knowledge that can be tapped and applied to national needs. According to Bush:

> The centers of basic research ... are the wellsprings of knowledge and understanding. As long as they are vigorous and healthy and their scientists are free to pursue the truth wherever it may lead, there will be a flow of new scientific knowledge to those who can apply it to practical problems in Government, in industry, or elsewhere. (Bush 1945, p. 12)

Implicit in Bush's metaphor is a linear model of the relationship between science and the rest of society: basic-applied-development-societal benefit. This model posits that societal benefits are to be found *downstream* from the reservoir of knowledge. Others have described the liner model as a *ladder,* an *assembly line,* and a *linked-chain* (Gomory 1990, Wise 1985, Kline 1985).

The linear/reservoir model is a metaphor explaining the relationship of science and technology to societal needs. It is used *descriptively* to explain how the relation actually works and *normatively* to argue how the relation ought to work. The linear model appears in discussions of both science policy, where it is used to describe the relation of research and societal needs (Brown 1992), and in technology policy, where it is used to describe the relation of research and innovation (Branscomb 1992). The linear model was based on assumptions of efficacy, and not comparisons with possible alternatives. In 1974 Congressman Emilio Daddario (D-CT), a member of the Science Committee of the U.S. House of Representatives (Science Committee), observed that members of Congress defer to the claims of scientists that basic research is fundamental to societal benefits "and *for that reason, if for no other*, they have supported basic research in the past" (Daddario 1974, p. 140; emphasis added). So long as policymakers and scientists felt that science was meeting social needs, the linear model was unquestioned.

The notion of basic research and the linear model of which it was a part has been tremendously successful from the standpoint of the values of the scientific community. Indeed the terms basic and applied have thus become fundamental to discussions of science and society. For example, the National Science Foundation (NSF) in its annual report *Science and Engineering Indicators* uses precisely these terms to structure its taxonomy of science. Not only did the basic-applied distinction present a compelling, utilitarian case for government support of the pursuit of knowledge, it also explicitly justified why pure research "deserves and requires special protection and specially assured support" (Bush 1945, p. 83). The special protections included relative autonomy from political control and standards of accountability determined through the internal criteria of science. In a classic piece, Michael Polanyi (1962) sketched in idealized fashion how a *republic of science* structured according to the values of pure science provides an *invisible hand* pushing scientific progress toward discovering knowledge which would have inevitable benefits for society.

Seeds of Conflict: Freedom versus Accountability

From the perspective of the scientific community, from the prewar to postwar periods, the concepts of pure research and basic research remained one and the same: the unfettered pursuit of knowledge. For the community of policymakers, however, there was an important distinction—pure research had little to do with practical benefits but basic research representing the "fund from which the practical applications of knowledge must be drawn" (Bush 1945, p. 19). From the perspective of policymakers, there was little reason to be concerned about science for the sale of knowledge alone; they had faith that just about all science would prove useful.

TABLE 1

Four Definitions of Basic Research

By product:	Basic research refers to those activities that *produce* new data and theories, representing an increase in our understanding and knowledge of nature generally rather than particularly (National Science Board 1996, Armstrong 1994).
By motive:	Basic research is conducted by an investigator with a *desire* to know and understand nature generally, to explain a wide range of observations, with no thought of practical application (National Science Board 1996).
By goal:	Basic research *aims* at greater knowledge and mastery of nature (White 1967, Bode 1964).
By standard of accountability:	Basic researchers are free to follow their own intellectual interests in order to gain a deeper understanding of nature, and are *accountable* to scientific peers (Polanyi 1962, Bozeman 1977).

SOURCE: Courtesy of Roger A. Pielke, Jr.

The different interpretations by scientists and policymakers of the meaning of the term basic research have always been somewhat troubling (Kidd 1959). A brief review of the use of the term basic research by the scientific community finds at least four interrelated definitions of the phrase, as summarized in Table 1.

From the standpoint of policymakers, basic research is defined through what it enables, rather than by any particular characteristic of the researcher or research process. These different interpretations of basic research by policymakers and scientists have coexisted largely unreconciled for much of the postwar era, even as for decades observers of science policy have documented the logical and practical inconsistencies. René Dubos (1961) identified a *schizophrenic attitude* among scientists, succinctly described as follows: "while scientists claim among themselves that their primary interest is in the conceptual aspects of their subject, they continue to publicly justify basic research by asserting that it always leads to 'useful' results" (Daniels 1967, p. 1700) It is this schizophrenia that has allowed postwar science policy to operate successfully under the paradigm of the linear model, apparently satisfying the ends of both scientists and politicians. *Basic research* was the term used to describe the work conducted in that overlap. The situation worked so long as both parties—society (patron) and scientists (recipient of funds)—were largely satisfied with the relationship.

The Changing Context

In the 1990s both scientists and politicians began to express dissatisfaction with the science policy of the post-World War II era. For instance, in 1998 the Science Committee undertook a major study of U.S. science policy under the following charge:

> The United States has been operating under a model developed by Vannevar Bush in his 1945 report to the President entitled *Science: The Endless Frontier*. It continues to operate under that model with little change. This approach served us very well during the Cold War, because Bush's science policy was predicated upon serving the military needs of our nation, ensuring national pride in our scientific and technological accomplishments, and developing a strong scientific, technological, and manufacturing enterprise that would serve us well not only in peace but also would be essential for this country in both the Cold War and potential hot wars. With the collapse of the Soviet Union, and the de facto end of the Cold War, the Vannevar Bush approach is no longer valid. (U.S. Congress 1998)

While the congressional report acknowledged the need for a new science policy, it did not address what that new policy might entail. However an understanding of the tensions leading to calls for change point in various directions.

These tensions have been long recognized. George Daniels (1967) sketches those underlying contemporary science policy: "The pure science ideal demands that science be as thoroughly separated from the political as it is from the religious or utilitarian. Democratic politics demands that no expenditure of public funds be separated from political ... accountability. With such diametrically opposed assumptions, a conflict is inevitable" (Daniels 1967, p. 1704) Such tensions were recognized even earlier, in 1960, by the Committee on Science in the Promotion of Human Welfare of the American Association for the Advancement of Science (AAAS): "Science is inseparably bound up with many troublesome questions of public policy. That science is more valued for these uses than for its fundamental purpose—the free inquiry into nature—leads to pressures which have begun to threaten the integrity of science itself" (AAAS 1960, p. 69). For many years under growing budgets in the context of the Cold War, postwar science policy successfully and parsimoniously evaded this conflict. Given pressures for accountability and more return on federal spending, conflict is unavoidable.

Why, more specifically, did postwar science policy remain largely unchallenged for a half century? From the point of view of society, it solved problems. First, science and technology were key contributors to victory in World War II. Infectious diseases were *conquered*.

Nuclear technology ended the war and promised power *too cheap to meter*. From the point of view of the scientific community, most good ideas received federal funding. The U.S. economy dominated the world. In such contexts, there was less pressure from the public and its representatives on scientists for demonstrable results; there was less accountability. Scientists, policymakers, and the broader public were largely satisfied with national science policies.

But at the beginning of the twenty-first century new challenges arose. Some infectious diseases rebounded through resistance to antibiotics, and new diseases, such as severe acute respiratory syndrome (SARS), threatened health. For many, the cost of healthcare made world-leading medical technologies unaffordable. The events of September 11, 2001, demonstrated the risks to modern society at the intersection of fanaticism and technology. The availability of weapons of mass destruction makes these risks even more significant. New technologies, in areas such as biotechnology and nanotechnology, created new opportunities but also threatened people and the environment. Many problems of the past have been solved, but new ones are emerging, and science and technology are often part of both the problem and possible solutions. The question of how to govern science and technology to realize their benefits is thus increasingly important.

In addition, many scientists were unhappy as budgets failed to keep pace with research opportunities: As the scientific community has grown and as knowledge has expanded, more research ideas are proposed than there is funding to support. Strong global competition and demands for political accountability create incentives for policymakers to support research with measurable payoffs on relatively short timescales, while within the scientific community competition for tenure and other forms of professional recognition demand rigorous, long-term fundamental research. As the context of science changes, scientists share anxieties with others disrupted by global economic and social changes.

New Science Policy Debates

While scientists perceive their abilities to conduct pure research constrained by increasing demands for practical benefits, policymakers simultaneously worry that basic research may not address practical needs. Insofar as postwar science policy has weakened, discussion of science policy has moved beyond the partial overlap of motives that helped sustain postwar science policy. Scientists now speak of their expectation of support for pure research, and policymakers increasingly ask for direct contributions to the solution of pressing social problems.

In this situation the differing views of scientist and policymaker can create conflict as the shared misunderstanding of the term basic research threatens to become pathological. In the words of Donald Stokes:

> The policy community easily hears requests for research funding as claims to entitlement to support for pure research by a scientific community that can sound like most other interest groups. Equally, the scientific community easily hears requests by the policy community for the conduct of "strategic research" as calls for a purely applied research that is narrowly targeted on short-term goals. (Stokes 1995, p. 26)

For their part, scientists seek to demonstrate the value of research to the public, often through increasing skill in public relations and contracting with consultants to provide cost-benefit studies that show the positive benefits of research investments. With few exceptions, the result of such concerns has not been constructive change, but rather defense of the status quo. In 1994 the National Research Council (NRC) convened scientists and informed members of the broader community to begin a constructive dialogue on the changing environment for science. The group found the public policy problem to be primarily the amount of federal funds devoted to research. A later National Academy report, *Allocating Federal Funds for Science and Technology* (1995), recommended that U.S. science should be at least world-class in all major fields, in effect recommending an entitlement for research. Similarly the 1998 "Science Policy Study" of the Science Committee similarly concluded, "The United States of America must maintain and improve its pre-eminent position in science and technology in order to advance human understanding of the universe and all it contains, and to improve the lives, health, and freedom of all peoples" (U.S. Congress 1998 Internet site)

Other approaches relate research and national needs. The Government Performance and Results Act of 1993 legislates formal accountability by requiring all government programs, including research, to quantitatively measure progress against established goals. Yet experience shows that asking for performance measures and actually developing and applying meaningful measures can be difficult. Daniel Sarewitz offers a penetrating critique of current policy and general steps that would pull research closer to society without sacrificing critical values of science. In particular he recommends research on research: "how it can be directed in a man-

ner most consistent with social and cultural norms and goals, and how it actually influences society" (Sarewitz 1996, p. 180). Donald Stokes (1995) resolves the dichotomy between research driven by purely scientific criteria and research responsive to societal needs by changing the single basic-versus-applied axis into a two-dimensional plane, with one dimension indicating the degree to which research is guided by a desire to understand nature, and the other indicating the degree it is guided by practical considerations. This conceptual advance demonstrates that *good science* can be compatible with practical application, but does not point to specific policy-relevant steps.

There is great potential for nations that have followed the Bush model, such as the United States, to learn from the experiences of those nations that have implemented differing science policies. What change will entail is not entirely clear, however, some trends are apparent. First, overall investments in science and technology show no signs of stagnation. If anything the world is investing more in science and technology, an amount that will in the near future exceed $1 trillion per year. These substantial investments are accompanied by increasing demands for accountability, relevance, and practicality. Such demands increasingly shape the context and practice of science in society. How science will shape and be shaped by these trends will undoubtedly mark a critical transition in science policy in the United States, and perhaps in the world.

ROGER A. PIELKE, JR.

SEE ALSO *Lasswell, Harold D.; Public Policy Centers; Social Theory of Science and Technology.*

BIBLIOGRAPHY

American Association for the Advancement of Science (AAAS). (1960). "Science and Human Welfare: The AAAS Committee on Science in the Promotion of Human Welfare States the Issues and Calls for Action." *Science* 132: 68–73.

Armstrong, John A. (1994). "Is Basic Research a Luxury Our Society Can No Longer Afford? *The Bridge* 24(2): 9–16.

Bode, Hendrik W. (1965). "Reflections on the Relation between Science and Technology." *Basic Research and National Goals: A Report to the Committee on Science and Astronautics, U.S. House of Representatives by the National Academy of Sciences*, pp. 71–76. Washington, DC.: Government Printing Office.

Bozeman, Barry. (1977). "A Governed Science and a Self-governing Science: The Conflicting Values of Autonomy and Accountability." Chapter 6 in *Science and Technology*

Policy, ed. Joseph Haberer. Lexington, MA: Lexington Books.

Branscomb, Lewis. (1992). "Does America Need a Technology Policy." *Harvard Business Review* 70: 24–31.

Brooks, Harvey. (1964). "The Scientific Advisor." In *Scientists and National Policymaking*, ed. Robert Gilpin and Christopher Wright. New York: Columbia University Press.

Brown, George E. (1992). "The Objectivity Crisis." *American Journal of Physics* 60(9): 776–863.

Bush, Vannavar. (1945). *Science—The Endless Frontier*. Washington, DC: Government Printing Office. Although its institutional recommendations were never fully adopted, this report is often referred to as setting the intellectual framework for thinking about science and its relationship to society for the post-World War II era.

Daddario, Emilio. (1974). "Science Policy: Relationships Are the Key." *Daedelus* 103: 135–142.

Daniels, George H. (1967). "The Pure-Science Ideal and Democratic Culture." *Science* 156: 1699–1706.

Dubos, René. (1961). "Scientist and Public: Why is the Scientist, Once a 'Natural Philosopher,' Now Considered a Barbarian by Many Educated Laymen?" *Science* 133: 1207–1211.

Dupree, A. Hunter. (1957). *Science in the Federal Government: A History of Policies and Activities to 1940*. Cambridge MA: Belknap Press of Harvard University Press.

"The Future of American Science." (1883). *Science* 1(1): 1–3. This is the first editorial that appeared in the journal *Science*, which in the early twenty-first century is one of the leading scientific journals in the world.

Gomory, Ralph. (1990). "Essay of Ladders, Cycles and Economic Growth." *Scientific American* June: 140.

Huxley, Thomas H. (1882). *Science and Culture, and Other Essays*. New York: D. Appleton and Company.

Kidd, Charles V. (1959). "Basic Research—Description versus Definition." *Science* 129: 368–371.

Kline, Stephen J. (1985). "Innovation Is Not a Linear Process." *Research Management* 28(4): 36–45.

National Science Board. (1996). *Science and Engineering Indicators*. Document NSB 96-21.Washington, DC: Government Printing Office. The U.S. National Science Board publishes data on a wide range of characteristics of science and engineering (e.g., funding, employment, disciplinary distributions of students, etc.) and as such provides an invaluable resource for characterizing the science and engineering communities and how they change over time.

Office of Technology Assessment. (1991). *Federally Funded Research: Decisions for a Decade*. OTA-SET-490. Washington, DC: Government Printing Office.

Pielke, Jr., Roger A., and Michele M. Betsill. (1997). "Policy for Science for Policy: Ozone Depletion and Acid Rain Revisited." *Research Policy* 26: 157–168.

Pielke, Jr., Roger A., and Radford Byerly, Jr. (1998). "Beyond Basic and Applied." *Physics Today* 51(2): 42–46.

Polanyi, Michael. (1962). "The Republic of Science." *Minerva* 1: 54–73. A classic piece that describes how the world would look if it operated under the norms of science.

Sarewitz, Daniel. (1996). *Frontiers of Illusion: Science, Technology and Politics of Progress*. Philadelphia: Temple University Press.

Stokes, Donald. (1995). "Sigma XI." In *Vannevar Bush II: Science for the 21st Century: Why Should Federal Dollars Be Spent to Support Scientific Research? Forum Proceedings, March 2–3*, ed. Kate Miller. Research Triangle Park, NC: Sigma XI Publications.

Stokes, Donald. (1997). *Pasteur's Quadrant: Basic Science and Technological Innovation*. Washington, DC: Brookings Institution Press. Offers a penetrating critique of post-World War II science policy and offers a clear conceptual alternative.

Weinberg, Alvin. M. (1970). "The Axiology of Science." *American Scientist* 58(6): 612–617.

White, Jr., Lynn. (1967). "The Historical Roots of Our Ecologic Crisis." *Science* 155: 103–1207.

Wise, George. (1985). "Science and Technology." *Osiris*, 2nd series 1: 229–246.

INTERNET RESOURCES

Guston, David H. (2000). "Retiring the Social Contract for Science." *Issues in Science and Technology Online* (Summer). A publication of the National Academy of Sciences, National Academy of Engineering, and Ida and Cecil Green Center for the Study of Science and Society. Available from http://www.nap.edu/issues/16.4/p_guston.htm.

U.S. Congress. (1998). Unlocking Our Future: Toward a New National Science Policy: A Report to Congress by the House Committee on Science, September 24, 1998. Available from http://www.house.gov/science/science_policy_report.htm.

SCIENCE SHOPS

• • •

Science shops provide independent, participatory research support in response to concerns experienced by civil society (Gnaiger and Martin 2001). Science in this context refers to all organized investigation, including the social and human sciences and arts, as well as the natural, physical, engineering, and technological sciences.

The concept of science shops was developed by students at universities in the Netherlands during the 1970s. This development was assisted by faculty and staff seeking to *democratize* the disciplinary hierarchies of the traditional university system. But arguably science shops are a manifestation of a movement stemming at least as far back as Thomas Jefferson's defense of the principle that "ideas should freely spread from one to another over the globe" (Jefferson 1813, Internet page).

The science shop concept spread worldwide in two waves. The first, in the late-1970s and early-1980s, was triggered by articles in *Nature* (Ades 1979) and *Science* (Dickson 1984) and led to initiatives in Australia, Austria, Belgium, Denmark, Northern Ireland, France, and Germany. The mid-1990s saw a resurgence based in large part on fast, inexpensive, and reliable communication technologies, such as the Internet. This growth led to new activities in England, Israel, South Korea, Malaysia, and New Zealand. Similar types of organizations have also been founded in Australia, Canada, South Africa and the United States but are referred to by other terms—Community-University Research Alliances, Community-based Research Centers, or Tecknikons.

There is significant variation in organizational structure among science shops, although three models dominate. The first is the university department model, where the science shop is attached to a disciplinary framework such as chemistry, biology, law, or physics. The second, most common model is the independent civil society organization, housing technical experts or brokering relationships with university or government researchers. The third model is the virtual alliance between partners in public, private, and not-for-profit sector institutions that jointly work on issues of mutual concern and benefit.

Despite differences in structure, Andrea Gnaiger and Eileen Martin point to six common elements found in all science shops. These include providing civil society with knowledge and skills through research and education; providing services on an affordable basis; promoting and supporting public access to and influence on science and technology; creating equitable and supportive partnerships with civil society organizations; enhancing understanding among policymakers and education and research institutions regarding the research and education needs of civil society; and enhancing the transferable skills and knowledge of students, community representatives, and researchers.

Science shops are closely associated with social justice, environmental, and community activist movements. The dominant research methodologies used include research mediation, participatory research, and participatory action research. The strengths of these approaches allow for the inclusion of the unique understanding of individuals and communities of their own local contexts, which helps establish causality of problems in a complex and diverse framework rather than in a reductionist manner. There is great adaptability and flexibility that allows for quick turnaround in problem identification and solving. The methods give people strong influence over both policy and practice at the local level. Local to global focus allows for scaling up of issues, providing grounded perspectives for national and international policies.

The principle weaknesses of the science shop methods are fourfold. Despite being a cost effective way of generating research, science shops suffer from chronic funding and resource shortfalls. With very few exceptions, unless funded through a philanthropic organization, government agency, or university, they spend almost as much effort on raising funds as they do performing research and advocacy work. Second, given their strong social justice tendencies, there appears to be institutional prejudice against working with corporations, governments, and intergovernmental agencies, or other organizations perceived to have a large *foot print*. This gap results in the absence of community partner and science shop perspectives in policy negotiations. Third, with the exception of the Netherlands, the lack of coordination among science shops and their relative absence from the dominant scientific communication streams means that there is a lack of comparability and a failure to generate commensurable information. This is currently being addressed by the creation of an International Science Shop Network, funded largely by the European Union. Finally, science shops have been accused of producing biased science, constructed to support the arguments of the clients they serve, a critique which is also aimed at scientists performing research for corporate clients. This criticism has been met by submitting research outputs to the same peer-review firewall that all scientific publication undergoes.

Science shops have proven to be an efficient and effective model for generating small-scale scientific and technological knowledge on issues of immediate and local concern. They provide a gateway for communities in gaining access to specialized data, information, and knowledge at a relatively low transaction cost. There are high residual effects within participating communities, leading to better understanding of science and technology as well as a critical capacity to assess the impact of scientific and technological issues on local social, economic, cultural, and environmental circumstances.

PETER LÉVESQUE

SEE ALSO *Global Climate Change; Governance of Science.*

BIBLIOGRAPHY

Ades, T. (1979). "Holland's Science Shops for 'Made To Measure' Research." *Nature* 281(18): B10.

Dickson, David. (1984). "Science Shops Flourish in Europe." *Science* 223: 1158–1160.

Gnaiger, Andrea, and Eileen Martin. (2001). *SCIPAS Report Nr.1, Science Shops: Operational Options.* Study financed by the European Commission DG XII Programme, "Improving the Human Research Potential and the Socio-Economic Knowledge Base (IHP), Strategic Analysis of Specific Political Issues (STRATA)" (HPV1-CT-1999-00001). Utrecht, Netherlands: Science Shop for Biology, Utrecht University.

Leydesdorff, Loet, and Peter Van Der Besselaar. (1987). "What We Have Learned from the Amsterdam Science Shop." In *The Social Direction of the Public Sciences, Sociology of the Sciences Yearbook*, Vol. 11, ed. Stuart Blume, Joske Bunders, Loet Leydesdorff, and R.D. Whitley. Dordrecht, Netherlands: Reidel.

Mulder, Henk; A.J. Thomas auf der Heyde; Ronen Goffer; and Carmen Teodosiu. (2001). *SCIPAS Report Nr. 2: Success and Failure in Starting Science Shops.* Study Financed by the European Commission-DG XII Programme, "Improving the Human Research Potential and the Socio-Economic Knowledge Base (IHP), Strategic Analysis of Specific Political Issues (STRATA)" (HPV1-CT-1999-00001). Utrecht, Netherlands: Science Shop for Biology, Utrecht University.

Sclove, Richard E. (1995). *Democracy and Technology.* New York: Guilford Press.

INTERNET RESOURCE

Jefferson, Thomas. (1743–1826). "Letter to Isaac Mcpherson Monticello, August 13, 1813" In *The Letters of Thomas Jefferson: 1743–1826: From Revolution to Reconstruction.* An .Html Project, Department of Alfa-Informatica, University of Groningen, The Netherlands. Available from http://odur.let.rug.nl/ ∽ usa/P/tj3/writings/brf/jefl220.htm.

SCIENCE, TECHNOLOGY, AND LAW

• • •

Law plays a growing critical role in the regulation of science and technology, including the ethical consequences of scientific research and new technologies. The relatively new field of *law, science, and technology* seeks to study systematically the diverse ways law interacts with science and technology. Law, science, and technology has been defined as "the discipline that deals with how our legal system can and must adjust to accommodate the problems created by the ever more urgent and ubiquitous impact of technology on society" (Wessel 1989, p. 260), and as seeking "to determine how the various processes of law—primarily judicial and legislative—respond to changes brought about by scientific advances" (Green 1990, p. 375).

Few law schools or legal scholars focused on the intersection of law with science and technology before the later part of the twentieth century. With advances in the computer, the Internet, biotechnology, genomics, telecommunications, and nanotechnology, technology has assumed an ever-increasing role in economic and daily life, and the law has struggled to keep pace. In the words of U.S. Supreme Court Justice Stephen Breyer, "[s]cientific issues [now] permeate the law" (Breyer 1998, p. 537). This has led to a proliferation in the study of law, science and technology interactions, including academic centers, textbooks (Sutton 2001, Areen et al. 1996), courses, specialized journals, conferences, and bar association sections (Merges 1988). There is also a growing awareness of the importance of scientific and technological developments by legal practitioners and scholars, with increased recognition among those outside the legal profession for the central importance of law in mediating the risks, benefits, and ethics of technology.

The field of law, science, and technology is premised on the belief that "[s]cience is a distinctive institution worthy of distinctive treatment by lawyers" (Goldberg 1986, p. 380). Despite increased awareness that science and technology present unique issues for the law, different formulations exist for examining law, science, and technology interactions. Here the field is divided into three primary strands. The first concerns the role of the law in managing the impacts of science and technology, including controlling the risks, promoting the benefits, and addressing ethical implications. The second concerns the institutions of law and science, examining how law affects the practice of scientific research, as well as the reciprocal relationship of how science and technology influence the law. The third involves a more generic inquiry into the problems and tensions that arise from the intersection of law with science and technology.

The Role of Law in Managing the Impacts of Science and Technology

Law plays a primary role in managing the impacts of science and technology. In the words of one prominent jurist, "[l]aw is the only tool that society has to tame and channel science and technology" (Markey 1984, p. 527). The impacts of science and technology that law seeks to manage can be subdivided into (a) risks, (b) benefits, and (c) ethical implications.

CONTROLLING RISKS OF NEW TECHNOLOGIES. New and existing technologies create many known and potential health, safety, environmental, and socioeconomic risks. Law is the principal societal institution for controlling these risks, through legislatures, regulators, and the judiciary (Jasanoff 1995). In developing such controls, the law relies on science to assess the relevant risks. Risk regulation thus involves two levels of science-law interactions: the role of law in regulating risks from science and technology; and the use of science by law to assess risk from new and existing technologies.

Legislation and regulation seek to address and reduce risks *ex ante*, before the risks are imposed. Most industrialized nations have comprehensive statutory and/or regulatory schemes in place to prospectively regulate potential risks from technologies such as pesticides, industrial chemicals, pharmaceuticals, natural resource extraction, genetically modified foods, and automobiles. *Ex ante* legislation and regulation by agencies statutorily empowered to do so presupposes the capability to adequately predict potential harms, a challenging undertaking for most risks. Indeed much of the complexity and controversy in *ex ante* risk regulation relates to uncertainties in the identification and quantification of potential risks. Nevertheless, given the preventive purpose of *ex ante* risk regulation, regulators are generally given considerable leeway in assessing risks, including the use of *conservative* (or plausible worst case) assumptions, requiring only substantial evidence and not necessarily the weight of evidence to support risk findings, and broad judicial deference to regulators' technical expertise.

One ongoing tension in *ex ante* regulation is the respective roles of legislators and regulators. The legislature in most jurisdictions has plenary power, and typically delegates to regulatory agencies the authority to regulate, subject to the substantive and procedural requirements included in the legislation. Regulatory agencies generally have greater technical expertise, available resources, and familiarity to address most risks associated with science and technology, and in that respect are the superior institution to make most risk regulatory decisions.

The legislature may take the lead when distrust between the legislature and regulatory agencies, or an issue itself, becomes so politically controversial that the greater legitimacy and accountability of the legislature is required (Goldberg 1987). A major concern is that legislation is usually more refractory to revision and updating than regulation, and thus inflexible statutory

risk requirements can quickly become obsolete in areas of rapid technological change. An example is the so-called Delaney clause (1958) in the United States, which banned all food additives found to cause cancer in animals or humans based on a 1950s-vintage *all or nothing* view of carcinogenicity that had been scientifically outdated for many years before the law was finally repealed in 1996 (Merrill 1988).

Ex ante regulation of risks associated with science and technology thus presents some unique issues and tensions in institutional choice. Given the pace of technological change and the complexity of the subject, legislatures are likely to be at a greater disadvantage compared to regulatory agencies in determining risks associated with science and technology. By contrast the fundamental social, policy, and ethical issues raised by many new scientific and technological advances call for the greater accountability and plenary power elected legislatures offer.

The other major legal mechanism for regulating risks from science and technology is *ex poste* litigation and liability. Individuals injured by technologies may bring tort or product liability lawsuits seeking compensation, and science plays a critical role in providing proof of causation in such cases. Based on concern that such litigation was vulnerable to expert testimony of dubious scientific credibility, courts have focused on ensuring that scientific evidence presented to juries is sound. A leading development in this regard is the U.S. Supreme Court's 1993 decision in *Daubert* v. *Merrell Dow Pharmaceuticals, Inc.* that requires federal courts to perform a *gatekeeping* function to ensure that scientific evidence and testimony is reliable and relevant before it can be admitted. This opinion has resulted in judges being proactive and knowledgeable in screening prospective scientific testimony, and has generated an enormous body of scholarly commentary on how judges should evaluate scientific evidence (Black et al. 1994, Beecher-Monas 2000). It has also stimulated professional scientific organizations such as the American Association for the Advancement of Science (AAAS) to seek to educate judges about science and to provide lists of qualified experts.

Unlike *ex ante* regulation that evaluates whether a particular product, process, or technology may present risks, *ex poste* regulation is directed more specifically at whether the technology caused a specific type of injury in a particular individual or group of individuals. The scientific obstacles and uncertainties in demonstrating *specific causation* are even more complex than those faced in demonstrating general causation in the regulatory context. The judicial system uses presumptions, burdens of proof, and standards of proof in reaching decisions under conditions of uncertainty.

PROMOTING THE BENEFITS OF NEW TECHNOLOGIES. The law also plays a critical role in fostering innovation and promoting the development of new technologies through several legal mechanisms and doctrines. Perhaps the most important of these relates to intellectual property, by which the law gives inventors and creators a time-limited exclusive right to commercially exploit the output of their work. Intellectual property is protected through a number of legal forms, including patents, copyright, trademarks, and trade secrets. The underlying rationale for protecting intellectual property is to promote innovation, by giving researchers and authors economic incentives to create new inventions and works. Intellectual property protection is particularly important in high technology industries such as computer software and biotechnology where ideas and innovations rather than infrastructure and machinery are primary company assets.

New technologies present fundamental challenges to traditional intellectual property doctrines. For example, digital information may not be adequately protected by traditional copyright enforcement procedures, which require the copyright owner to bring a lawsuit alleging infringement. Because unlimited numbers of perfect digital copies can be made at almost zero marginal cost by simply uploading the material onto the Internet, legislatures and courts have extended greater copyright protections for digital data. This is exemplified by the *notice and take-down* provision of the U.S. Digital Millennium Copyright Act (1998) that compels Internet service providers (ISPs) to promptly remove information that copyright holders claim is infringing their copyright.

The rapid growth and use of peer-to-peer file exchange likewise challenges the capability of copyright law to protect copyrighted digital works, and has resulted in a renewed interest in using data protection technologies such as encryption instead of, or in addition to, the law to protect copyright. This trend, in turn, has created the need for legal restrictions on anti-circumvention measures that could be used for unauthorized bypassing of data protection technologies. However restrictions on anti-circumvention technologies have also been criticized for extending copyright beyond its traditional limits, including by undermining the *fair use* of digital data and unduly restricting scientific research (Samuelson 2001).

There are similar challenges in adapting patent law to genetic discoveries. Patenting genes has raised many scientific, legal, ethical, and practical complexities that established patent law is not equipped to address. For example, the traditional distinction between non-patentable products of nature and patentable human inventions and discoveries has been blurred by technology that permits the isolation of genes (often in a slightly different form) from living organisms. How should ethical and moral concerns about patenting genes and living organisms be considered in patent decisions, if at all? Should there be exceptions from patent enforcement for patented genes and organisms used for research or clinical applications? Might gene patents actually impede research and slow innovation, contrary to the very purpose of patenting, due to overlapping and stacked patent rights that make the administrative costs of licensing prohibitive (the so-called *tragedy of the anticommons*) (Heller & Eisenberg 1998)?

In addition to its efforts to protect intellectual property, the law encourages advances in technology through antitrust doctrine. Antitrust law promotes innovation by preventing companies from exercising monopoly power or colluding together to block new market entrants and innovations. Technology industries present unique antitrust issues. On the one hand, increased antitrust concerns and scrutiny may be warranted because of the potential for network effects to result in path dependency. Specifically the positive externalities of having other users with a compatible system may create an entry barrier to new competitors that can result in a de facto monopoly for the early industry leader, because users will be reluctant to adopt a new, better technology if it is not compatible with other users. The high initial costs of creating and introducing a new product combined with the low marginal cost of many knowledge-intensive industries heavily favors superior market power for the already-established player.

On the other hand, there are factors to suggest that antitrust issues might be of less concern in high technology industries. Rapid technological progress in high-technology sectors can result in rapid changes in market position, even for a market leader. For example, WordStar was an early market leader in word processing software, but was quickly replaced by new market entrants with superior attributes. Given these conflicting factors, the role of antitrust law in regulating high technology industries and promoting technological innovation remains a major area of academic and policy debate (Hart 1998–1999, Liebowitz and Margolis 1996).

Antitrust actions brought in the United States and Europe against the Microsoft Corporation in the late 1990s and early 2000s illustrate these conflicting antitrust considerations. Government authorities claimed that Microsoft, by virtue of its Windows computer operating system, had a monopoly power with respect to other such operating systems that allowed Microsoft to suppress innovation in potentially competing products. Microsoft contended that it should be permitted to improve its products to include new functionalities (that is, a web browser), and that the antitrust enforcement actions were restraining such advances.

There are also other legal instruments for promoting innovation and advancing technology. Direct governmental funding of scientific research and development, as well as indirect subsidization through legal mechanisms such as research and development tax credits, are important stimulants. Technology-forcing regulations, such as motor vehicle emission standards, prompt technological progress in specific industries. Other standards that provide for uniformity of new technology formats, such as digital television, likewise are intended to facilitate technological development.

ADDRESSING ETHICAL IMPLICATIONS OF TECHNOLOGY. The law is the primary vehicle by which society seeks to resolve controversies raised by scientific research and new technologies. Whether the issue is surrogate motherhood, voluntary euthanasia, human cloning, genetic engineering, privacy in the workplace, online security, or any other technological advance with potential ethical consequences, society relies on legislatures and courts to develop and apply appropriate legal principles. The bioethicist Daniel Callahan has described this tendency to translate moral problems into legal problems as *legalism*, but he himself identifies a vacuum of societal institutions other than the law to resolve moral issues in a satisfactory manner (Callahan 1996). Indeed the failure to legally proscribe an activity carries an implicit message that the activity is morally acceptable.

In some cases, courts have restricted their own authority to consider the ethical aspects of controversial technological developments. For example, the U.S. Supreme Court held that living, engineered organisms such as the OncoMouse could be patented, and refused to address ethical arguments raised by such patenting, finding that those ethical objections were best addressed to the legislative arm of the government. Even when courts exclude ethical considerations, they often remain the primary motivation for litigation, which is then fought on surrogate legally-cognizable grounds.

Institutional Issues

The second major strand in the study of law-science interactions is the impact of science and technology on the practice of law, and the reciprocal effect of law on the practice of science.

EFFECTS OF SCIENCE AND TECHNOLOGY ON THE PRACTICE OF LAW. Scientific and technological advances have both substantive and procedural effects on the law. On the substantive side, new scientific evidence and techniques can change the way legal claims are resolved, including their outcomes. For example, forensic DNA evidence has fundamentally changed criminal law and paternity disputes by greatly improving the veracity of legal fact finding, while creating a plethora of new legal, ethical, and social issues (Imwinkelried and Kaye 2001). In criminal cases, forensic DNA has helped identify and convict guilty persons who might have otherwise escaped prosecution, and exonerated innocent persons accused or convicted. But this powerful forensic tool raises new issues, such as how and from whom DNA samples should be collected and stored, how genetic information may be used, and when convicted criminals should be permitted to reopen cases based on *new* DNA evidence.

Advances in technology are further revolutionizing the procedural aspects of law. The practice of law has historically been influenced by new technologies, including the printing press, telephone, photocopier, and fax (Loevinger 1985). In the early twenty-first century, digital evidence has improved the quality and availability of trial evidence, while raising concerns about tampering with digital photos and recordings. On-line databases, digital document repositories, electronic discovery, new graphics and presentation technologies, and *digital courtrooms* are changing the ways lawyers research, prepare, and present their arguments (Arkfeld 2001). On-line filing and availability of court records is increasing the convenience and availability of judicial proceedings, yet creating new privacy concerns.

EFFECTS OF LAW ON THE PRACTICE OF SCIENCE. According to Justice Breyer, "science depends on sound law—law that at a minimum supports science by offering the scientist breathing space, within which he or she may search freely for the truth on which all knowledge depends" (Breyer 1998, p. 537). Until recently, law rarely intruded into the inner sanctum of the space it created for science. Beginning in the 1980s, however, the law has steadily intruded into the practice of science. Investigations of claims of science misconduct have become more frequent and legalistic, as government investigators adopt adversarial and formal procedures approaching those used by criminal prosecutors. Individuals claiming to have been aggrieved by scientific misconduct or allegedly false claims of scientific misconduct frequently seek judicial remedies. Attorneys have even served non-party subpoenas on scientists who are doing research potentially relevant to a pending lawsuit, even if the subpoenaed scientists have no relationship to the litigation or any of the parties. This imposes a costly burden on scientists, and exposes them to intrusive searches and disclosures about their research activities.

Legislatures are also subjecting scientists to new legal requirements. Governmentally-funded researchers have long been subject to a number of requirements that are conditions of federal funding, such as requirements for human subject protection. But in 1998, the U.S. Congress passed the so-called Shelby Amendment that subjects researchers funded by the federal government to the Freedom of Information Act (FOIA), under which citizens can request and inspect all relevant documents not protected by limited exemptions. The Office of Management and Budget subsequently narrowed this legislation to federally-funded research directly relied upon in federal rulemaking, but even under such a constricted (and challengeable) interpretation, this legislation represented an unprecedented legal intrusion into the laboratory. In 2000 the U.S. Congress enacted the Data Quality Act, which imposes a series of substantive and procedural requirements on scientific evidence used by regulatory agencies. These developments indicate a trend of growing legal intrusion into the science, which was once perceived as a *self-governing republic* generally impervious to legal interventions (Goldberg 1994).

Tensions Between Law and Science

The third strand of law, science, and technology examines the tensions and conflicts that occur when law and science are juxtaposed in decision making. These tensions and conflicts generally flow from the fact that law and science have different objectives and procedures. One frequently mentioned difference is that the law focuses on process, whereas science is concerned with progress (Goldberg 1994). While both law and science are evidence-based systems for finding the truth (Kaye 1992a, Jasanoff 1995), the law is concerned with normative considerations such as fairness and justice, considerations generally outside the scientific framework. Given this difference, otherwise relevant evidence is inadmissible in law if its use or the way it was obtained is unfair, whereas the concept of excluding pertinent data is foreign to science (Loevinger 1992, Foster and Huber 1997). One U.S. federal judge described science as "mechanical, technical, value-free, and nonhumansitic," while law is "dialectical, idealistic, nontechnical, value-laden and humanistic" (Markey 1984, p. 527). Another difference is that "[c]onclusions in science are always probable and tentative," whereas "[c]onclusions in law are usually certain and dogmatic" (Loevinger 1985, p. 3). Given these and other contrasts, it is not surprising that tensions such as the following have developed.

TECHNICAL COMPETENCE. Most legal decision makers (for example legislators, judges, and juries) have very little scientific training and expertise, and yet are called upon to decide highly complex technological matters (Bazelon 1979, Faigman 1999). The result is that "amateurs end up deciding cases argued by experts" (Merges 1988, p. 324). There is therefore concern that legal decision makers will fail to reach scientifically credible decisions (Angell 1996) and will be improperly misled by *junk science* (Huber 1988).

The legal system has instituted a number of procedural and substantive innovations in an attempt to enhance the scientific merits and credibility of its decisions. One major change has been a systematic shift of decision-making authority from juries to judges, presumably because judges have greater capability and experience in distinguishing valid from invalid scientific testimony. Thus, as previously noted, judges in U.S. federal courts are required to perform a *gatekeeping* function to screen proposed scientific testimony for its reliability and relevance before it can be presented to a jury (*Daubert* v. *Merrell Dow Pharmaceuticals, Inc.* [1993]). Similarly, in patent infringement cases, the critical issue of interpreting the scope of a patent has been taken from juries and given to the trial judge pursuant to a 1996 U.S. Supreme Court decision.

Another innovation is the use of *neutral* or third party experts, appointed by the court rather than the contending parties to assist a judge or jury in understanding the scientific issues in a case. Some jurisdictions have also experimented with specialized courts better able to handle technological disputes, such as the digital court implemented by the State of Michigan. The increased use of pretrial conferences to narrow the scientific issues in dispute and the appointment of specially trained law clerks and *special masters* are other techniques courts employ to better handle complex scientific and technological cases (Breyer 1998).

In the legislative context, there is a growing recognition of the need for legislatures to have their own scientific and technological advisory bodies (Faigman 1999), with some pressures in the United States to replace the Office of Technology Assessment which was abolished in 1995. Most European governments and the European Union have established technology advisory bodies for their legislators.

LEGAL VS. SCIENTIFIC STANDARDS. Another area of dispute is whether the law should apply scientific standards and methods of proof, or apply its own standards to scientific evidence. An example is the concept of statistical significance, where the standard scientific convention is that a result will be considered statistically significant if the probability of the result being observed by chance alone is less than five percent (i.e., $p < 0.05$) (Foster and Huber 1997). Some legal experts argue that the law should apply a more lenient standard, specially in civil litigation where the standard of proof is the preponderance of the evidence (i.e., $p > 0.5$), because while science focuses primarily on preventing false positives, the law is equally if not more concerned about false negatives (Cranor 1995, Shrader-Frechette 1991). Other experts caution against equating the scientific standard of statistical significance with the legal standard of proof, because the two measures perform different functions and are like comparing *apples and oranges* (Kaye 1992b, Kaye 1987).

Judge Howard Markey, while sitting as Chief Judge of the U.S. Court of Appeals for the Federal Circuit, wrote that "[n]o court ... should base a decision solely on science if doing so would exclude the transcendental ethical values of the law" (Markey 1984, p. 525). He warned that "juriscience might displace jurisprudence" as a result of the tendency to "scientize the law" (Markey 1984, p. 525). In contrast, the U.S. Supreme Court's *Daubert* decision held that courts must ensure that scientific testimony have a "grounding in the methods and procedures of science," that is, be "derived by the scientific method" before it can admitted, which imports scientific standards of evidence into the law (*Daubert v. Merrell Dow Pharmaceuticals, Inc.* [1993], p. 590). Similarly Justice Breyer has argued "an increasingly important need for law to reflect sound science" (Breyer 1998, p. 538). Yet "some courts remain in the prescientific age" unless and until they "embrace the scientific culture of empirical testing" (Faigman 2002, p. 340).

TIMING OF DECISIONMAKING. Science and technology are progressing at increasing rates (Carlson 2003). A classic example of the rapid acceleration of technology is Moore's law, which predicts that the number of transistors on microchips will double every two years. The law is much slower to evolve, with case law advancing incrementally and gradually, and legislation advancing only sporadically. Statutes, in particular, can quickly become outdated as legislatures are limited, as a practical matter, to revisiting most issues every few years at best, and for some issues every few decades. Case law is also slow to adapt to advances in science and technology due to the binding effect of past precedents (*stare decisis*), something that does not impede science and technology. The result is that the law is often based on outdated scientific assumptions or fails to adapt to new technologies or scientific knowledge. Many experts argue that more flexible and adaptive legal regimes are needed to keep pace with advancing technological systems (Green 1990).

By contrast, there are situations where the law must address a question prematurely, before adequate scientific data are available (Faigman 1999). Science is in no rush to come to a final decision on any specific issue, and can afford to suspend judgment until *all the evidence is in*, even if that takes decades or centuries. Law does not always have the luxury of waiting (Goldberg 1994, Jasanoff 1995). When a defendant is charged with a crime, or a product manufacturer is sued for allegedly harming a citizen, the court must reach a final decision promptly without waiting for additional research to further clarify the issues. The bounded timeline of the law increases the risk of the legal system reaching decisions that may later be deemed scientifically invalid.

NEW TECHNOLOGIES VS. OLD LAWS. Another issue is whether new technologies require new laws or can be addressed by existing legal frameworks. One colorful articulation of this issue is the debate about whether there is any more need for the *law of cyberspace* than for the *law of the horse* (Easterbrook 1996, Lessig 1999). The analogy refers to the fact that there were no major legal doctrinal changes introduced to address the horse as it became a major part of commerce in earlier times, but rather existing doctrines were applied to the horse with only minor modifications. Thus there is a question about the need for new legal doctrines to address the Internet on issues such as privacy, copyright, pornography, and gambling. The passage of specialized laws such as the Digital Millennium Copyright Act and the Child Online Protection Act (1998) indicate a pattern of adopting new laws to address at least some cyberspace issues.

The same general issue arises in other technological contexts. One major debate in the regulation of genetically modified organisms is whether such products should be governed by existing environmental and food safety laws, or alternatively whether a new statutory regime created specifically for biotechnology products is required (Marchant 1988). Existing laws have generally been applied in the United States, while new enactments have been promulgated in Europe and other jurisdictions.

Another example is patent law, where to date existing patent rules have been applied to new technologies such as genes and other biomedical discoveries. Some commentators have argued that new laws, in particular new approaches that move away from the *one-size-fits-all* approach of current law, are needed to provide optimal patent protection for certain new and emerging technologies (Thurow 1997, Burk and Lemley 2002).

LEGAL INTERVENTION VS. MARKET FORCES. A final recurring issue is the respective roles of law and market in regulating new technologies. Specifically, under what circumstances is legal intervention (in the form of legislation or liability) appropriate, and when should the law pull back and leave the market to operate? Major disagreements on this fundamental issue exist. For example, there are conflicting views on whether government should restrict science funding to basic research, or also fund more applied research and development of new technologies.

This same basic tension between legal intervention and market forces underlay disagreements about whether Microsoft should have been subjected to antitrust enforcement because of its Windows operating system or whether market forces were adequate to prevent the company from unfairly exploiting its near monopoly. Another example is Internet privacy, where some commentators assert that technology and the market can provide adequate assurances of privacy, while others argue that a regulatory approach is needed. A third example is whether the government should set standards for technologies such as digital television and wireless communications, or leave it to the market to develop a de facto standard. These disputes rest on conflicting economic and political perspectives that are unlikely to be resolved in the foreseeable future.

Conclusion

The law interacts with science and technology in diverse ways. These interactions will proliferate in the future with advancing technologies that present novel risk, benefit, and ethical scenarios. The nascent legal field of law, science, and technology seeks to provide a

systematic treatment of these actions, and will grow and evolve in parallel and apace with its subject matter.

GARY E. MARCHANT

SEE ALSO *Aviation Regulatory Agencies; Building Codes; Communications Regulatory Agencies; Crime; Death Penalty; Environmental Regulatory Agencies; Expertise; Evidence; Food and Drug Agencies; Human Rights; Information Ethics; Intellectual Property; Internet; Justice; Just War; Misconduct in Science; Natural Law; Police; Regulation.*

BIBLIOGRAPHY

Angell, Marcia. (1996). *Science on Trial.* New York: W. W. Norton & Company. Critical analysis of law's treatment of science by medical researcher and former journal editor.

Areen, Judith; Patricia A. King; Steven Goldberg; et al. (1996). *Law, Science and Medicine,* 2nd edition. Westbury, NY: Foundation Press. Legal casebook for law and science courses.

Arkfeld, Michael R. (2001). *The Digital Practice of Law,* 5th edition. Phoenix, AZ: Law Partner Publishing.

Bazelon, David L. (1979). "Risk and Responsibility." *Science* 205: 277–280. Widely cited.

Beecher-Monas, Erica. (2000). "The Heuristics of Intellectual Due Process: A Primer for Triers of Science." *New York University Law Review* 75: 1563–1657.

Black, Bert; Francisco J. Ayala; and Carol Saffran-Brinks. (1994). "Science and the Law in the Wake of *Daubert*: A New Search for Scientific Knowledge." *Texas Law Review* 72: 715–802.

Breyer, Stephen. (1998). "The Interdependence of Science and Law." *Science* 280: 537–538. Influential analysis by a member of the U.S. Supreme Court.

Burk, Dan L., and Mark A. Lemley. (2002). "Is Patent Law Technology-Specific?" *Berkeley Technology Law Journal* 17: 1155–1206.

Callahan, Daniel. (1996). "Escaping from Legalism: Is It Possible?" *Hastings Center Report* 26, no. 6 (November–December): 34–35.

Carlson, Robert. (2003). "The Pace and Proliferation of Biological Technologies." *Biosecurity and Bioterrorism* 1(3): 1–12.

Cranor, Carl F. (1995). "Learning form the Law for Regulatory Science." *Law and Philosophy* 14: 115–145.

Daubert v. *Merrell Dow Pharmaceuticals, Inc.,* 509 U.S. 579 (1993). U.S. Supreme Court decision adopting new standards for admissibility of scientific evidence.

Diamond v. *Chakrabarty,* 447 U.S. 303 (1980). U.S. Supreme Court decision allowing patenting of living organisms.

Easterbrook, Frank H. (1996). "Cyberspace and the Law of the Horse." *University of Chicago Legal Forum* 1996: 207–216.

Faigman, David L. (1999). *Legal Alchemy: The Use and Misuse of Science in the Law.* New York: W. H. Freeman & Co. Excellent analysis of recent interactions between law and science.

Faigman, David L. (2002). "Is Science Different for Lawyers?" *Science* 297: 339–340.

Foster, Kenneth R., and Peter W. Huber. (1997). *Judging Science: Scientific Knowledge and the Federal Courts*. Cambridge, MA: MIT Press.

Goldberg, Steven. (1986). "The Central Dogmas of Law and Science." *Journal of Legal Education* 36: 371–380.

Goldberg, Steven. (1987). "The Reluctant Embrace: Law and Science in America." *Georgetown Law Journal* 75: 1341–1388.

Goldberg, Steven. (1994). *Culture Clash: Law and Science in America*. New York: New York University Press. Among the best book-length introductions.

Green, Harold P. (1990). "The Law-Science Interface in Public Policy Decisionmaking." *Ohio State Law Journal* 51: 375–405. Extensive, early overview of law-science relationship.

Hart, David M. (1998–1999). "Antitrust and Technological Innovation." *Issues in Science and Technology* Winter: 75–81.

Heller, Michael A., and Rebecca S. Eisenberg. (1998). "Can Patents Deter Innovation? The Anticommons in Biomedical Research." *Science* 280: 698–701. Influential article on how patents might impede innovation.

Huber, Peter W. (1988). *Liability: The Legal Revolution and Its Consequences*. New York: Basic Books, Inc.

Imwinkelried, Edward J., and D. H. Kaye. (2001). "DNA Typing: Emerging or Neglected Issues." *Washington Law Review* 76: 413–474.

Jasanoff, Sheila. (1995). *Science at the Bar: Law, Science, and Technology in America*. Cambridge, MA: Harvard University Press. Important overview of law-science relationship from social studies of science perspective.

Kaye, D. H. (1987). "Apples and Oranges: Confidence Coefficients versus the Burden of Persuasion." *Cornell Law Review* 73: 54–77. Clear, definitive analysis of relationship of legal and scientific standards of proof.

Kaye, D. H. (1992a). "Proof in Law and Science." *Jurimetrics* 32: 313–322.

Kaye, D. H. (1992b). "On *Standards and Sociology*." *Jurimetrics* 32: 535–546.

Lessig, Lawrence. (1999). "The Law of the Horse: What Cyberlaw Might Teach." *Harvard Law Review* 113: L501–549.

Liebowitz, S. J., and Stephen E. Margolis. (1996). "Should Technology Choice Be A Concern of Antitrust Policy?" *Harvard Journal of Law and Technology* 9: 284–318.

Loevinger, Lee. (1985). "Science, Technology and Law in Modern Society." *Jurimetrics* 26: 1–20. A pioneering work on the relationship between law, science and technology.

Loevinger, Lee. (1992). "Standards of Proof in Science and Law." *Jurimetrics* 32: 323–344.

Marchant, Gary E. (1988). "Modified Rules for Modified Bugs: Balancing Safety and Efficiency in the Regulation of Deliberate Release of Genetically Engineered Microorganisms." *Harvard Journal of Law and Technology* 1: 163–208.

Markey, Howard T. (1984). "Jurisprudence or *Juriscience*." *William & Mary Law Review* 25: 525–543. Provocative early critique of law's deference to science.

Markman v. Westview Instruments, 517 U.S. 370 (1996).

Merges, Robert P. (1988). "The Nature and Necessity of Law and Science." *Journal of Legal Education* 38: 315–229.

Merrill, Richard. (1988). "FDA's Implementation of the Delaney Clause: Repudiation of Congressional Choice or Reasoned Adaptation to Scientific Progress?" *Yale Journal on Regulation* 5: 1–86.

Samuelson, Pamela. (2001). "Anticircumvention Rules: Threat to Science." *Science* 293: 2028–2031.

Shrader-Frechette, K. S. (1991). *Risk and Rationality*. Berkeley: University of California Press.

Sutton, Victoria. (2001). *Law and Science: Cases and Materials*. Durham, NC: Carolina Academic Press. Legal casebook on law and science.

Thurow, Lester C. (1997). "Needed: A New System of Intellectual Property Rights." *Harvard Business Review* September–October: 95–103.

Wessel, Milton R. (1989). "What Is *Law, Science and Technology* Anyway?" *Jurimetrics* 29: 259–266.

SCIENCE, TECHNOLOGY, AND LITERATURE

• • •

The ethical implications of science and technology found in literaturre are varied and often implicit as well as explicit. A beginning survey may reasonably include the following non-exhaustive set of topics: the content of narratives that make asseissments of science and technology; orality, writing , printing, and electronic communication as technologies involving certain cultural contexts; and scientific theaories, experiments, and practices as sociocultural influences on literature. (Assessment of the stylistic and rhetorical strategies of science and technology, while also related, are treated in a separate entry.) Scholars in traditional disciplines have often touched on these topics, but only in the 1970s did interdisciplinary fields—the history of the book, science and technology studies, literature and science studies, and cultural studies—begin to give such concerns extensive attention. Tracing ethical aspects of science, technology, and literature calls for examining oratory, writing, printing, and electronic communication as technologies developed in cultural contexts; studying scientific theories, experiments, and practices as sociocultural influences on literature; assessing stylistic and narrative strategies in scientific discourse, including histories and philosophies of science, and elucidating how literary works and theories interpret and reconfigure science and technology as human endeavors. Scholars in traditional disciplines have touched on these topics for many years, but only in since the late 1970s have interdisciplinary fields—the history of the book, science and technology studies (STS), literature and science studies, and cultural studies—flourished to focus on such concerns.

Ancient and Early Modern Myths of Science and Technology

European classical representations of science and technology invoking ethical dilemmas appear in dramatic and didactic poetry. Greek and Roman myths describe Prometheus creating humans with Athena's consent and stealing fire for mortals from Zeus, actions that inspired John Ferguson's characterization of Prometheus as a master inventor and trickster whose rebellious intelligence helps humans rise above animals. Aeschylus's fifth-century *Prometheus Bound* posits that Zeus grew angry at human achievements and at Prometheus's theft, punishing the latter by chaining him to a rock. Hesiod's *Theogony* (c.700 B.C.E.) notes that Prometheus's brother Epimetheus married the beautiful Pandora, who was created as a punishment by Zeus. Pandora opens a container, releasing a host of miseries on humanity; however her curiosity inhibits human progress instead of encouraging innovation and invention. Biblical accounts imputing ethical aspects of science and technology include Genesis 6, which details the building of an ark by Noah, under God's direction, to protect animal species, including Noah's family, from the flood. Genesis 11, in the story of the Tower of Babel, relates how people built a tower and a city, thus prompting God to create different languages in order to constrain human achievement. These classical and Biblical texts represent scientific and technical projects as enhancing human life at the risk of alienating God.

Modern cautionary tales about Faust and the Sorcerer's Apprentice further consider the dangers of human meddling with science and technology. The Faust Chapbook of 1587 describes Dr. Faust as a master of science and sorcery who conjures the Devil and enters into a pact with him: The Devil promises to serve Faust and in exchange the doctor gives up his soul and renounces his Christian faith. Faust is celebrated for his ability to cast horoscopes but becomes increasingly debauched. The impropriety of Faust's aims and actions has inspired a range of European literary texts, including tragedies, narratives, and poetry by Christopher Marlowe, Johann Wolfgang von Goethe, Heinrich Heine, Paul Valéry, and Thomas Mann, and a number of musical works by Hector Berlioz, Charles Gounod, and Franz Liszt. Goethe's 1779 poem "The Sorcerer's Apprentice"

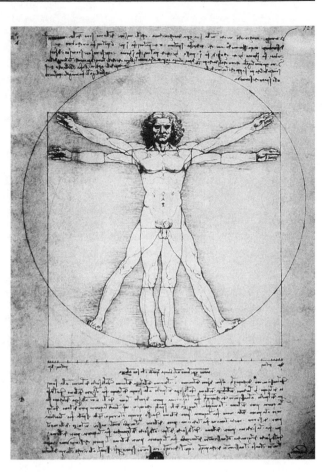

"The Vitruvian Man," 1490 drawing by Leonardo da Vinci. Made as a study of the proportions of the human body, the drawing is often used as an implied symbol of the essential symmetry of the human body, and by extension, to the universe as a whole. (© *Corbis.*)

("*Der Zauberlehrling*") interpreted through Paul Dukas's symphonic scherzo "*L'apprenti sorcier*" (1897) served as a source for the segment of Walt Disney's film *Fantasia* in which Mickey Mouse borrows the Sorcerer's magic broom and causes chaos before he is called to account for the mess. These legends suggest that human desire to know more about the world and control nature might be hubristic and selfish. The narratives imagine how endeavors motivated by extreme ambition inevitably lead to catastrophe. A bug in a computer protocol is commonly known by the term *sorcerer's apprentice mode*, as detailed in a number of websites linked to the Google search engine.

Linking themes of egotism and passion for new knowledge with contemporary theories about electricity in *Frankenstein, or The Modern Prometheus* (1818), Mary Shelley imagined how aspirations to conquer science and ancient alchemy inspire and destroy Dr. Victor Frankenstein. Frankenstein creates life only to turn his back on the creature he belatedly recognizes as a monster. Invoked often in fiction and film, the Frankenstein myth of creation gone awry retains potency for many in the age of bioengineering. Newspapers reporting on deliberations by the U.S. Congress and President's Council on Bioethics to ban cloning and restrict fetal tissue research invoked Shelley's novel (along with Aldous Huxley's *Brave New World*). Activists employ the term *Frankenfood* to denote food modified by processes of genetic transplantation.

Referring to Pygmalion rather than Prometheus, Nathaniel Hawthorne outlines the dangers of scientific ambitions and technological tinkering in stories such as "Rappaccini's Daughter" (1846) and "The Birth-Mark" (1846), whose plots explore how male scientists used their wives or daughters as subjects for their experiments. Villiers de L'Isle-Adam's mechanical fantasy *L'Eve future* (1880) follows a modern Pygmalion character who applies scientific knowledge to engineer a Galatea, only to find that even an artificial woman's needs surpass his scientific and technological ingenuity. Given the saliency of myths pointing up the dangers of science and technology, it is not surprising that themes of hubris, technology run amok, and scientific arrogance are common in science fiction, postmodern realist literature, and expository prose.

Printing and the Reading Revolution

Although the Sumerians created clay books as early as 3000 B.C.E. and the Chinese developed printing techniques in the early-second century C.E., accounts of modern printing technology usually begin with the importation of paper from Asia to Europe (Graff 1991). Early experiments with xylography and metallographic printing were *disappointing* (Havelock 1976). Johannes Gutenberg (1390–1468), who is credited with inventing typography, also is generally understood to be the first printer to use movable type in 1436. Metal type represented an advance on woodcuts, which were time-consuming to produce and of limited use. At the end of the sixteenth century, the printing industry was well established in many European cities even though printing remained a tedious process. While most books dealt with religious subjects, dramas and fictions were also published. Censorship and political restrictions curtailed some printers; in seventeenth-century England the government limited the number of printers.

After the Renaissance, advances in type and the use of paper covers decreased the cost of books while promoting a diversity of written materials. At the end of the eighteenth century, the invention of lithography and innovations in the power press advanced the printing industry, while improvements in papermaking and stereotyping decreased costs in the early-nineteenth century. By then reading had become a necessary part of everyday life for North Americans and Western Europeans in that work, worship, and social relations encouraged the activity and education became a fundamental goal of democracy (Graff 1991). In the United States during the antebellum period, children, prisoners, and freed slaves were taught to read as a means of socialization and economic empowerment, principles enunciated in didactic literature (Colatrella 2002).

Oral-Literacy Transformation

Developing scientific schema and philosophical theories, post-Enlightenment scholars demonstrated wide-ranging interests in linguistic, rhetorical, and narrative forms associated with oral and written texts. Linguists and philologists in the late-eighteenth and nineteenth centuries traced connections among Indo-European languages, studied classical rhetorical modes, and collected folktales from various regions. Romantics, who had an interest in ordinary people and their texts, celebrated the vernacular; James McPherson in Scotland, Thomas Percy in England, Jacob and Wilhelm Grimm in Germany, and Francis James Child in the United States collected examples from the oral traditions of those countries (Ong 1982). The work of these writers influenced twentieth-century formalists and structuralists, who melded textual and cultural analyses in their work on the periphery of the social sciences, notably in the fields of psychology and anthropology.

In the early-twentieth century, Andrew Lang demonstrated that oral folklore offered *sophisticated verbal art forms* (Ong 1982). Lang's work encouraged others to analyze techniques employed in classical poetry, particularly Homer's *Iliad* and *Odyssey*, and reinvigorated a debate begun in the seventeenth century concerning evolution and authorship of these works. In the twentieth century, Milman Parry viewed each Homeric epic as the culmination of orally delivered formulaic phrases used by bards. Building on Parry, Albert Lord hypothesized that "the idea of recording the Homeric poems, and the Cyclic epics [the Epic of Creation and the Epic of Gilgamesh], and the works of Hesiod, came from observations of or hearing about similar activity going on further to the East," specifically early versions of the Old Testament in ninth-century Palestine (Lord 1978, p. 156). Eric Havelock claimed the written versions of the *Iliad* and the *Odyssey* were the first products of the new Greek alphabet developed around 700–650 B.C.E. (Ong 1982).

Parry made phonographic recordings of working poets in 1930s Yugoslavia as a means of studying the composition of oral poems that might shed light on the development of the Homeric epics. After Parry's death, Lord continued the project, publishing *The Singer of Tales* in 1960, a book based on recordings and transcriptions. He argued that the Yugoslavian poets, who were generally illiterate, typically composed their songs during their performances according to mechanisms likely used in formulating the Homeric epics. Novice poets were able to create new songs because they had learned stories and formulaic phrases by watching the performances of others, a prerequisite for developing the special technique of composing by combining well-known formulas. Building on Parry, Lord argued that Homer composed oral narrative poetry through the same method, based on "intricate schematization of formulas" in Greek hexameter (Lord 1978, p. 142).

At the end of the twentieth century, the orality-literacy distinction drew the attention of theorists such as Jacques Derrida, J. L. Austin, John Searle, and Mary Louise Pratt, whose arguments influenced post-structuralist theories about literature. Derrida questioned the privileging of orality over writing, calling the practice phonocentrism and connecting it to logocentrism. He provoked speech act theorists Austin and Searle in pointing out that "the uses of language could not be determined as exclusively either normal or parasitic" (Halion 2003, Internet site). Suggesting the possibility of a unified theory of discourse, Pratt argued against the idea that the discourse of literature is functionally distinct from other verbal expressions.

Media Literacy

Contemporary interest in literacy shifts peaked in the the twentieth century as a transformation from print to new media developed. A number of non-fiction writers, including Marshall McLuhan, Ivan Illich, and Alvin and Heidi Toffler, addressed social issues concerning electronic media. The Tofflers conceived a popular theory of history describing three successive eras—the agricultural age, the age of the Industrial Revolution, and the Information Age, becoming famous as consultants to Newt Gingrich, who served as Speaker of the House in the U.S. Congress in the early 1990s. The Tofflers's work celebrates technological advances as progress. In contrast, Illich's writings question the assumed superiority of industrialized nations, the centralization of political authority, and faith in technology. He analyzed issues in medicine that denaturalize human control for the sake of technology.

Recognizing that consumers are bombarded with hundreds of advertisements, Illich criticized the reversal of the relation of needs and wants by materialist culture and argued that more technology does not produce greater leisure, freedom, or satisfaction; that what many think of as schooling is more properly termed *deschooling*; and that literacy can constrain rather than enable one's prospects in a culture. Some late-twentieth-century writers were inspired to apply Illich's theories in books such as *ABC: The Alphabetization of the Popular Mind* (1988) and *In the Vineyard of the Text* (1993), to projects associating literacy with technological change in the *convivial society*. Illich's concept of the convivial society in which technologies serve individuals rather than managers might have helped convince Lee Felsenstein, a founder of *Community Memory*—regarded by many as the world's first public computerized bulletin board system—to use the computer, which had been primarily promoted as having industrial applications, for artistic expression. English teacher Allan Luke positively characterizes literacy as *a communications technology* engaging individuals with real and fantastic worlds, creating a *simultaneous universe*, akin to McLuhan's *global village*, while Howard Rheingold describes *smart mobs* of individuals linked by electronic technologies.

McLuhan described his argument in *The Gutenberg Galaxy*, published in 1962 as *complementary* to those of Parry and Lord in dealing with cultural shifts affected by changing media; whereas their work accounted for the orality-literacy transformation, his provided trenchant analysis of the transformation from print to digital literacy. McLuhan resisted evaluating cultural change, instead concentrating on delineating connections among sociopolitics, culture, and media. In an interview, he explained how printing influenced nationalism: "Nationalism didn't exist in Europe until the Renaissance, when typography enabled every literate man to *see* his mother tongue analytically as a uniform entity. The printing press, by spreading mass-produced books and printed matter across Europe, turned the vernacular regional languages of the day into uniform closed systems of national languages ... gave birth to the entire concept of nationalism" (McLuhan 1995, pp. 243–244). McLuhan recognized that while technologies and media inevitably produce changes, such shifts could often be uncomfortable for those experiencing them and ought to be considered critically, as Illich and Neil Postman argue.

McLuhan's work allusively comments on cultures, texts, and media technologies, often through aphorisms attesting to diverse influences. His celebrated statement

"The medium is the message" from *Understanding Media* published in 1964, described technological consequences as continuous: "the personal and social consequences of any medium—that is, of any extension of ourselves—result from the new scale that is introduced into our affairs by each extension of ourselves, or by any new technology" (McLuhan 1995, p. 151). He recognized differences among media, distinguishing *cool* and *hot* media as media requiring engagement (telephone) or passivity (radio) on the part of the user. He described the inevitable constraints associated with technological progress; for example, that the alphabet can "alter the ratio among our senses and change mental processes" as "an aggressive and militant absorber and transformer of cultures" (McLuhan 1995, pp. 119, 144).

Digital Literacy

Many language and technology theorists have developed McLuhan's insights, extending them to other technical developments and evaluating their applicability to revisionist histories of literacy and cognition. Adopting some of McLuhan's ideas about the power of media to influence human perceptions in *Orality and Literacy* (1982), Walter Ong characterizes writing as a technology that changes human consciousness. Investigations in cognition formed the basis for the development of electronic communication media. In *How We Became Posthuman* (1999), Katherine Hayles describes Norbert Wiener's cybernetics, Claude Shannon's information theory, and the fictional contributions of Philip K. Dick to ideas of distributed consciousness and thereby offers a history of disembodiment in cybernetics. Brian Massumi reviews philosophies of perception, including those of Henri Bergson, William James, Gilles Deleuze, Felix Guattari, and Michel Foucault, to argue that new ways of reading are necessary to understand the body and media (film, television, and the Internet) as cultural formations.

Janet Murray argues that late-twentieth-century forms of media changed storytelling conventions to require interactivity. She acknowledges earlier narrative forms and strategies that provide precedents and points of comparison for such media, especially the epic, the picaresque, and the drama of Shakespeare, forcefully arguing that movies, computer games, and hypertext novels are new narrative forms requiring new ways of appreciating a story. Hypertext fiction, poetics, and history, and new media criticism by Michael Joyce, Stuart Moulthrop, George Landow, and Jay Bolter also proffer the argument that hypertextual narrative forms revise notions of interactivity and change perception in representing reality in new, perhaps dangerous, ways. In their joint work, Bolter and Richard Grusin detail changes in Internet media reflecting the remediation of different media forms and their effects on users, particularly in the way that the Internet has become another, albeit more interactive (cool), medium. Greg Ulmer considers electronic communication in teaching composition in universities, arguing that students accustomed to interactive technologies benefit from a constructivist rather than instrumentalist approach.

Authorship, Technology, and Ethics in the Information Age

Post-structuralists theorists Roland Barthes, Derrida, and Foucault questioned traditional notions of authorship. Their critiques suggest that it is impossible for anyone, even another author, to divine a writer's intentions and that readers provide intertextual and contextual information that expands the text. Barthes acknowledges in "The Death of the Author," which first appeared in 1968, that the plurality of voices in the text inevitably produce many possible meanings for readers. Foucault also questioned to what extent biographical information should affect consideration of an author's literary output in "What Is an Author?, first published in 1969, positing *the author function* and emphasizing the value of studying discourse rather than biography. The Internet complicates ideas of authorship. Each search produces a list of sites that could be one person's work, that of a group, or the official page of a company or institution, while many web pages have no identified authors. Contributors to an electronic forum collaborate as multiple authors to a boundless text.

In this way, electronic writing further reduces the distance between reader and text (a shift previously noted by Walter Benjamin), and increases the ephemerality of a text. The fixity of the printed text has transformed into the fluidity of electronic content. Scholars present electronic archives of canonical writers such as Emily Dickinson, Herman Melville, and Walt Whitman that incorporate all versions of particular texts, while hyperlinks organize text to present fluid documents with multiple reading pathways. Electronic sites also recuperate once-popular writers whose works appear on the Internet along with those never-before-published.

Although Internet communication enhances many aspects of social life, its boundlessness also creates ethical problems. Free speech advocates resist filtering information. Satisfactory technical solutions preventing electronic mail spam, plagiarism, identity theft, and pornography aimed at juveniles have not yet been developed. Free electronic distribution of music and film appeals to many users but chips away at intellectual property rights, as is argued by artists and producers in the recording and film industries. Ethical standards regarding authorship, as cases of plagiarism and false documentation of sources suggest, call into question the name on the book or the claims within it, but generally the production process appears to be opaque to a reader, who could easily assume, for instance, that a biography was researched and written by the author noted on the cover or that a reporter whose byline appears on an article witnessed an event, while there may in fact have been contributions from numerous research assistants or virtual research may have substituted for an on the scene account.

Critical Paradigms of Taste and Technology

Literary criticism has a long history of valuing some genres, writers, or works over others for ethical reasons. Plato characterized poetry as too dangerous to exist in the ideal republic because it inspired political critique, and Jonathan Swift satirized the seventeenth-century Battle of the Ancients and the Moderns that provoked many French and English critics to debate the merits of classical versus contemporary literature. Training in modern languages and literatures is a product of the post-Romantic age. Earlier education in liberal arts was dominated by study of classical texts; but by the early-twentieth century, ideas of canonicity transformed to include certain modern texts. Cultural tastes change over time; for example, the novels of Herman Melville gained popular attention in the late 1840s and 1850s, but his critical reputation then diminished before critics in the 1920s rediscovered his work. In the late-twentieth century the literary canon of Great Books expanded to include works from non-European or North American cultures and by women and minorities. Thus, while the high versus popular culture distinction has had particular resiliency, it has been applied to shifting sets of literary works.

The effects of technology on standards of literary taste have primarily concerned issues of reproduction associated with electronic media. In "The Work of Art in the Age of Mechanical Reproduction" (1936), Benjamin argues that advances in printing changed the status of art in making woodcut graphics reproducible in lithography, thereby enabling "graphic art to illustrate everyday life" (Benjamin 1985, p. 219). Benjamin notes the inverse relation of accessibility and quality of works of art that accounts for the popularity of a Chaplin film versus "the reactionary attitude toward a Picasso paint-

ing" (p. 234): "The greater the decrease in the social significance of an art form, the sharper the distinction between criticism and enjoyment of the public" (p. 234). His essay ends by suggesting the dangerous capacities of film to support totalitarianism.

Frederick Kittler also analyzes how the functions of literature depend upon contextual shifts of discourse systems and on changing technical capacities of media. Like Foucault, he organizes history into eras based on paradigms of how literature is read in relation to other discourses, and, like Benjamin, he is concerned about determining effects of technology on literature. Saul Ostrow references McLuhan's idea that technology extends the human body in remarking that "Kittler is not stimulated by the notion that we are becoming cyborgs, but instead by the subtler issues of how we conceptually become reflections of our information systems" (Kittler 1997, p. x). In an essay considering Bram Stoker's *Dracula* (1982), as a commentary on the reproducibility of technology, Kittler notes that communication systems determine modern interpretations and forecast the death of literature: "Under the conditions of technology, literature disappears ..."(Kittler 1997, p. 83).

Building on elements of Jacques Lacan, Foucault, and Derrida, Kittler theorizes about the discourse networks of 1800 and 1900. He identifies the classical romantic discourse network of 1800 according to its fundamental formulation of mothers socializing children through phonetic reading (*universal alphabetization*) and that of the modernist discourse network of 1900 by the influence of technologies such as the typewriter on writing and reading (*technological data storage*). Kittler recalibrates literary works and theories by representing them as media: "literature ... processes, stores, and transmits data" (Kittler 1990, p. 370). He argues that a transformed literary criticism ought to understand literature as an information network, thereby classifying literary study as a type of media studies. In representing literature as technology, Kittler's theories encourage literary criticism that connects works of art to scientific practices and theories.

Futurism

Agreeing with progressive thinkers who argued the benefits of modern technology, the early-twentieth-century Futurism movement recognized literature to be a form of imaginative anticipation of and stimulation toward scientific and technological change. Futurists reacted against Romantic conceptions of literature as a sentimental retreat from technology. In a 1909 manifesto,

Italian futurists such as Filippo Tommaso Marinetti proposed that products of the machine age might be celebrated alongside nature: "We will sing of the vibrant nightly fervour of arsenals and shipyards blazing with violent electric moons; greedy railway stations that devour smoke-plumed serpents; factories hung from clouds by the crooked lines of their smoke; bridges that stride the rivers like giant gymnasts ... adventurous steamers that sniff the horizon; deep-chested locomotives whose wheels paw the tracks like the hooves of enormous steel horses ..." (Tisdall and Bozzola 1978, p. 7). Marinetti excelled in performing manifestoes, designed to incite the crowd, at Futurist evenings; his arguments characterized "man as the conqueror of the universe, destined to impose change with the aid of science" (Tisdall and Bozzolla 1978, p. 89). Futurist painters concentrated on depicting dynamic forces, especially those of urban life. Photographers and filmmakers applied principles of Photodynamism to integrate light and line into action. Futurism encouraged poets, dramatists, and other writers to describe the life of matter without imposing versions of Romantic or pantheistic ego on material conditions.

Composers, architects, and activists were similarly drawn to the utopian promise of futurism. Antonio Gramsci, co-founder of the Italian Communist party, expressed sympathy for the Futurist attempts to destroy the foundations of bourgeois civilization because "they had a precise and clear conception that our era, the era of big industry, of the great workers' cities, of intense and tumultuous life, had to have new forms of art, philosophy, customs, language ..." (Tisdall and Bozzolla 1978, p. 201). In contrast, in "The Work of Art in the Age of Mechanical Reproduction," Benjamin pointed to how such radicalism, encouraged by technological change and promoting self-alienation, aestheticized destruction and contributed to Fascism.

Literature, Science, Technology, and Culture

Matthew Arnold in "Literature and Science" (1882) outlined a distinction between the disciplines later represented by C. P. Snow as the two cultures in his 1959 Rede lecture. Literary and cultural critics in the late-twentieth century changed the terms of such classification schemes in interpreting a range of texts—written, dramatized, ritualized, and so on—as cultural products. Clifford Geertz, Raymond Williams, and Victor Turner contributed fundamental concepts supporting the linguistic, or narrative, turn in anthropology and cultural studies. Geertz and Turner unpacked social events as cultural texts affecting individuals as community rituals,

while Williams looked at the symbolism of ordinary life that had previously been excluded from scholarly consideration. Sociologists Bruno Latour and Sharon Traweek examined laboratory life and scientists's networks and discourse. Their work, along with that of Stuart Hall and Frederic Jameson, among other cultural critics, effaced previously set boundaries dividing high and low culture, linked art and life, and blurred disciplinary divisions concerning methodologies.

Like writers and artists, scientists and technologists are subject to cultural ideologies and conditions, and they produce literature as well as a body of knowledge. Cultural critics understand literature and science as discursive, epistemological practices with reciprocal influence. Tracing the representations of scientists and scientific ideas in literature can be a critical step in confronting scientific theories and practices because literary genres entertain and educate. Scientific hypotheses and inventions in fictions and ethical issues represented in literature inspire scientists. Given the increasing imbrication of science and technology in everyday life, it is not surprising that many literary and artistic works weave such references into their discourse and offer some ethical commentary on their development and implementation.

Just as science and technology are constructed out of and influence social values, literary works reflect and refract cultural ideas and events, as Maurice Agulhon noted of the Rougon-Macquart novels by Emile Zola and their Darwinian intertexts. But the forms of engagement are not formulaic, with writers using literature to offer ethical arguments about science and technology. Romantic works privilege nature over technology, yet they inspire the individual to become a close observer of the natural world and thereby give some impetus to scientific study. Nineteenth-century campaigns against hunting for leisure and fashion and anti-vivisection movements, along with an appreciation for species developed post-Darwin and support for women's suffrage, inspired British women to write about nature (Gates 2002). U.S. writers such as Ralph Waldo Emerson and Henry David Thoreau promoted scientific observation of nature and reacted against the dehumanizing effects of technology. Melville's Moby-Dick (1851) describes the tools and techniques of whaling in telling the story of the doomed Ahab, who is willing to sacrifice his life and his crew to pursue the white whale. In his journals Household Words (1850–1859) and All the Year Round (1859–1870) and in a number of novels published serially in the mid-nineteenth century, Charles Dickens stimulated ethically inspired social reforms associated with technological changes of the Industrial Revolution; for example, he criticized how utilitarianism associated with factories crushes the human spirit in *Hard Times* (1854), how bureaucratic selfishness results in unjust incarceration in *Little Dorrit* (1855–1857), and how the law inexorably grinds on while ignoring human need in *Bleak House* (1852–1853).

Some feminist tales of science and technology suggest that ethical motivations inspire the creation of scientific knowledge and demonstrate how technology can be applied to effect social improvement. In the short story "Hilda Silfverling: A Fantasy" (1845), Lydia Maria Child depicts a conflict between scientific knowledge and domesticity but optimistically resolves it by technological means when the title character is preserved by a chemist experimenting with cryogenics rather than being executed for a crime she did not commit. Stories by Charlotte Perkins Gilman written between 1890 and 1916 in various magazines celebrate similar examples of women who escape from painful domestic situations by working, often by entrepreneurially employing an innovative management technique or adopting a new technology (Colatrella 2000). Gilman's utopian novel *Herland* (1915) imagines a matriarchal society that can alleviate psychic and social problems for women.

As scientists, particularly defenders of Charles Darwin from T. H. Huxley to Stephen Jay Gould, have appreciated, fiction and non-fiction literature helps people comprehend, digest, and accept scientific principles and applications. Although professional discourse in some fields can be too esoteric for non-scientists to appreciate, essays in newspapers and journals aimed at a broad range of scientists and/or the general public accessibly convey technical information, disseminating new ideas and articulating ethical issues of significance to scientists, technologists, and the public. Literary works of fiction, poetry, and drama also contextualize ethical dilemmas in pointed ways. Recent medical examples of how public understanding can influence scientific and technological processes include efforts to maintain ethical standards in testing AIDS vaccines in Africa, to speed up the drug review process for orphan diseases, and to administer treatment and research studies in a humane manner; in these cases, press reports and literary works (dramas, films, and novels) contributed to informing the public about science in public policy. The fiftieth anniversary of the atomic bombing of Japan inspired a number of books, novels, and films representing the scientific researchers and politicans involved. The fiftieth anniversary of the discovery of DNA also brought historical reconsiderations in film and in print, in this case docu-

menting Rosalind Franklin's contributions to James Watson's and Francis Crick's double helix model. While some considerations of science suggest the limitations of scientists and engineers, others verge on the hagiographical in representing their heroic dimensions. Whether one adopts Gould's ideal of literature as assisting in the process of scientific dissemination or Arnold's assumption that literature has an obligation to criticize science, almost everyone accepts that while researchers pursue knowledge for its own sake, it is impossible to disentangle scientific theory and practice and technological applications from morality and culture.

In conclusion, the interrelationships of ethics, science, and technology have often been represented in literature and other discursive media. Scientific and technical means have also sometimes been utilized to analyze literature, whether as tools of reproduction or as specific cultural circumstances affecting the production and reception of texts. While many literary works explore unpredictable and dangerous outcomes of scientific and technological experimentation, others consider the optimistic potentials of such work. Similarly, the enabling possibilities for humanity offered by computing and information technologies in recent decades have been invoked alongside constraints and problems that harm individuals and society. In studying technologies of representation such as writing, scholars connect humanistic study with scientific and technical research. Some critics and artists bring ethical perspectives to bear on representations of scientific and technology, while cultural historians and critics consider the scientific and technical mechanisms utilized in studying types of language and discourse forms such as the orally composed epic. In the Information Age, we recognize that media forms help structure our understanding and that out culturally constructed assumptions help develop and deploy technologies. Yet as questions concerning fetal tissue research and assisted reproduction testify, we have difficulty in believing that science and technology inevitably lead to progressive outcomes and that they are always ethically motivated and directed. We struggle to make sense of which historical representation of science and technology appears more accurate, while aiming to reduce the risks associated with current technologies and to design new and better ways of doing science and innovating technologies.

CAROL COLATRELLA

SEE ALSO *Asimov, Isaac; Brecht, Bertolt; Brave New World; Communication Ethics; Cybernetics; Foucault,*

Michel; Frankenstein; Huxley, Aldous; Hypertext; Illich, Ivan; Information; Information Ethics; Internet; Levi, Primo; McLuhan, Marshall; Morris, William; Movies; Rhetoric of Science and Technology; Science Fiction; Science, Technology, and Society Studies; Shelley, Mary Wollstonecraft; Thoreau, Henry David; Tolkien, J. R. R.; Utopias and Dystopias; Video Games; Wells, H. G; Zamyatin, Yevgeny Ivanovich.

BIBLIOGRAPHY

Aeschylus. (1956). "Prometheus Bound." In *Aeschylus II. (The Complete Greek Tragedies)*, ed. David Grene and Richmond Lattimore. Chicago: University of Chicago Press

Arnold, Matthew. (1882). "Literature and Science." *Nineteenth-Century* 12: 216–230. The first published version of Arnold's Cambridge Rede lecture, discussing whether natural science or literature and humanities, which is favored by him, ought to predominate in the curriculum.

Barthes, Roland. (1977). "The Death of the Author." In *Image Music Text*, trans. Stephen Heath. New York: Hill and Wang.

Benjamin, Walter. (1985). "The Work of Art in the Age of Mechanical Reproduction." In his *Illuminations*. New York: Schocken Books.

Bolter, Jay David, and Richard Grusin. (1999). *Remediation: Understanding New Media*. Cambridge, MA: MIT Press.

Child, Lydia Maria. (1997). *The Lydia Maria Child Reader*, ed. Carolyn Karcher. Durham, NC: Duke University Press.

Colatrella, Carol. (2000). "Work for Women: Recuperating Charlotte Perkins Gilman's Reform Fiction." *Research in Science and Technology Studies* 12: 53–76.

Colatrella, Carol. (2002). *Literature and Moral Reform: Melville and the Discipline of Reading*. Gainesville: University Press of Florida.

Dickens, Charles. (1850–1859). *Household Words*. London: Bradbury & Evans.

Dickens, Charles. (2003 [1852–1853]). *Bleak House*, ed. Nicola Bradbury. London and New York: Penguin.

Dickens, Charles. (1859–1870). *All The Year Round*. New York: J. M. Emerson & Co.

Dickens, Charles. (1969 [1854]). *Hard Times*. New York: Penguin.

Dickens, Charles. (1855–1857). *Little Dorrit*. London, Bradbury & Evans.

Ferguson, John. (1972). *A Companion to Greek Tragedy*. Austin: University of Texas Press.

Foucault, Michel. (1984). "What Is an Author?" In *The Foucault Reader*, ed. Paul Rabinow. New York: Pantheon.

Gates, Barbara T., ed. (2002). *In Nature's Name: An Anthology of Women's Writing and Illustration, 1780–1930*. Chicago: University of Chicago Press.

Gilman, Charlotte Perkins. (1979). *Herland*. New York: Pantheon.

Gilman, Charlotte Perkins. (1995). *The Yellow Wallpaper and Other Stories*, ed. Robert Shulman. Oxford: Oxford University Press.

Graff, Harvey. (1991). *The Legacies of Literacy: Continuities and Contradictions in Western Culture and Society*. Bloomington: Indiana University Press. Surveys the history of literacy in Western Europe and America. Examines philosophical, sociological, and anthropological issues associated with its development.

Havelock, Eric. (1963). *Preface to Plato*. Cambridge, MA: Harvard University Press.

Havelock, Eric A. (1976). *Origins of Western Literacy*. Toronto: Ontario Institute for Studies in Education.

Hawthorne, Nathaniel. (1987). *Selected Tales and Sketches*. New York: Penguin.

Hayles, M. Katherine. (1999). *How We Became Posthuman*. Chicago: University of Chicago Press.

Hesiod. (1983). *The Poems of Hesiod*, trans. with introduction and commentary by R. M. Frazer. Norman: University of Oklahoma Press.

Homer. (1963). *Odyssey*, trans. Robert Fitzgerald. New York: Doubleday.

Homer. (1975). *Iliad*, trans. Robert Fitzgerald. New York: Doubleday.

Illich, Ivan. (1978). *Toward a History of Needs*. New York: Pantheon.

Kittler, Friedrich. (1990). *Discourse Networks, 1800/1900*, trans. Michael Metteer, with Chris Cullens. Stanford, CA: Stanford University Press, 1990. Describes nineteenth-century romanticism and early twentieth-century modernism as constituting two discourse networks. Demonstrates that the meanings of literary texts are embedded other contemporaneous discourses.

Kittler, Friedrich. (1997). *Literature, Media, Information Systems: Essays*, ed. John Johnston. Amsterdam, The Netherlands: G & B Arts International.

Kittler, Friedrich (1999). *Gramaphone, Film, Typewriter*, trans. Geoffrey Winthrop-Young and Michael Wutz. Stanford, CA: Stanford University Press. Provides history of three technologies—gramaphone, film, and typewriter—that emerged in the nineteenth century. Analyzes sociopolitical and cultural discourses surrounding them, including contemporaneous appraisals and/or elaborations by Rainer Rilke, Franz Kafka, Martin Heidegger, Thomas Edison, Alexander Graham Bell, and Alan Turing. Considers computational media and information systems in relation to Jacques Lacan's psychological discourse and literary works such as Bram Stoker's *Dracula* and Thomas Pynchon's *Gravity's Rainbow*.

Lord, Albert. (1978). *The Singer of Tales*. New York: Atheneum.

Luke, Allan. (2003). "Literacy Education for a New Ethics of Global Community." *Language Arts* 81(1)(September): 20–22.

Massumi, Brian. (2002). *Parables for the Virtual*. Durham, NC: Duke University Press.

McLuhan, Marshall. (1995). *Essential McLuhan*, ed. Eric McLuhan and Frank Zingrone. New York: Basic Books.

McLuhan, Marshall. (1995). "Excerpt of 'The Gutenberg Galaxy.'" In *Essential McLuhan*, ed. Eric McLuhan and Frank Zingrone. New York: Basic Books.

Melville, Herman. (1988). *Moby-Dick, or The Whale*, ed. Harrison Hayford, Hershel Parker, and G. Thomas Tanselle. Chicago: Northwestern University/Newberry Library.

Murray, Janet H. (1997). *Hamlet on the Holodeck*. Cambridge, MA: MIT Press.

Ong, Walter J. (1982). *Orality and Literacy: The Technologizing of the Word*. New York: Routledge.

Shelley, Mary. (1991). *Frankenstein*. Oxford: Oxford University Press.

Tisdall, Caroline, and Angelo Bozzolla. (1978). *Futurism*. New York: Oxford University Press. A survey account of the development of the movement and the work of its practicioners.

Snow, C. P. (1964). *The Two Cultures; and, A Second Look*. London: Cambridge University Press. A later edition of Snow's 1959 Rede lecture, which argues that there is a divide between scholars in science and the humanities.

Villiers de L'isle Adam, Auguste. (1982). *Tomorrow's Eve*, trans. Robert Martin Adams. Urbana: University Of Illinois Press.

INTERNET RESOURCES

"Faust Legends." University of Pittsburgh. Available from http://www.pitt.edu/∼dash/faust.html#15/87.

Goethe, Johann Wolfgang von. *Der Zauberlehrling*. Available from http://german.about.com/library/blgzauberl.htm.

Halion, Kevin. "Deconstruction and Speech Act Theory: A Defence of the Distinction between Normal and Parasitic Speech Acts." e-anglais. com. Available from http://www.e-anglais.com/thesis.html. Ph.D. dissertation.

Pierre de la Mare. "Printing." Dotprint. 6 parts. Available from http://www.dotprint.com/fgen/history1.htm

SCIENCE, TECHNOLOGY, AND SOCIETY STUDIES

● ● ●

Science, Technology, and Society Studies, or STS, is an interdisciplinary field of academic teaching and research, with elements of a social movement, having as its primary focus the explication and analysis of science and technology as complex social constructs with attendant societal influences entailing myriad epistemological, political, and ethical questions. As such it entails four interlinked tenets or concepts that transcend simple disciplinary boundaries and serve as a core body of STS knowledge and practice. Several useful introductions to the STS field are available (Sismondo 2004, Cutcliffe and Mitcham 2001, Volti 2001, Cutcliffe 2000, Hess 1997, Jasanoff, et al. 1995).

Basic Themes

The field of Science, Technology, and Society Studies covers several basic themes.

CONSTRUCTIVISM. First and foremost, STS assumes scientific and technological developments to be socially constructed phenomena. That is, science and technology are inherently human, and hence value-laden, activities that are always approached and understood cognitively. This view does not deny the constraints imposed by nature on the physical reality of technological artifacts, but it does maintain that knowledge and understanding of nature, of science, and of technology are socially mediated processes.

CONTEXTUALISM. As a corollary to the notion of constructivism, it follows that science and technology are historically, politically, and culturally embedded, which means they can only be understood *in context*. To do otherwise would be to deny their socially constructed nature. This does not contradict *reality*, but does suggest that there are different contextualized ways of knowing. Likewise any given technological solution to a problem must be seen as contextualized within the particular socio-political-economic framework that gave rise to it.

PROBLEMATIZATION. A view of scientific knowledge and especially technological development as value-laden, and hence non-neutral, leads to the *problematization* of both. In this view science and technology have societal implications, frequently positive, but some negative, at least for some people. Thus it is not only acceptable, but, indeed, necessary to query the essence of scientific knowledge and the application of technological artifacts and processes with an eye toward evaluative and ethical prescription.

DEMOCRATIZATION. Given the *problematic* natures of science and technology, and accepting their *construction* by society, leads to the notion of enhanced democratic control of technoscience. Due to the inherent societal and ethical implications, there need to be more explicit participatory mechanisms for enhancing public participation in the shaping and control of science and technology, especially early in the decision-making process, when the opportunity for effective input is greatest. The ultimate goal is to structure science and technology in ways that are collectively the most democratically beneficial for society.

In adopting such a theoretical framework for the descriptive analysis and prescriptive evaluation of technoscience, STS serves as a location for discussing key societal and ethical issues of interest and concern to a democratic public. As such STS offers a set of conceptual tools and insights, themselves continually open to reflexive analysis and further evolution as scholars and activists gain ever more experience in understanding science and technology.

Historical Development

STS as an explicit academic field of teaching and research emerged in the United States in the mid-1960s, as scholars and academics alike raised doubts about the theretofore largely unquestioned beneficence of science and technology. Public concerns relating to such areas as consumerism, the environment, nuclear power, and the Vietnam War began to lead to a critique of the idea of technoscientific progress that many people had generally come to believe. Marked by such popular works as Rachel Carson's *Silent Spring* (1962) that raised questions about the hazards associated with chemical insecticides such as DDT and Ralph Nader's automotive industry expose, *Unsafe at Any Speed* (1965), STS reflected a widening activist and public engagement with technoscientific issues and concerns.

At approximately the same time this social movement was emerging, parallel changes within a number of traditional disciplinary academic fields were occurring. Evolving out of the work of scholars such as Thomas Kuhn, whose *The Structure of Scientific Revolutions* (1962), was tremendously influential, traditional philosophers, sociologists, and historians of science and technology, more or less independently of each other, began to move away from internalist positivist-oriented studies to reflect a more complete and nuanced understanding of the societal context of science and technology. Common to the intellectual analysis in each of these fields was criticism of the traditional notions of *objectivity* within scientific and technological knowledge and action, an examination that emphasized the value-laden contingent nature of these activities. As these fields evolved, they increasingly borrowed conceptual models and drew on case examples from each other, such that by the mid-1980s a clearly interdisciplinary academic field of study, replete with formalized departments and programs, professional societies, and scholarly journals, had emerged. Reflecting the more intellectual focus of their work, these scholars and their organizations began to use the term S&TS—Science and Technology Studies—to distinguish themselves from the more activist STS wing.

A third element or subculture within STS involves the more practice-oriented science and technology or engineering management and policy fields. Often referred to by the acronym STPP (Science, Technology and Public Policy) or SEPP (Science, Engineering, and Public Policy), this group is particularly interested in the practical policy issues surrounding science and engineering and in exposing scientific and engineering managers to the broader sociopolitical context they are likely to encounter. It too conducts research and scholarship and offers graduate education programs, but generally as part of a focused mission.

Collectively then this interdisciplinary group of scholars and sub-fields constitutes what has become known as STS or sometimes S&TS Studies. Together they examine the relationships between scientific ideas, technological machines and processes, and values and ethics from a wide range of perspectives. Independent of their specific motivations, approaches, and concerns, however, is a common appreciation for the complexities and contextual nature of science and technology in contemporary (and historical) society. Drawing on a strong base of empirical case studies by academic sociologists and historians of technoscience, more activist STSers and the STTP-oriented policy and management groups have since the 1990s been in a position to take a modest "turn toward practice" (Bijker 1993, p. 129) that should in principle, even if not always in practice, allow a more *democratic* public role in the ethical shaping and control of technoscience.

The STS Controversy

One result of this intellectual theorizing about the socially constructed nature of technoscience has been a strong, often polemical, backlash from certain quarters of the scientific community. This was unfortunate because much of the debate in what became known as the *Science Wars* appeared to miss, or ignore, the central focus and insights of STS, and was often polemical because of comments by participants on both sides. Many scientists hold tightly to the traditional ideal of objective knowledge based on reason and empirical evidence. For such individuals relativist claims that scientific knowledge is *socially constructed* and not to be found in an objective autonomous nature, but rather as the result of a set of historically and culturally elaborated set of conventions, was unsettling and struck more than a discordant note. Combined with widespread evidence of scientific illiteracy among school children and widely held pseudoscientific beliefs on the part of the general public, some scientists came to view much of STS as *anti-science* and indicative of a postmodern cultural decay.

Arguing in support of the objective nature of scientific evidence and science as a special way of knowing, a number of such individuals led by Paul Gross and Norman Levitt (1994) and Alan Sokol (1996a, 1996b,

1998) took issue with some of the more relativist-oriented STS scholars, such as Bruno Latour (1987), and launched a series of sharp attacks in print and at academic conferences. A spirited debate ensued, supposedly over the epistemological nature of scientific knowledge, but it veered into the social dynamics and political implications of science, and by association tended to indiscriminately taint all STS scholars as anti-science and engaged in a *flight from reason*.

Among the skirmishes Sokol, a physicist, wrote an article consisting of complete gibberish, but cast in postmodern constructivist language, that was published in the cultural studies journal, *Social Text* (Sokol 1996a), ironically in an issue intended as a response to the earlier work of Gross and Levitt (1994). Sokol was motivated by what he considered to be the "nonsense and sloppy thinking" that "denies the existence of objective realities" (Sokol 1996b, p. 63) and sought to expose it through his parody article, with the end result of adding fuel to the already hot fire of debate.

Without replaying the whole debate, which also included a bizarre invitation by Sokol for anyone who did not believe in scientific objectivity to come to his upper story office where they could test the law of gravity by stepping out the window, much of the dialog missed the common core of agreement that actually bound the combatants more closely together than perhaps at least science defenders realized. That is to say, most scientists, including Gross, Levitt, and Sokol, readily accept a *moderate constructivism*, one that views scientific knowledge of the natural world and its associated processes, and most certainly technological creations, to be *socially constructed* phenomena. Few moderate STS scholars or members of the public would deny the obdurate reality of nature, nor do they seek to control the underlying scientific epistemology, but it certainly is within reason for them to both understand and seek to control the sociopolitical implications of contemporary technoscientific advances. In the end then, it would appear there was probably more in common between the scientific combatants and that their *war* reflected much ado about little. Yet, at the same time, it does suggest just how difficult it may be for STS, either as a group of investigative scholars or as a social movement, to play an ethically and politically responsible role in the shaping and control of science and technology as the twenty-first century unfolds.

The Problem of Ethics

To say that incorporating an ethical awareness and normative framework into society's control and shaping of contemporary science and technology will be difficult, is not to say that it should not be attempted, nor that such attempts from within the STS community are not already occurring. Indeed that has been much of the raison d'etre of STS right from the beginning, even of those more intellectual scholars most interested in revealing the epistemological underpinnings of scientific knowledge. Thus it has been the case that STS social constructivists have often revealed the underlying values and ethical choice decisions made in scientific research and discovery, while those analyzing technological decision making, such as that surrounding the launch of the space shuttle Challenger (Vaughan 1996), similarly revealed the ethics of the decision to go forward that chilly Florida morning, even in the face of admittedly mixed evidence regarding the viability of O-rings at reduced temperatures. Other more specifically focused philosophers and ethicists have analyzed case studies of technoscientific failures or near failures, ranging from DC-10 aircraft landing gear to the San Francisco BART transportation system to the collapse of the Kansas City Hyatt Regency walkway, for what they reveal about the ethics and values subsumed in such technoscientific endeavors. Other scholars have examined such issues as the siting of toxic waste and hazardous manufacturing facilities because of what they show about environmental justice inequities.

Out of such analyses has come increased attention to the need to make scientists, engineers, and corporate managers much more socially and ethically attuned to the implications of their work. To that end, engineering education programs focus more attention on the ethics of engineering through required coursework, while organizations and groups such as the American Association for the Advancement of Science (AAAS), which established a Committee on Scientific Freedom Responsibility in 1975, and the computer science community, which created the ethics-oriented Computer Professionals for Social Responsibility in 1983, concentrate specific resources toward the effort to raise awareness of ethical issues.

Beyond this institutional level of response, increasing numbers of STS academic scholars have come to recognize and focus on normative concerns as an integral part of their work. In part this has been a response to the gauntlet thrown down by the political philosopher of technology, Langdon Winner (1993), who finds much of the largely *descriptive* constructivist analysis wanting in terms of *human well-being* and the *social consequences of technological choice*. One significant measure of the barometric shift in such matters has been the work of Wiebe Bijker, a leading constructivist scholar and the 2001–2003 President of the Society for the Social Studies of Science. In a number of works, including his 2001 pre-presidential address, Bijker explicitly argued the need for greater *political engagement* in matters technoscientific on the part of citizens and scholars alike, each drawing on the constructivist insights of STS. Such engagement in his view would entail much greater democratic participation in the technoscientific decision-making process on the part of the public and a larger role for STS scholars as *public intellectuals* who, by drawing on their STS insights, might contribute normatively to the civic enhancement of our modern technoscientific culture (Bijker 2001, 2003).

Summary

As the foregoing analysis suggests, STS, as an intellectual area of research and teaching, as applied policy analysis, and as a social movement, is not only a field well suited to explain the nature of science and technology (historically and in the contemporary world), but one that also holds out great promise for the normative and democratic enhancement of today's technoscientific society. STS both provides an analytical framework and serves as a locus of debate. Such is the potential of STS and the greatest opportunity for its application.

STEPHEN H. CUTCLIFFE

SEE ALSO *Interdisciplinarity; Merton, Robert; Scandinavian and Nordic Perspectives; Science, Technology and Law; Science, Technology, and Literature; Sokol Affair.*

BIBLIOGRAPHY

Bijker,Wiebe. (1993). "Do Not Despair: There is Life after Constructivism." *Science, Technology, & Human Values* 18 (Winter): 113–138.

Bijker, Wiebe. (2001). "Understanding Technological Culture through a Constructivist View of Science, Technology, and Society." In *Visions of STS: Counterpoints in Science, Technology, and Society Studies,* eds. Stephen H. Cutcliffe and Carl Mitcham. Albany: State University of New York Press.

Bijker, Wiebe. (2003). "The Need for Public Intellectuals: A Space for STS." *Science, Technology, & Human Values* 28 (Autumn): 443–450.

Carson, Rachel. (1962). *Silent Spring.* Boston: Houghton Mifflin.

Cutcliffe, Stephen H. (2000). *Ideas, Machines, and Values: An Introduction to Science, Technology, and Society Studies.* Lanham, MD: Rowman and Littlefield.

Cutcliffe, Stephen H., and Carl Mitcham, eds. (2001). *Visions of STS: Counterpoints in Science, Technology, and Society.* Albany: State University of New York Press.

Gross, Paul R., and Norman Levitt. (1994). *Higher Superstition: The Academic Left and Its Quarrels with Science.* Baltimore, MD: Johns Hopkins University Press.

Hess, David. (1997). *Science Studies: An Advanced Introduction.* New York: New York University Press.

Jasanoff, Sheila; Gerald E. Markle; James C. Petersen; and Trevor Pinch, eds. (2004). *Handbook of Science and Technology Studies.* Thousand Oaks, CA: Sage.

Kuhn, Thomas. (1962). *The Structure of Scientific Revolutions.* Chicago: University of Chicago Press.

Latour, Bruno. (1987). *Science in Action: How to Follow Scientists and Engineers through Society.* Cambridge, MA: Harvard University Press.

Nader, Ralph. (1965). *Unsafe at Any Speed: The Designed in Dangers of the American Automobile.* New York: Grossman.

Sismondo, Sergio. (2004). *An Introduction to Science and Technology Studies.* Malden, MA: Blackwell.

Sokol, Alan D. (1996a). "Transgressing the Boundaries: Toward a Transformative Hermeneutics of Quantum Gravity." *Social Text* 46/47(Spring/Summer): 217–252.

Sokol, Alan D. (1996b). "A Physicist Experiments with Cultural Studies." *Lingua Franca* 6(May/June): 62–64.

Sokol, Alan D., and Jean Bricmont. (1998). *Fashionable Nonsense: Postmodern Intellectuals' Abuse of Science.* New York: Picador.

Vaughan, Diane. (1996). *The Challenger Launch Decision: Risky Technology, Culture, and Deviance at NASA.* Chicago: University of Chicago Press.

Volti, Rudi. (2001). *Society and Technological Change.* New York: Worth.

Winner, Langdon. (1993). "Upon Opening the Black Box and Finding It Empty: Social Constructivism and the Philosophy of Technology." *Science, Technology, & Human Values* 18(Summer): 362–378.

SCIENTIFIC ETHICS

• • •

The term *scientific ethics* may refer to the ethics of doing science (Is one free to inject unwilling subjects with a pathogen so as to gain valuable scientific insights? or What role should animal experimentation play in biology?). In that sense, scientific ethics is a branch of applied ethics. The term may also refer to whether or not the methods and assumptions of science can be applied to the subject matter of ethics. The present entry is concerned with scientific ethics in the second sense—Can there be a science of norms?

Scientific ethics in this sense is often argued to be an oxymoronic term. Science deals in empirical facts, discovering what *is* the case, while ethics deals in normative matters, uncovering what *ought* to be the case. A scientific ethics would thus commit the naturalistic fallacy of confusing what is with what ought to be. Historically speaking, however, this distinction is as much the exception as the rule. Premodern ethical systems, such as the virtue theories of Plato and Aristotle, did not couch the debate about what ought to be done in a way that made facts and norms *non-overlapping magisteria* (Gould 2002). To understand the relationships between science and ethics, it is useful to begin with some working definitions.

Defining Ethics and Science

Ethics is divided into descriptive, normative, and metaethics. *Descriptive* ethics is the study of empirical facts related to morality, such as how people think about norms, use norms in judgment, or how the norms themselves evolve. There is a rich tradition of organizing knowledge about these things scientifically, ranging from the field of moral psychology (focusing on how people reason about norms) to some forms of sociobiology (studying how norms arose on evolutionary timescales).

Normative ethics is an attempt to organize knowledge about what human beings ought to do or intend, or what kind of people they ought to be—it provides guidance and advice. The three major versions of normative ethics are virtue theory, utilitarianism, and deontology. A virtue theoretic approach, such as found in Aristotle, focuses on the nature of persons or agents. Are they flourishing—functioning effectively as human beings—or failing to flourish? Virtue theorists focus on states of character (virtuous or vicious) and how they affect the ability to live the best human life. Utilitarians, such as Jeremy Bentham (1748–1832) or John Stuart Mill (1806–1873), focus instead on the consequences of an action, rather than the character of the person committing it. Specifically they look at the amount of happiness caused (or unhappiness prevented), with the happiness of all counting equally. Deontologists, such as Immanuel Kant (1724–1804), focus on the nature of the action itself rather than its consequences. Certain actions express appreciation for, and are done in accordance with, the demands of duty, respecting that which is the foundation of morality: rationality and autonomy.

Metaethical questions consider the scope and nature of moral terms. Do ethical terms such as good and bad refer to facts about the world, or merely to states of emotion in people making judgments? Does ethics constitute

knowledge or not; is ethical knowledge illusory? What is the structure of ethical arguments? It is less controversial that science may influence metaethical positions (although that position is also debated) than that there can be a science of normative ethics.

Science likewise comes in three forms. In the weakest sense, a science is an organized body of knowledge. If this is what is meant by science in relation to ethics, then a science of ethics certainly exists. The major moral theories just mentioned are attempts to bring some organization to what is known about morality.

Normally, though, science means something stronger and refers to a set of epistemological canons that guide inquiry. In one form, these canons are called *methodological naturalism*: the methods of inquiry used by an empirical science such as physics or biology. These include observation of the world, hypothesis formation, intervention and experiment, iterative formation and improvement of a theory, and more. Such activities are constitutive of the scientific method. If such methods can produce knowledge about norms, then a science of ethics is possible.

An even stronger form of science is *ontological naturalism*: Only those entities, events, and processes countenanced by the existing sciences may be used in theory construction. Methodological naturalism is a weaker form of science than an ontological naturalism. Consequently the possibility of an ethics grounded in ontological naturalism is more controversial.

In the weakest sense, ethics is a science if it can be organized into a coherent body of knowledge; in the moderate sense, ethics is a science if it can use the traditional epistemological canons of science to gain moral knowledge; and in the strongest sense ethics is a science if in addition to using the methods of science it also makes reference only to the entities and processes accepted by the extant, successful natural sciences. Only nihilists or radical moral particularists (those who contend that moral theory is so situation driven that general principles are impossible) would deny that there could be a science of norms in a weak sense. The moderate position is more controversial. Some would contend that moral knowledge is not gained using the empiricist methodology of the scientific method. For example, Kant's deontological theory does not require that humans reason empirically about morality; rather he maintains that they can know what they must do a priori independent of any particular experience. The strong position is the most controversial: Whether a normative theory can exist that differs neither in scope or content from the empirical sciences is debatable.

Naturalistic Fallacy

The argument offered most often against the possibility of scientific ethics in the moderate or strong senses is the naturalistic fallacy. First articulated by David Hume in *A Treatise of Human Nature* (1739), the naturalistic fallacy occurs when one moves from a list of empirical premises to a conclusion that contains a normative component. Hume is "surprised" when authors writing about ethics who were previously reasoning in the "usual way" suddenly begin to substitute "oughts" in places where before only the copula "is" had been present (Hume, Book III, Part I, Section I, Paragraph 24). Hume appears to point out a flaw in attempts to reason from the empirical to the normative—one will make reference to an unexplained term in a conclusion that was nowhere present in the empirical premises of the argument. Such an argumentative structure is invalid; the truth of the premises does not guarantee the truth of the conclusion. G. E. Moore advanced a similar argument early in the twentieth century when he argued that naturalized ethical systems fall prey to the *open question argument*. After one has identified normativity with a natural property such as avoidance of pain, for example, one can still meaningfully ask whether it is good to avoid pain. This means that utilitarians have not successfully reduced goodness to the natural property *avoiding pain*.

Whether or not the naturalistic fallacy and the open question argument provide *in principle* rationales against a moderate or strong scientific ethic is itself an open question. There are several possible responses. For example, both arguments rely on an analytic/synthetic distinction (a distinction between sentences true by definition and sentences true because of the way the world is), and many philosophers think no such distinction exists (see Casebeer 2003a). In addition, Hume's argument applies only to traditional deductive and inductive arguments. It may well be, though, that the relationship between natural ethical facts and the norms they deliver is *abductive*; one may best explain—abduction is often called *inference to the best explanation*—patterns of certain facts by assuming that they are also natural norms. Finally the open question argument probably does not generalize; it really amounts to saying that the two ethical systems Moore examines (Spencerian evolutionary ethics and hedonism) are not good natural ethical theories, and all but partisans would agree.

Why Scientific Ethics

Given disagreements about whether a scientific ethics in the moderate or strong sense is possible, why might people want such a thing? There are four possibly inter-related reasons. First science seems to some to have undermined traditional ethics, and hence human beings should use science to re-create ethics on firmer foundations. Second scientific ethics might be driven by concerns about the coherence of worldviews. Third scientific knowledge is the only real kind of knowledge. Fourth the sciences provide a prestige model, and in a highly scientific society people always try to imitate that which is of greatest prestige.

The first rationale may reflect a praiseworthy desire to reconsider long-standing issues in ethics from the perspective of contemporary science; for instance, what does contemporary cognitive science say about the existence of a free will, and what impact might this have on the conception of ethics? As another example, sociobiologists sometimes veer towards eliminativist extremes about the subject matter of ethics (morality is an illusion fobbed off on people by their genes). Strong scientific ethics thus might be a path to reconstruct what is purportedly illusory, whether it be a notion of agency compatible with the sciences or a scientific defense of the genuine objectivity of ethics.

The second rationale is closely related: Researchers may hold out hope that human knowledge can be unified. At the very least, they may ask that it be consistent across spheres of inquiry. Concerns about *consilience* can thus drive scientific ethics (Wilson 1975). The third and fourth rationales are strongly linked: If scientific knowledge is on a firmer footing than folk knowledge or nonempirical inquiry, then it is no wonder that funding and prestige would attach to scientific pursuits rather than not. Researchers in ethics may thus be attracted to the epistemic roots of science and the research support flowing from them. Sometimes this attraction leads to pseudoscientific ethics (just as it leads to pseudoscience), as in, for example, the work of Madam Vlabatsky's *theosophical scientific ethics* or in the eugenics movement. A thoughtful scientific ethics rejects pseudoscience and the pseudoethics that might follow.

Of course science advances, changing as time passes. Will attempts to connect science and ethics undermine the certainty some strive for in morality? They may, but this is no objection to the enterprise; it might be that the best one can hope for even in ethics is something like the *best guess hypothesis* offered by the practicing scientist.

Examples of Scientific Ethics

What might a moderate or strong scientific ethics look like? Herbert Spencer (1820–1903) claimed to offer such a theory in his work; he derived an evolutionary account of morality that is basically utilitarian in nature: If humans but allow the mechanisms of nature to do their work, there will be *natural social evolution* toward greater freedom. This will in turn lead to the greatest possible amount of happiness. While widely acclaimed during its time, Spencer's theory was ultimately rejected owing in part to its scientific inaccuracies, and to attacks upon it by Henry Sidgwick, Thomas Huxley, and G. E. Moore. At its worst, Spencer read repugnant norms into evolution; for example, here is what he said about Great Britain's Poor Laws, which mandated food and housing for the impoverished: "... there is an habitual neglect of the fact that the quality of a society is lowered morally and intellectually, by the artificial preservation of those who are least able to take care of themselves ... the effect is to produce, generation after generation, a greater unworthiness" (Spencer 1873 [1961], p. 313).

What might a more plausible scientific ethic look like? Such a theory might resemble that offered by the Greek philosopher Aristotle or the pragmatic philosopher John Dewey (1859–1952).

Aristotelian ethics is prescientific in the sense that the scientific revolution had not yet occurred; nonetheless, his method is empirical. For Aristotle, human flourishing is the *summum bonum* of existence; to say that an action is ethical or that a person is good is just to say that the action or the person contributes to or constitutes proper functioning. Contemporary ethicists have pursued this line of reasoning; for example, Larry Arnhart (1998) argues for a naturalized, Aristotelian ethical framework, and William Casebeer (2003a, b) argues that moral facts can be reduced to functional facts, with functions treated as an evolutionary biologist would (that is, as being fixed by evolutionary history). Leon Kass (1988) raises questions for such approaches; there are things that human passions and gut reactions say about the morality of certain actions that can never be captured with reason or the scientific method alone.

A related merging of science and ethics occurs in the work of the classic American pragmatists, such as Charles Pierce (1839–1914) and Dewey. Pierce argues that science itself is a form of ethics—it expresses respect for the values that underpin effective inquiry, and is subordinate to ethics insofar as it is human concerns about the efficacy of ideas that cause people to pursue science to begin with. Relatedly Dewey argues in his *Ethics* (1932) that the process of regulating ideas effectively—which is what science does in essence—enables human beings to become better able to express

values and act upon them. This approach of replacing *preexisting value* with the creation of value and understanding what genuinely follows from that positing of value is called axiology (Casebeer 2003a).

Even if moderate and/or strong versions of scientific ethics seem implausible, almost everyone admits that scientific results may limit the possible space of normative moral theories. Only the most trenchant antinaturalist would think that facts about human beings and how they reason have absolutely no bearing on moral concerns. These facts should, at the very least, constrain moral theorizing. For instance, Owen Flanagan advocates the *principle of minimal psychological realism*, which states that the moral psychologies required by moral theories must be *possible* for humans: "Make sure when constructing a moral theory or projecting a moral ideal that the character, decision processing, and behavior prescribed are possible ... for creatures like us" (Flanagan 1991, p. 32). So the scientific study of the genesis, neurocognitive basis, and evolution of ethical behavior *is* relevant to normative moral theory *even if* the moderate and strong versions of scientific ethics are misguided or fail.

Contemporary Developments and Future Possibilities

There are five general areas in which scientific research has the potential to constrain moral theory: moral psychology, decision theory, social psychology, sociobiology, and artificial modeling of moral reasoning. Moral psychologists focus on the psychological processes involved in moral thought and action. They study such phenomena as *akrasia* (weakness of the will), moral development, the structure of moral reasoning, and the moral emotions. Some of the best known work in this area revolves around moral cognitive development; Lawrence Kohlberg, for example, has formulated an empirically robust theory of moral development whereby people progress through three stages of moral reasoning, each broken into two levels. In the first stage, one reasons by asking, What's in it for me? In the second, one asks, What does culture or society say? In the third, one asks, To what contract would I be a party? What do universal moral principles demand? Progress through these stages or schema is universal and (with some exceptions) invariant. If Kohlberg is right, then perhaps a normative moral theory that takes issues of justice seriously is more viable than one that does not (although his research has been criticized for this very reason; see Lapsley 1996 for a summary).

Other moral psychologists have been exploring the relationship between reason and moral emotions such as guilt or shame. One longstanding debate in moral theory has involved the relationship between having a moral reason to do something and whether that reason necessarily motivates an individual to take action. Internalists (such as Plato or Kant) argue that moral reasons necessarily motivate: If, morally speaking, one ought not to do something then one will, *ceteris paribus*, be motivated not to do that thing. Externalists (such as Aristotle) argue that a moral reason must be accompanied by an appropriate motivational state (such as an emotion) in order to spark action. If certain normative moral theories require either an internalist or externalist psychology in order to be plausible, then results from empirical research may constrain moral theory. For example, Adina Roskies (2003) argues persuasively that neurobiological data about the relationship between emotion and reason rules out internalism and makes a Kantian psychology implausible. Other issues in moral psychology will stand or fall with progress in the cognitive sciences; for instance, moral cognitive development and moral concept development may both be subsumed by research into cognitive and concept development in general.

Decision theorists study the determinants of human choice behavior. Traditional rational actor assumptions (such as possessing unlimited time and computational power, a well-ordered preference set, and indifference to logically equivalent descriptions of alternatives and choice sets) usually inform decision theory. Whether or not these assumptions apply to human reasoning when it is done well may affect whether normative moral theories must be essentially rational and hence whether they must respond to the same norms as those of reason traditionally construed. Much work in decision theory has revolved around either extending the predictive power of traditional rational actor assumptions, or in articulating alternative sets of rational norms to which human cognition should be responsive. For instance, Amos Kahneman and Daniel Tversky's (1982) heuristics and biases research program explores the shortcuts human beings take to achieve a reasonable result when under time pressure or when working with incomplete information. It may very well be that normative moral theories constitute sets of heuristics and biases.

Gerd Gigerenzer and the Adaptive Behavior and Cognition Research Group (2000) focus on ecological rationality, demonstrating that traditional rational canons can actually lead people astray in certain environments. While there is a rearguard action to shore up traditional rational actor driven decision theory, in all likelihood, progress on this front will require articulating a new conception of rationality that is ecologically valid and cognitively realistic. The results of this program may, in turn, affect the structure of normative moral theory in much the same way that the structure of normative rational actor theory has been and will be affected.

Social psychologists study human cognition and emotion in the social domain. Given that moral judgments are paradigmatically about how people ought to treat others, work in this area usefully constrains normative theorizing. One controversy regards whether or not the *fundamental attribution error* (the human tendency to undervalue the situational influences on behavior and overvalue the internal character-driven causes) undermines traditional approaches to virtue theory. If, as some social psychologists argue, there is no such thing as *bravery* as a general trait, but rather only such fragmented virtue-theoretic traits as *brave while standing in the checkout line at the grocery store*, then it may very well be that virtue theory will have to become much more sophisticated if it is to be plausible (see Doris 2002 for a comprehensive discussion, as well as Harman 2000; Doris and Stich 2003 also offer a useful survey). The social nature of moral reasoning means that the latest studies of social psychological behavior can, on the weakest view, usefully constrain normative theorizing, and on a stronger view can usefully coevolve with it.

Sociobiologists such as E. O. Wilson study the origin and evolution of (among other things) moral norms. They argue that genes keep moral culture on some sort of leash: At the very least, the capacities human beings use to reason about morality are evolved capacities and need clear connections to the environments in which these capacities evolved; maximally moral norms may be *nothing more* than norms that have enabled organisms and groups of organisms to increase their genetic fitness. Sociobiological approaches to human social behavior have been controversial, but have nonetheless shed much light on how both the capacity to reason morally and the structure of some moral norms came to be (Boehm 1999, for example, discusses the evolution of egalitarian norms). Game-theoretic work on the evolution of the social contract and other moral norms has illuminated aspects of ethical behavior ranging from the propensity to be altruistic to the temptation to defect on agreements in certain instances. Sociobiological study reinforces the notion that any accepted normative theory should have a describable evolution and a discernable way of maintaining its existence (see Binmore 1994).

Computer models at both the micro and macro level have usefully informed all these fields of research. Changes in technology have influenced what philosophers make of the possibility of scientific impact on ethics. For example, Rene Decartes's inability to reconcile how mental states could be identical to brain states drove, at least in part, his dualism. The advent of in vitro methods for identifying the neural machinery of cognitive activity, such as Positron Emission Tomography (PET) and functional Magnetic Resonance Imaging (MRI), may have headed off dualism at the philosophic pass if such technologies were available during his time. The spread of inexpensive and powerful computing technology has made possible everything from the simulation of artificial societies (and hence has influenced sociobiological approaches) to the simulation of moral reasoning in an individual (and hence has influenced moral psychology). On the social simulation front, promising work by Jason Alexander and Bryan Skyrms (1996) on the evolution of contracts has usefully informed moral theorizing. On the individual level, work by cognitive modelers such as Paul Thagard (2000) and Paul Churchland (2001) has highlighted areas where normative moral theory can intersect with cognitive modeling.

Assessment

Is scientific ethics possible? Appropriately enough, this is an empirical matter. Should the promise held out by the rapidly progressing cognitive, biological, and evolutionary sciences be realized, there is reason to be sanguine about the moderate and strong programs for a scientific ethic. Science could reaffirm some of the pre-scientific insights into the nature of morality. But even if this very possibility is a misguided hope, scientific insights into human nature and cognition can usefully constrain the possible space of normative moral theory, and in this sense the existence of scientific ethics is a foregone conclusion. Science and ethics are indeed both magisterial, but they are, ultimately, overlapping.

WILLIAM D. CASEBEER

SEE ALSO Aristotle and Aristotelianism; Berlin, Isaiah; Decision Theory; Deontology; Emotion; Hume, David; Kant, Immanuel; Levi, Primo; Mill, John Stuart; Sociobiology; Spencer, Herbert.

BIBLIOGRAPHY

Aristotle. (1985). Nichomachean Ethics. Indianapolis, IN: Hackett. Aristotle's best-known work in normative ethics.

Aristotle was a student of Plato, another foundational figure in virtue ethics.

Arnhart, Larry. (1998). Darwinian Natural Right: The Biological Ethics of Human Nature. Albany: State University of New York Press. Defends a contemporary version of Aristotelian ethics using evolutionary biology.

Bacon, Francis. (1605 [2001]). The Advancement of Learning, ed. Stephen J. Gould. New York: Modern Library. Arguably the humanities/natural science split first occurred with Francis Bacon; a useful introduction to his thought.

Bacon, Francis. (1620 [1994]). Novum Organum. Chicago: Open Court Publishing Company. A good statement of the canons of empirical science early in their developmental history.

Binmore, Ken. (1994). Game Theory and the Social Contract, Vol. 1: Playing Fair. Cambridge, MA: MIT Press. Using assumptions from game theory, argues against the possibility of a Kantian reconstruction of John Rawls's theory of justice.

Boehm, Christopher. (1999). Hierarchy in the Forest: The Evolution of Egalitarian Behavior. Cambridge, MA: Harvard University Press. Discusses the evolution of egalitarian tendencies in apes and hominids.

Caplan, Arthur L., ed. (1978). The Sociobiology Debate: Readings on the Ethical and Scientific Issues Concerning Sociobiology. New York: Harper and Row. While dated, nonetheless serves as an excellent introduction to the sociobiology debate.

Casebeer, William D. (2003a). Natural Ethical Facts: Evolution, Connectionism, and Moral Cognition. Cambridge, MA: MIT Press. Argues for a strong form of scientific ethics, recapitulating a neo-Aristotelian virtue theory using resources from evolutionary biology and cognitive neuroscience.

Casebeer, William D. (2003b). "Moral Cognition and Its Neural Constituents." Nature Reviews: Neurosciences 4: 841–847. Discusses the interplay between neuroscience and normative moral theory while reviewing the neural mechanisms of moral cognition.

Churchland, Paul M. (2001). "Toward a Cognitive Neurobiology of the Moral Virtues." In The Foundations of Cognitive Science, ed. João Branquinho. Oxford: Oxford University Press. Reprint of Churchland's seminal 1998 article; discusses the affinity between virtue theory and neural network conceptions of cognition.

Dewey, John. (1932 [1989]). Ethics. Carbondale: Southern Illinois University Press. Dewey's best known text in moral theory.

Doris, John M. (2002). Lack of Character: Personality and Moral Behavior. Cambridge, UK: Cambridge University Press. Surveys results from contemporary social psychology, arguing that they make implausible the kinds of traits required by virtue theory.

Doris, John M., and Stich, Stephen. (2003). "As a Matter of Fact: Empirical Perspectives on Ethics." In The Oxford Handbook of Contemporary Analytic Philosophy, eds. Frank Jackson, and Michael Smith. Oxford: Oxford University Press. Offers an excellent introduction to empirical issues in ethics.

Elman, Jeff L., et al. (1996). *Rethinking Innateness: A Connectionist Perspective on Development*. Cambridge, MA: MIT Press. Argues that neural network accounts of cognition can account for multiple facts about the development of language in humans.

Flanagan, Owen. (1991). *Varieties of Moral Personality: Ethics and Psychological Realism*. Cambridge, MA: Harvard University Press. Discusses the relationship between psychology and normative moral theory.

Gigerenzer, Gerd; Peter M. Todd; and the ABC Research Group. (2000). *Simple Heuristics That Make Us Smart*. Oxford: Oxford University Press. Discusses the heuristics and biases at work in human thought, including what impact this has for the human conception of rationality.

Gould, Stephen J. (2002). *Rocks of Ages: Science and Religion in the Fullness of Life*. New York: Ballantine Books. Discusses the relationship between ethics and science, arguing that the two are nonoverlapping magisteria.

Harman, Gilbert. (2000). "Moral Philosophy Meets Social Psychology: Virtue Ethics and the Fundamental Attribution Error." In *Explaining Value and Other Essays in Moral Philosophy*. Oxford: Oxford University Press. Argues that the fundamental attribution error demonstrates that there are no such things as character traits.

Hume, David. (1739 [1985]). *A Treatise of Human Nature*, ed. Lewis A. Selby-Bigge. Oxford: Clarendon Press. Clearly articulates the naturalistic fallacy in this defense of noncognitivism about ethics.

Kant, Immanuel. (1785 [1993]). *Grounding for the Metaphysics of Morals*, trans. James Ellington. Indianapolis, IN: Hackett. A concise summary of Kant's views on ethical issues and one of the best-known works in deontology.

Kahneman, Daniel; Paul Slovic; and Amos Tversky, eds. (1982). *Judgment Under Uncertainty: Heuristics and Biases*. Cambridge, UK: Cambridge University Press. The seminal work in the heuristics and biases research program.

Kass, Leon. (1988). *Toward a More Natural Science: Biology and Human Affairs*. New York: Free Press. Argues that science can go too far if it is not appropriately regulated by the wisdom contained in emotional reactions to certain technological advances.

Lapsley, Daniel K. (1996). *Moral Psychology*. Boulder, CO: Westview Press. An excellent survey of issues in moral psychology.

Moore, G. E. (1902 [1988]). *Principia Ethica*. Amherst, NY: Prometheus Books. Articulates the open question argument against strong scientific ethics.

Roskies, Adina. (2003). "Are Ethical Judgments Intrinsically Motivational?: Lessons from Acquired Sociopathy." *Philosophical Psychology* 16(1): 51–66. Seminal article tackles the internalism/externalism debate (do moral reasons necessarily move people to act?) using neurobiological evidence.

Skyrms, Brian. (1996). *Evolution of the Social Contract*. Cambridge, UK: Cambridge University Press. Brings iterated game theory to bear on the evolution of the norms of justice and exchange; useful work in this area has also been accomplished by Jason Alexander, who studied under Skyrms.

Spencer, Herbert. (1873 [1961]). *Study of Sociology*. Ann Arbor: University of Michigan Press. One of Spencer's best known works in the field of scientific ethics.

Thagard, Paul. (2000). *Coherence in Thought and Action*. Cambridge, MA: MIT Press. Uses computer models to explain how concerns about coherence affect moral reasoning.

Wilson, Edward O. 1975 (2000). *Sociobiology: The New Synthesis*, 25th Anniversary edition. Cambridge, MA: Harvard University Press. Single-handedly established the field of sociobiology (or, in contemporary terms, evolutionary psychology).

SCIENTIFIC INTEGRITY

SEE *Research Integrity*.

SCIENTIFIC REVOLUTION

• • •

In the first half of the twentieth century it became a commonplace notion that modern science originated in a seventeenth-century "revolution" in thought precipitated by a new methodology for studying nature. In the last third of the twentieth century, a consensus developed among historians, philosophers, and sociologists of science that the emergence of modern science was more evolutionary than revolutionary. Furthermore, while modern science for 300 years claimed that its methodology generated value-free, objective knowledge, the late-twentieth-century consensus was that, implicitly and explicitly, the practice of science incorporated moral, ethical, and social value judgments.

The Seventeenth-Century Achievement

A fundamentally new approach to the study of nature did indeed emerge in seventeenth-century western Europe. The first herald of this development was Francis Bacon (1561–1626), who argued for a renovation in the human conception of knowledge and of knowledge of nature in particular. Especially in his *Novum Organum* (1620; New instrument [for reasoning]), Bacon formulated a radically empirical, inductive, and experimental-operational methodology for discovering laws of nature that could be put to use to give humankind power over nature. Bacon was primarily a social reformer who believed that knowledge could become an engine of national prosperity and power, improving the quality of life for all. To that end, he championed widespread education for all classes of society, featuring a strong mechanical-technical component that would assure widespread ability to create and maintain technological innovations. (The island of Laputa episode in Jonathan Swift's novel *Gulliver's Travels* (1726) mocks the Baconian faith in science-based innovation as improving the quality of life.)

Bacon was strongly opposed to mathematical accounts of natural phenomena, seeing in them a continuation of Renaissance magical nature philosophy and an erroneous commitment to deductive reasoning. René Descartes (1596–1650) by contrast, especially in his *Rules for the Direction of the Mind* (written 1628, but not published until 1701) and *Discourse on Method* (1637), roughly contemporary with Bacon's *Novum Organum*, articulated a mathematical and rigorously deductive, hence rational methodology for gaining knowledge of nature that employed experiment only to a limited degree and cautiously, because experimental results are ambiguous and subject to multiple interpretations. Descartes's own theory of nature was mechanistic, materialistic, and mathematical, hence deductive and deterministic. It became the basis for the mechanical worldwiew that was incorporated into enlightenment thinking and epitomized the view of nature as a clockwork world. Unlike Bacon, Descartes was a practicing researcher and a mathematician. He introduced analytical geometry—enabling algebraic solution of geometric problems—developed a materialistic cosmology in which the solar system and Earth formed naturally, discovered the reflex arc in his anatomical researches, developed a mechanical theory of life and biological processes, and wrote influentially on mechanics and optics, formulating his own theory of light.

Galileo Galilei (1564–1642), in his *Dialogues Concerning Two New Sciences* (1638), presented a deductive mathematical-experimental methodology that he attributed to Archimedes (c. 287–212 B.C.E.), several of whose treatises were translated into Latin and circulated widely beginning in the second half of the sixteenth century. In this work Galileo founded engineering mechanics and the mathematical theory of strength of materials, and he also extended and corrected earlier contributions to the science of mechanics (while perpetuating the mistaken notion that circular motion was "natural" and hence force-free). This work supplemented his more famous discoveries in astronomy based on his pioneering application of the telescope to the study of the moon and planets, and his defense of Copernicanism, the Sun-centered cosmological theory of Nicolaus Copernicus (1473–1543).

The Newtonian Triumph

Galileo's methodology probably comes closest to what people mean when they refer to "the scientific method"

and its invention in the seventeenth century. It reached its mature form in the hands of Isaac Newton (1642–1727) in the last third of the century. In all of his work, but especially in his majestic *Mathematical Principles of Natural Philosophy* (1687), considered the single most influential scientific text ever, and in *Optics* (1704), Newton synthesized induction and deduction, mathematics, and experimentation into a powerful methodology capable of revealing, in his view, the hidden "true causes" responsible for the phenomena of empirical experience. Like Descartes, whose methodology (and theories) he dismissed contemptuously, Newton made major contributions to mathematics, inventing, independently of Gottfried Wilhelm Leibniz (1646–1716), the calculus; to optics, inventing the reflecting telescope, discovering the phenomenon of diffraction and the seven-color composition of sunlight, and formulating a corpuscular, or particle, theory of light that would be dominant until the wave theory of light gained ascendance in the nineteenth century; to mechanics, in his famous three laws of motion; and to a theory of the universe based on his universal theory of gravitation, which provided a full account of the planetary orbits, confirming the validity of the earlier, scattered insights of Johannes Kepler (1571–1630).

Contrary to Descartes, who believed that matter was infinitely divisible, Newton favored an atomic theory of matter, and based physics and chemistry on a variety of forces acting nonmechanically and/or at a distance, rather than basing it only on mechanical contact forces. Newton's scientific style and his accomplishments represent the peak achievement of the seventeenth-century Scientific "Revolution." Until the mid-eighteenth century, many Continental natural philosophers—the term *scientist* was invented only in the 1830s—remained committed to Descartes's strictly mechanical model of scientific explanation while rejecting Descartes's particular theories. After that, Newtonianism effectively defined "modern" scientific study of nature until the early twentieth century and the rise of relativity and quantum theory.

By the end of the seventeenth century, then, modern science was firmly established, not only in mathematical physics and astronomy, but as a comprehensive philosophy of nature that was deterministic and materialistic, though explanations incorporated immaterial forces—such as gravity, electrical and magnetic attraction/repulsion, and selective chemical affinity—that acted according to strictly mathematical laws. This materialistic-deterministic approach to nature was broadly applied to biological and medical phenomena, especially in Italy and at the University of Padua, as reflected in William Harvey's (1578–1657) demonstration in 1628 of the closed circulation of the blood pumped by the heart and by the Galileo-influenced work of Giovanni Borelli (1608–1679) and others on the mechanics of the human skeletal and skeletal-muscular systems.

Even more than the telescope, the mid-seventeenth-century invention of the microscope by Antoni van Leeuwenhoek (1632–1723) revealed the existence of new worlds. The demonstration by Blaise Pascal (1623–1662) and Evangelista Torricelli (1608–1647) of the mechanical pressure exerted by the atmosphere using a simple barometer, which also showed that a vacuum could be created, strongly reinforced the mechanical conception of nature. A critical contribution to the new philosophy of nature was Christiaan Huygens's (1629–1695) midcentury demonstration that circular motion required a force to maintain it, contrary to the previous 2,000 years of Western thought. Descartes and Galileo both misunderstood this fact, which became a cornerstone of modern mechanics in Newton's principle of inertia. By the rise of the enlightenment in the second half of the eighteenth century, an amalgam of Descartes's mechanical worldview Cartesian mechanism and Newtonian deterministic mathematical physics was applied to society and its institutions, for example, by the Baron de Montesquieu (1689–1755), Anne-Robert-Jacques Turgot (1727–1781), and the Marquis de Condorcet (1743–1794) in France, and even to the human mind, for example, by David Hume (1711–1776) and Étienne Bonnot de Condillac (1715–1780).

Newtonianism Dethroned

In the nineteenth century, Newtonianism was severely challenged, and in the twentieth century it was displaced. The relationship between increasingly abstract mathematical models of nature and "reality" became an issue. The models worked empirically, but did they also provide a picture of reality? Meanwhile, the wave theory of light overthrew Newton's corpuscular theory and when incorporated by James Clerk Maxwell (1831–1879) into an electromagnetic field theory of energy led to attributing causal efficacy to space-filling immaterial entities. The introduction of the concept of energy on a par with matter diluted the deterministic materialism of modern science, while the new science of thermodynamics revealed that Newton's conception of time was flawed. Finally, with the kinetic theory of gases, statistical explanations were introduced into physics, which called determinism into question. With relativity and quantum theory, from 1905 on, Newtonian conceptions of space, time, matter, force, cause, and explanation, and Descartes's deductive model of rationality would all be replaced, and a fundamentally new form of science and a new, statistical conception of reality would emerge.

Seventeenth-century nature philosophy had presented itself as a body of impersonal knowledge, as simply descriptive of the way things were "out there," independent of personal, social, and cultural values. Given the religious wars of the first half of the seventeenth century, and the explicitly values-steeped character of Renaissance nature philosophy, this was a major epistemological innovation. The value-free character of the knowledge was guaranteed, it was thought, by a methodology employed in acquiring it that eliminated the influence of the subject on knowledge. However attractive such a conception of knowledge was then and continued to be through the nineteenth century, it created a gulf between facts and values, between knowledge and its applications, that in principle could not be bridged by reason, which increasingly came to be defined as reasoning in the scientific (hence objective) manner.

Bacon tacitly assumed that people would know what to do with the new mastery of nature that scientific knowledge would give them. But already by the mid-seventeenth century, the educational reformer John Amos Comenius (1592–1670) was warning that the new science was as likely to create a hell on Earth as a manmade heaven if application-relevant values were not explicitly linked to knowledge. In fact, right through the twentieth century and into the twenty-first, modernism, first in the West and then globally, has borne witness to the accuracy of Comenius's warning. While the scope and explanatory/predictive power of science in the nineteenth and twentieth centuries increased dramatically and became the basis of life-transforming technological innovations, there was no commensurate increase in conceptual "tools" for identifying which innovations to implement or how to implement them. Elimination of any influence on knowledge of the values held by the subject of knowledge eliminated any influence of knowledge on the values held by subjects!

As a result, even as science and technology became, after 1800, the primary agents of social change around the world, scientists and engineers remained outsiders to the terms of that change, which was driven overwhelmingly by scientifically nonrational political and market values. Both government funding of scientific research, especially in the United States after World War II, and industry dependence on science for technological innovations blurred the distinction between pure and applied science, reinforcing the post-1960s critique of science as in fact a value-laden ideology and not objective knowledge.

STEVEN L. GOLDMAN

SEE ALSO *Enlightenment Social Theory; Industrial Revolution; Modernization; Secularization.*

BIBLIOGRAPHY

Hankins, Thomas L. (1985). *Science and the Enlightenment.* Cambridge, UK: Cambridge University Press. Outlines the spread of Newtonianism in the eighteenth century.

Harman, P. M. (1982). *Energy, Force, and Matter.* Cambridge, UK: Cambridge University Press. Traces the growing conceptual complexity of nineteenth-century science.

Nye, Mary Jo. (1996). *Before Big Science: The Pursuit of Modern Chemistry and Physics, 1800–1940.* New York: Twayne. Excellent account of physics and chemistry at the dawn of their connection to government.

Shapin, Steven, and Simon Schaffer. (1985). *Leviathan and the Air-Pump.* Princeton, NJ: Princeton University Press. Classic study of the sociocultural context of the seventeenth-century Scientific Revolution.

Webster, Charles. (2002). *The Great Instauration: Science, Medicine, and Reform, 1626–1660,* 2nd edition. Oxford, UK: Peter Lang. Detailed account of the social context of Bacon's ideas and their influence on modern science.

Westfall, Richard S. (1971). *The Construction of Modern Science.* New York: Wiley. Excellent short history of seventeenth-century science.

Westfall, Richard S. (1993). *The Life of Isaac Newton.* Cambridge, UK: Cambridge University Press. Revised version of classic biography.

SCIENTISM

• • •

Scientism is a philosophical position that exalts the methods of the natural sciences above all other modes of human inquiry. Scientism embraces only empiricism and reason to explain phenomena of any dimension, whether physical, social, cultural, or psychological. Drawing from the general empiricism of the Enlightenment, scientism is most closely associated with the positivism of Auguste Comte (1798–1857), who held an extreme view of empiricism, insisting that true knowledge of the world arises only from perceptual experience. Comte criticized ungrounded speculations about phenomena that cannot be directly encountered by proper observation, analysis, and experiment. Such a doctrinaire stance associated with science leads to an abuse of reason that transforms a rational philosophy of science into an irrational dogma (Hayek 1952). It is this ideological dimension that is associated with the term *scientism.* In the early twenty-first century the term is used with pejorative intent to dismiss substantive arguments that appeal to scientific authority in contexts in which science might not apply. This overcommitment to science can be seen in epistemological distortions and abuse of public policy.

Epistemological scientism lays claim to an exclusive approach to knowledge. Human inquiry is reduced to matters of material reality. We can know only those things that are ascertained by experimentation through application of *the* scientific method. And because *the method* is emphasized with such great importance, the scientistic tendency is to privilege the expertise of a scientific elite who can properly implement the method. But the science philosopher Susan Haack (2003) contends that the so-called scientific method is largely a myth propped up by scientistic culture. There is no *single* method of scientific inquiry. Instead, Haack explains that "scientific inquiry is contiguous with everyday empirical inquiry" (p. 94). Everyday knowledge is supplemented by evolving aids that emerge throughout the process of honest inquiry. These include the cognitive tools of analogy and metaphor that help to frame the object of inquiry in familiar terms. They include mathematical models that enable the possibility of prediction and simulation. Such aids include crude, impromptu instruments that develop increasing sophistication with each iteration of a problem-solving activity. And everyday aids include social and institutional helps that extend to lay practitioners the distributed knowledge of the larger community. According to Haack, these everyday modes of inquiry open the scientific process to ordinary people and they demystify the epistemological claims of the scientistic gatekeepers.

The abuse of scientism is most pronounced when it finds its way into public policy. A scientistic culture privileges scientific knowledge over all other ways of knowing. It uses jargon, technical language, and technical evidence in public debate as a means to exclude the laity from participation in policy formation. Despite such obvious transgressions of democracy, common citizens yield to the dictates of scientism without a fight. The norms of science abound in popular culture, and the naturalized authority of scientific reasoning can lead, if left unchecked, to a malignancy of cultural norms. The most notorious example of this was seen in Nazi Germany where a noxious combination of scientism and utopianism led to the eugenics excesses of the Third Reich (Arendt 1951). Policy can be informed by science, and the best policies take into account the best available scientific reasoning. Lawmakers are prudent to keep an ear open to science while resisting the rhetoric of the science industry in formulating policy. It is the role of science to serve the primary interests of the polity. But government in a free society is not obliged to serve the interests of science. Jürgen Habermas (1978) warns that positivism and scientism move in where the discourse of science lacks self-reflection and where the spokespersons of science exempt themselves from public scrutiny.

MARTIN RYDER

SEE ALSO *Conservatism; Technicism; Technicization.*

BIBLIOGRAPHY

Arendt, Hannah, (1973 [1951]). *The Origins of Totalitarianism.* New York: Harcourt Brace. A student of philosopher and Nazi Martin Heidegger, Arendt recounts the social and intellectual conditions that gave rise to totalitarianism in Germany.

Haack, Susan. (2003). *Defending Science—Within Reason: Between Scientism and Cynicism.* Amherst, NY: Prometheus Books. A critical assessment by a logician and philosopher of science.

Habermas, Jürgen. (1978). "The Idea of the Theory of Knowledge as a Social Theory." In his *Knowledge and Human Interests,* 2nd edition, trans. Jeremy J. Shapiro. London: Heinemann Educational.

Hayek, Friedrich A. (1952). *The Counter-Revolution of Science: Studies on the Abuse of Reason.* Glencoe, IL: Free Press.

Midgley, Mary. (1992). *Science as Salvation: A Modern Myth and Its Meaning.* London: Routledge. A general criticism of scientism by a moral philosopher.

SECULARIZATION

• • •

Secularization is a concept important to science, technology, and ethics, because it encapsulates influential general theories about how moral influence may be exercised over and by science and technology under different historical and social conditions.

Most societies incorporate practices, beliefs, and institutions that correspond roughly to the domain of religion in modern Western cultures. These religious features presuppose the existence of non-human entities with powers of agency (i.e., gods) or the existence of impersonal powers endowed with moral purposes (i.e., karma). Moreover they generally assume that these non-human agents or powers have an impact upon human affairs. Secularization is a process by which religion comes to have decreasing importance in society along several dimensions.

First there is a decline in the status, prestige, and power of persons, practices and institutions associated primarily with religion. Second there is a decline in the importance of religion for the exercise of non-religious roles and institutions, including those associated with politics and the economy. Third there is a decline in the number of persons who take religion seriously and the degree of seriousness with which those involved in religion continue to take it. Secularization is highly correlated with the extent of industrialization in a society and with the development of scientific practices and institutions. But there is serious disagreement regarding whether secularization is largely a consequence of the growth of science and industry; whether science, industrialization, and secularization are relatively independent features of a more general process of modernization; or whether secularization is a prerequisite rather than a consequence of the growing importance of science in a society.

Three Theories of Secularization

Though he did not use the term, Auguste Comte (1798–1857) offered the first major theory of secularization in articulating what he called his law of three stages in his Positive Philosophy, developed in the 1820s. According to Comte every domain of knowledge passes through three progressive stages—a religious phase in which aspects of the universe are anthropomorphized (that is, human attributes including will and agency are projected onto non-human entities), a metaphysical phase in which impersonal forces (such as gravitational or electrical forces) are presumed to cause effects in the world, and a positive or fully scientific stage in which abstract causal explanations of events are abandoned in favor of general descriptive laws. Within Comte's system the rise of more reliable scientific knowledge drives out inferior religious belief; so secularization is a natural and necessary consequence of the rise of science. Even some sociologists of religion at the end of the twentieth century, such as Rodney Stark, retain a strong element of this positivist vision.

A near mirror image of the positivist view combines elements from the works of Early Modern historians such as Stephen McKnight and modern historians such as Howard Murphy. In their view Christian Humanism in the Renaissance focused Christian concerns on the amelioration of the human condition, encouraging the growth of science for the purpose of manipulating nature to serve human ends. Such views were strongly supported by Tomasso Campanella (1568–1639) in Italy, Johann Andreae 1586–1654> in the Germanies, and by Francis Bacon (1561–1626) in England. Later, when many intellectuals became disillusioned with organized religion because of the religious wars on the continent or because of the failure of institutionalized religion to promote causes of social justice, they turned to science as an alternative source of values that could improve peoples lives. From this perspective, science in Europe was nurtured within a religious context and then became the beneficiary of secularizing trends that emerged first within the Christian community itself.

A third relatively simple explanation of secularization derives from an evolutionary understanding of religion prominent among anthropologists such as Roy Rappaport and David Sloan Wilson. From this perspective religions serve primarily to establish group cohesion and social solidarity by promoting altruistic rather than individualistic behaviors. The growth of commercial economies tended to break down cooperative tendencies within societies, to promote in-group competition and individualism, and simultaneously to encourage inter-group cooperation and culture contact. As a consequence the local authority of religion was undermined both internally, as egoistic, liberal, ideology increasingly governed forms of behavior, and from the outside, as it became clear that many varieties of religion existed in other societies without subverting the functioning of those societies.

Twenty-First Century Perspectives on Secularization

Most social scientists at the beginning of the twenty-first century accept variants of a more complex account of secularization developed by Peter Berger and David Martin that grew out of the ideas of Max Weber (1864–1920). Within this account there are at least three interacting strands. One is a rationalizing trend that seems to emerge in monotheistic religions, especially those which, like Christianity, incorporate a transcendent God and therefore encourage attempts to understand the natural world without reference to specific instances of divine agency, and likewise grant human agency a predominant role in human affairs. Science and technology thus become consequences of the impli-

cit rationality of transcendent monotheism. This rationalizing strand would not necessarily by itself significantly reduce the authority of religion, but interacting with the others it does.

The second strand is a socioeconomic strand that begins from the Weberian claim that the protestant ethic promoted the rise of industrial capitalism. Industrial capitalism in turn encouraged the division of labor and promoted social differentiation into classes, breaking down the social homogeneity of pre-modern society and creating social and cultural diversity. The division of labor also transformed many social roles, which had once had important religious components, into specialized secular roles. Thus educators, health care professionals, government functionaries, and other professional groups developed specialized knowledge and institutions, creating new and non-religious sources of power and authority. Furthermore the breakdown of social homogeneity undermined the sense of communally shared values inculcated by religious practices and institutions.

Finally the Protestant Reformation promoted a sense of individualism that created a tendency for religious schism, the proliferation of competing sects, and a sense of religious relativism that was only exacerbated by culture contact with non-Christian cultures. One consequence of this relativism was the separation of Church and State, which found its most explicit separation in the first amendment to the U. S. Constitution. All of these tendencies—toward rationalization, science, and technological development; toward social differentiation and diversity; and toward religious pluralism—promoted the declining importance of religion relative to secular factors in promoting and controlling human activities. That is they all contributed to secularization.

In spite of such theories of secularization, it is clear that many issues associated with twenty-first century science and technology—from abortion to cloning, from nuclear weapons to internet piracy—are subject, even in such ostensibly secular societies as that of the United States, to religious interest-group influence. Thus the extent to which secularization adequately describes the general trend that shapes the context in which scientific, technological, and ethical interactions occur remains open to debate. There are even some proponents of cultural diversity and advocates of alternatives to modern European and North American industrial culture, who admit the importance of secularization, but who oppose the hegemony of the modern science and technology of those cultures and argue for a *re-enchant-*

ment or re-sacralization of the world. These persons point to such earth-centered spiritual traditions as those of Native Americans, as models that might promote a healthier and ultimately a more sustainable science and technology.

RICHARD OLSON

SEE ALSO *Comte, Auguste; Modernization; Urbanization; Weber, Max.*

BIBLIOGRAPHY

Berger, Peter. (1969). *The Social Reality of Religion.* London: Faber and Faber.

Berman, Morris. (1981). *The Reenchantment of the World.* Ithaca, NY: Cornell University Press. Influential updated Weberian account of secularization.

Bruce, Steve. (2002). *God Is Dead: Secularization in the West.* Oxford: Blackwell. Argues that science is more consequence than a cause of secularization, the chief driving force of which is cultural contact and a consequent awareness of the relativity of values.

Martin, David. (1978). *A General Theory of Secularization.* Oxford: Blackwell.

McKnight, Stephen A. (1989). *Sacralizing the Secular: The Renaissance Origins of Modernity.* Baton Rouge: Louisiana State University Press. Strong argument for the religious sources of modern science and, indirectly, of secularization.

Murphy, Howard. (1955). "The Ethical Revolt Against Christian Orthodoxy in Early Victorian England." *American Historical Review* (July): 800–817. Makes the argument that many nineteenth-century intellectuals turned to science as a source of values only as a consequence of a crisis of religious faith, rather than as a prelude to religious crisis.

Rappaport, Roy A. (1999). *Ritual and Religion in the Making of Humanity.* Cambridge, UK: Cambridge University Press. One of many evolutionary accounts of the survival value of religion. This one emphasizes the role of religion in encouraging in-group truth telling.

Stark, Rodney. (1963). "On the Incompatibility of Religion and Science." *Journal for the Scientific Study of Religion* 3: 3–20. Updated version of the traditional positivist argument that superior scientific knowledge drives out inferior religious faith.

SECURITY

• • •

Security has many dimensions, depending on the situation. People secure boats by tying them to a dock, secure loans from financial institutions, or secure promises with a handshake. People feel less secure, or insecure, when they doubt their own abilities, when they lose their privacy, when a thief steals their wallet or purse. Thus, security is a psychological as well as a physical state of feeling—as well as being—protected from loss, breach of trust, attack, or any real or perceived threat.

The word *security* is widespread and appears in many contexts, from the United Nations Security Council and the nuclear and environmental security councils worldwide to national security, social security, and neighborhood security watch groups formed to keep homes safe from burglars. The term has become enshrined as well in the Department of Homeland Security, which describes itself as working "to keep America safe" with one program slogan of "Don't be afraid, be ready." Closely related terms include *safety* and *fear*. Fear is a feeling, not always rational, of agitation and anxiety caused by the perception of danger. In the United States, in 2001, about 1,000 people died from airliner accidents, including those who died in the crashes of September 11, 2001, while in the same year, more than 42,000 people died in automobile crashes. Yet after the September 11 attacks, many people refused to fly and opted to drive. They no longer felt secure in airliners, even though they faced greater risk on the roads.

Pursuing Security

In between self-reliance and the appeal to religion (which places ultimate "security" in the divine), the most general efforts to enhance security involve science, technology, and politics. Many scientists, for instance, argue that insofar as fear arises from ignorance, scientific explanations of phenomena reduce superstition and increase understanding, thus promoting security through knowledge.

From earliest times human beings have also depended for their very existence on the technologies of food gathering, production, and preparation, as well as those that provide clothing and shelter. Technology, especially in the form of medicine, has a long history of combating the insecurity of disease. Virtually all forms of engineering propose to render human productivity and products more secure.

To protect technological gains, however, provisions for political security are a further requirement. The rise of the first civilizations was closely associated with the development of technologies of military security. In order to obtain civil security, people have even given their allegiance and surrendered their rights to emperors, kings, and governments. According to the English philosopher Thomas Hobbes in the *Leviathan* (1651), this compact between people and leaders is necessary because people naturally lack traits that would ensure mutual security. For Hobbes, people are essentially self-ish creatures with no concern for or connection to one another. Because humans are largely unsuccessful and constantly warring, they trade away their freedom and individuality in order to gain stability, law and order, a predictable future, leisure, and enjoyment. While other philosophers take a less dim view of human nature, all agree that security is essential for society, production, trade, and culture.

Hobbes and other early modern philosophers also argued that state security would not only protect tech-nological achievement but also promote it, and that security could be enhanced by turning those desires for material welfare that might otherwise lead to warfare between nations to a general warfare against scarcity. Although the pursuit of security thus plays important roles in virtually all modern technologies, the more explicit appeals to security are undoubtedly found in the discussion of computers and the military.

Computer and information professionals are at the front line of ensuring the confidentiality, integrity, oper-ability, and availability of information systems and data. Under the umbrella of those words come physical threats stemming from floods, hurricanes, sandstorms, and other natural disasters, as well as unintentional harm from careless use, and of course intentional harm from thieves, hackers, or terrorist attack. The focus of computer and information security often narrows to the means, such as encryption, passwords, and biometrics, rather than examining the motivations and goals of security. Among the many dimensions of this broader field are various levels of security, false senses of secur-ity, intrusive burden of security, and much more.

It is particularly important to differentiate between the *ordinary* and the *national* levels of security (Nissen-baum, Friedman, and Felten Internet article). The ordinary level comprises assurance of safety from the threats mentioned above, such as natural disasters, human error, or unwanted trespass. Computer and infor-mation professionals take what measures they can to protect from ordinary threats.

The national level, however, includes more extraor-dinary measures of action. In the name of national secur-ity, nations pursue extreme measures. As Helen Nissen-baum, Batya Friedman, and Edward Felten described it,

> The cause of national security can be parlayed into political measures as well: a lifting of typical

restraints on government activities and powers, especially those of security agencies. We may see also a curtailing of certain freedoms (e.g. speech, movement, information), a short-circuiting of certain normal democratic processes (e.g. those in the service of openness and accessibility), and even the overriding of certain principles of justice.

Thus, in some instances, ordinary security is trumped by national security, and the individual is left with fewer rights and feeling less, not more, secure. For example, national identity cards have only limited potential to enhance security but also entail an array of serious risks and other negative characteristics (Weinstein and Neu-mann 2001). Governments might impose national iden-tity cards and people might agree to them out of fear, rather than out of a rational need.

Specific Issues of Computer and Information Security

In most areas, governments, institutions, and manufac-turers give people visual reassurance that they are pro-tected from harm. Security is signified by armed guards standing at a checkpoint, childproof tops on pharmaceu-tical products, and locks on doors, windows, and cars. Banks are often solid structures, giving depositors the reassurance that their funds are safe. Screen savers can be password protected, although breaking through such protection is trivial. Whether effective or not, these measures calm and reassure people.

In the realm of computers and information, the physical and psychological aspects of security are more elusive, because the digital world is often devoid of the visual cues that lead people to feel secure. How can a user know that a document has not been altered, that no one has eavesdropped on a conversation, that an order comes from a real customer? Challenges include authenticating data and users, maintaining data integ-rity, and ensuring the confidentiality of communication.

The lack of transparency of technological devices easily renders end users both insecure and dependent. Although this is a problem associated with many tech-nological appliances such as radios, refrigerators, and air conditioners—devices that few can repair or even explain—the lack of "transparency" is peculiarly salient in computers, which are themselves increasingly inte-grated into other devices—to make the DVD player, car, or toaster "smart," but leaving the users feeling powerless and "dumb." When devices make people feel dumb, they also make them feel less secure.

What about the security threats of private spyware products? Not only do people have to be worried about governments or corporations spying on them, increasingly individuals have available sophisticated technologies for spying (spouses on each other, parents on kids, and so forth).

Another (closely related) issue: False security is provided by deleting computer documents, as some criminals have discovered to their chagrin. Computer professionals can recover many deleted files, even of non-criminals.

Security measures themselves can become burdensome, as when users have too many passwords to remember. Fear focused on one area may leave another more vulnerable. Indeed, professionals who concentrate too narrowly on the machine and wires and airwaves may overlook the danger of a disgruntled employee or an electromagnetic weapon. Research by Rebecca Mercuri into the dangers of electronic voting provides a cautionary tale, for this perceived cure for election errors and interference may result in the potential for even greater fraud.

Thus computer and information security are elusive goals that professionals aim to attain through technological fixes such as encryption, firewalls, and restricted networking. Sometimes these efforts are undertaken because of actual attacks and interference, and sometimes they are applied to allay fear or provide users with a sense of security.

Basic Issues of National and Military Security

The second most common area in which questions of security play a prominent role is that of national and military security. During the Cold War (1945–1990) the primary national security issue was nuclear weapons, and spies were sent into countries to learn more about them. Attempts to enhance nuclear weapons security and safety involved both controlling scientific knowledge that might be of use to an enemy, especially by means of secrecy, and engaging scientists and engineers in the development of technologies thought to enhance national security, technologies that ranged from "failsafe" command and control techniques to monitoring and surveillance devices. The demand for secrecy in some scientific research was nevertheless often argued to be a distortion of the scientific ideal, insofar as this ideal is committed to the production of shared knowledge. Indeed, some scientists argued that secrecy was actually counterproductive, and that greater security could be had through more openness in science.

As for spies, in the United States there were witchhunts and other wide-ranging and over-reaching investi-

gations by government that ruined the careers of innocent people and left many feeling insecure and vulnerable. The McCarthy hearings of the early 1950s involved telephone wiretaps and other intrusive acts used on innocent people.

With the end of the Cold War, the promotion of secrecy in science in the name of national security became less pronounced, but was sometimes replaced with the promotion of secrecy in science and technology in the name of corporate security and economic competitiveness. Then, with the advent of the so-called war on terrorism (2001–), needs for secrecy and control in science for national security reasons again became a prominent issue.

One specific example concerns biodefense and the boom in building high-security "hot labs" where the deadliest germs and potential bioterrorist weapons can be studied. Although the need for level 3 and level 4 biosafety labs and associated security measures are real, scientists such as David Ozonoff at the Boston University School of Public Health worry that there may be insufficient safeguards "to prevent work that violates the ethical standards of the scientific community" (Miller 2004). Stanley Falkow of Stanford University has even decided to destroy his own plague cultures rather than work under the new security regulations, pointing out the danger of security driving away talent (Miller 2004).

As these and other examples show, security needs will not abate, for they are deep in the human psyche and are built into the contract between people and their governments. Keeping security measures in balance with other values, such as freedom of speech and the pursuit of knowledge, poses a continuing challenge.

For more extensive discussion of this issue, see "A Difficult Decade: Continuing Freedom of Information Challenges for the United States and its Universities," available at http://www.murdoch.edu.au/elaw/issues/v10n4/woodbury104.html.

MARSHA C. WOODBURY

SEE ALSO Aviation Regulatory Agencies; Biosecurity; Building Destruction and Collapse; Computer Ethics; Computer Viruses/Infections; Freedom; Hobbes, Thomas; Information Ethics; Police; Privacy; Telephone; Terrorism.

BIBLIOGRAPHY

Miller, Judith. (2004). "New Biolabs Stir a Debate over Secrecy and Safety." New York Times, February 10.

Weinstein, Lauren, and Peter G. Neumann. (2001). "Risks of Panic" (Inside Risks column 137). *Communications of the ACM* 44(11): 152.

INTERNET RESOURCES

Mercuri, Rebecca. "Electronic Voting." Notable Software. Available from http://www.notablesoftware.com/evote.html.

Nissenbaum, Helen; Batya Friedman; and Edward Felten. "Computer Security: Competing Conceptions." Available from http://arxiv.org/html/cs.CY/0110001.

Woodbury, Marsha. "A Difficult Decade: Continuing Freedom of Information Challenges for the United States and its Universities." Available from http://www.murdoch.edu.au/elaw/issues/v10n4/woodbury104.html.

SELFISH GENES

• • •

Evolutionary biologists increasingly accept that genes are *selfish*. But what does this mean? Clearly genes do not have personal motivations, and even if they did, they could not achieve their designs without cooperation of the bodies in which they reside. In the most general sense, genes are merely blueprints, or, better, recipes, for the production of proteins. As such they influence the anatomy and physiology of living things including not only structural proteins but also enzymes and other factors that underlie the functioning of organisms. Genes ultimately affect the structure of kidneys, as well as the structure of nervous systems. Genes thus influence kidney function, just as they influence central nervous system function. When the central nervous system functions, behavior results. In this sense, genes are intimately connected to behavior, no less than they are to the physiology and structure of our internal organs.

Organisms are typically rather short-lived. Although they occupy the most obvious stage of the ecological and evolutionary theater, and natural selection appears to act on organisms whenever some reproduce differentially relative to others, the fact remains that natural selection among organisms is only important in the evolutionary sense insofar as it results in the disproportionate replication of some genes relative to others. Individual bodies themselves do not persist in evolutionary time; genes do. In fact genes are potentially immortal whereas bodies are not.

Selfish Genes and Modern Genetics

At the time of Charles Darwin (1809–1882), genetics was unknown, and so the focus of early evolutionary biology was on bodies. With the rise of Mendelian genetics and, subsequently, the field of population genetics, it became possible to trace the consequences of differential reproduction on their ultimate units, the genes themselves. Recognition of DNA as the genetic material, along with identification of its structure and the rise of modern genomic technology, has enhanced our understanding and also clarified the importance of focusing on these crucial units. When a hippo or a human being has a certain fitness, this means that his or her DNA is projected into the future with a given degree of success.

The term selfish, in relation to genes, is no more than a useful verbal short-hand. Selfishness simply refers to success in contributing to a particular gene's own replication. Natural selection rewards those genes that produce a *successful body* by causing more of the genes that influence the production of that body to be projected into the future. In this regard a successful body is one that metabolizes efficiently, that pumps blood successfully, that regulates its internal environment in a way conducive to life, and that also behaves in a manner that maximizes its success in reproducing, and/or in contributing to the reproduction of its component genes in the other major way available to it: by contributing to the success of genetic relatives, with the importance of each relative devalued in proportion as it is more distantly related (i.e., in direct proportion as a gene in a subject individual is likely to be present, by shared descent, in the body of another).

A key event in the development of selfish gene thinking was the recognition by British geneticist William Hamilton (1936–2000) that reproduction itself is only a special case of the more general phenomenon whereby genes contribute to their own replication. In a sexual species, reproduction occurs at some cost to the parent—in time, energy, risk—for which the sole evolutionary payoff is that each of the parent's genes has a 50 percent probability of being present in each offspring, and thereby are given a boost into succeeding generations. Hamilton observed that although reproduction is not normally considered selfish, in fact it is, at the level of genes. Moreover it is only because of the selfish payoff to the genes in question that reproduction is favored by natural selection in the first place!

Unlike the usual, negative implication of the word selfish, when applied to the attributes of genes, the term has no direct ethical implications. Living things are considered to behave in a manner that maximizes their *inclusive fitness*, which is simply the net effect of an act on identical genes present in other bodies. As a result selfish gene theory suggests that behavior that is selfish at the gene level typically involves actions that are *altruistic* at the level of bodies.

Hamilton effectively demonstrated that much seemingly altruistic behavior can be explained by this gene-centered perspective. Individual genes can promote their evolutionary success not only by helping produce offspring—new bodies within which some of these genes will reside—but also by contributing to the success of other individuals that have a probability of containing the genes in question. These other individuals are genetic relatives; indeed, a genetic relative is *defined* as an organism with an above-average probability of containing genes already present in a designated individual. For example, alarm-calling, whereby individuals who sense an approaching predator announce their discovery, that is directed preferentially toward genetic relatives. This can be selected for even if it reduces the likely survival of the alarm-caller so long as it increases the prospects that these relatives—and the alarm-calling genes within them—will survive and reproduce.

British biologist Richard Dawkins has been especially successful in explaining and popularizing this perspective, notably through his highly influential book, *The Selfish Gene* (1989). Dawkins argued that genes are essentially *replicators* whose biological role is to make additional copies of themselves. Those that succeeded in doing so went on to write the continuing history of life. Whereas early in evolutionary history replicators presumably floated freely in an *organic soup*, as natural selection continued, some discovered—quite by chance—that they were more successful by surrounding themselves with cell membranes and eventually, by aggregating together into multicellular bodies. Accordingly these bodies served, and still serve, as mere survival vehicles for the replicators.

This view is counter-intuitive because human beings subjectively experience themselves as the center of their own worlds, and therefore assume that their bodies—and not their genes—are equally the center of evolutionary concern. But bodies do not persist through evolutionary time. Although bodies can be selected for in the very short term, in that certain individuals are more reproductively successful than others, in the long term, these bodies are only vehicles for the differential success of their constituent genes, which replicate by virtue of the actions of the bodies in which they are enclosed.

Selfishness versus Altruism: A False Dichotomy

Critics of sociobiology and evolutionary psychology—both of which disciplines have been strongly influenced by the concept of selfish genes—often assume that this perspective implies that selfishness is more natural than altruism. The assumption has two significant flaws. First it suggests that identifying a trait as natural means that it is necessarily good, a view that was criticized by English philosopher David Hume (1711–1776), and, in the twentieth century, by philosopher George Edward Moore (1873–1958), who emphasized that *is* does not necessarily imply *ought*. Moore called this the *naturalistic fallacy*, and he argued that it is not philosophically or ethically defensible. Although many biologists—including Darwin—have maintained that morality is rooted in a natural moral sense, it is one thing to see morality as somehow deriving from one's biological heritage, quite another to validate behavioral tendencies simply because they are *natural*. It may be natural to respond violently to frustration, or in certain situations of competition, but is debatable whether in such cases, *naturalness* confers any ethical legitimacy.

Second, the suggestion that selfishness is somehow more natural than altruism ignores the crucial recognition that underlies all of selfish gene theory: the biological reality that genes cannot and do not behave in a vacuum, but only in the context of bodies. As such when a gene predisposes its body to behave selfishly (from the perspective of the gene), it often does so by inclining that *self* to act altruistically at the level of bodies. When parents provide food for their offspring, defend them against predators, or invest time and energy in their training, they may well be acting selfishly at the level of shared genes between parent and child, but altruistically insofar as individuals are behaving benevolently toward one another. Accordingly selfish genes need not behave selfishly!

The technology of cloning, stem cell research, and allied genomic sciences—including the identification of the human genome—has made considerations of human genes increasingly real. When developmental geneticists or evolutionary theorists speak of genes, they are increasingly able to speak authoritatively about specific DNA sequences, on identifiable chromosomes. It nonetheless does not seem likely that technology will permit the isolation of specific selfish or altruistic genes because selfish behavior does not exist as such, but rather, as a constituent of other characteristics and tendencies. For example, as discussed above, alarm-calling, which is a common textbook example of animal altruism, enhances the likely survival of others but at some increased risk to the alarm-caller. Alarm-calling need not be a result of generalized altruistic tendencies; rather it could derive from enhanced watchfulness due to anxiety, or even more acute eyesight, or a greater tendency to scan the surroundings for any number of reasons. Neither altruism nor selfishness per se, isolated as a generalized behavior trait, need be involved. The likelihood, therefore, is that advances in genetic technology will continue to elaborate genetic influences on behavior (just as they will with respect to proclivities for disease), without teasing out selfish genes as such. This, however, would not negate the scientific cogency of the concept, or even its genuine reality, because genes are selfish whenever they contribute to their own evolutionary success, without necessarily inducing their bodies to behave in an overtly self-aggrandizing manner.

Ethical Considerations Regarding Selfish Genes

Traditionally selfish behavior is considered unethical and its alternative, altruism, has been lauded as highly ethical. When biologists speak of selfish and altruistic behavior, they are simply defining these actions by their fitness consequences, and are not implying moral judgments. At the same time, one can speculate that the widespread, cross-cultural valuing of altruism and derogation of selfishness may itself derive from recognition that the living world inclines toward selfishness (at least at the level of genes) to a degree that may make exhortations to the contrary especially worthwhile.

Based on this cynics might point out that social and ethical systems may emphasize the desirability of altruism because of the payoff such behavior confers on others: Most people would be better off if others could be persuaded to be more altruistic, while they themselves remain comparatively selfish! Similarly biologists might point out that, as argued above, the boundaries between selfishness and altruism are unclear and often interpenetrating. Ethicists might emphasize that whereas evolutionary phenomena are crucially important to learn *about*, they are not suitable for learning *from*: Insofar as natural selection has produced human beings, along with other organisms, as the survival vehicles for selfish genes, the evolutionary process simply promotes whatever works. It is the responsibility of human beings to decide how they choose to assess such inclinations, and how, if at all, they elect to be influenced by that knowledge.

DAVID P. BARASH

SEE ALSO *Altruism; Dominance; Ethology; Sociobiology.*

BIBLIOGRAPHY

Alcock, John. (2001). *The Triumph of Sociobiology.* New York: Oxford University Press. An explanation of the scientific underpinnings as well as the successes of sociobiology and its selfish genes approach.

Barash, David P. (2001). *Revolutionary Biology: The New, Gene-Centered View of Life.* New Brunswick, NJ: Transaction Publishers. A nontechnical but scientifically accurate account of how a selfish genes perspectives helps illuminate human behavior in particular.

Dawkins, Richard D. (1989). *The Selfish Gene.* New York: Oxford University Press. A modern classic, which first brought the selfish genes perspective to popular attention.

De Waal, Frans B. M. (1996). *Good Natured.* Cambridge, MA: Harvard University Press. An account of various prosocial behaviors in nonhuman primates, showing that a selfish genes perspective is not incompatible with animal benevolence.

Hamilton, William D. (2002). *The Narrow Roads of Gene Land.* New York: Oxford University Press. A compilation of the important technical papers written by the founder of selfish gene thinking, with useful commentary.

Trivers, Robert L. (2003). *Natural Selection and Social Theory.* Cambridge, MA: Harvard University Press. A compilation of the important technical papers by one of the most original practitioners of selfish gene thinking, along with personal commentary.

Williams, George C. (1998). *The Pony Fish's Glow: And Other Clues to Plan and Purpose in Nature.* New York: Basic Books. An exceptionally lucid treatment of the benefits and limits of selfish gene thinking.

Wilson, Edward O. (1975). *Sociobiology: The New Synthesis.* Cambridge, MA: Harvard University Press. The now-classic summarization of sociobiology, which generated much of the controversy, and also brought together many important ideas, previously isolated.

SEMIOTICS

• • •

OVERVIEW

Semiotics (from the Greek root *sema* [sign]) proposes to be a science of signs and symbols and how they function in both linguistic (human and culture) and nonlinguistic (natural and artificial) systems of communication. In both instances the science has ethical dimensions. With regard to language and culture, some traditions of semiotics seek to expose what they argue are illegitimate uses of signs and symbols. With regard to nature and machines, questions arise about the legitimacy of conceiving interactions between noncultural phenomena in the same terms as cultural phenomena.

LANGUAGE AND CULTURE

Linguistic and cultural semiotics investigates sign systems and the modes of representation that humans use to convey feelings, thoughts, ideas, and ideologies. Semiotic analysis is rarely considered a field of study in its own right, but is used in a broad range of disciplines, including art, literature, anthropology, sociology, and the mass media. Semiotic analysis looks for the cultural and psychological patterns that underlie language, art, and other cultural expressions. Umberto Eco jokingly suggests that semiotics is a discipline for "studying everything which can be used in order to lie" (1976, p. 7). Whether used as a tool for representing phenomena or for interpreting it, the value of semiotic analysis becomes most pronounced in highly mediated, postmodern environments where encounters with manufactured reality shift humans' grounding senses of normalcy.

Historical Development

That human thought and communication function by means of signs is an idea that runs deep in Western tradition. Prodicus, one of the Greek Sophists of the fifth century B.C.E., founded his teachings on the practical idea that properly chosen words are fundamental to effective communication. Questioning this notion that words possess some universal, objective meaning, Plato (c. 428–347 B.C.E.) explored the arbitrary nature of the linguistic sign. He suggested a separateness between an object and the name that is used to signify that object: "Any name which you give, in my opinion, is the right one, and if you change that and give another, the new name is as correct as the old," (*Cratylus* [384d]). Aristotle (384–322 B.C.E.) recognized the instrumental nature of the linguistic sign, observing that human thought proceeds by the use of signs and that "spoken words are the symbols of mental experience" (*On Interpretation* [1, 16a3]). Six centuries later Augustine of Hippo (354–430 C.E.) elaborated on this instrumental role of signs in the process of human learning. For Augustine, language was the brick and mortar with which human beings construct knowledge. "All instruction is either about things or about signs; but things are learned by means of signs" (*On Christian Doctrine* 1.2).

Semiotic consciousness became well articulated in the Middle Ages, largely because of Roger Bacon (c. 1220–1292). In his extensive tract *De Signis* (c. 1267), Bacon distinguished natural signs (for example, smoke signifies fire) from those involving human communication (both verbal and nonverbal). Bacon introduced a triadic model that describes the relationship between a sign, its object of reference, and the human interpreter. This triad remains a fundamental concept in modern semiotics. John Poinsot (John of St. Thomas, 1589–1644) elaborated on the triad, laying down a fundamental science of signs in his *Tractatus de Signis* (1632). Poinsot observed that signs are relative beings whose existence consists solely in presenting to human awareness that which they themselves are not. It was the British philosopher John Locke (1632–1704) who finally bestowed a name on the study of signs. In his *Essay Concerning Human Understanding* (1690), Locke declared that *semiotike* or doctrine of signs should be one of the three major branches of science, along with natural philosophy and practical ethics.

Modern Semiotics

There are two major traditions in modern semiotic theory. One branch is grounded in a European tradition and was led by the Swiss-French linguist Ferdinand de Saussure (1857–1913). The other branch emerged out of American pragmatic philosophy through its primary founder, Charles Sanders Peirce (1839–1914). Saussure sought to explain how all elements of a language are taken as components of a larger system of language in use. This led to a formal discipline that he called *semiology*. Peirce's interest in logical reasoning led him to investigate different categories of signs and the manner by which humans extract meaning from them. Independently, Saussure and Peirce worked to better understand the triadic relationship.

Saussure laid the foundation for the structuralist school in linguistics and social theory. A structuralist looks at the units of a system and the rules of logic that are applied to the system, without regard to any specific content. The units of human language comprise a limited set of sounds called phonemes, and these comprise an unlimited set of words and sentences, which are put together according to a set of simple rules called grammar. From simple units humans derive more complex units that are applied to new rules to form more complex structures (such as themes, characters, stories, genres, and style). The human mind organizes this structure into cognitive understanding.

The smallest unit of analysis in Saussure's semiology is the *sign*, made up of a *signifier* or sensory pattern, and a *signified*, the concept that is elicited in the mind by the signifier. Saussure emphasized that the signifier does not constitute a sign until it is interpreted. Like Plato, Saussure recognized the arbitrary association between a word and what it stands for. Word selection becomes a matter, not of identity, but of difference. Differences carry signification. A sign *is* what all other signs *are not* (Saussure 1959).

Peirce shared the Saussurian observation that most signs are *symbolic* and arbitrary, but he called attention to *iconic* signs that physically resemble their referent and *indexical* signs that possess a logical connection to their referent (Peirce 1955 [1898]). To Peirce, the relationship of the sign to the object is made in the mind of the interpreter as a mental tool that Peirce called the *interpretant*. As Peirce describes it, *semiosis* (the process of sign interpretation) is an iterative process involving multiple inferences. The signifier elicits in the mind an interpretant that is not the final signified object, but a mediating thought that promotes understanding. In other words, a thought is a sign requiring interpretation by a subsequent thought in order to achieve meaning. This mediating thought might be a schema, a mental model, or a recollection of prior experience that enables the subject to move forward toward understanding. The interpretant itself becomes a sign that can elicit yet another interpretant, leading the way toward an infinite series of *unlimited semioses* (Eco 1979). By this analysis, Peirce shifts the focus of semiotics from a relational view of signs and the objects they represent to an understanding of semiosis as an iterative, mediational process.

Charles Morris (1901–1979) was a semiotician who adapted Peirce's work to a form of behaviorism. For Morris, semiotics involves "goal-seeking behavior in which signs exercise control" (Morris 1971 [1938], p. 85). Morris identified four aspects within the process of semiosis:

(1) the sign vehicle that orients a person toward a goal;

(2) the interpreter, or the subject of the semiotic activity;

(3) the designatium, or the object to which the sign refers;

(4) the interpretant, which is the cognitive reaction elicited in the mind of the interpreter.

Morris attempted to subdivide the field of semiotics into three subfields. *Semantics* studies the affiliations between the world of signs and the world of things. *Syntactics* observes how signs relate to other signs. *Pragmatics* explains the effects of signs on human behavior (Morris 1971).

Russian Influences

Saussure's abstraction of language as a self-contained system of signs became the target of criticism by those who saw language as a socially constituted fabric of human interchange. Language is highly contextual and humans acquire language by assimilating the voices of those around them. Language is not a fixed system but it changes as it is used through interaction with peers in modes of discourse. This philosophy, known as *dialogics*, was the outgrowth of intellectual development in Soviet Russia by a group whose work centered on the writings of Mikhail Bakhtin (1895–1975). The Bakhtin Circle, which included among its members Valantine Voloshinov (1895–1936), addressed the social and cultural issues posed by the Russian Revolution and its degeneration into the Stalin dictatorship. The group dissolved in 1929 after members faced political arrest. Bakhtin himself was not a pure semiotician, but he engaged with others, most notably Voloshinov, in the investigation of how language and understanding emerges in the process of dialogue.

Voloshinov argued that all utterances have an inherently dialogic character. According to Voloshinov, dialogue is the fundamental feature of speech. In his view, signs have no independent existence outside of social practice. Signs are seen as components of human activity, and it is within human activity that signs take on their form and meaning (Voloshinov 1986).

Another Russian, Lev Semenovich Vygotsky (1896–1934), applied the instrumental notion of semiotics toward cognition and learning (the relationship suggested much earlier by Aristotle and Augustine). Vygotsky identified the pivotal role language plays during the exercise of complex mental functions. In *Mind in Society* (1978 [1930]), Vygotsky observes how planning abilities in children are developed through linguistic mediation of action. "[The child] plans how to solve the problem through speech and then carries out the prepared solution through overt activity" (p. 28). He observed the similarity between physical tools and verbal artifacts as instruments of human activity. From his extensive and detailed observations of child development, Vygotsky concluded that higher-order thinking transpires by means of what he called "inner speech," the internalized use of linguistic signs (Vygotsky 1986).

Rhetorical Techniques and Ethical Implications

Roland Barthes (1915–1980) is probably the most significant semiologist to assume the mantle of Saussure. Barthes developed a sophisticated structuralist analysis to deconstruct the excessive rhetorical maneuvers within popular culture that engulfed Europe after World War II. Anything was fair game for Barthes's structuralist critique including literature, media, art, photography, architecture, and even fashion. Barthes's most influential work, *Mythologies* (1972 [1957]) continues to have an influence on critical theory in the early twenty-first century.

Myths are signs that carry with them larger cultural meanings. In *Mythologies*, Barthes describes myth as a well-formed, sophisticated system of communication that serves the ideological aims of a dominant class. Barthes conceived of myth as a socially constructed reality that is passed off as natural. Myth is a mode of signification in which the signifier is stripped of its history, and the form is stripped of its substance and then adorned with a substance that is artificial but appears entirely natural. Through mythologies, deeply partisan meanings are made to seem well established and self-evident. The role of the mythologist is to identify the artificiality of those signs that disguise their historical and social origins.

Barthes was critical of journalistic excesses that justified the French Algerian War (1954–1962). Skillfully, he deconstructed French journalism that had perfected the art of taking sides while pretending airs of neutrality, claiming to express the voice of common sense. Barthes observes that the myth is more understandable and more believable than the story that it supplants because the myth introduces self-evident truths that conform to the dominant historical and cultural position. This naturalization lends power to such myths. They go without saying. They need no further explanation or demystification.

American journalism is no less rich with its own mythical contributions to journalistic history. Examples include the Alamo (1835–1836), the sinkings of the *Maine* (1898) and the *Lusitania* (1915), the Gulf of Tonkin incident (1964), and Iraqi weapons of mass destruction (2003). In each case, the respective signifier was stripped of its own history and replaced with a more "natural" and believable narrative. These examples underscore the ethical implications of mythologies, because each was specifically instrumental in recruiting popular support behind an offensive war by making it appear to be a defensive war.

Mythologies are not limited to the realms of journalism, advertising, and the cinema, but find their way into all aspects of modern society. Science is no exception. The science educator Jay L. Lemke (1990) speaks of a "special mystique of science, a set of harmful myths that favor the interests of a small elite" (p. 129). Lemke believes that airs of objectivity and certainty in scientific discourse lend themselves to an authoritarian culture that serves to undermine student confidence. He describes linguistic practices that place artificial barriers between the pedagogy of science and common experience. He asserts that "a belief in the objectivity and certainty of science is very useful to anyone in power who wants to use science as a justification for imposing the policy decisions they favor. Science is presented as *authoritative*, and from there it is a small step to its becoming authoritarian" (Lemke 1990, p. 31).

George Lakoff and Mark Johnson (1980) describe a "myth of *objectivism*" in science writing that portrays a world of objects possessing inherent properties and fixed relations that are entirely independent of human experience. Objectivist writing emerged in the seventeenth century and now assumes the dominant position in modern discourses of science, law, government, business, and scholarship. Postmodern critics point to objectivism's failure to account for human thoughts, experience, and language, which are largely metaphorical. Metaphors are pervasive and generally unrecognized within a culture of positivism. Highlighting the use of metaphors is a useful key to identifying whose realities are actually privileged in academic writing (Chandler 2002).

Barthes's role as France's supreme social critic has been taken over by the French cultural theorist Jean Baudrillard (b. 1929). Baudrillard argues that postmodern culture, with its rich, exotic media, is a world of signs that have made a fundamental break from reality. Contemporary mass culture experiences a world of *simulation* having lost the capacity to comprehend an unmediated world. Baudrillard coined the term *simulacra* to describe a system of objects in a consumer society distinguished by the existence of multiple copies with no original. People experience manufactured realities—carefully edited war footage, meaningless acts of terrorism, and the destruction of cultural values.

In an age of corporate consolidation in which popular culture is influenced by an elite few with very powerful voices, semiotic analysis is deemed essential for information consumers. Semiotics informs consumers about a text, its underlying assumptions, and its various dimensions of interpretation. Semiotics offers a lens into human communication. It sharpens the consumer's own consciousness surrounding a given text. It informs consumers about the cultural structures and human motivations that underlie perceptual representations. It rejects the possibility that humans can represent the world in a neutral fashion. It unmasks the deep-seated rhetorical forms and underlying codes that fundamentally shape human realities. Semiotic analysis is a critical skill for media literacy in a postmodern world.

MARTIN RYDER

SEE ALSO *Peirce, Charles Sanders; Postmodernism; Rhetoric of Science and Technology.*

BIBLIOGRAPHY

Barthes, Roland. (1972). *Mythologies*, trans. Annette Lavers. New York: Hill and Wang. Originally published 1957. Fifty-four short critical reflections on mass culture in France during early 1950s. A classic work using semiotics to reveal the practices and artifacts of society as signifiers of the surface meanings and deep structures of contemporary life.

Baudrillard, Jean. (1988). *Selected Writings*, ed. Mark Poster. Stanford, CA: Stanford University Press. Provocative and controversial, Baudrillard describes a culture of people disenfranchised by the impotency of politics, media, and the consumer society.

Chandler, Daniel. (2002). *Semiotics: The Basics*. London: Routledge. A comprehensive introduction to semiotic theory for students of popular culture and mass communications.

Eco, Umberto. (1976). *A Theory of Semiotics*. Bloomington: Indiana University Press. In this classic text, Eco offers a theory of sign production that centers on the process of interpretation and the relationship between sign vehicles and the reality they portray. Eco's constructivist philosophy places the interpreter of signs on equal footing with the sign producer in the process of meaning construction.

Eco, Umberto. (1979). *The Role of the Reader: Explorations in the Semiotics of Texts*. Bloomington: Indiana University Press. Nine essays that explore the differences between "open" and "closed" texts, those that hold the reader at bay and those that actively engage the reader in the co-production of meaning.

Lakoff, George, and Mark Johnson. (1980). *Metaphors We Live By*. Chicago: University of Chicago Press. Linguist George Lakoff and philosopher Mark Johnson argue that metaphor is central to language and understanding.

Lemke, Jay L. (1990). *Talking Science: Language, Learning, and Values*. Norwood, NJ: Ablex Publishing. Lemke portrays science as language, suggesting that to learn science, one must learn the language of science; and to learn the language, one must engage with others in active dialog about science.

Morris, Charles. (1971 [1938]). *Foundations of the Theory of Signs*. Chicago: University of Chicago Press. This classic monograph proposed three divisions of semiotic theory: syntactics, semantics, and pragmatics.

Peirce, Charles Sanders. (1955). "Logic as Semiotic: The Theory of Signs." In *Philosophical Writings of Peirce*, ed. Justus Buchler. New York: Dover. Essay originally published in 1898. Considered the founder of pragmatism, Peirce introduced a logical model which he termed "abduction", the iterative process of formulating inferences through the interpretation of signs and testing those inferences with other signs as a means of advancing an investigative inquiry.

Saussure, Ferdinand de. (1959). *Course in General Linguistics*, ed. Charles Bally and Albert Sechehaye; trans. Wade Baskin. New York: Philosophical Library. Originally published 1916. This is a summary of Saussure's lectures at the University of Geneva from 1906 to 1911. In this seminal work, Saussure examines the relationship between speech and the development of language as a structured system of signs.

Voloshinov, V. N. (1986). *Marxism and the Philosophy of Language*, trans. Ladislav Matejka and I. R. Titunik. Cambridge, MA: Harvard University Press. Originally published 1929. Good introduction to the ideas of the Bakhtin Circle. In a series of articles written between 1926 and 1930, Voloshinov emphasizes the social essence of language and he tracks the development of ideology and consciousness at the level of discursive practice.

Vygotsky, L. S. (1978). *Mind in Society*, ed. Michael Cole, Vera John-Steiner, Sylvia Scribner, and Ellen Souberman. Cambridge, MA: Harvard University Press. Originally published 1930. A pioneer in developmental psychology, Vygotsky argued that language is central to learning, and that the workings of the human mind can best be explained in terms of its linguistic and cultural tools.

Vygotsky, L. S. (1986). "The Genetic Roots of Thought and Speech." In *Thought and Language*, trans. and ed. Alex Kozulin. Cambridge, MA: MIT Press. Russian edition originally published 1934, then translated in 1962 to become a classic foundational work in cognitive science. Vygotsky analyzed the role of speech in the development of human consciousness, and the relationship of language to complex thinking in humans.

NATURE AND MACHINE

Semiotics (from the Greek word for sign) is the doctrine and science of signs and their use. It is thus a more comprehensive system than language itself and can therefore be used to understand language in relation to other forms of communication and interpretation such as nonverbal forms. One can trace the development of semiotics starting with its origins in the classical Greek period (from medical symptomatology), through subsequent developments during the Middle Ages (Deely 2001), and up to John Locke's introduction of the term in the seventeenth century. But contemporary semiotics has its real foundations in the nineteenth century with Charles Sanders Peirce (1839–1914) and Ferdinand de Saussure (1857–1913), who, working independently of each other, developed slightly different conceptions of the sign. The development of semiotics as a broad field is nevertheless mostly based on Peirce's framework, which is therefore adopted here.

Ever since Umberto Eco (1976) formulated the problem of the "semiotic threshold" to try to keep semiotics within the cultural sciences, semiotics—especially Peircian semiotics—has developed further into the realm of biology, crossing threshold after threshold into the sciences. Although semiotics emerged in efforts to scientifically investigate how signs function in culture, the twentieth century witnessed efforts to extend semiotic theory into the noncultural realm, primarily in relation to living systems and computers. Because Peirce's semiotics is the only one that deals systematically with nonintentional signs of the body and of nature at large, it has become the main source for semiotic theories of the similarities and differences among signs of inorganic nature, signs of living systems, signs of machines (especially computer semiotics, see Andersen 1990), and the cultural and linguistic signs of humans living together in a society that emphasizes the search for information and knowledge. Resulting developments have then been deployed to change the scope of semiotics from strictly cultural communication to a biosemiotics that encompasses the cognition and communication of all living systems from the inside of cells to the entire biosphere, and a cybersemiotics that in addition includes a theory of information systems.

Biosemiotics and Its Controversies

Semiotics is a transdisciplinary doctrine that studies how signs in general—including codes, media, and language, plus the sign systems used in parallel with language—work to produce interpretation and meaning in human and in nonhuman living systems as prelinguistic communication systems. In the founding semiotic tradition of Peirce, a *sign* is anything that stands for something or somebody in some respect or context.

Taking this further, a sign, or *representamen,* is a medium for communication of a form in a triadic (three-way) relation. The representamen refers (passively) to its object, which determines it, and to its *interpretant,* which it determines, without being itself affected. The interpretant is the interpretation in the form of a more developed sign in the mind of the interpreting and receiving mind or quasi mind. The representamen could be, for example, a moving hand that refers to an object for an interpretant; the interpretation in a person's mind materializes as the more developed sign "waving," which is a cultural convention and therefore a symbol.

All kinds of alphabets are composed of signs. Signs are mostly imbedded in a sign system based on codes, after the manner of alphabets of natural and artificial languages or of ritualized animal behaviors, where fixed action patterns such as feeding the young in gulls take on a sign character when used in the mating game.

Inspired by the work of Margaret Mead, Thomas A. Sebeok extended this last aspect to cover all animal species–specific communication systems and their signifying behaviors under the term *zoösemiotics* (Sebeok 1972). Later Sebeok concluded that zoösemiotics rests on a more comprehensive *biosemiotics* (Sebeok and Umiker-Sebeok 1992). This global conception of semiotics equates life with sign interpretation and mediation, so that semiotics encompasses all living systems including plants (Krampen 1981), bacteria, and cells in the human body (called *endosemiotics* by Uexküll, Geigges, and Herrmann 1993). Although biosemiotics has been pursued since the early 1960s, it remains controversial because many linguistic and cultural semioticians see it as requiring an illegitimate broadening of the concept of code.

A code is a set of transformation rules that convert messages from one form of representation to another. Obvious examples can be found in Morse code and cryptography. Broadly speaking, code thus includes everything of a more systematic nature (rules) that source and receiver must know a priori about a sign for it to correlate processes and structures between two different areas. This is because codes, in contrast to universal laws, work only in specific contexts, and interpretation is based on more or less conventional rules, whether cultural or (by extension) biological.

Exemplifying a biological code is DNA. In the protein production system—which includes the genome in a cell nucleus, the RNA molecules going in and out of the nucleus, and the ribosomes outside the nucleus membrane—triplet base pairs in the DNA have been translated to a messenger RNA molecule, which is then read by the ribosome as a code for amino acids to string together in a specific sequence to make a specific protein. The context is that all the parts have to be brought together in a proper space, temperature, and acidity combined with the right enzymes for the code to work. Naturally this only happens in cells. Sebeok writes of the genetic code as well as of the metabolic, neural, and verbal codes. Living systems are self-organized not only on the basis of natural laws but also using codes developed in the course of evolution. In an overall code there may also exist subcodes grouped in a hierarchy. To view something as encoded is to interpret it as-*sign*-ment (Sebeok 1992).

A symbol is a conventionally and arbitrary defined sign, usually seen as created in language and culture. In common languages it can be a word, but gestures, objects such as flags and presidents, and specific events such as a soccer match can be symbols (for example, of national pride). Biosemioticians claim the concept of symbol extends beyond cultures, because some animals have signs that are "shifters." That is, the meaning of these signs changes with situations, as for instance the head tossing of the herring gull occurs both as a precoital display and when the female is begging for food. Such a transdisciplinary broadening of the concept of a symbol is a challenge for linguists and semioticians working only with human language and culture.

To see how this challenge may be developed, consider seven different examples of signs. A sign stands for something for somebody:

(1) as the word *blue* stands for a certain range of color, but also has come to stand for an emotional state;

(2) as the flag stands for the nation;

(3) as a shaken fist can indicate anger;

(4) as red spots on the skin can be a symptom for German measles;

(5) as the wagging of a dog's tail can be a sign of friendliness for both dogs and humans;

(6) as pheromones can signal heat to the other sex of the species;

(7) as the hormone oxytocin from the pituitary can cause cells in lactating glands of the breast to release the milk.

Linguistic and cultural semioticians in the tradition of Saussure would usually not accept examples 3 to 6 as genuine signs, because they are not self-consciously intentional human acts. But those working in the tradition of Peirce also accept nonconscious intentional signs in humans (3) and between animals (5 and 6) as well as between animals and humans (4), nonintentional signs (4), and signs between organs and cells in the body (7). This last example even takes special form in *immunosemiotics*, which deals with the immunological code, immunological memory, and recognition.

There has been a well-known debate about the concepts of primary and secondary modeling systems (see for example Sebeok and Danesi 2000) in linguistics that has now been changed by biosemiotics. Originally language was seen as the primary modeling system, whereas culture comprised a secondary one. But through biosemiotics Sebeok has argued that there exists a zoösemiotic system, which has to be called primary, as the foundation of human language. From this perspective language thus becomes the secondary and culture tertiary.

Cybersemiotics and Ethics

In the formulation of a transdisciplinary theory of signification and communication in nature, humans, machines, and animals, semiotics is in competition with the information processing paradigm of cognitive science (Gardner 1985) used in computer informatics and psychology (Lindsay and Norman 1977, Fodor 2000), and library and information science (Vickery and Vickery 2004), and worked out in a general renewal of the materialistic evolutionary worldview (for example, Stonier 1997). Søren Brier (1996a, 1996b) has criticized the information processing paradigm and second-order cybernetics, including Niklas Luhmann's communication theory (1995), for not being able to produce a foundational theory of signification and meaning. Thus it is found necessary to add biosemiotics ability to encompass both nature and machine to make a theory of signification, cognition and communication that encompass the sciences, technology as well as the humanities aspect of communication and interpretation.

Life can be understood from a chemical point of view as an autocatalytic, autonomous, autopoietic system, but this does not explain how the individual biological self and awareness appear in the nervous system. In the living system, hormones and transmitters do not function only on a physical causal basis. Not even the chemical pattern fitting formal causation is enough to explain how sign molecules function, because their effect is temporally and individually contextualized. They function also on a basis of final causation to support the survival of the self-organized biological self. As Sebeok (1992) points out, the mutual coding of sign molecules from the nervous, hormone, and immune systems is an important part of the self-organizing of a biological self, which again is in constant recursive interaction with its perceived environment *Umwelt* (Uexkull 1993). This produces a view of nerve cell communication based on a Peircian worldview binding the physical efficient causation described through the concept of energy with the chemical formal causation described through the concept of information—and the final causations in biological systems being described through the concept of semiosis (Brier 2003).

From a cybersemiotic perspective, the *bit* (or basic difference) of information science becomes a sign only when it makes a difference for someone (Bateson 1972). For Pierce, a sign is something standing for something else for someone in a context. Information bits are at most pre- or quasi signs and insofar as they are involved with codes function only like keys in a lock. Information bits in a computer do not depend for their functioning on living systems with final causation to interpret them. They function simply on the basis of formal causation, as interactions dependent on differences and patterns. But when people see information bits as encoding for language in a word processing program, then the bits become signs for them.

To attempt to understand human beings—their communication and attempts through interpretation to make meaning of the world—from frameworks that at their foundation are unable to fathom basic human features such as consciousness, free will, meaning, interpretation, and understanding is unethical. To do so tries to explain away basic human conditions of existence and thereby reduce or even destroy what one is attempting to explain. Humans are not to be fitted and disciplined to work well with computers and information systems. It is the other way round. These systems must be developed with respect for the depth, multidimensional, and contextualizing abilities of human perception, language communication, and interpretation.

Behaviorism, different forms of eliminative materialism, information science, and cognitive science all attempt to explain human communication from outside, without respecting the phenomenological and hermeneutical aspects of existence. Something important about human nature is missing in these systems and the technologies developed on their basis (Fodor 2000). It is unethical to understand human communication only in the light of the computer. Terry Winograd and Fernando Flores (1987), among others, have argued for a more comprehensive framework.

But it is also unethical not to contemplate the material constraints and laws of human existence, as occurs in so many purely humanistic approaches to human cognition, communication, and signification. Life, as human embodiment, is fundamental to the understanding of human understanding, and thereby to ecological and evolutionary perspectives, including cosmology. John Deely (1990), Claus Emmeche (1998), Jesper Hoffmeyer (1996), and Brier (2003) all work with these perspectives in the new view of semiotics inspired by Peirce and Sebeok. Peircian semiotics in its contemporary biosemiotic and cybersemiotic forms is part of an ethical quest for a transdisciplinary framework for understanding humans in nature as well as in culture, in matter as well as in mind.

SØREN BRIER

SEE ALSO *Peirce, Charles Sanders*.

BIBLIOGRAPHY

Andersen, P. B. (1990). *A Theory of Computer Semiotics: Semiotic Approaches to Construction and Assessment of Computer Systems*. Cambridge, UK: Cambridge University Press.

Bateson, Gregory. (1972). *Steps to an Ecology of Mind: Collected Essays in Anthropology, Psychiatry, Evolution, and Epistemology*. San Francisco: Chandler Publishing. Based on Norbert Wiener's classical cybernetics, a foundational work that develops the view toward second-order cybernetics and semiotics.

Brier, Søren. (1996a). "Cybersemiotics: A New Interdisciplinary Development Applied to the Problems of Knowledge Organisation and Document Retrieval in Information Science." *Journal of Documentation* 52(3): 296–344.

Brier, Søren. (1996b). "From Second Order Cybernetics to Cybersemiotics: A Semiotic Re-entry into the Second-Order Cybernetics of Heinz von Foerster." *Systems Research* 13(3): 229–244. Part of a Festschrift for von Foerster.

Brier, Søren. (2003). "The Cybersemiotic Model of Communication: An Evolutionary View on the Threshold between Semiosis and Informational Exchange." *TripleC* 1(1): 71–94. Also available from http://triplec.uti.at/articles.php.

Deely, John. (1990). *Basics of Semiotics*. Bloomington: Indiana University Press.

Deely, John. (2001). *Four Ages of Understanding: The First Postmodern Survey of Philosophy from Ancient Times to the Turn of the Twenty-First Century*. Toronto: University of Toronto Press. Integrates semiotics as a central theme in the development of philosophy, thereby providing a profoundly different view of the history of philosophy.

Eco, Umberto. (1976). *A Theory of Semiotics*. Bloomington: Indiana University Press. Foundational interdisciplinary work of modern semiotics.

Emmeche, Claus. (1998). "Defining Life as a Semiotic Phenomenon." *Cybernetics and Human Knowing* 5(1): 3–17.

Fodor, Jerry. (2000). *The Mind Doesn't Work That Way: The Scope and Limits of Computational Psychology*. Cambridge, MA: MIT Press.

Gardner, Howard. (1985). *The Mind's New Science: A History of the Cognitive Revolution*. New York: Basic.

Hoffmeyer, Jesper. (1996). *Signs of Meaning in the Universe*, trans. Barbara J. Haveland. Bloomington: Indiana University Press. The first and foundational book of modern biosemiotics.

Krampen, Martin. (1981). "Phytosemiotics." *Semiotica* 36(3/4): 187–209.

Lindsay, Peter, and Donald A. Norman. (1977). *Human Information Processing: An Introduction to Psychology*, 2nd edition. New York: Academic Press.

Luhmann, Niklas. (1995). *Social Systems*, trans. John Bednarz Jr. and Dirk Baecker. Stanford, CA: Stanford University Press.

Sebeok, Thomas A. (1972). *Perspectives in Zoosemiotics*. The Hague, Netherlands: Mouton.

Sebeok, Thomas A. (1992). "'Tell Me, Where Is Fancy Bred?' The Biosemiotic Self." In *Biosemiotics: The Semiotic Web, 1991*, ed. Thomas A. Sebeok and Jean Umiker-Sebeok. Berlin: Mouton de Gruyter.

Sebeok, Thomas A., ed. (1994). *Encyclopedic Dictionary of Semiotics*, 2nd edition. 3 vols. Berlin: Mouton de Gruyter. One of the most comprehensive and authoritative works on all aspects of semiotics.

Sebeok, Thomas A., and Marcel Danesi. (2000). *The Forms of Meaning: Modeling Systems Theory and Semiotic Analysis*. Berlin: Mouton de Gruyter.

Sebeok, Thomas A., and Jean Umiker-Sebeok, eds. (1992). *Biosemiotics: The Semiotic Web, 1991*. Berlin: Mouton de Gruyter.

Stonier, Tom. (1997). *Information and Meaning: An Evolutionary Perspective*. Berlin: Springer-Verlag. An important extension of the information processing paradigm, showing its greatness and limitations.

Uexküll, Thure von; Werner Geigges; and Jörg M. Herrmann. (1993). "Endosemiosis." *Semiotica* 96(1/2): 5–51.

Vickery, Brian C., and Alina Vickery. (2004). *Information Science in Theory and Practice*, 3rd edition. Munich: Saur. A comprehensive and important presentation, conceptualization, and use of the information processing paradigm in library and information science.

Winograd, Terry, and Fernando Flores. (1987). *Understanding Computers and Cognition: A New Foundation for Design*. Reading, MA: Addison-Wesley.

SENSITIVITY ANALYSES

• • •

Technically, a sensitivity analysis is a calculation or estimation, quantitative or not, in which all variables except one are held constant. This allows for a clear understanding of the effects of changes in that variable on the outcomes of the calculation or estimation. The methodologies of sensitivity analysis are well established in some areas of research, particularly those that employ methods of risk assessment and computer modeling (Satelli, Chan, and Scott 2000). However, the concept of sensitivity analysis has considerable potential for policy research, especially for understanding the role of different types of knowledge as factors contributing to particular value or ethical outcomes related to scientific research or technological change.

Potential use in Policy Making: Some Examples

In the context of research intended to support policy making a sensitivity analysis can help identify and frame the dimensions of a problem and thus clarify the potential efficacy of possible interventions. Consider a hypothetical example. There is a city in a desert that continually faces stress on its water resources. City officials invariably face finite time and budgets but have to make decisions about the community's water use. It is likely that they will hear from advocates proposing the development of new water projects such as dams and reservoirs as well as advocates who call for a reduction in water use in the community. Inevitably a question will arise: To what degree should the city consider limiting the use of water, for example, through conservation, versus increasing supply, for example, by building a new dam?

A sensitivity analysis can help policy makers understand the source of stresses on the community's water resources. Specifically, does stress result primarily from a growing population or from limited storage of water? From drought and climate? From a combination? If so, to what degree? The following idealized example shows how a sensitivity analysis might be organized in this case.

(1) A valued outcome is identified. In this instance the variable is water availability as measured by reservoir storage. Of course, other valued outcomes might be selected, and other measures might be selected.

(2) The existing literature is surveyed to assess the range of factors expected to influence the valued outcome over a period of time that is relevant to the decision context. For water resources the period of concern might be the upcoming decade. The two factors identified to be the most important influences affecting water availability might be rainfall and municipal water usage.

(3) With the two factors identified, the next step is to return to the literature to identify the distribution of views on the effects of rainfall and water use on water availability. The goal here is to identify the range of perspectives on the independent influence of (a) rainfall and (b) municipal water use on water availability.

(4) With a quantitative understanding of 3(a) and 3(b), it will be possible to compare the sensitivity of water availability to each of the two factors, with possible implications for decision making.

For example, if a sensitivity analysis showed that water use was expected to grow faster than variations in existing storage related to climate, policy makers might consider managing water use. Similarly, if a sensitivity analysis showed that reservoir storage was largely insensitive to accumulated rainfall, perhaps because there was far more rainfall than storage capacity, policy makers might consider building new reservoirs. A sensitivity analysis cannot determine what means and ends are worth pursuing, but it can shed some light on the connection of different means and ends.

The point of a sensitivity analysis is to identify factors that may be influenced by decision making in order to make desired outcomes more likely than undesired outcomes. Because the process of framing a problem (for example, using too much water versus not having enough water) necessarily implies some valued outcomes, a sensitivity analysis can help make those values explicit and demonstrate the prospects that different policy interventions might lead to desired outcomes.

More generally, in light of the multicausal nature of most phenomena that are of interest to policy makers (for instance, all the factors implicated in the supply of and demand for water in a large urban setting) and the large uncertainties typically associated with efforts to quantify the relationships between a particular cause (such as the challenges associated with projecting water supply over a period of decades) and an impact (for example, the difficulties of understanding who will be affected the most by water shortages and oversupply decades in the future), one obvious approach to guiding policy decisions is to look for areas of relative strength in relationships between causes and impacts and focus research to support decision making in those areas.

In a somewhat less idealized example Pielke et al. (2000) show that in light of scientific understanding as reported by the Intergovernmental Panel on Climate Change, demographic and socioeconomic change will be twenty to sixty times more important than climate change in contributing to economic losses related to tropical cyclones over the next fifty years. This sensitivity analysis suggests that (1) even if all losses resulting from climate change were prevented, the overall benefit would be dwarfed by increasing losses caused by the growth of populations and economies, and (2) research priorities relevant to the tropical cyclone threat could reflect those relationships by focusing on issues of preparation, planning, infrastructure, development, and resilience. The order-of-magnitude difference between these two sources of tropical cyclone impacts strongly suggests that more research on the sensitivity of tropical cyclones to climate changes is not likely to change the implications for decision making.

In another example one might consider the changing incidence and impacts of tropical diseases such as malaria to understand how predictions of the influence of climate change compare with other causal factors, such as growth in resistance to antibiotics, changes in healthcare delivery systems, migration and growth of populations, and annual-to-interannual climate variability.

Goals of Sensitivity Analyses

The goal is not to predict but to provide information about the relative sensitivity of impacts to various causal factors. That information can enhance the bases for effective decision making in the context of values and ethics as well as decisions about science priorities intended to support the generation of knowledge useful in pursuing desired outcomes without additional reduction in or characterization of scientific uncertainty.

In a policy setting sensitivity analysis does not attempt to resolve scientific disputes about causes of societal impacts but to compare and assess existing quantified predictions and observations of the multiple causes of such impacts to identify strong causal links. As the examples of water resources and tropical cyclones show, a sensitivity analysis approach can lessen the perceived need for reduction of uncertainty about future behavior as a prerequisite for decision making and point toward research avenues that can provide knowledge that can be useful in addressing high-priority sources of environmental change and societal vulnerability. Thus, sensitivity analysis can be an important tool for science policy decision makers in their attempt to enhance the societal value of their portfolios.

ROGER A. PIELKE, JR.

SEE ALSO *Science Policy*.

BIBLIOGRAPHY

Pielke, Roger A., Jr., Roberta A. Klein; and Daniel Sarewitz. (2000). "Turning the Big Knob: Energy Policy as a Means to Reduce Weather Impacts." *Energy and Environment* 11(3): 255–276.

Satelli, Andrea; Karen Chan; and E. Marian Scott, eds. (2000). *Sensitivity Analysis*. New York: Wiley. Focuses on the methodologies of sensitivity analysis and with emphasis on computer models. Written for a reader with a high level of expertise.

Saltelli, Andrea; Stefano Tarantola; Francesca Campolongo; and Marco Ratto. (2004). *Sensitivity Analysis in Practice: A Guide to Assessing Scientific Models*. New York: Wiley. Targeted at the non-expert, this book discusses the selection of sensitivity analysis methods and their applications.

SEX AND GENDER

• • •

Questions about the degree to which concepts of sex and gender influence science and engineering or are appropriate subjects for scientific research and technological manipulation are fundamental ethical issues. This entry discusses those issues and describes the genesis of the development of sex and gender discussions related to science and technology. The focus then shifts to the role of sex and gender in scientific knowledge and issues of inequity and their implications.

Historical Background

Gayle Rubin (1975) described the sex and gender system, distinguishing the biology of sex from the cultural and social construction of gender and revealing the male-centered social processes and practices that constrain and control women's lives. Rubin extended the implications of *The Second Sex* by Simone de Beauvoir (1947), who initiated the intellectual, theoretical foundations for the second wave of the women's movement, which itself built on the nineteenth-century first wave and took an activist turn in the United States in the context of protests and the civil rights movement of the 1960s. De Beauvoir provided the philosophical basis for existentialist feminism by suggesting that women's "otherness" and the social construction of gender rest on a social interpretation of biological differences (sex).

Rubin articulated the connection between biological sex and the social construction of masculinity and femininity that resulted in superiority being attached to what was labeled masculine and discrimination against what was defined as feminine across various societies. Although the definition of the tasks, roles, and behaviors that were considered masculine or feminine varied among societies, the lower status ascribed to the feminine and to femininity remained consistent. Rubin's articulation of the operation of the sex/gender system in a variety of contexts within a society and across societies provoked ethical questions about unequal treatment based on sex/gender in all arenas, including science and technology. That explication of the sex/gender system led to questions about whether sex/gender biases had permeated science and engineering on a variety of levels.

Sex and Gender in Scientific Knowledge

Inaccurate use of definitions and terms for sex and gender may lead to causal links that go beyond what the data warrant. As Londa Schiebinger (1993) documents,

human, particularly male, interest in certain anatomic features, such as mammary glands, has even influenced the taxonomic divisions and biological definitions of animal species. Moreover, aware of the fluidity in biological sex among a variety of species in the animal kingdom, including humans, biologists have explored the definition of biological sex and inappropriate extrapolations from the simplistic binary categories of biological male and female to the gender identities of masculine and feminine as well as inappropriate assumptions of their links with particular sexual orientations.

Indeed, although at the time of birth attendants categorize newborns into the binary category of male or female, numerous clinical examples demonstrate that biological sex can be disaggregated into genetic, hormonal, internal anatomic, and external anatomic components. Typically a genetic male (XY) produces some testosterone prenatally that causes an undifferentiated fetus to develop internal organs such as testes and external structures such as the penis that normally are associated with males. Breakdowns or changes at any level may cause development to take a different path. For example, individuals who are genetic males (XY) with androgen insensitivity (testicular feminization) have testes but have female external genitalia; individuals with Turner's syndrome (genetic X0) at birth have the anatomy of females (although their genitals may remain immature after puberty and they may or may not have ovaries) but do not have the XX sex chromosomes associated with "normal" females.

It once was assumed that after birth an individual categorized as male produces increased levels of testosterone at puberty that lead to the development of secondary sex characteristics such as facial hair and a deep voice, whereas a female develops breasts and begins menstruating in the absence of testosterone and in the presence of estrogen and progesterone. Clinical conditions such as congenital adrenal hyperplasia (CAH) demonstrated further breakdown in the uniformity of biological sex. The absence of the enzyme C-21-hydroxylase in individuals with CAH results in genetic females (XX) with female internal genitalia but male external genitalia.

These breakdowns demonstrating that being a genetic male does not always result in an individual with functioning male anatomy and secondary sex characteristics not only weakened the binary sex categories of male and female but also led scientists to question biologically deterministic models that linked the male sex with male gender identity, male role development, and heterosexuality. Statistical and interview data from the Kinsey Reports, coupled with clinical studies, revealed difficulties with the use of binary categories and assumptions of causality. For example, the studies of John Money and Anke Erhardt (1972) explored so-called ambiguous sex, or babies born with external genitalia "discrepant" with their sex chromosomes and internal genitalia, that is, genetic females (XX) with ovaries but with an elongated "penoclitoris" and genetic males (XY) with testes and androgen insensitivity.

Many of the babies in those studies were genetic females who had ambiguous external genitalia at birth because their mothers had been given synthetic progestins to prevent miscarriage. Money and Erhardt concluded from those studies that operations and hormone treatments that were intended to remove ambiguity would not prevent the "normal" development of gender identity congruent with the assignment of sex based on the construction of external genitalia, regardless of genetic or internal anatomic sex, as long as that reassignment occurred before eighteen months of age. At the time of those studies some ethical questions were raised about surgical attempts to construct "normal, appropriate" external genitalia, especially in the case of male identical twins in whom an accident during circumcision resulted in the amputation of the penis in one of the twins and the surgical reconstruction of genitalia for reassignment of that twin to the female sex.

Some people questioned the assumptions that Money and Erhardt made about appropriate gender identities and roles, such as whether exposure to androgens had resulted in the higher IQ of those genetic females and whether the parents of sexually reassigned individuals treated them in ways that would influence the children to develop an "appropriate" gender identity. In recent years more emphasis has been placed on the ethics of using surgery and hormones to provide conformity between biological sex and socially constructed gender roles. As adults the patients have raised questions about who made the decision to do sexual reassignment, who decided what was appropriate gender identity, and in many cases why they had not been told that those medical and psychological interventions had been performed on them.

Described as a solution for individuals who always felt that they were trapped in a body of the wrong sex, transsexual surgery became popular in the 1970s to make the socially constructed gender identity of individuals congruent with their biological sex. Although large numbers of "dissatisfied" or "problematic cases" of individuals who had undergone transsexual surgery surfaced almost immediately, realization by the broader medical and mainstream community that sex and gender are not the same and that binary categories of male and female, as well as masculinity and femininity, may be too limited and constraining, took longer.

John Money's treatment of Bruce/Brenda Reimer, as analyzed in a study by John Colapinto (2001), was instrumental in casting doubts on Money's social constructionist theories. Although the philosopher Janice Raymond (1979) pointed out that transsexual surgery would not be needed in a society that did not force people to conform to constricted, dichotomous gender roles based on their sex, not until the late 1990s did the transgender movement begin. Leslie Feinberg (1996) discussed how the social construction of gender allows her to assume a male gender role/identity without intending to undergo transsexual surgery; Feinberg understood and wanted to challenge the notion that biological sex determines gender, which is a social construction.

Inequitable Access to Science and Engineering on the Basis of Sex/Gender

Statistical data demonstrate a dearth of women in the physical sciences and engineering, suggesting that the sex/gender system prevents equitable access to education and employment in science and engineering for women and girls. The data document that legal actions in the late 1960s and early 1970s to remove the quotas (usually set at around 7 percent) on qualified women applicants to law, medical, and graduate schools have increased the percentages to parity in most fields. The physical sciences, computing, and engineering are major exceptions.

Although the number of women majoring in scientific and technological fields increased since the 1960s to reach 49 percent in 1998, as Table 1 demonstrates, the percentage of women in computing, the physical sciences, and engineering remains low. The percentage of graduate degrees in these fields earned by women is even lower. The small number of women receiving degrees in the sciences and engineering results in an even smaller percentage of women faculty members in those fields: For example, in 2000 only 19.5 percent of science and engineering professors at four-year colleges and universities were women. Outside academia the percentage of women in the scientific and technical workforce, which includes the social sciences, hovered at approximately 23 percent.

The Dearth of Women and a Gendered Science

Evelyn Fox Keller (1982, 1985) explored whether the dearth of individuals of one sex has led to the construction of a gendered science. Keller coupled work on the

TABLE 1

Women as a Percentage of Degree Recipients in 1996 by Major Discipline and Group									
	All Fields	All Science and Engineering	Psychology	Social Sciences	Biology	Physical Sciences	Geosciences	Engineering	Computer Science	Mathematics
Percentage of bachelor's degrees received by women	55.2	47.1	73.0	50.8	50.2	37.0	33.3	17.9	27.6	45.8
Percentage of master's degrees received by women	55.9	39.3	71.9	50.2	49.0	33.2	29.3	17.1	26.9	40.2
Percentage of doctoral degrees received by women	40.0	31.8	66.7	36.5	39.9	21.9	21.7	12.3	15.1	20.6

SOURCE: National Science Foundation. (2000). *Women, Minorities, and Persons with Disabilities in Science and Engineering*. Washington, DC: National Science Foundation, pp. 119, 170, 188.

history of early modern science by David Noble (1992) and Carolyn Merchant (1979), who demonstrated that women were excluded purposely and not permitted to be valid "witnesses" to scientific experiments, with theories of object relations for gender identity development. Keller applied the work of Nancy Chodorow (1978) and Dorothy Dinnerstein (1977) on women as primary caretakers of children during gender role socialization to suggest how that might lead to more men choosing careers in science, resulting in science becoming a masculine province that excludes women and causes women to exclude themselves. Science is a masculine province not only because it is populated mostly by men but because that situation causes men to create science and technology that reflect masculine approaches, interests, and views of the world.

Biases in Research in Science and Technology

The gendered nature of science has led to biases on several levels that are best illustrated by citing examples in science and technology that have led to ethical dilemmas.

EXCLUSION OF FEMALES AS EXPERIMENTAL AND DESIGN SUBJECTS. Cardiovascular diseases are an example of the many diseases that occur in both sexes from which women were excluded from studies until androcentric bias was revealed. Research protocols for large-scale studies of cardiovascular diseases failed to assess sex differences. Women were excluded from clinical trials of drugs because of fear of litigation resulting from possible teratogenic effects on fetuses. Exclusion of women from clinical drug trials was so pervasive that a meta-analysis published in September 1992 in the *Journal of the American Medical Association* that surveyed the literature from 1960 to 1991 on clinical trials of medica-

tions used to treat acute myocardial infarction found that women had been included in less than 20 percent and the elderly in less than 40 percent of those studies (Gurwitz, Col and Avorn 1992).

Dominance of men in engineering and the creative design sectors may result in similar bias, especially design and user bias. Shirley Malcom, in a personal communication to this author, suggests that the air bag fiasco in the U.S. auto industry is as an excellent example of gender bias reflected in design. Female engineers on the design team might have prevented the fiasco, recognizing that a bag that implicitly used the larger male body as a norm would be flawed when applied to smaller individuals, killing rather than protecting children and small women.

ANDROCENTRIC BIAS IN THE CHOICE AND DEFINITION OF PROBLEMS. Some subjects that concern women receive less funding and study. Failure to include women in studies of many diseases that occur in both sexes, such as cardiovascular disease, suggested that women's health had become synonymous with reproductive health. After a 1985 U.S. Public Health Service survey recommended that the definition of women's health be expanded beyond reproductive health, in 1990 the General Accounting Office criticized the National Institutes of Health (NIH) for inadequate representation of women and minorities in federally funded studies (Taylor 1994). This resulted in the establishment of the Women's Health Initiative (Healy 1991), which was designed to collect baseline data and look at interventions to prevent cardiovascular disease, breast cancer, colorectal cancer, and osteoporosis.

Having large numbers of male engineers and creators of technologies often results in technologies that are useful from a male perspective in that they fail to address important issues for women users. In addition the military origins for the development and funding of much technology makes its civilian application less useful for women's lives (Cockburn 1983). Men who design technology for the home frequently focus on issues that are less important to women users. For example, an analysis of "smart houses" reveals that those houses do not include new technologies; instead of housework they focus on "integration, centralised control and regulation of all functions in the home" (Berg 1999, p. 306). As Ruth Schwartz Cowan (1981) suggested, the improved household technologies developed in the first half of the twentieth century increased the amount of time housewives spent on housework and reduced their role from general managers of servants, maiden aunts, grandmothers, children, and others to that of individuals who worked alone doing manual labor with the aid of household appliances.

ANDROCENTRIC BIAS IN THE FORMULATION OF SCIENTIFIC THEORIES AND METHODS. Theories and methods that coincide with the male experience of the world become the "objective" theories that define the interpretation of scientific data and the use of technology. A 1996 study that included all prospective treatment and intervention studies published in the *New England Journal of Medicine*, the *Journal of the American Medical Association*, and the *Annals of Internal Medicine* between January and June in 1990 and 1994 revealed that only 19 percent of the 1990 studies and 24 percent of the 1994 studies reported any data analysis by gender despite the fact that 40 percent of the subjects were female (Charney and Morgan 1996).

Excessive focus on male research subjects and definition of cardiovascular diseases as male led to underdiagnosis and undertreatment of those diseases in women. A 1991 study in Massachusetts and Maryland by John Z. Ayanian and Arnold M. Epstein demonstrated that women were significantly less likely than men to undergo coronary angioplasty, angiography, or surgery when admitted to the hospital with a diagnosis of myocardial infarction, angina, chronic ischemic heart disease, or chest pain. A similar study (Steingart et al. 1991) revealed that women had angina before myocardial infarction as frequently as and with more debilitating effects than men, yet women were referred for cardiac catheterization only half as often.

These and other similar studies led Bernadine Healy, a cardiologist and the first woman director of the NIH, to characterize the diagnosis of coronary heart disease in women as the Yentl syndrome: "Once a woman showed that she was just like a man, by having coronary artery disease or a myocardial infarction, then she was treated as a man should be" (Healy 1991, p. 274). The use of the male as norm in research and diagnosis was translated into bias in treatments for women: Women had higher death rates from coronary bypass surgery and angioplasty (Kelsey et al. 1993).

In equally direct ways androcentric bias has excluded women as users of technology. The policy decision by Secretary of Defense Les Aspin (1993) to increase the percentage of women pilots uncovered the gender bias in cockpit design that excluded only 10 percent of male recruits by dimensions as opposed to 70 percent of women recruits. The officers initially assumed that the technology reflected the best or only design possible and that the goal for the percentage of women pilots would have to be lowered and/or the number of tall women recruits would have to be increased. That initial reaction, representing the world viewpoint of men, changed. When political conditions reinforced the policy goal, a new cockpit design emerged that reduced the minimum sitting height from 34 to 32.8 inches, thus increasing the percentage of eligible women (Weber 1999).

Implications of the Social Construction of Gender and of Science and Technology

Awareness and understanding of sex/gender biases raise the fundamental question of the way in which androcentric biases in scientific methods and theories occur. Should biological sex simply be termed essentialist and set aside, leaving the body to be viewed as a "coatrack" on which all that is cultural hangs, as suggested by Linda Nicholson (1994)? This interpretation implies that gender and all aspects of science and technology are socially, culturally constructed and nonobjective. Can scientists and engineers be objective? More important, is good science objective and gender-free? Or, as the title of Londa Schiebinger's 1999 book asks, *Has Feminism Changed Science?*

Most scientists, feminists, and philosophers of science recognize that no individual can be entirely neutral or value-free. To some "objectivity is defined to mean independence from the value judgments of any particular individual" (Jaggar 1983, p. 357). Scientific paradigms also are far from value-free. The values of a culture both in the historical past and in the present society heavily influence the ordering of observable phenomena into a theory. The worldview of a particular society, time, and person limits the questions that can be asked and thus the answers that can be given. Acceptance of a particular paradigm that appears to cause a "scientific revolution" within a society may depend on the congruence of the theory with the institutions and beliefs of the society (Kuhn 1970).

Scholars suggest that Darwin's theory of natural selection ultimately was accepted by his contemporaries, who did not accept similar theories proposed by the naturalist Alfred Russel Wallace (1823–1913) and others, because Darwin emphasized the congruence between the values of his theory and those held by the upper classes in Victorian Britain (Rose and Rose 1980). In this manner Darwin's data and theories reinforced the social construction of both gender and class, making his theories acceptable to the leaders of English society.

The current ideas of Darwinian feminists and feminist sociobiologists such as Patricia Gowaty (1997) and Sarah Blaffer Hrdy (1981) provide a biological explanation for female-female competition, promiscuity, and other behaviors practiced in modern society. Evolutionary psychologists carry this work a step further by positing biological bases for differences in the psychology of men and women. These biological differences, such as the ability of women to experience pregnancy, birth, and lactation, may give women different voices in ethical experiences, as has been suggested by Sara Ruddick (1989).

Not only what is accepted but what is studied and how it is studied have normative features. Helen Longino (1990) has explored the extent to which methods employed by scientists can be objective (not related to individual values) and can lead to repeatable, verifiable results while contributing to hypotheses and theories that are congruent with nonobjective institutions and ideologies, such as gender, race, and class, that are socially constructed in a society: "Background assumptions are the means by which contextual values and ideology are incorporated into scientific inquiry" (Longino 1990, p. 216). The lens of the sex/gender prism reveals how the dominance of men and masculinity in Western society has masked the androcentrism and ethical bias of many scientific experiments, approaches, theories, and conclusions.

SUE V. ROSSER

SEE ALSO Feminist Ethics; Homosexuality Debate; Sex Selection.

BIBLIOGRAPHY

Aspin, Les. (1993). Policy on the Assignment of Women in the Armed Forces. Washington, DC: U.S. Department of Defense.

Ayanian, John Z., and Arnold M. Epstein. (1991). "Differences in the Use of Procedures between Women and Men Hospitalized for Coronary Heart Disease." New England Journal of Medicine 325: 221–225.

Berg, Anne-Jorunn. (1999). "A Gendered Socio-Technical Construction: The 'Smart House.'" In The Social Shaping of Technology, 2nd edition, ed. Donald MacKenzie and Judy Wacjman. Philadelphia: Open University Press.

Charney, P., and C. Morgan. (1996). "Do Treatment Recommendations Reported in the Research Literature Consider Differences between Men and Women?" Journal of Women's Health 5(6): 579–584.

Chodorow, Nancy. (1978). The Reproduction of Mothering: Psychoanalysis and the Sociology of Gender. Berkeley and Los Angeles: University of California Press.

Cockburn, Cynthia. (1983). Brothers: Male Dominance and Technological Change. London: Pluto.

Colapinto, John. (2001). As Nature Made Him: The Boy Who Was Raised as a Girl. New York: Perennial.

Cowan, Ruth S. (1981). More Work for Mother: The Ironies of Household Technology from the Open Hearth to the Microwave. New York: Basic Books.

De Beauvoir, Simone. (1947). The Second Sex, ed. and trans. H. M. Parshley. New York: Vintage Books.

Dinnerstein, Dorothy. (1977). The Mermaid and the Minotaur: Sexual Arrangements and Human Malaise. New York: Harper Colophon.

Feinberg, Leslie. (1996). Transgender Warriors: Making History from Joan of Arc to Dennis Rodman. Boston: Beacon Press.

Gowaty, Patricia, ed. (1997). Feminism and Evolutionary Biology: Boundaries, Intersections, and Frontiers. New York: Chapman and Hall.

Gurwitz, Jerry H.; Col, N. F.; and Avorn, J. (1992). "The Exclusion of the Elderly and Women from Clinical Trials in Acute Myocardial Infarction." Journal of the American Medical Association 268(8): 1417–1422.

Healy, Bernadine. (1991). "Women's Health, Public Welfare." Journal of the American Medical Association. 266: 566–568.

Hrdy, Sarah Blaffer. (1981). The Woman That Never Evolved. Cambridge, MA: Harvard University Press.

Jaggar, Alison. (1983). Feminist Politics and Human Nature. Totowa, NJ: Rowman & Allanheld.

Keller, Evelyn F. (1982). "Feminism and Science." Signs 7(3): 589–602.

Keller, Evelyn F. (1985). Reflections on Gender and Science. New Haven, CT: Yale University Press.

Kelsey, S. F., et al. (1993). "Results of Percutaneous Transluminal Coronary Angioplasty in Women: 1985–1986 National Heart, Lung, and Blood Institutes Coronary Angioplasty Registry." Circulation 87(3): 720–727.

Kuhn, Thomas S. (1970). *The Structure of Scientific Revolutions*, 2nd edition. Chicago: University of Chicago Press.

Longino, Helen. (1990). *Science as Social Knowledge: Values and Objectivity in Scientific Inquiry*. Princeton, NJ: Princeton University Press.

Malcom, Shirley. Personal Communication, 1998.

Merchant, Carolyn. (1979). *The Death of Nature*. New York: Harper & Row.

Money, John, and Anke Erhardt. (1972). *Man and Woman, Boy and Girl: The Differentiation and Dimorphism of Gender and Identity from Conception to Maturity*. Baltimore: Johns Hopkins University Press.

National Science Foundation. (2000). *Women, Minorities, and Persons with Disabilities in Science and Engineering: 2000*. NSF 00-327. Arlington, VA: National Science Foundation.

Nicholson, Linda. (1994). "Interpreting Gender." *Signs: Journal of Women in Culture and Society* 20(1): 79–105.

Noble, David. (1992). *A World without Women: The Christian Clerical Culture of Western Science*. New York: Knopf.

Raymond, Janice. (1979). *The Transsexual Empire: The Making of the She-Male*. Boston: Beacon Press.

Rose, Hilary, and Stephen Rose. (1980). "The Myth of the Neutrality of Science." In *Science and Liberation*, ed. Rita Arditti, Pat Brennan, and Steve Cavrak. Boston: South End Press.

Rubin, Gayle. (1975). "The Traffic in Women: Notes on the 'Political Economy' of Sex." In *Toward an Anthropology of Women*, ed. Rayna Reiter. New York: Monthly Review Press.

Ruddick, Sarah. (1989). *Maternal Thinking: Toward a Politics of Peace*. Boston: Beacon Press.

Schiebinger, Londa. (1993). *Nature's Body: Gender in the Making of Modern Science*. Boston: Beacon.

Schiebinger, Londa. (1999). *Has Feminism Changed Science?* Cambridge, MA: Harvard University Press.

Steingart, R.M., et al. (1991). "Sex Differences in the Management of Coronary Artery Disease: Survival and Ventricular Enlargement Investigator." *New England Journal of Medicine* 321: 129–135.

Taylor, C. (1994). "Gender Equity in Research." *Journal of Women's Health* 3: 143–153.

Weber, Rachel. (1997). "Manufacturing Gender in Commercial and Military Cockpit Design" *Science, Technology and Human Values* 22: 235–253.

SEX SELECTION

• • •

Sex selection is an ancient and persistent practice. At some times and in some places, parents have selected the sex of their children by killing newborns or neglecting babies of the undesired sex, almost always female.

In the twenty-first century, technological developments and marketing practices are bringing new attention to sex selection, and raising an array of new concerns about it.

Some bioethicists and others defend sex selection as a matter of parental choice or "procreative liberty" (Robertson 2001). Others are highly critical, arguing that sex selection reflects and reinforces misogyny and gender stereotypes, undermines the wellbeing of children by subjecting them to excessive parental disappointment or expectations, and sets the groundwork for the future accessorizing and commodifying of children. The spread of prenatal screening for sex selection has caused alarm because of increasingly skewed sex ratios in some areas. Newer technologies now being used for sex selection also raise the prospect of a high-tech "consumer eugenics," in which other traits of future children are also chosen or "engineered."

Contemporary Sex Selection Methods

The development during the 1970s of prenatal testing technologies made it possible to reliably determine the sex of a fetus developing in a woman's womb. These procedures were initially intended to detect, and usually to abort, fetuses with Down Syndrome and other genetic anomalies, some of them sex-linked. But the tests were soon being openly promoted and widely used as tools for social sex selection, especially in South and East Asian countries where a cultural preference for sons is widespread. At the turn of the twenty-first century, prenatal screening followed by abortion remained the most common sex selection method around the world.

However, newer methods of sex selection are also coming into use. Unlike prenatal testing, these procedures are applied either before an embryo is implanted in a woman's body, or before an egg is fertilized. They do not require aborting a fetus of the "wrong" sex. In the United States, these pre-pregnancy methods are being promoted for social sex selection, as ways to satisfy parental desires, and are being marketed as forms of "family balancing" or "gender balancing."

EMBRYO SCREENING. Preimplantation genetic diagnosis (PGD), introduced in 1990, is an embryo screening technique. About three days after fertilization, a single cell is removed from each embryo in a batch that has been created using in vitro fertilization (IVF). Technicians test the cells for particular chromosomal arrangements or genetic sequences; then one or more embryos that meet the specified criteria—in the case of sex selection for a boy, those with both X and Y chromosomes—

are implanted in a woman's body. As a sex selection method, PGD is fairly reliable.

Like prenatal screening, PGD was presented as a way for parents to avoid having a child affected by certain genetic conditions (a motivation that has been strongly questioned by disability rights activists, whether involving prenatal tests or PGD). Before long, some assisted reproduction practitioners and bioethicists began suggesting that PGD should be made available to parents who want to fulfill their wish for a boy or a girl.

As of 2005, about 2,000 children have been born worldwide following the use of PGD, but no one knows how many of these procedures were undertaken for purely social sex selection reasons. In fact, the notoriously minimal regulatory environment for assisted reproduction facilities means that there is no firm data on the total number of PGD procedures conducted worldwide, or even on the exact number of clinics offering them. The risks of PGD to women who must undergo the hormone treatments and egg extractions required for all IVF procedures, and to the children born from screened embryos, are likewise unclear, both because of the small numbers involved so far and because of inadequate follow-up studies.

SPERM SORTING. Separating sperm that carry X chromosomes from those with Y chromosomes is the basis for a sex selection method that is less reliable, but that can be used without in vitro fertilization. A sperm sorting technique known as MicroSort® has been available since 1995. It relies on the fact that sperm with X chromosomes contain slightly more DNA than those with Y chromosomes, and uses a process called "flow cytometry," whereby X-chromosome-carrying sperm is separated from Y-chromosome-carrying sperm. The Genetics & IVF Institute (GIVF), the company that markets this technology for the "prevention of X-linked diseases and family balancing," claims that as of 2004, about 500 babies had been born after MicroSort® procedures. The company claims success rates of 88 percent for girls and 73 percent for boys. It reports that about 15 percent of its customers say they are trying to avoid the birth of a child who has inherited a sex-linked disease from the parents; the rest just want a boy or a girl.

Sex Selection as a Global Issue

In 1992 Nobel Prize-winning economist Amartya Sen (b. 1933) estimated the number of "missing women" worldwide—lost to neglect, infanticide, and sex-specific abortions—at one hundred million. Similarly shocking figures were confirmed by others. In areas of the world where sex-selection is most widespread, sex ratios are becoming increasingly skewed. In parts of India, for example, the sex ratio of young children is as low as 766 girls per 1,000 boys.

Some observers in the global North who express distress about the pervasiveness of sex-selective abortions in South and East Asia are untroubled by sex selection in countries without strong traditions of son preference. But politically and ethically, this double standard rests on shaky grounds.

As women's rights and human rights groups point out, an increased use and acceptance of sex selection in the United States would legitimize its practice in other countries, and undermines efforts there to oppose it. A 2001 report in *Fortune* magazine recognized this dynamic, noting that "[it] is hard to overstate the outrage and indignation that MicroSort® prompts in people who spend their lives trying to improve women's lot overseas" (Wadman 2001).

In addition, large numbers of South Asians now live in European and North American countries, and sex selection ads in publications including *India Abroad* and the North American edition of *Indian Express* have specifically targeted them (Sachs 2001). South Asian feminists point to numerous ways in which sex selection reinforces and exacerbates misogyny, including violence against women who fail to give birth to boys.

SOCIAL SEX SELECTION AS CONSUMER CHOICE AND COMMERCIAL ENTERPRISE. In North America and Europe, sex selection seems driven less by preference for boys than by a consumer ideology of "choice." In fact, anecdotal evidence suggests that of North Americans trying to determine the sex of their next child, many are women who want daughters.

However, a preference for girls does not necessarily mean that sex selection and sexism are unrelated. One study found that 81 percent of women and 94 percent of men who say they would use sex selection would want their firstborn to be a boy. Another concern is whether sex selection will reinforce gender stereotyping. Parents who invest large amounts of money and effort in order to "get a girl" are likely to have a particular kind of girl in mind.

The new sex selection methods have also been criticized as a gateway to consumer eugenics, both by public interest groups and by some practitioners in the assisted reproduction field. When the American Society for Reproductive Medicine seemed to endorse using PGD for social sex selection, the *New York Times* reported that this "stunned many leading fertility specialists." One fertility doctor asked, "What's the next step? As we learn more about genetics, do we reject kids who do not have superior intelligence or who don't have the right color hair or eyes?" (Kolata 2001).

Such concerns are exacerbated by the recognition that social sex selection constitutes a potential new profit center for the assisted reproduction industry. It would open up a large new market niche of people who are healthy and fertile, but who nonetheless could be encouraged to sign up for fertility treatments. Since about 2003, several assisted reproduction facilities have begun aggressively going after that market, running ads for social sex selection on the Internet, on radio, and in mainstream publications including the *New York Times* and the in-flight magazines of several airlines. If the parents of 5 percent of the four million babies born each year in the United States were to use MicroSort® sperm sorting at the current rate of $7,500 each, annual revenues would be $1.5 billion.

PROSPECTS FOR POLITICAL AND POLICY ENGAGEMENT. In India women's rights groups have long been at the forefront of efforts to enact laws prohibiting sex-selective abortion. As early as 1986 the Forum Against Sex Determination and Sex Pre-Selection began a campaign to enact legislation to regulate the misuse of embryo screening technology. Though laws have been on the books in India since 1994, they are often not enforced. China banned "non-medical" sex selection in 2004. The Council of Europe's 1997 Convention on Human Rights and Biomedicine also prohibits it, as do a number of European countries including the United Kingdom and Germany, with no adverse impact on the availability or legality of abortion. In 2004 Canada passed comprehensive legislation regulating assisted reproduction that includes a ban on sex selection. The United States currently has no federal regulation of sex selection.

In many parts of the world, even feminists who are deeply uneasy about sex selection have been reluctant to challenge it out of fear that to do so would threaten abortion rights. However, the emergence of pre-pregnancy sex selection methods makes it easier to consider sex selection apart from abortion politics, and may encourage new political and policy thinking about it.

MARCY DARNOVSKY
SUJATHA JESUDASON

SEE ALSO *Assisted Reproduction Technology; Eugenics; Sex and Gender*.

BIBLIOGRAPHY

Belkin, Lisa. (2000). "Getting the Girl." *The New York Times Magazine* July 25. A sympathetic account of women in the United States seeking sex selection techniques in order to have daughters.

Darnovsky, Marcy. (2004). "High-Tech Sex Selection: A New Chapter in the Debate." *GeneWatch* 17(1): 3–6. Also available at http://www.gene-watch.org/genewatch/articles/17-1darnovsky.html

Holmes, Helen Bequaert. (1995). "Choosing Children's Sex: Challenges to Feminist Ethics." In *Reproduction, Ethics, And The Law: Feminist Perspectives*, ed. Joan C. Callahan. Bloomington: Indiana University Press. A co-founder of the International Network Feminist Approaches to Bioethics discusses diverse feminist views on sex selection.

Kolata, Gina. (2001). "Fertility Ethics Authority Approves Sex Selection." *The New York Times* September 28: A16.

Mahowald, Mary. (2001). "Cultural Differences and Sex Selection." In *Globalizing Feminist Bioethics: Crosscultural Perspectives*, ed. Rosemarie Tong, with Gwen Anderson and Aida Santos. Boulder, CO: Westview Press. A philosopher argues that under an egalitarian theory of justice, sex selection is morally objectionable if its impact or intent is sexist.

Paul, Diane. (2001). "Where Libertarian Premises Lead." *American Journal of Bioethics* 1(1): 26–27. A response to John Robertson's defense of sex selection.

Robertson, John A. (2001). "Preconception Gender Selection." *American Journal of Bioethics* 1(1): 2–9. A defense of pre-pregnancy sex selection for social reasons by a bioethicist and legal scholar known for his theory of "procreative liberty."

Sachs, Susan. (2001). "Clinics' Pitch to Indian Émigrés: It's a Boy." *The New York Times* August 15: A1.

Talbot, Margaret. (2002). "Jack or Jill? The Era of Consumer-Driven Eugenics Has Begun." *The Atlantic Monthly* March.

Wadman, Meredith. (2001). "So You Want a Girl?" *Fortune* February.

Warren, Mary Anne. (1985). *Gendercide: The Implications of Sex Selection*. Totowa, NJ: Rowman & Allanheld. Despite her deep misgivings about sex selection, a philosopher argues that the risks of endangering other aspects of reproductive freedom would make any efforts against the practice counter-productive.

Wertz, Dorothy, and John Fletcher. (1992). "Sex Selection through Prenatal Diagnosis: A Feminist Critique." In *Feminist Perspectives in Medical Ethics*, eds. Helen Bequaert Holmes and Laura M. Purdy. Bloomington: Indiana University Press.

INTERNET RESOURCES

Gupte, Manisha. (2003). "A Walk Down Memory Lane: An Insider's Reflections on the Campaign Against Sex-Selective Abortions." An Indian feminist discusses efforts against sex selection and lessons learned. Available at http://www.genetics-and-society.org/resources/background/200308_gupte.pdf.

Human Fertilisation & Embryology Authority. (2003). "Sex Selection: Choice and Responsibility in Human Reproduction." Available from http://www.hfea.gov.uk/AboutHFEA/ Consultations.

Mallik, Rupsa. (2002). "A Less Valued Life: Population Policy and Sex Selection in India." *Newsletter for Center for Health and Gender Equity* October. Available from www.genderhealth.org/.

SHELLEY, MARY WOLLSTONECRAFT

• • •

Mary Wollstonecraft Godwin Shelley (1797–1851), author of *Frankenstein* (1818), often considered the first science fiction novel and source of the universal modern image of science gone awry, was born in London on August 30 and died there on February 1. Her father, William Godwin (1756–1836), to whom *Frankenstein* is dedicated, was an important liberal reformer now best known for *An Enquiry Concerning Political Justice, and Its Influence on General Virtue and Happiness* (1793). Her mother, Mary Wollstonecraft (1759–1797), who died four days after her daughter's birth, was an important early feminist now best known for *A Vindication of the Rights of Woman* (1792). In 1814 young Mary eloped to the European Continent with Percy Bysshe Shelley (1792–1822), considered one of the greatest Romantic poets. Two years later, having already produced two children and begun *Frankenstein*, Mary married Percy after the suicide of his first wife. They had four children before Percy drowned, but only Percy Florence survived into adulthood. Mary never remarried, devoting herself to motherhood, writing, and editing her husband's works.

Mary treated science less as a solution to practical problems or an intellectual discipline than as a means to "afford a point of view to the imagination for the delineating of human passions more comprehensive and commanding than any which the ordinary relations of existing events can yield" (Shelley 1969, p. 13) Her consistent philosophical position, expressed in science fictions, historical romances, travel books, and essays, was staunchly democratic, based on her belief that while genius must be encouraged, when the discoveries of genius impinge on others, there must be responsibility to the wider community. Frankenstein's murderous monster represents the escape of untempered genius into the world.

Her novel *The Last Man* (1826) is the first in English of the subgenre of works that imagine a global cata-

Mary Wollstonecraft Shelley, 1797–1851. Shelley is best known for her novel *Frankenstein; or, The Modern Prometheus*, which has transcended the Gothic and horror genres and is now recognized as a work of philosophical and psychological resonance. (*Source unknown.*)

strophe. In this case the Percy-like protagonist, Lionel Verney, moves from England to a progressively depopulated Europe, apparently the only human with a natural immunity to a new plague. In this situation science is encouraged to tame rampant Nature. Soon after the deaths begin, a character remarks to Verney that should "this last but twelve months . . . earth will become a Paradise. The energies of man were before directed to the destruction of his species: they now aim at its liberation and preservation" (Shelley 1965, p. 159).

Science always raises social and moral problems in Mary Shelley's writing. In her philosophical satire "Roger Dodsworth: The Reanimated Englishman" (1826), the fact that someone is brought back from frozen suspended animation to live out a 209 year life span, raises fundamental questions of authenticity. Was he *alive* while frozen? Is his even one *life*?

In her fiction Mary Shelley consistently articulates ethical issues related to science and technology that have since become major themes of public discussion. In Percy Bysshe Shelley's poem "Queen Mab" (1813), we see the cleft stick implicit in the progress of science: "Power, like a desolating pestilence, / Pollutes whate'er it touches; and [yet] obedience, / Bane of all genius, virtue, freedom, truth, / Makes slaves of men, and, of the human frame, / A mechanized automaton." Mary Shelley contributes to ethical thinking about science and technology by calling on society to consider how the power of scientific genius might be limited by the moral claims of the human community. Mary Shelley asks humans, by pursuing science within a community, to do better than they—and her characters—have.

ERIC S. RABKIN

SEE ALSO *Enlightenment Social Theory; Frankenstein; Science, Technology, and Literature.*

BIBLIOGRAPHY

Bennett, Betty T., and Charles E. Robinson, eds. (1990). *The Mary Shelley Reader.* New York: Oxford University Press..

Muriel Spark. (1987). *Mary Shelley.* New York: Dutton.

Shelley, Mary. (1965). *The Last Man,* ed. Hugh J. Luke, Jr. Lincoln: The University Press of Nebraska. Text originally published in 1826.

Shelley, Mary. (1969). *Frankenstein, or The Modern Prometheus,.* ed. M. K. Joseph. New York: Oxford University Press. M. K. Text originally published in 1818 and 1831.

SHINTŌ PERSPECTIVES

• • •

The indigenous religion of Japan, Shintō describes human existence much like the popular singer, Sting: as spirituality in the material world. This worldview is the foundation of Japanese civilization and has endured and adapted for centuries. While Shintō recognizes spirit over materiality as the basis of life, it shares something compelling with the perspective of science: the human propensity to identify that which is most powerful in nature and to harness that power for a comfortable and happy human life. Both are able to channel the raw potential of nature toward specific human aims on all levels of society, from the domestic to the national, and both regulate human control over nature through ethical standards that rely on an unquestioning belief in the value system upon which they are built.

Traditional Teachings

Some of the earliest forms of science and religion sought to answer the question of the origins of living things.

Practitioners of both looked to the sun for clues and based their theories and myths on its primordial role in sustaining life on Earth. The sun is the most reliable source of technology. It regulates time. Its proximity to the Earth allows life to flourish. The sun is the gravitational center of the solar system and causes all the planets to orbit it in precise yearly progressions. Hence many ancient cultures regarded the sun as a great celestial king, embodied as a human sovereign on earth.

Shintō, similarly, reveres the sun as the source of all forms of power in the world, both divine and temporal, and as the animating life force behind objective reality. The ancient Japanese personified the sun as a goddess, Amaterasu, who provided life-sustaining technologies— the cultivation of rice and wheat, the knowledge of harvesting silk from silkworms, and the invention of weaving. The goddess also allowed her grandson, Jimmu Tenno, to incarnate as the first historical mikado (emperor) of Japan. His descent to the sacred Japanese islands in 660 B.C.E. began an unbroken line in a divine solar dynasty. The mikado's chief role was to administer the life-giving force of the sun and its associated technologies within the conduct of Japanese life and ethics.

Shintō acknowledges the connection between fundamental natural processes, such as the live-giving, maintaining, and destructive nature of the sun, and the smooth function of human life lived in harmony with them. Nature is tangible power. Certain natural occurrences and objects possess more potency than others, such as the celestial bodies, mountains, rivers, fields, oceans, rain, and wind. These centralized embodiments of natural power, including also special people such as heroes and leaders, were divinized as *kami* (nature spirits) and worshipped.

Nature is very delicate; it can be disrupted easily. Of all living creatures, human beings have the unique propensity to consciously become disjointed from the balanced flow of nature. Its creative and destructive powers (*musubi*) and those objects (*kami*), both active and inert, that harness it rest on a fragile hinge. If nature's power is unleashed without a conduit, its destructive force can inhibit human happiness and survival. If the objects that house nature's power become contaminated, the creative functions of life stall or halt. The ancient Japanese regarded such obstructions as *pollution* (*tsumi*), overcome only through ritual ablution and lustration (*misogi harai*), likened to the polishing of tarnished silver. To overcome obstructions to nature's inherent balance caused by pollution, Shintō presents a threefold solution: conscious invocation of the power within a *kami,* ritual cleansing as the manner in which

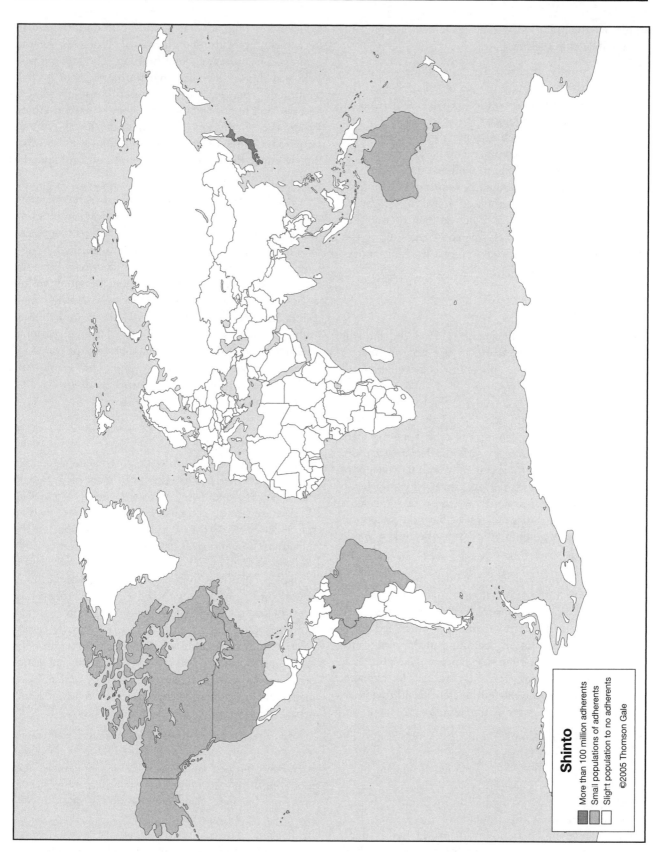

Shinto

More than 100 million adherents
Small populations of adherents
Slight population to no adherents

©2005 Thomson Gale

to remove the pollution, and ethical conduct to prevent such pollution in the first place.

The Shintō tradition of the divine emperor together with the living presence of *kami* relies on the complete integration of politics, science, and religion, with Shintō, the *shen* (spirit) *tao* (the way of), as the unbroken thread connecting these three societal divisions. Even after shogun temporal authority resigned the *tenno*, the *heavenly god-king*, to symbolic status, the divinity of the emperor remained powerful in the cultural mind of Japan. The emperor would always be regarded as the true ruler of Japan, so much so that the tradition was reinstituted in 1868, ending the feudal rule of the shogun and beginning the *taikyo* (great teaching) movement of 1870 to 1884.

Modern Shintō

The Great Teaching Movement (1870–1884) brought Shintō into the modern world in the same manner as many other neoreligious and political movements—in the guise of an ancient tradition. Even though the divinity of the emperor was considered the basis of all civic and devotional duty, the ideology of the modern Western nation-state was beginning to take shape in Japan. Shintō became synonymous with the Japanese nation. The notion that Shintō, specifically with its concept of the divine emperor, was the exclusive religion of Japan made the Japanese a unique race, a belief successfully promoted through the national education system. It remained Japan's guiding ethos until the end of World War II.

Japan's entrance into the modern world involved much more than the reassertion of traditional values in a foreign governmental model. For the first time, Japan was exposed to Western technology, which led to its own industrial revolution beginning in the nineteenth century. At the same time that Japan was adopting new technologies, the emperor was restored to temporal power—achieving the modern-ancient blend that characterizes all non-Western nation-states.

Before Japan's contact with the West, Shintō did not have a code of ethics comparable to those of Western religions. Humans were regarded as fundamentally good because positive forces of nature, the gods, had created them. There is no original sin in Shintō. Salvation is deliverance from the troubles of the world, which often means the malfunction of the world. Evil is simply the lack of harmony between spirit and matter, which can be restored through ritual appeasement of the disturbed *kami*. Ethics based on the strict division between good and evil did not emerge in Shintō until the seventeenth century with the influence of Confucian dualism expressed in the war code of Bushido. The samurai who followed this code contributed the qualities of loyalty, gratitude, courage, justice, truthfulness, politeness, reserve, and honor to Shintō's system of natural ethics. From the Confucian Teachings of Kogzi, Shintō acquired its three central insignia: the mirror to symbolize wisdom, the sword to symbolize courage, and the jewel to symbolize benevolence.

By the 1890s observance of Shintō's reverence to the emperor became the secular obligation of every Japanese citizen and not a matter of personal piety. As a result, a threefold code of ethics distinguished Japan's national identity: loyalty to the country; harmony within the family; and, by extension, harmony within society as a whole through modesty, fraternity, and intellectual development. After World War II, Shintō influence was no longer part of the Japanese national identity because the post-war constitution provided for strict separation of religion and state. There is no official government support for Shintō in early twenty-first century Japan.

Contemporary Issues

Shintō beliefs continue to undergird Japanese popular culture, particularly in its relation to technology, a field that Japan has dominated since the end of World War II. Because Shintō recognizes an unseen force behind the machinery of the world, its application to the numerous human-made devices that provide conveniences to humankind is obvious. The most notable example of Shintō's interaction with modern technology was in connection with the Apollo 11 moon mission. Before the launch of Apollo 11, Shintō purification rites were offered to placate a potentially restive *kami*, the moon-brother of the sun, Amaterasu. The rites aimed to secure two goals: to avert the imbalance of the moon's natural rhythms affected by human-made machinery landing on its virgin soil, and to assure a successful journey for the spacecraft and its crew.

In the early-twenty-first century, the Japanese increasingly rely on machines to make life easier. However many unseen factors can cause mechanical malfunction. With computer viruses and their consequences rampant, Japanese high-tech businesses often invoke the favor of Shintō *kami* to prevent the damage caused by hackers. The nation's computer network sustains 35,000 cyber attacks each month and many companies believe that antiviral software will not solve the problem. From playing a role in the development of technology and the resolution of its associated problems to averting domestic disharmony by presiding over wedding unions, Shintō continues to maintain the spirit behind the material world.

KATHERINE J. KOMENDA POOLE

SEE ALSO *Environmental Ethics; Japanese Perspectives.*

BIBLIOGRAPHY

Breen, John and Mark Teeuwen, eds. (2002). *Shintō in History: Ways of the Kami.* Honolulu: University of Hawaii Press.

Hardacre, Helen. (1989). *Shintō and the State: 1868–1988.* Princeton, NJ: Princeton University Press.

Holtom, D. C. (1943). *Modern Japan and Shintō Nationalism: A Study of Present-Day Trends in Japanese Religions.* Chicago: University of Chicago Press, 1943.

Littleton, C. Scott. (2002). *Shintō: Origins, Rituals, Festivals, Spirits, Sacred Places.* New York: Oxford University Press.

Muraoka Tsunetsugu. (1988). *Studies in Shintō Thought,* trans. Delmer M. Brown, and James T. Araki. New York: Greenwood Press.

Ono Sokyo. (1999). *Shintō: The Kami Way.* Singapore: Tuttle Publishing.

Philippi, Donald L. (1990). *Norito: A Translation of the Ancient Japanese Ritual Prayers.* Princeton, NJ: Princeton University Press.

SHIPS

• • •

Ships were invented before the beginning of recorded history. The Egyptians developed true sails by 3500 B.C.E., and the first sail-only boats were being used by 2000 B.C.E. For almost 4,000 years the leading technological developments involved refinements in sails and the design of larger and more powerful ships. The nineteenth century brought the development of steam power; after that time ships driven by electricity, fossil fuels, and even nuclear energy were developed.

Humans have used ships in warfare for almost the entire period of their development, first as a means of transporting soldiers and supplies, later as tactical vehicles for raids and looting expeditions, and then for strategic control of the seas. During the cold war era nuclear-equipped ships and submarines that were dispersed across the oceans to render them less vulnerable played a significant role in the nuclear deterrence strategy known as mutually assured destruction (Till 1984).

Today, in a world where loose aggregations of terrorist organizations are considered the enemy, the role of a navy is being redefined again in light of incidents such as the 2000 suicide attack on the U.S.S. *Cole* by men in a small, innocuous motorboat packed with explosives.

Commerce

Throughout history ships have served as unifying forces, promoting multilateralism and cultural diversity through trade. However, ships also were used as tools of colonialism and exploitation. Some analysts have observed that the more contact Europeans made with African culture, the more contempt they manifested and the more violence they committed (Scammell 1995). Ships also served as unwitting vectors of diseases such as smallpox, which decimated the native population of the Americas. Chartered shipping companies often acted as proxies of government, carrying out policies of ruthless exploitation that went well beyond what governments could do in the face of public opinion (Jackson and Williamson).

Safety

The most common type of ship collision involves two ships heading toward each other on a course that would lead them to pass each other without incident. At the last moment one of the ships turns into and collides with the other. These accidents always involve a classic misinterpretation of visual data: The captain of one ship assumes that the other ship is going away from his or her vessel and is turning to set a course landward of the first ship (Perrow 1984).

Technology, usually improperly used, can make captains complacent and careless. Studies of ship groundings have revealed that officers did not take soundings even though they knew they were in shoal water, failed to monitor the tide and current, did not keep a proper record of bearings, did not recheck the radar, and failed to adjust a magnetic compass, which in one disastrous case deviated 20 percent from true north (Moody 1948).

Design Issues

Huge ships, like skyscrapers, present safety issues that are implicit in their design. "[L]uxury passenger liners constitute the most serious fire risk afloat. Superimpose a hotel, a cinema, and a pleasure pier onto a very large cargo vessel…" with all of the possibilities for chaos that would entail (Sullivan 1943).

After the *Titanic* disaster in 1912 it was revealed that the ship did not carry enough lifeboats to accommodate every passenger and crew member. The *Titanic* had twenty boats that could carry only a third of its total passenger and crew capacity (Jim's Titanic Website 2004). When the *Andrea Doria* sank in 1956, it listed an angle greater than that envisioned by the designers, and so the lifeboats on the uphill (port) side could not be launched ("Andrea Doria: The Life Boats" 2004).

The Environment

Ships have a significant environmental impact. They act as a vector for invasive species such as hydrilla weed and zebra mussels, which arrive attached to a ship's hull or in the ballast and are released into local environment, where they drive out native species. Ships sometimes accidentally hit and damage fragile coral reefs such as those in Pennekamp State Park, Florida, and marine mammals such as whales, dolphins, and manatees frequently are maimed or killed after colliding with ships' propellers.

The public consciousness long retains the names of ill-fated oil tankers that dump their cargoes into the marine environment. On the evening of March 23, 1989, the *Exxon Valdez*, as a result of navigational errors, grounded in Prince William Sound, Alaska, with more than 53 million gallons of oil aboard. Approximately 11 million gallons of oil were spilled, resulting in the deaths of 250,000 seabirds, 2,800 sea otters, 300 harbor seals, 250 bald eagles, up to 22 killer whales, and billions of salmon and herring eggs (Exxon Valdez Oil Spill Trustees Council 2004).

However, the quiet dumping of engine oil during normal operations accounts for a majority of the oil that pollutes marine environments (Boczek 1992). A variety of treaties provide an international regime that governs dumping and oil spills. Those treaties include the United Nations Convention on the Law of the Sea, four 1958 Geneva conventions, the 1969 Brussels Convention passed in response to the Torrey Canyon disaster, another 1969 Convention on Civil Liability for oil spills, and a December 1988 annex to the Marpol agreement that established strict controls over garbage disposal from ships at sea (Boczek 1992).

Dangerous cargoes sometimes explode in port, as occurred in the July 17, 1944, incident in Port Chicago, California, when a Pacific-bound navy ship being loaded with explosives by a work crew consisting mostly of black sailors exploded, killing 320 men. Concerned about another explosion, 258 black sailors refused an order to load ammunition on another ship and were court-martialed ("A Chronology of African-American Military Service" 2004). Later large-scale peacetime ship explosions include the April 16, 1947, explosion of the S.S. *Grandcamp* at the pier in Texas City, Texas, killing 576 people (Galvan 2004), and the May 26, 1954, explosion aboard the carrier U.S.S. *Bennington* at sea, which killed 100 sailors (Hauser 1954).

Status of Seafarers

Contrary to popular belief as reflected in movies such as *Ben Hur*, most oared ships in antiquity were not operated by slaves. Citizen rowers were less expensive because they were paid only when aboard ship and their deaths did not cost the state anything. However, Athens turned to the use of slaves at a point in the Peloponnesian War when it ran out of available citizens (Casson 1994).

In 1598 the chronicler Hakluyt wrote of sailors: "No kinde of man of any profession in the commonwealth passe their yeres in so great and continuall hazard . . . and . . . of so many so few grow to gray haires" (quoted in Scammell 1995, p. 131). Sailors faced a high mortality rate from disease, accidents, and combat. Unable to recruit enough sailors, the British government began the impressment, and essentially enslavement, of unwilling agricultural and industrial workers in the 1500s, a policy that would continue for almost three centuries (Scammell 1995). However, the sea was one of the few careers that allowed people of humble rank to move up to positions of status and power (Scammell 1995). A significant path out of the working class was blazed by engineers (Dixon 1996).

Today the lives of itinerant seamen on cargo ships are still dangerous, grindingly hard, and poorly compensated (Kummerman and Jacquinet 1979).

JONATHAN WALLACE

SEE ALSO *Roads and Highways*.

BIBLIOGRAPHY

Boczek, Boleslaw A. (1992). "Global and Regional Approaches to the Protection of the Marine Environment." In *Maritime Issues in the 1990's: Antarctica, Law of the Sea and the Marine Environment*, ed. Dalchoong Kim. Seoul: Institute of East and West Studies. Summary of international law pertaining to the marine environment and shipping.

Casson, Lionel, (1994). *Ships and Seafaring in Ancient Times*. London: British Museum Press. Overview of ships in antiquity.

Dixon, Conrad. (1996). "The Rise of the Engineer in the 19th Century." In *Shipping, Technology and Imperialism*, ed. Gordon Jackson and David Williams. Hants, UK: Scolar Press. An account of the engineering profession as a pathway out of the working class.

Kummerman, Henri, and Robert Jacquinet. (1979). *Ship's Cargo, Cargo Ships*. London: McGregor Press. An overview of issues pertaining to modern cargo shipping.

Moody, Lieutenant Alton P. (1948). *Why Ships Ground*. Address to the fourth annual meeting of the Institute of Navigation. In the collection of the New York Public Library. An interesting brief analysis of the reasons why ships hit obstacles at sea.

Perrow, Charles. (1984). *Normal Accidents: Living with High Risk Technologies*. New York: Basic Books. Fascinating analysis of the elements of human error which contribute to high technology accidents.

Scammell, G. V. (1995). *Ships, Oceans and Empire: Studies in European Maritime and Colonial History, 1400–1750*. Brookfield, VT: Ashgate. Includes essays on ships as tools of colonialism.

Sullivan, A. P. L. (1943). *Ships and Fire: Suggested in Outline by the Deputy Chief of Fire Staff*. Liverpool and London: C. Birchall & Sons. In the collection of the New York Public Library. A brief summary of the reasons ships catch fire.

Till, Geoffrey. (1984). *Maritime Strategy and the Nuclear Age*, 2nd edition. London: Macmillan. The continuing relevance of warships in the age of nuclear weapons.

INTERNET RESOURCES

"Andrea Doria: The Life Boats." (2004). Available at http://www.andreadoria.org/TheLifeboats/Default.htm. Why half the Andrea Doria lifeboats weren't available after the collision.

"A Chronology of African-American Military Service from WWI through WWII." (2004). Available at http://www.africanamericans.com/MilitaryChronology4.htm. Includes a summary of the port Chicago incident.

Hauser, Don. "A Personal Account of Explosion Aboard *U.S.S. Bennington*, May 26, 1954." Available on the Crew Stories website, http://www.uss-bennington.org/stz-explosion-dhauser54.html.

Exxon Valdez Oil Spill Trustees Council. (2004). "Exxon Valdez Q&A." Available at http://www.evostc.state.ak.us/facts/qanda.htm. Useful summary of facts relating to the causes and impact of the Exxon Valdez grounding..

Galvan, Jose. (2004). "The Texas City Disaster." Available at http://www.useless-knowledge.com/articles/apr/july107.html.

Jim's Titanic Website. (2004). "Titanic Facts and Figures." Available at http://www.keyflux.com/titanic/facts.htm. Contains a description of the inadequacy of the Titanic lifeboats.

SIERRA CLUB

• • •

The Sierra Club is one of the leading non-governmental organizations that influence science, technology, and ethics relations from the environmental perspective.

Origins

The oldest environmental organization in the United States, the Sierra Club was founded in 1892 by a Scotsman, John Muir (1838–1914), who did not become a U.S. citizen until 1903. By 1892, however, he was already known to presidents and writers (including Ralph Waldo Emerson (1803–1882) as one of the country's most passionate advocates for the protection of wilderness.

Muir arrived in San Francisco, California, from Wisconsin in 1868 and headed to Yosemite Valley in the Sierra Nevada Mountains, which the avid outdoorsman had read about in a magazine. He spent the next seven years there, exploring, collecting plants, writing about his discoveries, and urging others to visit the high country. Those writings helped convince President Benjamin Harrison to create the Yosemite National Park in 1890.

In 1892 Muir became the first president of The Sierra Club, an association whose purpose as listed in its Articles of Incorporation was "To explore, enjoy, and render accessible the mountain regions of the Pacific Coast; to publish authentic information concerning them; and to enlist the support and cooperation of the people and government in preserving the forests and other natural features of the Sierra Nevada Mountains."

The Sierra Club-sponsored hiking and camping outings, called High Trips, that were fun but also meant to make members aware of and articulate about the preservation challenges facing the Sierra Nevadas. The education of such *activists* was important, for almost as soon as Yosemite National Park was established, efforts began to shrink it, strip it of federal protection, build a private railroad through it, and drown its beautiful Hetch Hetchy Valley behind a dam.

The park was shrunk and the proposal to build the dam passed in 1913, but all these fights—and especially the tragedy of the Hetch Hetchy defeat—helped transform the Sierra Club from a politically naive hiking club into a formidable and politically astute environmental organization. Its leaders now understood how the government worked and how important it was to win over public opinion to its causes. Outings and conservation were still integral to the Sierra Club, but so was political clout.

Contemporary Work

In the early twenty-first century, the Sierra Club is headquartered in San Francisco. With more than 750,000 members, it has lobbyists in Washington, DC, and a nationwide volunteer grassroots network striving to influence public policy on a variety of environmental issues.

Over the years, the club focus widened as environmental threats increased. Air and water pollution, urban sprawl, unsustainable logging, and the promotion of renewable energy—in addition to the protection of wilderness areas such as those in Yosemite—have emerged as some of the organization's top priorities. In recent years scientific pursuits in the areas of biotechnology—particularly as this new science relates to genetically modified organisms in agriculture and forestry—have been challenged by the club.

With regard to genetically engineered organisms, the club subscribes to a hard version of the Precautionary Principle and calls for a moratorium on the planting of all genetically engineered crops and the release of all genetically engineered organisms (GEOs) into the environment. It urges that where there are safer alternatives to the use of GEOs, these technologies should be given preference. On this topic the Sierra Club represents citizen science in action. Its biotechnology committee is all-volunteer. Some of its members are scientists but others are merely concerned citizens, worried about an unproven technology, who have researched the issue and feel compelled to act. Sierra Club committees make recommendations to the board of directors, which then formulates the club's *official* stand.

In the areas of energy conservation and renewables, the Sierra Club advocates for public transportation systems, energy efficient buildings and fuel efficient automobiles, and the use of renewable energy sources such as solar, wind, and geothermal power. The club has urged the U.S. Congress to provide for the expenditure of at least 2 billion dollars per year for at least five years for federal research and development—with emphasis on geothermal, solar, and fusion power; energy conservation and more efficient utilization of energy; and strip-mining reclamation. In 2001, when the U.S. government announced an energy plan that privileged oil, gas, and nuclear power interests, the Sierra Club sued to gain access to Vice President Dick Cheney's notes of meetings in which the energy policy was developed.

Following founder John Muir's statement that "Everybody needs beauty as well as bread, places to play in and pray in, where nature may heal and give strength to body and soul alike" (Muir 1912, p. 260), the Sierra Club has made an effort to broaden its preservation ethic to include what have come to be called *environmental justice* issues. Whether it is the threat to the Gwichin people's subsistence hunting from drilling in the Arctic National Wildlife Refuge or dioxin-spewing power plants in poor neighborhoods of Detroit or San Francisco, the Sierra Club attempts to reach out to communities not usually associated with the environmental movement and assist them in their struggles.

In the early 2000s the Sierra Club continues to promote outings, where hikers can explore and enjoy the wild places of the earth. But in a political and corporate environment that increasingly compromises the quality of water, air, and soil in pursuit of economic gain, organizations such as the Sierra Club have become essential advocates for the responsible use of the earth's ecosystems and resources. The Sierra Club's catalog of coffee table nature books and environmental literature can be accessed at http://www.sierraclub.org/books.

MARILYN BERLIN SNELL

SEE ALSO *Alternative Energy; Deforestation and Desertification; Ecological Restoration; Ecology; Environmental Ethics; Environmental Justice; Environmentalism; Genetically Modified Foods; Nature; Nongovernmental Organizations; Rain Forest; Sustainability and Sustainable Development; Water.*

BIBLIOGRAPHY

Muir, John. (1912). *The Yosemite.* New York: Century.

Turner, Tom. (1991). *The Sierra Club: 100 Years of Protecting Nature.* New York: Harry N. Abrams in association with the Sierra Club.

INTERNET RESOURCE

Sierra Club. Home page at http://www.sierraclub.org.

SIMON, HERBERT A.

• • •

Herbert Alexander Simon (1916–2001) was born in Milwaukee, Wisconsin, on June 15. He received his

Herbert Simon, 1916-2001. The study of decision-making behavior, especially in large organizations, led Simon to develop new theories in economics, psychology, business administration, and other fields. He was awarded the Nobel Prize in economics in 1978. He was also the first social scientist elected to the National Academy of Sciences. (AP/Wide World Photos.)

Ph.D. in political science from the University of Chicago in 1943, and taught at the Illinois Institute of Technology (1942–1949) before going to Carnegie Mellon University in 1949, where he remained until his death on February 9. Simon received major awards from many scientific communities, including the A.M. Turing Award (with Allen Newell; 1975), the Nobel Prize in Economics (1978), and the National Medal of Science (1986). During his career, Simon also served on the National Academy of Science's Committee on Science and Public Policy and as a member of the President's Science Advisory Committee. Simon made important contributions to economics, psychology, political science, sociology, administrative theory, public administration, organization theory, cognitive science, computer science, and philosophy. His best known books include *Administrative Behavior* (1947), *Organizations* (with James G. March 1958), *The Sciences of the Artificial* (1969), *Human Problem Solving* (with Newell 1972), and his autobiography, *Models of My Life* (1991). Having advanced the scientific analysis of decision-

making, Simon's thought also has evident implications for bringing ethics to bear on science and technology.

A New Theory of Decision-Making

Decision-making was the core of Simon's work. It was the heart of his dissertation, later published as *Administrative Behavior,* and it became the basis of his other contributions to organization theory, economics, psychology, and computer science. Decision-making, as Simon saw it, is purposeful, yet not rational, because rational decision-making would involve a complete specification of all possible outcomes conditional on possible actions in order to choose the single best among alternative possible actions. In challenging neoclassical economics, Simon found that such complex calculation is not possible. As a result, Simon wanted to replace the economic assumption of global rationality with an assumption that was more in correspondence with how humans actually make decisions, their computational limitations, and how they access information in a current environment (Simon 1955), thereby introducing the concepts of *bounded rationality* and *satisficing.*

Satisficing is the idea that decision makers interpret outcomes as either satisfactory or unsatisfactory, with an aspiration level constituting the boundary between the two. In neoclassical rational choice theory decision makers would list all possible outcomes evaluated in terms of their expected utilities, and then chose the one that is rational and maximizes utility. According to Simon's model, decision makers face only two possible outcomes, and look for a satisfying solution, continuing to search only until they have found a solution that is good enough. The ideas of bounded rationality and satisficing became important for subsequent developments in economics.

Simon used this view of decision-making to create (together with March and Harold Guetzkow) a propositional inventory of organization theory, which led to the book *Organizations* (1958). The book was intended to provide the inventory of knowledge of the (then almost nonexistent) field of organization theory, and also a more proactive role in defining the field. Results and insights from studies of organizations in political science, sociology, economics, and social psychology were summarized and codified. The book expanded and elaborated ideas on behavioral decision-making, search and aspiration levels, and the significance of organizations as social institutions in society. "The basic features of organization structure and function," March and Simon wrote, derive from the characteristics of rational human choice. Because of the limits of human intellective capacities in comparison with the complexities of the problems that individuals and organizations face, rational behavior calls for simplified models that capture the main features of a problem without capturing all its complexities." (p. 151)

The book is now considered a classic and pioneering work in organization theory.

Interdisciplinary Contributions

Simon also incorporated these views into his contributions to psychology, computer science, and artificial intelligence. For example, in his work with Newell, Simon attempted to develop a general theory of human problem solving that conceptualized both humans and computers as symbolic information processing systems (Newell and Simon 1972). Their theory was built around the concept of an information processing system, defined by the existence of symbols, elements of which are connected by relations into structures of symbols. The book became as influential in cognitive science and artificial intelligence as Simon's earlier work had been in economics and organization theory.

During his amazingly productive intellectual life, Simon worked on many projects, yet essentially pursued one vision—understanding how human beings make decisions. He contributed significantly to many scientific disciplines, yet found scientific boundaries themselves to be less important, even unimportant, *vis-à-vis* solving the questions he was working on. Even as Simon sought to develop the idea that one could simulate the psychological process of thinking, he tied his interest in economics and decision-making closely to computer science and psychology. He used computer science to model human problem solving in a way that was consistent with his approach to rationality. He implemented his early ideas of bounded rationality and means–ends analysis into the heart of his work on artificial intelligence.

MIE-SOPHIA AUGIER

SEE ALSO *Economics and Ethics; Management.*

BIBLIOGRAPHY

Augier, Mie-Sophia, and James G. March. (2002). "A Model Scholar." *Journal of Economic Behavior and Organization* 49(1): 1–17.

March, James G., and Herbert A. Simon. (1958). *Organizations.* New York: Wiley; 2nd edition, Cambridge, MA: Blackwell, 1993.

Newell, Allen, and Herbert A. Simon. (1972). *Human Problem Solving.* Englewood Cliffs, NJ: Prentice-Hall.

Simon, Herbert A. (1947). *Administrative Behavior.* New York: Macmillan; 4th edition, New York: Free Press, 1997.

Simon, Herbert A. (1955). "A Behavioral Model of Rational Choice." *Quarterly Journal of Economics* 69(1): 99–118.

Simon, Herbert A. (1969). *The Sciences of the Artificial.* Cambridge, MA: MIT Press; 3rd edition, 1996.

Simon, Herbert A. (1991). *Models of My Life.* New York: Basic.

SIMON, JULIAN

• • •

An economist who brought reams of evidence to bear against the conventional wisdom about the dangers of population growth and resource consumption, Julian Lincoln Simon (1932–1998) was born in Newark, New Jersey, on February 12; he attended Harvard University. After service in the Navy and work in advertising, Simon earned an MBA in 1959 and a Ph.D. in business economics in 1961, both from the University of Chicago. Although initially adopting the conventional Malthusian view that rapid population growth was a primary obstacle to economic prosperity in both the developed and developing worlds, his own research soon convinced him otherwise. Instead, science and technology, products of inexhaustible human ingenuity, have improved human welfare in nearly every measurable way and will continue to do so indefinitely into the future. He served as professor of business administration at the University of Maryland and distinguished senior fellow at the libertarian Cato Institute until his death from a heart attack in Maryland on February 8.

Against the Doomsayers

Simon had been fairly successful in the business and marketing fields during the mid-1960s. He operated a mail-order firm that was so lucrative he wrote the popular *How to Start and Operate a Mail-Order Business* (1965). But economic research led him to become critical of the grim Malthusian outlook on resource use and population growth popularized by Paul Ehrlich's *The Population Bomb* (1968) and *The End of Affluence* (1974), which argued that population growth was threatening human and environmental health. Simon replied that data from economists such as Simon Kuznets

FIGURE 1

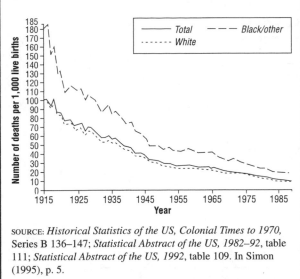

Infant Mortality Rate, Total and by Race, United States, 1915–1989

SOURCE: *Historical Statistics of the US, Colonial Times to 1970,* Series B 136–147; *Statistical Abstract of the US, 1982–92,* table 111; *Statistical Abstract of the US, 1992,* table 109. In Simon (1995), p. 5.

(1901–1985) and Richard Easterlin (b. 1926) showed there was no general negative correlation between population growth and living standards (Regis 1997).

Simon began his much maligned public crusade against the conventional wisdom "doomsayers" with a 1980 article in *Science,* which declared that false bad news about resources, population, and the environment was being widely published in the face of contrary evidence. Tellingly, the article was written in the form of a statement followed by facts, because Simon believed that sound science revealed unequivocal facts about the state of the world. As he wrote in the preface to *The Ultimate Resource 2* (1996), "Indeed, the facts and my new conclusions about population economics altered my wider set of beliefs, rather than the converse" (p. xxxi). Here he implies that his adversaries are poor scientists because they allow preconceptions to trump empirical evidence. His major books and articles elaborating a positive view of the state of humanity are notoriously crammed with trend data in hopes that the weight of the facts will persuade readers of the doomsayers' errors.

Two trends that he saw as most convincing are declines in infant mortality and rises in life expectancy (see Figures 1 and 2). He also presented data on decreasing pollution, rising agricultural productivity, increasing standards of living, and the declining prices of natural resources and commodities. All of these figures detail the overarching story of human progress and affluence made possible by the ultimate resource, the human mind. Indeed, his central premise was that human ingenuity is boundless, creating unlimited resources to "free humanity from the bonds in which nature has kept us shackled" (Simon 1995, p. 23).

The Dialectic of Scarcity and Abundance

For Simon, the problems of scarcity and the achievements of abundance are not so much fundamental opposites as they are different moments in an ongoing process.

> The process goes like this: More people and increased income cause problems in the short run. These problems present opportunity, and prompt the search for solutions. In a free society, solutions are eventually found, though many people fail along the way at cost to themselves. In the long run the new developments leave us better off than if the problems had not arisen. [Indeed, human beings now have in their hands] the technology to feed, clothe, and supply energy to an ever-growing population for the next seven billion years. (Myers and Simon 1994, p. 65).

The evident hyperbole of this rhetoric should not be used to portray Simon as a Pollyanna. Problems do arise, people are harmed, and people often fail in trying to solve them. But the larger perspective reveals that the process produces ultimate benefits for human welfare, which Simon insists are best measured by long-run trends. There is a sense of theodicy in Simon's vision.

With regard to long-run measurements, absolute trends comparing present and past states of affairs are more important than relative trends comparing two contemporary variables. Simon also argues that broad aggregate measures should emphasize effects on people rather than phenomena themselves. For example, he measures life expectancy rather than occurrences of AIDS, or agricultural productivity rather than global warming.

Moreover, the dialectic between scarcity prediction and abundance production highlights Simon's core belief that liberty is the most important precondition for progress. Free markets, free institutions, and even the free flow of immigrants are necessary for long-term material progress. Most centrally, people ought to be free to have as many children as they desire, in part because children, through their own inventiveness, will add to human welfare. A better future does not happen automatically, but requires free and well-informed decisions.

FIGURE 2

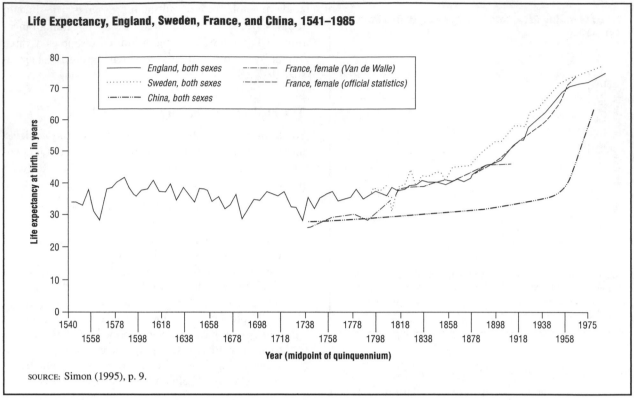

Life Expectancy, England, Sweden, France, and China, 1541–1985

SOURCE: Simon (1995), p. 9.

Finally, warnings about scarcities have a role to play in human welfare production. Unlike his opponents, who find his position detrimental, Simon actually grants critics an important if limited role in progressive developments. Simon's worldview partially depends on doomsayers to spark the impetus that steers humanity toward a better future.

Nonetheless, Simon believed that the "false bad news" of doomsayers is often overstated and can become counterproductive if not shamelessly self-promotional. With Herman Kahn (1922–1983) he co-edited *The Resourceful Earth* (1984) to discredit one such pessimistic volume, the *Global 2000 Report to the President* issued by the Global 2000 Study in 1980. More famously, Simon engaged in a highly publicized bet with Paul Ehrlich (b. 1932) in 1980. Ehrlich wagered that at least five of ten non-renewable resources (of his choosing) would be more expensive ten years later. Simon won the bet. In 1990, every one of the resources had declined in price by an average of forty percent. (When offered an opportunity to renew the wager for the next ten-year period, Ehrlich declined.)

As a result of his advocacy, Simon's ideas have won many converts to the idea that the status quo with some modest incremental adjustments will be sufficient for continued improvement in human well-being (e.g., Bailey 1993, Wildavsky 1995). His last major book, *The State of Humanity* (1995), was written with more than sixty collaborators. But despite the increased respectability accorded to Simon's views, they remain contentious and do not represent the mainstream in resource and population economics.

Science, Values, and the Hermeneutics of Data

From his very first article, Simon has been attacked by those who disagree with his views. Ehrlich called him an "imbecile," others considered his ideas simpleminded and dangerous, while most in the mainstream tried to refute the validity of his statistics (Regis 1997). But if the facts tell an unequivocal story, why is there so much disagreement? And if the facts corroborate Simon's analysis, why were his views so unpopular? Simon often felt that he was being ignored due to "a vast Malthusian population-environment-resources conspiracy of crisis" (1999, p. vii). In the posthumously published *Hoodwinking the Nation* (1999), he took up the question of why so much "false bad news" persists. He cited academic and media incentives and vested interests, psychological factors, strategies of change based on the assumption that crises mobilize action, racism, the non-intuitive nature of some of Simon's arguments, and widespread misunderstanding of resource creation and population economics. In all cases, he argued that what is at issue is the discrepancy between dominant, misguided beliefs and the facts of the matter.

On this level of psychological and sociological analysis, Simon undoubtedly presents some accurate findings. Yet a deeper level of analysis opens up beyond this limited argument that Simon has the true science and the absolutely correct data while others are just misled or willfully distorting the truth. For example, a graph may demonstrate that forest cover is increasing, but the reason for this may be the rise in forest plantations rather than recovery of more natural systems. Thus, the fact of increased forest cover leaves room for interpretation about its meaning and whether it is a good or a bad sign. Furthermore, some may find fault in Simon's anthropocentric view. They may regard global climate change as a problem even if humans are able to adapt to it, or they may object to his idea that genetic engineering and seed storage are reasonable responses to species extinction (1995, p. 15). Finally, some may argue that his categories miss the most important trends as he substitutes "what can be easily counted" for "what really counts." For example, in *The State of Humanity*, Simon admits that his trends describe only material and economic welfare but not emotional or spiritual welfare.

Unfortunately the underlying values differences between Simon and his adversaries are not often explicitly addressed. This held true of a similar controversy surrounding one of Simon's protégés, Bjørn Lomborg (b. 1965), author of *The Skeptical Environmentalist* (1998). Like Simon, Lomborg attacked the conventional wisdom and was in turn rebuked in a passionate series of exchanges with other scientists. Although disputants often claimed to be debating the facts, in reality the issues were much larger.

Despite his often zealous reliance on facts, Simon was perhaps aware of this dynamic to a greater extent than Lomborg. Whereas Lomborg concludes that we need to base decisions "not on fear but on facts" (p. 327), Simon concludes *The Ultimate Resource 2* with a section titled "Beyond the Data," including a subsection titled "Ultimately—What Are Your Values?" In this latter section he argued: "Whether population is now too large or too small, or is growing too fast or too slowly, cannot be decided on scientific grounds alone. Such judgments depend upon our values, a matter on which science does not bear" (p. 548). Measuring the real state

of humanity or the world involves normative as well as scientific considerations.

CARL MITCHAM
ADAM BRIGGLE

SEE ALSO *Environmental Ethics; Science Policy.*

BIBLIOGRAPHY

Bailey, Ronald. (1993). *Ecoscam: The False Prophets of the Environmental Apocalypse.* New York: St. Martin's Press. Deconstructs the conventional wisdom about resources and population growth in much the same way as Simon.

Ehrlich, Paul R. (1968). *The Population Bomb.* New York: Ballantine.

Ehrlich, Paul R., and Anne H. Ehrlich. (1974). *The End of Affluence: A Blueprint for Your Future.* New York: Ballantine.

Lomborg, Bjørn. (2001). *The Skeptical Environmentalist: Measuring the Real State of the World.* New York: Cambridge University Press. Full of trend data to support Simon's basic position that problems are mostly getting smaller rather than larger.

Myers, Norman, and Julian L. Simon. (1994). *Scarcity or Abundance? A Debate on the Environment.* New York: Norton.

Simon, Julian. (1965). *How to Start and Operate a Mail-Order Business.* New York: McGraw-Hill.

Simon, Julian. (1980). "Resource, Population, Environment: An Oversupply of False Bad News." *Science* 208(4451): 1431–1437.

Simon, Julian. (1996). *The Ultimate Resource 2*, rev. edition. Princeton, NJ: Princeton University Press.

Simon, Julian. (1999). *Hoodwinking the Nation.* New Brunswick, NJ: Transaction Books. Explains why the false litany of environmental bad news persists despite evidence to the contrary.

Simon, Julian, ed. (1995). *The State of Humanity.* Cambridge, MA: Blackwell.

Simon, Julian L., and Herman Kahn, eds. (1984). *The Resourceful Earth: A Response to Global 2000.* Oxford, UK: Basil Blackwell.

Wildavsky, Aaron. (1995). *But is it True? A Citizen's Guide to Environmental Health and Safety Issues.* Cambridge, MA: Harvard University Press. Case studies explore relationships between knowledge and action in environmental policy to argue that informed participation is a possible and necessary part of democratic citizenship. Concludes by rejecting the precautionary principle.

INTERNET RESOURCE

Regis, Ed. (1997). "The Doomslayer." *Wired Magazine* vol. 5, no. 2, pp. 136-140 and 193-198. Available from http://www.wired.com/wired/archive/5.02/ffsimon_pr.html. Recounts the confrontations between Ehrlich and Simon and clarifies Simon's basic points.

SIMPLICITY AND SIMPLE LIVING

• • •

The term *simple living* is generally used to refer to a voluntarily chosen way of life that is significantly less frenetic, and significantly less focused on "getting and spending," than life in the mainstream. Simple living traditions exist in a wide array of cultures, and date back thousands of years. But they take on special salience in highly affluent societies dependent on science and technology for their patterns of production and consumption.

The term *simplicity* is sometimes used synonymously with *simple living*, but this can lead to confusion as one of the potential uses of high levels of income is to purchase solutions to the burdens of everyday life. Thus, the very wealthy can afford to have personal assistants to take care of their finances, assist in childrearing, and manage the household, vastly simplifying their existence.

Basic Arguments

A theme common to many diverse simple living traditions is that too great an involvement with money is deeply problematic. A classic presentation of this thesis is found in Aristotle's *Politics* (4th century B.C.E.), which opens with a critique of excessively commercialized civilization. Aristotle (384–322 B.C.E.) distinguishes between what he terms natural and unnatural ways of life. Among the natural ways are hunting, fishing, and farming. What is distinctly unnatural is commerce, whose hallmark is that the pursuit of money takes on a life of its own, knowing no bounds.

Aristotle offers two critiques. The first anticipates the economic theorists of the nineteenth century: Aristotle argues for the diminishing marginal utility of money, maintaining that beyond a limited sufficiency, additional money does not contribute to human happiness. His second thesis is yet more radical, arguing that the unbridled absorption in attaining money results in the misuse of human capabilities and the distortion of the personality. When elevated to the social level, this produces a society in which all social roles have been corrupted. Doctors no longer pursue the health of the patient; jurists no longer seek justice. All activities are ultimately undertaken in pursuit of financial gain.

The two issues Aristotle raises, distortion of the personality and corruption of social roles, are two of a number of concerns that have motivated proponents of simple living. An example of the first is Henry David Thoreau (1817–1862), who wrote in *Walden* (1854) that wealth is a curse because it enslaves us. "I see young men, my townsmen, whose misfortune it is to have inherited farms, houses, barns, cattle and farming tools; for these are more easily acquired than got rid of." And, "The finest qualities of our nature, like the bloom on fruits, can be preserved only by the most delicate handling. Yet we do not treat ourselves nor one another thus tenderly" (Thoreau 1965, p. 4 and p. 6).

An example of the second concern, the health of the society, can be found in what has been called *Republican Simplicity* by historian David Shi. In the mid 1700s prior to the American Revolution, many of the leaders of that Revolution looked to the history of ancient Rome and Greece for guidance in their democratic venture. The lesson that they drew was that public virtue was necessary for the success of a republic, and that it could be undermined by excessive commercialism. John Adams (1734–1826) and Thomas Jefferson (1743–1826) corresponded about how to build a non-materialist society, and Jefferson looked to state-supported schools and value education as a foundation.

In the writings of the Quaker theorist John Woolman (1720–1772), one finds two lines of thought, both of interest. First, in contrast to the Puritans, Woolman suggested that the simple life also involved limitations on the amount of work one would do. This would later be expanded on by Thoreau, who suggested that we should have one day of work and six days of Sabbath. Secondly, Woolman argued that most of the ills of the world—poverty, slavery, war—could be traced to luxurious desires. He urged that we examine our own lives and see whether, unwittingly, we are part of the problem. He said we should "look upon our treasures, and the furniture of our houses, and the garments in which we array ourselves, and try whether the seeds of war have nourishment in these our possessions or not." The contemporary application of this outlook is the suggestion that war in the Middle East, and perhaps terrorism as well, have their roots in our excessive consumption of oil.

Benjamin Franklin (1706–1790), another American advocate of simple living, came to it from a rather different direction. Franklin argued the importance of the individual's liberation from the demands of onerous labor. "Employ thy time well, if thou meanest to gain Leisure." But Franklin argued for sharply limiting our consumption, so that we may save. His message was that we could all become wealthy if we learned to discipline ourselves, limited our desires, and earned more than we consumed.

Assessment and Application

These various examples make clear that simple living can be advocated for a wide variety of reasons. It represents no single philosophy of life. And while there are some exceptions—perhaps Franklin is one—what they have in common is the view that the good life, both individually and socially, is to be found largely outside the economic realm. Human happiness is obtained not by consuming more and more of what the economy has to offer, but by satisfying core economic needs, and then turning away from the economic to other realms of importance, whether they be religion, science, literature, service to others, or friends and family.

While much of the simple living literature is directed at the individual, offering advice and suggestions for how to live, simple living at times emerges as a politics of simplicity. Here it looks to social policy to offer the framework within which it becomes feasible for the average person to opt for a simple life. Such a politics offers a different paradigm for understanding the relationship between a technological economy and the good life. Economic performance is assessed not in terms of growth, but in terms of success in meeting core needs of the entire population. Technological and economic progress is measured more in terms of the expansion of leisure than the growth of gross domestic product (GDP). And work, rather than being seen as one productive input within the production process, is seen, potentially, as a realm within which personal growth and meaning can be achieved.

JEROME M. SEGAL

SEE ALSO *Consumerism*.

BIBLIOGRAPHY

Shi, David. (2001). *The Simple Life: Plain Living and High Thinking in American Culture*. Athens: University of Georgia Press. Shi provides a comprehensive and highly readable account of the different forms that simple living has taken, in practice and thought, throughout the American experience.

Segal, Jerome M. (2003). *Graceful Simplicity: The Philosophy and Politics of the Alternative American Dream*. Berkeley: University of California Press. Segal challenges the standard view within the simple living literature that, except for pockets of poverty, Americans have sufficient income to meet core economic needs. Rather than a "how to" book on simple living, he calls for a new approach to social and economic policy. He emphasizes the importance of beauty within the good life.

Thoreau, Henry David. (1965). *Walden and Other Writings*, ed. Brooks Atkinson. New York: The Modern Library.

SINGAPORE

• • •

Small states, like small businesses, often serve as the incubators of new forms of government. Perhaps no state has been so carefully and deliberately managed as Singapore, a multi-ethnic island city-state of 4 million inhabitants in an area of 250 square miles, or about the size of Guam. Because of the ways its management has sought to utilize science and technology to achieve certain social values, which has itself influenced some of these values, Singapore provides a useful case study in the possible relations between science, technology, and ethics.

Background

Located on the southern tip of the Malay Peninsula and separated from Indonesia, the largest Muslim country in the world, by the Straits of Malacca, Singapore was colonized by the British in the early 1820s due to its strategic location (for the British, it was the Gibraltar of the East). Important because it served as both a submarine port and had a major airfield, the Japanese captured Singapore during World War II. After the war it evolved toward independence in phases: It elected its first legislature in 1955 and was granted internal self-government in 1959. In 1963 Singapore joined the Federation of Malaysia, but separated in 1965 and has been fully independent since.

The People's Action Party (PAP), founded and dominated by Lee Kuan Yew (b. 1923), a British-educated lawyer, has led the country since the mid-1950s, creating a single-party state dedicated to the pursuit of economic growth through social order and efficiency under the guidance of a technocratic ideology. The result has been one of the most globalized entities in the world, measured in terms of foreign trade, investment, information inflows, and immigration. Between 1971 and 2003, Singapore's economy expanded at an average annual gross domestic product (GDP) growth rate of 7.2 percent. It enjoys one of the highest standards of living in Asia and was ranked sixth in the Growth Competitive Index conducted by the World Economic Forum in 2003.

From Stability to Creativity

Constant technological upgrading has been vital to the economic ascendancy of Singapore, and social policies have been reflexively monitored and implemented—whether in the streaming policies of the educational system, the level of civil liberties, or the value system of society—to ensure Singapore's global economic relevance. The political elite's Hobbesian view of national and international politics underpins Singapore's broad ethical approach to economic and technological development. The dominant image widely propagated in Singapore is that of a vulnerable city-state, lacking both natural resources and the cultural homogeneity of a Japan or a Korea (Singapore's ethnic composition is 76.8 percent Chinese, 14 percent Malay Muslim, and 8 percent Indian), and surrounded by potentially volatile Malay Muslim neighbors. The Singaporean leadership has used "survival" to justify the hierarchical management of society. The resulting political system has been dubbed by Chan Heng Chee (1989) as "the administrative state," a term that captures the depoliticization of the citizenry and the central place of a powerful bureaucracy in managing society. The political elite sees itself as practicing a pragmatic style of governance, understood as the ability to act rationally in the interest of the collective good without getting bogged down by moral and democratic excesses (Chua 1995).

The value framework has varied with the technological challenges facing Singapore. From the mid-1960s to the mid-1990s, technocratic planners invited multinationals from around the world to invest and manufacture consumer goods, and later highly sophisticated engineering components, for the global market. Founding leader Lee Kuan Yew, with strong eugenics views (Barr 2000), did not believe that Singapore's small population could produce a critical mass of creative individuals doing cutting-edge research. Instead, science and technology policies focused on producing highly competent citizens who could absorb and perhaps re-engineer products and processes from existing technology. Huge investments were made in tertiary education to supply technicians and engineers for the multinational sector at cheaper costs than in Western countries. Generous tax incentives, a highly controlled labor movement, and the sheer predictability of politics attracted some 7,000 well-known global companies to invest in the economy. These included such names as Philips, Honeywell, Hewlett-Packard, Seagate, Motorola, Exxon-Mobile, NEC, Siemens, and Sony.

In this phase, the ethical framework laid out by the government for technology development was a broad, society-wide one rather than a set of specific policies applied to particular industries or sectors. Singaporeans were expected to be socially disciplined, to comply with the technocratic goals of the government, and to refrain from excessive individualism and political expression (Quah 1983). They were asked to subscribe to a stereo-typical notion of Asian values, which the leaders believed would help the population ward off pernicious Western practices, such as weak commitment to the family, a propensity for contention over consensus, and a disrespectful youth culture. Singapore became famous for harsh punishments for behaviors such as littering, failing to flush public toilets, and small-scale drug dealing. The government expected conformity and in turn promised order, prosperity, integrity, and dedication to the collective good.

In the 1990s, however, new competitive pressures led to a major shift in the government's approach to technological development, and in almost cybernetic fashion, adjustments in social regulation policies. Countries previously outside the global capitalist system, such as China, India, and Central Europe, were now entering the global market. The Asian crisis that began in 1997 saw multinationals changing locations in the region. Gripped by concerns of national survival, planners saw the need to go beyond using multinationals for economic development and technology transfer, and undertook to produce original knowledge and technology. The planners hoped to build on existing educational and scientific infrastructures, such as the Institute of Molecular and Cell Biology (IMCB), which had been set up in 1987, to embark on original research.

The sectors targeted to spearhead the knowledge-based economy were bioscience and biomedical research, with foci in tissue engineering, stem cell research, immunology, and cancer research. Through these efforts, Singapore hoped to become a major player in pharmaceuticals, medical equipment, and health services. More than a billion U.S. dollars was committed toward creating an integrated medical and biotechnological park, Biopolis, and huge funds were earmarked for strategic investments in local and foreign biotechnology companies.

Framework for Policy and Ethics

The key question was how Singapore, without a long history of broad-based original research, would make the transition from being a technology-recipient to technological innovator. This challenge was met with a two-pronged approach. The planners mapped out a research process in which innovation would be carried out and directed by global research stars drawn to Singapore by alluring financial terms, including generous research funding. The other tack, and an important further inducement for researchers, was the creation of a stable and predictable milieu for long-term research, particularly in the biomedical area, unencumbered by moral and religious obstacles. Some technologically sophisticated nations, especially the United States, were putting restrictions on research involving living embryos, so Singapore's ability to provide a liberal moral climate allowing for such research would place it in a comparative advantage. Singapore's technocrats now had to use skills that had provided the high degree of economic, social, and political predictability during the technology-receiving phase to lay the requisite financial and ethical predictability for these new research and technological goals.

The challenge in creating a liberal moral climate involved coming to terms with local religious groups, particularly those from the growing Christian population among the upper stratum of Singaporeans. In addition, to gain legitimacy from the international community of researchers and regulators, Singapore had to demonstrate that it was not a morally renegade society but was committed to socially responsible research. This led the government to set up the Bioethics Advisory Council (BAC) in late 2000 to make recommendations for bioscience and biomedical research in Singapore. The committee, which was chaired by the former Vice-Chancellor of the National University of Singapore, stated that it would consult civil society groups, professional associations, and religious organizations in carrying out its charge, and promised to proceed with caution "so our findings and recommendations will be acceptable to society" (Straits Times, February 7, 2001).

Civil society in Singapore was generally quiescent (Tamney 1996), but on this morally sensitive issue involving the use of human embryos for research, religious groups freely gave their opinion. (Singapore is 42.5 percent Buddhist, 15 percent Muslim, 14.5 percent Christian, 8.5 percent Daoist, 4 percent Hindu, and 15 percent claiming no religion.) Most professional groups went along with embryonic stem cell research, but there was consternation among the religious representatives. Muslim representatives, believing that ensoulment of the human being begins forty days after conception, were amenable to early stage embryonic research. The same was true of the Buddhist groups, which view genetic research as helping humankind. By contrast, Protestant and Catholic bodies, as well as Hindu and Daoist representatives, objected to any destruction of embryos to obtain stem cells. Daoists argue it was against nature's way, Christians define life as beginning at conception, and Hindus see the destruction of the embryo as short-circuiting the karmic cycle. The deontological ethical position of these groups was at variance with the BAC, whose desire was to see bioscience develop in Singapore. As far as the BAC had an ethical position, it was a consequentialist one, proffering the benefit to humankind of finding cures to terrible diseases as a result of bioscience research. The Council subsequently ruled that its recommendations would not be dictated by religious positions, and argued, in typical pragmatic language, that research had to move ahead because "Singapore is a small place" (Straits Times, December 28, 2001).

Its recommendations, which were incorporated in the Biomedical Research Act of 2003, allowed for stem cells to be obtained from human embryos less than fourteen days old, the age just before the neurological system developed (Bioethics Advisory Committee 2002). Embryos less than fourteen days could be cloned but there would be no cloning of embryos for reproductive purposes. As if to underscore its ethical concerns, the Council stressed that all researchers and doctors required the consent of patients and embryo donors. In addition, the BAC was keen to point out that its recommendations were no more lax than legislation in other democracies such as the United Kingdom, Australia, Japan, and Sweden. In short, it was acting well within international norms. Despite some religious misgivings, resulting legislation is likely to preempt any future religious or moral objections, because both the government and the regulatory bodies can claim that society had been fully consulted in the decision-making process, and most groups went along with the final recommendations.

Singapore's liberal moral climate and weak civil society has earned the praise of many top scientists. A number of U.S. scientists, responding to the Bush administration's banning of embryonic research and its strict control over the use of existing stem cell lines, have found Singapore to be a more hospitable climate for their research. Dr. Philippe Taupin, a renowned biologist previously at the Salk Institute, gave the following reason for his move to Singapore in 2003: "I came here because I want to jumpstart my career. There are fewer ethical and political minefields than in the West, and Singapore has pledged a strong commitment to stem cell biology" (Straits Times, February 17, 2004).

Prospects

Singapore's strategy of bringing in experts from abroad has been impressive. Generous funding, which makes it unnecessary to apply constantly for research grants, and an uncritical climate, which extends to the plentiful supplies of laboratory mice undisturbed by animal rights activists, have been major draws. An influx of high-profile researchers would help both to leapfrog into cutting edge research and attract younger scientists the world over by establishing a prestigious and reputable climate. In 2003, 30 percent of the 3,600 Ph.D.s working in the biomedical sector were foreigners. Global stars such as Edison Liu, formerly at the National Cancer Institute in the United States, Alan Colman of "Dolly the sheep" fame, and Yoshiaki Ito from Kyoto University have given Singapore overnight attention as a global research center. Whole research teams from Japan and France have immigrated and been generously funded. The administrative coordination of education, immigration, and the health sector to support the advancement of bioscience has greatly impressed foreign researchers. Liu, who came to Singapore in 2001 to head the Genome Institute of Singapore (GIS), marveled at the integrative approach of the leaders and planners: "They are strategic thinkers, and are smart enough to view this as a whole. It is the most astounding social engineering I have seen in my life" (*Far Eastern Economic Review*, October 9, 2003).

The top-down control of society has not prevented the pragmatic relaxation of social controls from helping to realize the leaders' economic goals. Departing selectively from its previous preoccupation with social discipline and conformity, the government now asks Singaporeans to become creative individuals willing to take entrepreneurial risks. Activities such as bungee jumping, bar-top dancing, and street busking, once banned and frowned upon, are now being permitted to foster an adventurous spirit among the population. The most dramatic reversal has been to allow the lesbian and gay population to join the civil service. The tolerance of homosexuality, once derided as contrary to Asian values, is now seen as consistent with the pursuit of creativity—as argued by Richard Florida (2004).

Ethical Ambiguity

It would not be surprising if the urgency of meeting national economic goals in conjunction with the pragmatic design of the ethical framework for research should leave some ambiguity about the moral boundaries of research. A test case occurred in 2002, involving a world-famous British researcher, Dr. Simon Shorvon, a neurologist who had done pioneering work in epilepsy and Parkinson's disease. After being courted by Singapore authorities, he took up the position of Director of the National Neuroscience Institute. Shorvon's research into the role of genetic mutations in Parkinson's required patients to go off their medications while he studied the effects of administering various doses of L-Dopa and traditional Chinese herbs. The research

design required 1,500 Parkinson patients, but only twelve volunteers were available as of July 2002.

To secure more subjects, Shorvon retrieved records from the databases of pharmacies, deliberately bypassing the patient's doctors, and then led patients to believe that they had their physicians' approval for their research participation. In his experiments, Shorvon and his co-workers sometimes administered drugs at dangerously high levels, causing a few serious complications. When Singapore neurologists learned of his research and complained, he dismissed their concerns by saying that his methods were sensible and efficient, and claimed he had the backing of the various hospital review boards. Many of the Singapore doctors, including established professors, were torn between their commitment to patient rights and research ethics and the presumed importance of Shorvon's research. None of his peers and fellow neurologists made an official complaint.

Consistent with the top-down system of control in Singapore, it took a member of the inner circle of the elite to highlight and publicize the wrongdoing. Dr. Lee Wei Ling, a neurologist and (then) Deputy Director of the National Neuroscience Institute, is also the daughter of Lee Kuan Yew and sister to the current prime minister. When she was hospitalized for a neurological problem, her fellow neurologists mentioned the activities of Dr. Shorvon (*Straits Times*, April 4, 2003). Dr. Lee then reported him to the relevant authorities, leading to his removal. The interesting point about this case is not the lack of ethical standards in Singapore's research setting, but the fact that individual doctors and researchers did not feel sufficiently empowered by the hierarchical ethical system to take it upon themselves to expose wrongdoing. It took a member of the elite, who fortuitously happened to personally object to the egregious activities, to give weight to the ethical framework already in place.

JAMES V. JESUDASON
HABIBUL H. KHONDKER

SEE ALSO *Modernization; Political Economy; Political Risk Assessment; Science Policy.*

BIBLIOGRAPHY

Barr, Michael D. (2000). *Lee Kuan Yew: The Beliefs Behind the Man.* Washington, DC: Georgetown University Press. An informative book on the ideas behind Singapore's social engineering policies.

Bioethics Advisory Committee of Singapore (BAC). (2002). *Report on Ethical, Legal, and Social Issues in Human Stem Cell Research, Reproductive and Therapeutic Cloning.* Singapore: Author.

Chan Heng Chee. (1989). "The PAP and the Structuring of the Political System." In *Management of Success: The Moulding of Modern Singapore,* ed. Kernial Singh Sandhu and Paul Wheatley. Singapore: Institute of Southeast Asian Studies. A good distillation of this prominent author's works on Singapore politics.

Chua, Beng Huat. (1995). *Communitarian Ideology and Democracy in Singapore.* London: Routledge. An influential book on Singapore's political ideology.

Florida, Richard. (2004). *The Rise of the Creative Class.* New York: Basic Books. A controversial book arguing that a culture of diversity and tolerance in cities accounts for high economic growth.

Quah, Stella R. (1983). "Social Discipline in Singapore." *Journal of Southeast Asian Studies* 14(2): 266–289.

Tamney, Joseph B. (1996). *The Struggle Over Singapore's Soul: Western Modernization and Asian Culture.* Berlin: W. de Gruyter. A careful survey of Singapore's blend of Asian and Western values.

Magician James "the Amazing" Randi. Randi's media presence has brought the skeptical movement into the public consciousness. (© *Jeffery Allan Salter/Corbis.*)

SKEPTICISM

• • •

Skepticism has a long history that includes multiple meanings and in the early twenty-first century has complex ethical implications for science and technology. It plays an important role *within* science and technology but also can be *applied to* the same areas. In the former case skepticism may serve as a means to reject mistaken or false claims, limit fraud and misconduct, and produce evaluations of engineering designs and the safety of technologies. In the latter case skepticism may help the public place the benefits of science and technology in a larger perspective, although it also may deprive the public of certain real benefits.

Antecedents

The roots of skepticism can be traced back at least 2,500 years to the ancient Greeks. The historian of skepticism Richard Popkin states: "Academic scepticism, so-called because it was formulated in the Platonic Academy in the third century, B.C.E., developed from the Socratic observation, 'All I know is that I know nothing'" (Popkin 1979, p. xiii). In fact, the philosopher Pyrrho and his followers doubted the possibility of real knowledge of any kind, a viewpoint that led to a form of nihilism. Skepticism in this sense is a positive assertion about knowledge and thus cannot be held seriously if it is turned on itself: If one is skeptical about everything, one also has to be skeptical about one's own skepticism. Like a decaying subatomic particle pure skepticism uncoils and spins off the viewing screen of the mind's intellectual cloud chamber.

A more pragmatic meaning of the word *skeptic* can be found in the Greek word *skepsis*, which means "examination, inquiry, consideration." The *Oxford English Dictionary* gives this historical usage: "One who doubts the validity of what claims to be knowledge in some particular department of inquiry; one who maintains a doubting attitude with reference to some particular question or statement," along with "a seeker after truth; an inquirer who has not yet arrived at definite convictions." Skepticism is not "seek and ye shall find" but "seek and keep an open mind." In this context having an open mind means finding the essential balance between orthodoxy and heresy, between a total commitment to the status quo and the blind pursuit of new ideas, between being open-minded enough to accept radical new ideas and being so open-minded that one's brain cannot function.

Since the time of the ancient Greeks skepticism has evolved along with other epistemologies. On one level the Enlightenment was a century-long skeptical movement because there were few beliefs or institutions that did not come under the critical scrutiny of thinkers such as Voltaire (1694–1778), Denis Diderot (1713–1784), Jean-Jacques Rousseau (1712–1778), John Locke (1632–1704), and Thomas Jefferson (1743–1826). David Hume (1711–1776) in Scotland and Immanuel Kant (1724–1804) in Germany were skeptics' skeptics in an age of skepticism, and their influence continues to be felt in the early 2000s. In the twentieth century Bertrand Russell (1872–1970) and Harry Houdini (1874–1926) stood out as representatives of skeptical intellectuals and activists, respectively. Martin Gardner's *Fads and Fallacies in the Name of Science* (1952) launched the contemporary skeptical movement.

The Contemporary Skeptical Movement

Starting in the 1970s, the magician James "the Amazing" Randi's psychic challenges and media appearances pushed the skeptical movement to the forefront of public consciousness. In 1976 the philosopher Paul Kurtz (born 1925) founded an international skeptical organization called the Committee for the Scientific Investigation of Claims of the Paranormal (CSICOP), and in 1991 Michael Shermer cofounded the Skeptics Society and *Skeptic* magazine. This has led to the formation of a burgeoning group of people calling themselves skeptics who conduct investigations, hold monthly meetings and annual conferences, and provide the media and the general public with natural explanations for apparently supernatural phenomena.

Although intellectual skepticism flourishes in academia, skeptical activism has emerged as a powerful force in the application of science to all claims. In fact modern skepticism is embodied in the scientific method, which involves gathering data to formulate and test naturalistic explanations for natural phenomena. A claim becomes factual when it is confirmed to an extent where it would be reasonable to offer temporary agreement. However, all facts in science are provisional and subject to challenge, and skepticism thus is a method that leads to provisional conclusions.

Some claims, such as water dowsing, extrasensory perception (ESP), and creationism, have been tested and have failed the tests often enough that they may be rejected provisionally as false. Other claims, such as hypnosis, near-death experiences, and neurological correlates of consciousness, also have been tested, but the results have been inconclusive. Finally, there are claims, such as string theory, inflationary cosmology, and multiple or parallel universes, that are theoretically possible but have not been tested empirically. The key to skepticism is to apply the methods of science continuously and vigorously to make it possible to navigate the straits between "know nothing" skepticism and "anything goes" credulity. In this sense skepticism is the ethical component of science. It is the attitude that keeps the scientific method honest, the canary in the scientist's mine.

Ethical Issues

In regard to ethical concerns it is important to recognize the fallibility of science and skepticism. Although scientific skepticism is well suited for identifying certain kinds of mistakes and errors in thinking, such as what are called type I errors, or false positives, its standards are so high that it occasionally leads to the commission of a type II error, or false negative, failing to identify, for example, potential lifesaving medicines.

However, within this fallibility there are opportunities for self-correction. Whether mistakes are made honestly or dishonestly, whether a fraud is perpetrated unknowingly or knowingly, in time it will be recognized. The cold fusion fiasco in the late 1980s was a classic example of how organized skepticism can identify hype and error. Because of the importance of this self-correcting feature, there is in the profession what the Nobel laureate physicist Richard Feynman called "a principle of scientific thought that corresponds to a kind of utter honesty—a kind of leaning over backwards." As Feynman explained: "If you're doing an experiment, you should report everything that you think might make it invalid—not only what you think is right about it: other causes that could possibly explain your results" (1988, p. 247). Of course, not all scientists live up to this ideal.

What separates skepticism and science from other human activities is the tentative nature of all conclusions: There are no final absolutes, only varying degrees of probability. Skepticism is not the affirmation of a set of beliefs but a process of inquiry that leads to the building of a testable body of knowledge that is open to rejection or confirmation. In skepticism, knowledge is fluid and certainty is fleeting. That is the heart of its limitation and its greatest strength.

MICHAEL SHERMER

SEE ALSO *Libertarianism; Locke, John; Merton, Robert; Pseudoscience; Tocqueville, Alexis de; Wittgenstein, Ludwig.*

BIBLIOGRAPHY

Feynman, Richard P. (1988). *What Do YOU Care What Other People Think?* New York: Norton. A collection of first-person accounts by a Nobel laureate on how science is done, including his investigation into the Space Shuttle *Challenger* disaster.

Gardner, Martin. (1982 [1952]). *Fads and Fallacies in the Name of Science.* New York: Dover. The book that launched the skeptical movement. Still current after fifty years, it covers unidentified flying objects, scientology, alternative medical claims, and ways to detect pseudoscience.

Hecht, Jennifer Michael. (2003). *Doubt: A History: The Great Doubters and Their Legacy of Motivation, from Socrates and Jesus to Thomas Jefferson and Emily Dickinson.* San Francisco: Harper. A comprehensive history of skepticism from the ancient Greeks to modern atheists; includes both intellectuals and activists.

Popkin, Richard H. (1979). *The History of Scepticism from Erasmus to Spinoza.* Berkeley: University of California Press. An excellent history of intellectual skepticism by a philosopher and skeptic.

Randi, James. (1985). *Flim Flam! Psychics, ESP, Unicorns and Other Delusions.* Buffalo, NY: Prometheus Books. The classic work of a master investigator of psychics and scam artists; includes his many personal investigations into the world of the paranormal.

Shermer, Michael. (1997). *Why People Believe Weird Things: Pseudoscience, Superstition, and Other Confusions of Our Time.* New York: Freeman. A skeptical and scientific manifesto that includes analyses of creationism, Holocaust denial, immortality, near-death experiences, cults, and the nature of pseudoscience and superstition.

B. F. Skinner, 1904–1990. The American experimental psychologist became the chief exponent of that form of behaviorism known as operationism, or operant behaviorism. *(The Library of Congress.)*

SKINNER, B. F.

• • •

The reinventor and foremost champion of behaviorist psychology, Burrhus Frederic Skinner (1904–1990) was born in Susquehanna, Pennsylvania on March 20, and died at age 86 in Cambridge, Massachusetts on August 18. Building on the work of Ivan Pavlov (1849–1936), Edward Thorndike (1874–1949), and J. B. Watson (1878–1958), B. F. Skinner made unique contributions to the science of human behavior and intended for his work to serve as the basis for technologies by which human beings could control themselves and others for the benefit of all.

Life and Achievements

Graduating from Hamilton College, New York, with a bachelor's degree in English, Skinner initially wanted to become a writer. This vocation eluded him, and after a period of time in Greenwich Village he enrolled for graduate studies at Harvard University, where he earned his doctorate in psychology in 1931. In 1936 he went to teach at the University of Minnesota, where he met and married Yvonne Blue. In 1945 he became chair of the psychology department at Indiana University, but three years later returned to Harvard as a professor, where he remained for the rest of his academic career.

Skinner's work centered on the idea of *operant conditioning.* Unlike classical behaviorism, operant conditioning is the idea that as living organisms move about in their environments, behaviors that meet with reinforcing stimuli will be promoted, and other behaviors will not. Imagine saying "Hello" to associates at work, to which they give cheerful and friendly replies, leading to increased greetings; in the absence of any response, greetings will likely diminish or cease. Skinner elaborated this insight into diverse schedules of reinforcement (fixed and variable ratio and interval schedules) in order to investigate empirically their various degrees of effectiveness in behavior modification. Anthony Burgess's novel, *A Clockwork Orange* (1962) and the Stanley Kubrick film of the same title (1972) misrepresent behavior modification as using aversive reinforcement or stimuli (punishment) to discourage behavior, which Skinner regarded as ineffective.

Skinner was a fervent advocate of the application of operant conditioning. He even publicized that he applied his theories to his children, especially his younger daughter, who was in part raised in an air crib designed by Skinner. As a result of Skinner's work, operant conditioning became popular among therapists; some remained devotees into the twenty-first century.

But some problems with operant conditioning have led to skepticism. Among these are the underlying assumption of determinism and the dismissal of human consciousness. Skinner also proposed awkward ways for understanding emotions and thinking—the latter he dubbed "probability of verbal behavior"—so they would conform to the requirement of being observable (in Skinner's mind, a general requirement for all experimental sciences).

It is also unclear how some reinforcing stimuli become reinforcing in the first place. Suppose one hopes that saying "Hello" will encourage associates to leave one alone. Instead, they become intrusively friendly. The condition thus backfires. Ordinarily it is not difficult to tell a welcome response, but with complex actions this is no longer simple. Some critics argue that Skinner was openly ambivalent about whether human conscious life exists (Baars 2003), but others find in Skinner the most advanced way to apply modern science to human life and human society (Woodward and Smith 1996).

Controversies

Skinner thought that his insights into the technology of behavior ought to be used to cure sociopolitical problems. His presentation of this view in a utopian novel, *Walden II* (1948), and in such applications as *The Technology of Teaching* (1968), drew extensive criticism. Many charged him with proposing an anti-democratic technocracy that would extinguish human liberty and morality.

His response to this criticism was his most famous book, *Beyond Freedom and Dignity* (1971). Here he argued that "freedom" and "dignity" are pre-scientific concepts, and shifting to scientific terminology and applications would advance human life and society better than rhetoric. For Skinner, the scientific approach is the most dependable, reliable way to understand the world, and the implications of this approach are so significant as to render it imperative to follow it in all spheres of human concern. Religion, morality, free will, and even feelings are to be purged from an objective (that is to say, empirical) scientific conception of relationships to the world and each other. Indeed, Skinner thought that the more humans adopted his recommendations, the more likely they would be to achieve the goal of peace.

As to the overall success of Skinner's ideas, on some fronts his views have triumphed. His ideas that humans and other animals are pretty much the same have been well received in the burgeoning animal rights or liberation movement, for example. In applied psychology, however, Skinner has lost much appeal. Cognitive psychology, for example, has eclipsed his behaviorism. Skinner remains, however, one of the twentieth century's most prominent theorists about human behavior, next, perhaps, only to Sigmund Freud.

TIBOR R. MACHAN

SEE ALSO *Genetics and Behavior; Psychology; Utopia and Dystopia.*

BIBLIOGRAPHY

Baars, Bernard J. (2003). "The Double Life of B.F. Skinner: Inner Conflict, Dissociation and the Scientific Taboo Against Consciousness." *Journal of Consciousness Studies* 10(1): 5–25(21). This issue of the journal contains the lead essay by Baars and several replies, including one from Skinner's daughter, Julie S. Vargas.

Bjork, Daniel W. (1993). *B. F. Skinner: A Life.* New York: Basic Books. This is a comprehensive guide to B. F. Skinner's life.

Skinner, B. F. (1948). *Walden Two.* New York: Macmillan. Skinner's novel is his illustration of the kind of society his technology of behavior would produce.

Skinner, B. F. (1953). *Science and Human Behavior.* New York: Macmillan. Skinner's most accessible scientific book.

Skinner, B. F. (1968). *The Technology of Teaching.* New York: Appleton-Century-Crofts. The focus here is the deployment of Skinner's technique for teaching.

Skinner, B. F. (1971). *Beyond Freedom and Dignity.* New York: Viking. Skinner's most popular book on his understanding of human life and society.

Skinner, B. F. (1974). *About Behaviorism.* New York: Knopf. Skinner's explanation of his work in the later stages of his career.

Woodward, William R., and Laurence D. Smith, eds. (1996). *B. F. Skinner and Behaviorism in American Culture.* Bethlehem, PA: Lehigh University Press. A good collection of reflections on Skinner's work.

SLIPPERY SLOPE ARGUMENTS

● ● ●

"*Partial-birth abortion* bans are not themselves that bad. But you should oppose them because, if they are enacted, much broader bans on abortion will become more likely." "Letting dying people cut off their lifesaving treatment may seem proper on its own. But if we allow that, it may lead to dying people getting help in actively killing themselves, and then over time to involuntary killing of the comatose or even of the disabled." "Embryonic stem cell research might be OK in itself, but it may lead to people getting pregnant just to get abortions." Such arguments are commonplace in debates on many ethical topics: abortion, euthanasia, genetic engineering, gun control, free speech, privacy, and more.

All these arguments express concern about the *slippery slope*: the risk that implementing a seemingly modest and worthwhile decision A now will increase the likelihood of a much broader and more harmful decision B later. The arguments are sometimes made by political liberals and sometimes by political conservatives. They sometimes relate to judicial decisions and sometimes to legislative ones. But they are all prudential arguments about long-term consequences.

The slippery slope is not just a form of argument. It is also an asserted real-world phenomenon—the tendency of one decision to increase the likelihood of others. If this phenomenon is real, people may want to consider it when deciding where to stand on policy questions: After all, if a decision today does make likelier other decisions tomorrow, it is prudent to consider this risk when making the first decision.

Analyzing Slippery Slope Arguments

There is no well-established definition for what constitutes a slippery slope. Some limit it to situations where A and B are separated by a long series of incremental steps: first one restriction on gun ownership, then another, then a third, and eventually all guns are banned. Others limit slippery slopes to situations where A and B cannot be easily logically distinguished. Some philosophers define the slippery slope as a form of purely logical argument, that enacting A will logically require the enactment of B.

Still others look to the reason that people worry about slippery slopes. Voters, legislators, judges, and others often face the question, Should I support proposal A, or should I oppose it for fear that it might help bring

about B? To answer this, one must consider all the possible ways that A can help lead to B—whether sudden or gradual, logical or political. This entry will therefore use this broad definition: A slippery slope happens whenever one narrow judicial or political decision now (for instance, banning Nazi or Communist speech) increases the likelihood that another, broader decision will be enacted later (for instance, censorship of more speech).

Not *We*, but *They*

Why would slippery slopes ever happen? Say that we think gun registration (A) is good but gun confiscation (B) is bad. Why would decision A make decision B more likely? If we dislike gun confiscation now, would we not dislike it as much even after gun registration is enacted?

Social decisions are made by groups composed of individuals—voters, legislators, judges, and so on—who have different views. The slippery slope concern is that our support for decision A today will lead *other* people to support decision B tomorrow.

For instance, gun registration may make gun confiscation easier because the police will know where the guns are. It may also make confiscation more defensible legally because the police will be able to get warrants to search the homes of those people who have the guns. The cheaper a policy is, the more likely people are to support it. This year a swing group of voters may help enact gun registration because they like registration but not confiscation. But next year the same group might find itself outvoted by another group of voters who conclude that, because guns are now registered, confiscation is cheaper and thus more appealing.

The first group of voters will have fallen victim to the slippery slope: They voted for a modest step A, which they liked, but as a result got outcome B, which they loathe. They may then wish that they had considered the slippery slope dangers before making the first decision.

Different Slippery Slope Mechanisms

How can one evaluate the likelihood that supporting A will indeed lead others to support B? The metaphor of the slippery slope, unfortunately, will not help, precisely because it is just a metaphor. It is necessary to identify the mechanism behind it: How exactly will the first decision change the conditions under which others will evaluate the second proposal? There are several such mechanisms, all of which can be called slippery slopes,

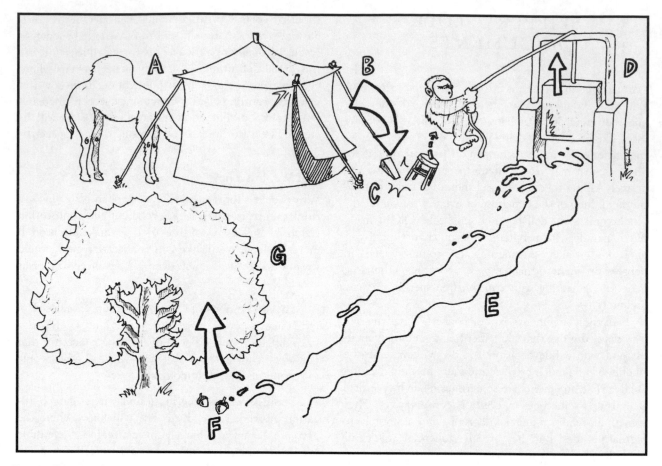

Drawing illustrating the concept of a slippery slope argument. Camel (A) sticks his nose under the tent (B), which collapses, driving the thin end of the wedge (C) to cause monkey to open floodgates (D), letting water flow down the slippery slope (E) to irrigate acorn (F) which grows into oak (G). (*Drawing by Eric Kim. Courtesy of Eugene Volokh.*)

but which are analytically different. Here are just a few examples.

COST-LOWERING SLIPPERY SLOPES. The gun registration example is one scenario. If decision A makes decision B cheaper, then it makes B more likely.

EQUALITY SLIPPERY SLOPES. Decision A may lead some people to feel that decision B must be enacted as well for equality reasons. For instance, some people argue that it is unfair to allow the dying to commit assisted suicide while refusing to permit the same release to those who are in great psychological pain but are not dying. The first step A may push some voters, legislators, or judges to support B, not because they like B as such, but because they oppose discrimination between A and B.

ENFORCEMENT NEED SLIPPERY SLOPES. When a modest restriction A—for instance, a mildly enforced prohibition on some drug—is often violated, some peo-

ple may come to support a much more severe restriction B (for instance, a war on drugs, with harsh punishments and intrusive searches) because they do not like to see the law being flouted. The intermediate position A thus becomes politically unstable, and slippage to B more likely.

ATTITUDE-ALTERING SLIPPERY SLOPES. Thus far this entry has discussed slippery slopes that operate without changing anyone's underlying attitudes. People might have the same attitudes about equality or cost as they did before A—but once A is enacted, those very attitudes lead them to support B, because of changed real-world circumstances.

Some slippery slopes, though, do operate by changing people's attitudes. Many voters, and even some legislators and judges, feel that they know little about certain issues. For instance, if they are asked whether they support some restriction on privacy, they might realize that privacy questions are very difficult, and that they have no good theory about which restrictions are good and which are not. Because they are thus *rationally ignorant*—they know the necessary limitations of their own knowledge—they may defer to the judgment of other authoritative institutions, such as courts and legislators. So if some kind of surveillance is legally permitted, many voters may therefore conclude that it is also morally proper.

This means that when proposal A is being considered, one must try to predict not only what A will do on its own terms, but also how it will change public attitudes. Will it, for instance, lead voters to alter their views to the point that they will also start supporting broader proposals like B? Will stem cell research on human embryos, for instance, change people's attitudes about the propriety of harvesting older fetuses or even babies for medical purposes? Would it lead people to think of abortions as a good rather than a necessary evil, and thus legitimize (for instance) people's getting pregnant just to harvest the resulting embryos? This sort of psychological prognostication is difficult, but it often has to be done if people are to decide whether the benefits of A indeed exceed its costs.

LEGISLATIVE-LEGISLATIVE, LEGISLATIVE-JUDICIAL, JUDICIAL-LEGISLATIVE, AND JUDICIAL-JUDICIAL SLIPPERY SLOPES. All these slippery slopes may in some measure operate whether decisions A and B are legislative decisions or judicial ones. Slippery slopes are often associated with judicial decision making, in which the doctrine of precedent helps accelerate the slide chiefly by strengthening the equality slippery slope and the attitude-altering slippery slope. But as some of these examples show, slippery slopes can operate even without any formal rule of precedent.

The Slippery Slope Inefficiency

None of these arguments, of course, always carry the day—nor should they. Sometimes we must make decisions even if there is a risk that the decisions will lead others to enact laws of which we disapprove. And yet some policy proposals that may be good on their own do end up being blocked because of eminently reasonable slippery slope concerns; one might call this the *slippery slope inefficiency*. Some people think this is true of gun registration, which has been blocked by concerns over a slippery slope to gun confiscation. Others think it is true of moderate assisted suicide proposals, which may be blocked by concerns that assisted suicide will become the norm for more and more patients.

Identifying this inefficiency suggests, perhaps surprisingly, that constitutional rights might sometimes *enable* modest regulation even while they *disable* broader prohibition. If gun right supporters feel that their right to own guns is constitutionally secure and, thus, that gun confiscation would be struck down by the courts, many of them might well drop their opposition to gun registration—an opposition that may be largely driven by slippery slope risks. If a trustworthy barrier against slippage is erected, then people may be more willing to take the first step out onto the slope.

EUGENE VOLOKH

SEE ALSO *Choice Behavior; Decision Theory.*

BIBLIOGRAPHY

Rizzo, Mario J., and Douglas Glen Whitman. (2003). "The Camel's Nose Is in the Tent: Rules, Theories, and Slippery Slopes." *UCLA Law Review* 51(2): 539–592.

Schauer, Frederick. (1985). "Slippery Slopes." *Harvard Law Review* 99(2): 361–383.

Volokh, Eugene. (2003). "Mechanisms of the Slippery Slope." *Harvard Law Review* 116(4): 1026–1137.

Walton, Douglas. (1992). *Slippery Slope Arguments.* New York: Oxford University Press.

SMITH, ADAM

• • •

Although Adam Smith (1723–1790) was not the originator of many of the ideas that became modern economics, his synthesizing treatise, *An Inquiry into the Nature and Causes of the Wealth of Nations* (1776), was so influential that he is generally considered the founder of the discipline. He effectively elaborated the concept of unplanned, spontaneous order, a feature of his economics that later played a part in other sciences such as evolutionary biology and cybernetics. Smith treated economic behavior as part of an entire ethical system, which he set out in his other major work, *The Theory of Moral Sentiments* (1759). Born in Kirkcaldy, Scotland, and baptized on June 5, he attended Glasgow and Oxford Universities and then returned to Glasgow as professor of moral philosophy. He died in Edinburgh on July 17.

Self Interest and Public Benefit

For Smith and contemporary practitioners, economics is in large measure the study of the outcome for society of

Adam Smith, 1723–1790. The Scottish economist and moral philosopher believed that in a laissez-faire economy the impulse of self-interest would work toward the public welfare. (*The Library of Congress.*)

individuals acting in their own interest without a view to public benefit. Smith thought the outcome was generally good. Unregulated, self-interested behavior could produce greater material wealth for society than could a system of policies designed by authorities to achieve wealth. Economists, historians, philosophers, and ethicists have debated his argument from his day to the present.

In support of his notion that beneficial order, not destructive chaos, can result from persons acting in their own interest, Smith repeatedly shows how desirable features of society are the unintended outcome of actions taken for other reasons. For example the division of labor, to which he attributed national wealth, was not the effect of human wisdom that intended the resulting material well being. Smith argues that humans, unlike animals that fawn to obtain favors, learn to divide tasks and specialize in producing goods and services that they can exchange for what they want. The division of labor, therefore, was the effect of the tendency of humans to barter in order to get what they want from others. It produces wealth because it saves time, develops specialized

skills, and prompts workers to invent technologies to ease their tasks.

Being aware of the productive advantages of specialization, authorities may presume that they can plan the division of labor. Smith traces the steps involved in producing a simple item and makes it clear that a planner would be incapable of assessing people's desires, devising tasks to satisfy them, and assigning the tasks to various workers. Even if people made their desires known in any one place, no person or group could imagine the skills and resources required to provide for any one desire. The division of labor functions most effectively if individuals learn from market prices the best way to employ their own time and abilities to satisfy the desires of others, thereby offering productive resources of which a planner would be unaware. When entrepreneurs seek the most profitable employment of their capital and workers go where wages are highest, the result, which neither intended, is that they unintentionally supply the desires of others in the cheapest way. Individuals do not have to have benevolent motives to produce social benefits.

Smith, who did not romanticize business, thought that employers always try to conspire to keep wages down and that sellers in the same trade always conspire to raise prices. Accordingly he admonished governments never to take actions that would make it easier for members of the same trade to cooperate. Self-interest leads to public benefit, but only if competition prevails.

Unregulated markets, when competitive, harness self-interested behavior to produce public benefit. Smith understood, however, that the authorities did not deliberately institute a market system to achieve this end. On the contrary, history taught him that the system emerged when landlords used the produce of their agricultural estates to buy luxuries rather than to maintain hundreds of tenants, soldiers, and servants. When they were no longer bound to their landlords, these individuals became freer to exchange their services for market-determined wages.

Smith's understanding of how the pursuit of individual interest produces the wealth of all led him to advocate the system of *natural liberty* in which the government's role, while indispensable, is confined to providing national defense, law and order, and goods that are unprofitable for private persons to produce, even though their benefits exceed their costs. Attempts by government to fix prices, encourage particular technologies, or subsidize certain industries for the benefit of society would be useless if not pernicious.

The Moral Basis of Markets

Smith devotes much of *The Wealth of Nations* to working out the implications of individuals being able to pursue their own interests, but he was aware that his system of natural liberty had a moral foundation. Markets not only had to be free from improper government interference and monopoly; legal and moral rules also had to protect them from injustice—murder, theft, and broken promises. In *The Theory of Moral Sentiments* (1759), Smith contended that orderly society was possible because the Author of Nature endowed humans with resentment of injustice and a desire to see it punished. For Smith society is possible because people passionately desire to punish injustice, not because they reason that their group will suffer if crimes against its members go unpunished. In his treatment of the social support for justice, as in his explanation of the emergence and functioning of markets, Smith emphasizes unintended outcomes. Individuals do not seek a wealthy society; they pursue their own interest and national wealth results. Similarly individuals do not strongly desire orderly society; their resentment of malice provides the basis for order.

It is easy enough to see that humans would resent malice toward themselves, but what of hurtful actions toward others? Humans are self-interested, but, as Smith claims in the opening line of *The Theory of Moral Sentiments*, they also care about the fortunes of others. By imagining what they themselves would feel in a similar situation, humans sympathize with the resentment of sufferers of injustice.

Smith does not limit the role of sympathy to ensuring that members of society will punish perpetrators of injustice. He uses the term sympathy to mean the human capacity to experience, to some degree, all the passions of others. When people share the passions that prompt others to act in ways they themselves would act in similar circumstances, they consider the acts of others just and proper. Similarly people approve of their own conduct if they feel that an *impartial spectator* would sympathize with the passions that influenced it. The impartial spectator acts as a constraint on self-interest. It approves of such self-regarding virtues as prudence, industry, and temperance, but recoils at *malevolence or sordid selfishness*.

Thus although Smith recognized the power of self-interestedness, he understood and celebrated other motives as well. According to his figure of speech, if the pillar of justice prevails, a society of the merely self-interested can exist, but without the ornaments of friendship, generosity, gratitude, and charity, people live a less happy, agreeable, and comfortable life. In his words, "to restrain our selfish, and to indulge our benevolent, affections, constitutes the perfection of human nature" (Smith 1969, p. 71).

Relevance to Current Policy

Smith's system of natural liberty does not provide guides for policies for the contemporary problems of poverty, environmental degradation, or for the alleviation of the stultifying effects of specialization. In these areas, later developments in specialized fields of economics have surpassed Smith's approach. At the same time, his understanding of human behavior and the sources of national wealth is still pertinent. The human tendency to regard first self-interest and that of family and friends has a basis in nature and is not entirely the consequence of education or culture. Therefore persons who make laws and policies must acknowledge it. It is fruitless to hope that authorities can persuade humans to provide for each other's needs out of benevolence. Self-interested individuals, however, will serve each other as they pursue their own interests, if competition exists and there are rules that punish violators of personal and property rights. Moreover authorities, as compared with the public, are no less self-interested and no more able to judge which industries or technologies will provide the greatest future social benefits. One lesson from Smith, then, is that governments should forgo planning and concentrate on promoting wealth and happiness by having legal systems that protect property rights and by encouraging ethical standards that honor following the rules of justice.

Another lesson is that markets do not become free because of the vision of some well-meaning and enlightened group. In the case of England, Smith observed that the market system resulted when landlords lost power. This historical observation is in keeping with his understanding of the limited effect of beneficial intent.

The twentieth-century failure of planned economies relative to those with freer markets lends support to Smith's free-market policies for the growth of national wealth. Even so, international agencies and national governments should be careful about promoting free markets by financially supporting authorities that promise to create them. Smith's historical perspective suggests that markets become freer when power changes hands, not when powerful leaders purport to make them free.

WILLIAM O. SHROPSHIRE

SEE ALSO *Capitalism; Cybernetics; Enlightenment Social Theory; Libertarianism; Market Theory; Political Economy.*

BIBLIOGRAPHY

Hayek, Friedrich. (1988). *The Fatal Conceit.* Chicago: University of Chicago Press. Includes an exposition of Smith's contribution to the concept of spontaneous order and references to his influence on Darwin and cybernetics.

Heilbroner, Robert. (1986). *The Essential Adam Smith.* New York: W. W. Norton, Inc. Excerpts from most of Smith's works, an excellent introduction to each of them, and a brief biography.

Marx, Karl. (1963). "Economic and Philosophical Manuscripts." In *Karl Marx: Early Writings,* ed. and trans. Tom B. Bottomore. New York: McGraw-Hill Book Company. Marx's view of the damage done to humanity by Smith's market-dominated society.

McCloskey, Deirdre. (1998). "Bourgeois Virtue and the History of P and S." *Journal of Economic History* 58: 297–317. An explanation of Smith's system of ethics as a combination of the prudent and social virtues.

Niebuhr, Reinhold. (1953). "The Christian Faith and the Economic Life of Liberal Society." In *Goals of Economic Life,* ed. A. Dudley Ward. New York: Harper Brothers. The perspective of a Christian theologian who came to a more moderate view of market society after being a socialist critic in his earlier years.

Samuels, Warren. (1977). "The Political Economy of Adam Smith." *Ethics* 87:189–207. An interpretation of Smith as a synthesizer of natural theology, ethics, jurisprudence, and expediency.

Schumpeter, Joseph. (1954). *History of Economic Analysis.* New York: Oxford University Press. A monumental history of economic thought that assesses Smith's importance and traces the background of many of his ideas, including the concept of spontaneous order.

Smith, Adam. (1969 [1759]). *The Theory of Moral Sentiments.* Indianapolis, IN: Liberty Classics Fund, Inc. Liberty Classics edition with introduction by Edwin G. West.

Smith, Adam. (1981 [1776]). *An Inquiry into the Nature and Causes of the Wealth of Nations,* eds. Russell Campbell and Andrew Skinner. Indianapolis, IN: Liberty Fund, Inc. Introduction by R. Campbell and A. Skinner.

West, Edwin. (1976). *Adam Smith: The Man and His Works.* Indianapolis, IN: Liberty Press. An accessible and sympathetic biography.

SOCIAL CONSTRUCTION OF SCIENTIFIC KNOWLEDGE

• • •

The leading research orientation in contemporary science and technology studies—the social construction of scientific knowledge (SSK, or social constructivism)—has been controversial since its inception in the 1970s. It primarily consists of a set of methodological imperatives for the study of science and technology that focus on the means by which people, ideas, interests, and things are organized in specific places and times to produce knowledge that has authority throughout society, especially among those not originally involved in the process of knowledge production. Thus, social constructivists tend to stress the diversity of interpretations and applications of knowledge across social contexts. However, in areas where philosophers and scientists might interpret that diversity as different representations or instantiations of an already established form of knowledge, social constructivists treat that variety as part of the ongoing core process of knowledge production.

Social constructivists therefore do not recognize a sharp distinction between the production and the consumption of knowledge. Thus, social constructivism has a "democratizing" effect on epistemology by leveling traditional differences in the authority granted to differently placed knowers. To a social constructivist a technologist using a scientific formula is "constructing" that formula as knowledge in exactly the same sense as did the scientist who originated the formula. Each depends on the other to strengthen their common "cycle of credibility" or "actor-network," in the words of Bruno Latour, perhaps the leading social constructivist. In contrast, most philosophers and scientists would raise the epistemic status of the original scientist to that of a "discoverer" and lower the status of the technologist to that of an "applier."

Basic Attitudes and Origins

In philosophical terms social constructivism is a form of antirealism: Social constructivists do not presuppose the existence of a reality independent of the procedures available to the examined agents for deciding the truth value of their assertions. In this respect social constructivism has affinities with idealism, pragmatism, phenomenology, and even logical positivism. The proponents of all those movements agree that aspects of the world that traditionally have been cited as evidence for "external reality" are in significant respects the intended and unintended products of human practices. However, this common insight has led to rather different philosophical responses. For example, positivists and phenomenologists strive to design criteria that can command universal assent, whereas idealists and pragmatists regard the resolution of conflict in the application of such procedures as the basis of future epistemic developments.

Social constructivists differ from earlier antirealists by challenging their common fundamental assumption of a centralized decision-making environment, whether it is a unified self or society.

In contrast, social constructivists presuppose that the social world in which construction occurs is highly dispersed. This implies that different decisions are taken across many places and times. This often is considered a "postmodern" feature of social constructivism. However, despite the lip service paid to French poststructuralist thinkers such as Michel Foucault (1926–1984) and Gilles Deleuze (1925–1995), social constructivists originally derived this characterization from the social phenomenologist Peter Berger (Berger and Luckmann 1967), from his Viennese teacher Alfred Schutz (1899–1959), and ultimately from Schutz's mentor, the neoliberal political economist Friedrich Hayek (1899–1992).

Just as Hayek had argued in the 1930s, against the socialists, that no central planner can determine fair prices more efficiently than can the spontaneous self-organization of buyers and sellers, social constructivists deny that a single philosophical method can determine the course of science more efficiently than can the spontaneous self-organization of scientific practitioners. Hayek grounded his argument on the unique knowledge possessed by people differently placed in the market. Thus, the social construction of scientific knowledge can be seen historically as an extension of a market mentality into an aspect of social life—science—that for much of the twentieth century tied its legitimacy to the control mechanisms of the state.

Despite often being portrayed as antiscientific, social constructivism has precedents in the history of science, starting with Aristotle's view of matter as an indeterminate potential that is given form through human intervention. In the nineteenth and early twentieth centuries the constructivist position was represented most clearly by chemists who contested the idea of an ultimate form of physical reality as defined by, say, "atoms." Instead, chemists appealed to "energy" as an updated version of Aristotelian potential. Current versions of constructivism further "socialize" this perspective by invoking concepts such as work and practice as the media through which scientific objects are brought in and out of existence. According to its proponents, social constructivism is the spontaneous philosophy of the working scientist, who is concerned more with making things happen in the laboratory, as well as in society at large, than with completing a philosophically inspired picture of ultimate reality. Not surprisingly, Latour and other leading social constructivists have flourished in engineering schools rather than in pure science faculties.

The Trajectory of Social Constructivist Research

The social construction of scientific knowledge normally is described in terms of its opposition to two familiar, although extreme, views that might be called philosophical rationalism and sociological determinism. Philosophical rationalism implies that science ultimately is driven by a concern for the truth, perhaps even a desire to provide a comprehensive and unified picture of reality. From that standpoint the social dimension of science functions as either a facilitator or an inhibitor of this quest. Sociological determinism implies that the science of a particular time and place is an ideological reflection of the social conditions that sustain it. From that standpoint the development of science is dependent on its larger societal functions. Social constructivism differs from those two perspectives by denying a strong ontological distinction between the "cognitive" (or "natural") and "social" (or "cultural") dimensions of science. Both dimensions are coproduced in any episode of scientific activity. As a result social constructivists see science as much more subject to agency and contingency than either philosophical rationalism or sociological determinism allowed.

David Bloor's *Knowledge and Social Imagery* (1976) was the first book to put forward the social constructivist case against both philosophers and sociologists. Bloor, a mathematician and psychologist, was influenced by Ludwig Wittgenstein's (1889–1951) later writings on rule following. Wittgenstein implied that there is no correct way to continue a number series (for example, 2, 4, 6 ...) except to abide by the judgement of the community engaged in the counting because any arithmetic series is open to an indefinite number of continuations (such as 8, 10, 12 ... or 7, 8, 9 and then 10, 12, 14 ...), depending on what is taken to be the rule underlying the number series. Bloor generalized that insight in the name of a thoroughly naturalistic approach to the study of knowledge that he called the "Strong Programme in the Sociology of Scientific Knowledge." That approach involved suspending all external normative judgements about the validity or rationality of knowledge claims. (In contrast, the "Weak Programme" would use sociology only to explain episodes of scientific dysfunction, because the canons of rationality were presumed to explain science's normal operation.) Bloor would look only to the standards of reasoning and evidence available to those who must live with the consequences of what they do. That approach encouraged what Bloor called a "symmetrical" attitude toward the various competing beliefs or courses of action in a particular situation. In other words the inquirer is to treat those beliefs or actions as seriously as the situated agents treat them, suspending any knowledge the inquirer might have about their likely or, in the case of historical cases, actual consequences. The import of this approach was to neutralize specifically philosophical appraisals of knowledge claims, which typically appeal to standards of rationality and validity that transcend the interests or even competence of the involved agents.

Whereas Bloor, along with his Edinburgh colleague Barry Barnes (1975), mapped out the conceptual terrain defined by social constructivism, the 1980s and 1990s brought a plethora of historical and sociological case studies inspired by that position. Constructivist historical studies characteristically reinterpret landmark scientific debates so that what traditionally was seen as an instance of truth clearly triumphing over falsehood came to appear as a more equally balanced contest in which victory was secured at considerable cost and by means that were specific to the contest. Attached to these reinterpretations is a view, traceable to Thomas Kuhn (1922–1996), in which every scientific success entails a rewriting of history to make it appear inevitable. In this respect social constructivist history of science aims to "deconstruct" the narratives of scientific progress typically found in science textbooks and works of science popularization.

Stephen Shapin and Simon Schaffer's *Leviathan and the Air-Pump* (1985) is perhaps the most influential work of this sort. It deals with Robert Boyle's (1627–1691) successful blocking of Thomas Hobbes's (1588–1679) candidacy for membership in the Royal Society. This episode normally is told in terms of Hobbes's persistent metaphysical objections to the existence of a vacuum long after it was found to be scientifically reasonable. However, it turns out that Hobbes was defending the general principle that experimental demonstrations are always open to philosophical criticism even if the philosopher could not have designed such an experiment. Hobbes's failure on this score set a precedent for the competence required for judging experiments that began to insulate science from public scrutiny.

Constructivist case studies typically draw on the sociological method of grounded theory, according to which the inquirer introduces a theoretical concept or perspective only if the agents under study also do so. Grounded theory originally was used to oppose structural functionalism, the leading school of U.S. sociology, which was associated with Talcott Parsons (1902–1979) and Robert Merton (1910–2003). Proponents of that school postulated that deviance is a well-defined

role that performs specific functions in the social system. In contrast, for grounded theorists the deviant role, say, in the context of asylums and hospitals, had to be constructed from moment to moment because generally speaking there was no clear observable difference between the behavior of so-called normals and that of deviants.

Achievements and Weaknesses

The groundbreaking, albeit perverse, insight of Latour and Steve Woolgar (Latour and Woolgar 1986), Karin Knorr-Cetina (1981), and the other early constructivist sociologists was to imagine that "deviance" may apply to people on the positive extreme as well as the negative extreme of a normal distribution curve. Thus, in their daily laboratory tasks scientists do not sound or look especially different from people working in an industrial environment subject to an intensive division of labor. Nevertheless, scientists are socially constructed as exceptionally rational, producing knowledge that commands authority throughout society. How is this possible? For a constructivist sociologist the answer lies in the "made for export" language scientists use to describe their activities and the specific distribution channels in which that language, as expressed in journal articles, preprints, and press releases, circulates. This produces a forward momentum, involving many other people, laboratories, interests, and so forth, that eventually turns a unique set of events into a universally recognizable fact.

There is little doubt that social constructivism has provided an important challenge to standard historical, philosophical, and sociological accounts of science. The question is its implications for science itself. The steadfast adherence of constructivism to the symmetry principle has been both a strength and a weakness.

The strengths of constructivism extend beyond intellectual insight to the ease with which it can be used in science policy research, especially in a time when constrained budgets and skeptical publics demand that science be evaluated in terms of its actual consequences rather than its professed norms. In this respect social constructivism has been a success in the marketplace, proving especially attractive to the increasing proportion of academic researchers who depend on external contracts for their livelihood. However, beneath that success lies a weakness: Constructivism lacks a clear normative perspective of its own. This lack largely reflects its decentralized vision of social life. Although constructivists excel in revealing the multiple directions in which science policy may go, they refuse to pass judgment on any of them or even on the means by which their differences might be resolved. In this respect social constructivism is indifferent to the future of science and the role of science as the vanguard of rationality and progress in society at large.

The program of "social epistemology" has attempted to redress this imbalance in social constructivism. It argues that social constructivism can provide the basis for a science policy that is both genuinely democratic and experimental. Conventional science policy tends to be problem-centered without evaluating the relevant discipline-based knowledge. Indeed, science policy analysts rarely think of themselves as *constructing* problems the problems they address—they are simply treated as given. In contrast, social epistemology moves science policy toward constructivism by critically examining the maintenance of institutional inertia: Why don't research priorities change more often and more radically? Why do problems arise in certain contexts and not others? These questions are addressed on the basis of three presumptions that take seriously the normative implications of the social constructivism (from Fuller and Collier 2003):

- *The Dialectical Presumption:* The scientific study of science will probably serve to alter the conduct of science in the long run, insofar as science has reached its current state largely through an absence of such reflexive scrutiny.

- *The Conventionality Presumption:* Research methodologies and disciplinary differences continue to be maintained only because no concerted effort is made to change them—not because they are underwritten by the laws of reason or nature.

- *The Democratic Presumption:* The fact that science can be studied scientifically by people who are themselves not credentialed in the science under study suggests that science can be scrutinized and evaluated by an appropriately informed lay public.

STEVE FULLER

SEE ALSO *Science, Technology, and Society Studies; Sokal Affair.*

BIBLIOGRAPHY

Barnes, Barry. (1975). *Scientific Knowledge and Sociological Theory.* London: Routledge & Kegan Paul.

Berger, Peter, and Thomas Luckmann. (1967). *The Social Construction of Reality.* Garden City, NY: Anchor Doubleday.

Bloor, David. (1976). *Knowledge and Social Imagery.* London: Routledge & Kegan Paul.

Collins, Harry, and Trevor Pinch. (1993). *The Golem: What Everyone Needs to Know about Science.* Cambridge, UK: Cambridge University Press.

Fuller, Steve. (1988). *Social Epistemology.* Indianapolis: Indiana University Press.

Fuller, Steve, and James Collier. (2003). *Philosophy, Rhetoric, and the End of Knowledge,* 2nd edition. Hillsdale, NJ: Erlbaum. (Originally published in 1993.)

Gilbert, Nigel, and Michael Mulkay. (1984). *Opening Pandora's Box.* Cambridge, UK: Cambridge University Press.

Glaser, Barney, and Anselm Strauss. (1967). *The Discovery of Grounded Theory.* Chicago: Aldine.

Golinski, Jan. (1998). *Making Natural Knowledge: Constructivism and the History of Science.* Cambridge, UK: Cambridge University Press.

Knorr-Cetina, Karin. (1981). *The Manufacture of Knowledge.* Oxford: Pergamon.

Latour, Bruno. (1987). *Science in Action.* Cambridge, MA: Harvard University Press.

Latour, Bruno, and Steve Woolgar. (1986). *Laboratory Life: The Social Construction of Scientific Facts,* 2nd edition. Princeton, NJ: Princeton University Press. (Originally published in 1979.)

Shapin, Steven, and Simon Schaffer. (1985). *Leviathan and the Air-Pump.* Princeton, NJ: Princeton University Press.

SOCIAL CONSTRUCTION OF TECHNOLOGY

• • •

The phrase *the social construction of technology* is used in at least two different, though overlapping, ways. Broadly it refers to a theory about how a variety of social factors and forces shape technological development, technological change, and the meanings associated with technology. More narrowly, the phrase refers to a specific account of the social construction of technology; the acronym SCOT is used to refer to this version of the broader theory (Pinch and Bijker 1987). According to Ronald Kline and Trevor Pinch (1999), SCOT uses the notions of *relevant social groups, interpretive flexibility, closure* and *stabilization;* the concept of interpretive flexibility is its distinguishing feature. To claim that technology has interpretive flexibility is to claim that artifacts are open to radically different interpretations by various social groups; that is, artifacts are conceived and understood to be different *things* to different groups.

Contra Technological Determinism

The starting point for understanding both the broad theory of the social construction of technology and the

SCOT version of that theory is to compare them with another view of technology referred to as *technological determinism*. Technological determinism has two basic tenets: (1) that technology develops independently from society; and (2) that when a technology is taken up and used, it has powerful effects on the character of society. According to the first tenet, technological development either follows scientific discoveries—as inventors and engineers *apply* science—or it follows a logic of its own, with new inventions deriving directly from previous inventions. Either way, technological development is considered to be separate from social forces; engineers and inventors work in an isolated domain in which all that matters is discovering and manipulating nature.

According to the second tenet of technological determinism, when technologies are adopted by societies or particular social groups, the adoption brings about—determines—social change and patterns of social behavior. In one formulation, technological change is said to create a *cultural lag* until culture catches up. One specific determinist argument proposed by historian Lynn White (1962) is that feudal society evolved from the invention of the stirrup. Another example is Langdon Winner's (1986) claim that society cannot have nuclear power without hierarchical organization. Winner's broader claim is that technologies necessitate particular forms of political organization. This principle of technological determinism leads to the commonly held view that technology determines society; that is, when technologies are adopted and used, they change the character of society.

The broad theory on the social construction of technology denies the first tenet of technological determinism entirely but makes a more nuanced response to the second tenet. In denying the claim that technology develops independently from society and follows science or its own logic of development, social constructivists argue that technological development is shaped by a wide variety of social, cultural, economic, and political factors. Nature does not reveal itself in some necessary or logical order. Scientists and engineers look at nature through lenses of human interests, theories, and concepts; engineers invent and build things that fit into particular social and cultural contexts. Technologies are successful not by some objective measure of their goodness or efficiency; rather, technologies are taken up and used because they are perceived to achieve particular human purposes and to improve a particular social world or to further the interests of individuals and social groups.

Broad theory proponents respond similarly to the second tenet of technological determinism: They claim the theory misses the fact that technology is being shaped by social factors and forces. But here the social constructivist does not wholly deny the technological determinist claim that technology affects society; rather, constructivists argue that forces may move in both directions. Technology shapes society *and* society shapes technology. Social constructivists claim that the theory of technological determinism gives an inadequate and misleading picture of the technology-society relationship in leaving out the powerful social forces at work in shaping the development, adoption, use, and meanings associated with technology. Social constructivists have also gone further in claiming that shaping does not just work in both directions but that technology and society are mutually constitutive; they cocreate one another.

Specific Theories of Social Construction

The critique of technological determinism and the emergence of the theory of the social construction of technology began and gained momentum in the 1980s along with other activities contributing to the development of a new field of study sometimes labeled *science and technology studies* (STS) and other times *science, technology, and society* (also STS). Within this field of study, two theoretical approaches are often distinguished: the version of social constructivism referred to as SCOT and actor-network theory (ANT). Both theories seek to explain why and how particular technologies are adopted while others are rejected or never developed. Both SCOT and ANT are concerned with how technological designs are adopted and become embedded in social practices and social institutions.

Actor-network theory takes as its unit of analysis the systems of behavior and social practices that are intertwined with material objects. This is the *network* part of actor-network theory. The *actor* part of actor-network theory emphasizes the presence of many actors, human and nonhuman. For instance, nature plays an important role in determining which technologies come to be adopted, and nature can be described as one of the actors in shaping the technologies that succeed in becoming embedded in the social world. Technologies and artifacts can themselves also be actors. Humans, nature, and artifacts collectively are referred to in actor-network theory as actants.

Resistance to Social Constructivism

Two issues often get in the way of understanding social constructivism in the broad sense. The first is an issue about which social constructivists disagree, the extent to which nature is real or *merely* socially constructed. *Realists* claim that nature is real and has an inherent or fixed character that scientists and engineers must manipulate to succeed. The hard character of nature shapes what engineers can do and what technologies are developed. Nevertheless while constructivist-realists claim that there is something real or *hard* about nature, they generally acknowledge that the only way humans have access to nature is through human meaning, human constructs, and human theories, all of which are social. Thus nature can be represented in different ways, in different knowledge systems. At the same time, *antirealists* claim that there is nothing hard or real about nature around which ideas and meanings can be constructed; at least, there is nothing real to which people have access. There are only ideas and meanings constructed by humans, and ideas and meanings are social.

While the chasm between realists and antirealists is wide, many social constructivists simply sidestep or bracket the issue without taking sides. For many social constructivists who seek to understand the cocreation of technology and society, it does not make a difference whether nature is real, because all concede that nature is viewed through the lenses of human beings, which are interested and social.

The second issue is the principle that new technologies build on older technologies. Technological determinists contend, for example, that computers could not have been developed if electricity and transistors and many other devices had not already been developed. Thus technology influences technology; later technology builds on prior technology. Social constructivists agree. What social constructivists reject, however, is that technological change and development follows a predetermined, linear path, a path necessitated by some nonhuman reality. Social constructivists argue that social factors influence the pace and direction of technological development and that development is often nonlinear.

How Social Construction Works

What does it mean to say that technology is socially constructed? As already mentioned, the theory referred to as SCOT makes use of the notions of relevant social groups, interpretative flexibility, stabilization, and closure.

THE BICYCLE STORY. Wiebe Bijker (1995) and Pinch and Bijker (1987) give an account of the development of the design of the bicycle—the design that has been used since the early-twentieth century. They argue that the path of development was complex, with various designs being tested and rejected by various groups in a nonlinear order. Relevant social groups—including sports enthusiasts, men and women who spent leisure time in public parks, bicycle makers, bicycle repair people, and more—responded to various models differently and found different advantages and disadvantages as well as meaning in them.

Development moved in many directions aimed at solving a variety of problems for riders, manufacturers, and those who repaired the bicycles; the problems included safety, ease of manufacture and repair, speed, ability to manage the roughness of roads, and so on. Designs had varying cultural meanings (was the bicycle macho or ladylike?), facilitated or constrained various social activities in public parks, and served the interests of various groups, including sports enthusiasts and manufacturers.

Design of the bicycle first took hold when the relevant social groups coalesced around one design because it solved problems for each group. This is the point Bijker refers to as stabilization. Once this happens small design changes may continue to be made, but tend to presume the overall design; designers tinker within that framework. In this way, Bijker shows that the design of the bicycle was socially constructed in the sense that the design that succeeded (that is, was adopted and pervasively used) was not the *best* in some objective sense, such as most efficient or elegant; rather, it was the one that the relevant social groups agreed upon because they were convinced it fit their needs.

The broader theory of the social construction of technology does not refute the SCOT theoretical apparatus; rather, the broader theory remains open to the use of alternative concepts, frameworks, and tools to study the cocreation of technology and society. Because social constructivism emphasizes the social shaping of technology, it may be useful to consider a few areas where social factors have a powerful influence on the technologies that are developed and what those technologies look like.

ECONOMICS. Perhaps the most obvious place to see the workings of society is in funding for the development of new technologies. Companies and government agencies invest large amounts of money, space, time, and effort in technological endeavors that seem promising. When enormous resources are put into an area of scientific or technological development, that area is much more likely to yield results. Thus, contrary to the inherent logic of development suggested by technological determinism, technology develops, at least in part, in an order that is determined by investment choices and other human decisions, and not by logic alone.

REGULATION. While governments often invest heavily in technological development, funding is not the only aspect of government that shapes technology. Governments often regulate technological domains and when they do so, the regulation affects future development. Consider, for example, the vast array of regulatory standards that automobiles must meet. Whether they are aimed at safety or clean air or decreasing dependence on fossil fuels, when governments set standards for automobiles, automobile manufacturers must design within the confines of those specifications. Hence regulation promotes development in a certain direction and forecloses development in other directions.

CULTURE. Yet another way that technological development is socially shaped is by the cultural meanings that influence the design of artifacts. Perhaps the best place to see this is in cross-cultural studies of technology. Such studies reveal how cultural meanings strongly influence technological development. Think, for example, of the lack of development in rail transportation in the United States where individualism and many other historical factors promote the use of automobiles, whereas in many European countries, this mode of public transportation has been successfully developed and enhanced for more than a century.

Ethics and Social Construction

Technological determinist theories such as that of Jacques Ellul (1964) seem to imply that technological development is autonomous and unstoppable; that is, individuals and even social movements can do nothing to change the pace or direction of development. Social constructivism can be seen as, at least in part, a response to the pessimism of technological determinism. Many social constructivist scholars see themselves as providing an account of technological development and change that opens up the possibility of intervention, the possibility for more deliberate social control of technology. Wiebe Bijker (1993), for example, describes the field as being rooted in critical studies. He claims that science and technology studies of the 1980s were "an academic detour to collect ammunition for struggles with political, scientific, and technological authorities" (Bijker 1993, p. 116). Thus, social constructivist theories might be seen as having an implicitly critical, and perhaps even a moral, perspective. However, social constructivist theories have been developed primarily by historians and social scientists, and scholars in these fields have traditionally understood the task of their scholarship to be that of description, not prescription. Hence, social constructivist theorists generally deny that their perspective is ethical.

Nevertheless, in bringing to light many of the otherwise invisible forces at work in shaping technology and society, social constructivist analysis often reveals the ways in which particular social groups wield power over others through technology. Knowledge of this aspect of technology opens up the possibility of deliberate action to counter the unfair use of power and the undesirable social patterns being created and reinforced through technology. A good example here is the work on gender and technology by such scholars as Judy Wajcman (1991) and Cynthia Cockburn and Susan Omrud (1993). By drawing attention to the ways in which technology reinforces gender stereotypes and more broadly, how gender and technology are co-created, these scholars make it possible for those involved with technological development to avoid reinforcing prevailing stereotypes or patterns of gender inequality. In this respect social constructivism has important ethical implications.

While social constructivism has significantly furthered the social analysis of science and technology, social constructivism is still relatively new. Perhaps the most serious criticism of social constructivism is that it consists only of a few theoretical concepts and a wide-ranging set of case studies. Hence, it still needs a more comprehensive theoretical foundation. Nevertheless, social constructivism has been influential and is likely to continue to be important in understanding the relationships among science, technology, and society.

DEBORAH G. JOHNSON

SEE ALSO *Science, Technology, and Society Studies.*

BIBLIOGRAPHY

Bijker, Wiebe E. (1993). "Do Not Despair: There is Life after Social-Constructivism." *Science, Technology, & Human Values* 18: 113–138. Bijker reviews four problems surrounding socio-historical technology studies: relativism, reflexivity, theory, and practice. He argues for scholars to break away from overly academic perspectives and to return to political issues.

Bijker, Wiebe E. (1995). *Of Bicycles, Bakelites, and Bulbs.* Cambridge, MA: MIT Press. Bijker provides three detailed case studies of the development of a technology. His sociological analysis of these cases makes use of and illustrates a theoretical, social constructivist framework.

Cockburn, Cynthia and Susan Omrud. (1993). *Gender and Technology in the Making.* London: Sage Publications.

Explains the evolution of "microwave technology" that became the "microwave oven." It is a story of a technology that begins as masculine and eventually becomes a feminine kitchen appliance.

Ellul, Jacques. (1964). *The Technological Society*, trans. John Wilkinson. New York: Knopf. This book is considered a classic. Ellul argues that the internal logic of technological development, that is, instrumental rationality and efficiency, transforms domains of human life and is not in control of humans.

Kline, Ronald, and Trevor J. Pinch. (1999). "The Social Construction of Technology." In *The Social Shaping of Technology*, 2nd edition, ed. Donald MacKenzie and Judy Wajcman. Buckingham, UK, and Philadelphia: Open University Press. This is an excerpt from a longer piece that appeared in 1996 in *Technology and Culture* 37: 763–795. Kline and Pinch concisely present the advantages and weaknesses of the specific version of the social construction of technology known as SCOT and developed by Bijker and Pinch.

Pinch, Trevor J., and Wiebe E. Bijker. (1987). "The Social Construction of Facts and Artifacts: Or How the Sociology of Science and the Sociology of Technology Might Benefit Each Other." In *The Social Construction of Technological Systems*, ed. Wiebe E. Bijker, Thomas P. Hughes, and Trevor J. Pinch. Cambridge, MA: The MIT Press. Using the conceptual apparatus of the SCOT model, this piece provides a historical account of the development of the bicycle.

Wajcman, Judy. (1991). *Feminism Confronts Technology*. University Park: Pennsylvania State University Press. Explores the relationship between technology and gender in western society. Wajcman uses the lens of feminism to examine the traditional association of technology with masculinity. She argues that this relationship is not natural but intertwined with socially and cultural processes. Notions of gender and technology social reinforce one another.

White, Lynn T. (1962). *Medieval Technology and Social Change*. New York: Oxford University Press. White examines the effects of technology on societies of medieval Europe. He attributes the collapse of feudalism to the development of machines and tools.

Winner, Langdon. (1986). *The Whale and the Reactor*. Chicago: University of Chicago Press. Consists of essays by Langdon Winner, which together constitute his unique perspective as a political scientist examining technology. As a whole the essays can be thought of as a critical philosophy of technology. Several of the essays had been published earlier and are seminal pieces in the field.

SOCIAL CONTRACT FOR SCIENCE

• • •

The social contract for science is an evocative ideological construct used to describe the relationship between the political and scientific communities. Participants in science policy debates often invoke the social contract for science uncritically and flexibly, ritually referring to Vannevar Bush as its author and *Science, The Endless Frontier* (1945) as its text. The term, however, has no explicit connection to Bush, but explaining its history and usage is enlightening.

Historical Origins and Decline

There are two helpful hypotheses for origin of the phrase. One focuses on what Don K. Price called the "master contract" that formed the "basic charter" of the postwar relationship between the U.S. government and the scientific community (Price 1954, p. 70). This relationship "gives support to scientific institutions that yet retain their basic independence" (Price 1954, p. 67–68). A second hypothesis holds that the social contract for science is related to a social contract for scientists, which describes how the profession of science is bound as a community to uphold behavioral norms and to "rely on the trustworthiness" of each other (Zuckerman 1977, p. 113).

Harvey Brooks polished the promise of the social contract for science as "widely diffused benefits to society and the economy in return for according an unusual degree of intellectual autonomy and internal self-governance to the recipients of federal support" (Brooks 1990, p. 12). Brooks's definition takes into account both hypotheses of origination by relying on the overall structure of Price's formulation and on the rationale of Zuckerman's formulation as why the *unusual degree* of autonomy and self-governance could be offered to science. That is, science could be granted autonomy because its members maintain their integrity by upholding group norms (Merton 1973).

In addition to evoking the contractual nature of the relationship between the public patron and the scientific community and the tacit trustworthiness of scientists to one another, the social contract for science has additional descriptive power. As with more formal social contracts from political philosophy, it offers an account of the provision of a public good, and it suggests the conditions of an original consensus against which change can be measured and evaluated (Guston 2000). Some scholars and policy makers, relying on a tacit understanding of the social contract for science, argue variously that science has been faithful to it but politics not particularly so (Press 1988); that the contract died in the late 1960s with a decline in research funding, only to be resuscitated in the 1980s (Smith 1994); and that the contract crumbled in the 1990s through various policy changes (Stokes 1997).

Using the Social Contract in Policy and Ethics

Guston (2000), however, argues that to serve as a baseline for historical change, the social contract for science must have its tenets elaborated in clear historical detail and have criteria for change derived from there. Thus although there is a consensus that any such agreement dates to the immediate post-World War II period, *Science, The Endless Frontier* is not the sole articulation of postwar science, and John Steelman's report, *Science and Public Policy* (Steelman 1947) must also be taken into account. Although these two analyses differed on how they imagined the organization and funding of postwar science, they both held—along with much theoretical writing of the period—that the political community would provide resources to the scientific community and allow the scientific community to retain its decision-making mechanisms and in return expects forthcoming but unspecified technological benefits. Such a contract was premised on the automatic provision of scientific integrity and productivity, which thus becomes the central criterion against which to measure change.

There were many potential challenges to the social contract for science, thus specified, over the postwar period in the United States, including inquiries into the loyalty of scientists in the 1950s, the changes in financial arrangements and funding in the 1960s, and greater emphasis on applied research and questions about the limits of scientific inquiry in the 1980s. But no challenges altered the presumption of the automatic provision of scientific integrity and productivity until the conflicts over scientific (or research) misconduct and over technology transfer in the late 1970s and early 1980s. Political perceptions in this period held that scientists might have broken the contract through the failure to control misconduct and to produce sufficient economic benefits. But scientific perceptions held that politicians might have broken the contract through meddling. Neither perspective is completely right (or wrong), but it was through their instigation of organizational innovation—the creation of the Office of Research Integrity and of offices of technology transfer—that these issues marked the end of the social contract for science and its assumption of the automatic provision of scientific integrity and productivity. The political and scientific communities collaborated over the creation of these institutions, and they ushered in a new era in which the political and scientific communities engage in a *collaborative assurance* of integrity and productivity instead. Scholars have traced similar transitions in science policies in European nations as well.

DAVID H. GUSTON

SEE ALSO *Research Integrity; Science Policy.*

BIBLIOGRAPHY

Brooks, Harvey. (1990). "Lessons of History: Successive Challenges to Science Policy." In *The Research System in Transition*, eds. Susan E. Cozeens; Peter Healey; Arie Rip; and John Ziman. Boston: Kluwer Academic Publishers.

Guston, David H. (2000). *Between Politics and Science: Assuring the Integrity and Productivity of Research.* New York: Cambridge University Press.

Merton, Robert K. (1973 [1942]). "The Normative Structure of Science." In *The Sociology of Science: Theoretical and Empirical Investigations*, ed. Norman W. Storer. Chicago: University of Chicago Press.

Press, Frank (1988). "The Dilemma of the Golden Age." In *The Presidency and Science Advising*, Vol. 6, ed. Kenneth W. Thompson. New York: University Press of America.

Price, Don K. (1954). *Government and Science: Their Dynamic Relation in American Democracy.* New York: New York University Press.

Smith, Bruce L. R. (1994). "The United States: The Formation and Breakdown of the Postwar Government-Science Compact." In *Scientists and the State: Domestic Structures and the International Context*, ed. Etel Solingen. Ann Arbor: University of Michigan Press.

Steelman, John R. (1947). *Science and Public Policy.* 5 Vols. The President's Scientific Research Board. Washington, DC: U.S. Government Printing Office.

Stokes, Donald E. (1997). *Pasteur's Quadrant: Basic Science and Technological Innovation.* Washington, DC: Brookings Institution Press.

Zuckerman, Harriet. (1977). "Deviant Behavior and Social Control in Science." In *Deviance and Social Change*, ed. Edward Sagarin. Beverly Hills, CA: Sage.

SOCIAL CONTRACT THEORY

• • •

The idea of a social contract can have broad and narrow meanings. In the broad sense a social contract can simply be short hand for expectations in relations between individuals or groups. In the narrow, more technical sense social contract theory has a long and venerable history that in the present has been rhetorically adapted to assess general expectations between science and society. A review of various theoretical perspectives nevertheless raises questions about the adequacy of such adaptations.

Social Contracts in General

Contracts in the strict sense are agreements between two parties that establish mutual obligations and are enforceable by law. The idea of a social contract is more fundamental, and argues that society comes into existence as a kind of contract. In the classical or premodern views that are sometimes identified as anticipations of social contract theory, the social contract is not so much an originating action as one that implicitly exists between a preestablished order and individuals within it. This is, for instance, the view argued by Socrates in Plato's *Crito*. The modern view, by contrast, is that individuals come first, and through their agreement establish a new phenomenon called the state.

For most modern theorists this contract is not a historical event, much less an actual legal document, but an ideal construct to aid in postulating how things should be. It depends on two key assumptions: (a) that human beings as individuals are in some sense prior to any established social order, so that their obedience to the state has to be justified; and (b) that the condition of human beings outside the socially constructed state, or in what is called the *state of nature*, is ultimately unsatisfactory, thus providing humans a reason to escape such a condition by social contract. From these assumptions Thomas Hobbes (1588–1679) and John Locke (1632–1704), without using the term, developed social contract theory to examine the status of a monarch. When Jean-Jacques Rousseau (1712–1778) subsequently coined the term *social contract*, he used the same kind of theory to defend a notion of democratic equality. Later John Rawls (1921–2004) adapted social contract theory to defend a system of distributive justice.

From Hobbes through Kant

Early modern versions of social contract theory were justifications for overthrowing tyrants who had overstepped the bounds allotted them, failing therefore to meet their obligations. Manegold of Lautenbach (c. 1030–c. 1112), Englebert of Volkersdorf (fl. c. 1310), Mario Salamonio (c. 1450–1532), and Junius Brutus (fl. 1572) all argued that a sovereign was bound by an implicit contract to act in the interest of his subjects. If he abused these obligations, the population had the right to take up arms.

Hobbes used contract theory for the exact opposite reason than most of his predecessors when he argued that a ruler should never be overthrown. Heavily influenced by the destruction of the English Civil War (1639–1651) and the resulting social upheaval, his version presented an appeal against such atrocities. In his *Leviathan* (1651), Hobbes pictured the original state of nature for prepolitical humans was one of constant war, which he argued any rational person would want to end. In their desire for peace, individuals would forfeit their natural liberty. Hobbes's contract between individuals rather than between subjects and sovereign establishes an obligation on all to obey the sovereign as a *rule of reason*, which he also calls a *law of nature*. Thus, for Hobbes, subjects never have the right to oppose their sovereign. Likewise Hobbes sees no contractual constraint on the sovereign, because only the sovereign can preserve a state of peace.

Unlike Hobbes, Locke in his *Two Treatises of Government* and *A Letter Concerning Toleration* (both 1689) argued that an absolute monarchy is inconsistent with civil society. For Locke, the prepolitical state of nature is a peaceful yet moral society where humans are bound by divinely commanded natural law. Social problems develop insofar as they lack a common judge with authority over all. In the absence of this common judge, individuals strive for power to exert wills and attempt to seize each other's property. This situation calls for someone with the authority to act as judge in order to protect life, liberty, and estate. The lack of a state prevents enforcement of the laws of nature, so citizens create one. As with Hobbes, the contract is between individuals rather than between governed and ruler. But citizens who institute a government to prevent people from occasionally violating natural law and showing partiality do not give up their liberty in the contract. They simply grant the state the right to judge and punish offenders of natural law. The state, therefore, has very limited authority based on its contractual powers. Its primary duty is to protect property. The contract is dissolved and resistance is justified if the government commits any breach of trust.

During the eighteenth century, a time of monarchial excess in much of Europe, social contract theory moved away from just overthrowing the king to arguing for a more equitable political system. The most notable theorist in this regard was Rousseau, whose treatise on *The Social Contract* (1762) foreshadowed both the American and French revolutions. These theories were no longer concerned with the status of a monarch, but with the idea that monarchy was itself a suspect political system. The social contract was no longer between the people and a sovereign; now the people have become sovereign.

Rousseau discussed the idea of social contract on two separate occasions. In neither case did he claim that the contract was an actual historical event. Instead he offered a theoretically ideal contract concerned with the origin of government. He did not write about how it actually happened but how it ought to have happened. He believed that the state of nature was one of individual liberty where each person was free and equal and none had by nature any legitimate authority over any other. The prepolitical state was also a presocial state. The result of the establishment of social relations was the rise of inequalities in social and economic forums. It is this that leads to conflict between individuals, because only social individuals could begin to acquire wealth and hence have reasons for war. The rich end up controlling the masses because they manipulate society in order to protect it from the ravages of war. Hence there is a need for an ideal contract that should be established to preserve equality. This contract between citizens establishes a government that is ruled by the general will or what is best for all. Rousseau's ideal contract creates not a sovereign person but a sovereign people. The government can only be an agent of the people's will. It is an exchange of natural liberty for civil liberty, where each member has an equal share in the expression of a general will.

More systematically than Rousseau, Immanuel Kant (1724–1804) extended social contract theory by presenting the contract as a regulatory ideal. Kant's contract was not so much what people would have agreed to as what they should have agreed to in such a hypothetical situation. For Kant, the social contract was that ideal to which individuals would agree if they were ideal moral beings. In his view all laws should be framed so that everyone would consent to them if given the choice.

The social contract theories of the eighteenth century provide a justification for a political system based on the equality of all citizens. The emergence of republican democracies at the same time is no coincidence. The idea that citizens were equal was not particularly novel, because earlier contract theory began with a prepolitical state of nature in which all were equal. But the idea that individuals in the political state should retain their equality creates a whole new conception of government.

It is important to note that social contract theory not only arose in historical association with the rise of modern democracy, but also in association with the rise of modern science and technology. Indeed the theories of the state of nature in both Hobbes and Locke provide justifications for the pursuit of technology. With Hobbes the justification is one of necessity, in order to escape the oppression of nature. With Locke the justification is more that of seizing opportunities for advancement. Moreover the social order within science is not unlike that elaborated by Rousseau and Kant: one of free and equal members in a well-ordered body politic. Indeed the scientists of the Enlightenment often referred to the *republic of letters* and the *republic of science*—and saw this democracy in science as a model for that to be established outside science. The term republic of science has continued to be used by such defenders of science as Michael Polanyi (1962) and Ian Jarvie (2001).

John Rawls and a Theory of Justice

Interest in social contract theory declined in the nineteenth century and was displaced by utilitarianism, the theory that actions are right when they produce more benefit than harm for society. But in the mid-twentieth century, social contract theory reemerged as a theory for justice, first in economics and then in philosophy.

Economist James Buchanan, for instance, has developed an argument derived out of rational choice theory dealing with the distribution of wealth in society. Like others, Buchanan is not talking about a historical event but rather suggests a contract theory that could be used to propose changes in political institutions. For him, the optimum decision making rule is to minimize the cost of collective action and promote what is advantageous to utility-maximizing citizens.

Philosopher Rawls, however, has altered the overall emphasis of the social contract by using it to promote a theory of justice. The social contract ensures that all people's interests are properly protected. The problem of justice arises because individuals make competing claims to the same goods produced through social cooperation. Unlike earlier versions of contract theory, Rawls sees social contract theory as a means for addressing this problem of conflicting interests. The distribution of social goods is just if and only if it would be acceptable to all parties prior to any party knowing which goods he or she might receive. In order to meet this requirement Rawls imagines a *veil of ignorance* behind which "no one knows his place in society, his class position or social status" (Rawls 1971, p. 12), a condition from which any social order could be constructed.

Michael Lessnoff's *Social Contract* (1986) argues that Rawls's theory of justice is the culmination of social contract theory. Although he believes that the problem of justice is the correct subject for contract theory, he nevertheless proposes a reformulation of Rawls. First, all must enjoy equal basic liberties unless an unequal distribution would improve the total basic liberty of those with less. Second, a fair and equitable opportunity must exist for all to achieve their desired social and economic positions, unless the inequality improves the lives of those with fewer opportunities. Third, inequalities of various social and economic goods must be to the benefit of those who have less of them.

Thus in the twentieth century social contract theory moved from a theory of governance to one of distributive justice. As such it has been used to question some of the situations brought about by science and technology. For instance, there are questions of justice regarding the practices of the United States that, with about 4 percent of the world's population, uses more than 20 percent of the world's resources. Distributive justice questions also come into play in assessing access to science and science education on the basis of economic class, gender, or ethnicity. Finally from the perspective of Rawls' veil of ignorance, one can ask whether the contemporary distribution of governmental funding for science is just. Instead of defending particular governmental funding policies for science from the perspective of particular scientific interest group politics, would it not be more just to ask how physicists, chemists, and biologists would distribute societal support for science, before knowing which kind of scientists they were going to become?

Science, Technology, and the Social Contract

The idea of a social contract has appeared in a number of different forms when discussing science and science policy. Classic sociology of science, such as that found in the work of Robert Merton (1973) and Joseph Ben-David (1984), although they do not use the term, might well be read as describing how a social contract among scientists leads to the creation of a distinctive scientific ethos. Studies of the history of engineering as a profession (Layton 1971) point in the same direction: that engineering as a profession was self-defined in part by means of a social contract among engineers. (It might also be interesting to note the special situation among social scientists, who both study and are constituted by such contracts.) In the broad sense, a social contract between science, technology, and society may also simply refer to common expectations in the relations between professional representatives in each of these

three sectors: scientists, engineers and technologists, and politicians, respectively.

In this second sense of a social contract between scientists and the body politic, discussions have been at once more explicit and less well-grounded in social contract theory. As with social contract theory, a social contract for science need not refer to any specific historical agreement in a prepolitical period between the scientific community and the state or government. Instead it may be argued to be a logical extension of a desire on the part of individuals to better their condition, insofar as any such desire can itself be argued to benefit from scientific progress.

The whole concept of government spending on items such as science, technology, and medicine can thus be derived both from the original idea of individuals giving up their freedom to secure life, liberty, and property and from Rawls's idea of justice as directing resources to science and technology so as to increase benefits for all. Because the government is obligated by the social contract to improve its citizens' welfare, and insofar as science and technology are seen as having the potential to improve citizens' lives, the government invests in science, technology, and medicine.

Most explicitly science policy analysts in the United States have argued that Vannevar Bush's *Science—The Endless Frontier* (1945) established a social contract between the scientific community and government. In this case the public was left out of the agreement or at best represented by the government. In this contract, scientists promised to eliminate disease, feed the world, increase national security, and increase jobs in return for government funding and the right to maintain their autonomy. One description of this contract as a military-industrial complex became a focus for liberal political criticism during the 1960s. Antitechnology criticism of science as the cause of environmental pollution was a further spur to such criticism. In the 1980s and early 1990s with the downturn in the U.S. economy and the end of the Cold War, policy analysts began to question this social contract as well. They argued that the scientific community had failed to live up to its end of the bargain or was no longer as crucial to national welfare as it had been previously, and that public funding of science should be reexamined. With the reemergence of the U.S. economy in the mid-1990s and the rise of global terrorism in the 2000s such concerns tended to disappear.

The previous analysis assumes a kind of symbiosis between science and technology in what is often called technoscience. But in fact it can be argued that the situation with technology, especially that form of technology known as engineering, needs to be distinguished. For engineers, at least in the United States, any presumed social contract is mostly manifested in the marketplace. Industrial or market success substitutes for the social contract. When it comes to engineering, the problem is that there is no social contract—and yet the technologies that are developed and commercialized often have a social impact that consumers are not able intelligently to anticipate and governmental regulation is not sufficient to control.

The idea of a social agreement or contract continues to be invoked by politicians. For instance, in 1974 the British Labor Party proposed to save the United Kingdom by means of a social contract with the trade union movement. In 1994 the Republican Party in the United States ran its political campaign based on a *Contract with America*. The usefulness of social contract theory is its ability to ask what rational individuals would do if given a choice, and then to critique a system based on an argument about what is best for everyone. Even in Hobbes's defense of the monarchy, he begins with the assumption of what is best for all and not just a minority. Rawls extends this idea to justice and the distribution of resources to criticize any historical situation. Both approaches have been indirectly appealed to in discussions of a social contract for science, but it remains to be shown that such rhetoric has drawn at all deeply on the social contract theory tradition.

FRANZ ALLEN FOLTZ

SEE ALSO *Hobbes, Thomas; Kant, Immanual; Locke, John; Rousseau, Jean-Jacques; Social Contract for Science.*

BIBLIOGRAPHY

Ben-David, Joseph. (1984). *The Scientist's Role in Society: A Comparative Study,* with a new introduction. Chicago: University of Chicago Press. First published 1971.

Buchanan, James M. (1975). *The Limits of Liberty.* Chicago: University of Chicago Press.

Bush, Vannevar. (1945). *Science—The Endless Frontier.* Washington, DC: Government Printing Office.

Cozzens, Susan E.; Peter Healey; Arie Rip; and John Ziman. (1989). *The Research System in Transition.* Dordrecht, The Netherlands: Kluwer Academic Publishers.

Hobbes, Thomas. (1968). *Leviathan.* Harmondsworth, UK: Penguin.

Jarvie, Ian C. (2001). *The Republic of Science: The Emergence of Popper's Social View of Science.* Amsterdam: Rodopi.

Kant, Immanuel. (1977). *Kant's Political Writings,* ed. Hans Reiss. Cambridge, UK: Cambridge University Press.

Layton, Edwin T., Jr. (1971). *The Revolt of the Engineers: Social Responsibility and the American Engineering Profession.* Cleveland, OH: Press of Case Western Reserve University.

Lessnoff, Michael. (1986). *Social Contract.* London: Macmillan. A collection of edited selections from eleven of the most influential writings on social contract theory.

Lessnoff, Michael. (1990). *Social Contract Theory.* New York: New York University Press.

Locke, John. (1966). *Two Treatises of Government* and *A Letter Concerning Toleration.* Oxford: Blackwell.

Merton, Robert K. (1973). *The Sociology of Science: Theoretical and Empirical Investigations,* ed. Norman W. Storer. Chicago: University of Chicago Press. A collection of more than twenty important papers, from 1935 to 1972, by the founder of the sociology of science.

Nozick, Robert. (1974). *Anarchy, State, and Utopia.* Oxford: Basil Blackwell.

Polanyi, Michael. (1962). "The Republic of Science." *Minerva* 1: 54–74.

Rawls, John. (1971). *A Theory of Justice.* Cambridge, MA: Harvard University Press.

Rousseau, Jean-Jacques. (1968). *The Social Contract and Discourses.* London: Penguin Books.

SOCIAL DARWINISM

• • •

Social Darwinism was a prominent ideology in the late-nineteenth and early-twentieth centuries that emerged when biologists and social thinkers tried to apply the biological theories of Charles Darwin (1809–1882) to human society. Social Darwinists believed that humans were subject to scientific laws, including Darwinian natural selection and the struggle for existence. They viewed human competition as a beneficent force bringing progress. However serious differences emerged among those who tried to formulate social theories based on Darwinism. One of the most controversial disputes among social Darwinists was whether humans should model their societies on nature or use scientific knowledge to vanquish nature. Specifically the question was whether humans should sharpen or soften the struggle for existence. Though most social Darwinists never admitted it, this fundamental question was not tractable scientifically, but depended on one's ethical perspective, because Darwinian processes could not predict future outcomes nor provide moral guidance. Not all Darwinists embraced social Darwinism, of course, and some promoted eugenics as a way to evade the human struggle for existence.

From Malthus to Darwin

Tracing the origins of social Darwinism is complicated, because many ideas associated with social Darwinism—such as laissez-faire economics, militarism, and racism—predated Darwin and influenced the formulation of his biological theory. Probably the most important of the forerunners of social Darwinism was Thomas Robert Malthus (1766–1834), whose population principle claimed that human populations tend to expand faster than the food supply. This population imbalance, according to Malthus, inevitably produces human misery, famine, and death. Darwin forthrightly incorporated Malthus's ideas, along with other concepts from nineteenth-century economics, into his biological theory. However he also gave a new twist to Malthus that would be important in the rise of social Darwinism. While Malthus considered the human misery caused by overpopulation entirely harmful and lamentable (though inevitable), Darwin construed it as beneficial and progressive, because it drove the evolutionary process, producing new species. The rise of Darwinian theory in the late-nineteenth century gave greater currency to Malthus's ideas, which became prominent in social Darwinist circles.

Darwin was clearly a social Darwinist, because he believed that the Malthusian population principle demonstrated the necessity of a struggle for existence among humans, leading to competition both within and between human societies. However these two levels of competition could work at cross-purposes, presenting Darwin (and other social Darwinists) with a dilemma. Which was more important: individual or group competition? Most social Darwinists—including Darwin—insisted that both operated simultaneously, though they did not always agree on which was more important. Darwin believed that individual competition among humans manifested itself primarily as peaceful economic competition, while group competition often brought warfare and racial conflict.

Another important plank of social Darwinism that Darwin propagated was human inequality. Natural selection could only function if there were significant differences between organisms. Also, in order to make their theory of human evolution more plausible, Darwinists had to emphasize the tremendous diversity within the human species, while showing the proximity of humans to other species. This led them to stress the differences between races, and the proximity of "primitive" races to primates. Darwin specifically claimed that "savage" races were biologically inferior to Europeans. He believed their intellectual prowess was far below that of Europeans, and because he considered moral character a hereditary trait, he also accused them of being biologically inferior in their moral character.

In most of his writings Darwin confined himself to *describing* the process of human evolution. However at times he became *prescriptive*, proposing public policy based on his theory. He generally supported laissez-faire economics, because it would promote competition among individuals, allowing the "fittest" to succeed. In a private letter he expressed concern that labor unions were deleterious, because they opposed individual competition. He also used his theory to justify national and racial competition, which was reflected in British and other European attempts to dominate the globe through imperialism. In *The Descent of Man* Darwin stated, "At some future period, not very distant as measured by centuries, the civilised races of man will almost certainly exterminate and replace throughout the world the savage races" (Darwin 1981, vol. 1, p. 201). Darwin, however, did make it clear that despite his view that wars have played a crucial role in human evolution, he hoped they would cease in the future.

Classic Social Darwinism

While justifying and supporting human competition as biologically beneficial, Darwin did not believe that the human struggle for existence was completely ruthless. He thought that human morality—which he explained as a product of the struggle for existence—tempered the struggle, at least within societies. Herbert Spencer (1820–1903), whom Darwin and many of his contemporaries considered a great philosopher, but whose star has waned since, likewise argued that ethics was the pinnacle of human evolution. However, like Darwin, he thought that too much altruism would be detrimental to humanity, because it would diminish human competition.

Spencer's role in the development of social Darwinism has been hotly debated, because before Darwin published his theory, Spencer already believed in biological evolution and embraced a competitive ethos and laissez-faire economics. However Spencer's pre-Darwinian ideas about evolution were shaped by Lamarckism, which taught that organisms passed acquired traits on to their offspring. Spencer's pre-Darwinian view of competition was not really social Darwinism. After 1859 Spencer integrated natural selection and the struggle for existence into his social views, thus espousing a form of social Darwinism. Like Darwin, he did not think the human struggle for existence had to be violent. On the contrary, he thought the struggle was becoming more and more peaceful as society progressed.

Not all social Darwinists thought warfare was becoming obsolete, as Spencer did. William Graham Sumner (1840–1910), a prominent American sociologist who pioneered social applications of Darwinism, claimed that Darwinism proved the inevitability of war. He even stated that "nothing but might has ever made right" (Hawkins 1997, p. 117), a position that Darwin rejected, but that several social Darwinists embraced. Even so, Sumner advised avoiding war if possible, so he was far from being a rabid militarist. However some social Darwinists, including the German general Friedrich von Bernhardi (1849–1930), author of the best-selling book, *Germany and the Coming War* (1912), used social Darwinism to promote militarism.

Racial competition was an even more prominent and widespread theme in social Darwinist thought than was national competition. Ernst Haeckel (1834–1919), the leading Darwinian biologist in Germany in the late-nineteenth century, was even more racist than Darwin. He argued that the distinctions between the human races were so great that humans should be divided into twelve separate species, which he placed in four separate genera. These races, he claimed, were in a competitive conflict that would only end with the extermination of the least fit races. Ludwig Gumplowicz (1838–1909), a law professor at the University of Graz in Austria, published one of the most extensive treatments of this theory in *The Racial Struggle* (1883), a term that became popular among social Darwinists in the 1890s and first decades of the twentieth century. Gumplowicz did not consider races a biological entity at all, however, as did most later racial thinkers, but rather he stressed their cultural construction. Nonetheless he argued that races are locked in an ineluctable Darwinian struggle for existence, and he believed that the ethnic conflicts within the Austro-Hungarian Empire were part of this universal struggle.

Another influential social Darwinist in the late-nineteenth and early-twentieth centuries who emphasized the racial struggle for existence was Georges Vacher de Lapouge (1854–1936), who exerted greater influence in Germany than in his native France. Lapouge was worried that certain "inferior" European races were displacing the "superior" forms. He wanted to supplement the racial struggle with eugenics. He hoped to replace the slogan of the French Revolution—liberty, equality, fraternity—with a more "scientific" triad—determinism, inequality, and selection. He warned in 1887, "In the next century people will be slaughtered by the millions for the sake of one or two degrees on the cephalic index [i.e., cranial measure-ments]. ... the superior races will substitute themselves by force for the human groups retarded in evolution, and the last sentimentalists will witness the copious extermination of entire peoples" (Hecht 2000, p. 287).

Social Darwinist racism also found much support in Britain and the United States. Walter Bagehot (1826–1877), one of the first writers in Britain to apply Darwinism to politics, thought racial competition was a blessing to the human race, stimulating progress. He asserted that even though some races may not accept the superiority of the European race, "we need not take account of the mistaken ideas of unfit men and beaten races" (Hawkins 1997, p. 70). Karl Pearson (1857–1936), a leading British biologist, wanted to mitigate individual competition to increase national and racial vitality. He promoted eugenics as a way to give the British a competitive advantage in the racial struggle, and he supported the extermination of other races to make room for British settlement. In 1916 Madison Grant (1865–1937), a well-connected lawyer who served as president of the New York Zoological Society, published *The Passing of the Great White Race*. The preface to his book was written by one of the leading scientists of his time, Henry Fairfield Osborn (1857–1935), who was both a professor at Columbia University and president of the American Museum of Natural History. In his book Grant proposed using immigration restrictions and eugenics to restore the vitality of the "Great White Race," which was threatened with biological decline. Pearson and Grant were by no means idiosyncratic in supporting eugenics within their countries to strengthen their nation or race to compete successfully in the wider national or racial struggle for existence.

Conflicting Perspectives

One of the striking things about nineteenth-century social Darwinism was the variety of political positions that could use social Darwinist arguments to buttress their positions. British liberals—like Darwin—could use the theory to support laissez-faire economics and imperialism. But some non-Marxian socialists thought social Darwinism was on their side. For example, the physician Ludwig Büchner (1824–1899), one of the earliest and most famous Darwinian popularizers in Germany, argued that individual competition was essential for human advancement. However, he denied that the capitalist system was best in promoting competition. Capitalism, he thought, skewed the struggle for existence, because those who inherited capital would have an unfair advantage over those from poor families. Büchner suggested eliminating the inheritance of capital to level the playing field, so one's biological traits and abilities would be the only factors determining success or failure. Similar arguments were advanced by prominent Fabian socialists in Britain, such as Sidney Webb (1859–1947), and by the Labour Party leader, Ramsey MacDonald (1866–1937), who both promoted their socialist ideas as the logical outcome of Darwinian theory.

Though appropriated by scholars and politicians embracing a wide variety of political positions, social Darwinism would have its greatest impact on the world stage through the political power exerted by a fanatical social Darwinist whose racist brand of social Darwinism would drive him to unleash World War II in Europe. In *Mein Kampf* (1925–1927) Adolf Hitler argued that racial competition was a part of the universal struggle for existence, which destroys the weak and unfit. Hitler believed that morality consisted in cooperating with nature in destroying the weak, so the healthy, "superior" individuals could triumph.

Social Darwinism declined in popularity in the mid-twentieth century, and not only because of its association with the Nazis. Biological explanations for human behavior gave way in the mid-twentieth century to environmental explanations. Behaviorism dominated psychology in the 1950s, cultural relativism dominated anthropology, and Marxism and other non-Marxist forms of economic and environmental determinism displaced biological determinism in the social sciences. By the 1960s biological determinism had almost completely disappeared from serious scholarly work. After Richard Hofstadter wrote the first major historical work titled *Social Darwinism in American Thought* (1944), the term social Darwinism was generally used disparagingly.

In the 1970s a new movement within the scientific community emerged that reinvigorated biological determinism. Edward O. Wilson provoked intense controversy with the publication of his book, *Sociobiology* (1975). Many accused Wilson of resurrecting social Darwinism, but he and supporting colleagues denied the charge. Indeed Wilson did embrace some of the positions of earlier social Darwinists (for example, his stress on biological determinism, the importance of Darwinian selection on human behavior, and so on), but he did not embrace the crude nationalism and racialism that Hofstadter identified as leading characteristics of social Darwinism.

RICHARD WEIKART

SEE ALSO *Christian Perspectives; Darwin, Charles; Dominance; Eugenics; Holocaust; Population; Race; Sociobiology; Spencer, Herbert.*

BIBLIOGRAPHY

Bannister, Robert. (1979). *Social Darwinism: Science and Myth in Anglo-American Social Thought.* Philadelphia: Temple University Press. Bannister challenges Hofstadter's thesis by claiming that the influence of social Darwinism was minimal. His unconvincing argument is that social Darwinism was merely a straw man created by progressives who opposed a competitive ethos.

Clark, Linda. (1984). *Social Darwinism in France.* Tuscaloosa: University of Alabama Press. Clark's work shows the varieties of social Darwinism in France, especially emphasizing what she calls reform Darwinism.

Darwin, Charles. (1981). *Descent Of Man.* 2 Vols. Princeton, NJ: Princeton University Press.

Degler, Carl N. (1991). *In Search of Human Nature: The Decline and Revival of Darwinism in American Social Thought.* New York: Oxford University Press. Degler provides an excellent survey of the nature-nurture debate in the human sciences in the twentieth-century United States, focusing especially on the role of Darwinism in this debate. He is sympathetic to the rise of sociobiology.

Hawkins, Mike. (1997). *Social Darwinism in European and American Thought, 1860–1945.* Cambridge, UK: Cambridge University Press. This is the best single volume on social Darwinism available. Hawkins provides in-depth analysis of key social Darwinist thinkers in various countries.

Hecht, Jennifer Michael. (2000). "Vacher De Lapouge and the Rise of Nazi Science." *Journal of the History of Ideas* 61 (2000): 285–304.

Hofstadter, Richard. (1955). *Social Darwinism in American Thought,* revised edition. Boston: Beacon Press. This is the pioneering work that introduced the term social Darwinism into historical discourse. While providing excellent explanations and examples of social Darwinism, he wrongly equated it with conservative political thought.

Jones, Greta. (1980). *Social Darwinism and English Thought: The Interaction between Biological and Social Theory.* Sussex, UK: Harvester. Jones shows the prominence (and variety) of social Darwinist thought in Britain during the late-nineteenth and early-twentieth centuries.

Weikart, Richard. (2004). *From Darwin to Hitler: Evolutionary Ethics, Eugenics, and Racism in Germany.* New York: Palgrave Macmillan. This work shows the impact of Darwinism on ethical and moral thought in Germany during the late-nineteenth and early-twentieth centuries. It explores eugenics, euthanasia, racism, militarism, and the impact of Darwinism on Hitler's thought.

Wilson, Edward O. (1975). *Sociobiology: The New Synthesis.* Cambridge, MA: Belknap Press of Harvard University Press. This is the classic work by a Harvard biologist that brought sociobiology into prominence in the late twentieth century.

Weindling, Paul. (1989). *Health, Race and German Politics between National Unification and Nazism, 1870–1945.* Cambridge, UK: Cambridge University Press. A long chapter on social Darwinism in Germany shows the predominance of political liberalism in social Darwinist discourse. Much of the rest of the book examines eugenics.

Young, Robert. (1985). "Darwinism *Is* Social." In *The Darwinian Heritage*, ed. David Kohn. Princeton, NJ: Princeton University Press. Young cogently explains how social thought impacted Darwin and influenced the formulation and expression of his theory.

SOCIAL ENGINEERING

• • •

Social engineering occurs in two forms: large scale and small scale. The debate surrounding these two approaches to the design of social institutions constitutes a fundamental issue in the ethics of science and technology. To what extent is it possible and legitimate for scientific expertise to serve as the basis for social policy and action? Can humans use science to rationally design and successfully implement an enduring society? Different concepts of scientific knowledge and technological action supply different answers to these questions and variously support large scale versus small scale engineering efforts.

Large Scale Social Engineering

Large scale efforts to improve the human condition are a modern phenomenon. Such endeavors require technical knowledge, political muscle, and economic resources. In supporting these claims, James Scott (1998) characterizes the rise of high modernism in social-political, agricultural, industrial, and architectural contexts during the last two centuries. High modernism encompasses a quest for authoritarian control of both human and nonhuman nature, a belief that carefully crafted social order surpasses happenstance, and a confidence in science as a means to social progress. Once the improvement of humanity becomes a plausible state goal, the convergence of rising social science, state bureaucracy, and mass media undergirds five-year collectivist plans, colonial development schemes, revolutionary agricultural programs, and the like, often under the control of a single planning entity.

In urban planning, for example, Scott details the designs of the Swiss architect, Charles-Edouard Jeanneret, (1887–1965), known professionally as Le Corbusier. For Le Corbusier, urban design expresses universal scientific truths. His geometric symmetries often structured human activity, as inhabitants conformed to the design rather than vice versa. This approach applied to entire cities as well as individual homes ("machines for living"). Le Corbusier's formulaic concatenation of single function components produced simplicity via widely separated spaces for living, working, shopping, and recreating. Defining the good of the people, often the working poor, in terms of detached, scientific principles and their authoritarian imposition is, according to Scott, emblematic of high modernist, large scale attempts at social engineering.

Small Scale Social Engineering

In conceiving the perfect, nondecaying state, Plato envisions a radical departure from existing society. Marxists, too, as self-described social engineers, use historical interpretation in aiming for revolutionary, holistic change. The Anglo-Austrian Philosopher, Karl Popper (1902–1994) contrasts these utopian endeavors with "piecemeal social engineering." When society needs reforming, the piecemeal engineer

> does not believe in the method of re-designing it as a whole. Whatever his ends, he tries to achieve them by small adjustments and re-adjustments which can be continually improved upon. . . . The piecemeal engineer knows, like Socrates, how little he knows. He knows that we can learn only from our mistakes. Accordingly, he will make his way, step by step, carefully comparing the results expected with the results achieved, and always on the look-out for the unavoidable unwanted consequences of any reform; and he will avoid undertaking reforms of a complexity and scope which make it impossible for him to disentangle causes and effects, and to know what he is really doing. (Popper 1957, pp. 66–67)

These claims resonate with Camus's (1956) distrust of ideologically calculated revolution and his preference for limited but inspired rebellion. In Popper's view, mistakes are inevitable, and more radical innovations produce more mistakes. Because foolproof social forms are unattainable, some mechanism for identifying needed improvements must be an integral part of a necessarily gradual implementation process. This view contrasts with that of large scale social engineering on several dimensions and highlights multiple points of contention.

Spontaneous versus Consciously Controlled Change

Popper's (1972) concept of evolutionary epistemology supports not only the idea that advances are slow and piecemeal but also that they are guided by no overarching plan. This view resembles that of the twentieth-century British economist Friedrich Hayek (Nishiyama and Leube 1984). Hayek (1967) emphasizes the view that significant social phenomena emerge spontaneously via the unintended effects of individual actions, and he finds support for the benefits of this process in the ideas of the British political economist, Josiah Tucker (1711–1799), and especially the Austrian economist Karl Menger (1840–1921), that social institutions compete with one another in a kind of survival of the fittest. Because knowledge required for large-scale planning is widely distributed among many minds and cannot be narrowly concentrated, Hayek rejects centralized planning. Popper (1963a) advocates "negative utilitarianism," the view that proposals for reform should be judged by how little suffering is caused. Government should thereby ameliorate enduring social ills (such as poverty and unemployment) and leave efforts to increase happiness to individual enterprise. These views shape the method (monitored, incremental change) and the goals (amelioration) of social engineering.

The nature of social reform is also examined by the American philosopher and educator John Dewey (1859–1952). But when Dewey speaks about the need for liberalism to advance beyond its early gains in securing individual freedom, his vision is incongruent with that of Hayek and Popper. For Dewey, liberalism should advance a social order that "cannot be established by an unplanned and external convergence of the actions of separate individuals, each of whom is bent on personal private advantage" (Dewey 1963 [1935], p. 54). This social reform must be thoroughgoing in its quest for institutional change.

> For the gulf between what the actual situation makes possible and the actual state itself is so great that it *cannot be bridged by piecemeal policies undertaken ad hoc.* The process of producing the changes will be, in any case, a gradual one. But "reforms" that deal now with this abuse and now with that without having a *social goal based upon an inclusive plan,* differ entirely from efforts at reforming, in its literal sense, the institutional scheme of things. (p. 62)

Dewey sees the necessity of early planning in his thinking about social reform (Geiger 1971 [1939]), and while it is clear that Popper restricts not planning per se but only its scope and method, Dewey projects a wider, more vibrant use of planning in achieving social renovation. Education, science (the method of intelligence), and well-designed government policy are keys to social improvement.

The Nature of Scientific Knowledge

Any call for social engineering requires some clarification of the relationship between science and engineering. Popper differentiates natural and social science in ways that Dewey does not. In natural science, Popper's realist perspective dictates that theories make claims about unobservable realities responsible for observed regularities. These claims are tested by means of controlled experiments. In contrast, Popper construes social science as producing low-level empirical laws of a negative sort ("you cannot have full employment without inflation"), which are tested through practice in social engineering. This amounts to a narrow view of social science and contributes to the contrast between his scientific radicalism, which focuses on natural science, and his engineering conservatism, which is linked to social science. The contrast between Dewey the pragmatist and Popper the realist is instructive here. From Dewey's pragmatic perspective, "the ultimate objects of science are guided processes of change" (Dewey 1958 [1929], p. 160). Both natural science and social science provide an illustration of this concept (Dewey 1947). Popper's general aversion to abstract theories in social science may be linked to his desire to reject certain theories, such as that of the Austrian psychiatrist Sigmund Freud, on the basis of unfalsifiabilty. Dewey's acceptance of a wider range of theory plus empirical law in social science allows for testing to occur in a greater range of circumstances, not only in practice (which is often problematic: even piecemeal change simultaneously introduces multiple causal factors) but also in controlled, even laboratory, settings. Contemporary studies in social science embrace such methods, including those of simulation (Liebrand, Nowak, and Hegselmann 1998; Ilgen and Hulin 2000). Moreover, when guided by theory and experimental tests, changes introduced into practice need not be small scale. Large-scale changes may be introduced for larger scale problems (such the Great Depression or disease epidemics). Linking Science to Practice Popper and Dewey differ when relating science to social engineering. In disputes with the American philosopher Thomas Kuhn (1922–1996), Popper emphasizes the value of critical and revolutionary action (bold conjectures and severe tests) over and above the uncritical plodding of normal science (Popper 1970). This contrasts with his recommendations for social engineering where action should be piecemeal. This contrast, acknowledged by Popper (1976) himself, may arise from the use of the scientific community as a model for society at large. Nevertheless, the degree of openness and fruitfulness of criticism differs significantly within these two realms (Burke 1983). Robert Ackermann proposes that an explanation "of the relative isolation of theoretical scientific knowledge from practical concerns is required to explain how a form of social conservatism can be held consistently with a form of theoretical radicalism" (Ackermann 1976, p. 174).

Such concerns are related to Scott's analysis of why large scale schemes have often failed to improve the human condition. Scott sees knowledge of how to attain worthwhile, sustainable solutions as being derived not from scientific theory, nor from the low level empirical laws cited by Popper, but by a form of know how (*metis*, from the ancient Greek) rooted in localized, cultivated practice. Like Dewey's conception, which builds an inherent normative element ("*guided* processes") into knowledge itself, there is no need to search for means of effective "application." The implication is that useful knowledge springs from contextualized activities, not from using local conditions to fill in the variables of general principles. This view raises serious doubts about the practical relevance of scientific expertise, in the modern sense, and its ability to produce sustainable solutions to social problems. Indeed, some have suggested that such limitations exist not only in large scale enterprises but also in small scale efforts involving more narrowly focused problems (Hamlett 1992, Winner 1992). A narrow focus can undermine the need to address larger issues and long run concerns and can mire the political process in gridlock. From these considerations, it should be clear that small scale engineering offers no panacea and that different concepts of small scale enterprise point the way in somewhat different directions.

Impact of the Social Engineering Issues

Questions concerning appropriate scale and the interaction of social science and social engineering have wide impact. An entire school of social scientists use Popper as a guide in trying to design effective social policy. The works of the incrementalist Charles Lindblom (*The Intelligence of Democracy*; *Usable Knowledge: Social Science and Social Problem Solving*; *Inquiry and Change: The Troubled Attempt to Understand and Shape Society*; etc.) provide, by title alone, some measure of the impact of Popper and Dewey and of social scientists' pursuit of social engineering. Moreover, differences between planned, rule-governed (top-down) versus unplanned, evolutionary (bottom-up) approaches inform methodologically diverse explorations within social science itself (Banathy 1996, Read and Miller 1998). Whether or not humans can effectively design social systems is essentially a question concerning human intelligence, and efforts to build automated intelligent systems confront the same methodological controversy concerning rule-governed versus connectionist, evolutionary designs

("Sackler Colloquium" 2002). Finally, controversies over the promises of planned societies continue to echo the dispute between Popper and Marxists over the true nature of social engineering (Cornforth 1968, Marquand 2000, Notturno 2000, Postrel 2001).

MARVIN J. CROY

SEE ALSO Dewey, John; Incrementalism; Popper, Karl; Plato.

BIBLIOGRAPHY

Ackermann, Robert. (1976). *The Philosophy of Karl Popper*. Amherst: University of Massachusetts Press. Provides analyses of several central concepts, including Popper's account of the social sciences.

Banathy, Bela. (1996). *Designing Social Systems in a Changing World*. New York: Plenum. Explains design as a collective human activity and system design as a process for addressing a wide variety of problems from a holistic, large scale perspective.

Burke, T. E. (1983). *The Philosophy of Popper*. Dover, NH: Manchester University Press. Critically assesses Popper's connection of epistemological concepts to issues of freedom and values.

Cornforth, Maurice. (1968). *The Open Philosophy and the Open Society: A Reply to Dr. Karl Popper's Refutations of Marxism*. New York: International Publishers.

Camus, Albert. (1956). *The Rebel: An Essay on Man in Revolt*. New York: Alfred A. Knopf. Articulate and inspired statement of Camus's rejection of ideology, including Marxism, and his view of where that leaves any effort to improve the human condition.

Dewey, John. (1947). "Liberating the Social Scientist: A Plea to Unshackle the Study of Man." *Commentary* 4: 378–385.

Dewey, John. (1958 [1929]). *Experience and Nature*, 2nd edition. New York: Dover Publications.

Dewey, John. (1963 [1935]). *Liberalism and Social Action*. New York: Capricorn Books. Classic statement on the nature of Liberalism and its promise for an improved social order.

Geiger, George. (1971 [1939]). "Dewey's Social and Political Philosophy." In *The Philosophy of John Dewey*, ed. Paul Arthur Schilpp. La Salle, IL: Open Court.

Hamlett, Patrick. (1992). *Understanding Technological Politics: A Decision Making Approach*. Englewood Cliffs: Prentice Hall. Combines the strengths of theoretical and empirical approaches by analyzing the framework and consequences of particular decisions concerning technology development and implementation.

Hayek, Friedrich. (1967). *Studies in Philosophy, Politics, and Economics*. Chicago: University of Chicago Press. Impressive collection of twenty five articles authored in two decades that illustrate the range and genius of Hayek's thinking.

Ilgen, Daniel, and Charles Hulin. (2000). *Computational Modeling of Behavior in Organizations: The Third Scientific Discipline*. Washington, DC: American Psychological Association.

Liebrand, Wim B. G.; Andezej Nowak; and Rainer Hegselmann, eds. (1998). *Computer Modeling of Social Processes*. Thousand Oaks, CA: Sage. This collection explores simulations, neural networks, and data analysis methods in explaining and predicting complex social processes.

Marquand, David. (2000). "A Tale of Three Karls: Marx, Popper, Polanyi, and Post-Socialist Europe." In *Philosophy and Public Affairs*, ed. John Haldane. Cambridge, UK: Cambridge University Press.

Nishiyama, Chiaki, and Kurt R. Leube, eds. (1984). *The Essence of Hayek*. Stanford, CA: Hoover Institution Press, Stanford University.

Notturno, Mark Amadeus. (2000). *Science and the Open Society: The Future of Karl Popper's Philosophy*. New York: Central European University Press.

Popper, Karl. (1957). *The Poverty of Historicism*. New York: Basic. Contains the kernel of Popper's argument against large scale social engineering, particularly the Marxist variety, plus the characterization of piecemeal social engineering that stimulated decades of incrementalist thinking in the social science.

Popper, Karl. (1963a). *Conjectures and Refutations: The Growth of Scientific Knowledge*. New York: Harper and Row.

Popper, Karl. (1963b). *The Open Society and Its Enemies*, 4th edition. Princeton, NJ: Princeton University Press.

Popper, Karl. (1970). "Normal Science and Its Dangers." In *Criticism and the Growth of Knowledge*, ed. Imre Lakatos and Alan Musgrave. Cambridge, UK: Cambridge University Press.

Popper, Karl. (1972). *Objective Knowledge: An Evolutionary Approach*. Oxford, UK: Oxford University Press.

Popper, Karl. (1976). "Reason or Revolution." In *The Positivist Dispute in German Sociology*, ed. Theodor W. Adorno et al. London: Heinemann.

Popper, Karl. (1983 [1956]). *Realism and the Aim of Science*. Totowa, NJ: Rowman and Littlefield.

Postrel, Virginia. (2001). "The Future and Its Enemies: Dynamism versus Stasis." In *Competition or Compulsion? The Market Economy versus the New Social Engineering*, ed. Richard Ebeling. Hillsdale, MI: Hillsdale College Press.

Read, Stephen, and Lynn Miller. (1998). *Connectionist Models of Social Reasoning and Social Behavior*. Mahwah, NJ: Erlbaum.

"Sackler Colloquium on Adaptive Agents, Intelligence, and Emergent Human Organization: Capturing Complexity through Agent-Based Modeling." (2002). *Proceedings of the National Academy of Sciences* 99(suppl. 3): 7187–7316.

Schilpp, Paul Arthur, ed. (1974). *The Philosophy of Karl Popper*. La Salle, IL: Open Court.

Scott, James C. (1998). *Seeing Like a State: How Certain Schemes to Improve the Human Condition Have Failed*. New Haven, CT: Yale University Press. Clearly delineates the differences between small scale and large scale social engineering, with accessible examples of each, and provides helpful analyses of the differences.

Winner, Langdon, ed. (1992). *Democracy in a Technological Society*. Dordrecht: Kluwer Academic Publishers.

SOCIAL INDICATORS

• • •

The historical tradition of social indicators may be traced back to Jeremy Bentham's (1789) ideas about a *felicific calculus* that would allow decision makers to calculate the net pleasure or pain connected to everyone affected by an action, with evidence-based public policy choices made to get the greatest net pleasure or least net pain for the greatest number of people. From a consequentialist moral point of view, the aim of government should be to increase the pleasure or happiness, broadly construed, of the maximum number of persons.

This approach is similar to the naturalist tradition in American pragmatism as argued in work by William James (1909), Ralph Barton Perry (1926, 1954), John Dewey (1939), and C. I. Lewis (1946), but more complicated. It is similar in the sense that pragmatism, like Bentham, naturalizes ethics by basing it in subjective preferences. It is more complicated in that most early-twenty-first century social indicators researchers believe the relatively objective circumstances of people's lives merit at least as much attention as how people assess those lives. The argument is that a morally complete assessment of people's lives, or a full assessment of people's lives from a moral point of view, requires a thorough examination of the nature or being as well as the value or good of those lives. In philosophical jargon, social indicators rest on an ontological answer to the question, What is its nature?, and an axiological answer to the question, What is its value?

Basic Concepts

The term *social indicator* denotes a statistic that has significance for measuring the quality of life. The term *social report* designates an organized collection of social indicators, and *social accounts* names a balance sheet in which costs and benefits are assigned to the indicators in a social report. Briefly the main difference between social reports and accounts is that the former answers the question, How are we doing?, and the latter answers the question, At what price?—where price may be measured in dollars, energy, personal satisfaction or dissatisfaction, or some other applicable metric.

From a linguistic perspective, social indicators usually consist of a term denoting a subject class and a term denoting some *indicator property*. For example, the second term of the phrase *infant mortality* denotes the indicator property mortality and the first term denotes a particular class of things, namely infants that may possess that property. By replacing the subject term infant by *one-year-old, two-year-old*, or more, one can routinely generate (social) mortality indicators for as many age groups as desired. Similarly by replacing the subject term by *male, Indian*,or others, one can routinely generate mortality indicators for as many kinds of groups as one likes.

Social indicator phrases are like variable names in logic and mathematics, and social indicators are like the variables themselves. Furthermore just as one speaks of the values of variables in logic and math, one may speak of the indicator-values of social indicators. For example, the annual percent of undergraduate degrees awarded to females in engineering in the United States in the 1990s was about 16 percent. So one may say that this variable (annual percent of undergraduate degrees awarded to females in engineering in the United States in the 1990s) had an indicator-value of 16 percent.

Social indicators that refer to personal feelings, attitudes, preferences, opinions, judgments, or beliefs of some sort are called *subjective indicators*, for example, satisfaction with one's health, attitudes toward science or scientists, and beliefs about the dangers of some new technology. Social indicators that refer to things that are observable and measurable are called *objective indicators*, for instance, the height and weight of people, numbers of automobiles manufactured or sold each year, and numbers of people employed in research and development.

Positive indicators are those for which most people equate an indicator-values increase with quality of life improvement, such as elderly citizens incomes and minority-group educational attainment. The female engineering degrees indicator mentioned above would be regarded as positive by those who think that the quality of women's lives tends to improve as their access to the full range of professional occupations improves. *Negative indicators* are those for which most people equate an indicator-values increase with quality of life deterioration, namely, infant mortality rates and murder rates. (Notice that an indicator is regarded as positive or negative not in virtue of whether or not its values in fact increase or decrease, but only in virtue of whether or not most people would like its values to increase or decrease. What is relevant is not the *fact* but the *desirability* of an increase or decrease in its values.)

Unclear indicators are such that either (a) most people will not be willing or able to say whether higher indicator-values indicate a better or worse state of affairs, for instance, welfare payments, or (b) there is serious disagreement about whether higher indicator-values indicate a better or worse state of affairs, namely, divorce rates. In the case of welfare payments, it is difficult to say, because as the values increase there may be an increase of people in need of such assistance, which is bad; while, at the same time, there is an increase in the amount of assistance given, which is good. In the case of divorce rates, many people know exactly what they want to say, and they happen to disagree with what some other people want to say.

Input indicators indicate some sort of inputs into a process or product, such as numbers of people engaged in research and development. *Output indicators* indicate some sort of output of a process or product, such as numbers of articles published or patents awarded per 1,000 people employed in research and development. Unlike the previous indicator classifications, what counts as an input or output indicator depends on the purposes of the classification. For example, from the point of view of a teacher, the amount of time a student spends studying could be regarded as an output indicator measuring the effects of a student's own need for achievement as well as from advice, admonitions, and threats given to the student. However from the point of view of a student, time spent studying could be regarded as an input indicator measuring the necessary investment made in the interest of obtaining such important measurable outputs as university degrees, good jobs, and higher income. In some contexts it is useful to talk about *intermediate output indicators* (for example, that count the machines that make consumer products), *throughput indicators* (for instance, that assess choices people make for certain consumer goods) and *outcome indicators* (such as those that measure longer-term net results of inputs).

When people use the phrase *quality of life*, they sometimes intend to contrast it with quantities or numbers of something. There are, then, two different things that one might reference when using the phrase *quality of life*. First, one might want to refer to sorts, types, or kinds of things, rather than to mere numbers of things. For example, one might want to know not merely how many people received bachelors degrees majoring in mathematics, but also something about who they were, male or female, in public or private institutions, with or without scholarship aid, and so on. When the term *qual-*

ity in the phrase *quality of life* is used in this sense, one may say that it and the phrase in which it occurs is are intended to be primarily *descriptive*.

Second, one might want to refer to the value or worth of things when using the term quality in the phrase *quality of life*. For example, one frequently hears of people making a trade-off between a high salary and better working or living conditions. Presumably the exchange here involves monetary and some other value. That is, one exchanges the value of a certain amount of money for the value of a certain set of working or living conditions. When the term *quality* in the phrase *quality of life* is used in this sense, one may say that it and the phrase in which it occurs is intended to be primarily *evaluative*.

Both senses of the phrase *quality of life* are important. It is important to be able to describe human existence in a fairly reliable and valid fashion, and it is important to be able to evaluate human existence in the same way. In the early years (1960s) of social indicators research, people asked, Should researchers measure the nature and value of life with objective or subjective indicators, or both? In the early twenty-first century, nearly everyone agrees that both kinds of measures should be used.

Uses and Abuses

It cannot be emphasized too strongly that social reporting is an essentially political exercise and that its ultimate success or failure depends on the negotiations involved in creating and disseminating the reports. Every opportunity to use social indicators is equally an opportunity to abuse their use. For examples, indicators:

(1) provide convenient numerical summaries of important features of society, but also encourage commission of The Number-Crunchers' Fallacy, which is this: Anything that cannot be counted is unimportant and anything that can be counted is important.

(2) can be used to predict and alter future behavior, for better or worse depending on the nature of the behavior and the alterations.

(3) can give visibility to problems, and also create them by focusing attention on them, or by hiding some in the interest of emphasizing others.

(4) can help obtain balanced assessments of conditions against mere economic assessments, and can distort appropriate assessments by assuming that everything valuable can be given a price in monetary terms.

(5) can help in the evaluation of current public policy and programs, and also contribute to perverse evaluations because the statistics routinely collected may not allow decision makers to control for important contaminating variables when they are trying to decide what has caused what.

(6) can help determine alternatives and priorities, but also allow an elite corps of statisticians and other experts to unduly influence the public agenda by providing the *official version* of the state of the world.

(7) can facilitate comparisons among nations, regions, and cities, and service providers, but also encourage invidious comparisons, raising aspirations and hopes too high or not high enough.

(8) can suggest areas for research to produce new scientific theories and more knowledge about the structures and functions of systems, but also retard action because people may be unwilling to act in the absence of a perfect theory or model.

(9) can provide an orderly and common framework for thinking about social systems and social change, perhaps so orderly and common that alternatives from different points of view might be perceived as unrealistic, unthinkable, totally radical, and incredible merely because they are different.

(10) can stimulate thinking about new polices and programs, or stifle such thought as a result of massive *group-thinking*.

Critical Issues

Anyone constructing social indicators with the aim of integrating them into a social reporting or accounting system to monitor changes in the quality of people's lives will have to address the following thirteen issues, which collectively yield more than 200,000 possible combinations representing at least that many different kinds of systems.

1. *Settlement/aggregation area sizes:* For example, best size to understand air pollution may be different from best size to understand crime.

2. *Time frames:* For example, optimal duration to understand resource depletion may be different from optimal duration to understand impact of sanitation changes.

3. *Population composition:* For example, analyses by language, gender, age, education, ethnic background, and income, among others, may reveal or conceal different things.

4. *Domains of life composition:* For example, different domains such as health, job, family life, and housing give different views and suggest different agendas for action.

5. *Objective versus subjective indicators:* For example, relatively subjective appraisals of housing and neighborhoods by actual dwellers may be very different from relatively objective appraisals by *experts.*

6. *Input versus output indicators:* For example, expenditures on teachers and school facilities may give a very different view of the quality of an education system from that based on student performance on standardized tests.

7. *Measurement scales:* For example, different measures of perceived subjective well-being provide different views of people's well-being and relate differently to other measures.

8. *Report writers:* For example, different stakeholders often have very different views about what is important to monitor and how to evaluate whatever is monitored.

9. *Report readers:* For example, different target audiences need different reporting media and/or formats.

10. *Quality-of-life model:* For example, once indicators are selected, they must be combined or aggregated somehow in order to get a coherent story or view.

11. *Distributions:* For example, because average figures can conceal extraordinary and perhaps unacceptable variation, choices must be made about appropriate representations of distributions.

12. *Distance impacts:* For example, people living in one place may access facilities (hospitals, schools, theatres, museums, and libraries) in many other places at varying distances from their place of residence.

13. *Causal relations:* Prior to intervention, one must know what causes what, which requires relatively mainstream scientific research, which may not be available yet.

In the presence of the potential abuses and the great variety of reports that might be produced as people make different choices regarding the thirteen critical issues, the general rule to be used is to try to have a development process that is maximally inclusive and transparent. William James came close to capturing the appropriate aim in 1891.

> That act must be the best act ... which makes for the *best whole,* in the sense of awakening the least sum of dissatisfactions. In the casuistic scale, therefore, those ideals must be written highest which prevail at the least cost, or by whose realization the least possible number of other ideals are destroyed. ... The course of history is nothing but the story of men's struggles from generation to generation to find the more and more inclusive order. (James 1977, p. 623)

ALEX C. MICHALOS

SEE ALSO *Science and Engineering Indicators.*

BIBLIOGRAPHY

Berger-Schmitt, Regina, and B. Jankowitsch. (1999). *Systems of Social Indicators and Social Reporting: The State of the Art.* Mannheim, Germany: Centre for Survey Research and Methodology (ZUMA).

Dewey, John. (1939). *Theory of Valuation.* Chicago: University of Chicago Press.

Hagerty, Michael R., et al. (2001). "Quality of Life Indexes for National Policy: Review and Agenda for Research." *Social Indicators Research* 55(1): 1–96.

James, William. (1909). *The Meaning of Truth: A Sequel to Pragmatism.* New York: Longmans, Green and Co.

James, William. (1977). "The Moral Philosopher and the Moral Life." In *The Writings of William James: A Comprehensive Edition,* ed. J. J. McDermott. Chicago: University of Chicago Press. Originally published in 1891.

Lewis, Clarence I. (1946). *Analysis of Knowledge and Valuation.* LaSalle, IL: Open Court.

Michalos, Alex C. (1997). "Combining Social, Economic and Environmental Indicators to Measure Sustainable Human Well-Being." *Social Indicators Research* 40(1–2): 221–258.

Michalos, Alex C. (1980–1983). *North American Social Report,* 5 vols. Dordrecht, The Netherlands: D. Reidel.

National Science Board. (2002). *Science and Engineering Indicators.* Washington, DC: Government Printing Office.

Nordhaus, William D., and E. C. Kokkelenberg, eds. (1999). *Nature's Numbers: Expanding the National Economic Accounts to Include the Environment.* Washington, DC: National Academy Press.

Perry, Ralph Barton. (1926). *General Theory of Value.* Cambridge, MA.: Harvard University Press.

Perry, Ralph Barton. (1954). *Realms of Value.* Cambridge, MA: Harvard University Press.

INTERNET RESOURCES

Bentham, Jeremy. (1907 [1789]). *An Introduction to the Principles of Morals and Legislation.* Oxford: Clarendon Press. Available from http://www.econlib.org/library/Bentham/bnthPML.html

London Group of Environmental Accountants. (2002). "SEEA 2000 Revision." Available from http://www4.statcan.ca/citygrp/london/publicrev/pubrev.htm.

SOCIAL INSTITUTIONS: OVERVIEW

• • •

Ethics is involved not only with personal decisions and the assessments of individual behavior but also with social institutions, especially, in the contemporary world, with those institutions constituted by scientific and technical professions as well. Classic sociology—as developed by social scientists considered in entries on "Durkheim, Émile," "Marx, Karl," and "Weber, Max," among others—identified a number of basic social institutions such as the family, religion, state, economy, and education. Social institutions in this sense are defined by persons acting in concert to address distinctive human interests; as such they are characterized by social roles that people accept when acting, for instance, in relation with those to whom they have biological links (the family), in relation to that which is seen as sacred (religion), in relation to the exercise of group power (state), and so on. Each social institution is thus defined by and defines a sphere of human behavior, and the roles woven into these institutions traditionally constitute both descriptive or empirical (and in this sense scientific) and prescriptive or normative (and thus ethical) phenomena. Roles both describe and prescribe human behavior within the contexts of social institutions.

Science and technology, while acquiring the status of social institutions, have likewise influenced and altered other social institutions and social roles in at least three overlapping ways. First, technological change over the long sweep of human history has shifted the relative weights or balances between different roles. For thousands of years, during the preliterate period of human history, when humans were primarily hunters and gatherers, the institution of the family occupied the dominant position with only the most modest autonomy granted to religion and even less to those activities now associated with the state, economy, and education. With the domestication of plants and animals, however, divisions of labor arose that in turn gave rise and increasing prominence to religion, state, economy, and education, while also transforming the institution of family (as is considered, for example, in the entry on "Family").

Second, over the course of written history science or the systematic pursuit of knowledge in its various permutations altered fundamental ideas about these basic social institutions and their justifications. Mythical narratives of the gods and relations between gods and humans as the original behavior patterns to be differentially imitated by different social institutions were supplemented by accounts that appealed to patterns in nature. The science of nature slowly introduced alternative understandings of social institutions and social roles, as can be seen, for instance, in Plato's *Republic*, with its rational account of the need for myths or likely stories about the differences between the social roles of the three basic classes (or social institutions) of artisans, soldiers, and rulers.

Finally, in the modern period, new unifications of science and technology in both the "Scientific Revolution" (sixteenth century) and the "Industrial Revolution" (eighteenth century) intensified the proliferation of social institutions and social roles through the development of scientific disciplines and industrial divisions of labor. These historical changes altered anew the balances between institutions (giving both science and economy, for instance, a weight previously unknown in human history), granted each institution more autonomy or independence, and ultimately relativized the power of particular social roles through their very proliferation. Beginning in the second half of the twentieth century, the growing multiplicity and complexity of roles began to be linked and networked in synchronic hybrids of interdisciplinarity and diachronic career changes. (Entries on "Education" and "Interdisciplinarity" are especially relevant in regard to such changes.)

Beyond entries already mentioned, others in the *Encyclopedia of Science, Technology, and Ethics* break out social institution–related issues in different ways. The perspective of the basic institution of religion finds expression in a series of entries on "Buddhist Perspectives," "Christian Perspectives," "Hindu Perspectives," and more. The basic institution of the state is engaged with entries on "International Affairs," "Military Ethics," "Police," "Science Policy," and "Science, Technology, and Law." Entries on such basic social institutions are complemented by ones on more fine-grained social organizations and agencies (professional societies such as the "American Association for the

Advancement of Science"), on related processes (such the emergence of "Professions and Professionalization"), and on ethical questions that repeatedly challenge and are challenged by social institutions (such as "Justice").

CARL MITCHAM

SEE ALSO *Aristotle and Aristotelianism; Bell, Daniel; Civil Society; Ethics: Overview; Modernization; Nongovernmental Organizations; Plato; Polanyi, Karl; Professional Engineering Organizations; Regulation and Regulatory Agencies; Science, Technology, and Society Studies; Work.*

SOCIALISM

•••

Socialism has been one of the most popular political ideas in history, rivaling in some ways even the great religions. By the late 1970s, a mere 150 years from the time the term *socialism* was coined, roughly 60 percent of the world population was living under governments that called themselves "socialist," although these varied widely in their institutions and were often violently at odds with one another.

Socialism drew impetus from the rise of industry in Europe in the nineteenth century. The new wealth generated by new methods of production encouraged the belief that now it would be possible to assure a comfortable standard of living for every member of society. The uneven distribution of this new wealth was seen to pose ethical questions that were less often asked about long-entrenched class disparities prevalent in the countryside. Socialism was seen by many of its advocates as not only an ethical but also a scientific response to these new circumstances. Drawing on the Enlightenment critique of religion, socialism offered an image of the ideal life as something to be achieved in the here and now rather than in the great beyond.

Five Types of Socialism

The myriad forms of socialism that were actually put into practice might be grouped into five broad categories: communism, social democracy, Third World socialism, fascism, and communal socialism. (There were others, such as anarcho-syndicalism, that remained forever in the realm of speculative thought.) Each of these five requires a note of explication.

In the early decades of socialist thought the terms *socialism* and *communism* were often used interchange-

ably, and while some writers attempted to define the distinction between the two, no such distinction ever achieved widespread acceptance. When Vladimir Ilich Lenin (1870–1924) led his group of Bolsheviks to power in Russia in 1917, he announced that they would henceforth call themselves communists. Until then, they had been merely the *bolshevik* (meaning majority) segment of Russia's Social-Democratic movement. (This had been a single party, at least formally, until 1912, when Lenin's faction announced it was a party in itself. Still they were all social democrats.)

In the years following 1917, as parties modeled after Lenin's appeared in dozens of countries, a clear distinction emerged between social democracy and communism. There were countless points of dispute and differences, but probably the most profound was that social democrats sought parliamentary means to power and adhered to the principle that political systems should have multiple parties, whereas communists envisioned a revolutionary path to power and believed that communist parties, as the only true representatives of the working class, were the only legitimate ones. This made for such a wide gulf that thereafter social democrats never called themselves communists, and communists never called themselves social democrats. The distinction, however, continued to be clouded by the fact that both sides claimed the term *socialism* for themselves. Thus the country Lenin created was called the Union of Soviet Socialist Republics, and at the same time the international federation that brought together the world's social democratic parties (the British Labour Party, the German Social Democrats, etc.) called itself the Socialist International.

Third World socialism is a loose category comprising "African socialism," "Arab socialism," and various cognate forms that appeared elsewhere in poorer countries after World War II. These were usually dictatorial in their political practice (although not in all cases: India offers a dramatic counterexample), but rarely was the state as all controlling as in communist systems. Some of these states (for example, Tanzania under Julius Nyerere [1922–1999]) elaborated complex blueprints of economic development, whereas in others "socialism" probably served as little more than a popular label for a hodgepodge of policies of a military dictator or a rationalization for strengthening the power of the central government (for example, Somalia under Mohammed Siad Barre [c. 1919–1995]).

To include fascism as a subset of socialism invites controversy because fascist movements often made their appeal on the promise to protect society from socialists or communists, because they were almost always part of the Right (whatever that may mean) rather than the Left, and because their inclusion may be taken as a polemical device to tar socialism with the odium attached to fascism. Yet the historical basis for their inclusion is strong. Adolf Hitler's party called itself National Socialist (as did some similar groupings in other countries, such as Hungary, some of which thought up the name independently of, and even prior to, Hitler). In Italy, Benito Mussolini formed his fascist movement as a leftist pro-war breakaway from the Socialist Party, of which he was a top leader. Each of these movements attempted to retain some of the elements of socialism while substituting the nation (or in Hitler's case the German *volk*) for the working class that had been seen as the main engine and beneficiary of socialism in traditional theory. Once in power, both Mussolini's party and Hitler's continued to preserve some of the accouterments of their socialist heritage. Mussolini himself probably captured best the relationship between these isms when he declared that fascism was a "heresy" of socialism, suggesting something that had sprung from the same premise but turned to challenge some of socialism's integral tenets.

Communal socialists differ from all the others in that they do not focus on trying to gain power (whether by vote or violence) in order to establish a socialist system over an entire country. Rather they are groups of individuals whose primary goal is to live a socialist life themselves by organizing communities operating on socialist principles. (No doubt many commune members also hope that their example might inspire emulation.) Usually such communities have numbered a few hundred members, although some have measured only in the tens and others in the low thousands. In the United States, a few hundred such societies were founded over the course of the nineteenth century, some by people whose driving belief was socialism, per se, others by devotees of religious sects, such as Shakers, for whom sharing property was but a facet of their sense of spirituality. Israeli kibbutzim are another important example of this form.

Historical Origins

Except for the communal, all of these forms grew from the same acorn: the French Revolution of 1789, with its ethos of "liberty, equality, fraternity." Although the Revolution itself did not aim for socialism, and although the term socialism was not coined until decades later, it was in pursuit of this inspiring triad of goals that socialism came to be conceived and then popularized. How

can there be equality, it was asked, with vast disparities between rich and poor? How can there be brotherhood in a context of heartless economic competition? How can there be liberty if most people are enslaved to material necessity?

These questions presented themselves with greater urgency as the Industrial Revolution took hold. Although the poor of the factories were not poorer than the poor of the farms, their poverty, concentrated in urban slums, was more visible. Moreover, the Industrial Revolution entailed new ills such as industrial accidents and work environments devastating to human health. The labor of young children in factories offered a spectacle more heartbreaking than work of children on farms, which seemed a natural part of rural life from time immemorial.

The solution, it was argued, was to be found in collective ownership of property and the egalitarian distribution of the goods of society. These twin principles were to remain at the heart of socialism, although each of them, as well as many lesser points of doctrine, were to be disputed, refined, and amended repeatedly. Collective ownership in an individual commune was easy to envision. Collective ownership of the economic assets of an entire society was more difficult to conceptualize. It might mean ownership by the central government, but in other versions it might mean something less centralized—for example, that individual enterprises would be owned by the people who worked in them or by local communities. Egalitarian distribution did not necessarily mean exactly equal shares. The most fetching socialist slogan was "from each according to his abilities; to each according to his needs," which implied a measure of inequality but raised the question of how such needs would be determined. In Israeli kibbutzim, one place where an earnest effort was made to implement this principle, special committees existed to which kibbutz members could bring their special needs or abilities (a medical condition, an artistic calling, a family emergency abroad that required travel, and the like), and these committees were empowered to distribute resources accordingly.

Relation to Science and Technology

The connection between socialism and science originates in the claim of Karl Marx (1818–1883) and Friedrich Engels (1820–1895), the most influential of all socialist thinkers, to have discovered "scientific socialism." By this they meant to distinguish themselves from such early-nineteenth- (or in a few cases, late-eighteenth-) century visionaries as Henri de Saint-Simon

(1760–1825), Robert Owen (1771–1858), Charles Fourier (1772–1837), and Étienne Cabet (1788–1856), who had inspired the founding of various communes. Marx and Engels ridiculed the idea that a group of individuals could move the world toward socialism by creating model communities to demonstrate socialism's benefits. They saw this as naive because they doubted that political forms or even political ideas emerged simply from the free play of the human mind. To believe this, they said, is to be "utopian."

This term *utopian* itself is misleading because Marx and Engels were not objecting to the fancifulness of some of the early socialist visions. (Fourier's socialism, for example, envisioned that lions and whales would be tamed so as to free humans from physical labor and that each citizen would be entitled not only to a "social minimum" of economic rewards but also a "sexual minimum" of carnal satisfaction.) The fleeting glimpses Marx and Engels offered of life under socialism were pretty idyllic in themselves: People would do only those activities that they find intrinsically gratifying, say, hunting in the morning, fishing in the afternoon, writing poetry in the evening. Rather, what Marx and Engels found unrealistic, hence "utopian," about the earlier thinkers were their ideas about how socialism could be brought about. "Life is not determined by consciousness but consciousness by life," they wrote (*The German Ideology* part 1A,1845).

What they meant by this was that socialism could not come about until the objective conditions—which meant a certain level of wealth and technology—were right. Nor would it be brought about by individuals who happened upon the idea of socialism through reading or contemplation; rather its engineers would be people impelled to fight for socialism by the very conditions of their daily lives. Specifically, they held that socialism had not been possible in rural society but that the advent of industrialization laid open a new era. For one thing, the new technologies generated unprecedented abundance, making it possible for every member of society to enjoy a high standard of living. (Of course, what seemed a high standard in 1850 would be considered quite low by twenty-first-century standards, a wry comment perhaps on the elasticity of human need.) Moreover, the character of industrial production, depending on highly collective human effort, was conducive to collective ownership, making socialism a natural choice.

For the first time, because of this change, socialism had become a realistic possibility. Indeed its appearance had become likely, perhaps even inevitable. This was because industrialization brought the flowering of capitalism. Capitalist competition forced manufacturers to cut costs, including labor costs, thus driving down rates of pay. As a result, the very individuals whose sweat was providing the new abundance were left with too little income to share in it themselves. Eventually, driven in part by a sense of injustice but even more by the whip of destitution, they would rise up to abolish the system of private capitalism and replace it with socialism. This would not be because anyone had persuaded or taught them to do so but because bitter circumstances would impel them to do it.

In sum, Marx and Engels believed that they had discovered the processes that drive social and political change, and that these were rooted in the march of technology rather than in anything as arbitrary as individual will or cognition. They believed that this revelation of the laws of social evolution was analogous to the recent sensational revelation of the principle of the evolution of species. As Engels put it in his graveside eulogy to Marx in 1883: "Just as Darwin discovered the law of development of organic nature, so Marx discovered the law of development of human history."

Relation to Science and Ethics

Science, itself, as it is now understood, was not as clearly demarcated in their time, and from the perspective of the early twenty-first century it is easy to see the flaws in Marx and Engels's claims to science. To start with, they did an injustice to those they invidiously compared to themselves as "utopian." Owen among others also considered himself a man of science. Like Marx and Engels, Owen sought to draw generalizations about human behavior from his observations. His most cherished belief was that persons' characters are formed by the circumstances of their lives rather than by inner moral convictions or any other factors that they can control themselves. This notion, that one's thoughts and actions are shaped by forces larger than oneself, is very akin to Marx and Engels's central scientific claim and anticipated them by a full generation.

Moreover, in their approach to socialism, a good case can be made that the "utopians" were more scientific than Marx and Engels. Having hit on the idea that socialism would furnish a cure for society's ills, they set out to demonstrate its efficacy by attempting socialist experiments. Insofar as experimentation lies at the heart of the scientific method, the "utopians" were more genuinely "scientific socialists" than Marx and Engels, who discounted any such attempt. The latter duo claimed they could see where history was heading, but it is hard to imagine how this counts as more scientific than any other exercise in prophecy.

Beyond the absence of experimental method, Marx and Engels never stated any testable proposition nor did they betray any doubts inspired by the failure of specific details of their prophesies. Tellingly, they never treated their own forecasts as if they did amount to "science," at least as the term has come to be understood. As the decades passed they poured forth an endless stream of commentary, much of it arresting, on unfolding political events. But they rarely displayed any sense of needing to examine whether and in what way these new events comported with their larger theories. That is, they conducted themselves as what today are sometimes called "public intellectuals" or as activists, not as people who thought of themselves as scientists.

Still, it is difficult to dismiss Marx and Engels's claim to "science" without conceding that their method of attempting to distill systematic generalizations from the study of contemporary history constitutes a main building block of contemporary social science. There may be room to debate about how "scientific" social science is, falling as short as it does from the methodological rigor of "hard science," but insofar as its scientific legitimacy is accepted, then Marx and Engels must be given credit as pioneers, however imperfect their methodology.

In terms of its relationship to ethics, socialism presents an ambiguous picture. By claiming that they were doing no more than divining historical laws that showed that socialism was due to triumph, Marx and Engels shifted the argument in favor of socialism from the realm of "ought" to "is" (or, more precisely, to "will be"). And they specifically denied the possibility of absolute or universal moral principles, as opposed to principles that merely served the interests of a particular class. "Law, morality, religion are to [the proletarian] so many bourgeois prejudices, behind which lurk in ambush just as many bourgeois interests," wrote Marx and Engels in *The Communist Manifesto* (1848).

At the same time, it would be hard to deny that the force of Marx and Engels's indictment of capitalism is the sense of moral indignation that flows through it. Despite their own militant atheism, they decried capitalism as a system under which "all that is holy is profaned." A similar ambiguity can be found in various non-Marxist socialists. To take Owen, his fervent assertions that people's characters were molded for them seemed to negate any sense of moral responsibility. Yet he was very interested in discovering methods to mold characters to some kind of proper moral standard. He was for this reason a pioneer in early childhood educa-

tion, and the organization of his followers in the 1830s called itself the Society for the New Moral World.

Owen, like Marx and Engels, was a vituperative opponent of revealed religion. (They called it an "opiate"; he called it one of the "three great evils" afflicting humanity.) In contrast, however, there have always been some religious socialists. As already mentioned, various socialist communes rested on religious bases, and a broader movement of Christian socialism made a strong appearance during the twentieth century. These adherents saw socialism as an expression of the biblical precept to love thy neighbor as thyself and of the Christian emphasis on spiritual rather than material values.

This points toward another aspect of the ambiguity of the relationship between socialism and ethics. On the one hand, socialist ideas aim to create a society that will fulfill certain moral goals, such as liberty, equality, and brotherhood. On the other hand, the emphasis on politics and policy has meant that many socialists have made little use of traditional notions of individual moral agency. The socialists who have most fully avoided this dilemma are the communal socialists who aim to carry out socialism in their own lives rather than to engineer larger political changes.

Their great emphasis on improving the world through political and economic changes rather than uplifting individual behavior has also brought socialists into a fraught confrontation with the question of whether, or to what extent, ends justify means. In the main, communists (as well, of course, as fascists, if one counts them under the socialist umbrella) have been ruthless in their means and ruthless in justifying this. As Leon Trotsky (1879–1940) once put it: "Only that which prepares the complete and final overthrow of imperialist bestiality is moral, and nothing else. The welfare of the revolution—that is the supreme law!" ("The Moralists and Sycophants Against Marxism," essay in his *Their Morals and Ours* [1936])

Social democrats and other noncommunist socialists have ordinarily rejected such claims, and they have often chastised the communists on moral grounds for their deceptive or violent tactics. Yet the force of such condemnations in intrasocialist debates was often vitiated by the emphasis on social change as the preeminent path to improving the world. If social change bulks so much larger than individual behavior, then might not unsavory tactics be justified in pursuit of the necessary policies?

The Legacy of Socialism

By the twenty-first century, much of the body of socialism has wasted away. Fascism, if it ever deserved to be counted here, is little more than a grim memory—although the term continues to be applied to various violent authoritarian movements. Communism has disappeared from the large majority of once-communist states. The remaining communist states all seem either to be following China in gradually shedding their distinctly communist features or to be living on borrowed time, awaiting the demise of a powerful dictator. Communal socialist societies are few and far between. Even their most triumphant exemplars, the kibbutzim, have mostly transformed themselves into miniature market economies.

What remains strong, however, is the legacy of social democracy. Social democratic parties justly claim most of the credit for various forms of worker protection and a wide variety of services and benefits that every developed democratic society provides. And these parties continue as powerful forces throughout the democratic world. None of them aim any longer to displace capitalism; rather their program is to continue to tame or modify it. Although markets have, to most minds, proven their superiority over the socialist dream of "economic planning," there still are social values—protection of the weak or of the environment or the provision of certain public services, for example—that unfettered markets do not serve. Social democracy has found an enduring niche as the advocate of these values—which have been put into practice through such programs as social security and socialized medicine.

If this is a dilute residue of socialism, so, too, do the scientific and ethical issues that have long surrounded socialism endure in dilute form. Contemporary protests against "globalization" echo earlier ones against capitalism itself. While there are few remaining believers in "scientific socialism" or in Marx and Engels's economic determinism, the question of the degree to which individual behavior should be attributed to free will as opposed to external or biological influences continues to be hotly debated in such policy areas as criminal justice and the rights of homosexuals. And the deep discourse over whether it is more efficacious to improve society by uplifting individuals or to improve individuals by reforming the society seems certain to endure.

JOSHUA MURAVCHIK

SEE ALSO *Arendt, Hannah; Communism; Critical Social Theory; Fascism; Marxism; Marx, Karl; Mondragón Cooperative Corporation; Morris, William.*

BIBLIOGRAPHY

Harrington, Michael. (1970). *Socialism*. New York: Bantam. A brief for socialism by the leading American socialist writer of the late twentieth century, who was both a social democrat and a Marxist.

Kolakowski, Leszek. (1978). *Main Currents of Marxism*, trans. P. S. Falla. 3 vols. Oxford: Clarendon Press. A comprehensive account of the many varieties of thought flowing from socialism's most influential thinker. Includes also Marx's antecedents.

Lichtheim, George. (1970). *A Short History of Socialism*. New York: Praeger. A concise history of Western socialism by a leading European intellectual of socialist bent.

Muravchik, Joshua. (2002). *Heaven on Earth: The Rise and Fall of Socialism*. San Francisco: Encounter Books. An overview of the history of socialism in all its forms told through biographical sketches of its key figures.

SOCIAL THEORY OF SCIENCE AND TECHNOLOGY

• • •

The idea of social theories of science and technology initially seems counterintuitive, because commonsense notions of science and technology separate them from the social world, and place them instead into the world of nature and fact. But closer scrutiny reveals a number of relevant aspects of social theory that can assist in understanding the development of science and technology, and the ethical and political aspects of such changes.

Social Theory: Scale, Structure, Agency, and Critique

Social theory is a body of scholarly work that describes and explains the social world. While ordinary people use workable models of social interaction and causality to get through the day, these folk sociologies, psychologies, and economic theories are not carefully articulated as testable models, and are often limited in scale and scope.

The idea of scale—or the size, duration, and level of complexity at which phenomena occur—is one of the first dimensions of variation in all social theory. One expects, and finds, different mechanisms and patterns to explain the behavior of small groups in comparison with large, complex societies. The disciplines themselves mirror this issue of scale, in which psychology, for example, is mostly concerned with individuals and small-group processes while sociology, anthropology, or economics examine the behaviors of whole populations or cultures. Moreover within each discipline of the social sciences and humanities are specialties that focus on different scales or levels of analysis. In sociology, this is the distinction between micro- and macro-sociologies, between models of small-group interactions and explanations of whole social systems.

With distinctions based on issues of scale, questions arise concerning scope—to articulate models appropriate at one scale with those of a larger or smaller level of analysis, or of a longer or shorter duration in time. What are the relationships among small groups and larger social institutions? How do social forces, historical trends, and cultural formations impact individuals? This remains a challenge for interdisciplinary social theory, and points to a related set of questions regarding the relationship between individual agency and social structure as well as relations between ethics and politics. How, and in what ways, are individual thoughts and actions, including ethical assessments, influenced by preexisting cultural, social, and economic conditions? If individual actions are strongly determined by social structure, where does social, scientific, or technological innovation come from—not to mention ethical criticism? If individuals freely innovate and criticize, why do social structures and belief systems persist over time? Issues of scale, structure, and agency link very closely to long-standing issues in the study of science and technology, particularly concerning questions about the balance between society determining technology (social constructivism) and technology determining society (technological determination)

The issue of social criticism is particularly important to science, technology, and ethics. Much social theory includes some assessment (positive or negative) of the social world. For example, Karl Marx (1818–1883) articulated his theory of the means of production determining the social structure and belief system of a society, while witnessing the devastating poverty of rapid industrialization and urbanization in Manchester, England in the mid-nineteenth century. Twenty-first-century authors are concerned with an array of issues, such as explaining new technologies and their effects on indigenous cultures, often with an implied concern that these societies are threatened by technological change. Others focus on the way common work and language practices of science shape how experiments are conceived and interpreted, or how social power influences what research is prioritized for funding. Focusing on how technology affects work and employment often leads to concern with systems of wealth and social stratification, with the unequal distribution of goods and harms. Social theory, then, always intersects with the political and ethical sides of science and technology because it is concerned ultimately with the human dimensions, both causes and consequences, of change.

Approaches to Science and Technology Studies

Science and technology studies, like economic theory, can be read as an argument with the ghost of Marx. In his voluminous writings, Marx articulated a model of the constitution of society literally from the ground up. In this model, the productive relationships of a society, meaning economy and agriculture, determined the basic social organization, in terms of classes and the structure of the state. Society then determined the cultural formations and basic ideologies, including science as an explanatory system. This model implies a degree of technological determinism in which social relations are determined by technology. The first generations of scholars concerned with science and technology wrestled with this issue, with Lewis Mumford (1895–1990), Jaques Ellul (1912–1994), and Ivan Illich (1926–2002) leading the way in developing critical theories of contemporary society adopting and criticizing Marx's insights. In the early-twenty-first century, Langdon Winner (1986, 1977) continued this tradition.

Focused on science, Robert Merton was also influenced by another founding social scientist, Max Weber (1864–1920), to formulate a theory of science as a modern institution based on the Protestant work ethic and the development of capitalist economic systems. Merton's *normative structure of science* articulated formally what had been a set of assumptions and values governing science as it emerged in sixteenth-century Europe. The values of *communalism*, in which knowledge is to be shared; *disinterestedness*, against personal or economic gain from knowledge acquisition; *universalism*, in which the identity of the author of scientific statements is not to be taken into account; and *organized skepticism* to provide the mechanisms for self-correction in science continue to be upheld and are presented to science and technology students as the primal values governing good science. Writing in the mid-twentieth century, Merton was concerned with demonstrating that democracy needed science, and science needed democracy, to avoid the distortions of Stalinist and Nazi influence he saw occurring early in his career.

Scholarship on science and technology struggled, however, with whether or not the social structure affected merely the social organization of these activities or the content and details of scientific and technological change as well. Within historical scholarship on technology this led to two major streams of thought: the internalist, which focused on the internal logic of development, seeing it as resistant to all social influences, and the externalist, which focused on the pervasiveness of social influences and impacts on scientific and technological change. This parallels questions of whether internal professional ethics or external political pressures should be granted priority in the governing of science and engineering.

Toward the latter third of the twentieth century the opposition between social and technological determinism was partially resolved with the development of the social construction conjecture. Social construction is based, in part, on insights derived from Thomas Kuhn's (1962) work in the history and philosophy of science, especially his notion of paradigm and paradigm shift, which spread quickly through the scholarly world, influencing studies of both science and technology. Focus on moments of change and controversy allowed scholars to see how both the social and natural are always present in shaping science and technology. The first generation of scholarship (Mulkay and Knorr-Cetina 1983) articulated what would come to be called the empirical program of relativism that generated the *symmetry principle*, which proposes that both true and false beliefs should be amenable to the same kind of social analysis. (In the past, true propositions were explained as reflecting the way nature is, false ones as reflecting the distorting interests of scientists or society.) Symmetry models have been further refined over time, for example by scholars such as Bruno Latour, who with colleague Steve Woolgar articulated the term *technoscience* to represent the confluence of technology and science as organized ways of interacting with the material world.

Technology studies applied these insights in its own way, and the editors of *The Social Construction of Technological Systems* (1989) presented a collection of works for what would become the *SCOT* model. This model describes how the *working* of a technology is primarily dependent on the social processes leading to its manufacture and the decisions of various end user groups as to whether or not it meets their needs as they decide how to employ the new technology. A technology whose material parts are in functioning order may still, and is often, deemed to be *not working* or a failure because it does not meet people's needs. In effect this appeared to constitute an ethical and political assessment of the adequacy of the status quo, a position criticized by Winner (1993) and generating further scholarly discussion.

What the initial constructivist studies of science and technology focused on was the microsocial processes of laboratory and workbench activities, such as the socially-grounded work of the interpretation of experiments. Negotiations among different groups in the design processes followed quickly, eventually moving up in scale to study organizational and bureaucratic contexts for generating models of change. Studies of cultural ideas, language, and values can generate explanations for the general trends of development in science and technology, but not the strong causal explanations aspired to by prior generations of scholars. Despite the advantages of having concrete artifacts and well-defined scientific ideas to trace, the shift from context to context and across different scales of social action remains challenging for social theorists of science and technology.

Similarly the question of determinism and the relationship between individual agency and social structure still challenge explanatory models. Rather than strong causal laws, heuristics outlining the applicability of models and propositions guide social studies of science and technology. For example, while a strongly deterministic model of the origins of new science and technology cannot be true, because that would be to ignore all evidence of the work, politics, and economic choice leading up to the new technoscience, it often *feels* true to consumers of science and technology to whom all of those prior social relations are invisible. Wiebe Bijker (1997) has developed a theory that helps to explain this by noting that people with low inclusion in the construction process often face a *take-it-or-leave-it* choice with new science and technology. Technoscience seems determined, to them, while those with high inclusion in the process see much of the construction. This interpretation of the construction of technoscience raises important ethical and political issues related to levels of participation in scientific and technological processes.

Indeed the roles of end users and stakeholders in science and technology have gained increased attention, in research on the public understanding of science, vernacular design, and consumer analyses. In the first instance, it has been pointed out that users are strongly dependent on technological *scripts*—that is, cultural and behavioral frameworks for understanding and interacting with technologies (Bijker and Law 1992). But users also create opportunities to rewrite scripts, and to modify not only the meaning, but the materiality and affordances of new technologies. End users can be creative appropriators of technology: "Low-riders" are transformations of automobile suspension systems by Hispanic urban culture for cultural self-expression; artisans use old tools in new ways to produce new effects; cell phones can be used to organize "smart mobs" and synchronize political action.

Contemporary Issues and Elaborations of the State of the Art

John Staudenmaier (1989) has cataloged the major historical themes in the history of technology since the inception of the Society for the History of Technology (1958), such as work and labor, military, aerospace, and gender. Recent scholarship on science and technology continues and expands these topics. For example, technology, labor, and work receive attention from sociologists such as Steven Vallas (2001), particularly in the roles that information technology and computerization have in different kinds of industries and organizations. Older models of technology, as always deskilling workers and centralizing power in organizational leadership, have given way to more nuanced models of context- and work-dependent implementations of new technology.

Computerization has become a major topic in social theories of technology. Much work is focused on the emergence of information and telecommunication technologies, their contexts of production, and the impacts of their use and adoption. A second, also revolutionary area of inquiry is the transformation of the life sciences, producing the emerging biotechnology industry, in which distinctions between pure and applied research or fundamental understanding of life processes and product development are increasingly blurred. These two areas come together in interesting ways in *cyborg theory*. Developed by Donna Haraway (1991), this is the treatment of human beings and the material world as interconnected and interdependent, with humans seen as biological, social, and information-based beings that obscure traditional boundaries between nature and culture, human and machine.

Some level of constructivism in both science and technology is well-argued consensus within the field, although its counter-intuitive elements often provoke commentary and criticism from those outside the social studies of science and technology. Finer distinctions among models and theories have been generated, for example between SCOT and its sibling, actor-network-theory (Law and Hassard 1999). Actor-network theory analyzes the networks of humans and material objects to generate specific explanations for the success or failure of ideas or artifacts. It is perhaps a methodology rather than a theory, per se, but nonetheless has value in gen-

erating detailed analysis of the various components of technoscientific projects. Such a method may also offer resources for analyzing the influence or failure of various ethical or political responses to technoscience.

Various social movements have picked up insights from social theories of science and technology. A first heuristic derived from constructivism is that things might have been otherwise. Designs could have turned out differently; the pursuit of scientific knowledge prioritized on different values would lead in new directions.

A second heuristic is that scientific and technological change generally follows the lines of power and resources already prevalent within a society. This does not mean that technoscience cannot have revolutionary effects on social relations, but that it is more likely that people will use technoscience to attempt to preserve power and privilege that already exists.

From these insights, environmentalists, social justice organizations such as feminist and anti-racist groups, and critics of development and globalization can make better informed interventions in the formulation, conduct, and effects of scientific and technological change. Feminists and racial or ethnic minorities, for example, point to the potential benefits of increasing the diversity of formal scientific and technological involvement because diverse backgrounds can be resources for new ideas, and for different values to motivate practice. They also point to the inventive and problem-solving activities of ordinary people, and take into account the moral and cultural values that might have bearing on the products of technoscience and their consequences for diverse communities.

Environmentalists point to the unequal distribution of the harms of technoscience, for example that poor communities and nations often face far greater harm from industrial pollution, and conduct research to help ameliorate those problems. There is also an evident tension between improving the economic and health circumstances of people in non-industrialized countries and preserving important ecological and cultural configurations. Social theories of science and technology may help anticipate the related consequences of technological change, and design interventions to minimize their negative outcomes.

Formal policy-making has taken up social theories of science and technology unevenly. One of the most concise models of science and society from a policy perspective is indirectly informed by social theories of technoscience. Backing away from a traditional linear model that privileges basic research leading directly to development and application, Donald Stokes (1997) proposes a more complex model in which different kinds of technoscientific problem formulation and research processes are supported and managed in different ways.

Whether broadly or narrowly defined, social theories of science and technology are as dynamic as technoscientific change itself. The connection is both strength and weakness. There is always a lot to do; new questions emerge daily. But there are too few resources or people to do all the work. Cutting edge analysis of technoscience easily becomes a quaint historical account of a forgotten technology or discredited science. More seriously, with rapid change and diverse topics, it is often difficult to see commonalities across fields of inquiry, and to develop generalizations about scientific and technological processes that are independent of specific contexts and thus subject to general ethical assessment. Integrating research across different scales of interaction, from individuals and identity formation processes to macroeconomic changes in global economic activity, remains a daunting task for all forms of social theory.

The final challenge for social theories of science and technology is one faced by all disciplines: to remain relevant to a diverse public audience and policy professionals. All disciplines face the possibility of becoming too focused on internal, scholastic issues, rather than seeking to develop broad heuristics that can be of benefit to those seeking to understand the important questions all social theories engage: How do I know? Why did this happen? Is it a good thing? What can be done about it?

JENNIFER L. CROISSANT

SEE ALSO Autonomous Technology; Ellul, Jacques; Illich, Ivan; Kuhn, Thomas; Marx, Karl; Merton, Robert; Mumford, Lewis.

BIBLIOGRAPHY

Bijker, Wiebe. (1997). Of Bicycles, Bakelites and Bulbs: Toward a Theory of Technological Change. Cambridge, MA: MIT Press.

Bijker, Wiebe, and John Law, eds. (1992). Shaping Technology, Building Society: Studies in Sociotechnical Change. Cambridge, MA: MIT Press.

Bijker, Wiebe; Thomas P. Hughes; and Trevor Pinch, eds. (1989). The Social Construction of Technological Systems: New Directions in the Sociology and History of Technology. Cambridge, MA: MIT Press.

Haraway, Donna. (1991). Simians, Cyborgs, and Women: The Reinvention of Nature. New York: Routledge.

Kuhn, Thomas. (1962). The Structure of Scientific Revolution. Chicago: University of Chicago Press.

Latour, Bruno, and Steve Woolgar. (1986). Laboratory Life: The Construction of Scientific Facts. Princeton, NJ: Princeton University Press.

Law, John, and John Hassard, eds. (1999). Actor-Network Theory and After. London: Blackwell.

Mulkay, Michael, and Karin Knorr-Cetina, eds. (1983). Science Observed: Perspectives on the Social Study of Science. Thousand Oaks, CA: Sage Publications.

Staudenmaier, John. (1989). Technology's Storytellers: Reweaving the Human Fabric. Cambridge, MA: MIT Press. A basic history of the history of technology as represented by the Society for the History of Technology and its journal Technology and Culture.

Stokes, Donald. (1997). Pasteur's Quadrant: Basic Science and Technological Innovation. New York: The Brookings Institute.

Vallas, Steven. (2001). The Transformation of Work. Greenwich, CT: JAI Press.

Winner, Langdon. (1977). Autonomous Technology: Technics-out-of-Control as a Theme in Political Thought. Cambridge, MA: MIT. Press.

Winner, Langdon. (1986). The Whale and the Reactor: A Search for Limits in an Age of High Technology. Chicago: University of Chicago Press.

Winner, Landgon. (1993). "Upon Opening the Black Box and Finding it Empty: Social Constructivism and the Philosophy of Technology." Science, Technology and Human Values 18(3): 362–378.

SOCIOBIOLOGY

• • •

Sociobiology denotes the attempt to provide a biological explanation for the social behavior of animals, including humans, although the focus is more often on social insects such as ants and honey bees. Because ethics is also concerned with social behavior among human beings, achievements in sociobiology may also have implications for a possible science of ethics.

The Darwinian Background

As a term the word sociobiology first appears in Principles of Animal Ecology (1949) by Warder C. Allee, Alfred E. Emerson, et al., but the subject matter is much older. In On the Origin of Species (1859), Charles Darwin argued that there is constant population pressure brought on by the fact that numbers of organisms always outstrip food and other resources. There is therefore a constant struggle for existence. Some organisms have features enabling them to better succeed in the struggle, and thus there is a natural selection of the winners over the losers. This leads to evolution, but evolution of a special kind. Selection produces and perfects features useful in the struggle—organisms have adaptations such as the hand and the eye that aid them in survival (and, even more importantly, to reproduce).

Darwin realized that behavior is as much part of an animal's repertoire in the struggle for existence as are any physical adaptations. He was particularly interested in social behavior such as that of the hymenoptera (the ants, bees, and wasps). His interest was spurred not only by the phenomenon itself but because (in Darwin's opinion) such behavior seems to go against the workings of selection. Darwin believed that the struggle for existence pits every organism against every other organism, and hence selection can only promote adaptations that are valuable to the individual. (In contemporary language, Darwin was an individual selectionist rather than a group selectionist.) How then do organisms develop social features that seem to help the nest, perhaps even at the cost of total sacrifice of the interests of the individual? Sterile workers apparently spend their whole lives looking to the needs of their mothers and siblings. Eventually Darwin came to believe that the nests of social insects should be regarded as one large superorganism, rather than a group of individuals working together. In that way, the individuals in a nest are more parts of the whole (like the heart and liver are parts of the human body) rather than organisms existing in their own right with their own interests.

For a number of reasons, in the century after *On the Origin of Species* was published, the study of behavior by biologists lagged behind other areas of evolution. First behavior is much more difficult to record and measure than are physical characteristics. Experimentation is particularly difficult, for it is notoriously true that animals change their behaviors in artificial conditions. Secondly practitioners of the new social sciences thought that they exclusively should examine behavior, and that biology had no place in their endeavors. Unfortunately there existed a strong ideology that experience and training are the cause of most, if not all, behavior, and hence evolutionary factors tended to be discounted before any research was done. Continental students of behavior known as ethologists were a notable exception to this indifference to evolutionary theory, although their work was (as judged by twenty-first century standards) hampered by unjustified assumptions about the significance of group selection.

Breakthroughs in the 1960s

Major breakthroughs occurred in the 1960s, due, in large part, to the work of William Hamilton (1936–2000) in England. Promoting the theory now known as *kin selec-*

tion, Hamilton, then a graduate student, pointed out that in modern terms, selection is equivalent to passing on a particular individual's genes (or rather copies of those genes) more effectively than competitors. However when a person's close relatives reproduce, because they share copies of that person's genes, they also pass on those same copies: reproduction by proxy as it were. Normally it is biologically most efficient to reproduce oneself because (except for identical twins) an individual cannot be genetically more closely related to any other being. Hamilton argued there are some exceptions to this general rule. The hymenoptera particularly have an unusual reproductive system, with females having both mothers and fathers and males having only mothers. Queens get all the sperm they will ever use on the nuptial flight. To produce a female, the queen releases a sperm; in contrast, in producing male offspring, no sperm is released. Thus sisters (50% $\male$ + 50% $\female$ x ½ = 75%) are more closely related than mothers and daughters (50% $\female$ x ½ + 50% $\male$ x ½ = 50%), and so, from an evolutionary perspective, a nest member is better off raising fertile sisters than fertile daughters. From an individual selection perspective, sociality is advantageous.

After Hamilton others proposed theories using an individualistic perspective. One important contribution was Robert Trivers's notion of *reciprocal altruism*, based on the *you scratch my back and I'll scratch yours* principle, which holds that some forms of sociality succeed because organisms gain more through cooperation than through conflict. Also significant were insights based on the use of game theory, particularly the idea of an Evolutionarily Stable Strategy (ESS). A whole group is sometimes less well adapted than it could be because self-interest is paramount. Sex ratios are a case in point. Females do not need a large number of males for fertilization. But because 50:50 seems to be a more stable balance in the population, the group maintains a surplus of males, instead of a more efficient 10:90 male to female ratio. Building on ideas like this, the study of evolution started to change dramatically, and by the 1970s the study of social behavior, in theory and in practice, became one of the most advanced and exciting areas of evolutionary inquiry. The ideas were presented in popular form by British biologist Richard Dawkins in his *The Selfish Gene* (1976), and in what became the bible of the movement and gave the field its name, *Sociobiology: The New Synthesis* (1975), by the American scholar of the study of social insects, Edward O Wilson.

Controversies

These works, Wilson's in particular, were highly controversial, mainly (although not exclusively) because they extended to humans. Much like Darwin himself, having surveyed social behavior in the animal world from the most primitive forms to the primates, Wilson argued that Homo Sapiens is part of the evolutionary world in its behavior and culture. Although he did allow that experience and training can have some effects, Wilson believed that genes are the real key to understanding human thought and behavior. In male-female relationships, in parent-child interactions, in morality, in religious yearnings (a very important phenomenon for Wilson, a Southerner), in warfare, in language, and in much else, biology matters crucially.

Social scientists and left-leaning biologists (especially Richard Lewontin and Stephen Jay Gould), and philosophers (especially Philip Kitcher in a witty attack, *Vaulting Ambition*), accused sociobiologists—particularly human sociobiologists—of a multitude of sins. Epistemologically these detractors judged the work of sociobiologists to be false, and then (not entirely consistently) charged them with producing ideas and theories that are not falsifiable. One particularly effective rhetorical charge was that sociobiologists's claims are akin to the *Just So* stories by Rudyard Kipling, in which a fantastical tale is created (i.e., that of how the elephant's nose is long because it was pulled by a crocodile) and then is alleged to be fact. Sociobiologists were also found to be wanting ethically. Their work was attacked as sexist, racist, homophobic, capitalist, and in short, guilty of every possible transgression that exists in a patriarchal, unjust society. They were accused of supporting the status quo in Western societies, and of pretending to give genuine scientific answers to bolster what were really ideological convictions.

There was undoubtedly some truth to all of these claims. Yet some change can be progress, and there is little doubt—at the animal level particularly—that evolutionists have taken full note of critics' complaints and worked hard to address them. Modern techniques, particularly those that employ the insights of molecular biology, have been of great help here. For instance many sociobiological claims concern parenthood. If males are competing for females, for instance, and (as in birds) males are also contributing to childcare, one expects efforts to be tied to reproductive access and success. But while it is difficult if not impossible to determine paternity with traditional methods, that Gordian Knot is cut as soon as one starts using genetic fingerprints. Not only are the scientific claims testable but in many cases they have been found to be correct. Animal sociobiology is no more tentative than other scientific fields. It can be persuasively argued that in science bold conjectures are

needed in abundance. However when those conjectures are accepted as fact without being tested, there is a problem. Science requires continual, rigorous challenge.

Human sociobiologists argue that they too have theories that can be, and are, put to the test, such as theories about infanticide, showing that this occurs when and generally only when it is in the biological interests of parents not to have all of the children to which they (or sometimes, rivals) have produced. One well-known theorem (with much support in the animal realm) asserts that females who are more fit will tend to skew birth rates toward males, and less fit females toward females. The reason for this is that even unfit females generally get impregnated, whereas if there is competition among males—and there usually is—the fitter male tends to get the prize. Hence because fit mothers are more likely than unfit mothers to have fit offspring, for fit mothers having males is a good strategy, whereas for unfit mothers having females is a good strategy. There is incidentally no necessary presumption that this always requires conscious intention—fluctuating hormones, for instance, might be the proximate causes. Human sociobiologists argue that this also occurs in human societies, with the members of upper classes tending to dispose of daughters, either physically by allowing them to die or giving them away shortly after birth, or through other methods that effectively prevent reproduction even without killing (for instance, by forcing daughters into religious orders that require celibacy thereby effectively preventing them from reproducing). In recent years, human sociobiology has changed into what is now called *evolutionary psychology*. The emphasis is less on behavior and more on the mental traits that lead to behavior. This view is still philosophically controversial, with much debate about how and whether one can talk of psychological characteristics as being *innate* (and how one would test the theory).

Sociobiologists have countered vigorously against the social and ethical charges levied against them. Almost without exception, human sociobiologists have not had significant social agendas and are greatly concerned by the misuse that can (and sometimes is) made of their work. They repudiate strongly the charge that they are crypto-nazis or subscribers to other vile doctrines, and deplore the fact that sometimes people favorable to these ideas invoke the authority of sociobiology in support. They stress that differences between races, for instance, are far less than similarities, and in any case differences in themselves do not necessarily spell superiority or inferiority. Although their work has been much criticized by feminists, human sociobiologists respond that pointing out differences between males and females is not in itself sexist. Indeed one might argue that not to recognize differences can be morally wrong. If boys and girls mature at different rates, insisting that they all be taught in the same ways could be detrimental to both sexes. In more specific issues also sociobiology is not necessarily erroneous or promoting an immoral agenda. To hypothesize that something such as sexual orientation is dictated by an individual's genes (and that there is a pertinent underlying evolutionary history to explain it) could be a move toward recognizing that all people are equally worthy of moral tolerance and respect.

There is ongoing philosophical debate over all of these issues. In the early twenty-first century there is renewed interest in the possible evolutionary underpinnings of religion. It is clear that sociobiology—animal and human, and by whatever name the field is known—is not about to disappear, and is in fact a thriving area of inquiry.

MICHAEL RUSE

SEE ALSO Aggression; Animal Tools; Darwin, Charles; Ethology; Evolutionary Ethics; Game Theory; Nature versus Nurture; Scientific Ethics; Selfish Genes; Social Darwinism; Sociological Ethics.

BIBLIOGRAPHY

Darwin, Charles. (1859). On the Origin of Species. London: John Murray.

Dawkins, Richard. (1976). The Selfish Gene. Oxford: Oxford University Press.

Hamilton, William D. (1996). Narrow Roads of Gene Land: The Collected Papers of W. D. Hamilton. New York: W.H. Freeman/Spektrum.

Kitcher, Philip. (1985). Vaulting Ambition. Cambridge, MA: MIT Press.

Ruse, Michael. (1979). Sociobiology: Sense or Nonsense? Dordrecht, The Netherlands: Reidel.

Trivers, Robert L. (1971). "The Evolution of Reciprocal Altruism." Quarterly Review of Biology 46: 35–57.

Wilson, Edward O. (1975). Sociobiology: The New Synthesis. Cambridge, MA: Harvard University Press.

Wilson, Edward O. (1978). On Human Nature. Cambridge, MA: Cambridge University Press.

SOCIOLOGICAL ETHICS

• • •

Sociology, or the scientific study of society, social institutions, and social relationships, is one of the most important social sciences and may include in its concerns anthropology, economics, history, political science, and psychology. As a field of study it is inherently intertwined with ethics. Because any society is dependent on common assumptions about what is acceptable and unacceptable behavior among its members, sociological analysis has to include descriptions of those ethical beliefs and practices. Indeed, the society constituted by sociologists may be defined by its internal ethical commitments. At the same time, insofar as sociologists do research in and on society, they produce knowledge about moral values and their social functions, and questions arise about the proper guidelines for their work, especially when that work may conflict in various ways with accepted social norms.

The Sociology of Ethics

Early in the formation of sociology morals and values entered into the picture and influenced sociological thought and practice. A specific concentration such as a "sociology of moral values" may not exist (Durkheim 1993, p. 14), but morality has played a central role in the prevailing concepts that have shaped and molded sociology. This ideology can be seen in the works of individuals such as Karl Marx (1818–1883), Max Weber (1864–1920), and Emile Durkheim (1858–1917). These classical sociologists agreed on issues surrounding industrial capitalism and how values and morals worked to keep a society together; however, they nonetheless differed in their views of the function these elements have and how they change over time.

Although Marx is credited for playing a key role in establishing the field, Weber is the one considered to be the father of sociology. Marx's challenging social criticism was replaced by Weber's value-neutral sociology, which nevertheless stressed, as in The Protestant Ethic and the Spirit of Capitalism (1904), the ethical foundations of social orders. Marx was intrigued by the interaction between science and society, whereas Weber examined social structure and focused more on the notion of value-free science. Weber believed people acted on their own accord and emphasized the importance of the individual rather than the role of society as a collective whole. He also emphasized the notion that people should not expect science to tell them how to live their lives.

Durkheim's theories are considered by some sociologists to be even more applicable today than they were at the time he formulated them (Turner 1993). His primary contribution to sociology was his stance on social solidarity, social roles, and the division of labor. Morality and the connection between science and society also influenced Durkeim's work on professional ethics. Durkheim touted the importance of moral education on everyday life and emphasized its inclusion in the study of sociology. Marx, Weber, and Durkheim may have developed their theories in a different academic era, but they continue to influence and impact the field of sociology today.

Works by Weber and Durkheim were the precursors to those by Robert Merton, the first sociologist to win the National Medal of Science and the founder of the sociology of science. Merton's focus was on the functional analysis of social structures, and he discounted subjective dispositions, such a motives and aims. Things Merton is best known for are coining the terms "self-fulfilling prophecy," "deviant behavior," and implementing the focus group concept in a research setting.

The Ethics of Sociology

The first attempt to promote international cooperation and professionalize the field of sociology can be seen in the formation of the International de Sociologie by Rene Worm in 1893. In 1905 a number of well-known sociologists across the United States met to create an entity to promote the professionalization of the field of sociology. This organization was called the American Sociological Society and later evolved into what is known today as the American Sociological Association. Today, the ASA is the largest organization of sociologists and its membership is not only made up of students and faculty, but 20% of its membership is comprised of individuals who represent government, business, and non-profit groups. In the spring of 1997, the ASA membership approved its current version of the Code of Ethics. It includes an introduction, a preamble, five general principles, and specific ethical standards. Rules and procedures for handling and investigating complaints are also noted.

As time went on, more organizations such as the International Sociological Association were formed to support sociologists and advance knowledge about this field of study. Like ASA, these entities have also developed and established codes of ethics for their membership to follow. ISA, an organization founded in 1949, drafted its own code of ethics and the current version was approved by their Executive Committee in the fall of 2001. Other groups, such as the North Central Sociological Association, have preferred to base their codes on those outlined by ASA.

New and exciting research opportunities often bring unforeseen scenarios, many of which revolve around the sociologist's relationship with subjects. Dilemmas involving the applicability of informed consent, the use of deception, and the protection of privacy and confidentiality are common in social science research. A conflict between the desire to protect human subjects and the goal of obtaining data may not be easy to rectify even if guidelines are followed.

Research misconduct and authorship violations are also concerns that face social scientists. Abuses vary in severity and may encompass plagiarism, data fabrication, and falsification of data and results. The ethical dilemmas encountered in sociology are not unique. As science and technology become intertwined further with society, these ethical questions will become even more complex.

Sociological Issues Related to Science and Technology

Problems that occurred during the 1960s and 1970s, such as the thalidomide drug tests (1962) and the Tuskegee syphilis study (1932–1972) emphasized the fallibility and injustices of scientific research and added momentum to appeals for more regulations and guidelines. Scientific investigations, especially those in biomedicine, often are considered high-risk and life-threatening, but the social sciences also have encountered less obvious but not necessarily less dangerous situations. One case that is discussed frequently in social science circles is Stanley Milgram's work on obedience to authority in 1963. Milgram found that a majority of the individuals participating in this series of studies were willing to administer what they believed to be harmful electrical shocks to their victims. Laud Humphreys's tearoom trade in 1970 also sparked controversy. Humphreys studied homosexual encounters in a St. Louis park restroom without revealing the true nature and intention of his research. Philip Zimbardo's Stanford prison experiment in 1973 is another example of an infraction that sent up red flags to those involved in protecting human subjects (Sieber 1982). Zimbardo's study, which ended early due to concerns about its effects on the subjects, used role playing to determine what happens when good people are put in an environment that fosters evil.

Informed consent is a key component of human subjects research, but it can be controversial in disciplines such as sociology. Regulations require that in most cases informed consent be obtained before research can commence, but consent often is seen as an unrealistic obstacle in the social sciences. Research conducted by social scientists often involve the use of ethnographic methods, the collection of oral histories, and survey procedures, which do not readily lend themselves to the written informed consent process. Obtaining written consent may be problematic for researchers working in situations where language and cultural differences pose as a barrier. This may occur in situations where the individuals are illiterate or merely speak a different language. Some cultures consider the signing of a document taboo or an act reserved for certain situations such as the signing of legal documents. Evidence also indicates that subjects who sign consent forms, like those who participated in Milgram's study, do not always comprehend the full extent of the project (Mitchell 1993). Many social science initiatives include individuals involved in illegal activities where anonymity is essential. In these situations the informed consent document may compromise confidentiality by being the only link to the subject.

Steps taken to protect the privacy of the subject and ensure the confidentiality of the data may instill a false sense of security in the researcher and the subjects. A researcher may code identifiers, destroy data after project completion, use pseudonyms to mask identity, and avoid gathering personal information altogether in an attempt to provide protection. These measures are not infallible, and violations are evident in numerous cases. The use of thinly disguised pseudonyms that provoked the "Springdale" controversy can be seen in Arthur Vidich's *Small Town in Mass Society* (Vidich and Bensman 2000). Sociologist Arthur Vidich and anthropologist Joseph Bensman conducted a study of small town life and assigned the pseudonym "Springdale" to the upstate New York community. It didn't take long for the community's true identity to be revealed, which caused Vidich's and Bensman's research practices to be called into question. Other infractions have involved the subpoena of data, as in the case of Rik Scarce, who underwent 159 days of incarceration for refusing to release his field notes (Scarce 1995). Even with protections in place the subject's privacy and confidentiality may be at risk.

All researchers wrestle with similar issues of research misconduct. A survey published in *American Scientist* (November–December 1993) that measured perceived rather than actual misconduct examined some of those concerns. Doctoral candidates and faculty members representing the fields of chemistry, civil engineering, microbiology, and sociology were asked questions about scientific misconduct, questionable research practices, and other types of wrongdoing. Several conclusions were extracted from the data results, including

reports that scientific transgressions occurred "less frequently than other types of ethically wrong or questionable behavior by faculty and graduate students in the four disciplines" surveyed (Swazey, Anderson, and Lewis 1993, p. 552). Other entities, such as the media, chose to concentrate on practices that painted a dire picture of academic integrity.

Funding and sponsor involvement constitute other factors that can create serious ethical dilemmas for researchers. Certain departments, such as sociology, often struggle for financial support and rely heavily on government and corporate sponsorship. Project Camelot, which has been regarded by some as "intellectual prostitution," was used to "predict and influence politically significant aspects of social change in developing nations of the world, especially Latin America" (Homan 1991, p. 27). Warnings by critics like Derek Bok, the former president of Harvard and author of the book "Universities in the Marketplace: The Commercialization of Higher Education" (Princeton University Press) indicate that pressure by academia to attract industry involvement is a precarious undertaking that can lead to the "commercialization of higher education" (Lee 2003, p. A13). These relationships also may result in pressure on researchers to skew results to favor the sponsor. In the end stiff competition for research funding and pressure to attract industry involvement may compromise ethical and professional standards (Homan 1991).

Changes in Science and Technology That Affect Sociology

Regulations and guidelines based on a biomedical model have had a dramatic impact on sociology. After the atrocities that occurred during World War II a series of codes were implemented to focus on the protection of human subjects in research. Some of the more noted ones include the Nuremberg Code, the Declaration of Helsinki, and the 1971 guidelines published by the U.S. Department of Health, Education, and Welfare (DHEW).

The Nuremberg Code, a set of ten principles designed to protect human subjects in research, was a ruling announced in 1947 by the war crimes court against Nazi doctors who conducted experiments on their prisoners. The Declaration of Helsinki was approved by the Eighteenth World Medical Assembly in 1964 and was designed to assist physicians in biomedical research involving human subjects. The continuation of ethical infractions invoked calls for additional regulations. Guidelines published in 1971 by the DHEW were one response to those demands and would prove to be the inspiration for the development of institutional review boards (IRBs) for federally funded research initiatives.

Another instrumental document resulted from the formation of the National Commission for the Protection of Human Subjects of Biomedical and Behavioral Research. The Belmont Report elaborated on the ten points outlined in the Nuremberg Code and placed the emphasis on respect for persons, beneficence, and justice. Those regulations were revised in 1981, and Title 45, Code of Federal Regulations, Part 46, became known as the Common Rule.

Professional codes of ethics are a relatively recent phenomenon. The codes that existed before World War II were found primarily in the major professions of that time, such as medicine and law. Most modern organizations have developed codes based on those in the sciences, but the codes used in the social sciences often lack the power to impose sanctions for noncompliance. Unlike the case in some professional associations, participation in an organization such as the American Sociological Association (ASA) is not necessary for a person to be a sociologist or to conduct social science research. The lack of an enforcement mechanism for ethical violations also weakens the power of codes such as that of the ASA. The notion that professional codes of ethics are merely symbolic has been attributed to the government's decision to implement regulations (Dalglish 1976).

Contributions to Science, Technology, and Ethics Discussions by Sociology

A debate has been brewing among scientists and social scientists who submit research protocols for approval. The DHEW declared on July 12, 1974, that to obtain federal funding for a research project an IRB had to be in place to review projects that involved human subjects in biomedical and behavioral research. Today IRBs apply one set of rules, based on a biomedical format, to review all project submissions. Those requirements have proved to be inapplicable to numerous social science proposals and are next to impossible to carry out in all research settings. Sociologists and other social scientists have joined forces to form alliances, such as the Social and Behavioral Sciences working group, to improve the IRB process for social science researchers. In some cases, however, IRBs continue to interpret "the requirements of the Common Rule in a manner more appropriate to high risk biomedical research, ignoring the flexibility available to them in the Common Rule" (Sieber, Plattner, and Rubin 2002, p. 2).

Sociologists also have collaborated with researchers in science and technology on a number of ethics initiatives. Joint facilities and centers have helped facilitate those efforts by encouraging cross-curriculum dialogue and research. The Hastings Center was founded in 1969 to "examine the different array of moral problems engendered by advances in the biomedical, behavioral, and social sciences" (Abbott 1983, p. 877). The Center for Applied Ethics at the University of Virginia, also founded in 1969, has worked on integrity issues that span various fields and subject matters. Another interdisciplinary effort is the Ethical, Legal and Social Implications Research Program (ELSI). Founded in 1990, ELSI has focused on a number of issues, including informed consent, public and professional education, and discrimination, by bringing together experts from multiple, diverse disciplines and conducting workshops and orchestrating policy conferences to discuss these pertinent issues.

Education is imperative to promote academic integrity, and students in all disciplines should be instructed on matters that may have an adverse effect on their research. Acceptable academic behavior can be conveyed through formal methods such as workshops and symposia or through the use of informal techniques such as discussions with advisers, mentors, and classmates. Conversations that introduce possible solutions to the ethical predicaments encountered in research also can be beneficial. Teaching new researchers how to act in an ethical manner will help reduce the number of violations and will create research professionals dedicated to upholding the morals that are valued in society.

The Future

Ethical dilemmas will continue to plague researchers whether they are in the sciences or the social sciences. A state of risk-free research is not foreseeable, and steps will continue to be taken to minimize the severity and frequency of these problems. Changes in the regulations will be felt most heavily in the biomedical and science fields, but the social sciences will not be spared from increased scrutiny. Some efforts may prove to be worthy and circumvent or minimize ethical quandaries, whereas others may violate personal rights and academic freedom in the process. Cooperation among disciplines is essential to communicate the importance of ethics and create researchers who conduct their work with integrity. In the words of Johann Wolfgang von Goethe, "Knowing is not enough; we must apply; willing is not enough, we must do."

SHARON STOERGER

SEE ALSO *Codes of Ethics; Durkheim, Émile; Human Subjects Research; Informed Consent; Institutional Review Boards; Merton, Robert; Misconduct in Science: Social Science Cases; Privacy; Research Ethics; Sociobiology; Tuskegee Experiment; Weber, Max.*

BIBLIOGRAPHY

Abbott, Andrew. (1983). "Professional Ethics." *American Journal of Sociology* 88(5): 855–885.

Dalglish, Thomas Killin. (1976). *Protecting Human Subjects in Social and Behavioral Research: Ethics, Law, and the DHEW Rules: A Critique.* Berkeley: Center for Research in Management Science, University of California, Berkeley.

Durkheim, Emile. (1993). Ethics and the Sociology of Morals. Translated by Robert T. Hall. Buffalo, NY: Prometheus Books.

Federman, Daniel; Kathi E. Hanna; and Laura Lyman Rodriguez, eds. (2002). *Responsible Research: A Systems Approach to Protecting Research Participants.* Washington, DC: National Academies Press.

Gouldner, Alvin. (1971). *The Coming Crisis of Western Sociology.* New York: Avon Books.

Homan, Roger. (1991). *The Ethics of Social Research.* New York: Longman.

Lee, Felicia R. (2003). "The Academic Industrial Complex." *New York Times,* September 6.

Mitchell, Richard G., Jr. (1993). *Secrecy and Fieldwork.* Newbury Park, CA: Sage.

Report and Recommendations of the National Commission for the Protection of Human Subjects of Biomedical and Behavioral Research. (1978). "The Belmont Report." Washington DC: U.S. Government Printing Office.

Scaff, Lawrence A. (1984). "Weber before Weberian Sociology." *British Journal of Sociology* 35(2): 190–215.

Scarce, Rik. (1995). "Scholarly Ethics and Courtroom Antics: Where Researchers Stand in the Eyes of the Law." *American Sociologist* 26: 87–112.

Sieber, Joan E., ed. (1982). *The Ethics of Social Research: Surveys and Experiments.* New York: Springer-Verlag.

Sieber, Joan E.; Stuart Plattner; and Philip Rubin. (2002). "How (Not) to Regulate Social and Behavioral Research." *Professional Ethics Report* 15(2): 1–4.

Silber, Joan E. (1992). *Planning Ethically Responsible Research: A Guide for Students and Internal Review Boards.* Newbury Park, CA: Sage.

Silber, Joan E. (2001). *Summary of Human Subjects Protection Issues Related to Large Sample Surveys.* Washington, DC: U.S. Department of Justice, Bureau of Justice Statistics. Available at http://purl.access.gpo.gov/GPO/LPS16482.

Stanley, Barbara H.; Joan E. Sieber; and Gary B. Melton, eds. (1996). *Research Ethics: A Psychological Approach.* Lincoln: University of Nebraska Press.]

Swazey, Judith P.; Melissa S. Anderson; and Karen Seashore Lewis. (1993). "Ethical Problems in Academic Research." *American Scientist* 81: 542–553.

Turner, Stephen P., ed. (1993). *Emile Durkheim: Sociologist and Moralist.* London: Routledge.

Vidich, Arthur J., and Joseph Bensman. (2000). *Small Town in Mass Society: Class, Power and Religion in a Rural Community.* Urbana: University of Illinois Press.

SOFT SYSTEMS METHODOLOGY

• • •

Soft systems methodology provides a framework for structuring, analyzing, and solving problems in systems that involve people. It integrates logical, cultural, and political analyses of a problem situation in order to imagine, discuss, and then implement actions to improve the situation, with the consensus of the participants. Soft systems methodology is used primarily by managers and consultants working on technical or organizational problems; it has proved particularly useful in the Information Technology/Information Systems sector.

Peter Checkland developed soft systems methodology because classic systems engineering and systems analysis (*hard* systems methodologies), which work excellently in many engineering situations, often disappoint in management situations. Hard systems methodologies are well-suited for designed systems where the task of the analyst is to find the most efficient means of reaching a well-defined goal, but they cannot deal with the cultural and social dimensions in what Checkland terms human activity systems, which are systems that include human self-consciousness and freedom of choice. One of the characteristics of human activity systems is the wide range and importance of world-views, or *Weltanschauungen*, held by the participants in the system, and the consequent lack of clearly defined or agreed goals within such a system. Soft systems methodology is designed to deal with human activity systems where "in the complexity of human affairs the unequivocal pursuit of objectives which can be taken as given is very much the occasional special case" (Checkland 1999, p. A6).

There are four main activities in Checkland's methodology:

1. Finding out about a problem situation, including its cultural and political dimensions;

2. Formulating relevant purposeful activity models (devising scenarios of possible future actions and outcomes);

3. Debating the situation with participants, using the models, seeking from that debate both

a) changes that would improve the situation and are regarded as both desirable and (culturally) feasible, and

b) the accommodations between conflicting interests that will enable action-to-improve to be taken;

4. Taking action in the situation to bring about improvement. (Checkland 1999, p. A15).

Soft systems methodology provides practitioners with almost the same analytical techniques and many of the same conceptual approaches as Harold D. Lasswell's policy sciences, but laced with more pragmatism and less idealism. Soft systems methodology focuses on business and industry applications, it seeks agreed solutions, and is based in management science and engineering. The policy sciences are concerned with representative democracy and public policy, they are rooted in the social sciences, and they emphasize a moral rather than consensual basis for decision making. Both approaches agree that the analyst becomes involved in the system under examination; that the viewpoint of the analyst must be made explicit; that there are non-rational elements in human behavior; and that history, perception, relationships, and culture are important factors in human activity systems.

Peter Checkland, the founder of soft system methodology, was born in Birmingham, England in 1930. He studied chemistry at Oxford University in the 1950s, then worked at ICI Ltd. as a technologist and manager. He moved to the Department of Systems at the University of Lancaster in 1969, and in the early twenty-first century is Professor of Systems, Management Science, in the Lancaster University Management School.

MAEVE A. BOLAND

SEE ALSO *Engineering Method; Lasswell, Harold; Systems.*

BIBLIOGRAPHY

Checkland, Peter. (1999). *Systems Thinking, Systems Practice: Includes a 30-Year Retrospective.* Chichester, UK: John Wiley & Sons. Reissue of the first book on soft systems methodology, with a major retrospective essay reviewing thirty years of developments in the field.

Checkland, Peter, and Sue Holwell. (1998). *Information, Systems, and Information Systems.* Chichester, UK: John Wiley & Sons. Focuses on the place of information technology in human affairs, with an emphasis on IT in organizations.

SOFTWARE

SEE *Free Software; Hardware and Software.*

SOKAL AFFAIR

• • •

The Sokal Affair was the central and most highly publicized episode of the "Science Wars," a fracas that roiled the academic atmosphere throughout the 1990s. The main point at issue in these conflicts was the accuracy and indeed the legitimacy of critiques of science and technology propounded by scholars committed to or influenced by postmodern thought and identity politics. The hoax itself, as well as the volume of *Social Text* (no. 46/47, Spring/Summer 1996) in which it appeared, arose chiefly in response to an earlier science wars salvo, the book *Higher Superstition: The Academic Left and Its Quarrels with Science* by Paul R. Gross and Norman Levitt (1994), which aggressively criticized the "science studies" movement that had emerged from poststructuralist and social-constructivist doctrines.

The squabbles ignited by *Higher Superstition* alerted Alan Sokal, a mathematical physicist at New York University, to the controversy. Further research nullified his initial suspicions that the book might merely be yet another "culture wars" diatribe from the right. He concluded, despite his own leftist sympathies, that postmodern and relativistic views of science epitomized the weaknesses he had already discerned in some versions of contemporary left-wing thought. It struck him that a parody article satirizing the pretensions of science studies might provoke useful debate around this issue. The resulting essay, *Transgressing the Boundaries: Toward a Transformative Hermeneutics of Quantum Gravity*, mischievously combined references to arcane physics and mathematics with laudatory citations of major postmodern theorists, ostensibly to support the thesis that postmodern dogma accords with advanced ideas in foundational physics.

The essay was submitted to *Social Text* just as that journal was planning its own rejoinder to *Higher Superstition.* Editor Andrew Ross, himself a prominent target of Gross and Levitt, had recruited a number of well-known proponents of science studies as contributors. When Sokal's Trojan-horse manuscript arrived, its Swiftian character escaped detection and the piece was promptly accepted because of the author's physicist credentials, as well as his authentic leftist pedigree and his feigned detestation of the enemy camp.

The "Science Wars" number of *Social Text* appeared in May 1996. Within days, Sokal unmasked his own hoax in the magazine *Lingua Franca,* and the episode quickly made its way into the mass media. Subsequent denunciations of Sokal by *Social Text*'s editors and supporters did little to staunch the widespread glee that erupted from some quarters.

The greatest significance of the affair lies, indeed, in the very fact that it became so widely known and evoked such intense responses. In itself, Sokal's piece was intentionally sophomoric, a transparently silly joke. It "proved" little more than that a handful of academics had been overeager to recruit a "real" scientist to their side of an acrimonious dispute. Why, then, the enormous uproar?

The answer lies in the hostility that had been building for a decade or more in response to the pretensions and what many saw as the monopolistic ambitions of the postmodern left. Such resentment was hardly limited to scholars of conservative bent: It was widely shared by liberals and leftists who had come to view postmodern academic culture as bizarre and overbearing. Consequently, the Sokal Hoax became the symbolic center of an intellectual firestorm whose stakes extended well beyond anything directly connected to the prank itself. It brought into the open long-brewing anxieties over scholarly priorities and their effect on the academic pecking order. The myopia of *Social Text* came to stand, rightly or wrongly, for the pretensions of postmodern scholarship per se. Sokal's success emboldened many long-suffering professors to decry at last the impostures of a subculture that had long cowed them with its self-ascribed sophistication. Most scientists were understandably amused by the spectacle, but in regard to what was really at issue, they were bystanders. This was, at heart, a battle fought by non-scientists.

In the early twenty-first century, the postmodern left seems to have declined, at least as the hegemonic trendsetter of the academy. For good or ill, many of its social precepts remain central to university culture, but with diminished stridency. "Theory," as postmodernists were wont to use the term, has lost much of its power to intimidate. At the same time, many humanist scholars who once employed the vaunted insights of science studies to disparage science now affect to admire it deeply. Postmodernism and the political style linked to it certainly endure, but in a more subdued mode. The Sokal Affair was by no means the sole or even the most important catalyst for these changes, but it was timely and amazingly effective.

NORMAN LEVITT

SEE ALSO *Science, Technology, and Society Studies.*

BIBLIOGRAPHY

Editors of *Lingua Franca.* (2000). *The Sokal Hoax: The Sham that Shook the Academy.* Lincoln: University of Nebraska Press. Reprints Sokal's "confessional" piece, along with a wide array of comments and observations about the Hoax.

Koertge, Noretta, ed. (1998). *A House Built on Sand: Exposing Postmodern Myths About Science.* New York: Oxford University Press. A collection of essays sharply critical of some of the best-known work in postmodern science studies.

Levitt, Norman. (2000). "Confessions of a Disagreeable Man." *Skeptic* 8(3): 93–97. An insider account of the genesis of the Hoax, with some opinions on its significance.

Ross, Andrew, ed. (1996). *The Science Wars.* Durham, NC: Duke University Press. A book version of *Social Text* 46/47, with some additional pieces included, but with the Sokal hoax article itself deleted.

Sokal, Alan. (1996a). "Transgressing the Boundaries: Toward a Transformative Hermeneutics of Quantum Gravity." *Social Text* 46/47 (Spring/Summer): 217–252. The Sokal Hoax article.

Sokal, Alan. (1996b). "A Physicist Experiments with Cultural Studies." *Lingua Franca* 6(4): 62–64. Sokal's "confession" of the hoax.

Sokal, Alan, and Jean Bricmont. (1998). *Fashionable Nonsense: Postmodern Intellectuals' Abuse of Science.* New York: Picador. A critique by Sokal and another physicist of the misuse of scientific notions by a number of postmodern theorists.

Segerstrale, Ullica, ed. (2000). *Beyond the Science Wars: The Missing Discourse about Science and Society.* Albany: State University of New York Press. Essays on the significance of the "science wars" and the hopes for replacing those squabbles with a less recriminatory dialog.

SPACE

• • •

To the question, "Where are you in this moment?" a pilot would answer, "At longitude x, latitude y, altitude z." But if one asks, "Where do you live?", the answer may instead evoke neighborly relations weaved through the years, a climate, old stones, the freshness of water. Depending on who is asked about what, the *where* question can be answered by *space* determinations or by the memories of a concrete *place*. Space and place are two different ways of conceiving the "where" or, using the Latin word for "where" as a *terminus technicus*, two answers to the *ubi* question.

Place and Space

Place is an order of beings vis-à-vis the body. This order (*kosmos* in Greek) always mirrors the great cosmos. This vis-à-vis or mirroring is the essence of what has been called *proportionality* (Illich and Rieger 1996). According to Albert Einstein, the concept of space disembedded itself from the "simpler concept of place" and "achieve[d] a meaning which is freed from any connection with a particular material object" (Einstein 1993, p. xv). Yet Einstein insisted that space is a free creation of imagination, a "means devised for easier comprehension of our sense experience" (Einstein 1993, p. xv). In pure space, however, the body would be out of place and in a state of perceptual deprivation.

The focus here is on the radical monopoly that space determinations exert on the *ubi* question. Wheels and motors seem to belong to space as feet do to places. And just as the radical monopoly of motorized transportation on human mobility leaves some freedom to walk, space determinations leave remnants of placeness to linger in perception and memory. Ethics, then, can only be rebuilt by a recovery of placeness.

Origins of Space

A general conception of space is conspicuously absent from ancient mathematics, physics, and astronomy. The Greek language, so rich in locational terms, had no word for space (Bochner 1998). *Topos* meant place, and when Plato in the *Timaeus* (360 B.C.E.) located the demiurge in an uncreated *ubi* in which one can have no perception because it does not *exist*, he called it *chôra*, fallow land, the temporary void between the fullness of the wild and cultivation. According to Plato, the demiurge's chôra could only be conceived "by a kind of spurious reason," "as in a dream," in a state in which "we are unable to cast off sleep and determine the truth about it" (passage Timaeus 52). In hindsight, one may say that this was a first intuition of the antinomy between place and what is has come to be called "space." In the fourteenth century, Nicolas d'Oresme imagined an incorporeal void beyond the last heavenly sphere, but still insisted that, in contrast, all real places are full and material. Space, still only a pure logical possibility, became a *possibile realis* between the times of d'Oresme and Galileo (Funkenstein 1986, p. 62).

Following the canons of Antiquity and medieval cartography, a chart summarized bodily scouting and measuring gestures. Pilgrims followed *itineraria;* sailors, charts of ports; and surveyors consigned ritually performed acts of mensuration on marmor or brass plates. These were not maps in the modern sense, because they did not postulate a disembodied eye contemplating a land or a sea from above. The first maps in the modern sense were contemporary with early experimentations in

central perspective and, like these, construed an abstract *eye* contemplating a distant grid in which particulars could be relatively situated. In 1574, Peter Ramus wrote a *lytle booke* in which he exposed a *calculus of reality* where all topics were divided in mental spaces that immobilized objects in their definitions precluding the understanding of knowledge as an act (Pickstock 1998). Cartesian coordinates and projective geometry gave the first mathematical justification to the idea of an immaterial vessel, unlimited in extent, in which all material objects are contained.

Non-Euclidean Space

Had space been invented, as Einstein contended, or discovered? In the eighteenth century, Immanuel Kant announced that space was an *a priori* of perception. For him, Euclidean geometry and its axioms were the mathematical expression of an entity—space—that cannot be perceived, but, like time, underlies all perceptions. The first attempt to contradict Euclidean geometry was published in Russian in 1829 by Nicolay Lobachevsky (1792–1856), whose ideas were rooted in an opposition to Kant. For him, space was an *a posteriori* concept. He sought to prove this by demonstrating that axioms different from Euclid's can generate different spaces. In light of Lobachevsky's—and then Georg Riemann's (1826–1866)—non-Euclidean geometries, Euclidean geometry appears ex post facto as just another axiomatic construct. There is no a priori space experience, no *natural*, or *universal* space. Space is not an empirical fact but a construct, an arbitrary frame that *carpenters* the modern imagination (Heelan 1983).

Einstein occupies an axial and simultaneously ambiguous position in the history of this understanding. In order to express alterations of classical physics that seemed offensive to common sense, he adopted a mathematically constructed *manifold* (coordinate space) in which the space coordinates of one coordinate system depend on *both* the time and space coordinates of another relatively moving system. On the one hand, like Lobachevsky and Riemann (1854), Einstein insisted on the constructed character of space: Different axioms generate different spaces. On the other, he not only came to consider his construct as ruling the unreachable realms of the universe, but reduced earthly human experience to a particular case of it. In Einstein's space, time can become extension; mass, energy; gravity, a geometric curvature; and reality, a distant shore, indifferent to ethics. This view of space has reigned over the modern imagination for a century. Yet the idea that the realm of everyday experience is a particular case of this general construct has not raised fundamental ethical questions.

Ethics in Space

The subsumption of the neighborhood where one lives into the same category as distant galaxies transforms neighbors into disembodied particularities. This loss of the sense of immediate reality invites a moral suicide. Hence, ethics in the early-twenty-first century requires an epistemological distinction that evokes that of d'Oresme in the fourteenth: Contrary to outer space, the perceptual milieu is a place of fullness. According to its oldest etymology, *ethos* means a place's gait. Space recognizes no gait, no body, no concreteness, and, accordingly, no ethics. The *ubi* question must thus be ethically restated.

Body historians and phenomenologists provide tracks toward an ethical recovery of placeness in the space age. Barbara Duden (1996) argues that one can only raise fundamental ethical questions related to pregnancy by relocating the body in its historical places. For their part, phenomenologists, those philosophers who cling to the *primacy of perception* in spite of tantalizing science-borne and technogenic *certainties*, restore some proportionality between body and place. For Gaston Bachelard (1884–1962), for instance, there is no individual body immersed in the apathetic void of space, but an experience of *mutual seizure* of the body and its natural *ubi*. Maurice Merleau-Ponty (1908–1964) further articulates the complementarity of these two sides of reality. These can be steps toward a recovery of the sense of the vis-à-vis without which there is no immediate reality, and hence no ethics.

JEAN ROBERT

SEE ALSO *Cyberspace; Einstein, Albert; Foucault, Michel.*

BIBLIOGRAPHY

Bachelard, Gaston. (1983). *Water and Dreams. An Essay on the Imagination of Matter.* ed. and trans. Edith R. Farrell. Dallas: Dallas Institute of the Humanities and Culture. Originally published in French in 1956.

Bochner, Salomon. "Space." (1998). In *Raum und Geschichte*, Kurseinheit 4, comp. Ludolf Kuchenbuch and Uta Kleine. Hagen: Fernuniversität.

Coxeter, H. S. (1965). *Non-Euclidean Geometry.* Toronto: University of Toronto Press.

Duden, Barbara. (1996). *The Woman Beneath the Skin.* Cambridge, MA: Harvard University Press. Originally published in 1991.

Einstein, Albert. (1993). "Foreword." In *Concepts of Space: The History of Theories of Space in Physics*, by Max Jammer. New York: Dover. Einstein's text was originally published in 1954.

Funkenstein, Amos. (1986). *Theology and the Scientific Imagination from the Middle Ages to the Seventeenth Century.* Berkeley: University of California Press.

Heelan, Patrick A. (1983). *Space-Perception and the Philosophy of Science.* Berkeley: University of California Press.

Illich, Ivan, and Rieger, Matthias. (1996). "The Wisdom of Leopold Kohr." In *Fourteenth Yearly E. F. Schumacher Conference.* New Haven, Connecticut, October 1994. Reprinted in French as "La sagesse de Leopold Kohr," in *Ivan Illich, La Perte des sens*, trans. Pierre-Emmanuel Dauzat. Paris: Fayard. The French version is the most recent authorized version; the English version is not generally available.

Lobachevsky, Nikolay Ivanovich. (1886). *Geometrische Untersuchungen zur Theorie der Parallellinien.* Kasan, Russia: Kasan University. Originally published in Berlin in 1840.

Merleau-Ponty, Maurice. (1964). *The Primacy of Perception.* Chicago: Northwestern University Press.

Pickstock, Catherine. (1998). *After Writing: On the Liturgical Consummation of Philosophy.* Malden, MA: Blackwell.

Ramus, Peter. (1966). *Logike.* Leeds: The Scholar Press. Originally published in 1574.

Riemann, Bernhard. (1876). "Über die Hypothesen, welche der Geometrie zu Grunde liegen" [On the hypotheses that are the base of geometry.]. In *Gesammelte mathematische Werke und wissenschaftlicher Nachlaß* [Collected mathematical works and legacy], ed. Richard Dedekind and Heinrich Weber. Leipzig, Germany: Teubner. Originally published in 1854.

SPACE EXPLORATION

• • •

Space exploration is the investigation of the cosmos beyond the upper regions of the Earth's atmosphere using telescopes, satellites, space probes, spacecraft, and associated launch vehicles.

Background

The desire to explore space is nearly primal for Homo sapiens. Early humans quickly spread out of Africa to every region on the planet, then came to speculate that the stars and planets were yet other material places worthy of exploration. The idea to travel to these other worlds was inevitable.

However for thousands of years, humans commonly drew fundamental distinctions between the Earth and non-Earth environments. In the formulation of Aristotle taught that the laws of nature that applied on Earth did not necessarily apply beyond the Earth, thus severely restricting the very possibilities for human space exploration.

During the great age of European exploration of the Earth, astronomers such as Galileo Galilei (1564–1642) and his contemporary, Johannes Kepler (1571–1630), began the modern observational exploration of the heavens, in fact of space, using new techniques and instruments of science. A result of this exploration of space was the scientific revolution itself. Science was now seen as applicable to understanding the entire world, to both heaven and Earth. Civilization was transformed.

It now seems natural that Kepler's "Somnium," about a journey to the Moon, includes a realistic description of the lunar surface and how a traveler might physically survive such a trip. But this pioneering story began a long tradition of science fiction literature examining ethical and political issues of space exploration and scientific enterprise.

Twentieth-Century Developments

Planning and experiments to develop the science and technology of physical space exploration began with Konstantin E. Tsiolkovsky (1857–1935) in Russia and Robert Goddard (1882–1945) in the United States. Both of these inventors considered the long-term implications of their work for humanity. Application of their technology to weapons of war soon became evident. Although Goddard helped the U.S. military with rocket-assisted take off of conventional aircraft, it was the Germans who made extensive use of Goddard's published rocket development during World War II.

As the war ended, the *space race* began in earnest between the Soviet Union and the United States. Efforts were made by both countries to enlist German scientists, who had worked on the Nazi rocket program. Many Americans were shocked when, on October 4, 1957, the Soviet Union launched the first artificial earth satellite, Sputnik I. Some Americans viewed the Soviet triumph as an indication of U.S. weakness in science and technology, and considered it a political imperative to match and surpass Soviet accomplishments. Many voiced concern about the threat presented by the combination of nuclear weapons with ballistic missiles.

At the same time, some saw a great potential for peaceful exploration and development of the space environment. Ethical issues were debated about both the commercial and military aspects of this new human

Sputnik I. Launched by the Soviet Union on October 4, 1957, it was the first artificial earth satellite. (*AP/Wide World Photos.*)

enterprise. The National Aeronautics and Space Administration (NASA) was created by Congress in 1958, at the height of the Cold War. It is remarkable that the NASA charter specifically states that the agency is restricted from military activity. (Nonetheless NASA would not always adhere to the charter. For instance, design of the space shuttle was driven significantly by military requirements at a time when Congressional support for NASA was waning.)

Space Law

Despite international competition, there was early agreement that space and celestial bodies were open to peaceful use by all nations, and that principles of international law would be followed in this new realm. Parallels with, and precedents set by, maritime law guided the formulation of space law and regulation. On December 13, 1963, the U.N. General Assembly adopted the Declaration of Legal Principles Governing Activities of States in the Exploration and Use of Outer Space. Further work by the United Nations resulted in the Outer Space Treaty, first signed by sixty-three nations in 1967, and adopted by most countries in the early twenty-first century.

Although much progress has been made in space law, there are challenging near-term issues. For example, the orbital location and radio frequency allocation of communication satellites is a type of territorial issue. At bottom, these resources are limited. Humans have the ancient challenge, in new guise, of how to share these resources peacefully and wisely. The information content of direct-broadcast satellite transmissions is also a complex issue involving national sovereignty on the one hand, and freedom of expression on the other. Observation or spy satellites bring issues of privacy versus freedom of inquiry and information. The United States, Russia, and others have entered into more than 100 treaties and agreements regarding issues of orbit and frequency allocation, as well as launching, tracking, monitoring, and recovery of satellites and space vehicles.

Human Exploration

The first human to orbit the earth, Soviet cosmonaut Yuri Gagarin, returned safely from space in April of 1961. The U.S. astronaut John Glenn followed with a similar mission the next year. These flights, and the many that followed, helped to transform human perspective of the earth and its place in the universe, just as the unmanned missions were doing. Only eight years after Gagarin's flight Neil Armstrong stepped onto the lunar surface on July 20, 1969.

Following the first earth orbit missions, both nations continued without a reported loss of human life until 1967 when three astronauts were lost during a ground test of Apollo 1 and a cosmonaut was lost during return from a Soyuz space mission. Nonetheless, manned space exploration has had a remarkably good safety record. Any space mission must balance goal, schedule, and budget, as well as recognizing risk and the unknown. In achieving this balance in space missions, it is important to keep in mind Richard Feynman's remarks about the loss of the space shuttle *Challenger*: "For a successful technology, reality must take precedence over public relations, for nature cannot be fooled" (Feynman 1986, F5).

Over the last several decades launch failures have been on the order of 1 percent. The space shuttle record, with a total of 112 successful flights and the loss of shuttle *Challenger*, reflects this value. *Columbia*, on the other hand, was the first loss of an American crew on reentry. In both cases the loss appears to be due to schedule and mission demands taking priority over safety.

An astronaut moves along the Space Shuttle Discovery. *(NASA.)*

The loss of human life in space flight development has been relatively low compared to the pioneering days of aviation. This may in part be due to risk-benefit and budget considerations. Experimental airplanes were relatively inexpensive to create and pilots were willing to take considerable risks. It was cost effective to risk pilot and plane to develop the new technology. This is not the case with spaceflight development and exploration. The loss of one mission costs billions of dollars and results in untold costs in schedule slippage and decreased political support. It is remarkable though that during the Mercury, Gemini, and Apollo spaceflights, there was no loss of human life. The Russian space effort has also been relatively free from loss of human life. The most well known soviet accident, Soyuz 11 in 1971, resulted in the death of three crew members as they returned to earth. Overall the loss of life in the U.S. and Russian programs has been similar, if one includes the unannounced Soviet losses of perhaps twelve.

The live coverage loss of *Challenger* and *Columbia* reminded the world that spaceflight is not yet routine. Exploration at the frontier must always remain riskier than day-to-day experience. There is, however, reasonable expectation that near-earth spaceflight will become safer in the foreseeable future. It remains to be seen how the advent of commercial spaceflight will change the equation, but the long term effect should be for improved safety.

Ethical Issues

Although certainly chartered upon a wider canvas, the challenges in space development are, in the first instance, those related to the ongoing challenges faced by the nation states. These issues are mostly of increased degree, rather than entirely new for humans. The ethics of space exploration from this perspective are addressed in such documents as the ESA-UNESCO report, *The Ethics of Space Policy* (Pompidou 2000).

Beyond these issues are those prompted by questions about the impact on human civilization of asteroid and comet orbit modification, space elevators-to-orbit development, or planetary, space, and asteroid colonization. Such endeavors could have impact on civilization beyond that of *normal* human activity.

Also of importance are issues such as interplanetary contamination, the terraforming of planets, and contact with extraterrestrial intelligence. These issues center on questions about the effect of the universe on human beings, and their effect on it.

An elementary case of this sort is the detection of primitive extraterrestrial life in the form of microbes or microfossils. Because the nature of such life is not known, one can only make informed speculation about what the effects might be on civilization and on life on Earth. Or, indeed, what effect humankind might have on such life.

Space exploration may result in the detection of extraterrestrial life or even other civilizations. A scientific Copernican-Darwinian worldview suggests the likelihood of finding evidence of this sort. In any case, it appears likely that people will continue to look for such evidence.

Several outcomes of the detection of life elsewhere in the universe have been suggested: a mostly harmless event, with gain in the knowledge that other life exists in the universe; a major change in life itself or civilization; the loss of civilization; the change or loss of dominant species; loss or change of all *higher order* species; loss of the planetary biosystem; or some unpredicted transformation of life and civilization. These changes are not necessarily in only one direction.

Several decades prior to the physical exploration of space the British ethicist and philosopher Olaf Stapledon (1886–1950) and the crystallographer J. D. Bernal (1901–1971), wrote about some of these wider issues of space exploration. Their pioneering efforts influenced later thinkers from the futurist and novelist Arthur C. Clarke (b. 1917) to the British-American physicist Freeman Dyson (b. 1923).

Responding to Ethical Issues

Humans have attempted to develop some approaches for dealing with the new ethical issues presented by space exploration. Prevention of potential contamination to the Earth's biosphere was practiced during the first lunar expeditions. Astronauts, spacecraft, lunar samples, and equipment were isolated upon their return to Earth from the Moon. The Lunar Receiving Laboratory is in operation to this day, protecting lunar rocks and soil, even though there is now no risk to life on this planet. Space probes are decontaminated prior to leaving the Earth in most cases. Considerable care of this sort was taken with spacecraft, such as Viking (1975) and Sojourner (1996) that would land on the Martian surface. The trajectory of Galileo (1989) was purposely changed, at the end of its mission, in order to send the spacecraft to fiery destruction in the upper atmosphere of Jupiter to insure no contamination of the Jovian moons with terrestrial microorganisms.

In 1991 the Declaration of Principles Concerning Activities Following the Detection of Extraterrestrial Intelligence was drafted by the International Academy of Astronautics (Billingham 1994). The Board of Directors of the International Institute of Space Law approved the declaration. This document is an effort to outline a responsible and orderly set of activities for scientists and others to follow after the detection of extraterrestrial intelligence. An obvious objective of this protocol is to protect life and civilization on Earth.

One can optimistically view the development of portions of the agreements regarding space exploration as the emergence of a principle of non-interference with extraterrestrial life. In a sense humankind seems to be developing a sort of *prime directive* rule of space exploration, which was once only addressed in science fiction. The prime directive restricts human beings from interfering with any extraterrestrial life that is *less developed* than they are.

Carl Sagan (1934–1996) and others have argued that the sort of extraterrestrial life that is likely to be detected will either be of an elementary sort, or a civilization well beyond our imagination. If this turns out to be the case, then the proper conduct, in either of these situations, will not be the sort fancied in the popular space operas of interstellar diplomacy and conflict. Humans would be either the fortunate caretakers of a wholly new primitive life system or the subjects of scientific interest, perhaps protected or transformed beyond recognition.

The American biologist and essayist Stephen Jay Gould (1941–2002), pointed out that the revolution of Copernicus and Galileo was about *real-estate*, but that the Darwinian revolution was about *essence* and thus had much the greater impact. This situation is reflected in questions about the present and future exploration of space. Presently human explorers are experiencing the Galilean, or real estate, phase of the space enterprise. But soon the essence, or Darwinian, phase may commence. Beginning in the mid-1990s, many planets, orbiting other stars, were found by astronomers. Space-born experiments directed at trying to detect some tell-tale signs of life on planets of other solar systems are planned for the first half of the twenty-first century. Even in the Earth's home-system there is hope for detecting life: The oceans that may exist below the ice surface of the Jovian satellite Europa are currently of prime interest to astrobiologists.

The nation states of Earth have created many agreements for the peaceful exploration of space. Space law is now an active field. Humankind has made a start in constructive and peaceful conduct during the early stages of space exploration.

Space exploration is not a one-way enterprise. The "pale blue dot" vision of earth in space, the close-up images of the many worlds of this solar system, returned samples from space, and the countless Hubble space telescope vistas, are transforming the human mind. This transformation is playing a key part in the evolution of the ethics of space exploration—an evolution that may now be at a stage where there is a need to develop a preliminary "prime directive," in order to define conduct with other life in the galaxy. The need may be closer than imagined.

MICHAEL GILMORE

SEE ALSO *Apollo Program; Galilei, Galileo; National Aeronautics and Space Administration; Space Shuttle* Challenger *and* Columbia *Accidents; Space Telescopes.*

BIBLIOGRAPHY

Bernal, J. D. (1969 [1929]). *The World, the Flesh, and the Devil*. Bloomington: Indiana University Press.

Billingham, John, et al., eds. (1994). *Social Implications of the Detection of Extraterrestrial Civilization*. Mountain View, CA: SETI Press.

Chaikin, Andrew. (1994). *A Man on the Moon*. New York: Penguin Books.

Crowe, Michael J. (1986). *The Extraterrestrial Life Debate 1750–1900*. New York: Cambridge University Press.

Dyson, Freeman. (1979). *Disturbing the Universe*. New York: Harper & Row, Publishers.

Feynman, Richard P. (1986). *Appendix F, Personal Observations On The Reliability Of The Shuttle*. Report of the Presidential Commission on the Space Shuttle Challenger Accident, vol. 2, p. F5. U.S. Government Printing Office.

McDougall, Walter A. (1985). *...the Heavens and the Earth: A Political History of the Space Age*. New York: Basic Books

Pompidou, Alain, Co-ordinator. (2000). *The Ethics of Space Policy*. New York: UNESCO.

Shklovskii, Iosif S., and Sagan, Carl. (1966). *Intelligent Life in the Universe*. San Francisco: Holden-Day, Inc.

Stapledon, Olaf. (1968). *"Last and First Men" and "Star Maker": Two Science Fiction Novels*. New York: Dover Publications, Inc.

SPACE SHUTTLES CHALLENGER AND COLUMBIA ACCIDENTS

• • •

The losses of the space shuttles *Challenger* in 1986 and *Columbia* in 2003 dramatically illustrated the risks involved in the human exploration of space, and provide starkly instructive case studies in the ethics of science and technology.

A central mission of the National Aeronautics and Space Administration (NASA) is human exploration of space. Given this legitimate political commitment to human space exploration, the space shuttle program is ethically and politically acceptable insofar as the agency in charge, NASA, promotes careful and honest examination of the human risks and, in reaching the compromises unavoidable in balancing safety against performance, involves those most subject to the risks and those making the political commitment.

The careful, honest examination of risk cannot be done once; it must continue as flight experience accumulates. In balancing safety and performance the shuttle's design both represents NASA's understanding of the system and predicts that the shuttle's flight will safely meet performance requirements. To count as a success, a shuttle flight must perform as the design predicts, not merely return "safely" to Earth. As long as flight does not conform to design, that is, has "anomalies," the design remains provisional; it is not fully understood; and the system is "developmental" not "operational." Both disasters revealed that NASA truncated the examination of risk by deeming the shuttle "operational"; by treating as "successful" flights that did

FIGURE 1

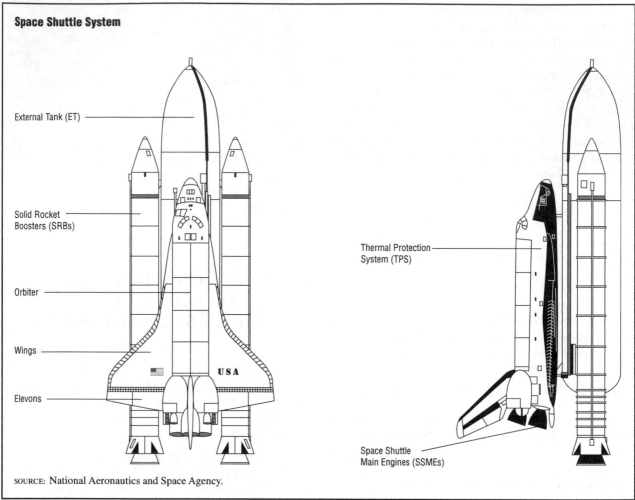

Space Shuttle System

External Tank (ET)

Solid Rocket Boosters (SRBs)

Orbiter

Wings

Elevons

Thermal Protection System (TPS)

Space Shuttle Main Engines (SSMEs)

USA

SOURCE: National Aeronautics and Space Agency.

The figure shows the launch configuration of the main elements of the Space Shuttle System, in "top" and starboard views: The winged Orbiter, which sits atop the large External Tank. On each side of the external tank are the two Solid Rocket Boosters.

not perform as predicted; and by "accepting" risks inherent in anomalous performance. Continuing instances of anomalies signaled the existence of inexplicable risks, which, accepted, culminated in the disasters.

Shuttle History and Design

After Apollo NASA needed a large program to justify its size and budget. It ambitiously planned a shuttle, a space station, and planetary exploration, but budgetary constraints limited the post-Apollo program to the space shuttle. To secure approval of the shuttle, NASA promised to launch all U.S. payloads. Also the reusable orbiter was presented as a means of long-run cost savings: With regularly scheduled, once-per-week operational launches promised by the mid- to late 1980s, the shuttle was to pay for itself. To develop fifty shuttle payloads every year, however, would have required a space

budget ten times as large as NASA's actual budget. There was clearly an unrealistic presentation of feasibility on the part of NASA and uncritical thinking on the part of the U.S. Congress. The promises remain a root cause of pressure to launch the shuttle on schedule.

As Figure 1 shows, the shuttle consists of two solid rocket boosters (SRBs) to provide major thrust at launch, an external tank that carries fuel for the orbiter's main engines, and the orbiter, which carries the crew, payload, and main engines. The burnt-out SRB casings drop into the ocean where they are retrieved and later reused. The orbiter returns to Earth for servicing and reuse. The external tank is taken nearly to orbit before separation from the orbiter, and burns up on reentry. The official investigative reports, cited below, describe the shuttle, normal operations, and each disaster.

Shuttle development presented many design problems. One of the most challenging was a "thermal protection system" to protect the orbiter from the heat of reentry, when temperatures may exceed 5,000 degrees Fahrenheit. Another was providing a reliable seal between SRB segments.

Disasters Compared

The two disasters were very different superficially. The *Challenger* disaster occurred in the first moments of launch on an unusually cold January 28, 1986. Because of the cold weather, an O-ring seal between SRB segments leaked hot combustion gas, which quickly triggered the explosion that destroyed the vehicle. The dynamics of launch cause the joints between SRB segments to flex, and to prevent leaks the O-rings must be resilient enough to "follow" this flexure and maintain their seal. The cold O-rings were too stiff to follow the joint flexure.

The *Columbia* disaster culminated during reentry on February 1, 2003, after completion of the mission's on-orbit tasks. During launch the external tank had shed a large piece of foam insulation, which struck the orbiter's left wing, damaging its thermal protection system. Because of this unknown damage to the wing during launch, the heat of reentry destroyed the wing, leading to the breakup of the orbiter.

Similarities between the cases in three areas—no-return decisions, misunderstood anomalies, and overridden concerns from engineers—reveal the common ethical issues.

NO-RETURN DECISIONS. In both cases an explicit no-return decision left no chance to avoid disaster: For *Challenger* this occurred at launch—specifically, the ignition of the SRBs. For *Columbia* this came at initiation of reentry—the firing of the retro-rockets. Between the identification of an anomaly and this no-return decision there was time to have averted the disaster.

Regarding *Challenger*, the danger of a cold launch was suspected from heat damage to SRB seals—anomalies—in previous flights over several years. But the analysis of trends of seal damage as related to temperature omitted flights suffering no seal damage, all of which occurred at warm temperatures. This omission obscured the relationship of damage to temperature. If the many no-damage, warm launches had been considered, the significance of the few high-damage, cold launches would have emerged and convinced engineers that cold launches were unsafe (Vaughan 1996).

With respect to *Columbia*, occurrences of shedding of foam—anomalies—were known even before the *Challenger* accident. Foam strikes were "accepted" because efforts to prevent foam shedding were unsuccessful but flights were "successful." If NASA can fix the shedding problem in the halt in shuttle flights that followed the *Columbia* accident, so it could have during the similar halt after *Challenger*. This would have caused minimal (if any) delay and would have prevented the second disaster.

MISUNDERSTOOD ANOMALIES. The root cause of both disasters was misunderstanding anomalies. The 2003 *Columbia* disaster report quotes the 1986 *Challenger* report to show that the causes were identical. In effect, anomalies in performance—if followed by a successful landing—were considered evidence of safety instead of what they really were, evidence that the shuttle did not perform as designed. Thus safely landing after foam shedding or seal erosion reinforced the conviction of safety. This "normalization of deviance" violates the trust given NASA to accomplish human spaceflight safely (Vaughan 1996).

OVERRIDEN CONCERNS FROM ENGINEERS. In both cases working-level engineers most familiar with the relevant systems expressed timely concerns that could have averted the disaster, and their concerns were overridden. Regarding *Challenger*, engineers at the SRB contractor wanted to postpone the launch for a few hours or for a day for warmer weather, and were heard by company management in last-minute "readiness-to-launch" reviews, but management overrode them after NASA officials expressed frustration and desire to launch. They were overridden in part because of the inadequate trend analysis mentioned above. Warmer conditions could have averted the disaster. Desire to launch prevailed. With respect to *Columbia*, because the impact seemed more significant than the many previous instances of foam striking the orbiter, NASA engineers reviewing launch videos were alarmed. They requested a damage assessment but were overridden by management without a hearing. Had management honored the request, the disaster might have been prevented—the crew rescued but the orbiter lost (CAIB 2003).

The engineers did not push their arguments because of fear for their careers. Deciding to launch a shuttle had changed from a process requiring agreement that the system is safe to launch, per the design, to a process assuming launch and requiring anyone asking for delay to prove it unsafe. As "accepted" risks, damage to seals and strikes by foam were no longer an issue. This acceptance meant that a major foam strike on a launch shortly before *Columbia* (on October 7, 2002) was not declared an anomaly (CAIB 2003). Consistent with NASA's 1982 declaration of the shuttle as "operational," insulation strikes and seal damage became normal, while raising questions about these issues became deviant. William Langewiesche (2003) shows the depth of NASA managers' belief that insulation striking the orbiter was not a risk; he shows that only seeing an experimental demonstration of damage to a mock wing could destroy their belief, and that the demonstration left them in shock. Raising questions about foam shedding to such managers would damage one's career.

A healthy organization provides an environment and information conducive to decisions that advance the organization's goals within ethical constraints. Clearly, pressure to launch biased decisions by overemphasizing the partial, short-term goal of launching on schedule, reified in a lack of substantive, ethical discussion preceding the fatal no-return decisions. Astronauts, those most at risk, were not represented in the discussions. As the official reports reveal, typical predecision discussions were formal and procedural and laden with acronyms, emphasized the need to launch, and lacked ethical substance.

RADFORD BYERLY, JR.

SEE ALSO *Apollo Program; Engineering Ethics; National Aeronautics and Space Administration; Space Exploration.*

BIBLIOGRAPHY

Langewiesche, William. (2003). "*Columbia*'s Last Flight." *Atlantic Monthly* 292(4): 58–87. An accessible but rigorous analysis of the accident and NASA's reaction to it.

U.S. *Columbia* Accident Investigation Board (CAIB). (2003). *Report of the Columbia Accident Investigation Board*, Vol. 1. Arlington, VA: Author. The official report on *Columbia*; comprehensive.

U.S. House. Committee on Science and Technology. (1986). *Investigation of the* Challenger *Accident*. 99th Cong., 2nd sess. The congressional report on *Challenger*, essentially a supplement to executive branch *Challenger* report immediately below.

U.S. Presidential Commission on the Space Shuttle *Challenger* Accident. (1986). *Report of the Presidential Commission on the Space Shuttle Challenger Accident*. 5 vols. Washington, DC: Author. Official executive branch report on *Challenger*; comprehensive.

Vaughan, Diane. (1996). *The* Challenger *Launch Decision: Risky Technology, Culture, and Deviance at NASA*. Chicago: University of Chicago Press. How NASA's culture contributed to the accident.

INTERNET RESOURCES

U.S. *Columbia* Accident Investigation Board (CAIB). "The CAIB Report." Available from http://www.caib.us/news/report/.default.html.

U.S. Presidential Commission on the Space Shuttle *Challenger* Accident. "Report of the Presidential Commission on the Space Shuttle *Challenger* Accident." Available from http://history.nasa.gov/rogersrep/51lcover.htm.

SPACE TELESCOPES

• • •

The idea of a space-based telescope dates back to a proposal by R. S. Richardson in a 1940 issue of *Astounding Science Fiction*, but Richardson thought the moon would be a suitable venue. The U.S. proposal to put a telescope in orbit around the earth was made by Lyman Spitzer in "Astronomical Advantages of an Extra-Terrestrial Observatory," a paper written for a project for the Rand Corporation in 1946. In 1958, after a call for proposals by the Space Science Board of the National Academy of Sciences, the National Aeronautics and Space Administration (NASA) Space Sciences Working Group began developing proposals for orbiting astronomical observatories. The idea of an orbiting observatory received support at the highest government levels on the basis of arguments for national prestige, which was in need of shoring up after the launch of *Sputnik I* in 1956 by the Soviet Union.

Project Development

In 1960 and 1961 NASA initiated the process that eventually led to the Hubble Space Telescope (HST). It issued several calls for proposals for launch vehicles and astronomical hardware. By separating the two issues NASA created the grounds for serious planning problems because the limitations of the launch vehicle would have serious implications for the size and design of the observatory. By not insisting on coordinating the two from the start, NASA was, perhaps unknowingly, preparing the ground for later arguments about the constitution of the observatory.

In 1969 after debates among a variety of interest groups, the National Academy of Sciences clearly backed the proposal for a space-based telescope. NASA soon bought into the idea. However, NASA always has been and continues to be a management enterprise of considerable complexity with a myriad of problems that lead to difficulties in making decisions. Much decision

The Hubble Space Telescope, attached to a space shuttle. Named after Edwin Hubble, the telescope was launched into orbit in 1990 as a joint project of NASA and the European Space Agency. Initial optical errors were corrected in 1993, and high-quality imaging began in 1994. HST is projected to continue operating until 2009. (© 1996 Corbis.)

making at NASA is influenced strongly by politics. The many and often competing interests NASA managers felt they had to satisfy ranged from internally competing science groups to contractors, politicians, public interest groups, regional NASA facilities, and national priorities, along with international considerations. In addition, there was always competition from other NASA projects. Funds were limited, and the demands were many. The space telescope, as was the case with many other projects, stalled.

Among other activities under way at NASA at the time when the space telescope was being debated was the planning of a space shuttle program, which was approved in 1972. To restart the stalled planning for the space-based telescope, NASA proposed that the launch vehicle for the telescope be the shuttle. That proposal had serious design implications for the telescope, which would have to fit into the baggage bay of the shuttle.

The Large Orbiting Observatory project was beset by arguments that delayed its completion. There were arguments over where the central control would be: The Goddard Space Flight Center at Beltsville, Maryland, or The Marshall Space Flight Center at Hunbtsville, Alabama.. There were arguments over who would have authority over what; what kinds of instruments should be built; how much money was available; which con-

tractor would build the instruments; how much existing technology, such as military spy satellite technology, could be appropriated; and eventually, who would be blamed for the big mistake of the spherical aberration of the primary mirror and how it would be fixed.

The Large Orbiting Observatory, by now called the HST, was completed in 1986, shortly before the *Challenger* disaster. The grounding of the shuttle program forced a four-year delay in launching the HST. When the HST finally orbited in 1990, it was discovered immediately that its primary mirror had a spherical aberration: The images it sent to earth were blurry. After a number of investigations, including congressional hearings, it was concluded that the mistake was due to a failure of both the engineering team at the contractor for the mirror, Perkin-Elmer, and its management. Perkin-Elmer agreed to repay the government $25 million.

The problems with the Hubble eventually were fixed, and the HST has been instrumental in revolutionizing scientists' conception of the universe. It allowed astronomers to look deeper into space than ever before, revealing features of the universe that confirmed some theories and made others doubtful.

Reflections

When one reflects on the history of the HST, the variety of factors that played a role in its development, and its impact on astronomical understanding, several themes emerge. First, the building of a large and expensive scientific instrument is not a simple process. Furthermore, instruments with the size and complexity of the HST require such vast resources that only a national government or another entity capable of putting together a conglomerate of considerable size can undertake a project of such magnitude. Second, in a world of limited resources the commitment to undertake one project of that size means that other projects will suffer. Thus, not only was there considerable tension between advocates of earth-based telescopes and advocates of space-based ones, directing funds toward the HST meant that less money was available for new and larger earth-based telescopes. Third, most of the conflicts involving the HST were clashes of values that often were multidimensional.

The initial battle over launching a large space telescope as opposed to several smaller, more specialized telescopes was not just an argument about whether the project was feasible. In a 1983 symposium sponsored by the Smithsonian Institution the physicist Freeman Dyson (b. 1923) argued against the idea of doing science with instruments with the size and scale of the HST and for a smaller, diversified kind of science employing specialized, smaller, and much cheaper instruments. Dyson was arguing against big science, which had become a distinctive characteristic of the U.S. physics community.

Dyson may have had a point. The U.S. physics community had continued to rely on large instrument projects to a risky extent. The lesson was learned the hard way when the Super-Conducting Super Collider (SSC) project was canceled fifteen years after it had been proposed and billions of dollars had been spent. The physics community reacted as if it had received an amputation: It had no visible capacity to do microphysics at the cutting edge.

This episode shows the flaw inherent in insisting on a hegemony in a science. The lesson to be learned from the Hubble, however, actually goes in the other direction. The turn to big science/technology need not limit the scientists to one large project; it also can generate small science projects in its wake. Smaller and less expensive types of telescopes, such as an infrared telescope, are being placed in orbit to discover what the Hubble could not reveal. What was missing from the thinking about the SSC were ideas about what would follow from it by way of subsidiary projects such as smaller more specialized experimental devices.

The HST illustrates other value clashes as well. Many people argue against this kind of project while people are suffering from hunger, disease, and lack of education. Big science/technology, it is claimed, is a luxury at a time when many millions are living in misery. This is a hard argument to refute, and it is not clear that one should try. It is important to be reminded of the human cost of science and technology. At the same time it is possible also to consider another human dimension to big science/technology that although it does not refute the argument from human physical need speaks to a different form of human need.

In the Middle Ages there was much misery. In Europe most of the population lived in squalor, disease was rampant, and ignorance was the norm. However, despite those circumstances, people in that era gave of their time, labor, and meager belongings to build some of humankind's most magnificent edifices: Gothic cathedrals. The cathedrals of Europe present a statement of humanity's commitment to seek more than it can find on earth. Projects such as the HST may be considered a continuation of that quest.

JOSEPH C. PITT

SEE ALSO *National Aeronautics and Space Administration; Space Exploration*.

BIBLIOGRAPHY

Chaisson, Eric J. (1998). *The Hubble Wars*. Cambridge, MA: Harvard University Press. A personal account of some of the internal battles in the Hubble saga.

Needell, Allan A., ed. (1983). *The First 25 Years in Space: A Symposium*. Washington, DC: Smithsonian Institution Press.

Pitt, Joseph C. (2000). *Thinking about Technology*. New York: Seven Bridges Press. Formulates the author's philosophical position behind the observations at the end of this article concerning value conflicts.

Smith, Robert W. (1989). *The Space Telescope*. New York: Cambridge University Press. The most detailed and definitive history of the Hubble project.

SPECIAL EFFECTS

• • •

Special effects (which typically refers to visual effects in live-action moving-image media but also includes audio effects and other possibilities) are the methods used to produce on-screen (or on-air) events and objects that are physically impossible or imaginary, or too expensive, too difficult, too time-consuming, or too dangerous to produce without artifice. The ethics of the related technologies are seldom discussed but are nevertheless significant.

Origins

Cinematic special effects grew out of trick photography and began with the trick film tradition popularized by early filmmakers such as Georges Méliès (1861–1938), a special effects pioneer who was the first to develop many in-camera techniques. Silent films used a variety of special effects techniques, particularly in the genres of science fiction and horror. Many new special effects technologies became possible after the invention of the optical printer in 1944, resulting in a new generation of science-fiction films in the 1950s that used the new techniques, as well as more realistic-looking effects in other films. Finally, the late 1980s and 1990s saw another advance in effects technology: the rise of digital special effects created in computers, which allowed live-action footage to be combined with anything that could be rendered in computer graphics.

Special effects are a large part of the film industry in the early twenty-first century, with a number of companies such as Industrial Light & Magic and Digital Domain specializing in the production of special effects.

Jim Carrey as Stanley Ipkiss in a scene from the 1994 film *The Mask*.

Special effects can be found in almost every genre of filmmaking, in both big-budget and low-budget films, as well as on television, most notably in advertising, where high budgets and short formats allow filmmakers to experiment with expensive new techniques.

Types of Special Effects

Special effects can be divided into four types: practical effects, in-camera effects, optical effects, and digital effects. Practical effects, also known as physical effects, are those that occur in front of the camera, such as rigged explosions, pyrotechnics, animatronics figures or puppetry, makeup effects, and so forth. Practical effects have the advantage of occurring on the set where they appear directly in the scene and the action of the shot, and require no postproduction processes.

In-camera effects are achieved through forms of trick photography and are made in the camera at the time of shooting. Such effects include shots taken at different camera speeds, shots using lens filters, and day-for-night shooting, all of which change the kind of image being recorded. Superimpositions and multiple-exposure matte shots require the film to be exposed, rewound, and exposed again, adding two or more images together onto the same piece of film before it is developed (this combining of imagery is also called *compositing*). Foreground miniatures, glass shots, and matte paintings make use of the monocular nature of the camera by falsifying perspective and making small objects close to the camera look as if they are part of larger objects farther away from the camera. Buildings can be extended and other large set pieces can thus be made inexpensively through the use of detailed models and paintings done with the correct perspective. Front projection and rear projection processes combine foreground sets and actors with backgrounds made from projected imagery (most typically as moving background imagery placed behind an actor driving a car).

Optical effects involve the use of an optical printer, a device invented by Linwood Dunn in 1944 that allows images on developed pieces of film to be rephotographed and composited together onto a single piece of film. An optical printer is basically a camera and a projector (or multiple projectors, in some cases) set up with a camera in such a way that film frames can be rephotographed directly from another strip of film. Optical processes allow frame-by-frame control and greater preci-

A special-effects artist signs autographs near a model of the character Gollum from the *Lord of the Rings* film trilogy. The groundbreaking CGI character was built around an actor's voice, movements, and expressions by using a motion capture suit which recorded his movements and applied them to the digital character. (© *Reuters NewMedia Inc./Corbis.*)

sion in spatially positioning elements than is possible with in-camera compositing. Perhaps the most common form of optical compositing is the matte shot, wherein a foreground element is combined with a background, without the background visible through the foreground element (as would be the case with superimposition). To achieve this, keying processes are used for the production of foreground elements, and the most typical of these, blue screening and green screening, place the actor in a solid-color background, which is later optically removed from the shot. A holdout matte is made from the foreground element, which leaves a part of the rephotographed background plate unexposed, and the foreground element is later exposed onto the same plate, fitting into the unexposed area. Traveling mattes also make this technique possible for moving objects and moving camera shots.

Digital effects are all done in a computer. Images are either shot with digital cameras or scanned from film into a computer, where they are edited and composited digitally. Digital effects avoid the generational loss (the loss

that occurs when film images are rephotographed onto another piece of film) that happens during optical rephotography, and the computer makes matteing much easier and faster and gives the effects technician greater control over the image. Digital effects technology also allows computer-generated imagery to be combined with live-action footage, and allow images to be controlled down to individual pixels. Light, shadow, and color can all be adjusted, and digital grading can replace color correction and matching that was previously done during the color timing (the matching of colors from shot to shot during postproduction) of prints in postproduction. Digital effects were experimented with during the 1980s and came into common use during the mid-1990s as techniques were developed and computer systems became powerful enough to make digital effects work affordable.

Some special effects (such as dinosaurs, space battles, monsters, and so forth) are obviously special effects no matter how well they are done, because the objects or events they portray clearly do not or no longer exist. Other effects, known as "invisible effects," are less noticeable because they portray objects and events (for example, background buildings, smoke, and building extensions) that do not call attention to themselves and that usually could have been done conventionally had the budget allowed it. Another type of invisible effects are effects in which something is erased or removed from the image. One example is wire removal, in which the wires used to fly an actor or object are digitally erased during postproduction.

Ethics

The alteration and faking of photographs has existed as long as photography itself. Whether or not the use of special effects is ethical depends on the intentions and truth claims of the work in which they appear. By altering, combining, or fabricating images, special effects work reduces or removes the correspondence, or indexical linkage, that an image may have to its real-world referent. Thus, while special effects may be acceptable in films that are fictional or are clearly re-creations of events, one would not expect to find them in news or documentary footage that claims to be a record of actual events. Even when they are used in an entirely fictional film, how special effects are used can still greatly determine how a film is received by an audience. For example, Jackie Chan's earlier films, in which he actually does all his own stunts, are more impressive than his later films in which some of his stunts are the result of wire work and special effects. Likewise, while the digital crowd scenes in *The Lord of the Rings: The Two Towers* (2002) and *Star Wars, Episode II: Attack of the Clones* (2002) are impressive, one is still aware that they are special effects, unlike the massive crowd scenes in older movies such as *Gandhi* (1982) and the Russian version of *War and Peace* (1966–1967), which were all done using actual crowds. At the same time, not only are special effects used to create spectacle, but their creation itself has become a spectacle, as witnessed by "making of" featurettes often found among the DVD extras. For many, knowing how an effect was made can enhance the viewing experience rather than spoil the effect.

Advances in special effects have made fantastic ideas possible and allowed filmmakers to give them concrete expression. The fact that many effects in the early twenty-first century are photo-realistic and seamlessly integrated into live-action footage also means that a discerning viewer will need a certain degree of sophistication. Combined with unlikely storylines, the use of special effects, which makes unlikely or impossible events appear possible and plausible, may help to erode the ability of younger or unsophisticated viewers to distinguish between what is plausible and what is not. Despite the fact that the films in which special effects appear are often clearly fictional, seeing photo-realistic representations of what look like actual events can make an impression on some viewers, particularly in a culture in which so much of what people see of the world is mediated through film and television imagery. At the same time, because of magazines, books, and DVD extras detailing special effects techniques and technology, contemporary viewers often are more aware of how special effects are done and how they are incorporated into a film.

MARK J. P. WOLF

SEE ALSO *Computer Ethics; Entertainment; Movies; Video Games*.

BIBLIOGRAPHY

McAlister, Michael J. (1993). *The Language of Visual Effects*. Los Angeles: Lone Eagle Publishing. A dictionary of special effects terminology.

Rickitt, Richard. (2000). *Special Effects: The History and Technique*. New York: Watson-Guptill. A good overview of the history and techniques of special effects.

Vaz, Mark Cotta, and Craig Barron. (2002). *The Invisible Art: The Legends of Movie Matte Painting*. San Francisco: Chronicle Books. A book on a specific effect, matte paintings, which are the background images used in wide shots and combined with live action.

Wolf, Mark J. P. (2000). "Indexicality." In *Abstracting Reality: Art, Communication, and Cognition in the Digital Age*. Lanham, MA: University Press of America. A look at how digital technology has changed art, communication, and how people view their world and environment.

SPEED

• • •

The word *speed* is derived from the Middle English *spede* (good luck), which in turn originated from older roots meaning to prosper or succeed. In its contemporary usage, speed refers to a rate of change. It commonly denotes the time it takes to travel a certain distance (e.g., a rate of 60 miles per hour), but it is also used to describe the time needed to perform certain tasks or operations, especially in information processing (e.g., a computer with a 500-megahertz processing speed). Individual artifacts such as cars, airplanes, and computers are achieving ever-greater speeds, which has effectively decreased and in some cases nearly eliminated distance. The speed of modern travel and communication has shrunk the world and radically altered the experience of time and place for individuals, corporations, and nations. Increased speed at this level of analysis presents several important safety and ethical issues.

The Technological Singularity and Other Analyses

But even more profound implications derive from the speed at which the very processes of technological innovation and knowledge creation occur. Moore's law (holding that growth in the number of transistors per integrated circuit will be exponential) was generalized to all technologies by Raymond Kurzweil in his "law of accelerating returns." Some futurologists claim that this acceleration will lead to a "technological singularity." This denotes the point in the development of a civilization at which technological change accelerates beyond the ability of present-day humans to fully comprehend, guide, predict, or control it. It derives mostly from the use of the term *singularity* in physics to indicate the failure of conventional models to predict change as one approaches a gravitational singularity—an event or location of infinite power such as a black hole, where matter is so dense that its gravity is infinite. When a black hole absorbs nearby matter and energy, an event horizon separates this region from the rest of the universe, constituting a rupture in the structure of space and time. Vernor Vinge (1993) developed the concept of technological singularity and applied it more specifically to the advent of greater-than-human intelligence. Beyond the technological singularity lies a fundamentally transformed world, perhaps one dominated by machines that have goals inconsistent with those of humanity. Vinge concluded that if the singularity can happen it will, because the competitive advantage afforded by advances in technology assures their implementation.

Many other analyses of modernity have noted this acceleration and described its personal and social consequences. Theodore Kaczynski, the Unabomber, warned of its actual and impending dehumanizing effects. Alvin Toffler (1970) summed up this wider rendition of speed with his coinage "future shock," as the overwhelming rate of change transforms institutions, shifts values, and undermines cultural and personal foundations. Toffler argued that the rate of change can be even more important than the direction of change in terms of psychological and social impacts. With his concept of "cultural lag," William F. Ogburn (1922, revised 1950) focused more on differential rates of change between interde-

pendent parts of society. For example, science and technology usually operate at a much faster—though in his 1950 revised version, Ogburn admitted it might not be an ever increasing—rate than cultural beliefs and social institutions. Deborah G. Johnson (2001) argued that this differential speed creates "policy vacuums" as social decisions lag behind technological innovation. The French essayist and urbanologist Paul Virilio (1995) similarly claimed that immediacy and instantaneity present the most pressing challenges and ethical concerns at the personal, economic, political, and military levels.

Perception and Experience

In a psychological and even existential sense the perception of relative speeds is rooted in the workings of human consciousness. Oliver Sacks (2004) noted how early psychologists used developments in cinematography to elucidate the perception of time. Late-nineteenth-century innovations in cinecameras allowed photographers to register larger or smaller numbers of events over a given period by adjusting the frames exposed per second. This allowed them to capture the frenzied flapping of bees' wings or the slow unfurling of fern crosiers and re-present them at the rate of normal human perception.

In his *Principles of Psychology* (1890), William James (1842–1910) used the metaphor of altering the frames per second exposed to light to explain the human perception of time. If we were able to process 10,000 events per second instead of the usual ten, then time (measured, as it must be, by our experiences or sense impressions of the world) would slow down. So too, if we were able to process only one-thousandth of the sensations per second than normal, then time would speed up. In the former case, the sun would stand still. In the latter case, mushrooms would spring up and shrubs would rise and fall like restlessly boiling water. Human consciousness is a roll of film spinning at such a rate as to expose a certain number of frames per second, thus giving rise to normal perceptions of time and the speed, as it were, of human awareness or being.

Later sensory psychologists have examined cases of aberrant time perception. For example, several subjects have reported a tremendous slowing of time when suddenly threatened with mortal danger. The metaphorical explanation often proposed for these phenomena is that the human brain, in moments of extreme stress, is able to reduce the duration of individual frames and expose more of them per second. This accelerates thought and increases the speed of decision-making capabilities. From a physiological perspective, such instances may result from a flood of excitatory or a relaxation of inhibitory neurotransmitters.

Certain drugs also provide departures from normal time. Hashish makes events appear to slow down, whereas mescaline and amphetamines accelerate them. Indeed the latter drug is commonly referred to as "speed," indicating the subjective, phenomenological quality of time as a function of brain chemistry and consciousness. Sacks notes there are persistent disorders of neural speed, some of which can be caused by encephalitis lethargica and Parkinson's disease. Some patients can experience radical slowing of thought and movement, which can sometimes be reversed by reducing dopamine deficiencies with the drug L-dopa.

On another experiential level, Virilio states that the primary consequences of the increasing speed of modern life are personal, amounting to disorientation concerning reality. He argues that the globalized, instantaneous flows of information in cyberspace undermine the deep-seated spatial and temporal anchors of the human experience. His views find support, for example, in the way that some virtual relationships have led to tragic decisions by teenagers who become victims of sexual predators on the web. The lightning speed of cyberspace communications has undoubtedly altered fundamental human experiences such as love and intimacy. In many urban areas the Internet is reshaping dating and courtship. Love at hyperspeed brings conveniences by matching supply and demand in a more systematic fashion than haphazard meetings, but it also shifts the meaning of relationships in ways that require personal and social adjustment.

The speed of Internet and satellite communication provides the benefit of instantaneously connecting loved ones separated by great distances. Cyberspace, however, may give only a false sense of closeness. For example, the members of a suburban family in the United States usually have hectic schedules that scatter them significantly in physical space and, when by means of a cell-phone family plan, computer messaging, or both, they succeed in communicating mostly on-the-go, this form of communication eclipses more traditional ones occurring in such shared places as the dinner table or the living room.

It is nevertheless important not to romanticize the past. At least since the 1950s in industrialized countries, family time and communication between fathers and their children were infrequent in many households. The increase of dual-income families and the rise of television viewing have further undermined family intimacy. Nonetheless, the experience of cyberspace communica-

tions is qualitatively different in that the interlocutors' bodily presences and languages are absent from voice or text messages.

Despite variances in the range of speeds at which human thought can operate, there are basic neurological determinants that limit human cognitive capacities (e.g., serial computations, recognition, and associations). These limits are frequently tested by the accelerated flows of information and technical change in modern life, but drugs, supplements, and perhaps even neural human–computer interfaces may be able to expand cognitive processing speeds. Cognitive prostheses can improve human cognition, much as eyeglasses improve vision. For example, an airplane cockpit display has been developed that shows crucial information so that a pilot can understand what the aircraft is doing in a fraction of a second instead of the usual few seconds (Bower 2003). Such technologies are based in human cognitive studies research on information processing and visual tracking.

There is, however, controversy about whether such mind-expanding devices are a blessing or a curse, because they bring about even greater pressures by increasing the speed of information processing. This raises the stakes in case of human or machine error. Beyond concerns of safety, however, these actions raise profound issues about how humans synchronize with nature and society. Toffler (1970) echoed the sentiments of many critics of modernity by suggesting that there is something dangerous and even alienating about the rapid tempo of change. Individuals and society are maladapted to such breakneck speeds, and we require social and personal mechanisms to regulate change and decelerate it to a more human pace.

Economic Consequences

At least since Karl Marx's critique of industrial capitalism in the mid-nineteenth century, many theorists and workers alike have disparaged some of the effects of greater speed introduced into manufacturing processes by automated production equipment. They argue that these devices should conform to the humans operating the equipment, not the other way around; otherwise, increased speed jeopardizes the physical and mental health of workers. Critics also point out that these changes often involve exploitation by decreasing bargaining power, pay, status, and/or self-esteem. The increased speed and efficiency of machines has also caused unemployment as human workers become less profitable. Tracking the economic consequences of technological innovation is difficult, however, because it often creates new employment opportunities elsewhere.

The increased speed of financial and economic activity raises more concerns than just competitiveness versus risks to physical and mental health. Indeed, on a larger scale, it could be argued that the competitive profit motive driving capitalism is a major cause of the accelerating pace of modern life. Internet transactions have globalized financial markets as investments can be made at the speed of light and funds shuffled between countries at the press of a button. Transnational businesses are able to create information networks that bypass the traditional power of the nation-state. Toffler (1980) noted the rise of "third wave" societies based on information, communication, and technologies operating at rapid speeds. Not only does this shift power in the sense that nonstate actors make more and more major decisions, but it also increases the interconnectivity of third wave countries because communication linkages and knowledge have largely replaced industrial processes as their economic lifelines.

Interconnectivity brought about by increased reliance on swift, automated information technologies allows for a more fluid and responsive economy and greater specialization of production. It also, however, increases volatility and vulnerability to shocks anywhere in the system. This had led some (e.g., Siegele 2002) to propose the need for economic "circuit breakers" to protect global markets from cascading failures. Such precautionary measures and restrictions, however, need to be balanced against the benefits of free flows of global capital. Furthermore, even if the speed and integration of information flows may lead to more sudden downturns, they can increase the rate of economic recovery as well. Nonetheless, economic laws, regulations, and institutions are forced to globalize at the same speed as the technology in order to secure and harmonize economic activities.

The instantaneity of communication has generated the real-time economy, which has large macroeconomic effects and impacts at the level of individual companies. Real-time enterprises, ideally, will be able to monitor internal and external conditions in order to react to changes instantaneously. Through increased communication with customers, they will also be able to rapidly offer new products and services, thus more tightly coupling demand and supply. The flood of information threatens to overload companies, which have responded by developing software to optimize supply chains and automate certain responses to real-time cues.

Rapidly changing markets and technologies increase competitive pressures for firms to increase integration and flexibility, which can lead to organizational problems. The emergence of the real-time economy more directly pins economic vitality on the smooth functioning of integrated technologies. A software virus, for example, could cause massive economic collapse. Ludwig Siegele (2002) offers the conclusion that such drawbacks are not inherent in the technologies, but arise from the way they are used. But he adds, "it is worth asking to what extent we want computers to run our lives" (p. S20).

Cultural benefits are also generated by the speed of new communication technologies. For example, the time gap between the release of a Hollywood movie in the United States and its debut elsewhere in the world has been drastically cut, symbolizing the free flow of art and culture made possible by these new speeds. In some cases this may foster greater cross-cultural understanding and tolerance. Some, however, perceive this as a threat to local economies and cultures, which now must accelerate to keep up with foreign competition. Cultural homogenization may result.

Social and Political Consequences

Economic consequences of increased technological speed spill over into social changes. Harriet B. Presser (1999) noted that the use of rapid communication technologies is one factor in the widespread prevalence of nonstandard work schedules. The globalization of markets and the ability to be "on call" all the time require expanded hours of operation. This affects the family lives of workers and requires social institutions such as daycare to adapt to changing needs. The increased reliance on rapid communication technologies by the military also carries social and political consequences. Such advances in the U.S. military have tested the limits of telecommunication capacity, or bandwidth, which is expensive to expand. Although the real-time information gained can help protect both troops and civilians, politicians face trade-off dilemmas concerning the best investment of public funds.

Real-time politics has brought both beneficial and detrimental effects to democratic processes. The immediacy of citizen participation in government may contribute to political accountability and strengthen civic commitments. For example, the Internet has sparked a new wave of social responsibility by organizing protestors around the world (McPherson and Schapiro 2001). It allows like-minded activists to communicate, build consensus, coordinate activities and information, and provide mutual moral support. Campaigns against "sweatshop" labor have been primarily organized via the Internet. Such forms of communication may even help foster democracies in nations controlled by tyrants. One drawback, however, is that passions unleashed at the speed of the Internet often outstrip facts and evidence, which can delegitimize well-meaning social reformers.

Other negative effects can result from real-time politics. Virilio argues that representative democracy is undermined by the virtualization of government and the rise of opinion democracy patterned on viewer counts and opinion polls. Political leaders may pander to public opinion rather than make unpopular, but perhaps better, decisions. Public opinion polls often reflect short-term interests, whereas leaders must balance these with long-term common-interest goals. The greater speed of communication often undermines careful deliberation and reasoned judgment, but it can also better inform such deliberation. But referendum reforms were altering the balance of participatory and representative democracy before the real-time computerization of politics. So, cyber-speeds may aggravate more than cause this dilemma.

Increasing speed of information flows can exacerbate the complexity and multiplicity of policy issues, leading to issue overload. This is a situation in which the multitude and complexity of issues exceeds what individuals can understand and societies can handle through the courts (leading to court-case overload), legislation (producing tunnel-vision laws), or executive or other institutional channels (Breyer 1993).

On a larger scale, Stewart Brand (1999) argued that the accelerating pace of technological change, the short-term perspective of consumerist lifestyles, and the short-term focus of political election cycles have all eroded the concept of long-term responsibility. The acceleration of experiential time effectively reduces the timescale of interest, thus shrinking the horizon of felt obligation. Brand writes, "Our ever hastier decisions and actions do not respond to our long-term understanding, or to the gravity of responsibility we bear" (p. 8). In order not to be doomed by speed, we must slow down enough to allow time to apply the brakes in case of emergencies.

He proposed a balancing corrective to this short-sightedness to help us accept our long-term responsibilities to nature and future generations. In cooperation with others, Brand founded the Long Now Foundation in 1996 and began to design the Clock of the Long Now, a giant mechanical clock to be set somewhere in the U.S. desert to record time for 10,000 years. The goal is to embody deep time in a way that counterbalances the shrinking timescales experienced by those caught up in the speed of modern life.

A. PABLO IANNONE
ADAM BRIGGLE

SEE ALSO Artificial Intelligence; Communication Systems; Cyberculture; Cyberspace; Economics and Ethics; Information Overload; Internet; Turing Tests; Work.

BIBLIOGRAPHY

Bower, Bruce. (2003). "Mind-Expanding Machines: Artificial Intelligence Meets Good Old-Fashioned Human Thought." Science News 164(9): 136–138.

Brand, Stewart. (1999). The Clock of the Long Now: Time and Responsibility. New York: Basic.

Breyer, Stephen. (1993). Breaking the Vicious Circle: Toward Effective Risk Regulation. Cambridge, MA: Harvard University Press.

Johnson, Deborah G. (2001). Computer Ethics, 3rd edition. Upper Saddle River, NJ: Prentice Hall.

McPherson, Michael S., and Morton Owen Schapiro. (2001). "When Protests Proceed at Internet Speed." Chronicle of Higher Education, 23 March, B24.

Ogburn, William F. (1922). Social Change with Respect to Culture and Original Nature. New York: Huebsch. Ogburn's first systematic formulation of cultural lag as a part of his theory of social change, which was presented in the chapter "The Hypothesis of Cultural Lag." Later editions, including the 1950 revision of the book, were published by Viking.

Presser, Harriet B. (1999). "Toward a 24-Hour Economy." Science 284(5421): 1778–1779.

Sacks, Oliver. (2004). "Speed: Aberrations of Time and Movement." New Yorker 80(23): 60–69.

Siegele, Ludwig. (2002). "How About Now? A Survey of the Real-Time Economy." Economist, 2 February, S3–S20.

Toffler, Alvin. (1970). Future Shock. New York: Random House. Argues that there can be too much change in too short of a time, and therefore we need to improve our ability to wisely regulate, moderate, and apply technology to serve human ends.

Toffler, Alvin. (1980). The Third Wave. New York: Morrow. Develops the thesis that information societies are a qualitative shift beyond industrial societies and explains several consequences of this shift.

Vinge, Vernor. (1993). "Technological Singularity." Whole Earth Review, no. 81: 88–95.

INTERNET RESOURCE

Virilio, Paul. (1995). "Speed and Information: Cyberspace Alarm!" trans. Patrice Riemens. CTHEORY.net, 27 August. Available from http://www.ctheory.net/text_file.asp?pick=72.

SPENCER, HERBERT

• • •

British philosopher and sociologist, Herbert Spencer (1820–1903) was born in Derby, England, on April 27, and became well known for developing and applying evolutionary theory to sociology, philosophy, and psychology. Following an informal education in the anti-establishment views of his father, he briefly trained as a civil engineer before becoming a journalist and political writer. Spencer began writing books in the early 1850s, and presented a systematic and comprehensive account of his views on ethics, sociology (government, politics, and education), and biology in the nine-volume *A System of Synthetic Philosophy* (1862–1893). Although his ideas were influential during the last few decades of the nineteenth century, his reputation subsequently waned. Spencer died in Brighton, England, on December 8.

Basic Ideas

Spencer's scientific and empirical method exhibits affinities with Auguste Comte's positivism. Central to his approach was the synthetic practice of deriving fundamental principles from disparate phenomena in many sciences and then demonstrating how the principles of one science interact with and affect the other fields of inquiry. Using Charles Darwin's evolutionary theory, Spencer thus constructed a general account of human progress that came to be known as "Social Darwinism."

For Spencer, natural progress was the necessary process of evolution from simple to more complex and heterogeneous forms, but this was not, he insisted, teleological or purpose-driven. Spencer coined the phrase "survival of the fittest," which Darwin employed in later editions of *On the Origin of Species* (first published in 1859), but neither thinker addressed the ambiguity (that is, are individuals, groups, or species the relevant unit of selection?) and near tautology of this phrase ("fitness" is often defined in terms of survival, so that survival of the fittest is akin to saying survival of that which survives the best). Although Darwin admired Spencer, the two disagreed on several aspects of evolutionary theory including the possible inheritance of acquired characteristics.

Human life is on a continuum with the evolutionary unfolding of the natural world, and, because progress toward complexity and individuation are necessary, human nature cannot be thought of as stable and unchanging. Rather, humans are collections of instincts and sentiments that must continually adapt to the chan-

Herbert Spencer, 1820–1903. Spencer was an English philosopher, scientist, engineer, and political economist. In his day his works were important in popularizing the concept of evolution and played an important part in the development of economics, political science, biology, and philosophy. (*The Library of Congress.*)

ging societal context. Society is likewise an extension of the organic human body and nature. Finally Spencer argued that society too expresses evolutionary laws or principles that can serve as the foundation of morality and law. Evolutionary science, then, serves as the base of his comprehensive natural law philosophy of morality and politics and explains how *The Principles of Biology* (1864, 1867) flows naturally into the conclusions reached later in *The Principles of Sociology* (1882, 1898) and *The Principles of Ethics* (1892).

Spencer believed that modern evolutionary science had weakened traditional beliefs in ethics as a supernatural code of divine commandments. Science could fill this ethical vacuum left by religion, by providing the principles from which to deduce a naturalistic ethics of rational egoism. Science ought, therefore, to command the dominant position in education, displacing art and the humanities (1861). Spencer reconciled the apparent contradiction between his naturalized, a-teleological laws of society and morality, on one hand, and human freedom and purpose, on the other, by arguing that it is precisely individual freedom that alone can guarantee continued evolutionary progress. Indeed for Spencer, individual liberty is primary and relations with others are largely contractual, made from the realization that social life is necessary to reach certain individual goals.

Furthermore, in a move that is similar to John Stuart Mill and the logical commitment implied in Alan Gewirth's "principle of generic consistency," Spencer claimed that morality contains a "law of equal freedom." This law states that individuals must recognize the individuality of others and curtail their freedom so as not to infringe on the freedom of others. This sort of minimalist, contractual view of society underpins his *laissez faire* political philosophy from *Social Statics* (1851) to *Man versus the State* (1884). The state's function is condensed to dispensing justice, which amounts to protecting individual rights. These rights follow naturally from the law of equal freedom, because the recognition of others' individuality immediately implies the duty to recognize their rights.

Decline and Continuing Influence

Spencer's decline can be attributed to several inconsistencies in his work, growing social unease with founding society on evolution, social rejection of his strongly libertarian principles, and the demise of any residual scientific belief in the inheritance of acquired characteristics. Yet some of Spencer's voluminous thoughts continue to be of influence. His work on intellectual and physical education has left deep imprints on modern curricula. His political thought, especially his defense of natural rights, has been invoked by libertarian philosophers such as Robert Nozick. And Spencer's idea that nature shows a progressive trend toward increased complexity of organization has been revived by some biologists and social theorists. Robert Wright (2000) argues that evolution tends to produce ever more complex forms of life, because cooperation through expanded forms of organization produces selective advantages. In human social evolution, this explains the move from primitive hunting-gathering tribes to large states and finally to global systems. New technologies—such as the agricultural production of food or the transmission of information through computer networks—make possible wider forms of social cooperation.

The evolutionary theorist Stephen Jay Gould (1989) nevertheless rejected Spencer's idea of progressive evolution and argued instead that the history of life is a random process that could have turned out differ-

ently. By contrast paleobiologist Simon Conway Morris (2003) sees evidence for evolutionary patterns inclined to produce intelligent life. If Gould is right, then the human sense of purpose has no ontological support. If Conway Morris is right, human purposefulness might fulfil an end inherent in the universe from the beginning. The fundamental issue—with deep moral and religious implications—is whether the universe is pointless or purposeful. This was also the central tension underlying Spencer's lifelong attempts to bridge the natural and the human worlds.

ADAM BRIGGLE

SEE ALSO *Evolutionary Ethics; Scientific Ethics; Social Darwinism.*

BIBLIOGRAPHY

Conway Morris, Simon. (2003). *Life's Solution: Inevitable Humans in a Lonely Universe.* Cambridge, UK: Cambridge University Press. Contra Gould, Morris argues that the historical course of evolution is strongly selection-constrained and thus progress toward intelligent beings is inevitable.

Gould, Stephen Jay. (1989). *Wonderful Life: The Burgess Shale and the Nature of History.* New York: W.W. Norton. Details the science behind different interpretations of the Burgess Shale formation in order to put forward his notion of biological contingency or the non-directedness of life's history, namely, that small initial differences could have made subsequent evolution radically different.

Spencer, Herbert. (1851). *Social Statics.* London: Chapman. Argues that civilization is a natural, continual process of humans adapting to changing circumstances and that progress toward perfection is the same as the achievement of a perfect adaptation to surroundings.

Spencer, Herbert. (1861). *Education: Intellectual, Moral, and Physical.* London: Williams and Norgate. Outlines Spencer's educational theory.

Spencer, Herbert. (1862–1893). *A System of Synthetic Philosophy.* 10 vols. London: Williams and Norgate. Contains all of Spencer's major works, including *The Principles of Psychology* (1855), *First Principles* (1862), *The Principles of Biology* (1864, 1867), *The Principles of Sociology* (1882, 1898), and *The Principles of Ethics* (1892).

Spencer, Herbert. (1884). *Man versus the State.* London: Williams and Norgate. Champions a *laissez-faire* state and free market toward the ultimate goals of freedom, peace, and justice.

Wright, Robert. (2000). *Nonzero: The Logic of Human Destiny.* New York: Pantheon. Uses game theory to argue that cultural evolution leads to higher levels of complexity just as biological evolution does.

SPENGLER, OSWALD

• • •

Oswald Spengler (1880–1936) was born in Blanken-
burg, Germany, on May 29, and attended the universities of Munich, Berlin, and Halle, where he studied
mathematics and the natural sciences, which led to his
becoming a secondary school teacher of mathematics in
Hamburg. He abandoned teaching in 1911 to work on
his magnum opus—*The Decline of the West* (1918–
1922)—which he did steadily during the World War I.
He intentionally published the first volume to coincide
with the German military defeat and industrial collapse
of 1918, and the second four years later. From this time
until his death in Munich on May 8, he wrote other,
shorter books and pamphlets on social and political subjects, including *Man and Technics* (1931).

Despite his marginal status in the German academic world and the controversy with which his ideas
were greeted, Spengler's influence on social science was
far greater than that of those who tried furiously to
refute him. His impact derives from the fact that in
examining the nature of Western Europe and North
America he makes predictions about its future, drawing
inferences based on a metaphysical reading of history
during a period of serious crisis.

The key to Spengler's philosophical anthropology
and accompanying philosophy of history is his use of the
Faustian legend in popular German literature to interpret modern technology. According to him, humans are
the only predators able to select and design weapons for
attacking nature and each other. At some point around
the tenth century this ability developed to such an
extent in Western European culture that humans seized
for themselves the prerogatives of domination over nature. This inexorable destiny is a radical break with earlier periods of thought, in which humans saw themselves
as subject to nature; yet it was a destiny made possible
by nature, when nature gave human beings both mental
superiority and hands. The hands are fundamentally
weapons. More than a *tool of tool*, as described by Aristotle, the hand perfects itself in conflict more than manufacture. Indeed just as Spengler interprets the plough
as a weapon against plant life, so he sees instruments of
worship as arms against the devil. But Spengler does not
confuse technology with tools or technological objects.
Technology is a set of procedures or practical means for
producing a particular end in view. In Spengler's words,
technology *is the tactics of living*, a conception that goes
beyond human life. Following Friedrich Nietzsche, he
identifies life with struggle, a fierce and merciless struggle that springs from the will to power, with the
machine being the *subtlest of all possible weapons*.

Oswald Spengler, 1880–1936. The German philosopher is famous for
his *Decline of the West*. He held that civilizations, like biological
organisms, pass through a determinable life cycle and that the
modern West was approaching the end of such a cycle. (© *Corbis-
Bettmann*.)

Having placed the origin of Faustian culture in the
Nordic countries, Spengler interprets the Enlightenment as the moment when the machine replaced the
Creator. The machine became a god, with factories for
temples and engineers for priests, whose mysteries were
the esoteric features of mechanization. Nineteenth-century machine age industrialization imposed itself on nature with standardized, inert forms that are hostile to the
natural world and the precursors of decline. But in order
to feed the *technological-machinist army* Western Europe
and North America furthered the destruction of nature
across the globe, creating an untameable monster that
threatens to conquer humans themselves and lead culture to a grandiose suicide. The tragedy of humanity lies
in humans raising their hands against their own
mother—nature. All the great cultures defeats. The
struggle against nature is a struggle without hope, even
though people pursue it to the end.

Contrary to the views of Enlightenment theorists
such as Henri de Saint-Simon or Auguste Comte, the
domination of nature by Faustian technology does not

seek human emancipation, but is the manifestation of a blind will to power over the infinite. As Hermínio Martins (1998) argues, Spengler rejects the rationality of technological history. The history of Western European and North American technology is simply human tragedy because the infinite is always greater than efforts to tame it. Inspired also by Nietzsche's cyclic vision of history, Spengler sees culture, rooted in the soil, being replaced by civilization, in which the intellect prevails, decaying again eventually into culture.

The significance that Spengler attributes to technology, his defense of science-as-technology, his cultural pessimism, and his hostility to liberal, democratic values and institutions were commented on by Max Weber, and influenced thinking during the Nazi regime, despite the fact that he rejected national socialism completely in 1934. Many of his insights and expressions regarding the essentially non-transferable character of Western European and North American technological culture as a destiny, the will to power as the foundation of technology, and the conceptual and ontological dependency of science on technology are further echoed in Martin Heidegger and Ernst Jünger, as well as in some members of the first generation of the Frankfurt school.

JOSÉ LUÍS GARCIA

SEE ALSO *Faust; German Perspectives.*

BIBLIOGRAPHY

Hughes, H. Stuart. (1952). *Oswald Spengler: A Critical Estimate.* New York: Charles Scribner's Sons.

Martins, Hermínio. (1998). "Technology, Modernity, Politics". In *The Politics of Postmodernity,* eds. James Good, and Irving Velody. Cambridge, England: Cambridge University Press.

Spengler, Oswald. (1918 [1922]). *Der Untergang des Abendlandes* [The Decline of the West], 2 vols. Munich: C. H. Beck. English translation by Charles Francis Atkinson (London: George Allen & Unwin [1926/1928]).

Spengler, Oswald. (1931). *Der Mensch und die Technik: Beitrag zu einer Philosophie des Lebens* [The Man and Technics: A Contribution to a Philosophy of Life]. Munich: C. H. Beck. English translation by Charles Francis Atkinson (London: George Allen & Unwin [1932]).

SPORTS

• • •

Ethical issues related to science and technology in sports only began to attract critical attention during the sec-

ond half of the twentieth century. This paralleled the increasing scientific study of sports and the creation of sports science, as well as the discovery and development of performance enhancing drugs and technological transformations in sports equipment. The latter two influences have been especially problematic, and have played a central role in the emergence of critical studies in the field.

Modern Sports Development

This scientizaton reflects a shift in values concerned with sports. Allen Guttmann describes, in *From Ritual to Record* (1978), how the development of timing technology introduced the possibility of records, now a dominant feature of modern sports. The late-nineteenth century British public school games, which championed *muscular Christianity,* repositioned physical exertion as central to the development of a productive and civil society. It also led to the politicization of sports and, along with the revived modern Olympic movement, which began in 1896, steadily became a focus of international political propaganda. With a philosophy that champions humanistic virtues of peace, culture, and education, the modern Olympic movement is less about sports contests than about ideology. It occupies an ambiguous social position as an organization that has devalued amateurism and embraced commercialization, while maintaining that there is something philosophically and socially meaningful about the games.

Ethical discussions concerning technology in sports generally focus on establishing what constitutes just or fair competition. The limited accessibility of a technology is often used as a reason for prohibiting its use in competition. In addition if the use of a particular innovation contravenes the agreed upon rules, that use may also be unethical. However because disputes exist as to what rules have been agreed to, the ethical issues are often blurred.

Drugs and Sports

During the 1980s, concerns about technology in sports focused largely on technologies of doping and drug use. This was prompted by a series of doping incidents in international sports, some of which resulted in death or serious injury for a number of athletes (Brown 1980, Houlihan 2002). The situation was accentuated by high-profile cases, for example that of the Canadian runner Ben Johnson who was stripped of the gold medal he won at the 1988 Olympics in Seoul after testing positive for anabolic steroids. Discussions about doping continue, accentuated by the emergence of new technologies, such as genetic modification, that challenge the ability of anti-doping authorities to detect *cheaters* (Miah 2004). Gene doping could challenge ethical theories in sports: Are genetically enhanced athletes cheats if they are altered before birth (embryogenesis)? Also if the genetic technologies at issue are not harmful to athletes, there is no persuasive health argument to support a ban on their use.

Sports Artifacts

Beyond doping, the increased use of technology and technologically advanced artifacts in sports raises a number of ethical questions (Miah and Eassom 2002, Gelberg 1998). Innovative techniques have radically changed some sports or events, such as the Fosbury flop in high jumping or the O'Brien shuffle in shot put. These have been seen as ethically contentious, though legitimate, because they increased the demands placed on athletes in competition.

Since the late-twentieth century, events in the sporting world have clearly illustrated the ethical implications that arise from the use of technology in sports. A few examples are the development of running shoe technology; lighter and stronger implements, such as golf clubs, cricket bats, and tennis rackets; and innovations such as the Fast-Skin swimming suit, which was used for the first time at the 2000 Olympics in Sydney. Many new sports technologies have been accepted. Technologically advanced running shoes, tennis rackets, bicycles, golf clubs, and others have been identified as beneficial improvements to sports because they enhance the safety of an activity or allow athletes to perform without interference from inadequate, cumbersome technology.

Technology has even democratized participation in sports to some extent, with the mass production of equipment permitting more people to play sports with the same kind of equipment used by elite athletes. However, this has also carried a burden of making elite sports subservient to the public or more specifically, sport spectators. Television audiences often dictate scheduling for competitions, which raises problems for sports federations, because so-called prime-time television schedules can conflict with the time of day when it is most desirable for athletes to compete.

One of the central components of these ethical discussions is the degree to which technologies are replacing the athlete in performance or are dehumanizing sports (Hoberman 1992). For example, double-stringed (so-called spaghetti strung) tennis racquets were banned

in the 1980s because they offered too much performance enhancement by enabling athletes to exert an unusually high amount of spin on the ball. There is an ethical concern about the *means* that allow athletes to achieve high levels of performance: An *undeserved enhancement* is considered unethical. Yet it can be argued that sports performances are necessarily technological and athletes must embrace their cyborgian identities by recognizing technology as a valued aspect of their performance.

When technology appears to make a sport easier for athletes, thus seemingly undermining or *devaluing* the performance, there are also ethical issues raised. Of key importance is what is meant by devaluing sports, because it is possible that technology could also be described as removing performance inhibitors, which is desirable when such inhibition is athletically irrelevant. For example, highly sophisticated running shoes might appear to enhance performance, or alternatively can be said to reduce inhibitions caused by the natural weakness the human foot.

This argument requires determining the factors that are *athletically relevant* to specific sports, an often contentious issue that can appeal to definitions of the goals of sports (Suits 1973). Do piezoelectric circuits in skis remove a performance inhibitor or make the activity unacceptably easier? The technology is designed to reduce the vibrations felt by skiers, thus giving them better control. It can certainly be argued that the new technology has made the activity easier because athletes no longer have to deal with the same degree of vibration as before. However it can also be argued that vibration is an irrelevant aspect of skiing—skiing does not test the ability of athletes to cope with vibration—and thus that the technology is not ethically suspect. Breaking records in the wake of technological advances in a particular sport suggests that an activity has become easier as a result of the innovation or that the advances have contributed to enhanced performance. It is, however, sometimes more accurate to conclude that the new technology has enabled a more representative measure of athletic performance.

Other ethical discussions involve whether technology changes the nature of the sport. For example, despite having sanctioned many changes to the construction of competitive bicycles, the International Cycling Union (ICU) banned Graeme Obree's *superman* design, in which one rides with arms stretched out in front of the body (like Superman), crouched over the handlebars. The ICU justified the ban by arguing that the new design would be generally unavailable, and thus the competitive sport would actually be different than the *normal* cycling experienced by the average rider. The ban seems to have been imposed because the innovation created a new concept of what constituted cycling, which conflicted with some kind of traditional, ideal form.

However some technological changes are beneficial to sports and disallowing them because they change traditional concepts is wrong. Changes to the construction of the javelin in the 1980s paved the way for a new type of successful participant, as opposed to the athletes who had been traditional winners in the event. However without such changes the natural progress of the sport would have resulted in athletes throwing the javelin into the audience, possibly requiring elimination of the activity from track and field competitions.

Conclusion

Alasdair MacIntyre's (1985) articulation of practice communities, which discusses the intrinsic good of sports and the distinction between novice and expert, is a useful retheorization of sports values (Morgan 1994). William Morgan's thesis is an explanation of the political economy of sports and the problematic hierarchical structures that have marginalized specific voices within specific practice communities. According to Morgan, there are two possibilities when sports are altered through technological developments. Society must either redescribe the activity—such as in the case of the javelin throw when the sport changed to sustain its character. Or society must accept the emergence of cyborg-athletes, which entails a redefinition of what it means to be a human being. By offering a subtle shift in the perception of humanness, sports provide an arena in which what it means to be human, as a living being and as an athlete, is ambiguous, liberated, and technologized.

ANDY MIAH

SEE ALSO *Drugs*.

BIBLIOGRAPHY

Brown, W. Miller. (1980). "Ethics, Drugs and Sport." *Journal of the Philosophy of Sport* 7: 15–23. One of the early papers on the doping issue; asks what should be against the rules in the first place.

Gelberg, J. Nadine. (1998). "Tradition, Talent and Technology: The Ambiguous Relationship between Sports and Innovation." In *Design for Sport*, ed. Akiko Busch. London: Thames and Hudson. A cross-disciplinary analysis of technology in sport, considering the aesthetics and tradition of technology as binding factors in ethical discussions about progress in sport.

Guttmann, Allen (1978). *From Ritual to Record: The Nature of Modern Sports*. New York: Columbia University Press. This book draws attention to the evidence that records and measurement have not always been grounding values in elite sport.

Hoberman, John M. (1992). *Mortal Engines: The Science of Performance and the Dehumanization of Sport*. New York: Free Press. A socio-historical analysis of doping in elite sport. Hoberman argues that the situation in elite sport is intractable.

Houlihan, Barrie. (2002). *Dying to Win: Doping in Sport and the Development of Anti-Doping Policy*. Strasburg, France: Council of Europe Publishing. A comprehensive analysis of anti-doping policy, Houlihan's work reveals the inadequacy of ethical theorizing within applied sports settings.

MacIntyre, Alaisdair. (1985). *After Virtue: A Study in Moral Theory*, 2nd edition. London: Duckworth.

Miah, Andy. (2004). *Genetically Modified Athletes: Biomedical Ethics, Gene Doping, and Sport*. London: Routledge. The first book length study of ethical issues arising from genetic modification in sport. Argues for genetic exceptionalism in sport and for anti-doping organizations to reconsider characterizing genetic modification as just another form of doping.

Miah, Andy, and Simon B. Eassom, eds. (2002). *Sport Technology: History, Philosophy & Policy*. Oxford: Elsevier Science. Provides a foundation for theorizing technological issues in sport, building upon themes in cultural studies of the cyborg, otherness and gender.

Morgan, William J. (1994). *Leftist Theories of Sport: A Critique and Reconstruction*. Urbana: University of Illinois Press. Questions the legitimacy of sporting structures and their capacity to enable democratically sensitive, ethical decision making.

Suits, B. (1973). "The Elements of Sport." In *The Philosophy of Sport: A Collection of Original Essays*, ed. Robert G. Osterhoudt. Illinois: Charles C. Thomas Publishers. Suits is a founding father of philosophy of sport. His theoretical insights on defining games, play, and sports continue to underpin discussions about the ethics of sport technology.

STAKEHOLDERS

• • •

The stakeholder concept derives from a simple premise: Organizations and technologies exist in constellations of relationships. Organizations operate in a network of market and nonmarket relationships with other organizations, groups, and individuals. Likewise technologies emerge and exist in a network of suppliers, end users, and others who bear the impact of the technology. Generally with reference to both organizations and technologies, these related parties are termed *stakeholders*, meaning that they hold a stake in the outcomes of the organization or technology.

Stakeholder research has important implications for science, technology, and ethics, as stakeholder thinking concerns itself both with the distribution of benefits among stakeholders and the procedures by which stakeholders work together toward desirable ends. After a brief history of the concept, this entry summarizes the distributive and procedural aspects of stakeholder thinking, particularly as they apply to three areas: corporate decision making, technology assessment, and environmental regulation.

History of the Concept

The stakeholder concept has its origins in the study of corporations and how they make decisions. R. Edward Freeman's *Strategic Management: A Stakeholder Approach* (1984), is regarded as seminal in the study of stakeholders, though Freeman attributes the term to scholars at the Stanford Research Institute in the 1960s. Farther back still, the premise that organizations must concern themselves with the demands of multiple constituencies traces back to classic management studies by Chester Barnard and Mary Parker Follett.

Contemporary discussions of stakeholders address three main questions. Social scientists have examined two. First, what are the consequences of different approaches to managing stakeholder groups? For example, Thomas Jones (1995) argues that a corporation's ethical treatment of its stakeholders has demonstrable financial implications. Second, why do stakeholder groups behave the way they do? For example, Tim Rowley and Mihnea Moldoveanu (2003) trace collective action by stakeholder groups to both the interests and the collective identity of group members. Put simply, the first question concerns the *instrumental* value of managing stakeholders effectively; the latter is a *descriptive* question aimed at helping decision makers to understand the environment in which they operate (Donaldson and Preston 1995).

Philosophers have concentrated on a third and equally important question: How should corporations behave toward stakeholders? This inquiry reflects the essentially normative nature of the concept—the term *stakeholder* itself serves as a counterpoint to the claim that corporations are responsible only to their stockholders—and has given rise to the search for a so-called *normative core* for stakeholder theory, a fundamental set of principles governing the ethical treatment of stakeholders (Donaldson and Preston 1995). Drawing on a host of ethical theories, ethicists have developed Kantian, feminist, rights-based, and Rawlsian arguments, among others.

The Distributive Dimension

In practical terms, much stakeholder research (especially in the third, normative, stream) addresses the issue of distribution: how corporations, public policy makers, and technology managers allocate rights and values across multiple stakeholders. Normative stakeholder arguments offer ways to assess the moral quality of these distributive patterns, and these arguments have important implications for ethical issues in the realm of science and technology.

For example, the question of who should benefit from emergent technologies—nanotechnology, pharmaceutical advances, and the human genome, among others—is, at its core, a question of distribution (Singer and Daar 2001) that stakeholder theory helps to resolve. Specifically the principle of stakeholder fairness developed by Robert Phillips (2003) derives from a widely accepted notion of reciprocity and holds that obligations accrue to participants in a cooperative scheme in proportion to contributions by stakeholder groups.

This logic also applies to the less tangible benefits and costs of technology. An emerging issue concerns the steps technology managers take to prevent employees from inappropriately using information technology resources such as e-mail and the Internet. The conflict is not over material resources but rather the tension between the privacy rights of employees, who seek to use these resources for personal reasons without the threat of invasive monitoring, and the property rights of stockholders, who would bear the cost of lawsuits if inappropriate technology use results in hostile work environment lawsuits. An exclusive emphasis on stockholder interests might advocate a total ban on the use of these technologies for nonbusiness purposes, whereas stakeholder theory would suggest a moderate position, allocating rights proportionally and allowing, for example, some personal use of information technology resources along with unobtrusive forms of monitoring to protect stockholder interests.

The Procedural Dimension

Stakeholder research also addresses procedural concerns that are central to the application of stakeholder theory to science and technology. Evan and Freeman (1993) draw on a Kantian perspective to spell out principles specifying how corporations should engage with stakeholders. They suggest, in part, that stakeholders have a right to participate in decisions that affect them. This concern for procedural justice extends to decisions in the realm of science and technology, where technologies, development paths, and potential science-related policies must be evaluated in light of stakeholder interests. Consequently one finds frequent reference to the procedural aspects of stakeholder theory in the areas of technology assessment and environmental regulation. Here stakeholder theory maintains that those groups with a vested interest in a technology, action, or organization should have an opportunity to express those interests and, in some cases, to participate in decision making. As some have argued, this participation should take the form of comprehensive dialogue among various stakeholder groups .

As diverse development agencies, corporations, and government regulators (from the United Nations to the World Bank to Motorola Corporation) apply these procedural principles by initiating dialogue with stakeholders concerning new technologies and environmental policies, they discover that the procedural aspect of stakeholder management is not only ethically desirable but highly practical. As stakeholder thinkers have long maintained, sharing information, ongoing dialogue, and meaningful participation in decision making enables better collaboration, reduces conflict, and ensures smoother implementation of policies and technologies (Freeman 1984, Johnson-Cramer, Berman, et al. 2003).

In sum, the value of stakeholder theory in resolving ethical issues in science and technology lies, to date, in offering prescriptions (a) that answer the distributive questions arising from development, utilization, and marketing of new technologies by businesses, and (b) that guide the procedural treatment of stakeholders in diverse areas such as technology assessment and environmental regulation. Ultimately amidst efforts to develop general principles and insights, stakeholder researchers have done little to apply their insights to specific questions about science and technology. The potential is clear, but much work remains to be done to demonstrate the usefulness of stakeholder theory in this domain.

MICHAEL E. JOHNSON-CRAMER
ROBERT PHILLIPS

SEE ALSO *Georgia Basin Futures Project; Management: Models; Participation; Science Policy.*

BIBLIOGRAPHY

Donaldson, Thomas, and Lee Preston (1995). "The Stakeholder Theory of the Corporation: Concepts, Evidence, and Implications." *Academy of Management Review* 20(1): 65–91. A review of the first ten years of stakeholder research, this article identifies three streams of relevant work: descriptive-empirical, instrumental, and normative.

The authors also advance a normative argument, rooted in the theory of property, for attending to multiple stakeholders.

Evan, William, and R. Edward Freeman. (1993). "A Stakeholder Theory of the Modern Corporation: Kantian Capitalism." In *Ethical Theory and Business*, ed. Tom Beauchamp and Norman E. Bowie. Englewood Cliffs, CA, Prentice Hall.

Freeman, R. Edward. (1984). *Strategic Management: A Stakeholder Approach.* Boston: Pitman. The seminal work in stakeholder research, this book advances the idea that organizations have stakeholders and organizational performance depends on an organization's ability to satisfy its multiple stakeholders.

Johnson-Cramer, Michael; Shawn Berman; et al. (2003). "Reexamining the Concept of Stakeholder Management." *Unfolding Stakeholder Thinking*, Vol. 2., ed. Jorg Andriof, Sandra Waddock, Sandra Rahman, and Bryan Husted. London: Greenleaf. This chapter offers more specific analysis of what it means to manage stakeholders. It outlines a model based on both procedural concerns (how policies toward stakeholders are formulated) and substantive concerns (the content of those policies).

Jones, Thomas. (1995). "Instrumental Stakeholder Theory: A Synthesis of Ethics and Economics." *Academy of Management Review* 20(2): 404–437. Jones argues that business firms that are dishonest or opportunistic will incur higher costs in contracting with stakeholders. This logic underlies the instrumental argument that firms that do not satisfy stakeholders perform poorly.

Phillips, Robert. (2003). *Stakeholder Theory and Organizational Ethics.* San Francisco: Berrett-Koehler. Phillips's book represents the most recent advance in normative stakeholder theory, offering a response to the stakeholder identification problem (that is, who actually count as stakeholders?).

Rowley, Timothy J., and Mihnea Moldoveanu. (2003). "When Do Stakeholders Act? An Interest and Identity-based Model of Stakeholder Mobilization." *Academy of Management Review* 28(2): 204–219.

Singer, Peter, and Abdallah Daar. (2001). "Harnessing Genomics and Biotechnology to Improve Global Health Energy." *Science* 294: 87–89.

STATISTICS

•••

Basic Concepts of Classical Inference
History, Interpretation, and Application

BASIC CONCEPTS OF CLASSICAL INFERENCE

Statistics may be defined as the study and informed application of methods for drawing conclusions about the world from fallible observations. It has three distinct components: (1) It is based on the mathematical theory of probability, (2) as inductive inference it belongs to the philosophy of science, and (3) its subject matter is any of a wide range of empirical disciplines.

Humanity has been counting, measuring, and recording from antiquity, but the formal history of statistics dates to the first systematic analyses of official registries in the seventeenth century. The origin of the name is from the eighteenth century, the German *Statistik*, meaning "study of the state" or political science (generally qualitative). It was appropriated in the 1780s for use in English as *statistics*, an unusual new name for the quantitative analysis of conditions in a country (replacing *political arithmetic*), in order to attract public attention (Pearson 1978). Applied subsequently to measurement error in astronomy, the statistical approach using probability spread in the nineteenth century to social phenomena, to physics, and then to biology. Formal statistical inference came into being around the turn of the twentieth century, motivated in large measure by the study of heredity and evolution.

Intensive developments of theory and methodology, with the enormous impact of the electronic computer, have made statistics the most widely used mathematical discipline, applied to virtually every area of human endeavor. Analysis and interpretation of empirical results is basic to much of modern technology and the controversies surrounding its use. Statistical methodology, readily available in computer software packages, is easy to apply but not so easy to understand. Lack of professional competence, conflicts of interest, and oversimplified reporting by the media pose real dangers of abuse. Yet intelligent participation in the shaping of public policy requires the insights of a thoughtful, well-informed electorate.

There is a vast and constantly growing body of statistical methods, but the most commonly reported results employ the classical, or Neyman-Pearson, theory of statistical inference. Presented herein are the basic concepts of the classical theory in concise form. Further details, with many examples, can be found in textbooks on various levels of mathematical sophistication.

Descriptive versus Inferential Statistics

Statistics can be understood as descriptive or inferential. *Descriptive statistics* are methods for organizing, summarizing, and communicating data, or the data themselves. The resulting tables and graphs may represent complete information on a subject or only a selected sample. *Inferential statistics*, the subject here, refers to methods for reaching conclusions extending beyond the observations actually made, to statements about large classes of potential observations. It is inference from a sample, beyond its description.

From Sample to Probability

Statistics begins with data to explore a question about some large *target population* (of people or objects) that can be expressed in quantitative form. It is often impossible to observe the entire population of interest, and therefore a sample is selected from the best available, sometimes called the *sampled population*, to distinguish it from the target population.

RANDOM SAMPLE. The sample, on which the inference will be based, should be representative of the population, and thus be selected at random. This means that each member of the population should have an equal chance of being selected—an aim that in real-life situations can at best be approximately met. For example, to determine what proportion of patients with a certain type of cancer would benefit from a new treatment, the outcome of interest could be the proportion surviving for one year after diagnosis, with the study sample drawn from patients being seen in a particular hospital. The representativeness of the sample is always a key question in statistics.

STABLE RELATIVE FREQUENCY. It is known from experience that the observed proportion of a characteristic of a population becomes stable with increasing sample size. For example, the relative frequency of boys among the newborn fluctuates widely when studied in samples of size 10, and less so with samples of size 50. When based on samples of size 250, it is seen to settle just above .5, around the well-established value of .51. It is the observed stability of frequency ratios with increasing sample size that connects statistics with the mathematical concept of probability.

FREQUENTIST DEFINITION OF PROBABILITY. Classical statistical inference uses the *frequentist* definition of probability: The probability of an event denotes the relative frequency of occurrence of that event in the long run. This definition is reflected in a fundamental principle of probability, the *law of large numbers*: In the long run, the relative frequency of occurrence of an event approaches its probability. The probability may be known from the model, such as obtaining a six with a balanced die, namely 1/6. This is an example of the classical definition of probability, pertaining to a finite number of equally likely outcomes. Otherwise by definition the probability is whatever is obtained as long-run relative frequency. The size of the sample is of central importance in all applications.

The frequentist definition is embedded in the axiomatic approach to probability, which integrates statistics into the framework of modern mathematics. There are three basic axioms, using concepts of the theories of sets and measure. Expressed simply, the axioms state that: (1) the probability of any event (set) in the sample space of events is a number between 0 and 1, (2) the probability of the entire sample space is 1, and (3) if two events are mutually exclusive (only one of them can occur), then the probability that one or the other occurs is the sum of their probabilities.

RANDOM VARIABLES AND THEIR DISTRIBUTIONS. The numerical or coded value of the outcome of interest in a statistical study is called a *random variable*. The yes/no survival status of a cancer patient one year after diagnosis is a *binary* random variable. In a sample of size n, the number of patients surviving is some number S_n between 0 and n, called a *binomial* random variable. S_n/n is the relative frequency of surviving, and $1 - S_n/n$ the relative frequency of not surviving one year. The distribution of S_n, to be discussed below, is the binomial distribution showing the probabilities of all possible outcomes between 0 and n. An example of a *continuous* random variable X is the diastolic blood pressure (in millimeters of mercury) of patients treated for hypertension, at a given point of treatment. The relative frequency of different values assumed by X is the observed distribution of the random variable.

The concrete examples of a random variable and its distribution have direct counterparts in the mathematical theory of probability, and these are used in the development of methods of inference. A random sample of size n in statistics is considered a sample of n independent, identically distributed random variables, with independence a well-defined mathematical concept. These are abstract notions, often omitted in elementary presentations that give only the computational formulas. But they are the essential link for going from an observed set of numbers (the starting point of statistics) to mathematical entities that are the building blocks of the theory on which the methods of statistics are based.

PARAMETERS OF A DISTRIBUTION. The *probability distribution* of a random variable X describes how the probabilities are distributed over the values assumed by X along the real line; the sum of all probabilities is 1. The distribution is defined by *parameters*, constants that specify the location (central value) and shape of the distribution, often denoted by Greek letters. The most commonly used *location* parameter is the *mean* or *expected value* of X, $E(X)$, denoted by μ ("mu"). $E(X)$ is the weighted average of all possible outcomes of a random variable, weighted by the probabilities of the respective outcomes. A parameter that specifies the *spread* of the distribution is the variance of the random variable X, Var(X), defined as $E(X - \mu)^2$ and denoted by σ^2 ("sigma square"). It is the expected value of the squared deviations of the observed values from the mean of the distribution. The square root of the variance, or σ, is called the *standard deviation* of X.

THE BINOMIAL DISTRIBUTION. An important distribution deals with counting outcomes and computing proportions or percentages, often encountered in practice. Independent repetition of an experiment with a binary outcome and the same probability p of success n times yields the *binomial distribution* specified by the parameters n and p. The random variable X, defined as the number of successes in n trials, can have any value r between 0 and n, with probability function

$$P(X = r) = C(n, r)p^r(1 - p)^{n-r},$$

where $C(n, r)$ is the combination of n things taken r at a time and has the form

$$C(n, r) = \binom{n}{r} = \frac{n!}{r!(n - r)!}.$$

($n!$, called "n factorial," is the product of integers from 1 to n, with $0! = 1$. For example, $4! = 1 \times 2 \times 3 \times 4 = 24$.) It can be shown that for a binomial random variable, $E(X) = np$, and $Var(X) = np(1 - p)$. As the sum of n outcomes coded 0 or 1, X is also denoted by S_n.

THE NORMAL DISTRIBUTION. The most basic distribution in statistics is the *normal* or *Gaussian distribution* of a random variable X, defined by the probability density function

$$f(x) = \frac{1}{\sqrt{2\pi}\sigma} e^{-\frac{(x-\mu)^2}{2\sigma^2}},$$

where μ is the mean and σ is the standard deviation. The formula includes the constants $\pi = 3.142$ and $e = 2.718$, the base of the natural logarithm.

One reason for the importance of this equation is that many variables observed in nature follow an approximate normal distribution. Figure 1 shows frequency histograms of two samples, of height and diastolic blood pressure, with the corresponding normal distribution. The smoother fit in Figure 1a is the result of the

FIGURE 1

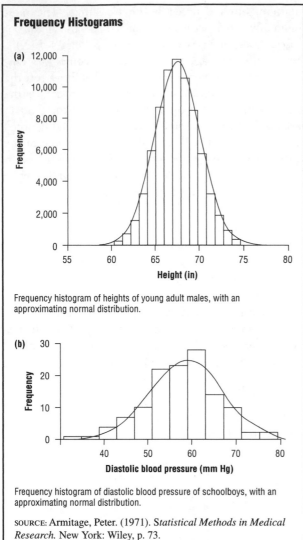

Frequency Histograms

(a)

Frequency histogram of heights of young adult males, with an approximating normal distribution.

(b)

Frequency histogram of diastolic blood pressure of schoolboys, with an approximating normal distribution.

SOURCE: Armitage, Peter. (1971). *Statistical Methods in Medical Research.* New York: Wiley, p. 73.

far larger sample size as compared with the number of observations used in Figure 1b.

THE STANDARD NORMAL DISTRIBUTION. An important special case of the normal distribution is the *standard normal*, with mean 0 and standard deviation 1, obtained by the transformation

$$Z = \frac{X - \mu}{\sigma}.$$

Any normal variable can be transformed to the extensively tabled standard form, and the related probabilities remain the same. Figure 2 shows areas under the normal curve in regions defined by the mean and standard deviation, for both the X-scale and Z-scale. It is useful

to remember that for a normally distributed random variable, about 95 percent of the observations lie within two standard deviations of the mean.

THE SAMPLE MEAN. Statistical inference aims to characterize a population from a sample, and interest is often in the *sample mean* as an estimate of the population mean. Given a sample of n random variables $X_1, X_2, \ldots, X_n$, the sample mean is defined as

$$M = \bar{X} = \frac{X_1 + X_2 + \cdots + X_n}{n}.$$

If the variables are independently distributed, each with mean μ and variance σ^2, then the *standard error of the mean* is

$$SE = SE(\bar{X}) = \frac{\sigma}{\sqrt{n}}.$$

For simplicity of notation, the symbols M and SE are used below.

THE CENTRAL LIMIT THEOREM. The normal distribution plays a special role in statistics also because of the basic principle of probability known as the *central limit theorem*: In general, for very large values of n, the sample mean has an approximate normal distribution. More specifically, if $X_1, X_2, \ldots, X_n$ are n independent, identically distributed random variables with mean μ and variance σ^2, then the distribution of their standardized mean

$$\frac{M - E(M)}{SE} = \frac{\bar{X} - \mu}{\sigma/\sqrt{n}}$$

tends to the standard normal distribution as $n \to \infty$. Nothing is said here about the shape of the underlying distribution. This principle, observed empirically and proved with increasingly greater precision and generality, is important to much of statistical theory and methodology.

APPLICATION TO THE BINOMIAL DISTRIBUTION. In the case of the binomial distribution, where $X = S_n$ is the sum of n independent random variables with outcomes 0 or 1,

$$M = \frac{S_n}{n} \quad \text{and} \quad E(M) = p,$$

$$\text{Var}(M) = \frac{p(1-p)}{n} \quad \text{and} \quad SE = \sqrt{\frac{p(1-p)}{n}}.$$

FIGURES 2–3

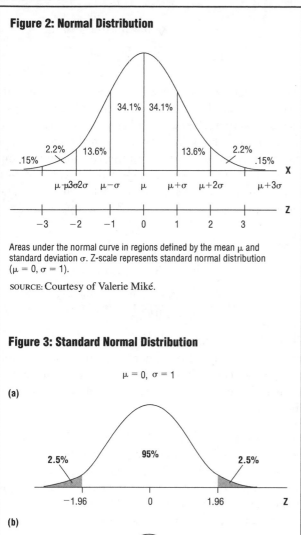

Figure 2: Normal Distribution

34.1% 34.1%

2.2% 13.6% 13.6% 2.2%

.15% .15%

$\mu-3\sigma$ $\mu-2\sigma$ $\mu-\sigma$ μ $\mu+\sigma$ $\mu+2\sigma$ $\mu+3\sigma$ **X**

−3 −2 −1 0 1 2 3 **Z**

Areas under the normal curve in regions defined by the mean μ and standard deviation σ. Z-scale represents standard normal distribution ($\mu = 0$, $\sigma = 1$).

SOURCE: Courtesy of Valerie Miké.

Figure 3: Standard Normal Distribution

$\mu = 0$, $\sigma = 1$

(a)

2.5% 95% 2.5%

−1.96 0 1.96 **Z**

(b)

5% 90% 5%

−1.645 0 1.645 **Z**

Rejection regions for tests at 5% significance level
 a. Two-sided test (both tails of distribution)
 b. One-sided test (either upper or lower tail)

SOURCE: Courtesy of Valerie Miké.

By the central limit theorem, the distribution of the standardized mean

$$\frac{M - E(M)}{SE} = \frac{S_n/n - p}{\sqrt{p(1-p)/n}}$$

tends to the standard normal distribution as $n \to \infty$. (The approximation can be used if both $np > 30$ and

$n(1 - p) > 30$. A so-called continuity correction of $-1/2n$ in the numerator improves the approximation, but is negligible for large n.)

Inference: Testing Statistical Hypotheses

Performing tests of statistical hypotheses is part of the scientific process, as indicated in Table 1, ideally with the professional statistician as member of the research team. The conceptual framework of subject matter specialists is an essential component, as is their close participation in the study, from its design to the interpretation of results.

FORMAL STRUCTURE. The formal steps of testing, summarized in Table 2, involve defining the null hypothesis, denoted H_0, to be tested against the alternative hypothesis H_1. The aim is to reject, or "nullify," the null hypothesis, in favor of the alternative, which is typically the hypothesis of real interest. The test may be two-sided or one-sided. For example, if the mean of a distribution is μ_0 under the null hypothesis, one may use the *two-sided test*, usually displayed as follows:

$$H_0 : \mu = \mu_0 \quad \text{vs.} \quad H_1 : \mu \neq \mu_0.$$
$$\text{Reject } H_0 \text{ if } |z| > z_{\alpha/2} = c,$$

that is, if the absolute value of the *test statistic z*, calculated from the observations, is outside the *critical value c*, determined by the significance level α ("alpha"). The corresponding *one-sided test* would be one of the following:

$$H_0 : \mu \leq \mu_0 \quad \text{vs.} \quad H_1 : \mu > \mu_0.$$
$$\text{Reject } H_0 \text{ if } z > z_\alpha = c.$$

$$H_0 : \mu \geq \mu_0 \quad \text{vs.} \quad H_1 : \mu < \mu_0.$$
$$\text{Reject } H_0 \text{ if } z < z_\alpha = c.$$

An outcome in the *rejection region*, the tail(s) of the distribution outside c, is considered unlikely if the null hypothesis is true, leading to its rejection at significance level α. The form of the test used, one- or two-sided, depends on the context of the problem, but the actual test used should always be reported.

AN EXAMPLE IN TWO PARTS. A senator, running for reelection against a strong opponent, wants to know his standing in popular support. An eager volunteer conducts a survey of 100 likely voters (Case #1) and reports

TABLES 1–2

Table 1: Testing a Statistical Hypothesis: the Scientific Context

1. Conceptual framework or paradigm
2. Formulation of testable (falsifiable) hypothesis
3. Research design, including selection of sample
4. Data collection
5. Data analysis
6. Interpretation of results
7. Generalization to some population: Inference
8. Follow-up in further studies

SOURCE: Courtesy of Valerie Miké.

Table 2: Testing a Statistical Hypothesis: the Procedure

1. Set up *null hypothesis* vs. *alternative hypothesis*.
2. Collect data in accordance with research design.
3. Analyze data for overall patterns, outliers, consistency with theoretical assumptions, etc.
4. Compute the *test statistic,* to be compared with the *critical value,* which divides the distribution of the test statistic under the null hypothesis into "likely" and "unlikely" regions, determined by the *significance level* α. The conventional division is 95% and 5%, for $\alpha = .05$.
   a. If the test statistic is in the 95% region, considered a "likely" outcome, do not reject the null hypothesis.
   b. If the test statistic is in the 5% region, considered an "unlikely" outcome, reject the null *hypothesis*. The result is said to be *statistically significant* at P = .05.
5. Review analysis with subject matter specialist, for possible implications and further studies.

SOURCE: Courtesy of Valerie Miké.

back that 55 plan to vote for the senator. Meanwhile, a professional pollster retained by the campaign manager takes a sample of 1,100 likely voters (Case #2), and also obtains a positive response from 55 percent. What can they conclude?

Each may choose a two-sided test of the null hypothesis that the true proportion p of supporters is .5, at significance level $\alpha = .05$:

$$H_0 : p = .5 \quad \text{vs.} \quad H_1 : p \neq .5.$$

By the central limit theorem for the binomial distribution each can use the test statistic z, assuming the standard normal distribution,

$$z = \frac{M - .50}{SE},$$

and carry out a z-test for Case #1 ($n = 100$) and Case #2 ($n = 1,100$). The sample mean M is .55 for each, but SE involves the sample size:

Case #1:
$$SE = \sqrt{p(1-p)/n}$$
$$= \sqrt{.5 \times .5/100} = .05$$
so that $$z = \frac{.55 - .50}{.05} = 1.0.$$

(To distinguish between a random variable and its observed value, the latter is often denoted in lower case, such as Z versus z.) As seen in Figure 3a, this test statistic is just one standard deviation from the mean under the null hypothesis, well within the likely region. Figure 3b shows that even a one-sided test would require a test statistic of at least $z = 1.645$ to reject H_0. The senator cannot be said to be ahead of his opponent.

Case #2:
$$SE = \sqrt{p(1-p)/n}$$
$$= \sqrt{.5 \times .5/1,100} = .015$$
so that $$z = \frac{.55 - .50}{.015} = 3.33.$$

Figure 3a shows that this test statistic is greater than the critical value 1.96, leading to rejection of the null hypothesis. The pollster can report that the senator is statistically in the lead, whereas the volunteer's result is inconclusive.

ERRORS ASSOCIATED WITH TESTING. Two types of error that may occur in testing a statistical hypothesis are shown in Table 3: Type I, rejecting H_0 when it is true, and Type II, not rejecting it when it is false. (The expression "accept" instead of "do not reject" H_0 is sometimes used, but strictly speaking the most that can be asserted is that the observed result is consistent with, or is a "likely" outcome under, the null hypothesis; it is always a tentative conclusion.) The Type I error means that when H_0 is rejected at $P = .05$ (or $\alpha = .05$, the significance level of the test), an outcome in the rejection region would occur by chance 5 percent of the time if H_0 were true. The Type II error, its probability denoted by β ("beta"), is not as well known; many users of statistical methods even seem unaware that it is an integral part of the theory. The complement of β, or $(1 - \beta)$, the probability of rejecting H_0 when it is false, is called the *power* of the test.

THE P-VALUE. In reporting the results of a study, statistical significance is usually indicated in terms of what has become known as the *P*-value, written as $P < .05$ or $P < .01$, referring to the significance level α. In analyses carried out by computer, the software typically also provides the actual value of P corresponding to the observed test statistic (properly doubled for two-sided

TABLES 3–4

Table 3: Errors Associated with Testing a Statistical Hypothesis

Conclusion of test	Null hypothesis true	Null hypothesis false
Do not reject H_0 "Not statistically significant"	No error	Type II error (β)
Reject H_0 "Statistically significant"	Type I error (α or P) Significance level	No error $(1-\beta)$ Power

SOURCE: Courtesy of Valerie Miké.

Table 4: Power of Test: Example of a Randomized Clinical Trial

Assume one-year survival rate with current treatment is 50% and with new treatment is

n	55%	60%	65%	75%	85%	95%
25	.06	.11	.19	.46	.78	.98
50	.08	.17	.33	.74	.97	*
100	.11	.30	.58	.96	*	*
250	.20	.61	.93	*	*	*
500	.35	.89	*	*	*	*
1,000	.61	.99	*	*	*	*
2,500	.94	*	*	*	*	*

First column shows n = number of patients in each treatment group. Entries in columns 2–7 represent power of test $(1-\beta)$ = probability of rejecting H_0 for different values of H_1; α = .05, two-sided test (arcsine transformation). For entries marked (*) the power is greater than .995.

SOURCE: Courtesy of Valerie Miké.

FIGURE 4

Relationship Between Significance Level and Power

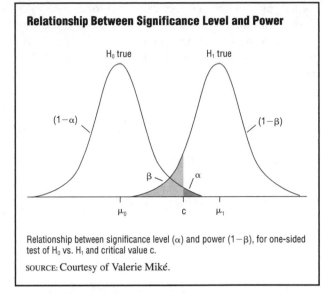

Relationship between significance level (α) and power $(1-\beta)$, for one-sided test of H_0 vs. H_1 and critical value c.

SOURCE: Courtesy of Valerie Miké.

Power as a function of sample size and alternative hypothesis is illustrated in Table 4. Assuming that a certain type of cancer has a one-year survival rate of 50 percent with the standard treatment, a randomized clinical trial is planned to evaluate a promising new therapy. The table shows the power of a two-sided test at α = .05 for a range of possible survival rates, with the new treatment and different numbers of patients included in each arm of the study.

For example, if there are 100 patients in each group, a new treatment yielding a one-year survival rate of 75 percent would be detected with probability (power) .96. "Detect" here refers to the probability that the observed difference in survival rates will be statistically significant. But if the improvement is only to 60 percent, the corresponding power is a mere .30. To detect this improvement with high power (.99) would require a sample size of 1,000. In any particular case, investigators have a general idea of what improvement can reasonably be expected. If the survival rate in the study arm is unlikely to be higher than 60 percent, then a clinical trial with just a few hundred patients is not a good research design and may be a waste of precious human and financial resources.

Inference: Estimating Confidence Intervals

An intuitive everyday procedure is point estimation, obtaining a summary figure, such as the sample mean, for some quantity of interest. But it is generally desirable to give an indication of how good—how precise—this estimate is, and this is done with the confidence interval.

tests). In Case #1 above, the value corresponding to $z = 1.0$ can be read off Figure 2 as $P = .32$. For Case #2, the value for $z = 3.33$ is seen as $P < .003$; it can be looked up in a table of the normal distribution as $P = .0024$. In results reported in the applied literature, at times only the observed P-value may be given, with no discussion of formal testing.

THE POWER OF THE TEST. Tests of the null hypothesis can be carried out without reference to the Type II error, but along with α and the sample size n, consideration of β is crucial in the research design of studies. The level of β, or equivalently, the power of the test, is always defined in terms of a specific value of the alternative hypothesis. The relationship between α and β for fixed n is shown in Figure 4 for a one-sided test of μ_0 versus μ_1. Changing the critical value c shows that as α increases, β decreases, and vice versa. A shift of μ_1 in relation to μ_0 indicates that the distance between them affects the power of the test.

THE FORMAL STRUCTURE. It is assumed here that the normal distribution is applicable, so that the terms already introduced can be used, with estimation of the population mean μ by the sample mean M. By definition, the following holds for the standard normal z-statistic

$$P\left(-z_{\alpha/2} < \frac{M - \mu}{SE} < z_{\alpha/2}\right) = 1 - \alpha.$$

As can be seen from Figure 3a, for $\alpha = .05$ this becomes

$$P\left(-1.96 < \frac{M - \mu}{SE} < 1.96\right) = .95.$$

Rewriting the expression inside the parentheses yields

$$P(M - 1.96SE < \mu < M + 1.96SE) = .95,$$

which is called a 95 percent confidence interval for the unknown population mean μ. It means that in a long sequence of identical repeated studies, 95 percent of the confidence intervals calculated from the sample would include the unknown parameter. There is always a 5 percent chance of error, but a larger sample size yields a smaller SE and narrower limits.

TWO-PART EXAMPLE CONTINUED. In the senator's reelection campaign, the point estimate M = .55 was obtained with different samples by both the volunteer and the pollster, and here the unknown parameter estimated by M is the true proportion p. Using the expression above yields

Case #1: $P(.45 < p < .65) = .95,$

for $n = 100,\ SE = .05.$

Case #2: $P(.52 < p < .58) = .95,$

for $n = 1,100,\ SE = .015.$

The critical value $c = 1.96$ for the standard normal (two-sided, $\alpha = .05$) is close to 2.0, and results are often presented in the form $M \pm 2SE$.

Case #1: $.55 \pm .10$

Case #2: $.55 \pm .03$

The latter expression may be reported by the media as "55 percent with a 3 percent margin of error," putting the senator clearly in the lead. What is omitted is that

this is a 95 percent confidence interval, with a 5 percent chance of error on the interval itself.

RELATIONSHIP BETWEEN TESTING AND ESTIMATION. Any value included in a $(1 - \alpha)$ confidence interval would in general be accepted (not rejected) as the null hypothesis in the corresponding test of significance level α, and values outside the interval would be rejected. In this example the null hypothesis of $p = .50$ was rejected in Case #2, but not in Case #1. The confidence interval is a useful, informative way to report results.

Overview

A statistical study may be *observational* or *experimental* and may involve one or more samples. The polls and the clinical trial were examples of a *one-sample* survey and a *two-sample* experiment, respectively. The methods of inference described a simple prototype of the Neyman-Pearson theory, using the binomial and standard normal distributions, but they are valid in a wide range of contexts. Other important probability distributions include two generated by a stable random process: the *Poisson*, for the number of events occurring at random in a fixed interval, and the *exponential*, for the length of the interval between the occurrence of random events. Radioactive decay, traffic accidents in a large city, and calls arriving at a telephone exchange are random processes that illustrate both distributions.

If the variance of a normal distribution is unknown and estimated from the sample (using a computational formula involving the observations), the z-test used above is replaced by the *t-test* for small samples ($n < 30$), with its own distribution. For larger samples the normal distribution is a close approximation. The *chi-square test*, perhaps the most widely used method in applied statistics, assesses the relationship between two categorical variables (each taking on a finite number of values, displayed in a two-way table), or the "goodness-of-fit" of observed data to a particular distribution. *Multivariate techniques* deal with inferences about two or more random variables, including their interaction; basic among these are *correlation* and *regression*. Important and central to the design of experiments is the *analysis of variance*, a method for partitioning the variation in a set of data into components associated with specific causes, in order to assess the effect of any hypothetical causes on the experimental result.

There are specialized techniques for *time series* and *forecasting*, for *sample surveys* and *industrial quality control*. *Sequential analysis* refers to procedures for repeated testing of hypotheses along the way, to minimize the sample size needed for a study. The class of *nonparametric methods* uses tests that do not assume a specific parametric form for the probability distributions, all within the classical theory. *Decision theory* formulates statistical problems as a choice between possible decisions based on the concept of utility or loss.

The same data can often be analyzed by different techniques, using different assumptions, and these may yield conflicting results. Statistical theory aims to provide the best methods for a given situation, tests that are most powerful across the range of alternatives, and estimates that are unbiased and have the smallest variance. Given an adequate model, statistics can control the uncertainty attributable to *sampling error*. But it cannot control *systematic error*, when the data are not even closely representative of the assumed population. Inference is based on an abstract logical structure, and its application to messy reality always requires the mature judgment of experienced investigators.

VALERIE MIKÉ

SEE ALSO *Biostatistics; Epidemiology; Meta-analysis; Probability; Qualitative Research.*

BIBLIOGRAPHY

Anderson, David R.; Dennis J. Sweeney; and Thomas A. Williams. (1994). *Introduction to Statistics: Concepts and Applications*, 3rd edition. Minneapolis/St. Paul: West Publishing. A textbook of applied statistics with many examples; no calculus required.

Cox, D. R., and D. V. Hinkley. (1974). *Theoretical Statistics*. London: Chapman and Hall. A graduate level text on the theory of modern statistics.

Cox, D. R., and D. V. Hinkley. (1978). *Problems and Solutions in Theoretical Statistics*. London: Chapman and Hall. Additional material pertaining to the text above.

Fisher, R. A. (1990). *Statistical Methods, Experimental Design, and Scientific Inference*, ed. J. H. Bennett. Oxford: Oxford University Press. Three classic works by a founder of modern statistics, published in a single volume.

Freedman, David; Robert Pisani; and Roger Purves. (1998). *Statistics*, 3rd edition. New York: Norton. An introductory text presenting concepts and methods by means of many examples, with minimal formal mathematics.

Kotz, Samuel; Norman L. Johnson; and Campbell B. Read, eds. (1982–1999). *Encyclopedia of Statistical Sciences*. 9 vols. plus supp. and 3 update vols. New York: Wiley.

Kruskal, William H., and Judith M. Tanur, eds. (1978). *International Encyclopedia of Statistics*. 2 vols. New York: Free Press.

Lehmann, E. L. (1959). *Testing Statistical Hypotheses*. New York: Wiley. A theoretical development of the Neyman-Pearson theory of hypothesis testing.

Miller, Irwin, and Marylees Miller. (2004). *John E. Freund's Mathematical Statistics with Applications*, 7th edition. Upper Saddle River, NJ: Prentice Hall. An intermediate-level introduction to the theory of statistics.

Pearson, Karl. (1978). *The History of Statistics in the 17th and 18th Centuries against the Changing Background of Intellectual, Scientific, and Religious Thought: Lectures by Karl Pearson Given at University College, London, during the Academic Sessions, 1921–1933*, ed. E. S. Pearson. London: Griffin. Includes documentation on the origin of the name *statistics*.

Snedecor, George W., and William G. Cochran. (1989). *Statistical Methods*, 8th edition. Ames: Iowa State University Press. A classic text of applied statistics.

Strait, Peggy Tang. (1989). *A First Course in Probability and Statistics with Applications*, 2nd edition. San Diego, CA: Harcourt Brace Jovanovich. Thorough presentation of mathematical concepts and techniques, with hundreds of examples from a wide range of applications.

HISTORY, INTERPRETATION, AND APPLICATION

Numerous jokes are associated with statistics and reflected in such caustic definitions as "Statistics is the use of methods to express in precise terms that which one does not know" and "Statistics is the art of going from an unwarranted assumption to a foregone conclusion." Then there is the time-worn remark attributed to the English statesman Benjamin Disraeli (1804–1881): "There are three kinds of lies: lies, damned lies, and statistics."

Statistics may refer to individual data, to complete sets of numbers, or to inferences made about a large population (of people or objects) from a representative sample of the population. The concern here is with inferential statistics. Its methodology is complex and subtle, and the risk of its abuse very real. There is no end in sight for the public being inundated with numbers, by the market and all kinds of interest groups. It has been estimated that children growing up in a pervasive television culture are exposed to more statistics than sex and violence combined. It was another Englishman, the novelist and historian H. G. Wells (1866–1946), who said: "Statistical thinking will one day be as necessary for efficient citizenship as the ability to read and write."

For those who understand, statistics is an exciting venture, a bold reaching out by the human mind to explore the unknown, to seek order in chaos, to harness natural forces for the benefit of all. Its development was integral to the rise of modern science and technology, its critical role recognized by the brilliant founders of new disciplines.

After a brief sketch of the history of statistical inference, this article offers a commentary on interpretations of statistics and concludes with a discussion of its applications that includes a case study of statistics in a scientific context.

Highlights of History

This quick survey of the history of statistics is presented in two sections, beginning with the emergence of statistical inference and then turning to the use of statistical concepts in philosophical speculation.

FROM STATISTICAL THINKING TO MATHEMATICAL STATISTICS. The normal distribution, which plays such a central role in statistics, was anticipated by Galileo Galilei (1564–1642) in his *Dialogue concerning the Two Chief World Systems—Ptolemaic and Copernican* (1632). He spoke of the errors in measuring the distance of a star as being symmetric (the observed distances equally likely to be too high as too low), the errors more likely to be small than large, and the actual distance as the one in which the greatest number of measurements concurred—a description of the bell-shaped curve. Discovered by Abraham de Moivre (1667–1754), the normal distribution was fully developed as the law of error in astronomy by Pierre-Simon de Laplace (1749–1827) and Carl Friedrich Gauss (1777–1855).

The statistical approach was applied to social phenomena by the Belgian astronomer Adolphe Quetelet (1796–1874), in what he called *social physics*, by analogy with *celestial physics*. He introduced the concept of the *average man* to show that observed regularities in the traits and behavior of groups followed the laws of probability. He strongly influenced Florence Nightingale (1820–1910), the British nursing pioneer and hospital reformer, who urged governments to keep good records and be guided by statistical evidence.

The fundamental contributions of the Scottish physicist James Clark Maxwell (1831–1879) to electromagnetic theory and the kinetic theory of gases would lead to communications technology and ultimately to Albert Einstein's special theory of relativity and Max Planck's quantum hypothesis. Having learned of Quetelet's application of the statistical error law to social aggregates, Maxwell theorized that the same law governed the velocity of gas molecules. His work in statistical mechanics and statistical thermodynamics foreshadowed a new conception of reality in physics.

The Austrian monk Gregor Johann Mendel (1822–1884) carried out plant crossbreeding experiments, in the course of which he discovered the laws of heredity.

Traits exist as paired basic units of heredity, now called genes. The pairs segregate in the reproductive cell, and the offspring receive one from each parent. Units corresponding to different traits recombine during reproduction independently of each other. Mendel presented his results at a scientific meeting in 1865 and published them in 1866, but they were ignored by the scientific community and he died unknown.

Statistical inference as a distinct discipline began with Francis Galton (1822–1911), a cousin of Charles Darwin, whose *On the Origin of Species* (1859) became the inspiration of Galton's life. The theory of evolution by natural selection offered Galton a new vision for humanity. He coined the term *eugenics* to express his belief that the conditions of humankind could best be improved by scientifically controlled breeding. He devoted himself to the exploration of human inheritance in extensive studies of variability in physical and mental traits, constructing what would become basic techniques of modern statistics, notably regression and correlation. In 1904 he established the Eugenics Record Office at University College, London, which in 1911 became the Galton Laboratory of National Eugenics, with Karl Pearson (1857–1936) appointed its director.

A man of classical learning and deep interest in social issues, Pearson was attracted to Galton's work in eugenics. Becoming absorbed in the study of heredity and evolution by the measurement and analysis of biologic variation, he developed a body of statistical techniques that includes the widely used chi-square test. In 1901 he founded the journal *Biometrika*. But he never accepted Mendel's newly rediscovered laws of inheritance involving hereditary units as yet unobserved, and engaged in a feud with Mendelian geneticists. Pearson was appointed the first professor of eugenics in 1911, with his Biometric Laboratory incorporated into the Galton Laboratory of National Eugenics, and the department became a world center for the study of statistics. When he retired in 1933, the department was split in two; his son Egon Pearson (1895–1980) obtained the chair in statistics, and Ronald A. Fisher (1890–1962) became professor of eugenics.

Trained in mathematics and physics, Fisher emerged as the greatest single contributor to the new disciplines of statistics and genetics and the mathematical theory of evolution. He did fundamental work in statistical inference, and developed the theory and methodology of experimental design, including the analysis of variance. Through his books *Statistical Methods for Research Workers* (1925), *The Design of Experiments* (1935), and *Statistical Methods and Scientific Inference*

(1956), he created the path for modern inquiry in agronomy, anthropology, astronomy, bacteriology, botany, economics, forestry, genetics, meteorology, psychology, and public health. His breeding experiments with plants and animals and his mathematical research in genetics led to the publication of his classic work, *The Genetical Theory of Natural Selection* (1930), in which he showed Mendel's laws of inheritance to be the essential mechanism for Darwin's theory of evolution.

Egon Pearson collaborated with the Russian-born mathematician Jerzy Neyman (1894–1981) to formulate what is now the classical (Neyman-Pearson) theory of hypothesis testing, published in 1928. This is the theory used across a wide range of disciplines, providing what some call the null hypothesis method. Neyman left London in 1937 to become a strong force in establishing the field in the United States. Another major contributor to American statistics was the Hungarian-born mathematician Abraham Wald (1902–1950), founder of statistical decision theory and sequential analysis.

STATISTICS AND PHILOSOPHY. Statistical developments in the eighteenth century were intertwined with natural theology, because for many the observed stable patterns of long-run frequencies implied intelligent design in the universe. For Florence Nightingale in the nineteenth century, the study of statistics was the way to gain insight into the divine plan.

Francis Galton had a different view. For him the theory of evolution offered freedom of thought, liberating him from the weight of the design argument for the existence of a first cause that he had found meaningless. Karl Pearson, author of *The Grammar of Science* (1892), was an advocate of logical positivism, holding that scientific laws are but descriptions of sense experience and that nothing could be known beyond phenomena. He did not believe in atoms and genes. For him the unity of science consisted alone in its method, not in its material. Galton and Pearson gave the world statistics, and left as philosophical legacy their vision of eugenics.

James Clark Maxwell was a thoughtful and devout Christian. He argued that freedom of the will, then under vigorous attack, was not inconsistent with the laws of nature being discovered by contemporary science. The statistical method, the only means to knowledge of a molecular universe, yielded information only about masses of aggregates, not about individuals. He urged recognition of the limits of science: "I have endeavored to show that it is the peculiar function of physical science to lead us to the confines of the incomprehensible, and to bid us behold and receive it in faith, till such time as the mystery shall open" (quoted in Porter 1986, p. 195).

In 1955 Fisher, by then Sir Ronald Fisher, said in a London radio address on the BBC: "It is one of the evils into which a nation can sometimes drift that, for about three generations in this country, the people have been taught to assume that scientists are the enemies of religion, and, naturally enough, that the faithful should be enemies of science" (Fisher 1974, p. 351). Scientists, he insisted, needed to be clear about the extent of their ignorance and not claim knowledge for which there was no real evidence. Fisher's advice remains sound at the start of the twenty-first century.

Interpretation: A Commentary

The following are comments on various aspects of statistics, painted of necessity in broad strokes, and concluding with some thoughts concerning the future.

STATISTICS AND THE PHILOSOPHY OF SCIENCE. Two distinct types of probability—objective and subjective—have been recognized since the emergence of the field in the seventeenth century. The classical (Neyman-Pearson) theory of hypothesis testing is based on the objective, frequentist interpretation. The subjective, degree-of-belief interpretation yields variations of so-called Bayesian inference. The latter involves combining observations with an assumed prior probability of a hypothesis to obtain an updated posterior probability, a procedure of enduring controversy. But the frequentist theory, as pointed out by its critics, does not provide any measure of the evidence contained in the data, only a choice between hypotheses. The American mathematical statistician Allan Birnbaum (1923–1976) did pioneering work to establish principles of statistical evidence in the frequentist framework, his two major related studies being "On the Foundations of Statistical Inference" (1962) and "Concepts of Statistical Evidence" (1969). Exploring the *likelihood principle*, Birnbaum reached the conclusion that some sort of confidence intervals were needed for the evaluation of evidence. A leading advocate of the subjective approach, of what he called personal probability, was another American statistician, Leonard J. Savage (1917–1971), author of the classic work *The Foundations of Statistics* (1954).

Statistics as commonly taught and used is that based on the frequentist theory. But there is lively interest in Bayesian inference, also the focus of serious study by philosophers (Howson and Urbach 1993). The entire subject has been engaging philosophers of science, giving rise to a new specialty called the philosophy of probability. An example is the edited volume *Probability Is the Very Guide of Life: The Philosophical Uses of Chance* (Kyburg and Thalos 2003), a collection of essays by philosophers of probability that explores aspects of probability as applied to practical issues of evidence, choice, and explanation—although without consensus on conceptual foundations. The title refers to a famous remark of Bishop Joseph Butler, one of the eighteenth-century natural theologians who saw statistical stability as a reflection of design and purpose in the universe (Butler 1736). Another edited volume, *The Nature of Scientific Evidence: Statistical, Philosophical, and Empirical Considerations* (Taper and Lele 2004), has contributions by statisticians, philosophers, and ecologists, with ecology used as the illustrative science. What remains clear is the persistent conflict between the frequentist and Bayesian approaches to inference. There is no unified theory of statistics.

STATISTICS IN THE FIELD. At the other end of the statistical spectrum is the approach expressed by the term *exploratory data analysis* (EDA), introduced by John W. Tukey (1915–2000), the most influential American statistician of the latter half of the twentieth century. Exploratory data analysis refers to probing the data by a variety of graphic and numeric techniques, with focus on the scientific issue at hand, rather than a rigid application of formulas. Tukey's textbook on EDA (1977) contains techniques that can be carried out with pencil and paper, but the approach is well suited to computer-based exploration of large data sets—the customary procedure. EDA is an iterative process, as tentative findings must be confirmed in precisely targeted studies, also called *confirmatory data analysis*. The aim is flexibility in the search for insight, with caution not to oversimplify the science, to be wary of pat solutions.

Practicing statisticians need to understand established theory, know the methods pertaining to their area of application, and be familiar with the relevant software. They must know enough about the subject matter to be able to ask intelligent questions and have a quick grasp of the problems presented to them. For effective communication they must be sensitive to the level of mathematical skills of the researchers seeking their assistance. It is easy to confuse and alienate with technical jargon, when the intention is to be of service. What is asked of them may range from short-term consultation—analysis of a small set of data, or help with answering a statistical reviewer's questions on a manuscript submitted for publication—to joining the research team of a long-range study that is being planned. Unless

otherwise agreed, it is understood that frequentist theory will be used, with routine preliminary exploration of the data. A statistician who strongly prefers the Bayesian approach may recruit investigators interested in collaborating on Bayesian analysis of suitable scientific problems.

Some points to remember: Statistics is a tool—more precisely, a collection of tools. Creative researchers know a lot of facts and have hunches and ideas; they may seek interaction with a compatible statistician to help sort things out, and that is where the tools come in. Which ones are actually used may not matter so much in the end. On occasion, the statistician's real contribution may not even involve formal analysis. A mind trained in mathematics views problems from a special perspective, which in itself may trigger insight for the scientist immersed in the material. Other situations require structured research designs with specification of proposed methods of analysis. These include cooperative studies, such as large multinational clinical trials involving hundreds of investigators. But in any case and even in the most masterful hands, statistics can be no better than the quality of the underlying science.

THE FUTURE OF STATISTICS. The explosive growth of information technology, with its capacity to generate data globally at a fast pace and in great volume, presents the statistical profession with unprecedented opportunity and challenge. The question is not that of either/or, of theory versus practice, but of perspective and balance: Continue exploration on every front, but make what is established widely available. Apply what is known, and do it well. Make sure that wherever statistics is potentially useful, it is at hand.

A promising development here is the Cochrane Collaboration, founded in 1993, an independent international organization dedicated to making accurate, up-to-date information about health care interventions readily available around the globe (Cochrane Collaboration). The organization promotes the search for evidence in the form of randomized clinical trials and provides ongoing summary analyses. By late 2004 there were twelve Cochrane centers worldwide, functioning in six languages, serving as reference centers for 192 nations, and coordinating the work of thousands of investigators. Such a vast undertaking must use objective criteria and uniform statistical methods that can be precisely communicated. That is the strength of the standard frequency approach.

In the realm of theoretical advances, some economic constraints may be cause for concern. Young graduates in academic positions, often struggling in isolation while carrying heavy teaching loads, are under great pressure to produce publications, any publications, to attain job security and professional advancement. This may not be the wisest use of their intellectual potential. A man of wit, Tukey would say that one should do theory only if it is going to be immortal. By contrast, those in a practical setting, such as a large biostatistics department, have to cope with the endless flow of data to be analyzed, under the constant pressure of immutable deadlines. The loss of major research grants may put many jobs in jeopardy, including their own. There should be other, readily available and steady sources of support that provide time for reflection, to find and explore areas of interest that seem to offer promise down the road. Such a path should include attention to what is happening in philosophy and close involvement with a field of cutting-edge empirical research. The great founders of statistics were widely read, hands-on scientists.

Application of Statistics

In the last decades of the twentieth century statistics continued its vigorous growth into a strong presence not only in the sciences but also in political and social affairs. Its enormous range of applications, with specialized methodology for diverse disciplines, is reflected in the thirteen-volume *Encyclopedia of Statistical Sciences*, published between 1982 and 1999 (Kotz, Johnson, and Read). The term *statistical science* refers to statistical theory and its applications to the natural and social sciences and to science-based technology. The best general advice in the application of statistics is to proceed with care and suspend hasty judgment. This is illustrated by a case study of the diffusion of neonatal technology.

STATISTICS IN CONTEXT: A CASE STUDY. The role of statistics in the interplay of forces affecting technological innovation was explored in a case study in neonatal medicine, a specialty created by technology (Miké, Krauss, and Ross 1993, 1996, 1998). It is the story of transcutaneous oxygen monitoring (TCM) in neonatal intensive care, introduced as a scientific breakthrough in the late 1970s and rapidly adopted for routine use, but abandoned within a decade. The research project included interviews with executives and design engineers of ten companies marketing the device, with investigators who had pioneered the technology, and with directors of neonatal intensive care units (NICUs).

Supplemental oxygen, essential for the survival of premature infants, had been administered since the 1930s, first via incubators and then by mechanically assisted ventilation. But in the 1940s an eye disease often leading to blindness, initially called retrolental fibroplasia (RLF) and later renamed retinopathy of prematurity (ROP), became the major clinical problem of surviving prematurely born infants. Over fifty causes were suggested, and about half of these were formally evaluated, a few in prospective clinical trials. When in the mid-1950s supplemental oxygen was identified as the cause of ROP in two large randomized clinical trials, the recommended policy became to administer oxygen only as needed and in concentrations below 40 percent. By this time more than 10,000 children had been blinded by ROP worldwide.

But subsequent studies noted higher rates of mortality and brain damage in surviving infants, as the incidence of ROP persisted and then rose, with many malpractice suits brought on behalf of children believed to have been harmed by improper use of oxygen. There was an urgent need for better monitoring of oxygen in the NICU.

Measurement of oxygen tension in arterial blood by means of the polarographic Clark electrode had been possible since the 1960s. The procedure was only intermittent, however, and the related loss of blood harmful to tiny, critically ill newborns. The new technology of TCM involved a miniaturized version of the Clark electrode that could monitor oxygen continuously across the skin, bypassing the need for invasive blood sampling. But the device was difficult to use, babies were burned by the electrode, and ROP was not eliminated. Within years TCM was being replaced by pulse oximetry, a still more recent technology with problems of its own.

A number of issues emerged. Subsequent review found serious flaws in the two randomized clinical trials that had implicated oxygen, and a series of methodological errors was noted in the early studies of other possible causes. The effectiveness of TCM in the prevention of ROP had not been shown before the adoption of the technology, and results of a randomized trial finally published in 1987 were inconclusive. It became clear that the oxygen hypothesis was an oversimplified view. ROP had a complex etiology related to premature physiology, even as the patient population itself was changing, with the survival of smaller and smaller infants.

A mistaken view of disease physiology, coupled with preventive technology advocated by its pioneers, heralded by the media, and demanded by the public—with industry only too eager to comply—led to the adoption of an untested technology that was itself poorly understood by those charged with its use. There was no special concern with statistical assessment, reliance on regulations of the Food and Drug Administration (FDA) being the norm. And there is no clear-cut way to assign ultimate responsibility. The study concluded with the overarching theme of complexity and uncertainty.

SUMMING UP Statistics is a powerful tool when in competent hands, one of the great intellectual achievements of the twentieth century. Ethical issues pertain to its misuse or lack of adequate use.

Elementary texts of applied statistics have traditionally been called "cookbooks," teaching mainly the "how" and not the "why." But in the present-day fast food culture hardly anyone cooks any more, and this applies equally to statistics. Computer software provides instant analysis of the data by a variety of techniques, allowing the user to pick and choose from the inevitable sprinkling of "significant" results (by definition of the meaning of P-value) to create a veneer of scientific respectability. Such meaningless and misleading activity, whatever the reason, can have harmful consequences. Another danger of abuse can come in the phrasing of questions in public opinion polls, known to affect the response, in a way that biases the results in favor of the sponsor's intended conclusion.

The ideal role of statistics is to be an integral part of the investigative process, to advise, assess, and warn of remaining uncertainties. The public needs to be informed and offer its support, so that the voice of statistics may be clearly heard in national life, over the cacophony of confusion and conflicting interests. This theme has been developed further in the framework of a proposed *Ethics of Evidence*, an approach for dealing with uncertainty in the context of contemporary culture (Miké 2003). The call for education and responsibility is its predominant message.

VALERIE MIKÉ

SEE ALSO *Biostatistics; Epidemiology; Galilei, Galileo; Galton, Francis; Meta-analysis; Nightingale, Florence; Probability.*

BIBLIOGRAPHY

Birnbaum, Allan. (1962). "On the Foundations of Statistical Inference (with Discussion)." *Journal of the American Statistical Association* 57(298): 269–326.

Birnbaum, Allan. (1969). "Concepts of Statistical Evidence." In *Philosophy, Science, and Method: Essays in Honor of Ernest Nagel*, ed. Sidney Morgenbesser, Patrick Suppes, and Morton White. New York: St. Martin's Press.

Box, Joan Fisher. (1978). *R. A. Fisher: The Life of a Scientist.* New York: Wiley. Biography written by Fisher's daughter who had also been his research assistant.

Butler, Joseph. (1736). *The Analogy of Religion, Natural and Revealed, to the Constitution and Course of Nature.* London: Printed for James, John, and Paul Knapton.

Fisher, R. A. (1930). *The Genetical Theory of Natural Selection.* Oxford: Clarendon Press. 2nd edition, New York: Dover, 1958.

Fisher, R. A. (1974). "Science and Christianity: Faith Is Not Credulity." In *Collected Papers of R. A. Fisher*, Vol. 5 (1948–1962), ed. J. H. Bennett. Adelaide, Australia: University of Adelaide. A 1955 talk given in London on BBC radio.

Fisher, R. A. (1990). *Statistical Methods, Experimental Design, and Scientific Inference*, ed. J. H. Bennett. Oxford: Oxford University Press. Three classic works by a founder of modern statistics, published in one volume.

Galilei, Galileo. (1967 [1632]). *Dialogue concerning the Two Chief World Systems—Ptolemaic and Copernican*, trans. Stillman Drake. 2nd edition. Berkeley and Los Angeles: University of California Press. Includes description of the bell-shaped curve of error, known subsequently as the normal distribution.

Gigerenzer, Gerd; Zeno Swijtink; Theodore Porter; et al. (1989). *The Empire of Chance: How Probability Changed Science and Everyday Life.* Cambridge, UK: Cambridge University Press. Summary of a two-volume work by a team of historians and philosophers of science, written for a general audience.

Hacking, Ian. (2001). *An Introduction to Probability and Inductive Logic.* Cambridge, UK: Cambridge University Press. Introductory textbook for students of philosophy, with many examples, explaining different interpretations of probability.

Howson, Colin, and Peter Urbach. (1993). *Scientific Reasoning: The Bayesian Approach*, 2nd edition. Chicago: Open Court.

Kotz, Samuel; Norman L. Johnson; and Campbell B. Read, eds. (1982–1999). *Encyclopedia of Statistical Sciences.* 9 vols. plus supp. and 3 update vols. New York: Wiley.

Kruskal, William H., and Judith M. Tanur, eds. (1978). *International Encyclopedia of Statistics.* 2 vols. New York: Free Press.

Kyburg, Henry E., Jr., and Mariam Thalos, eds. (2003). *Probability Is the Very Guide of Life: The Philosophical Uses of Chance.* Chicago: Open Court. Collection of essays by philosophers of probability.

Miké, Valerie. (2003). "Evidence and the Future of Medicine." *Evaluation & the Health Professions* 26(2): 127–152. Presents the *Ethics of Evidence* in the context of contemporary medicine and culture.

Miké, Valerie; Alfred N. Krauss; and Gail S. Ross. (1993). "Reflections on a Medical Innovation: Transcutaneous

Oxygen Monitoring in Neonatal Intensive Care." *Technology and Culture* 34(4): 894–922.

Miké, Valerie; Alfred N. Krauss; and Gail S. Ross. (1996). "Doctors and the Health Industry: A Case Study of Transcutaneous Oxygen Monitoring in Neonatal Intensive Care." *Social Science & Medicine* 42(9): 1247–1258.

Miké, Valerie; Alfred N. Krauss; and Gail S. Ross. (1998). "Responsibility for Clinical Innovation: A Case Study in Neonatal Medicine." *Evaluation and the Health Professions* 21(1): 3–26.

Pearson, Karl. (1991 [1892]). *The Grammar of Science*. Bristol, UK: Thoemmes Press. 3rd edition, London: Adam and Charles Black, 1911. A classic work on the philosophy of science.

Porter, Theodore M. (1986). *The Rise of Statistical Thinking, 1820–1900*. Princeton, NJ: Princeton University Press. A general history, considering scientific and economic currents that gave rise to the field.

Savage, Leonard J. (1954). *The Foundations of Statistics*. New York: Wiley. 2nd edition, New York: Dover, 1972. A classic work by the advocate of personal probability.

Stigler, Stephen M. (1986). *The History of Statistics: The Measurement of Uncertainty before 1900*. Cambridge, MA: Harvard University Press, Belknap Press. A thoroughly researched history, with detailed discussion of the origin of statistical methods.

Stigler, Stephen M. (1999). *Statistics on the Table: The History of Statistical Concepts and Methods*. Cambridge, MA: Harvard University Press. A collection of essays, a sequel to the author's 1986 work.

Tanur, Judith M.; Frederick Mosteller; William H. Kruskal; et al., eds. (1972). *Statistics: A Guide to the Unknown*. San Francisco: Holden-Day. A collection of forty-four essays describing applications of statistics in everyday life, written for the general reader.

Taper, Mark L., and Subhash R. Lele, eds. (2004). *The Nature of Scientific Evidence: Statistical, Philosophical, and Empirical Considerations*. Chicago: University of Chicago Press. A collection of essays, with ecology used to illustrate problems in assessing scientific evidence.

Tukey, John W. (1977). *Exploratory Data Analysis*. Reading, MA: Addison-Wesley. Textbook for an approach championed by the author, a prominent statistician, with many simple examples that do not require use of a computer.

Wald, Abraham. (1971). *Statistical Decision Functions*, 2nd edition. New York: Chelsea Publishing. Classic work of the founder of the field.

Wald, Abraham. (2004 [1947]). *Sequential Analysis*. New York: Dover. Another classic work by Wald, who also founded sequential analysis.

INTERNET RESOURCE

"The Cochrane Collaboration." Available from http://www.cochrane.org. Web site of the organization.

STEINMETZ, CHARLES

• • •

Electrical engineer and socialist Charles Proteus Steinmetz (1865–1923), born in Breslau, Germany, on April 9, was a public figure of the Progressive Era who tried to *engineer* a better society by creating an early code of engineering ethics, running for political office, and advocating a technocratic form of socialism. He died on October 26 in Schenectady, New York.

Trained in mathematics and physics, Steinmetz emigrated to the United States in 1889 to avoid being arrested for his socialist activities as a student in Germany. He became a leading researcher in the areas of magnetic hysteresis (a property of the metal cores used in transformers and electrical machines) and theories of alternating currents, electrical machinery, and high-voltage transmission lines. As chief consulting engineer of the newly formed General Electric Company (GE), which he joined in 1893, Steinmetz trained a generation of engineers in the use of advanced mathematics to design electrical equipment, established an engineering research laboratory, and published several books while teaching part-time at Union College in Schenectady, New York, the headquarters of GE. A dwarfed hunchback with a flair for publicity, he gained a national reputation as an electrical wizard for creating lightning in the laboratory and engaging in politics within and outside the engineering profession.

Steinmetz developed a distinct philosophy regarding the social responsibility of engineering. He argued that engineers should compromise with business interests in regard to ethical concerns within professional societies and address political issues on their own. In this way, engineers could maintain control over the profession against commercial interests and be able to promote political solutions in a wider arena.

Steinmetz carried out that philosophy in 1912 when he helped write the first code of ethics for the American Institute of Electrical Engineers (AIEE), the forerunner to the Institute of Electrical and Electronics Engineers (IEEE). Steinmetz was a president of the AIEE (1901–1902) and an active member of its first two ethics committees. The AIEE code, established in 1912, favored the interests of the employer over that of the engineer—up to a point. Rather than making engineers responsible for defective equipment, as the first draft of the code had done, for example, the revised code required engineers simply to report the problem, a common element in twenty-first century engineering codes of ethics. Inside GE, Steinmetz advised engineers in his group to keep silent rather than defend a company position with which they disagreed.

Steinmetz was active in politics at all levels. He served as president of the board of education under George Lunn, the socialist mayor of Schenectady in 1912, and was president of the city council in 1915. An evolutionary socialist who belonged to the conservative wing of the Socialist Party of America, Steinmetz drew on his corporatist experiences at GE, his work in local politics, his presidency of the AIEE, and as president of the National Association of Corporate Schools (NACS) to develop a theory of corporate socialism, which he expressed in some detail in *America and the New Epoch* (1916). In this form of technocracy, an enlightened industrial corporation, one that attended to the welfare of its workers, was the model for society. He proposed that the U.S. government be reorganized like an efficient corporation with democratic safeguards. The government would own and operate transportation and communication systems. An Industrial Senate, composed of leaders of large corporations, would coordinate and supervise industry. A democratically elected Tribunicate would set national and foreign policy, but could only veto the Senate.

Near the end of his life, Steinmetz acted on his belief that widespread electrification, by requiring cooperation to build networks and regulate consumption, would lead to socialism. He ran for New York state engineer in 1922 on a platform of harnessing the full power of Niagara Falls. The same year, he offered to help Vladimir Ilyich Lenin electrify Russia, in accord with Lenin's proposal text "Soviets + Electricity = Socialism."

To resolve the tensions he faced as a corporate engineer and a socialist, Steinmetz developed a patchwork of compromises that allowed agencies, such as the AIEE and NACS, and engineering colleges to retain autonomy by cooperating with industrial corporations. This would prepare corporations to become the model for the state and thus would be a step on the road to socialism. His ideas influenced President Woodrow Wilson's *war collectivism* and later proposals for the New Deal.

Steinmetz was able to promote his peculiar combination of conservative and radical views because of his public status as an electrical wizard, a new breed of scientific researcher that replaced cut-and-try inventors such as Thomas Edison. Steinmetz used his public position to demonstrate one way in which corporate engineers could address ethical and social issues in engineering.

RONALD KLINE

SEE ALSO *Engineering Ethics*.

BIBLIOGRAPHY

Jordan, John M. (1989). "Society Improved the Way You Can Improve a Dynamo: Charles P. Steinmetz and the Politics of Efficiency." *Technology and Culture* 30: 57–82.

Kline, Ronald R. (1992). *Steinmetz: Engineer and Socialist.* Baltimore, MD: Johns Hopkins University Press.

Layton, Edwin T., Jr. (1986 [1971]). *Revolt of the Engineers: Social Responsibility and the American Engineering Profession.* Baltimore, MD: Johns Hopkins University Press.

Steinmetz, Charles P. (1916). *America and the New Epoch.* New York: Harper & Brothers.

STEM CELLS

SEE *Embryonic Stem Cells*.

STRAUSS, LEO

• • •

Leo Strauss (1899–1973) was the most influential political philosopher of the twentieth century as well as its most extraordinary teacher. He was born into an Orthodox Jewish family in Kirchain, Hessen, Germany, on September 20. Strauss completed a doctorate at Hamburg in 1921 and immigrated to the United States in 1938. He taught at several American universities and attracted many gifted students. Their respect for his thought has led to those students being called disciples or *Straussians*. He died on October 18 in Annapolis, Maryland.

Philosophy and History

Like many scholars who left Germany in the 1930s, Strauss believed that a philosopher's work must be understood in the light of a political situation. Perhaps uniquely, he thought that all philosophers are in the *same* situation. Every regime, every society that sustains a government, is founded on certain shared opinions about what is noble and sacred, what is just, and what is in the common interest. Philosophers want to replace those cherished opinions with knowledge. This means that philosophy is by definition potentially subversive and is always likely to arouse the hostility of the regime. The story of Socrates' trial and execution is the best expression of this problem.

Strauss's view of philosophy is closely connected to his doctrine of esoteric writing, which is elaborated in

Persecution and the Art of Writing (1952). When philosophers write books, they must take pains both to protect philosophy from the hostility of citizens and to protect political life from subversion by philosophy. Their complete teachings can be communicated only by hints and clues. For example, a philosopher may write in one place that nothing should be taken seriously unless it is founded on experience and write in another place that religion is not founded on experience; only an attentive reader will be able to tell how seriously the author takes religion.

Because he read philosophy in this way, Strauss rejected the historicism that was prevalent in his time. According to historicists, person-to-person communication is not possible across historical boundaries; it is necessary to study past thinkers as objects in their historical context rather than as persons trying to talk to their later readers. Strauss taught that it is possible to understand Aristotle as he understood himself, for at least in the respect discussed above his situation is not fundamentally different from that of his modern readers. Strauss's most important book, *Natural Right and History* (1953), presents a sustained challenge to historicism. It is likely that the title implies a challenge to the philosopher Martin Heidegger's (1889–1976) *Being and Time* (1927). For Strauss, it is possible to arrive at a grasp of being that is not radically dependent on the flow of history.

Quarrels in Philosophy

Philosophy is the desire for wisdom, not the possession of wisdom. It may never amount to more than a clear grasp of the most fundamental questions. Strauss organized those questions into a number of historical quarrels. One of the most important is that between Athens and Jerusalem. Jerusalem stands for the concept of biblical revelation: Everything human beings must know is revealed to them in God's law. Athens stands for reason: Human beings can find out what they want to know by means of relentless questioning. Strauss taught that this quarrel was the most important source of intellectual vitality in Western civilization. However, although Strauss wrote extensively about Jewish philosophy and theology, his students disagree about how seriously he took biblical revelation.

With the power of revelation fading in modern civilization, Strauss sought to revive another quarrel: the one between the ancients and the moderns. The ancient thinkers, classical and medieval, looked to an authority higher than the human (nature or God) as the standard of truth and justice and based their political teachings on duties and virtues. The moderns began with a more or less explicit rejection of ancient thought. They viewed humankind as independent of any higher authority and based their teachings on rights rather than duties and on frank appraisals of human nature. Strauss argued, against the scholarly orthodoxy, that classical political philosophy had to be taken seriously as an alternative to the modern version. It is not clear whether he believed that ancient thought is superior on the whole.

Political Philosophy and Science

Strauss did consider classical social science to be manifestly superior to its modern counterpart. Social science in Strauss's time aspired to be "value-free." It sought to explain social facts the way a physicist explains the momentum of particles, without contaminating the explanation with historically conditioned expectation or judgment. However, the clarity the scientific method secures for physics induces a dangerous blindness when it is applied to human things: "A social science that cannot speak of tyranny with the same confidence with which medicine speaks of cancer cannot understand social phenomena for what they are" (Strauss 1991, p. 177). Classical social science recognized that human communities may flourish or fall victim to decay, and so it had something useful to say.

However, classical social science seems to rest on the strength of Strauss's analogy between the science of medicine and the sciences of politics and ethics. The physician not only can describe human biology but can prescribe remedies because medicine distinguishes what is naturally healthy from what is not. Can a knowledge of human nature similarly allow a philosopher to identify what is just and what is unjust, what saves and what destroys families and cities? The Platonists argued that it could, and Strauss refers to their teaching as classical natural right.

Classical natural right is concerned with articulating a hierarchy of natural ends. Thus, the perfection of human capacities, which the ancients called virtue, is primary and provision for survival, comfort, and freedom is secondary. Early modern political philosophy rejected the former and concentrated on the latter. That was largely a consequence of the rejection of Aristotelian teleology by modern science. Aristotle ascribed goals and purpose, or *teloi*, to nature. Modern thought recognizes only mechanical forces as natural; goals are products only of human will.

According to Aristotle, the issue between the mechanical and biological accounts of nature turns on how one interprets the motion of heavenly bodies. On this count the victory of modern science seems complete: There is no teleology on a cosmological scale. Because a value-free social science is useless, it becomes necessary to accept a dualism consisting of nonteleological physical sciences and social sciences that allow teleology. In a letter Strauss ascribes to Plato the view that this dualism cannot be reconciled. Strauss seems to have accepted this limitation for the most part, confining himself to political questions and largely ignoring not only modern natural science but classical biology and physics as well.

However, Strauss was choosing not the ancients over the moderns but Plato over Aristotle. Aristotle believed that biology could bridge the gap between "knowledge of inanimate [nature] and knowledge of man" (Strauss 1991, p. 279). His biology gives full weight to matter and momentum but recognizes a role for formal and teleological explanations. If Strauss had lived a bit longer, he would have witnessed some rehabilitation of Aristotle as a philosopher of biology, and that might have led him to reconsider the question.

Philosophy and Moderation

Although Strauss ignored contemporary science, he was attentive to its roots in modern thought. The early moderns proposed the unlimited conquest of nature for the purpose of the eventual satisfaction of all human desires. That project would include the conquest of human nature by some state, and that state would have to become universal and homogeneous if it were to eliminate all contradictions between states or between citizens. Such a state would need technologies of manipulation and coercion beyond any previously available to a government. Once the state accomplished its goal, perhaps it would whither away. Why would it be necessary to govern those whose every desire is satisfied?

However, if, as Strauss suspected, the complete satisfaction of human desires is impossible, the last state would in fact become a pervasive and immortal tyranny. This would mean the end of freedom and hence of philosophy. Strauss preferred Socratic philosophy to its modern counterpart at least insofar as it combined the pursuit of wisdom with moderation. It would be far better to settle for a decent form of government than to risk everything for one that is perfect. Of course, the philosopher will, because of the nature of this choice, be especially aware of its imperfections.

Accordingly, Strauss was both a supporter and a critic of modern liberal democracy. Although democracy is almost certainly the best viable form of government, Strauss had witnessed the weakness of the Weimar Republic in Germany and was concerned that a similar failure of nerve would affect Western democracies in their confrontation with communism. Moreover, democracy seemed problematic for philosophical reasons. Philosophers must stand apart from their fellow citizens and put more confidence in what reason tells them than in what the majority says. Philosophy is therefore elitist by necessity. Finally, because it is difficult to combine wisdom and political power, Strauss distrusted radical politics in any form. Anticommunism, elitism, and an insistence on political moderation have not endeared Strauss or the Straussians to their more orthodox colleagues in the universities.

KENNETH C. BLANCHARD, JR.

SEE ALSO *Aristotle and Aristotelianism; Democracy; Plato.*

BIBLIOGRAPHY

Deutsch, Kenneth L., and John Murley, eds. (1999). *Leo Strauss, the Straussians, and the Study of the American Regime.* Lanham, MD: Rowman and Littlefield.

Deutsch, Kenneth L., and Walter Soffer, eds. (1987). *The Crisis of Liberal Democracy: A Straussian Perspective.* Albany: State University of New York Press. Includes essays by students and other persons influenced by Strauss. Especially interesting are a number of arguments about the meaning and depth of Strauss's allegiance to political interests in general and modern liberal democracy in particular.

Drury, Shadia. (1988). *The Political Ideas of Leo Strauss.* New York: St. Martin's Press. This polemic is the best known critique of Strauss and his ideas. Drury sees no mysteries or ambiguities: Strauss was simply antiliberal.

Strauss, Leo. (1952). *Persecution and the Art of Writing.* Westport, CT: Greenwood Press.

Strauss, Leo. (1953). *Natural Right and History.* Chicago: University of Chicago Press. Strauss's best known work; includes a critique of historicism and a profound exploration of classical and modern natural right.

Strauss, Leo. (1959). *What Is Political Philosophy? and Other Studies.* Chicago: University of Chicago Press. A good general collection of Strauss's essays.

Strauss, Leo. (1991). *On Tyranny,* rev. and expanded edition, including the Strauss-Kojève correspondence, ed. Victor Gourevitch and Michael S. Roth. New York: Free Press. First published in 1963, this collection includes the famous debate between Strauss and the Hegelian philosopher Alexandre Kojève as well as their letters to each other. It may be the clearest introduction to Strauss's view of philosophy.

Strauss, Leo. (1997). *Jewish Philosophy and the Crisis of Modernity: Essays and Lectures in Modern Jewish Thought*, ed. Kenneth Hart Green. Albany: State University of New York Press. An excellent collection of essays on Jewish philosophy and theology, including the important "Jerusalem and Athens."

Strauss, Leo, and Joseph Cropsey, eds. (1987). *History of Political Philosophy*, 3rd edition. Chicago: University of Chicago Press. First edition, 1963; second edition, 1972. A collection of essays by Strauss and his students, with each essay focusing on a different political philosopher.

STRESS

• • •

Stress is an engineering concept that is applied metaphorically in the life sciences and social sciences. The ethical implications of stress in the social sciences lie in its perceived significance for work and health in technologically advanced societies. Stress provides an exemplary case for the interactions of science, technology, and ethics.

Origins

Although the word *stress* existed long before it became a technical term—it originally meant hardships and afflictions, as in "the stress of weather"—the earliest modern meanings of the term belong to engineering. In the nineteenth century considerations of stress in a modern sense took shape in several fields: strength of materials, thermodynamics, and medicine. William Rankine (1820–1872), who did pioneering work in civil engineering and thermodynamics, defined stress as the forces a material exerts in response to external forces applied to it. Those engineering developments applied not only in theory but also in practice as the steam engine, railroads, and heavy industry transformed the everyday world. If the resultant stresses are not taken into consideration, buildings and bridges collapse.

At that time physicians turned their attention to engineering aspects of the human body. In the eyes of nineteenth-century physicians, "overstrain" and "overpressure" of the nervous system and the heart produced serious and even fatal diseases. In part, "overstrain of the heart" and "neurasthenia" expressed people's anxiety over the "strange disease of modern life" (Arnold 1853 [1965]) with its harried pace and engineered infrastructure.

Twentieth-Century Developments

In the twentieth century the experimental psychologist Walter B. Cannon (1871–1945) developed the concept of homeostasis to call attention to an organism's response to emergency situations: the fight or flight syndrome. In "The Stresses and Strains of Homeostasis" (1935) Cannon reviewed the forces that lessen the efficiency of homeostatic processes in an organism. The physiologist Hans Selye (1907–1982) studied other endocrine responses to external threats, leading to his concept of stress as "a specific syndrome which consists of all the nonspecifically-induced changes within a biologic system" (Selye 1976, p. 64). Laboratory studies represented the intersection of clinical work in psychosomatic medicine and psychiatry, especially the work of the migraine identifier Harold G. Wolff (1898–1962) and others. Two military psychiatrists, Roy Grinker (1900–1993) and John Spiegel (1911–1991), who treated U.S. Army Air Corps crews published their findings in *Men under Stress* (1945). Through such investigations stress emerged as a central category to describe the effects of modern warfare and then was extended to include all of modern life. The meaning of stress was complicated by the fact that Selye's definition referred to the response, whereas in the other cases it referred to the stimulating cause of psychosomatic distress.

In the 1970s the related notion of trauma, or excessive stress, became a key to legitimating posttraumatic stress disorder as a diagnosis for American veterans of the Vietnam War. Stress as a cause of war neuroses later was extended backward to include puzzling illnesses that appeared during the American Civil War (irritable heart and nostalgia), World War I (shell shock, traumatic neurosis, neurasthenia), and World War II (combat fatigue). Trauma and stress became emblematic of the violence, productive and destructive, of technologically advanced societies.

After the 1950s stress became a key term in cybernetics and the social sciences. In cybernetics and systems theory the concept of stress was applied to all levels of organization, from the cellular to the global, organism and machine. One result has been vagueness in the meaning of the term, especially in the social sciences: Stress can refer to objective features of life events measured by psychological instruments such as the Social Readjustment Rating Scale of Thomas H. Holmes (1918–1988) and Richard H. Rahe (b. 1936), subjective features as in Richard S. Lazarus's (1922–2002) notion of the cognitive appraisal of threat as vital in the stress-coping process, and an interaction between situational and dispositional factors.

Stress as a category has had the most significant impact in the areas of health and work. A stress-diathesis model of illness causation proposes that excessive demands (stress) on adaptive capacities interact with psychosocial and biological predispositions (the diathesis), resulting in the breakdown of the weakest link in an individual's biopsychosocial systems. Thus, one person develops asthma, another depression, and a third cardiovascular disease. Although oversimplified, this suggests the thrust of contemporary thinking about possible causal links between stress and disease. Insofar as considerations of stress affect health, they affect work, and stress management has become important in the regulation of behavior in technologically advanced societies.

Ethics

The ethical implications of stress are twofold. First are the implications that arise from the experience of what is called stress. Stress plays a role in defining the limits of human performance: If demands are excessive, psychological or physical illness can result. Individual, corporate, and social responsibilities for minimizing stress and its effects have become significant. Excessive stress has become the basis for legal action. Although social inequalities are sources of stress, the emphasis in some societies, such as the United States, has been on individuals assuming increased personal responsibility for lifestyle choices that can result from and/or lead to stress and its deleterious effects.

Second are the implications that arise from the way that stress frames the trials and troubles of living. The construct of stress reframes the tribulations of living in rationalized or engineered terms: Stress is what individuals and organizations seek to manage. Ethical considerations thus appear in terms of efficiency and control. Management as the norm for dealing with stress reduces the ethical act to devising means to adjust to ends that may not be questioned.

ROBERT KUGELMANN

SEE ALSO Psychology; Social Indicators.

BIBLIOGRAPHY

Arnold, Matthew. (1853 [1965]). "The Scholar–Gipsy." In *The Poems of Matthew Arnold*, ed. Kenneth Allott. New York: Barnes & Noble.

Cannon, Walter B. (1935). "The Stresses and Strains of Homeostasis." *American Journal of the Medical Sciences* 189: 1–14.

Cooper, Cary L., and Philip Dewe. (2004). *Stress: A Brief History*. Oxford: Blackwell. This useful volume gives special attention to work-related stress and to Lazarus's concept of cognitive appraisal.

Grinker, Roy R., and John Spiegel. (1945). *Men under Stress*. Philadelphia: Blakiston. A study of American aviators in World War II and the psychosomatic illnesses many suffered as a result of the stress of war.

Kugelmann, Robert. (1992). *Stress: The Nature and History of Engineered Grief*. Westport, CT: Praeger. A phenomenological and historical study focusing on how it is that people describe life's difficulties as "stressful," and dealing with the ethical implications of stress management.

Selye, Hans. (1950). *The Physiology and Pathology of Exposure to Stress*. Montreal: Acta. Selye here summarized the physiological research on stress, the general adaptation syndrome, and diseases of adaptation.

Selye, Hans. (1976). *The Stress of Life*, revised edition. New York: McGraw-Hill. Selye's classic statement, written for a general audience. Includes what he saw as the ethical implications of his research.

INTERNET RESOURCE

Brown, Stephen D. (1997). "The Life of Stress: Seeing and Saying Dysphoria." Unpublished Ph.D. diss. Reading, UK: University of Reading. Available from http://devpsy.lbor-o.ac.uk/psygroup/sb/thesis.htm. The history of stress through a poststructuralist approach, looking at theories, practices, and discourses of stress, including the self-help literature. Examines the role of technology and the ethics involved in stress.

SURVEILLANCE

SEE *Monitoring and Surveillance*.

SUSTAINABILITY AND SUSTAINABLE DEVELOPMENT

• • •

The concept of sustainable development (SD) has been a part of the global ecological dialogue among scientists and governmental leaders for more than two decades. One outcome of the 1992 United Nations Conference on Environment and Development (UNCED, or the Earth Summit) was The Earth Charter, a policy statement about the ethics of international SD. The Charter opens, "We must join together to bring forth a sustainable global society founded on respect for nature, universal human rights, economic justice, and a culture of peace" (Earth Charter International Secretariat 2000).

This statement captures the ethical context in which policy-makers developed the SD concept.

The most commonly used definition of SD comes from the 1987 report prepared for the Earth Summit, *Our Common Future* (1987). SD is "Development that meets the needs of the present without compromising the ability of future generations to meet their own needs" (WCED 1987). The 178 heads of state that gathered at the Earth Summit sought to address both the *environmental problem* and the *socioeconomic development problem*. The SD concept presented a paradigm in which officials viewed environment and development as partners rather than adversaries. The WCED view of SD presumed that socioeconomic growth and environmental protection could be reconciled in an equitable manner.

The SD idea contrasts with development that focuses on socioeconomic gain often at the expense of the environment. Some natural resource extractive industries, such as mining and fishing, deplete resources in the name of promoting socioeconomic growth. Unsustainable development, however, can be devastating for the environment and society. In 1992, for instance, the northern cod fishery collapsed in Newfoundland due to overfishing. The government, in light of this natural resource drawdown, called for a two-year moratorium on cod fishing so that the stocks could recover. This action affected thousands of workers (Haedrich and Hamilton 2000). The tension between biological/ecological concerns and human socioeconomic concerns, in this case and others like it, highlights the importance of finding a balance between society and the environment.

While the WCED definition has the greatest international recognition, a range of definitions are associated with SD. David Pearce and colleagues, for example, present a thirteen-page annex of definitions of the term. What the WCED brief definition has in common with others is that it identifies three main, but not equal, SD goals: (a) socioeconomic growth; (b) environmental protection; and (c) social equity. Interest groups highlight different aspects of this three-part definition. The economic concerns of national and transnational industrialists are incorporated into the definition, as are the concerns of environmentalists, and the socioeconomic concerns of nongovernmental organizations and governments wishing to alleviate poverty and injustice.

While the WCED popularized the concept, the phrase *sustainable development* had already been around for at least ten years. The International Union for the Conservation of Nature used the term in *World Conser-vation Strategy* (1980). *World Conservation Strategy*, however, emphasizes ecological sustainability, not the integration of ecological, economic, and social sustainability. SD draws upon *limits to growth, appropriate and intermediate technologies, soft energy paths,* and *ecodevelopment* discourses of the 1970s and 1980s (Humphrey, Lewis, and Buttel 2002, Mitcham 1995).

For example, the limits to growth debate centers around the much-publicized *The Limits to Growth* (1972), a study produced by Donella Meadows and others for the Club of Rome (Humphrey and Buttel 1982, Mitcham 1995). The book presents evidence that severe biophysical constraints would impinge upon the growth and development of societies. *The Limits to Growth* predicts ecological collapse if current growth trends continued in population, industry, and resource use. The study provoked tremendous international debate, attention, and critique (Sandbach 1978). The limits to growth idea became politically unpopular in the less developed countries (the Global South) "on the grounds that it was unjust and unrealistic to expect countries of the [Global] South to abandon their aspirations for economic growth to stabilize the world environment for the benefit of the industrial world" (Buttel 1998, p. 263).

While the limits to growth debate asks whether environmental protection and continued economic growth are compatible, the mainstream SD discourse assumes that the two are complimentary and instead focuses on *how* SD can be achieved (Baker, et al. 1997). The SD discourse does not assume there are fixed limits to socioeconomic development; it is pro-technology, pro-growth, and compromise oriented. The WCED report clearly states, "The concept of sustainable development does imply limits—not absolute limits but limitations imposed by the present state of technology and social organization on environmental resources and by the ability of the biosphere to absorb the effects of human activities. But technology and social organization can be both managed and improved to make way for a new era of economic growth" (Ekins 1993, p. 91).

The discourse on SD presents a shift in thinking about human development. SD is presented as a solution to the problems of economic development and environmental degradation. International aid agencies, such as the U.S. Agency for International Development (USAID) and the World Bank, adopted the SD framework for the design of their development programs. The emergence of the concept came at the same time that environmental policymakers began framing environmental problems such as biodiversity loss, the greenhouse effect, and the thinning of the ozone layer, as *global problems*. No longer was it enough to *think globally, act locally*. In an era of globalization, the new interpretation of environmental problems suggested that people must *think globally, act globally*. SD ethically frames many of these actions.

The Definitional Problems of SD

While critics of SD come from many policy positions, they all agree on its lack of clarity. What should be *sustained* in SD: the economy, the environment, human welfare? Whose *needs* and whose *development* should be promoted? What should be *developed*? Is *development* the same as growth? Does development refer to production growth, as is typically indicated by growth of gross national product; does it refer to environmental growth, such as an improvement of environmental resources; or does development refer to growth in human welfare, including health, working conditions, and income distribution? (Ekins 1993). To deal with some of these problems, analysts and communities have begun constructing indicators for SD, such as those being created by "sustainable cities," such as Seattle (Portney 2003).

Some critics of the concept argue that it is old wine in new bottles in that it only requires slight modifications to existing modes of production, existing political structures, and existing values. New laws, international treaties, and better education, among others, will produce SD. Marxist interpretations, such as that put forward by Sharachandra Lélé, note that the concept "Does not contradict the deep-rooted normative notion of development as economic growth. In other words, SD is an attempt to have one's cake and eat it too" (Lélé 1991, p. 618). Fred Buttel, nonetheless, points out some of the advantages of the concept:

> SD still does focus our attention on the two great contradictions of the world today: The long-term compromising of the integrity of ecosystems (local as well as global ones) and the tendency toward reinforcement of the socioeconomic processes of social exclusion of billions of the world's people. Because of its relevance to spotlighting attention on these two great institutional failures of our epoch, SD allows a range of groups to contest structures and policies and to develop alternative visions of the future. (Buttel 1998, p. 265)

The treatment here assumes that there are three realms involved in SD that must be harmonized: ecological, economic, and social. Edward Barbier asserts that the objective of SD is "to maximize the goals across all these systems through an adaptive process of trade-offs (1987,

p. 104). In sum, for development to be sustainable, the environment should be protected; people's economic situation should be improved; and social equity should be achieved.

Alternative Theoretical Perspectives on SD

According to some social theorists and science policy analysts, the impending scarcity of oil, the carbon buildup in the atmosphere, and the potential for global climate change are among the leading ecological problems now facing the world. These problems do not speak well for the sustainability of western cultural traditions, such as the national and international expansion of free market capitalism. Yet modern social theorists and science policy analysts are not of one mind as to how science, technology, and society may deal with these ecologically critical, global sustainability issues in the twenty-first century. Three different models to approach a sustainable future are outlined: the conservative, ecological modernization model; the state-oriented, managerial model; and the radical, neo-Marxian model.

THE CONSERVATIVE, ECOLOGICAL MODERNIZATION PERSPECTIVE. Some theorists and science policy analysts foresee the twenty-first century as the period of ecological modernization. As the impending global ecological crisis gathers force, capitalists—the leaders of national and multinational business and industry—will reflect upon their vital predicament and, through the power of the market and innovative technologies, create sustainable societies throughout the world.

In 1997 Amory and Hunter Lovins of the Rocky Mountain Institute together with Ernst von Weizsacher, Director of the Wuppertal Institute (Germany), published *Factor Four: Doubling Wealth, Halving Resource Use*. Their work, in the spirit of ecological modernization, focuses on waging a worldwide *efficiency revolution—increasing energy savings by a factor of four*. They note that, historically, production efficiency improved through technological changes in labor practices: industrialization, automation, and robotics. For them, the new focus of the production efficiency revolution will be gains in the use of natural resources, notably energy. To wage this revolution, they propose harnessing the power of markets through price adjustments to create incentives for technological innovation.

The authors of *Factor Four* cast a wide net, focusing on how the efficiency revolution applies to transportation, design and building methods, natural resource conservation, agriculture, and energy. Common to these ways of using energy and natural resources more efficiently is the argument that "in many cases saving resources could cost less than buying and using them" (von Weizsacher, Lovins, and Lovins 1997, p. 146). Their examples include the Morro Bay, California, homebuilding program. In that program, builders were required to demonstrate that they reduced water consumption by twice what their next new home owners would consume by free installation of water efficient plumbing in already existing homes. Other examples include the use of more costly fluorescent lamps that last ten times longer than incandescent lamps; laptop computers that use one percent of the electricity consumed by desktop units; and more efficient air conditioning, in part through *superwindows* made to emit light, not heat.

Von Weitzsacher, Lovins, and Lovins identify former President Clinton's Partnership for a New Generation of Vehicles as a voice of the efficiency revolution. The hypercar is the centerpiece of this partnership between government and the Big Three U.S. auto makers—DaimlerChrysler, Ford, and General Motors. Capable of making a coast-to-coast trip on a single tank of fuel, the hypercar achieves fuel efficiency through the dual strategy of a streamlined, *slippery* body that is ultralight and a hybrid-electric/gasoline power unit. The hypercar also circumvents the problem of managing the waste build-up of engine batteries that could leak acid into the ground, water, or both.

The ecological modernization approach may contribute to economic growth and environmental protection, however, it is not clear whether it promotes or enlarges social equity. The model has been especially prevalent in Europe (Mol and Spaargaren 2002).

THE STATE-ORIENTED, MANAGERIAL PERSPECTIVE. A managerial approach seeks to reform, but not revolutionize, the existing political and legal structure of societies to achieve SD. Some recent programs undertaken by national governments and government-funded international development agencies exemplify managerial approaches. One such managerial effort is biodiversity conservation. Biodiversity protection addresses the goals of SD by preserving biological diversity and providing the potential for long-term social and economic benefits through sustained resource use and tourism. This effort at SD is exemplified by work on Ecuador done by environmental sociologist Thomas Rudel in 2003.

Esmeraldas, located in northwestern Ecuador, consists of tropical rain forests that contain an array of rarely seen biodiversity. It also has one of the highest deforestation rates in Latin America (between 2–4% annually). The rapid deforestation of this ecologically significant environment drives international efforts to make forestry sustainable in Ecuador. At least three social forces impel the rapid deforestation of Esmeraldas's lush tropical forests: It contains commercially valued hardwood; it is accessible to urban markets; and there is economic and population pressure to attain work logging the rain forest.

Over the last half of the twentieth century, the Ecuadorian government established an extensive set of national parks and forest reserves. Two reserves are located in Esmeraldas, the Cayapas-Mataje Reserve and the Cotacachi-Cayapas Reserve. A state-appointed forest service manages all of Ecuador's forest reserves. The forest service issues logging permits to the urban-based lumber companies and receives a stumpage tax for harvested trees in the reserves. The Ecuadorian government uses the tax receipts to pay forest service officers and to pay off government debt to international economic development agencies. Thus a fourth cause to deforestation in this area is that this state managerial arrangement encourages the exploitation of Ecuador's rain forests.

In spite of this state-induced system of tropical deforestation, increasingly influential national and international environmental groups and development organizations working in Ecuador have managed to promote sustainable forestry practices in the reserves. One such arrangement involves an economic development contract between the Ecuadorian government and USAID. The goal of this program is to form and develop Sustainable Use of Biological Reserves (SUBIR) in Ecuador. Using USAID funds, Ecuadorian officials fund ecologists to set the annual volume of rain forest harvesting equal to the annual rate of rain forest growth in the reserves and buffer zones adjacent to the reserves. In the rural community of Playa de Oro outside of the Cotacachi-Cayapas Reserve, village leaders are trying to take advantage of SUBIR by developing ecotourism. Thus the USAID program is leading to both sustainable forestry and economic growth for a rural village.

In another example, Deutche Gesellschaf fur Technische Zusammenarbeit (GTZ), the German equivalent of USAID, has organized a council of more than fifty Afro-Ecuadorian village leaders to practice sustainable forestry, to bargain collectively with the lumber companies, and to replant whatever trees are harvested. By practicing sustainable forestry, and by gaining a fairer return on the trees harvested in the reserves, Esmeraldas villages are an important, new experiment in sustainability in a highly diverse ecosystem.

These Ecuadorian SD efforts represent two of thirty-two working contracts involving international economic development agencies, national and provincial officials, village leaders, lumber companies, and environmental organizations. These efforts simultaneously attempt to alleviate problems of poverty, inequality, and biodiversity loss through land conservation. They are not without problems; however, they are a concrete attempt at reconciling the tensions between ecological, economic, and social systems.

THE RADICAL, NEO-MARXIAN PERSPECTIVE. Marxists or, in this designation, radicals, conceptualize environmental problems as inherent irrationalities in the capitalist mode of production (Humphrey, Lewis, and Buttel 2002). Radicals insist that economic expansion is the basic causal force by which capitalism resolves economic and social crises. The capitalist class and their allies, such as state officials, deflect discontent with social inequality by perpetuating economic growth necessary for the increased wages and rising material standards of living for the working class. Through this material, wage-based enfranchisement of workers, the capital class avoids the overt repression of workers, protects their own privileged relationship to private property, and garners monetary profit, at a substantial cost to the environment.

Anthropologist Ramachandra Guha's *The Unquiet Woods* (2000) illustrates the radical framework in the context of Badyargah. Located in the foothills of northern India's Himalayas, Badyargah is a cluster of homogeneous, egalitarian rural villages in the state of Tehri Garhwal. For centuries, the villagers of Badyargah, practiced a form of sustainable subsistence agriculture. Badyargah villagers lived well on fresh fish, rice, wheat, millet, and the meat of their lambs and sheep. The sustainability of Badyargah's agriculture began to decline following the first state-subsidized road building in the mid-1960s. At the time India's national government began boosting private capital expansion by awarding private logging contracts to outside lumber companies. Once a national forest surrounding a Badyargah village was harvested, Indian state foresters strictly excluded villagers from reentry to protect the regeneration of commercially valued trees.

Anticipating a particularly large commercial logging contract in 1979, Badyargah village leaders began planning rural, grassroots resistance. They contacted Sunderlal Bahuguna, a leading environmental activist in the Indian hill region. Bahuguna and his followers persuaded residents of forest-dependent villages to practice Chipko. To resist logging, the villagers hug trees.

The Chipko movement forces loggers to choose between sparing the trees or taking human lives. As part of this episode, Bahuguna went on a well-publicized hunger strike, and, day and night, 3,000 villagers guarded the site of the anticipated commercial logging. The government and contractor abandoned the logging plans.

This radical, grassroots resistance movement to protect local forests for use by the villagers was by no means an isolated episode in this part of rural India. Local, radical resistance to commercial logging in Tehri Garhwal became so prevalent that the government forestry department declared a fifteen-year, statewide moratorium on commercial logging beginning in 1982. Yet scholarly observers such as Guha do not anticipate the end of the Chipko movement in northern India. The modernization process, driven by capitalism, is bringing large dams, increased mining, and mountaineer tourism into the region. "The intensification of resource exploitation," Guha writes, "has been matched almost step by step with a sustained opposition, in which Chipko has played a crucial role, in catalyzing and broadening the social consciousness of the Himalyan peasantry" (Guha 2000, p. 179). Whether this radical environmentalism will bring back the sustainable rural economy of rural northern India remains to be seen.

Assessment

Beginning with the international debates over the implications of *The Limits to Growth* in the 1970s, scientists, environmentalists, and state officials have extensively engaged in global efforts to seek international consensus about the meaning and practice of SD. SD policies, ultimately, involve ethical decision-making about how science and technology can be applied in economic development efforts worldwide. The examples used to illustrate contemporary SD efforts highlight an important point. There is no one-size-fits-all model of SD.

Ecological modernization appears to be central to SD efforts in the Global North (the more developed, industrialized nations) in the early twenty-first century. Led by profit-oriented entrepreneurs trained in science and technology, ecological modernization aims to ecologize the economies of advanced industrial countries. Ecological modernization as an SD effort, exemplified by the hypercar, has a strong appeal to capitalists and mainline environmental groups. This form of modernization emphasizes ecological rationality in the use of natural resources for profit. Using the ethical criteria for SD, however, indicates that ecological modernization trades off social equity concerns for the sake of environmental and economic gains.

The grassroots, rural resistance movements against modernization in parts of the Global South—exemplified by the Chipko movement in northern India—is an oppositional struggle for SD. Reflecting the Gandhian tradition of nonviolent resistance that brought India to national independence in the mid-twentieth century, the Chipko movement brings sustainable rural subsistence traditions to SD efforts in India. The Chipko movement trades off economic growth for the sake of social equity and environmental integrity.

Rural development in the province of Esmeraldas, Ecuador, underscores the not-one-size-fits-all nature of SD. According to Rudel, forest-dependent organizations in Esmeraldas have initiated lobbying efforts to lift the national ban on timber exports. Because of the sustainable harvesting practiced by these Ecuadorian organizations, and because of the relatively high wages earned by the new logging cooperatives, Esmeraldas's export lumber could be ecologically approved by an international, third party certification agency. This potential certification could mean a higher demand for Ecuadorian tropical woods in the international lumber market. That potential development, in turn, could bring more wealth, sustainable forestry, and, possibly, more income equality among Esmeraldas workers—the three criteria needed for fully meeting the ethical standards for SD. Esmeraldas, thus, could become an exemplary SD model in the early-twenty-first century.

TAMMY L. LEWIS
CRAIG R. HUMPHREY

SEE ALSO *Change and Development; Development Ethics; Ecological Footprint; Ecology; Georgia Basin Futures Project; Mining; Modernization; Progress; Sierra Club; Waste.*

BIBLIOGRAPHY

Baker, Susan; Maria Kousis; Dick Richardsom; and Steven Young. (1997). *The Politics of Sustainable Development: Theory, Policy, and Practice Within the European Union.* London: Routledge.

Barbier, Edward B. (1987). "The Concept of Sustainable Economic Development." *Environmental Conservation* 14(2):101–110.

Buttel, Frederick H. (1998). "Some Observations on States, World Orders, and the Politics of Sustainability." *Organization & Environment* 11: 261–286.

Earth Charter International Secretariat. (n.d.). *The Earth Charter: Promoting Change for Sustainable Development.* San Jose, Costa Rica: Author. A document produced by the 1992 Earth Summit in Rio de Janiero.

Ekins, Paul. (1993). "Making Development Sustainable." In *Global Ecology*, ed. Wolfgang Sachs. London: Zed Books. One of Europe's leading ecological economists outlines three principles to achieve sustainable development in the north and south.

Guha, Ramachandra. (2000). *The Unquiet Woods: Ecological Change and Peasant Resistance in the Himalaya.* Berkeley: University of California Press. A leading environmental anthropologist analyzes the causes of a grassroots struggle to protect sustainable forestry practices in rural India.

Haedrich, Richard L., and Lawrence C. Hamilton. (2000). "The Fall and Future of Newfoundland's Cod Fishery." *Society & Natural Resources* 13: 359–372.

Humphrey, Craig R., and Frederick H. Buttel. (1982). *Environment, Energy, and Society.* Belmont, CA: Wadsworth Publishing.

Humphrey, Craig R.; Tammy L. Lewis; and Frederick H. Buttel. (2002). *Environment, Energy, and Society: A New Synthesis.* Belmont, CA: Wadsworth Publishing. An entire review of environmental sociology, the book includes a unique chapter on sustainable development.

International Union for the Conservation of Nature. (1980). *World Conservation Strategy.* Gland, Switzerland: Author. One of the earliest uses of the term *sustainable development.*

Lélé, Sharachandra M. (1991). "Sustainable Development: A Critical Review." *World Development* 19: 607–621.

Meadow, Donella N.,; Dennis L. Meadows; Jorgen Randers; and William W. Behrens III. (1972). *The Limits to Growth.* New York: Universe. Computer simulations of the world ecosystem suggest that severe biophysical constraints are impinging upon the growth of present-day societies.

Mitcham, Carl. (1995). "The Concept of Sustainable Development: Its Origins and Ambivalence." *Technology and Society* 17: 311–326. This paper, in the author's own words, considers "the historical and philosophical background of sustainability, the immediate origins of the concept of sustainable development, sustainable development and its near neighbors, and critiques of sustainability" (p. 312).

Mol, Arthur P. J., and Gert Spaargaren. (2002). "Ecological Modernization and the Environmental State." In *The Environmental State Under Pressure*, ed. Arthur P. J. Mol and Frederick H. Buttel. Two of Europe's leading social theorists outline the meaning of ecological modernization and its consequences for governments in environmental policy making.

Pearce, David. W.; Anil Markandya; and Edward B. Barbier. (1989). *Blueprint for a Green Economy.* London: Earthscan.

Portney, Kent E. (2003). *Taking Sustainable Cities Seriously: Economic Development, the Environment, and Quality of Life In American Cities.* Cambridge, Mass.: MIT Press.

Rudel, Thomas K. (2003) "Organizing for Sustainable Development: The Biodiversity Crisis and an Encompassing Organization in Esmeraldas, Ecuador." In *Environment, Energy and Society: Exemplary Works*, eds. Craig R. Humphrey; Tammy L. Lewis; and Frederick H. Buttel. Belmont, CA: Wadsworth Publishing. An environmental sociologist reports on a case from the tropical rainforest in Ecuador that has achieved some success in moving toward sustainable development.

Sandbach, Francis. (1978). " The Rise and Fall of the *Limits to Growth* Debate." *Social Studies of Science* 8: 495–520.

Von Weitzsacher, Ernst; Amory Lovins; and Hunter Lovins. (1997). *Doubling Wealth, Halving Resource Use*. London: Earthscan.

World Commission on Environment and Development. (1987). *Our Common Future*. New York: Oxford University Press. Contains the classic or standard definition of SD.

SYSTEMS AND SYSTEMS THINKING

•••

A system is defined by a set of distinctive relationships among a group of components that interact with one another and their environment through the exchange of energy, matter, and/or information. These relationships produce a new entity, the whole, that requires its own level of analysis. The technical use of the concept of a system in science and technology dates back to the 1950s. Systems thinking subsequently become a catchall term for different postwar developments in a variety of fields, such as cybernetics, information theory, network theory, game theory, automaton theory, systems science and engineering, and operations research. An underlying theme in these developments is a shift from reductionistic thinking and compartmentalized organization to holistic thinking aimed at understanding linkages among parts and increasing organizational communication. The rise of systems thinking has broad ethical and societal implications that range from practical changes in public decision making to the emergence of a worldview critical of some instances of scientific and technological hubris.

A Taxonomic History

During the second half of the twentieth century amalgams of the terms *system* and *systems* became ubiquitous. Computer and operating systems were joined by biological, business, and political systems. Systems science and systems engineering were complemented by systems management, systems medicine, and the practice of looking at the earth as a system. However, the systems thinking in all these cases can be divided into three basic types: systems theory, systems methodology, and systems philosophy. In the history of systems thinking each realm has followed its own path, with many overlaps and interactions.

SYSTEMS THEORY. The birth of systems theory took place in the technical sciences during World War II when the scientist Norbert Wiener (1894–1964) studied control problems with antiaircraft fire. Those studies concerning communication and control in particular technical systems inspired Wiener to more general reflections on what he came to call the science of cybernetics (Wiener 1948). Although Wiener did not stress the system concept, system, he argued in effect that any type of system can be understood with the help of general laws or principles. In Wiener's cybernetics two main ideas figure: feedback, with its regulating and stabilizing properties, and transmission of information, which helps transform the many parts of a complex system into a whole. A mathematical elaboration of the concept of information was developed by Claude E. Shannon (1916–2001).

The success of cybernetics and information theory created a fertile climate for a theoretical movement based on new principles and oriented toward concepts such as system, organization, and regulation. A leading figure in the rise and development of systems theory was the biologist-philosopher Ludwig von Bertalanffy (1901–1972), who attempted to overcome mechanistic reductionism, in biology in particular but also in scientific thought in general, and persistently opposed a machine view of the world. Although he agreed with Wiener that cybernetics can provide insights into the teleological behavior of systems, he argued that the principle of feedback adopts essentially a machine view.

For von Bertalanffy (1968) a machine is composed of durable components and therefore is primarily static in character. A characteristic of the cybernetic model is that fixed structures must be present to make regulation by feedback possible. An organism, however, is characterized primarily by a dynamic ordering and maintains its structures in a continuous process of building up and breaking down (e.g., human red blood cells are replaced at a rate of 2 million to 3 million per second). The organism is thus not a closed system with a static mechanical structure but an open system in flowing or dynamic equilibrium. Such systems also are characterized by emergent properties: characteristics that are not evident when one studies system components in isolation from one another. Systems theory often is seen as a way to retain holism and organicism without positing teleological or vitalist philosophies.

Opposing Wiener's claim that the cybernetic model is the basis for a universal science, von Bertalanffy argued that the open-system model has universal validity and provides the proper foundation for a "general system theory." In 1954 he and others, among them Kenneth Boulding, Anatol Rapoport, and Ralph Gerard, founded the Society for General Systems Research, which later was renamed the International Society for the Systems Sciences (ISSS). The ISSS brought together areas of research with dissimilar contents but similar structures or philosophical bases to enable researchers in various fields to develop a common language. Systems theory in this sense aspired to become a transdisciplinary science.

Systems theory and the quest for a general systems theory received a new impetus in the 1960s when Heinz von Foerster (1911–2002) introduced the concept of self-organization and later, in the 1970s, when Humberto R. Maturana and Francisco J. Varela (1980) proposed the concept of autopoiesis and developed the model of the organism as an autopoietic system. The term *autopoiesis* means "self-creation" and refers to the propensity of living and certain other nonequilibrium systems to remain stable for long periods despite the fact that matter and energy flow through them. Ilya Prigogine (1917–2003) further refined systems theory with the notion of dissipative systems: open systems that exchange energy, matter, and information with their environment; operate far from thermodynamic equilibrium; and display the spontaneous appearance of complex organization.

According to the social theorist Niklas Luhmann (1995), the concepts of self-organization and autopoiesis allow a further step, moving from a general systems theory based on the open-system model to a general theory of self-referential systems of social meaning and communication. Luhmann's application of systems theory to modern societies rejected the normative orientation of sociologists such as Émile Durkheim (1858–1917) and Talcott Parsons (1902–1979). He argued instead that systems theory has to drop all references to actors and their self-interpretations and focus on the ways in which complex social systems arise, much as living organisms do, through autopoiesis.

SYSTEMS METHODOLOGY. Systems methodology is concerned with the scientific method for approaching practical problems in technology and society. It may be defined as the theoretical study of practice-oriented methods in science and engineering, in which the notion of the system indicates an approach that is intended to be integrating and holistic. As with systems theory, systems methodology arose out of postwar developments in technology, in this case systems engineering and operations research. Although operations research usually is concerned with the operation of an existing

system, systems engineering investigates the planning and design of new systems.

The dominance of reductionistic and mechanistic thinking that was criticized by von Bertalanffy (1968) in his quest for a general system theory also became an important issue in systems methodology. As a leading representative, Russell L. Ackoff (1974) defended a systems approach to counter what he called "Machine Age" thinking. Together with C. West Churchman he founded one of the first systems groups in the United States at the philosophy department at the University of Pennsylvania shortly after World War II. Comparable developments took place in England at the University of Lancaster with the pioneering work of Geoffrey Vickers and Peter Checkland. Checkland observed that variants of systems thinking transferred from technology to the social domain were not especially successful. Following from that observation Checkland started to seek an alternative for the engineer's approach and tried to shift from what he called "hard systems thinking" (technical, quantitative models) to "soft systems thinking" (the incorporation of human values and perspectives).

A new impetus to the development of systems methodology came from the work of the social theorist and philosopher Jürgen Habermas (b. 1929). Habermas critiqued the dominance of technical categories in Luhmann's theory and the absence of human actors with conscious intentions in the development of modern society. In the 1980s this inspired a younger generation to work out a program termed critical systems thinking. Michael Jackson, Robert Flood, and Werner Ulrich became influential in this area.

In the late 1990s, inspired by the legacy of the Dutch philosopher and legal theorist Herman Dooyeweerd (1894–1977) an attempt was made in systems thinking to break with the Western idea of human autonomy and autonomous rationality. Fundamental to that research program was the notion of intrinsic meaning and the normativity of reality. Merging Dooyeweerd's theory of modalities and Stafford Beer's cybernetic theory of management, J. D. R. de Raadt launched "multi-modal systems thinking." Sytse Strijbos followed another more radical strategy by focusing on the underlying ontology and philosophical underpinnings of systems methodology. Borrowing from Dooyeweerd's notion of disclosure, Strijbos laid the foundations of "disclosive systems thinking." Industrial ecology and product life-cycle analyses are other versions of systems methodology that are used to make large-scale decisions with the goal of achieving sustainable energy and material flows (Graedel and Allenby 2003).

SYSTEMS PHILOSOPHY. Although systems philosophy was mentioned earlier in conjunction with systems theory (Wiener, von Bertalanffy, and others all attempted to develop the philosophical implications of their work) and systems methodology (for a while Ackoff and Churchman were based in an academic philosophy department), this approach merits independent recognition. In the 1970s, for instance, the Hungarian philosopher Ervin Laszlo tried to build on von Bertalanffy's ideas for a new scientific worldview, including a philosophy of nature, to develop a systems philosophy that would bring the latest developments in science to bear in conceptualizing the social problems of the emerging global society (Laszlo 1972). However, for clarity it is useful to distinguish at least four senses in which the terms *system* and *philosophy* have been connected.

First, there is the traditional sense in which philosophy aspires to be systematic, that is, to cover all the basic issues in a manner that properly subordinates and relates them. It is in this sense that one speaks of a philosophical system such as those of the philosophers Immanuel Kant (1724–1804) and Georg Friedrich Wilhelm Hegel (1770–1831). This is the oldest but in the current instance least significant connection.

Second, in the 1970s Laszlo aspired to formulate a systems philosophy keyed to the latest developments in science and to the urgent problems of contemporary global society. This type of systems thinking plays heavily into larger changes both in cultural norms and in social laws and institutions. Laszlo has been a prolific author whose books range from promotional work on systems philosophy to analyses of world modeling, sustainability, globalization, consciousness, and future studies. He is the founding editor of *World Futures: The Journal of General Evolution,* which began publication in 1980. Systems philosophers of this type often draw inspiration from process philosophy, especially the ideas of Alfred North Whitehead (1861–1947).

A more hard-nosed version of systems philosophy is found in the work of the Argentine-Canadian philosopher Mario Bunge (1979). For Bunge systems science is a research program for the construction of a "scientific metaphysics" built on well-defined, scientifically based concepts but having broad generality.

Third, systems philosophy deals with the philosophical issues of systems theory. Systems philosophy in this sense may be related to philosophical analyses of chaos and complexity and efforts to draw from those studies general implications for understanding nature and acting in the world. Chaos theory and complexity theory especially emphasize emergent properties and the self-organization of complex systems.

Fourth, systems philosophy concerns the philosophical foundations of systems methodology and thus deals with issues about human intervention in the world. It is a distinguishing feature of E. G. Churchman's work in management science that it closely connected with a philosophy of the systems approach. Management to Churchman has to deal with the ethical challenge to design improvement. But what constitutes an improvement and how can we design improvement without understanding the whole system?

Implications and Assessment

Systems thinking denotes the effort to define a nonreductive method for conceptualizing and explaining phenomena in both nature and society. As such it has a number of ethical and political implications that may be indicated roughly as follows.

First, systems thinking often claims to give a better account of the genealogy of ethics than did previous analyses. Ethics is described as an emergent property of complex living systems. Second, the opposition of systems thinking to nonsystems thinking almost always has a moral dimension. Systems thinking is said to be superior to nonsystems thinking in both theory and practice because it understands the world more accurately and provides better guidance for human action. Just as systems science yields better knowledge of the complexities of nature and artifice, systems engineering and systems management ground more effective interventions in nature, the construction of large-scale artificial systems, and the maintenance and management of their complex interactions.

These morally flavored claims can, however, cut two ways: to promote science and technology or to delimit them. On the one hand, systems thinking has played a large role in advancing scientific knowledge and technological development in the post–World War II era. It has done this both in the form of specific methodologies and theories and in the inculcation of a general receptivity to and awareness of interconnectivity in scientific and engineering communities. Some of its most significant impacts have occurred in biology, especially in the rise of ecology and in refinements of genomics. Institutional changes in the social structure of knowledge, especially increased interdisciplinarity, also have resulted from systems thinking.

On the other hand, systems thinking at times has criticized the modern scientific and technological project. In this critique of technological hubris, connections can be developed easily, for instance, between systems thinking and environmental thinking. Although he did not use the term, Aldo Leopold (1887–1948) essentially argued that the concept of the system forms the foundation of ethics: "All ethics so far evolved rest upon a single premise: that the individual is a member of a community of interdependent parts" (Leopold 1949, p. 203).

In a like manner Fritjof Capra (1997) has argued that new research on the organization of living systems promotes a reexamination of social policies. Systems thinking is both a scientific shift and a cultural paradigm shift away from mechanism and reductionism, but the relationship between those two shifts is complex and ethically charged. Capra, for instance, argues that systems research supports social egalitarianism, but that argument raises ethical questions about deriving political and moral conclusions from observations about nature. This is the same dilemma often raised by political conclusions drawn from the more reductionistic theories of sociobiology. The focus on wholeness, interconnectedness, and complexity thus has had an ambiguous impact on the larger realm of cultural and philosophical thought.

Thus, although it doubtlessly has been associated with some criticisms of technological and scientific hubris, systems thinking also has generated new versions of that hubris. For example, Luhmann's brand of systems thinking seeks to abstract a "grand theory" or a universal framework that is not concerned with individual humans, only the abstractions of information exchange. That led Habermas to label it as a version of "antihumanistic" sociology that denies the ability of individuals and institutions to guide social change consciously. Indeed, worldviews that stress holism always create a risk of losing lose sight of individual values such as dignity, freedom, and intentionality. In this case modern societies are seen as polycentric, and democratic participation and control as illusory, in the face of overwhelming complexity. However, it is difficult to conceive of justice and many other social values being realized by an autopoietic process devoid of intentional agency.

Similar two-sided features can be identified in proposals by Brad Allenby and others for the development of earth systems engineering and management. The bottom line is that systems and systems thinking remain ambivalent in their ethical import with regard to science and technology, but that ambivalence also may be their basic strength. Surely there is a sense in which science and technology need to be promoted and criticized at the same time.

SYTSE STRIJBOS
CARL MITCHAM

SEE ALSO *Complexity and Chaos; Reliability of Technology: Technical and Social Dimensions; Soft Systems Methodology.*

BIBLIOGRAPHY

Ackoff, Russell L. (1974). *Redesigning the Future: A Systems Approach to Societal Problems.* New York: John Wiley & Sons. The crux of the book is the provocative claim that "God's work is to create the future. Man must take it away from Him."

Beer, Stafford. (1959). *Cybernetics and Management.* New York: Wiley. The first of several books in which the author has developed his Viable System Model. Invited by President Salvador Allende, he applied this cybernetic model in the early 1970s in an ambitious project to regulate the whole social economy of Chile.

Bunge, Mario. (1979). *A World of Systems.* Boston: D. Reidel. This constitutes volume 4, part 2, of Bunge's *Treatise on Basic Philosophy.*

Capra, Fritjof. (1997). *The Web of Life: A New Scientific Understanding of Living Systems.* New York: Anchor.

Churchman, C. West. (1968). *The Systems Approach.* New York: Dell. A slightly revised and updated version appeared in 1979.

Flood, Robert L. (1990). *Liberating Systems Theory.* New York: Plenum. First volume in the series "Contemporary Systems Thinking," series ed. Robert L. Flood.

Graedel, T. E., and B. R. Allenby. (2003). *Industrial Ecology.* Upper Saddle River, NJ: Prentice-Hall. The field of industrial ecology has quickly developed since the 1990s as one of the latest separate branches of systems thinking and practice.

Laszlo, Ervin. (1972). *The Systems View of the World: The Natural Philosophy of the New Developments in the Sciences.* New York: George Braziller. See also the subsequent update of this book, *The Systems View of the World: A Holistic Vision for Our Time.* Cresskill, NJ: Hampton Press, 1996.

Leopold, Aldo. (1949). *A Sand County Almanac: And Sketches Here and There.* London: Oxford University Press.

Luhmann, Niklas. (1995). *Social Systems,* trans. John Bednarz, Jr. Stanford, CA: Stanford University Press. Originally published in German in 1984.

Maturana, Humberto R., and Francisco J. Varela. (1980). *Autopoiesis and Cognition: The Realization of the Living.* Boston: D. Reidel.

Strijbos, Sytse. (1988). *Het Technische Wereldbeeld: Een wijsgerig onderzoek van het systeemdenken* [The Technical Worldview: A philosophical investigation of systems thinking]. Amsterdam: Buijten and Schipperheijn. The study gives a critical discussion of important sources of systems thinking and of a fundamental philosophical critique. It focuses on the claim made by Von Bertalanffy and a "group" around him that systems thinking overthrows the technical worldview and establishes a new conception in place.

Von Bertalanffy, Ludwig. (1968). *General System Theory: Foundations, Development, Applications*. New York: George Braziller. A classic study in systems theory.

Wiener, Norbert. (1948). *Cybernetics: Or, Control and Communication in the Animal and the Machine. A classic study in systems thinking*. New York: Wiley.

TAYLOR, FREDERICK W.

• • •

Frederick Winslow Taylor (1856–1915), who believed that his system of scientific management provided the foundations for a scientific ethics, was born in Germantown, Pennsylvania, on March 20. His early education took place in private schools in Pennsylvania, Europe, and New Hampshire, and he was accepted for admission into Harvard University. But fascinated by the relationship among science, technology, and ethics, he decided on an apprenticeship at a steel company in Philadelphia, where, from 1878 to 1884, he advanced from common laborer to a supervisory mechanical engineer. In the process he became familiar with *soldiering,* when workers, to protect jobs and keep piece-rates high, increased output while bosses were watching and decreased it otherwise. An ardent believer in the Puritan work ethic, Taylor was troubled by this inefficient and unethical behavior, and came to believe that he had a solution not only for the Midvale Steel Company but for institutions throughout the world. He pursued this vision until his death in Philadelphia, Pennsylvania, on March 21.

Taylor's Studies

Taylor began by systematically studying machinery and human beings to discover precisely how much a diligent worker, using the best machines and procedures, could produce in a day. For example, his empirical analysis of metal-cutting machinery allowed him to more than double the machine's speed, and by analyzing the machinist's procedures into elementary motions, and timing them with a stop watch, he was able to minimize wasteful motions and optimize beneficial ones. This led to a

Frederick W. Taylor, 1856–1915. Taylor consolidated a system of managerial authority, often referred to as scientific management, that encouraged a shift in knowledge of production from the workers to the managers. (© *Bettmann/Corbis.*)

belief that all tasks, from the lowliest to the highest, could be made more efficient, and the resulting increase in productivity would optimize everyone's compensation and job satisfaction. He argued that a "single best way" existed for accomplishing every task, and that his scientific analysis of human technology interventions achieved an ethical goal: the resolution of the age-old conflict between labor and management.

After Taylor left Midvale in 1890, he spread the gospel of scientific management while occupying a series of positions from Maine to Wisconsin. He lived at a time when many Americans believed science and technology had the solution to many problems of humanity, but also during a time when bitter strikes sometimes resulted in the deaths of workers. Labor leaders and politicians criticized Taylor's claim that his system would end owner-worker hostility and render unions and strikes unnecessary. They pointed out that workers could not be treated in the same way as machines, and that several creative ways existed for accomplishing tasks rather than Taylor's one best way. Others questioned Taylor's yoking of productivity and morality. Taylor emphasized that wise work produced ethical workers, whereas others insisted that human morality motivated hard work.

During the final decades of Taylor's life, his obsession with efficiency deepened. Managers as well as laborers often resented his despotic attempts to change traditional methods of work and management. To those who said that scientific management was antidemocratic, he insisted that his techniques energized workers, promoted their self-reliance, increased their wages, and shortened their work week. To those who said that scientific management was unethical, he emphasized that his methods enhanced fellow feeling among workers and between workers and managers because he promoted true justice by encouraging the maximum efficiency and prosperity of all those involved in his system. But labor leaders and some politicians saw scientific management simply as a tool for maximizing production and profits to the neglect of the emotional and physical health of the workers. For them, Taylor's methods debilitated workers and increased accidents.

Taylor's Influence

In the early decades of the twentieth century, Taylor's ideas continued to generate both critics and advocates. In 1911 Taylor's disciples founded the Society to Promote the Science of Management (called, after his death, the Taylor Society) and he himself published *The Principles of Scientific Management*. In 1912 Taylor's system was debated at a Congressional hearing during which he defended his system as a force for good, but some committee members felt that he did not grasp the deep asymmetry between labor and management. Nevertheless, in its report the committee found some things to praise in scientific management—for example, standardization.

In the years after Taylor's death, Taylorism spread around the world. Taylor's disciples preached the gospel of efficiency to a wider audience than just businessmen—including housewives, teachers, even clergy. Like Taylor, his disciples viewed his doctrines as a means of transforming society, because the pivotal point differentiating civilized from uncivilized societies was productivity. Some of Taylor's disciples criticized their master—for example, Frank Gilbreth advocated replacing stopwatch studies with "micromotion" analyses in which each minute of a worker's activities was filmed and divided into a hundred units. Even Vladimir Lenin was influenced and thought Taylorism compatible with communism.

However, humanists such as Lewis Mumford (1895–1990) felt that Taylor's system got it backward: Humans come before and transcend systems. Some even saw Taylorism as deeply unethical, because its mechanistic treatment of workers was both an illusion and a delusion. During the twentieth century scientific management evolved and diversified, and it was no longer a unified and consistent body of thought. Although Taylor's goals of establishing social and economic justice and ending class conflict have not been achieved, his ideas, transformed and diversified, continue to influence various ideologies of science, technology, and ethics.

ROBERT J. PARADOWSKI

SEE ALSO *Management*.

BIBLIOGRAPHY

Copley, Frank Barkley. (1923). *Frederick W. Taylor: Father of Scientific Management*, 2 volumes. New York: Harper. For many years this was the standard biography, but it tends to be hagiographical.

Haber, Samuel. (1964). *Efficiency and Uplift: Scientific Management in the Progressive Era, 1890–1920.* Chicago: University of Chicago Press. An analysis of how some Taylor disciples created reform programs "without an appeal to conscience."

Kanigel, Robert. (1997). *The One Best Way: Frederick Winslow Taylor and the Enigma of Efficiency.* New York: Viking. The best critical biography of scientific management's founder.

Nadworny, Milton J. (1955). *Scientific Management and the Unions, 1900–1922.* Cambridge, MA: Harvard University Press. Historical study that emphasizes industrial relations, industrial management, and syndicalism in the United States.

Nelson, Daniel. (1980). *Frederick W. Taylor and the Rise of Scientific Management*. Madison: University of Wisconsin Press. A chronological account of an "unlikely revolutionary" that stresses Taylor's interaction with contemporary businesses and his relationship to workers.

Taylor, Frederick Winslow. (1947 [1911]). *The Principles of Scientific Management*. New York: Harp. Taylor's best and most influential account of his ideas.

TECHNICAL FUNCTIONS

• • •

One common way to describe artifacts is in terms of how they technically *function*. In a telephone sound is transformed into electronic signals that are then transmitted over some distance and transformed back into sound by another telephone. Such technical functions are strongly related to human uses. Telephones are designed and built so that they can be used for transmitting the human voice over distances well beyond its normal range. Because references to technical functions are often the basis for assessing human uses of artifacts, and insofar as such assessments express certain values, the relation between technical functions and uses is an issue for any ethics of technology.

Judging Actions and Artifacts

All intentional human behaviors or actions are subject to normative judgments. These judgments are of two sorts: deontic and evaluative. Deontic judgments express what one ought and ought not to do or what one has reasons for doing. Evaluative judgments describe something as good or bad. Using an artifact is subject to these types of judgments, in the first place because it is a form of action. It is generally wrong, for example, to hurt another person with a knife, which is merely a specification of the judgment that one ought, generally, not to hurt someone.

Additionally, however, the use of artifacts is subject to judgments that relate directly to the particular function of the artifact. For instance, one may say that it is wrong to use a Phillips screwdriver to open a paint can. Assuming that the attempt to open the can is itself perfectly in order, the *wrong* here is not morally wrong but *instrumentally* or functionally wrong: Using the Phillips screwdriver will not smoothly lead to the desired outcome. Typical for artifact use, such judgments may be translated, so to speak, to the artifacts themselves. An artifact is said to perform its function well or to function poorly or to malfunction. One can also say that a particular artifact, in the prevailing circumstances, *ought to do such-and-such a thing*. Even natural objects can, in a context of use, be subject to such judgments, for instance when one says that a particular stone is *a good stone to use as a hammer*.

Functions

The use of the term function in the previous paragraphs sets aside a considerable philosophical debate about the meaning of functions, one that has taken place largely in relation to the analysis of functions in biology (the function of the heart is to pump blood) and the social sciences (the function of religion is to create social cohesion).

Briefly there are two major competing concepts of functions: system functions as first stated by Robert Cummins, and proper functions as first stated by Larry Wright (1973) and further analyzed by Ruth Millikan, Karen Neander, and others. According to Cummins (1975), who is primarily concerned with biological systems, something has a function insofar as it contributes to the capacity of some system. According to Millikan, by contrast, the *proper* function of an organ or system is what helps to account for the survival and proliferation of its ancestors (1993). Millikan aims for a theory of functions that applies to artifacts as well as organisms.

Against these attempts to bring all uses of the notion of function under a single theory, Beth Preston (1998) argues for a *pluralistic theory of functions* that includes Cummins's system functions and Millikan's proper functions. Wybo Houkes and Pieter Vermaas (2004) hold that theories of artifacts are overly function-oriented and that a theory of artifact functions can be derived from a theory of artifact actions. For Preston, as well as for Houkes and Vermaas and for many others, functions often become the locus in both science and technology for the uniting of deontic and evaluative judgments.

Uniting Deontic and Evaluative Judgments

There is, however, no consensus of what precisely unites deontic and evaluative judgments insofar as they jointly comprise the realm of the normative. One account proposed by Joseph Raz (1999) and Jonathan Dancy (2005) holds that normative facts are facts expressing how other facts—natural or positive facts—matter to the question how to act. The deontic judgment that *To do X is right* then expresses the normative fact that there is the positive fact of X possessing certain features, *and* that these features are such that, in the circumstances at hand, the balance of reasons points toward the doing of X. The evaluative judgment that X *is good* similarly expresses the normative fact that the features of X are such that one has reason, perhaps even a compelling reason, to adopt a certain positive attitude toward X. What this positive attitude could be depends on the nature of X.

In contrast to a lack of clarity concerning the normative in general, there is wide agreement among philosophers that *instrumental* value should be sharply distinguished from *moral* or *ethical* value. To see the difference between these two forms of value, consider the statement: This is a good knife to kill Mrs. Robinson with. The knife is instrumentally good as a *means* to an end, but the *end*, the killing of Mrs. Robinson, is *morally bad*. One has reason to disapprove of Mrs. Robinson's violent death, and ought to prevent it. But *given* that the killing of Mrs. Robinson is sought, to do it with this knife may be considered a good and recommendable choice. Instrumental value is therefore in a sense conditional: It concerns the fitness of a particular means to the realization of an end once that end is given, whereas it is not concerned with any pros or cons regarding the end itself.

The distinction between moral value and instrumental value is closely related to a distinction among the sorts of reasons that back up an act or an attitude or a belief. If means M is fit to end E such that one ought to choose M or choose to do M, this concerns an ought on rational grounds. By contrast, if M is morally good or bad, such that one ought to approve or disapprove of M, this concerns an ought on moral grounds. This way of distinguishing rational grounds from moral grounds sees the notion of rationality exclusively as instrumental rationality. Not all philosophers will agree, however, that rationality should be viewed thus.

Designing and Using Artifacts

The design and use of artifacts is involved with both kinds of grounds for normative judgments, but in particular cases it is not always obvious whether one or the other kind is at issue. Malfunction judgments and judgments of poor or proper functioning certainly have a special relation to considerations of rationality. A statement such as *Artifact A malfunctions* expresses the positive fact that A does not or will not show the behavior it was designed to show. However, this positive fact does not exhaust the meaning of the statement. It also seems that when an artifact malfunctions or functions poorly, human beings *by definition* have a reason not to use it, or at least not to use it as designed, on rational grounds.

One cannot go as far as saying that the notion of malfunction or of poor functioning semantically implies that the item ought not be used. There may be reasons such that, on balance, it is rational to use the thing anyway. But if one applies the judgment of malfunction prior to any considerations of use, as a mere factual statement of the artifact's failure to show a certain behavior, it makes no sense to then ask whether that fact means anything about what one will do with the artifact. When an artifact is said to malfunction, one necessarily has at least *a* reason not to use it as designed.

Similarly to say that a particular artifact functions well is not just to say that it shows a certain behavior, as a positive fact, regardless of anything that one might do with it. This judgment implies that the item shows a particular behavior and that one has a reason to use it as designed. In this case, however, the conditionality of instrumental reason really has a bite: One has an overall reason to use something to produce the result that using the artifact in question produces. If one does not have a reason to use a car in the first place, because one is not going anywhere, then neither does one have a reason to use this particular car, which happens to be a very good car.

Whether one also has a reason *on moral grounds* not to use a malfunctioning or poorly functioning artifact, or even ought not to use such an artifact on moral grounds, is a question that raises different issues. The judgment might be motivated, for instance, by fear that the artifact's use would pose a hazard for other people. But such judgment often depends on the particular case at hand and thus is not covered in the *meaning* of malfunction. It is hardly worthwhile to discourage someone from using a Phillips screwdriver to open a paint can on moral grounds.

The rationality of artifact use depends critically on knowledge. To judge that the use of a particular object is the best means to achieve a certain goal requires an adequate knowledge of the object's properties and the effects of manipulating it in the prevailing circumstances. The use of the object can be rational only to the extent that the user's beliefs about the object are rationally justified. Rationally, in this sense, refers to *epistemic* rationality, and not *practical* rationality, which was the form of rationality relevant in the preceding considerations. In practical rationality, the issue is what it is best to *do*, or what one has a reason to do, given one's end of realizing a particular situation. For epistemic rationality the issue is what it is best to *believe*, or what beliefs one has a reason to adopt, given the end of holding as many true beliefs as possible or holding only true beliefs.

Proper Use and Good Design

When someone uses an artifact in disregard of its designed function, that is, according to some privately conceived use plan, reasons of epistemic rationality seem all that matter. (The concept of function as a use plan is developed at length by Houkes, Vermaas, Dorst and de Vries, 2002.) When the artifact's use fails to have the desired result, there is no one to blame. This is no longer true when an artifact is used for its designed function, in circumstances that are consistent with the artifact's use plan as explicated in the instructions for use. When handing over an artifact to a client who ordered it, or to the market, the designer/producer is committed to the veracity of the predictions made about the artifact's behavior. These predictions have the force of a promise, and the commitment accordingly has the character of a moral obligation. One could say that a designer ought, on moral grounds, to be epistemically rational. In practice the extent of this obligation is articulated in the form of standards that say how much research and testing is sufficient to vindicate the claims that are to be made about the artifact's performance.

It is part of the human condition that neither the criteria of epistemic and practical rationality nor the criteria of moral obligations can guarantee the realization of plans. One may be disappointed by fellow human beings as well as, metaphorically speaking, by nature. The ubiquity of uncertainty shows in the use of language when one says that a particular artifact *ought to do* something when handled in a certain way. This may express the idea that one is epistemically justified in one's belief that the artifact will perform as expected, given the amount of research and testing adopted in designing or in repairing the artifact, but at the same time there is a recognition that there is always the possibility that something was overlooked. The statement may also express the idea that one *has a right* to the artifact's performance, on the basis of a promise by a designer/producer, retailer, or repair service person, while there is at the same time the awareness that such promises are occasionally broken.

It seems natural that in what is summarily described as *good design* the grounds distinguished above play a role. An artifact that can be termed a good design must be instrumentally fit for its function in a range of plausible circumstances. However a well-designed artifact must also be one that it is morally vindicated to use. This can either mean that it is not likely to lead to outcomes of low moral value, for instance by being safe, or that it is likely to lead to outcomes of high value, which will often be a comparative matter.

Thus the features of a particular artifact may give rise to reasons, even compelling reasons, for its use in order to contribute to the realization of one's goals, by which such artifact is instrumentally good. It may also have features such that one has reasons, even compelling reasons, to approve and promote its use, by which it is morally or ethically good. Additionally, artifacts are often judged on the basis of a third criterion, previously not discussed, namely aesthetic appeal. Technical artifacts may have both instrumental and ethical value, or both instrumental and aesthetical value, or even all three. Some trash receptacle for public use may not only function perfectly as a trash receptacle, but it may also encourage people to use it to a larger extent than another type of trash receptacle, and on top of that be considered a beautiful object.

MAARTEN FRANSSEN

SEE ALSO *Engineering Design Ethics; Engineering Ethics.*

BIBLIOGRAPHY

Audi, Robert. (1989). *Practical Reasoning*. London: Routledge. An introductory text on practical rationality.

Cummins, Robert. (1975). "Functional Analysis." *Journal of Philosophy* 72(20): 741–765. The classical text for the theory that an object's function is its contribution to the overall behavior of a system of which it is a part.

Dancy, Joseph. (2005). "Ethical Non-Naturalism." In *The Oxford Handbook of Ethical Theory*, ed. David Copp. Oxford: Oxford University Press. Proposes, among other things, a general characterization of normativity.

Fransen, Maarten. (2005). "The Normativity of Artefacts." *Studies in History and Philosophy of Science* 36(x). Analyzes the interrelations between the various normative statements in which artifacts figure and their grounding on criteria of rationality and morality.

Grice, Paul. (1991). *The Conception of Value*. Oxford: Oxford University Press. A general text on the notion of value and evaluation.

Houkes, Wybo, and Pieter Vermaas. (2004). "Actions versus Functions: A Plea for an Alternative Metaphysics of Artifacts." *The Monist* 87(1): 52–71. Argues that use is a more basic notion to make sense of artifacts than function. Discusses (ir)rational and (im)proper use of artifacts.

Houkes, Wybo; Pieter Vermaas; Cees Dorst; and Marc de Vries. (2002). "Design and Use as Plans: An Action-Theoretical Account." *Design Studies* 23(3): 303–320. Develops the notion that the function of an artifact has to be understood from the role it plays in a use plan by which an individual seeks to produce a certain end.

Korsgaard, Christine. (1996). *The Sources of Normativity*. Cambridge, UK: Cambridge University Press. A general philosophical text discussing various conceptions of normativity.

Millikan, Ruth Garrett. (1993). *White Queen Psychology and Other Essays for Alice*. Cambridge, MA: MIT Press. This is a more accessible statement of a view first argued in Millikan's *Language, Thought, and Other Biological Categories: New Foundations for Realism*, Cambridge, MA: MIT Press, 1984, that all functional statements, including those concerning the function of language and the meaning of sentences, an be grounded in terms of evolutionary biology.

Preston, Beth. (1998). "Why is a Wing like a Spoon? A Pluralist Theory of Function." *Journal of Philosophy* 95(5): 215–254. Argues that only a pluralist theory of function is able to do justice to biological as well as artifact functions.

Raz, Joseph. (1999). *Engaging Reason: On the Theory of Value and Action*, 2nd edition. Oxford: Oxford University Press. An important text, first published in 1975, on the relation between value and reasons for acting.

Wright, Georg Henrik von. (1963). *The Varieties of Goodness*. London: Routledge & Kegan Paul; New York: The Humanities Press. Discusses the interrelations among a broad spectrum of evaluative terms.

Wright, Larry. (1973). "Functions." *Philosophical Review* 82(2): 139–168. First statement of the view that an object's function is what is causally responsible for its existence, which underlies later evolutionary accounts such as Millikan's.

TECHNICISM

• • •

The term *technicism* is parallel in construction to "scientism" and serves many of the same purposes, although it is less common. While closely associated with the process of "technicization," technicism, like all "isms," offers a special perspective on the world and its character. The belief in technology as central to the world can take different forms, but is most commonly manifest in what may be called ethical technicism.

Origins

In the *Gorgias* Plato (c. 428–347 B.C.E.) already identified the character of technicism, the belief in means as in some sense primary over ends. Gorgias, a sophist, has separated his rhetorical skills (*technai*) from any firm subordination to substantive social or cultural traditions, not to mention to the good. This is a position that Socrates (c. 470–399 B.C.E.) strongly criticizes, but according to Karl Polanyi (1886–1964), Lewis Mumford (1895–1990), and other historians, it is precisely such a project of separating culture into various components and then pursuing each on its own terms that is the foundation of modern technology. When technics is pursued in terms of its own logic it becomes technology.

According to Max Weber (1864–1920), in his posthumously published studies titled *Economy and Society* (1922), traditional societies contain "techniques of every conceivable types of action, techniques of prayer, of asceticism, of thought and research, of memorizing, of education, of exercising political or hierocratic domination, of administration, of making love, of making war, of musical performance, of sculpture and painting, of arriving at legal decisions" (vol. 1, p. 65). But in traditional societies these techniques are embedded in mores and counter-mores institutions. The planting of crops is done efficiently, but also in accord with certain religious rituals. The building of houses is done effectively, but also with respect for various craft traditions and social distinctions. Efficiency and effectiveness do not operate independently of other social, culture, religious, aesthetic, ethical, and political constraints.

In the German tradition Max Scheler (1874–1928) was among the first to use the term *Technizismus* (technicism) to name an attitude toward the world that takes the pursuit of material effectiveness in means as itself a fundamental ideal. The term appears in Scheler's 1926 book *Die Wissensformen und die Gesellschaft*, but was also used in papers as early as 1914 in which he provided phenomenological sketches of different types of persons and leaders. For Scheler, it is the historical development of modern technological civilization that gave rise to technicism as a form of discourse (Janicaud 1994) or consciousness (Stanley 1978) that chooses to privilege means over ends—that is, to center public life around the pursuit of ever more effective means, while relegating questions of ends to issues of personal or private choice and decision-making. From this perspective, technicism has become a pejorative term especially among nonbehavioral social scientists.

Ethical Technicism

Among the first philosophers to analyze the ethical implications of separating out means from ends was José Ortega y Gasset (1883–1955). In the English translation of his *The Rebellion of the Masses* (1929), Ortega identifies three principles as fundamental to the twentieth century: liberal democracy, scientific experiment, and industrialism. "The two latter may be summed up in one word: technicism" (1932, p. 56). In fact, insofar as liberal democracy is also committed to public policies that promote the maximization of means, leaving ends to be determined by individuals, technicism covers the first principle as well. (In Spanish Ortega actually used the word *técnica*, but the translation "technicism" is significant as one of the earliest English occurrences in a new sense. In the previous century "technicism" meant simply excessive reliance on technical terminology.)

The next decade, in *Meditación de la técnica* (1939), Ortega outlined a historical movement from the chance inventions that characterize archaic societies, through the trial-and-error techniques of the artisan, to the scientific technologies of the engineer. According to Ortega, the difference between these three forms of making lies in the way one creates the means to realize a human project—that is, in the way technicalness or technicity is manifest. In the first epoch technicity is hidden behind accidents, whereas in the second, technicity is cultivated and protected in craft traditions. In the third, however, the inventor has undertaken scientific studies of technics and, as a result, "prior to the possession of any [particular] technics, already possesses technics [itself]" (*Obras completas* V, p. 369). It is this third type of technicity that constitutes "modern technicism" (and here Ortega himself uses the term *tecnicismo*).

But technicism understood as the science of how to generate all possible means independent from any lives making and using context creates a unique existential problem. There is a temptation to pursue technical invention as a good in itself, to become lost in the technical means as exciting or valuable in their own right. Prior to the modern period human beings were limited by circumstances in which they at once acquired a way of life and the technical means to realize it. Now in liberal societies they are given in advance a plethora of technical means but no well-defined sense of the good other than personal choice. "To be an engineer and only an engineer is to be everything possibly and nothing actually" (*Obras completas* V, p. 366). In the midst of modern technicsm Ortega discovers a crisis of imagination and choice. Insofar as people can be anything at all, why should they be any one thing? What Ortega imagined has become real in the case of those who play with their avatars in cyberspace while failing to become something in the world.

Epistemological Technicism

The engineer Billy Vaughn Koen, however, proposes the engineering method as the fundamental way of knowing and acting in the world in a way that turns technicism from an ethical problem into an epistemological method. Koen does not use the term technicism, perhaps because of its negative connotations. But his argument is that engineering is the method that all human beings use, and indeed must use, whenever they solve problems. "*To be human is to be an engineer*" (2003,

p. 7; italics in original) whether one knows it or not. There is simply no alternative.

For Koen, "The *engineering method* is the use of heuristics to cause the best change in a poorly understood situation within the available resources" (p. 59). Heuristics are simply strategies based on some hunch, rule of thumb, or intuition, about what might work, that include both a rejection of any absolute sense or certainty and a willingness to revise in response to experience in order to make things better. In Koen's perspective, the engineering method is universal precisely because it does not claim to be universal. The engineering response to Ortega's problem is simply to try something. No situation of even apparently unlimited possibilities can remain that way forever. There is in the end an excitement about an epistemological technicism that sees necessity as unnecessary and is therefore willing to play with possibilities and see what happens.

WILHELM E. FUDPUCKER

SEE ALSO *Scientism*; *Technicization*.

BIBLIOGRAPHY

Janicaud, Dominique. (1994). *Powers of the Rational: Science, Technology, and the Future of Thought*, trans. Peg Birmingham and Elizabeth Birmingham. Bloomington: Indiana University Press. Originally published as *La puissance du rationnel*. (Paris: Gallimard, 1985).

Koen, Billy Vaughn. (2003). *Discussion of the Method: Conducting the Engineer's Approach to Problem Solving*. New York: Oxford University Press.

Stanley, Manfred. (1978). *The Technological Conscience: Survival and Dignity in an Age of Expertise*. New York: Free Press.

TECHNICIZATION

• • •

The challenges posed by modern science and technology to ethics include the challenge of technicization. Technicization is a process that some contend infects and thereby corrupts ethics. To understand this claim requires an understanding of the process of technicization (related terms: technicism, technization, technicalization, scientism, scientization, mechanization) in relation to the task of ethical reflection.

Technological civilization is made up not only of machines but also and more importantly the methods or "techniques" that produce machines. Technique is rooted in the human capacity for language that gives humans the ability to imagine ever-new goals and the means to achieve them. For most of human history techniques were embedded in a wider array of cultural beliefs and practices and passed on as part of the culture from one generation to the next. One did things in a certain way because that was how one's ancestors did them. Such techniques were not inherently related to science.

When modern science intersected with ancient technologies beginning in the seventeenth century, the result was the technicization of society. This occurred when scientific investigation systematically evaluated not only the array of techniques historically available from all cultures for accomplishing human ends but also systematically studied the process by which techniques come to be invented, so as to refine the efficiency and effectiveness of the invention process itself. The ultimate goal of the science of technical development is the creation of the most efficient techniques in all areas of human endeavor so that every aspect of life is shaped by technical norms of efficiency.

The application of science to technique transforms the way human beings understand themselves and human societies organize themselves and their tasks. In premodern societies "essence" was thought to precede "existence"—that is, human beings thought of their selves and their institutions as having a preordained natural course of development (their *telos*) as part of an unchanging sacred natural order established by the gods and ancestors and/or nature itself.

From the ancient Greeks right on through the Enlightenment, social, political and ethical theory was dominated by the assumption that there is either a supernatural (the Platonic tradition and its successors) or a natural (the Aristotelian tradition and its successors) telos or archetype that must be discovered and implemented in human society. Society, as the Greeks said, is the cosmos writ small and later thinkers such as Hobbes and Rousseau still gave assent to such a view although there understandings of "nature" certainly differed. Even when Kant split noumena from phenomena he still assumed a universal rational human nature. Hence, whereas there was comparative reflection on social organization and speculation as to the best order for society from the time of the ancient Greek philosophers, it was assumed that the best order could not be discovered in the practices of social convention (the artificial) but only through the discovering the right order of nature (its true essence and telos).

Society is not an empirical object. The awareness of society as a realm separate from nature is a work of the human imagination. It was only with the emergence of the comparative and cross-cultural studies of the social sciences in the nineteenth century that society came to be imagined as existing as a distinct realm apart from nature, an artificial or humanly made order that had no inner telos. Society came to be understood as a technological or artificial product, Existence precedes essence and society is what humans make of it.

In this way, with the emergence of the critical historiographical and ethnographical techniques of the social sciences in the nineteenth century, the mythic stories of "natural order" were demythologized and replaced with a technological understanding of society. This transformation came to be expressed in four new ways of thinking about self and society: (1) the *existential self*, (2) the *managerial society*, (3) *public policy*, and (4) *social ethics*. Because the order of society is not fixed and given with the order of nature, humans must (1) choose who they shall become individually and as a society (2) reorganize the structures of society to make such choices possible, (3) engage in public debate in order to make choices about what kind of society they want to create and (4) therefore engage in social ethics as the attempt to define the norms by which they shall make such choices and so invent themselves.

In premodern societies ethics is primarily the ethics of virtue and so is concerned with individual choices. The task of ethics is to actualize one's essential "human nature" in accord with one's *telos*, within the social order as the cosmos writ small. Once institutions are seen as human creations based on choice rather than being fixed and given as part of a sacred cosmic order of nature, ethics is forced to enlarge its horizons to engage in the critique of institutional behavior without reverting to the essentialist model of cosmological thinking. A technological civilization fundamentally transforms the understanding of the task of ethics by introducing the novel idea of social ethics as a post-essentialist critique of society as a technological artifact through those public policies or social choices that shape one's personal identity and institutional life.

For some (for example, Niklas Luhmann, 1927–1998) the technological civilization that emerges out of the new social scientific consciousness of the artificiality of society seems to promise greater freedom and control, and so a greater scope for ethics through managerial social policy. However, others (such as Jacques Ellul 1912–1994, Jürgen Habermas b. 1929) argue technicization threatens to undermine that freedom and the practice of ethics by producing the technobureaucratic rationalization and mechanization of society.

Indeed, a major motif among the giants of sociology (Karl Marx, Émile Durkheim, Max Weber) is the mechanization of society so as to create what Weber called "the iron cage" of technobureaucratic societies. In this view, managerial societies are dominated by bureaucracies of scientific-technical experts who identify and promote the most efficient ways to meet human needs in all areas of endeavor (that is, maximizing results while minimizing costs and energy expenditures), and technical efficiency eliminates choice. The focus shifts from ends to means. The less efficient society cannot compete with the more efficient society any more than the less efficient business can compete with the more efficient business.

This process of technicization threatens the human ability to think and act ethically. Insofar as ethics entails the Socratic question—*Is what people call good really the good?*—how can that question be raised and acted on in a society that defines efficiency as the ultimate good? How can ethicists expect to succeed in introducing nontechnical norms such as justice and compassion in a society that seems to make acting on nontechnical norms virtually impossible? And how can norms be asserted at all in a post-essentialist technological society?

The seriousness of this problem is evidenced by the technicization of ethics itself. In a technical civilization only people who have technical expertise command respect and are socially and financially rewarded. In response ethicists' reflections have become increasingly too technical and specialized to be understood by society at large and so must be left to the calculations of technobureaucratic experts. As a consequence the Socratic task of the "gadfly" who calls into question what people call "good" in order to introduce a broader (nontechnical) vision and practice of "the good life" is in danger of being neutralized as irrelevant. If the ethical task of the gadfly is to be possible, it will have to begin by calling into question the "technological bluff" of the adequacy of technical language and norms as sufficient for realizing the good life.

DARRELL J. FASCHING

SEE ALSO *Scientism; Technicism.*

BIBLIOGRAPHY

Ellul, Jacques. (1990). *The Technological Bluff*, trans. Geoffrey W. Bromiley. Grand Rapids, MI: Eerdmans. A critique of the ideological discourse of technology as the only solution to human problems.

Fasching, Darrell J. (1993). *The Ethical Challenge of Auschwitz and Hiroshima: Apocalypse or Utopia?* Albany: State University of New York Press. An analysis of the technological or artificial nature of human identity and the implications for developing a post-essentialist cross-cultural and interreligious global ethic to guide public policy.

Habermas, Jürgen. (1984–1987). *The Theory of Communicative Action*, trans. Thomas McCarthy. 2 vols. Boston: Beacon Press. A critique of the rationalization or technicization of society that calls the "technological bluff" into question and relocates ethical reflection in a utopian theory of communicative action.

Mumford, Lewis. (1970). *The Myth of the Machine*, Vol. 2: *The Pentagon of Power*. New York: Harcourt Brace Jovanovich. The second volume of Mumford's history of technology; focuses on the mechanization of society.

Rammert, Werner. (1999). "Relations that Constitute Technology and Media that Make a Difference: Toward a Social Pragmatic Theory of Technicization." *Techné: Journal of the Society for Philosophy and Technology* 4(3): 23–43. Also available from http://scholar.lib.vt.edu/ejournals/SPT/.

TECHNOCOMICS

• • •

Technocomics are illustrated narratives in which science and technology play a major role in the determination of character and action. Superhero comics are often good examples, insofar as many of their protagonists receive superpowers as an unexpected consequence of some scientific phenomenon. Peter Parker becomes Spider-Man, for instance, during a school outing to a science museum where he is accidentally bitten by a radiated spider; the X-Men all experience genetic mutations as a result of environmental contamination and thus confront problems of social prejudice and responsibilities between generations. Technocomics as a genre are thus closely related to science fiction and may serve to both mirror and shape popular reflection on questions related to science, technology, and ethics.

The comic book superhero first emerged from pulp fiction in the 1930s in what is known as the Golden Age of DC Comics and its protagonists such as Superman, Batman, and Wonder Woman, who were only marginally associated with science and technology. The post-World War II period saw a decline in the popularity of these figures. But in the 1960s, Marvel Comics brought about a Silver Age by creating a new pantheon of superheros including Spider-Man, the Incredible Hulk, and the X-Men, all of whom reflected a deeper concern for the ethical issues associated with science and technology in the nuclear age. The following analy-

Superman. *(AP/Wide World Photos.)*

tic introduction assumes some familiarity with this particular genre as it has developed in the United States, a genre that has also extended to movies, video games, and, in the early-twenty-first century, to some advanced simulations such as *Technocracy* (see Brucato, Long, and DeMayo 1999). For more general introductions to technocomics see the work of Mike Benton (1992), Richard Reynolds (1994), and Geof Klock (2002).

Radiation: Science as Savior and Scapegoat

Radiation has from the very beginning played a key role in technocomics, which perhaps reflects twentieth-century American societal fascination with, as well as aversion to, nuclear technology and its applications during times of both war and peace. Many superheroes of both the Golden and Silver ages of comics derived their special abilities from some type of radiation in one of three ways. The first, rarest, and perhaps most optimistic way is when the character comes to reside in a different environment and is exposed to a form of radiation that alters the physiology of his already existing anatomy. Superman, one of the earliest protagonists of the Golden Age of comics, is an example of this type of superhero. Origin-

ally the source of his special powers were unexplained; later, however, they were linked to the effects on his body of the light radiation from the Earth's yellow sun as opposed to that of the red sun of his home planet Krypton. Later comics involving Superman included a substance called Kryptonite (no relation to the element Krypton), whose green and other forms had various effects on him, including the nullification of his powers.

The second way in which radiation bestows superpowers in technocomics illustrates one of the most common fears of the nuclear age—mutation. This preoccupation with the unexpected, negative effects of radiation (which gave rise to a series of Godzilla movies in Japan), is manifested in such Silver Age technocomic protagonists as the X-Men, who are born with superpowers because ambient radiation from atomic bombs has changed their genetic codes.

Yet the third way in which superheroes derive their powers from radiation in these comics is the most prevalent—the alteration of an individual's genetic makeup through accidental or intentional exposure (such as nuclear accidents, atomic experiments, and others). Some of the most famous superheroes who have attained their powers in this way include Spider-Man, the Teenage Mutant Ninja Turtles, Dr. Manhattan, Daredevil, and Captain Atom. The most representative of this type of superhero, however, is the Incredible Hulk, whose alter ego, Dr. Bruce Banner, was a research scientist for the military-industrial complex who was attempting to develop a gamma bomb for the U.S. Army. During the first test of this bomb, Banner entered the testing area to save a civilian from the explosion, thus exposing himself to the gamma radiation that causes him, in a Jekyll-and-Hyde-like manner, to transform into the Hulk, a huge, immensely strong creature.

Human Response: Technology as Superpower

While superhuman characters such as Superman and Spider-Man experienced permanent changes that made their special powers innate, other characters have developed and employed technology in an attempt to achieve superhero status. Technological research and development organizations began to appear in superhero comics (such as Advanced Idea Mechanics [A.I.M.], a criminal organization in the Marvel universe, and Scientific Technological Advanced Research Laboratories (S.T.A.R.), a scientific organization in the DC universe), creating new technology to both the benefit and detriment of society. Devices such as ray guns, flying cars, and power armor appear in myriad forms in these comics, in which technological processes and pharmaceuticals such as cloning and *supersoldier sera* are also common. The list of technologically assisted superheroes and supervillains is long, and the majority of them utilize special suits and gear. The most famous of these characters include Iron Man, Green Arrow, the Punisher, Nick Fury and the Supreme Headquarters International Espionage Law-Enforcement Division (S.H.I.E.L.D.), the Atom, Hank Pym, Blue Beetle, Owl Man, Doctor Doom, Lex Luthor, Booster Gold, Captain America, the Engineer, and Batman (who nevertheless would be a force to reckon with even without his Bat Computer and infamous utility belt).

One of the most unique of these superheroes, Booster Gold, is of special interest because his origins illustrate ambivalent feelings toward the corporate technological complex. Booster, a twenty-fifth-century football player banished from professional sports for illegal betting, steals a force field belt, flight ring, and a time sphere which he and a robot named Skeets use to travel back in time. He then becomes the CEO of Booster Gold International, a monolithic holding company and tax shelter, as well as *America's Most Popular Super Hero*.

Still other superheroes use technology in the form of symbiotes (organisms, alien or otherwise, that grant abilities to their hosts), chemical alterations of their bodies, and even artificially intelligent constructs. Perhaps the most famous example of a chemically enhanced superhero is Captain America. According to the account of his origins, during World War II the United States developed an experimental supersoldier serum. It was first tested on Steve Rogers, a frail man unfit for combat, to whom it gave increased mental and physical capabilities. The doctor who created the formula was soon after killed by a Nazi spy, leaving Rogers as the first and only supersoldier—Captain America.

Still other superheroes and supervillains have obtained their powers through a combination of the effects of radiation and technological enhancement. A good example here is one of the X-Men, Wolverine, a born mutant who is later *improved* with technology. Wolverine's original mutations included animal senses and an amazing capacity for self-healing. This latter power enabled the Weapon X Program to implant the unbreakable metal adamantium into his bones without killing him, thus making him virtually indestructible.

The Ethics of Power

Ethical questions regarding science and technology make natural themes for technocomics, given the great number of technologically created superheroes and supervillains who serve as their protagonists. One of the most common of these questions is that regarding the limits of scientific experimentation. J. Robert Oppenheimer's concern about the atomic bomb finds its echo in technocomics: Does ability imply permission? Do humans have the right to use technology just because they have invented it? These questions are debated time and again in the pages of technocomics (for example, in the cases of the Weapon X Program, the origin of the Hulk, and Brainiac 5's creation of Computo). Such comics play an important ideological role, because they are often a young person's first introduction to these questions, and furthermore offer a safe, fictional representation that spurs critical thinking about the real dilemmas (such as human cloning) faced by contemporary society.

Many technocomic superheroes demonstrate the desire to use their powers ethically and strive to accept a responsibility to others that they believe accompanies their special gifts. For example, heroes such as the almost omnipotent Professor X and Spider-Man (whose message "With great power comes great responsibility" has become a mantra for generations of comics fans) seem to be always defending and disseminating their belief that those who possess special abilities must not exploit those who do not.

Homo Superior: Social Darwinism in Technocomics

Although Social Darwinism is a misapplication of a scientific theory, it generates many debates in technocomics, especially given their superhuman protagonists. Should the strongest, most talented, and most intelligent rule the world to the detriment of the weak? Perhaps the most important site of this debate in the technocomic world is found in the X-Men comics, in the conflict between Professor X and his archrival, Magneto. Magneto is a superpowerful mutant who survived life in a concentration camp during World War II, and has therefore experienced firsthand the horrors that humans are capable of inflicting on one another. He is convinced that mutantkind (human beings who have mutated and developed superior abilities) is the next step in human evolution and that mutants should therefore take their place as the new rulers of the world. Professor X, however, takes the stance that mutants—however different they may be—are still humans and must learn to live alongside less-gifted humans.

Technoscientific Authoritarianism

Ethical questions surrounding technoscientific authoritarianism are often present in technocomics, given

Spider-Man. (AP/Wide World Photos/Courtesy Marvel Comics.)

that absolute power is a goal that many technically enhanced supervillains strive for. A particularly relevant instance of this debate, albeit ultimately unresolved, appears in those Marvel comics dealing with Doctor Doom, the supreme ruler of a fictional country called Latveria. This country is described as being free from racism and social unrest; its inhabitants enjoy economic prosperity while remaining ecologically and physically safe and sound. But while the government of Latveria is considered to be an *enforced monarchy* by Doom and his subjects, all others consider it a dictatorship. The question of whether it is acceptable to give up democratic and personal freedoms to a technocrat in return for safety and security arises. At one point in the Marvel universe, Doctor Doom manages to take control of the entire world after which he eliminates disease and hunger and brings about world peace with an iron hand. Even the staunch defender of democracy, Captain America, has to admit that, while the method Doom uses is unacceptable, the changes he brings about are in the best interest of humanity. Nevertheless, at the end of the series, Doom is removed from power and the world reverts to its previous state, with relief food rotting on the docks in Africa, arguments breaking out in the United Nations, and the winds of war again stirring worldwide. Readers are left to decide for themselves which type of government is preferable.

Subsequently, in 2004, Captain America, the technologically enhanced supersoldier, was involved in a critique of the very military industrial complex that created him. He is sent to Guantanamo to oversee the treatment of the Taliban and Al-Qaeda prisoners being held there by the heavily armed, technologically superior U.S. soldiers, and is shocked at the human rights abuses he witnesses being committed by members of his own team.

Questionable Experimentation and Creation: Progress versus Safety

Questions surrounding the ethical ramifications of experimentation, especially experimentation on living beings, arise frequently in technocomics. Should experiments be done if they are not safe for the individuals involved? Is questionable scientific experimentation ethical if it causes human and/or animal suffering in pursuit of the alleviation of future suffering? Are technological processes that extend the quantity of life worth their possible toll in quality of life? The previously mentioned Weapon X Program in which Wolverine gains his adamantium skeleton, along with the ambivalent feelings many superheroes have toward their own powers, is only one of the ambiguous situations in technocomics that promote such ethical pondering.

Artificial intelligence (AI) plays a central role in many technocomics. The philosophical questions raised in this regard range from the ontological (Is a machine that can think a living creature?), to the epistemological (How does one recognize life?), to the ethical (Is it ethical to try to create a machine that can think? If a thinking machine has accidentally been created, should it be shut off? Should humans allow themselves to become so dependent on machines in general, and on artificial intelligence in particular?).

Not only does sentient AI life exist in the world of technocomics, but it is also often imbued with the theological categories of good and evil. One example can be found in the *Avengers* series of comics, in which the scientist/Avenger Hank Pym accidentally creates Ultron, an evil, artificially intelligent being who is able to remodel himself as well as to create other AI machines. The Vision, one of the machines modified by Ultron, using his newly acquired free will for more noble purposes, rebels against his programming, joins the Avengers, and even marries. Similarly, in the *Brainiac* series of stories, Brainiac 5 creates an AI machine named Computo, that ends up killing dozens of people before being turned off.

Using Technocomics

Technocomics have introduced many scientific and ethical questions into the minds of readers, and can be expected to continue to do so by incorporating into fiction new technologies and scientific theories as they emerge in the real world. Technocomics have been a source of entertainment for so long that their value as teaching tools are often overlooked. Nevertheless, in the early-twenty-first century, there is increasing awareness of the effectiveness of using technocomics to spark scientific and philosophical debate in the classroom. The Department of Chemistry at the University of Kentucky, for example, supports a web site linking science to technocomics that lists, in periodic table structure, the occurrences of elements in comic books, both in the form of facts and misconceptions. At times the superpowers portrayed in technocomics, as well as the scientific errors that they frequently entail, can be as useful as scientific facts for teaching purposes. James Kakalios, a professor at the University of Minnesota, incorporated such misconceptions in a course titled "Everything I Know of Science I Learned from Reading Comic Books," which compares and contrasts the science portrayed in technocomics with real-world physics, including thermodynamics and the material sciences. Kevin Kinney of DePauw University discusses many misconceptions of biology in comic books, such as those related to superpowers and the amount of nutrients that would be needed to fuel them.

Interestingly while science fiction in general has logically been appropriated by teachers of ethics as a springboard for debates about ethical issues in science and technology, the use of technocomics for these same purposes appears to have been overlooked. Nevertheless the success of film adaptations of such technocomics as *The Hulk* (2003), *Spiderman* (2002, 2004), and *X-Men* (2000, 2003) will almost certainly guarantee serious reconsideration as to how these works both reflect and mold popular opinions and conceptions about the nature—ethical or otherwise—of scientific investigation and technological innovation.

MATTHEW BINIEK
JAMES A. LYNCH

SEE ALSO *Movies; Popular Culture.*

BIBLIOGRAPHY

A.I.M (Advanced Idea Mechanics) *Strange Tales* #146. Marvel Comics Group, Marvel Entertainment Group, Inc.

"The Batman." *Detective Comics* #27, May 1939. DC Comics.

Benton, Mike. (1992). *Superhero Comics of the Golden Age: A History*. Dallas: Taylor.

"Booster Gold, Origin." *Boo$ter Gold* #7, August 1986. DC Comics. First appearance.

Brucato, Phil; Steve Long; and Tom DeMayo. (1999). *Guide to the Technocracy*. La Mesa, CA: White Wolf. A general introduction to the history and characters of the White Wolf role playing simulation game called *Technocracy*.

"Captain America. *Captain America Comics* #1, March 1941. Marvel Comics Group, Marvel Entertainment Group, Inc. First appearance and origin.

The Engineer. *The Authority* #1, May 1999. Image Comics. First appearance.

"The Hulk/Bruce Banner." *The Incredible Hulk* #1, May 1962. Marvel Comics Group, Marvel Entertainment Group. First appearance and origin.

Klock, Geof. (2002). *How to Read Superhero Comics and Why*. New York: Continuum. Applies Harold Bloom to superhero interpretation.

Reynolds, Richard. (1994). *Super Heroes: A Modern Mythology*. Oxford: University of Mississippi Press. Identifies seven laws for super heroes, including devotion to justice, secret identifies, lost parents, and accountability only to conscience, which are interpreted as key issues in male adolescent development.

"Superman." *Action Comics* #1, June 1938. DC Comics. First appearance and origin.

"Wolverine." *The Incredible Hulk* #180, October 1974. Marvel Comics Group, Marvel Entertainment Group, Inc. First appearance.

"X-Men." *X-Men* (First Series) #1, 1963. Marvel Comics Group, Marvel Entertainment Group, Inc. First appearance.

TECHNOCRACY

• • •

Technocracy may be generally described as an organizational structure in which decision makers are selected based on their specialized, technological knowledge, and/or rule according to technical processes. It has also been defined more simply as rule by experts. In all such cases technocracy constitutes a particular interaction between science, technology, and politics that has led to significant ethical debate.

Historical Development

The concept of technocracy needs to be qualified because the idea of rule by experts is at least as old as Plato's proposal for philosopher kings. Similarly, in his *New Atlantis* (1627), Francis Bacon envisaged an ideal society directed by scientists. But the contemporary meaning of technocracy presupposes the existence of complex industrial societies and the large-scale production and consumption processes that arose at the beginning of the twentieth century. It is only under these conditions that a class of experts in organization and production, namely engineers or technologists, could form. Technocracy, then, is rule by this particular type of expertise. Its advocates either assume or explicitly state that the efficient, rational production and distribution of goods for material abundance is the primary or even exclusive goal of society, because only in this way could they justify expert governance in these fields.

Early in the nineteenth century, the French writer Henri de Saint-Simon (1760–1825) foreshadowed calls for modern technocracy by arguing that the organization of production was more important to society than any other political end. By the 1890s, an emerging ambiguity in the social role of engineers led some to question their traditional subservience to employer goals. Unlike doctors, lawyers, and most other experts, engineers used their expertise to shape productive and technological systems, thereby transforming entire societies. Many began to feel that their power enabled or even obliged them to bring about social progress. With his idea that scientific laws would govern the efficient management of labor and use of resources, Frederick Taylor (1856–1915) provided a practical platform to extend the domain of engineering expertise into management and politics.

Henry Gantt (1861–1919) and James Burnham (1905–1987) further argued for the independence of engineers in their critiques of societal irrationalities and inefficiencies. Thorstein Veblen (1857–1929) critiqued wastefulness in the dominant political and economic system (i.e., the capitalist price system) and argued that engineers were best suited to direct society, because their objectivity was preferable to the short-sighted greed of business leaders. One of his disciples, Howard Scott (b. 1926), formed the Technical Alliance (in 1918) and later—rivalling with the "Continental Committee on Technocracy" (led by Harold Loeb and Felix Frazer)—Technocracy Inc. (in 1933). Members of Technocracy Inc. advocated a transition away from the price system and the establishment of a "governance of function," or a Technate, on the North American Continent. They argued that the scientific design of social operations would guarantee abundance for all.

Types of Technocracy

Analytically there exist at least seven variations on the technocracy theme. First, there is the notion of "expertocracy," or a conspiracy of experts who usurp decision making powers from democratically elected representatives. Second, technocracy can serve as a form of social engineering, where administrative procedures and organizational contrivances, rather than experts, gain power and form a "technological state." Third, there is a technocracy of work best articulated by Taylor's *Principles of Scientific Management* (1911). Fourth, the technological imperative of "can implies ought," in which means and feasibility determine goals, may create a technocracy that values the improvement of instrumentalities as a primary end. Fifth, there is the systems technocracy that may emerge from dynamic, interdependent systems engineering and by thereby administrating soci(et)al and political systems. Sixth, technocracy can refer to a situation in which laws are enforced by designing systems such that it is almost impossible to break them and that societal decisions and developments are totally streamlined by them and/or computerization. Finally, there is the technocratic movement spearheaded by Technocracy Inc. Additionally, the term has also been applied to a number of dictatorship governments and to a virtual reality game that claims to be based on "the inexorable advance of real-life technocracy" (see the web site at www.white-wolf.com/Games/Pages/MagePreview/technocracy.html).

Nevertheless, only four of these possibilities exhibit continuing viability. The idea of technocracy as expertocracy remains the most popular: a conspiracy of experts taking power through their personal, knowledge-based control of complex decision making. In the version promoted by Veblen (1925) this would involve rule by engineers especially in industrial corporations. But other alternatives might stress the intelligence and efficiency of more localized expertise, such as medical doctors to run health care systems. In all instances, expertocracies are argued to increase intelligence and efficiency in technical action—but threaten democracy.

A second widely discussed possibility focuses on the scientific optimization of social engineering through public administration. Here it is not experts as persons but administrative procedures and organizational structures that would exercise power. No individual or group would rule; individuals or groups would at most have a role in properly managing institutions and processes. This is the vision of technological politics presented by Jacques Ellul and others in which technological and administrative decisions replace political deliberation. Legislation by elected officials would wither under such an automated bureaucracy.

During the 1960s the idea of a technological imperative led to the articulation of another important version of technocracy, although one that has declined in intellectual salience. According to critical social theorist Herbert Marcuse (1898–1979) and science fiction writer Stanislaw Lem (b. 1921), there is a strong tendency for technical possibilities to determine social or political goals. Anything that can be done or produced will be done or produced, even becoming a matter of need. Means would determine ends; can implies ought. In a society established along these lines, improvement of instrumentalities becomes of singular value; the constant improvement of technology becomes the goal.

A fourth form of technocracy that continues to be examined conceives it in system terms. This is an important new variation on the technocracy theme. Systems engineering as well as systems analyses of the interconnections and complexities of society (as in the work of Niklas Luhmann) suggest a new kind of systems-technocracy. Discussions of systems-technocracy and the special case of "computerocracy" have emerged as serious issues in association with the rise of the so-called era of "information and systems technology" (Hans Lenk 1971, 1973).

Is systems-technocracy the wave of the future? There certainly are trends pointing in this direction, and the discussion should not be left to sociologists and politicians only. Instead, the single-focus framework of the social sciences should be combined with historical, engineering, and philosophical approaches to create an adequately interdisciplinary perspective. From such a perspective it can be argued that in a pluralistic technoscientific society the best way forward is to steer a pragmatic middle path between the extremes of an inhumanly efficient technocracy, a ruthless power politics, and a vulgar democracy devoid of intelligence.

Assessment

As Jean Meynaud (1964) summarized the issue, the decades-old debate on technocracy comes down to the fact that there is no conspiracy on the part of the technical community to usurp political power, though technical matters have taken on ever increasing importance.

Because the complexity of social, technological, economic, and ecological systems has increased, there is a progressive demand for technological, scientific, and organizational expertise. At the same time, narrow expertise calls forth a complementary needs for generalists, people with a broad view ("specialists of the general") of interdisciplinary complexes who can take a systems approach toward problems.

Historically speaking, the technocracy debate simply continued the social criticisms of technology from the early part of the twentieth century. Its dominant characteristic has been a pessimistic attitude that ignores the extensive ways technology has humanized the world. But the privileged position of experts in particular cases has not led to the demise of politics in the so-called "technolocal state" (Helmut Schelsky) or of the importance of its interplay between conflicting and overlapping interest groups and power structures. The opposite seems to be the case. The most significant outcome of the technocracy debate is thus an awareness that complex political decisions cannot be replaced by the technological or "computerocratic" procedures of optimization and maximization.

There are several explanations for this. Most significant is the fact that complex political decisions involve both information and the adjudication of a plurality of values. The inexplicable and undecidable character of political questions in contrast to technological answers, as was argued by Hans Lenk (1973), has largely been confirmed by experience. Society and the state are not machines with mere objective standards of performance, and there is no scientifically generated "one best way" (as Schelsky believed) to solve many technical, let alone political, problems. Attempts to apply science to societal problems with this intention often lead to interminable debates among competing experts, while the underlying values at stake remain unexamined.

Yet it remains true that technical matters have taken on ever increasing importance in the complex problems of modern societies and computerocracy as a virulent version of systems technocracy is an imminent danger in our hi-tech societies. The challenge for democratic governance is to integrate technical experts with non-expert participants to strike common interest solutions in contexts where many elements are beyond the comprehension of all but a few specialists. These interdisciplinary contexts may even demand generalists capable of integrating diverse sets of knowledge and perspectives.

HANS LENK

SEE ALSO Ellul, Jacques; Expertise; Participation.

BIBLIOGRAPHY

Akin, William E. (1977). *Technocracy and the American Dream: The Technocrat Movement, 1900–1941.* Berkeley: University of California Press.

Armytage; W. H. G. (1965). *The Rise of the Technocrats: A Social History.* London: Routledge.

Beville, G. (1964). *Technocratie Moderne.* Paris: Librairie générale de droit et de jurisprudence.

Boisdé, R. (1964). *Technocratie et Démocratie.* Paris: Plon.

Burnham, James. (1941). *The Managerial Revolution.* New York: John Day.

Elgozy, G. (1965). *Le Paradoxe des Technocrates.* Paris: Denoël.

Ellul, Jacques. (1954). *La Technique Ou L'enjeu du Siècle.* Paris: Economica. (The Technological Society, New York: Knopf 1964).

Elsner Henry. (1967). *The Technocrats: Prophets if Automatization.* Syracuse, NY: Syracuse University Press.

Fischer, Frank. (1990). *Technocracy and the Politics of Expertise.* Newberry Park, CA: Sage Publications.

Koch, Claus, and Dieter Senghaas, eds. (1970). *Texte Zur Technokratiedieskussion* [Materials on the Debate about Technocracy] Frankfurt: Europaeische Verlagsanstalt.

Lenk, Hans. (1971). *Philosohphie im Technologischen Zeitalter* [Philosophy in the Technological Era] Stuttgart: W. Kohlhammer.

Lenk, Hans. (1984). "Toward a Pragmatical Social Philosophy of Technology and the Technological Intelligentsia." In *Research in Philosophy and Technology,* ed. by Paul T. Durbin.

Lenk, Hans. (1991). "Ideology, Technocracy, and Knowledge Utilization." In *Europe; America, and Technology: Philosophical Perspectives,* edited by Paul T. Durbin. Dordrecht/ Boston/ London: Kluwer.

Lenk, Hans, ed. (1973). *Technokratie als Ideologie: SozialphilosophischeBeiträge zu einem politischen Dilemma* [Techncracy as Ideology: Social Philosophical Contributions to a Political Dilemma]. Stuttgart: W. Kohlhammer.

Marcuse, Herbert. (1964). *One-Dimensional Man: Studies in the Ideology of Advanced Industrial Society.* Boston: Beacon Press.

Meynaud, Jean. (1960). *Technocratie et Politique.* Lausanne: Études de science politique.

Meynaud, Jean. (1964). *Technocratie, mythe, ou réalité?* Paris: Paynot. English translation by Paul Barnes: *Technocracy* (New York: Free Press, 1969).

Radaelli, Claudio M. (1999). *Tecnocracy in the European Union.* London: Longmans.

Schelsky; Helmut. (1965). *Auf Der Suche Nach Der Wirklichkeit.* [In Search of Reality] Duesseldorf/Cologne: Eugen Diederichs.

Scott, Howard. (1929). "The Scourge of Politics in a Land of Manna." In *The One Big Union Monthly* Sept., pp. 14 ff.

Taylor, Frederick. (1911). *The Principles of Scientific Management.* New York: Harper and Brothers.

Veblen, Thorstein. (1921). *The Engineers and the Price System*. New York: Huebsch.

TECHNOETHICS

• • •

Technoethics is a term coined in 1974 by the Argentinian-Canadian philosopher Mario Bunge to denote the special responsibilities of technologists and engineers to develop ethics as a branch of technology. However, in 1971 the chemical engineer and theologian Norman Faramelli had used a word of only one less letter, *technethics*, to argue for a general ethics of technology from a Christian theological perspective. In 1973 the *Britannica Book of the Year* defined the same term, without referencing Faramelli, as indicating "the responsible use of science, technology and ethics in a society shaped by technology."

Bunge's use is the more significant and radical. For Bunge engineers and managers, because of their enhanced powers, acquire increased moral and social responsibilities. To meet these responsibilities they cannot rely on traditional moral theory; since moral theory itself is underdeveloped having "ignored the special problems posed by science and technology" (Bunge 1977, p. 101). Instead, engineers must adapt science and technology, tools that are foreign to most philosophers, to construct a new theory of morality.

According to Bunge, rational moral rules have exactly the same structure as technological rules. Technological rules come in two types: ungrounded and grounded. Ungrounded technological rules either are irrational or are based on empirical evidence that has not been systematized. Grounded technological rules are based on science. According to an earlier argument, Bunge (1967) sees technology as being constituted by scientific theories of action. Modern technology develops when the rules of prescientific crafts, which are based on trial-and-error learning, are replaced by the scientifically "grounded rules" of technological theories.

During the late 1990s and the early 2000s the term *technoethics*, especially in Spanish and Italian cognates, appeared anew in an effort to parallel another coinage from the 1970s: *bioethics*. However, the prefix *techno* has connotations that are at odds with *bio*, which references life and its nuances. Ethics is a living field. *Techno* denotes the hard-edged and loud, as in *technomusic*, *technoart*, and *technoeconomics*. Given these uses, *technoethics* fails to connote as readily the broad concerns that have been easy to include in bioethics.

Indeed, Bunge's use of the term seems more appropriate.

In the preparation of the *Encyclopedia of Science, Technology, and Ethics* there was some initial debate about making it an "Encyclopedia of Technoethics." The conclusion, however, was that such an alternative would have been inadequate in building bridges between a number of applied ethics fields ranging from computer and engineering ethics to research and environmental ethics, including history, literature, and philosophy along the way. The expansive if less catchy title *Encyclopedia of Science, Technology, and Ethics* defines in a more inclusive way the scope of a reference work that should appeal to scholars; professionals in the sciences, engineering, and the humanities; and general readers.

CARL MITCHAM

SEE ALSO *Chinese Perspectives; Science, Technology, and Society Studies.*

BIBLIOGRAPHY

Bunge, Mario. (1967). *Scientific Research II: The Search for Truth*. New York: Springer. See especially the chapter titled "Action," pp. 121–150.

Bunge, Mario. (1977). "Towards a Technoethics," *Monist* 60(1): 96–107. This is the first publication of a paper presented at the International Symposium on Ethics in an Age of Pervasive Technology, Technion-Israel Institute of Technology, December 21–25, 1974, with abridged proceedings published in Melvin Kransberg, ed., *Ethics in an Age of Pervasive Technology*, Boulder, CO: Westview, 1980.

Esquirol, Josep M., ed. (2002). *Tecnología, Ética y Futuro: Actas del I Congreso Internacional de Tecnoética*. Bilbao, Spain: Editorial Desclée de Brouwer.

Esquirol, Josep M., ed. (2003). *Tecnoética: Actas del II Congreso Internacional de Tecnoética*. Barcelona: Publicaciones Universitat de Barcelona.

Faramelli, Norman J. (1971). *Technethics: Christian Mission in an Age of Technology*. New York: Friendship Press.

TECHNOLOGICAL FIX

• • •

Technology is often couched in terms of solving problems such as curing disease, providing for reliable food production, or affording efficient means of transportation. Indeed, technology has proved powerfully effective for solving any number of problems, from the massive project of sending people into space to the minor chore of fastening pieces of paper together. But in a 1966 article, atomic physicist Alvin M. Weinberg raised the following question: Are there some types of problems that cannot—or should not—be *fixed* by technology? Weinberg coined the term *technological fix* to describe the use of technology to respond to certain types of human social problems that are more traditionally addressed via political, legal, organizational, or other social processes. Although Weinberg advocated the use of technological fixes in some cases, the term has come to be used frequently as a pejorative by people critical of certain uses of technology.

Writing during the cold war, Weinberg cites nuclear weapons as an example of a technological fix for war. The technological ability to unleash global devastation serves as a deterrent to international aggression. But critics argue that such a solution is at best tenuous, and at worst lessens people's resolve to work diplomatically at ameliorating the underlying clashes of ideology, economy, and culture that lead to war. Nuclear weapons also served as an alternative to maintaining a large standing army such as that of the Soviet Union, thus shifting social sacrifice from the less to the more democratically acceptable—from personal service to government investment in advanced technological weapons research and development. It is this aspect of technological fixes—their tendencies to mask the symptoms of complex social problems without addressing their causes or true costs—that generally evokes ethical concern.

For example, if large numbers of children are being disruptive or having trouble concentrating in school, is the liberal prescription of psychotropic drugs a viable technological way to ease the problem, or does this simply allow parents and teachers to abdicate their responsibilities for good parenting and maintaining discipline, respectively? If employees are using company computers for personal business or entertainment, is installing software to monitor and curb such behavior a viable technological solution, or does this simply foster an atmosphere of distrust without addressing the causes of the problem, perhaps poor morale or inefficient tasking?

These are difficult questions because there are surely some children who could benefit from psychotropic drugs, and there are arguably certain situations in which an employer has a legitimate need to monitor an employee's use of the computer. But once such technological fixes become available, they run the risk of proliferating into universal *easy ways out*. Or they may simply shift the locus of the problem; in the case of the work computers, spy software does not guarantee greater employee productivity, only that employees will not be unproductive in a particular way.

Despite these criticisms, sociologist Amitai Etzioni (1968) defended the use of what he called technological *shortcuts*. Etzioni argued that many of the concerns levied against such shortcuts were based on conjecture rather than hard evidence. For example, when better lighting is installed on city streets in an effort to discourage crime, critics claim that this approach treats only the symptoms and does not do anything to address the underlying motivations for crime, nor does it necessarily reduce crime overall; rather, they claim, it just shifts the criminal activities to other locations. But while sounding plausible, such criticisms are typically unsupported by any definitive data. The questions to be asked in this example are, where do criminals go, and what do they do, when their previous stalking grounds are illuminated? "No one knows," writes Etzioni, but "[t]he one thing we do know is that the original 'symptom' has been reduced" (p. 45).

Etzioni also pointed to the deep-seated and intractable nature of many social problems, which suggests the near impossibility of ever implementing any comprehensive solutions via social transformation, particularly given fervent political disagreement about the propriety of various transformation strategies. Thus stopgap shortcuts may be the only recourse. "Often," writes Etzioni, "our society seems to be 'choosing' not between symptomatic (superficial) treatment and 'cause' (full) treatment, but between treatment of symptoms and no treatment at all" (p. 48).

The fundamental difficulty with technological fixes—or shortcuts—is the inherent incompatibility between problem and solution. Technologies are most useful for solving specific, well-defined, and stationary problems, such as how to get cars from one side of a river to the other (for example, using bridges). In contrast, social problems, such as crime, poverty, or public health, are broad, ill-defined, and constantly evolving. Weinberg, like Etzioni, was not naïve about this difficulty, writing, "Technological Fixes do not get to the heart of the problem; they are at best temporary expedients; they create new problems as they solve old ones" (p. 8).

BYRON P. NEWBERRY

SEE ALSO *Science, Technology, and Society Studies.*

BIBLIOGRAPHY

Etzioni, Amitai. (1968). "Shortcuts to Social Change?" *Public Interest* 12: 40–51. A sociologist argues for the use of technological means to treat the symptoms of pressing social problems.

Weinberg, Alvin M. (1966). "Can Technology Replace Social Engineering?" *Bulletin of the Atomic Scientists* 12(10): 4–8. An atomic scientist discusses the merits of using technology to eliminate or attenuate social problems as an alternative to pursuing the more difficult strategy of changing prevalent social attitudes.

TECHNOLOGICAL INNOVATION

• • •

Technological innovation has been a leading agent of social change, worldwide, since the late 1700s, serving as the conduit into society of developments in science and technology. As such, it has been at the center of ethical issues ranging from the morality and justice of the early Industrial Revolution to the consequences of genetic engineering, nanotechnology, and artificial intelligence (AI). In spite of its extraordinarily high social visibility, however, innovation is almost universally misunderstood and misrepresented, typically as synonymous with invention. Invention, in turn, is presented as a value-free, hence ethically neutral, application of new or existing technical knowledge. Treating innovations as inventions implies that ethical issues associated with their implementation derive not from factors intrinsic to innovations, but from how society chooses to implement them. Such an interpretation frees innovators from moral responsibility for the ethically problematic consequences of their activities, as well as buffering these activities from public assessment.

What Innovation Is

Innovation is a social process in which technical knowledge and inventions are *selectively* exploited on behalf of (corporate or government) institutional agendas driven by marketplace values or political policies. Inventions, and more broadly scientific and engineering expertise, are merely raw materials for technological innovation, which is the value-laden, ethically provocative process that determines *whether* an invention is introduced into a society, the *form* in which it is introduced, and the *direction* of its subsequent development as society responds to the innovation. The introduction of the automobile, television, nuclear power plants, and the Internet are examples of the value-laden innovation process, including how societal responses feed back into the course of innovation developments over time.

Conceptual Emergence and Practical Engagement

The beginning of the twentieth century saw leading economists focused on determining the conditions for supply-demand equilibrium. For Austrian economic theorist Joseph Schumpeter (1883–1950), however, what needed to be analyzed was not equilibrium but the disequilibrium created by economic growth. Looking back over the nineteenth century and the first decade of the twentieth, Schumpeter argued that entrepreneurship in combination with technological innovation—that is, risking capital by creating new businesses that transform inventions into innovations—was the engine of economic growth in modern societies. This combination of innovation and entrepreneurship created new wealth, destroyed old wealth, and created new concentrations of social and political power. Schumpeter defended what he called the *creative destruction* that often accompanied implementing innovations. The creation of synthetic dye, electric power, and the automotive industries, for example, undermined established industries based on natural dyes, steam and water power, and horse drawn transportation. Businesses were indeed destroyed, jobs were lost, people suffered but, Schumpeter claimed, *better* businesses were created, employing more people in *better* jobs. Schumpeter eventually also defended the wasteful and often frivolous character of the combination of innovation and entrepreneurship in an industrial capitalist environment driven by opportunistic profit-seeking.

After World War I, individual thinkers, among them the American economist Thorstein Veblen (1857–1929) and future U.S. president Herbert Hoover (1874–1961), argued that technological innovation would be central to national security and industrial competitiveness. Only in Germany, however, was there a strong national commitment to an innovation-driven military and industrial agenda, initiated by Prince Otto von Bismarck in the 1860s and developed further by all subsequent German governments, especially the National Socialists. In the United States and Great Britain, by contrast, calls for such national commitments were repeatedly rejected. For example, George Ellery Hale (1868–1938), one of the world's leading astronomers and the person responsible for maintaining America's leadership in telescopy from 1897 into the 1980s, failed in his attempt to win government acceptance of his plan to harness academic scientists to the nation's war effort during World War I. He failed again in his postwar attempt to create a national research foundation to be cosponsored by the federal government and major corporations.

World War II changed all this. The role that technology and science played in waging and winning the war for the Allies, especially the role of the U.S. Office of Scientific Research and Development (OSRD) headed by Vannevar Bush (1890–1974), led if anything to an overestimation of the power of innovation in the postwar period. In his report titled *Science: The Endless Frontier* (1945), Bush argued that U.S. industrial prosperity and military security would in the future be critically dependent on continuous science-based technological innovation. The federal government needed to create mechanisms for government-subsidized basic research, primarily at universities, to *feed* the commercial innovation process. For Bush, this was the lesson of such OSRD accomplishments as the Manhattan Project, of the Massachusetts Institute of Technology's (MIT) Radiation Laboratory or RadLab that produced a constant stream of electronic warfare and counterwarfare technologies, and of mass-produced cheap antibiotics and blood products. Yet as Bush later acknowledged, this *push* or linear model, in which basic research leads to applied science, which then leads to commercial technological innovations, overestimates the dependence of innovation on basic science. This view was confirmed in Project Hindsight (1966), a Department of Defense study of twenty weapons systems, introduced since 1946, that concluded that basic science affected less than 10 percent of these systems. A follow-up study by the National Science Foundation (NSF), TRACES (Technology in Retrospect and Critical Events in Science [1968]), defended the basic research-driven model in the Bush report by looking back fifty years instead of twenty.

Since 1970 research by historians of technology has supported a version of the Project Hindsight conclusion. While basic research sometimes *pushes* innovation, innovation far more often *pulls* research, which may then enable further innovation. The exponential growth of innovation in the semiconductor and computer industries exemplifies this relationship.

Bush's report and its basic science push model nevertheless anchored postwar-U.S. science and technology policy. For the first time in U.S. history, there was a mandate for large-scale federal support of basic as well as applied scientific research. The ethics of giving scientists public funds to do research on subjects of their choice gave rise to contentious political debates that held up creation of the NSF in 1950. But the NSF budget for basic research was then and has remained modest compared to the budgets for applied research linked to innovation, which until 1989 was driven primarily by Cold War military agendas and secondarily by the evolving war on cancer, war on AIDS, and Human Genome Project agendas of the National Institutes of Health (NIH) and the U.S. space program.

In the 1960s leading political figures including Presidents John F. Kennedy, Lyndon B. Johnson and Richard M. Nixon promoted innovation as the key to U.S. economic growth. In 1962 President Kennedy explicitly identified industrial innovation as the source of new jobs and new wealth that would be shared by all. But it was only in the 1970s and after, in the wake of the Silicon Valley phenomenon and the astonishing pace of wealth creation in the semiconductor and computer industries, that a national consensus recognized the civilian economy as critically dependent on innovation for growth. It was in the 1960s and 1970s that Schumpeter's identification of innovation and entrepreneurship as engines of economic growth was rediscovered. It had sparked little interest when published in 1911 or even after Schumpeter's migration to Harvard University in the 1930s. Nor did University of Chicago economist Frank Knight (1885–1982) stimulate interest in the link between innovation and entrepreneurship with his pioneering 1921 study of the dynamic role played by risk in creating new businesses. Knight coupled a penetrating analysis of the economics of innovation-driven entrepreneurship to a stinging moral critique of the wastefulness of innovation in a capitalist economy. The importance of the ideas of Schumpeter and Knight would be appreciated only when innovation had engaged the general political consciousness and conscience. Early-twenty-first-century American economist Paul Romer is an influential neo-Schumpeterian, arguing that growth is generated by ideas of which innovation is a symptom and defending the virtues of the unmanaged U.S. innovation model over the managed innovation models in Japan and east Asia.

The Ethics of Innovation

Recognition of the scale and scope of innovation-enhancing policies provoked broad criticism of social and ethical implications of the dependence of society on innovation. Jacques Ellul in *The Technological Society* (1954), for instance, argued that such dependence reflected a gamble that would compel societies to transform themselves into vehicles for supporting continuous innovation at the expense of traditional personal and social values. Ellul's ethical and political critique of technology-based society attracted many followers who developed it further in the 1960s and 1970s, and were significantly responsible for the creation of university-based science, technology, and society (STS) studies programs as an academic response to the new institutionalization of innovation by government and industry. Alvin Toffler's *Future Shock* (1970) was a more popular caution against and criticism of the personal as well as social disorientation caused by continuous innovation. Its commercial success suggests a responsive chord of concern in the general public, which nevertheless embraced the flood tide of innovations affecting every aspect of personal and social life, locally, nationally, and globally, that poured into the marketplace during the last third of the twentieth century.

By the turn of the twenty-first century, that economic prosperity was keyed to continuous technological innovation in a global competitive environment was enshrined as an ineluctable fact, a principle of nature, a kind of categorical imperative. *Innovate or stagnate* not just economically, but culturally as well. Open to serious debate in principle were such questions as whether innovation-induced social change constituted true growth or was just change; whether such change was progressive, improving the quality of life, or just sound and fury busyness signifying nothing very deep. Yet public debates on such questions rarely took place. What was broadly recognized as inescapable, though, was that the innovation-driven economic growth process institutionalized after World War II and adopted globally by 2000 was characterized by a kind of positive feedback. Only *continuous* growth was possible; stasis, with the loss of the expectation of growth, threatened economic collapse.

Meanwhile the accumulated scholarship of the STS studies community generated new insights into the innovation process. Contrary to the inherited wisdom that technical knowledge was value-free, innovation is in fact ethically *preloaded*. Innovations enter the marketplace incorporating a broad range of value judgments primarily determined by the agendas of the commercial institutions and governmental agencies pursuing innovation on behalf of those agendas. The so-called *negative externalities* of innovation—including Schumpeter's *creative destruction* of superseded technologies along with their institutions, facilities, and people—also include negative environmental impacts, the introduction of new forms of personal and social life, and the creation of new vested economic, social, and political interest groups and power centers, each committed to perpetuating itself. All such concomitants of innovation raise ethical concerns that dwarf the public processes available for addressing them.

Organizational theorist and Nobel economics laureate Herbert Simon noted in the 1960s that complex systems are by definition ones whose behaviors include unpredictable outcomes. Technological innovations often result in the implementation by society of complex systems to support them. As a result, even with the best of corporate, governmental, and public intentions, it is impossible to predict in advance all of the consequences, negative or positive, of innovations in, for example, antibiotics, television, the Internet, and cell phones. Such unpredictability motivated Bill Joy—a cofounder of Sun MicroSystems Corporation, its chief scientist, and a cocreater of the Java programming language—to issue a passionate call in 2001 for a moratorium on innovation in biotechnology, nanotechnology, and robotics. Joy's argument was that these three technologies were converging and had the potential for unpredictable consequences that posed profound threats to human survival. Joy stumped the nation warning academic, industrial, and public audiences of the potential for catastrophic harm from continuing our postwar policy of unfettered innovation followed by catch-up attempts at regulation as problems arose.

A similar moratorium had been argued for in 1974 by Paul Berg, inventor of recombinant DNA technology. Berg's call, following a year-long cessation of research in his own lab, led to the 1975 Asilomar Conference, which substituted heightened laboratory safeguards for a moratorium, and subsequently sanctioned a biotechnology innovation free-for-all. In the 1980s, Jeremy Rifkin and others attempted to block innovation in genetically modified food crops and plants, to little if any avail. Joy's call did provoke a substantial response within the technology community. Raymond Kurzweil, an eminent engineer-inventor, debated Joy on a number of occasions, orally and in print, championing unrestricted innovation as both progressive and capable of containing any unanticipated harmful consequences of innovation. In spite of rapid commercial development of biotechnology and nanotechnology industries at the start of the twenty-first century, the public was not engaged in the ethical issues raised by innovations that were under research and development, in the prototype stage, or being introduced into the marketplace.

STEVEN L. GOLDMAN

SEE ALSO *Business Ethics; Invention; Science, Technology, and Society Studies.*

BIBLIOGRAPHY

Bush, Vannevar. (1980). *Science: The Endless Frontier*. New York: Arno Press. Originally published in 1946, the seminal U.S. science policy document.

Chandler, Alfred Dupont. (1980). *The Visible Hand*. Cambridge, MA: Belknap Press. A history of American industry's exploitation of innovation.

Ellul, Jacques. (1967). *The Technological Society*. New York: Vintage Books. The most influential attack on modern technology as antidemocratic and antihuman.

Erwin, Douglas H., and David C. Krakauer. (2004). "Insights into Innovation." *Science* 304: 1117–1118.

Galbraith, John K. (1998). *The Affluent Society*. Boston: Houghton Mifflin. Originally published in 1958, a still powerful critique of consumerism and greed.

Hughes, Thomas P. (1989). *American Genesis: A Century of Invention and Technological Enthusiasm, 1870–1970*. Chicago: University of Chicago Press. An excellent history of technological innovation American style.

Mowery David C., and Nathan Rosenberg. (1998). *Paths of Innovation*. Cambridge, UK, and New York: Cambridge University Press. An in-depth examination of four industries created by innovation.

Rosenberg, Nathan. (1983). *Inside the Black Box: Technology and Economics*. Cambridge, UK, and New York: Cambridge University Press. An economist's analysis of the relation between invention, innovation, commerce, and society.

Schumpeter, Joseph. (1983). *Theory of Economic Development: An Inquiry into Profits, Capital, Credit, Interest, and the Business Cycle*, trans. Redvers Opie. New Brunswick, NJ: Transaction Publishers. Expands Schumpeter's focus on innovation to a comprehensive theory of business cycles as inevitable in a growing economy.

TECHNOLOGY ASSESSMENT

SEE *Constructive Technology Assessment; Technology Assessment in Germany and Other European Countries; United States Office of Technology Assessment.*

TECHNOLOGY ASSESSMENT IN GERMANY AND OTHER EUROPEAN COUNTRIES

• • •

From its mid-1970s origins, technology assessment (TA) in Germany and in Western Europe has been presented as a methodical, ethical, and theological as well as natural-, engineering- and social-science-oriented reflection on the technological preconditions for the formation and design of modern societies and the impacts of technology on such societies. TA analyzes both the development of technologies and the entities that have the competence, resources, and strategic potential to create them. Using prediction procedures, decision-theory approaches, and model simulations—all of which resemble economic models—the goal is to raise awareness of the desired and undesired, synergetic, and cumulative consequences of new technologies, if possible before they become issues of public debate. TA further aims to reveal the basic values underlying any *assessment*.

Representative Institutions

Understood as a form of *political counseling*, a series of TA institutions were founded by some Western European parliaments. Among these institutions are the following:

Scientific and Technical Options Assessment (STOA), by the European Parliament (1985)

Office Parlementaire d'évaluation des choix scientifiques et technologiques (OPECST), France (1983)

Büro für Technikfolgen-Abschätzung beim Deutschen Bundestag (TAB) or Office of Technology Assessment at the German Parliament (1990)

Rathenau Institute, Netherlands (1986)

Parliamentary Office of Science and Technology (POST), United Kingdom (1989)

There are also parliamentary institutions in Denmark, Austria, Finland, Belgium, Greece, Norway, Switzerland, Sweden, and Spain, which in the near future will join this circle of parliamentary counselors in the cooperative European Parliamentary Technology Assessment (EPTA). Some Eastern European countries, in particular Poland, Hungary, and the Czech Republic, have also established independent TA institutions. Of the independent institutions founded in Germany, of particular interest is the Institut für Technikfolgenabschätzung und Systemanalyse (ITAS or Institute for TA and System Analysis) of the Karlsruhe Research Center (RZE), a member of the Helmholz-Gemeinschaft Deutscher Forschungszentren (Helmholz Association of National Research Centers), the largest scientific organization in Germany. ITAS is also the operating authority of TAB. ITAS publishes the only significant TA journal in Germany titled *TA in Theory and Practice*.

Two major research institutes in the Helmholtz Community Association of National Research Centers among those that conduct projects on sustainability research relating to TA, should be mentioned: Forschungszentrum Jülich (Juelich Research Institute) and the Deutsches Zentrum für Luft- und Raumfahrt (DRL) (German Center for Aviation and Space Flight), Cologne. Another national organization is the European Academy for Research on the Consequences of Scientific/Technical Development, which is located in Bad Neuenahr and primarily supported by the state of Rhineland Palatinate and by the DLR. It is less technology-transfer oriented than, for example, the ITAS or TAB because its research is focused more on basic questions concerning the acceptability of technology use as an element of forward-looking policies. The Academy for TA, founded in Stuttgart in 1991, was closed by the state of North Rhine-Westphalia at the end of 2003. This was a severe setback for TA research in Germany, in particular because the academy had an impressive public profile as a result of its efforts to link socially relevant discourse with areas of science, economics, and politics.

Research Themes

Among the important TA topics in Germany, sustainability dominates current research. Indeed efforts are aimed at institutionalizing the principles of sustainable development at all levels of national and transnational political systems.

In addition to biotechnology (as related to agriculture, pharmacy, textiles, and food), research into gene technology, diagnostics, and therapy are at the center of public interest. In Germany discussions have concentrated on the fields of biomedicine, and in particular on the ethical justification of research using human embryos and preimplantation diagnoses (PID). Stem-cell research is examined in terms of future application to tissue and organ regeneration. The acquisition of stem cells from embryos, or so-called therapeutic cloning, is the subject of numerous investigations. The compatibility of biomedical developments with the principle of human dignity as defined by the German basic law (or constitution) and the EU constitution is an especially important issue.

The development of nanotechnology is also of interest, especially because this field has frequently been presented as a key technology for the twenty-first century. Applications of nanotechnology are projected in the fields of space flight, agriculture, information processing, and medicine. The implementation of nanotechnology materials is discussed in relation to ecological and medical issues.

In the context of the process of globalization—especially in university research projects—there are TA

questions about the consequences and effects of the virtualization of social life—politics, economics, ecology, culture, and law. With regard to politics, studies have focused on e-government, electronic democracy, and the dismantling of nation-states. With regard to economics, TA has concerned itself mainly with the transformation of work. In addition, TA continues to address classic issues such as traffic, new energy sources (nuclear fusion), privatization of health systems, pharmacology, food technology, multimedia technology, and information or data processing.

Evaluation

The German and European TA landscape deserves evaluation on the basis of the following: Have the numerous TA activities had any influence? If so, what kind of influence have they had on technological developments and on related underlying decisions? *Technological Assessment in Europe: Between Method and Impact* (2003), a study by ITAS and the European Academy, is a useful guide in answering these questions. This study presents a typology of three types of impacts: the generation of knowledge; the alteration of opinions and forms of behavior; and the initiation of action.

The study concludes that: "Based on the typology of the impacts on TA it is shown that the impacts of TA present more than just the direct influences of political decisions ... TA—independent of whether it is more classically scientific or participatory—contributes in various ways to society's communication process and to the political decision process: Through the preparation of a balanced basis of knowledge, through the initiation of a new discussion in a gridlock situation, through the working out of new perspectives on a problem" (Decker and Ladikas 2004, p. 78).

Finally the report of the European Science and Technology Observatory (ESTO), an association of twenty European institutions, should be mentioned. In 2002 at the direction of the Institute for Prospective Technological Studies (JRC-IPTS) of the European Commission, ESTO produced an overview of technology-forecasting activities in Europe.

This working document arose within the frame of the ESTO project "Monitoring of Technology Forecasting Activities," funded by the Joint Research Center Institute for Prospective Technological Studies (JRC-IPTS) of the European Commission. This project was part of a larger ESTO monitoring activity, which ran from February 2000 until June 2001. The main results of this ESTO activity are published in "Strategic Policy

Intelligence: Current Trends, the State of Play Perspectives, IPTS Technical Report series, EUR 20137 EN.

RABAN GRAF VON WESTPHALEN

SEE ALSO *Discourse Ethics; German Perspectives.*

BIBLIOGRAPHY

Decker, Michael, and Miltos Ladikas. (2004). "EU: Procekt: Technology Assessment in Germany; Between Method and Impact (TAMI)."*Technikfolgenabschätzung* 1(13): 71ff.

European Commission, Joint Research Centre (JRC). (2002). *Monitoring of Technology Forecasting Activities—ESTO Project Report.* Düsseldorf, Germany: VDI Technology Center, Future Technologies Division.

Graf von Westphalen, Raban, ed. (1999). *Technikfolgenabschätzung als politische Aufgabe,* 3rd edition. Munich, Germany: Oldenbourg Verlag. An overview of methods and mode of operations based on case studies.

Grunwald, Armin. (2002). *Technikfolgenabschätzung—eine Einfülrung.* Berlin: Edition Sigma. The best German language introduction in the field of technical-impact assessment.

TECHNOLOGY LITERACY

SEE *Public Understanding of Science.*

TECHNOLOGY: OVERVIEW

• • •

Technology may be broadly defined as the making and using of artifacts. In its simplest forms, however, use will involve no more than natural objects, and in more abstract instances fabrication and use can both be of concepts—in which case logic may be described as a technology. The etymology of the word leads back to the Greek *techne,* from which is derived *technique* and *technics.* In the opening lines of *Nicomachean Ethics,* Aristotle (384–322 B.C.E.) observed that "Every *techne* and every inquiry, and similarly every *praxis* and pursuit, is believed to aim at some good" (1.1.1094a). Thus the centrality of human ends or intentions to technology makes ethical analyses vital. Ethical inquiry is made difficult, however, by the diversity of ways technology can be understood. According to one proposed analysis, technology may be distinguished into objects, knowledge, activities, and intentions (Mitcham 1994). Each of these types of technology constitutes a source and challenge for ethics.

Historical Dimensions

Before considering these different types of technology, which are covered in a plethora of entries in this encyclopedia, there are historical transformations from technics to technology to acknowledge. These transitions, which are also often described as shifts from ancient to modern or from prescientific to scientific technology, can be discussed in terms of artifacts and attitudes. In relation to artifacts, humans used lithic (or stone) tools from the early Paleolithic period (about 2.6 million years ago) up to the close of the Neolithic period around 5,000 years ago. The widespread control of fire occurred roughly 124,000 years ago and crops were domesticated around 10,000 years ago. Up until approximately 40,000 years ago, the interplay between human physiology and technics no doubt influenced the evolution of human cognitive and other physical capacities.

The development of bronze and iron tools marked the end of the Neolithic and the transition into the classical age, in which technological artifacts in the form of structures became increasingly significant. Premodern structures, initially in the early civilizations of Egypt, Mesopotamia, India, and China, then especially in China's Han dynasty (206 B.C.E.–220 B.C.E.) and the Greek and Roman periods in Europe, became interrelated with governance, and the works of architects began to influence daily life. In the European Middle Ages progressive developments in mechanics and the harnessing of nonhuman sources of power promoted further change in artifactual history.

The emergence of technology in a distinctly modern sense is correlated with the rise of modernity itself. Through the Industrial Revolution tools, machines, structures, industrial processes, and mass-produced consumer goods increased in complexity and number, acquiring an unprecedented societal influence. Additionally, during and after the Enlightenment, technology became progressively associated with accumulating scientific knowledge, to the point where, in the late twentieth century the connection was occasionally denominated with the term *technoscience.*

In relation to attitudes, which exhibit inherently ethical components, history may be broken out into a threefold taxonomy of arguments about technology and its proper role in the good life. Although partially historical, these basic attitudes (with countless gradations) nevertheless continue to coexist today. First, ancient or premodern attitudes about technology were generally skeptical, tending to view it as a necessary but dangerous turning away from God or the gods. Artifacts were judged to be less real than natural objects, technical information was not considered true wisdom, and technical affluence was thought to undermine higher goods such as individual virtue and political stability.

Second, modern Enlightenment attitudes about technology were optimistic, viewing it as a means of socializing individuals and creating public wealth. The will to technology was ordained by God or nature. Technical engagement with the world provided true knowledge, and nature and artifice were judged as operating by the same mechanical principles.

Finally, Romantic attitudes about technology reintroduced a degree of premodern uneasiness to constitute an ambivalence that tried to strike a middle ground between premodern skepticism and modern enthusiasm. Technology was viewed as one manifestation of human creativity, and thus to be affirmed, but also as manifesting a lamentable tendency to crowd out other forms of creativity. Technology engendered freedom but simultaneously alienated individuals from affective strength, weakened cultural bonds, and introduced new forms of social control. Artifacts expanded the processes of life, but imagination and vision deserved to be defended against the encroachments of technical knowledge.

Technology as Object

Technology is most commonly thought of in terms of artifacts, physical objects designed and produced by human beings. Ethical issues related to artifacts include the concerns of health and safety. These are especially illustrated by elements of risk and uncertainty, because it is often impossible to predict how objects will interact with the complex physiological, social, and ecological contexts in which they are deployed. Important work in engineering design seeks to integrate safety concerns throughout the process, but in some sense accidents and failures may be an inevitable part of complex modern artifacts.

Other ethical issues stem from justice and equity concerns that arise, for example, in cases of technology transfer and other manifestations of globalization. Matters of justice and equality are also involved in the representation of females and minorities in technology development and application policies. Freedom is a further important consideration in debates about technological determinism (in the thought of Jacques Ellul) or the liberating potential of technology (as argued by Julian Simon). Moreover, philosophers such as Langdon Winner have argued that artifacts have politics, in that they may be intentionally designed to limit the freedoms of certain groups. Other objects inherently lead to

different political systems of control along the spectrum from authoritarianism to democracy.

Technological objects raise additional ethical and phenomenological questions about how they influence individual and group self-identities. For example, the design of buildings and public spaces in urban environments, in addition to impacts on safety, health, and equity, influence community character and quality of life. Finally, there is a sense in which technological objects as consumer goods can alter both culture and, through pollution and waste, the natural environment.

Not only do many of the key themes just mentioned have their special entries, but sample encyclopedia entries on almost any technology—from "Airplanes" and "Biological Weapons" to "Movies" and "Television"—illustrate these issues. Entries on thinkers such as "Anders, Günther," "Ellul, Jacques," "Illich, Ivan," and "Simon, Julian" present particular arguments. Slightly more general discussions that emphasize structures and hardware can be found in "Architectural Ethics" and "Computer Ethics," respectively.

Technology as Knowledge

Much of the philosophical work on technology as knowledge has naturally been epistemological, but ethical issues have also received consideration. One of these concerns freedom of speech and censorship. For example, terrorist threats highlight the dual-use character of technical knowledge, which may often be used for beneficial as well as nefarious purposes. This raises age-old questions about whether some knowledge should be forbidden, or if not, how its production and exchange should be regulated. Because technoscientific knowledge is not easily separable from applications, it may not be feasible or wise to argue that ethical considerations need only take place after knowledge has been produced.

With advances in genetics and information technologies, the issue of intellectual property rights has sparked debate about the ethical and societal implications of the private ownership of technical knowledge. Pertinent topics in this area are open-source software and the patenting of genetic material. In agriculture, the latter area has raised difficult questions about the legal status of indigenous technical know-how. Another important topic is the increasing privatization of academia driven by incentives for university researchers to patent the technological products that result from their research. This raises ethical issues about the proper role of the academy and the value of open information exchange in science.

One last broad set of ethical issues is raised by the theme of expertise and the role of experts, especially engineers, in a democracy. Many problems in modern industrial societies require the specialized knowledge of engineers, but most would claim that a technocracy, or rule by experts, represents an undesirable departure from democratic ideals. (It is worth noting, however, that in some cases technocrats are praised because of their lack of attachment to fundamentalist political or religious ideologies; technical knowledge and competence has its virtues.) Although engineers have much to offer regarding management and policy decisions, many nontechnical or political issues tend to become unproductively debated as if they could be resolved by technical knowledge. Other issues related to the accumulation of specialized knowledge by experts are the deskilling of the workforce, equity concerns about access to education, and widespread technological illiteracy even in societies utterly dependent on the smooth functioning of technological systems. All of these issues raise important questions about knowledge as a form of power.

Encyclopedia entries that deal directly with technology as knowledge thus include those on "Expertise," "Intellectual Property," "Public Understanding of Science," and "Technocracy." Related questions are also addressed in more general entries on, for example, "Computer Ethics" and "Information Ethics."

Technology as Activity

Technology as activity shades from personal to institutional and social modes. It may conveniently be divided into the two broad themes of production and use. With regard to production, most of the ethical issues are internal to the various technical professions. They raise issues of professional, engineering, and management ethics, which are often formalized in codes of ethics and are being increasingly integrated with professional training and education programs. Different ethical issues arise along the spectrum of engineering functions from the initiating actions of inventing and designing to the subsequent processes of testing, constructing, and operating. But across the board one common theme is that of the social responsibility of engineers, managers, and the organizations in which they are embedded.

Technology as activity is nevertheless more complex than a one-way flow of products from invention to application or use. Not only are engineers influenced in subtle ways by cultural norms, their work is often con-

sciously informed and directed by formal and informal involvements of governments and publics. These take the broad form of technical standards, regulation, and technology policy, as various institutions and actors engage in decision-making procedures about which technologies to produce, ban, limit, or otherwise manage. Examples include regulatory bodies such as the Food and Drug Administration (FDA), advisory bodies such as bioethics commissions, and technology assessment agencies such as the Office of Technology Assessment (OTA) or tools such as environmental impact statements. Public decisions about the production and use of technology raise manifold ethical issues about who should be involved, how involvements should be structured, how risks, costs, and benefits should be measured, and what goals should drive the policymaking process. Broader debate occurs over the proper roles of market mechanisms and government control.

Ethical analyses of the use of technology flow naturally from the fact that such uses are subordinate to, or in the service of, some goal. Issues of use often raise the question of whether artifacts can be considered ethically neutral. For example, computer technology can be used to help researchers find cures for diseases, or it can be used to hack into financial systems and steal money. Although it is common to conceptualize technology in this way, there is significant evidence for the nonneutrality of technology.

Indeed technological changes fundamentally alter human experiences in ways that can be judged good or bad, but certainly not neutral. Such changes are best illustrated by work, the most prominent form of technology as activity. The large-scale production and use of modern technologies has brought about the transformation of craftwork into industrial labor, which is marked by division of labor, mass production standardization, and bureaucratic organization.

For more analysis of the ethical issues related to technology as activity it is thus useful to consider encyclopedia entries on "Professions and Professionalism," specific professional organizations such as the "Institute of Electrical and Electronics Engineers," and regulatory agencies such as the "Food and Drug Administration" and the "Federal Aviation Administration." Also relevant would be entries on the principles that are said to guide much technical activity such as "Efficiency," "Safety," and "Reliability."

On a philosophic note, it is also important to consider how technological activities or processes of a more impersonal sort alter human relationships and relationships between humans and nature. The entry on "Tools and Machines" makes suggestions with regard to human–human relationships. The entry on "Arendt, Hannah," provides further background to her argument about the ways traditional technics or premodern technology was limited by the materials and energy given in nature. The development of steam, electric, and nuclear power qualitatively changed this human–nature relationship. Finally, Arendt noted how technology as action is a deeply troubling contradiction. Traditionally, action was associated with the political realm and its qualities of plurality, indeterminacy, and choice. Modern mass society has subordinated this realm to the pursuit of scientific technology and technologically mediated work, an effort that seeks to replace the contingencies of nature and the polis with the control and certainty of technology. Ethical and metaphysical quandaries result about the modern attempt to control, manage, and even make nature. Much of the rhetoric around the notion of ecological sustainability, for example, is dominated by concerns of control and efficiency rather than political and ethical considerations of the meaning of the good life and humankind's proper relationship with other species. And contemporary worries about the uncertainty of much scientific and technical knowledge would arise only in a world that aspired to certainty in human affairs.

Technology as Intention

Technology as intention is at once the most basic yet the most difficult to consider. As Aristotle noted, neither technics nor technology can exist without the exercise of intentionality. Moreover, because ethics is itself so closely tied to the idea of intentions and their assessment, to think of technology as intention would seem to bring technology more closely into the ethical realm than to think of technology as object, knowledge, or perhaps even action. At the same time, the slipperyness of intentionality presents its own difficulties, especially in relation to technology. Is there any such thing as a distinctively technological intention in the same way there are technological objects, forms of knowledge, and activities? Is it possible, for instance, to distinguish between religious, political, and technological intentions—or between premodern and modern technology in terms of intentionalities? Or are intentions just mental states to which technical activities are necessarily subordinated? Is there one intention to procure food, which can then be achieved by, say, political or technological means? But surely the intentional selection of technological over political means constitutes a kind of technological

intention. (See, in this respect, the entry on "Technological Fix.")

The most common way in which intentionality has been invoked when examining the ethics of technology is in fact in relation to the idea of modern technology as emanating from a distinctive will or volition, a philosophical argument more common to phenomenological than to analytic traditions in philosophy. Discussions of technology as volition span the spectrum from technology as a creative life force to technology as a restricting urge to control. Technology can be celebrated in a Nietzschean aesthetics of self-making in the project to wrest control of life from the vagaries of nature and even achieve immortality. But there is a sense in which technologies have a "will of their own" and are not infinitely plastic to the impress of different human intentions. Perhaps it is not just human intentions or volitions that shape technology, but technologies that also influence human intentions. There are limits to what one can do with any particular technology: It is difficult to use a hammer to screw a nut onto a bolt.

To analyze technology as a form of intentionality further requires that ethical assessments of use be coupled with empirical work on the properties of technologies. One form this has taken is to conceptualize intending as a form of decision making, which may in turn be undertaken by rational analysis. More generally, the increasing powers unleashed by modern technology suggest a need for increased knowledge of what ends they are to serve and knowledge of the consequences before they are put into use. But such needs must themselves be translated into action. And failure to take action is a form of weakness of intention or will that recurs frequently in situations of public and personal decisions about technology.

Most discussions of the ethics of technology deal with specific technologies: biomedical technologies, computers, nuclear weapons, and more. But in a few instances philosophers working in the phenomenological tradition have sought to bridge technological divides and consider the parameters of technology as a whole. Here the contributions of such thinkers as "Anders, Günther" and "Jonas, Hans" as well as "Heidegger, Martin" are especially significant. Related discussions can be found in entries on such philosophical schools as "Existentialism" and "Critical Social Theory."

Generalization

The distinctions between ethical issues in technology as object, as knowledge, as activity, and as intention should not serve to excuse anyone from thinking about ethics and technology in other ways as well—or for seeking to integrate these four modes of the manifestation of technology. For instance, Albert Borgmann's provocative interpretation of modern technological objects as tending toward what he terms the "device paradigm" of supplying some commodity with minimal human engagement and contextual dependency at the same time depends on a unique form of (virtual) knowledge and sponsors a distinctive type of (unfocused) activity. Borgmann's ethical assessment of technological devices is coordinate with his ethical judgment regarding technological knowledge and activity. To distribute ethical issues across a spectrum of manifestations of technology may serve simply as a provisional means for appreciating the breadth of concerns that fall under the idea of relating technology and ethics. Similarly, Don Ihde's analysis of different forms of human engagement with technology—from embodied extension to perceptual transformation—crosses the boundaries of technology as object, knowledge, and action in ways that invite scientists, engineers, and the general public to ask broad ethical questions about the techno-lifeworld they are in the process of creating.

Finally, the breadth of concerns must not be thought of as one determined only by problems. The praise of technology that is distinctive of the modern project and Enlightenment aspirations invests technology with rich ethical promise for better goods and services, understanding, human health, and intentional fulfillment. From this perspective the ethical problems are addressed so that they can be negotiated with that distinctively human behavior that originally gave rise to all technology, ancient and modern, in order to pursue and promote true human flourishing. Problems need not be limitations; they can also be conceived as the stimulus to new achievements.

ADAM BRIGGLE
CARL MITCHAM
MARTIN RYDER

SEE ALSO *Architectural Ethics; Computer Ethics; Engineering Design Ethics; Engineering Ethics: Overview; Ethics: Overview; Expertise; Industrial Revolution; Professions and Professionalism.*

BIBLIOGRAPHY

Barbour, Ian G. (1993). *Ethics in an Age of Technology.* San Francisco: HarperSanFrancisco. This is the second of two volumes, the first of which dealt with science and religion.

Borgmann, Albert. (1984). *Technology and the Character of Contemporary Life: A Philosophical Inquiry.* Chicago: University of Chicago Press. A provocative interpretation of the special character of modern technology as object.

Chadwick, Ruth, ed. (2001). *The Concise Encyclopedia of the Ethics of New Technologies.* San Diego, CA: Academic Press. Thirty-seven articles on various technologies (biotechnology, genetic engineering, nuclear power) and related issues (brain death, intrinsic and instrumental value, precautionary principle) selected from the *Encyclopedia of Applied Ethics* (1998).

Ihde, Don. (1990). *Technology and the Lifeworld: From Garden to Earth.* Bloomington: Indiana University Press. A broad phenomenological analysis of human-technology interactions.

Jonas, Hans. (1984). *The Imperative of Responsibility: In Search of an Ethics for the Technological Age,* trans. Hans Jonas and David Herr. Chicago: University of Chicago Press. Combines two German books published first in 1979 and 1981.

Kaplan, David M., ed. (2004). *Readings in the Philosophy of Technology.* Lanham, MD: Rowman and Littlefield. Includes substantial sections on ethics and politics.

Mitcham, Carl. (1994). *Thinking through Technology: The Path between Engineering and Philosophy.* Chicago: University of Chicago Press.

Mitcham, Carl, and Robert Mackey, eds. (1983). *Philosophy and Technology: Readings in the Philosophical Problems of Technology.* New York: Free Press. The "Ethical-Political Critiques" section includes a number of classic texts; other articles in this early collection are also relevant.

Tavani, Herman T. (2003). *Ethics and Technology: Ethical Issues in an Age of Information and Communication Technology.* Hoboken, NJ: Wiley.

Winner, Langdon. (1986). *The Whale and the Reactor: A Search for Limits in an Age of High Technology.* Chicago: University of Chicago Press.

INTERNET RESOURCE

"The Online Ethics Center for Engineering and Science." Case Western Reserve University. Available from http://onlineethics.org.

TECHNOLOGY TRANSFER

• • •

Technology transfer is a complex and multi-faced process. Initially, transfer occurs from research laboratories such as universities to the market. Prior to 1980 when The Patent and Trademark Laws Amendment Act, more commonly know as the Bayh-Dole Act was passed, there was limited flow of government-funded inventions to the private sector. In 1980, the federal government held title to approximately 28,000 patents. Fewer than 5 percent of these were licensed to industry for development of commercial products (U.S. Government Accounting office, 1998). The Bayh-Dole Act permitted universities to retain title to inventions developed under government funding and encouraged universities to collaborate with companies to promote the utilization of invention arising from federal funding. Since the passage of this Act, partnerships between universities and industry have moved new discoveries from the laboratory to the market place for the benefit of society.

There is substantial evidence to suggest that the Bayh-Dole Act has promoted a considerable increase in the technology transfer from universities to industry, and ultimately to the people around the world. However, it is obvious that economic interests were the driving forces for the change in governmental policy. Licensing by universities, National Institutes of Health or other governmental agencies in life sciences has yielded substantial profits to pharmaceutical companies, sometimes at the cost of human suffering. If the public good is not served by, or is undermined by technology transfer, then it is ethically justified to change public policy.

Historically, and to a large extent even in the early twenty-first century, the transfer of technology occurs between and among developed nations. However, new forms of multi-national enterprise imply a dispersion of production tasks across globe. In the case of developing countries, the technology must meet the local needs and be socially accepted. If the technology is not appropriate it may cause negative economic, social, and environmental impacts. The chemical disaster in Bhopal, India, is a case in point. Methylisocyanate (MIC) leaking from a Union Carbide corporation pesticide plant immediately killed more than 2,000 people and injured or disabled more than 200,000 others. The death toll has reached 20,000 since December 3, 1984, when the accident occurred. Information about hazardous technologies was lacking, workers were poorly trained, and major safety equipment was inoperative because of poor maintenance. In this case the technology should have been modified to make it adaptable to the new environment.

Mechanisms of Technology Transfer

The most important legitimate channels for technology transfer are licensing, foreign direct investment, and joint ventures. Most technology transfer takes place in the form of licensing under specific terms and conditions agreed to by both suppliers and recipients. The suppliers gain monetary rewards, whereas the recipients expand their economic opportunities.

Foreign direct investment refers to a process by which multinational corporations (MNCs) transfer production operations to the developing countries through wholly owned subsidiaries. In this context, the transfer of technology takes place internally between parent MNCs and their branches and subsidiaries in different countries. This enables MNCs to retain technology within the corporations.

Joint ventures have emerged as an alternative to foreign direct investment because most developing countries have issued investment laws that regulate foreign investment. These laws promote joint ventures between local and foreign partners. Consequently, with greater emphasis on national participation and control by the developing countries, technology transfer has assumed a new meaning, although control over proprietary technology and know-how has remained with MNCs.

Technology Transfer and Ethical Issues

Given these basic mechanisms of technology transfer, one may nevertheless ask: Why technology transfer? Can technology transfer improve the economic conditions of people living in the developing countries? Can technology transfer create global equity?

Proponents of globalization have suggested that technology and its diffusion can improve living standards, increase productivity, generate employment opportunities, improve public services, and create competitive markets for products. Have these goals been achieved? There are two contending theories: the dependency theory and the bargaining theory.

DEPENDENCY THEORY. Proponents of this theory (Cardoso and Faletto 1979) claim that, because of the insistence of multinational corporations on foreign direct investment (which transfers technology from the parent companies to the foreign subsidiaries), developing countries are denied access to modern technologies. These theorists contend that technology is key to development and, if denied, developing countries will remain dependent on developed countries. This will create negative economic outcomes, such as increased inequality and wage stagnation. Consequently, the balance of trade between developed and developing societies will remain unequal and therefore exploitive. Sunil K. Sahu (1998) suggests that such technological dependence creates an enclave economy for the developing countries, and that it will be difficult for their economies to expand or even survive.

BARGAINING THEORY. This theory takes a view opposite that of dependency theory. Bargaining theory recognizes the potential benefit that MNCs can bring to their host countries. In other words, the technologies of the advanced countries do not have adverse effects on the economy of the developing societies. Raymond Vernon (1971), an advocate of this theory, has developed a concept known as "obsolescing bargaining" that explains the relationship between MNCs and host countries. The bargaining power of the developing countries tends to increase after a certain period, specifically when technology becomes stabilized and competition for the same technology by other developed countries intensifies. The competition among developed countries increases the choices available to the developing countries. Additionally, once the foreign investment is "sunken," the host country is in a much stronger position to negotiate a better deal, and at this point MNCs cannot credibly threaten to withdraw (Stepan 1978). Vernon also suggests that the monopoly of the innovator is not permanent because most products tend to pass through a transition from "monopoly to oligopoly to workable competition" (Vernon 1971, p. 91). This is also known as the product life-cycle theory.

Can Technology Transfer Create Global Equity?

Technology transfer has accelerated the process of globalization, and it is suggested that it may lift all people and raise their living standards. The Industrial Revolution brought new wealth first in Europe and then in the United States. Since the Industrial Revolution, the difference between the rich and the poor in the world has increased. It is estimated that the difference between the per capita incomes of the richest and poorest countries was 3 to 1 in 1820, 11 to 1 in 1913, 35 to 1 in 1950, 44 to 1 in 1973, and 72 to 1 in 1992 (UNDP 1999). The gap is further reflected in how the world's wealth is distributed. The wealthiest 20 percent of the world's people—all from developed countries—control 85 percent of global income. The remaining 80 percent of people share 15 percent of the world's income. Such disparity has led to greater poverty in the developing countries. Statistics show that the number of people who are living on less than $1 per day (a frequently used poverty line) was rising in the late twentieth and early twenty-first centuries. The number of these people grew from 1.2 billion in 1987 to 1.5 billion in 2000, and there could be nearly 2 billion poor people by 2015. In addition, approximately 45 percent of the world population live on $2 per day (World Bank 2000). Some countries such as South Korea, Taiwan, Singapore, Hong Kong,

and China have benefited from global economies, but others have not. The growth of proprietary technology, covered by patents and industrial property rights, has served as a major barrier to new entrants, and it will continue to do so unless proprietary rights are modified.

MURLI M. SINHA

SEE ALSO *Development Ethics; Technological Innovation.*

BIBLIOGRAPHY

Cardoso, Fernando Henrique, and Enzo Faletto. (1979). *Dependency and Development in Latin America,* trans. Marjory Mattingly Urquidi. Berkeley and Los Angeles: University of California Press.

Rosenberg, Nathan, and Claudio Frischtak, eds. (1985). *International Technology Transfer: Concepts, Measures, and Comparison.* New York: Praeger.

Sahu, Sunil K. (1998). *Technology Transfer, Dependence, and Self-Reliant Development in the Third World: The Pharmaceutical and Machine Tool Industries in India.* Westport, CT: Praeger.

Stepan, Alfred. (1978). *The State and Society: Peru in Comparative Perspective.* Princeton, NJ: Princeton University Press.

United Nations Development Programme (UNDP). (1999). *Human Development Report, 1999.* New York: Oxford University Press.

United States Government Accounting Office. (1998). "Technology Transfer." Report to Congressional Committees. Administration of the Bayh-Dole Act by Research Universities.

Vernon, Raymond. (1971). *Sovereignty at Bay: The Multinational Spread of U.S. Enterprises.* New York: Basic.

Vernon, Raymond. (1977). *Storm over the Multinationals: The Real Issue.* Cambridge, MA: Harvard University Press.

World Bank. (2000). *World Development Report, 1999–2000.* New York: Oxford University Press.

TECHNOSCIENCE

• • •

Technoscience refers to the strong interactions in contemporary scientific research and development (R&D) between that which traditionally was separated into science (theoretical) and technology (practical), especially by philosophers. The emphasis that the term *techno(-)science* places on technology as well as the intensity of the connection between science and technology varies. Moreover the majority of scientists and philosophers of science continue to externalize technology as *applications and consequences* of scientific progress.

Nevertheless they recognize the success and efficiency of technology as promoting realism, objectivity, and universality of science.

The prehistory of the concept of technoscience goes back at least to the beginning of modern science. Francis Bacon (1561–1626) explicitly associated knowledge and power; science provided knowledge of the effective causes of phenomena and thus the capacity for efficient intervention within them. The concept became clearer during the first half of the twentieth century. Gaston Bachelard (1884–1962) in *Le nouvel esprit scientifique* (1934; The new scientific spirit) places the *new scientific spirit* under the preponderant influence of the mathematical and technical operations, and utilizes the expression *science technique* to designate contemporary science. However the term techno(-)science itself was not coined until the 1970s.

The History of Techno(-)science

The first important occurrence of the term appears in the title of an article titled "Ethique et techno-science" by Gilbert Hottois, first published in 1978 (included in Hottois 1996). This first usage expresses a critical reaction against the theoretical and discursive conception of contemporary science, and against philosophy blind to the importance of technology. It associates technoscience with the ethical question, What are we to make of human beings? posed from an evolutionist perspective open to technical intervention.

Throughout the 1980s two French philosophers, Jean François Lyotard and Bruno Latour, contributed to the diffusion of the term in France and North America. For Lyotard technoscience realizes the modern project of rendering the human being, as argued from the work of René Descartes (1596–1650), a *master and possessor of nature*. This project has become technocratic and should be denounced because of its political association with capitalism. As a promoter of the postmodern, Lyotard thus facilitates diffusion of the term within postmodern discussions.

In *Science in Action* (1987), Latour utilizes the plural *technosciences* in order to underline his empirical and sociological approach. The technosciences refer to those sciences created by human beings in real-world socioeconomic-political contexts, by conflicts and alliances among humans and also among humans and nonhumans (institutions, machines, and animals among others). Latour insists on networks and hybrid mixtures. He denounces the myth of a *pure science*, distinct from technologies susceptible to good and bad usages. In reality it is less technology that Latour internalizes in the idea of science than society (and therefore politics), of which technologies are part in the same ways as other artifacts. He rejects any philosophical idea, whether ancient or modern, of a science that is supra- or extra-social and apolitical. The worldwide successes of the technosciences are a matter of political organization and will, and do not derive from some universal recognition of a rational and objectively true knowledge that progressively imposes itself. Latour has contributed to the success of the term technoscience in social-constructivist discussion since the 1990s.

The work of Donna Haraway illustrates well the diffusion of technoscience crossed with the postmodern and social-constructivist discussions in North America. Technoscience becomes the word-symbol of the contemporary tangle of processes and interactions. The basic ingredients are the sciences, technologies, and societies. These allow the inclusion of everything: from purely symbolic practices to the physical processes of nature in worldwide networks, productions, and exchanges.

In France, in continental Europe, and in the countries of Latin America, the use of the term technoscience has often remained closer to its original meaning that involves more ontological (as with German philosopher Martin Heidegger (1889–1976)), epistemological, and ethical questioning than social and political criticism. Indeed in a perspective that complements the one provided here, in *La revolución tecnocientífica* (2003; The technoscience revolution), Spanish philosopher Javier Echeverría provides an extensive analysis of technoscience as both concept and phenomenon. A political usage is not, however, rare, especially in France where there is a tendency to attribute to technoscience a host of contemporary ills such as technicism and technocracy, multinational capitalism, economic neo-liberalism, pollution, the depletion of natural resources, the climate change, globalization, planetary injustice, the disappearance of human values, and more, all related to U.S. imperialism. The common archetype of technoscience is Big Science, originally exemplified by the Manhattan Project, which closely associated science, technology, and the politics of power. In this interpretation, technoscience is presented from the point of view of domination, mastery, and control, and not from that of exploration, research, and creativity. It is technocratic and totalitarian, not *technopoiétique* and emancipating.

The Questions of Technoscience

What distinguishes contemporary science as technoscience is that, unlike the philosophical enterprise of science identified as a fundamentally linguistic and theoretical activity, it is physically manipulative, interventionist, and creative. Determining the function of a gene whether in order to create a medicine or to participate in the sequencing of the human genome leads to technoscientific knowledge-power-doing. In a technoscientific civilization, distinctions between theory and practice, fundamental and applied, become blurred. Philosophers are invited to define human death or birth, taking into account the consequences of these definitions in the practical-ethical plans, that is to say, in regard to what will or will not be permitted (for example, the harvesting of organs or embryonic experimentation).

Another example is familiar to bioethicists. Since the 1980s there has existed a line of transgenic mice (*Onco mice*) used as a model for research on the genesis of certain cancers. Here is an object at once natural and artificial, theoretical and practical, abstract and concrete, living and yet patented like an invention. Their existence and use in research further involves many different cognitive and practical scientific questions and interests: therapeutic, economic, ethical, and juridical. It is even a political issue, because transgenic mice are at the center of a conflict between the European Union and the United States over the patentability of living organisms.

The most radical questions raised by technosciences concern their application to the *natural* (as a living organisms formed by the evolutionary process) and *manipulated* (as a contingent creation of human culture). Such questions acquire their greatest importance when one takes into account the past and future (unknowable) immensity of biological, geological, and cosmological temporality, in asking, for example: What will become of the human being in a million years? From this perspective the investigation of human beings appears open not only to symbolic invention (definitions, images, interpretations, values), but also to techno-physical invention (experimentation, mutations, prosthetics, cyborgs). A related examination places the technosciences themselves within the scope of an evolution that is more and more affected by conscious human intervention. Both approaches raise questions and responsibilities that are not foreign to ethics and politics but that invite us at the same time to consider with a critical eye all specific ethics and politics because the issues exceed all conceivable societal projects.

GILBERT HOTTOIS
TRANSLATED BY JAMES A. LYNCH

SEE ALSO *Critical Social Theory*.

BIBLIOGRAPHY

Echeverría, Javier. (2003). *La revolución tecnocientífica* [The technoscience revolution]. Madrid: Fondo de Cultura Económica de España.

Haraway, Donna. (1997). Modest-Witness@Second-Millenium: Female-Man_Meets OncoMouse™. New York: Routledge.

Hottois, Gilbert. (1996). *Entre symboles et technosciences* [Between symbols and technosciences]. Seyssel and Paris, France: Champ Vallon and Presses Universitaires de France

Hottois, Gilbert. (2002). Species Technica [Species technica]. Paris: Vrin.

Latour, Bruno. (1987). Science in Action. Cambridge, MA: Harvard University Press.

Lyotard, Jean-François. (1988). Le postmodern expliqué aux enfants [The postmodern explanation of children]. Paris: Galilée.

Séris, Jean-Pierre. (1994). La technique. [The technique] Paris: Presses Universitaires de France.

TELEPHONE

• • •

Telephone technology allows a person to talk to nearly anyone in any place who has similar equipment. There are substantial ethical questions related to the uses and abuses of the telephone. Among other things, the telephone is a communication system that provides political leaders, pollsters, and social science researchers with some understanding of public attitudes and behaviors. It gives voice to the needs and wishes of citizens as they attempt to make their views known to governments and corporations. Additionally, the telephone is a conduit for the delivery of professional services. As a result of these aspects of what has been an everyday but rapidly changing technology, considerable attention has been devoted to the telephone from ethical, legal, and policy viewpoints.

Historical Development

The term *telephone* is based on the combination of the Greek words, *tele* ("distant" or "afar") and *phon* ("sound" or "voice"); it was first used in France in the 1830s to name a crude acoustic device. By the mid-1800s something akin to a pair of tin cans connected by a taut string was known in the United States as the "lover's telephone." In 1876 Alexander Graham Bell (1847–1922) won a patent for a device that has come to be known as the telephone.

Alexander Graham Bell testing his telephone invention in front of onlookers. Graham won a patent for the device in 1876. (*U.S. National Aeronautics and Space Administration.*)

The traditional telephone operates by converting the mechanical energy of sounds carried in the air (the speaker's voice) into electrical impulses for transmission to a receiver. The receiver reverses the process, changing the electrical impulses back into vibrations. Those vibrations are heard as sounds. The original telephones transmitted electrical impulses by wires. Radio and other portions of the electromagnetic spectrum subsequently supplemented or supplanted wires as digital forms replaced analog.

The uses of the telephone have expanded to include multiple forms of data transmission, including fax, photo, and video image formats. Ancillary services have been created and have been widely adopted, including answering machines, caller-ID boxes, and telephone-based security systems. The Internet owes much of its success to the ability of users to go online by means of telephone lines.

In the early period of the telephone myriad uses were explored, including the "broadcasting" of news, opera, weather reports, and religious services. Some con-templated services never materialized: Bell speculated that the telephone might be used to communicate with the deceased. Other services did not materialize because they were outdated before they could deployed: France's national telephone company conducted extensive research in the 1960s to see if the telephone touch-tone pad could be adapted to serve as a home calculator. Yet other services were initially innovative and popular, but then, as technology continued to advance, they were left in the backwater. The fax machine and the French Minitel system are examples of this phenomenon.

Ethical Issues

PRIVACY, SECURITY, AND SURVEILLANCE. Among the early ethical questions was the way the telephone was used to invade privacy in the household and give outsiders access to household members. In particular the telephone allowed outsiders to make social connections with the members of a household, thus violating rigid gender and class roles. Ethical questions relating to various roles in the household, along with the power relationships among those roles, have been exacerbated by the telephone. For example, teenagers and parents come into conflict over appropriate norms for telephone use.

The telephone often leads to disruption of household routines and may allow for social subversion through practical jokes and harassing or obscene phone calls. Women especially have been victimized by such calls, though a surprisingly large number of men have been as well. Although commentators see great net benefits arising from the telephone, they also recognize the moral dilemmas that result from the "distant presence" (a phrase popularized by Kenneth J. Gergen) the telephone allows. The American humorist Mark Twain (1835–1910) was an early acerbic critic of the way the telephone could disrupt trains of thought and ordinary social interaction. In addition, characteristic of early telephone technology was the large proportion of homes that shared local service party lines; this meant that neighbors could listen in to conversations and learn family secrets.

Larger questions of privacy surrounded systematic wiretapping conducted by both licit and illicit organizations. Only a few years after the telephone was invented numerous devices were built to allow not only tapping but also recording of telephone conversations. (Many of these microphone devices also can be used to listen in on in-room conversations.) A wide variety of practices legal and illegal, moral and immoral, could be identified and documented.

Police forces and other governmental agencies sometimes carried out large-scale wiretapping not only in pursuit of wrongdoers but also to monitor those perceived as opposing government policy. In what has become a well-established cycle of innovation, new ways to communicate were followed by new ways to penetrate those forms, followed by steps to enhance privacy. Often a variety of codes would be devised to hinder attempts to collect data and conduct surveillance. The question of the areas in which people had a "reasonable expectation" of privacy was brought to a head when the U.S. Supreme Court decided in *Katz* v. *United States* (389 US 347 [1967], docket number 35) that public phone booths were not eligible for systematic tapping by the police.

Although monitoring of workers has been a perennial workplace issue, the telephone gave that issue added impetus because it greatly expanded the ability of managers to tap into the conversations of employees. Telephone companies often have conducted extensive monitoring, sometimes to the point of abuse, when they have used their own technology to monitor employees' behavior and comments. Switchboard operators once were notorious for eavesdropping, though sometimes that allowed them to interrupt the execution of crimes. (Eavesdropping, as opposed to service monitoring or surveillance by officials, is generally prohibited everywhere.) Many companies, including especially telephone companies, have published rulebooks and etiquette guides directed to their employees and managers regarding eavesdropping. While these efforts presumably reduced the problem, they have not been sufficient to extinguish the practice.

TELEMARKETING AND RESEARCH. Telemarketing is the offering of goods or services through sales presentations on the telephone. Because it can be a low-cost, high-profit enterprise, its rapid proliferation has become a source of general annoyance to the targeted public. The Direct Marketing Association and the American Marketing Association instruct their members not to use approaches that might be considered illegal. Moreover, there are numerous laws that regulate telemarketing at the national and local levels. Major moral dilemmas are related to this situation.

On the one hand, there are the claimed rights of businesses to "freedom of commercial speech," which includes the freedom to communicate with potential customers and participation in "fair and efficient markets." (These rights are protected strongly in the United States.) These rights often are carried out with increasingly powerful telephone support technology and data-base-mining software. On the other hand, individuals have a right to be left alone and not to have information about them collected in secret and without their permission. (These rights are protected strongly in the European Union nations and not as well protected in the United States.) Despite such efforts on both the technological front (such as caller-ID and call blocking) and the legal front (such as the compilation of "do not call lists" and the regulation of times when sales calls may be made), this problem persists.

Social science research and public opinion surveys often are reliant on polling by telephone. Numerous agencies and associations, such as the American Sociological Association (ASA) and American Association for Public Opinion Research (AAPOR) have created codes of conduct for their members, and in some cases governments have stepped in to create regulations in this area. Criminal penalties can be imposed for collecting data improperly by telephone. Many institutional review boards (IRBs) at universities require that researchers demonstrate that they will protect the data and not cause psychological distress, and this applies to telephone surveys as well as to medical experimentation. In more extreme cases, such as at the University of Newcastle in Australia, researchers are required to notify the target population in advance with a written information sheet that warns that telephone contact will be made and includes complete contact information.

UNIVERSAL SERVICE, SOCIAL EQUITY, AND DEMOCRACY. An important ethical component of national and regional policies for telephone technology is equitable distribution. As Claude Fischer (1992) has noted, in its early years the telephone could be considered only a luxury. However, what was an expensive enhancement to lifestyle has in contemporary society become a near necessity for most people.

For much of the twentieth century national telecommunication policies were aimed at subsidizing low-income and rural populations by indirectly taxing (through higher rates) urban and nonpoor telephone subscribers. This was done under the rubrics of social equity and economic development. In fact, in the United States the promise of universal service at an affordable cost was accepted by the government in exchange for the granting of near-monopoly status to the American Telephone and Telegraph Company (AT&T). However, the initial moral clarity of those policies has been obscured as advanced telecommunication technologies have proliferated, especially in the case of the mobile phone.

Mobile phone. Mobile phones have a long and varied history that stretches back to the early 1970s. Due to their low establishment costs and rapid deployment, mobile phone networks have since spread rapidly throughout the world, outstripping the growth of fixed telephony. (© Leland Bobbe/Corbis.)

It is noteworthy that around the world hundreds of millions of subscribers have flocked to new mobile phone services. Those services allow subscribers to leapfrog the long waits and frequently high prices associated with wireline residential services. Moreover, cross-subsidization by ordinary telephone subscribers of low-cost services for schools and hospitals, as is the practice in the United States, means that many people with modest incomes are being penalized for the benefit of institutions in wealthy communities. (Mobile phone subscribers in the United States are exempt from these taxes.)

There can be little doubt that the telephone is an important adjunct to democracy on the level of political expression and as a bulwark against excessive governmental power. At the same time, terrorists and those seeking radical regime change can use the telephone to further their aims. In light of this situation many governments monitor telephone conversations and in some cases limit or prohibit mobile phone services. As instances, North Korea forbids civilian mobile phones on security grounds and Colombia's mobile phone networks were selectively turned off by the government in an effort to detect the location of cell phone-toting drug lord Pablo Escobar.

Public Use of Mobile Telephones

Each major advance in telephone technology has been accompanied by some social disruption. In most cases the disruptions have been transient. With the advent of the mobile telephone, however, high levels of conflict continue. These conflicts often may be understood in terms of what is known in psychology as the actor-observer paradox. The person who wishes to use the mobile phone (the actor) does so because he or she has good cause and with the expectation that others will understand and accept that necessity. However, the people around the user (the observers) view the situation differently. They feel that the mobile phone user is being selfish and self-indulgent and is failing to respect the conventions of polite society. The public use of mobile phones is likely to remain a source of normative conflict because the sources of irritation are not merely conventional but seem to go to the core of human cognitive processes. The result could be that as mobile phone users pursue the private pleasures of conversation there will be a reduction in civility and personal engagement in public places. Perhaps no better illustration of this process is the havoc wrought by drivers who are preoccupied by their mobile telephone conversations.

Provision of Professional Services

The ease and flexibility of telephone use have led many professional organizations to develop codes of conduct that allow their members to use the telephone, under appropriate conditions, to serve clients. This is the case with the many national and worldwide associations of lawyers, for instance. However, the potential for abuse also has led many organizations, such as the Legal Profession Advisory Council, to remind their members that whereas the telephone can be used to discuss and provide confidential information, both the professional and the client have to agree to this in advance. It further recommends that a scrambling device or other encryption technology be used. All advertisements for lawyers should bear the attorney's phone number prominently.

The question of recording telephone conversations is fraught with ethical and moral questions. In one instance (LEO 1738, 48/10 Va Lawyer Reg 23, April 13, 2000) the Virginia state bar association reexamined the subject of taping telephone conversations. That association concluded that all forms of wiretapping, along with one-party-consent recording of telephone conversations by lawyers, are prohibited. Although many people disagreed with that conclusion, it did arrive at the formulation that because wiretapping involves "deceit," the practice must be forbidden. This raises problems when, for instance, testers try to prove housing discrimination by pretending to be people other than who they are. The rules even make it unethical for an attorney who receives an obscene or threatening phone call to record it.

The American Medical Society counsels physicians that telephone advising and referral services should be used only to complement face-to-face interaction and that both the physicians and the clients should be well aware of the limitations of the medium. They urge that no physician make a clinical diagnosis or prescribe medications by telephone and at the same time be certain to elicit all-important information over the phone. They also should avoid generating large telephone bills that their patients or others have to pay.

Counseling by Telephone

Telecounseling has been defined as using the telephone for synchronous but distant interaction between counselors and clients for one-to-one conferencing. Obviously, such interactions are fraught with ethical issues. In response, the National Board for Certified Counselors (NBCC) says that its members should base the use of telecounseling on the needs and convenience of the client. The NBCC further stresses that telecounseling should only be a supplement to face-to-face counseling.

Confidentiality is an important consideration because it may be difficult to know precisely with whom one is speaking when one receives a telephone call. Thus, the American Psychological Association's guidelines warn counselors about privacy and confidentiality issues. The International Chiropractors Association of California has in its code of ethics the statement that its members "shall not discuss any patient information over the telephone with anyone without the patient's consent, preferably in writing." The International Association of Coaches instructs coaches to take precautions to ensure the confidentiality of telephone communications with clients.

In areas in which telephone counseling would be inappropriate professional codes of conduct underscore the importance of avoiding abuse. Thus, the Michigan Speech and Hearing Association urges that the telephone not be used for "diagnosis, treatment or re-evaluation of individual language, speech or hearing disorders." Medical and legal associations have guidelines that also are meant to avoid problems and underscore to their members that using the telephone may be construed as entering into a relationship with a client, with all the demands such a relationship entails.

More Complications Ahead

Because the telephone can obscure many of the ways in which people recognize each other or understand an evolving situation and can transcend distance, it opens new opportunities for ethically questionable or unethical behavior. In addition, as a result of the simplicity and power of the telephone, it has become a vital component of modern life. A variety of codes of conduct, laws, and corporate and governmental regulations have been developed to address these problems. However, these attempts have had incomplete success. Even as recent events are grappled with through norms and regulations, new telephone-based technologies that allow even more forms of use and abuse are complicating efforts to control telephone behaviors through technological countermeasures and moral and legal sanctions.

JAMES E. KATZ

SEE ALSO *Communication Ethics; Communication Systems; Monitoring and Surveillance; Networks; Privacy; Security.*

BIBLIOGRAPHY

Fischer, Claude. (1992). *America Calling: A Social History of the Telephone to 1940.* Berkeley: University of California Press. A profound analysis of technology's influence on social relationships and community. Fischer explores the integration of the telephone into society, thereby uncovering and illuminating important interplays between the communication processes and social structures.

Katz, James. (1999). *Connections: Social and Cultural Studies of the Telephone in American Life.* New Brunswick, NJ: Transaction. A detailed, empirical study of the ways telephone are used in ordinary (and extraordinary) circumstances. Explores norms and behaviors revolving around the telephone via methods including national surveys, interviews, and ethnographic observation.

Pool, Ithiel de Sola, ed. (1977). *The Social Impact of the Telephone.* Cambridge, MA: MIT Press. This path-breaking collection of historical and cultural studies on the telephone has become a classic and still has much to offer. It grew out of AT&T-sponsored lectures at MIT to celebrate the Centennial of the telephone's invention.

Ronnell, Avital (1989). *The Telephone Book: Technology-Schizophrenia-Electric Speech.* Lincoln: University of Nebraska Press. An eccentric but probing look at the deeper processes involved in the social and phenomenological construction of the telephone. The volume is a wry postmodern achievement in both form and content.

TELEVISION

• • •

Along with the radio, television has become the primary means for broadcast communication and entertainment. As such it calls for ethical and political assessment.

What follows will thus focus on such assessments, noting a spectrum of views running from positive to negative in relation to both content and practice.

Background

The word *television*, a hybrid compound of the Greek *tele* (distance) and the English *vision*, names a technological invention from the 1920s in which electromagnetic waves are used to control a beam of electrons scanning a cathode-ray tube so as to create an image. The initially distinctive feature of this technology was that, unlike motion pictures but like radio, it could be personalized for home or individual use. Over the course of more than half a century the electronics underwent continuous modification: Vacuum tubes were replaced with transistors and then integrated circuits; the black-and-white cathode-ray tube became colored and was then replaced by a high-definition, flat, liquid crystal display; and analog transmission was transformed to digital. The information transmitted thus became increasingly rich in a technical sense.

The commercial development and regulation of television followed the pattern established by radio: Television was initially promoted by the same corporations, and existing regulatory agencies and frameworks were adopted to distribute a limited transmission spectrum among competing private interests. Some countries established national broadcast operations independent of or complementing private operations. But in all cases television viewers received programs free of charge, except for advertising time or taxes. With the advent of video recording systems, cable, and satellite television, however, transmission resources were greatly enlarged, and fundamental shifts took place within the industry that created pay-for-view television. Yet this further increase in technical information delivery and in regulatory regime change failed to alter the basic content, which has remained of two sorts: information and entertainment. Indeed, the TV is the centerpiece of home entertainment systems.

Moral Promise and Threat

From its post–World War II appearance, the promise of TV has been at once praised and criticized. As a new, more vivid and pervasive form of mass communication than anything that had preceded it (magazines, newspapers, radio, and movies) it was subject to intensified versions of both the hype of modernity, which sees technological innovation as inherently beneficial, and mass culture criticism, which argues technology's dangers and debasements. A love–hate relationship was manifest in tensions between promises of increased democratic enlightenment and worries about the commercialization of culture.

On the one hand, television brings diverse quality dramas into the home, and international news programs depict a variety of countries, cultures, and perspectives in a single broadcast. On the other, pop-culture programs, such as those on MTV, present fragmented images that draw from a multicultural mix of music, fashion, sexuality, and ethnic traditions. The moral significance of numerous "high" and "low" cultural programs can be attributed to their ability to deconstruct monolithic images and ideologies: "Implicit in pluriculture is a kind of *bricolage* relativism. One may pick and choose culture fragments, multiply choices, and in the process reflectively find one's own standards provincial or arbitrary—certainly no longer simply *a priori* obvious" (Ihde 1995, p. 155).

While traditional cultures find themselves forced to confront modern secular images, so too are provincial U.S. (and other Eurocentric) audiences forced to question their own identities when confronted with traditional religious images. Television thus presents viewers with the opportunity to engage the global "community of those who have nothing in common" (in Alphonso Lingus's formulation) such that they may become more reflective about the arbitrary nature of their own cultural identity. Any particular cultural position is but one of many such perspectives within the wider cultural arena. For example, the multiperspectival international coverage of the "War on Terrorism" suggests that the conflict between East and West cannot be adequately explained by the partial metanarratives of either side.

Criticism

For present purposes television criticism may be distinguished into three types: those not influenced by Marshall McLuhan (1911–1980), those influenced by McLuhan, and those reacting against or going beyond McLuhan. As the typology suggests, the ideas of McLuhan, who argued the primacy not of television content but of its formal properties, have played a central role. "The medium is the message" was the sound-bite summary of his theory in *Understanding Media* (1964).

Prior to or subsequently ignoring McLuhan have been studies focused on issues related to the content of television and the social influence of this content. Does television advertising work? Do the attitudes and opinions expressed on TV influence or just represent those of the viewers? In the 1950s concern often emphasized the impact of television on leisure and culture. In the early 2000s the concern shifted to the political or cultural biases of television programming. This tradition of criticism also distinguishes different genres—news, cultural programming, sports, soaps, and so on. Most television criticism in the mass media has been of this type, which thus represents the most common critical approach. Studies by Cecelia Tichi (1991) and Lynn Spigel (1992) are scholarly contributions to this tradition.

Among criticisms that have been influenced by McLuhan's work are more intellectual studies, some of which have become classic references. Examples include Tony Schwartz's *Media: The Second God* (1981), Neil Postman's *Amusing Ourselves to Death* (1985), and Joshua Meyrowitz's *No Sense of Place* (1985). More thickly analytic than McLuhan, but in the same vein, Stanley Cavell (1984) contrasts the basic experience of movies as viewing with that of television as monitoring. All successful TV formats—from sitcoms and game shows to sports coverage and news—are forms of monitoring. For Cavell it is no accident that the television receiver is called a monitor, and that TV is used to monitor everything from banks to parking lots.

Most representative of the reaction to McLuhan is the work of Brian Winston (1998), who originally titled his work *Misunderstanding Media*. For Winston television is not the radically new medium envisioned by McLuhan, but simply another instance of technological performance based on progressively developing scientific competence. Moreover, "there is nothing in the histories of electrical and electronic communication systems to indicate that significant major changes have not been accommodated by preexisting social formations" (p. 2). Building on but transcending McLuhan is the teletheory of Gregory L. Ulmer (1989) and the concept of the televisual as developed by Tony Fry (1993).

Cutting across these three types of criticism are negative and positive assessments that focus either on the physical aspects of the technology or its content/form. Although there is no proof that a person can become physically ill from watching television, conclusive scientific evidence does not exist that details what challenges to health are likely to arise from exposure to extremely low doses of low-level radiation over long periods of time. Indeed, critics suggest that there is no threshold of exposure below which radiation may not harmfully affect humans. From an environmental perspective, critics further note not only that the process of manufacturing televisions generates toxic problems, but also that the level of electronic waste is growing rapidly.

This dilemma is exacerbated by the fact it is often less expensive and more convenient to replace rather than fix a malfunctioning television.

Negative assessments of the content of television programs vary. There are psychological worries about exposing children to violent and sexually charged programs, feminist and multicultural arguments about how television programs routinely stereotype women and other minorities in adverse ways, and sociopolitical concerns about the connection between television and political propaganda. Whereas the televised coverage of the Vietnam War in the 1960s facilitated a negative public reaction of the conflict because of its association with the "real" coverage of battlefield and civilian casualties, recent critical works that exemplify McLuhan's famous pronouncement that "the medium is the message," such as Jean Baudrillard's provocatively titled *The Gulf War Did Not Take Place* (1995) and Paul Virilio's *Strategy of Deception* (2000), suggest that the selective presentation of events during the Gulf War and the Kosovo conflict are indicative that people now live in a "hyper-real" time in which ever proliferating images are produced that are dissociated from reality. For example, during the Gulf War the impression that indiscriminate bombing and civilian causalities were minimized was fostered through the media's constant presentation of "smart bombs" that that destroyed only deliberately chosen and carefully delimited targets. A more recent argument, presented by Michael Moore in his Oscar-winning documentary *Bowling for Columbine* (2002), is that the media distortion of topics such as urban violence has produced a culture of fear in which American citizens routinely mistake deliberately sensationalized reporting for the presentation of unbiased facts.

In *Four Arguments for the Elimination of Television* (1978), Jerry Mander, a disillusioned advertising mogul, goes so far as to argue that because television is biased in favor of corporate interests and because it functions best when conveying simplified linear messages, it is beyond reform; the power of television to discipline people into accepting repressive control can be combated only by eliminating it completely. Mander also contends that television bolsters the tendency toward living in an artificial environment. This argument is given more in-depth philosophical examination by Albert Borgmann.

Ethical Criticism

Considering the sociological reports concerning how highly people esteem their televisions, Borgmann insists that the "telephone and television are the technological devices that have weakened literacy and impoverished the culture of the world" (1995, p. 90). Writing letters, telling stories, engaging in conversations, attending plays, reading to one another, and silently reading books and periodicals to oneself have all taken a backseat to watching television. Television routinely provides an alienating experience that disengages subjects from one another and inhibits genuine intersubjective connection by promoting self-oriented comportment. Whereas the scattered family once gathered around the "culture of the table," today TV dinners dominate. Not only is food reduced to a meal to be grabbed, but the festive and conversational context of dining—a focal practice—is lost. Seduced by the soothing presence of the television, people have come to experience engagement with others and with nature as exertion, as a cruel and unjust demand. When their favorite show is on, they do not want anyone to interrupt and pull them away from their passive contentment.

Borgmann grounds his negative assessment of television in an ontological distinction between two kinds of reality: *disposable devices* and *commanding things*. Disposable devices are readily available commodities that make technologically mediated experiences instantly available without the use of much skill. Indeed, learning to watch television requires little effort; young children ascertain how to do it, often without any formal instruction. Disposable devices thus belong to a world of pliable material; their emotional and moral significance is subjective and flexible. Their use, as Borgmann takes the example of television to illustrate, encourages a shallow life of distraction and isolation.

By contrast, commanding things are focal objects that express meaning on the basis of their own intrinsic qualities; the emotional and moral significance that people invest in them is largely based on the sense-bestowing capacity of the objects themselves. Commanding things direct one's attention because they require skill to use and we treat people who can adroitly operate them with respect. Whereas one does not value someone because they know how to operate a television, one admires musicians whose disciplined training allows them to create beautiful, memorable music. Furthermore, in contrast to the withdrawn and individualist behavior that disposable devices such as television encourage, commanding things further the end of communal engagement. One of the reasons why a person learns to use an instrument is to be able to extend the range of communication, to be expressive to others through the sounds that the instrument makes possible.

Assessment

Borgmann's criticisms, along with many others, have themselves been criticized as failures to appreciate the potential for enriching one's world through multivalent monitoring. From aesthetic installations of multiple television monitors to sports bars and space probe transmissions, television has the power to extend the human sensorium in ways not unlike the telescope and microscope. The ultimate promise of television may not be its utility to preexisting cultural ideals (such as democracy) but its performative presentation of scientific experience in ways that cannot help but insinuate science and technology ever more deeply into culture. To the extent to which science and technology may themselves be viewed as morally worthy projects, so too may television be viewed throughout its increasingly information-rich manifestations.

EVAN M. SELINGER

SEE ALSO *Advertising, Marketing, and Public Relations; Communications Ethics; Communication Systems; Entertainment; Globalism and Globalization; Popular Culture; Violence.*

BIBLIOGRAPHY

Baudrillard, Jean. (1995). *The Gulf War Did Not Take Place*, trans. Paul Patton. Bloomington: Indiana University Press.

Borgmann, Albert. (1984). *Technology and the Character of Contemporary Life: A Philosophical Inquiry.* Chicago: University of Chicago Press.

Borgmann, Albert. (1995). "The Moral Significance of the Material Culture." In *Technology and the Politics of Knowledge,* ed. Andrew Feenberg and Alastair Hannay. Bloomington: Indiana University Press.

Cavell, Stanley. (1984). "The Fact of Television." In his *Themes Out of School: Effects and Causes.* San Francisco: North Point Press.

Fry, Tony, ed. (1993). *RUA TV? Heidegger and the Televisual.* Sydney, Australia: Power Publications.

Heidegger, Martin. (1966). *Discourse on Thinking.* New York: Harper and Row.

Ihde, Don. (1995). "Image Technologies and Traditional Culture." In *Technology and the Politics of Knowledge,* ed. Andrew Feenberg and Alastair Hannay. Bloomington: Indiana University Press.

Mander, Jerry. (1978). *Four Arguments for the Elimination of Television.* New York: Morrow.

McLuhan, Marshall. (1964). *Understanding Media: The Extensions of Man.* New York: McGraw-Hill.

Mellencamp, Patricia, ed. (1990). *Logics of Television: Essays in Cultural Criticism.* Bloomington: Indiana University Press.

Meyrowitz, Joshua. (1985). *No Sense of Place: The Impact of Electronic Media on Social Behavior.* New York: Oxford University Press.

Newcomb, Horace, ed. (1976). *Television: The Critical View.* New York: Oxford University Press. A collection representative of early-generation TV criticism.

Postman, Neil. (1985). *Amusing Ourselves to Death: Public Discourse in the Age of Show Business.* New York: Viking.

Schwartz, Tony. (1981). *Media: The Second God.* New York: Random House.

Spigel, Lynn. (1992). *Make Room for TV: Television and the Family Ideal in Postwar America.* Chicago: University of Chicago Press. Cultural history.

Tichi, Cecelia. (1991). *Electronic Hearth: Creating an American Television Culture.* New York: Oxford University Press. Although the focus is on content, also influenced by McLuhan.

Ulmer, Gregory L. (1989). *Teletheory: Grammatology in the Age of Video.* New York: Routledge; 2nd edition, New York: Atropos Press, 2004.

Virilio, Paul. (2000). *Strategy of Deception*, trans. Chris Turner. London: Verso.

Winston, Brian. (1998). *Media Technology and Society: A History; From the Telegraph to the Internet.* London: Routledge. This is a "reworking and updating" of his *Misunderstanding Media* (1986).

Edward Teller, 1908–2003. The Hungarian-American physicist—sometimes called the "father" or the "architect" of the hydrogen bomb—was for decades on the forefront of the nuclear question and in the 1980s was an advocate of the Strategic Defense Initiative (SDI), also known as "Star Wars." (*The Library of Congress.*)

TELLER, EDWARD

• • •

Edward Teller (1908–2003) was born in Budapest, Hungary on January 15, emigrated to the United States in 1939, and became known publicly as the "father of the hydrogen bomb." From the late 1940s until his death, he defended the U.S. development of nuclear weapons and the ethics of nuclear deterrence; as a public policy adviser he argued for the peaceful use of nuclear power and advocated national missile defense. He died in Palo Alto, California (September 9).

Education and Hydrogen Bomb Development

Teller worked with many of the early physics greats in Europe between the two world wars, distinguishing himself first in atomic and molecular physics (the Inglis-Teller and the Jahn-Teller effects), and then in nuclear physics. After serving at several universities, he eventually established permanent residence at the Lawrence-Livermore National Laboratory, of which he was one of the principal founders. (Livermore was originally dedicated to military research and development, although its work is now more general.) Teller also served as a senior researcher at Los Alamos during World War II,

although his efforts were directed more toward development of fusion (hydrogen) bombs rather than fission (uranium and plutonium) devices, which were the highest priority.

In the early postwar years Teller became a principal advocate for the development of the hydrogen bomb by the United States, on the basis of strong belief in the deterrence concept, and distinctly conservative political views, which made him unpopular among many physicists. A centerpiece of his political ideology lay with his extremely strong antipathy to Communism. It was his fear that the Soviet Union would develop fusion weapons first and then use them to blackmail North American and Western European countries, especially the United States, that drove him into advocating their development. Along with Stanislaw Ulam (1909–1984), he is credited with coming up with the scheme that led to successful development of the H-bomb.

Teller's advocacy of the H-bomb placed him in direct disagreement, even confrontation, with many of the leading weapons scientists, most notably J. Robert Oppenheimer, who had been the scientific director at Los Alamos. The confrontation reached its climax during security hearings for Oppenheimer in Washington, DC, in 1954. Whereas most of Oppenheimer's contemporaries acted as friendly and supporting witnesses, Teller was a notable exception. He did not state categorically that he was in favor of denying Oppenheimer clearance, but he did say that he would be uncomfortable having Oppenheimer privy to important weaponry secrets. Partly as a result of Teller's testimony Oppenheimer was denied clearance. This act led to what amounted to a permanent ostracization of Teller by the mainstream U.S. physics community, although he remained friendly with a number of important, loyal friends, including Hungarian colleagues.

Later Work and Assessment

Teller was an innovative, energetic, talented individual, well liked on a personal level by most who knew him. He was the source of innumerable ideas concerning both military and peaceful uses of atomic energy, though many of these turned out to be impractical. He was a strong advocate of the deterrence concept and a principal spokesperson for the concept of strategic missile defense, although his advocacy was diluted by his unwarranted claims concerning its effectiveness. He was a leader in "Project Plowshare" during the late 1950s and 1960s, whose goal was to utilize nuclear explosions for peaceful purposes. For example, he proposed creating artificial harbors and canals by this means, which he termed "geological engineering." None of these schemes was realized, and the idea eventually died.

Despite the contrary opinions of many distinguished scientists, including Albert Einstein as well as Oppenheimer, there appears to be little if any doubt that the Soviet Union would certainly have proceeded to build its own hydrogen weapons. Without U.S. equivalency, the twenty-first century world would likely be very different. In hindsight Teller's strong advocacy seems to have been warranted.

BENJAMIN BEDERSON

SEE ALSO *Atomic Bomb; Missile Defense Systems; Nuclear Ethics.*

BIBLIOGRAPHY

Blumberg, Stanley A., and Gwinn Owens. (1976). *Energy and Conflict: The Life and Times of Edward Teller.* New York: G. P. Putman's Sons.

Rhodes, Richard. (1995). *Dark Sun: The Making of the Hydrogen Bomb.* New York: Touchstone. Good general history.

Sanders, Ralph. (1962). *Project Plowshare: The Development of Peaceful Uses of Nuclear Explosions.* Washington, DC: Public Affairs Press.

Teller, Edward. (1947). "The Two Responsibilities of Scientists," *Bulletin of the Atomic Scientists* 1(December).

Teller, Edward. (1987). *Better a Shield than a Sword.* New York: Free Press. An argument for nuclear missile defense.

Teller, Edward. (2001). *Memoirs: A Twentieth Century Journey in Science and Politics.* Cambridge, MA: Perseus Publishing. For a very different take on Dr. Teller see the review of this book by Richard Rhodes in the *New York Times*, November 25, 2001.

TERRORISM

• • •

Terrorism was first used to define a systematic policy of violence during the French Revolution and has since undergone important transformations that have been topics of both scientific investigation and efforts at technological control. What is now called terrorism is an old practice that has acquired new dimensions as a result of science and technology in at least three respects: rationale, publicity, and weapons (and other means). Any adequate ethical or policy assessment of terrorism requires consideration of all three aspects of the problem.

Historical Aspects

Terrorism is an ill-defined but ethically charged term, which generally refers to the highly public, calculated use of violence, destruction, or intimidation to gain political, religious, or personal objectives. Yet in this sense many wars and even some police actions might be described as terrorist insofar as they seek to induce or exploit fear. Some observers also argue that there is little principled difference between official U.S. definitions of terror and counterinsurgency measures described in U.S. armed forces manuals (Atran 2003).

From certain Roman emperors to the Spanish Inquisition (beginning in the fifteenth century) and the French Revolution's Reign of Terror (1793–1794), early forms of terrorism were primarily conducted by the state or other parties with high political power such as the Catholic Church. The nineteenth century, however, witnessed the development of complementary efforts by individuals or small groups such as the small band of Russian revolutionaries known as Narodnaya Volya (People's Will) who grew impatient with the slow pace of tsarist reforms. Members of this group are among the few to refer to themselves as terrorists, and, aided by the development of powerful and affordable explosives, they assassinated Tsar Alexander II in 1881. The Fenian Brotherhood, an Irish-American group, planted explosives around London in the mid-1800s to protest the British occupation of Ireland, thus demonstrating one of the main objectives of many terrorist organizations, namely, to attempt to reacquire territory that they feel is legitimately theirs. On June 28, 1914, Gavrilo Princip, a member of the Serbian nationalist terrorist organization called the Black Hand, assassinated Archduke Francis Ferdinand of the Austro-Hungarian Empire, thus triggering the social and political upheavals of World War I.

World War II witnessed the uses of state terrorism by both the Allied and Axis powers. After the war, terrorism continued to broaden beyond the assassination of political leaders. Terrorist movements developed in certain European colonies to both pressure colonial powers and intimidate indigenous populations into supporting a particular group. After colonialism had waned in the 1950s and 1960s, terrorism continued in several areas and for a variety of purposes. These attacks often targeted civilians, as in the case of the murder of eleven Israeli athletes at the Olympic Games in Munich in 1972.

Although suicide terrorism has deep historical roots (Atran 2003), it has played a major role in Middle East politics since the early 1980s. Since at least 1993, suicide attacks by groups such as the Islamic Resistance Movement (Hamas) have continually thwarted peace efforts between Israel and Palestine. Although Islamic religious extremism is involved in many of these terrorist attacks, it should be noted that other religious groups have committed acts of terror. The same holds true for secular groups, such as the Liberation Tigers of Tamil Eelam in Sri Lanka.

In the 1990s, Osama bin Laden, a member of a wealthy Saudi family, rose to prominence as the leader of al-Qaeda (the Base), an Islamist terrorist organization. Determined to resist Western influence in Muslim countries, members of this group killed hundreds in bombings of U.S. embassies in Africa in 1998. Al-Qaeda members have been able to create a complex, networked organization capable of transcending national borders. Such capabilities allowed them to hijack commercial airplanes and crash them into the World Trade Center towers in New York City and the Pentagon in Washington, DC, on September 11, 2001. Passengers onboard a fourth plane forced it to crash in a Pennsylvania field.

These attacks caused approximately 3,000 deaths and extensive social, psychological, and economic damage, and set off major political changes around the world, much of which bears on the use of science and technology both as potential security threats and as sources of counterterrorist measures.

Rationales

The justifications that terrorists give of their actions are perhaps even more difficult to consider than the definition of the actions themselves. It is easier—and initially appears more accurate—to describe terrorists as cowards or insane. But such a reaction runs the danger of misconstruing the phenomena and feeding into counterproductive responses.

Works by al-Qaeda and Theodore Kaczynski (the Unabomber) suggest that a major underlying rationale for some contemporary forms of terrorism is a condemnation of the dangers and depravity of modernity, including liberalism, capitalism, and a technological materialism divorced from spiritual or ethical guidance. Paul Berman (2003) traces much of the ideological impetus of al-Qaeda back to Egyptian Islamic fundamentalist groups and their "intellectual hero," Sayyid Qutb (1906–1966), who presented an extended critique of the modern world and the tyranny that technology holds over life. Qutb traced the source of error back to a split between the spiritual and material realms, which put humans out of touch with their own nature. He did not lament science but did decry the alienating effects of scientific "progress" (and the attendant consumerism) divorced from spirituality. The split between the secular and the sacred, he argued, was the fatal error that rendered the modern world inhospitable to a meaningful human existence and relationship with God.

Qutb's cultural critique also offered a revolutionary program to save humankind by calling for a small vanguard to establish sharia, the religious law of Islam, for all of society. Competing interpretations of the Koran and the meaning of Islam have created conflicts along the spectrum of liberal and extremist Muslims. For Berman Islamic terrorists are heirs to modern European fascism, with their ideals of submission, absolutism, and "the one instead of the many." William A. Galston (2003) suggests that such an interpretation erases key distinctions such as that between the meaningless self-annihilation of nihilists and the politically motivated acts of suicide terrorists. Furthermore, the thesis of liberalism versus totalitarianism reinforces the belief that terrorists "hate us for what we are, not what we do," which curtails critical scrutiny of policy decisions.

Kaczynski developed a related rationale in his manifesto, *Industrial Society and Its Future* (published by the *New York Times* and *Washington Post* in 1995). Whereas Qutb placed the problems of modernity in religious history and sought solutions in religious texts, Kaczynski appealed to human evolutionary history to explain modern social and psychological problems and relied on Western philosophers to buttress his critique. Nonetheless, both provided similar justifications for taking radical steps to undermine modern techno-industrial society. Alston Chase (2003) argues that (just as with Qutb) Kaczynski's writing cannot be simply dismissed as fringe lunacy or simple-minded Luddism. His ideas were shaped by real experiences as a mathematician at Harvard University in the late 1950s and early 1960s.

First, Kaczynski was subjected to dehumanizing psychology experiments at the hands of Henry A. Murray. Second, the climate of academia (and the wider culture) was saturated by the tenets of logical positivism, which held that ethical claims are meaningless, because science cannot prove them either true or false. Ethical and other values are purely matters of private emotion. As with Qutb, Kaczynski saw this separation of private (moral) and public (material) and other such fundamental schisms in modern industrial society as the root cause of unethical science and technology, vacuous consumerism, and massive human indignities and feelings of meaninglessness. Finally, Kaczynski held that science and technology had become servants of a military-industrial complex in ways that echoed the arguments of other critics such as the American mathematician Norbert Wiener (1894–1964). Such an argument justifies Kaczynski's rejection of the combatant/civilian distinction, because virtually all academic scientists and engineers could be perceived as caught up in a web of culpability. There is no doubt that acts of terror are objectionable, but this does not erase the possibility that their underlying rationale may at least be intelligible.

Although one major way to avoid considering the reasons that terrorists give for their actions is to reject terrorists themselves as irrational, another is to propose a sweeping historical thesis such as Samuel P. Huntington's "clash of civilizations" (1996). In response, Amartya Sen (2002) has argued that the "clash thesis" dangerously oversimplifies the heterogeneity of motives and objectives behind terrorist acts by reducing complex people and organizations to one dimension. Huntington's thesis paints a patina of coherence over the messy reality—that rationales for terrorism are diverse, complex, changing, and poorly understood. Context matters and terrorism cannot be reduced to a single "root cause" such as poverty, political conflict, or the intrusion of Western values on other cultures.

As an alternative to reliance on a large-scale historical thesis, it would perhaps be useful to undertake more detailed psychological and social scientific studies of terrorists and terrorist organizations. According to Scott Atran (2003), for instance, suicide terrorists have no appreciable psychopathology and are at least as educated and economically well-off as their surrounding populations, although there is a fairly strong negative correlation between civil liberties and suicide terrorism. In their studies attempting to uncover the causes of terrorism, Alan B. Krueger and Jitka Maleckova (2003) conclude that "any connection between poverty, education, and terrorism is, at best, indirect, complicated, and probably quite weak" (p. B10). They suggest that terrorism is a response to political conditions and feelings of indignity and frustration that are only weakly linked to economic circumstances. Marc Sageman (2004) similarly claims that people join terrorist organizations to escape a sense of alienation.

Atran also notes a correlation between U.S. involvement in international situations and terrorist attacks against the United States. Adolf Tobeña and Scott Atran (2004) suggest that understanding terrorists' motivations requires research both on social conditions and individual traits. Hector N. Qirko (2004) proposes a model from evolutionary psychology to explain suicide terrorism. He suggests that this non-kin altruistic behavior can be explained in terms of inclusive fitness, because institutions that train suicide terrorists essentially create "fictive kin."

Contemporary terrorists are usually young males who feel that they have no alternative path to influence and power and that their voice will otherwise be ignored. Humiliation, despair, and loss of economic or social advantage are factors that often play into motivations to join terrorist movements. In many Muslim areas, expanding youth populations cannot find opportunities because of rigidly authoritarian regimes. For many, the allure of martyrdom becomes a strong case for carrying out suicide missions. Indeed Nasra Hassan (2001) reports that there is an excess of young recruits hoping for martyrdom.

Publicity

A primary terrorist objective is the creation of fear in a targeted population in order to use the psychological impact of actual or threatened violence to effect political change. The capability to cause terror has been multiplied not just by more powerful weapons, but also by the expanded media coverage of terrorist acts made possible by innovations in communication technologies. Knowledge of terrorist acts is much more immediate, vivid, and widely disseminated than ever before.

Before the advent of mass media and modern communication technologies, acts of terror were committed in crowds in order to gain publicity. This led Brian M. Jenkins (1974) to describe "terrorism [as] theatre," which was vividly confirmed by the September 11 attacks, designed in part to provide billions of television viewers with images symbolizing the weakness of the United States. Timothy McVeigh, who bombed the Alfred P. Murrah Federal Building in Oklahoma City in 1995, chose that target for the open space surrounding it, which allowed for extensive television coverage. The Colombian leftist terrorist group known as the Revolutionary Armed Forces of Colombia (FARC) has its own radio broadcasts, and there are more than 4,000 terrorist websites (Wright 2004). Terrorists have adapted strategies with the emergence of satellite networks such as the Arabic news network Al Jazeera and the video capabilities of the Internet to expand their abilities to gain publicity.

Brigitte L. Nacos (1994) has explored the relationship between terrorism and the media, and suggested that the media unintentionally help terrorists achieve goals of publicity, recognition, instability, and respect. Focusing on the Iranian hostage crisis (1979–1981) and the downing of Pan Am Flight 103 (1988), Nacos argued that terrorists successfully manipulated the linkages between the news media, public opinion, and presidential decision-making by staging spectacles of terror. The opposite view is that media attention harms terrorist causes. Images of death and destruction focus attention not on the group's message but on its method, which can delegitimize its cause and alienate potential supporters.

What is not controversial, however, is the fact that media attention can and often has shaped the outcome of terrorist activities. It can disrupt counterterrorist operations and influence the dynamics of hostage situations. Terrorist groups increasingly target the media, which attracts attention and shapes coverage. The decision by managers of two U.S. newspapers (as urged by the Federal Bureau of Investigation) to publish the Unabomber's manifesto led to his identification and capture. Nacos argued, however, that this was a shameful act of government acquiescence to mass-media pressure, which might eventually encourage more terrorism. The mass media holds wider powers too, in the sense that its public representations partially define what counts as terrorism and what counts as legitimate acts of violence.

Such issues raise questions about the responsibility of the media in covering terrorism. Excessive coverage may further terrorist causes and encourage more attacks, but it is also true that too little coverage would not fulfill the media's goal of informing the public. One specific example of this dilemma is posed by the occasional audio and videotapes released by bin Laden. How much coverage should he be granted? An example of self-imposed limits on media coverage emerged in the aftermath of the 2003 U.S.-led invasion of Iraq, when the major media organizations declined to air images of beheadings performed by terrorists. But the explosion of media outlets, especially on the Internet, makes it easier for terrorists to publicize their message.

Media coverage of terrorism also raises the important ethical issue of tradeoffs between freedom of the press and security interests. Democratic governments must walk a fine line to find the proper balance for controlling media actions. In the 1980s the British government banned the broadcasting of statements by members of terrorist organizations and their supporters. Margaret Thatcher, the then prime minister of Britain, justified this policy by claiming that the surest way to stop terrorism was to cut off "the oxygen of publicity." Some argue that coverage of vulnerabilities in U.S. national security (e.g., the susceptibility of nuclear power plants to terrorist attacks) might also help terrorists prioritize future acts.

Finally, the publicity received by Islamic fundamentalist groups has given the impression that they commit the majority of suicide terrorist acts. But Robert A. Pape (2003), in a quantitative study of the 188 documented acts of suicide terrorism from 1980 to 2001, concluded that this impression was false. The leading instigator of suicide attacks was the Tamil Tigers in Sri Lanka, a secular Marxist-Leninist group that was responsible for seventy-five of the incidents.

Weapons and Other Means

As in many other areas of interaction between science, technology, and society, the most dramatic transformation in contemporary terrorism is new technological means. These means come in two forms: means of communication among terrorists that facilitate their planning and execution, and means in the form of weapons. The thousands of deaths resulting from the September 11 attacks signal terrorists' abilities to manipulate modern technologies to cause ever greater devastation. Contemporary terrorist attacks highlight the fact that not just the use of individual technological instruments is at stake. Developed societies' dependency on centralized, complex technological systems looms as a source of vulnerability that gives terrorists enormous power.

Lawrence Wright (2004) uses the March 2004 Madrid train bombings by al-Qaeda to detail the importance of the Internet to terrorist organizations. He argues that the Internet serves two interrelated purposes. First, it is a vehicle for strategic and tactical goals such as planning and organizing attacks, raising funds, and training recruits. The Internet and other communication technologies (e.g., cell phones and satellite phones) allow for highly coordinated international attacks. Al-Qaeda even publishes two online magazines that feature how-to articles on kidnapping and other terrorist tactics. Coded communications are used, and web sites are continually moved in order to avoid detection.

The second purpose served by the Internet is more fundamental. Muslim immigration in Europe is creating massive social and psychological disruptions. Many young Muslims have trouble adapting to their new situations and are confused about whether their adopted homelands are part of "the land of believers" or "the land of impiety." The Internet provides a virtual community and a compassionate, responsive forum that "stands in for the idea of the *ummah*, the mythologized Muslim community" (Wright 2004, p. 49). This virtual community strengthens feelings of common identity and provides mutual emotional support to combat feelings of alienation. Arabic satellite channels are being replaced by the Internet as the main conduit of information and communication among a growing global "jihadi subculture."

Marc Sageman (2004) sees further implications of the new Internet culture. Al-Qaeda, for example, is a nonhierarchical network, which increasingly uses bottom-up, self-selected recruitment strategies (rather than top-down selection) as a result of emerging Internet communities. Various levels of adherents form according to different interpretations of the ideology and purpose of al-Qaeda. Top-down control is diminished, as leaders no longer approve all attacks. After losing its Afghan sanctuary, the leadership of al-Qaeda is more reliant on such semi-independent cells in diverse regions. Sageman sees such local cells as the wave of the future, a theory supported by the Madrid bombings, which were carried out by a semi-independent cell. Because of the Internet, al-Qaeda is becoming a virtual community (not dependent on any one geographical locale) in the global space of the Internet. It is a "virtual

Islamist state that is trying to find a place for itself in the actual world" (Wright 2004, p. 53). The cohesiveness of this virtual community presents fundamental questions about its legitimate recognition in the international arena.

The use of the Internet and other communications technologies has sparked a technological arms race as government entities develop their own innovations to track and monitor terrorist activities. Government intervention such as shutting down web sites that are judged to support terrorism has sparked controversies about the proper limits to free speech (e.g., should instructions on bomb making be available online?).

Terrorists also use technologies in the form of weapons, which span the spectrum from simple to complex. Nasra Hassan (2001) explains that the materials used to build suicide bombs (nails, gunpowder, light switches, acetone, etc.) are not only readily available but so affordable that the most expensive part of some Palestinian suicide missions is the transportation to the site of the attack. Similarly, very little expertise or high-tech equipment is needed to make effective agricultural bioterrorist weapons (Wheelis, Casagrande, and Madden 2002). Timothy McVeigh used an ammonium nitrate and fuel oil (ANFO) bomb, which was composed of many simple and readily available components (e.g., fertilizer) but was most likely fairly complicated to construct. So-called dirty bombs (combinations of TNT or ANFO explosives with highly radioactive materials) are similar in that radioactive materials are relatively easy to procure (significant quantities have even been found in scrap yards), but constructing and deploying an effective dirty bomb capable of widely dispersing radiation is difficult (Levi and Kelly 2002).

Nuclear weapons are extremely difficult to build and nuclear material is rare and hard to refine, but political unrest in nations possessing them has increased fears that terrorists could acquire existing nuclear weapons. The term *loose nukes* refers to nuclear weapons, materials, or knowledge that could fall into terrorist hands. The black market in uranium and plutonium and poorly paid Russian scientists are of special concern. Al-Qaeda has repeatedly attempted to purchase highly enriched uranium, and states that sponsor terrorism continually try to build nuclear weapons. The threat of nuclear terrorism raises the old "nuclear dilemma" former U.S. President Dwight Eisenhower noted in the 1950s, namely, how to ensure atomic power is used to promote peace rather than threaten war. Fear surrounding these possibilities also spreads rumors of new weapons, such as "red mercury," which could make nuclear fusion weapons easier to build. Controversy surrounds the nature and very existence of red mercury, however (Edwards 1995).

Biological and chemical agents have also been used to kill and terrorize targeted populations. At least one British officer gave blankets used by smallpox patients to Native Americans during the French and Indian War (1754–1763), and reports exist of similar acts by land speculators and settlers. In 2001 an unidentified terrorist mailed letters laced with anthrax to U.S. senators and media icons. Five people died as a result. In the late 1980s Saddam Hussein used a combination of chemical agents including sarin, mustard gas, and possibly VX to kill as many as 5,000 and wound another 65,000 Kurds in northern Iraq.

In addition to both simple and more complex weapons, terrorists have adapted other technologies to serve as weapons. The most dramatic example is the use of commercial airplanes and skyscrapers by terrorists on September 11, 2001. It could also be argued, however, that terrorists even use television as a psychological weapon by creating images that induce fear.

Perhaps the most frightening reality raised by contemporary terrorist acts is the inherent vulnerability of complex sociotechnical systems. As Langdon Winner (2004) argues, life in modern civilization increasingly depends on large-scale, complex, geographically extended, and often centralized technological systems. The Y2K scare vividly raised the specter of vulnerability, as citizens, governments, and businesses alike realized how fragile such highly integrated and tightly coupled systems are. Examples include information and computer networks, dams and water purification systems, nuclear power plants, the energy transmission and distribution infrastructure, the communications infrastructure, chemical plants, gas pipelines, railroads, the mail system, food supply chains, huge fields of monoculture crops, and the containerized cargo system.

The human demands and material costs of policing these systems are, in the long term, unsustainable. Totalitarian societies have "hardened" their technologies to provide the necessary surveillance and protection, but this destroys civil freedom. Reliable engineering can solve only some of the problems. The only alternative left for free, democratic societies, Winner argues, is to embrace an attitude of trust. Citizens expect that key technologies will always work reliably. The relationship is reciprocal as it informs the structure and operation of technological systems themselves. The upshot is that "Many key components are built in ways that leave them open to the possibility of inadvertent or deliberate interference" (p. 156).

When this attitude of openness and trust is undermined by a sense of vulnerability and dread, rights and democratic institutions are threatened. Fears of cyber-, bio-, eco-, and other terrorist plots lead to a society that begins to treat all citizens as suspects, because anyone could potentially cause massive damage given the vulnerability of high-density populations dependent upon tightly integrated systems of all sorts.

Winner speculates that "Although seldom mentioned in the mass media, the ultimate fear driving public and private policies in the post 9/11 [era] is an awareness that seemingly secure, reliable structures of contemporary civilization are, taken together, an elaborate house of cards" (p. 167). This taps into our deepest fears about technology: that the powers we seek to control will come back to destroy us. Winner presents a suite of options based on the premise of designing technical systems that are more loosely coupled and "forgiving." Environmental design and bioregionalism provide models for shifting to locally available resources and decentralized systems.

The vulnerability of sociotechnical systems presents a curious reversal of the technological and power asymmetries in the relationship between terrorists and the groups they attack. The latter are generally regarded as privileged in terms of technology and power, whereas the former must take recourse to terrorist tactics precisely because of their position of weakness. Certainly, many of these groups are oppressed. But power in this dynamic is revealed as a two-way, nonhierarchical affair. The massive vulnerability of technological systems (and the fact that many technologies are becoming easier to manufacture on small scales partially because of the wide dissemination of knowledge) gives to individuals and small groups an inordinate amount of power to inflict damage and spread terror.

ADAM BRIGGLE
CARL MITCHAM

SEE ALSO *Aviation Regulatory Agencies; Biological Weapons; Building Destructions and Collapses; Chemical Weapons; Fire; Information Ethics; International Relations; Security; Terrorism and Science.*

BIBLIOGRAPHY

Atran, Scott. (2003). "Genesis of Suicide Terrorism." *Science* 299(5612): 1534–1539.

Berman, Paul. (2003). *Terror and Liberalism*. New York: Norton.

Chase, Alston. (2003). *Harvard and the Unabomber: The Education of an American Terrorist*. New York: Norton.

Edwards, Rob. (1995). "Cherry Red and Very Dangerous." *New Scientist* 146(1975): 4–5.

Elshtain, Jean Bethke. (2003). *Just War against Terror*. New York: Basic.

Galston, William A. (2003). "Why They Hate Us." *Commonweal* 130(11): 22–24.

Gilbert, Paul. (1994). *Terrorism, Security, and Nationality: An Introductory Study in Applied Political Philosophy*. London: Routledge.

Hassan, Nasra. (2001). "An Arsenal of Believers." *New Yorker* 77(36): 36–41.

Huntington, Samuel P. (1996). *The Clash of Civilizations and the Remaking of World Order*. New York: Simon and Schuster.

Jenkins, Brian M. (1974). *Terrorism and Kidnapping*. Santa Monica, CA: Rand.

Krueger, Alan B., and Jitka Maleckova. (2003). "Seeking the Roots of Terrorism." *Chronicle of Higher Education*, 6 June, B10–B11.

Levi, Michael A., and Henry C. Kelly. (2002). "Weapons of Mass Disruption." *Scientific American* 287(5): 76–81.

Nacos, Brigitte L. (1994). *Terrorism and the Media: From the Iran Hostage Crisis to the World Trade Center Bombing*. New York: Columbia University Press.

Pape, Robert A. (2003). "Dying to Kill Us." *New York Times*, 22 September, A17.

Qirko, Hector N. (2004). "'Fictive Kin' and Suicide Terrorism." *Science* 304(5667): 49–51.

Sageman, Marc. (2004). *Understanding Terror Networks*. Philadelphia: University of Pennsylvania Press.

Sen, Amartya. (2002). "Civilizational Imprisonments." *New Republic* 226(22): 28–33.

Tobeña, Adolf, and Scott Atran. (2004). "Individual Factors in Suicide Terrorism." *Science* 304(5667): 47–49.

Wheelis, Mark; Rocco Casagrande; and Laurence V. Madden. (2002). "Biological Attack on Agriculture: Low-Tech, High-Impact Bioterrorism." *BioScience* 52(7): 569–576.

Winner, Langdon. (2004). "Trust and Terror: The Vulnerability of Complex Socio-technical Systems." *Science as Culture* 13(2): 155–172.

Wolin, Richard. (2003). "Are Suicide Bombings Morally Defensible?" *Chronicle of Higher Education*, 24 October, B12–B14.

Wright, Lawrence. (2004). "The Terror Web." *New Yorker* 80(21): 40–53.

INTERNET RESOURCE

Kaczynski, Theodore. (1995). *Industrial Society and Its Future*. Available from http://www.unabombertrial.com/manifesto.

TERRORISM AND SCIENCE

• • •

When the U.S. Department of Homeland Security (DHS) was proposed in 2002, President George W. Bush (b. 1946) noted that "in the war against terrorism, America's vast science and technology base provides us with a key advantage." What he failed to mention is that science and technology are also major sources of vulnerability to terrorist attacks, requiring decisions about censorship of publication and restriction of access to sensitive areas and materials. Thus terrorism poses special problems for the scientific and technical community in two respects: how to limit terrorist access to sensitive knowledge and technology, and what scientific research and technological developments to pursue in the interests of countering terrorist threats. Although scientists and engineers must bring their professional ethical responsibilities to bear on both tasks, it is equally important that decision makers understand the related limitations of science and technology.

Limiting Terrorist Access

Because of their multiple use capabilities, scientific knowledge and technological devices can be used by terrorists for purposes other than those originally intended. Preventing such misuse presents policy makers and the scientific and engineering communities with two challenges. First, they must insure that knowledge and information are not inappropriately disclosed. Second, they must secure existing and proposed technologies (e.g., nuclear power plants) and research materials (e.g., pathogens). In general, policies in the first case involve restricting the availability of sensitive information by the government, scientists, or both. Actions in the second case generally involve containment, monitoring, and restriction of access. Both actions raise tensions between the goals of security and scientific freedom and openness in the creation and exchange of knowledge and products. Striking the proper balance between these competing goods has taken on heightened importance since the terrorist attacks of September 11, 2001, and the responses by the governments of the United States and other nations.

The situation is made more complex by the notion that some degree of scientific freedom is necessary for national security, because it facilitates the creation of new knowledge and artifacts that may be useful in preventing or responding to terrorist attacks. Especially in the biomedical field, circumstances are further complicated by the potential twin effects of secrecy and restricted access. In some cases, these effects may protect public health by preventing terrorist from acquiring sensitive information or dangerous pathogens. In others, they may harm public health by preventing the development of cures and vaccines or inhibiting the coordination of response efforts to disease outbreaks. In some cases, the potential benefits of researching pathogens to mitigate the effects of terrorist attacks may not be worth the risks. This has sparked controversies about the creation and siting of biosafety laboratories that handle dangerous pathogens.

The free creation and exchange of knowledge by scientists can present dangerous, unintended consequences for society. A paper by Ronald Jackson and other researchers found that the insertion of IL-4 genes into mousepox viruses resulted in near total immunosuppression (Jackson, Ramsay, Christensen, et al. 2001). This advanced valuable knowledge about immune system functioning, but it also evoked fears that terrorists could use such knowledge to engineer hyper-virulent viruses. Similarly, the journal *Science* published a paper in 2002 that showed how to assemble a poliovirus from readily available chemicals (Cello, Aniko, Eckerd 2002). The threat of terrorist acts has caused political leaders and members of the scientific community to question whether such knowledge should be created, and if so, how its publication and exchange should be regulated.

In *New Atlantis* (1627), Francis Bacon (1561–1626) imagined the self-censoring activity of scientists in recognition of the fact that politically authorizing experimental science entails societal risks. The twentieth century provided several examples of tradeoffs between security and openness in the pursuit of knowledge. The Manhattan Project that produced the first atomic bomb cultivated a culture of secrecy. A similar culture developed among researchers studying microwaves during World War II. During the Cold War, the U.S. government attempted to constrain information exchange in some areas of mathematics and the physical sciences that may have aided Soviet nuclear weapons development (Monastersky 2002). Physicist Edward Teller (1908–2003) and others eventually persuaded policy makers that openness, rather than secrecy, was the best tactic for security during the Cold War.

In 1975, an international group of scientists held the Asilomar conference to debate the proper use and regulatory oversight of recombinant DNA research. During the late 1970s, the National Security Agency (NSA) regulated cryptographers developing new algorithms, but the two groups eventually agreed to a system

of voluntary submission of papers for review. In 2002, the U.S. government began to withdraw from public release more than 6,600 technical documents dealing mainly with the production of germ and chemical weapons. In a controversial move, the U.S. national policy for the restriction of information that may threaten national security was altered in the wake of the September 11 attacks to include restrictions on publication of federally-financed research deemed to be "sensitive but not classified" (Greenberg 2002).

As these examples illustrate, limitations on research and the availability of technical knowledge can come in the form of self-imposed screening mechanisms by the scientific community or government regulation. The Asilomar conference, for example, led to a suite of self-policing mechanisms within the scientific community, including the decentralized system of Institutional Biosafety Committees (IBCs). This same mechanism has been proposed by the National Science Advisory Board for Biosecurity (NSABB) as a way to prevent the misuse of biological research by terrorists. The NSABB also works to develop codes of conduct for researchers and laboratory workers, which underscores the importance of ethical conduct by individuals, especially where no rules exist or where the precise meaning of rules is unclear. Some professional associations and journals, including *Science* and *Nature*, have instituted procedures to give special scrutiny to papers that raise security concerns (Malakoff 2003). Putting such control in the hands of journal editors has caused some to argue that an advisory group like the Recombinant DNA Advisory Committee (RAC) would be a better mechanism.

Mitchel Wallerstein (2002) points out that the dangers posed by terrorists acquiring sensitive science and technology information differ from the state-related threats that were of primary concern during World War II and the Cold War. Terrorists generally do not seek out and would not be able to use the results of most basic research, but states may possess the intellectual and financial capital necessary to turn basic research into weapons. Daniel Greenberg (2002) contends that terrorists do not rely on new science. Rather, readily accessible information that has long been available suffices to fulfill most of the goals of terrorist organizations.

Biological weaponry is the area of science that could most directly benefit terrorist organizations. Wallerstein writes, "Information that improves knowledge of dangerous pathogens, their safe handling, and their weaponization increases the likelihood that such weapons could be produced covertly on a small scale" (p. 2169). His general conclusion is that restrictions on scientific and technical communications need occur only on a much smaller scale than during the Cold War. In fact, many echo his conclusion that sensitive research is a very narrow slice of the scientific world, which allows for severe but highly targeted restrictions.

Restricting the publication of information deemed sensitive and controlling access to technologies and research materials can help achieve security goals, but not without costs (Knezo 2002a). Some impacts are relatively minor, such as new standards for the construction and management of laboratories. Other impacts are more severe, including the impact of national security policy measures on the research process. Tightened laboratory access policies, publication rules, and visa restrictions may reduce the number of applications by foreign students to U.S. universities and colleges. This could hamper cross-cultural understanding. According to State Department rules, consular officials may deny visas for study in the United States in sixteen categories specified on the Technology Alert List to students from countries listed as "state sponsors of terrorism." Additional exemptions to the Freedom of Information Act (FOIA) and the withdrawal of information from federal agency websites have also sparked concerns about constraints on legitimate scientific work and academic freedoms.

Economic losses are also a concern about some legislative responses to security risks posed by science and technology. Instituting security and tracking measures in academic laboratories entails additional costs for researchers. Restrictions on foreign researchers can damage technological developments and economic productivity. The U.S. Immigration and Customs Enforcement (ICE) agency operates "Project Shield America" to prevent the illegal export of sensitive munitions and strategic technology to terrorists. It is intended to prevent terrorism, but may also entail losses to economic competitiveness.

Science and Technology to Counter Terrorism

Since the September 11 attacks, science and technology have increasingly been advertised as ways to prevent terrorist attacks as well as reduce vulnerabilities and minimize impacts of such attacks (e.g., Colwell 2002). This is in part a response by scientists and engineers to the sizeable increases in homeland security and counterterrorism research and development (R&D).

The National Research Council's Committee on Science and Technology for Countering Terrorism issued a report in 2002 that described the ways in which science and engineering can contribute to making the nation safer against the threat of catastrophic terrorism. It outlined both short-term applications of existing technologies and long-term research needs. The report recommended actions for all phases in countering terrorist threats, which can be roughly ordered as awareness, prevention, protection, response, recovery, and attribution. Different threats pose different challenges and opportunities across these phases. For example, nuclear threats must be addressed at the earliest stages, whereas biological attacks are more difficult to preempt, but more opportunities exist for technological intervention to mitigate their effects.

Scientific research and technological innovations can improve performance of all phases, from threat analyses and vulnerability assessments to post-attack investigations and restoration of services. For example, the Bush administration established BioWatch, a nationwide system of sensors to detect the presence of certain pathogens, and a public-health surveillance system that monitors the databases of eight major cities for signs of disease outbreaks. Early warning systems can detect the presence of certain pathogens by utilizing computer chips and antibodies or pieces of DNA (Casagranda 2002). Explosives-detection technologies have also been spurred since September 11, 2001 in order to bolster airline security.

Other examples include the use of biometrics (e.g., fingerprints and retinal signatures) to develop national security identity cards. The shipping industry is slowly adopting new security measures such as sophisticated seals and chemical sensors. Other researchers are developing strategies for securing information systems. Military infrared countermeasures for surface-to-air missiles may be used on civilian aircraft. Technologies for decontamination, blast-resistant walls, and protective gear for first responders are other components of research programs. Increasing flexibility and innovating measures to isolate failing elements could increase security of more complex technical systems such as transportation and communication infrastructures. Researching and developing broader applications of renewable energy can harden the energy infrastructure. Social scientists and psychologists also provide research for understanding causes and motivations of terrorists as well as the dynamics of terrorist group formation. Some (e.g., Susser, Herman, Aaron 2002) have demonstrated that, because terrorists choose targets to maximize psychological impact, mental health must be considered a top response priority.

With all of these potential applications of science and technology, decision makers need to address questions about how to coordinate, organize, prioritize, and evaluate investments to serve the goals of security and public health. Genevieve Knezo (2002b) reported that prior to September 11, 2001, the Government Accountability Office (GAO) and other authorities had questioned whether the U.S. government was adequately prepared to conduct and use R&D to prevent and combat terrorism. Partially in response to the need to better coordinate counterterrorism efforts (including R&D), the cabinet-level Department of Homeland Security (DHS) was created by legislative act in 2002. This incorporated half of all homeland security funding within a single agency. In addition to legislative activity, new advisory bodies such as the NSABB have been formed to guide the creation of new rules and development of new institutions to maximize the benefits of science and technology while minimizing unintended negative impacts.

Since September 11, 2001, established institutions have benefited from significantly increased funding for homeland security and public health research. For example, in 2002 President Bush proposed a 2,000 percent budget increase for the National Institute of Allergy and Infectious Diseases (NIAID) from pre-September 11 levels. Other institutions and agencies have either received additional funding (especially the National Institutes of Health) or made attempts to restructure their priorities to take advantage of shifts in R&D funding priorities (Congressional Research Service 2002; American Association for the Advancement of Science 2004).

Investments in science to reduce terrorist threats raise several ethical issues. First, the scale of vulnerabilities outstrips resources to reduce them, which raises equity issues in the process of prioritizing investments. For example, bioweapons detectors are too expensive to deploy on every street corner, so locations must be prioritized. Likewise, not all areas pose equal risks from terrorist attacks, so efforts need to be targeted to match threats.

Second, Arthur Caplan and Pamela Sankar (2002) note the increase in "research protocols that call for the deliberate exposure of human subjects to toxic and noxious agents" (p. 923). Such dilemmas are not new, as many trials on U.S. Navy and Army crew members took place in the 1960s in an effort to document the effects of biological and chemical weapons. Many research subjects were neither informed of the details of the study nor issued protective gear (Enserink 2002). Such

research needs clear guidelines and unequivocal justification for its relevance to national security. Professional ethical issues also arise when unemployed scientists and engineers face financial incentives to aid terrorist organizations (Richardson 2002).

Third, the integrated nature of socio-technical systems raises considerations of equity and civil liberties. For example, forty percent of all containerized cargo that arrives in the Long Beach harbor in Los Angeles is destined for the U.S. interior. How should the burden of increased security costs be distributed? Furthermore, the process of hardening these systems can reduce access and curtail certain civil liberties (Clarke 2005).

Finally, several analysts have criticized dominant U.S. counterterrorism science policies as ineffective. Bruce Schneier, security technologist and cryptographer, argues that managers too often seek technological cure-alls and rarely consider the consequences of system failures (Mann 2002). For example, all security systems require secrets, but they should be the components that are most easily changed in case system integrity is breached. Biometric identity devices that use fingerprints can centralize so many functions that they create "brittle" systems that fail poorly in case they are stolen. New banking account numbers can be issued in case of fraud, but not new fingerprints. Schneier contends that in airline security the only effective measures are the low-tech solution of reinforcing cockpit doors and the non-technical fact that passengers now know to fight back against hijackers. Both measures pass Kerckhoffs' principle, which occurs when a system remains safe even when almost all of its components are public knowledge. Schneier also holds that security systems are at their best when final decision-making responsibility is given to humans in close proximity to the situation, not computers. Security systems should be ductile, small-scale, and compartmentalized to mitigate the effects of inevitable failures.

Stephen Flynn (2004) focused less on the inherent limitations of technology as a means of countering terrorism; rather he critiqued government R&D prioritizations. Flynn argued that some high-tech solutions such as digital photographs of container loading processes, internal emissions sensors in cargo containers, and GPS tracking devices can improve security, but they have not been given adequate funding.

The 2002 report by the Committee on Science and Technology for Countering Terrorism openly recognizes the fact that science and technology are only one part of a broad array of strategies for reducing the threat of terrorism that includes diplomacy, cross-cultural learning, and economic, social, and military policies. Furthermore, as the U.S. experience in the Vietnam War and the Soviet experience in the 1980s invasion of Afghanistan demonstrate, technological superiority does not guarantee victory. Success in the war on terror is measured by accomplishments, not R&D budgetary numbers.

From communism to environmental problems and the challenges posed by a globalizing economy, science and technology have often been put forward as ways to protect national interests and secure prosperity (Jenkins 2002). Scientists, engineers, and politicians often define problems in ways that call for technical solutions, but they must be held accountable for such problem definitions. Scientists and engineers especially must exercise ethical responsibility by not unduly exaggerating arguments that their research will serve societal goals.

Assessment

The two sections of this entry are interrelated in that increased scientific research on counterterror measures will create new knowledge and opportunities for terrorist exploitation, which will create new challenges for securing that knowledge. Given that security, health, and civil liberties are at stake in decisions about science and terrorism, it is important that measures be taken to involve and inform citizens. This entry has focused on actions by the U.S. government because it plays a leading role in matters of science and terrorism. But other countries and international coalitions face similar ethical dilemmas and policy choices. Private companies own many of the infrastructures that are targets for terrorist attacks, so regulations may be required to induce the private sector to invest in counterterrorism technologies that may not have commercial markets. Some scientific research, however, may have viable market applications, meaning that some of the R&D burden can be privatized, which raises other ethical issues that partially mirror those involved in the privatization of war.

ADAM BRIGGLE

SEE ALSO Biological Weapons; Building Destructions and Collapses; Chemical Weapons; Fire; Information Ethics; International Relations; Security; Terrorism.

BIBLIOGRAPHY

Caplan, Arthur L., and Pamela Sankar. (2002). "Human Subjects in Weapons Research." *Science* 298(5595): 923.

Casagrande, Rocco. (2002). "Technology against Terror." *Scientific American* 287(3): 83–87.

Cello, Jeronimo, Paul V. Aniko, and Wimmer Eckard. (2002). "Chemical Synthesis of Poliovirus cDNA: Generations of Infectious Virus in the Absence of Natural Template." *Science* 297(5583): 1016–1018.

Clarke, Richard A. (2005). "Ten Years Later." *Atlantic Monthly* 295(1): 61–77. Imagines future scenarios of terrorist attacks and government reactions.

Enserink, Martin. (2002). "Secret Weapons Tests' Details Revealed." *Science* 298(5593): 513–514.

Committee on Science and Technology for Countering Terrorism, National Research Council. (2002). *Making the Nation Safer: The Role of Science and Technology in Countering Terrorism*. Washington, DC: National Academy Press.

Flynn, Stephen. (2004). *America the Vulnerable: How Our Government Is Failing to Protect Us from Terrorism*. New York: HarperCollins.

Greenberg, Daniel S. (2002). "Self-Restraint by Scientists Can Avert Federal Intrusion." *Chronicle of Higher Education* October 11: B20.

Jackson, Ronald J., Alistair J. Ramsay, Carina D. Christensen, et al. (2001). "Expression of Mouse Interleukin-4 by a Recombinant Ectromelia Virus Suppresses Cytolytic Lymphocyte Responses and Overcomes Genetic Resistance to Mousepox." *Journal of Virology* 75(3): 1205–1210.

Jenkins, Dominick. (2002). *The Final Frontier: America, Science, and Terror*. London: Verso. Presents the argument that changes in technology have created a situation where all citizens are vulnerable to catastrophic terrorist attacks.

Malakoff, David. (2003). "Researchers Urged to Self-Censor Sensitive Data." *Science* 299(5605): 321.

Mann, Charles C. (2002). "Homeland Insecurity." *Atlantic Monthly* 290(2): 81–102.

Monastersky, Richard. (2002). "Publish and Perish?" *Chronicle of Higher Education* October 11: A16–A19. Focuses on the dilemma of scientific openness and national security.

Richardson, Jacques G. (2002). *War, Science and Terrorism: From Laboratory to Open Conflict*. London: Frank Cass. Sees the connection between science and terrorism largely as an outgrowth from the partnership between science and the military and asks to what extent science is promoted by actual or threatened armed conflict and whether war is an extension of science by other means.

Susser, Ezra S., Daniel B. Herman, and Barbara Aaron. (2002). "Combating the Terror of Terrorism." *Scientific American* 287(2): 70–77.

Wallerstein, Mitchel B. (2002). "Science in an Age of Terrorism." *Science* 297(5590): 2169.

INTERNET RESOURCES

American Association for the Advancement of Science. (2004). "Defense and Homeland Security R&D Hit New Highs in 2005; Growth Slows for Other Agencies." Available from http://www.aaas.org. A summary of AAAS estimates and analyses of appropriations in the FY 2005 omnibus bill and final FY 2005 appropriations bills for federal R&D.

Colwell, Rita R. (2002). "Science as Patriotism." Available from http://www.sciencecoalition.org/presskit/articles/puboped/colwell_jan302002.pdf. Paper presented at the Annual Meeting of the Universities Research Association, Washington DC.

Congressional Research Service. (2002). "Science and Technology Policy: Issues for the 107th Congress, 2nd Session." Available from http://www.ncseonline.org.

Knezo, Genevieve. (2002a). "Possible Impacts of Major Counter Terrorism Security Actions on Research, Development, and Higher Education." CRS Report for Congress. Available from http://www.fas.org.

Knezo, Genevieve. (2002b). *Federal Research and Development for Counter Terrorism: Organization, Funding, and Options* CRS Report RL31202. Abstract available from http://www.pennyhill.com/index.php?lastcat= 200&catname= Terrorism&viewdoc=RL31202.

THEODICY

• • •

Theodicy is a concept developed by Gottfried Wilhelm Leibniz (1646–1716) to justify the existence and absolute perfection of God despite the evil that exists in the world. The term appeared in 1710 in the title of Leibniz's work *Theodicy—Essays on the Goodness of God, of the Freedom of Man, and the Origin of Evil,* and with it he coined an optimistic variant *par excellence* on theories of evil. Insofar as science and technology are often interpreted as responses to evil, theodicy is related to their modern emergence.

Background and Emergence

Theories of evil have been developed by Plotinus (204–270), Augustine (354–430), and others in which evil is seen as necessary for universal harmony. Within the framework of the complex theological discussions on the origin of evil, Leibniz's theodicy denies both the idea of God as a malevolent creator of the world (a position taken by certain Gnostics) and the refutation of this theory by Origen (c. 185–254) and Augustine who, in postulating human freedom, attributed moral responsibility for all the evils of the world to human beings, in the form of sin.

Leibniz's particular approach was to interpret perfection as the state of a thing when it attains its highest level of being. This definition highlights God's perfection. From the quantitative point of view, God has all perfections; from the qualitative point of view, these perfections reach their highest form in him. God is therefore omniscient and omnipotent. Despite the impressions that evil, injustice, and suffering give us of the world, God's perfection is necessarily expressed in his creation.

This theory is, paradoxically, a key philosophical element of transition to modernity, a vital bridge to the new philosophies that emerged in the second half of the eighteenth century: the philosophy of history, philosophical anthropology, and aesthetic philosophy. The advance of these philosophies is tied to a new understanding of human nature that rejects the naturalism of seventeenth century thought, as well as traditional Christian theology. All the images of the human that developed in the eighteenth century were optimistic in ways reflecting theodicy—as can be illustrated in moral humanity (Anthony Ashley Cooper Shaftesbury [1671–1713]), rational humanity (Jean-Jacques Rousseau [1712–1778], Immanuel Kant [1724–1804]), economic humanity (Adam Smith [1723–1790]), and perfectible humanity (Condorcet [1743–1794]).

Although the idea of a human fall did not immediately disappear, a new concept began to replace it—not exactly of human greatness, but of the ability of humans to do what was necessary to make the world better for the human species. To understand this situation is to recognize the significance of Leibnizian theodicy for modern science and technology, as well as for ethics in the era of modernity. Leibniz's theodicy was both necessary for and representative of the modern world, insofar as it gave expression to a vision of the human condition as one which, aided by science and technology, was no longer characterized by powerlessness, suffering, and evil. These were henceforth looked at outside Leibniz's own metaphysical framework as being essentially surmountable.

Collapse and Continuity

With the Lisbon earthquake of 1755, Leibniz's justification of God in the face of worldly evil collapsed, in a complex historical context where science began progressively to replace religion as the cultural frame of reference. Nevertheless, the semantic core of Leibniz's arguments, that to compensate for evil is in fact the purpose the divine creator had before him, held firm. As Odo Marquard (1989, pp. 38–63) argued, Leibniz provided the teleological framework in which science and technology could become both means and ends. In Leibniz's theology that basic principle is "*malum* through *bonum*": God does not make up for evil with good, but evil is rehabilitated by the good it pursues. Tolerance in the face of evil is justified by having the highest good as the end in view, insofar as evil is the condition that makes the good possible.

In this sense, the principle of theodicy is that the ends justify the means. With the collapse of Leibnizian theodicy in its original form, human beings take the place left vacant by the omnipotent creative will and theodicy is transformed into anthropodicy or human progress. Humanity as an end in itself is free to use everything else as mere means, inheriting God's role in order to realize and complete theodicy in history. Every goal achieved became a new means toward another end.

As a result of this teleological sequence of means and ends, what came to predominate was not the possible uses of the means, but the very means themselves. The ends no longer justified the means, the means justified the ends. This logic is linked to the cost/benefit compensation criterion of utilitarianism: Every good has its price. As Thomas Robert Malthus (1766–1834) wrote in his *Essay on the Principle of Population* (1798): "There is evil in the world, not in order to produce despair, but rather activity." This idea is equally present in other modern thinkers such as Bernard Mandeville (1670–1733): "There are 'private vices' [*malum*], but they are 'public benefits' [*bonum*-through-*malum*]."

The Example of Cournot and Teilhard

Among those who developed philosophies of history guided by an optimistic approach or who believed in humanity's ascending progress to an ideal state were the Frenchmen Antoine-Augustin Cournot (1801–1877), a teacher of mathematics and author of several works on the philosophy of history, and Pierre Teilhard de Chardin (1881–1955), a Jesuit priest, paleontologist, and philosopher of nature. Though sometimes neglected, these two thinkers developed unusual and powerful syntheses that reflect the subtle and penetrating influence wielded by the Leibnizian idea of an omnipotent creative will. Their work had significant repercussions during their own lifetimes, and their theoretical constructs are still surprisingly topical in the twenty-first century: Cournot as a prophet of post-historical technological civilization, Teilhard as the prophet of transhumanism.

For the century in which he lived, Cournot was the thinker who developed with the greatest persistence a philosophy of history in which science and technology take pride of place. His philosophy of history is based on a series of binary opposites: chance and necessity, reason and instinct, passions and interests. With these concepts, his reading of history was finalistic, and he argued for the likelihood or even the inevitability of what has come to be called "the end of history," a partly Hegelian premise that was revived at the end of the twentieth century in a world that claimed the end of ideology, of utopia, of politics, of the human. Hermínio Martins (1998), who has emphasized the importance of Cournot for the philosophy of technology, argues that Cournot's "end of history" semantics do not imply a form of necessitarianism, in the sense of extinction or termination, but more correctly exhaustion, completion, fulfillment, or consummation.

Cournot's temporal interpretation of collective human existence is based on a system of three great time-phases, as found in the work of Auguste Comte (1798–1857) and Karl Marx (1818–1883), and closely related to different kinds of discourse. The first phase has been labeled "ethnological" and is characterized by the subordination of reason to instinct, of the planful to the unreflective; habit and custom predominate, and are accompanied by natural or human disasters. The second stage is the phase of history itself. This is defined by an increase in rationality in thought and action, and by a combination of passions and interests as the springs of action with sufficient power to give rise to colossal events, of which the French Revolution is an example. The third and terminal phase is the closest possible approximation to the ideal, which humanity will never be able to attain. In this phase, "political faiths" decline, as occurred during the French Revolution, and give way to the peaceable play of economic interest and the *doux commerce*.

This third stage establishes a post-historic society that conquers nature by systematic scientific discovery, technological invention, innovation, and economic growth. Cournot anticipates positions that were further developed in the twentieth century, such as Joseph Schumpeter's routinization of economic innovation and what Alfred North Whitehead calls the "invention of invention," but does not show any significant concern with the possible intrinsic limits of scientific progress, which might bar further fundamental technological advance.

Teilhard's approach to human history also embodies finalism, and the role of scientific and technological advance within it, although his vision embraces different domains from those of Cournot. Teilhard's arguments have roots in the philosophy of Henri Bergson (1859–1941), and are part of the new theology of history that seeks to protect theology from the temptation of rationalist hermeneutics. Nonetheless, it did not shy away from dealing with "earthly realities," such as the relationship between humans and nature, the carnal nature of human beings, scientific humanism, and the theology of science. Teilhard's thinking embodied these contributions, and added a lively intuition of the evolutionist and voluntarist scientific and technological type that aroused serious suspicions in Rome. Contravening some basic postulates of Christianity, he argued for the "spiritual value of matter," and developed a conception in which humankind, with its artistic achievements, technological artifacts, and religions, is part of an overall evolutionary scheme in which there exists a progressive manifestation of biochemical complexity on the path to a growing unified consciousness.

In the tradition of the omnipotent creative will, Teilhard argued that perfection lies in the progress not of individuals, but of humanity as a whole, on a path toward unification with God who, being in essence supernatural, is at the same time the natural outcome of evolution. In his main work, *The Phenomenon of Man* (1959), he develops a suggestive synthesis of science and religion, in the context of a view of the universe as a system that develops from one phase to another with ever-higher forms of consciousness.

Teilhard's speculations anticipated those who favor a transhuman future which appears possible and desirable. These transhumanists are convinced that the new computational technologies are creating a collective human intellect, a kind of cognitive and mental hyperextension of the human mind. Cournot, by contrast, thought that organic life would remain fundamentally inaccessible to mathematical and experimental science, while postulating that increasing knowledge of inanimate nature would be sufficient to ensure technical perfectibility and material progress.

JOSÉ LUÍS GARCIA

SEE ALSO *Leibniz, G. W.; Progress.*

BIBLIOGRAPHY

Anderson, Perry. (1992). *A Zone of Engagement.* London: Verso.

Cournot, Antoine-Augustin. (1861). *Traité de l'Enchaînement des Idées Fondamentales dans les Sciences et dans l'Histoire* [Treatise on the Chain of Fundamental Ideas in Science and History]. Paris: Hachette.

Cournot, Antoine-Augustin. (1875). *Matérialisme, Vitalisme, Rationalisme* [Materialism, Vitalism, Rationalism]. Paris: Hachette.

Leibniz, Gottfried Wilhelm. (1952 [1710]). *Theodicy—Essays on the Goodness of God, of the Freedom of Man, and the Origin of Evil*, ed. Austin Farrer, trans. E.M. Huggard. London: Routledge & K. Paul.

Marquard, Odo. (1989). *Farewell to Matters of Principle: Philosophical Studies*, trans. Robert M. Wallace with the assis-

tance of Susan Bernstein and James I. Porter. New York: Oxford University Press.

Martins, Hermínio. (1998). "Technology, Modernity, Politics." In *The Politics of Postmodernity*, ed. James Good and Irving Velody. Cambridge, UK: Cambridge University Press.

Ruyer, Raymond. (1930). *L'Humanité de l'Avenir d'après Cournot* [The Humanity of the Future according to Cournot]. Paris: Alcan.

Teilhard de Chardin, Pierre. (1959). *The Phenomenon of Man*, trans. Bernard Wall. New York: Harper.

THERAPY AND ENHANCEMENT

• • •

It is common, in classifying interventions, to sort them into those that are therapeutic, that is, directed at diminishing the harms suffered by a patient, and those that are enhancing, that is, directed at increasing the goods experienced by a patient. At least three independent but related questions can be raised about the therapy/enhancement distinction: (1) Can the two terms *therapy* and *enhancement* be defined clearly, reliably, and accurately? (2) Assuming they can be satisfactorily defined, under what circumstances is it morally justified for a physician to engage in either activity? (3) Assuming they can be satisfactorily defined, what implications does labeling an intervention as therapeutic or enhancing have on the issue of whether the cost of the intervention should be borne in part or in whole by third-party funding agencies?

Defining Therapy and Enhancement

The distinction between therapy and enhancement can be most clearly made by first having available a clear definition of a third term: *malady*. The following definition of a malady, adapted from Gert, Culver, and Clouser 1997 (p. 104) classifies all clear cases of maladies as maladies and does not classify as a malady any condition that is clearly not a malady.

> An individual has a malady if and only if (s)he has a condition that is not normal for a person in his (her) prime, other than his (her) rational beliefs or desires, such that (s)he is suffering, or is at a significantly increased risk of suffering, a nontrivial harm or evil (death, pain, disability, loss of freedom, or loss of pleasure) in the absence of a distinct sustaining cause.

Therapies are interventions whose intention is to reduce or eliminate the harms that are a defining characteristic of maladies. If an intervention is not directed toward reducing or eliminating the harms associated with a malady, then it is not a therapy. Enhancements are interventions directed toward increasing the personal goods experienced by another person, such as abilities (including knowledge), freedom, and pleasure. If an intervention is not directed toward increasing another's personal goods, then it is not an enhancement. These definitions seem to correctly classify all cases of therapies and enhancements.

An extensive project ("The Enhancement Project") sponsored by the Hastings Center concluded that the two terms could not be defined clearly and could thus serve only as "conversation starters." In the words of the project coordinator, "Like many distinctions, the treatment/enhancement distinction is permeable, unstable, and can be used for pernicious purposes" (Parens 1998, p. 25). In contrast, the present authors think the two terms can be defined clearly and that one advantage of clear definitions is that they decrease the likelihood of any pernicious applications of the terms defined.

There are inevitable borderline cases. For example, how should one classify the administration of growth hormone to a child destined to be very short but who shows no evidence of an endocrinopathy? There is disagreement about this question, but not because of any avoidable vagueness in the definitions given here. Instead the disagreement is about whether this condition is a malady. If it is not, then administering growth hormone is not a therapy; if it is, then it is a therapy. Both Eric T. Juengst (1998) and Norman Daniels (1994) also use the concept of malady in distinguishing between therapies and enhancements, although Daniels's definition of malady differs from the one given here.

The Moral Justifiability of Administering Therapies and Enhancements

There is a general consensus that it is ethically justified to administer interventions when certain conditions are met. First, the intervention must be a rational one for the patient to choose under his or her circumstances. Second, patients must give valid consent to an intervention: They must be given adequate information about the intervention, must not be coerced into consenting, and must be fully competent to consent. If these conditions are met, then it is ethically justified to administer an intervention. If one of them is not met, then it may or may not be ethically justified to administer the intervention.

If an intervention can be accurately predicted to cause only an increase in the personal goods experienced by an individual, and the individual gives a valid consent to the intervention, then there is nothing morally problematic about administering the enhancement. What often makes enhancements problematic is that there is uncertainty about whether there might be significant harms that will, sooner or later, accompany the enhancement. Breast augmentation surgery may result in abscesses or in later disfiguring and irreversible structural lesions. Exogenous growth hormone administration might result in later endocrinopathies or even tumors. Mood-altering drugs might result in short-term tranquility or euphoria but long-term deleterious psychic (or neurochemical) effects. Even in cases of enhancements with possible risks, unless it would be irrational for the adequately informed competent patient to choose to have the enhancement, it seems morally justified to administer the enhancement if the patient has validly consented to it.

One moral problem that arises concerning enhancements is not the moral acceptability of enhancing with valid consent, but whether the resources spent developing enhancements detract from the resources that are available for therapy. Except when the harms suffered are trivial and the goods involved are extraordinary, it is almost universally acknowledged that it is more important to prevent or relieve harms than to promote goods. Thus if the enhancements that are developed and marketed decrease the resources that are available for therapy, then it might be argued that it is not morally acceptable to develop and market such enhancements. It is very doubtful, however, that preventing the development of enhancements would increase the resources used for therapy, so that it is not clear how much force this argument would have.

Another moral problem concerning enhancement is that it is sometimes used to gain an unfair advantage over others, such as the case of athletes who take prohibited drugs to gain a competitive edge. The problem here, however, is not with enhancements themselves but with their use to gain an unfair advantage. It might be claimed that the existence of enhancing drugs provides such a strong temptation that merely making them available is morally problematic. But most enhancing drugs are also used therapeutically and, in fact, were originally developed for therapeutic use. That enhancing drugs are sometimes used unfairly is no more of an argument against their morally acceptable use than the fact that automobiles are sometimes used in committing a crime is an argument against their morally acceptable use. Other arguments against the use of enhancements, such as that they cause envy, create social pressure for their use, and increase the disparity between people, are also arguments against elite colleges, expensive cars, and personal trainers.

A rhetorically powerful but completely mistaken argument against enhancement is that it is not natural. This argument has no force because almost the entire world humans now live in is not natural, if by *natural* one means independent of human artifice. Even most of the trees and plants humans use are not natural. Medicine is not natural. Before abandoning traditional ways of acting and doing, whether natural or artificial, it is certainly important to ensure that the undesirable unintended consequences will not overwhelm the desired consequences. The larger the change the more caution is appropriate, especially if the desired consequences are not the prevention or relief of evils, but only the promotion of goods, such as germ-line genetic engineering that is used solely for enhancing.

HUMAN GENETIC THERAPY AND ENHANCEMENT.
An important application of the therapy/enhancement distinction occurs with genetic therapy and genetic enhancement, and examples of both processes may well proliferate in the future. It is important to distinguish germ-line genetic engineering from somatic-cell genetic engineering. Both involve directly altering the genetic structure of an organism, but somatic-cell genetic engineering, which is done by altering the somatic cells of an organism, is not intended to have any consequences for the descendents of that organism. Germ-line genetic engineering alters the genetic structure of an organism in ways that will or may have consequences for all of its descendents. Gene therapy is genetic engineering aimed at eliminating the genetic cause of (a) a serious malady or (b) a significantly increased risk of suffering that malady. Genetic enhancement is genetic engineering aimed at providing an organism with new or improved traits that are deemed useful or desirable by those doing the altering. Genetic engineering for plants and nonhuman animals is almost always genetic enhancement. Gene therapy is now being considered for human beings, but there is already talk of genetic enhancement for human beings.

If somatic-cell genetic engineering does not have any consequences for future generations, it is not considered controversial. Unlike the genetic engineering that is used in plants and animals, somatic-cell gene therapy alters only the genetic structure of the individual who receives the somatic-cell gene therapy; the altered genetic structure is not passed on to that indivi-

dual's offspring. Although it is possible for somatic-cell genetic engineering to affect the germ line, this is not yet considered a serious risk, and so its effects are thought to end with the individual treated. Unless some argument is provided to show that somatic-cell genetic engineering has serious risks, there is no stronger reason not to have somatic cell gene enhancement than not to have plastic surgery to improve the appearance of normal people. Indeed, it is hard even to imagine an argument against somatic-cell gene enhancement that is not also a general argument against any kind of technological enhancement.

The moral controversy that is the main subject here concerns whether there is any morally significant difference between germ-line gene therapy and germ-line gene enhancement with regard to human beings. In what follows, gene therapy and gene enhancement will always refer to germ-line gene therapy and germ-line gene enhancement. Gene therapy is regarded by some as the best way to correct severe genetic defects such as thalassemia, severe combined immunodeficiency, or cystic fibrosis. One argument is that because there is no nonarbitrary line between therapy and enhancement, acceptance of gene therapy, even to cure a serious genetic malady, makes it impossible not to accept gene enhancement as well.

This argument is used both by those who are opposed to genetic engineering of any kind, and those who favor gene enhancement. The former argue that because scholars are unable to draw a nonarbitrary line between gene therapy and gene enhancement, people should protect themselves against the latter by not even beginning with the former. The latter argue that because it is clear that one ought to accept gene therapy, one ought to also accept gene enhancement. Nevertheless, the objection that gene therapy will lead to gene enhancement presupposes that there is something intrinsically morally wrong with gene enhancement. No one has yet provided a strong theoretical argument that shows that genetic enhancement to produce greater size, strength, or intelligence, or increased resistance to toxic substances, is morally problematic. Yet neither is it clear that one ought to accept gene therapy or that there is no morally significant distinction between gene therapy and gene enhancement.

In fact, it is possible to draw a nonarbitrary line that distinguishes gene therapy and gene enhancement because there is an adequate definition of a genetic malady, related to the above general definition of a malady:

> An individual has a genetic malady if and only if
> (s)he has a genetic condition that is not normal

for a person in his (her) prime, other than his (her) rational beliefs or desires, such that (s)he is suffering, or is at a significantly increased risk of suffering, a non-trivial harm or evil (death, pain, disability, loss of freedom, or loss of pleasure) in the absence of a distinct sustaining cause.

Genetic conditions such as hemophilia, cystic fibrosis, and muscular dystrophy all share features common to other serious maladies, such as cancer, high blood pressure, and tuberculosis and so fit the definitional criteria of malady. Genetic conditions that do not meet the definitional criteria of a malady should obviously not be counted as a malady, and gene engineering for these constitutes gene enhancement. Examples of genetic nonmaladies might include blue eyes, widow's peak, freckles, O blood type, or curly hair.

Nonetheless, it is inevitable that there will be some genetic conditions about which there will be disagreement concerning their malady status. The number of such conditions is small, however, and the disagreement is based on the nature of maladies, not on vagueness in the malady definition. Borderline conditions, such as short stature or mild obesity, will be conditions about which people disagree on their malady status because it is not clear whether these conditions significantly increase the risk of suffering nontrivial harms. Because such borderline conditions are not very serious in the medical sense, they are quite unlikely to be candidates for gene therapy, at least initially. For all practical purposes gene therapy would be limited to the clear cases of genetic maladies. Indeed, the moral argument against gene enhancement, outlined below, is also an argument against genetic engineering for mild or borderline cases of genetic maladies.

The moral argument against gene enhancement is fairly straightforward. It is not morally acceptable to cause harm or a significant risk of harm to some people simply in order to create benefits for some other people. It is sometimes morally acceptable, however, to cause harm or a significant risk of harm to some people in order to prevent more serious or more certain harm to others. The government is allowed to quarantine people, that is deprive them of their freedom, even without their consent, if failure to quarantine would cause serious harm, as in the sudden acute respiratory syndrome (SARS) epidemic of 2003. This restriction of freedom, however, would not be justifiable simply in order to provide benefits to people. Gene enhancement does, at present, pose an unknown but possibly significant risk of harm to the descendants of the person who is being genetically enhanced. This genetic enhancement is not done to prevent a more serious or certain harm to this person. Therefore genetic enhancement is not morally justified. As noted, this same argument can be used against gene therapy for mild or borderline cases of genetic maladies. With regard to serious genetic maladies, this argument does not have the same force, for in these cases, the harm being prevented is more serious and certain than any harm that might be created. This does create a morally significant difference between gene therapy and gene enhancement.

Another completely different kind of argument can be given that leads to the same conclusion. Gene therapy simply aims to replace a defective gene with a nondefective allele of the same gene. If the technique for replacing genes is perfected, which at present it is not, then there is little or no chance that some unknown harmful side effect will result. The genetic structure of the organism will be identical in the relevant respect to the genetic structure of the majority of the human species. With gene enhancement, however, a new gene is being introduced with far greater chance of unknown harmful side effects. There are many genetic effects that do not show up for many generations. The identical gene inherited from the mother may have different effects when inherited from the father. There are expanding genes (triplet repeats) that do not have any effect until after several generations. Gene enhancement could create harms for the third or fourth generation, when it may not even be possible to track these individuals. This is another morally significant difference between gene therapy and gene enhancement.

Because preimplantation screening can eliminate almost all of the genetic maladies that would be eliminated by gene therapy, it seems clear that the primary reason for engaging in any kind of genetic manipulation is gene enhancement. Thus, although there is a morally significant difference between gene therapy and gene enhancement, given that the alternative of preimplantation therapy has less risks than gene therapy, it may be that there is at present no moral justification for engaging in either of these practices.

NONHUMAN GENETIC THERAPY AND ENHANCEMENT. As previously noted, genetic engineering is practiced on plants and nonhuman animals, and indeed has a long history in the nondirect forms of selective breeding and hybridization. In these cases what is almost always of interest is not genetic therapy for the good of the organism but genetic enhancement for the good of human users. On the basis of all the arguments already given, there is no reason to make a general objection to the genetic enhancement of plants and nonhuman animals.

Reimbursements for Therapies and Enhancements

Discussions of the therapy/enhancement distinction are sometimes linked to the question of third-party reimbursement for the two kinds of interventions. It may be assumed that therapies should be reimbursed and enhancements should not (see Parens 1998 for a discussion of these arguments). While there may be a societal consensus that most therapies should be reimbursed and that most enhancements should not, this is a contingent and not an invariant relationship.

Suppose two new managed-care companies start up and offer somewhat different ranges of benefits. Company A pays not only for essentially all therapies but also for most borderline cases whose therapy/enhancement status is a matter of dispute, and even pays for a few enhancements that are clearly specified in the terms of the contract. Company B pays only for therapies and states ahead of time that they will not reimburse for borderline conditions (which they might list) and will not reimburse for any enhancements whatsoever. Company A's premiums are higher, while company B is offering a lower cost, less-inclusive policy. Neither company is acting unethically or in an unjust fashion.

If, however, the issue concerns medical plans that are financed by taxes, then there may be an argument that only therapies, and not enhancements, should be covered. Yet even in this case, there is no obvious way to determine which, if any, borderline cases should be covered. In democratic societies decisions about government-financed medical treatments should reflect the prevailing public consensus, as determined through democratic political processes.

BERNARD GERT
CHARLES M. CULVER

SEE ALSO *Bioethics; Human Cloning.*

BIBLIOGRAPHY

Brock, Dan W. (1998). "Enhancements of Human Function: Some Distinctions for Policymakers." In *Enhancing Human Traits*, ed. Erik Parens. Washington, DC: Georgetown University Press. Discusses circumstances in which it is ethically justified for a health insurance company not to reimburse the treatment of a disease and to reimburse the giving of some enhancements.

Daniels, Norman. (1994). "The Genome Project, Individual Differences, and Just Health Care." In *Justice and the Human Genome Project*, ed. Timothy F. Murphy and Marc A. Lappé. Berkeley and Los Angeles: University of California Press.

Gert, Bernard. (2000). "Thinking about Huxley's *Brave New World:* Was It Wrong to Create a Genetic Hierarchical Society? Is It Wrong to Prevent One?" In *Etica ricerca biologica* [The ethics of biological research], ed. Cosimo Marco Mazzoni. Florence, Italy: Leo S. Olschki. Reprinted in *Ethics and Law in Biological Research*, ed. Cosimo Marco Mazzoni. Boston: Kluwer Academic, 2002. Explores the issue of whether genetic enhancement should be allowed.

Gert, Bernard; Charles M. Culver; and K. Danner Clouser. (1997). *Bioethics: A Return to Fundamentals*. New York: Oxford University Press. Contains an extensive discussion of the criteria for a valid consent and the criteria for justified paternalistic interventions.

Gert, Bernard; Edward M. Berger; George F. Cahill Jr.; et al. (1996). *Morality and the New Genetics*. Boston: Jones and Bartlett. Contains a detailed discussion of the concept of a genetic malady.

Juengst, Eric T. (1998). "What Does Enhancement Mean?" In *Enhancing Human Traits*, ed. Erik Parens. Washington, DC: Georgetown University Press.

Parens, Erik. (1998). "Is Better Always Good? The Enhancement Project." In *Enhancing Human Traits*, ed. Erik Parens. Washington, DC: Georgetown University Press.

THOMAS AQUINAS

• • •

Thomas of Aquino (ca. 1225–1274), a philosopher and theologian, was born into an aristocratic family at Roccasecca, near Naples, Italy. He joined the Dominican order in 1245, taking a *licentia docendi* at Paris in 1256. He later taught at Paris, Rome, Orvieto, and Naples. Thomas died at the Cistercian abbey of Fossa Nuova on March 7 and was canonized in 1323 by Pope John XXII. The *Summa contra Gentiles* was completed about 1264. His longest and most influential work, the *Summa Theologiae*, was unfinished at the time of his death.

Ethics and Politics

Thomas was the foremost contributor to the thirteenth-century recovery of Aristotle. His achievement in ethics lies chiefly in the application of a Christianized version of Aristotle to politics and law. In most respects he departs from the Augustinian orientation of previous generations that found the present world sin-laden and disordered and its politics harsh and coercive.

Thomas accepted the rational, humane, ordered world depicted by Aristotle. There is no tension between the acquisition of present goods on earth and the achievement of eternal ones in heaven so long as the former are directed toward and subordinated to the

Thomas Aquinas, ?–1274. Aquinas was an Italian theologian and philosopher of the Dominican Order of the Catholic Church and is regarded as one of the greatest and most influential thinkers of the Church. He had an important influence on the intellectual awakening that occurred in western Europe during and after his lifetime. (© *The National Gallery, London/Corbis.*)

latter. Human beings have a final ethical end—eternal blessedness—that transcends all earthly ends, but earthly happiness is also possible and desirable. God has equipped human beings with the rational capacity to pursue earthly as well as heavenly goods, and although sin has impaired the will, it has not obliterated reason. Thomas believes, as Augustine (354–430) did not, that humans are capable, under proper governance, of cooperating with one another to achieve a common good.

For Thomas human beings are by nature political animals; government is not merely a consequence of sin. Even if the Fall of Adam had not occurred, no individual would be able to acquire all the necessities of life unaided; only cooperation can secure the benefits of divisions of labor. However, there are many ways to achieve human ends, and so a community must be guided toward the common good by just and wise rule. The best government is a "mixed" constitution of the kind that Aristotle called *politeia*. Kingship may be the most efficient form of rule, but it is also the most likely

to deteriorate into tyranny. It therefore must be tempered by elements of democracy and aristocracy. A king should choose the best people as his counselors, and what he does should be ratified by the people. Thomas follows Aristotle in supposing that a government in which as many people as possible participate will be the most stable because it will commend itself to all sections of the community.

Law and Ethics

In the *Summa Theologiae* Thomas develops a typology of law as eternal, natural, human, and divine. This theory has a Platonic starting point insofar as law is defined as a rational pattern or form. In the political realm law thus serves as a "rule and measure" for citizens' conduct. When citizens obey the law, they "participate" in that order in the way a table "participates" in the rational pattern or form of a table.

Because God is the supreme governor of everything, the rational pattern or form of the universe that exists in God's mind is law in the most comprehensive sense: the law that makes the universe orderly and predictable. This rational pattern is what Thomas called eternal law, and to it everything in the universe is subject. The eternal law is similar in content to what science now calls the laws of nature.

Inasmuch as humankind is part of the eternal order there must be a portion of the eternal law that relates specifically to human conduct. This is the *lex naturalis*, the "law of [human] nature": an idea present in Aristotle to which Thomas gave extensive elaboration. In developing his natural law theory Thomas restored human reason to a central place in moral philosophy. For Thomas, as for Aristotle, human beings are preeminently reason-using creatures. The law or order to which people are subject by their nature is not a mere instinct to survive and breed. It is a moral law ordering people to do good and avoid evil, have families, live at peace with their neighbors, and pursue knowledge. It is natural in that humans are creatures to whom its prescriptions are rationally obvious. To all humans, pagans included, these precepts simply "stand to reason" by virtue of a faculty of moral insight or conscience that Thomas called *synderesis*.

However, humans act on the principles of natural law with the assistance of more particular and coercive provisions of what Thomas called human law. The natural law is too general to provide specific guidance. Part of this specific guidance can come from the moral virtues that equip people to achieve practical ends: prudence, justice, temperance, and fortitude. However, these personal guidelines are developed and reinforced by human or positive laws that that help cultivate such good habits. These particular, positive rules of behavior include civil and criminal laws of the state as formulated by practical reason, or what Aristotle called *phronesis*, in the light of the general principles of natural law and have a morally educative function. Human laws that are not based on natural law—laws that oppress people or fail to secure their good—have more the character of force than that of law. Obedience may be called for if disobedience would cause greater harm, but people are not obliged to obey unjust laws. Individuals may exercise independent moral judgment; they are not simply subjects but rational citizens.

The fourth kind of law—divine law—is part of the eternal law but, unlike human law, is not derived from rational reflection on more general principles and historical circumstances. It is a law of revelation, disclosed through Scripture and the Church and directed toward people's eternal end. Human law is concerned with external aspects of conduct, but salvation requires that people be inwardly virtuous as well as outwardly compliant. The divine law governs people's inner lives: It punishes people insofar as they are sinful rather than merely criminal.

Applying Thomism

The strongest implications of Thomas's thought for ethics, science, and technology are found in the doctrine of natural law and the underlying idea of human equality. For instance, Pope Leo XIII in his encyclical *Rerum Novarum* (1891) drew on law theory to criticize the conditions of labor under industrial capitalism. Insofar as it requires people to do good, avoid evil, pursue knowledge, and live at peace with their neighbors, the natural law suggests that governments should support scientific and technological research intended to have beneficial outcomes. By the same token, it supports the principle that governments should not sponsor such research when it involves the development of weapons of mass destruction or the exploitation of some human beings by others.

Natural law doctrine implies as well that governments should not harm, but seek to preserve, the physical environment of humankind: the natural world that God created and over which humans properly exercise dominion. In regard to biological and medical science, the idea of human nature as a repository of value implies a distinction between laudable biomedical research, which is a work of charity beneficial to the human race, and unacceptable research involving the manipulation or distortion of human nature. In this connection Thomas often is cited in support of the Catholic Church's prohibition of artificial (as distinct from natural) methods of contraception.

Finally, it may be noted that Thomas's insistence on citizen participation in government speaks against any suggestion that political decisions should be made by technocratic elites of scientists and engineers rather than by those who will be affected by those decisions. Thomas presided over a thorough revaluation of the capacity of human beings for autonomous moral action and hence for responsible political participation. In effect, he reinvented the Aristotelian ideal of citizenship after its long medieval eclipse, and that reinvention would apply today to scientific and technological decision making.

R. W. DYSON

SEE ALSO *Aristotle and Aristotelianism; Christian Perspectives; Just War; Natural Law; Virtue Ethics.*

BIBLIOGRAPHY

Chesterton, G. K. (1974). *Saint Thomas Aquinas: The Dumb Ox.* New York: Image. A popular and uncritical biography, but useful for purposes of orientation.

Dyson, R. W., ed. *Political Writings/Thomas Aquinas.* Cambridge, UK, and New York: Cambridge University Press. A comprehensive selection of passages in translation, with detailed notes and introduction.

Finnis, John. (1998). *Aquinas: Moral, Political and Legal Theory.* Oxford, UK: Oxford University Press. A comprehensive and recent scholarly treatment of Thomas's thought on ethics, politics and law.

McInerny, Ralph M. (1982). *Ethica Thomistica: The Moral Philosophy of Thomas Aquinas* Washington DC: Catholic University of America. A clear exposition and summary of Thomas's moral philosophy; still regarded as a standard textbook on the subject.

O'Connor, D. J. (1967). *Aquinas and Natural Law.* London: Macmillan.

THOMISM

• • •

Thomism is a philosophical system of thought based on the writings of Thomas Aquinas, from his death in 1274 to the present. As a philosophy Thomism may be viewed as a moderate realism developed within medieval and Renaissance scholasticism that has been in continuous dialogue with alternate systems of thought in the modern and contemporary periods. The focus here is on Thomism specifically as it relates to science, technology, and ethics in the present.

Notion and Relevance

Thomas of Aquino (1225–1274) was a Dominican who studied under Albert the Great (c. 1200–1280) in Paris and Cologne and then taught at the University of Paris and in various Italian cities. Thomas was a prolific writer, known in his own day as a commentator on Aristotle, who adapted his thought to explicating the Catholic faith. Thomas was himself competent in the science of nature in the Aristotelian sense, and owed much to Albert's knowledge of the biological and psychological sciences. The relevance of both Albert and Thomas to modern science and its problems has been explored extensively by three contemporary Dominicans, Benedict M. Ashley, William A. Wallace, and James A. Weisheipl (1923–1984).

Modern science differs from *scientia* as understood in the Thomistic tradition, where it is defined as true and certain knowledge acquired by demonstration through prior knowledge of principles and causes. Modern science makes a lesser epistemic claim, only to knowledge acquired by hypothetico-deductive reasoning yielding conclusions with a high degree of probability but that fall short of certitude. Mathematical logic is instrumental for science, but science itself remains fallible and revisable. For Thomists this is too pessimistic. They would say that philosophers of science should rediscover the epistemology of Aristotle's *Posterior Analytics*, and rather than basing their reasoning on logic alone, could also focus on concepts provided by the philosophy of nature developed within the Aristotelian tradition (Wallace 1996).

For Thomism's relevance to technology a balanced view is that of a former Dominican and Wallace student, Paul T. Durbin. Durbin insists, first, that technology in the present day is essentially related to science, and second, that an identifiable social group is the carrier of technology. Thus the term *technology* can be taken to cover this scientific and technical community, including its inner structure and functions, its products, its particular values, and its implicit view of human nature. The term *philosophy of technology* then means a set of generalizations or a systematic treatment, in philosophical language, of one or another or all of the above social phenomena.

With regard to ethics, of the three terms—*science*, *technology*, and *ethics*—the last has the most explicit and enduring relationship to Thomism. There ethics is seen as the philosophical study of voluntary human action, with the purpose of determining what types of activity are good, right, and to be done, or bad, wrong, and not to be done, so that human individuals might live well. As a philosophical study, ethics treats information derived from a person's natural experience of the problems of human living. The term ethics is etymologically connected with the Greek *ethos*, meaning customs or behavior, and is the same as moral philosophy, similarly connected with the Latin *mores*, also meaning customs or behavior. It is a practical science in the sense that its objective is not simply to know, but to know which actions should be done and which should be avoided, so as properly to translate knowledge into action. Thus understood, only one thesis on ethics is listed among various theses seen as essential to Thomism. This states that humans have by nature the right to cooperate with others in society in the pursuit of personal happiness in the common good, and that this pursuit of happiness is guided by conscience, laws both natural and positive, and virtues both private and public. Briefly, Thomistic ethics is a virtue ethics that infers from nature what humans ought to do or be to achieve their proper perfection.

Historical Overview

Albert and Thomas wrote in the medieval period of high scholasticism. Albert was the first to appreciate the importance of the newly imported Greek-Arabic learning for science and philosophy, and he set himself to making encyclopedic summaries for his students, which earned for him the title "the Great" in his own lifetime. He had many followers among German Dominicans, including Meister Eckehart (c. 1260–1327) and Theodoric of Freiberg (c. 1250–1310), the second of whom worked out the first correct theory of the rainbow. But Albert's work bore principal fruit in the monumental synthesis elaborated by his pupil Thomas. Called the "Angelic Doctor," Thomas brought natural philosophy and metaphysics into the heart of theology to develop the unique synthesis known as Thomism. Its major teachings are that first matter is pure potentiality and its first actuation is by substantial form; that the human rational soul is the unique substantial form of the human body, endowed with powers that are really distinct from it; that human knowledge originates with the senses but is capable of attaining universals; and that humans can reason to the existence of God and some of God's attributes from the visible things of the world.

In later scholasticism Thomism became the official doctrine of the Dominican Order, where it was championed by Harvey Nedellec (1250 or 60–1323), John of Naples (d. 1330), and Jean Capréolus (c. 1380–1444). The Renaissance was the period of great commentaries on Thomas known as "Second Thomism," when Dominicans exerted strong influence at Paris and Salamanca as well as northern Italy. The more famous of the figures of Second Thomism were Thomas de Vio Cajetan (1469–1534), who debated the German religious reformer Martin Luther on the Eucharist; Francisco de Vitoria (1486?–1546), who developed the theory of natural law during Spain's period of colonial expansion; and Vitoria's colleague Domingo de Soto (c. 1494–1560), whose work foreshadowed to a degree Galileo Galilei's law of falling bodies (Wallace 2004). The same period saw the foundation of the Jesuits, who were initially trained as Thomists but then developed their own versions of Thomism. Jesuits and Dominicans later entered into prolonged controversy over the efficacy of God's grace on human free will and God's foreknowledge of human free actions, and were convinced that many modern evils stem from false philosophy, to which Thomas's thought would supply a needed corrective.

Developmental Thomism

The period from the mid-sixteenth to the late nineteenth century saw little development within Thomism. The system itself had received strong endorsement by the Council of Trent (1545–1563), and, as what may be referred to as Scholastic Thomism, it was taught in Catholic seminaries as a philosophical preparation for the study of theology. It was often seen as the "perennial philosophy," an integrated system that gave enduring answers to central questions about reality and knowledge. And it was largely unaffected by the scientific revolution of the seventeenth century, which was mainly concerned with physical sciences that seemed to have little relevance to Catholic teaching.

This situation changed dramatically after the issuance in 1879 of the encyclical *Aeterni Patris* of Pope Leo XIII (1810–1903), which gave rise to a movement known variously as neo-scholasticism or neo-Thomism (or, among Dominicans, "Third Thomism.") The stimulus came from the labors of medieval historians such as Maurice De Wulf (1867–1947) and Martin Grabmann (1875–1949), who had recovered works of medieval thinkers and focused attention on Thomas's thought as containing answers to pressing contemporary problems.

With Pope Leo's endorsement, Thomism underwent extensive development in the twentieth century and came in dialogue with other philosophical movements. Arguably it is the most extensively developed systematic philosophy in the present day.

In this expanded sense, the term *Thomism* has itself undergone a change of meaning. An "ism" need not refer exclusively to an original system of thought. It might also refer to a system of thought that has taken on new meaning in light of developments that were unforeseen and unknown by its originator. In this alternate sense René Descartes could not be a Cartesian nor could Immanuel Kant be a Kantian. This sense would apply to those who came after them and assimilated new knowledge into their syntheses in ways consistent with the principles they had established, while rejecting matter that had been superseded in the interim. This is obviously a more speculative enterprise, but it is in this sense that one might speak of one or more developmental Thomisms.

Types of Thomism

The development of overriding importance is the growth of modern science in its classical and contemporary senses and how this affects Thomism as a whole. Allied to this are three subsidiary developments that may be characterized as different types of Thomism. Of these, two have already achieved the status of movements, namely, Existential Thomism, which arose from confrontation with existentialist thought, and Transcendental Thomism, which arose from the confrontation with Kantianism and other forms of idealism seen in the works of Continental philosophers. A third, resulting from the confrontation with Anglo-American philosophy, may be described as Analytical Thomism, though it is not yet regarded as a movement.

EXISTENTIAL THOMISM. The two philosophers most identified with this movement were the Frenchmen Jacques Maritain (1882–1973) and Étienne Gilson (1884–1978), both former students of Henri Bergson (1859–1941). Maritain became interested in the thought of Thomas after being converted to Catholicism. His most lasting achievements have been in the area of epistemology, in elucidating the different degrees of knowledge and their interrelationships, so as to constitute an integral, Christian humanism. He also made substantial contributions to social and political philosophy and to constructive critiques of modern culture and art. In his theoretical philosophy he stressed the authentic existentialism of Thomas, maintaining the primacy of existence in a realist philosophy of being, and seeing this as also providing the basis for an understanding of knowledge and of love.

Gilson did his early work on Descartes, which led him to a study of medieval philosophy and of Thomism in particular. He saw the philosophy of the Middle Ages as a Christian philosophy, one that, while keeping the orders of faith and reason distinct, considers Christian revelation as an indispensable auxiliary to reason. In Thomas he found a metaphysics of existence that conceives God as the very act of being (*Ipsum Esse*) and creatures as beings centered on the act of existing (*esse*). His disciples regarded his existential metaphysics as a corrective to the essentialism that had insinuated itself in Renaissance and rationalist versions of Thomistic thought.

TRANSCENDENTAL THOMISM. The roots of this movement can be traced to Désiré Mercier (1851–1926) and Maurice Blondel (1861–1949), and to the efforts of two Jesuits, Jean-Pierre Rousselot (1846–1924) and Joseph Maréchal (1878–1944), to rehabilitate critical philosophy in light of the teachings of Thomas. Maréchal's thought passed through several phases, but in a later formulation he proposed the act of judgment as an affirmation of absolute reality that objectifies the form or concept and so grasps it as being. Then, beyond the concept, the intellect is made aware of a further intelligibility by its own tending, in a dynamism unleashed by the concept itself, toward something infinite and absolute—actually the infinite act of existing that is God. The intellect thus "constitutes" its object as belonging, in a finite and participatory way, to the realm of the real.

Maréchal's innovative views gained new insights from dialogues with phenomenology by two German Jesuits, Karl Rahner (1904–1984) and Emerich Coreth (b. 1919), and by analyses of modern science by a Canadian Jesuit, Bernard Lonergan (1904–1984). From these have emerged a new metaphysics in which the being investigated is that which occurs in consciousness. So Coreth writes of an immediate unity of being and knowing in the very act of knowing, and Lonergan looks upon being as whatever is to be known by intelligent grasp and reasonable affirmation, and so extrapolates from the being of consciousness to the being of the cosmos. For Rahner an analysis of the performance of the human spirit discloses an innate drive to being as absolute and really existing, which itself is human nature as "spirit in the world" or finite transcendence. They elaborate these insights in various ways through the use of what is called a transcendental method.

ANALYTICAL THOMISM. Like phenomenology, analytical philosophy is more a method or way of doing philosophy than it is a philosophy itself. Bertrand Russell (1872–1970) was one of its pioneers, and after him came the logical positivists, with their anti-metaphysical programs, and finally a more relaxed conception of linguistic analysis, culminating in the work of Ludwig Wittgenstein (1889–1951). One of Wittgenstein's students, Elizabeth Anscombe (1919–2001), along with her husband Peter Geach (b. 1916) were the first analysts to attend to Thomism in their writings. A related thinker is Alasdair MacIntyre (b. 1929), whose work in Aristotelian politics and virtue ethics brought him to the study of Thomas. Also noteworthy is the work of John N. Deely, a former Dominican and student of Weisheipl, who recovered the work on semiotics of the early-seventeenth-century Thomist John Poinsot, known in the Dominican Order as John of St. Thomas. By the early twenty-first century, the most distinctive contributor to the emerging movement is John J. Haldane, of the University of Aberdeen, who has published extensively in the philosophy of mind and the philosophy of God from a Thomist perspective.

Areas of Continuing Research

Thomists in the United States seem more inclined to pursue the analytical route than the other two movements, and have two main areas of research. The first focuses on an analysis and critique of scientific concepts with reference to the Aristotelian-Thomistic heritage, particularly the latter's use of first matter and transient entities to develop a view of creation and evolution that concords with recent theories of cosmogenesis (the origin of the cosmos). The second focuses on problems in bioethics, particularly through a recovery of Thomas's teaching on delayed hominization as this relates to the study of homogenesis.

On the theme of cosmogenesis, this line of research associates God's creative act at the beginning of time with the "big bang" theory of cosmic origins (Wallace 2002). Time began some 13 billion years ago by the production by God, ex nihilo (out of nothing), of the primordial mass-energy of which the universe is now composed. Along with the act of creation, God as prime mover also initiated the "big bang," releasing the enormous energy of the primitive mass for the formation of the natures now found in the universe. These are, in order, transient natures, inorganic natures, plant natures, animal natures, and human nature. They correspond to the stages of evolution commonly accepted among scientists: the period of fundamental particles impelled at high energy; that of element and compound formation; the two periods of biogenesis, wherein first plants and then animals were generated; and finally that of hominization, when *Homo sapiens* first appeared. All of these stages except the last were accomplished by a natural process Thomas referred to as "the eduction of [substantial] form from the potency of first matter" (*Summa Theologiae* I, q. 90, a. 2).

The final stage of cosmic evolution would then be hominization, the appearance of humans with a special type of substantial form, an immaterial (and immortal) soul. Here there is a break in the line of causality extending back to creation, because, according to Catholic teaching, such a soul cannot be educed from the potency of matter. Up to this point the entire process of evolution can bring organisms to a level just below that of thought and volition, but they cannot progress to the final stage. Here God's creative act is again required. This second input of divine causality is the production, ex nihilo, of the immaterial souls of the first humans, tailored to match the ultimate disposition of first matter, as this has been prepared, over billions of years, for their reception.

With regard to bioethics, an important advance has been in the recovery of Thomas's teaching that the beginning of human life is a gradual process: that the human soul is not infused into the incipient organism at fertilization but rather is prepared for by a succession of substantial forms that dispose first matter for the reception of an intellective soul (Wallace 1995). Less well known is his speculation that the reverse process may occur at the ending of human life, namely, that the human soul may depart from the body well before all signs of life have disappeared from it. Both views are opposed to the notion of immediate hominization, commonly taught in Catholic circles, namely, that human life begins at fertilization, when the rational soul is infused by God into the body, and terminates at death, when the same human soul departs from the body.

With regard to human generation, Thomas followed Aristotle in holding that the conception of a male child was not completed until the fortieth day after intercourse, whereas that of the female child was not completed until the ninetieth day. The details of Thomas's treatment, now referred to as delayed hominization, were worked out on the basis of Aristotle's teaching as developed by medieval commentators, particularly Avicenna (980–1037). Little empirical evidence was available to support the various steps of the argument. In the early twenty-first century, however, the human reproductive process is being studied inten-

sively, and much evidence can be brought to bear on the problem of hominization.

Catholic theologians have advanced two lines of argument that generally favor Thomas's solution. The first, proposed by Norman M. Ford (1988), is based on the possibility of twinning in the formation of the fetus and is essentially an argument from individuation. This would propose that the definitive individuation of the human fetus does not occur until fourteen days after conception, and thus that the intellective soul, and so the human person, need not be present before that time. The second argument, advanced by Joseph F. Donceel (1970), is based on the organ systems required first for sensitive life and then for the exercise of reason, which would involve the senses, the nervous system, the brain, and especially the cortex. The time when such organ systems are present in the human fetus must be ascertained by embryology. This probably occurs somewhere between several weeks and the end of the third month after conception, and so it is possible, on this theory, that human animation does not occur before this time.

Both of these conclusions, if accepted, would have far-reaching implications for future work in human genetics. Because the Catholic Church has thus far not taken a definitive position on the precise time when the human soul is present in the developing organism, the question remains open to discussion.

WILLIAM A. WALLACE

SEE ALSO *Christian Perspectives; Thomas Aquinas; Virtue Ethics.*

BIBLIOGRAPHY

Ashley, Benedict M. (1985). *Theologies of the Body: Humanist and Christian.* St. Louis, MO: Pope John Center.

Deely, John N. (1990). *Basics of Semiotics.* Bloomington: Indiana University Press.

Donceel, Joseph F. (1970). "Immediate Animation and Delayed Hominization." *Theological Studies* 31(1): 76–105.

Durbin, Paul R. (1968). *Philosophy of Science: An Introduction.* New York: McGraw-Hill.

Durbin, Paul R. (1986). "Ferment in Philosophy of Science: A Review Discussion." *The Thomist* 50 (1986): 690–700.

Ford, Norman M. (1988). *When Did I Begin? Conception of the Human Individual in History, Philosophy, and Science.* Cambridge, UK: Cambridge University Press.

Furton, Edward J., and Louise A. Mitchell, eds. (2002). *What Is Man, O Lord? The Human Person in a Biotech Age.* Boston: National Catholic Bioethics Center.

John of St. Thomas. (1985). *Tractatus de Signis: The Semiotic of John Poinsot,* ed. John N. Deely. Berkeley and Los Angeles: University of California Press.

Kretzmann, Norman, and Eleonore Stump, eds. (1993). *The Cambridge Companion to Aquinas*. Cambridge, UK: Cambridge University Press.

Wallace, William A. (1989). "Nature and Human Nature as the Norm in Medical Ethics." In *Catholic Perspectives on Medical Morals*, eds. Edmund D. Pellegrino, John P. Langan, and John Collins Harvey. Dordrecht, Netherlands: Kluwer Academic.

Wallace, William A. (1995). "St. Thomas on the Beginning and Ending of Human Life." In *Sanctus Thomas de Aquino Doctor Hodiernae Humanitatis*. Studi Tomistici, 58. Libreria Editrice Vaticana: Pontificia Accademia di S. Tommaso, pp. 394–407.

Wallace, William A. (1996). *The Modeling of Nature: Philosophy of Science and Philosophy of Nature in Synthesis*. Washington, DC: Catholic University of America Press.

Wallace, William A. (2000). "Thomas Aquinas and Thomism." In *The History of Science and Religion in the Western Tradition: An Encyclopedia*, ed. Gary B. Ferngren. New York and London: Garland.

Wallace, William A. (2002). "*Fides et Ratio*: The Compatibility of Science and Religion." In *What Is Man, O Lord?* eds. Edward J. Furton and Louise A. Mitchell. Boston: National Catholic Bioethics Center.

Wallace, William A. (2004). *Domingo de Soto and the Early Galileo: Essays on Intellectual History*. Collected Studies Series, CS 783. Aldershot, UK: Ashgate Publishing Ltd.

Wallace, William A., and James A. Weisheipl. (1967). "Thomas Aquinas, St." In *New Catholic Encyclopedia*, Vol. 14.

Weisheipl, James A. (1967). "Thomism." In *New Catholic Encyclopedia*, Vol. 14, continued in "Scholasticism, 3," 12: 1165–1170.

Weisheipl, James A. (1974). *Friar Thomas d'Aquino: His Life, Thought, and Work*. Garden City, NY: Doubleday.

Weisheipl, James A., ed. (1980). *Albertus Magnus and the Sciences: Commemorative Essays, 1980*. Toronto: Pontifical Institute of Mediaeval Studies.

THOREAU, HENRY DAVID

• • •

Henry David Thoreau (1817–1862) was born in Concord, Massachusetts, on July 12, and died there of tuberculosis on May 6, two months shy of his forty-fifth birthday. He is best known as the author of *Walden* (1854), an account of the two years (1845–1847) he spent living in a cabin he built on the shores of Walden Pond (outside Concord), and "Civil Disobedience" (originally delivered as a lecture entitled "The Rights and Duties of the Individual in Relation to Government"), a polemical political essay describing the events surrounding, reasons for, and consequences of his arrest for nonpayment of taxes.

Henry David Thoreau, 1817–1862. Thoreau was an American writer, a dissenter, and, after Emerson, the outstanding transcendentalist. He is best known for his classic book, *Walden*. (*The Library of Congress.*)

Thoreau is often portrayed as an anti-modern romantic, placing him in strong opposition to the modernizing forces of science and technology. There is good evidence for this portrait scattered throughout his work. He wrote as an advocate of nature, and frequently suggested that the artifacts of civilization violated the goods and principles found in nature. For example, in his first book, *A Week on the Concord and Merrimack Rivers* (1849), he claimed that he would prefer to destroy the dams on the rivers and free the fishes; in a late essay, "Walking" he famously declared that "in Wildness is the preservation of the World" (Thoreau 1893, p. 275). He wrote in *Walden* of the need for people to simplify their lives ("Simplicity, simplicity, simplicity!" [Thoreau 1985, p. 395]), and many have interpreted this as an injunction to turn away from the world of modern science and technology in order to restore a more independent, even primitive lifestyle.

Despite the occasional evidence in support of this understanding of Thoreau's teaching, however, there is good reason to believe it is not a true picture of either his life or his intentions as an author. Any reader of Thoreau's books, essays, or fourteen volume *Journal* will be struck by his preoccupation with observing the natural world. He was a skilled, committed, and lifelong naturalist, and he provided field reports and specimens to the foremost biologist in the United States at the time, Louis Agassiz of Harvard University. He was also something of an archaeologist, gathering one of the most extensive collections of American Indian artifacts of his generation. Equally important *Walden* can be read as a philosophical commentary on modern economics, suggesting Thoreau's interest in social science. Thoreau was skilled as a surveyor and a carpenter, and proved his genius as a technologist by developing a new formula and manufacturing process for the graphite in the pencils manufactured by his family's business, which made these the highest quality pencils produced in the United States at the time. Thoreau's biography and writings reveal a man with a much more sophisticated view and knowledge of modern science and technology than is often acknowledged. While it is true that Thoreau often juxtaposed modern science and technology with what he took to be the wisdom or laws of nature, this does not preclude his being a serious natural and social scientist.

In fact Thoreau's complaint was not with science or technology in themselves, both of which he admired (and tried successfully to practice) in their proper place, but with the uncritical exercise and use of both. Although he was a skilled naturalist and technologist, he was most importantly a literary artist and a moralist. The message of *Walden* is not that modern science and technology are bad, but rather that they are bad as human beings currently practice them. This complaint is inspired by a concern for liberty, and is built on the fear that people are using science and technology to build wealth even if it costs them their freedom. He complained that people "have become the tools of their tools" (Thoreau 1985, p. 352) and that they would be more likely to learn "beautiful housekeeping" and "beautiful living" (p. 353) if they were willing to cultivate a more thoughtful poverty and independence. Ultimately Thoreau was a critic not of science and technology, but of the modern political economy and the way it employed these tools. His fear was that people were becoming morally ignorant about the cultivation of a good human life even as they were becoming scientifically and technically proficient.

As a social critic Thoreau has inspired many in the modern environmental movement who share his fear that society uses science and technology to war against nature rather than to learn to live in peace and harmony with it. Thoreau continues to be one of the most powerful literary voices in America. He is a reminder of the need to continually probe the purposes and ends to which science and technology are employed.

BOB PEPPERMAN TAYLOR

SEE ALSO *Environmental Ethics; Environmentalism; Freedom; Nature; Science, Technology, and Literature.*

BIBLIOGRAPHY

Buell, Lawrence. (1995). *The Environmental Imagination: Thoreau, Nature Writing, and the Formation of American Culture.* Cambridge, MA: Harvard University Press.

Harding, Walter. (1966). *The Days of Henry Thoreau.* New York: Alfred A. Knopf.

Neufeldt, Leonard N. (1989). *The Economist: Henry Thoreau and Enterprise.* New York: Oxford University Press.

Richardson, Robert D. (1986). *Henry Thoreau: A Life of the Mind.* Berkeley and Los Angeles: University of California Press.

Taylor, Bob Pepperman. (1996). *America's Bachelor Uncle: Henry Thoreau and the American Polity.* Lawrence: University Press of Kansas.

Thoreau, Henry David. (1893). *Excursions.* Boston: Houghton Mifflin.

Thoreau, Henry David. (1985). *A Week on the Concord and Merrimack Rivers; Walden, or Life in the Woods; The Maine Woods; Cape Cod.* New York: Library of America.

THREE GORGES DAM

• • •

The Three Gorges Multipurpose Water Control Project on the Yangtze River in China is one of the largest engineering projects in history. When complete, it will rival the Great Wall in technical and cultural significance. Unlike the Great Wall, however, and in accord with contemporary notions of scientific and technological decision making, the Three Gorges Dam has been the subject of considerable ethical and environmental assessment.

Historical Background and Description

The Yangtze River originates at 6,000 meters (20,000 feet) in the mountains of Tibet and then flows for 6,300 kilometers (3,900 miles) east through central China, passing through Nanjing, the capital of Jiangsu Province, before emptying into the East China Sea, through the port of Shanghai. From Shanghai, for the first 2,500 kilometers up the lower river is generally broad, calm,

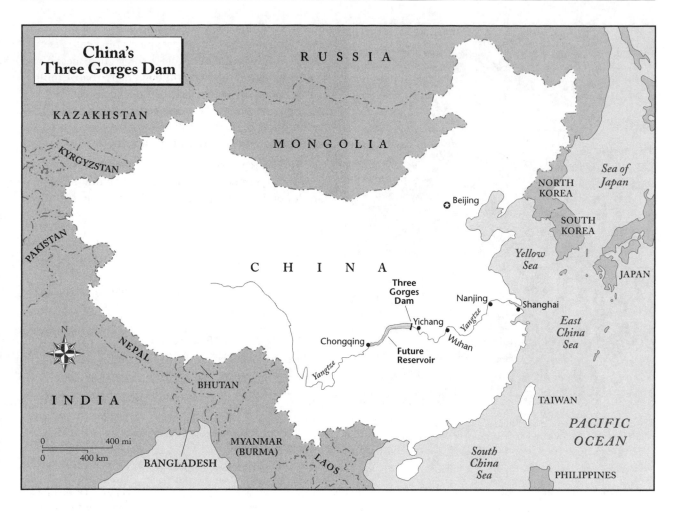

China's
Three Gorges Dam

KAZAKHSTAN

KYRGYZSTAN

RUSSIA

MONGOLIA

PAKISTAN

NORTH
KOREA

Sea of
Japan

Beijing

SOUTH
KOREA

CHINA

Yellow
Sea

JAPAN

NEPAL

Three
Gorges
Dam

Nanjing

Shanghai

Yichang

Wuhan

East
China
Sea

Chongqing

Yangtze

Future
Reservoir

Yangtze

INDIA

BHUTAN

TAIWAN

PACIFIC
OCEAN

N

0   400 mi
0   400 km

BANGLADESH

MYANMAR
(BURMA)

LAOS

South
China
Sea

PHILIPPINES

and navigable, serving as a major transportation artery as it flows through the traditional rice basket of China. At Yichang there is a series of three, sheer-cliffed gorges, Xiling, Wu, and Qutang, that stretch up river for another 1,000 meters.

The idea of building the dam was first proposed by Sun Yat-sen in 1919, but it was not until 1994, with the backing of Deng Xiaoping, that construction actually began.

The project consists of three parts: the dam itself, hydroelectric stations located on each side of the dam, and navigation locks on the left side of the dam. When finished, the dam at the mouth of Xiling Gorge will be 185 meters high, 3,035 meters long at the top, and will create a reservoir that stretches 600 kilometers through each of the gorges in turn, with a surface area of 10,000 square meters and a volume of 39.3 billion cubic meters. It will provide flood control, generate electric power, and improve navigation.

Ethical Issues

The Three Gorges Dam project has been subject to three basic criticisms. It has been judged by the World Bank as economically unsound, by many environmentalists as ecologically destructive, and by some social scientists as socially and culturally disruptive. All of these issues have been discussed at length, and efforts have been made to address the objections.

Because of the negative judgment of the international financial community, China has raised the money for construction from its own resources. At the same time, it has tried to structure the project so that in the long term the investment will benefit Chinese economic development.

The Three Gorges Dam will indeed have significant ecological and social consequences. The ecological impact is justified not only by great social but also significant environmental benefit. When completed, for instance, the dam will provide for extensive flood con-

The Three Gorges Dam on the Yangtze River at Yichang, in China's Hubei Province. (*AP/Wide World Photos. Reproduced by permission.*)

trol on a river that has caused major disasters on an average of every ten years in the past. It will also produce 18.2 million kilowatts of electricity, the equivalent of ten standard coal-fired power plants that would together burn more than 50 million tons of coal each year, create 2 million tons of sulphuric oxide, 10,000 tons of carbon monoxide, and 370,000 tons of nitrous oxide, which would severely pollute the environment there. The dam will use an otherwise wasted, and sometimes destructive, energy source, water, to supply clean electricity for industrial and economic development.

But the Three Gorges project is also a means for scientific and technological collaboration at both the national and international levels, and thus an opportunity to exercise human self-realization or achievement by bringing science and technology together to cause a beneficial transformation of nature. The project is in fact utilizing and developing advanced construction techniques, and will install the highest quality power generation equipment available. On site concrete formulation takes place in Japanese machines, the hydroelectric generators come from Europe, and so on.

Finally the design of the Three Gorges Dam has been the subject of extensive ethical discussion and aims to contribute to the contemporary ideal of sustainable development. Where possible, biological preserves have been established to protect threatened species and to preserve water quality. Although more than 1 million people along the river are being relocated, they are being provided with new and better housing than they had in the past. Additionally efforts have been made to preserve materials of archeological value.

The Three Gorges Dam project is thus a major learning experience in China. It is teaching an important lesson in relating science, technology, and ethics.

FAN CHEN
YINGHUAN ZHAO

SEE ALSO *Dams; Ecology; Environmental Ethics; Pollution; Water*.

BIBLIOGRAPHY

"A Brief Introduction to the Three Gorges Dam Project," 2nd edition. *People's Daily*, November 5, 1997.

Chetham, Deirdre. (2002). *Before the Deluge: The Vanishing World of the Yangtze's Three Gorges*. New York: Palgrave Macmillan. A historical and personal story by a Chinese scholar with more than twenty years experience on the Yangtze.

Hersey, John. (1956). *A Single Pebble*. New York: Knopf. A fictional story about a civil engineer from the West who travels up the Yangtze in order to develop a Three Gorges Dam proposal, in the process discovering the rich historical traditions that any such project would disrupt.

Winchester, Simon. (1996). *The River at the Center of the World: A Journey Up the Yangtze and Back in Chinese Time*. New York: Henry Holt.

INTERNET RESOURCES

"The historical review of the Three Gorge Project". Available from http://www.people.com.cn.

"Will Rare Underwater Plants Become Extinct?" Available from http: //www. jxnews.com.cn.

Ziyun, Fang, and Yongbai Chen. "The Practice and Research of Environmental Protections of the Three Gorge Project." Available from htpp://www.emcp.acca21.org.cn.

THREE-MILE ISLAND

• • •

On March 28, 1979, a series of events took place at the nuclear reactor at Three Mile Island, Unit 2 (TMI-2), near Harrisburg, Pennsylvania, that resulted in an accident in which a significant fraction of the nuclear reactor core melted and a small amount of radioactivity was released to the environment. After more than twenty years of government-stimulated development of the nuclear power industry and in the context of increasing public objections, that accident became the focus for an intensely polarized debate about the wisdom of further construction of nuclear reactors. The accident at the Three Mile Island nuclear power station has taken on a key historical role in discussions concerning science, technology, and ethics.

Reactor Design

Understanding the accident requires a general understanding of the way the TMI-2 reactor worked. TMI-2 was a pressurized water reactor. A simple diagram of the system is shown in Figure 1.

The fission process—splitting the atom, with the release of energy—occurs in the reactor core. This generates heat, and so the core is cooled with water under high pressure, which is needed to prevent the water from boiling. The reactor is contained in a thick (ten inches) steel-walled reactor vessel. Two loops circulate the water. The primary loop carries the pressurized water through the reactor, where it is heated, to a device called a steam generator. In the steam generator heat is transferred from the primary loop to water in a secondary loop, which is not under pressure, and thus is converted to steam. Water in the primary loop does not mix with water in the secondary loop. Radioactivity in the primary loop never mixes with water in the secondary loop. The cooled water in the primary loop then is pumped back to the reactor for reheating. The steam produced in the secondary loop is piped to a turbine, where it hits turbine blades and causes them to spin. The turbine is connected to a generator that produces electricity. The steam then condenses below the turbine and is pumped back to the steam generator for its own reheating.

The primary loop is contained inside a steel-lined, steel-reinforced concrete building in which the walls are three to five feet thick. This containment building, as shown in Figure 1, is designed to prevent or at least minimize radiation leakage to the environment in case of a serious accident. It is a requirement in the United States that all commercial reactors be built inside a containment building. This is part of the "defense-in-depth" philosophy that has been required from the beginning in the design of commercial nuclear power plants in the United States.

The Accident

The accident began at 4:00 A.M., when maintenance activities caused secondary loop pumps to shut down, leading to a buildup of heat in the primary loop. The reactor shut down automatically, but the pressure in the primary loop increased significantly. As is shown in Figure 1, a pressurizer outside the reactor vessel monitors the primary loop. If the pressure gets too high, a valve opens and radioactive water escapes to the drain tank below the reactor.

This is what happened at TMI-2. When the pressure returned to normal, the operator sent an electrical signal to the motor that closes the valve. An indicator light showed this action was taken, causing the operator to believe the valve had closed. Unfortunately the indicator did not show the actual valve position, which was partially stuck open. One of the changes resulting from the accident is an indicator that actually shows closure of the valves. Another sensor in the control room showed high pressure in the reactor drain tank, which indicated a leak, but this indicator was located behind a seven-foot-high instrument panel.

Alarms and warning lights began to go off in the control room, indicating problems in different systems. This confused the operators and made it difficult to diagnose the problem and choose the appropriate corrective action. Water continued to leak through the open valve from the primary loop to the basement, where it overflowed from the reactor drain tank onto the basement floor. It then was pumped to tanks in the adjacent auxiliary building. When those tanks overflowed, radioactive water spilled onto the floor of the auxiliary building, enabling the radioactive gas xenon, an inert gas that is not incorporated into the body tissue, to escape from the building through the ventilation system. This resulted in a low-level exposure to residents in surrounding communities.

Even when a reactor is shut down, residual radioactive fission products in the reactor core continue to produce heat that must be removed. An emergency cooling system turned on automatically and started pumping water into the primary loop. The operators, however, thinking that the valve on the pressurizer was closed and noting that the water level indicator in the pressurizer showed that the pressurizer was full, throttled back and then shut down the emergency cooling system because they feared that the primary loop would overfill with water and cause a dangerous overpressure in the loop.

FIGURE 1

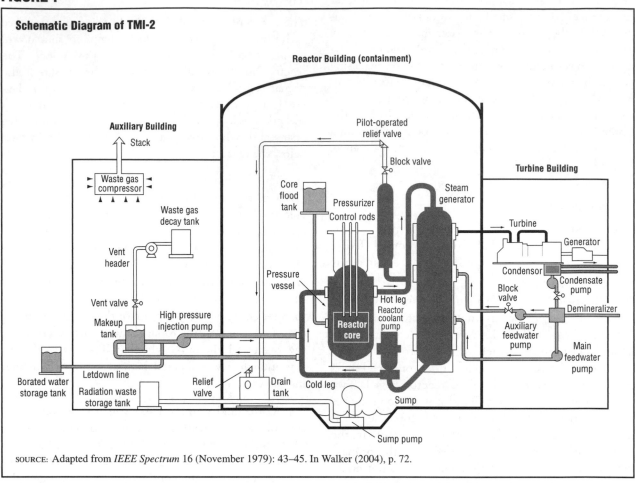

Schematic Diagram of TMI-2

SOURCE: Adapted from *IEEE Spectrum* 16 (November 1979): 43–45. In Walker (2004), p. 72.

Schematic diagram of TMI-2. (© *IEEE 2004.*)

Actually, the pressure was dropping in the primary loop because of the open valve, and boiling of the remaining water began to occur. The large pumps for the primary cooling water began to vibrate heavily because they were filling with steam from the boiling water. Those pumps were shut down to prevent them from being damaged. Although the primary loop water was boiling off, with the steam going through the open valve, serious damage to the core still would have been avoided if the emergency core cooling system had continued operating.

After about 100 minutes enough water had leaked from the core through the open pressurizer valve that the top of the core was no longer covered with cooling water. The temperature in the uncovered parts of the core began to rise. The fuel is contained in tubes called cladding made of a zirconium alloy, and the uncovered tubes began to react with the steam, releasing hydrogen. Some of that hydrogen escaped into the containment building and later underwent a rapid burn (mild explo-

sion) that caused some equipment damage. Some of the hydrogen accumulated in the top of the vessel that held the reactor and inhibited reactor cooling for several days. It also led to concern by some Nuclear Regulatory Commission (NRC) staff members that the hydrogen might explode. (It turned out that this was not possible because of an oxygen shortage in the system.) Because of uncertainty about the condition of the reactor two days after the accident began Pennsylvania Governor Richard Thornburgh advised pregnant women and pre-school-age children within a five-mile radius of the plant to evacuate.

After 142 minutes the cause of the leak was determined, and a backup valve for the pressurizer was closed, stopping the loss of water. However, by that time about one-third of the primary loop water had escaped. Because of concern that introducing cold water into the intensely heated core would cause the fuel elements to fracture, the emergency core cooling system was not restarted until four and a half hours after the accident began.

FIGURE 2

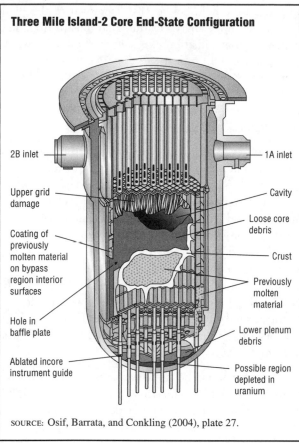

Three Mile Island-2 Core End-State Configuration

2B inlet

1A inlet

Upper grid damage

Cavity

Coating of previously molten material on bypass region interior surfaces

Loose core debris

Crust

Previously molten material

Hole in baffle plate

Lower plenum debris

Ablated incore instrument guide

Possible region depleted in uranium

SOURCE: Osif, Barrata, and Conkling (2004), plate 27.

Condition of the reactor core after the accident. (*Courtesy of the Pennsylvania State University Engineering Library.*)

As the core overheated and the cladding underwent chemical reactions as well as melting, the core structure began to lose strength and the top of the fuel elements collapsed into a pile, some of which heated to the melting temperature of the fuel, creating a large molten mass in the center. Some of that molten fuel eventually spilled over the side of the core and accumulated below the core. Altogether approximately 50 percent of the core melted. Fortunately, there was sufficient cooling water to prevent the molten fuel from rupturing the reactor vessel. Except for the radioactivity in the cooling water that leaked into the drain tank and then was pumped into the auxiliary building, from which there were small gaseous releases to the environment, almost all the radioactivity was contained within the containment building. The final state of the core at the end of the accident is shown in Figure 2.

Health Effects

The Nuclear Regulatory Commission, the Environmental Protection Agency (EPA), the Department of Health, Education and Welfare, the Department of Energy, and the state of Pennsylvania conducted studies on the health effects of the accident. All those studies concluded that the dose any member of the public received was far less than the natural background radiation. There was no increase in cancer in the surrounding communities.

Some nongovernmental groups and university researchers rejected those reports. Although the accident led to no generally accepted radiation injuries to the public or to workers, it did cause an emotional trauma to the local citizens and indeed to the nation. Without question it led to a loss of public confidence in nuclear power.

Lessons and Changes

Analysis of the accident revealed several significant operations problems in the industry as well as oversight problems at the NRC. Of particular importance was the finding that operator error had resulted from a lack of understanding of how the system behaved, a lack of information at the control panel to help operators make a correct diagnosis, and a control panel design that promoted confusion rather than understanding. Other issues in the accident included poor communication between the reactor site and NRC headquarters, ineffective communication with the public and the press, and an inadequate communication system for the NRC and industry to inform operators of safety problems identified at other plants. For example, the operators did not know that a similar stuck valve incident had occurred at another reactor eighteen months earlier.

In response the industry created an operations oversight organization called the Institute for Nuclear Power Operations (INPO). Among its activities are plant visits by expert teams on a regular basis (twelve to eighteen months), assistance to plant operators to improve their skills, and the creation of the National Academy for Nuclear Training, which accredits nuclear training programs in maintenance and operations to assure high standards. Simulators that replicate the behavior of the plant now exist at each site and are used to train operators on normal operations and accident scenarios. A key goal of INPO is the promotion of a culture at nuclear power plants that emphasizes "safety first" as the basis of decision making.

Finally, the NRC and industry used information from the accident to develop computer models that describe the progression of serious accidents. There are now emergency centers that conduct regular emergency exercises, including the use of local community response teams of emergency workers and fire fighters. All these efforts have transformed the U.S. nuclear industry and its regulation and have resulted in remarkable improvements in safe operations as well as economic performance, both of which were needed.

In the United States the nuclear power industry had developed rapidly in the 1960s and 1970s, with different companies involved and with diverse designs and changes in design with each new reactor. The power output of the reactors increased quickly from the early small reactors, with the belief that there would be an "economy of scale" with larger units. The regulatory process developed in parallel with industry growth, and changes in regulations were made as experience was gained and plants got larger. As a result each reactor was unique, and it was difficult to maximize learning in construction, operation, and maintenance. This contrasts with both the French and the Japanese nuclear power industries, which were initiated later and chose one or a small number of designs for their reactors, which contributed to facilitated learning in building and operations.

Accident Cleanup

Cleanup of the accident included the processing and storing of radioactive contaminated water in the auxiliary and reactor buildings and removal of contaminated building materials and the reactor core to a safe storage site at the Idaho National Engineering Laboratory (INEL). This was a lengthy, expensive, and contentious process. Numerous technical challenges, many of them first of a kind, had to be overcome. Those challenges included (1) building and operating systems to treat the radioactive water; (2) inspecting damage to the core, which revealed a collapse of the top five feet of the reactor material into a rubble bed, with a five-foot-thick section of solidified melted fuel below; (3) development and use of tools to break up the solidified section of the core so that it could be loaded into casks and shipped to INEL; (4) solving a biological growth problem that caused clouding of the water; and (5) the development and use of robotic equipment to decontaminate the reactor building basement. In addition to finding solutions to the technical problems, NRC approval was needed for each step in the cleanup. This often resulted in delays, partly because the NRC frequently sought general public input and acceptance.

Some of the contentious issues that arose delayed the cleanup. One was the venting of radioactive gas from the containment building to allow worker entry

and building cleanup to begin. Two raucous public meetings were held before NRC approval of the plan. The public was angry, fearful, and mistrusting, and assurances that radiation exposure to the public would be negligible fell on deaf ears. The venting took place from June 28 to July 11, 1980, and was monitored by the NRC, the EPA, a state agency, the utility company, and a citizen's group. Radiation exposure was determined to be negligible.

Another issue was more technical and involved the use of a crane above the reactor vessel to remove the vessel head to allow access to the fuel. The conditions inside the containment were junglelike, including high humidity and even rain. Extensive maintenance was performed on the crane to ready it for use, but one engineer, Richard Parks, wanted to do a full load test before attempting to lift the multiton vessel head. When management decided against this, Parks went directly to the NRC with his concern and was fired for whistle-blowing by the general contractor, Bechtel. The NRC sided with his concern, and testing was performed before the head was lifted.

Additional public concerns arose about shipping canisters of highly radioactive waste off-site to INEL and about the disposal of the decontaminated water after the rest of the cleanup had been completed. The simplest and least expensive solution would have been to release the water gradually to the river. This would not have presented any hazard to the public, but there was strong citizen opposition to putting the water into the Susquehanna River. In the end the utility agreed to evaporate the water. That operation was completed in August 1993 after a two-and-a-half-year process.

It took approximately eleven years to complete the cleanup and place the building in a monitored shutdown state. The cost was approximately $1 billion. This does not include the cost of replacement electricity or the cost related to TMI-1 being shut down for six years before it was allowed to restart. The cost to the industry was also substantial because the NRC required numerous modifications to the safety systems of all pressurized water reactors as well as changes to operating procedures. Although those changes did enhance plant safety in most cases, making changes in response to a crisis is generally more expensive and undoubtedly drove up the cost of nuclear power generation in the 1980s.

Ethical and Policy Issues

Several ethical and policy issues have arisen regarding the safety of nuclear power plants and whether another accident might occur. The first issue is whether electric power generation companies might put economics before safety. Although the industry has found that the safest plants are also the most economical, decisions to keep a plant operating even though conservative safety considerations suggest it should be shut down occasionally still occur. One example was the Davis-Besse plant in Ohio in 2002, where evidence of continuous corrosion of the reactor vessel was not investigated thoroughly until the corrosion completely penetrated the head. Fortunately, the steel liner was able to hold the reactor pressure until the problem was discovered. The public will have to judge whether the safety record of the industry and the oversight of the NRC are sufficient to justify the continued operation of nuclear power plants.

Second, and perhaps more significant in the early twenty-first century, is whether, in light of potential terrorist attacks against nuclear power plants, the nation should continue to use nuclear power, which in 2000 supplied approximately 20 percent of the electricity consumed in the United States. Could a group of terrorists breach all safety systems and cause a significant radiation injury to the public? After the terrorist attacks of September 11, 2001, security has been enhanced at each nuclear site, including the hiring of additional guards. Also, studies have been made on the effect of an airplane crash into the containment building and other parts of the plant. These studies suggest that the use of standard evacuation procedures would be sufficient to prevent any serious injury to the public. Nonetheless, some public officials and critics of nuclear power lack confidence in the results and believe nuclear power plants should be eliminated.

There are, however, national security and environmental benefits of nuclear power that must be considered. Nuclear power does not require the use of imported fossil fuels such as oil or future imports of natural gas. Furthermore, there are no emissions of sulfur oxides, nitrous oxides, or carbon dioxide as there are with the burning of fossil fuels. Indeed, nuclear power is already the dominant method of avoiding carbon dioxide emissions in the nation. Any replacement of the 20 percent of electricity generated by nuclear power could increase the cost of electricity generation, reduce the reliability of the electrical grid system, and/or increase pollutants emitted to the environment. Nuclear power may be critically needed to reduce the potential consequences of global warming. Also, as the price of natural gas rises and as it is recognized that natural gas may be able to serve as a substitute for oil in transportation, nuclear power may be the most cost-effective means for producing electricity, especially for electrical generation that has a minimum of environmental consequences.

EDWARD H. KLEVANS
DARRYL L. FARBER

SEE ALSO *Chernobyl; Nuclear Ethics; Nuclear Regulatory Commission.*

BIBLIOGRAPHY

Knief, Ronald Allen. (1992). *Nuclear Engineering: Theory and Technology of Commercial Nuclear Power*, 2nd edition. Washington, DC: Hemisphere. A technical account of the accident contained in a text on commercial nuclear power. Has a good summary of the lessons learned.

Osif, Bonnie A.; Anthony J. Baratta; and Thomas W. Conkling. (2004). *TMI 25 Years Later*. University Park: Pennsylvania State University Press. Written for the general public; provides a good summary of how nuclear power plants work as well as a description of the accident, its causes, and responses to it.

President's Commission on the Accident at Three Mile Island. (1979). *The Need for Change, the Legacy of TMI: Report of the President's Commission on the Accident at Three Mile Island*. Washington, DC: Author. (Also available in paperback and hardcover from Pergamon, Elmsford, NY, 1979.) The first and most important analysis of the accident, with particular emphasis on the failures of the NRC, along with ample discussion of problems in the nuclear industry. The authors identified needed improvements and made numerous specific recommendations to improve the industry and its regulation.

Rees, Joseph V. (1994). *Hostages of Each Other: The Transformation of Nuclear Safety since Three Mile Island*. Chicago: University of Chicago Press. The history of the Institute for Nuclear Power Operations. The primary emphasis is on the industry's response to the TMI-2 accident.

Walker, J. Samuel. (2004). *Three Mile Island: A Nuclear Crisis in Historical Perspective*. Berkeley: University of California Press. A well-documented book that describes the accident in great detail, summarizing what happened each day from Wednesday, when the accident started, to the Monday after the accident, when the most intensive crisis phase was over. Effectively captures the human dynamics involving the NRC, the governor of Pennsylvania, and state offices. Also continues the story of the accident through the reviews that followed.

TILLICH, PAUL

• • •

Born in Starzeddel, Germany, on August 20, Paul Johannes Tillich (1886–1965) explored the theological

Paul Tillich, 1886–1965. The American Protestant theologian and philosopher ranks as one of the most important and influential theologians of the 20th century. He explored the meaning of Christian faith in relation to the questions raised by philosophical analysis of human existence. (*Harvard University News Office.*)

and philosophical depths of contemporary culture. His experiences as a German army field chaplain in World War I shook Tillich's confidence in Western civilization, leading him to question its cultural and religious assumptions. In a series of professorships culminating in an appointment at the University of Frankfurt he spelled out his "theology of culture," exploring the unconscious, self-evident faith implicit in ostensibly secular social thought and structures. After he was dismissed from his professorship on April 13, 1933, by the Nazi government, on November 3 of that year Tillich arrived in the United States, where he held positions at Union Theological Seminary, Harvard University, and the University of Chicago. He died on October 22 in Chicago.

Tillich understood technology as an adjusting of means to an end. That process is present in animal behavior such as the building of a nest, but human technology transcends organic processes by making tools for unlimited use. Tillich called the technical forms closest to natural processes "unfolding" technologies, for example, cattle breeding; those technical forms conserve and develop the potentialities implicit in natural forms. "Realizing" technologies such as musical instruments represent the direct expression of spirit in symbolic productions. "Transforming" technologies, exemplified by machines, destroy living connections by imposing purposes that are not implicit in natural forms.

Tillich defined science (*Wissenschaft*) as any methodologically disciplined cognitive approach to reality. In the subject-object structure of knowing, science separates itself from its object. For Tillich modern science is also a form of controlling knowledge or technical rationality because of its intimate connection to technological application.

Science and technology are "ambiguous," Tillich argued, both creative *and* destructive. They provide liberation from superstition and debilitating work but are enslaving in other ways. This shadow side of science and technological development arises not from their essential structures but from their isolation from wider contexts of meaning and their domination (what Tillich calls imperialism) over other ways of knowing and acting. In this fallen state of autonomy they achieve a quasi-religious status as "scientism" and "technicism." Along with capitalism they form a trinity of social forces that determine the religious situation of modernity.

The fulfillment of scientific and technological possibilities cannot come from their subjection to political or religious authority, however. That would constitute the imposition of "heteronomy," or determination from outside. Science must be free to question every presupposition, Tillich argued, or it loses its character as science. The creative potential of science and technology must proceed though an autonomy aware of its own depth to become "theonomous," or transparent to the ground of being (God), and thus reunited with broader conceptions of the meaning of life.

Ambiguity as the mixture of creativity and destructivity pervades technological production as the tools that liberate humanity also subject humankind to the rules of the making of those tools. Ambiguity is manifest in humanity's limited ability to adapt itself to limitless technical productivity, including atomic weapons. It is revealed in the emptiness created by the production of gadgets, which represent means that become their own end. It is manifest in an objectification of both natural objects and persons that transforms both into things. Neither the external restrictions of heteronomy (including religious determination) nor the fallen autonomy of running ahead indefinitely in a meaningless world is adequate to overcome these ambiguities.

Scientism and technicism must be overcome by what Tillich calls *theonomy*. Theonomy does not prescribe particular technological objects but instead calls for the creation of technical *Gestalten* (wholes) that people can love for the form and meaning embodied in them. It does this through production that follows rather than precedes human needs and maintains the intrinsic power in things. It would not halt scientific inquiry into the nature of the atom, for example, but would ban the destructiveness of inventions such as the atomic bomb by limiting the desire to create such devastation. Theonomy demands that people be treated as a ends rather than means, overcoming technological structures of dehumanization. It resists the attempt to control knowledge or monopolize the cognitive function, influencing science indirectly by determining the attitude and style of scientific creations.

Science is ambiguous in that the observer remains estranged from objects, examining them for the sake of domination. It proceeds through observation and conclusion. However, the observed changes, in the process of being observed, result in the discovery not of the "real" but of an encountered reality. Science carries unexamined assumptions into arguments that may influence its discoveries, with every statement about an object adopting concepts that require further definition, ad infinitum.

Autonomous reason, without the depth of reason (the true-itself), is driven to solve its dilemmas by combating relativism with absolutism, formalism with emotionalism, and subjectivism with objectivism. In theonomy, however, reason is grounded in the depth of reason, leading toward a more inclusive pattern of participation and insight, delving not only into the nature but also into the ultimate meaning and existential significance of things. Science tends toward a nominalistic form of methodological reductionism that is manifest in empiricism and positivism. Cut off from the depth of reason, scientism creates its own quasi-religious myth of a meaningless universe that swallows everything, including scientific passion. Theonomy, however, rejects an "objective" approach that loses its objectivity by grasping only one element of an object and not the whole, reducing reality to its own terms.

Contemporary technological society is ambiguous, Tillich states, just like the technological era that brought it into being. The task of a theonomous technological society would be to move autonomy to its own depth, making things and structures transparent to the ground of their being, thus making them not only useful but significant components of a meaningful world.

Few modern theologians have attempted the broad and deep conversation Tillich carried on with political, social, economic, and cultural phenomena. His distinctively neoclassical style of thought, however, is more intelligible to those steeped in the European intellectual traditions than to those grounded in pragmatic American thought. The theologian and ethicist Reinhold Niebuhr (1892–1971), in contrast, is more accessible to readers in the United States. For those who can negotiate his prose, however, Tillich provides a systematic and comprehensive ethical, philosophical, and theological assessment of modernity, from art and architecture to space travel.

J. MARK THOMAS

SEE ALSO *Christian Perspectives*.

BIBLIOGRAPHY

Thomas, J. Mark. (1987). *Ethics and Technoculture*. Lanham, MD: University Press of America. This book reviews the evaluation of technological society in the thought of Talcott Parsons, Herbert Marcuse, Martin Heidegger, and Paul Tillich, arguing that the presuppositions of each thinker determine the terms of his social and ethical judgment.

Thomas, J. Mark. (1990). "Are Science and Technology Quasi-Religions?" In *Research in Philosophy and Technology*, vol. 10, ed. Frederick Ferré. Greenwich, CT: Jai Press. This essay uses thought within the "civil religion" debate and the work of Paul Tillich to analyze the proposition that science and technology constitute quasi-religions. It concludes that they are not quasi-religions but that scientism and technicism are.

Tillich, Paul. (1956). *The Religious Situation*. New York: Meridian. Originally written in 1926, this book explores the religious situation of modernity as it has been determined by the divine trinity of science, technology, and capitalism.

Tillich, Paul. (1963). *Systematic Theology*, vol. III. Chicago: University of Chicago Press. The third and final volume of Tillich's system explores the ways in which theonomy overcomes the ambiguities present in every dimension of finite life, including technological and scientific endeavors.

Tillich, Paul. (1988). *The Spiritual Situation in Our Technical Society*, ed. J. Mark Thomas. Macon, GA: Mercer University Press. This collection of essays brings together Tillich's writings on science and technology from his earliest German period (1927) until almost the end of his career (1963). It organizes those essays in five parts, describing the situation in technical society; the structure and meaning of science and technology; science, technology and human self-interpretation; dehumanization in technical society; and the symbols and ambiguities of a technical society.

TOCQUEVILLE, ALEXIS DE

• • •

Politician and author Alexis de Tocqueville (1805–1859), who was born in the village of Tocqueville in France on July 29 and died on April 16, is best known for his two politically minded books, *Democracy in America* (1835–1840) and *The Old Regime and the Revolution* (1856). Tocqueville was born into an aristocratic family and lived as an aristocrat. He had no children and no strong desire to perpetuate his family's noble name. His passion was to promote human liberty in democratic times, to keep alive what was best about the old aristocracies in societies devoted to the democratic understanding of justice. Tocqueville's political career was undistinguished, but he deserves to be remembered for his literary legacy.

Democracy in America, the outgrowth of an extended visit to the United States from May 1831 until February 1832, remains the best single book written on democracy and the best book written on America. It has in many ways become more true over time, as America has become more democratic. Tocqueville presents democracy not just as a form of government but as a way of life; the democratic ways of thinking, feeling, and acting, he correctly thought, had infused and would gradually continue to infuse themselves into every aspect of American and modern life.

Tocqueville's explicit discussion of democratic science, technology, and ethics occurs in Part 1 of *Democracy* Volume 2, where his subject is the democratic mind. There he describes Americans as Cartesians without ever having read a word of Descartes. They are habitual skeptics; they view all claims of personal authority as nondemocratic claims to rule. Skeptical of the soul, Americans act feverishly on behalf of the body and its enjoyments. So they prize scientific knowledge far less for its own sake than for its applications or technological effects. The Americans dismiss the proud and pure desire to know characteristic of theoretical science as an aristocratic prejudice. Democratic peoples subordinate pleasures of the mind to those of the body.

Tocqueville himself embraces neither the aristocratic nor democratic views of science, but adopts the position of an umpire determining what is true and false about each partial or extreme view. The pride associated with the ruling class in an aristocracy leads scientific inquirers to confine themselves to the haughty and sterile pursuit of abstract truths. All scientific advances find their roots in such fundamental inquiry, but aristocrats inconsiderately or unethically neglect what applied science might do to improve ordinary human life.

Alexis de Tocqueville, 1805–1859. Tocqueville was a French political thinker and historian who championed liberty and democracy. *(The Library of Congress.)*

Democrats, Tocqueville adds, are so selfishly enthralled with the benefits of technology that they neglect to provide for pure or theoretical inquiry. Democracies characteristically do not have a class that possesses the leisure required for the theoretical sciences; the mind needs relatively calm or unagitated social circumstances to achieve its possible perfection. The theoretical life is rarely possible for members of a merely middle class, for free beings who must work to earn a living.

For minds in democratic times, the most magnificent products of human intelligence are methods that quickly produce wealth and machines that reduce the need for human labor and the cost of production. Those who direct democratic nations, Tocqueville contends, must use their influence and power to go against the democratic grain by raising those minds on occasion "to the contemplation of first causes," to elevate them sometimes with the magnificence of the theoretical life. Their failure to do so might mean the near disappearance of scientific geniuses such as Blaise Pascal (1623–1662) and even the gradual decline of scientific progress itself. A nation with no theoretical passion at all might end up wallowing in the scientific stagnation characteristic of the China that Europeans discovered. The technical genius of America finally depends on the perpetuation of a way of life that disdains mere technology in the name of truth.

Tocqueville also worried about the effect of a democratic technological orientation on the souls of most human beings. He writes that if he had lived in an unjust, poor, and otherworldly aristocratic age, he would have attempted to turn people toward the study of physical science and the pursuit of material wellbeing. But in a democracy, people are readily pushed by social circumstances in that technological direction; there is no longer any need to promote applied science. Instead, the need is to raise souls in the direction of heaven, greatness, a love of the infinite, and the love of immaterial pleasures. The democratic danger is that "while man takes pleasure in [the] honest and legitimate search for well-being, he will finally lose the use of his most sublime faculties, and that by wishing to improve everything around him, he will finally degrade himself" (*Democracy in America*, Volume 2, Part 2, Chapter 14). So any comprehensive scientific claim for the truth of materialism—for the idea that there is no truth at all to claims for the soul's immortality—should be condemned by thoughtful human beings in democratic times as probably untrue and certainly pernicious.

Tocqueville was also a critic of the effect of applied science on language in democratic times. Language becomes progressively more vague and impersonal; human action is described using words more appropriate to mechanical motion. Precise personal distinctions and assertions become suspect, and metaphysics and theology slowly lose ground. Instead of saying, "I think," those who aim to influence democratic opinion say, "studies show." Having rejected personal authority, people in democratic times are far less skeptical concerning impersonal scientific claims about the various *forces* that shape their lives. Having freed themselves from aristocratic tyranny, people are seduced by the expertise of *schoolmasters* whose despotism is milder but exceedingly meddlesome. A democratic danger is the loss of any conception of free will or personal liberty; people will too easily be governed both by the claims of impersonal expertise and public opinion determined by no one in particular.

Tocqueville's significance is his account of all of modern life in terms of democracy. Many of his observations and fears anticipate, for instance, Martin Heidegger's account of all of modern life in terms of *technology*, and certainly modern democracy would be impossible without the liberation of technological progress for the most part from moral and political concerns. But Tocqueville emphatically refuses to equate technological progress with human progress. His judgments about democratic progress are friendlier to democracy and more judicious than Heidegger's. Democratic thought is partly true and partly not, and there is no reason to believe that people will not be able to correct some of its excesses in the directions of truth and liberty.

PETER AUGUSTINE LAWLER

SEE ALSO *Democracy; Freedom; Skepticism*.

BIBLIOGRAPHY

de Tocqueville, Alexis. (2000). *Democracy in America*, trans., ed., and introduced by Harvey C. Mansfield and Delba Winthrop. Chicago: University of Chicago Press.

Lamberti, John-Claude. (1999). *Tocqueville and the Two Democracies*, translated by Arthur Goldhammer. Cambridge, MA: Harvard University Press.

Lawler, Peter Augustine. (1993). *The Restless Mind: Alexis de Tocqueville on the Origin and Perpetuation of Human Liberty*. Lanham, MD: Rowman and Littlefield.

Manent, Pierre. (1996). *Tocqueville on the Nature of Democracy*, trans. John Waggoner. Lanham, MD: Rowman and Littlefield.

TOLKIEN, J. R. R.
• • •

Born in Bloemfontein, South Africa on January 3, fantasist, philologist, and critic John Ronald Reuel Tolkien (1892–1973) served in France during World War I and saw action at the Battle of the Somme. He completed his undergraduate studies at Exeter College, Oxford, in 1915, and from 1920 until 1924 was Reader and Professor of English Language at Leeds University. In 1925 Tolkien was elected Rawlinson and Bosworth Professor of Anglo-Saxon at Oxford University and Fellow of Pembroke College. In 1945 he was elected Merton Professor of English Language and Literature at Oxford. He published *The Lord of the Rings* in three volumes from 1954 to 1955 and retired from his professorship in 1959.

Man and Nature vs. Technology

In a 1951 letter to an editor, Tolkien commented that *The Lord of the Rings* and *The Silmarillion* (1977) were primarily concerned with "the Fall, Mortality, and the Machine." He explained that *the Machine* (or *magia*, magic) were plans or devices that dominated, either by

J. R. R. Tolkien, 1892–1973. Tolkien gained a reputation during the 1960s and 1970s as a cult figure among youths disillusioned with war and the technological age; his continuing popularity evidences his ability to evoke the oppressive realities of modern life while drawing audiences into a fantasy world. (*AP/Wide World Photos.*)

destroying the environment or by controlling the wills of people (Carpenter 2000, pp. 145, 146). His Middle-earth writings (*The Hobbit* [1937], *The Silmarillion*, *The Lord of the Rings*, the posthumously published *Unfinished Tales* [1980], and the twelve-volume *History of Middle Earth* [1982–1996]), can be understood as at least a partial response to a modern world that was embracing industry and technology. Tolkien believed the Machine (technology) was destroying his beautiful, rural, Edwardian countryside (represented in *The Hobbit* by the peaceful Shire) with wars, factories, cars, railroads, and pollution, and he saw no end in sight. He passed on his distaste for mechanization to his hobbits in the prologue of *The Lord of the Rings*: "They [hobbits] do not and did not understand or like machines more complicated than a forge-bellows, a water-mill, or a hand-loom . . ." (Tolkien 1994, p. 1). His two major villains in the story, Saruman and Sauron, are dependent on machines and use them to dominate and destroy the countryside. His descriptions of the realm of Mordor, with its desolate, scarred plains and history of being a stronghold of evil, were taken from his experiences on the battlefield.

Tolkien was not opposed to technology in itself, but he despaired of the motives behind it, which he saw as primarily concerned with speed, *immediacy*, and the desire for power and control. He compared the Machine with art, which created new worlds of the mind and imagination, and complained that labor-saving machines only added more and less effective work. He lamented that the *infernal combustion* engine had ever been invented, and expressed doubts that it could ever be put to rational use. He also disliked the fact that the Machine was increasingly associated with English daily life. He once owned a car, but found it difficult to drive in Oxford's traffic congestion, and commented that the *spirit of Isengard* (the evil Saruman's fortress) had led planners to destroy the city in order to accommodate more cars and traffic. Near the end of World War II he sarcastically suggested the war had been conducted by bureaucrats (*the big Folk*) who viewed most of it in *large motor-cars*.

Some critics suggested that *The Lord of the Rings* was an allegory and protest of atomic power and the dangers inherent in nuclear warfare. Tolkien emphatically denied this, saying that the story (which predated the nuclear age) was not about atomic power, but power exerted for domination. In his view nuclear physics could be used for domination, but it should not be used at all, and he further emphasized that the story was really about *Death and Immortality*. But he was stunned and outraged when he learned of the dropping of the atomic bomb on Hiroshima. He called the scientists who developed the bomb *lunatic physicists* and raged that it was idiocy to "consent to do such work for war-purposes, calmly plotting the destruction of the world!" (Carpenter 2000b, p. 116).

Tolkien's conservative Christian (Roman Catholic) beliefs contributed substantially to his attitudes about technology. In his seminal essay "On Fairy Stories" (1939, originally a lecture at the University of St. Andrews), he stated that human beings were *subcreators* who were created by God in his image to use their gifts wisely and in accordance with his wishes. The inclination of modern society toward domineering technology was, for Tolkien, a denial of God as creator. He called *The Lord of the Rings* a "fundamentally Christian and Catholic work" (Carpenter 2000b, p. 172), and his view of Christianity saw the universe as a place of conflict between good and evil.

Translation of *The Lord of the Rings* Into Film

In late 1957 Tolkien was approached by a group of American businessmen who gave him drawings and a story-line for a proposed animated film version of *The Lord of the Rings*. He wrote a member of the group a scathing letter of denunciation, explaining that the proposal and script, in whole and detail, was totally unacceptable, and that he did not want his story *garbled*. The early twenty-first century film versions of *The Lord of the Rings* have received generally favorable notices, particularly on the Internet and from young people. But several Tolkien scholars have written of their displeasure at the crass commercialization of the films, and the many liberties taken with characters and events. The films have been marketed by deploying the latest technology to sell to younger fans, and Tolkien's complex fantasy has been simplified into a visually stunning, character-driven action story with emphasis on spectacle rather than content.

Tolkien's son Christopher, the literary executor of his father's estate, did not disapprove of the film, but voiced doubts about the transformation of *The Lord of the Rings* into dramatic form. Tolkien, no doubt, would voice his displeasure over the films, and contend that technology has been used to reproduce and garble his narrative. He was resigned to the use of the Machine as a self-destructive tool of the modern world, which desired, in his view, to eliminate tradition and the past. He expressed his resignation in 1956, just a year or so after the publication of the final volume of *The Lord of the Rings*: "If there is any contemporary reference in my story at all it is to what seems to me the most widespread assumption of our time: that if a thing can be done, it must be done" (Carpenter 2000b, p. 246).

PERRY C. BRAMLETT

SEE ALSO *Anglo-Catholic Cultural Criticism; Christian Perspectives; Science Fiction; Science, Technology, and Literature*.

BIBLIOGRAPHY

Carpenter, Humphrey. (2000a). *J. R. R. Tolkien: A Biography*. Boston: Houghton Mifflin. The standard biography of Tolkien. The author was given unrestricted access to all Tolkien's papers and interviewed his friends and family.

Carpenter, Humphrey, ed. (2000b). *The Letters of J. R. R. Tolkien*. Boston: Houghton Mifflin.

Curry, Patrick. (1998). *Defending Middle-Earth—Tolkien: Myth & Modernity*. London: HarperCollins. Defends Tolkien's work from escapist and reactionary charges and maintains that *The Lord of the Rings* addresses the global realities and problems associated with the misuse of technology and destruction of the environment.

Purtill, Richard L. (1984). *J. R. R. Tolkien: Myth, Morality and Religion*. San Francisco: Harper & Row. Examines the religious and ethical ideas in Tolkien's work, with particularly trenchant chapters on the nature and role of myth and the art of storytelling.

Schick, Theodore. (2003). "The Cracks of Doom: The Threat of Emerging Technologies and Tolkien's Rings of Power." In *The Lord of the Rings and Philosophy*, ed. Gregory Bassham and Eric Bronson. Peru, IL: Open Court Publishing.

Tolkien, J. R. R. (1994). *The Lord of the Rings*. Boston: Houghton Mifflin.

OTHER RESOURCES

J. R. R. T.—A Film Portrait of J. R. R. Tolkien. (1992). Produced by Helen Dickinson. Directed by Derek Bailey. 110 minutes. Visual Corporation Limited. Videocassette. This is a video, the first made on Tolkien, with valuable contributions from Tolkien himself, his son Christopher, and noted Tolkien scholar Tom Shippey.

TOLSTOY, LEO
• • •

Lev Nikolaevich (Leo) Tolstoy (1828–1910) was born at Yasnaya Polyana, the Tolstoy family estate a hundred miles south of Moscow on August 28. He died on November 20 at a nearby railroad station, having fled in the night from an increasingly contentious marriage and a set of familial relationships that had been hardened in large part by Tolstoy's attempts to apply his radical moral beliefs to his own life. In the intervening eighty-two years Tolstoy became perhaps the most prominent novelist in an age and place of great authors as well as a vociferous critic of science and modernization.

Tolstoy's international fame rests primarily on two novels, *War and Peace* (1865–1869) and *Anna Karenina* (1875–1877). His fictional works also include short masterpieces such as "The Death of Ivan Ilyich" (1886), "The Kreutzer Sonata" (1889), and "Master and Man" (1895). In addition he wrote autobiographical accounts of his childhood (*Childhood, Boyhood, Youth* [1852–1857]) and his experiences as a soldier in the Crimean War (*Sevastopol Sketches* [1855]). With regard to issues of science, technology, and ethics Tolstoy's most relevant writings include a variety of short, passionate nonfiction works, particularly "What I Believe" (1884), "What Then Must We Do?" (1887), "On the Significance of Science and Art" (1887), "What Is Art?" (1898), and "I Cannot Be Silent" (1908), all of which address a confluence of moral and intellectual errors he perceived in modern life and thought at the turn of the twentieth century.

Leo Tolstoy, 1828–1910. Tolstoy was a Russian novelist, reformer, and moral thinker, notable for his influence on Russian literature and politics. (*The Library of Congress.*)

Tolstoy directed his most trenchant criticisms at the insensitive intellectuality of the urban elites, which he considered distant from the natural values of the land and its laborers; the modern Western adherence to science and its methods; and thinkers such as Auguste Comte (1798–1857), Georg Hegel (1770–1831), and simplistic interpreters of the philosopher Immanuel Kant (1724–1804) who built positivist historical and scientific doctrines on what he considered rickety evidence.

Despite his turn toward the simplicity of peasant agricultural values and the teachings of the Gospels, Tolstoy's commitment to a questioning, empirical worldview was deep. Tolstoy was never interested in a vague and disconnected mysticism. Those who consider themselves capable of circumscribing the infinite multiplicity of the world with their "scientific" theories were deluding themselves, he argued. People are not incapable of knowing or perceiving many of the causes or influences on which the natural and human world has been founded; it is simply that there are far too many influences, causes, and effects for people to remember and record, and to be able to integrate the available material in a scientifically conclusive manner. Positivis-

tic science rests on a lack of respect for the multiplicity of the natural and human worlds. Assuming too much about human capabilities to know and understand is, in the world of social action and belief, morally dangerous.

Like his contemporary Fyodor Dostoevsky (1821–1881), whom he never met, Tolstoy was broadly concerned with the spiritual future of the human race. He attempted to confront the gradual movement away from traditional values with an almost Aristotelian emphasis on the permanent relationships of things, promoting the universality of natural and religious values of love and labor to which he believed the human heart responds. Although the West now knows him as the writer of large and perhaps infrequently read novels, his influence on writers and political dissidents such as Mohandas Gandhi (1869–1948) and Alexander Solzhenitsyn (b. 1918) has been enormous, and his thought provides resources for ethical assessments of science and technology that have not yet been explored fully.

GLENN R. WILLIS

SEE ALSO *Russian Perspectives.*

BIBLIOGRAPHY

Bayley, John O. (1967). *Tolstoy and the Novel.* New York: Viking Penguin. Regarded by some as the finest, most incisive one-volume study of Tolstoy and his literary legacy.

Berlin, Isaiah. (1978). "The Hedgehog and the Fox" and "Tolstoy and the Enlightenment." In *Russian Thinkers.* New York: Penguin. Berlin's syntheses of historical and intellectual currents in nineteenth century Russia are accessible and brilliant. "The Hedgehog and the Fox" may be his most famous essay. Berlin's analyses of Tolstoy's influence on his own epoch, and the epoch's influence on Tolstoy, are unsurpassed.

Troyat, Henri. (2001). *Tolstoy,* trans. Nancy Amphoux. New York: Grove/Atlantic. Troyat's biography of Tolstoy is a new translation from the French. The dominant Tolstoy biography in Europe, it is an aspiring work of literature in itself.

Wilson, A. N. (2001). *Tolstoy: A Biography.* New York: Norton. Written by a prolific English literary biographer and novelist; more analytic in tone than Troyat's biography.

TOOLS AND MACHINES

• • •

Tools and machines are almost universally thought of as beneficial, which would make their invention morally praiseworthy. Indeed, without tools it is difficult to see how human beings could survive, and the increasing adoption of machines shows that most people see them as salutary contributions to human affairs. Although isolated tools or particular machines may on occasion be criticized for their negative impacts, this is done mostly to improve technological implements or to reform their uses. Nevertheless, one may note important distinctions between tools and machines as such, and how these distinctions, independent of any particular uses, may be ethically significant.

Distinctions

What is the difference between a tool and a machine? This question is complicated by lexicographic shifts over time. The Greek and Latin words for machine (*mechane* and *machina*) name a kind of tool (*organum* or *instrumentum*) for lifting heavy weights. Classical mechanics identified six basic types of such machines: the lever, wedge, wheel and axle, pulley, screw, and inclined plane. Machines, unlike other tools, presented a conundrum: How do they enable human users to lift weights that would otherwise be beyond their power to move?

Unlike with a stick used for poking or scratching, which serves as a straightforward extension of some human operation, determining how machines work is more difficult. Aristotle's *Mechanical Problems* was an early attempt to solve the mystery concerning how machines do what they do, that is, how they work or operate. What happens is that all six simple machines function as machines by transforming a smaller force exerted over a longer distance into a greater force exerted over a shorter distance by means of a structured redirection of the force in question.

But machines in this premodern sense are just one kind of tool. All tools, even simple machines, require two types of direct human inputs: energy and guidance. The hammer is swung with the arm and guided by hand–eye coordination. By contrast, machines in a modern sense require only one type of direct human input: guidance. The difference is that between a human-powered and -guided bicycle and a human-guided car; a person does not pedal a car, but simply drives it.

After human beings have constructed them or found natural objects with properties such that they can be used as tools, any use will involve some energy and guidance from a user. The guidance, precisely because it constitutes the introduction of intelligence, involves skill. In this sense the skillful use of tools is different from the more passive use of other artifacts such as baskets, chairs, and houses. The coordination of human power inputs, as when a group of men operates a battering ram, and the substitution of animal and other nonhuman sources of power such as wind for human power, foreshadow the development of machines in the modern sense.

The standard definition for the modern machine is: "a combination of resistant bodies so arranged that by their means the mechanical forces of nature can be compelled to do work accompanied by certain determinate motions." Alternatively, a machine is an "assemblage of resistant bodies, connected by movable joints, to form a closed kinematic chain with one link fixed and having the purpose of transforming motion." (Both definitions are from Franz Reuleaux, who in the late 1800s formulated the modern science of mechanics.) Mechanics, or the science of machines, analyzes the ways forces are compelled and transformed to do work in terms of their structures (statics) and functional operations (dynamics).

Functions and Uses

Tools and machines have internal operations or workings that can be used for many different purposes. These operations are commonly analyzed in modular terms: Gears slow down or speed up motion. A cam transforms reciprocal into rotary motion. Although how tools and machines operate or function does not fully determine their uses, they place boundary conditions on or for possible uses. Indeed, when an inventor applies for a patent on a new machine, the inventor is required to specify both its (external) use and how (internally) it is designed to operate or function so as to make possible the intended use. Engineering design thus considers both extrinsic use and internal structure and operation, and is successful when it unites the two.

But just as with the tool–machine distinction, so that between function and use is difficult to nail down. In many instances the word *function* can be replaced by the words *working*, *operation*, or even *use*. One must be careful in speaking about functions not to create an imaginary ontological substance that is nothing more than projected use. But to say that the machine operations or functions of pounding, drilling, or rotating are the uses of pounding, drilling, or rotating shifts attention from the structure of the machine and how it works to the intentions or purposes of the user.

For engineers who focus on machines, then, machines and their component parts are as often distinguished by operations or functions as by uses. Indeed, it is precisely in this sense that classical machines are distinguished from tools. The machine works to increase force across decreasing distance in ways that other tools do not. Moreover, the working or functioning of tools as tools depends on human energy and skillful guidance; modern machines work or function with only human guidance. Because of this, using machines requires less human work and, by placing greater and greater power in human hands, makes consciousness or forethought an ethical imperative. One does not have to be nearly as conscious about what is going on when riding a bicycle as when driving an automobile.

In general the experience of using machines is different from that of using tools in terms of the decline in human energy input and a corresponding increase in human mental input. This transformation of the use experience is of ethical significance and is independent of any particular use. It is true no matter what kind of machine one is operating and what one is producing with it or where one might be traveling in it. No matter what kind of machines are involved, machine users are morally obligated to think more than tool users about what is going on. To some extent this shift in the character of the use experience may also be described as setting the pattern for living in a machine-dominated technological world.

CARL MITCHAM
ROBERT MACKEY

SEE ALSO Animal Tools.

BIBLIOGRAPHY

Mitcham, Carl. (1994). *Thinking through Technology: The Path between Engineering and Philosophy.* Chicago: University of Chicago Press. See especially chapter 7, "Types of Technology as Object."

Reuleaux, Franz. (1876). *The Kinematics of Machinery: Outlines of a Theory of Machines,* trans. and ed. Alex B. W. Kennedy. London: Macmillan. Translation of *Theoretische Kinematik: Grundzüge einer Theorie des Maschinenwesens,* 1875.

TOTALITARIANISM

• • •

Totalitarianism is defined as a political system or regime in which the government seeks total control of society. This requires breaking down all the intermediate associations of civil society or turning them into agencies of the government, so that all that exists are, on the one hand, atomistic individuals and, on the other, the unity of the state.

Totalitarian systems have significant implications for science, technology, and ethics. Totalitarian governments rely on communications technology to spread an official ideology and to monitor subjects, while totalitarian control of the economy creates major hurdles to technological invention and innovation. Scientists face numerous ethical challenges in totalitarian systems, from ideological conditions often imposed on their research (a rejection of *Jewish science* in Germany and the promotion of Trofim Lysenko's genetics of the inheritance of acquired characteristics in the Soviet Union) to the kinds of projects on which they may be required to do research.

Features of Totalitarianism

The two classic scholarly examinations of totalitarianism are Hannah Arendt's *The Origins of Totalitarianism* (1951) and Carl Friedrich and Zbigniew Brzesinski's *Totalitarian Dictatorship and Autocracy* (1956). Friedrich and Brzesinski identify totalitarianism as a unique political order, opposed to democracy yet distinct from authoritarianism and dictatorship, and characterized by six key features. The first is an official ideology. In totalitarian systems, this ideology includes a blueprint for remaking society, either in ethnic or racial terms (as in the case of *fascism*) or in class terms (as in the case of communism) as well as justification for the monopoly of political power.

The second basic feature is a single mass political party, usually with a single leader, with a monopoly of political power. This group is part of the total penetration of society by the rulers. Other rival group identities in society—religious organizations, voluntary associations, other political parties—are either destroyed or brought under the control of the party.

The third characteristic is the existence of a secret police force and rule through the development of terror in the population. Because the leaders of the political system seek to penetrate and remake society, they are ruthless in dealing with political and cultural opponents. Any autonomous organization of activities is seen as a threat and all who are not active in their support of the ruling party are possible targets of harassment by the secret police. Even active, and loyal, party members are not immune, however. The purges of the Communist Party under Stalin, for example, were aimed at party members who were deemed not diligent enough in their identification and condemnation of potential threats to the system.

The fourth feature is its monopoly over the means of communication. Although it is impossible to control all forms of communication, totalitarian regimes seek to limit the autonomous flow of information. Control over information is a crucial component in solidifying the ideology in the minds of the population—facilitating the establishment of legitimacy for the leaders, justifying its monopoly of political power, and creating support for its social blueprint.

The fifth characteristic, highlighted by Friedrich and Brzezinski, is the monopoly over weapons in society. This is not a feature unique to totalitarian systems (many democracies control access to weapons by the general population). It is, however, a necessary feature of totalitarian control.

The final feature is a centrally controlled economy. Control over the economy serves three purposes. First it assures the social blueprint; economic development can be structured in the way most supportive of the plan for remaking society, and the workplace can be used as an arena for socializing the masses in support of the system. Second it assures access by the state to the resources it requires to maintain power at home and expand its influence abroad. Finally, and perhaps most important, a centrally planned economy makes people dependent on the state. Thus, while arguably economically inefficient, a planned economy is politically efficient.

Arendt proposed a similar description of totalitarianism, emphasizing its ability to *atomize* the population (controlling the ability of the population to engage in group activities autonomous from the party or the state) and its effective use of ideology. The development of a mass adherence to official ideology is essential for the formation of legitimacy in totalitarian systems. Control over communications—particularly the educational system and mass media—made the development of such adherence theoretically possible.

Totalitarianism in Practice

In practice totalitarianism has never achieved the complete penetration and control of society. Although people were careful in public, and often went through the motions of participating in state-sponsored mobilization efforts, they led separate public and private lives. Terror crept into the private lives of individuals—one had to be extremely leery of speaking ill of the government even among one's good friends—but people also partook in the activities of normal life: shopping, attending the ballet, walking in the park, and so on.

Because the ideal differed from the reality of totalitarian life, some political scientists and many social historians (see, for example, writings by Sheila Fitzpatrick and Stephen Cohen) criticized the totalitarian model for overemphasizing politics, underemphasizing the role of society, and assuming a system of tight, top down control devoid of political and social conflict. The totalitarian model of politics assumed that everyone was completely controlled and atomized, and that leaders never responded to society. But in the Soviet case, leaders sometimes appealed to constituencies, and policies were, at times, sparked by initiatives from below.

The three examples in the real world that came closest to approaching the totalitarian ideal have been Adolph Hitler's Nazi Germany (1933–1945), Joseph Stalin's Soviet Union (1929–1953), and, more recently, the Taliban-run system in Afghanistan. None of these, however, achieved full realization of the totalitarian ideal. These three cases provide helpful examples of three forms of totalitarianism: fascism, communism, and Islamism. Friedrich and Brzezinski argue that fascist and communist dictatorships were basically alike, though one can identify different points of emphasis between the two forms of government. Fascism is a form of totalitarianism that emphasizes racial and/or ethnic superiority, engages in militarism, and argues for the need for a dominant state to develop the capacity of the superior race and/or ethnic group. According to Barrington Moore (1966), fascism develops as the result of an alliance among the state, the land-owning elite, and the industrial bourgeoisie. Communism emphasizes the remaking of society to eliminate economic exploitation through state control of the means of production. Moore argues—ironically, given Karl Marx's prediction of workers' revolutions in the most economically developed countries—that communism developed where the lack of a middle class and the presence of a large and disgruntled peasantry allowed revolutionary leaders to seize control of the government in the name of destroying the old economic order. In both forms of totalitarianism in practice, increasing control over the economy and society were justified through the claim that one or more groups (for example, capitalists or Jews) were *enemies* of the people.

Islamism is a more recent variant of totalitarianism. Its ideology is anti-western, critical of modernization, and emphasizes the dominance of Islamic law—as interpreted by the leaders—over society.

Science, Technology, and Ethics

The totalitarian goal to penetrate and remake society completely has significant implications for science and technology. The control and monitoring that characterized totalitarianism shaped the practice of science dramatically. In the ideal totalitarian system, scientists are less free than in any other type of system to pursue their research as they see fit. Scientific research and related technological advances become the property of the party-state. This situation poses ethical dilemmas for scientists. On the one hand, the likelihood that the fruits of their labor could be used in unpleasant ways by the state creates a disincentive for scientists. On the other hand, working through the official scientific channels is the only way for such scientists to conduct their research. Thus, although in practice scientists in systems with totalitarian features conducted pioneering research, such scientists were limited both by the imperatives of the totalitarian ideology and by their personal ethical concerns about the consequences of their research.

Technology is a necessary tool in the transformation of tyranny into totalitarianism. Friedrich and Brzezinski emphasize technology in their discussion of totalitarianism, arguing that this type of political system could only have arisen in an era of modern technology. They highlighted the role of technology in allowing control over communications and making possible large scale economic planning, as well as in facilitating the monitoring of everyday life by the secret police. Totalitarian governments direct scientists to develop such technology.

Though technology is a necessary part of a modern totalitarian state, technology was not easily absorbed into the totalitarian system in practice. Not all technological products of scientific research found a receptive audience in the party-state bureaucracy. The economic planning approach that was a feature of the Soviet system, for example, made it difficult to incorporate technology. Many economic planners feared the introduction of new technology because of the uncertainty that accompanied the introduction. As a result, when there was a clear goal to increase production, and when this increase could be achieved through the addition of more inputs into the system (extensive growth), the totalitarian planning system worked fairly well. As the global economy moved in the direction of growth resulting from technology-driven improvements in efficiency (intensive growth), the Soviet planning system lagged behind.

Finally technologically-conditioned improvements in communication posed serious problems for totalitarian systems. While technology made monitoring of large numbers of citizens possible in the middle part of the twentieth century, the growth of fax machines, personal computers with printers, cellular telephones, and Internet connections by the early twenty-first century, provided citizens in dictatorial countries with access to information from outside the country and enabled them to compose and spread antigovernment messages quickly and relatively anonymously. Technology may allow Big Brother more ways to monitor citizens, but it also provides citizens more opportunities to engage in subversive activities.

LOWELL W. BARRINGTON

SEE ALSO *Arendt, Hannah; Authoritarianism; Conservatism; Fascism.*

BIBLIOGRAPHY

Arendt, Hannah. (1951). *The Origins of Totalitarianism.* New York: Harcourt, Brace.

Cohen, Stephen. (1985). *Rethinking the Soviet Experience: Politics and History since 1917.* New York: Oxford University Press.

Easton, David. (1965). *A Framework for Political Analysis.* Englewood Cliffs, NJ: Prentice-Hall.

Fitzpatrick, Sheila. (1986). "New Perspectives on Stalinism." *Russian Review* 45(October): 357–373.

Friedrich, Carl, and Zbigniew Brzezinski. (1956). *Totalitarian Dictatorship and Autocracy.* Cambridge, MA: Harvard University Press.

Moore, Barrington, Jr. (1966). *Social Origins of Dictatorship and Democracy: Lord and Peasant in the Making of the Modern World.* Boston: Beacon Press.

TOURISM

• • •

The ancient Greek philosophers thought that leisure was a necessary component of human flourishing even though freedom from the demands of necessity was possible only for a few people. Modern industrialized countries have achieved economies that for many of their members facilitate leisure, or, as Thorstein Veblen (1857–1929) suggested, the "non-productive consumption of time" (Veblen 1994 [1899], p. 43). In this context tourism is a form of unproductive consumption that is peculiar to the technologically advantaged. Tourism, however, also has become a major stimulus to economic production.

Purpose and Effects

Tourism is travel based on desires to relax, sightsee, appease curiosity, satisfy a sense of adventure or an adventurous self-image, compete with one's peers or colleagues, re-create images of paradise or luxury or the exotic, and escape. Tourism affects the economies and cultures of destination sites in both positive and negative ways. Those locales may organize their production activities around the satisfaction of tourists' demands for leisure, fantasy, adventure, or knowledge, activities that may operate to the detriment of local cultures.

As with any human relations involving production and consumption, even an activity centered on leisure, tourism thus calls for ethical and philosophical reflection. Only recently, however, has the phenomenon of tourism become a subject of ethical consideration, largely through its connection to other concerns, such as environmental degradation (to which "ecotourism" is one response), economic development, and cultural impacts.

Distinctions

The word *tourism* is derived from the Latin *tornus* and before that the Greek *tornos*, referring to a tool for making a circle (the word *turn* comes from the same root). Taking a tour thus implies circumnavigating, and the term *tourism* initially had depreciatory connotations of superficiality. In the early twenty-first century the connotations are more complex.

Tourism must be distinguished from other kinds of and motivations for travel. Economic and political migration, for example, is not new, but its increased extent is considered a significant element of globalization (Held, McGrew, Goldblatt, and Perraton, 1999). Contemporary economic migration includes the journeys made by migrant laborers and travel for business purposes in a postindustrial age of transnational corporations and labor markets, prompted also by international disparities in wealth and movement, especially between less developed countries and Organization for Economic Cooperation and Development (OECD) countries. Political migration includes refugees from crisis or conflict areas. Tourism, in contrast, has no material imperative, although one could argue that advertising and the media create a perceived necessity for tourism.

Flâneurism, a form of consumption activity that is much closer to tourism, is leisurely and detached urban promenading among the crowds, allowing spontaneous perceptual encounters to determine the directions of one's movements and thoughts. Although the expression came from the poet Charles Baudelaire (1821–1867), the philosopher Walter Benjamin (1892–1940) is perhaps the preeminent exponent of flâneurism through his writings on walking in Paris (Benjamin 1999). Voyeurism suggests a disengaged onlooker without a commitment to the local environment and thus overlaps with many common tourist practices. At its most benevolent voyeurism is the observation or immersion experience of other cultures in a way that allows one to extricate oneself when the experience becomes uncomfortable or problematic. The observation or experience, though perhaps immersive, allows a relatively easy exit from the situation, unlike the case for members of the local culture. There is a fine line between "authentic," engaged traveling and voyeurism.

Tourism is first and foremost an industry. It is one of the largest modern industries, accounting for hundreds of billions of dollars per year, and is the most significant industry for many countries. According to the World Tourism Association, which became an executing agency of the United Nations Development Programme in 1976, tourism grew from 456 million international travelers in 1990 to more than 700 million in 2002. Tourism appeared as both a word and a phenomenon in the early 1800s in association with increases in the means of transportation brought about by the construction of roads and highways, advances in carriage technology, and the building of the railroads. The current growth in tourism is due largely to the same processes and technologies that drive and constitute globalization and its consequences, including ease and frequency of transport and the growth of information and communications technologies. Economically advantaged people increasingly seek more far-flung and diverse destinations for vacation and pleasure.

The idealized motivation driving some forms of tourism is, as the Spanish-American philosopher George Santayana (1863–1952) suggested, that "there is wisdom in turning as often as possible from the familiar to the unfamiliar: it keeps the mind nimble, it kills prejudice, it fosters humor" (Santayana 1968, p. 15). Arguably, however, tourism today is much more epistemically ordered even when it takes on authenticity-seeking or adventurous forms.

Varieties of tourism or leisure travel have been distinguished in regard to the authenticity of the experience of other cultures and places (see Boorstin 1961). Dean MacCannell (1999 [1976]) suggests that actual gradations in the search for authenticity resist polar categorizations of tourism as authentic or inauthentic. Rather, destination places, tourist objectives and perceptions, local expectations and dependencies, "staged authenticity" (MacCannell 1999 [1976], title of chapter 5), and the dynamic nature of cultural activities and artifacts render such categories indistinct. The global journeying that through the years has created "backpacker meccas" in places such as Goa (India), Kathmandu (Nepal), and Lamu (Kenya) may seem a more authentic quest for rich cultural experiences in comparison to sheltered resort vacationing (enclave tourism), in which the actual place or culture is insignificant. Authenticity, however, is framed by the tourist's cultural expectations as much as it is a property of the experience of foreign destinations. The "inauthentic," moreover, may involve a relatively benign mutual exploitation or exchange between tourists and locals.

Ethical Issues

The paradox of the authenticity-seeking traveler is that the more tourists vacation in a particular place, the more a tourism infrastructure is developed and the more that place comes to resemble the tourist's home, causing local cultural and environmental deterioration. Pico Iyer (1989) has written about the unusual juxtapositions and hybrids of different cultures one finds across the globe as a result of the forces of globalization and tourism. This paradox creates a dilemma regarding whether to visit a place or to tour at all. The question for anthropologists and environmentalists is whether it is appropriate to visit a fragile culture or a pristine environment when one's visitation contributes to its alteration. Furthermore, as a tourist destination becomes more developed and attracts increasing numbers of visitors, many tourists may look elsewhere for less-traveled destinations. As a consequence they may perpetuate the same cycle, and some overdeveloped areas ultimately may witness a decline in the visits on which their economies depend.

From the perspective of those who welcome the local tourist industry may provide much-needed income and infrastructure development, but the cycle of unmanaged tourism development ultimately places those economic benefits at risk. Although income is generated locally from the industry, the distribution of benefits is uneven, and there may be severe damage to local cultures, other parts of local economies, and the natural environment. Such considerations have generated antitourism and protourism positions, with the former generally concerned with the environmental

and cultural impact and the latter with economic development.

Tourism raises specific and clear ethical and cultural concerns in regard to some of its manifestations, for example, sex tourism and reality tourism, with the latter involving poor or oppressed people inviting visitors to observe and experience their living conditions (an example of voyeurism). Opponents of tourism point to increased child labor, greater crime rates, and increased prostitution.

Tourism may contribute indirectly to resource conflicts and tensions with traditional land-use practices in addition to eroded cultural values and commodification of traditional practices. Economically it can lead to increased prices for basic goods for local people and higher costs for infrastructural development, diverting resources from other critical social sectors. Environmentally tourism may lead to the depletion of natural resources and pollution (air pollution, sewage, solid waste) in addition to problems such as coral reef anchoring, trampling, construction and deforestation, and disruption of ecosystem processes. Other common foci of criticism include the large amounts of fuel burned by airliners transporting tourists to and from their destinations, the construction of golf courses in environmentally fragile areas, and the aesthetic pollution of overdevelopment.

Proponents of tourism point to new infrastructure development for residents, greater civic participation, and reinvigoration of cultural traditions in addition to the mutual understanding and respect that may result from cultural exchange. Tourism may contribute to state revenues and foreign exchange earnings, increase employment opportunities, and help local economies grow. Environmentally tourism may contribute to new investments in conservation efforts, lead to regulatory measures and improved management practices, and provide new forms of employment. It also may indirectly involve the development of better technologies for conservation programs through technology transfer and the growth of science-based programs for environmental management.

The distinction between negative and positive effects depends principally on the specific contexts, rendering the prospects of a global management program extremely challenging. Environmental impacts, however, can have a far-ranging effect beyond the particular tourism context. This contributes another dimension to already complex ethical questions of obligations beyond borders, especially in a globalizing era.

The expansion of ecotourism is a major response to such concerns over environmental and cultural degradation and an attempt to invigorate local economies that otherwise are dependent on environmentally unsustainable practices. In some cases such practices are directly related to the tourism industry (for example, deforestation in the Himalayas for wood-fire cooking); in others the practices may be the sole (and sometimes illegal) source of income (such as rain forest logging).

Ideally, the goals of ecotourism are to combine ecological and cultural awareness with sustainable local economies and resource use and preserve local cultural identities and values. Ecotourism may include what is sometimes referred to as "scientific tourism." This form of tourism may range from volunteer fieldwork in the collection of scientific data to tourism accompanied by an ecologically informed guide. The growth of ecotourism in some areas, however, often represents a superficial assuaging of tourists' environmental concerns and expectations rather than an actual advance in conservation practices. Cheating on the ecotourism designation is common in some areas in the form of advertising regular activities, accommodations, or management practices as "eco-friendly" to attract unsuspecting tourists concerned about ecological impact. This has prompted efforts to certify and monitor ecotourism companies. Nevertheless, genuine ecologically benign tourism, even if it is possible, seeks to attract tourists to fragile places, thus re-creating the paradox mentioned above.

More recently these collective considerations have found expression in international forums. The World Tourism Organization (WTO), which is affiliated with the United Nations, has drafted a "Global Code of Ethics for Tourism" (1999). The code consists of ten general principles intended to guide "stakeholders" and supplement the tourist industry's emphasis on the market and private enterprise aspects of tourism. The WTO seeks to encourage "sustainable tourism," encompassing some of the considerations raised above. The United Nations Environmental Programme also attempts to integrate tourism considerations with international agreements such as the United Nations Convention on Biological Diversity.

The intersection of facilitating technologies, economics, and culture, along with environmental impacts, generates ethical considerations and dilemmas involving tourism. The new directions of tourism remain to be seen as globalization proceeds. Some places focus on the tourist industry to boost economies whose other industries may be stagnating or nonexistent. However, as a result of the fickle nature of tourism and its poten-

tial for the destruction of local environmental and cultural resources there is urgent cause for concern over dependency on tourism, particularly in developing countries. Ecotourism may provide only a temporary answer to economic and ecological realities without a more closely regulated and monitored industry or different global economic arrangements. If tourism is inevitable, perhaps the best option is the development of a global regime of "sustainable tourism." The Kingdom of Bhutan may provide an educative example, as it limits the numbers of visitors per year in the name of sustainable environmental and cultural considerations while trying to sustain economic well-being.

These issues perhaps may be overcome through shared, direct experience of places such as the Nepalese Himalayas, the biodiverse rain forests in Costa Rica, and the coral reefs of the South Pacific or of the peoples of New Guinea, Lapland, and central Africa. Perhaps what is needed is an ethics of tourism that is attentive to character, obligations, equity, and rights so that the benefits of tourism may flourish without doing harm. Perhaps there is also a need for a practical ethics of tourism that can admit that sometimes it is better not to be a tourist at all.

THOMAS C. HILDE

SEE ALSO *Benjamin, Walter; Consumerism; Science, Technology, and Society Studeis.*

BIBLIOGRAPHY

Benjamin, Walter. (1999). *The Arcades Project*, trans. Howard Eiland and Kevin McLaughlin. Cambridge, MA: Belknap Press.

Boorstin, Daniel J. (1961). *The Image: A Guide to Pseudo-Events in America*. New York: Harper and Row.

Fussell, Paul. (1980). *Abroad: British Literary Traveling between the Wars*. Oxford: Oxford University Press.

Held, David; Anthony McGrew; David Goldblatt; and Jonathan Perraton. (1999). *Global Transformations*. Stanford, CA: Stanford University Press.

Iyer, Pico. (1989). *Video Night in Kathmandu and Other Reports from the Not-So-Far East*. New York: Vintage.

MacCannell, Dean. (1999 [1976]). *The Tourist: A New Theory of the Leisure Class*. Berkeley: University of California Press.

Santayana, George. (1968). "The Philosophy of Travel." In *The Birth of Reason and Other Essays*, ed. Daniel Cory. New York: Columbia University Press.

Smith, Mick, and Rosaleen Duffy. (2003). *The Ethics of Tourism Development*. New York: Routledge.

Veblen, Thorstein. (1994 [1899]). *The Theory of the Leisure Class*. New York: Penguin.

World Tourism Organization. (2002). *Tourism Market Trends*. Madrid: Author.

INTERNET RESOURCE

World Tourism Organization. (1999). "Global Code of Ethics for Tourism." Available from http://www.world-tourism.org/projects/ethics/principles.

TOXIC CHEMICALS

SEE *Arsenic; Chemical Weapons; Regulatory Toxicology.*

TOXIC METALS

SEE *Heavy Metals.*

TOXICOLOGY

SEE *Regulatory Toxicology.*

TOYS

SEE *Robot Toys.*

TRADEOFFS

• • •

Tradeoffs occur under constraints similar to zero-sum games in which one participant's gain (or loss) is balanced by another's loss (or gain). A tradeoff is an exchange that occurs as a compromise, giving up one set of interlocked advantages and disadvantages in order to gain another, more desirable set. The benefits that are foregone in a particular case are often referred to as the opportunity-costs of that decision. Many personal and policy decisions regarding scientific research, technological development, and the use of technological products, processes, or systems depend either consciously or unconsciously on accepting tradeoffs. In many cases so-called ethical criticisms of science and technology are themselves criticized as ignoring the need for tradeoffs. Analysis of the concept of tradeoffs is thus an important feature of any general appreciation of relations between science, technology, and ethics.

Examples in Science and Technology

Human life is saturated with tradeoffs because time is a limiting resource. People can only perform a limited number of activities and thoughts in a given period of time. Usually, the routine of life masks the tradeoffs made and opportunity costs incurred.

ECONOMICS AND SCIENCE. People are perhaps most aware of tradeoffs in financial choices because money is another limiting resource. For example, with the money I have, I can choose between buying a car and taking a vacation. As Kenneth Arrow (1974) noted, much of economics involves saying "this or that, not both" (p. 17).

Budget allocation scenarios present important instances of tradeoffs in science as well. For example, the National Institutes of Health (NIH) experienced annual increases of fifteen percent between 1998 and 2003. Such large growth was justified by the potential health benefits of new advances in biomedical research, but some complained that the physical sciences and engineering suffered as a result of this prioritization. Other tradeoffs occur further downstream in the allocation of these funds through competitive grant processes. At the NIH, for example, decisions must be made about which diseases to prioritize and which researchers and facilities are most qualified to carry out that research. Indeed, this illustrates a more general point that prioritization is one way of dealing with tradeoffs, and the failure or inability to set priorities is a failure or inability to appreciate the reality of tradeoffs.

ENGINEERING. Tradeoffs are essential to both the internal operations of engineering and architecture as well as their social interactions. According to Edward Wenk (1986), "The most demanding skill in engineering design may ... be the *acute weighing of tradeoffs*" (p. 53). Different materials have different advantages and disadvantages for a given project and competing goals such as beauty, efficiency, responsiveness, and durability must be traded off against one another.

But tradeoffs in the design and implementation of technology are not an insular affair, limited only to considerations of material and design constraints. Another important factor in engineering tradeoffs is the public perception of risk. Engineers must incorporate safety margins and/or redundancy into their designs in order to reach socially acceptable levels of risk. These extra measures impose additional costs and other constraints, which can lead to declines in efficiency or functional performance.

Wenk demonstrated how political and financial aspirations can be traded off against safety in the use of technology. In the 1980s several highway bridges collapsed, but the problem was not poor design or age. Rather, political leaders caved into the pressure from trucking lobbies to permit greater truck weights by relaxing load limits. Citing the costly but failed U.S. federal bailout of railroads and the persistent pursuit of the Strategic Defense Initiative despite signs of systematic problems, he wrote, "the more massive a technology, the greater seems to be the political momentum for implementation and the greater the difficulty in identifying the tradeoffs occasioned by its accomplishment" (p. 38). Wenk also speculated about the influence of political concerns on the ill-fated *Challenger* shuttle. It was launched on the morning of the State of the Union address, which may have affected the managerial decisions about how to treat warnings of a possible failure of the O-rings. These cases point out the ethical responsibility of engineers when political considerations are traded off against safety concerns.

APPLICATIONS. Yet Wenk's most important point is that every choice involving science and technology presents tradeoffs because technological innovation and implementation are not unqualified goods. There are disadvantages to go along with the advantages and costs to go along with the benefits. This symmetrically implies that forgoing or somehow altering the pursuit and application of knowledge presents benefits as well as costs. For example, participants in the lengthy Environmental Impact Statement (EIS) process concerning the construction of a wind farm in Nantucket Sound, Massachusetts, were weighing many tradeoffs, including the one between clean energy and the beauty of a relatively pristine seascape.

The case of chlorofluorocarbons (CFCs) is another example. Industrial and political leaders were at first unaware that the use of CFCs involved tradeoffs between human and environmental health and the conveniences of widespread and cheap refrigeration. The international decision to phase out the use of CFCs was another tradeoff between the costs of such a large-scale economic transition and improved human and environmental health. Companies that produce hazardous wastes face tradeoffs between the costs of containing and storing that waste and the potential liability for damages to human and environmental health. As they attempt to minimize costs, the risks to health usually increase (Sewall 1990). Another example stems from the threat of terrorist attacks and the resulting tradeoffs between national security and scientific freedom of inquiry. In these cases, decisions must be made by public leaders, but many tradeoffs involving the use of technology are made by individuals. For example, those who choose general over commercial aviation accept the tradeoff of increased cost and risk for greater convenience.

RISKS. John Graham and Jonathan Wiener (1995) argued that as technology has come to saturate modern life, government has increasingly adopted the role of reducing risks to environmental and human health. They point out that risk tradeoffs often confound these efforts, as well-intentioned efforts to reduce some risks can turn out to increase others. Efforts to counter a "target risk" can generate "countervailing risks," which are commonly known as side effects (medicine), collateral damage (military tactics), or unintended consequences (public policy). If decision makers are well informed, they may be able to reduce overall risk by choosing "risk-superior" options, but sometimes risk tradeoffs are unavoidable.

Risk tradeoffs occur at both personal and societal levels. For example, a woman dealing with menopause can take hormonal replacement therapies to ward off the risk of osteoporosis and chronic pain, but in so doing she may increase the risk of uterine and breast cancer. Similarly, visiting a hospital can reduce risks from trauma and illness, but it can also lead to other illnesses. On a social level, decision makers must choose when to chlorinate drinking water, which kills harmful microbes but may add a cancer risk. Spraying hot water on the beaches of Prince William Sound, Alaska, after the 1989 *Exxon Valdez* oil spill reduced risks to nearby otters and birds, but may have harmed the longer-term ability of the ecosystem to recover by killing certain marine organisms and microbes. Grahm and Wiener proposed a risk tradeoff analysis framework to help decision makers grasp the entire portfolio of risks that science and technology can present within a given decision.

Tradeoffs as an Explanatory Concept

The notion of tradeoffs is important not only in decision making but as an explanatory term in several scientific disciplines, including economics and evolutionary biology. British economist Lionel Robbins called economics the study of human behavior as a relationship between ends and scarce means that have alternative uses. Indeed, microeconomics rests largely on the math of constrained maximization (for example, Lagrange multipliers). Robbins' definition of economics shows its close connection to ethics and politics as all involve the assessment of social institutions and the consequences of alternative decisions. The ethics of political-economics derives from the fundamental tradeoffs posed by scarcities of land, labor, and capital. Even social programs that do achieve their goals leave society with fewer available resources to further values in other policy areas. Steven Rhoads (1985) stated "spending and regulatory decisions that use scarce resources ... incur costs in terms of forgone alternatives (that we no longer have the capacity to undertake) elsewhere" (p. 11). But economic activity is not entirely a zero-sum game. For example, comparative advantage can increase overall output and welfare if countries specialize their production processes and engage in trade. Similarly, although many tradeoffs exist between environmental protection and economic growth, there are several cases where environmentally friendly practices are also most cost-effective.

Rhoads (1985) noted that economists and engineers often clash in their understanding of opportunity-costs and tradeoffs. Engineers, he argued, have a narrower conception that revolves around materials selection, whereas economists account for all social costs. The former ask about tradeoffs between using steel and reinforced concrete in building projects, whereas the latter consider ways to solve the problem without building at all. Their differences also point out contrasts in the meaning of efficiency. Engineers push for the implementation of the latest technological innovations, whereas economists account for the tradeoffs involved in replacing older technologies. The former is the path to increasing technological efficiency, whereas the latter implies that economic efficiency takes wider social costs into account.

Although it is true that economic transactions are not always zero-sum games, there can be a tendency by some to underemphasize the importance of tradeoffs in some areas. The broken window fallacy, for example, states that when a child breaks the baker's window, he or she actually spurs economic activity. After all, the baker must buy a new window, which gives money to the window-maker to spend on new shoes, etc. However, "hidden costs" are ignored in this calculus. The money spent by the baker on a new window would have been spent on shoes. Now, for the same cost, instead of a window and shoes the baker only has a window.

The Panglossian attitude of the broken window fallacy has also been attacked in evolutionary biology. Stephen Jay Gould and Richard Lewontin (2001) critiqued the dominant adaptationist program, which atomizes an organism into its traits. It then explains that an organism cannot optimize each trait without imposing expenses on others: "The notion of 'trade-off' is introduced, and organisms are interpreted as best compromises among competing demands" (p. 77). Organisms

are presented as the result of an optimization problem, where "each trait plays its part and must be as it is" (p. 77). Gould and Lewontin borrowed the metaphor of spandrels to argue that organisms must be analyzed as integrated wholes with "Baupläne," or phyletic and developmental constraints. These constraints, they contended, are more important in explaining evolutionary change than selective forces. The plurality of tradeoffs between selective pressures, random forces, and various constraints, rather than strictly between selective forces, expands the relevant foci of analysis.

Ethical Analysis

Tradeoffs can be abstracted into a taxonomy of competing goods, including equity, efficiency, freedom, and security (see Okun 1975). Indeed public policy, by virtue of being public, tends to require tradeoffs due to a plurality of views and interests. Science and technology play major roles in several policies that make tradeoffs among social priorities, between costs and risks, between various sectors of the population, and between long- and short-term timescales (Wenk 1986).

The latter tradeoff has become increasingly important as technological capacities have increased our power to create negative consequences deep into the future. This tradeoff is often posed as one between short-term gains and obligations to future generations, although the degree to which this is an ethical concern in any given circumstance is usually contested. Technology-induced displacements of the workforce also seem to create tradeoffs between long-run, aggregate gains and short-term, localized losses.

The development, use, and regulation of technologies pose many other ethical dilemmas in the form of tradeoffs. Some of the most charged issues involve tradeoffs between economic growth and human health and safety. For example, regulations on pollution emissions and synthetic chemicals protect health and safety, especially of workers who come in close contact with those pollutants and chemicals. Similarly, traffic laws and regulations on automobiles ensure some measure of safety. Theoretically, banning pollution, chemicals, automobiles, and other dangerous technologies could save millions of lives annually. Yet even marginally increasing restrictions on certain emissions (let alone banning them) can bring major tradeoffs that pose the difficult question of how much a human life is worth. Rhoads (1985) cited a proposed 1980 benzene emission standard by the U.S. Environmental Protection Agency (EPA) that would have imposed large costs on industry but would not prevent a case of leukemia until 37,000 years had passed. The estimated cost of saving one life was $33 billion. Rhoads argued that decision makers can minimize opportunity-costs by investing money in other areas (for example, traffic safety) where saving lives costs much less.

Cases such as this raise the question of how risks should be measured (for instance, what toxicological dose-response model) and how they are perceived by different elements of society. They also highlight the fact that the tradeoff concept itself depends upon a consequentialist ethic. One must be willing to base a decision on the consequences of alternative course of action to even participate in the logic of tradeoffs. A deontologist who believes it to be immoral to jeopardize human life no matter what the consequences will not accept the tradeoffs mentioned above. They would argue that $33 billion is not too much to pay to save a human life, because protecting human life is considered an inviolable duty.

Another important insight is that individuals may make different decisions about tradeoffs depending on how they encounter information. For example, Norman Augustine (2002) presented his students a hypothetical opportunity of investing in a new product that would create millions of jobs and enhance the quality of life for most people. He received an enthusiastic response, but then he adds that the product would kill a quarter of a million people every year. None of the students remained interested in investing, and most said the product should be banned. He then tells them that he is referring to the automobile. Tradeoff decisions clearly depend on cultural norms, personal experiences, and the socio-psychology of risk perception as much as they do on a rational tabulation of relative costs and benefits (see Slovic 2000).

Whether performed consciously or unconsciously, every time new knowledge is sought and new technologies are applied, a tradeoff has been made. In many cases, the bundle of benefits and costs chosen is obviously more desirable than the forgone alternatives. However, in other instances there may be considerable disagreement on whether and how to proceed. These cases pose challenging questions of who should make such decisions and how they should be made.

Decision makers have several tools for making tradeoff decisions. On the technical end, a tradeoff calibration can be used, which involves filling lookup tables by balancing different objectives. For example, this tool can help an engineer who wishes to increase torque while restricting nitrogen oxide emissions. Economic tools include risk-cost-benefit analyses, revealed preferences, and expressed preferences (for example, contingent valuation and willingness-to-pay surveys). Psychological tradeoff analyses show cross-cultural differences in the interactions between an individual's moral reasoning and the consequences of decisions (see for example Swinyard et al. 1989). More strictly governmental tradeoff analysis techniques include advisory panels and institutions dedicated to assessing decisions and assigning accountability for successes and failures. Decision makers can be guided through the oftentimes high-stakes tradeoffs presented by science and technology by specialized assessment institutions such as the U.S. Office of Technology Assessment (OTA), which existed from 1972 to 1995.

Decision making is inherently forward-looking, so one of the biggest challenges posed by many tradeoffs involving science, technology, and society is uncertainty about likely future outcomes of alternative decisions. Increasing information is often a worthwhile means to reduce uncertainties and increase foresight, but this must also be accompanied by decision-making structures capable of synthesizing that information. Furthermore, uncertainties will remain. For example, regulating toxic chemicals involves tradeoffs between costs and acceptable risks. But the situation is complicated by uncertainties in modeling dose-response functions, ecological interactions, and economic impacts. Eliminating these uncertainties is often impossible, at least on the time-scales required by decision makers.

Therefore, many tradeoff decisions must be made not between two (or more) well-characterized competing bundles of advantages and disadvantages, but rather between two (or more) dimly understood future scenarios. Partially for this reason, Edward Wenk (1986) argued that tradeoffs require anticipatory governments capable of assessing different alternatives and their probabilities. He also insisted that tradeoffs involving science and technology call for participation by an "attentive public" not just political, commercial, and scientific elites. Such assessments raise the fundamental question of which alternative will make us better off. Thus, they are the responsibility of all citizens, not the domain of any particular expertise.

ADAM BRIGGLE

SEE ALSO *Consequentialism; Double Effect and Dual Use; Risk Ethics; Unintended Consequences.*

BIBLIOGRAPHY

Arrow, Kenneth. (1974). *The Limits of Organization.* New York: Norton.

Augustine, Norman R. (2002). "Ethics and the Second Law of Thermodynamics." *The Bridge* 32(3): 4–7. Also available as a publication from the National Academy of Engineering, Washington, DC.

Gould, Stephen Jay, and Richard C. Lewontin. (2001). "The Spandrels of San Marco and the Panglossian Paradigm: A Critique of the Adaptationist Programme." In *Conceptual Issues of Evolutionary Biology*, 2nd edition, ed. Elliot Sober. Cambridge, MA: MIT Press.

Graham, John D., and Jonathan Baert Wiener, eds. (1995). *Risk versus Risk: Tradeoffs in Protecting Health and the Environment*. Cambridge, MA: Harvard University Press. A collection of nine case studies in risk tradeoffs with an introduction and conclusion by the editors, in which they abstract out the main issues, critically analyze them, and offer some ways to resolve risk tradeoffs.

Okun, Arthur M. (1975). *Equality and Efficiency: The Big Tradeoff*. Washington, DC: The Brookings Institution. Explores the question of to what extent governments should pursue economic equality and notes four major tradeoffs involved in this pursuit, which are weighed against the tradeoffs of allowing the market to function based solely on efficiency.

Rhoads, Steven E. (1985). *The Economist's View of the World: Governments, Markets, and Public Policy*. London: Cambridge University Press. Contains sections on opportunity cost (pp. 11–24) and benefit-cost analysis (pp. 124–142) that are most relevant to tradeoffs.

Sewell, Kenneth S. (1990). *The Tradeoff between Cost and Risk in Hazardous Waste Management*. New York: Garland. Provides a conceptual framework to evaluate these tradeoffs and applies it to a case study in Massachusetts.

Slovic, Paul (2000). *The Perception of Risk*. Sterling, VA: Earthscan Publications.

Swinyard, William R., Thomas J. Delong, and Peng Sim Cheng. (1989). "The Relationship between Moral Decisions and their Consequences." *Journal of Business Ethics* 8: 289–297. Proposes a tradeoff analysis framework to elucidate the way in which these decisions are made.

Wenk, Edward Jr. (1986). *Tradeoffs: Imperatives of Choice in a High-Tech World*. Baltimore: Johns Hopkins University Press. One of the best single overviews of the topic. Examines the decisions that must be made as technology intersects with culture, politics, and individual lives.

TRANSACTION-GENERATED INFORMATION AND DATA MINING

• • •

The term *transactional information* was first employed by David Burnham (1983) to describe a new category of information produced by tracking and recording individual interactions with computer systems. Unlike most human interactions, those processed by computer systems are easily recorded and aggregated to yield knowledge about individual behaviors that would have otherwise been more difficult to acquire and often less complete. Known as transactional-generated information (TGI), it is information acquired from commercial and noncommercial transactions involving individuals in many increasingly computerized day-to-day activities. Examples of commercial transactions include withdrawing money from an ATM machine or credit-card shopping; examples of noncommercial transactions include checking books out of a library or participating in an online educational program. TGI can be contrasted with but does not exclude more traditional information such as a person's age, place of birth, education, work history, and so forth.

The Special Character of TGI

The practice of collecting information about persons is hardly new. Governments have collected census data since the Roman era. But through the twentieth century, the few records that existed about individuals contained information about when and where they were born, married, worked, or owned property. Information about the day-to-day transactions of individuals was rarely, if ever, collected and stored. Even if it had been collected, it would have been difficult to process and store. Armies of clerks would have been needed to sort through this information and huge warehouses or repositories would have been required to store the physical records. Those conditions changed, of course, with the advent of computers and electronic databases.

Additionally much traditional information about persons is gathered in ways that require conscious acts of disclosure on the part of those providing it. When individuals fill out census forms, they are generally aware of providing information about themselves to a government agency. By contrast, with TGI data subjects are not always consciously aware they are providing information about themselves to some data collector. When motorists use the convenience of an Intelligent Highway Vehicle System, such as E-ZPASS, they seldom realize that a transaction occurs each time they pass a toll plaza. Not only is a motorist's pre-paid account with E-ZPASS debited, but the exact time of passing through the toll booth is electronically recorded and stored.

Cookies

Next consider a kind of *on-line* transaction involving typical Internet users, who may have no knowledge that TGI is being collected. Via programs called *cookies*, TGI is routinely gathered about users who visit web sites. Cookies technology enables web site owners to collect certain kinds of data about users who access their sites, including information about the user's Internet Protocol (IP) address and Internet Service Provider (ISP). This information is stored in a text file placed on the hard drive of the user's computer and then retrieved from that computer and resubmitted to the web site the next time the user accesses it. It provides the operator of a web site with information about a user's on-line browsing preferences. Transactions involving the use of cookies to exchange data between users and web sites typically occur without the knowledge and consent of users.

Since their implementation on the web in the 1990s, the use of cookies technology has been controversial. The owners and operators of on-line businesses and Web sites, who defend the use of cookies, claim that they are performing a service for repeat users of a web site by customizing a user's means of information retrieval. For example, they point out that cookies technology enables them to provide a user with a list of preferences for future visits to that Web site. Defenders of cookies also note that users can elect to disable cookies via an option provided on their web browsers.

Privacy advocates, on the other hand, argue that because cookies technology involves the monitoring and recording an individual's activities while visiting a Web site, as well as the subsequent downloading of that information onto a user's PC (without informing the user), the use of cookies clearly cross the privacy line. They also point out that many web sites do not permit users to disable cookies, and they note that users must first be aware of cookies before they can opt out (i.e., reject cookies) on web sites that allow them to do so. Some privacy advocates also worry that information gathered about a user via cookies can eventually be acquired by on-line advertising agencies, which could then target that user for on-line ads.

Merging and Mining TGI

Because TGI exists in the form of electronic records, it can be easily exchanged between databases in a computer network; these records can also be *merged. Computerized merging* is the technique of extracting information from records about individuals (or groups of individuals) that reside in two or more databases, which are often unrelated, and then integrating that information into a composite file.

Information gathered about an individual's on-line activities and preferences via Internet cookies can also be merged with information about an individual's transactions in off-line activities in physical space to construct a general profile. In 1999 DoubleClick.com, an on-line advertising firm that used cookies technology to amass information about Internet users, proposed to purchased Abacus, an off-line database company. DoubleClick's pending acquisition of Abacus was criticized by many privacy advocates who feared that the on-line ad company would combine the information it had already acquired about Internet users (via cookies) with the records of some of those same individuals that resided in the Abacus database.

DoubleClick would have been able to merge web profiles with off-line transactional data about consumers. In January 2000, however, DoubleClick was sued by a woman who complained that her right to privacy had been violated by that company. The woman filing the suit claimed that DoubleClick's business practices were deceptive because the company had quietly reversed an earlier policy in which it provided only anonymous data about Internet users (acquired from cookies files) to businesses. Because of public pressure, DoubleClick backed off its proposal to purchase Abacus. However, because of the controversy surrounding the DoubleClick incident, many realized for the first time the kinds of privacy threats that can result from the merging of electronic data. And even though the DoubleClick-Abacus merger did not materialize, the danger of future mergers of this type remain.

In addition to being merged, TGI can also be *mined. Data mining* is a computerized technique used to reveal non-obvious patterns in data that otherwise would not be discernible. Data-mining technology also generates new classifications or categories (of individuals), which are not always obvious to the individuals who populate them. Some of these newly discovered/created categories or groups suggest *new facts* about individuals who constitute these groups. For example, a young executive with an impeccable credit history could, as a result of data-mining technology, end up being identified as a member of a (newly generated) category of individuals who are perceived to be high-credit risks because of certain patterns found in aggregated data, despite the fact that the particular person's credit history is unblemished. That is, a data-mining program might associate the young executive with a group of individuals who are likely to start their own businesses in the next three years and then file for bankruptcy within the next five years.

Because of concerns about the ways in which electronic records can be exchanged between two or more databases, various privacy laws have been enacted at the federal and state levels. For example, the Health Insurance Portability and Accountability Act (HIPPA) of 1996, enacted into law on April 14, 2003, provides protection for personal medical records. And the Video Protection Act (also known as the "Bork Bill" because it was passed through the U.S. Congress in the aftermath of Judge Robert Bork's nomination to the U.S. Supreme Court) protects consumers from having records of their video rentals from being collected and exchanged. However, these laws primarily aim at protecting personal information that is: (a) *explicitly* identifiable in electronic records, and (b) considered *intimate* or *confidential*.

Information acquired via data mining fits neither category. First, as noted, it is derived from *implicit* patterns in data, which without data-mining technology, would not be accessible to data collectors. Second the kind of personal information generated in the data-mining process is often considered non-intimate or non-confidential because it is derived from information acquired through transactions in which individuals engage openly and in public places.

The use of courtesy cards in supermarket transactions might initially seem innocuous from the perspective of personal privacy. The items purchased are typically transported in an open shopping cart that is visible to anyone in the store so there is nothing confidential or intimate about the activity. However a record of courtesy card purchases can be used to generate a consumer profile. This profile reveals patterns that identify, among other things, the kinds of items purchased and the time of day/week an individual typically shops. Such information is useful to *information merchants* who use it to target consumers in their advertising and marketing campaigns. Furthermore information in a consumer profile can be used to make judgments about personal lifestyles, health, spending habits, and more. Indeed such a profile may be created even when the aggregated data on which it is based is inaccurate because the courtesy card was loaned to another person.

The new forms of information produced by TGI and data mining thus present special challenges to privacy. First individuals may not be aware of the degrees to which their activities are being tracked by a constellation of computer system interactions and their interactions analyzed by data mining techniques. The lack of knowledge in these regards is itself an ethical issue that deserves to be addressed by general education and disclosure statements associated with the particular computer systems. Second because it is easy for such TGI and

data mining products to include inaccuracies that may have substantial if subtle impacts, it may be necessary to consider possibilities for personal review or disclosure when TGI is used to influence decision making.

HERMAN T. TAVANI

SEE ALSO *Computer Ethics; Internet; Privacy.*

BIBLIOGRAPHY

Burnham, David. (1983). *The Rise of the Computer State.* New York: Random House. One of the earliest accounts of how computer databases could be used by government agencies to store and exchange information gained about individuals from their electronic transactions.

Fulda, Joseph S. (2004). "Data Mining and Privacy." In *Readings in CyberEthics,* 2nd edition, ed. Richard A. Spinello and Herman T. Tavani. Sudbury, MA: Jones and Bartlett Publishers. This anthology comprises fifty readings; Fulda's article succinctly describes how privacy concerns involving data mining differ from other computer-related privacy issues.

Johnson, Deborah G. (2004). "Computer Ethics." In *Academy and the Internet,* ed. Helen Nissenbaum and Monroe E. Price. New York: Peter Lang. This anthology comprises twelve readings; Johnson's article provides an overview of computer-ethics issues, including a discussion of specific privacy concerns involving TGI.

Tavani, Herman T. (1999). "Informational Privacy, Data Mining, and the Internet." *Ethics and Information Technology* 1(2): 137–145. Examines how privacy issues arising from data mining differ from those associated with traditional data-retrieval techniques; also illustrates how mining data from the Internet differs from data mining in off-line contexts such as "data warehouses."

Tavani, Herman T. (2004). *Ethics and Technology: Ethical Issues in an Age of Information and Communication Technology.* Hoboken, NJ: John Wiley and Sons. Comprises eleven chapters that cover a wide range of computer-ethics issues, including an extensive discussion of data mining and cookies in Chapter 5: "Privacy and Cyberspace."

TRANSPLANTS

SEE *Organ Transplants.*

TREAT, MARY

• • •

Accomplished amateur botanist and entomologist, Mary Treat (1830–1923), born in Trumansville, New York, on September 7, was a popular chronicler of the plant, insect, and bird life that shared her small Vineland,

New Jersey, home. Treat, who was considered a peer and valued correspondent by countless scientists (including Asa Gray (1810–1888), Charles Darwin (1809–1882), Gustav Mayr (1830–1908), and Auguste Forel (1848–1941)), was widely acknowledged as an authority on insectivorous plants, harvesting ants, and burrowing spiders. She is credited with discovering two species of spider, as well as rare fern and plant species. The recognition she received for her scientific research distinguishes her in the history of women in the sciences. It is her investigations into the nest-making actions of birds and insects, however, that illuminates her concern with ethics and the effects of human action in the natural world.

Treat's scientific nature essays, first published in *Harper's* and the *Atlantic Monthly*, then collected in *Home Studies in Nature* (1885), reflect the shift in scientific investigation prompted by the publication of Darwin's *Origin of Species* (1859). Treat described a world in which the landscape of morality changed significantly, where humans no longer resided securely at the apex of creation. Treat agreed with Darwin's notion of nature "red in tooth and claw"; she saw instances of struggle, violence, chance, and adaptation all around her. Yet Treat, unlike many other American intellects of the time, refused to see nature exclusively in these terms. Instead, she advocated a sophisticated brand of Darwinian evolution—one that incorporated ideas expressed in Darwin's *Descent of Man* (1871) and *The Expression of Emotions in Man and the Animals* (1872)—to explain how animals and insects construct their domestic spaces in the face of their struggle to survive.

Treat revised the model of nature she inherited from the tradition of women nature writers preceding her—nature is not simply a model for human behavior, nor is it something that exists solely for humans to control. Instead, as she learned from her reading of Darwin, nature is composed of separate but interrelated communities; the moral sense, as Darwin notes, comes into being with the social instincts that animals develop as they learn to live in a community. Treat focused her scientific studies on how birds and insects build their nests and observed that they, like humans, exercise reason in the construction of their homes. These observations led her to question the supposed difference between human and non-human, and she used nest construction to demonstrate kinship through reason. Humans, or at least those whom Treat called "good observers" of nature, cannot deny this kinship with non-human communities and are, as a result, obligated to act in an ethical way toward nature.

Treat did not escape the anthropocentric observer position common to many women writing about nature in the nineteenth century, but like her mid-twentieth century counterpart Rachel Carson, she used what she saw (and how she saw it) to justify her call for the ethical treatment of all inhabitants of nature.

TINA GIANQUITTO

SEE ALSO *Darwin, Charles; Environmental Ethics; Sex and Gender*.

BIBLIOGRAPHY

Bonta, Marcia Myers. (1995). *American Women Afield: Writings by Pioneering Women Naturalists*. College Station: Texas A&M University Press.

Creese, Mary. (1998). *Ladies in the Laboratory? American and British Women in Science, 1800–1900*. Lanham, MD: The Scarecrow Press.

Norwood, Vera. (1993). *Made from This Earth: American Women and Nature*. Chapel Hill: University of North Carolina Press.

Treat, Mary. (1885). *Home Studies in Nature*. New York: American Book Company.

TRUST

• • •

Trust of science and technology and of the people who conduct research and invent, design, develop, manufacture, operate, maintain, and repair technology is essential to the development of science and technology. When the trust proves unwarranted, however, the result can be disaster in forms varying from harm to health and safety, to persistent distortions of knowledge, to theft of credit or property that cripples cooperation necessary to support the growth of knowledge and development of technology. A deeper question is what it means for science and technology, and the people responsible for them, to be trust*worthy*.

The Concepts of Trust and Trustworthiness

Although Sissela Bok (1978) discussed trust as a moral resource beginning in the 1970s, the question of the morality of trust relationships—the conditions under which, from a moral point view, one ought to trust—was not explicitly discussed until a decade later by Annette Baier (1986). Two earlier essays were important in laying the foundation for this major turn in the discussion. Ian Hacking (1984) provided a devastating assessment of game theoretic approaches to solving problems of trust, such as the Prisoner's Dilemma. Baier (1985) herself previously had argued for broadening the focus in ethics from obligations and moral rules to the subject of whom one ought, as a moral matter, to trust and when. As Kathryn Addelson (1994) points out, Baier's change of focus establishes a general perspective on ethical legitimacy that is shared by all, rather than privileging the perspective of those who make, instill, and enforce moral rules.

As Baier (1986) argues, trust involves both confidence and reliance. If people lack other options, they may continue to rely on something, such as the water supply, even when they no longer trust it. Similarly, people may have confidence in something, or confidence in their expectations concerning it, without relying on it. To rely only where one can trust is a fortunate circumstance.

Baier's general account of the morality of trust illuminates the strong relation between the trustworthy and the true. A trust relationship, according to Baier, is *decent* insofar as it stands the test of disclosure of the premises of each party's trust. For example, if one party trusts the other to perform reliably only because the truster believes the trusted is too timid or unimaginative to do otherwise, disclosure of these premises will give the trusted party an incentive to prove the truster wrong. Similarly if the trusted party fulfills the truster's expectations not through trustworthiness but only through fear of detection and punishment, disclosure of these premises will lead the truster to expect that the trusted would defect, if able to do so undetected.

Although explicit discussion of moral trustworthiness is relatively recent, both professional ethics and the philosophy of technology have given considerable attention to the concept of (prospective) responsibility. Because being trustworthy is key to acting responsibly in a professional capacity, or to being responsible in the virtue, as contrasted with causal, sense, the literature on responsibility provides at least an implicit discussion of many aspects of trustworthiness.

Niklas Luhmann (1979) has shown how trust simplifies human life by endowing some expectations with assurance. It is prohibitively difficult and time-consuming to consider all possible disappointments, defections, and betrayals by those persons or circumstances on which one relies; all possible consequences of those disappointments; and all actions that might prevent those disappointments or change their effect. Trust reduces that burden. In a later work, Luhmann urged a different distinction between confidence and trust: that *trust* be

used only when the truster has considered the alternatives to trusting. Luhmann's discussion of the distinction between trust and confidence highlights the element of risk in trusting. Risk or vulnerability does characterize situations in which trust is necessary, in contrast to those in which one's control of the situation makes trust unnecessary. However the notion of reliance in Baier's definition of trust as confident reliance does capture the sense of vulnerability. One's vulnerability in reliance does not require consideration of the alternatives to such reliance.

The risk taken in trusting does leave the truster liable to disappointment (or worse), whether that trust is of persons, objects, or circumstances (such as, that the temperature will go below freezing overnight). However only if one's trust is in agents capable of recognizing intention, can one be let down. Furthermore, although one may disappoint without intending to, one must at least be aware of behaving in the way that turns out to disappoint in order to be said to have let someone down. So if Alice does not know she is waking Bob each morning by closing the garage door, she cannot be said to have let him down by not waking him today. Because science and technology do not arise except through human intervention, the focus of this entry is on trust in people, individually or acting as a group.

Trustworthy Professionals

For the professionals behind science and technology to behave in a responsible or trustworthy manner requires both technical competence and moral concern—specifically a concern to achieve a good outcome in the matter covered, which is sometimes called their fiduciary responsibility, the responsibility of a person in a position of trust. The moral and technical components of professional responsibility led sociologist Bernard Barber (1983) to speak of these as two *senses* of trust. However if the public is to trust the members of the science and engineering professions, it is not in two senses. Rather the public trusts the professional to achieve some outcome for which both competence and concern are required. For researchers, the outcome typically centers on the accurate report of the methods and results of research, a report that fairly acknowledges any contributions of others. For engineers, it typically centers on the provision of a technology that performs its function and does not pose unnecessary threats to safety.

For engineers the competence and concern are engineering competence and concern for such social goods as public safety, confidentiality of information, fairness in competition, the public understanding of science and technology, protection of the environment, and the quality and performance of the technology in question. Engineering codes of ethics enjoin engineers not to take on work beyond their competence, so at least for engineers technically incompetent performance is also recognized to be a moral failing. In contrast, researcher investigators generally do not regard undertaking research beyond one's competence as a *moral* failing, although certain incompetencies, such as those that result in harm to experimental subjects or to public health, might be.

Because the exercise of professional responsibility characteristically draws on a body of specialized knowledge that is brought to bear on the promotion or preservation of another's welfare, to trust a person to fulfill a professional responsibility is to trust that professional to perform in a way that someone outside of the profession cannot entirely specify, predict, or often even recognize. The point is not captured in the frequent suggestion that trust is necessary because the trusting party cannot control or monitor the trusted party's performance. It would do the layperson little good to have full knowledge of the plans for a medical device or an experiment, or even the ability to guide the actions of the science and engineering professionals. Although laypeople might be able to recognize some acts of gross negligence, they would not know the implications of most of what they saw or how to improve the professional's performance. For this reason, from the point of view of the public, there are no good alternatives to having trustworthy professionals. In her 2001 Gifford Lectures, Onora O'Neill (2002) makes the same point that nothing can guarantee trustworthiness and emphasizes the burden that what she calls the *culture of suspicion* places on officials and professionals, such as medical researchers.

The question of whether scientists and engineers are responsible for the ultimate uses of the knowledge and technology they create is sometimes called the *end use question*. Caroline Whitbeck (1998) has argued that for scientists and engineers to be entrusted to prevent evil end uses of their products and discoveries those uses must be intended as well as foreseeable, because, for example, it would be impossible to forego the creation of all the many useful tools from hammers to pokers, to kitchen knives to hatchets that one can foresee can also be used as weapons.

It is arguably unreasonable to say that scientists and engineers are untrustworthy (more specifically, negligent), if they fail to consider *unforeseeable* uses and consequences. Indeed National Academy of Engineering President William Wulf (2004) draws attention to technological systems, such as computer systems, that are so complex that failures in them are inherently unpredictable, so it would not be possible for engineers to predict them. Criteria for trustworthy behavior or policies regarding such systems have yet to be settled.

The application of standards of professional responsibility in science and engineering is complicated by the fact that not all scientific and engineering professions have the same developed understanding of themselves as professions. Although U.S. engineering societies formulated ethical codes and guidelines from the early decades of the 1900s, attention to the professional responsibilities of research investigators has only received broad attention since the mid-1980s. However trust and trustworthiness became a central theme in those discussions in the 1990s (Whitbeck 2004). An international perspective provides even greater variation although the so-called Washington Accord, an agreement that recognizes equivalency of accredited engineering education programs in participating countries, is leading to more uniformity.

Trustworthy Policies

Some questions about the ethical implications of science and technology are policy questions, sometimes called *macro issues*. Although people can and do praise and blame particular individuals for formulating, adopting, implementing, or carrying out policies regarding science and technology, in a democracy these are societal decisions. Policy decisions run the gamut from decisions about what research and development should be given public support or even legally permitted, to what can be used as research material (for example, embryonic stem cells), to how and when to prevent or clean up toxic and nuclear contamination, to whether and how to control the social consequences of new technologies, such as privacy invasions on the Internet.

Typically such policy questions must be decided under conditions of significant uncertainty. Often the nature of the possible outcomes as well as the likelihood of various outcomes are unknown. Such uncertainties lead to misgivings about the pace of innovation and discovery. Technology is said to create new options, but technological advance also forecloses options. For example, after the introduction of the automobile, one could no longer choose to keep a horse and buggy in the city. Furthermore its consequences may contradict expectations. For example, historian Ruth Cowan (1983) found that household appliances did not reduce housework but raised the standards for that work. The relationship between science and technology and the societies in which they develop is extremely complex. Therefore the extent to which the frequent criticism of modern life in technologically developed societies is most properly directed at science and technology (or at least the pace of their development); at social factors, such as market forces affecting their development; or at human tendencies to use and abuse power in general is likely to continue to be disputed.

CAROLINE WHITBECK

SEE ALSO *Virtue Ethics.*

BIBLIOGRAPHY

Addelson, Kathryn. (1994). *Moral Passages: Notes toward a Collectivist Ethics.* New York: Routledge. Addelson presents a collectivist account of responsibility and ethics.

Baier, Annette. (1986). "Trust and Antitrust." *Ethics* 96: 232–260. Reprinted in Baier's *Moral Prejudices.* Cambridge, MA: Harvard University Press, 1994. Baier discusses the subject of trust and the curious neglect of this subject in most of moral philosophy. She gives an account of what it is to trust and a partial account of when it is immoral to expect or fulfill trust.

Barber, Bernard. (1983). *The Logic and Limits of Trust.* New Brunswick, NJ: Rutgers University Press. This work examines trust in various contexts including professional practice and business. As a sociologist, Barber delineates professions from non-professions in terms of their social status, rather than in terms of their claim to provide ethical guidelines for their practice.

Bok, Sissela. (1978). *Lying.* New York: Vintage Books. Bok addresses the subject of trust, one of the few philosophers to do so before Baier and provides a model of what was to become known as practical (as contrasted with "applied") ethics.

Cowan, Ruth Schwartz. (1983). *More Work for Mother: The Ironies of Household Technology from the Open Hearth to the Microwave.* New York: Basic Books. Cowan's history of household technology demonstrates supposed labor-saving household technology raised the standards for housework, rather than reducing the time spent performing it.

Hacking, Ian. (1984). "Winner Take Less: A Review of *The Evolution of Cooperation* by Robert Axelrod." *New York Review of Books,* June 28, pp. 17–21. Hacking praises *The Evolution of Cooperation* and criticizes twentieth-century modes of philosophizing that seeks to force-fit human life to logical models that fail to do justice to notions such as cooperation.

Ladd, John. (1979). "Legalism and Medical Ethics." In *Contemporary Issues in Biomedical Ethics,* ed. John W. Davis, Barry Hoffmaster, and Sarah Shorten. Clifton, NJ: Humana Press. Ladd here develops his philosophical position on the primacy of the notion of responsibility in practical ethics and the secondary status of rights and obligations/duties.

Luhmann, Niklas. (1979). *Trust and Power.* New York: John Wiley & Sons. Luhman's principle work on trust examines the relationship between trust and confidence and discusses the function of trust in human life.

O'Neill, Onora. (2002). *A Question of Trust: 2002 BBC Reith Lectures.* Cambridge, UK: Cambridge University Press. O'Neill examines what she regards as the current climate of suspicion and argues for a better appreciation of the evidence of trustworthiness.

Whitbeck, Caroline. (1998). *Ethics in Engineering Practice and Research.* New York: Cambridge University Press. Examines the notions of responsibility, trust, and trustworthiness in engineering and science.

Whitbeck, Caroline. (2004). "Trust and the Future of Research." *Physics Today* 57(11): 48–53.

Wulf, William A. (2004). "Keynote Address." In *Emerging Technologies and Ethical Issues in Engineering: Papers from a Workshop, October 14–15, 2003.* Washington, DC: National Academies Press. Wulf, President of the National Academy of Engineering, reviews the accomplishments of engineers in the twentieth century and their challenges in twenty-first century, including the moral and ethical responsibilities associated with the impact of engineering on society.

INTERNET RESOURCE

Washington Accord. Available from http://www.washingtonaccord.org. Internet site of the Washington Accord.

TURING, ALAN

• • •

Alan Mathison Turing (1912–1954), the founder of modern computer science and an important World War II cryptanalyst, was born in London on June 23. He died near Manchester, England, on June 7. His short life illustrates the ethical conflicts and ambiguities of scientific and technological aspirations.

Basic Creativity

Turing's early life was characterized by an intense enthusiasm for science that was only weakly supported by his upper-middle-class family. In 1931 he became an undergraduate at Cambridge University and read mathematics, demonstrating a rapidly emerging originality. At age twenty-four he settled an important problem in the foundations of mathematics, using a method that had much wider implications. Turing developed a precise way to characterize the concept of the "effectively calculable." This consisted of the "Turing machine," as the logician Alonzo Church immediately dubbed Turing's construction when reviewing it in 1937.

A Turing machine is an imaginary device with a finite number of possible configurations, a finite table of instructions for moving from one configuration to another, and the capacity to read, erase, and write a set of finitely many different symbols on a tape. With this structure Turing captured the idea of a finite mechanism, which he compared with the finite capacity of the human mind. By allowing unlimited space and time for working out the machine's operations, Turing was able to argue that such a device could encompass everything that could be achieved by a human calculator following a definite rule. Church endorsed Turing's argument that the concept of "effectively calculable" had been given a natural and convincing definition in terms of being computable by a Turing machine, a proposition now known as the Church-Turing thesis.

More recently there has been discussion of whether there could be, in the real universe or an imaginary one machines capable of operations beyond the scope of a Turing machine, and this debate has generated controversy about the correct interpretation of the Church-Turing thesis (Floridi 2003). At the time, however, Church simply characterized "computable" by reference to what could be done by any kind of machine of a finite size, and Turing similarly referred to that term as being synonymous with *mechanical*.

What is not in dispute is the fact that the Turing machine is still definitive as the foundation for computer science. By attacking an abstruse problem in the most rarefied and philosophical aspects of mathematics Turing arrived at the principle behind the dominant technology of the late twentieth century. Indeed, it was Turing who, seeing the practical potential of his ideas in 1945, was a leading designer and promoter of the electronic computer and its software.

However, this was possible only because of world events between 1938 and 1945 that gave Turing unique insight into practical computation and the promise of digital electronic technology. During the World War II Turing was the chief scientific figure in the successful British effort to decipher coded German communications, a project that became a joint Anglo-American operation after 1941. Turing's ingenious logical methods and theory of information measuring were used throughout the communications war, especially in the section he personally headed, which was responsible for reading U-boat signals.

By 1945 Turing thus possessed unrivaled theoretical and practical experience in the emergent field of information processing. He was disappointed by the practical progress of his plans at the National Physical Labora-

Alan Turing, 1912–1954. The British mathematician was noted for his contributions to mathematical logic and to the early theory, construction, and use of computers. *(Photo Researchers, Inc.)*

tory, the British government establishment to which he was appointed. He soon left to take up another, also disappointing, position at Manchester University. However, those short-term setbacks illustrated the fact that Turing's interest was never in the economic potential of computers but only in the long-term scientific question of what he called intelligent machinery, now usually referred to as artificial intelligence.

Is the computer in principle capable of rivaling human thought? That question was hinted at even in Turing's prewar references to human memory and states of mind but became much more prominent after 1945. In that period Turing went much further than he had in 1936, arguing that the computer could emulate all aspects of human thought, not merely those corresponding to a human being following a definite method. At that time he also spoke frequently about the physical basis of mental operations and informally described his work as "building a brain."

Contested Issues

A crucial element in Turing's argument is that the computer is a practical form of a universal machine that is capable of performing any algorithm. According to this argument, if the function of the brain can be described as any sort of definite process, in principle a computer can simulate it. It is not suggested that the architecture of the brain should resemble that of a digital computer. Another vital part of Turing's argument is that programs that modify themselves can be considered as learning from experience. He expected them to show the features of surprise and originality that characterize the apparently "nonmechanical" aspects of human thought. Turing's famous 1950 paper (reprinted in Boden 1990) introduced the "imitation game," now called the Turing test, in an attempt to make an objective comparison between computational and human processes.

Interest in these issues has never flagged. The arguments of Roger Penrose (1989) have supplied important new ingredients. It is noteworthy that the interpretation of Gödel's theorem and the quantum-mechanical nature of matter, which are central to Penrose's arguments, are also issues that Turing found important and difficult to address.

The Turing test for intelligence can be accused of having been set up to evade questions of consciousness and responsibility: It is the problem of mind made into a game perhaps in the way codebreaking made it possible to think of World War II as a fascinating and exciting but bloodless game. In real life Turing struck everyone as a person of great integrity, not as a superficial or insensitive person. However, he did not offer an ethical view in his writing on mind and machines. It was the same with the war in which he played so important a role: Turing never spoke about motivation or political allegiance, though his actions showed a strong commitment to the defeat of Nazi Germany. His moral speech was generally directed against anything "phony." In this he was like G. H. Hardy, the Cambridge champion of pure mathematics, but whereas Hardy hated war and rejoiced if his work was "useless" for it, Turing applied mathematics to more effect in war than perhaps anyone else ever had.

After 1950 Turing devoted himself mainly to a mathematical theory of biological growth and form, a quest roughly parallel with the elucidation of DNA. This time he stated a motivation: to defeat the religious "argument from design" and vindicate the power of scientific explanation. However, in 1952 Turing was arrested as a homosexual, and after the ensuing trial he was sentenced to receive injections of estrogen, which was the advanced "scientific" treatment of that period.

Turing rose to the crisis with a staunch defense of his personal liberty and equality that has become a standard of European human rights but in his time was an isolated position. He was even more isolated because of his unique access to sensitive Anglo-American military secrets. At the height of Cold War paranoia in June 1954 Turing found his life impossible. He died by taking cyanide.

That period has been dramatized for the stage and television (Whitemore 1986) in scenes in which a fictional Turing gives speeches to an audience, but the real person left his life without a word about the major ethical conflicts he faced. Although Turing was a farsighted and original thinker on fundamental scientific questions and an extraordinary personality, in his silence and unwillingness to pontificate he bore witness to a particular view of scientific practice.

ANDREW P. HODGES

SEE ALSO *Artificial Intelligence; Computer Ethics; Turing Tests; von Neumann, John.*

BIBLIOGRAPHY

Boden, Margaret A. (1990). *The Philosophy of Artificial Intelligence.* Oxford, UK: Oxford University Press. Classic papers, beginning with Turing's.

Davis, Martin. (2000). *The Universal Computer: The Road from Leibniz to Turing.* New York: Norton. A vivid exposition of Turing's 1936 work as the culmination of symbolic logic.

Floridi, Luciano. (2003). *Blackwell Guide to the Philosophy of Computing and Information.* Oxford, UK: Blackwell.

Hardy, G. H. (1940). *A Mathematician's Apology.* Cambridge, UK: Cambridge University Press. A classic essay on the ethics of mathematics from Turing's Cambridge background.

Hodges, Andrew. (1983). *Alan Turing: The Enigma.* London: Burnett; New York: Simon & Schuster. New editions: London, Vintage 1992; New York, Walker 2000.

Penrose, Roger. (1989). *The Emperor's New Mind: Concerning Computers, Minds, and the Laws of Physics.* Oxford and New York: Oxford University Press. A serious challenge to Turing's theory of mind that shares Turing's materialist assumptions and base in mathematics.

Whitemore, Hugh. (1986). *Breaking the Code.* New York: Samuel French. A stage dramatization of Turing's life.

INTERNET RESOURCE

Hodges, Andrew. Alan Turing Website. Available at http://www.turing.org.uk. Covers all aspects of Turing's life and work.

TURING TESTS

● ● ●

Turing tests are procedures to test the functional equivalence of people and computers. They generalize the thought experiment proposed by the British mathematician Alan M. Turing (1912–1954) in his pioneering 1950 paper, "Computer Machines and Intelligence," to answer the question, Can machines think?:

> [T]he "imitation game" ... is played with three people, a man (A), a woman (B), and an interrogator (C) who may be of either sex. The interrogator stays in a room apart from the other two. The object of the game for the interrogator is to determine which of the other two is the man and which is the woman.
>
> In order that tones of voice may not help the interrogator the answers should be written, or better still, typewritten. The ideal arrangement is to have a teleprinter communicating between the two rooms.
>
> We now ask the question, "What will happen when a machine takes the part of A in this game?" Will the interrogator decide wrongly as often when the game is played like this as he does when the game is played between a man and a woman? These questions replace our original, "Can machines think?" (Turing 1950, pp. 433–434)

Turing's proposed test has been very influential in the philosophy of mind and cognitive science. Variations apply as well to some issues in the ethics of technology. Consider four cases.

A Moral Problem

First, Turing's original test had a moralizing aspect. A tricky game is needed to arrive at a fair test of human versus machine ability because humans are prejudiced against machine intelligence. Turing's blind test combats this prejudice. Note that in Turing's imitation game, computers serve two roles: as potential artificially intelligent interlocutor and as filtering media. This second role has become more significant with the spread of networked computer-mediated communication. As Turing noted, when people communicate only by typing, many cues drop out, and it is not immediately obvious who is male or female. Indeed, in a long-running Internet implementation based on Turing's original male–female game gender turns out to be very difficult to detect (Berman and Bruckman 2001). The spread of the Internet has made this filtering and uncertainty, which

might be termed a "Turing effect" of computer-mediated communication, practically important. Its equalizing and liberating aspect is summed up by Peter Steiner's 1993 *New Yorker* cartoon caption: "On the Internet, nobody knows you're a dog." So too do age and rank in organizations drop away in chat rooms and e-mail, creating one of the moral risks of Internet anonymity: adults posing as children and vice versa. Indeed, the recent winners of the annual Loebner metals for best Turing test performance have been chatbots (Loebner Prize Internet site).

Machines with Moral Standing?

Second, and more speculatively, were a computer program to pass Turing's original test for intelligence, this success might have moral implications. For Roger Penrose, ownership of a device that passed the test "would involve us in *moral responsibilities* [because] to operate [such a] computer to satisfy our needs without regard to its sensibilities would be reprehensible." This could be morally equivalent to slavery. "Turning off the computer, or even perhaps selling it, when it might have become attached to us, would present us with moral difficulties" (Penrose 1989, p. 8). Of course, this argument assumes human-level intelligence sufficient for moral standing. A broader account of moral standing leads to an extension of Turing's test.

Third, there is the direct ethical extension of the Turing test. Instead of testing for intelligence, one could test for moral standing itself. Arguably, a computer program that could discuss ethically complex issues indistinguishably from a person should be granted moral standing (Allen, Varner, and Zinser 2000). Variations on this theme of testing for moral personhood via indistinguishability is common in science fiction. For example, in Ridley Scott's 1982 film *Blade Runner*, humans and computer-based "replicants" are indistinguishable by any nonphysical (invasive) Turing test.

Problems with Turing Tests

These Turing test applications disclose some of its problems: (a) The original version tests for communicative ability, but ethics (and perhaps intelligence) arguably requires the ability to *act* as well as to communicate. (b) Turing tests make playing a game (the imitation game) the criteria for intelligence or ethics, respectively. But the ability to deceive is neither necessary (think of naive but intelligent agents) nor sufficient (think of programmed con artists) for moral considerability. (c) More generally, experience with computers because Turing makes it obvious that people tend to overestimate the abilities of computer programs. Notwithstanding such problems, the Turing test remains ethically salient, invoking core moral ideals of fairness and the equivalence of the indistinguishable to challenge prejudice about the unique status of human abilities.

Human versus Machine: Chess

Fourth and again quite practically, there are indirect ethical questions about the human values challenged by machine performance of activities once thought to be open only to humans. The most noted example is the game of chess and the victories of IBM's Deep Blue computer system over grandmaster Gary Kasparov in 1996 and 1997. This can be considered, loosely, a real-world Turing test, whereby master level chess ceased to be a realm in which humans could be distinguishable from machines.

Predictably Deep Blue's success led to a strategic retreat, distinguishing easily (we say now!) mechanizable formal games such as chess from "really difficult" tasks embedded in thick human contexts. Subsequently the Internet search engine Google introduced automated news editing, and reviewers claimed that its editing service was indistinguishable from that of normal human editors. It remains open whether people will view these tests as raising the value of what machines can now do or lowering it. The initial reaction to Deep Blue's victory suggests the latter.

PETER DANIELSON

SEE ALSO *Artificial Intelligence; Robots and Robotics; Turing, Alan.*

BIBLIOGRAPHY

Allen, Colin; Gary Varner; and Jason Zinser. (2000). "Prolegomena to Any Future Artificial Moral Agent." *Journal of Experimental and Theoretical Artificial Intelligence* 12(3): 251–261. Thorough consideration of ethical aspects of the Turing Test.

Berman, Joshua, and Amy Bruckman. (2001). "The Turing Game: Exploring Identity in an Online Environment." *Convergence* 7(3): 83–102. Reports on a research platform for experimenting with online Turing Tests.

Millican, P & A. Clark, eds. (1996). *Machines and Thought: The Legacy of Alan Turing.* Oxford, UK: Clarendon Press. A collection of leading articles on the Turing Test emphasizing cognitive science and philosophy of mind.

Penrose, Roger. (1989). *The Emperor's New Mind: Concerning Computers, Minds, and the Laws of Physics.* Oxford: Oxford

University Press. A well-known scientist considers one ethical aspect of the Turing Test.

Turing, Alan M. (1950). "Computing Machinery and Intelligence." *Mind* 59: 433–460. Turing's original article. See also under Internet Resources below.

INTERNET RESOURCES

The Loebner Prize. Available at http://www.loebner.net/Prizef/loebner-prize.html.

Turning, Alan M. (1950). "Computing Machinery and Intelligence." Available from http://www.loebner.net/Prizef/TuringArticle.html. Online reproduction of Turning's original article.

The Turning Test Page. Available from http://cogsci.ucsd.edu/∼asaygin/tt/ttest.html#onlineref. Contains links to many Turning Test papers.

TUSKEGEE EXPERIMENT

• • •

From 1932 to 1972 the U.S. Public Health Service (PHS) tracked the nonmedicated course of syphilis, a disease that is caused by the bacterium *Treponema palladium*, among 399 patients and 201 controls at Tuskegee Institute (now Tuskegee University). In the region around Tuskegee in Macon County, Alabama, the PHS, in conjunction with the county health department and the Rosenwald Foundation, initially began a survey and small treatment program for African-Americans with syphilis.

The study goals and research methods soon shifted in response to financial limitations, and the project became the longest nontherapeutic observational study on human beings in medical history, manifesting major violations of basic human rights and ethical precepts. The legacy of government-sanctioned refusal to treat syphilis continues to influence the reluctance of African-Americans and other ethnic minorities to participate in government-funded clinical trials, contribute to organ and tissue donation campaigns, support biomedical research initiatives, and be involved in routine preventive medical care programs.

Throughout forty years of untreated observations infected poor rural African-American men intentionally were denied effective therapy as their disease progressed. Indeed, the premise of the study entailed nontreatment until the participating men died and could be autopsied to document the effects of syphilis on their tissues and organs. U.S. government health professionals withheld the standard treatment for syphilis in the early years of the project, injections of arsenic-based salvarsan and

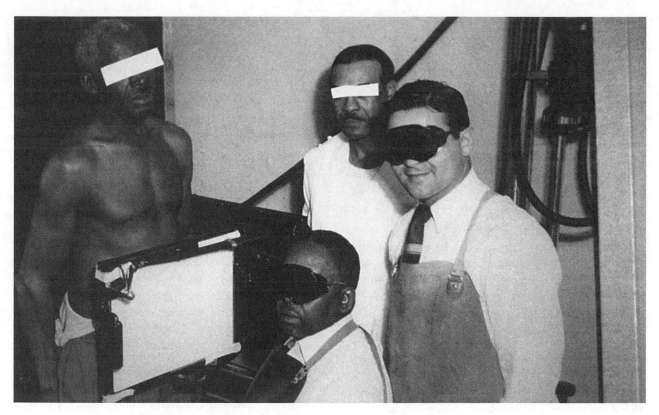

Doctors taking an x-ray of a Tuskegee subject. (© *Corbis Sygma*.)

topical applications of mercury or bismuth ointments; study participants never received clear advice about their disease state. When penicillin became the therapeutic agent of choice, study participants continued to be denied access to this known cure and their unremediated infections progressed.

Ongoing participation in the study by the men and their families was secured through the deception that they were receiving valuable medical care. Although the PHS provided the bulk of the medical personnel for this study, participant's primary contact throughout the years was with the Tuskegee-trained, PHS employed African-American nurse.

Permission for the study was obtained from key officials, including the U.S. surgeon general, the president of Tuskegee Institute, the medical director of Tuskegee Institute's John A. Andrew Hospital, and public health officials of Macon County. However, at no point were the basic human rights of the study participants protected. There was no voluntary, informed consent of the men under study and no opportunity to end the experiment at will, and the participants continued to be deceived throughout the study. The project, often called America's Nuremberg, reflected the convergence of scientific insensitivity and arrogance, racial injustice

and dehumanization, and socioeconomic class–based duplicity in the victimization of the study participants.

Target participants in the study were syphilitic African-American men in the later stages of the disease. In these less contagious stages untreated syphilis still causes serious cardiovascular abnormalities, neurological disorders, blindness, and death in infected individuals. Lack of treatment through participation in the study caused 28 to 100 men to die, and it has been estimated that the withholding of medical care adversely affected 22 wives, 17 children, and 2 grandchildren who subsequently contracted syphilis. The impact of intentional nontreatment of the men who were studied on rates of offspring miscarriages, stillbirths, infant mortality, and infants born with serious syphilis-related mental and physical problems remains unknown. Additionally, the degree of infertility among women sexually affiliated with the study's untreated syphilitic men has not been quantified.

The study was continued at a time when Jim Crow racism and segregation dominated interethnic interactions in the American South and when patients with sexually transmitted diseases faced social and medical discrimination. In the United States syphilis was both a medical problem and a metaphor for immorality and indecency. The PHS study focused on a nonrepresentative cohort of poor, uneducated African-American men residing in a remote location. Their selection was compatible with the emergence of U.S. eugenic programs.

Syphilis historically had been a significant social scourge in much of the Western world; the development of effective treatments for treponemal disease increased public confidence in the capacity of science to develop innovative technological solutions for persistent social problems and suggested that this dreaded sexually transmitted disease could become rare. Tracking its natural history in an expendable group was for some people a "tolerable" breach of ethics.

The government study was exposed publicly in 1972. In 1997 U.S. President Bill Clinton apologized on behalf of the nation to the few surviving victims. Ten million dollars in lawsuit-generated reparations was distributed among six hundred study participants and their descendants in partial compensation for their suffering.

FATIMAH L.C. JACKSON

SEE ALSO *Human Rights; Race; Sociological Ethics.*

BIBLIOGRAPHY

Jones, James H. (1993 [1981]). *Bad Blood: The Tuskegee Syphilis Experiment.* New York: Simon and Schuster.

TWO CULTURES

• • •

The term *two cultures* refers to a failure of scientists and humanists to comprehend the content, nature, and implications of each other's intellectual activities. An issue that goes back at least to the rise of modern science as a distinct practice and the romantic criticism of some of the results of the scientific worldview, it received international attention when Charles Percy Snow (1905–1980) considered the breakdown in a 1959 lecture, "The Two Cultures and the Scientific Revolution."

Snow, who had experience as a novelist and a scientist, coined the phrase to deplore a widening *gulf of mutual incomprehension* between *literary intellectuals* and natural scientists. The division between cultures represented a dilemma over the role of science and technology in human affairs and led to the failure to address the three *menaces* of nuclear weapons, overpopulation, and the gap between rich and poor. Although he recommended broadening education for both groups, Snow ultimately implied that solving these problems simply required more science and technology. Accordingly Snow accused literary intellectuals of being anti-scientific: While scientists held *the future in their bones,* the literati (whose ideas, Snow believed, unduly influenced western policy makers) were *natural Luddites.*

Critics, notably Frank R. Leavis, criticized Snow for being anti-cultural: In reducing humanistic knowledge to the equivalent of factual information, Snow undermined the capacity for *reflexive* ethical inquiry. In "A Second Look" (1964), Snow acknowledged that his phrase ignored the emergence of a *third culture* of social scientists that studied *the human effects of the scientific revolution.* Snow's phrase, imprecise in excluding *third* groups and in reducing *culture* to a set of conditioned responses, nevertheless calls attention to the problem of specialization and the disagreements about the proper function of science and technology that have persisted to this day.

Exchanges such as the *science wars* demonstrate that in many respects intellectual chasms have only continued to widen. Moreover, public policy debates over the relations among science funding, technology development, and the common good are often indicative of clashing worldviews reminiscent of Snow's two cultures. Within academe, most often in engineering and science curricula, occasional multidisciplinary and interdisciplinary programs do allow students to analyze and even synthesize humanistic and scientific paradigms; these offset to some extent the trends of increased specialization and balkanization. Public science agencies have likewise paid increasing attention to the ethical and societal implications of their research and development activities.

Efforts to integrate the two cultures can potentially balance technological goals with humanistic ones, but they can also be superficial and even counter-productive if they treat humanistic contributions as afterthoughts. Moreover the problem is not simply one of social groups; engineers, for example, tend to be the main advocates of appropriate technology. The gap, however, will continue to widen as specialized knowledge continues to be valued over broader, more integral understanding. A modern educational grounding in the fundamental concepts and practices of technical and humanistic traditions would be ideal. At the very least, interdisciplinary efforts that critically engage values and assumptions on both sides are indispensable if there is to be communication, understanding, and collaboration across the various intellectual divides.

ERIK FISHER

SEE ALSO *Governance of Science; Interdisciplinarity; Science Policy.*

BIBLIOGRAPHY

Brockman, John, ed. (1995). *The Third Culture.* New York: Simon and Schuster.

Leavis, Frank R. (1963). *Two Cultures? The Significance of C. P. Snow.* New York: Pantheon. Leavis, a literary critic, maintains the primacy of the humanities, not science; contains "An Essay" by Michael Yudkin.

Snow, Charles P. (1998). *The Two Cultures.* Cambridge, UK: Cambridge University Press. This reprint includes Snow's original lecture "The Two Cultures and the Scientific Revolution" (1959) and his "Second Look" (1964); the introduction by Stefan Collini presents the Snow-Leavis debate in light of precursors such as that between T. H. Huxley and Matthew Arnold.

U

UNCERTAINTY

• • •

The privative concept of uncertainty is more important in science, technology, and ethics than its positive root, certainty. (There is no entry in the encyclopedia on certainty.) This is the case for two reasons: Uncertainty is more common than certainty, and the implications of uncertainty for human action are more problematic than certainty. Uncertainty in science or engineering appears to call for an ethical assessment; uncertainty in ethics is a cause for moral concern. Nevertheless before discussing uncertainty, it is useful to begin with some considerations of certainty, the positive notion from which it is derived.

Certainty and Uncertainty in History

Concern for certainty as a distinct issue emerges at the same time as modern natural science. In premodern philosophy and science, it is difficult to find any term or concept that is strictly analogous. The Latin *certus*, the etymological root of certainty, is from the verb *cernere*, meaning to decide or determine; the Greek cognate *krinein* means to separate, pick out, decide, or judge. This sense remains in English when speaking of *a certain X*, indicating one item picked out from a group.

The concept of certainty in something approaching the modern sense is first given extended analysis in relation to religious faith. Faith, according to Augustine, is more certain than other forms of knowledge. Thomas Aquinas replies (*Quaestiones disputatae de veritate*, q. 14) that faith is psychologically but not epistemologically more certain than knowledge. Falling between knowledge and opinion in its degree of certainty, faith is defined as "an act of the intellect assenting to divine truth at the command of the will moved by the grace of God" (*Summa theologiae* II-II, q. 1). Moreover the certainty of faith provides a basis for moral judgment that is more secure than any provided by natural knowledge. Through faith, ethics takes on obligations of a stronger character than would otherwise be possible.

From theology, certainty becomes an issue for science when philosophers such as Francis Bacon and René Descartes argue for seeking cognitive certainty not through faith but through new methodologies. As interpreted by John Dewey in *The Quest for Certainty* (1929), "The quest for certainty is a quest for peace which is assured, an object which is unqualified by risk and the shadow of fear that action casts" (Dewey, p. 7). But the effort to secure such certainty and security that was originally undertaken through religious acceptance or propitiation of the gods is, in the early twenty-first century, commonly sought by means of technology and science. Extending Dewey, it is noteworthy that significant worries about lacks of certainty only became prominent as the new methods began to succeed so as to raise expectations of still further achievement. Thus has the pursuit of certainty through science and technology acquired a sense of ethical obligation.

The quest for certainty implies the presence of uncertainty, so that although this could not have been said prior to the modern period, it is now common to describe all human action as taken in the context of uncertainty. Insofar as this is the case, uncertainty is a locus of ethical discourse and conflict. Yet there are two forms of uncertainty in the modern sense that are most basic. Although often thought of as incomplete knowledge and applied to propositions, uncertainty can also be a psychological state. This distinction is important because perceived uncertainties may or may not reflect the actual state of incompleteness in knowledge. Perceptions of uncertainty may themselves be uncertain.

Uncertainty in Science

Characterizing and quantifying uncertainty is a core activity of science. Uncertainty emerges from research methodologies themselves, from the inherent characteristics of the processes and phenomena being studied, from incomplete or imperfect understanding, and from the contexts within which human beings seek to understand their surroundings. These sources of uncertainty may be understood, but they can never be eliminated. Uncertainty is always present to some degree in scientific knowledge, and in our formal knowledge of the world. This phenomenon is most famously embodied in Heisenberg's Uncertainty Principle, which states that the location and momentum of subatomic particles—the fundamental components of existence—can never simultaneously be known with complete accuracy.

Uncertainty is conceptually and practically distinct from fallibilism, or the notion that all scientific knowledge may turn out to be false. While both uncertainty and fallibility are attributes of knowledge, uncertainty refers to the accuracy of knowledge; fallibility to the provisional nature of knowledge. As Heisenberg's Uncertainty Principle illustrates, even if some knowledge (in this case, the uncertainty principle itself) were not provisional, uncertainty would still exist.

If, by contrast, the world were largely deterministic—that is, if its behavior could be explained through comprehensible and invariant cause and effect relations—then uncertainty could be eliminated, at least in theory. In practice, determinism can be approximated in some important human activities. Engineered systems, for example, can be designed as closed systems whose functional behavior is dictated by well-tested, scientific laws (laws of gravity, thermodynamics, and more), tested in laboratories, and supported by experience. Thus, for example, a bridge, or electronic circuit, or nuclear reactor, may operate with high reliability for decades. Eventually, however, the apparently closed system is breached—by corrosion, contamination, earthquake, or terrorism, among others—and the behavior of the system can no longer be thought of as deterministic or certain. The embeddedness of all engineered systems in larger social and natural systems dictates that uncertainty will eventually be introduced into engineering.

Uncertainties can be known with accuracy in closed systems that display random, or aleatory, behavior. Once the laws governing such system behavior are well elucidated, aleatory uncertainties cannot be further reduced. The obvious example is a game of dice or cards, where probabilities of particular outcomes can be determined from relatively simple statistical methods due to the known behavior of six-sided dice or fifty-two-card decks. Random behavior, and thus aleatory uncertainty, also exists in nature (for example, radioactive decay, Brownian motion), and can be approximated by some living systems (such as growth of bacteria in a medium) over limited periods of time, and often described by simple mathematical relations. Aleatory uncertainty is a property of random behavior in closed systems; it is inherent in the system itself.

For open systems whose governing laws cannot be fully elucidated, which includes all social and many technological and natural systems, uncertainty is said to be epistemic—a consequence of incomplete knowledge about cause-and-effect relations. In such cases—that is, most of the real world—uncertainty is a characteristic of both the system itself, and the psychological state of those who are assessing the uncertainty. Most problems at the interface of science, uncertainty, and ethics, are problems of epistemic uncertainty.

Epistemic uncertainties are most typically measured and expressed in probabilistic terms. Probabilities may be determined through frequentist approaches based on statistical analysis of past events or phenomena, or through subjectivist approaches, such as eliciting expert opinions, or surveying the scientific literature on a given subject. It is important to keep in mind that probability distributions derived from subjectivist approaches are distributions of beliefs about events, not of actual event occurrences.

Epistemic uncertainties also may be expressed in qualitative terms (such as *likely*, *unlikely*, and *doubtful*), or nonprobabilistically as ranges in values (for example, as error bars on a graph). Quantitative, nonprobabilistic uncertainties can also be derived from a comparison of the differences among outputs from different mathematical models ("model uncertainty").

Uncertainty in some complex systems or problems can be successfully addressed with frequentist approaches, because observational experience is sufficient to allow rigorous statistical treatment. Insurance companies, for example, set premiums using population-based data on life expectancy, morbidity, and frequency of auto accidents, among others. Engineers use data from tests and historical performance to estimate probabilities of failures in technological systems. Weather forecasts take advantage of a long history of careful observation of meteorological events. In such cases, uncertainty estimates can be refined and sometimes reduced on the basis of ongoing experience. It is important to recognize, however, that frequentist estimates of uncertainty are not necessarily accurate indicators of future probabilities, because in open systems, past behavior, however well documented, does not necessarily foretell future behavior. For example, 100-year flood levels, which are based on historical records and used in the United States for planning and insurance purposes, derive from the false assumption that climate behavior does not vary on time scales of more than a century (Pielke 1999).

Contextual Origins of Uncertainty

Uncertainty is a crucial concept in human affairs because knowledge of the future is always imperfect, and decisions are therefore always made in the face of uncertainty about their outcomes. From this perspective, the word uncertainty refers most generally to the disparity between what is known and what *actually is* or *will be*. Uncertainty, that is, reflects an incomplete and imperfect characterization of current conditions relevant to a decision, and the incomplete and imperfect knowledge of the future consequences of the decision. Logically, then, one way to improve the success of a decision should be to characterize, and if possible reduce, the uncertainty relevant to that decision, and considerable resources in science are devoted to this task. But significant obstacles stand in the way of this goal.

Many, perhaps most, of the important decisions faced by society have one or more of the following attributes: (1) the problem cannot be characterized in terms of easily measured outcomes in a well-defined population; (2) sufficient or relevant historical data are not available to allow frequentist approaches; (3) the dynamics of system behavior are incompletely and imperfectly understood; (4) the system is open; (5) numerous disciplines can contribute relevant understanding; and (6) different interests or values define the problem in different ways. For these reasons, most uncertainties in human affairs are epistemic, and most must be assessed through subjectivist methods. In all such cases, estimates of uncertainty are themselves both uncertain and strongly conditioned by the social context within which they are generated and used.

Less uncertainty can be an attribute of less knowledge. Continual research into and experience with complex, open systems should be expected to reveal new questions and new intricacies that may add to uncertainty over time. New knowledge does not necessarily translate into a greater ability to make well-constrained

statements about cause-and-effect relations relevant to human decisions. The archetypal example of this phenomenon is the climate change controversy, where ongoing research into the operations of the earth system and its interactions with human activities is continually introducing new variables and parameters, new appreciation of existing complexities, and new areas of scientific disagreement. While the observation of global warming is robust, and the rising impact of climate on society well documented, continued investigation into the causal relations between these two observations yields an ever expanding array of possible causal agents, and growing intricacy in the relations among agents.

A conventional view of this problem describes a *cascade of uncertainty*, where the more modest uncertainties embodied in the understanding of relatively simple systems or phenomena are introduced into and magnified at the next level of complexity, which in turn introduces its own, perhaps greater, uncertainties (Schneider and Kuntz-Duriseti 2002). The importance of this notion lies especially in the fact that simpler systems are generally farther away from real world problems. Thus it is hard enough to understand and reduce the uncertainties surrounding greenhouse gas behavior in the atmosphere, but if the concern is the impacts of those gases on society via changes in regional climate, then uncertainties cascade beyond comprehension or control.

This view of the problem locates uncertainty in the complexity of natural and social systems being studied, but uncertainty also arises from the conduct of these studies. Science is not a unitary activity; multiple disciplinary approaches often yield multiple perspectives that do not fit together to yield a seamless picture of nature, but rather create multiple and sometimes even conflicting pictures (Dupré 1993). For example, plant geneticists and those in related fields commonly evince greater certainty than ecologists that genetically modified crops will be beneficial to humanity and the environment. These differences derive in part from different ways of understanding nature. Plant geneticists, employing reductionist approaches to crop engineering, are thus confident about their ability to control crop behavior. Ecologists, in contrast, study complex systems where small variations in conditions are often seen to have large and unpredictable impacts.

Lying beneath these epistemological differences are likely to be ethical tensions between one worldview where control of nature yields human benefit and another where pretensions to control can be futile and dangerous. For complex issues where relevant knowledge comes from multiple disciplines, estimates of

FIGURE 1

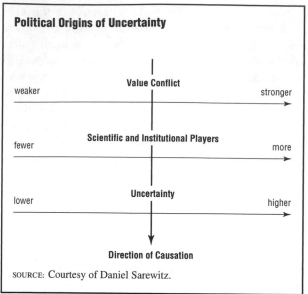

Political Origins of Uncertainty

Value Conflict

weaker   stronger

Scientific and Institutional Players

fewer   more

Uncertainty

lower   higher

Direction of Causation

SOURCE: Courtesy of Daniel Sarewitz.

uncertainty may thus partly be a reflection of competing disciplinary perspectives, and the ethical commitments entailed in those perspectives. These relations are likely to be reinforced by behavioral attributes of scientists. In particular, experts typically underestimate uncertainty in their own area of expertise (Kahneman et al. 1982) while locating the sources of uncertainty in disciplines other than their own (Pinch 1981).

Uncertainty estimates may strongly reflect institutional and political context. Consider, for example, that the U.S. National Aeronautics and Space Administration (NASA) initially estimated the reliability of its space shuttle fleet at 0.9997, or one failure every 3,333 launches (Pielke 1993). Since then two shuttles out of 112 total launches have self-destructed during flight, yielding a historical reliability of 0.98—thirty times less than the initial estimate. High certainty about shuttle reliability could exist when experience with shuttle flights was small, and knowledge was limited. Yet high certainty was also consistent with the political interests of NASA, and with the institutional incentives in the agency, which rewarded launching shuttles, not grounding them. Another illustration comes from medical science, where a number of studies have shown that clinical trials directly or indirectly supported by pharmaceutical companies often yield more favorable assessments of new therapies—greater certainty about positive results—than trials that are not tied to the private sector in any way (Angell 2000). The point here is not that scientists are engaging in fraudulent research in an effort to bolster desired conclusions, but experimental design and interpretation of data are partly matters of judgment, and judgment may be influenced by the incentives, priorities, and culture of one's work environment.

Additional examples from such areas as climate change science (van der Sluijs et al. 1998), earthquake prediction (Nigg 2000), oil and gas reserve estimates (Gautier 2000), and nuclear waste disposal (Metlay 2000) show that uncertainty estimates are strongly dependent on institutional and political context, and that opening up the research process to additional scientific and institutional perspectives often leads to significant changes in perceived uncertainty.

Uncertainty and Values

Important decisions in human affairs create winners and losers relative to the status quo ante, and thus implicate competing interests and values. In areas of decision making that include a significant scientific component, such as the environment, public health, and technological risk, uncertainty provides the space for disputes between competing interests and values to play out, because those who hold contesting positions can make conflicting or disparate science-based claims about the consequences of particular courses of action. Thus, for example, supporters of genetically modified foods can point to the potential for gains in crop productivity, and opponents can point to the threat of diminished crop genetic diversity. This is a self-reinforcing process: As value disputes grow more heated, they bring out the latent uncertainties associated with a problem or decision by expanding the realm of phenomena, disciplinary perspectives, and institutional and political players relevant to the problem. These relations are schematically illustrated in Figure 1.

So long as uncertainty is understood simply in terms of the incomplete but ever-improving knowledge of the world, reduction of uncertainty will be prescribed as a path toward resolving political disputes. But when uncertainty is also recognized as an outgrowth of the contexts within which scientific inquiry is structured and carried out, the path begins to look Sisyphean. Indeed the contextual diversity of science is the manifestation of, not the solution to, the conflicting values that underlie political debate. These observations suggest that the taming of uncertainty must depend not on the capacity of science to characterize and reduce uncertainty, but on the capacity of political processes to successfully resolve value disputes that underlie the choices that humans face.

DANIEL SAREWITZ
CARL MITCHAM

SEE ALSO *Precautionary Principle; Reliability of Technology: Risk; Technical and Social Dimensions; Unintended Consequences.*

BIBLIOGRAPHY

Angell, Marcia. (2000). "Is Academic Medicine for Sale?" *New England Journal of Medicine* 342: 1516–1518. How interests and uncertainty interact in clinical medicine.

Dewey, John. (1984). *John Dewey: The Later Works, 1925–1953;* vol. 4: (1929) *The Quest for Certainty: A Study of the Relation of Knowledge and Action.* Carbondale: Southern Illinois University Press.

Dupré, John. (1993). *The Disorder of Things: Metaphysical Foundations of the Disunity of Science.* Cambridge, MA: Harvard University Press. Why science does not provide a unified explanation of nature.

Gautier, Donald L. (2000). "Oil and Gas Resource Appraisal: Diminishing Reserves, Increasing Supplies." In *Prediction: Science, Decision Making, and the Future of Nature,* eds. Daniel Sarewitz; Roger Pielke Jr.; and Radford Byerly Jr. Covelo, CA: Island Press. Uncertainty and predictions of hydrocarbon reserves.

Kahneman, Daniel; Paul Slovic; and Amos Tversky. (1982). *Judgment under Uncertainty: Heuristics and Biases.* New York: Cambridge University Press. Classic treatment of uncertainty and human decision making.

Metlay, Daniel. (2000). "From Tin Roof to Torn Wet Blanket: Predicting and Observing Groundwater Movement at a Proposed Nuclear Waste Site." In *Prediction: Science, Decision Making, and the Future of Nature,* eds. Daniel Sarewitz; Roger Pielke Jr.; and Radford Byerly Jr. Covelo, CA: Island Press. Uncertainty and nuclear waste storage.

Nigg, Joanne. (2000). "Predicting Earthquakes: Science, Pseudoscience, and Public Policy Paradox." In *Prediction: Science, Decision Making, and the Future of Nature,* eds. Daniel Sarewitz; Roger Pielke Jr.; and Radford Byerly, Jr. Covelo, CA: Island Press. Uncertainty and earthquake prediction.

Pielke, Roger A., Jr. (1993). "A Reappraisal of the Space Shuttle Program." *Space Policy* 9: 33–157.

Pielke, Roger A., Jr. (1999). "Nine Fallacies of Floods." *Climatic Change* 42: 413–438. How misunderstanding uncertainty can lead to flawed public policies.

Pinch, Trevor J. (1981). "The Sun-Set: The Presentation of Certainty in Scientific Life." *Social Studies of Science* 11: 131–158. How scientists perceive uncertainty.

Schneider, Stephen H., and Kristin Kuntz-Duriseti. (2002). "Uncertainty and Climate Change Policy." In *Climate Change Policy: A Survey,* eds. Stephen H. Schneider; Armin Rosencranz; and John O. Niles. Washington, DC: Island Press. Natural science perspective on uncertainty in climate change science.

van der Sluijs, Jeroen; Josee van Eijndhoven; Simon Shackley; and Brian Wynne. (1998). "Anchoring Devices in Science for Policy: The Case of Consensus around Climate Sensitivity." *Social Studies of Science* 28(2): 291–323. Social science perspective on uncertainty in climate change science.

UNINTENDED CONSEQUENCES

• • •

Human activities often produce consequences very different from those intended. Indeed this is a theme of classical tragedy and much premodern argument about the indeterminacy of human affairs. Sociologist Robert K. Merton was one of the first to subject "The Unanticipated Consequences of Purposeful Action" (1936) to systematic analysis, noting the influences of the need to act in spite of uncertainties, the allocation of scarce resources such as time and energy, and how personal interests shape perspectives and decisions. Advances in science and technology seem particularly likely to change the world in unanticipated ways. Innovations are by definition something new and are likely to involve unknowns. Innovations may be used in unplanned ways that trigger surprising results. The more complex a system, the harder it is to anticipate its effects. Unintended consequences can shift the cost-benefit analysis of a new technology, theory, or policy; distribute costs and benefits inequitably; or lead to other direct or indirect social problems. Such consequences raise questions of responsibility and liability; decision making under uncertainty; equity and justice; and the role of individual citizens, corporations, universities, and governments in managing science and technology.

Types of Unintended Consequences

Unintended consequences occur in many forms, although the categories are neither entirely discrete nor universally recognized. *Accidents* are usually immediate and obvious, and result from problems such as mechanical failure or human error, such as the disastrous 1986 explosions, fires, and releases of radiation at the nuclear rector in Chernobyl, Russia.

Side effects are additional, unanticipated effects that occur along with intended effects, such as gastrointestinal irritation resulting from aspirin taken to relieve pain. *Double effects,* meaning simply two effects, often refer to simultaneous positive and negative effects, as in the aspirin example. Many medical side effects are well documented, such as the devastating effects of diethylstilbestrol (DES) and thalidomide and the ability of bacteria to develop resistance to antibiotics (Dutton et al. 1988).

Surprises could apply to any unintended consequence, but the term is more specifically used, along with *false alarms,* to describe errors in prediction. A false alarm is when a predicted event fails to occur, such as the millennium computer bug, whereas a surprise is an unexpected event, such as the 2004 Indian Ocean tsunami (Stewart 2000).

Henry N. Pollack (2003) refers to *inadvertent experiments*, in which human actions unwittingly allow and sometimes force society to consider the effects of its actions. He cites the hole in the ozone layer and climate change as classic examples. Historians of science and technology also have noted the occasional benefits of *serendipity* in both discovery and invention.

More provocatively science and technology sometimes have the reverse of their intended effects. In the 1970s Ivan Illich (1973) among others argued that scientific and technological development, after crossing a certain threshold, may exhibit a counterproductivity, producing new problems even as it solves old ones. Extending this notion into political theory, Ulrich Beck (1986) argues that unintended consequences in the form of *boomerang effects* are transforming politics into a concern for the just distribution not of goods but of risks.

With a more individualist focus, Edward Tenner identifies *revenge effects* as the "ironic unintended consequences of mechanical, chemical, biological, and medical ingenuity" or, more anthropomorphically, as "the tendency of the world around us to get even, to twist our cleverness against us" (Tenner 1997, p. 6). He further divides revenge effects into *rearranging effects*, that shift the locus or nature of a problem, such as urban air-conditioning making the outside air hotter; *repeating effects*, that have people "doing the thing more often rather than gaining time to do other things"; *recomplicating effects* such as the annoying loops of voice mail systems; *regenerating effects*, in which a proposed solution such as pest control makes a situation worse; and *recongesting effects*, such as the human ability to clog space with debris from space explorations (Tenner 1997, p. 10).

Direct effects are those that occur fairly quickly, with no intervening factors. *Indirect effects* are likely to take longer to develop and may involve interactions with other factors; *latent side effects* also refer to impacts that occur later in time. *Secondary effects* are the next level of impacts resulting from direct effects; they generally impact people or places other than those a product or activity is intended to affect; these may also be called *ripple effects*. The secondary effects of smoking on nonsmokers have been well documented. *N-order effects* are even more removed from the direct effects. *Cumulative effects* are additive. Combinations of substances, particularly pesticides or medicines, are sometimes called *cocktail effects*, especially in the United Kingdom. *Interaction effects* are those resulting from a combination of two or more factors that act on or influence each other to produce a result different from either acting alone.

The military uses the term *collateral damage* to describe injuries to people and property other than intended targets, such as the destruction of the Chinese Embassy during the 1999 North Atlantic Treaty Organization (NATO) bombing campaign in Yugoslavia. Civilian casualties are often framed as collateral damage because ethical principles of noncombatant immunity proscribe the deliberate injury of civilians.

Economists often refer to unintended consequences as *externalities*, "An action by either a producer or a consumer that affects other producers or consumers, yet is not accounted for in the market price" (Pindyck and Rubinfeld 1998, p. 696). Pollution is usually considered an externality, as its effects on human health, safety, and quality of life are often not factored into industrial costs. Externalities may require management such as government imposed regulations, subsidies, or market-based mechanisms to prevent economic inefficiencies. Externalities such as pollution or hazardous wastes often impose unequal burdens on the poor or powerless, raising questions about equity and environmental justice.

Unintended consequences are different from *unanticipated consequences*, in which effects may be suspected or known to be likely but are not part of the intended outcome. Some anticipated consequences may be ignored if they interfere with the interests of decision makers or seem relatively minor; cumulative or interactive effects may make them more serious. Knowledge about effects, or effects that should have been anticipated, may be important in deciding who, if anyone, should be held legally, politically, or morally responsible for unintended outcomes.

Causes and Effects

Unintended consequences of science and technology can have many causes. Design flaws may lead to project failure. Materials may not meet expectations. Assumptions may prove incorrect.

Human factors frequently trigger unintended consequences. Human errors, sometimes interacting with technical failures and environmental stresses, often cause accidents, such as the 1984 release of poisonous gas from Union Carbide's pesticide plant in Bhopal, India (Jasanoff 1994). People often use science and technology in unexpected ways. What appears to be operator error may be the result of an overly complex or inherently unsafe technology. Additionally safety measures such as seat belts sometimes may actually increase hazards as people compensate by taking more risks, illustrating a phenomenon known as *risk homeostasis*.

Unintended consequences may have social, economic, or behavioral as well as physical causes and impacts, especially when transferred from one culture to another. Anthropologists, for instance, have well documented the often unintentionally destructive outcomes of technology transfer across cultures (Spicer 1952). The movie *The Gods Must Be Crazy* (1981) depicts a comic version of this phenomenon. Effects may be catastrophic, even when the transfer is only from laboratory to market place.

Richard A. Posner (2004), for instance, distinguishes four types of catastrophe, all but one resulting from the unintended consequences of science and technology. The exception is a natural catastrophe. The other categories are accidents from the products of science and technology, such as particle accelerators or nanotechnology; unintended side effects of human uses of technology, such as global climate change; and the deliberate triggering of destruction made possible by dangerous innovations in science and technology, which can be considered technological terrorism. Posner also notes "the tendency of technological advance to outpace the social control of technology" (Posner 2004, p. 20), an instance of cultural lag.

Not all unintended consequences are bad; many innovations have beneficial side effects, and effects can be mixed. For example, 2004 studies on some pain relievers, such as Vioxx or Celebrex, suggest that they may reduce cancer risks while enhancing risks of heart attacks. From the perspective of social scientist Michel de Certeau creative, unintended uses may actually serve as a means for the assertion of human autonomy; using products in ways unintended by the designer is a way of resisting technological determination. Some writers see occasional benefits even in negative unintended consequences. Fikret Berkes and Carl Folke suggest that in some cases, "breakdown may be a necessary condition to provide the understanding for system change," although crisis cannot be allowed to reach the point where it imperils the survival of the system (Berkes and Folke 1998, p. 350). Complexity theorists have even argued the emergence of new forms of spontaneous order from unintended chaotic situations.

Managing Unintended Consequences

How should unintended consequences be managed? Some impacts may be avoided with more careful planning in the design and implementation of innovations, but many writers assume that unexpected negative consequences are inevitable, *normal accidents* (Perrow 1984),

and advocate systems that either minimize such effects or try to manage them.

Unintended consequences often cross temporal and spatial boundaries. When effects cross physical or political barriers, unintended consequences raise questions about responsibility. Indeed, one ethical response to such technological changes in the scope and reach of human action is to argue for the articulation of a new *imperative of responsibility* (Jonas 1984). How does one country hold another responsible when pollution or other effects cross borders? This is a major question in climate change, where industrialized countries have been the major human source of greenhouse gases but developing countries will suffer the most severe impacts expected, such as sea rise and increased and prolonged regional droughts. In some limited cases national tort law provides compensation for injuries caused by actions taking place outside the borders of the sovereign state. International law is even more problematic, since there is no sovereign providing enforcement, and countries must rely on their ability to reach international agreements to deal with novel and intractable problems such as the hole in the ozone.

Conventional methods of dealing with risk, such as insurance, legal remedies, and emergency procedures, were not designed to deal with the current spread of side effects. When effects occur much later in time they affect future generations, raising issues of *intergenerational* equity. Is it fair to leave a seriously degraded and hazardous world for future generations?

Three types of errors may be made at the more mundane level of managing unintended consequences (Tenner 1997). *Type I errors* are those where unnecessary preventive measures are taken, such as keeping a safe and effective product off the market. *Type II errors* occur when an important protective measure is not taken, such as allowing the use of a very harmful product. *Type III errors* involve displaced risks, new risks created by protective measures, such as the economic effects of unnecessary environmental regulations.

David Collingridge describes the essential problem with technology, the *dilemma of control*: "Attempting to control a technology is difficult, and not rarely impossible, because during its early stages, when it can be controlled, not enough can be known about its harmful social consequences to warrant controlling its development; but by the time those consequences are apparent, control has become costly and slow" (Collingridge 1980, p. 19) He proposes "a theory of decision making under ignorance" to make decisions more "reversible, corrigible, and flexible" (p. 12). He works within the *fallibilist tradition*, which "denies the possibility of justification, and sees rationality as the search for error and the willingness to respond to its discovery" (p. 29). Collingridge advocates a decision process that allows errors to be identified quickly and managed inexpensively. Options should be kept open so that changes can be made as new information becomes available, but this becomes more difficult the longer a technology is in use.

Others have suggested similar systems. Aaron Wildavsky talks about the *resilience* of systems and advocates a gradual system of response as new information becomes available. Steve Rayner (2000) also stresses the importance of developing resilience to improve society's ability to deal with surprises. Sheila Jasanoff (1994) advocates planning in both the anticipation of and the response to disasters. Kai Lee (1993) and Berkes and Folke (1998) propose using *adaptive management* to build resilience into the management of natural resources.

Arguing that science and technology themselves can play multiple roles, not only as a source of risks but as means to help identify and prevent problems, as well as to develop adaptation measures to ease negative impacts, Posner (2004) recommends the use of *cost-benefit analysis* to evaluate risks, saying it is an essential component of rational decisions. He also recognizes that uncertainties create many ethical, conceptual, and factual problems and suggests several methods for coping. Some application of the *precautionary principle*, or the *better safe than sorry* approach to decisions, may be appropriate as a variation of cost-benefit analysis in which people choose to avoid certain risks.

John D. Graham and Jonathon B. Wiener (1993) describe the *risk tradeoffs* that are inevitably faced in protecting human health and the environment; minimizing one risk may actually increase other *countervailing risks*. In some cases, reducing one risk will cause other *coincident risks* to decrease, as well. The authors propose a *risk trade-off analysis* to reveal the tradeoffs likely in any decision, and examine ethical as well as scientific issues. Factors to be considered in evaluating risks include "magnitude, degree of population exposure, certainty, type of adverse outcome, distribution, and timing" (Graham and Wiener 1993, p. 30). Consideration of these factors before making a decision may make it possible to reduce but not eliminate surprise effects.

Corporations, think tanks, universities, or other private institutions may not consult the public about their scientific and technological decisions. Even government-sponsored research and regulation typically involve little public participation. Yet the public is usually the intended user of innovations and bears many of the benefits and burdens of both intended and unintended consequences. Questions for a democratic society include whether the public should play a larger role in decisions regarding science and technology, how meaningful public involvement can be achieved, and how public opinions should be balanced with scientific expertise. Greater public involvement would increase the diversity of interests and values brought to an analysis of and debate about the risks and benefits of innovations in science and technology.

Science and technology funding raise questions about the optimal allocation of public and private funds. Funding rarely is devoted to assessing risks of innovations. Funding to develop solutions to one problem may end up creating other unintended consequences. Should funding agencies require more analysis of possible consequences of funded projects, and should the agencies be held partially responsible for consequences?

Conclusion

The unintended consequences of science and technology are ubiquitous and complex in the contemporary world. They raise important questions about the kind of society in which humans choose to live in, including issues relating to allocation of scarce societal resources; the types and levels of risks society is willing to tolerate; the attribution of responsibility and liability; the right to compensation for injury, the equitable distributions of societal costs and benefits; and the role of individuals, corporations, governments, and other public and private institutions in the control of science and technology.

MARILYN AVERILL

SEE ALSO *Enlightenment Social Theory; Normal Accidents; Precautionary Principle; Uncertainty.*

BIBLIOGRAPHY

Beck, Ulrich. (1986). *Riskogesellschaft: Auf dem Weg in eine andere Moderne* [Risk society: Towards a new modernity]. Frankfurt am Main: Suhrkamp. English version, translated by Mark Ritter (London: Sage Publications, 1992). Beck maintains that the changing nature of risks is transforming the developed world from an industrial society to a risk society in which the side effects of industrialization increase in frequency and severity and force society to focus more on risk production than on wealth production.

Berkes, Fikret, and Carl Folke, eds. (1998). *Linking Social and Ecological Systems: Management Practices and Social Mechanisms for Building Resilience.* Cambridge, UK: Cambridge University Press. This edited volumes presents case studies of the interactions of social and ecological systems in response to environmental change. Authors focus on the nature of the problems faced, responses to those problems, and lessons to be learned.

Collingridge, David. (1980). *The Social Control of Technology*. New York: St. Martin's Press. Focusing on the social effects as well as the social control of technology, Collingridge describes the problem of control when undesirable effects are not known until change has become more difficult. He proposes a theory of decision making under ignorance so that decisions can be modified as new information becomes available.

De Certeau, Michel. (1984). *The Practice of Everyday Life*, trans. Steven F. Rendall. Berkeley: University of California Press.

Dutton, Diana D.; Thomas A. Preston; and Nancy E. Pfund. (1988). *Worse than the Disease: Pitfalls of Medical Progress*. Cambridge, UK: Cambridge University Press.

Graham, John D., and Jonathon Baert Wiener. (1993). *Risk vs. Risk: Tradeoffs in Protecting Health and the Environment*. Cambridge, MA: Harvard University Press. This book describes the many tradeoffs that must be made in attempting to manage risks, including risks relating to unintended consequences. It provides case studies of tradeoffs that have been made, analyzes the problems inherent in such tradeoffs, and suggests ways to manage risks more effectively and democratically.

Illich, Ivan. (1976). *Medical Nemesis: The Expropriation of Health*. New York: Pantheon Books. In this work, Illich criticized the medical establishment as a major threat to health, claiming that even though more money is being spent on healthcare, fewer benefits are being realized.

Jasanoff, Sheila. (1994). "Introduction: Learning from Disaster." In *Learning from Disaster: Risk Management after Bhopal*, ed. Sheila Jasanoff. Philadelphia: University of Philadelphia Press. Authors from a variety of disciplines seek to identify lessons to be learned from the 1984 lethal gas leak from a plant in Bhopal, India. The book discusses the problems with transplantation of technologies across cultures and the way culture, politics, and other human factors shape approaches to risk management.

Jonas, Hans. (1984). *The Imperative of Responsibility: In Search of an Ethics for the Technological Age*. Chicago: University of Chicago Press. Viewed as Jonas' most important work. Here he reflects on the challenges brought to society by nuclear weapons, chemical pollution, and biomedical technologies.

Lee, Kai. (1993). *Compass and Gyroscope: Integrating Science and Politics for the Environment*. Washington, DC: Island Press. Sustainable development; civic science, and adaptive management are discussed within the context of the controversies over salmon populations in the Columbia River Basin. Lee recommends that science be coupled with public debate in addressing environmental problems.

Merton, Robert K. (1936). "The Unanticipated Consequences of Purposive Social Action." *American Sociological Review* 1(6): 894–904.

Perrow, Charles. (1999). *Normal Accidents: Living with High Risk Technologies*. Princeton: Princeton university press.

Pindyck, Robert S., and Daniel L. Rubinfeld. (1998). *Microeconomics*, 4th edition. Upper Saddle River, NJ: Prentice Hall.

Pollack, Henry N. (2003). *Uncertain Science ... Uncertain World*. Cambridge, UK: Cambridge University Press.

Posner, Richard A. (2004). *Catastrophe: Risk and Response*. Oxford: Oxford University Press. Posner Addressed Risks that could actually threaten the future of human life on earth; several of these Potential disasters are the unintended outcomes of innovations in science and technology. He calls for interdisciplinary policy responses to address these potential catastrophies.

Rayner, Steve. (2000). "Prediction and Other Approaches to Climate Change Policy." In *Prediction: Science, Decision Making, and the Future of Nature*, ed. Daniel Sarewitz, Roger A. Pielke Jr., and Radford Byerly Jr. Washington, DC: Island Press.

Spicer Edward H., (1952). *Human Problems in Technological Change: A Cashbook*. New York: John Wiley. Includes Lauriston Sharp's widely referenced case study on "Steel Axes for Stone Age Australians."

Stewart, Thomas R. (2000). "Uncertainty, Judgment, and Error in Prediction." In *Prediction: Science, Decision Making, and the Future of Nature*, ed. Daniel Sarewitz, Roger A. Pielke Jr., and Radford Byerly Jr. Washington, DC: Island Press.

Tenner, Edward. (1997). *Why Things Bite Back: Technology and the Revenge of Unintended Consequences*. New York: Vintage Books. Unintended consequences are at the center of this book, which describes how and why technology, when combined with human behavior and institutions sometimes seems to turn on the society that produced it. Extensive examples are provided relating to medicine, the environment, computers, and the spread of pests.

UNION OF CONCERNED SCIENTISTS

• • •

The Union of Concerned Scientists (UCS) is a nonprofit alliance of more than 100,000 scientists and citizens that works to promote environmental and global security solutions based on sound science. UCS scientists, engineers, and analysts collaborate with colleagues across the country to conduct technical studies on renewable energy options, cleaner cars and trucks, the impacts of and solutions to global warming, the risks of genetically engineered crops, deforestation, invasive species, nuclear power plant safety, missile defense, the security of nuclear material, and other issues. Research results are shared with policy makers, the news media, and the public in order to shape public policy, corporate practices, and consumer choices.

Founding and Finances

UCS was founded in 1969 out of a movement at the Massachusetts Institute of Technology, where an ad hoc group of faculty and students joined together to protest the misuse of science and technology. They put forth a Faculty Statement—the genesis of UCS—calling for greater emphasis on the application of scientific research to environmental and social problems, rather than military programs.

UCS derives approximately 50 percent of its operating revenue from foundations, 40 percent from membership, and 10 percent from planned giving and other sources. Member and foundation support has grown steadily over the years, and in the early twenty-first century UCS has an operating budget of nearly $10 million. More than 75 percent of the operating budget is applied directly to program work.

Historical Development

In its early work, the UCS focused on nuclear weapons, weapons-related research, and nuclear power plant safety. In April 1969, it released its first report, *ABM ABC*, criticizing President Nixon's proposed Safeguard anti-ballistic missile system. UCS's ongoing opposition helped build public support for the ABM Treaty, signed by the United States and the Soviet Union in 1972. In 1979, when Three Mile Island Unit II experienced a near meltdown, UCS provided crucial independent information to the media and the public seeking to understand the accident and the risks to neighboring communities.

In the early 1980s, when the Reagan administration proposed a missile defense program called the Strategic Defense Initiative (SDI), also known as "Star Wars," UCS mobilized swift and sweeping opposition in the scientific community to the SDI program, and analyzed its technical and strategic drawbacks, providing a crucial counterweight to the claims and promises of its proponents.

In 1987, UCS successfully sued the Nuclear Regulatory Commission to strengthen safety enforcement at nuclear power plants. Four years later, UCS forced the shutdown of the Yankee Rowe nuclear plant in Massachusetts due to safety concerns.

UCS kicked off its new climate change campaign in 1990, when 700 members of the National Academy of Sciences signed UCS's Appeal by American Scientists to Prevent Global Warming. In 1992, some 1,700 scientists worldwide, including a majority of Nobel laureates in the sciences, issued the World Scientists' Warning to Humanity. UCS Chair Henry Kendall, a Nobel laureate in physics, wrote and spearheaded the statement, an unprecedented appeal from the world's leading scientists on the destruction of the earth's natural resources.

In 1993, UCS pioneered new analytical techniques to demonstrate the breadth of renewable energy resources in twelve Midwestern states. The attention and commitment to clean energy that the research generated continues into the twenty-first century. UCS also launched a new program the same year, focusing on sustainable agriculture and biotechnology. The program's first report, *Perils Amidst the Promise,* analyzes the ecological risks of the commercialization of transgenic (genetically engineered) crops. Two years later, in response to grassroots pressure generated by UCS, the U.S. Environmental Protection Agency imposed new transgenic crop standards.

The UCS Clean Vehicles program, which was launched in 1991, had a number of major policy victories in the mid- to late-1990s. UCS led the successful campaign to open the market to clean, nonpolluting cars in California in 1996. The state's low-emission vehicle (LEV) standards, which include zero-emission vehicle (ZEV) production requirements, have been adopted by several states in the northeastern United States. In 1998, UCS helped convince California to require SUVs, light trucks, and diesel cars to meet the same tailpipe emissions standards as gasoline cars. *Greener SUVs,* a 1999 report demonstrating numerous "off the shelf" technologies available to automakers to cost-effectively increase the gas mileage of their cars and trucks, has provided a technical basis for the environmental community's efforts to raise national fuel economy standards.

Success and Shortcomings

UCS has secured some major policy victories in the early twenty-first century. Its 2000 report *Countermeasures,* which demonstrated that the proposed national missile defense system could be defeated by missiles equipped with simple countermeasures, convinced President Clinton not to deploy the system. In 2001, UCS issued the first-ever analysis of antibiotic use in livestock feed, demonstrating that widespread overuse threatens the efficacy of drugs used in human medicine. And UCS continues to play a key role in shaping California environmental policy; in 2002, the state passed the first global warming emission rules for cars and light trucks, and the nation's strongest renewable energy standard (20% by 2017). The U.S. Senate also passed a 10 percent renewable energy standard in 2002, the first-ever renewable energy legislation of its kind in Congress.

USC's advocacy of forward-thinking solutions on environmental and arms control issues have prompted some national media label to UCS a "liberal" group and has also made it a target of criticism of various groups invested in the status quo. Despite these challenges, UCS has forged relationships with leaders, on both sides of the aisle, who understand that independent scientific analysis has an important role to play in the decisions about public health, safety and the environment.

Since its inception, however, the Union of Concerned Scientists has played an influential role in environmental and security policy development. It has brought independent scientific analysis to pressing issues facing the global society and effectively communicated these findings to the public and policy makers to demonstrate their meanings at the national, regional, and community level. UCS believes scientists can and should play an important role informing public policy choices. As long-time UCS board chair Henry Kendall put it, "If scientists do not speak out, significant opportunities are lost" (Kendall 2000, p. 1).

SUZANNE SHAW

SEE ALSO Federation of American Scientitsts; Nongovernmental Organizations; Professional Engineering Organizations.

BIBLIOGRAPHY

Kendall, Henry W. (2000). A Distant Light: Scientists and Public Policy. New York: AIP Press/Springer.

INTERNET RESOURCES

Union of Concerned Scientists. "History." Available from http://www.ucsusa.org/ucs/about/page.cfm?pageID=767.

UNITED NATIONS EDUCATIONAL, SCIENTIFIC AND CULTURAL ORGANIZATION

• • •

The United Nations Educational, Scientific and Cultural Organization (UNESCO) was conceived within the United Nations (UN) Charter, which was ratified on October 24, 1945. In the view of its founders, it was to revive within the new UN system, the International Institute of Intellectual Cooperation (IIIC), created, in 1924, by the League of Nations' International Committee on International Cooperation (ICIC). The institute had counted among its members such eminent world personalities as Albert Einstein, Henri Bergson, Sigmund Freud, Marie Curie, Gabriela Mistral, Aldous Huxley, Miguel de Unamuno, Paul Valéry, and Rabindranath Tagore. The UNESCO Constitution was adopted on October 24, 1945, by thirty-seven countries, By October 2003, it was composed of 190 Member States and six Associate Members.

At the outset, some of its more influential members were of the opinion that UNESCO should be the world organization in which "intellect would be allowed to have more scope and real power in the things of this world" (an expression used by Valéry, a leading member of the first French delegation to the new organization, who had also represented France in the old IIIC). It was thought that this approach could better protect the institution from excessive dependence on changing political pressures. The same concerns could explain why, at the outset, the members of the Executive Board were conceived to be more than just representatives of their respective governments; they would be chosen by the General Conference (the highest organ of UNESCO) on the basis of their personal qualifications and independence of mind, as had been the case with the IIIC. But because of political considerations, the practice moved in a different if not opposite direction. Accordingly, the UNESCO Constitution was amended in 1992 to make it clear that the representatives on the Board would always follow the instructions of their respective governments.

UNESCO's five original fields of competence were placed under the headings of education, exact and natural sciences, social sciences, culture, and communication. To these were later added intersectoral activities that embrace both the sciences and culture as well as fundamentally multidisciplinary projects such as the Protection of the World and Cultural Heritage and collaboration with other organizations of the UN system and with international nongovernmental organizations (NGOs).

Education

UNESCO's first publication was a report titled Fundamental Education: Common Ground for All Peoples (1946). Although, ten years later, a working party of the General Conference proposed a new definition of this concept ("to help people who have not obtained such help from established educational institutions to understand the problems of their environment and their rights and duties to acquire a body of knowledge and skills for the progressive improvement of their living conditions and to participate more effectively in the economic and social development of their community"), the term *fundamental* or *basic* education stopped being used, on the ground that it was liable to confer official status on a "cut-rate" educational goal that would run counter to the goal of universal primary education.

The pursuit of a world model of schooling based on the experience of industrially developed countries had often exacerbated the difficulties of the poorer populations in developing their vernacular modes of learning. Therefore, at the World Education Forum (held in Dakar, Senegal, in 2000), UNESCO adopted a new approach under the name Education for All. This program was designed to reach six goals by the year 2015: (1) expand early childhood care and education, (2) improve access to and complete free schooling of good quality for all children of primary school age, (3) greatly increase learning opportunities for youth and adults, (4) improve adult literacy rates by 50 percent, (5) eliminate gender disparities in schooling, and (6) improve all aspects of educational quality (UNESCO, "World Education Forum").

Since 1964 UNESCO has taken a similar approach in working toward its goal of eradicating literacy in the world. At the 1965 World Conference of Ministers of Education in Tehran, the organization introduced the notion of "functional literacy," a conception in which learning to read and write was no longer regarded as an end in itself, but was more closely linked to the exercise of rights, responsibilities, and aptitudes in the professional, social, civic, and cultural fields. Despite some technically impressive results, these massive interventions did not succeed in absorbing the residual number of some 900 million "illiterate" persons who live in the world. Even the UN Literacy Decade Program, launched in 2003, seems to have accepted that despite the intensification of the efforts aimed at accelerating the literacy campaigns, the number of the illiterate will still be of the order of 820 million by 2010.

Natural Sciences

The International Hydrological Programme and the Man and the Biosphere Programme are two of the most important UNESCO programs in the field of natural sciences.

INTERNATIONAL HYDROLOGICAL PROGRAMME (IHP). IHP aims to provide technical training and policy advice required to manage water resources efficiently,

fairly, and in an environmentally sound manner. The program is also involved in developing tools and strategies to prevent water conflicts from erupting between and within states.

UNESCO hosts the secretariat of twenty-three UN partners, which constitute the World Water Assessment Programme. The *U.N. World Water Development Report* (WWDR) provides a comprehensive, up-to-date overview of this resource. The first edition of the report, *Water for People, Water for Life* was launched on World Water Day, May 22, 2003, at the Third World Water Forum in Kyoto, Japan.

MAN AND THE BIOSPHERE (MAB) PROGRAMME. MAB is a most innovative program. In 1968, four years before the UN Conference on the Human Environment in Stockholm, UNESCO held the Conference on the Biosphere in Paris with a view to reconciling the environment and "development." The term *biosphere* was used to designate all living systems covering Earth and the processes allowing them to function. MAB got underway in 1971 as an intergovernmental interdisciplinary activity aimed at developing scientific knowledge about the rational management of natural resources and their conservation in the light of the different types of human activity and the world's different land systems. More than 10,000 researchers from some 110 countries participated in this worldwide effort. More than 400 "biosphere reserves" have also been created that work as "living laboratories," each testing ways of managing natural resources while fostering economic development.

OTHER MAJOR ACTIVITIES IN THE NATURAL SCIENCES. The list of UNESCO's other activities in the natural sciences includes the following:

Intergovernmental Oceanographic Commission (IOC): This coordinating body of UN agencies and institutes monitors ocean conditions to improve weather forecasts, predict the onset of El Niño, and provide early warnings of tsunamis and storm surges. IOC also helps build the Global Ocean Observing System, which weaves together data from special buoys, ships, and satellites to better understand the links between ocean currents and climate.

International Geoscience Programme: Formerly called the International Geological Correlation Programme (IGCP), this joint endeavor of UNESCO and the International Union of Geological Sciences (IUGS) was launched in 1972. It maintains active interfaces with disciplines such as water, ecological, marine, atmospheric and biological sciences. As an international forum for multi-disciplinary geo-environmental research, it is designed to help scientists in more than 150 countries assess energy and mineral resources, while expanding the knowledge base of Earth's geological processes and reducing the risks of natural disasters in less-equipped countries.

Environment and Development in Coastal Regions and Small Islands (CSI): The CSI platform for intersectoral action was initiated in 1996 to contribute to environmentally sustainable, socially equitable, culturally respectful and economically viable development in small islands and coastal regions. The program is based upon three complementary and mutually reinforcing approaches: field-based projects on the ground; UNESCO chairs and University Twinning (UNITWIN) arrangements; and a multi-lingual, Internet-based forum on "wise coastal practices for sustainable human development."

The CSI platform has generated two cross-cutting projects: the Local and Indigenous Knowledge Systems (LINKS) project and the Small Islands Voice (SIV) project. The LINKS project focuses on this interface between local and indigenous knowledge and the Millennium Development Goals of poverty eradication and environmental sustainability. It addresses the different ways that indigenous knowledge, practices and worldviews are drawn into development and resource management processes.

Social and Human Sciences

Often perceived as the conscience of the United Nations, UNESCO is further mandated to develop ethical guidelines, standards, and legal instruments in the field of science and technology—specifically bioethics. The ongoing revolution in science and technology has indeed given rise to some fears that unbridled scientific progress poses a threat to the culturally established ethics of world societies in dealing with their life and their human and natural environment. UNESCO's Programme on the Ethics of Science and Technology was designed to place such progress in the framework of ethical reflection rooted in the cultural, legal, philosophical, and religious heritage of the various human communities. This program includes the Bioethics Programme, the International Bioethics Committee (IBC), the Intergovernmental Bioethics Committee (IGBC), and the World Commission on the Ethics of Scientific Knowledge and Technology.

BIOETHICS PROGRAMME. Created in 1993, this program has been a principal priority of UNESCO since 2002. With its standard-setting work and the multicultural and multidisciplinary forums it has helped to organize, the program has played a leading institutional role at the international level. The Bioethics Programme oversees the activities of the IBC and the IGBC.

UNIVERSAL DECLARATION ON THE HUMAN GENOME AND HUMAN RIGHTS. The first major success of the Bioethics Programme came in 1997, when the General Conference adopted the Universal Declaration on the Human Genome and Human Rights. The only international instrument in the fields of bioethics, this landmark declaration was also endorsed by the UN General Assembly in 1998. Adopted unanimously and by acclamation by the twenty-ninth session of the General Conference, the declaration serves as a legal reference and a basis for reflection on such critical issues as human cloning. In the early twenty-first century, work was underway to evaluate the impact of the declaration worldwide, in accordance with the Guidelines for the Implementation of the Declaration (1999), and to develop a new international declaration on human genetic data.

INTERNATIONAL BIOETHICS COMMITTEE. Created in 1993, this body, composed of thirty-six independent experts named by UNESCO's Director General, follows progress in the life sciences and its applications in order to ensure respect for human dignity and freedom. As the only internationally recognized global body for in-depth bioethical reflection, the IBC acts as a unique forum for exposing the issues at stake. It does not pass judgment on one position or another. Instead, it invites each country, and particularly the lawmakers therein, to decide between the different positions and to legislate accordingly.

INTERGOVERNMENTAL BIOETHICS COMMITTEE. The IGBC, created in 1998, comprises thirty-six member states whose representatives meet at least once every two years to examine the advice and recommendations of IBC. It informs the IBC of its opinions and submits these opinions along with proposals for follow-up of the IBC's work to the Director General for transmission to member states, the Executive Board, and the General Conference.

WORLD COMMISSION ON THE ETHICS OF SCIENTIFIC KNOWLEDGE AND TECHNOLOGY (COMEST). Also created in 1998, this commission formulates the ethical principles that provide noneconomic criteria for decision makers concerning sensitive areas such as sustainable development; freshwater use and management; energy production, distribution, and use; outer space exploration and technology; and issues of rights, regulations, and equity related to the rapid growth of the information society.

From the 1999 World Conference on Science, COMEST also received a mandate to pursue research and come up with recommendations on instilling ethics and responsibility into science education. As a first step toward fulfillment of this mandate, COMEST organized a Working Group on the Teaching of Ethics. This group was asked to give the necessary advice on how to integrate awareness and competence in the field of ethics and responsibility of scientific education and research in the training of every young scientist. The report of the group, endorsed in December 2003, includes a survey of existing programs, an analysis of their structure and content, and detailed curriculum advice on how to integrate into scientific education both ethics and training in the history, philosophy, and cultural impact of science.

MANAGEMENT OF SOCIAL TRANSFORMATION (MOST) PROGRAMME. The list of the programs started by UNESCO with a view to setting ethical frameworks for the advancement of scientific discoveries cannot be completed without mentioning MOST, a program aimed at extending UNESCO's new ethical approach to the larger social transformations linked to globalization. Through this program, which was created in 1993, UNESCO seeks to conduct studies on issues such as urban development and governance through a range of grassroots projects, consultations, and academic networks. MOST increasingly focuses on research to help national and local governments develop appropriate governance policies and structures in multicultural societies, even addressing such issues as social inclusion and the eradication of poverty.

A Critical Assessment of UNESCO's Activities

UNESCO has often been criticized for having failed to act as "the conscience" of the people composing the United Nations and, in the particular field of science and technology, to fully implement its mandate to contain their unbridled development within internationally accepted ethical principles. Such criticisms need to be assessed against philosophical, structural, and institutional limits to UNESCO actions.

A first limit is the fact that the "conscience" attributed to UNESCO is nothing but a metaphor. It represents, at best, the hopes placed by the world populations in the performance of its organizational mandate. In practice, however, an insurmountable gap exists between these populations and the politicians, experts, and economists who often act in their name in the way that each side perceives how science, technology, education, and communication affect their lives. For the latter side, composed of the dominant groups of power and knowledge, ethics have seldom had the same meanings as have been conferred to it by the overwhelming number of humans suffering from the so-called fallouts of modern economic and technological development.

A second serious limit to attempts by UNESCO—or any other similar organization—to humanize science and technology or to curtail their unbridled advancement, stems from the very nature of these institutions. Ethics, by definition, poses questions of morality and of adherence to a set of humanly and socially defined *moral* values, whereas the advancement of science and technology remains solely defined by the state of the art in knowledge and performance. As Jacques Ellul (1954) has argued, technology, in particular, is not neutral. It tends to colonize the very behavior and worldview of the subjects it serves. The same way that an unbridled economy tends to "dis-embed" itself from the society that needs it, technologies such as human cloning or genetic engineering create for themselves an autonomous or transcending "ethics" that tends to defy that of a historically defined culture.

The twin set of reasons mentioned above have been quite detrimental to the hopes raised by UNESCO in the implementation of its "grand design" to act as the "conscience of the world." In their greatest majority, the delegates composing its General Conference and its Executive Board were led, more or less, to defend the passing interests of their respective governments rather than uphold the spirit of its constitution. Some of the more politically or financially powerful members of the organization did not even hesitate to openly impose on it their particular views, regardless of their obligations. On the other hand, the power of experts and specialists defending the dominant discourse in the fields of governance, development, market economy, science, and technology have had a steady repressive effect on the growth of different forms of resistance to that power. The result has been that, despite the fact that UNESCO can be credited with some important technical and legal achievements in its fields of competence, these have fallen far short of fulfilling the hopes that the people of the world had placed in its potentialities.

MAJID RAHNEMA

SEE ALSO *Education; Human Rights; International Relations; United Nations Environmental Program.*

BIBLIOGRAPHY

Bose, Frédéric. (2002). *Tell Me About: UNESCO.* Paris: UNESCO/Nouvelle Arche de Noé Editions.

Conil-Lacoste, Michel. (1994). *The Story of a Grand Design: UNESCO 1946–1993: People, Events, and Achievements.* Paris: UNESCO Publishing.

Dutt, Sagarika. (2002). *UNESCO and a Just World Order.* New York: Nova Science Publishers.

Ellul, Jacques. (1964). *The Technological Society,* trans. John Wilkinson. New York: Knopf, 1964. Revised edition, New York: Knopf/Vintage, 1967.

Stenou, Katérina. (2000). *UNESCO and the Issue of Cultural Diversity: Review and Strategy, 1946–2000: A Study Based on Official Documents.* Paris: UNESCO Publishing.

INTERNET RESOURCE

United Nations Educational, Scientific and Cultural Organization (UNESCO). "World Education Forum." Available from http://www.unesco.org/education/efa/wef_2000.

UNITED NATIONS ENVIRONMENTAL PROGRAM

• • •

As "the voice for the environment" in the United Nations system, the United Nations Environment Programme (UNEP) speaks on behalf of generations not yet born and acts as a clearinghouse for scientific information. It works with other U.N. entities as well as international organizations, national governments, nongovernmental organizations, and the private sector, reporting on the changing state of the world environment, tracking the causes of change, and working collaboratively to develop responses to those changes. Based in Nairobi, Kenya, UNEP has six regional offices and centers, including the Global Resource Information Database and the World Conservation Monitoring Center. Its Division of Technology, Industry, and Economics is headquartered in Paris. UNEP also hosts several secretariats that were formed in response to the passage of international treaties, conventions, and protocols relating to the environment.

Origins

At the time of the first Earth Day celebration in 1970 there was growing awareness of the transnational threats posed by pollution but no international body to advocate for global environmental health. That void was filled in 1972 in Stockholm, Sweden, at the United Nations General Assembly Conference on the Human Environment, which established UNEP.

Delegates to the conference, which was convened to examine the relationship between the environment and development, agreed that humankind had the fundamental right to "freedom, equality, and adequate conditions of life, in an environment of a quality that permits a life of dignity and well-being" and that human beings bear "a solemn responsibility to protect and improve the environment for present and future generations" (*Declaration of the United Nations Conference on the Human Environment,* Stockholm, June 1972).

However, agreement was not forthcoming about how to balance concern for the environment and development to achieve those ends. Officials from developing countries worried, for example, that the resource-protection policies suggested by many of the delegates would hinder economic development in poor nations. At the urging of developing-world leaders such as Indira Gandhi, philosophical statements about "loyalty to the earth" were displaced by practical considerations of economic growth.

From Stockholm onward UNEP has tried to set a course that both is visionary and grapples with the realities of life. To that end it has underscored the fact that poverty, hunger, and misery in the developing world must be addressed if an environmental agenda is to be successful, emphasizing the need for economic growth that would allow developing countries to make progress without repeating the environmentally disastrous mistakes of the industrialized world. The term *sustainable development* came into use in the 1980s to describe that approach.

Areas of Concern

In the 1980s UNEP defined several areas of environmental concern, including climate change and atmospheric pollution; pollution and the shortage of freshwater resources, along with the deterioration of coastal areas and oceans; and land degradation, including desertification and the loss of biological diversity.

AIR. Since the first book on air pollution was written in the seventeenth century, the situation has gotten decidedly worse. There is still urban air pollution, and with it concern about sulfur dioxide, nitrogen oxides, carbon monoxide, ozone, and suspended particulate matter. Every day nearly one billion people in urban areas breathe air with unacceptable levels of pollution. The problem has widened to include the depletion of the stratospheric ozone layer caused by chloroflourocarbons, acid rain that burns forests with heavy doses of sulphur dioxide and nitrogen oxides, and climate change, which is melting glaciers and raising sea levels at an alarming rate and has been studied in a concerted way since 1980 by UNEP's World Climate Programme. UNEP's efforts to improve understanding about the sources of atmospheric pollution and climate change helped bring about the entry into force of a Global Convention (Vienna 1985) and a Global Protocol (Montreal 1988) for the protection of the ozone layer. UNEP collaborated with other groups in the development of the Climate Change Convention that was signed in 1992.

WATER. As far back as 1977 at the United Nations Water Conference delegates were alarmed about rising levels of water consumption and pollution. The conference's Mar del Plata Action Plan challenged the international community to create an integrated long-term plan for water management. The first step was taken in 1985, when UNEP launched the Programme for Environmentally Sound Management of Inland Waters (EMINWA) in an effort to protect the world's supplies of fresh water.

Oceans and seas, which cover 70 percent of the earth's surface, are another area of concern, particularly with regard to coastal development, discharges of municipal and industrial waste, and the overexploitation of water through the use of long-line and drift nets. In the early twenty-first century more than 120 countries take part in UNEP's Regional Seas Programme, which encourages research, monitoring, and the control of pollution and the development of coastal and marine resources.

LAND. The degradation of drylands, which is known as desertification, is an increasingly severe problem that affects more than a sixth of the world's population. Caused mostly by agricultural and grazing practices that ignore the fragility and productive limits of the land, desertification also is brought about by prolonged drought. More humid areas are at risk of degradation as a result of urbanization, unsustainable agriculture, and deforestation, which clears more than 11 million hectares (27.2 acres) of forest per year. In 1977 UNEP was designated to coordinate the United Nations Plan of Action to Combat Desertification. With regard to deforestation and habitat loss, agreement on a set of nonbinding principles for forest conservation was reached in 1992. UNEP also initiated a series of in-depth country-by-country studies of biodiversity that led in 1992 to the Convention on Biological Diversity.

Results and Successes

The success of UNEP-instigated treaties, conventions, and protocols has resulted from the agency's effective use of scientific and expert advice to inform decision makers about complex environmental problems. For instance, the Global Biodiversity Assessment of 1995, which led to the Convention on Biodiversity, involved roughly 1,500 scientists. UNEP helped develop ways to produce, synthesize, and legitimize the expert knowledge of those scientists and then to provide reliable and accessible scientific advice on environmental policy options. In 1988 the World Meteorological Organization and UNEP set up the Intergovernmental Panel on Climate Change (IPCC). Since that time the 2,500 scientists associated with IPCC have produced a series of reports that have been highly influential in the debate about climate change. UNEP is not an environmental protection agency as such but more of a scientific advisory institution.

Historical Development

Twenty years after the Stockholm conference UNEP continued to explore the relationship of the environment and development at the 1992 United Nations Conference on Environment and Development (Earth Summit) in Rio de Janeiro, Brazil. One hundred seventy countries came together in Rio and adopted by consensus a common global strategy for environmental protection called Agenda 21. Among other things, Agenda 21 laid the groundwork for the 1997 Kyoto Protocol of the Framework Convention on Climate Change. Some participants felt that the recommendations of the Earth Summit favored development over environmental protection. Examples include state sovereignty over resources (and environmental and development policies), the promotion of global free trade and open markets, and a "polluter pays" approach in which market instruments and not strict regulatory mechanisms are used to curb environmental degradation.

It was also at the Earth Summit that the Precautionary Principle received a global hearing. The delegates agreed in Principle 15 of the Rio declaration on environment and development that "where there are threats of serious or irreversible damage, lack of full scientific certainty shall not be used as a reason for postponing cost-effective measures to prevent environmental degradation." The Precautionary Principle became a cornerstone for the 2000 Cartagena Protocol on Biosafety of the Convention on Biodiversity and has been used in additional forums to argue against genetically modified agricultural products and other forms of biotechnology. The European Union calls the Precautionary Principle a "principle of common sense" and uses it in judging food safety; San Francisco was the first American city to adopt the principle for its purchases and building projects.

In 2002 at the United Nations World Summit on Sustainable Development in Johannesburg, South Africa, development again seemed to occupy center stage. UNEP's executive director, Klaus Toepfer, diplomatically called the summit "satisfactory," but many delegates were angered by efforts, most notably those of the United States, to derail timetables and targets for environmental policies such as the use of renewable energy. Nevertheless, UNEP continues to be the best hope for international cooperation and global governance on life-threatening issues that know no boundaries.

MARILYN BERLIN SNELL

SEE ALSO   Biodiversity; Deforestation and Desertification; Ecology; Environmental Ethics; Environmentalism; Environmental Justice; Environmental Regulatory Agencies; Global Climate Change; Nongovernmental Organizations; Pollution; Rain Forest; Waste; Water.

BIBLIOGRAPHY

Atchia, Michael, and Shawna Tropp, eds. (1995). *Environmental Management: Issues and Solutions.* New York: Wiley.

Brown, Noel, and Pierre Quiblier, eds. (1994). *Ethics and Agenda 21: Moral Implications of a Global Consensus.* New York: United Nations Environment Programme.

Toldba, Mostafa K.; Osama A. El-Kholy; E. El-Hinnawi; et al, eds. (1992). *The World Environment 1972–1992: Two Decades of Challenge.* London: Chapman and Hall. 1992.

Worldwatch Institute, in cooperation with the United Nations Environment Programme. (2002). *Vital Signs 2002: The Trends That Are Shaping Our Future.* New York: Norton.

UNITED STATES ENVIRONMENTAL PROTECTION AGENCY

SEE *Environmental Regulation.*

UNITED STATES FEDERAL AVIATION ADMINISTRATION

SEE *Aviation Regulatory Agencies.*

UNITED STATES FEDERAL COMMUNICATION COMMISSION

SEE *Communications Regulatory Agencies.*

UNITED STATES FOOD AND DRUG ADMINISTRATION

SEE *Food and Drug Agencies.*

UNITED STATES GEOLOGICAL SURVEY

SEE *National Geological Surveys.*

UNITED STATES NATIONAL ACADEMIES

SEE *National Academies.*

UNITED STATES NATIONAL SCIENCE FOUNDATION

SEE *National Science Foundation.*

UNITED STATES NUCLEAR REGULATORY COMMISSION

SEE *Nuclear Regulatory Commission.*

UNITED STATES OFFICE OF TECHNOLOGY ASSESSMENT

SEE *Office of Technology Assessment.*

URBANIZATION

• • •

Urbanization is a historical phenomenon closely linked to changes in technology and to some extent science that also influences and is influenced by ethical ideals. Both technology and science develop with more intensity in cities, in part promoted by urban models of human behavior, which in turn may be reinforced by notions of technological instrumentalism and scientific objectivity.

Urbanization, Ancient and Modern

The term *urbanization* refers to the increasing concentration of people in cities. The first cities appeared after the development of plant cultivation and animal domestication. Formerly nomadic tribes settled in fertile river valleys and became increasingly dependent on agriculture.

The ancient cities of Mesopotamia were established between about 4000 and 3000 B.C.E. The cities of ancient Egypt appeared around 3300 B.C.E. and were closely linked to the increasing power of the pharaohs, who were both secular and spiritual leaders who could use their power to create new cities. By about 2500 B.C.E. urban societies had developed in other parts of the world, such as the Indus River Valley in India and Pakistan and the Yellow River Valley of China. Subsequent urban developments of a classical form occurred in Athens, Rome, and other parts of the eastern Mediterranean. Despite urbanization in these ancient forms most people continued to live outside cities.

The modern city is linked closely to the development of industrialization, especially in Europe and North America. Before the Industrial Revolution cities were primarily centers for trade, political power, and religious authority. The rise of the machine in the late 1700s in both Europe and North America led to new city forms characterized by larger numbers of people living in areas with greater population density. As machines were developed and manufacturing increased, people began to migrate to cities from rural areas as laborers and consumers.

Technological change is not exclusive to the post–Industrial Revolution era. What distinguished that historical period was the unprecedented rapid increase in the number, kind, and effects of technological innovation and associated increases in urbanization. About 3 percent of the world population lived in urban areas in 1800, a number that rose to 13 percent in 1900 and more than 40 percent in 2000.

The Modern City

The rise of the modern city had significant economic, social, and cultural impacts. Urbanization changed many of the traditional institutions, values, and human experiences that characterized preindustrial cities. For example, while cities grew in importance in economic terms, they also became centers of poverty. Cities also brought together people of different cultures with different worldviews, traditions, and values. In addition, the concentration of people in urban areas created a host of ethical issues related to living together closely.

In 1905 the German social theorist Max Weber (1864–1920) observed that industrialization represented a fundamental process of social change that was embedded in the development of rationality and scientific knowledge. According to Weber, "demystification" challenged traditional religious ideas by providing an alternative basis of knowledge. Weber concluded that this brought about a notable decline in the acceptance of the spiritual explanations that are at the heart of religious beliefs and practices. As a result human activities that previously had been dominated by religious authority were controlled by an appeal to scientific and rational thinking.

In 1965 the Harvard professor of divinity Harvey Cox observed a close interconnectedness between the rise of urban civilization and the collapse of traditional religion. "Urbanization," Cox stated, "constitutes a massive change in the way men live together, and became possible in its contemporary form only with scientific and technological advances which sprang from the wreckage of traditional views" (Cox 1965, p. 1). Cox argued that that epochal change in worldviews resulted directly from the changing nature and character of cities. As cities became more cosmopolitan and as technology fostered greater interconnectedness through travel and communications, religion, Cox argued, lost its centrality in the hearts and minds of people. Nonreligious perspectives on the human condition replaced Christian religious norms and standards for conduct.

Urbanization in a Global Context

The patterns of economic, social, and cultural changes caused by rapid urbanization in the nineteenth and twentieth centuries are observable in modern cities. In general terms the world population is becoming predominantly urban. Industrialized or more developed countries were more than 75 percent urbanized in 2000, compared with 39 percent for less developed countries. To a certain extent economic gain and higher incomes are associated with urbanization. The expansion of production, communication, knowledge, and trade helped raise standards of living in the more developed countries.

In developing countries the urbanization experience has been vastly different: Industrialization accounts for a much lower proportion of the national economy, and these countries also have significantly lower income per capita. The concern in developing countries is the rate at which increases in the numbers of people living in urban areas are occurring. According to the United Nations Center for Human Settlements (2001), 40 percent of the population of developing countries was living in urban areas in 2001. By 2020 that number is expected to increase to 52 percent.

In 2001 three-quarters of global population growth occurred in urban areas in developing countries, posing significant problems associated with rapid growth in the parts of the world least capable of accommodating it. Most of the projected growth will occur in megacities:

cities with a population of ten million or more. These areas already face increasing difficulties in providing their inhabitants with adequate water, food, shelter, employment, sanitation, and basic services. Poverty has become increasingly urbanized as more people migrate from rural to urban areas. The United Nations Center for Human Settlements (2001) estimates that more than a billion people live in crowed slums in inner cities or in squatter settlements on the periphery of large urbanized areas. Not only does this result in strained local conditions, the rapid growth and concentration of poverty in urban areas in the developing world often leads to adverse consequences for national economies.

Although modern cities are part of a highly interdependent global network fostered by new information, communication, and transportation technologies, one significant characteristic of cities in the twenty-first century is the growth of disparities between the rich and the poor. The United Nations calls this the "divided city," and it is characteristic of urban areas in both developed and developing countries. Some researchers predict a new wave of rapid technological change in urban areas driven by information and communications technologies, which reinforce urban polarization and cause further erosion of traditional economic, social, and cultural activities. New technologies, they observe, reinforce and extend the reach of the economically and culturally powerful. Those who already have access to new technologies and most able to benefit from the potential of new technologies will use them to their advantage to assure their place as the principal beneficiaries of the "information revolution."

Another phenomenon closely linked to the modern city, especially in North America and parts of Europe, is suburbanization. Driven by advances in transportation and communication technologies, sprawl patterns of urbanization from central cities to suburbs began to emerge after 1945. By 1960, 60 million people in the United States were living in suburbs, compared with only 45 million in cities. Since 1980 suburban populations have grown ten times faster than have central-city populations.

In response to the problems associated with the rapid rise of modern urbanization and its attendant problems, urban planning emerged in the United States around the end of the nineteenth century. Although examples of planned cities date back several thousand years, urban planning developed from demands for social reform in both England and the United States. In the early twenty-first century urban planners are part of a distinct occupational skill group that applies a specified body of knowl-edge and techniques addressing land use, city functions, and a wide variety of other urban characteristics.

M. ANN HOWARD

SEE ALSO *Industrial Revolution; Modernization; Secularization.*

BIBLIOGRAPHY

Callahan, Daniel, ed. (1966). *The Secular City Debate*. New York: Macmillan. This collection of essays, critiques, and book reviews by a diverse group of authors was compiled in response to Harvard professor Harvey Cox's book *The Secular City*. The editor notes that the response to Cox's work was surprising, with more than 225,000 million copies sold in the first printing. An afterword by professor Cox is included.

Cox, Harvey. (1965). *The Secular City*, rev. edition. New York: Macmillan. Explores the theological significance of the modern city and includes an interpretation of the relationship between urbanization and modern secular society. Cox contrasts the modern city (technopolis) with traditional forms of human communities, particularly with regard to the influence of religion and religious institutions.

Ginsburg, Norton. (1966). "The City and Modernization." In *Modernization*, ed. Myron Weiner. New York: Basic Books.

Hetzler, Stanley A. (1969). *Technological Growth and Social Change*. London: Routledge & Kegan Paul. This collection of essays was originally prepared as lectures for *Forum*, an educational radio program sponsored by Voice of America. Twenty-five scholars explore how modernization, defined as technological development, occurs and how it can be accelerated. Ginsburg's essay reviews the history of cities and their historical functions. He describes the distinctly different characteristics of the modern city.

Lebebvre, Henri. (2003). *The Urban Revolution*, trans. Robert Bononno. Minneapolis: University of Minnesota Press. Henri Lefebre (1901–1991) was an influential French philosopher and sociologist. This work was originally published in 1970, but not translated into English until this edition. Highly theoretical, connecting urban research with social theory and philosophy, Lefebre's work marked a new view of "urbanism." Lefebre posited that "urban society" more aptly describes modern societies, rather than the term "postindustrial society," arguing that all forms of human settlements have been altered due to industrialization and urbanization. He noted that agricultural activity is inextricably linked to "industrialization." Even traditional forms of village life around the globe have been permanently transformed by industrial production and consumption.

Mumford, Lewis. (1961). *The City in History: Its Origins, Its Transformations, and Its Prospects*. New York: Harcourt Brace Jovanovich. An extensive exploration of the history of the city, beginning with Ancient Mesopotamia and Egypt, through the modern. In addition to an historical overview, Mumford critiques many of the historical urban forms. He is especially critical of the modern manifestation of urban communities and the negative influence of capitalism resulting in resource depletion. The annotated bibliography is extensive and very helpful to anyone interested in in-depth material.

Prud'Homme, Remy. (1989). "New Trends in Cities of the World." In *Cities in a Global Society*, ed. Richard V. Knight and Gary Gappert. Newbury Park, CA: Sage. Collection of essays that details the phenomenon of rapid global urbanization and the forces influencing the extraordinary changes in modern human settlements. Prud'Homme observes that the cities will have increasingly distinct "global" as opposed to "national" roles to play as economic forces become more global.

United Nations Center for Human Settlements. (2001). *Cities in a Globalizing World*. London: Earthscan. Compilation of work by more than eighty international researchers. The report reviews the status of the world's cities and summarizes global trends that will impact the cities of the future. Especially notable is the observation regarding the increased isolation of the urban poor in both developed and developing countries.

UTILITARIANISM

SEE *Consequentialism*.

UTOPIA AND DYSTOPIA

• • •

Part of being human is the ability to dream of a better (or worse) life, either in this world or the next. Some dreams have led to the study of nature and humans, from the deep mysteries of the atom and the gene, to the even deeper challenges of individual and collective sanities—all with an understanding that how one acts can be as important as why, especially when studies of nature (science) and how to transform nature (technology) confer ever greater powers and responsibilities on human beings. Some of humanity's best thinkers and artists have, for 2,500 years, created moral compasses by distilling human wisdom (and folly) into imaginative works called utopias and dystopias (sometimes called anti-utopias). These compasses are neither timeless nor universal; instead, their poles are constantly aligned and realigned by the forces of history, economics, politics, and aesthetics. Messages from these explorers of science, technology, and ethics have long had the potential to both frighten and enlighten. Indeed, they have been doing so at least since the hero escaped from that allegorical cave of shadows in Plato's classic utopia, *The Republic* (360 B.C.E.)—a parable clearly revisited and updated in the film *The Matrix* (1999).

Utopia Defined: Thomas More's Pun and the Myth of Utopianism

The word *utopia* originated in December 1516, when Thomas More published a book with that one word, capitalized, as its title. More wrote his text in Latin. Its complete, twenty-seven word title—*De optimo reipublicae statu deque nova insula utopia libellus vere aureus, nec minus salutaris quam festivus, clarissimi disertissimique viri Thomae Mori inclutae civitatis Londinensis civis et Vicecomitis*—features not only a latinizing of his own name and city but also a brand-new word coined as a trilingual pun. In Latin and English, utopia minimally disguises its truncated roots in two made-up, latinized homophones from the Greek words for a *good place* (*eutopos*) and for *no place* (*ou-topos*). Hence, "Utopia: the good place which is no place" (Sargisson, p. 1). Since 1516 More's readers and translators alike have wrestled with the many puns and ambiguities of this multi-voiced dialogue that is, in Vita Fortunati's words, "a bewildering mixture of reality and fiction" (Fortunati and Trousson 2000, p. 153).

The full title of More's book, in its first English translation by Ralph Robinson in 1551, was *On the best State of a Commonwealth and on the new Island of Utopia A Truly Golden Handbook, No Less Beneficial than Entertaining by the Most Distinguished and Eloquent Author Thomas More Citizen and Undersheriff of the Famous City of London*. This language—especially *best* and *Handbook*, *Commonwealth* and *Beneficial*—evokes the common understanding of *utopia* and *Utopia* as a blueprint for a perfect society. Such an initial reading makes it easy to dismiss utopian arguments as just unrealistic. Since the late twentieth century, scholars such as Ruth Levitas, Tom Moylan, Lyman Tower Sargent, Lucy Sargisson, and W. Warren Wagar have challenged this colloquial, negative view of utopian texts, thoughts, and theories.

The recorded usages of utopia expose a long history of undervaluing the impulse for social dreaming, for collectively desiring a better way of being. Denotations for utopia show a sustained effort to disempower minority reports from the critics of the dominant ideologies that have sustained (mostly premodern) heads of state and (mostly modern) captains of capital. A distinction between *imaginary* and *imaginative* is helpful here. After first asserting, "Utopian thought is imaginative," Northrop Frye observes that "The word *imaginative* refers to hypothetical constructions, like those of literature or mathematics. The word *imaginary* refers to something that does not exist" (Frye 1957, p. 193). More's island is a new *no place* that people can hold in their hands and in their minds; it is imaginative, not imaginary.

Another, less nuanced point is raised by the adjective *perfect* being applied to this system depicted by More. There is a figure—it is tempting to call him a character—in More's *Utopia* called "More," who spends much of his time listening to the exploits of Raphael Hytholoday, a sailor and scholar who has been to Utopia. As Hytholoday (an imagined figure whose name means *peddler of nonsense*) tells his tale, "More," the character, expresses several reservations. For "More"—and, one could surmise, for More, the man,—many of the Utopians' laws and customs "were really absurd" (More 1995, p. 110). Then, when Hytholoday has finished, "More" says, "Meanwhile I can hardly agree with everything he has said (though he is a man of unquestionable learning and enormous experience of human affairs), yet I freely confess that in the Utopian commonwealth there are many features that in our own societies I would like rather than expect to see" (More 1995, p. 110–111). *Utopia* depicts, not a perfect social, legal, and political system, but instead a complex debate, enriched by humor, between More's earned political realism of low expectations and his cautious optimism of higher desires for society.

Utopian Studies: Modern Scholarship Challenges Utopian Stereotypes

The debilitating myth of utopianism as unrealistic perfectionism comes, in part, from concentrating on the content and form of utopias—on what is held and what is doing the holding—rather than on the function of utopia. Some important work has been done with the content and form approaches, most significantly the magisterial tome by Frank and Fritzie Manuel, *Utopian Thought in the Western World* (1979). But as Ruth Levitas notes, "to focus on the function of utopia is already to move away from colloquial usage, which says nothing about what utopia is for, but implies that it is useless" (Levitas 1990, p. 5).

Turning attention to how utopias function, scholars, led by Lyman Tower Sargent (1988), have challenged the dominant commonplace understanding of utopia by reexamining the history of utopian expressions, locating many newly-discovered and rediscovered resources. Other scholars, following the example of Ernst Bloch (1970), have expanded utopia by finding it "immanent in popular culture, in the fashion industry, dance, film, adventure stories, art, architecture, music, and even medical science" (Sargisson, p. 12). Even a Parisian graffito from May 1968—"Be realistic. Demand the impossible"—becomes fodder for utopian analysis, with its second command serving as the apt title for Tom Moylan's 1986 study of science fictional treatments of the critical utopian impulse by Joanna Russ, Ursula K. Le Guin, Marge Piercy, and Samuel R. Delany.

This new wave of utopian studies operates not as a small, monolithic cabal but rather as a growing international community. For example, Fortunati and Raymond Troussan's 700-page *Dictionary of Literary Utopias* (2000) has ninety-nine contributors from more than a dozen countries. This key reference work offers a thorough comparative and interdisciplinary perspective on literary utopias and dystopias, yet even it cannot claim anything approaching complete coverage of utopian and dystopian thought. For a sweeping overview, historian and novelist W. Warren Wagar, contends, "At least two great rivers of utopian dreaming flow through the history of ideas, corresponding to the two great families of world-views, the naturalist and the idealist, which have contended with one another for thousands of years in every philosophical arena in the world" (Wagar 1991, p. 56). Furthermore "Since the seventeenth century, most blueprints for good societies have emanated from the naturalist family, as represented by the classic texts of Bacon, Condorcet, Comte, Cabet, Marx, Bellamy, Wells, and Skinner. But not all. Many utopian visions are grounded in such members of the idealist family of world-views as Platonism, mysticism, orthodox religious piety, and modern and postmodern irrationalism" (Wagar 1991, p. 56). Key writers, for Wagar, in this second tradition include William Morris, George Bernard Shaw, Herman Hesse, Aldous Huxley, Teilhard de Chardin, C. S, Lewis, William Burroughs, and Doris Lessing. In their idealist works, "utopia is not a bustling city registering worldly progress but a community of spirit earning grace" (Wagar 1991, p. 56).

Naturalistic Utopias: Bacon and Science

For present purposes, the name at the head of Wagar's naturalist tradition should be highlighted, Francis Bacon. His *New Atlantis* (1627) brings the politically responsible use of science and technology to the forefront of utopianism by way of its House of Salomon, a grand research institution that, historically speaking, serves as the prototype for modern laboratory science. Writing in 1665, Joseph Glanville affirms, "Salomon's House in the *New Atlantis* was a prophetic scheme of the Royal Society" (Fortunati and Trousson, p. 448). Before detailing its personnel, equipment, and methods, an Elder of the House of Salomon first explains its underlying goals: "The end of our foundation is the knowledge of causes, and the secret motions of things; and the enlarging of the bounds of human empire, to the effecting of all things possible" (Bacon 1627, p. 240).

Their division of labor anticipates such ventures as the Massachusetts Institute of Technology (MIT), the Manhattan Project, and Bell Laboratories. One subgroup of Elders functions, in Bacon's words, as *interpreters of Nature*, whose role foreshadows the modern scientific method itself. These protoscientists "raise the former discoveries by experiments into greater observations, axioms, and aphorisms" (Bacon 1627, p. 240). That is to say, two hundred years before the word *scientist* was coined, Bacon divided practitioners into the experimenters and the theorists. Moreover, his *New Atlantis* initiates the major model of modern utopias, ones that imagine liberating humanity through enhanced production and consumption, including Louis-Sebastien Mercier's *The Year 2440* (1770), Etienne Cabet's *Voyage en Icarie* (1840), and, after *Uncle Tom's Cabin* (1852) by Harriet Beecher Stowe, the most popular nineteenth-century American novel, Edward Bellamy's *Looking Backward* (1888).

Idealist Utopias: Morris and Community

Idealist utopias are quieter than their naturalistic cousins. A sense of community is earned in them not by way of technology but through the avenues of spirit in Hermann Hesse's *The Glass Bead Game* (1949). In the naturalistic utopias (and in their dark avatars, the naturalistic dystopias), communication is enhanced (or thwarted) through the agency of faster and better telephones, telegraphs, and computers, among others, while in the idealist utopias (and their avatars) communication honors its root in *communing*, in the fullest sense of a people sharing life. (Tom Moylan [2000] provides an analysis of key examples of these science fictional utopias and dystopias from the 1980s and 1990s.) Idealist utopias are often explicit responses to naturalistic texts, as in Morris's *News from Nowhere* (1890) as a pastoral reply to Bellamy's *Looking Backward* (1888) and its shiny vision of an industrial army circa 2000. On rare occasions, a naturalist dystopia and its paired idealist utopia are written by the same author— for example, Aldous Huxley's *Brave New World* (1932) and *Island* (1962). Taken together, these major utopian streams engage in a complex critique of science and technology, especially in the twenty-first century science fiction short story, novel, and (to a lesser degree) film.

Charting Wilde's Map of the World and Beyond

In "The Soul of Man Under Socialism" (1891), Oscar Wilde poeticizes the positive utopian impulse, saying, "A map of the world that does not include Utopia is not worth even glancing at, for it leaves out the one country at which Humanity is always landing. And when Humanity lands there, it looks out, and seeing a better country, sets sail. Progress is the realization of Utopias." (Wilde 1891, p. 34). Yet while many anticipated and welcomed the rise of modern industry, science, and technology, a minority questioned their impact, wondering not about the feasibility but the wisdom of utopian schemes. Utopias and dystopia are asymmetrical concepts, akin to health and disease, whereby one person's hopeful dream is another's dyspeptic nightmare. One key example is behaviorist B. F. Skinner's *Walden Two* (1948), written as a positive, naturalistic utopia, yet often read as a dystopia—and one Henry David Thoreau would not have warmed to.

Overall, the miscoupling of science (natural and psychological) and power (political and economic) found its most compelling expressions in the great twentieth-century dystopias, especially Yevgeny Zamyatin's *We* (1920), Huxley's *Brave New World* (1932), and George Orwell's *1984* (1949). *We* is especially germane because of its moral calculus. That is, in Zamyatin's hyper-rational world, ethical values are literally, not metaphorically, based on mathematical calculations. Even more disturbing is Huxley's prophetic extrapolation of modern consumerism. He invented the perfect narcotics—soma and the *feelies*—for the dystopian year of our Ford, 632; in the twenty-first century, both can be found at the local mall. Lastly, *1984*'s impact on the understanding of power and politics, language and truth, and banality and desire are difficult to underestimate. After all, not every writer has his name become a ubiquitous adjective—Orwellian.

ROBERT SHELTON

SEE ALSO *Bacon, Francis; Brave New World; Critical Social Theory; Huxley, Aldous; More, Thomas; Morris, William; Posthumanism; Science Fiction; Science, Technology, and Literature; Wells, H. G; Zamyatin, Yevgeny Ivanovich.*

BIBLIOGRAPHY

Bacon, Francis. (1627). *New Atlantis. Famous Utopias of the Renaissance*, ed. Frederic R. White. Putney, VT: Hendricks House, 1981. Utopian and early modern scientific visions powerfully merged.

Bloch, Ernst. (1970). *Philosophy of the Future*. New York: Herder & Herder. A modern European philosopher reinvigorates utopian thinking.

Fortunati, Vita, and Raymond Trousson, eds. (2000). *Dictionary of Literary Utopias*. Paris: Champion. Insights and commentary by a hundred leading scholars of literary utopias and dystopias.

Frye, Northrop. (1957). *Anatomy of Criticism: Four Essays*. Princeton, NJ: Princeton University Press. A seminal text in modern literary theory.

Levitas, Ruth. (1990). *The Concept of Utopia*. Syracuse, NY: Syracuse University Press. Utopia examined by way of its form, function, and content.

Manuel, Frank E., and Fritzie P. Manuel. (1979). *Utopian Thought in the Western World*. Cambridge, MA: Harvard University Press. A sweeping overview of the main thinkers in utopian fiction and non-fiction.

More, Thomas. (1516). *Utopia*, eds. George M. Logan, and Robert M. Adams. New York: Cambridge University Press, 1995. The work that gave humanity the word *utopia*.

Moylan, Tom. (1986). *Demand the Impossible: Science Fiction and the Utopian Imagination*. New York: Methuen. Utopian and dystopian themes in contemporary science fiction.

Moylan, Tom. (2000). *Scraps of the Untainted Sky: Science Fiction, Utopia, Dystopia*. Boulder, CO: Westview Press. Still more recent science fiction seen through utopian and dystopian lenses.

Sargent, Lyman Tower. (1988). *British and American Utopian Literature, 1516–1985: An Annotated, Chronological Bibliography*. London: Garland Publishing. The major bibliography of utopian literature, annotated with insight and wit.

Sargisson, Lucy. (1996). *Contemporary Feminist Utopianism*. New York: Routledge. Feminist reexamination of utopian thought, literature, and criticism.

Wagar, W. Warren. (1991). "J. G. Ballard and the Transvaluation of Utopia." *Science-Fiction Studies* 18: 53–70. Wagar is an important utopian scholar, novelist, and teacher.

Wilde, Oscar (1891). "The Soul of Man Under Socialism." In *Selected Essays and Poems*, ed. Owen Dudley Edwards. London: Penguin, 1954. Wilde was indeed more than an amusing playwright.

V

VACCINES AND VACCINATION

• • •

Nowhere is the effort to use science and technology for human benefit clearer than in the development of vaccines against serious infectious diseases. Although vaccinations seldom have unintended consequences they can, on occasion, pose complex ethical issues.

Historical Developments

One of the key figures in Western vaccine history is the English physician Edward Jenner, who developed an effective prophylactic against smallpox in 1796. The protective material that Jenner used came originally from a cow infected with cowpox. When infected cows were milked, Jenner noted, as others had before him, that the milkmaids who did the milking developed pustules or sores on their hands that disappeared in time and did not harm the girls. These girls, it was further observed, did not get smallpox, which is a disease that left unsightly pockmarks on the face and skin where the pustular sores had been, and in about 30 percent of the cases was lethal. On May 14, 1796, Jenner took liquid from the developing sores of the hand of a milkmaid named Sarah Nelms and injected it into the skin of eight-year-old James Phipps. After about six weeks, when the sore had resolved and disappeared, Jenner injected Phipps's skin with virulent human smallpox. Phipps did not get the generalized disease and was protected against the widespread eruption of sores. After many such experiments, Jenner called the material he injected to achieve protection against virulent human smallpox, a vaccine deriving the term from the Latin *vacca* (cow).

In the decades following Jenner's discovery many other vaccinators took up the procedure, and gradually the number of smallpox outbreaks started to decrease noticeably. By the time Louis Pasteur was developing his vaccines (for fowl cholera, rabies, and anthrax) in the late nineteenth century, smallpox was so much on the wane that Jenner's contribution to human well-being was much lauded in all the countries that had adopted his methods.

The continued use of the vaccine, made for mass distribution by collecting the pustular material from the skin of deliberately infected calves, led eventually to a campaign spearheaded by Donald Henderson, working with the World Health Organization (WHO), to completely eradicate human smallpox from planet Earth. In this they were successful, as attested by a declaration of the chief WHO scientists on December 9, 1979. This was the first time that a serious and socially debilitating human infectious disease had been entirely eradicated, and is one of the highest achievements of humankind. A second disease targeted for elimination through the use of vaccines is polio. By 2004, the disease notification returns showed that there were but several tens of cases in fewer than five countries per annum. It is held that elimination will have occurred when there will have been zero notified cases worldwide for at least two years, and this is confidently expected to occur before 2010.

The consequence of these pioneering results in vaccine use has been that many infectious diseases afflicting humans throughout history are now so much in the past that people have forgotten how dangerous, damaging, and deadly they can be. Human diseases that are in this category include diphtheria, tetanus, pertussis or whooping cough, polio, measles, mumps, rubella, rabies, hepa-

FIGURE 1

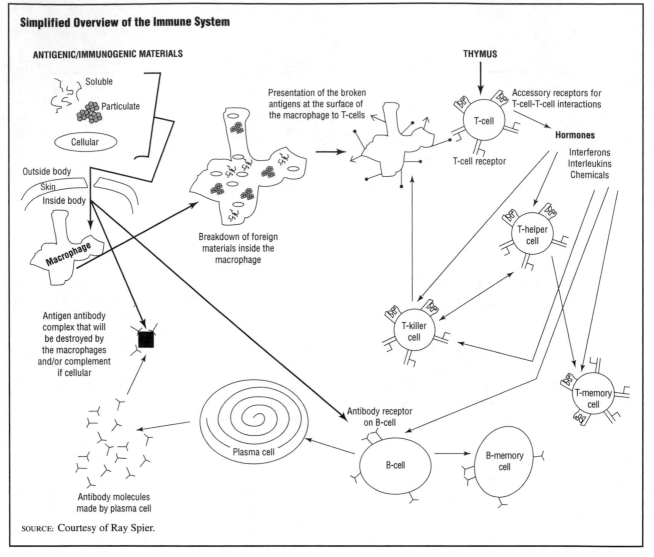

Simplified Overview of the Immune System

SOURCE: Courtesy of Ray Spier.

titis A and B, *Haemophilus influenzae* b, and yellow fever. A similar list for veterinary diseases includes foot-and-mouth disease, Newcastle disease, Marek's disease, anthrax, and canine distemper.

Vaccine Science

Vaccines that are intended to prevent infectious diseases are normally made from the organisms or close relatives of the organisms that cause the disease. In the former case, before the organism is used as a vaccine, it is killed or inactivated by a variety of techniques that include heating, treating with inactivating agents such as formaldehyde or acetyl-ethylene-imine, exposing to ultraviolet light or gamma radiation, and using denaturing agents such as urea and/or proteolytic enzymes. A further option when using an inactive vaccine is to use a part or subunit

of the pathogenic agent. Here the bacterium or virus is disrupted and one or several of its component parts are used in the vaccine to which is added an adjuvant. This latter material is a nonspecific stimulator of the immune system that greatly potentiates the killed organism or its components as used in the vaccine.

When a live organism is used for a vaccine it is chosen for its relatedness to the disease-causing organism plus its inability to cause disease, normally as a result of its attenuated or weakened nature. Both types of vaccines benefit from stabilizers and preservatives, but while the killed vaccines are more likely to survive at room temperature the live vaccines must be either held in refrigerated conditions or freeze-dried, in which state they will withstand limited exposure to more elevated temperatures.

The actual materials of the killed or live organisms that are active in the achievement of the vaccination effect are either the proteins or complex carbohydrates that exist on the outer skins or envelopes of the pathogenic microorganisms. While most carbohydrates are generally inert, they can be made into powerful and active agents when they are covalently joined to proteins. For the proteins themselves, scientists can identify immunogenic sites that are either dependent on a linear string of not less that six amino acids or a region of a protein molecule wherein an amino acid from one part of the linear chain comes into proximity with other amino acids to make up an immunogenically active area or site; these are called conformational immunogens. Normally two or three amino acids would be involved in such a conformational determinant. Lipids and nucleic acids are not normally involved in the immunogenicity of pathogenic organisms except in that they participate in creating an environment in which proteins and carbohydrates take up their active three-dimensional configurations.

These immunogenic materials can be found on all types of viruses and bacteria, and some protozoa. Whole animal cells that express proteins or glycoproteins (proteins to which carbohydrate groups are attached) at the exterior surface have also been used as vaccines, such as the anticancer vaccine for chickens that utilizes infection by the herpes virus that causes Marek's disease in poultry. There have been experiments demonstrating the feasibility of this approach to vaccinate against some human cancers, but more testing needs to be done.

Vaccines work by stimulating the human or animal immune system to produce glycoprotein antibodies and specialized cells that seek out and kill body cells that contain infecting organisms or materials they do not recognize as belonging to the normal body. In this way they are able to recognize cancer cells because such cells make unique molecules that are exposed at the surface of the cancerous cell and label that cell as one of the abnormal cells of the body that may be killed by the killer T cells of the immune system. Many such cancerous cells are made in the lifetime of a human and are dealt with in this way, thus protecting the human from the uncontrolled and killing effects of a rapidly expanding cancer. While much is known about these reactions, scientists are yet uncertain about many of the details that relate the invading or foreign microorganism to the response it evokes and the consequences of that response. Some of the complexity of the immune system may be gleaned from the overview diagram of Figure 1.

When a microorganism penetrates the skin barrier and survives the antimicrobial agents in the skin and tissues, it is ingested into a macrophage-dendritic cell. These cells (of which there are more than twenty tissue-specific types) engulf foreign particles and then break them down to smaller molecules that are released to become expressed on the outer surface of the cell. From this exposed site they attract another kind of cell of the immune system known as the T cell. This cell, found in the white cells of the blood, is formed when an undifferentiated white cell passes through the thymus gland.

There are many different and specialized T cells. T cells sport receptors on their surfaces that interact with specialized molecules on the surface of the macrophage cells, which proffer the broken-down piece of the invading organism to the T-cell receptor system. When this interaction occurs, the T cell excretes a number of locally acting hormones (parachrines) that cause other cells of the immune system, such as the antibody-producing B cells, to reproduce and differentiate to plasma cells. These cells excrete antibody molecules that bind to the foreign invading organism or foreign molecule, forming an antibody-antigen complex. Several such antibody molecules with differing binding specificities may bind to a single invading organism or complex molecule. The consequence of these attachments is that the foreign molecule or organism is marked for destruction by either the other specialized killer T cells or scavenger macrophages. Other T cells retain a "memory" of the immunogenic components of the foreign organisms, so that when the body is invaded at a later date (which may be many years later) the body is primed to respond in a more rapid and vigorous way to the invader. There are two main processes involved here. Each has its own cells, cell receptors, and parachrine hormones; each has its specialized cells with their own unique growth and differentiation responses. The resulting complexities have, so far, prevented the design of a new vaccine based solely on knowledge of how a vaccine works.

Notwithstanding these complexities, many new diseases are being targeted for control or elimination by vaccination. Among these, many new types of vaccine are being tested for acquired immunodeficiency syndrome (AIDS), caused by the human immunodeficiency virus (HIV), as well as novel vaccines that may protect children against malaria infection. Diarrhea and pneumonia are other killer diseases affecting young children and neonates (resulting in 2 million deaths per year). These bacterial diseases are preventable by vaccination, but the means for the inexpensive and safe delivery of

Incredulous people grouped around Dr. Edward Jenner as he administers the first vaccine, 1796. (*The Library of Congress.*)

the vaccine materials is still under investigation. Vaccines for herpes simplex, papilloma virus cancers of the uterus, and staphylococcal infections are under development, as are the new techniques of DNA vaccines and powerful adjuvants such as CpG (multiples of the dinucleotide cytosine-guanine).

Ethical Issues

Vaccines are generally given to people and neonatal infants in good health. As with any medical treatment, it is possible that, as a side effect, serious illness or disease may result from the administration of a vaccine. This raises ethical questions: How much harm should be incurred to achieve a benefit that is expressed as an increase in the well-being of a population or society? How much individual suffering justifies a particular social gain? Because the suffering has been inflicted by an individual vaccinator, this might be thought less acceptable than the natural suffering that would otherwise afflict an unvaccinated population. One or two seriously diseased children may be the result of a vaccination campaign that has prevented several hundred deaths and thousands of diseased and disabled people. This ethical issue can be approached on the basis of a calculus of suffering. The chance that any one individual will experience harm, pain, or loss (with no advance knowledge of who will be so affected) has to be set against the thousands of people who would almost certainly suffer if they were not vaccinated. This utilitarian calculus tends to hold sway in most parts of the world, but there will be individuals in advanced as well as developing societies with dissenting views.

Those of a fundamentalist persuasion might argue that preventing people from becoming diseased is preventing God from exacting a punishment by causing a disease on those that have turned to idols or otherwise misbehaved by disobeying God's commandments. Another similar statement might be that by taking action to prevent disease, humans are acting unnaturally. Counterarguments to these statements is that one of God's commandments is "Therefore choose life" (*Deut* 30:19), so vaccines are acceptable in that they preeminently save lives. The argument about unnaturalness turns on the definition of the natural or that which obeys the laws of thermodynamics. Vaccines are in the latter category and should, thereby, be both natural and acceptable.

In the early days of the smallpox vaccination campaigns, and before Jenner, the argument from the pulpit was that as there was a small chance that the vaccinee would catch smallpox from the vaccination (then called variolation), thus creating a way that an individual could commit suicide (albeit inefficiently), which was forbidden by both religious and secular authorities. Clearly the intent of the vaccinee is normally to avoid death so the commitment of a crime that involves an evil mind or intent is not applicable in this area.

A case can be made that by vaccinating all the young children of the developing world against neonatal infections there will be increases in population numbers that will eventually lead to starvation and further suffering. But as developing populations advance and more of the female population receive some education, birthrates decline—a situation aided by the increased probability that newly born children will survive the hurdles of the infectious diseases of childhood.

Some vaccines are expensive; for example, a three-dose course of vaccination for hepatitis B when it first became available was about $1,000. The U.S. Food and Drug Administration, or a comparable agency in another country, must license the marketing and widespread use of a vaccine. To obtain this license, a company may spend anywhere from $300 million to $800 million testing the vaccine's safety and efficacy and ensuring consistency of production, and this cost must be recouped within the remaining lifetime of whatever patent was taken out when the mere possibility of a vaccine was recognized. The poor or the people of the developing world clearly cannot afford expensive new vaccines. In a decade or so, however, the price of most new vaccines come down to affordable levels, and agencies such as WHO, charitable foundations such as the Bill and Melinda Gates Foundation, and local governments find the funds to buy vaccines purchased at special low prices. For people in the developing world these vaccines are generally free at the point of use.

It clearly costs less to test a vaccine in a developing country where the prevalence of the disease is at a higher level, and so the challenge level (the level of the virus in the population that can constitute a cause of disease infection against which the vaccine generates a protective response) is higher; fewer people therefore have to be enrolled in the efficacy tests. But it is clear that from a safety point of view, the people in the test are exposed to the risks of harmful side effects. Why should people in the developing world accept the risks of harm from a vaccine that is intended to decrease the risk of disease in the advanced or developed world? To obviate this disproportion, arrangements are often made so that those who have participated in the trial and others in their society may obtain preferential supplies of vaccine. But this is not always the case.

Finally, some argue that vaccines both promote more risky sexual behavior and obviate the need for the development of self-control by the use of a technical fix—the vaccine. This latter argument is parried by the contention that without the vaccine the disease situation in the society would be considerably worse and that a person's self-control is a matter for their personal determination and conscience. That a vaccinated person would behave in a way that would increase the chances of becoming infected is a real issue. If, however, the herd effect is to apply, then the increase in risky behavior will come to naught as the herd is so well protected that, no matter the risky behavior, the chances of getting the disease are drastically reduced.

In an era of heightened threats of terrorist attacks, it is important to realize that one such threat is the deliberate release of pathogenic microorganisms. To prevent such an event becoming a disaster it would be important to have available the necessary vaccines to limit the spread of contagious disease. This in turn could lead to further developments of pathogens that are not affected by the vaccines. An escalatory process is thus engendered. Determining how much of a society's resources will be devoted to these contingencies will require much skill in deliberation and adaptation to current conditions and future potential developments. Nevertheless, the relative importance of alternative personal and social expenditures will need to be continually reevaluated.

Notwithstanding the ethical issues, vaccines remain one of the most effective and powerful tools for controlling, reducing, or eliminating debilitating diseases. They

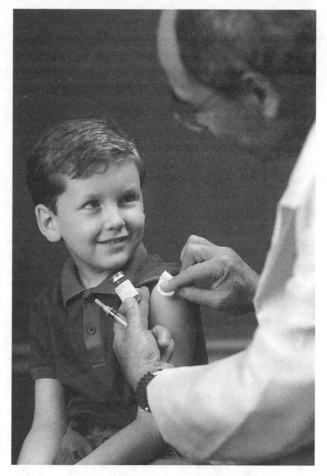

Elementary-school student receiving a vaccine. (© Bob Krist/Corbis.)

also point the way ahead for the development of medicine in that more effort should be expended on the development of methods for the prevention of disease rather than the cure of diseases that could have been prevented.

R. E. SPIER

SEE ALSO Antibiotics; Bioethics; HIV/AIDS.

BIBLIOGRAPHY

Baxby, Derrick. (1981). Jenner's Smallpox Vaccine. London: Heinemann Educational.

Bazin, Hervé. (2000). The Eradication of Smallpox, trans. Andrew and Glenise Morgan. San Diego: Academic Press.

Coico, Richard; Geoffrey Sunshine; and Eliezer Benjamini. (2000). Immunology: A Short Course, 5th edition. Hoboken, NJ: Wiley-Liss. Useful and clear introduction.

De Kruif, Paul. (1927). Microbe Hunters. London: Jonathan Cape. An exciting review of the early work on vaccines written by an enthusiastic journalist.

Gregoriadis, Gregory; Anthony C. Allison; and George Poste, eds. (1989). *Immunological Adjuvants and Vaccines.* New York: Plenum Press. Includes most current developments.

Lederberg, Joshua, ed. (1999). *Biological Weapons: Limiting the Threat.* Cambridge, MA: MIT Press. A useful overview.

Plotkin, Stanley A., and Walter A. Orenstein, eds. (2003). *Vaccines,* 4th edition. Philadelphia: Saunders. Covers mainly human vaccines.

Radostits, O. M.; D. C. Blood; and C. C. Gay. (1994). *Veterinary Medicine,* 8th edition. London: Baillière Tindall. Covers animal diseases, treatments, and vaccines.

Vallery-Radot, René. (1927). *The Life of Pasteur,* trans. R. L. Devonshire. Garden City, NY: Doubleday.

VALUES AND VALUING

• • •

The concept of value is more complex than it might initially appear. Values can range across personal preferences as indicated by pleasures, desires, wants, and needs to more objective goods such as health, efficiency, progress, truth, beauty, and more. Values can also be negative as well as positive, and in the former case they are commonly termed "disvalues," with examples being pain or illness. Values in all these senses both influence and are influenced by science and technology.

However, what precisely makes each of these diverse phenomena into values is more difficult to indicate. The concept of value, its manifestation in values, and the process of valuing (and evaluation) have been subject to diverse economic, social scientific, and philosophical analyses, each of which introduces numerous distinctions of relevance to any description and assessment of values in and resulting from science, engineering, and technology. Because of such difficulties, the present review attempts no more than some general introductions to three areas of discussion and includes a briefly annotated bibliography to mostly philosophical references.

Economic Perspectives

The term *value* is derived from the Latin *valere,* to be worthy or strong, the root as well of *valiant, valor,* and *valid.* It can be used as a noun ("Science is one of the primary *values* in modern culture") or verb ("We *value* modern technology"), or turned into a modifier ("Engineering is a *valuable* activity"). The term first emerged during the rise of the modern period to refer to the monetary worth of some commodity. Eighteenth-century economists conceptualized value as dependent on humans, and as such value was subtly opposed to premodern notions of goodness as a transcendental manifestation (along with truth and beauty) of being as such.

In the labor theory of value, commonly referenced to the English philosopher John Locke (1632–1704), value is created by humans when they technologically transform nature. In classical economics the market price of a commodity was thought to reflect the objective value contributed to it by human labor. But critics of this view argued in favor of price reflecting almost wholly the values that consumers attribute to products in a competitive marketplace. Exchange value replaced use value as the primary form of value. In economic science the basic concern has thus become to analyze interactions between human values and market behavior.

Social Scientific Perspectives

A different analysis of values developed in the social sciences, where the concern was more with how values are rooted in or related to the self and how they constitute society or influence political behavior. One mid-twentieth century effort to promote the scientific study of social values was advanced by the pragmatist philosopher Charles Morris (1901–1979). Extending earlier work, Morris (1956) distinguished between operative, conceived, and object values; did an empirical, cross-cultural analysis of value preferences among college students in Canada, China, India, Japan, Norway, and the United States who completed a "ways to live" inventory; and then speculated about the social, psychological, and biological determinants of values. The results of this psychometric research, which revealed both stability in structures among thirteen different ways of life and differences between national samples, were not especially profound, but they nevertheless promoted the idea that values are amenable to empirical investigation. This was in opposition to any assumption that the fact/value distinction would exclude values from scientific examination.

On a more personal level, one of the most widely referenced psychological analyses of value is that of Abraham H. Maslow (1908–1970). According to Maslow (1971) human beings try to satisfy needs or pursue values in the following order of priority: physiological needs (air, water, food), safety (security, stability), needs of belongingness and love, esteem needs, and self-actualization. The need for self-actualization was further associated by Maslow with the pursuit of what he called B(eing)-values such as truth, goodness, and beauty.

An observation by Langdon Winner bears on the implications for science and technology of many psychological (and even some economic) approaches to values. Once values are subjectivized, "[r]aising the question of value is no longer so much an occasion to think about the qualities of things or conditions outside us [as it is] an opportunity to look within, to perform an inventory of emotions" (Winner 1986, p. 158). Persons no longer purchase objects because the objects themselves have value as they are likely to purchase objects to realize their own values.

In sociology and anthropology values are described not so much in individual or personal terms as dimensions of culture. Shared values create collective identity and solidarity in culture and society. Socialization is a process of inculcating values from one group or generation to another. Sociologists of science analyze what particular values are shared within communities of technical professionals and how the inculcation and reinforcement of such values takes place. Values are both expressive and functional more than cognitive.

It should also be noted that within modern societies as a whole, one of the features that defines them as modern is the shared value placed on science and technology. Some critics of technological society in turn argue that this shared commitment to and/or acceptance of science and technology may undermine other socializing values such as religion. Questions thus arise about the absolute value of scientific knowledge—and about the possibility of technologies configured by alternative values.

Philosophical Perspectives

In philosophy the examination of values is closely linked to ethics. The philosophical examination of values and valuing as distinct from ethics came of age in the mid-twentieth century in different ways in the pragmatic, analytic, and phenomenological traditions.

PRAGMATIC TRADITION. In the pragmatic tradition, work by John Dewey (1859–1952), Ralph Barton Perry (1876–1957), Stephen C. Pepper (1891–1972), and C. I. Lewis (1883–1964) has been central. For Perry, value is defined as "any object of any interest" (1926, p. 115), so that to say that X is valuable means that Y takes an interest in X. Pepper sees Perry's definition as too narrow and argues more generally that values are constituted by "all selections by a selective system that are relevant to human decisions" (1958, pp. 690–691).

Dewey and Lewis continued the pragmatic empiricism of Perry and Pepper by arguing the foundational character of the human creative act of valuing. For Dewey, values are *ends-in-view*, that is, always provisional and able to become means to another end-in-view. Going beyond sheer animal impulses or appetites that produce effects, human interest, desire, "having ends-in-view, and hence involving valuations, is the characteristic that marks off human from nonhuman behavior." Moreover, when science is put to "distinctively human use" its knowledge about the nonhuman world is utilized to assess such ends-in-view in terms both of whether they are likely to be achievable by the proposed means or capable of becoming means themselves for further provisional ends. "In this integration not only is science itself *a* value (since it is the expression and the fulfillment of a special human desire and interest) but it is the supreme means of the valid determination of all valuations in all aspect of human and social life" (1939, p. 66).

Like Dewey, Lewis sees evaluations as forms of empirical knowledge related to courses of human action. Values have empirical content, although this content bears solely on personal preferences and courses of action, which makes values subject to democratic choice and scientific assessment. The general study of values, which can involve more than ethical values, is for pragmatists more properly termed theory of value or axiology than ethics.

ANALYTIC TRADITION. In the analytic tradition, the early leaders were Charles L. Stevenson (1908–1979), A. J. Ayer (1910–1989), and R. M. Hare (1919–2002). According to Ayer, the philosophical analysis of values is better described as *metaethics* than as ethics, because its goal is more the clarification of the meaning of terms than normative argumentation. Adopting a positivist interpretation of science as the paradigm of knowledge, Ayer and Stevenson argued that ethical and value statements were simply noncognitive expressions of likes and dislikes. Hare subsequently merged metaethical analysis with ordinary language philosophy to undertake a critical examination of the "language of morals." Linguistically, value statements were argued to entail a universalization of likes and dislikes.

Another even more abstract metaethical approach to values can be found in the work of G. H. von Wright (1916–2003), a student of Ludwig Wittgenstein. Von Wright (1963) subjects a particular value, goodness, to extended conceptual analysis. For von Wright it is not so much the value of goodness that is a creative projection of human action as a human commitment to a specific value that establishes that value as a norm. Von Wright and others such as Sven Ove Hansson (2001) have further sought to develop a formalized logic of values and norms reasoning.

PHENOMENOLOGICAL TRADITION. In the phenomenological tradition the defining work was that of Max Scheler (1874–1928). Whereas pragmatism focused on the process of valuing and analytic philosophy on the meaning and logic of value propositions, Scheler sought a conceptual elucidation and critical assessment of the substantive value feelings people experience. Scheler undertook his phenomenological descriptions of experienced values in opposition to Kantian formalism and universalism—a formalism echoed in metaethical formalism. For Scheler, prerational or intuitive preferences are at the basis of substantive ethics. These feelings can be grouped into five basic types: sensible values, pragmatic values, life values, intellectual values, and spiritual values. For Scheler (and most subsequent phenomenologists) technology is constituted by pragmatic values and science by intellectual ones.

Implications

The philosophical study of values yields a number of distinctions used in reflecting on relations between science, technology, and values. Such distinctions include those between instrumental and final values (means and ends), between extrinsic and intrinsic values, and subjective and objective values. Although related, these distinctions are subtly different. For instance, instrumental or use values may be extrinsic or designed into technological artifacts so as to become intrinsic values that have subjective and objective dimensions.

In relation more specifically to science and technology there are three interrelated issues with regard to values: What sort of property is involved with having a value or being valuable? (That is, are values primarily aspects of things or of knowers and users?) Is this property subjective or objective? (That is, to what extent is value subject to scientific study?) How might this property be designed into products, processes, or systems? (That is, can values be part of engineering design and technological invention?)

By and large values are taken in economics and in philosophy to be second-order properties that arise in interactions among human beings (markets) or depend on human beings (their interests). Values are thus not determined by science though they are certainly manifested in science, and science can study values in at least three ways: inventorying what values people express, analyzing structural relations among values, and criticiz-

ing specific values as likely or not to be able to be realized given the way the world is. The engineering design of products, processes, or systems is always undertaken with some values in view both with regard to process and project termination. That is, questions are increasingly asked about whether certain values such as user-friendliness, gender equity, or democratic participation can be designed into technologies. But the degree to which such a question can be answered in any systematic manner remains problematic.

The problematic character of the values–science relation is another continuing issue. One of the most persistently defended distinctions in science and technology is that between facts and values. Although widely criticized—because it is not clear whether the distinction is itself a fact or a value or both—one of the most persistent difficulties is to figure out how best to relate the two once distinguished. Even those who want to defend the difference also want to argue that values should have some bearing on what kind of science gets done and how it is done, and on which kind of technology gets created and how it should be used.

One general effort to address such questions is Loren R. Graham's *Between Science and Values* (1981), in which the author distinguishes between restrictionist and expansionist relationships. In the restrictionist view, science and values are strongly separated, and science is argued to be autonomous with no univocal influence on values. According to Graham, this is a view that is more defensible in physics than in biology, especially when the biology involves research on human beings. In the expansionist view, science is argued to have either direct or indirect implications for values and vice versa. This is the view that Graham thinks is most reasonable, but also one that he admits is both difficult to determine the boundaries for and dangerous. Indeed, as his historical case studies in physics and biology across the twentieth century reveal, almost any effort to deal with the science–values relation has weaknesses as well as strengths. Values and valuing are as much a challenge to science as science is to values.

In conclusion, it is worth observing that discussions of science, technology, and values in the 2000s have become less central than in the 1950s or 1960s. Were Jacob Bronowski's widely read *Science and Human Values* (1956) to have been published in the 1990s it would more likely have been titled something like "Science and Ethics."

CARL MITCHAM

SEE ALSO *Axiology; Critical Social Theory; Ethical Pluralism; Existentialism; Neutrality in Science and Technology.*

BIBLIOGRAPHY

Bronowski, Jacob. (1956). *Science and Human Values.* New York: Julian Messner.

Dewey, John. (1939). *Theory of Valuation.* Chicago: University of Chicago Press. An introduction to a pragmatic theory of values or axiology.

Graham, Loren R. (1981). *Between Science and Values.* New York: Columbia University Press. A historical examination of the interaction of science and values in the twentieth century using cases from physics and biology.

Grice, Paul. (1991). *The Conception of Value.* Oxford: Clarendon Press. A theory of value drawing on both Aristotle and Immanuel Kant.

Hansson, Sven Ove. (2001). *The Structure of Values and Norms.* Cambridge, UK: Cambridge University Press. Presents a formal system for the representation of values and norms.

Jonas, Hans. (2000). *The Genesis of Values,* trans. Gregory Moore. Chicago: University of Chicago Press. Originally published 1997. A historicophilosophical analysis of the work of Friedrich Nietzsche, William James, Émile Durkheim, Georg Simmel, Max Scheler, John Dewey, and Charles Taylor arguing that "values arise in experiences of self-formation and self-transcendence" (p. 1).

Lewis, Clarence Irving. (1946). *An Analysis of Knowledge and Valuation.* La Salle, IL: Open Court. Argues a naturalist theory of evaluations as a form of empirical knowledge.

Maslow, Abraham H. (1971). *The Farther Reaches of Human Nature.* New York: Viking. A humanistic psychologist's argument for a hierarchy of needs-values.

Morris, Charles. (1956). *Varieties of Human Value.* Chicago: University of Chicago Press. An early psychometric study of different value preferences.

Pepper, Stephen C. (1958). *The Sources of Value.* Berkeley: University of California Press. One of the most systematic and expansive pragmatic theory of values.

Perry, Ralph Barton. (1926). *General Theory of Value: Its Meaning and Basic Principles Construed in Terms of Interest.* New York: Longmans, Green. Reprint, Cambridge, MA: Harvard University Press, 1950. Defines value as simply the object of an interest.

Perry, Ralph Barton. (1954). *Realms of Value: A Critique of Human Civilization.* Cambridge, MA: Harvard University Press. Extends the 1926 interest theory of value into a critical assessment of interests in society, ethics, politics, law, economics, science, art, and more.

Scheler, Max. (1973). *Formalism in Ethics and Non-formal Ethics of Values: A New Attempt toward the Foundation of an Ethical Personalism,* trans. Manfred S. Frings and Roger L. Funk. Evanston, IL: Northwestern University Press. Originally published 1913–1916. Distinguishes five basic feeling values: physical, useful, vital, mental, and religious.

von Wright, Georg Henrik. (1963). *The Varieties of Goodness.* London: Routledge and Kegan Paul. A conceptual analysis of goodness as a type of value.

Winner, Langdon. (1986). *The Whale and the Reactor: A Search for Limits in an Age of High Technology.* Chicago: University of Chicago Press. A collection of influential essays.

VEBLEN, THORSTEIN

• • •

Economist, sociologist, and a founder of institutional economics, Thorstein Bunde Veblen (1857–1929) was born in Manitowoc County, Wisconsin, on July 30. He studied under the economist John Bates Clark at Carleton College in Minnesota, then at Johns Hopkins University before earning his doctorate in philosophy at Yale University in 1884. After a career of teaching at the University of Chicago, Stanford University, the University of Missouri, and the New School for Social Research, he died near Menlo Park, California, on August 3.

Veblen was an iconoclast. During the early twentieth century he was the foremost critic of the business establishment and its effects on culture and society. He alienated other academics by challenging their acquiescence to business interests. He was a prolific writer whose most famous work earned both popular success and intense academic scrutiny.

As one of the first institutional economists, Veblen's writings were often diametrically opposed to classical or neoclassical economics. For Veblen neoclassical economics relies on static notions of individually determined self-interests. In contrast, institutional economics maintains that social institutions, arising from individual economic behavior, influence that behavior in return. This approach views the economy as an evolving system and places a strong emphasis on dynamics, changing structures (including technologies, institutions, and ethics), and shocks to the system arising from technological innovation.

His most famous work, *The Theory of the Leisure Class: An Economic Study of Institutions* (1899), was a scathing sociocultural commentary. Veblen provides both a dynamic theory of class movement and a theory of consumption. He paints a picture of the business class as evolving from an earlier stage of "savagery," in which people peacefully went about their daily lives without any notion of private property and with relatively little material wealth. Culture then evolved from this primitive state to one of "barbarianism" characterized by private ownership and a leisure class that did not have to

Thorstein Veblen, 1857–1929. The American political economist, sociologist, and social critic wrote about the evolutionary development and mounting internal tensions of modern Western society. (© Corbis-Bettmann.)

work, but instead derived its wealth from the exploitation of other human beings through technology. Members of the leisure class gained their status through control and knowledge of technology. Veblen maintained that the leisure class would remain in power and receive the economic benefits of being in power as long as they could appropriate technological skills, tools, and labor. This appropriation depends mainly on private property and the profits derived from ownership of economic resources. This ability to remain in power and to maintain a dominant class position depends in turn on creation of institutions through business and government to protect the property rights of the leisure class at the expense of everyone else.

Veblen argued that the concentration of technology and power would often lead to the accumulation of wealth in the hands of a small leisure class at the expense of those at the other end of the economic spectrum. In the absence of institutions, effective property rights, and cultural norms the majority of the population would have access to neither capital nor the means to secure it. This has proven to be the case in many developing countries, where the absence of well-defined and enforceable property rights makes capitalism prone to inequitable outcomes.

Veblen's theory of consumption, especially the idea of consuming something beyond basic necessities, was unique. Conspicuous consumption provides the basis for twentieth-century consumerism in which consumption of goods and services serves not only as a tool to meet basic needs but also as a symbol of status.

Veblen recognized both the importance of science and technology in the creation of wealth and tensions between scientific technology and commercial enterprise. In *The Theory of Business Enterprise* (1904) and again in *The Engineers and the Price System* (1921) he analyzed the tensions between technological efforts to create good products and commercial interests in making money. Because of his praise of the "instinct of workmanship" (in his 1914 book published under that title) in ways that would eventually be echoed by Samuel C. Florman's *The Existential Pleasures of Engineering* (1976), Veblen's analysis inspired the technocracy movement and its effort to place engineers in positions of political power.

Veblen was one of the great thinkers of the twentieth century. Whether it was jealousy of his publishing success or because of his aloof nature, Veblen was shunned by his colleagues during most of his career. Ironically, near the end of his life, the American Economic Association offered him one of the highest honors in the field, the presidency of the association. He declined as he was unconcerned with either fame or recognition by his peers. Instead Veblen focused his efforts on writing and cofounding the New School for Social Research in New York. The posthumous rediscovery of Veblen's ideas has lead to renewed interest in both institutional and evolutionary economics and a new appreciation for and interpretation of Veblen's ideas. His legacy in the creation of social and economic theory continues to grow in importance.

WILLARD DELAVAN

SEE ALSO *Engineering Ethics; Management.*

BIBLIOGRAPHY

Dorfman, Joseph. (1934). *Thorstein Veblen and His America.* New York: Viking Press. One of the best biographies of Veblen, comprehensive and well written.

Florman, Samuel C. (1976). *The Existential Pleasures of Engineering.* New York: St. Martin's Press.

Jorgensen, Elizabeth Watkins, and Henry Irvin Jorgensen. (1999). *Thorstein Veblen: Victorian Firebrand.* Armonk,

NY: M.E. Sharpe. A wonderful biography replete with personal letters and interpretations of the letters.

Mouhammed, Adil H. (2003). *An Introduction to Thorstein Veblen's Economic Theory*. Lewiston, NY: Edward Mellen Press. Comprehensive guide to Veblen's economics and his place relative to other great thinkers of the twentieth century.

Veblen, Thorstein. (1899). *The Theory of the Leisure Class: An Economic Study in the Evolution of Institutions*. New York: Macmillan. His most cited and popular work; a critique of society and business practices.

Veblen, Thorstein. (1904). *The Theory of Business Enterprise*. New York: Scribners. A thoroughly convoluted attempt to define the modern corporation and explain why as an institution it is failing society

Veblen, Thorstein. (1914). *The Instinct of Workmanship, and the State of Industrial Arts*. New York: Macmillan. Veblen himself said this was his only important book. A synthesis of most of his important arguments.

Veblen, Thorstein. (1921). *The Engineers and the Price System*. New York: B. W. Huebsch. A collection of essays attacking the control of industry by investment bankers and outlining the role of engineers and the common worker in changing the inefficiencies of the existing power structure.

VEGETARIANISM

• • •

Vegetarianism is a traditional ethical stance and practice that has been influenced around the turn of the twenty-first century by science and technology. Strictly speaking, vegetarianism is a way of life in which one abstains from eating meat including fowl and fish. The *vegan* (pronounced "veegan") diet excludes all animal products, including eggs and milk. *Lacto-vegetarians* include milk products in their diet, and *lacto-ovo-vegetarians*, both milk and eggs. In the techno-scientific culture, a vegetarian diet may also be conscientious in other ways, such as by taking into account agricultural and food production methods, transportation distances, and the fairness of trade.

History of Vegetarianism

The history of vegetarianism began around the same time in the Mediterranean area and India. In Greece, Pythagoras (circa 569–475 B.C.E.) and his group were the first known to profess vegetarianism programmatically. Later the philosophers Epicurus, Plutarch, and some Neoplatonists recommended a diet without meat.

In India, the newly born Jain and Buddhist religions initiated the practice of vegetarianism in the fifth century B.C.E. Soon their idea of nonviolence (*ahimsa*) spread to Hindu thought and practice. In Buddhism and Hinduism, vegetarianism is still an important religious practice.

The religious reasons for vegetarianism vary from sparing animals from suffering to maintaining one's spiritual purity. In Christianity and Islam, vegetarianism has not been a mainstream practice although some, especially mystical, sects have practiced it. Monasticism in both East and West has often promoted vegetarianism.

In European and North American culture, vegetarianism witnessed a revival beginning in the seventeenth and eighteenth centuries and especially during the nineteenth century in part as a protest against some aspects of the scientific and industrial revolutions. Well-known vegetarians include Leonardo da Vinci (1452–1519), Mary Wollstonecraft Shelley (1797–1851), Richard Wagner (1813–1883), Henry David Thoreau (1817–1862), Leo Tolstoy (1828–1910), George Bernard Shaw (1856–1950), Mohandas Gandhi (1869–1948), and Albert Einstein (1879–1955).

Contemporary Issues

In contemporary culture, individuals have various reasons for pursuing vegetarianism. Although religious and spiritual arguments continue to be made, scientific research has also provided new justifications for vegetarianism. First, there is clear evidence that, contrary to early modern scientific theories that animals were like machines, animals in fact feel pain, anxiety, and other forms of stress. Thus it appears that breeding and killing animals for food causes them suffering. Moreover, some nutritional research indicates that a vegetarian diet is healthier than a carnivorous one. Finally, meat is ecologically more expensive to produce for food than vegetables: On average, the input ratio of units of proteins and energy fed to livestock to produce one unit of meat is ten to one.

Technologically enhanced food production has raised other concerns. For instance, it is highly questionable whether animals live in sufficiently humane conditions on contemporary farms. Indeed, the movement to promote the humane treatment of animals in the 1970s was extended from pets to other animals, and has had an influence on contemporary vegetarianism, as well as on the treatment of laboratory animals. Additionally, pesticides, hormones, and antibiotics involved in raising livestock have caused uneasiness. Similarly, the huge transport distances and the questions of global justice have encouraged people to think about what they eat, since food often is produced in Third World countries for wealthier nations.

The most common rejoinders to such vegetarian arguments are as follows: The ills of meat production do not directly imply any moral obligation for vegetarianism; meat has been a traditional part of human diet for thousands of years, hence it is not clear whether a vegetarian diet really suits everyone; and it is possible to arrange farms so that animals do not suffer unnecessarily. Moreover, often vegetarians have been accused of fanaticism and moralism; one common view is that they are just unbalanced people. In fact, it has also been noted that Adolf Hitler was a vegetarian.

The question of animal rights may also be related to vegetarianism. Just as *racism* involves one race oppressing another, it can be argued that *speciesism* involves one species oppressing another. Those who argue for the existence of animal rights commonly use their view to support vegetarianism. However, acceptance of the idea of animal rights immediately raises problems of the depth and extension of these nonhuman rights. Do animals have more than rights to life? Do all living creatures, including bacteria, have such rights? Usually only moral agents have rights, and duties as well; how does this apply to animals?

TOPI HEIKKERÖ

SEE ALSO *Animal Welfare; Food Science and Technology; Nutrition and Science; Organic Foods.*

BIBLIOGRAPHY

Dombrowski, Daniel. (1984). *The Philosophy of Vegetarianism.* Boston: University of Massachusetts Press. Provides a historical perspective into the philosophical ideas behind vegetarianism.

Singer, Peter. (1990 [1975]). *Animal Liberation,* 2nd edition. New York: Random House. A contemporary classic on the animal rights (first edition was published in 1975). Has had an important role in both vegetarian and animal liberation movement.

Spencer, Colin. (2002). *Vegetarianism: A History,* 2nd edition. New York: Four Walls Eight Windows. A historical account of vegetarianism, both from practical and theoretical points of view.

VEGETATIVE STATE

SEE *Persistent Vegetative State.*

VEREINS DEUTSCHER INGENIEURE

SEE *Professional Engineering Organizations.*

VERNE, JULES

• • •

The French novelist and playwright Jules Verne (1828–1905) was born in Nantes on February 8 and died in Amiens on March 24. He is best known for a series of novels published under the inclusive title *Voyages extraordinaires* (Extraordinary journeys). Some of these works have been interpreted, especially in English-speaking countries, as early science fiction, or used to stimulate discussion of ethical and political issues related to developments in science and technology—views that are at best only partial appreciations of his achievement.

Verne earned his *licence en droit* (master's degree in law) in Paris in 1850. After twelve years producing plays, opéras comiques, operettas, and short stories, he became famous in 1863 for his first published novel, *Five Weeks in a Balloon.* Verne subsequently published some fifty-three novels, among the best-known titles being *Journey to the Center of the Earth* (1864), *From the Earth to the Moon* (1865), *Twenty Thousand Leagues under the Seas* (1870), *Mysterious Island* (1870), *Around the World in Eighty Days* (1872), and *Michael Strogoff* (1876). After Verne's death, Hetzel, continuing the *Voyages extraordinaires* collection, published several novels, still under the name of Jules Verne, but all modified by his son Michel Verne, who added new chapters and new characters; Michel Verne even wrote a complete novel, *L'Agence Thompson and Co.* (The Thompson Agency and Company), which was edited under his father's name.

The objective of Verne's novels was primarily to teach geography, history, and the sciences to the French family. To make such dry disciplines attractive, Verne created initiatory stories happening in different geographies, such as: a judicial error and the innocence of the supposed culprit demonstrated through a cryptogram during the descent of the river Amazon in *The Jangada* (1881; also known as *The Giant Raft*); the cryptogram opens the novel, but cannot be solved during the whole story, because the key was considered as lost (after the discovery of the key, Jules Verne ends the novel with the readable message hidden in the cryptogram at the opening of the story); a *jeu de l'oie* (goose game, kind of snakes and ladders), allowing the reader to discover the United States in

The Will of an Eccentric (1899); or even a search for the missing link of human evolution in the African jungle in *The Aerial Village* (1902; also known as *The Village in the Treetops*). Writing for the French middle-class family did not prevent Verne from putting into his novels his views about colonialism, politics, and the society of his time. Antimilitarist and against the death penalty, Verne also denounced the misdeeds of slavery. He condemned British Victorian imperialism in such novels as *The Kip Brothers* (1902). During his lifetime he became known as a writer for children and was considered a scientific prophet. These two erroneous opinions continue to persist in the early twenty-first century. In reality, Verne was a writer of his time, using a style in which wordplay and hidden meanings were abundant; his work nevertheless heralded the structure of the modern novel.

Well into the twentieth century, Verne's works were so badly translated in the Anglo-Saxon countries that his readers could appreciate only his rare "futuristic" views, supported by a few extraordinary machines used to support the novelistic intrigue. Since the early 1960s, however, new translations by Walter James Miller, Edward Baxter, and William Butcher have allowed English-language readers to appreciate Verne as a true writer—a precursor of surrealism and other literary movements of the twentieth century such as the Collège de Pataphysique. (Pataphysics, an absurdist concept coined by the French writer Alfred Jarry, is the idea of a philosophy or science dedicated to studying what lies beyond the realm of metaphysics. It is a parody of the theory and methods of modern science and is often expressed in nonsensical language. A practitioner of pataphysics is a pataphysician.) Many scholarly studies in Europe and the United States show the modernity in Verne's novels, where irony and cold humor are always present.

Verne's many plays, usually written in collaboration with other authors, such as Charles Wallut and Adolphe d'Ennery, and most of his vaudeville works, operettas, and so on have grown old and would fail to have appeal in the early twenty-first century. *Journey through the Impossible* (1882), however, is a modern masterpiece, written at the juncture of the optimistic and pessimistic periods of Verne's life. This three-act play, cowritten with d'Ennery and inspired by *The Tales of Hoffmann*, a grand opera by Jacques Offenbach (1819–1880), is one of the main peaks in Verne's output. For the first and only time, the heroes do the impossible, when in the novels they did only what was extraordinary: The heroes from Verne's novels, including Nemo, Ox, and Ardan, meet onstage and go to the center of the earth in the first act, to the bottom of the Sea in the second, and to

Jules Verne, 1828–1905. The French novelist was the first authentic exponent of modern science fiction. The best of his work is characterized by intelligent predictions of technical achievements actually within man's grasp at the time Verne wrote.

the far planet Altor in the third. The principal hero is the son of Captain Hatteras, who was the first discoverer of the North Pole. During the three acts, his fiancée Eva shares his adventures and difficulties—an unusual fact in the *Voyages extraordinaires*—and he hesitates between love and knowledge, the same way Hoffmann hesitates between love and art.

Verne's work has provided scenarios for more than four hundred films and television programs, not only in Hollywood but also in countries as far away as China. In many instances they have continued to provide a popular introduction to the wonders of science and technology, propagating the image of Jules Verne as science fiction author. Jules Verne wrote his novels during the time when steel and steam engines became popular, when electrical power was used more and more, and when the Eiffel Tower was built, and he uses all these new technologies in his novels to be an integral part of the adventures he was telling his readers.

There are two ways in reading Jules Verne: the first level is the initiatory story with an adventure and sometimes more or less unusual and fantastic machines. Because of the bad English translations, it was the only way English-speaking readers could enjoy Jules Verne. The second level is appreciating the use of technology and science as narrative tools, enjoying the imaginary solutions of problems and desperate situations of an adventure happening in a world where war, confrontations and intolerance exist.

JEAN-MICHEL MARGOT

SEE ALSO Science Fiction; Science, Technology, and Literature.

BIBLIOGRAPHY

Evans, Arthur B. (1988). *Jules Verne Rediscovered.* New York: Greenwood Press.

Taves, Brian, and Stephen Michaluk Jr. (1996). *The Jules Verne Encyclopedia.* Lanham, MD: Scarecrow Press.

Verne, Jules. (2002 [1898]). *The Mighty Orinoco,* trans. Stanford L. Luce, ed. Arthur B. Evans, introduction and notes by Walter James Miller. Middletown, CT: Wesleyan University Press. Originally published 1898. This novel is one of the most modern novels; the subject of the story is transgender, the transformation of a man into a women.

Verne, Jules. (2003). *Journey through the Impossible,* trans. Edward Baxter, introduction and notes by Jean-Michel Margot. Amherst, NY: Prometheus Books.

VIDEO GAMES

• • •

Video games may be defined as games involving electronic technology in which real-time interactive game events are depicted graphically on a screen through pixel-based imaging. Elements one would expect to find in a *game* are *conflict* (against opponents or circumstances), *rules* (determining what can or cannot be done and when), use of some *player ability* (skill, strategy, or luck), and some kind of *valued outcome* (winning vs. losing, highest scores, or fastest times, among others). All are usually present in video games in some manner, albeit to varying degrees. In video games, the scoring of points, adherence to the rules, and display of the game's visuals are all monitored by a computer, which also can control the opposing characters within a game, becoming a participant as well as referee. Most arcade video games, home computer games, and home video games using a television would qualify as video games.

The development of the video game was shaped by film, television, and computer technology, and its influences include pinball, arcade games, science fiction, sports, and table-top games. Video games appeared during a time in which interactive art, minimalism and abstraction, and electronic music were developing, and these provided an important part of the cultural context in which the video game evolved.

Modes of Exhibition

Video games have appeared in a number of different modes of exhibition, including mainframe games, coin-operated arcade video games, home video game systems, hand-held portable games and game systems, and home computer games.

The games created on the giant mainframe computers were limited to the large mainframe computers found only in laboratories and research centers. These games were experiments and were neither sold commercially nor generally available.

Coin-operated arcade games come in several forms: stand-alone consoles; cocktail consoles; and sit-inside or ride-on games. A stand-alone console, the most common, is a tall boxlike cabinet that houses the video screen and the control panel for the game. The game controls can include joysticks, track-balls, paddles (round, rotating knobs), buttons, and guns with triggers. Occasionally there are controls for more than one player, although single-player games are the most common.

The *cocktail* console is designed like a small table, with the screen facing upward through a glass tabletop. Often the game is designed for two players, with a set of controls on each end of the table and the screen between them. This type of console is popular in bars or restaurants where patrons can sit and play a video game, while setting their drinks on the tabletop (hence the name cocktail).

Sit-inside or ride-on consoles hold or contain the player's body during play. They may even involve physical movement, usually to simulate the driving or flying of a vehicle in the game, typically with a first-person perspective. In driving and racing games, foot pedals and stick shifts are sometimes included. Other games involve bicycle pedaling, skis, skateboards, and simulated horses.

Home video game systems typically use a television or computer monitor for their graphic displays, although some systems come with their own screens. Home game systems that display their graphics on a television can be console-based, cartridge-based, or use laserdiscs, CD-

Two youngsters play video games at an arcade. (*AP/Wide World Photos.*)

ROMs, or DVD-ROMs (home computer games also appeared on cartridges, floppy disks, diskettes, and audio tape). Console-based systems have their games hardwired into the console itself, while cartridge-based game systems have their games hardwired into cartridges or cards that are plugged into the game console, allowing new games to be sold separately. CD-ROMs and DVD-ROMs are used for most contemporary game systems, because they can contain far more data than traditional cartridges.

Hand-held portable games and game systems that run on batteries can be carried along with the player. They are usually small enough to fit in the palm of one's hand, and typically have small LCD screens with buttons and controls around the screen. Some of these systems are cartridge-based as well.

Networked games involve multiple participants connected via the Internet to a video game world on a server, where they interact with the world and with each other's characters. Some of these games have hundreds or thousands of players and run twenty-four hours per day, with players logging on and off whenever they want. Players in these on-line worlds meet, converse, and form alliances and friendships without ever meeting face-to-face. Because real people control the player-characters, the social interaction is real, albeit in a more limited bandwidth than in-person interaction.

Ethics

Like film and television, video games have been criticized for having excessive violence, explicit sex, occasional racism, stereotypical characters, and an overall lack of edifying content. As graphics develop toward photo-realism, games grow more concrete in their visual representations and more like the images produced in other media, including those through which the player receives real world information (for example, television) and interacts socially (for example, the Internet). Combined with a simulated world in which players can act, video games can subtly influence players' behavior, beliefs, and outlook in real-life.

Most narrative media embody world-views through the ways in which characters' actions are linked to consequences, while video games link consequences to the player's own actions. Instead of merely watching and identifying with a character, the video game player is an active participant in the action seen on-screen. Whereas watching martial films does not help one develop physical skills, a video game *can* sharpen the player's hand-eye coordination skills and reflex responses, and stimulate aggression. The speed at which game action occurs often requires players to develop reflex responses at the expense of contemplation, sometimes resulting in a kind of repetitive stimulus-response training in which reaction speed is crucial. These responses can vary, from abstract figure manipulation, strategic thinking, and problem solving, to the hair-trigger automatic killing in fast-action games. While games can be designed to develop a variety of skills, shooting and killing are unfortunately among the most common.

On a larger scale, ethical worldviews can also be affected as successful game play often encourages or requires players to think in certain ways, and game narratives may link actions to outcomes and consequences that reinforce certain types of behavior. Thus it is a question of how the medium is used, how games are designed, and what values those designs embody. On-line role-playing games, for example, differ greatly from other forms of video games in that they are played by vast numbers of people in *persistent* (twenty-four-hour-per-day) game worlds, and games are ongoing and cannot be restarted. Some players invest a great deal of time and money in such games, building up their characters' powers and possessions, so there is often more at stake during game play, and ethics takes on greater importance as consequences within the game begin to extend into the real world.

While most people can clearly distinguish between video games and real life, ideas learned through the games can spill over to other behaviors in either positive or negative ways. Clearly there is a difference between real-world morality and that of the on-line game world. *Killing* another player's character may be considered an act of aggression, however the behavior falls within the established rules of play, and players whose characters are killed often come back with new characters. Yet the metaphor of killing remains, as does the fact that many people consider pretend killing to be fun. Likewise the goal-oriented nature of video games focuses more on what a player does and achieves rather than on what a player becomes. Additionally the malleability and repeatability of most video game experiences can lead to both experimentation and desensitization through repetition, because nothing is final or irreversible when a game can be restarted or when a player has multiple *lives*.

Other potential effects involve the player's default assumptions and ways of analyzing the world. For example, in most games everything is structured around the player and is present to produce an experience for the player. Other characters are there to either help or hinder the player-character, and often they speak in direct address to the player-character. Game objects exist for the player to use, take, or consume. The overall effect can be to promote a self-centered, utilitarian point of view in which players consider everything in the game world according to how it will affect or be of use to them.

At the same time, video games can have a positive influence, enhancing problem-solving skills, powers of observation, and patience. Completing an adventure game's objective, for example, usually requires goal-oriented behavior and often single-minded pursuit. Even when laden with puzzles and ambiguity, most adventure game problems and goals are clear-cut and simple relative to the problems and goals encountered in real life. The video game may remove the player momentarily from the complex problems of real life and offer solvable, simplified conflicts and goals that can be solved in a few hours (or days) and for which solutions already exist. In either case, these effects may be subtle, but repeated exposure to situations in which one is required to think a certain way can have gradual, long-term effects. Some values may find affirmation outside the games, such as overcompetitiveness and the accruing of personal wealth and goods.

In order to regulate games and hold game makers accountable, professional codes, such as that of the Association of Computing Machines (ACM) have been created. Additionally, the Entertainment Software Ratings Board (ESRB) provides a series of ratings (Early Childhood (EC), age 3 and up; Everyone (E), age 6 and up; Teen (T), age 13 and up; Mature (M), age 17 and up; and Adults Only (AO), age 18 and up), although these ratings are not always enforced in stores, where games might be sold to underage players.

While it is true that many games in the early twenty-first century are graphically violent and sexually explicit, it should be remembered that some of the best-selling games of all time (*The Sims, Myst,* and *Pac-Man,* for example) have been nonviolent, indicating that it is good game design, not sex or violence, that sells.

MARK J. P. WOLF

SEE ALSO *Computer Ethics; Entertainment; Science, Technology, and Literature; Special Effects; Violence.*

BIBLIOGRAPHY

DeMaria, Rusel, and Johnny Lee Wilson. (2002). *High Score!: The Illustrated History of Electronic Games.* New York: McGraw-Hill Osborne Media. An illustrated history covering the games, the hardware, and the people behind the industry.

Kent, Steven L. (2001). *The Ultimate History of Video Games: The Story Behind the Craze that Touched Our Lives and Changed the World.* Roseville, CA: Prima Publishing. A journalistic account of the history of video games compiled from dozens of interviews conducted by the author.

Wolf, Mark J. P., ed. (2001). *The Medium of the Video Game.* Austin: University of Texas Press. The first scholarly book devoted to examining the video game as an artistic medium.

Wolf, Mark J. P., and Bernard Perron, eds. (2003). *The Video Game Theory Reader.* New York: Routledge Press. A collection of scholarly essays examining the video game from a multitude of perspectives.

VIOLENCE

• • •

One of the multiple battlefields of environmental determinists versus biological determinists relates to the causes of violence. The former see violence as a primarily culturally rooted phenomenon, whereas the latter see it as being biologically determined This controversy, however, may be due to a failure to distinguish between aggressiveness and violence.

Aggressiveness and Violence

Aggressiveness is an instinct and therefore is a product of bioevolution. However, nature has not selected for the trait of aggressiveness alone but together with a set of inhibiting factors that are activated in certain circumstances, for instance, when two individuals who belong to the same group fight with each other and the life of one of them is threatened. As Irenäus Eibl-Eibesfeldt (1984) argues, a widely obeyed commandment in nature is "thou shalt not kill thy neighbor." Not even animals with as bad a reputation as wolves are an exception to this law.

In humans aggressiveness is linked primarily to the brainstem and the so-called limbic system or emotional brain (Sanmartín 2002). This part of the brain contains the structures that appear to be responsible for the responses (autonomous, somatic, hormonal, and neurotransmitter) that make up aggressive behavior. These automatic responses are triggered unconsciously by certain stimuli and coordinated by the amygdala, a structure in the inner region of the temporal lobe of both brain hemispheres.

The amygdala sets off the chain of effects that constitute the acting out of aggressive behavior in response to a stimulus. It also is responsible for stopping those effects when it receives inhibiting stimuli such as the emotional expression of fear shown by victims.

If humans were only the product of bioevolution, their aggressiveness would be regulated by the amygdala exclusively. However, humans are much more than a product of biological evolution. Indeed, the amygdala is connected to certain brain regions that are considered the seat of consciousness and that experienced extraordinary growth approximately 1.5 million years ago (Damasio 1994). These regions are in the frontal part of the brain cortex, the so-called prefrontal cortex. Their functions appear to be linked closely to the abilities that traditionally have been considered humankind's noblest: imagination, thought, and feeling. Ideas, thoughts, and feelings make up the framework that analyzes emotions and decides whether to reinforce or extinguish them. If, for instance, one of the emotions that constitute aggressiveness is reinforced, aggressiveness may go out of control and its natural inhibitors may be rendered inoperative. Soldiers and terrorists usually undergo a process of cognitive restructuring in which they learn to view their victims not as persons, but as things or symbols. Once victims are not seen as persons, it is impossible to empathize with them; consequently, their facial expressions have no inhibitory capacity.

Strictly speaking, violence is what occurs when the interaction between the expression and the inhibition of aggressiveness is disrupted in a way that hypertrophies aggressiveness and adds the intention to cause damage knowingly, as in the case of soldiers and terrorists. This disruption is influenced by ideas, thoughts, and feelings acquired over the course of a lifetime. Of equal importance are some of the products of the mind and in particular certain technical products. All of them are cultural elements. Violence therefore can be said to be primarily the result of the effects exerted by certain cultural elements on natural aggressiveness, hypertrophying it and conveying intentionality. The adverb *primarily* is used here because in certain cases (around 20 percent) the alteration of natural aggressiveness is caused by biological pathologies.

Technological Change as a Source of Violence

In most cases violence is born out of culture. Culture in turn is shaped by technology, as Ortega y Gasset (1939) argued, because humans are *a nativitate* (from birth) technical animals. Human beings change and survive because of bioevolution. However, humans do not worry as much about surviving as they do about the quality of life, always striving to achieve higher standards of well-being. This goal is achieved, but at the cost of creating a sort of supranature that consists of instruments or tools, machines, various forms of social organization, and instruments that apparently free humans of all the elements in nature that make them needy beings: cold, food, and the like.

Human beings also have directed technology toward themselves. On the one hand, they have constructed external prostheses that have modified their natural appearance. On the other hand, technology has penetrated so deeply into humans that they can in principle alter even their genetic information and therefore reconstruct themselves by following preestablished patterns and desires.

These technical interventions have had some negative effects, such as the conversion of innate aggressiveness into violence. Technology (not entirely by itself, of course) has upset the balance between natural aggressiveness and its natural mechanisms of control.

In addition, this technical supranature regularly experiences strong convulsions. At these times there take place the great technological changes that seem to drive historical transformations with increasingly greater speed. The mechanical clock, the fifteenth-century arts of navigation, railways, airplanes and spaceships, nuclear weapons, computers, the Internet, gene technology, and cloning are all technical inventions that have shaken traditions and compelled humans to adapt quickly to new situations. Human beings appear to be forced to adapt themselves to the changes in their technical supranature, not the other way round; this process often is described as social progress.

The consequent demand for adaptation generates a certain amount of stress that is becoming increasingly difficult to control. Uncontrolled stress usually degenerates into violence. In this sense, then, technology in general may become a source of violence (Sanmartín 2000).

Television and Violence

Two technologies are especially linked to violence: the mass media (especially visual media) and weapons.

In the early twenty-first century, not even the industry denies that exposure to violent images in television, video games, or the Internet has effects on the audience and, particularly, on children and adolescents. What is under discussion, however, is the type and degree of these effects. Albert Bandura (1977) stressed the idea that children learn violent behavior not only by imitating the real violence that is present in their environments, but also by emulating the violence (fictitious or not) broadcast on television. This correlation was confirmed by a longitudinal study started in 1960 by Leonard Eron, Monroe Lefkowitz, Leopold Walder, and L. Rowell Huesmann (Huesmann and Eron 1986) that used a sample of 800 eight-year-old children. Jeffrey Johnson (2002) published another longitudinal study showing that seeing violence on television at age fourteen correlated significantly with later aggression (assault and battery, violent or armed robbery). According to Johnson and associates (2002), if exposure to television was one hour per day at age fourteen, 5.7 percent of the individuals at a mean age of sixteen or twenty-two exhibited violent behaviors, and if exposure was increased to three or more hours per day, violent behavior went up to 25.3 percent. Craig Anderson and Brad Bushman (2002), in a related meta-analysis of longitudinal studies, cross-sectional studies, field experiments, and laboratory experiments carried out to that date concerning the possible influence of violence on television, demonstrated that all studies supported the existence of a significant correlation between exposure to violence on television and violent behavior.

Other studies have provided more clues to this problem. Foremost among them that of Jo Groebel (1999), which showed that the relationship between real violence and screen violence is interactive: Violent people use audiovisual media to reinforce their beliefs and attitudes, becoming even more violent. This study dealt with a large sample: 5,000 twelve-year-old children from twenty-three different developed and developing countries with social environments containing high or low rates of real violence. One interesting discovery was that 88 percent of the children had seen the move *The Terminator* (1984). An even more interesting discovery was that in high-violence environments half the children wanted to be like the Terminator, whereas in low-violence environments the number was only 37 percent. In other words, the influence of screen violence on real violence depends on the amount of real violence surrounding a child.

Before blaming television for violence in society, especially among children and adolescents, one must consider carefully the social environments of those children. When children live in homes in which they suffer or witness abuse, where there is alcohol or drug abuse, where parents and children do not get along, where the homes have cramped or unhealthy living conditions, without the support of other family members or friends, the result may be an environment in which the spark of television violence has little difficulty causing a fire by adding to preexistent violent attitudes and behaviors.

In the early twenty-first century, this environment often also contains videogames, either on game consoles or computer. Many authors, such as Degaetano and Grossman (1999) and Anderson (2004), state that the effects of violent images on video games are even worse than other kinds of images, for several reasons. Firstly, as opposed to films and television programs, in violent video games the player is forced to identify with the main character (the aggressor). Secondly, violent video games require active participation, and active participation promotes learning. Thirdly, video-game violence is directly rewarded. Finally, the level of violence in video games is far superior to that in films or on television.

Computers, and in particular the Internet, are connected to violence, especially violent crime, in different ways. Rather than generating new forms of crime, the Internet has revolutionized some traditional forms of crime by accelerating their transnationalization. If there is one thing that has rapidly globalized, it is the criminal activities of mafias and extremist organizations. Sexual exploitation—and pornography, in particular (Von Feilitzen and Carlsson, 2000)—drugs trafficking, the smuggling of chemical, nuclear and radioactive material, and especially the money-laundering business, have benefited from the globalizing effect of the World-Wide Web. In fact, every year more than 600 billion dollars (slightly more than 2% of global gross domestic product) are laundered world-wide, practically cost-free.

Weapons and Violence

One particular type of technology is especially linked to violence: weapons. From a naturalistic point of view, bioevolution has poorly equipped the human animal for causing severe damage and especially for killing other humans. Human beings do not have fangs, sharp claws, or pointed horns. In order to kill they have to use their feet or fists with great force and skill or put their hands on a victim's neck for several minutes. In such cases, killing takes place at close quarters and a victim's aggression-inhibiting signals are quite effective.

These inhibitors are bypassed, however, when weapons are used. From knives and swords to guns and bombs, weapons have evolved to increase the distance between users and victims, until the victims have disappeared from direct view. This is not a coincidence. Once distance has blurred facial expressions, postures, and other aggression inhibitors, victims cease to be seen as persons and become things. One cannot empathize with a thing, stop one's destructive actions, or even feel sorry afterward. In this way one of the beings most ill-equipped by nature for killing has become, by virtue of technology, one of the most effective killers.

JOSÉ SANMARTÍN

SEE ALSO Aggression; Entertainment; Just War; Military Ethics; Movies; Television; Video Games.

BIBLIOGRAPHY

Anderson, Craig A. (2004). "An Update of the Effects of Violent Video Games." Journal of Adolescence 27, 122-133.

Anderson, Craig A., and Brad J. Bushman. (2002). "The Effects of Media Violence on Society." Science 295: 2377–2379.

Bandura, Albert. (1977). Social Learning Theory. Englewood Cliffs, NJ: Prentice-Hall.

Damasio, Antonio R. (1994). Descartes' Error: Emotion, Reason and the Human Brain. New York: Putnam.

Degaetano, Gloria, and Dave Grossman. (1999). Stop Teaching Our Kids to Kill: A Call to Action against TV, Movie and Video Game Violence. New York: Crown.

Eibl-Eibesfeldt, Irenäus. (1984). Krieg und Frieden [War and peace]. Munich: R. Piper & Co. Verlag.

Groebel, Jo. (1999). "Media Access and Media Use among 12 Year Olds in the World." In Children and Media: Image, Education, Participation, ed. Cecilia von Feilitzen and Ulla Carlsson. Göteborg: The UNESCO International Clearing House on Children and Violence on The Screen.

Huesmann, L. Rowell, and Eron, Leonard D. (1986). Television and the Aggressive Child: A Cross-National Comparison. Hillsdale, NJ: Erlbaum.

Johnson, Jeffrey G.; Cohen, Patricia; Smailes, Elizabeth M.; et al. (2002). "Television Viewing and Aggressive Behaviour during Adolescence and Adulthood." Science 295: 2468–2471.

Ortega y Gasset, José. (1939). Meditación de la técnica. English Version: "Thoughts On Technology," in Philosophy and Technology: Readings in the Philosophical Problems of Technology, ed. Carl Mitcham and Robert Mackey. New York: The Free Press, 1972.

Sanmartín, José. (2000). La violencia y sus claves [The clues to violence]. Barcelona: Ariel.

Sanmartín, José. (2002). La mente de los violentos [The violent mind]. Barcelona, Ariel.

Von Feilitzen, C., and Carlsson, U. (2000). Children in the New Media Landscape: Games, Pornography, Perceptions.

Göteborg: The Unesco International Clearing House on Children and Violence on the Screen.

VIRTUAL REALITY

• • •

Virtual reality (VR) technology emerged in the 1980s, with the development and marketing of systems consisting of a *head mounted display* (HMD) and *datasuit* or *dataglove* attached to a computer. These technologies simulated three-dimensional (3-D) environments displayed in surround stereoscopic vision on the head-mounted display. The user could navigate and interact with simulated environments through the datasuit and dataglove, items that tracked the positions and motions of body parts and allowed the computer to modify its output depending on the recorded positions. Other types of VR that arose subsequently included *projection virtual reality*, in which users who wear special glasses interact with three-dimensional virtual models that are projected in a room and can be perceived from different angles, and *desktop virtual reality*, in which users stereoscopically view a virtual environment represented on a computer screen (using special stereo glasses) and interact with it using datagloves, or, more commonly, a mouse.

VR is used to simulate real environments, such as existing buildings or city areas, or to visualize imaginary ones, for instance spaceships or battlegrounds. VR is a technique with great possibilities for training, visualization, and entertainment. Applications are found in computer-aided design, construction, computer gaming, education, military exercises, aviation training (flight simulators), surgical training, therapy, and art.

Meanings of Virtual Reality

As Howard Rheingold (1991) notes, VR merges overlapping interests from the military for more realistic but risk-free training, of the science fiction imagination, and of entertainment industry efforts to intensify the vividness of various media. Although the term "virtual reality" most often refers to systems of the type just described, it is also used in a wider sense, to denote not fully realized virtuality, as in lesser forms of three-dimensional computer-simulated environments that are engaged from a first-person perspective. The most common example is first-person 3-D computer games. Such games are varieties of desktop virtual reality minus the stereo glasses. Wider still, VR sometimes denotes any

interactive computer-generated environment, including those represented through two-dimensional graphics or through texts or symbols. In fact, the term *virtual* may be attached to any kind of object, event, or environment that is not realized physically but electronically, as in virtual money, virtual casinos, or virtual doctors (medical doctors that can be consulted over the Internet). In such cases, *virtual* may mean no more than "computer-simulated," or "on the Internet," or "in cyberspace," as opposed to "in physical space." This broad use of the term points to the fact that for many people, the term "virtual reality" and "virtual" are interpreted metaphysically as denoting a new, fictional kind of reality.

Mostly, however, the term *virtual reality* is used more narrowly, to refer to 3-D computer-simulated environments incorporating a first-person perspective that includes some degree of immersion, meaning that users feel that they are situated in an environment. Immersion can be enhanced through such means as realistic graphics and sounds, surround and stereo vision, surround sound, position tracking, and force and tactile feedback.

A distinction can be made between single-user and multi-user or networked VR. In single-user VR, there is only one user, whereas in networked VR, there are multiple users who share a virtual environment and appear to each others as *avatars*, which are graphical representations of the characters played by users in VR. A special type of VR is *augmented reality*, in which aspects of simulated virtual worlds are blended with the real world that is experienced through normal vision or a video link, usually through transparent glasses on which computer graphics or data are overlaid. Related to VR are *telepresence* and *teleoperator systems*, systems that extend a person's sensing and manipulation capability to a remote location by displaying images and transmitting sounds from a real environment that can (optionally) be acted on from a distance through remote handling systems such as robotic arms.

Ethical issues in virtual reality

VR has been the subject of speculation and critique in both academic circles and mass media. Popular culture portrays futures in which immersive VR is routinely used in society, as in science fiction movies such as *The Matrix* (1999), *Lawnmower Man* (1992), *Existenz* (1999), and the Star Trek series (with the Holodeck), and in novels such as William Gibson's *Neuromancer* (1984) and Neal Stephenson's *Snow Crash* (1992). VR is portrayed both positively, as a medium that offers endless possibilities for learning, entertainment, social interaction, and self-experimentation; and negatively, as a medium that causes users to flee from or deny everyday reality, that is used by evil minds to manipulate and gain control over others, and that dissolves any distinction between reality and fiction.

In the academic literature, authors have mainly tried to come to grips with the questions of how VR will transform people's conception of reality and how it will transform social life. As for the former question, authors tend to agree that VR will change the concept of reality and cause the distinction between reality and fiction to blur. However, some authors, such as Michael Heim (1993) and Sherry Turkle (1995), have argued that a distinction between physical and virtual reality will always exist because people are biological human beings that are born and die in the physical world and retain their roots there, whereas others, such as Philip Zhai (1998) have argued that such biological background facts are irrelevant and that VR can offer us a limitless world as rich and detailed as physical reality and can even replace the physical world as one's primary habitat.

As for social and ethical aspects of VR, most discussion has focused on the question of how the blurring of reality and fiction in VR may affect its users, on how reality is (mis)represented in VR, and on what forms of immoral behavior may occur in virtual environments. These issues will now be discussed in turn.

VR AND THE REAL WORLD. Some authors who hold that the extensive use of VR applications induces a blurring of the boundary between the real and the imaginary worry about negative social consequences. They worry that the idealized, vacuous and consequenceless worlds of VR come to serve as a model by which people comprehend the real (that is, physical) world, and conversely, that the attention and care that people attach to real-world people, animals, and things is also attached, inappropriately, to virtual things and personae. Another worry is that people may come to prefer the freedom and limitlessness of virtual reality and cyberspace over the limitations of physical existence and invest most of their time and energy in their virtual life, to the neglect of the real people and affairs in their physical lives. Proponents of VR argue instead that most people will be able to maintain a good sense of reality and will strike a healthy balance between their virtual life (which is, in part, also real life) and their physical life.

REPRESENTATION IN VR. VR environments that are intended to simulate actual realities may misrepresent these realities, according to expected standards of accu-

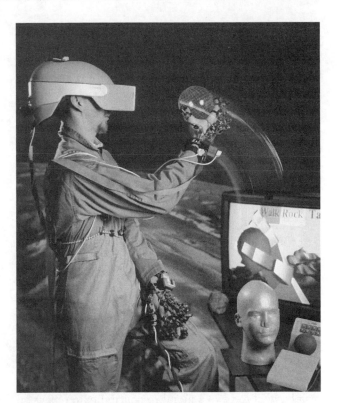

A man demonstrates a virtual reality device by lifting a virtual rock on a simulated Martian surface, wearing a video helmet and virtual reality gloves. (© *Roger Ressmeyer/Corbis.*)

racy. This may cause their users to make false decisions or act wrongly, with potentially serious consequences, especially in areas in which life-or-death decisions are made, such as medicine and military combat. When VR is used for education and training, therefore, high standards of accuracy and realism should be expected, and developers have a responsibility to adhere to such standards. VR simulations may also contain biased representations that are not necessarily false, but that contain prejudices about people or situations. For example, a surgery training program may only practice surgery on young white males, a VR game may represent women and minorities in stereotypical ways, or a combat simulation program may only simulate combat situations in which civilians are absent. Like other media, VR may also break taboos by depicting morally objectionable situations, including violent, blasphemous, defamatory, and pornographic situations.

BEHAVIOR IN SINGLE-USER VR. Most moral issues regarding representation in VR are not unique to it, and also apply to other types of simulations and pictorial representations. What is unique about VR, however, is the possibility to interact with environments that look real but are not. Because virtual environments are not real, any consequences of one's actions in VR, specifically in single-user VR, are not real-life consequences. It is therefore possible to perform actions in VR that would be cruel and immoral in the real world because they do harm, but can be performed without retribution in VR because no real harm is done. But is it morally defensible for people to act out graphic and detailed scenarios of mass murder, torture, and rape in VR, even when done in private? Are there forms of behavior that should not be encouraged or allowed even in VR, either because of their intrinsically offensive nature, or because such simulations desensitize individuals and may facilitate immoral behavior in the real world? Or is it the case that the possibility to act out fantasies in VR keeps some people, such as sex offenders or people prone to violence, from acting out this behavior in the real world, so that VR may actually prevent crime?

The interactivity made possible by VR developers also raises moral questions. VR applications may invite or discourage, require or prohibit, reward or punish behaviors. They may cheer users who go on killing sprees, or may instead voice moral outrage. Developers may be held to have a moral responsibility to reflect on the way in which they deal with immoral behavior by users, and whether and how they signal approval or disapproval of such behavior, or remain neutral.

INTERACTIONS IN MULTI-USER VR. In multi-user VR, users may engage in immoral or illegal behaviors such as theft, vandalism, murder, sexual assault, and adultery. What is confusing is that some of these behaviors may be real while others are imaginary. A user may harm or kill another user's avatar, but cannot harm or kill another user. Yet a user may also cause real harm to another user, by deeply insulting that user, stealing an identity, or wreaking havoc in a virtual apartment. Such actions are thought of as real and may even lead to criminal prosecution. Sometimes, however, it is not so clear what actions mean. Does genuine sexual assault occur when one user fondles another user's avatar against his or her will? What if such behavior is performed by a programmed avatar (a *bot*) that has been programmed to do so by its owner? Very different moral intuitions may exist about these and many other actions in multi-user VR, and more broadly in cyberspace.

Another issue that plays in multi-user VR and cyberspace is identity. As has been argued extensively in academic studies, VR avatars and role-playing in cyberspace enable people to experiment with identities and to experience otherness more vividly than ever before. A man can learn what it is like to be a woman, a white person can have the experience of a black person,

and so forth. Negatively, such role-playing can be used to deceive others about one's true identity. But as psychologist Sherry Turkle (1995) has argued, such experiences may help users expand and develop their own identities and may deepen a distinctly human form of self-awareness.

PHILIP BREY

SEE ALSO *Cyberspace; Information Ethics.*

BIBLIOGRAPHY

Brey, Philip. (1999). "The Ethics of Representation and Action in Virtual Reality." *Ethics and Information Technology* 1: 5–14. Considers ethical issues of representation and single-user behavior in VR.

Ford, Paul. (2001). "A Further Analysis of the Ethics of Representation in Virtual Reality: Multi-User Environments." *Ethics and Information Technology* 3: 113–121. A follow-up to Brey's essay, this essay considers ethical issue or representation and behavior in multi-user VR.

Heim, Michael. (1993). *The Metaphysics of Virtual Reality.* New York: Oxford University Press. A series of essays on the history and future of contemporary people's computerized lives, including reflections on the nature and appeal of cyberspace and virtual reality.

Rheingold, Howard (1991). *Virtual Reality.* New York: Summit. An early but still significant study of VR, its history, and its promises for the future.

Turkle, Sherry (1995). *Life on the Screen: Identity in the Age of the Internet.* New York: Simon and Schuster. An important psychological study of how the Internet and its virtual environments are impacting identity formation and our way of thinking about reality, relationships, politics, and sex.

Zhai, Philip (1998). *Get Real: A Philosophical Adventure in Virtual Reality.* Lanham, MD: Roman & Littlefield. Extensive study of VR, virtuality, and its meaning for humanity. Zhai argues that the whole empirical world can and should be created in virtual reality.

VIRTUE ETHICS
• • •

The prominence of rules, consequences, rights, and duties is a relatively recent phenomenon in moral thought. For Plato, Aristotle, Laozi (or Lao-tzu), Confucius, the Buddha, and Jesus, the primary focus of the good life was on cultivating virtues and battling vices. Yet among these diverse traditions moral character and its significance for personal and social good have been subject to considerable debate—which continues in the early twenty-first century by drawing on the thought and research in sociology, anthropology, film studies, folklore, religion, biology, neurophysiology, pedagogy, medicine, and other disciplines. Both ancient reflection and contemporary scientific inquiries seek to identify the principle virtues and vices and how they develop or weaken. Adversaries debate whether the virtues (and vices) are intertwined, whether they exist independently, or whether there is a chief virtue (or vice). Such inquiries easily lead to more general questions of human flourishing and distinctiveness, so that ultimately at issue are basic questions concerning the nature of human happiness and the good society.

From the perspective of virtue ethics, science and technology are arguably enduring components of the good life. Aristotle (384–322 B.C.E.), for instance, describes virtue as a kind of human excellence or striving for perfection. (The Greek word for virtue is *arête*, which encompasses both moral capability and specific talents. A musician, for example, might give a *virtuoso* performance.) In this sense both *episteme* (knowledge or science) and *techne* (craft, art, skill, know-how) are forerunners to the modern notion of technology and involve human *arête*. Controversies about the responsibility of scientists and engineers evoke this twofold sense of virtue, insofar as they address the special types of knowledge they pursue as well as their moral positions regarding the results and applications.

Scientific discoveries and technological products also pose challenges to understanding and embodying a virtuous life. Studies of animal and human behavior raise questions about possible similarities between animals and humans in promoting cooperation or fostering competition. In place of proposals for political utopias and personal desires for posthumanist transformations, can advanced technosciences be limited or guided by the values found in folk wisdom, venerable sages, or sacred texts? Or do many technical inventions thwart the search for a virtuous life by zealously promoting and catering to ordinary vices? Instead of assisting with the cultivation of temperance, justice, courage, love, or charity, do they perhaps tempt humans with vanity, sloth, anger, lust, and greed?

Background

A virtue-based ethics is agent centered, presumes a telos or purpose for human life, and encompasses both personal and public goods. In light of ecological problems growing out of the human use of technology, critics have charged virtue ethics with being anthropocentric. It neglects or devalues the welfare of animals, natural entities, and the environment. Defenders of virtue ethics respond that a fundamental virtue such as humility promotes recognition of human limits and asks humans to view themselves as simply parts of a larger cosmic whole. Moreover, the concept of virtue as a perfection applies to nonhuman as well as human entities. While the idea of a telos or purpose in nature is problematic for science and technology, the topic remains a source for lively discussion among philosophers of biology who study possible adaptations to ethical theory. Indeed, even in the philosophy of technology, analyses of the role of functions is a research issue of potential relevance to virtue ethics.

As such issues indicate, despite the tendency to portray virtue ethics as a settled tradition of strong consensus and enduring narratives, there have always been lively debates about the scope of a virtuous life, the relative strengths or weaknesses of specific virtues and vices, and the best vision of human happiness. For example, three classic representatives of virtue ethics emphasize contrary views on pride. Aristotle considered it a principal virtue. One should attain a proper sense of self—one's accomplishments and contributions. A proud individual is not driven by vanity or boasting, for these lead to excesses of indulgence that would be unworthy of a free and rational person. The proud individual is courageous—the most fundamental virtue, for without courage one can hardly embody other cardinal virtues such as justice or prudence. A model is the citizen whose democratic participation is free of destructive vices such as envy or rancor.

Augustine of Hippo (354–430) and Thomas Aquinas (1225–1274), though, were among many Christians and religious thinkers who believed pride to be indicative of an exaggerated sense of self, involving vanity or, worse, the temptation to view oneself in godlike or superhuman terms. Pride was the queen of vices, for it spawned the decline toward deadlier ones, such as envy, anger, and lust. Such vices corrupt one's moral character and undermine efforts to become a just person. This distortion of self brings about neither happiness nor salvation, only ruin or damnation.

Buddhism and Daoism, meanwhile, taught that true virtue seeks the no-self or personal transcendence. This involves overcoming the drive for individuality in which satisfying the desires and needs of one's physical self is primary. Intellectual nitpicking may derail this goal. But reflection on the nature of this goal remains essential, and can generate parables and paradoxes that are potential guides to enlightenment (see Saeng 1991, Chuang Tzu 1996). Enlightenment is realized not when one becomes a dutiful citizen or achieves self-esteem, but is moved by compassion for another. This is an experience of insight and joy.

Disciples and pedagogues have continually debated the nature and prominence of the virtues. The Western tradition that featured the seven cardinal virtues—courage, justice, temperance, and prudence among pagans, and love, hope, and charity as the Christian additions—is hardly carved in stone. Seven has been a magical number, but other virtues have also been considered essential to the good life. Aristotle, for instance, devotes more attention to friendship than any other single virtue, and friendship may be considered the basis of scientific and technical communities.

At the same time the underpinnings of virtue have been extensively debated. For example, pagans focus on the meanings and demands of individual courage or the extent of its relation to political justice, whereas monotheists anchored a moral life not in self and society, but primarily in God. The contentiousness of these disputes and their failures to successfully promote virtue eventually led to a radical challenge of virtue ethics that nevertheless did not eliminate its relevance. Rather, according to the historian of modern moral philosophy J. B. Schneewind (1998), these disputes relegated the virtues and vices to secondary status. Displacing them as the primary focus of ethics were duties, happiness or pleasure, autonomy rather than character, and the right rules or laws for gauging ethical conduct.

After more than 200 years of rationalism and emotivism in moral theory, toward the end of the twentieth century virtue ethics underwent a revival. Dissatisfied with the inability of prominent moral theories to address human well-being, resolve concerns about justice in an increasingly technological world, and inform or guide individuals toward the good, philosophers began reassessment of the centrality of virtue. A key contributor to this was Alasdair MacIntyre, author of *After Virtue* (1981) and *Whose Justice? Which Rationality?* (1988). Invoking the wisdom of Aristotle and Thomas Aquinas, along with the lessons of contemporary social and political thought, MacIntyre espoused an enriched view of the integral and narrative self that challenged rival notions of the self as little more than a utility maximizer or logical servant to duty.

MacIntyre's learned eloquence and sharp critique of his own intellectual and moral times spawned a veritable industry. Responses ranged from best-selling children's books on the virtues to theoretical and scientific inquiries into the nature of moral character, whether or how it can be taught, and the relation of individuals to others: other humans, other species and life-forms, even deities. Some scientists have contended that, contrary to MacIntyre's emphasis on human identity as flourishing in cultural and historical storytelling, human morality should more sensibly emulate animals. Monkeys and chimpanzees, birds and elephants, according to zoologists, illuminate more accessible and realistic moral guidance than the (less realistic) heroes and saints who permeate human literature.

Such disputes—interweaving disciplines, incorporating historical and cultural contexts, responding to calls for justice or courage and to temptations of anger or lust—underscore the lasting appeal of virtue ethics. Unable to resolve all the philosophical questions put to it in journals and seminars, nor ready to dictate every moral situation (which theories can?), virtue ethics highlights controversies as vigorously as any other moral theory. Nowhere is this more clearly illustrated than in relation to science and technology.

Sloth, Leisure, Efficiency

Medieval Christians learned the seven deadly sins through the mnemonic device of an acronym—*s-a-l-i-g-i-a*. Each letter represented a deadly sin in order from the queen of vices (*superbia* being pride) to the deadliest (*acedia* or sloth). In between are situated *avaritia* (greed or covetousness), *luxuria* (lust), *invidia* (envy), *gula* (gluttony), and *ira* (anger). Warnings against sloth—from the Benedictine rule concerning the dangers of idleness to popular jokes about couch potatoes—represent it as the death of the soul as well as the spirited body. Sloth is more than laziness or lethargy; it constitutes a lack of purpose, an indifference to others and the goings-on of the world. In his *Pensees* (c. 1660), Blaise Pascal frequently remarked how people fill their time with diversions, such as games, chatter, and sensual delights. These prevent contemplation of more defining matters that include the meaning of one's death or a believer's relation to God.

By contrast, leisure is upheld as a sign of independence and accomplishment. What Pascal denigrates as diversions can be praised as just desserts. In leisure individuals explore their potential, be it in time of play, hobby, volunteer work, or even, as G. K. Chesterton (1874–1936) wryly noted, "the time to do nothing at all." To this end inventions promise to lessen arduous chores while opening more opportunities for whatever one desires. Household gadgets save on cleaning and organizing; robotics and assembly lines spare the sweat and blood of labor; sophisticated weaponry produce greater damage with risk to fewer personnel. Leisure relies on the promises of efficiency. These promises, however, can be misleading insofar as they exchange one set of difficult expectations for another. For example, the historian Ruth Schwartz Cowan (1983) has demonstrated the deceptive attractions of household technologies. The washing machine cuts down the once-a-week ardor of washing by hand and wringers, but introduces the everyday demand for a clean set of clothes, hence making laundry a daily chore. The invention of the four-burner stove with oven shifts family expectations from variations of a pot of stew to a five-course meal, hence the popularity of the cookbook. The overall result is that technologies tend to reduce the physical pressure of housework while increasing the solitariness and frequency of household tasks. The promise of more leisure, concludes Cowan, is often illusory.

From a virtue ethics perspective, however, there remain additional concerns with leisure. While *scholar* derives from the Latin word for leisure—implying both an individual and cultural good—leisure nevertheless poses considerable danger. As studied by Sissela Bok in *Mayhem* (1998), many leisure technologies involve decadent forms of play. Video games, television shows, and movies featuring callous and malicious regard for human (and animal) life have gradual effects on participants and audiences that can be just as pernicious as the tortures of ancient spectacles. This danger prevents individuals from seeking or realizing their potential as genuine human beings.

From the perspective of what might be called a technological virtue—if not technological duty—of efficiency, which emphasizes cost–benefit analysis, convenience, speed, and reliability, these gradual and pernicious effects are difficult to assess. A consequentialist or utilitarian option might consider measurable and substantive results enjoyable in the near future or negative influences on other virtues. Indeed, there are kinds of leisure that mask opportunities for sloth. This is not free time as envisioned by those who endorse human flourishing, but an appeal to vanity that plants the seeds for a slow death of one's humanity and moral character. Worse, humans become less focused on other virtues, such as justice, care, or loyalty.

Pride, Vanity, Control

As noted, Aristotle and his adherents view pride as a positive value, whereas Christian philosophers see it as vicious. Though numerous moral traditions and religions challenge pride, often what they have in mind is hubris or vanity. Hubris involves a kind of arrogance, boasting, or overweening confidence. Vanity involves an undue or unrealistic sense of one's self. Hubris is portrayed in one who fails because of unwarranted sense of self-worth. Vanity is depicted in one who wants to look younger, richer, more powerful, or more knowledgeable than one really is. Boasting, begrudging, and being envious are some of the cravings of vanity. These cravings are often driven by a technological fix, the unshaken belief that a device will always arise—such as diet pills, cosmetic surgery, or transplants—that helps to overcome the effects of aging or unwanted anatomical features. The vain person thus hopes others see a version of oneself that one does not quite believe. That is why medieval moralists pictured the vainglorious person staring into the mirror.

Pride is ambiguously presented in the human trait that desires control. Humans are increasingly adept at withstanding or overcoming natural forces. Protecting themselves from the whims of weather, rechanneling water sources so they can dwell in deserts, or regulating their own predatory or procreative tendencies, they find in science and technology the powers to explain and control the forces of nature. Humans also attempt to extend this control to human domains that were previously resolved in terms of freedom, wisdom, upbringing, or environment. For example, by reclassifying a vivacious or imaginative child as one with attention deficit disorder or disciplinary problems, the child shifts from a subject in need of a certain kind of pedagogy to a candidate for Ritalin.

Determining when technological control should yield to a moral approach is a perennial concern for virtue ethics, particularly for those who support Aristotle's notion that part of a virtuous life is striving for the means between the extremes. With increasing capabilities brought by a variety of technologies, humans still need to strike a balance between turning nature into a managed artifact and resigning themselves to all the challenges and threats nature presents.

The desire for control can nevertheless be another form of vanity. That the world, nature, or other people act without any regard for one's wishes or well-being—indeed, that they seem oblivious to one's very existence—insults a person's own (inflated) sense of self-worth. Symptomatic of this inflation is the ubiquity of cell phones. Owners insist they carry them for possible emergencies. But this claim is betrayed by its omnipresent use. Is the desire to be always and immediately accessible to anyone a symptom of vanity, justified pride, or unending control?

Honesty, Loyalty, Responsibility

Honesty is often described as an intellectual and a moral virtue. The ability to understand things clearly, to know one's own motives and aspirations, and to comprehend circumstances and other humans involves intellectual abilities that precede and accompany moral deliberations and actions. Yet the temptation to deceive others and manipulate the truth also makes honesty a moral issue.

This temptation is especially pronounced in professional ethics. Given their expertise, authority, and the confidence ordinary humans have in them, scientific and technical professionals have a distinctive responsibility to understand and articulate the possible effects of their research. The details of this responsibility can be overshadowed by conflicting loyalties. According to the American philosopher Josiah Royce (1855–1916), loyalty is a virtue essential to the good life. Though its etymology comes from law (*lex* in Latin), Royce views loyalty more in terms of love, purpose, and commitment. Individuals find meaning in their lives when anchored by the object of their loyalty; moreover, this attitude generates respect for the loyalties that give others a purpose.

In professional circles, however, loyalties are not always unified. Among researchers and engineers, for example, there can be obligations to one's employer, the sponsor of a research grant, colleagues and the principles of the discipline, families, and of course the general public. A notable exemplar is the scientist Joseph Rotblat (b. 1908). He was a contributor to the Manhattan Project, in which the United States developed the atomic bomb during World War II. After the defeat of Germany, Rotblat concluded that the project was no longer justified by the danger of Nazi bomb development and left the project. His is a difficult example to follow. Often researchers and even college professors can elucidate the lofty principles that they are supposed to adopt, but when millions of dollars from a grant are at stake, their loyalty to truth can be compromised by loyalty to the research momentum. Some moralists believe the virtues of integrity, self-respect, and honesty can overcome conflicts of loyalty and corruptible compromises. In complex enterprises, however, the notion of personal responsibility can be overshadowed by demands of the workplace or a competitive climate in which one sticks to the proverbial rules and goals of the game rather than challenging the legitimacy of the rules and goals. In such a context, the virtue of responsibility may be torn between courageous criticism and loyal adherence to the team, group, or community.

Justice, Greed, Progress

According to Plato and Aristotle, justice is a virtue that involves harmony or analogy between perfections in citizens and in the state. Modern political philosophy has been skeptical of this view and questioned whether the virtues of individual and society need to reflect one another. In his famous *The Fable of the Bees* (1705) Bernard Mandeville contended that a society can flourish in spite of—and often because of—the vices of its citizens. With appropriate constraints—such as a competitive market or constitutionally separated powers—the natural impulses and selfish appetites of the populace can be harnessed to yield social benefits. As Mandeville poetically noted: "Thus every part was full of vice / Yet the whole mass a paradise." This attitude persists insofar as economists claim that even though gas-guzzling sport-utility vehicles (SUVs) fuel vanity and greed, the Internet indulges lust, and fast food sates gluttony, economic growth and the general welfare are assured.

Virtue ethics theorists nevertheless question such an assessment. For instance, John Casey (1990) argues that justice is first and foremost a disposition within individuals, and defends the traditional view of the truly just person as one who leads a balanced life, recognizing the claims and goods of others. From this perspective, economic greed threatens justice. Though often associated with tycoons, royalty, and celebrities, greed is a temptation in nearly everyone. This is why, A. F. Robertson (2001) writes, stories and concerns about greed cross all ages, and are manifest in everything from children's tales such as "Puss in Boots" to intergenerational squabbles over property and controversies about professionals who appear more devoted to income and prestige than family or service to society. Daniel Callahan (1987) has further argued that with the advances of medical technology, the question needs to be raised whether humans have become greedy for life, attempting to live in excess of a natural life cycle, when they can no longer function or contribute, and at the expense of the well-being of younger generations.

From a virtue perspective, it is essential to ask whether greater affluence spawns generations of more just individuals (and more just societies) or creates more possibilities for vices to thrive (and injustice to grow). How often have parents and grandparents not lamented that increases in the number and glamour of toys among children are not easily correlated with any increases in willingness to share? To what extent does the example of the United States, whose abundance is historically unprecedented, but whose level of government-sponsored foreign aid is not particularly impressive, bear on assessments of political justice? According to Leo Marx (1987), in eighteenth-century America, philosophers such as Benjamin Franklin and Thomas Jefferson saw both personal and social justice as essential measures for assessing national progress. In the nineteenth century, however, the meaning of progress shifted from rights, equality, and personal freedom to material gain and industrial growth, a change that continued across the twentieth century and into the twenty-first.

Scholars such as Dinesh D'Souza contend that many critics miss the central issue on the debates over the meaning or evidence of progress. Instead of seeing wealth as a potential obstacle for the establishment or expansion of justice, D'Souza sees wealth as the key to increasing global well-being. While he acknowledges that enormous increases in scientific knowledge, technological power, and material prosperity characteristic of the 1980s and beyond have carved new gaps between the world's rich and the poor, he points out that in absolute terms the poor and the rich today live much more comfortable lives than did the poor and the rich 500 years ago. Whereas in 1500 only the most wealthy had indoor plumbing and well-heated homes, today even the traditional poor—such as students, seasonally employed, or those too feeble to work—possess cars, reside in secure surroundings, and rely on pricey media such as the Internet, cable TV, and cell phones. Interpreting Thomas Jefferson as a defender of class hierarchies based on a natural aristocracy of individual merits D'Souza believes capitalism has been a gift rather than curse to human life. The desire and search for wealth tames the destructive potential of greed and envy. Guided by a virtue of prosperity, capitalism embodies the prudence to use science and technology that, according to D'Souza, "... has in practice done more to raise the standard of living of the poor than all the government and church programs in history" (D'Souza 2000, p. 240).

This systematic effort towards greater wealth can also be the basis for an essential social virtue—namely, trust. Trust involves a common and cooperative regard for norms or mutual self-interest. In the view of social scientist Francis Fukuyama, this regard is most effective in communities where social capital and ethical values are most prominent. These communities are not, however, rooted in traditional units such as the family. They are instead found in associations that transcend kinship, such as businesses and companies. The benefits of these associations are most notably seen in three advanced technological and capitalist societies: the United States, Germany, and Japan. Here, according to Fukuyama, one

understands the basis of other social virtues and their relation to a life of prosperity.

The estimated benefits of capitalism's virtues are not readily supported by research. Contrary to those who assume a millionaire's summer palace that perilously rests on the ledge of a shore cliff is the spark to global justice, demographers and ecologists find that prosperity's recipients are segmented rather than universal. That is, pockets of great wealth often have negligible or negative influences on the range of human (and non-human) suffering, starvation, or disease. Moreover, excesses of fortune foster a sense of obliviousness to the conditions of others. Such obliviousness—a potential vice insofar as it is interpreted as willful ignorance—turns a blind eye to human threats to the climate. It overlooks human causes of continual increases of pollution, thus jeopardizing the traditional lifestyles of native peoples. It downplays the continued emphasis on consumption of natural resources that generate droughts and scarcities among the world's poorer populations. Obliviousness becomes vicious when it pooh-poohs scientific claims that drastic changes in weather patterns brought on by human pollutants—in the year 2000 each American produced 4.5 pounds of garbage per day—endanger the lives of animals and fish throughout the planet (See, for example, De Souza, Williams, and Meyerson 2003, Post and Forchhammer 2004).

Character, Self, Other

Proponents of virtue ethics emphasize the development of moral character. This development assumes that there is an integral person, a core to an individual that is definitive. Moral pedagogy is directed to this core. The lessons about courage, loyalty, justice, or compassion found in traditional narratives, folktales, sacred texts, honest dialogue, or exemplars help form one's true or genuine identity. These sources reside in other humans, those who spin the narratives, relay the tales and texts, or are admired exemplars. Despite Voltaire's quip that character is so inborn humans could no more change it than wolves could lose their instincts, proponents of virtue ethics generally argue that moral character can be developed, taught, changed, and practiced.

This assumption has three challenges. The first is biological. Paul M. Churchland (1998), for one, proposes that human virtues can be more thoroughly understood from a neurophysiological perspective. Pedagogy and environment obviously have some influence, but they play a secondary role to identifying and treating malfunctioning synapses or chemical imbalances that might prevent the moral agent from successfully cooperating in the well-being of the group. Zoologist and ethnologist Frans de Waal (1996) contends humans have much to learn from animals who exhibit uncanny methods for establishing justice, tolerance, and compassion, and resolving conflicts, without resorting to massacres and war.

Second is a scientific and creative challenge. This challenge stems from the ambiguous human disposition of curiosity. Humans want to know, a desire that seems unquenchable. Curiosity is a likely culprit behind the original sin of Adam and Eve. The French philosopher Jean-Paul Sartre (1905–1980) describes the attempt to know another as a form of capture. At the same time, inventions give humans radical new ways for seeing, hearing, and learning about the world and the universe. Anyone with a stereo can hear Beethoven indefinitely more times than residents of nineteenth-century Europe. The depths of the oceans and dark abysses of the universe are as impossible for human curiosity to resist as exploring their own genetic material or the chemical charges that drive their urge to mate. And under the rubric of transhumanism, researchers are exploring how such fields as genetics and nanotechnology can reinvent the human forms of intelligence, emotion, physiology, and communication. This curiosity does not have to lead to identification of a true self; it can introduce possibilities for creating new selves. With the advent of cyberspace, according to Allucquère Rosanne Stone (1995), humans have found ever-more ways of experimenting and playing with a variety of identities. The face-to-face encounter is not the ideal, just one of many options. It has its own limitations, from which cybercommunities can be valued as liberating rather than alienating.

Third is a philosophical and pedagogical challenge. The idea of a core self is neither self-evident nor coherent. For example, Alphonso Lingis (2004) describes an array of virtuous deeds—of illiterate mothers, gallant youths, mute guerillas, compassionate prisoners, free-spirited nomads—that cannot be attributed to an integral or holistic self. The realization of a virtuous capacity seldom springs from proper habits, one's internal biology, or the narratives of ancestors or cybercommunities. Instead, humans learn about courage, justice, or love as imperatives from contact with others—in their physical or embodied presence. Science and technology should expand rather than displace the possibilities for face-to-face encounters. Such possibilities suspend the insistence on control and self-respect by emphasizing respect for and openness to others, regardless of whether or not they are neighbors, friends, strangers, or aliens. This respect is not grounded in or preceded by under-standing or knowledge of shared values. Instead, writes Lingis, it involves courage rather than caution to trust another insofar as trust dissipates one's own projects and identities. "Trust is a force that can arise and hold on to someone whose motivations are as unknown as those of death.... There is an exhilaration in trusting that builds on itself" (Lingis 2004, p. 12).

Such challenges recognize an ambiguity in the human relation to science and technology. Whether this ambiguity demonstrates progress or regress in ethical life is subject to debate. From a virtue-ethics angle, this debate must include the relative strengths of the virtues and vices, their personal and social significance, whether or how they can be taught, and to what extent science and technology primarily guide humans to realization of their true selves or invite them to devise or create other ways of being.

ALEXANDER E. HOOKE

SEE ALSO *Aristotle and Aristotelianism; Augustine; Buddhist Perspectives; Christian Perspectives; Confucian Perspectives; Jewish Perspectives; Islamic Perspectives; Pascal, Blaise; Plato; Shintō Perspectives; Thomas Aquinas; Thomism.*

BIBLIOGRAPHY

Berrigan, Daniel. (1989). "In the Evening We Will Be Judged by Love." In *Sorrow Built a Bridge: Friendship and AIDS*. Baltimore: Fortkamp Publishing. Memoirs by priest and political activist, who recalls with sorrow and wit his experiences in accompanying those who are dying from fatal diseases.

Bok, Sissela. (1998). *Mayhem: Violence as Public Entertainment*. Reading, MA: Addison-Wesley. Lucid and balanced presentation on potential social and moral hazards emerging from a culture entertained by everyday violence, particularly in the form of news, television dramas, and video games.

Callahan, Daniel. (1987). *Setting Limits*. New York: Simon and Schuster. Thoughtful account of how traditional morality and ongoing changes in medical technology have produced a tragic dilemma for contemporary life.

Casey, John. (1990). *Pagan Virtue: An Essay on Ethics*. Oxford: Clarendon Press. A thorough account of the principal virtues—particularly courage, temperance, justice, and prudence—that preceded Christian moral thought.

Chuang Tzu. (1996). "The Sign of Virtue Complete." In *Chuang Tzu: Basic Writings*, ed. and trans. Burton Watson. New York: Columbia University Press.

Churchland, Paul M. (1998). "Toward a Cognitive Neurobiology of the Moral Virtues." *Topoi* 17(2): 83–96.

Cowan, Ruth Schwartz. (1983). *More Work for Mother: The Ironies of Household Technology from the Open Hearth to the Microwave*. New York: Basic. Fascinating survey of promises and pitfalls encountered among household gadgets and utilities.

Crisp, Roger, and Michael Slote, eds. (1997). *Virtue Ethics*. Oxford: Oxford University Press.

Darling-Smith, Barbara, ed. (1993). *Can Virtue Be Taught?* Notre Dame, IN: University of Notre Dame Press. Scholars discuss the extent to which virtue ethics and specific ethics can be central feature of education.

Darling-Smith, Barbara, ed. (2002). *Courage*. Notre Dame, IN: University of Notre Dame Press.

De Souza, Roger-Mark; Williams, John; and Meyerson, Frederick. (2003). "Critical Links: Population, Health, and the Environment." *Population Bulletin* 58(3): 3–43.

D'Souza, Dinesh. (2000). *The Virtue of Prosperity: Finding Values in an Age of Techno-Affluence*. New York: Free Press. Articulate explanation and defense of the rise of capitalism and its positives influences on moral progress and human welfare.

Etzioni, Amitai, ed. (1995). *New Communitarian Thinking: Persons, Virtues, Institutions, and Communities*. Charlottesville: University Press of Virginia. Collection of essays by academic writers, with central focus on defending and reviving the sense of a traditional community as the basis for citizens leading a virtuous life.

Flanagan, Kieran, and Peter C. Jupp, eds. (2001). *Virtue Ethics and Sociology: Issues of Modernity and Religion*. New York: St. Martin's Press.

Fukuyama, Francis. (1995). *Trust: The Social Virtues and the Creation of Prosperity*. New York: The Free Press. Historical and sociological account of differentiations between high-trust and low trust societies, with contemporary industrial and capitalist countries triumphing over kinship and clan-driven countries in terms of generating greater trust and prosperity.

Gordon, Mary. (1993). "Anger." In *Deadly Sins*, ed. Thomas Pynchon et al. New York: Morrow.

Grayling, A. C. (2002). *Meditations for the Humanist: Ethics for a Secular Age*. Oxford: Oxford University Press.

Hooke, Alexander E., ed. (1999). *Virtuous Persons, Vicious Deeds*. Mountain View, CA: Mayfield Publishing. Presents a wide range of virtues and vices as discussed by philosophers and nonphilosophers, and historical and contemporary writers.

Horowitz, Maryanne Cline. (1998). *Seeds of Virtue and Knowledge*. Princeton, NJ: Princeton University Press. Historical account of two enduring themes in moral upbringing: one, virtues are like seeds, and need to be planted in young children early and nourished constantly; two, virtues and vices are in a battle over the human soul, and virtues must be ever vigilant in defeating the vices.

Hursthouse, Rosalind. (1999). *On Virtue Ethics*. Oxford: Oxford University Press. Systematic reflection on the nature of virtue ethics and discussion of relation of specific virtues to the good life.

Jordan-Smith, Paul. (1985). "Seven (and More) Deadly Sins." *Parabola* 10(4): 34–45.

Kupfer, Joseph. (1999). *Visions of Virtue in Popular Film*. Boulder, CO: Westview Press. Highlights a variety of films that depict virtues in plots and main characters.

Lingis, Alphonso. (2004). *Trust*. Minneapolis: University of Minnesota Press. Elaborate descriptions of various encounters involving humans having the courage and trust to rely on one another regardless of a common language or culture.

Lomborg, Bjørn. (2001). *The Skeptical Environmentalist: Measuring the Real State of the World*. Cambridge, UK: Cambridge University Press.

MacIntrye, Alasdair. (1981). *After Virtue*. Notre Dame, IN: University of Notre Dame Press. 2nd edition, 1984.

MacIntrye, Alasdair. (1988). *Whose Justice? Which Rationality?* Notre Dame, IN: University of Notre Dame Press.

Martin, Mark W., and Roland Schinzinger. (2005). *Ethics in Engineering*, 4th edition. Boston: McGraw-Hill.

Marx, Leo. (1987). "Does Improved Technology Mean Progress?" *Technology Review* 90(1): 33–41, 71.

Meilaender, Gilbert C. (1984). "It Killed the Cat: The Vice of Curiosity." In *The Theory and Practice of Virtue*. Notre Dame, IN: University of Notre Dame Press.

Post, Eric; and Forchhammer, Mads. (2004). "Spatial Synchrony of Local Populations Has Increased in Association with the Recent Northern Hemisphere Climate Trend." *Proceedings of the National Academy of the Sciences* 101(25): 9286.

Richards, Norvin. (1992). *Humility*. Philadelphia: Temple University Press. Extended analysis of a virtue that at first seems quaint, but here proposed to be central to living a good life.

Robertson, A. F. (2001). *Greed: Gut Feelings, Growth, and History*. Cambridge, UK: Polity Press. Anthropological study of historical and contemporary attitudes and practices that indulge and battle the temptations of greed.

Royce, Josiah. (1924). *Loyalty*. New York: Macmillan.

Saeng, Chandra N. (1991). "Insight-Virtue-Morality." In *Buddhist Ethics and Modern Society*, ed. Charles Wei-Hsun Fu and Sandra A. Wawrytko. Westport, CT: Greenwood Press.

Schneewind, J. B. (1998). *The Invention of Autonomy*. Cambridge, UK: Cambridge University Press. Scholarly investigation of the disputations of early modern moral philosophers, and how Immanuel Kant's idea of autonomy attempts to resolve the philosophical difficulties of formulating a secular and rational ethic.

Schwartz, Nancy L. (2004). "'Dreaded and Dared': Courage as a Virtue." *Polity* 36(3): 341–366.

Segal, Lore. (1996). "My Grandfather's Walking Stick; or, The Pink Lie." *Social Research* 63(3): 931–941.

Shaw, Bill. (1997). "A Virtue Ethics Approach to Aldo Leopold's Land Ethic." *Environmental Ethics* 19(1): 53–67.

Stone, Allucquère Rosanne. (1995). *The War of Desire and Technology at the Close of the Mechanical Age*. Cambridge, MA: MIT Press.

Taylor, Gabriele. (1996). "Deadly Vices?" In *How Should One Live?* ed. Roger Crisp. Oxford: Clarendon Press. Concentrated analysis of the meaning of a deadly vice, and whether specific virtues can confront or overcome the vices.

Waal, Frans de. (1996). *Good Natured: The Origins of Right and Wrong in Humans and Other Animals*. Cambridge, MA: Harvard University Press.

VON NEUMANN, JOHN

• • •

One of the most brilliant mathematicians of the twentieth century, John von Neumann (1903–1957) was born in Budapest, Hungary on December 28. He died February 8 in Washington, DC, having created the mathematical foundation for quantum mechanics, one of three competing theories of the physics of the universe, a theory of mathematical economics, the process for creating an implosion atomic bomb, and the theory of automation.

Von Neumann studied at the University of Budapest, the University of Berlin, and the prestigious Technische Hochschule in Zurich. While in Zurich, he worked with two outstanding mathematicians, Hermann (1885–1955) Weyl and George Polya (1887–1985). In 1926, von Neumann was awarded a Ph.D. in mathematics from the University of Budapest and a diploma in chemical engineering from the Zurich University.

Von Neumann lectured at the University of Berlin (1926–1929) and the University of Hamburg (1929–1930). During this later period he also held a Rockefeller fellowship that enabled him to do postdoctoral study with one of the mathematical giants of the time, David Hilbert (1862–1943), at the University at Göttingen. By 1927, von Neumann was acknowledged worldwide as a young mathematical genius, and in 1929, Oswald Veblen (1880–1960) invited him to Princeton University to lecture on quantum theory. In 1930 he became a visiting lecturer at Princeton and in 1931 was appointed a professor. In 1933, the Institute for Advanced Study was formed, and he became one of the first six full time members of the School of Mathematics. Von Neumann held this position for the remainder of his life.

Von Neumann published 130 articles and books during his career, evenly split between pure and applied mathematics, as well as twenty articles and books that made significant contributions to physics.

His 1932 book *Mathematische Grundlagen der Quantenmechanik* created a firm mathematical foundation for quantum mechanics. Quantum theory assumes that energy is not absorbed or radiated continuously, but rather discontinuously and only in multiples of definite

John von Neumann, 1903–1957. The Hungarian-born American mathematician was the originator of the theory of games and an important contributor to computer technology. (© UPI/Corbis Bettmann.)

invisible units called quanta. Quantum mechanics is a physical theory that describes the motion of objects using the principles of quantum theory. In this work, he also introduced a new form of algebra that he named *rings of operators*. In his monograph *Algebras of Operators in Hilbert Space*, von Neumann extended this algebra to group representation as well as to quantum mechanics. This part of mathematics is now called von Neumann algebras.

Von Neumann's 1937 paper "A Model of General Economic Equilibrium" has been repeatedly cited as the greatest paper in mathematical economics ever written. The paper provided a theory of capitol and economic growth based upon a mathematical foundation.

Von Neumann created the entire field of game theory. His 1944 book (written with Oskar Morgenstern), *Theory of Games and Economic Behavior*, not only completed the theory but also introduced several other sets of axioms in other fields of economics.

During the Second World War, von Neumann worked with the scientists and administrators at Los

Alamos on the development of the atomic bomb. His two principal contributions to the Los Alamos project were the introduction of mathematical decision making and refinement of the implosion or plutonium bomb. He did not originate the idea of an implosion, but he did develop the correct density of explosives required to achieve the correct implosion.

Von Neumann's development of MANIC—an acronym for Mathematical Analyzer, Numerical Integrator, and Computer—enabled the United States to produce and test the world's first hydrogen bomb in 1952. Von Neumann spent much of his later life working in automata theory, a field that attempts to understand multiple automation applications working together to form a process or perform a task. He was also an early advocate of stored programs within a computer. His computer architecture is common to all personal computers and has come to be known as von Neumann architecture.

HENRY H. WALBESSER

SEE ALSO *Decision Theory; Turing, Alan; Wiener, Norbert.*

BIBLIOGRAPHY

Aspray, William. (1990). *John von Neumann and the Origins of Modern Computing.* Cambridge, MA: MIT Press. Describes the contributions of von Neumann to the development of von Neumann computer architecture.

Heims, Steve J. (1980). *John von Neumann and Norbert Wiener: From Mathematics to the Technologies of Life and Death.* Cambridge, MA: MIT Press. A dual biography, this is a work that traces the academic contributions of von Neumann and Norbert Wiener to the numerous areas of study each influenced as well as their relationship to one another.

Macrae, Norman. (1992). *John von Neumann.* New York: Pantheon. A general bibliography of the life, person, and contributions of the man.

von Neumann, John. (1937). "A Model of General Economic Equilibrium." In *Ergebnisse eines mathematischen Kolloquiums* [Reports of a mathematical colloquium], ed K. Menger. Notre Dame, IN: University Press of Notre Dame. Foretells what is to become the theory of games.

von Neumann, John, and Oskar Morgenstern. (1944). *Theory of Games and Economic Behavior.* Princeton, NJ: Princeton University Press. Offers a theory to explain certain classes of economic behavior.

von Neumann, John. (1958). *The Computer and the Brain.* New Haven, CT: Yale University Press. A work that stimulated the development of artificial intelligence as a field of study in computer science.

von Neumann, John. (1966). *Theory of Self-Reproducing Automata.* Urbana: University of Illinois Press. Published posthumously. A theory that explains the underlying structures needed to create automated systems capable of reproducing themselves.

VON WRIGHT, GEORG HENRIK

• • •

Philosopher and inventor of deontic logic, Georg Henrik von Wright (1916–2003), who was born in Helsinki, Finland, on June 14, was also a cultural critic of techno-scientific progress. In philosophy, von Wright is best known as Ludwig Wittgenstein's successor in the chair of philosophy at Cambridge (1948–1951), and for participating in the publishing of Wittgenstein's papers posthumously. Von Wright was also a major contributor to the rebirth of modal logic in 1950s. Among his most important academic works are *Norm and Action* (1963), *Varieties of Goodness* (1963), and *Explanation and Understanding* (1971). The last had a distinctive role in efforts to bridge the gap between the Anglo-American and continental European traditions in philosophy.

Apart from his work within academic philosophy, von Wright was an important public intellectual in Finland and Scandinavia. Throughout his career he wrote philosophical essays in which he dealt extensively with the questions of the effects of science and technology on human life. He presented his cultural analysis in *Vetenskapet och förnuftet. Ett försök till orienteering* (Science and reason: An attempt at orientation), published in 1986.

Von Wright's cultural philosophy focuses on the critical situation of modern Western civilization, seen as threatening the whole globe. Many of the most serious problems of the modern world can be understood as direct consequences of techno-scientific advance. Von Wright wrote about the ecological crisis, the existence of weapons of mass destruction capable of devastating all human life, the ethical vacuum that has followed secularization and collapse of traditional value systems, and the expansion of instrumental reason in all areas of human life.

Von Wright sought the origins of these problems in the history of ideas. He located the roots of modern science and technology in the objectification of nature, the inclination toward mechanistic and deterministic causal explanations, and reductionism. The manipulative ethos of modern natural science is explicit in the writings of the pioneering philosophers of science, such as Francis Bacon (1561–1626) and René Descartes (1596–1650). It is clear that this conceptual framework has produced impressive results. However, von Wright asserted that the cost has been high.

Furthermore, von Wright noted how science is becoming an ever more important force for production. This development is problematic for science itself. The crucial question concerns what will happen to truth as the goal of science, if science becomes dependent on demands for profit, and if new discoveries are kept secret for commercial and military purposes. Von Wright also doubted the ability of modern science to provide a culturally understandable and meaningful worldview.

Although von Wright arrived at his conclusions independently, his analysis of techno-scientific progress has predecessors. Cultural critics such as Oswald Spengler, Lewis Mumford, Jacques Ellul, and the thinkers of Frankfurt School developed similar themes. Von Wright's achievement is the sobriety and transparency of his analysis. His background in analytical philosophy makes his argument especially interesting, because this tradition has usually been very optimistic concerning modern natural science.

TOPI HEIKKERÖ

FSEE ALSO Scandinavian and Nordic Perspectives; Wittgenstein, Ludwig.

BIBLIOGRAPHY

Schilpp, Paul Arthur, and Lewis Edwin Hahn, eds. (1989). *The Library of Living Philosophers*, Vol. 19: *The Philosophy of Georg Henrik von Wright*. La Salle, IL: Open Court.

von Wright, Georg Henrik. (1963). *The Varieties of Goodness*. London: Routledge & Kegan Paul.

von Wright, Georg Henrik. (1986). *Vetenskapen och förnuftet. Ett försök till orienteering* [Science and reason: An attempt at orientation]. Borgå, Sweden: Söderström.

von Wright, Georg Henrik. (1993). *The Tree of Knowledge and Other Essays*. Leiden: Brill.

W

WASTE

• • •

Advanced industrial societies produce enormous quantities of waste. People know it when they see it, yet waste does not admit of any strictly physical definition. Moreover what is at one point waste can at another point easily be resource. Examples include archaeological digs in archaic trash dumps, artistic creations of *objets trouvés* co-generation plants, and recycling centers.

However waste is defined and measured, it is safe to say that never before have humans produced and thrown away as much as they do in the early twenty-first century. Mass production through industrialization, extensive packaging (to facilitate both shipping and sales), and rapid obsolescence (whether planned or as an accidental effect of technological progress) in a free market economy, driving the compulsion to make things and consume them, have formed a world in which artifacts are produced, consumed, and discarded to an historically unprecedented extent.

Indeed there is a tendency for the lifetime of *durable* products to be shortened to that of *consumables*, and for non-renewable natural resource stocks to be consumed in the same way as renewable production flows, which some critics ascribe to the inability of *free market forces* to distinguish between them. Given the size of the phenomenon and its potential damaging effects on public health, the environment, and future generations, waste is one of the fundamental problems facing the techno-scientific and consumer society.

Regulations

The rapid growth, diversification, and toxicity of waste production have been accompanied, though not matched, by legislation, the development of regulatory institutions, and new methods of treatment and control. Waste has become a priority of environmental risk politics for national and international authorities (for example, the European Union [EU], U.S. Environmental Protection Agency [EPA], Organization for Economic Co-Operation and Development [OECD], World Health Organization [WHO], United Nations Environment Programme [UNEP], and so on), and one of the crucial concerns of social and ecological movements (such as Greenpeace).

The roots of this politicization go back to the nineteenth century and the earliest public health reforms spearheaded by medical scientists and advocates of public hygiene (Melosi 1981). This process is related to growing feelings of repugnance and the formalizing of new rules of conduct, discipline and self-control. Waste, which was increasing as the population of urban areas grew, was synonymous with chaos, disorder, and contagion, and had to be put out of sight. The concept of *matter out of place*, used by Mary Douglas (1966) in an anthropological study of dirt and pollution, offers a vision of waste as something that intrudes on ordered arrangements where everything has its rightful place.

Another impulse for the politicization of waste came in the 1970s with the emergence of ecological movements and environmental ethics. Rachel Carson's pioneering book, *Silent Spring* (1962), was a decisive influence in these developments. In it she denounced the harmful effects on human and animal health of the massive application of DDT and other chemical pesticides in agriculture. Consciousness of ecological frailty and feelings of ambiguity in relation to the unexpected consequences of technological advances were later reinforced by environmental accidents in the fields of technology and energy (for example, Times Beach and Love Canal in the United States, and the Seveso dioxin-contaminated waste drums in Europe).

Waste Policy

Waste policy is formed as part of a wider strategy, either to decrease pollution and protect the environment, or to bring about technological and industrial change and innovation. In each of these aims, there is remarkable ambivalence regarding the technological implications.

On the one hand, technology itself is responsible for much waste production and global pollution. Each technical development, despite its many benefits, has brought an increase in the amounts and types of waste. After the non-degradable waste produced by the steel and iron industries of the early industrial era, plastic, chemical, and pharmaceutical products have given rise to even more waste products that are more toxic and difficult to treat, control, and dispose of. On the other hand, technology is also absolutely necessary for waste prevention and the disposal of pollutants. All the principles of current international waste management strategy—minimization, recycling, reuse, and improving final disposal and monitoring—depend, in general, on techno-scientific solutions. For example, the ability to recycle is built into some products at the design stage; and some technological innovations are created specifically to improve the treatment or recovery of waste.

So-called ecological or green strategies are made difficult by the many sources of waste—domestic, commercial, industrial, medical, agricultural, construction, and so on—and its physical and chemical nature, comprising (among other materials) metals, plastics, glass, paper, and vegetable matter, often in complex and hard-to-separate combinations as in batteries, cartons, and cars. When waste cannot be recycled or reused, it is usually burnt ("incinerated") at high temperatures or dumped into landfill sites. However each of these methods may cause air, water, and soil pollution, and may have harmful effects on human, plant, and animal health.

Hazardous and Radioactive Waste

Hazardous waste, and especially radioactive waste, requires extra care in its treatment and disposal. Because of their potential harmful effects—and the political, social, and ethical questions they raise—hazardous and radioactive wastes are generally the most studied. Most international policies and treaties deal with waste of these types, whose environmental problems are global

in scope and indifferent to national, generational, or class boundaries. Yet despite similarities, entirely separate legislation governs the two types, and they have different regulatory institutions and interest groups.

The contents of hazardous wastes may cause serious damage to human health and/or the environment, when improperly treated, stored, transported or disposed of. There are differing definitions and systems of classification in different countries and even between states and regions of the same country. It is symptomatic that there is little agreement on the definition of *hazardous*, on who is responsible for this definition, and on what substances are considered as hazardous waste.

According to Brian Wynne (1987), a sociologist who has addressed environmental issues and in particular the problem of waste, the lack of consensus between countries over hazardous waste is the main difficulty for international regulation. Furthermore this type of waste is usually taken to be not dispersed and diluted in the environment, but *packaged* for further treatment before eventual destruction, containment, and/or dispersal, and is thus more liable to have concentrated and harmful effects. In their life cycles these wastes not only change in physical and chemical terms, but also pass through the control of various human agents. A complex *behavioral-technical system* therefore underpins hazardous waste, bringing together natural processes and human interaction in an unpredictable and imprecise way. This happens all over an industrial network, whose entire infrastructure—for collection, transport, storage, treatment, and disposal of waste—requires extensive regulation.

In general this type of waste is identified in three ways: (a) by reason of certain properties, detected by test procedures such as flammability (may cause or prolong fire), corrosiveness (may destroy live tissue that comes into contact with it), toxicity (inhaling, swallowing or penetration through the skin may involve serious risk or even death), etc.; (b) by the presence of toxic chemical elements or abnormal concentrations of these, also detectable by tests; and (c) by listings of specific categories of waste identified as being hazardous and for which no tests are necessary. Radioactive waste contains substances which emit ionizing radiation. Proper management and safe and environmentally sustainable storage are vital but complex tasks. Nuclear waste, depending on the source, its levels of radioactivity, longevity and hazard, may be classified in two broad categories: "high-level" (from the reprocessing of spent nuclear fuel) and "low-level" (generally in the form of radioactively contaminated industrial or research waste). Other categories are transuranic radioactive waste and uranium mill tailings. One may identify two key problems with this classification: first, "low-level" waste contains some elements that are more radioactive than some of those contained in "high-level" waste; second, the public tends to perceive all radioactive waste as being "high-level."

Regardless of whether the risks are great or small, citizens typically fear toxic products and their carcinogenic effects in general, and nuclear radiation in particular. Despite accusations of irrational "chemophobia," the concerns of ordinary people are based on the impact of accidents such as those at Three Mile Island, Chernobyl, and Bhopal. In addition to these accidents, and compounding the potential threat of chemical products, each year several hundred synthetic chemical products are brought to market without being subjected to any prior tests. This underlies the phenomenon of "bioaccumulation," whereby all substances that are resistant to degradation, whether tested or not, gradually build up in successive stages of the food chain.

Ethical Issues

The regulation of waste raises four key ethical and political issues. The first derives from the need for integrated waste management involving a range of actors on different levels. In addition to international responsibility—which is necessary, for example, to control exports of waste and to avoid illegal dumping in the oceans—the following are also key elements:

(a) the model of economic development, for example one in which recycling and waste reduction activities are encouraged, leading to the idea of sustainable development;

(b) scientific research that can salvage traditional technologies that are less harmful to the environment, invent alternative technologies, and develop products with an ecologically friendly design;

(c) attitudes and incentives in business, where new designs and technologies can be used to minimize the environmental impact of a product;

(d) the civic consciousness of citizens, who may demand environmentally friendlier ("greener") products, less packaging, and access to reliable information through, for example, labeling (such as the "eco-label" – a flower logo in Europe).

A second issue concerns the ethical dilemmas raised by the risks associated with waste technologies. Given the rational impossibility of a *zero-risk* society, the debate about the threshold of acceptable risk and how it ought to be distributed generally swings between utilitarian and egalitarian ethical perspectives. Problems arise because no standard threshold provides all citizens with equal protection from harm. Moreover that threshold, which is an average annual probability of fatality linked to some hazard, may not protect the basic rights of all individuals with their specific characteristics and needs.

For Kristin Shrader-Frechette (1991), a leading investigator of the ethical dilemmas associated with nuclear waste, it is essential to obtain the free and informed consent of those who are exposed or put at risk. Those who impose societal risks on others should compensate them in order to obtain their consent. Informed and freely-given consent and compensation are guidelines which are appropriate for avoiding popular hostility. This arises frequently in discussions on where to site waste treatment facilities, reflecting syndromes known as NIMBY (*not-in-my-backyard*), NIABY (*not-in-anybody's-backyard*), or LULU (*locally-undesirable-land-use*).

A third issue is the link to the methodology used in technological assessment and analysis of environmental impact. A socially acceptable study of these problems cannot be reduced to simple cost-benefit analysis based on calculations of mathematical probabilities while ignoring moral values such as equality, equity, social justice, and common well-being.

Apart from examining the magnitude, risks, and benefits, any assessment should also weigh the moral acceptability of technology, because the issues involved cannot be reduced to factual terms. To fail to recognize this is to commit a version of the *naturalistic fallacy* (Moore 1903) by deducing and justifying ethical conclusions from technical considerations (Shrader-Frechette 1980). This error is even more serious when found in studies used to support policy decisions relating to matters of public interest.

A final ethical consideration is that a significant number of waste-related activities, from collection to recycling, are very profitable. Indeed wastes are a vital part of the capitalist economy: consumerism and an active throwaway mentality encourage constant production and fuel ever-expanding human needs.

However the fact that an entirely new industry has developed, on a for-profit basis, to deal with the waste problem, gives rise to a conflict between public and private interests. The involvement of private groups in matters of public interest may create conflict, even though a strong public sector can encounter problems with excessive bureaucracy and consequent distortions. To avoid exacerbating such conflicts, citizens are often given access to full information on each case and/or committees of experts are appointed to give scientific opinions on the regulation of waste management.

Modern society strives for a balance between economic development and environmental protection, finding a threshold that reconciles the inevitable production of waste with a commitment to ecological sustainability. The depletion of natural resources that may not be renewable, and the (often related) by-production of hazardous waste, is an increasingly important focus of long-running debates regarding conflict between state regulation and market forces, between individual action and collective consequences, and between the practical and the ethical impact of new or newly mass-consumed technologies.

HELENA MATEUS JERÓNIMO

SEE ALSO Carson, Rachel; Consumerism; Ecology; Environmental Ethics; Environmental Impact Assessment; Environmental Regulatory Agencies; Hazards; Nuclear Waste; Pollution; Risk; Sustainability and Sustainable Development; United Nations Environmental Program.

BIBLIOGRAPHY

Carson, Rachel. (1962). Silent Spring. Boston: Houghton Mifflin.

Douglas, Mary. (1966). Purity and Danger. London: Routledge & K. Paul.

Dowling, Michael, and Joanne Linnerooth. (1987). "The Listing and Classifying of Hazardous Waste." In Risk Management and Hazardous Waste: Implementation and the Dialectics of Credibility, ed. Brian Wynne. Berlin and New York: Springer-Verlag.

Melosi, Martin V. (1981). Garbage in the Cities: Refuse, Reform, and the Environment, 1880–1980. College Station: Texas A & M University Press.

Moore, George Edward. (1903). Principia Ethica. Cambridge, UK: Cambridge University Press.

Shrader-Frechette, Kristin. (1980). Nuclear Power and the Public Policy: The Social and Ethical Problems of Fission Technology. Boston: D. Reidel Publishing Company.

Shrader-Frechette, Kristin S. (1991). Risk and Rationality: Philosophical Foundations for Populist Reforms. Berkeley: University of California Press.

Strasser, Susan. (1999). Waste and Want: A Social History of Trash. New York: Metropolitan Books.

Wynne, Brian, ed. (1987). Risk Management and Hazardous Waste: Implementation and the Dialectics of Credibility. Berlin and New York: Springer-Verlag.

WATER

• • •

Water is the liquid of life and is crucial to every type of organism, from simple bacteria to megafauna, as well as to many of the physical processes that shape the planet, as in the weathering of mountains and valleys. For life in all forms, water is more important than even oxygen, because there exist anaerobic bacteria that can live without air but no anhydroxic bacteria that can exist without water. When astrobiologists seek to determine the possibility of life on other planets, their first question concerns the presence of water. Throughout human history, however, water has had as much a symbolic as biological significance, and human beings have adapted to environments both abundant and scarce in water, through different technological, ethical, and political engagements. Water is so rich in metaphor that it cannot be reduced to merely H_2O, nor to a fluid circulated in pipes, metered, and then distributed by authorities. The duality of meaning that water embodies includes the fact that it can be both deep and shallow, life-giving and destructive, a blessing and a curse, and something that cleans the surface and also purifies the inner soul.

Water in Science

As a chemical compound water is composed of one atom of oxygen and two of hydrogen. Because acids are characterized by hydrogen ions (H^+) and bases by hydroxide ions (OH^-); water (H_2O) may be described as neither acidic nor basic, rather equally both:

$$H_2O \rightarrow H^+ + OH^-$$

The structure of water is:

$$H : \ddot{O} : H$$

Oxygen is attached to two hydrogen atoms with two covalent bonds leaving two nonbonding pairs of electrons. Hydrogen bonding is particularly important in biochemical systems, because biochemical molecules contain many oxygen and nitrogen atoms that participate in hydrogen bonding. Hydrogen bonds between water molecules are responsible for the interesting physical properties of water that made it the solvent of life. Together with the extended temperature range between its solid (ice) and gaseous (steam) states, that makes liquid water able to serve as the foundation for those extremely complex carbon formations that constitute living organisms.

When present at a depth of at least two meters (six feet), pure water is a pale blue, odorless, tasteless, and transparent liquid. Other observed colors are due to various impurities, nonliving and living. It is mostly "blue water" that flows in rivers and into lakes and aquifers. "Green water" refers to the precipitation that is directly used by nonirrigated agriculture, pasture, and forests, and to evapotranspiration.

In its liquid and solid forms, water covers 71 percent of the surface area of the globe. Humanity's anthropocentric worldview explains why this mostly "blue" planet was (mis)labeled Earth. Of all the water on the planet, only 3 percent is freshwater, a figure that includes glacial ice and other hard-to-reach water sources. Of this, only 0.003 percent of the surface and subsurface water is usable by humans.

Hydrology is the science of the properties, distribution, and circulation of surface and subsurface water. In hydrologic terms, water that collects in rivers, lakes, or reservoirs is called surface water. That which seeps into the shallow or deeper layers of Earth is called aquifer. The gaseous, solid, or liquid phases of water affect both the element's chemistry such as its bonding and its physics such as its density. Water is an excellent solvent, and hence it has many constituents that are dissolved or suspended in it. These facilitate chemical interactions, which aid complex metabolisms. This explains why water is critical for all life-forms.

Pure water can be obtained through painstakingly and costly mechanical processes. Water is then the most benign of all chemical compounds known to humans. Water that contains dissolved carbonates such as calcium and magnesium is known as hard water. People notice this because it suppresses the formation of lather with soap, and when boiled, it leaves a "lime scale" that is seen in cookware. Soft water is free of such carbonates.

Water circulates from the ocean and surface of Earth to the atmosphere and then gravitates back in various forms including snow, rain, and fog. Human activities affect this hydrological cycle, most prominently through the building of physical barriers such as dams and through modifications of watersheds. Most water resources are renewable except for fossil (or connate) water that is laid down in sedimentary rocks and sealed off by overlying beds. Nevertheless, human contamination of groundwater stock, and alterations of watersheds (or, in British parlance, "drainage basins") through, for example, deforestation or paving over hydrologically critical areas can reduce aquifer recharge, alter flow characteristics, and, in severe cases, deplete a formerly renewable resource.

Many large watersheds lack time series data, and scientists in riparian states (those who study watersheds) often use different methodologies for collecting their data, which makes data sharing among water basin states ineffective and integrated management of the river system difficult.

Technologies of Water

Natural water is managed through a system of wells, dams, artificial reservoirs, conveyance systems, and human-made ponds. Humans withdraw untreated water from surface sources and pump it from aquifers. The water is treated and then pumped into carefully laid-out distribution systems such as water mains, which are connected to underground networks and sometimes to (elevated) storage facilities.

The geographical setting of the source of water, water treatment, its distribution, return flow collection, and return flow treatment—each requires a unique technological approach in order that people can access and use the resource. Economic considerations and regulations regarding human health and environmental protection also affect the choice of technology.

Easy-to-tap water sources were the first to be developed. Growing water needs require new and innovative technologies because water is increasingly extracted from deeper wells and piped in from further and further locations; furthermore, in a growing number of countries that have exhausted their supplies, freshwater is obtained by removing the salts and other contaminants from sea water (desalination). Growing water scarcity is inducing the development of water-efficient technologies. Given that agriculture is by far the largest consumer of water, drip irrigation techniques offer huge water savings, especially when compared to sprinkler irrigation or the traditional, but low-cost, flood irrigation.

Historically, the water wheel, a wheel with paddles or buckets attached to the outside, was first used to lift water from a river onto irrigation channels. Eventually, a water-powered wheel was developed and used in the Middle Ages for extracting power from a flow of water. Its applications included milling flour and machining and pounding linen for use in paper. Similarly, the steam engine contributed to Europe's economic development especially during the Industrial Revolution. This engine coverts the potential energy of the pressure in steam to mechanical work.

Water systems have been targets in warfare, and the threat of terrorism is requiring new technologies and strategies to protect water supply systems, especially in large metropolitan centers, and in countries where in which the majority of the population depends on a few desalination plants. Efforts are afoot to develop remote but real-time water-quality monitoring systems that not only encompass the traditional water-quality parameters but also can detect currently unmonitored biological agents that could threaten freshwater supplies, such as bacteria, viruses, and protozoa.

Ethics of Water

Water is central to the health of the ecosystem, central to the beliefs and customs of many religious communities, and vital to the maintenance of the economic well-being of modern and traditional lifestyles. Allocating water across competing users must thus be tempered by extensive stakeholder participation and weighed against any adverse social or ecological impacts that a solely economic approach may cause.

The increasing demand for freshwater is related to population growth, trends toward more protein-based diets, and overall improvements in the quality of life. Countries typically tap their lowest cost and most reliable sources of water first. As these sources become fully utilized, the development of new sources carries with it heavier financial costs and environmental consequences.

The equitable allocation and sustainable use of water require good governance that is rooted in policies that are scientifically, culturally, and economically sound; in institutional structures that are community friendly and invite public participation; and in decision makers who are competent and fair, and have the support of the political forces. It also requires employing modern technologies that have been adopted in many Western countries but are beyond the reach of poorer ones.

In 2002 the United Nations Committee on Economic, Social and Cultural Rights declared water a human right. It stated that the human right to water entitles everyone to sufficient, safe, acceptable, physically accessible, and affordable water for personal and domestic uses. An adequate amount of safe water is necessary to prevent death from dehydration, reduce the risk of water-related disease, and provide for consumption, cooking, and personal and domestic hygienic requirements. The signatories to the International Covenant on Economic, Social and Cultural Rights are required to progressively ensure access to clean water, equitably and without favoritism.

Politics of Water

Negotiating water-sharing agreements on for international rivers tends to be complex. Allocation agreements among competing users often involve a combination of geoclimatic factors as well as legal, historical, technological, demographic, political, and ethical considerations. In the case of international rivers, upstream states are generally seen as having leverage in influencing the allocation process simply because they control the "water tap."

Water allocation arguments include the largely discredited view that a country has an absolute sovereignty over resources that originate inside its political boundaries. Prior appropriation agreements state that the earliest users of water have rights to it. This convention is widely used in the American Southwest and by a few other countries, such as Iraq in connection with its share of the waters of the Tigris and Euphrates rivers. Before a balanced allocation formula can be reached, several factors need to be carefully considered and fairly weighted for every riparian country. These factors include a country's contribution to the total flow of the river, current and projected population size, area of arable land, and the extent to which the health of the national economy is dependent on water. A sustainable and ethical management strategy must also consider and protect the needs of aquatic life, upstream habitats (especially forests), wetlands, and floodplains, as well as the water needs of future generations. International agreements make the integrated (and sustainable) management of river systems easier.

Acute and protracted water scarcity is likely to be a source of violent conflict especially in countries where the agricultural sector is a vital contributor to national economic health. This danger has helped place water scarcity high on the world's political agenda. Globally, the overwhelming majority of water is consumed by the agricultural sector. There has been a gradual and continuing shift away from supply management to demand management of water, whereby people are asked to make the most out of their existing resources. Communities try to maximize their crop yields per unit of water (more "crop per drop") and their financial returns by planting suitable, lucrative crops. Similarly, a few arid and semiarid countries are gradually shifting away from water thirsty crops such as citrus to ones that are more suited to their own climatic and physical environments such as wheat, lentils, and chickpeas.

Immense amounts of water are wasted through leakage from antiquated urban supply networks and unsustainable irrigation strategies. Existing technologies such as the efficient, water-saving drip irrigation technique and microsprinklers have been around for decades but used on only around 1 percent of all irrigated lands. Even relatively small improvements in efficiency through the transfer of appropriate irrigation technologies and the implementation of various policy incentives and/or disincentives will result in substantial water savings.

One proposed strategy would involve governments gradually charging farmers the real and full cost of water. Progressively higher charges per unit of water consumed would induce most users to think before they turn on the water. Water quality can be protected by raising people's awareness about the adverse effects of pollution, making it prohibitively expensive to pollute, and by building sanitation infrastructures and wastewater treatment plants. This will minimize pollution levels and provide the public with recycled water to be used in nonhuman ways that do not directly affect food production, such as car washes and irrigation of lawns.

When national sources are exhausted, countries seek alternatives such as importing water, usually from nearby countries. Globalization and the opening of international markets are likely to encourage large-scale trading of freshwater across international borders. This is a controversial because of the likely environmental impacts and the political implications that a dependency on imported water may create.

Desalination, however, is an increasingly promising water-augmentation method. This process entails removing soluble salts from water to make it suitable for various human uses. Technological advances have been steadily decreasing its unit price, which is inducing more countries and facilities to use it. A growing number of countries have been increasingly adopting desalination technologies to augment their national or area-specific freshwater supplies.

HUSSEIN A. AMERY

SEE ALSO *Acid Mine Drainage; Air; Dams; Deforestation and Desertification; Earth; Environmental Ethics; Fire.*

BIBLIOGRAPHY

Al-Jamal, M. S.; S. Ball; and T. W. Sammis. (2001). "Comparison of Sprinkler, Trickle, and Furrow Irrigation Efficiencies for Onion Production." *Agricultural Water Management* 46(3): 253–266.

Amery, Hussein A., and Karen B. Wiley. (2003). "Resolution of International Water Conflicts: From the U.S. Southwest to the Euphrates River Basin States." *International Journal of Global Environmental Issues* 3(2): 226–239.

Amery, Hussein A., and Aaron T. Wolf, eds. (2000). *Water in the Middle East: A Geography of Peace.* Austin: University of Texas Press.

Birkett, James D. (1984). "A Brief Illustrated History of Desalination: From the Bible to 1940." *Desalination* 50: 17–52.

Faruqui, Naser I.; Asit K. Biswas; and Murad J. Bino, eds. (2001). *Water Management in Islam.* Ottawa: International Development and Research Centre; Tokyo: United Nations University Press. A unique anthology that outlines the potential role culture and religion could play in the protection and efficient management of water resources.

Gleick, Peter H. (2004). *The World's Water, 2004–2005.* Washington, DC: Island Press. A biannual report that contains impressive current data on various aspects of water resources from around the globe, and a few state of the art articles.

Illich, Ivan. (1985). *H2O and the Waters of Forgetfulness.* Dallas: Dallas Institute of Humanities and Culture.

Olivera, Marcela, and Jorge Viaña. (2003). "Winning the Water War." *Human Rights Dialogue* 2(9). Also available from http://www.cceia.org/viewMedia.php/prmID/951

Postel, Sandra. (1999). *Pillar of Sand: Can the Irrigation Miracle Last?* New York: Norton. The book examines the role irrigation played in the rise and fall of ancient civilizations. It draws on this to examine how the current growing reliance on irrigation, the mounting water scarcity, salinization of soils, and the rising tensions between riparian states affects countries and could impact them in the twenty-first century.

Stumm, Werner, and James J. Morgan. (1996). *Aquatic Chemistry*, 3rd edition. New York: Wiley.

Swan, C. H. (1978). "Middle East: Canals and Irrigation Problems." *Quarterly Journal of Engineering Geology* 11(1): 75–78.

United Nations. World Water Assessment Programme. (2003). *Water for People, Water for Life.* New York: UNESCO Publishing.

U.S. National Research Council, Committee on Sustainable Water Supplies for the Middle East; Israel Academy of Sciences and Humanities; Palestine Academy for Science and Technology; and Royal Scientific Society, Jordan. (1999). *Water for the Future: The West Bank and Gaza Strip, Israel, and Jordan.* Washington, DC: National Academy Press. Examines the need to enhance and protect sustainable water supplies in the West Bank and Gaza Strip, Israel and Jordan. It proposes criteria needed for developing sustainable water resources while maintaining environmental support systems. The collaborative study brought scientists and engineers from the U.S. National Academy of Research and from other countries in the region, and it marked unprecedented cooperation among Israel, Jordan and the Palestinian organizations.

World Commission on Dams. (2000). *Dams and Development: A New Framework for Decision-Making.* London: Earthscan.

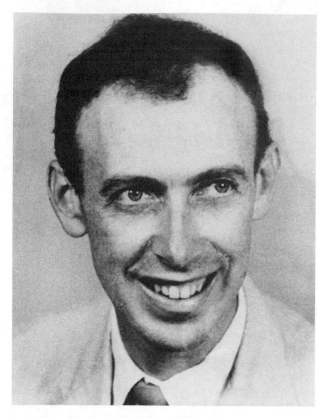

James Watson, b. 1928. The American biologist was a discoverer of the double-helical structure of the deoxyribonucleic acid molecule. (*The Library of Congress.*)

WATSON, JAMES

• • •

Co-discoverer of the molecular structure of DNA, James Watson (b. 1928) was born in Chicago on April 6, and became a controversial figure in debates about the social and ethical implications of genetic research. Watson received his Ph.D. in zoology from Indiana University in 1950. His partnership with Maurice Wilkins and Francis Crick led to the 1953 discovery of the complementary double-helix configuration of the DNA molecule, for which the three researchers shared the 1962 Nobel Prize in physiology and medicine. In 1968 Watson was named director and in 1994 president of Cold Spring Harbor Laboratory, which he shaped into a leading center of research on the genetic basis of cancer. In 1988 Watson was appointed Associate Director for Human Genome Research at the National Institutes of Health (NIH), where he initiated the Ethical, Legal, and Social Implications (ELSI) program as part of the Human Genome Project (HGP).

Although Watson continued his research, including important work on the function of messenger RNA (mRNA), his career shifted toward administration and the promotion of science (McElheny 2003). In these capacities, he confronted some of the political and ethical dilemmas born of his co-discovery of "the key to life." The subsequent revolution in genetics raised questions about the proper use of this new knowledge. Indeed, Watson on occasion made controversial and sometimes-contradictory statements on several of these issues, including recombinant DNA (rDNA) research, reproductive rights, and germline genetic therapy (see Watson 2000).

During congressional testimony in 1971, Watson expressed strong concerns about genetic engineering and reproductive technologies, and in the mid-1970s he played a role in establishing a moratorium on certain kinds of rDNA research. However, he later came to regret this position and even called critics of the research "a bizarre collection of kooks, sad incompetents, and down-right shits" (Beckwith 2003, p. 357). Watson defended a cornucopian attitude about the promises of genetic technologies to solve societal problems and dismissed public fears as irrational, Luddite paranoia.

In this regard, two of his strongest convictions about the use of genetic technologies were his libertarian ideology and a desire to engineer the human genome. First, he argued that society should not impose rules on individuals concerning their use of genetic knowledge. People should be allowed to make those decisions in private, especially women who are faced with difficult reproductive choices. Second, he maintained that germline gene therapy, despite its similarity to morally reproachable governmental eugenics programs, deserves serious consideration as a personal option because of the potential for human betterment. In other words, "If we could make better human beings by knowing how to add genes, why shouldn't we?" (Wheeler 2003). For Watson, the genome is a cruel limitation on the vast possibilities that scientists could create by manipulating human DNA.

Watson's most lasting legacy in the realm of the politics of science is his creation of the ELSI in the HGP carried out by the National Center for Human Genome Research Institute (NCHGI). In an "unprecedented experiment in American science policy," Watson unilaterally set aside 3 to 5 percent of the HGP budget to support ELSI studies of new advances in genetics with the goals of identifying and defining major issues and developing initial policy options (Juengst 1996).

It is difficult to decipher Watson's intentions in creating the ELSI program. He was quoted as saying,

"I wanted a group that would talk and talk and never get anything done" (Andrews 1999, p. 206). Yet he also claimed, "Doing the Genome Project in the real world means thinking about [social impacts] from the start, so that science and society can pull together to optimize the benefits of this new knowledge for human welfare and opportunity" (Watson and Juengst 1992, p. xvi).

Most likely, Watson viewed the ELSI program as a form of enlightened scientific self-interest. It could create a social environment conducive to genetics research by aiding in the development of policies that prevent people from being harmed by the use of genetic information and technologies. In Watson's view, genetics research produces inherently valuable knowledge. As Juengst explains, "The question that the ELSI program addresses is the virtuous genome scientist's professional ethical question: 'What should I know in order to conduct my (otherwise valuable) work in a socially responsible way?'" (1996, p. 68). The societal buffer that the program creates may explain why Watson referred to the creation of the ELSI program as one of his top accomplishments. Although Watson created it on a whim, the ELSI program has had a lasting impact on the practice of science as similar programs are becoming common aspects of scientific research.

ADAM BRIGGLE

SEE ALSO *Genetic Research and Technology; Human Genome Organization.*

BIBLIOGRAPHY

Andrews, Lori B. (1999). *The Clone Age: Adventures in the New World of Reproductive Technology.* New York: Henry Holt. Memoir touches on several of the most pressing issues in reproductive technology.

Beckwith, Jon. (2003). "Double Take on the Double Helix." *American Scientist* 91: 354–358. Review of *DNA: The Secret of Life* and *Watson and DNA: Making a Scientific Revolution.*

Juengst, Eric T. (1996). "Self-Critical Federal Science? The Ethics Experiment Within the U.S. Human Genome Project." *Social Philosophy and Policy* 13(2): 63–96. Comprehensive analysis of the ELSI program; surveys various critiques of the program and makes a case that its strength as a policy mechanism lies in its decentralized "un-commission" design.

McElheny, Victor K. (2003). *Watson and DNA: Making a Scientific Revolution.* Cambridge, MA: Perseus. Authoritative biography of Watson, covering his scientific accomplishments, his impact on the scientific community, and his political and ethical views about the use of genetics in society.

Watson, James D., and Eric T. Juengst. (1992). "Doing Science in the Real World: The Role of Ethics, Law, and the Social Sciences in the Human Genome Project." In *Gene Mapping: Using Law and Ethics as Guides,* ed. George Annas and Sherman Elias. New York: Oxford University Press. The editors introduce the book by surveying the questions raised by new knowledge in the field of genetics and articulates the role of the ELSI program in the HGP.

Watson, James D. (2000). "Genome Ethics." *New Perspectives Quarterly* 17(4): 48–50. Outlines Watson's views on the ethics of genome research and demonstrates his strong materialist account of human origins and the implications of this philosophy for decisions regarding reproduction and the use of genetic technologies.

Wheeler, Timothy. (2003). "Miracle Molecule, 50 Years On." *Baltimore Sun* February 4, p. 8A. Surveys the implications of the discovery of DNA and Watson's life since then.

INTERNET RESOURCE

Watson, James D. (2000). "Genome Ethics." *New Perspectives Quarterly* 17(4): 48–50. Available from http://www.digitalnpq.org.

WEAPONS OF MASS DESTRUCTION
• • •

The phrase *weapons of mass destruction* (WMDs) was first used in the *London Times* in 1937 to describe Germany's blanket-bombing—using conventional weapons—of the city of Guernica, Spain (Mallon 2003). During the Cold War, the Soviet Union adapted the phrase to describe, collectively, nuclear, biological, and chemical (NBC) weapons (Norris and Fowler 1997). The U.S. Department of Defense defines WMDs as "weapons that are capable of a high order of destruction and/or being used in such a manner as to destroy large numbers of people," including high explosives, nuclear, chemical, biological, and radiological weapons. WMDs, however, often refer primarily to nuclear weapons.

History

Historical accounts of WMDs include the use of toxic smoke during the Peloponnesian War and during the Sung Dynasty in China (Hersh 1968); the Tartars catapulted plague-infected corpses into walled cities. Use of a *scorched earth policy* (Langford 2004) was also a common battle tactic in which retreating armies would destroy crops, burn villages, and poison wells and water supplies.

Large-scale production and deployment of nonnuclear WMDs was not possible until the beginning of the twentieth century (Hersh 1968), at which time scientists developed a more comprehensive understanding of how various chemicals functioned and of the manufacturing technologies necessary to synthesize large quantities of toxins. Advances in science thus led to the proliferation and stockpiling of numerous chemical agents such as mustard gas, phosgene, and chlorine. Chemical-weapons use during World War I resulted in the death of at least 90,000 people with more than 1.3 million additional casualties (Hersh 1968). Germany was the first nation to use poison gas during the war, but Great Britain, France, and the United States also used chemical weapons.

During World War II Germany and Japan conducted numerous chemical and biological weapon experiments on civilian and prisoner populations, yet such weapons were not used during combat. The United States was the first nation to use nuclear weapons when it bombed Hiroshima and Nagasaki in 1945. Many historians suggest that the incendiary bombing of Tokyo and Dresden by the United States during World War II, which killed thousands of civilians, also constituted use of WMDs. Use of chemical and biological weapons by several nations continued in the latter half of the twentieth century. One example is the defoliant Agent Orange that was used extensively by the United States in Vietnam to destroy vegetation. Iraq illegally used poison gas against the Iraqi Kurds killing tens of thousands of civilians. Although an exact accounting is impossible, the Federation of American Scientists indicates that dozens of nations possess, are developing, or are capable of developing WMDs.

The September 11, 2001, terror attacks that caused mass destruction and loss of life, however, were not perpetrated with NBC weapons, leading some experts to push for a more expansive definition of WMDs. Everett Langford describes WMDs as "those things which kill people in more horrible ways than bullets or trauma, or which cause effects other than simply damaging or destroying buildings and objects, with an element of fear or panic included" (Langford 2004, p. 1). Using this definition, WMDs would also include the airplanes used in the 2001 terror attacks; fungi used to destroy specific crops; defoliants; large scale incendiary devices; pathogens that kill agricultural animals; and other nonlethal agents. Sohail Hashmi and Steven Lee, however, argue that WMDs are different from conventional weapons because, "when used in war, [they are] inherently indiscriminate, meaning that their use ... would almost certainly result in the deaths of many civilians" (Hashmi and Lee 2004, p 10).

Ethics

For several reasons WMDs, especially NBC weapons, fall into different moral and ethical categories than conventional weapons. Over millennia, humans developed ethical guidelines and rules for *just war*. But Michael Walzer argues that nuclear weapons "are the first of mankind's technological innovations that are simply not encompassable within the familiar moral world" (Hashmi and Lee 2004, p 5).

Unlike more conventional arms, WMDs do not stay in the location in which they were deployed; detonation of NBC weapons invariably produce plumes of radiation and toxins that can travel hundreds of miles, well beyond the boundaries of the battlefield. The plume could kill innocent civilians within the country and in neighboring countries not involved in the conflict. Use of WMDs could also render large tracts of land uninhabitable, not only affecting the short term ability of a nation to feed itself after hostilities cease, but also that of future generations.

With conventional weapons, large numbers of people are needed to deploy enough bombs in order to cause widespread damage, so that there is at least some level of checks and balances in the decision process. WMDs, by contrast, may require just a handful of people whose actions can cause large-scale devastation, and thus WMDs are inherently less democratic than conventional weapons. The strongest ethical argument against using WMDs is quite simply that their use could destroy the world, killing billions of innocent people in *mutually assured destruction* (Hashmi and Lee 2004).

Politics

The world community made several attempts to control WMDs after World War I. The most important treaties are the Geneva Protocol (1925), which prohibits the use of both biological and poison gas methods in warfare; the Nuclear Non-Proliferation Treaty (1968), which prohibits states from acquiring nuclear weapons if they had not already detonated a nuclear weapon by January 1, 1967; the Biological and Toxin Weapons Convention (1972), which prohibits the development, stockpiling, and acquisition of biological weapons; and the Chemical Weapons Convention (1993), which prohibits the use, development, and stockpiling of chemical weapons.

Proliferation of WMDs during the twentieth century was characterized by the activities of large nation-states that possessed the financial resources, infrastructure, and intellectual capital necessary to research, test, and produce such weapons. Rapid technological advances in biological and chemical science coupled with readily accessible how-to information via the Internet and the collapse of the Soviet Union have markedly increased the risk of proliferation of WMDs. Individuals and small groups now have the capability of producing WMDs such as ricin, anthrax, and radioactive *dirty bombs*, without state support.

Through even more rapid technological advances in the years to come, the world may see a future with even more dangerous WMDs capable of being produced and deployed by just a few talented individuals, using genetic engineering, nanotechnology, and robotics (Joy 2000). Unlike the *old WMDs* of the twentieth century that required significant state support to produce, and thus could be controlled to some degree through international treaties, *new WMDs* pose entirely new problems of control, not to mention ethical and moral considerations that have yet to be fully addressed by the scientific community.

A first attempt in this direction is the "Statement on Scientific Publication and Security" produced by a group of scientific journal editors, scientists, and government officials at a National Academy of Science (NAS) meeting in January 2003. In the statement the authors acknowledge that some scientific information "presents enough risk of use by terrorists that it should not be published" (Journal Editors and Authors Group 2003, p. 1149). Rather than establishing strict guidelines for censorship, however, the authors leave such decisions up to the journal editors, who must weigh the possible security threats against the scientific merit and potential societal benefits of publishing the article. There are many more questions to ask, and actions to take, however, if society is to adequately address the threat of WMDs in the twenty-first century.

ELIZABETH C. MCNIE

SEE ALSO *Atomic Bomb; Baruch Plan; Biological Weapons; Chemical Weapons; Limited Nuclear Test Ban Treaty; Just War; Military Ethics; Nuclear Ethics.*

BIBLIOGRAPHY

Hashmi, Sohail H., and Steven P. Lee, eds. (2004). *Ethics and Weapons of Mass Destruction: Religious and Secular Perspectives.* Cambridge, UK: Cambridge University Press. An excellent resource with views on the ethics of WMDs from numerous religious and other and perspectives.

Hersh, Seymour. (1968). *Chemical and Biological Warfare: American's Hidden Arsenal.* Indianapolis, IN: Bobbs-Merrill Company.

Journal Editors and Authors Group. (2003). "Statement on Scientific Publication and Security." *Science* 299(5610): 1149.

Joy, Bill. (2000). "Why the Future Doesn't Need Us." *Wired* 8(4): 238–262. Also available from http://www.wired.com/wired/archive/8.04/joy.html.

Langford, R. Everett. (2004). *Introduction to Weapons of Mass Destruction: Radiological, Chemical and Biological.* Hoboken, NJ: John Wiley. A thorough review of all technical, health-related and medical issues concerning WMDs in the early twenty-first century.

Norris, John, and Will Fowler. (1997). *NBC: Nuclear, Biological and Chemical Warfare on the Modern Battlefield.* Cambridge, UK: Cambridge University Press.

INTERNET RESOURCES

Federation of American Scientists. "States Possessing, Pursuing or Capable of Acquiring Weapons of Mass Destruction." Available from www.fas.org/irp/threat/wmd_state.htm.

Mallon, Will. "WMD: Where Did the Phrase Come From?" History News Network. Available from http://hnn.us/articles/1522.html.

WEBER, MAX

•••

Max Weber (1864–1920) was arguably the most important social and political theorist of the twentieth century, as well as the unwilling father of modern sociology (a role he unknowingly shared with Èmile Durkheim). The eldest of six children (with a brother Alfred, who also became a famous sociologist and cultural analyst), Max Weber was born in Erfurt, Prussia, on April 21, grew up in a suburb of Berlin, and spent his entire adult life in German university towns. He pursued law, economics, and philosophy at Heidelberg, Strassburg, Berlin, and Göttingen (1882–1886), served in the army reserve for two years during college, returned home, and studied law in Berlin, graduating in 1889. He won academic appointments in Berlin and Freiburg, but was forced to retire from teaching after suffering a nervous breakdown that immobilized him between 1897 and 1903—an almost pure example of what Sigmund Freud at precisely the same time had labeled the *Oedipus complex.* Finally recovered enough to take an extended, transformative trip to the United States in 1904, and freed of teaching duties by an inheritance, Weber spent the next sixteen years producing an unrivalled body of sociocultural, economic, and sociological analyses that

Max Weber, 1864–1920. The German social scientist was a founder of modern sociological thought. His historical and comparative studies of the great civilizations are a landmark in the history of sociology. (*The Library of Congress.*)

is second to none in the history of modern social science. He died unexpectedly on June 14 at the age of 56, a victim of the global influenza pandemic. Weber had married his cousin, Marianne Schnitger, in 1893, and it was her tireless work between 1920 and 1924 as editor of his many posthumous books that fixed Weber's rightful place in the social science pantheon, because during his life he had published only a small percentage of what he wrote.

Weber's common fame rests on his *Protestant Ethic and the Spirit of Capitalism* (1904–1905), originally published as two articles in a scholarly journal. Here he demonstrated why northern European Protestant behavior was more conducive to the formation of early capitalism than were southern European Catholic beliefs and practices, a hypothesis that has given rise to thousands of commentaries and critiques. But he also contributed fundamental works to the sociology of law (which he virtually invented), the sociology of music (also a first), the sociology of the economy, the philosophy of social science method, the comparative sociology of religion (also his creation), social stratification, the sociology of

bureaucracy, and of power and *charisma* (his term), and so on. His major work is *Economy and Society* (1922), a massive study assembled by his wife (herself an important feminist public intellectual), and translated into English for the first time in 1968. Weber's importance grows with time, and he is the only classic social theorist for whom in the early twenty-first century an entire scholarly journal is named. A recent bibliography of works in English concerning Weber numbers more than 4,900 items, and as Karl Marx and Freud become increasingly less tenable as the major analysts of the modern world, Weber's ideas become ever more pertinent and revealing.

Weber's thoughts about science and ethics are neatly summarized in two of the most famous lectures ever given by a social scientist, "Science as a Vocation" (November 1917) and "Politics as a Vocation" (January 1919). Both were delivered at the University of Munich before large audiences of returning veterans and other students (among them, Rainer Maria Rilke) in a highly politicized atmosphere, with Weber expected to take a strongly nationalistic stance similar to many of his colleagues. Instead he spoke in contrarian terms by insisting that science requires objectivity and *value-freedom* from its practitioners, who must be motivated by a selfless *Beruf* (vocational calling) dedicated solely to the discovery of truth, and never by mundane self-aggrandizement or political values. He warned against the *cult of personality* and the seductive weakness for *selling a worldview* that interferes with proper scientific work. Weber drew on Friedrich Nietzsche, Leo Tolstoy, the Sermon on the Mount, Charles-Pierre Baudelaire, Immanuel Kant, and his young friend, Georg Lukács (1885–1971) in making a strong case for scientific research as a single-minded search for the unprettified truth, and nothing else.

In the companion lecture, "Politics as a Vocation," Weber continued in this vein, introducing one of his most famous distinctions, between *an ethic of ultimate ends* and *an ethic of responsibility*. The former defines the bailiwick of scientists, while the latter belongs to politicians and other activists, whose *raison d'être* is the strategic furthering of an ideological program. Weber warned that when these two ethics are joined within a single person, they inevitably lead to the degeneration of both roles, and to cultural calamity. As Weber explained in one of his most famous and controversial paragraphs:

> We must be clear about the fact that all ethically oriented conduct may be guided by one of two fundamentally differing and irreconcilably opposed maxims: conduct an be oriented to an "ethic of ultimate ends" or to an "ethic of responsibility." This is not to say that an ethic of ultimate ends is

identical with irresponsibility, or that an ethic of responsibility is identical with unprincipled opportunism. Naturally nobody says that. However, there is an abysmal contract between conduct that follows the maxim of an ethic of ultimate ends—that, is in religious terms, "the Christian does rightly and leaves the results with the Lord"—and conduct that follows the maxim of an ethic of responsibility, in which case one has to give an account of the foreseeable results of one's action ("Politics as a Vocation" in *From Max Weber*, p. 120).

Within a very few years, the scientists and ethicists of Nazi Germany experienced the dire consequences of ignoring the thrust of Weber's speeches—which accounts in part for the Nazi government's interest in discrediting the memory of Weber after his death. Interestingly Weber is one of few German intellectuals of the twentieth century whose reputation was never threatened by world memory of the Third Reich.

ALAN SICA

SEE ALSO *Axiology; Durkheim, Émile; Ethical Pluralism; Marx, Karl; Secularization; Sociological Ethics; Spenser, Herbert.*

BIBLIOGRAPHY

Eliaeson, Sven. (2002). *Max Weber's Methodologies.* Cambridge, UK: Polity Press.

Käsler, Dirk. (1988). *Max Weber: An Introduction to His Life and Work*, trans. Philippa Hurd. Oxford: Polity Press; Chicago: University of Chicago Press.

Sica, Alan. (2004). *Max Weber: A Comprehensive Bibliography.* New Brunswick, NJ: Transaction Publishers.

Weber, Max. (1930). *The Protestant Ethic and the Spirit of Capitalism*, trans. Talcott Parsons. London: Allen and Unwin; New York: Charles Scribner's Sons.

Weber, Max. (1946). "Science as a Vocation"; "Politics as a Vocation." In *From Max Weber*, trans. and eds. Hans Gerth and C. Wright Mills. New York: Oxford University Press.

Weber, Max. (1968). *Economy and Society: An Outline of Interpretive Sociology*, 3 vols., ed. Guenther Roth and Claus Wittich. New York: Bedminster Press. Reissued, University of California Press, 1978, in 2 volumes.

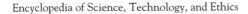

WEIL, SIMONE

• • •

French philosopher, mystic, and social critic Simone Weil (1909–1943) was born in Paris on February 3 and

Simone Weil, 1909–1943. The French thinker, political activist, and religious mystic was known for the intensity of her commitments and the breadth and depth of her analysis of numerous aspects of modern civilization. *(AP/Wide World Photos.)*

died in Ashford, Kent, in England on August 24. Though raised in a prosperous bourgeois family and classically educated at the Ecole Normale Supérieure, Weil sympathized from an early age with the plight of the poor, the oppressed, and the afflicted.

Before the age of twenty Weil identified herself as an anarcho-syndicalist. She was attracted to the philosophy of Marx but refused to join the communist party. Her earliest sustained social analysis, "Reflections Concerning the Causes of Liberty and Social Oppression," provided a critique of Marxism that Albert Camus (1913–1960) judged the most profound of the twentieth century. This critique focused on what Weil thought was the inadequacy of Marx's optimistic view that technological progress would lead inevitably to the liberation of the proletariat. For her, technological development gave humanity more control over nature only at the expense of greater dependency on what she called the *collectivity.* The collectivity includes the bureaucratic structure of the state (political and legal authority, including the government and the police) as well as the private corporations that produce the goods and services of the economy.

Weil argued that labor is not in itself the cause of oppression. For her, genuine human freedom meant freedom from the illusions that, in industrial society, take the form of ideologies and myths, of which the idea of progress is the preeminent example. In order to be free of the tyranny of illusion, human beings must come to know themselves as limited beings. Their finitude is revealed through methodical, thoughtful engagement with necessity; in other words, through work. Work is therefore a good that is not to be eliminated but ought to be the spiritual center of civilization. Weil argued that the problem with modern technology is that methods (mechanical or bureaucratic) are built into machines or organizations, thereby eliminating the need for thinking. A method, once developed, can be applied indefinitely, without ever being understood by the person who applies it. Generally, there is method in the motions of work, but none in the minds of the workers who tend automatic machines. They are reduced to slavery; they have lost their freedom.

This analysis formed the basis of Weil's critique of the industrial system that, in her view, dedicated itself to the maximization of the productivity of the worker rather than the maximization of freedom in the work process. In her two years of factory work (1934–1935), she saw that workers usually cannot understand the techniques they apply and this fact undermined their thinking relationship to reality. Due to the division and coordination of labor which in turn is a function of the techniques of production, there is a virtually complete divorce between thought and action. The manual laborers on a production line are not free, are dehumanized and reduced to slaves, not because they perform physically laborious tasks but because their tasks are so structured as to exclude the possibility of thought. Mental workers, those who make up the essential bureaucratic structure by which the activity of the workers is brought into coordinated relation, may be as enslaved as the manual laborers themselves because their thinking is ordinarily divorced from any direct action or work, and does not involve a dialogue with those whose lives they order. They too have lost touch with necessity.

Weil's critique of modern industry led her to analyze modern science as itself having become a thoughtless collective enterprise that relies on specialization for its advancement. No single mind can grasp even a subdiscipline of physics or chemistry. Researchers take over not only the results but the methods developed by their predecessors without understanding them or their relation to the whole. Weil concluded that the scientist can be crushed by science in much the same way that the workers are crushed by their work.

Toward the end of her life when her most profound religious thinking and social analysis was done, Weil contrasted modern (or, as she called it, classical) science, developed after Galileo and Newton between the sixteenth and the nineteenth centuries, with ancient Greek science. She concluded that modern science had emancipated the study of nature, first understood on the analogy of work (that is, in terms of energy), from the idea of the good, and then from the idea of necessity. In the 1940s, Weil predicted that the incomparable technical achievements of science would become divorced from any ordering principle and destroy human scale, as complexity was piled on complexity and society became uprooted.

Weil died prematurely in England at the age of 34. The significance of her posthumously published writings on religion as well the social and political crises of her times are only beginning to be appreciated for their depth and originality.

LAWRENCE E. SCHMIDT

SEE ALSO *Freedom; Humanization and Dehumanization; Marxism.*

BIBLIOGRAPHY

McLellan, David. (1990). *Utopian Pessimist*. New York: Poseidon Press. A biography by the renowned scholar of Marxism that emphasizes her links to the labor movement in France between the wars.

Pétrement, Simone. (1976). *Simone Weil: A Life*. New York: Pantheon Books. A carefully documented biography by a dear personal friend and renowned scholar of ancient Gnosticism.

Weil, Simone. (1968). "Classical Science and After." In her *On Science, Necessity and the Love of God*. Oxford, UK: Oxford University Press. Written in Marseille in 1941, this essay takes up the critique of classical (modern) science adumbrated in "Reflections Concerning the Causes of Liberty and Social Oppression."

Weil, Simone. (1971). *The Need for Roots*. New York: Harper Colophon. Written at the end of 1942 for De Gaulle who was leading the Free French in London, this report outlines Weil's proposal for the reconstruction of France after Hitler's defeat.

Weil, Simone. (2001). "Reflections Concerning the Causes of Liberty and Social Oppression." In her *Oppression and Liberty*. Translated by Arthur Wills and John Petrie. New York: Routledge. Written in 1933–1934 before her year of factory work, this lengthy essay outlines Weil's reasons for rejecting the liberal and Marxist optimism about the liberating potential of technological progress.

WELLS, H. G.

• • •

Herbert George Wells (1866–1946) was born in Bromley, Kent, United Kingdom, on September 21, to servants turned shopkeepers. After a poor education in local private schools he was apprenticed to the drapery trade at age fourteen. After a spell as a pharmacist's assistant Wells became a student-teacher in Midhurst, where he won a scholarship to study for a degree under the biologist Thomas Henry Huxley (1825–1895) at the Normal School of Science in South Kensington. After initially failing to earn a degree, he became a schoolteacher and completed his bachelor of science degree in zoology at the University of London in 1890. He died in London on August 13.

Although eventually Wells became world famous as the author of *The Time Machine* (1895), *The Invisible Man* (1897), *The War of the Worlds* (1898), and other novels, his first two books were science textbooks published in 1893. Throughout the 1890s Wells was a regular contributor to scientific periodicals and wrote popular science articles for the mainstream press. Even after becoming famous as a writer of fiction, Wells maintained an interest in science as a Fellow of the Zoological Society after 1890 and joined the Sociological Society (on its foundation in 1904). He debated eugenics with the scientist Francis Galton (1822–1991) and others and published scientific works such as *Anticipations of the Reaction of Mechanical and Scientific Progress upon Human Life and Thought* (1901), *The Science of Life* (1930), and *Science and the World-Mind* (1942).

Wells's contribution to science, technology, and ethics was considerable. He recognized from his university days that although human progress was not inevitable, science would play a key role in human achievement. From Huxley he adopted the notion of ethical evolution: humankind's responsibility to influence the biological destiny of humans and other species positively. That notion ultimately led Wells to promote, at the micro level, a welfare state based on negative eugenics and state provision of a "basic minimum" and, at the macro level, a cosmopolitan world state based on education, cooperation, and socialist planning.

Eugenics was an important subject for Wells during much of his career. He first considered it in *Anticipations*

H. G. Wells, 1866–1946. The English author began his career as a novelist with a popular sequence of science fiction that remains the most familiar part of his work. He later wrote realistic novels and novels of ideas.

(1901) before analyzing it more closely in works such as *Mankind in the Making* (1903), *A Modern Utopia* (1905), *Men Like Gods* (1923), *The Science of Life* (1930), and *The Work, Wealth and Happiness of Mankind* (1931) and finally rejecting it outright in *The Rights of Man* (1940) and *'42 to '44* (1944). During the Edwardian period Wells believed that negative eugenics could be a viable means of preventing the procreation of "the people of the abyss": the incurably diseased, habitual criminals or drunkards, and those unable to adapt to the rapidly changing modern world. Gradually he tempered his position, seeing welfare provision, education, and medical science as more important factors for improving the quality of successive generations. With the rise of Nazi eugenics after 1933, Wells distanced himself from general eugenic theory, declaring that any form of compulsory or state eugenics would be a fundamental breach of human rights in *The Rights of Man* (1940).

According to Wells, human progress rests on technological advancement, and he predicted that in the twentieth century humanity would either destroy itself or create material abundance and cosmopolitan unity. His 1935 film *Things to Come* is a marvel of invention, with ultramodern architecture, highly skilled workers, scientific population control, space flight, moving footpaths, and more. However, the society it portrayed was brought about only by generations of warfare, and in this lies the tension that existed between Wells's vision of a technological future and the means to achieve it.

Although Wells preached disarmament and world peace throughout his life, his futuristic utopian societies founded on the power of science consistently had to go through devastating wars to be achieved. Humankind had to learn a severe lesson before it would apply the gifts of science to its destiny. Thus, in *The War in the Air* (1909), powered flight leads to aerial combat; in *The World Set Free* (1914), harnessing the atom leads to nuclear war; and in *The Shape of Things to Come* (1933), material progress leads to global conflict and an "air dictatorship." All these stories end with global human fellowship and peace, but they are achieved at a high price.

Wells's legacy in terms of science, technology, and ethics lies in his imaginative application of science to invention, his hopefulness about what science may produce for humanity, but also his warnings about what the abuse of science may mean for the human race. In his nonfiction writings Wells was ambiguous throughout his life, never able to offer a peaceful route to the achievement of his predicted scientific utopias. Although Wells was never certain in his hope or despair for the future, his ultimate mood on the subject is aptly characterized in the title of his final work, published a few months before the dropping of atomic bombs on Japan, *Mind at the End of Its Tether* (1945).

JOHN S. PARTINGTON

SEE ALSO *Eugenics; Progress; Science Fiction; Science, Technology, and Literature; Utopia and Dystopia.*

BIBLIOGRAPHY

Haynes, Roslynn D. (1980). *H.G. Wells: Discoverer of the Future: The Influence of Science on His Thought.* New York: New York University Press. A detailed study of Wells's application of scientific concepts in his literary art; an alternative reading to Reed (see below).

Partington, John S. (2003). *Building Cosmopolis: The Political Thought of H.G. Wells.* Aldershot, UK: Ashgate. A comprehensive analysis of Wells's political thought, considering the influence of Huxleyan notions on Wells's worldview, and detailing Wells's support and later rejection of eugenics.

Reed, John R. (1982). *The Natural History of H.G. Wells.* Athens: Ohio University Press. An alternative study of the influence of science on Wells's literary work; an alternative reading to Haynes (see above).

Smith, David C. (1984). *H.G. Wells: Desperately Mortal: A Biography.* New Haven, CT: Yale University Press. The outstanding biography of Wells.

Wagar, W. Warren. (1961). *H.G. Wells and the World State.* New Haven, CT: Yale University Press. The first thorough study of Wells's world-state ideas.

Wells, H.G. (1984). *Experiment in Autobiography: Discoveries and Conclusions of a Very Ordinary Brain (Since 1866).* London: Faber and Faber. Wells's outstanding autobiography, first published in 1934.

WHISTLEBLOWING

• • •

The origin of the term *whistleblowing* is uncertain. It may refer to English policemen blowing whistles to alert others to an illegal act or to sports referees stopping a game due to a rule infraction. The term began to be used in a way relevant to science, technology, and ethics in the 1960s and became part of the common vocabulary as a result of Ralph Nader's investigative activities during the 1970s. *The American Heritage Dictionary* defines a whistleblower as "one who reveals wrongdoing within an organization to the public or to those in positions of authority," but a more detailed analysis of the term is appropriate.

Analysis of the Concept

Based on the above definition, it is possible to distinguish between internal and external whistleblowing. Internal whistleblowing occurs when the hierarchical chain of command within an organization is violated, so that one's immediate superiors are bypassed, perhaps because they have refused to act or are themselves involved in the wrongdoing. The whistleblowing is internal, however, because it stays within the organization. External whistleblowing refers to going outside the organization, possibly to a regulatory agency, the press, or directly to the public. The philosophical literature often restricts the use of the term to external whistleblowing, but the media typically use it in both senses.

A further distinction may be made between open and anonymous whistleblowing. The former means that the identity of the whistleblower is known, while such identity remains unknown in the latter. Anonymous whistleblowing is generally considered to be less effective, because it is more easily ignored and because no follow-up with the whistleblower is possible. Organizations have also shown themselves willing to devote significant resources to discover the identity of anonymous whistleblowers, a task made easier by the limited number of individuals who typically have access to the information being revealed. At the same time, whistleblowers often want to hide their identities because of possible reprisals from their organizations or colleagues.

The idea of *wrongdoing* also requires clarification. Generally not every wrongdoing is considered to be a legitimate subject for whistleblowing. The wrongdoing must entail serious harm, whether physical, psychological, or financial. Depending on the particular philosophical perspective, the notion of harm might be extended to situations where human beings are only indirectly affected, such as through damage to the environment. Serious harm is considered to be the appropriate criterion in that whistleblowing itself is an act which that tends to harm the parties involved and thus requires a balancing of outcomes.

Finally, although omitted in the popular definition, whistleblowers need to be insiders, that is, either currently or formerly associated with the organization on which they are blowing the whistle. Outsiders might be considered spies, investigative reporters, or moles, but not whistleblowers. Whistleblowing must involve a conflict of loyalties, between the duty of loyalty to an organization and duties to the public or to a principle. For an infiltrator, no such duty of loyalty to the organization exists. It should be noted, however, that whistleblowers often do not perceive themselves as being disloyal, especially in instances of internal whistleblowing, but believe they are working for the long-term organizational good.

Ethical Perspectives

From the perspective of ethics, whistleblowers are faced with deciding whether breaking the bond of loyalty is justified in a particular circumstance. The philosopher Richard DeGeorge proposed the classic criteria for justifying whistleblowing; most other criteria are a reaction to his formulation. DeGeorge argues that external whistleblowing is morally permissible if three conditions are met: (a) substantial harm will be done to persons; (b) the immediate superior is made aware of the problem; and (c) the chain of command of the organization is exhausted. DeGeorge contends that whistleblowing is morally obligatory if two additional conditions are met: (d) enough documented evidence is available to the whistleblower to convince an impartial individual; and (e) the whistleblower has a justified belief that the wrongdoing will be corrected as a result of going public.

A number of critiques have been leveled against DeGeorge's criteria, including questions about the extent to which a future rather than a past harm must be involved, immediacy of the harm, lack of consideration for the fate of the whistleblower, and importance of the motives governing the action. Fundamentally these debates reflect the divergence between consequentialist and deontological approaches to the issue. Consequentialist thinkers emphasize the costs to the institution and to the whistleblower and the detrimental results of mistaken or malicious whistleblowing, while deontological thinkers tend not to distinguish as significantly between degrees of harm and are more concerned with justice being done.

For engineering, in particular, the issue of whistleblowing has been a major focus of ethical discussions because of the potential impact of engineering activities on public safety. Most codes of engineering ethics follow the lead of the Accreditation Board for Engineering and Technology (ABET) by emphasizing that "engineers shall hold paramount the safety, health and welfare of the public in the performance of their professional duties." Preventing physical harm to people is seen as a special professional responsibility of engineers based on their technical expertise. Many codes even obligate engineers to blow the whistle by requiring notification of *the proper authority* when public safety is endangered.

Whistleblowing in science generally has a different justification. Most often it is related to the research process and falsification of data, although research can also directly harm the human subjects involved or the public at large through the introduction of products based on falsified data. The difference in emphasis between science and engineering whistleblowing can be traced to the fundamental emphasis given to truth and accuracy in science, as opposed to the need to protect the public from harmful technologies in engineering.

Due to the serious consequences associated with whistleblowing, most analyses stress that it should be an avenue of last resort. Many discussions have emphasized ways that organizations can avoid whistleblowing, including creating an internal ethics office, fostering open door practices, having clear organizational policies, or appointing an ombudsperson. One reason to highlight such alternatives is that the whistleblower often becomes the target of subsequent investigations, directing attention away from the misconduct that was revealed.

In fact, the consequences for whistleblowers are so universally negative, including shunning by colleagues and organizational reprisals, that whistleblowing is legitimately an act of moral heroism. Commentators such as Kenneth Alpern argue that engineers, given their special responsibility for the public safety, should be required to be moral heroes. Others, such as Mike Martin, believe that whistleblowing is a supererogatory act whose obligatory nature must be evaluated on a case-by-case basis, taking into account both professional duty and personal considerations. Whether certain individuals should be singled out and required to suffer grave consequences for the common good will continues to be a matter of debate.

HEINZ C. LUEGENBIEHL

SEE ALSO *Engineering Ethics.*

BIBLIOGRAPHY

Alpern, Kenneth D. (1983). "Moral Responsibility for Engineers." *Business and Professional Ethics Journal* 2(2): 39–47. Argues that engineers have a higher duty than others to blow the whistle.

Davis, Michael. (1996). "Some Paradoxes of Whistleblowing." *Business and Professional Ethics Journal* 15(1): 3–19. Critiques DeGeorge's analysis of whistleblowing and substitutes a complicity theory.

DeGeorge, Richard T. (1999). *Business Ethics*, 5th edition. Upper Saddle River, NJ: Prentice Hall. Provides the most commonly cited treatment of whistleblowing. Also distinguishes between responsibilities of business people and professionals.

Johnson, Roberta Ann. (2003). *Whistleblowing: When it Works—and Why.* Boulder, CO: Lynne Rienner. Emphasizes the legal dimensions of whistleblowing.

Lubalin, James S., and Jennifer L. Matheson. (1999). "The Fallout: What Happens to Whistleblowers and Those Accused But Exonerated of Scientific Misconduct?" *Science and Engineering Ethics* 5(2): 229–250. An empirical study of consequences associated with whistleblowing in science.

Martin, Mike W. (1992). "Whistleblowing: Professionalism, Personal Life, and Shared Responsibility for Safety in Engineering," *Business and Professional Ethics Journal* 11(2): 21–37. Argues against DeGeorge that deciding whether whistleblowing is obligatory should be contextually determined based on an all-things-considered judgment.

Martin, Mike W., and Roland Schinzinger. (1996). *Ethics in Engineering*, 3rd edition. New York: McGraw-Hill. The most widely used text on engineering ethics. It extends DeGeorge's analysis of whistleblowing.

Whitbeck, Caroline. (1998). *Ethics in Engineering Practice and Research.* New York: Cambridge University Press. Focuses on ethical issues in scientific research.

INTERNET RESOURCE

National Society of Professional Engineers. "NSPE Code of Ethics for Engineers," rev. January 2003. Available at http://www.nspe.org/ethics/eh1-codepage.asp.

WIENER, NORBERT

• • •

Born in Columbia, Missouri, on November 26, Norbert Wiener (1894–1964) gained prominence as a world-famous mathematician who founded the interdisciplinary field of *cybernetics*, questioned its social implications, and encouraged scientists and engineers to consider the social consequences of their work. He died in Stockholm, Sweden, on March 18.

A child prodigy, Wiener earned a B.S. from Tufts University at the age of fourteen and a Ph.D. from Harvard at eighteen. As a professor of mathematics at the Massachusetts Institute of Technology (MIT), he made his mark in the areas of statistical theory, harmonic analysis, and prediction and filtering. While doing research on an antiaircraft system during Word War II, Wiener developed the key idea behind cybernetics: Humans and machines could both be studied using the principles of control and communication engineering. Both were information-processing entities that interacted with the environment through feedback mechanisms to pursue goals.

The atomic bombings of Japan in August 1945 brought the issue of social responsibility to the fore for Wiener. He wrote a resignation letter to the president of MIT that fall, stating that he intended to leave science because scientists had become the armorers of the military and had no control over their research. Although Wiener may have never sent the letter, he stopped doing military work. He became well-known for this stance in 1947 when the press reported his refusal to attend a military-sponsored symposium on computers and to share his war-time research with a company developing guided missiles. Wiener reasoned that "the bombing of Hiroshima and Nagasaki, has made it clear that to provide scientific information is not a necessarily innocent act, and may entail the gravest consequences" (Wiener 1947, p. 46).

Wiener expressed his views on the ethical and social aspects of science and technology in *Cybernetics* (1948), *The Human Use of Human Beings* (1950), and *God and Golem, Inc.* (1964). All three books warn about the potentially dangerous social consequences of the very field he had founded. Wiener claimed that cybernetics had "unbounded possibilities for good and evil" (Wiener 1948, p. 37). Electronic prostheses would benefit humans, and automated factories, the basis of a *second industrial revolution*, could eliminate inhuman forms of labor. If, however, humans "follow our traditional worship of progress and the fifth freedom—the freedom to exploit—it is practically certain that we shall face a decade more of ruin and despair" in implementing this technology (Wiener, 1950, p. 189). He also criticized game theory and military science for viewing the world as a struggle between good and evil.

Wiener considered whether to stop working on cybernetics because of its dangers. But it belonged "to the age, and the most any of us can do by suppression is to put the development of the subject into the hands of the most irresponsible and most venal of our engineers," namely, those doing military work. He recommended educating the public about the social implications of his field and confining research to areas, "such as physiology and psychology, most remote from war and exploitation" (Wiener, 1948, p. 38–39). Near the end of his life, Wiener said scientists and engineers should stop being amoral *gadget worshipers* (Wiener 1964) and imagine the consequences of their work well into the future. In regard to growing concerns about the dehumanizing effects of computerization, he recommended a cybernetic division of labor: Humans should perform functions best suited to them, computers those best suited to computers.

Cybernetics has led a life of its own outside of Wiener's control. In the late 1940s, philosophers in the Soviet Union criticized Wiener for attacking dialectical materialism, then did an about-face in the 1950s and adopted cybernetics wholeheartedly. In Western Europe and North America, in the 1960s, cybernetics lost prestige among scientists who questioned its rigor and universal claims. Beginning in the 1980s, some humanists praised Wiener's antimilitarism, while others criticized cybernetics for creating a philosophy of nature and a computer-based material culture that turns humans into *cyborgs* (cybernetic organisms). At the same time, historian and philosopher Donna Haraway co-opted Wiener's cybernetic vision to create an ironic cyborg epistemology with which to critique the global corporate-military-university complex and the technosciences that sustain it.

RONALD KLINE

SEE ALSO *Automation; Cybernetics; von Neumann, John.*

BIBLIOGRAPHY

Galison, Peter. (1994). "The Ontology of the Enemy: Norbert Wiener and the Cybernetic Vision." *Critical Inquiry* 21(Autumn): 228–266.

Heims, Steve J. (1980). *John von Neumann and Norbert Wiener: From Mathematics to the Technologies of Life and Death.* Cambridge, MA: MIT Press.

Norbert Wiener, 1894–1964. The American mathematician studied computing and control devices. Out of these studies he created the science of cybernetics. (*The Library of Congress.*)

Wiener, Norbert. (1947). "A Scientist Rebels." *Atlantic Monthly* 179(1): 46. Reprinted in *Bulletin of the Atomic Scientist,* February 1947, 31.

Wiener, Norbert. (1950). *On the Human Use of Human Beings: Cybernetics and Society.* Boston: Houghton Mifflin.

Wiener, Norbert. (1961). *Cybernetics: Or Control and Communication in the Animal and the Machine,* 2nd edition. New York: MIT Press.

Wiener, Norbert. (1964). *God and Golem, Inc.: A Comment on Certain Points Where Cybernetics Impinges on Religion.* Cambridge, MA: MIT Press.

WILDERNESS

• • •

Few currents in literature, the arts, and religion run deeper than the cultural fascination with wildness, and its locational concomitant, wilderness—places where primordial reality dominates and the artificialities of humans, including their sciences and technologies, are not apparent. Marks of the depth of the idea are its universality and flexibility. Appeals to wilderness can be found in cultures as diverse as China and North America. In its many intellectual guises and emotional overlays it has proven adaptable and meaningful across great historical divides. Although here the emphasis will be on its Euro-American manifestations, it is important to recognize that wilderness is not an idea exclusive to that culture.

Euro-American Context

In the Euro-American context the idea of wilderness is associated with the view that humans by nature separate themselves from nature, which then provides the backdrop for most considerations of ethics. Yet throughout western history, ethical principles have been formulated to apply only on the human side of the human-wildness divide. This separation of humans from wildness is especially important normatively, because it shapes the context in which new technologies are evaluated, including technologies that radically alter nature and irreversibly destroy wildness in the process of *development* and *progress*. Against the backdrop of wilderness, science has sometimes been judged both tame and distorting. Appeals to wilderness are often the basis for criticizing technologies, especially technologies that radically alter nature or irreversibly destroy wildness in the process of *development* and *progress*.

Max Oelschlaeger (1991) hypothesizes that Mediterranean cultures, especially at the eastern end where agriculture was taking hold, began developing in their mythology a separation of human culture from nature as early as 10,000 B.C.E. in the Yahwist tradition. In later Hebrew history, sojourns in the wilderness became symbols for the spiritual purification of prophets; at the same time, wild lands were treated as wastelands awaiting transformation into *productive* farmland. Oelschlager attributes this ambivalence to residual tensions between settled agriculturalists and nomadic, *wilder* tribes of herders and gatherers. These two themes—wildness and civilization through cultivation—are entwined throughout the Judeo-Christian tradition. Christ's sojourn in the wilderness, in keeping with the Hebrew tradition of seeking purification by retreating from society into wilderness, is portrayed as a time of spiritual strengthening in preparation for a future ministry.

The idea of wilderness as an obstacle to the human will, which grew out of the earlier tendency of agriculturalists to distinguish their works—the domain of their physical control—from wild nature lying beyond civilization, took on a renewed meaning with the discovery of the New World. In this context, the fascination with wilderness was expressed as a struggle between the Enlightenment view of human perfectibility through science and technology, and romanticism as expressed in the French philosopher Jean-Jacques Rousseau's idea of the *noble savage*. According to this view, pre-civilized humans, not yet corrupted by the affectations of society, have a purity not found in contemporary society. This fascination with wildness was inspired by the discovery of *primitive* cultures, and reinforced the romantic critique of the overly rational and mechanical world of the Enlightenment.

Once imported into the New World, the wilderness versus civilization theme took on new vitality as colonists came in direct contact with wilderness and with *wild* tribes they saw as *savages*. Jonathan Edwards—despite his reputation as a brimstone orator—preached benevolence toward the whole of God's nature (Miller 1967, p. 283). More concretely, the battle between civilization and wilderness was fought again each day at the advancing edge of colonial development of lands formerly inhabited by Native American tribes. In a classic analysis of U.S. history, Frederick Jackson Turner (1920) emphasized the importance of the frontier in the identity of the United States, predicting that a huge transformation in consciousness would ensue as the frontier closed. Turner saw the existence of an open frontier, and the idea of *manifest destiny* associated with it, as definitive of the American experience. Accordingly the closing of the frontier was thought to usher in a new era in American life.

Two Views of Wilderness

It is useful to separate two aspects of the wilderness idea as it has developed in American thought. First there was the indicated experience of wildness as a countervailing force resisting the daily transformation of wild lands into farmland and cities in the path of westward expansion. This process of *civilizing* lands that had before been the habitat of nomadic tribes of hunters and gatherers represents a replay of the growth of agricultural societies across the Middle East and Europe in the original expansion of agriculture in the Old World. In this conflict, wilderness was cast as one pole in a dialectic between human culture and wild nature.

The reality of these day-to-day struggles to transform wilderness into productive land may be contrasted with a second, emergent idea of wilderness, an idea—one might say an idealization—of wildness and wilderness that has evolved within academic and intellectual circles, especially in North America and in Australia. The works of Perry Miller (1967), Leo Marx (1967), Roderick Nash (1982) and the philosophers Mark Sagoff (1974) and Max Oelschlager have all articulated and emphasized the importance of the idea of wilderness in the American identity and self-perception. These authors, whose careers correspond to a growing academic interest in environmental studies all brought new dimensions to a vital strain in American intellectual life, as exemplified, for example, in the writings of Ernest Hemingway, Wallace Stegner, Annie Dillard, and many others. These authors reprise a longstanding theme—as exemplified in the Leatherstocking Tales of James Fenimore Cooper and his hero, Natty Bumppo, of associating life at the edge of wilderness as symbolic of freedom, self-reliance, and character.

The emphasis in the United States on the idea of wilderness led to a re-shaping of the related concept, "nature," which came to mean "primordial nature," whereas in Europe—where most land had been altered by humans long ago—people enjoyed the "countryside," with farms, homes, and businesses distributed across the landscape, as "natural." The assimilation of the idea of nature to that of primordial nature, and referring only to lands where humans have no presence, has contributed to the polarization of thought about nature in the United States. Whereas Europeans enjoy mixed landscapes, Americans distinguish wilderness from "the working landscape," and there are bitter disagreements about what activities are appropriate in wilderness areas. Advocates of wilderness thus try to eliminate activities, such as motorized recreation, from wilderness areas, considering such uses inappropriate and damaging to the primordial quality of wilderness.

The complex, often conflicting theme of nature versus culture has been important in environmental thought and action. Henry David Thoreau, the transcendentalist, said "in Wildness is the preservation of the World" (Thoreau 1998 [1862], p. 37), and his ideas are echoed in the work of John Muir (founding president of the Sierra Club) and many other wilderness advocates. Muir's reverence for forests and wild nature clashed with the ideas of Gifford Pinchot, the first Forester of the National Forest Reserves, who argued that all resources should be developed to improve the material lot of humans. So reverence—and passion—for wildness exists in sharp contrast to another, opposing theme: the need to control and civilize nature for human use. This tension in the environmental movement, it could be argued, reflects the broader ambivalence of Euro-American culture toward wildness and civilization.

Wilderness Policy

Muir's respect for wilderness also motivated Aldo Leopold, the philosophical forester who worked tirelessly to protect wild areas from development, from within and, later, outside the U.S. Forest Service. Leopold convinced the Forest Service to set aside the Gila Wilderness in 1922, and he co-founded the Wilderness Society—an activist group that advocates for wilderness protection—in 1935. Leopold advocated for preservation of the wilderness on several bases; he countered the utilitarians and materialists by noting that wilderness backpacking and hiking are uses, too, and that some land has more utility for back-country recreation than for development. He also argued that humans need wild, natural systems as *models* of healthy systems if they are ever to become intelligent managers of the modified systems that are their immediate habitats. Leopold, however, at his most passionate, argued for wildness and wilderness as a *cultural* necessity, and as a matter of *intellectual humility.* "The shallow-minded modern" must, he thought, learn to appreciate wilderness as a symbol of our "untamable past," and "giving definition and meaning to the human enterprise" (Leopold 1949, p. 200–201, 96).

In 1964 the U.S. Congress passed the Wilderness Act, which gave wilderness areas considerable protections. This act, which provides for the designation and protection of wilderness areas, defined wilderness "in contrast with those areas where man and his own works dominate the landscape, is hereby recognized as an area where the earth and its community of life are untrammeled by man, where man is a visitor who does not remain." Far from resolving the conflict between wilderness advocates and advocates of economic development, however, the passage of the act has resulted in a series of political struggles regarding which, and how much, U.S. government land would be designated as wilderness, and what kinds of activities would be allowed on designated land.

A New Wilderness Debate

In the 1990s, a new wilderness debate broke out, as philosophers, historians and scientists all called into question the truth and efficacy of the wilderness *myth.* An early salvo in this new war came from the philosopher J. Baird Callicott, who criticized the entrenched idea/myth of wilderness because it supports an inaccurate view of humans as separate from nature; and because the myth has colonialist overtones, treating members of the cultures who lived there as less than human, whereas these peoples *managed* the land, albeit less intensively than the European colonists. Callicott also thinks the myth confuses policy by emphasizing exclusion of all humans from wilderness. If the emphasis were shifted to protecting *wildness,* protection would only forbid the intrusion of modern, industrial uses, Callicott argues, and one might encourage people to live with nature in unobtrusive ways in order to cohabit with wildlife.

Subsequently this debate was rekindled in two contexts. First the historian William Cronon, who had implicitly raised some of Callicott's issues in his 1983 book, *Changes in the Land*, published a book in which he and his co-authors emphasized that the idealized, mythical idea of wilderness is very much an American construction, a culturally relative idea that should be recognized as very particular to the United States, and prone to hide rather than illuminate the reality of European settlement and colonial land transformations (Cronon 1995).

The wilderness debate also shaped a subsequent debate in conservation biology, as conservation biologists suggested that, whatever the original rationale for wilderness, the wilderness areas in the early twenty-first century are indispensable reserves to protect biological diversity. This idea has since been criticized by Callicott, who argues against the assumption that wilderness areas must be depopulated in order to protect wild species, arguing that conservation biologists requiring wilderness simply perpetuates the old dichotomy between humans and nature. Further the philosopher of biology, Sahotra Sarkar, has argued persuasively that the goals of biodiversity protection and wilderness preservation often conflict; this debate shows signs of continuing well into the twenty-first century. (Sarkar 1999).

The idea of wilderness has been, and remains, both seminal and controversial in ongoing discussions of the American character. Further this idea provides an attitudinal backdrop for explorations in environmental ethics and environmental thought, and also for debates about environmental policy. Given this central role in European and North American—especially U.S.—thought and action, it is not surprising that the idea deeply affects the ways humans understand—and evaluate—new and emerging technologies.

BRYAN G. NORTON

SEE ALSO *Earth; Environment; Environmental Ethics; Wildlife Management.*

BIBLIOGRAPHY

Callicott, J. Baird, and Michael P. Nelson. (1998). *The Great New Wilderness Debate*. Athens: The University of fGeorgia Press.

Cronon, William. (1983). *Changes in the Land: Indians, Colonists, and the Ecology of New England*. New York: Hill and Wang.

Cronon, William, ed. (1995). *Uncommon Ground: Toward Reinventing Nature*. New York: Norton.

Leopold, Aldo. (1949). *A Sand County Almanac and Sketches Here and There*. New York: Oxford University Press.

Marx, Leo. (1967). *The Machine in the Garden: Technology and the Pastoral Ideal in America*. London: Oxford University Press.

Miller, Perry. (1967). *Nature's Nation*. Cambridge, MA: Harvard University Press.

Nash, Roderick. (1982). *Wilderness and the American Mind*, 3rd edition. New Haven, CT: Yale University Press. First edition was published in 1967.

Oelschlaeger, Max. (1991). *The Idea of Wilderness*. New Haven, CT: Yale University Press.

Sagoff, Mark. (1974). "On Preserving the Natural Environment." *Yale Law Journal* 81: 205–267.

Sarkar, Sahotra. (1999). "Wilderness Preservation and Biodiversity Conservation—Keeping Divergent Goals Distinct." *Bioscience* 49: 405–412.

Thoreau, Henry David. 1998 (1862). "Walking." In *The Great New Wilderness Debate*, ed. J. Baird Callicott and Michael P. Nelson. Athens: University of Georgia Press.

WILDLIFE MANAGEMENT

• • •

In some sense, wildlife management is not new. Wildlife was managed for subsistence hunting—by burning fields to create grass for ungulates, for example—by early humans and even perhaps by protohumans. *Game management*—management of animals for *sport* hunting, in particular—has been traced at least as far back as ancient Egyptian civilizations. Large game fields, managed for sport, were maintained for the recreation of Egyptian royalty. Hunting restrictions—which can be thought of as the precursors of modern wildlife management—can be traced back to early tribal customs and taboos. Typically game management involved few species—mostly for food and sport, but also for aesthetics in some cases—and was practiced over relatively small areas in a decentralized manner.

Since the twentieth century, due mainly to a confluence of developments in ecology and society, game management has been supplemented by more comprehensive wildlife management in most developed countries. Game management programs often dominate government wildlife management departments because of their political popularity and because they have, in hunting and fishing license fees, a strong source of revenue. Beginning in the 1920s with the pioneering work of Aldo Leopold, wildlife management took its place next to game programs. Eventually many governments reconceptualized game management as one specialization in the broader field of wildlife management, and in the early-twenty-first century most governments include agencies that accept some responsibility for maintaining healthy populations of almost all indigenous species.

Leopold and Evolution of Wildlife Management in the United States

Leopold, working with his more field-oriented friend, Herbert Stoddard, provided both the intellectual and practical leadership in shifting government agencies, at least in the United States, toward a more holistic approach toward wildlife management. As a consultant and researcher on game populations in the early 1930s, Leopold met and became friends with the British ecologist Charles Elton, an advocate of the empirical study of whole ecosystems. Elton and Leopold both recognized the implication of ecology: It is very difficult to manage single species in isolation without upsetting important ecological processes over time. This insight was driven home to Leopold by the abnormal fluctuations in deer populations in the southwestern United States, where he was director of operations, and sometimes game manager, over national forest holdings. Leopold had employed predator eradication as a means to create an artificially large herd of deer for hunting. During an especially bad winter, more than 60 percent of the deer died because they had eaten all available browse, causing a population crash, destroying vegetation, and encouraging soil erosion. In areas where top predators have been removed and there are no natural checks on wildlife population increases, there are often disagreements about the ethical treatment of animals, including conflict with private hunters and with government management agencies over policies involving culling of wildlife populations. Reducing populations of species whose natural predators have been eliminated is a great challenge. Agencies charged with controlling wildlife populations are sometimes strongly criticized by the public, which has become increasingly concerned with animal welfare and animal rights (Dizard 1999, Sharpe et al. 2001).

Leopold, years later in his classic book of essays *A Sand County Almanac and Sketches Here and There* (1949), included a brief but elegant mea culpa. He said he had mismanaged the land, creating starving deer and eroding hillsides, because he had not yet learned "to think like a mountain" (Leopold 1949, p. 130). Leopold treated his conversion as a revelation and also as a metaphor that must guide the future of wildlife management. Haunted by the "fierce green fire" that he saw in the eye of a dying she-wolf—a wolf shot by his group of forest rangers—Leopold realized, he said, that "there was something new to me in those eyes—something known only to her and the mountain" (Leopold 1949, p. 131). Leopold gradually rejected predator eradication programs and eventually advocated protection of wolves in wilderness areas. He devoted his remaining career to advocating and practicing holistic wildlife management, applying ecological principles to whole ecosystems. He was learning to think on the timescale significant to mountains—and was accepting moral responsibility for the long-term results of his short-term thinking about wolves and deer.

After leaving the U.S. Forest Service in 1928, Leopold became, first, a private consultant on game and sport hunting, and eventually the first professor of game management at the University of Wisconsin. He concluded that predators were an essential element in a healthy ecosystem, and shifted emphasis in his managerial theory and practice toward more holistic habitat management and away from management for single species (Leopold 1939, Flader 1994). By the late 1960s and early 1970s, some states had initiated *nongame wildlife programs*, and since then wildlife programs have flourished in response to strong public support and coexist, more or less easily, with game management programs. Demographic changes also had an impact as more of the population moved to the suburbs and the exurbs. These changes corresponded with an increase in leisure time and an increased demand for opportunities to interact with wildlife in nonconsumptive ways, for example during popular activities such as hiking, camping, and birdwatching. By appealing to this growing interest, governmental and nongovernmental agencies built a political constituency that supported parks, reserves, and wilderness. (Hays 1987).

Leopold, following Henry David Thoreau and John Muir—other *holists* who were very influential in conservation—thus shifted the focus of management from species to systems, and departed from his *resource management* approach. He moved toward a *biotic* view, which sets out to protect the integrity of ecological systems.

Leopold's evolution began with his belief that the goal of management is to maximize game availability; by the time he published his landmark book, *Game Management* in 1933, Leopold had also begun to emphasize the *quality* of game, arguing that quality is inversely related to artificiality. He advocated minimizing interference in the hunter/prey relationship to the greatest extent possible. Leopold believed sportsmanship was enhanced—and moral and aesthetic values supported—

when the sportsman interacts directly with wild game, without the interference of wildlife managers. Leopold realized, however, that growth and dispersion of human populations increases the need for more invasive management. Thus he saw game management as a negotiation between demand for quantity of game for increasing populations and the continuing threat to the quality of game and the hunting/fishing experience.

Leopold also argued that the same methods that he and others had applied to game management should be employed to maximize wildlife more generally, and closed the 1933 book by arguing that managers should apply similar methods to all wildlife. He stated that the goal of wildlife management was "to retain for the average citizen the opportunity to see, admire and enjoy, and to challenge to understand, the varied forms of birds and mammals indigenous to his state" (Leopold 1933, p. 403). Leopold advocated use of agricultural tools to produce more wildlife, claiming that the goals of the profession were not just to keep all life forms in existence, but also to ensure "*that the greatest possible variety of them exist in each community*" (Leopold 1986, p. 403).

By 1939 Leopold had become less optimistic regarding the possibility of managing for particular species, recognizing that ecological relationships are so complex that manipulation of systems to maximize some species will always have unforeseen consequences; species are so intertwined that only habitats can be protected. Leopold advocated protection of whole habitats and argued that society should value whole communities of plants and animals, and stop trying to value and favor some species inordinately. Leopold continued, until his death in 1949, to advocate holistic management, and registered many successes in protecting natural areas. He recognized, however, that truly holistic management remained mostly a dream. His influence, nevertheless, continues, as many wildlife managers follow Leopold's principles and emulate his method of integrating ecological science and management.

Issues in the Twenty-first Century

Since Leopold's time, and especially since the 1980s, concern with wildlife management has been supplemented with attempts to save *biological diversity*, which is a very broad and complex concept that includes wildlife. In the United States, biodiversity policy has been shaped by the Endangered Species Act of 1973, which restricts activities that threaten species of concern, and also mandates species and habitat restoration for species that are listed because of risk of extinction or extirpation from regional habitats. Although the act is

politically controversial, protection of species remains a high priority for large majorities of the public. The act has also been criticized for retaining a bias toward *single-species management*, and there have been many efforts to reshape wildlife management to protect ecosystems and habitats. In this broader effort, endangered species protection is an important element, and the act, with its emphasis on single species management, nevertheless protects many species and their habitats through its designation and protection of critical habitats for listed species, which are of course shared with other plants and animals.

One important ethical controversy arises over the treatment of wild animals in captivity. While zoos have since late in the twentieth century shifted their message from purely recreational enjoyment of animals toward a conservation emphasis, animal rights organizations attack zoos as *animal prisons*, and question the holding of wild animals in captivity as a way to supplement or shore up sagging wild populations. Critics of invasive management of specimen animals ask: What gives humans—who have already disrupted animal communities all over the world—the right to capture and hold animals for conservation breeding purposes? (Norton et al. 1995)

Since 1970, as wildlife management and biodiversity protection policies have become more scientific by incorporating ecology and many other physical and social sciences into the management process, several important consensuses regarding both goals and methods have emerged. One important consensus is that large parks and preserves are necessary, but usually not sufficient, to protect all varieties of wildlife, because even large parks often lose significant numbers of mammal species (Newmark 1995). Accordingly there is increasing interest in managing *the matrix* of private lands that embeds reserves. This may involve creating buffer zones of lighter use around reserves, and creation of protected riparian corridors to connect various reserves and populations of animals (Harris 1984).

Gap analysis has emerged as the state of the art method for protecting biological diversity. According to this technique, ecosystem and habitat conservation programs are judged by comparing biodiversity priorities with existing and proposed reserves. By identifying *gaps*—important ecological communities that have no protection—conservation efforts can be concentrated on saving all community types and, in the process, the species of wildlife that depend upon them (Church et al. 1996, Scott and Csuti 1996). The goal of international conservation is to protect representative samples of all the biological communities in the world (McNeely 1989). Efforts are underway to restore some whole ecological systems and to reintroduce predators in some areas, such as the Greater Yellowstone Ecosystem in the western United States. Restoration of wildlife populations and protection of their habitat is praised not only for its ecological benefits, but also as a means to involve communities in local conservation projects, thereby building community leadership and making citizens more aware of environmental values.

The future of wildlife management—and of wildlife itself— in the early twenty-first century is uncertain. As cities expand into countryside, it becomes more difficult to maintain populations of many species, especially large predators. Scientific experts fear that species such as wolves, mountain lions, and bears will become increasingly hard to protect. As areas not dominated by human uses shrink, wildlife will have to be managed more invasively to protect the diverse biological heritage each generation has inherited. Such management, however, undermines the wildness of wildlife and affects, as Leopold stressed, the quality of the human experience of wild creatures.

Learning to protect truly wild populations will be a challenge for the future. Rapidly accelerating rates of extinction demonstrate that humans have not learned these protection methods yet. As the pressures of expanding populations and cities continue though the twenty-first century, much wildlife will be lost as ubiquitous species that easily cohabit with humans take over the remaining, fragmented habitats. Only a concerted effort to understand and to act decisively can avoid a drastic simplification of the biological context in which humans evolved. Such an effort would involve unprecedented cooperation among scientists, governments, private land-owners, and wildlife management agencies, and could only achieve success if techniques are developed to manage whole regions to maintain adequate reserves and other protections to form a complex matrix of human and natural communities.

BRYAN G. NORTON

SEE ALSO Environment; Environmental Ethics; Management; Wilderness.

BIBLIOGRAPHY

Church, Richard L.; David M. Stoms; and Frank W. Davis. (1996). "Reserve Selection as a Maximal Covering Location Problem." Biological Conservation 76: 105–112. A review of scientific approaches to reserve design.

Dizard, Jan. E. (1999). Going Wild: Hunting, Animal Rights, and the Contested Meaning of Nature. Amherst: University of Massachusetts Press. A thoughtful discussion of the problems of managing wild populations in close proximity to human communities.

Elton, Charles. (1927). Animal Ecology. New York, NY: Macmillan. Elton led the way toward a more systems-approach to studying animal species.

Elton, Charles. (1958). The Ecology of Invasions of Plants and Animals. New York, NY: John Wiley. The classic ecological treatment of migrations of plants and animals.

Flader, Susan L. (1994) Thinking Like a Mountain: Aldo Leopold and the Evolution of an Ecological Attitude toward Deer, Wolves and Forests. Madison: University of Wisconsin Press. Originally published in 1974. A detailed study of Leopold's changing views on wolf management.

Harris, Larry. (1984). The Fragmented Forest. Chicago: University of Chicago Press. An analysis of the causes and remedies of habitat fragmentation.

Hays, Samuel. (1987). Beauty, Health, and Permanence: Environmental Politics in the United States, 1955–1985. Cambridge, UK: Cambridge University Press. A historical look at evolving goals of environmental management.

Leopold, Aldo. (1939). "A Biotic View of Land." Journal of Forestry 37: 113–116. A brief essay that marked a key milestone in Leopold's evolution toward holistic management.

Leopold, Aldo. (1949). A Sand County Almanac and Sketches Here and There. Oxford: Oxford University Press. Leopold's much-read and much-loved reminiscences of a life in conservation science.

Leopold, Aldo. (1986). Game Management. Madison: University of Wisconsin Press. Leopold's classic (1933) "textbook" of management science.

McNeely, Jeffrey A. (1989). "Protected Areas and Human Ecology: How National Parks Can Contribute to Sustaining Societies of the Twenty-first Century." In Conservation for the Twenty-First Century, ed. David Western and Mary Pearl. New York: Oxford University Press.

Newmark, William D. (1995). "Extinction of Animal Populations in Western North American National Parks." Conservation Biology 9: 512–527. An important scientific paper often cited to show the limitations of reserves in the protection of all species.

Norton, Bryan G.; Michael Hutchins; Elizabeth Stevens; and Terry Maple, eds. (1995). Ethics on the Ark: Zoos, Animal Welfare, and Wildlife Conservation. Washington, DC: Smithsonian Institution Press. An anthology of differing viewpoints on zoos and their role in conservation.

Scott, J. Michael, and Blair Csuti. (1996). "Gap Analysis for Biodiversity Survey and Maintenance." In Biodiversity II: Understanding and Protecting Our Biological Resources, ed. Marjorie L. Reaka-Kudla, Don E. Wilson, and Edward O. Wilson. Washington, DC: John Henry Press. A summary of the goals and techniques of gap analysis.

Sharpe, Virginia A.; Bryan Norton; and Strachan Donnelly. (2001). Wolves and Human Communities: Biology, Politics, and Ethics. Washington, DC: Island Press. An anthology of viewpoints on the possible re-introduction of wolves to upstate New York.

Thoreau, Henry David. (1960). *Walden*. New York: New American Library. Thoreau was a pioneer in seeing wildlife as a key to living a better life. *Walden* was first published in 1854.

Thoreau, Henry David. (1998). "Walking." In *The Great New Wilderness Debate*, ed. J. Baird Callicott and Michael P. Nelson. Athens: University of Georgia Press. Essay originally published in 1862.

WITTGENSTEIN, LUDWIG

• • •

Engineer, architect, and one of the most influential analytic and linguistic philosophers of the twentieth century, Ludwig Wittgenstein (1889–1951) was born in Vienna, Austria, on April 26 and died a few days after his sixty-second birthday in Cambridge, England, on April 29. Although seldom emphasized in works about the philosopher, Wittgenstein's life was deeply engaged with technology. He studied mechanical engineering in Berlin and aeronautical engineering in Manchester, England, securing the patent for a propeller in 1911. He also conducted combustion chamber research and his ideas were used for helicopter engines after World War II. Even after abandoning his engineering career, Wittgenstein's engineering education continued to exercise an influence on his philosophical work.

Wittgenstein began his career as a philosopher in 1912 after reading Bertrand Russell and Alfred North Whitehead's *Principia Mathematica* (Volume I, 1910). The logical foundations of mathematics was one of the most important philosophical issues of the day, and between 1914 and 1918 Wittgenstein wrote one of his *Tractatus Logico Philosophicus*. Its spare hundred pages contain a philosophy of logic, of language and meaning, of science, an ontology, and by implication, ethics. Language is the basis for all thought, so that the first philosophical task must be to understand its relation to the world in order to clarify its meaning. Many philosophical problems rest on confusions about the meaning of language; when these confusions are revealed, the problems vanish. Only scientific problems are real and thus able to be truly solved.

Wittgenstein's work was a fundamental influence on the philosophical program of the logical empiricists of the Vienna Circle, including Otto Neurath (1882–1945), Moritz Schlick (1882–1936), and Rudolf Carnap (1891–1970). This program argued that metaphysics,

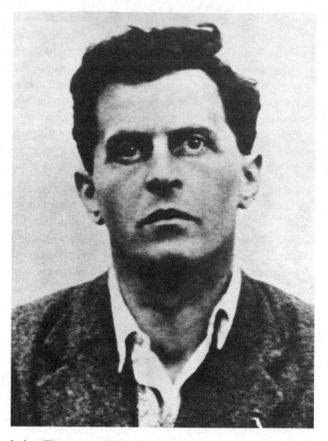

Ludwig Wittgenstein, 1889–1951. After making important contributions to logic and the foundations of mathematics, the Austrian philosopher Wittgenstein moved away from formalism to an investigation of the logic of informal language. (*Hulton Archive/ Getty Images.*)

ethics, and religious beliefs were *non-scientific* and therefore beyond serious philosophical enquiry. Ethical values themselves were sometimes presented as no more than expressions of personal or social emotions. This positivist interpretation of Wittgenstein's thought remained influential even into the 1980s. In the *Tractatus* itself, however, Wittgenstein maintained that although only scientific problems are real, what really matters for human beings are unsolvable questions about right and wrong, good and bad, the meaning of life and so on (Wittgenstein classified these as mystical questions). To be unable to give acceptable scientific answers to such questions did not imply their meaninglessness.

After his death the publication of Wittgenstein's *Philosophical Investigations* revealed a very different Wittgenstein than that associated with the *Tractatus*. To some extent, Wittgenstein turned away from a logical, scientific clarification of language, because diversity of language uses demonstrates the futility of the effort. Language does not function as a scientific mirror of the world but as a profound social phenomenon, as a practice among people. The meanings of words are found in their uses in different contexts, as they are used in *language games*, which belong to specific *ways of life* or *forms of life*, and mistakes arise when philosophers try to find essential meanings in words, because such meanings do not exist. Language is also a learned technique, and to some extent all techniques, even scientific ones, are similar: They all have a deeply social element. There is no *super-game* of philosophy or science that could subsume all other games.

Wittgenstein neither considered himself a scientist nor accepted the idea of technological progress, and he departed clearly from the standard interpretations of scientific development as articulated first by the logical empiricists and then in revised form by Karl Popper (1902–1994) and his followers. In scattered remarks, such as those found in *Culture and Value* (1980), Wittgenstein expressed distrust of modern science and technology and considered them, along with industrialization, as the main causes of war. "Man has to awaken to wonder—and so perhaps do peoples. Science is a way of sending him to sleep again," he once wrote (*Culture and Value*, p. 5e). In his view, science not only fails to deal with the most significant issues but also tends to homogenize the world. The scientific age is associated with a decline in culture, and attempts to popularize science are, according to Wittgenstein, largely mistakes.

Influenced by Viennese cultural and artistic critics such as Karl Kraus (1874–1936), Wittgenstein was sensitive to the negative effects of modern science and technology. Skeptical of progress, he wrote, "It isn't absurd, e.g., to believe that the age of science and technology is the beginning of the end for humanity; that the idea of great progress is a delusion" (*Culture and Value*, p. 56e). The experience of both World Wars and the disappearance of a whole way of life help explain Wittgenstein's critical distrust of scientific and technological development alone as inherently beneficial. In response to the use of the atomic bomb, he actually considered the possibility that modern technology might destroy the whole human race. His pessimism was similar to that of many other intellectuals, including his mentor Bertrand Russell (1872–1970). However, Wittgenstein did not pursue these concerns in any rigorous way.

Many of Wittgenstein's ideas are key features in subsequent criticisms of science and technology. The political theorist Langdon Winner (1986) uses the *form of life* concept to explain how technology becomes a part of one's humanity, as a kind of second nature. As a consequence, technological artifacts often acquire a political character. From an epistemic point of view, sociologist David Bloor (1983) also draws on Wittgenstein to develop a critical assessment of the social nature of scientific knowledge. The so-called "strong program" of the Edinburgh school in the sociology of scientific knowledge uses Wittgenstein's ideas as a basis for their research. Wittgenstein's influence is pervasive and his thinking leaks out into many different fields, including discussions of values in science and technology. For instance, John Searle used Wittgensteinian techniques to attack claims for artificial intelligence (1986). Wittgenstein's main contribution to science and technology criticism consists of a heightened sensitivity to "bewitchment" (Wittgenstein's term) in technological discourse.

ANDONI ALONSO

SEE ALSO *Logical Empiricism; Progress; Skepticism; von Wright, Georg Henrik.*

BIBLIOGRAPHY

Alonso Puelles, Andoni. (2002). *El arte de lo indecible (Wittgenstein y las vanguardias)*. Cáceres, Spain: Universidad de Extremadura. An examination of mutual influences between Wittgenstein and art.

Bloor, David. (1983). *Wittgenstein: A Social Theory of Knowledge*. New York: Columbia University Press.

Glock, Hans-Johann. (1996). *A Wittgenstein Dictionary*. Oxford: Blackwell. The most useful general reference to the many aspects of Wittgenstein's thought and the controversies it has generated.

Kenney, Anthony. (1988). *Wittgenstein*. Harmondsworth, UK: Penguin. The best English introduction to Wittgenstein's philosophy as a whole.

Monk, Ray. (1990). *Ludwig Wittgenstein: The Duty of Genius*. New York: Penguin. The best general biography.

Searle, John R. (1986). *Minds, Brains and Science*. Cambridge, MA: Harvard University Press.

Winner, Langdon. (1986). *The Whale and the Reactor: A Search for Limits in an Age of High Technology*. Chicago: University of Chicago Press.

Wittgenstein, Ludwig. (1922). *Tractatus Logico-Philosophicus*, trans. C.K. Ogden and F.P. Ramsey. London: Routledge.

Wittgenstein, Ludwig. (1953). *Philosophical Investigations*, ed. G.E.M. Anscombe and R. Rhees. Oxford: Blackwell.

Wittgenstein, Ludwig. (1961). *Tractatus Logico-Philosophicus*, trans. D.F. Pears and B.F. McGuinness. London: Routledge.

Wittgenstein, Ludwig. (1980). *Culture and Value*, ed. G.H. von Wright in collaboration with H. Nyman; trans. Peter Winch. Oxford: Blackwell.

Wittgenstein, Ludwig. (1993). *Philosophical Occasions: 1912–1951*, ed. James C. Klagge and Alfred Nordmann.

Indianapolis, IN: Hackett. An extensive collection of shorter texts, including "A Lecture on Ethics."

WORK

• • •

Work done by human beings is purposive action guided by intelligence; work that is repetitive or arduous is often called labor. Both purpose and intelligence may originate in persons other than those actually doing the work. Associated with the basic definition are many related usages including effort expended (also called toil); the result of that effort (a work of art); and one's job or employment, workplace, trade, occupation, or profession. In all these senses work is subject to technological modification, scientific and literary study, and ethical reflection.

Historical Background

In early civilized societies, the kind of work people did depended on their class: The elite had slaves do whatever they considered demeaning, notably if it involved unrewarded physical exertion. Certain religious attitudes perpetuated this devaluation. Some Buddhist and Christian monks, for instance, have associated physical inactivity with the highest spiritual states.

By contrast, in medieval Europe, a combination of prayer and work (*ora et labora*) came to be viewed as a more fully human expression of spirituality. Government despoliation of monasteries during the reformation reduced the feasibility of a life devoted primarily to prayer. But comparable lifestyles are still possible. These aside, the Industrial Revolution tied most workers' survivability to remunerative employment.

Throughout recorded history societies have adopted various attitudes and expectations regarding work. Knowledge of this history lends perspective to contemporary attitudes and expectations. Over time, though, vast technological changes have been made in the production, marketing, and distribution of goods and services, so that past arrangements may not pertain to contemporaneous circumstances. The young Karl Marx (1818–1883) thought history pointed toward an egalitarian society in which every worker would freely choose which activities to engage in. Hannah Arendt (1958) preferred instead a socially stratified society, as in ancient Athens, where a knowledgeable few engage in (political) action, while others work (produce something) or labor (exert themselves physically).

History aside, work-related matters are now routinely viewed in economic terms. In particular, all types of paid activity are identified as labor (skilled and unskilled), and labor costs are largely determined by supply and demand. The supply of labor is, in turn, increasingly a function of globalization; and labor is sought mostly for tasks that technology has not mastered. In this context, work is conceptualized as remunerative employment and is commodified.

Indeed, in the early twenty-first century most people associate work with earning a living and, for the career-oriented, enhancing social status. Frequently, though, personal career aspirations exceed what is attainable under the prevailing economic system—whence arise a number of ethical issues.

These ethical issues include the following: Is the character of work determined solely by the market? Who is obliged to work? Under what circumstances? Should remuneration provide a decent living for the worker (a living wage) and for the worker's family (a family wage)? Which if any institution(s) should provide and/or assure employment, humane working conditions, meaningful and satisfying work? Are those unable to find employment entitled to subsistence? The social effects of scientific and technological change increase the salience of these issues.

Established Ways to Think about Work

Practical approaches to such questions involve both ethical determinations and public policies. These, in turn, draw on research findings in such disciplines as history, economics, sociology, psychology, and jurisprudence, most of which have tended to reinforce socially favored attitudes toward work.

Work is now commonly treated as something bought and sold—typically, a service or product. Employers decide which services or products to offer or generate in a given locale and employ workers accordingly. Workers' remuneration is a function of their productivity in their economic environment. This productivity, in turn, is measured by subtracting overhead—that is, the expenses incurred by conducting a business on-site—from revenue received for services or products. Because a large part of overhead is labor costs, management strives to keep these to a competitive minimum, and may therefore resort to workforce downsizing, technological displacement, and/or workplace relocation. From these practices arise many ethical issues directed to fostering cross-cultural fairness in every aspect of the employment relationship, but especially those having to do with hiring, retention, remuneration, and working conditions. Also important, especially to employees, is finding in work both personal fulfillment and conformity with a socially promoted work ethic (Gini 2000).

A work ethic involves making work a key measure of personal success (Rose 1985, Beder 2000). Industrial-era capitalists fostered a work ethic to maintain a sufficient supply of willing workers. But workplace rationalization and globalization (see below) have rendered the work ethic an unreliable incentive. Some theorists nonetheless still call for meaningful work (Schwartz 1982, Byrne 1990) and a right to work. The latter expression sometimes signifies individualist opposition to unionization (Dickman 1987) and sometimes, gainful employment as such (Harvey 2004, Skopcol 1990). In either sense it is stymied by cost-cutting strategies that replace higher- with lower-paid workers and human beings with machines.

Since the Great Depression in the 1930s governments have assumed some responsibility for this problem by funding systems of unemployment compensation (UC): twenty-two countries had done so by 1949, and sixty-eight countries by 2004. Some scholars argue that any structurally unemployed person is entitled to subsistence income. But governments increasingly require a claimant for UC (as distinguished from generic welfare support) to have been employed and/or be actively seeking employment. Thus in the 1990s even Nordic countries, long noted for their generous UC programs, made these programs less accessible and their benefits less supportive.

Many factors enter into the amount of compensation a person receives for work done. These include the level of development and/or indebtedness of the economy within which one works, one's social and political affiliations, and one's gender, race, national origin, and so on. For example, work done by women is sometimes labeled differently from men's work to justify paying women less (Wright et al. 1987; Mohanty 2003). A society may also set ethical limits on the time a worker may devote to play (Byrne 1990). This pro-work mindset (manifested even in the career-oriented way parents view their children's preschool activities) seeks to maintain an abundance of available labor, now on a global scale.

Large corporations increasingly dominate worldwide employment without assuming responsibility for the negative consequences of their decisions regarding workforce size or location. Even in the face of automation (Byrne 1990) and globalization (Goudzwaard 1979), though, less socially disruptive strategies are possible. These are supported by calls for decent working conditions and a living wage (for example, the United Nations' Universal Declaration of Human Rights, arts. 23 and 24; John Paul II, 1981). Such declarations, though, forestall few if any downsizing decisions. Moreover, the unemployed are often stigmatized and considered personally responsible for their situation even as governments dismantle programs that would mitigate the effects of unemployment (Beder 2000). These conflicting attitudes about work show that the ways in which work has been viewed are no longer adequate to the challenges now emerging.

Finding New Ways to Think about Work

The premodern fusion of work and life associated with primitives and studied by cultural anthropologists is now rare. The modern fusion of work and compensation is coming undone as the availability of jobs depends less on individual skills or dedication than on strategic workplace and/or workforce selections that contribute to profit maximization. In short, the industrial-age problem of worker displacement engendered by rationalization of process is now being compounded by globalization. So earlier analyses of work-related problems need to be reviewed through new lenses if a humane approach to work is to be restored.

Already in the eighteenth century some theorists began speculating about the future of work in view of the inroads of mechanization. Building on earlier utopian visions, some social planners proposed founding communes that would use technologies selectively (Manuel and Manuel 1979). But classical economists, including Adam Smith (1723–1790) and David Ricardo (1772–1823), believed that an unfettered market would achieve "full employment equilibrium." As explained by the French economist Jean-Baptiste Say (1767–1832), for example, supply creates its own demand and this engenders full employment. This "law of markets," or Say's law, predicts that as laborsaving devices replace workers more products become available at prices more consumers can afford, thereby creating a need for additional workers. On this theory, unemployment is not structural (inevitable given system priorities) because a machine-challenged workforce will accept lower wages, which in turn diminishes the need for more expensive machinery (Gini 2000). The mature Marx predicted instead that capitalists' continued recourse to laborsaving devices would engender a great mass of marginalized and potentially insubordinate poor. Proving this prediction incorrect has been a priority for theorists and politicians ever since.

In the nineteenth and early twentieth centuries laborers were assumed to have minimal intelligence, which Taylorization and Fordism sought to exploit. But such workplace strategies destroy job satisfaction, lowering productivity. So during much of the twentieth century social scientists were recruited to improve workplace *human relations* and *quality of work life*, in large part to forestall unionization. In this vein, industrialist Henry Ford once raised his workers' wages above then-current rates so his employees could afford to buy his automobiles. Still others, from John Stuart Mill (1806–1873) to Franklin D. Roosevelt, worried about what the British economist John Maynard Keynes (1883–1946) called *technological unemployment* (Gini 2000, Goudzwaard 1979). Contemporary defenders of Say's law do not share these concerns. Their *trickle-down economics*, however, do not address the emerging phenomenon of companies "churning" a literally global workforce to cut costs. So this survey of work-related issues must, finally, take note of recent attempts to evaluate these new approaches to workforce dynamics.

The problem, in brief, is how to accommodate the tendency (a) of employers to pursue the least costly means of production and (b) of employees to seek the most advantageous compensation. In the age of discovery made possible by the development of reliable ships, employers combined on-site production with slave labor. In the industrial era, employers welcomed wage laborers to their fixed-site factories. Now in the age of computers and electronic telecommunications it is possible to locate supplies, employees, equipment, product, and vendors in whatever mix most favors a given business. Enslavement is now a violation of human rights under international law. It still occurs, however, and in other ways as well. Transnational corporations exploit Third World workers and will continue doing so until prohibited under international law (Moran 2002). They will do so because they gain monetary, trade, tax, and other advantages by locating facilities and employees so as to minimize total labor costs and maximize return on capital. Adding these strategies to automation, capitalist management strives to control workers, as did communist managers (Shaiken 1985). Control of the work process now depends, however, not just on routinizing a task but on where and by whom that task is most profitably carried out.

Most workers need to use tools, including highly complex machines that sometimes replace workers. Thus the availability of employment depends in part on what technology and operators are available. With this in mind, contemporary experts, like their forebears, debate whether introducing new technologies expands or contracts job opportunities (Aronowitz and DiFazio 1994, Bix 2000). In fact, it does both, either by requiring additional workers, as did the assembly line, or by rendering skills previously in demand obsolete, as has containerization and automated manufacturing processes, or both eliminating some jobs and creating others, as has the computer. The U.S. Department of Defense's funding of science and engineering since World War II has severely skewed educational and hiring priorities in many technical fields (Standler 2004). And computer-based network technology generally reduces complex layers of jobs to comparatively few, thereby rendering many employees superfluous. Some laid-off workers can be retrained for new jobs (hence the U.S. Workforce Investment Act of 1998). These jobs, however, are often temporary and/or part-time with no employer-provided benefits. In this context employers no longer stress company loyalty but promise their employees heightened skills for placement elsewhere. But those seeking reemployment may be deemed *overqualified*, in part because they are in a labor pool that includes many others, some no less skilled, in or from countries where compensation is substantially lower. Partly because of this migration of work unemployment is much lower in many developing countries, especially in the Asia-Pacific region, than in some developed countries, especially in Europe. This situation remains subject, however, to profit-maximizing strategies, which are ever under review. So however work is distributed around the world, it will enhance a globalized buyer's market that primarily benefits corporate executives and investors.

This noted, economic growth does tend to lower unemployment, albeit not precisely in accord with Okun's law (a 1% increase in the rate of economic growth lowers the unemployment rate by 0.3%). Lower unemployment, though, is not inconsistent with job obsolescence. Individuals with advanced degrees, especially in technical and business-related fields, do have better marketability than do those less or less appropriately educated. And it is true that in developed countries, especially in Europe, new jobs are being created mainly in the service sector. This sector, though, is itself being transformed by the same network technology that has reduced the number of jobs in manufacturing.

Workers' Rights in a Global Workplace

The global marketplace raises pressing ethical issues regarding workers' rights. But workers' rights are difficult to enforce in many countries. So business ethicists recommend codes of ethics that can be applied cross-culturally. These have tended to favor management, but public awareness of corporate executives' malfeasance and disproportionate compensation has generated support for tighter external regulation of business practices. The decades-old debate about corporate responsibility now takes into account stakeholder theories, which extend property rights to groups other than shareholders and management, such as plant-location cities, suppliers, and customers. But such theorizing is difficult to apply to structurally consolidated professional services, such as in health care, or to transnational combinations in industries such as finance, telecommunications, and retail groceries. Government and corporate leaders extol the resulting increases in productivity, even as they blame the unemployed for not having jobs (Beder 2000). Such politically motivated problem skimming, however, does not address people's growing sense that the globalized marketplace is limiting their employment opportunities.

Global employment strategies that are advantageous to an employer disadvantage some potential employees more than others. Protective tariffs may be imposed to safeguard jobs tied to goods not produced at competitive costs. But the availability of substantially cheaper labor in or from developing countries disfavors retention of higher-paid employees in developed countries. Thus by the year 2015 the U.S. electronics industry will have transferred some three million jobs to India, and possibly as many as that to China. Comparable moves are planned in Europe, even in non-English-speaking countries. Meanwhile, China now produces four times as many apples as the United States so that only growers in the state of Washington can still compete without tariff protections. And if U.S. tariffs on orange juice are abolished under a proposed free trade agreement, Brazil's product will capture the U.S. market and Florida orange growers will no longer hire Mexican migrant workers. Changes of this magnitude in job markets cannot be neutralized by extolling the rewards of adhering to a work ethic. A better response might be to somehow apply Marx's maxim: from each according to ability, to each according to need. This ideal, however, is not easily introduced into the corporation-dominated global economy.

Economists who study the effects of globalization disagree about their ultimate ramifications. Some retain the optimism of Say's law by arguing that the global economy as a whole improves whenever something is produced where it can be done efficiently and at a substantially lower cost than elsewhere. This thesis, which economists explain in terms of *comparative advantage*, needs to be modified to take into account both international monetary exchange rates and the losses incurred by displaced workers. Moreover, if the comparative advantage in question depends on exploiting workers (for example, in sweatshops) or engaging in illegal activities (such as laundering money), it is subject to additional ethical objections. To address such distortions of global fairness both the International Labour Organization and its parent body the United Nations (UN) have identified certain core labor standards with which all employers should comply. Subscribed to by many UN member nations, these standards favor workers' right to organize and condemn forced or compulsory labor, child labor, and discrimination in employment or occupation. Much debated is whether the inclusion of these core labor standards in trade agreements would mostly benefit Third World workers or First World corporations (Basu et al. 2003).

Work in the Future

In short, the ethical problems associated with a globalized and technologically challenged workforce involve not only economic but social and political considerations as well, especially because their solution requires moving beyond the modern tendency to base people's income eligibility almost exclusively on their work. This is rarely considered in the United States, where job responsibilities (such as being "on call 24/7") are blurring the line between work and leisure. Meanwhile in the United Kingdom programs are being developed precisely to achieve better "worklife balance." In some places, such as Alaska and Saudi Arabia, resource-based wealth has been distributed to all citizens, even those not participating directly in the generation of that wealth. Expanding such arrangements and devising others not dependent on the market is desirable (Offe and Heinze 1992) but unlikely so long as such traditional capitalist values as property rights and the work ethic remain dominant. For the foreseeable future, then, few besides the independently wealthy will be able to live decent lives without engaging in wage work. Where, then will this work be found?

This question is often answered by extending historical precedents into the future, namely, by viewing past transitions (from agricultural to industrial to service and to information sectors) as an evolutionary indication of another major employment sector to come. This may be so, but present data fail to reveal this new source of work in the world.

In the United States alone, some three-fourths of all workers deal with—create, collect, or use—information. The complexity of their involvement, and thus of their compensation over a lifetime, is partly a function of their education. But only 25 percent of the U.S. workforce have had four years of college, and only 5 percent have advanced degrees. The growth rate of researchers in the United States is a third less than the rate for all OECD countries. A fourth of the scientists in the United States are foreign-born, as are a third of doctorate-level scientists and engineers. Moreover, the United States has in recent years been attracting fewer foreign students to its technical programs. Indeed, it has been lowering annual ceilings for high-skilled (H-1b) visas even as the Japanese have greatly increased theirs. Meanwhile, more than 10 percent of all U.S. workers, mostly women, do not have regular full-time jobs. Given such indications of the present situation, what is needed is surely neither utopian nor anti-utopian scenarios but all the social inventiveness Americans can muster.

EDMUND F. BYRNE

SEE ALSO *Affluence; Automation; Business Ethics; Capitalism; Class; Critical Social Theory; Efficiency; Entrepreneurism; Industrial Revolution; Globalism and Globalization; Levi, Primo; Management: Models; Marx, Karl; Money; Poverty.*

BIBLIOGRAPHY

Arendt, Hannah. (1958). *The Human Condition.* Chicago: University of Chicago Press. Argues that the political sphere should address intellectual concerns, not the life-maintenance concerns of laborers.

Aronowitz, Stanley, and William DiFazio. (1994). *The Jobless Future: Sci-Tech and the Dogma of Work.* Minneapolis: University of Minnesota Press. Affirms that as technology diminishes job opportunities society must replace free market mythology with more egalitarian policies.

Basu, Kaushik, Henrik Horn, Lisa Román, and Judith Shapiro, et al., eds. (2003). *International Labor Standards: History, Theory, and Policy Options.* Oxford: Blackwell. Examines the history and economic implications of the controversial call for international labor standards.

Beder, Sharon. (2000). *Selling the Work Ethic.* Carlton North, Victoria, Australia: Scribe. Discusses the idea that the capitalist-fostered work ethic is both factually unfounded and detrimental to human well-being.

Bix, Sue. (2000). *Inventing Ourselves Out of Jobs? America's Debate over Technological Unemployment.* Baltimore and London: Johns Hopkins University Press. Diverse interested parties over three generations support technological advances, disagree about responsibility for structural unemployment.

Byrne, Edmund F. (1990). *Work, Inc.: A Philosophical Inquiry.* Philadelphia: Temple University Press. Asserts that defense of workers' rights requires community constraints on corporate hegemony.

Dickman, Howard. (1987). *Industrial Democracy in America.* La Salle, IL: Open Court. Claims organized labor is an unjustifiable hindrance to *laissez-faire* capitalism.

Gini, Al. (2000). *My Job, My Self: Work and the Creation of the Modern Individual.* New York: Routledge. Maintains that work is essential to maintenance of a human being's sense of self-worth.

Goudzwaard, Bob. (1979). *Capitalism and Progress: A Diagnosis of Western Society,* trans. Josina Van Nuis Zylstra. Grand Rapids, MI: Eerdmans. Argues that supportive economic theories notwithstanding, people's faith in progress is in crisis.

John Paul II, Pope. (1981). *On Human Work: Encyclical "Laborem Exercens."* Washington, DC: United States Catholic Conference. Has value primarily as manifestation of the worker's intrinsic dignity and vocation.

Manuel, Frank E., and Fritzie P. Manuel. (1979). *Utopian Thought in the Western World.* Cambridge, MA: Harvard University Press, Belknap Press. A comprehensive survey of utopian writings in the Western intellectual tradition.

Mohanty, Chandra Talpade. (2003). *Feminism Without Borders: Decolonizing Theory, Practicing Solidarity.* Chapel Hill, NC: Duke University Press. Describes how diverse local cultures regarding gender are used to exploit women workers.

Moran, Theodore H. (2002). *Beyond Sweatshops: Foreign Direct Investment and Globalization in Developing Countries.* Washington, DC: Brookings Institution Press. Advocates more equitable distribution of foreign investment in developing countries.

Offe, Claus, and Rolf G. Heinze. (1992). *Beyond Employment: Time, Work, and the Informal Economy,* trans. Alan Braley. Philadelphia: Temple University Press. Examines the emergence of an informal economy that replaces waged labor with non-monetary exchanges of services.

Rose, Michael. (1985). *Re-working the Work Ethic.* London: Batsford. Examines changes in workplace values and responsiveness to rewards and controls.

Schwartz, Adina. (1982). "Meaningful Work." *Ethics* 92(4): 634–646. Argues that a more democratic division of labor is better suited to enhancing individual autonomy.

Shaiken, Harley. (1985). *Work Transformed: Automation and Labor in the Computer Age.* New York: Holt, Rinehart, and Winston. Affirms that workers should have more input into the planning, programming and control of computerized automation.

Skocpol, Theda. (1990). "Brother, Can You Spare a Job? Work and Welfare in the "United States." In *The Nature of Work: Sociological Perspectives,* ed. Kai Erikson and Steven Peter Vallas. New Haven, CT: Yale University Press. Presents the idea that employment assurance, as recommended by New Deal planners, needs to be reconsidered now that the era of selective welfare is over.

Wright, Barbara Drygulski, et al., eds. (1987). *Women, Work, and Technology: Transformations.* Ann Arbor: University of Michigan Press. Feminist responses to gender-biased accounts of women's involvement in workplace technologies.

INTERNET RESOURCES

Harvey, Philip (2004). The Right to Work and Basic Income Guarantees: Competitive or Complementary Goals? Available at http://www.etes.ucl.ac.be/BIEN/Files/Papers/2004Harvey.pdf. Assesses alternative ways proposed to redress the failure of economies to assure a decent standard of living.

Standler, Ronald B. (2004). "Funding of Basic Research in Physical Science in the USA." Available at http://www.rbs0.com/funding.pdf. Argues that science in the United States is too dependent on military funding priorities.

WORLD BANK

· · ·

The World Bank (Bank) or International Bank for Reconstruction and Development was created at a meeting of the forty-four World War II allied nations in Bretton Woods, New Hampshire, in 1944. Because of its promotion of economic development, the Bank is also an international institution involved to some extent with issues relevant to science, technology, and ethics.

Historical Emergence

At its inception, the Bank's mission was to make long term capital loans to countries harmed by World War II and, more generally, to undeveloped countries worldwide. Sister organizations founded at the same time, with overlapping missions, include the International Monetary Fund (IMF), the International Development Association, the International Finance Corporation, and the Multilateral Investment Guarantee Agency. Through an agreement signed in November 1948, the Bank acts as a specialized UN agency.

Surprisingly the Bank was largely irrelevant to the process of rebuilding Europe after World War II; the majority of the huge financial commitment came through the United States' Marshall Plan. In the first twenty-five years after its creation, the World Bank made only a handful of loans to European states (albeit large ones), including loans for the reconstruction of the steel industry in France, Belgium, and Luxembourg (McLellan 2003). With money in hand collected from its subscribing members, the Bank nevertheless felt an intense *pressure to lend*, and fell back to a secondary mission, that of lending to economically underdeveloped countries.

The Bank's charter contained language militating in favor of project-based lending, and in the early years most of its loans were for the finance of specific projects such as the development of mines or dams (Skogly 2001). The Bank, which experienced a failure rate of as much as 70 percent of its loans in the poorest countries (McLellan 2003), soon noticed that local conditions did not support the success of these projects. Among the factors cited by the Bank for project failure in poor countries are ineffective government, corruption, and *lack of transparency* (World Bank 1994).

To respond to these problems, the Bank began a program of so-called structural adjustment loans or SALs, which represented a movement away from its original project-based lending. SALs involve money advanced for a variety of projects and efforts, and are explicitly conditioned on the implementation of structural and economic changes by the borrowing country, including decentralization, privatization, cost-cutting, and discontinuance of tariffs and supports for its own currency.

In 2002 the Bank made $11.5 billion in loans in support of ninety-eight projects in forty countries. It currently has a total of about 1,800 projects in almost every developing country (McLellan 2003).

Evaluation

The main charge leveled against the Bank is that its ideological approach to lending actually creates the poverty it is intended to combat. Most critics focus on the SALs with their attendant mandatory conditions. Vikas Nath says that the Bank reduced many Third World nations to even greater poverty and dependence on Western aid. Countries often have to borrow from other sources to repay the Bank. Borrowing countries "gradually lost their ability to shape their own future...." (McLellan 2003, p. 62). In the poorest countries, government employment arguably provides a social safety net when jobs in private industry are unavailable. Critics argue that, by forcing cuts in government employment, the Bank throws people into poverty, since the predicted growth in private employment does not materialize soon enough, or with salaries high enough, to pick up the slack.

For many years, the Bank rarely assessed the environmental or social impact of projects it funded. The Sardor Sarovar dam project in India, projected to displace 1 million people, was canceled because of local protests. The Bank admits that under its current portfolio of projects, some 26 million people have been evicted, lost land, or lost livelihoods. As a result, in the early-twenty-first century the Bank conducts environmental reviews of all projects, and lending for environmentally beneficial projects makes up 10 percent of its portfolio. (McLellan 2003).

Critics also question whether a for-profit institution can carry out a not-for-profit mission in the Third World. "The World Bank focuses on economic growth until it is distracted by other issues like hunger, women, health, the environment, etc. The World Bank tries to adapt itself to these considerations without giving up its basic goal" (Danaher 1994, p. ix).

Such critics contend that the SALs in particular lead to the repression of democratic rights in poor countries, without reducing poverty. "Structural adjustment is a policy to continue colonial trade and economic patterns developed during the colonial period. . . . [Third World countries] are more dependent on the ex-colonial countries than we ever were" (Danaher 2003, p. 4). Thirty out of forty-seven African governments have been in SALs for many years—yet by 1992, rather than being reduced, their external debt had more than doubled (to $290 billion) (Danaher 2003).

Shakrukh Rafi Kahn studied the impact of Bank lending in Pakistan over a twenty-year period. Though some initiatives, such as privatization of state-owned banks, were somewhat successful, he noted the greatly disproportionate impact of the Bank's SAL policies on the nation's poor: "They have been hurt many times over. Not only have they borne a disproportionate burden of the cuts in employment, cuts in subsidies and the rise in prices, but they also have started bearing more of the tax burden" (Kahn 1999, p. 120).

The Bank, in more guarded language, seems to be aware of the problems with its programs. In a publication on governance in developing countries, the Bank notes that the form of government (democratic or autocratic) is not one of its concerns. In reviewing its SALs around the world, the Bank concedes that things have not gone well in Africa: "Bank assistance to Africa is dominated by the collapse of public sector capacity in many countries, brought about by a combination of state over-extension, delayed adjustment to changed external economic circumstances, natural events, and poor governance" (World Bank 1994, p. 9). It recognizes that Western solutions to problems cannot always be transferred wholesale to countries with very different traditions. The Bank concludes "Performance in sub-Saharan Africa has been disappointing" (World Bank 1994, p. 11).

In a more overtly self-critical document, water expert George Keith Pitman (2002) argues that the Bank is poorly organized to implement its own water resources management strategy. Knowledge and leadership on water issues is seriously fragmented within the Bank's management structure, while budget cuts have eroded the knowledge function. Pitman also quotes certain nongovernmental organizations (NGOs) that believe "the pressure to lend . . . has not been removed and continues to work against aspects of the water policy that recommend greater attention to smaller and cheaper alternatives" (Pitman 2002, p. 39).

The Poverty Action Lab, a Massachusetts Institute of Technology project, has begun randomized evaluations of the impact of Bank projects. Its researchers agree that success cannot be measured only by concrete achievements; assessments must include the impact of Bank projects on the lives of the poor (Dugger 2004). For example, hiring additional teachers for rural Indian schools did not improve test scores, but treating debilitating intestinal worms in Kenyan students raised attendance at a cost of only $3.50 per treated person per year.

Economists at the Poverty Action Lab say that the Bank's culture led to a certain complacency in the past, preventing the Bank from rigorously evaluating its own projects. The Bank is beginning to pay attention, organizing its own randomized studies.

Columbia professor Joseph Stiglitz believes that the Bank has been more successful than the international monetary fund in undertaking sweeping reforms of its own structure and approach: "the bank has always been less hierarchical than the IMF and more accepting of alternative views. . . . [by 1997] the bank had begun to seriously address the fundamental criticisms levied at it" (Stiglitz, p. 122).

Conclusion

The Bank is a well-funded, powerful Western institution with the mission of aiding developing countries. Many of its good intentions may be wasted due to its attempt to apply free market solutions in countries with very different traditions, or that are simply not ready for these approaches.

JONATHAN WALLACE

SEE ALSO *Development Ethics*; *Modernization*; *Money.*

BIBLIOGRAPHY

Danaher, Kevin, ed. (1994). *Fifty Years is Enough: The Case Against the World Bank and the I.M.F.* Boston: South End Press. A collection of essays from a left-wing and critical perspective.

Dugger, Celia. (2004). "Letter From Washington: World Bank Challenged: Are the Poor Really Helped?" *The New York Times*, July 28, p. A4.

Kahn, Shakrukh Rafi. (1999). *Do World Bank and I.M.F. Policies Work?* New York: St. Martin's Press. A study of the impact of twenty years of structural adjustment loans in Pakistan

McLellan, Elisabeth P., ed. (2003). *The World Bank: Overview and Current Issues*. New York: Nova Science Publishers. An overview of the Bank's history and impact on the third world

Pitman, George Keith. (2002). *Bridging Troubled Waters: Assessing The World Bank Water Resources Strategy*. Washington, DC: World Bank. A World Bank analyst argues that the Bank's strategy on water resources is flawed

Skogly, Sigrun. (2001). *Human Rights Obligations of the World Bank and the International Monetary Fund*. London: Cavendish Publishing. An analysis of the impact of the Bank under the International Law of Human Rights.

Stiglitz, Joseph. (2003). "Democratizing the International Monetary Fund and the World Bank: Governance and Accountability," in *Governance: An International Journal of Policy, Administration and Institutions*, 16 (1) 111–139. A defense of the World Bank as an institution responsive to its critics.

World Bank. (1994). *Governance: The World Bank's Experience*. Washington, DC: Author. The Bank's critique of the impact of its structural adjustment loans, placing some of the blame on local corruption and conditions.

WORLD COMMISSION ON THE ETHICS OF SCIENTIFIC KNOWLEDGE AND TECHNOLOGY

• • •

The World Commission on the Ethics of Scientific Knowledge and Technology (COMEST) mirrors at the international level numerous national commissions on science, technology, and ethics. In the early 1990s, the General Conference of the United Nations Educational, Scientific, and Cultural Organization (UNESCO) formally recognized that the changes wrought by science and technology raise questions that demand ethical reflection (Pompidou 2000). In 1997 it approved the creation of COMEST to institutionalize this growing awareness that ethical reflection must become an integral part of scientific research and its technological applications. COMEST and the Bioethics Programme at UNESCO comprise its Programme on the Ethics of Science and Technology, which is designed to further the mission of UNESCO to serve as the conscience of the United Nations.

In addition to advising UNESCO on its program concerning the ethics of scientific knowledge and technology, COMEST is mandated to: (a) be an intellectual forum for the exchange of ideas and experience; (b) identify the early signs of risk situations; (c) advise decision makers on such issues; and (d) promote dialogue between scientific communities, decision makers, and the general public. COMEST is composed of eighteen members and eleven ex-officio members diversified by expertise, nationality, and culture. The operating budget of COMEST for the 2002–2003 biennial was $3.8 million.

By mid-2004, COMEST had held three main sessions. The first was in Oslo in April 1999, which focused on analysis of the ethical aspects in the fields of energy and freshwater resources and the information society. The second, in Berlin in December 2001, was devoted to assessing the progress of COMEST and its influence. In addition, a youth forum on the ethics of science and technology was held and a statement about space policy developed. The third session was held in Rio de Janeiro in December 2003 with a significantly wider agenda incorporating the ethics of biotechnology and nanotechnology. In addition, the Rio de Janeiro Declaration on Ethics in Science and Technology, signed by representatives from Portugal and several countries in South America and Africa, committed the party nations to pursue ethical approaches to scientific and technological advance. The sessions were attended by policy makers, scientists, and representatives from various organizations and member nations.

Complementing these three sessions are various COMEST subcommissions and working groups. Four subcommissions have focused on the ethical aspects of freshwater resources, space policy, energy use, and the information society. These research topics were designed to follow up on the efforts of the World Conference on Science hosted by UNESCO and the International Council for Science (ICSU) held in Budapest in 1999. Working groups have addressed issues such as ethics education and ethics and responsibility in research training.

The principle outputs of the subcommissions and working groups have been a series of publications. Each report surveys an issue and highlights the ethical questions involved. The main focus is on recommendations to COMEST or other decision-making bodies regarding alternative courses of action. For example, the subcommission on the ethics of freshwater created the Research and Ethical Network Embracing Water (RENEW) to identify and endorse examples of best ethical practices of freshwater use. It also produced a report, "Some Examples of Best Ethical Practices in Water Use," that used five case studies to highlight fundamental ethical principles to guide water use policies. Other subcommissions have outlined considerations that could point the way toward more just and sustainable policies (for example Pompidou 2000, Kimmins 2001). The principle of precaution and the concepts of sustainable development and environmental responsibility underpin these recommendations.

COMEST has grown over its short history. Its budget increased 50 percent for the 2002–2003 sessions, and it has broadened its scope of topics. The global scale of COMEST provides its three main strengths. First the internationalization of ethical issues involving science and technology necessitate a global forum such as COMEST to foster communication and mediate conflicts. Its multicultural and interdisciplinary analyses have contributed to better identification of the ethical issues involved in areas such as freshwater, space, and energy. They have also detected early signs of possible risks to society and articulated guidelines for decision makers in the public and private sectors. Second its global reach allows COMEST to promote the development of ethical reflection on these issues in countries that do not have such institutions. Third the scope of COMEST allows it to formulate universal norms to guide the wise use of science and technology.

The global scale of COMEST is also a weakness because it can distance its analyses from the site-specific considerations necessary for formulating ethical policies. Invoking universal standards and ethical principles in concrete situations presents COMEST with its biggest challenge. Several subcommission reports recognize the need to tailor solutions to local conditions (Kimmins 2000). Yet this means that COMEST may be out of synch with its intended audiences and must strive to reconcile its global reach with local needs. Toward this end, COMEST must establish more objective assessments of its work in order to evaluate its efforts.

ADAM BRIGGLE

SEE ALSO *United Nations Educational, Scientific, and Cultural Organization.*

INTERNET RESOURCES

Brelet, Claudine. (2004). "Some Examples of Best Ethical Practice in Water Use." United Nations Educational, Scientific, and Cultural Organization. Available from http://portal.unesco.org/shs/en/ev.php-URL_ID=4382&URL_DO=DO_TO-PIC&URL_SECTION=201.html. A 54 page report by the COMEST subcommission on the Ethics of Freshwater Use.

Kimmins, James P. (2001). "The Ethics of Energy: A Framework for Action." United Nations Educational, Scientific, and Cultural Organization. Available from http://unesdoc.unesco.org/images/0012/001235/123511eo.pdf. A 55 page report by the COMEST subcommission on the Ethics of Energy, covering the current situation of energy throughout the world and the ethical challenge of energy, with recommendations for future action.

Pompidou, Alain. (2000). "The Ethics of Space Policy." United Nations Educational, Scientific, and Cultural Organization. Available from http://portal.unesco.org/shs/en/ev.php-URL_ID=4382&URL_DO=DO_TOPIC&URL_SECTION=201.html. A 137 page report by the working group on the Ethics of Space Policy addressing ethical implications of space activities for different sociocultural contexts.

World Commission on the Ethics of Scientific Knowledge and Technology. United Nations Educational, Scientific, and Cultural Organization. Available from http://portal.unesco.org/shs/en/ev.php-URL_ID=1373&URL_DO=DO_TOPIC&URL_SECTION=201.html. Link to the COMEST page of the UNESCO web site.

WORLD HEALTH ORGANIZATION

• • •

The World Health Organization (WHO) is one of the sixteen United Nations (UN) specialized agencies, with a mission to promote world health. The organization's broad conception of health as including politicized issues such as poverty, apartheid, and environmental quality has aroused controversy over the years.

Organization and History

The WHO was conceived at the 1945 San Francisco conference at which the United Nations was formed. It came into being on April 7, 1948, after its constitution was ratified by twenty-six of the original sixty-one members. WHO is based in Geneva and has six regional offices: Africa, Europe, Southeast Asia, Americas, Eastern Mediterranean, and Western Pacific. Governance is provided by the World Health Assembly, with representatives from (as of 2005) 192 member states. The assembly selects an executive board, which in turn nominates a director general, who is elected by the assembly for a five-year term.

The original top WHO priorities in 1948 were malaria, maternal and child health, tuberculosis, venereal disease, nutrition, and environmental sanitation.

Subsidiary concerns included public health administration, parasitic and viral diseases, and mental health.

WHO is the successor to a series of international Sanitary Commissions, beginning in the nineteenth century, that concentrated on the containment of infectious diseases. Whereas the philosophy of those earlier organizations was to keep infectious diseases out of nations or regions, the philosophy of WHO was to eradicate those diseases wherever they were found, a "total change of perspective" from that of its predecessors (Beigbeder 1998, p. 13).

In the early twenty-first century WHO fields emergency teams of medical professionals who respond to the outbreak of new infectious diseases such as severe acute respiratory syndrome (SARS) and avian flu. WHO also helps member countries create or improve medical schools and services.

Concept of Health

The WHO definition of health is very broad. According to the organization's constitution, health is "a state of complete physical, mental and social well-being and not merely the absence of disease or infirmity." WHO conceives of health as a fundamental human right and cornerstone of world peace.

In line with its broad definition of health, WHO has been a pioneer in environmental concerns. It was concerned as early as the 1950s about the effects of the eradication of insect species and the peaceful uses of nuclear power.

Although WHO has been most effective as a detail-oriented technical organization concentrating on medical and scientific problems such as smallpox eradication, its broad mandate has opened the door to numerous efforts to politicize it. From the beginning the WHO assembly has debated and voted on resolutions introduced by its members on political topics such as the effect on Palestinian physical and mental health of the Israeli occupation or on Nicaraguan health of U.S. sanctions. From the date Israel joined WHO the Eastern Mediterranean group always held its meetings in Arab capitals to which the Israeli delegates were not permitted to travel, effectively keeping Israel from playing a role in WHO regional activity. This situation was not resolved for more than thirty years, when Israel was invited to join the European region.

The U.S. ambassador William Scranton said in 1976 that "the absence of balance, the lack of perspective and the introduction by the World Health Organization of political issues irrelevant to the responsibilities of the World Health Organization do no credit to the United Nations" (Siddiqi 1995, p. 8).

Smallpox Eradication: A WHO Success

WHO played a lead role in one of the more dramatic medical victories of modern times: the worldwide elimination of smallpox. The organization announced its smallpox campaign in 1966 and was able to declare victory in 1979, at a cost of about $313 million. WHO acted as a clearinghouse for strategy, knowledge, and vaccine and coordinated a worldwide volunteer effort. To date smallpox is the only infectious disease that WHO or any other organization has succeeded in eradicating. Unlike malaria, one of the most visible failures of WHO, smallpox was an easier target because it is transmitted from human to human with no animal vectors, has a low rate of transmission and develops slowly, is easy to diagnose, and is easy to contain with small doses of vaccine.

Malaria: A WHO Failure

In 1955 WHO announced the ambitious goal of worldwide elimination of malaria; by 1960, sixty-five countries and territories had antimalarial programs. Those programs relied primarily on spraying the walls of houses with DDT. In 1966 WHO announced that 813 million people, 52 percent of the at-risk population, had been insulated from the disease. From 1959 to 1966 almost 11 percent of the organization's annual budget was devoted to the malaria campaign. However, things began to backslide soon afterward as malaria cases began to increase in some countries; for example, Pakistan, which had only 9,500 cases in 1968, had 10 million in 1974 (Siddiqi 1995).

By 1969 WHO recognized that the eradication program had failed. Many mosquitoes lived, bred, and bit their victims far away from the house walls that were being sprayed; some forms of shelter did not have walls; some species were becoming resistant to DDT or otherwise had changed their behavior; WHO had failed to account for population migratory patterns; and many countries did not have the infrastructure necessary to support the program. In 1969 WHO acknowledged that the eradication program did not "adequately take into account economic and social factors" in malaria-ridden countries (Siddiqi 1995, p. 163). Subsequently, WHO changed its focus from eradication to control of malaria. The disease continues to be the world's most lethal parasite-borne ailment and the second most important killer after tuberculosis in more than 190 countries inhabited by 40 percent of the world's population (Beigbeder 1998).

Despite its failure to conquer malaria, WHO has continued to attempt the worldwide eradication of infectious diseases. It vowed to eliminate polio by 2005. However, the September 11, 2001, attacks and the perceived intentions of al Qaeda to use any biological weapon available to attack the West led to renewed consideration of whether disease eradication will ever be possible (Roberts 2004).

Infant Formula: A Controversial Initiative

In December 1969 WHO began to focus on the decline in breast-feeding in Third World countries, which it believed might have been attributable to the aggressive promotion of formula substitutes. Many highly political nongovernmental organizations (NGOs) had seized on this issue as an important one, symbolic of the continuing fallout from colonialism and the exploitation of the Third World by multinational companies. In October 1979 WHO and the United Nations International Children's Emergency Fund (UNICEF) cosponsored a conference that was attended by NGOs and the formula industry. WHO, which had accepted a mandate to mediate between the opposing sides, adopted a working document that appeared to the companies to adopt many NGO grievances without citing supporting data. This led to collisions with "important commercial interests" (Beigbeder 1998, p. 76). The conference resulted in no compromises, and the NGO-industry dialogue was discontinued. WHO and UNICEF pressed on, in 1981 adopting nonbinding recommendations to member states relating to the marketing of substitutes for breast milk.

During the formula debate WHO was seen by critics as intervening in an ideological debate without citing firm scientific evidence for the proposition that babies were being harmed or killed by the use of formula instead of breast milk. WHO also was accused by the industry of disregarding social and even medical factors that contributed to the use of formula, such as its use by women with inadequate production of breast milk (Beigbeder 1998).

The Normative Role of WHO

WHO has three different modes of action under its constitution: It can adopt conventions, make regulations, or issue nonbinding recommendations. Whereas the first two actions bind its members to act, the third does not.

In practice most of the work done by WHO has been an exercise of its nonbinding recommendation power. The organization has been extremely reluctant to exercise its normative powers to make binding international law or rules. This is partly attributable to the initial reluctance of the United States to ratify the WHO charter, fearing that its actions would dictate the passage of domestic legislation: "Clearly, WHO's more influential member states have no intention to convert the Organization into a World Ministry of Health, no more than they wish to create a world government" (Beigbeder 1998, p. 15). WHO has proposed a single convention on tobacco that was never adopted. Even its nonbinding recommendations are a "starkly limited tool" (Koplow 2003, p. 143). Some commentators believe that WHO's reluctance to exercise its normative powers is a product of "organizational culture established by the conservative medical professional community that dominates the institution" (Taylor 1992, p. 303). David Koplow has noted that the WHO "has no power to enforce compliance, to mandate any particular resolution of a dispute, or to impose sanctions upon recalcitrant states" (Koplow 2003, p. 145).

Organizational Effectiveness

The organization's executive director Halfdan Mahler asked in 1987 whether WHO was to be "merely a congregation of romanticists talking big and acting small" (Beigbeder 1998, p. 191). His successor, Hiroshi Nahajima, appointed in 1988, said that "in the past, we have tended to be rigid and doctrinaire, when, in fact, the utmost flexibility is called for" (Beigbeder 1998, p. 28).

In a 1991 report the Danish government evaluated the effectiveness of WHO programs in Kenya, Nepal, Sudan, and Thailand and found "weak analytical capacity," a lack of prioritization, and failure to delegate authority (Beigbeder 1998, p. 191). Member nations often lack the resources to pay for the measures recommended by WHO or do not have the infrastructure or commitment necessary to implement them.

In the early years of the twenty-first century WHO, like other UN agencies, experienced a struggle for dominance between its First World and Third World members. While the United States continued to pay 25 percent of the organization's budget, the WHO executive board, only 42 percent of whose members came from Third World nations in 1950, by that time had an overwhelming majority of Third World representation (68 percent) (Siddiqi 1995). The United States and its allies frequently exercised behind-the-scenes influence on the outcome of WHO deliberations in a way that contradicted the apparently democratic and majoritarian structure of the organization. For example, the United States and Russia, the holders of the last publicly known smallpox stocks, were able to set WHO policy on the destruction of those stocks.

Evaluation

When it concentrates on technical cooperation, WHO sometimes has been extremely effective, as it was in eliminating smallpox from the world. However, like its sister UN agencies it has expended a large proportion of its resources and credibility in political and ideological disputes that have detracted from its technical mission.

JONATHAN WALLACE

SEE ALSO Bioethics; Health and Disease.

BIBLIOGRAPHY

Beigbeder, Yves. (1998). *The World Health Organization*. The Hague, Netherlands: Martinus Nijhoff.

Koplow, David A. (2003). *Smallpox: The Fight to Eradicate a Global Scourge*. Berkeley: University of California Press.

Roberts, Leslie. (2004). "Polio: The Final Assault." *Science* 303: 1960–1968.

Siddiqi, Javed. (1995). *World Health and World Politics: The World Health Organization and the United Nations System*. London: Hurst.

Taylor, Allyn Lise. (1992). "Making the WHO Work." *American Journal of Law and Medicine* XVIII(4): 101–346.

WORLD TRADE ORGANIZATION

• • •

The World Trade Organization (WTO) is the largest, most powerful international organization dealing with global rules of trade among nations. It was formed in 1995 following the so-called Uruguay Round of negotiations under the General Agreement on Tariffs and Trade (GATT), the previous multilateral trading system established in 1948. Whereas GATT was primarily concerned with trade in goods, the WTO covers trade in goods and services, banking and finance, intellectual property, dispute settlement, and trade policy reviews. The purpose of the WTO is to provide a negotiating forum for nations to form agreements to lower trade barriers to ensure that trade flows as freely, fairly, and predictably as possible. The WTO regulates trade by administering and negotiating trade agreements, resolving trade disputes, reviewing national trade policies, providing technical assistance and training programs in developing nations, and cooperating with other international organizations. All WTO trade agreements are the result of a consensus among representatives of member governments, ratified by the parliaments of the participating nations. These binding agreements guarantee nations their trade rights and responsibilities. For the 147 member nations, the WTO is the most influential institution of international commerce.

WTO Agreements and Organization

Under WTO agreements, countries should neither discriminate among their trading partners nor should they discriminate between foreign and domestic products and services. Every government should be given "most-favored-nation" status whereby any favor granted to one nation must be granted to every other nation, thus ensuring that all trade partners be treated equally. The WTO aims to make trade more free and more fair by lowering trade barriers such as customs duties (tariffs), eliminating import bans or quotas, and limiting the nontariff trade barriers that nations may implement and enforce, such as domestic laws regulating product standards and liability, environmental protections, use of tax revenues for public services, and other domestic laws regulating investment and trade. The WTO limits the nature of tariffs a nation may impose, as well as what kind of nontariff barriers to trade nations may implement and enforce. Through the WTO Dispute Settlement Process, nations can challenge each other's laws on behalf of their commercial interests if they believe barriers to trade exist. If member nations do not conform to WTO regulations they face possible economic sanctions.

Six main agreements comprise the WTO: the umbrella agreement establishing the WTO, the General Agreement on Tariffs and Trade, the General Agreement on Trade in Services (GATS), and the agreements on Trade-Related Aspects of Intellectual Property (TRIPS), Dispute Settlement, and Trade Policy Reviews. The highest authority is the Ministerial Conference, where delegates from member nations meet every two years to reach consensus on multilateral agreements. The second level of authority, responsible for decisions between Ministerial Conferences, has three branches: the Dispute Settlement Body, the Trade Policy Review Body, and the General Council. The General Council is divided into three more councils, each handling a different area of trade: the Goods Council, the Services Council, and the TRIPS Council. Numerous specialized committees and working groups work on the details of individual agreements, as well as issues relating to the environment, development, finance, and regional trade agreements. The WTO Secretariat is based in Geneva, Switzerland, headed by a director-general with limited authority. The Secretariat's main duties include providing legal and technical support to the various councils and ministerial conferences, conducting research, and performing public affairs activities.

Relation to Science and Technology

Many WTO agreements affect the science and technology laws and practices of member nations. One example is the Sanitary and Phytosanitary Measures Agreement (SPS), which sets food safety and animal and plant health standards, including quarantine, inspection, and testing requirements. The aim of the SPS agreement is to establish standards based on accepted science to allow countries to set reasonable health and safety regulations but only to the extent necessary to protect human, plant, or animal life or health. The SPS agreement prevents countries from using higher sanitary and phytosanitary measures in order to protect domestic producers. WTO members can challenge each other's food health and plant and animal safety regulations if they exceed mandated limits.

The Agreement on Technical Barriers to Trade (TBT) ensures that product standards, regulations, testing, and certification for all goods, including industrial and agricultural products, do not become obstacles to trade. The TBT agreement sets limits on the standards governments may enforce to achieve social, environmental, consumer, or public health objectives. The aim is to prevent technical regulations and industrial standards from being used for protectionism. The WTO recognizes the rights of nations to protect the environment and public welfare but not if standards give domestically produced goods an unfair advantage or so far exceed the standards of other nations that they become an obstacle to trade. The TBT agreement subjects national product standards and regulations to scrutiny under WTO Trade Policy Reviews and challenges in Dispute Settlement Court.

The Agreement on Trade-Related Aspects of Intellectual Property establishes the levels of protection governments have to give the intellectual property rights of other governments. The agreement covers copyright (including computer programs, music recordings, and film), trademarks (signs and slogans), geographical indications (place-names that indicate where a product is from and what it is, such as champagne or tequila), industrial designs (for large-scale technologies), patents (protecting products and processes lasting for twenty years), trade secrets (and other undisclosed information with commercial value), and integrated circuit layout designs. TRIPS extends intellectual property rights to include pharmaceuticals, plant varieties, human and plant cell lines, microorganisms, and genes. The agreement defines what counts as intellectual property, how governments should enforce rights, and how to settle disputes over rights between member nations.

Criticisms of the WTO

The WTO has been dogged by controversy from its inception. It continues to be on the defensive against criticism that its agreements privilege corporate interest goals over public interest goals. Critics maintain that the WTO illegitimately dictates the policies of sovereign nations, promotes free trade at any cost, and gives commercial interests priority over development, the environment, health, safety, and worker rights. They further claim that it eliminates both job security and food security, favors developed nations over underdeveloped nations, and fosters a dispute resolution process that is undemocratic and unaccountable. The WTO maintains that through lowering import tariffs and "harmonizing" the international rules of commerce trade should become more predictable, more competitive, and more beneficial for all nations, especially less-developed nations.

DAVID M. KAPLAN

SEE ALSO *Intellectual Property; Political Economy; Property.*

BIBLIOGRAPHY

Wallach, Lori, and Michelle Sforza. *Whose Trade Organization?: Corporate Globalization and the Erosion of Democracy: An Assessment of the World Trade Organization.* Washington, DC: Public Citizen, 1999.

World Trade Organization, ed. *The Legal Texts: The Results of the Uruguay Round of Multilateral Trade Negotiations.* Cambridge, UK, and New York: Cambridge University Press, 1999.

INTERNET RESOURCE

World Trade Organization Website. Available from http://www.wto.org.

WRIGHT, GEORG HENRIK VON

SEE *von Wright, Georg Henrik.*

Z

ZAMYATIN, YEVGENY IVANOVICH

● ● ●

Yevgeny Ivanovich Zamyatin (1884–1937), who was born in Lebedyan, Tamov district, Russia on February 1, is best known for having written *We* (1920), the archetypal anti-utopian novel. The son of a Russian Orthodox priest and a mother who had received a liberal education, he was a constant critic, siding with the Bolsheviks before the revolution and chiding the new government after their victory.

Zamyatin's critical posture was not limited to Russia. Although he was a naval architect by training, when he was in Great Britain (1916–1917) to supervise the building of Russian icebreakers, he published *The Islanders*, a satire of the English. Over the course of his career Zamyatin wrote about forty books, a few of which were quite influential in their time, but he is remembered primarily for one he could not publish. When the Soviets began to censor literature in 1922, the first manuscript banned was *We*, which then appeared in English in the United States (1924) and in Russian in Prague (1927). After 1929 Zamyatin could not publish at all at home. In 1931 at Zamyatin's request, Stalin allowed him to emigrate to Paris, where he lived, unsupported by the local Russian community, until his impoverished death on March 10.

We is the forty-record journal of D-503, an engineer supervising the building of *The Integral*, a spaceship intended to impose the philosophy of the totalitarian One State on other planets: "If they will not understand that we are bringing them a mathematically faultless happiness, our duty will be to force them to be happy" (p. 3). The fundamental contradiction between mechanism and individualism defines the novel. People are "Numbers": The higher the number, the higher the rank; there are vowels and even numbers for females, consonants and odd numbers for males. The "Lex Sexualis" states, "A Number may obtain a license to use any other Number as a sexual product" (p. 22)

Everyone lives according to a Table of Hours. All residences are made of glass. Curtains may be drawn only during Sexual Hours. Despite his role and self-conscious desire to be a good citizen, D-503 develops a soul. The first, unexamined symptom is his desire to express himself, to write the book that is before the reader. The second is a complex passion he feels for I-330, a bold woman revealed as a revolutionary who is trying to use D-503 to gain control of *The Integral* but also may have fallen in love with him.

The development of his soul subtly changes D-503's viewpoint: "As I crossed the avenue, I turned around. Here and there in the huge mass of glass penetrated by sunshine there were grayish-blue squares, opaque squares of lowered curtains, the squares of rhythmic, Taylorized happiness" (p. 41). The reference to Frederick Winslow Taylor (1856–1915), the inventor of time-motion studies and "industrial engineering," Zamyatin's high priest of dehumanizing technology, suggests why D-503 says, "Love = f(D), love is the function of death" (p. 127).

I-330 does seduce D-503, but an assistant prevents a takeover of the ship. I-330 is killed publicly, and D-503, like every other citizen of One State, undergoes a new procedure to remove the imagination, after which he concludes, with horrible happiness, "Reason must prevail" (p. 218).

Perhaps it must. In 1988, under glasnost, when the Soviet Union began to "rehabilitate" banned literature, *We* was on the very first list.

The fundamental contradiction between mechanism and individualism that Zamyatin explored has resonated ever since in discussions of science, technology, and ethics. As societies, by adopting modern science and technologies, have come to possess increasingly potent tools for individual action, those tools often have resulted in the conscious imposition or spontaneous emergence of machinelike social orders. For good and bad, after all, railroads make people run on time. This dilemma echoes through powerful and popular works ranging from Edgar Rice's play *The Adding Machine* (1923) to monitory novels such as Aldous Huxley's *Brave New World* (1932), Ayn Rand's *Anthem* (1938), George Orwell's *Nineteen Eighty-Four* (1948), and William Gibson's *Neuromancer* (1984) as well as potent sociological analyses such as Jacques Ellul's *The Technological Society* (1964) and touchstone movies such as *Blade Runner* (1982).

ERIC S. RABKIN

SEE ALSO *Science Fiction; Science, Technology, and Literature; Utopia and Dystopia.*

BIBLIOGRAPHY

Cooke, Brett. (2002). *Human Nature in Utopia: Zamyatin's We.* Evanston, IL: Northwestern University Press. A literary critical and sociobiological analysis of the meanings and place of *We*.

Richards, David John. (1962). *Zamyatin: A Soviet Heretic.* London: Bowes and Bowes.

Zamiatin, Eugene [Zamyatin, Yevgeny]. (1959 [1924]). *We,* trans. Gregory Zilboorg. New York: Dutton. (Original Russian Ms., 1920.)

Zamyatin, Evgeny [Yevgeny]. (1978 [1918]).*The Islanders,* trans. T. S. Berczynski. Ann Arbor, Michigan: Trilogy Publishers.

ZOOS

• • •

Evidence suggests that humans first domesticated animals beginning about 10,000 B.C.E., but collecting wild and exotic animals did not begin until about 3,000 B.C.E. During the next few millennia, gardens, animal collections, parks, and animal reserves grew in numbers and range. But it was not until the development of the nation-state in the sixteenth century that organized menageries, zoos, and aquaria emerged and proliferated (Kisling Jr. 2001). In the early twenty-first century visiting zoos is one of the most popular activities in many countries, yet keeping animals in zoos—particularly large mammals such as elephants and whales—raises important ethical questions that pit the interests of science and conservation against those of animal rights.

History

The first recorded examples of animal collections were found in the great civilizations of Mesopotamia, such as Assyria, Sumeria, and Babylon. Animal collections were the privilege of the wealthiest people, usually royalty, who could afford to capture or purchase, and maintain, exotic animals. Early collections often included falcons, deer, exotic birds, fish, gazelle, apes, monkeys, ostriches, lions, and elephants. Falcons and lions were often used in royal sport for hunting and fighting, and some parks and preserves were created for this very purpose (Kisling Jr. 2001). Animal collections, gardens and parks also existed in ancient Egypt, Asia, India, Greece, North and South America, and later in Europe, but continued to be a hobby enjoyed primarily by royalty.

In medieval Europe, animal collectors grew in number to include monasteries and municipalities, although collecting was still an expensive practice. As these collections grew in size during the Renaissance, particularly with the addition of exotic animals captured in the new world, they were referred to as menageries. With the onset of the industrial revolution, more people had extra spending money and leisure time in which to indulge in various interests, including the financial support of menageries. In the late-eighteenth and early-nineteenth centuries, private collections evolved into publicly supported menageries (Kisling Jr. 2001). The shift from menagerie to *zoological garden,* or simply zoo, also occurred in the early-nineteenth century. In 1825 the Zoological Society of London suggested creating a zoological garden in which living animals with "their nature, properties and habits may be studied" (Kisling Jr. 2001, p 37), indicating a shift to a more scientifically grounded purpose in collecting animals.

Human knowledge of animal husbandry has improved significantly since the mid-1800s. In early zoos it was not uncommon to see animals kept in small cages with dirt or concrete floors and in generally poor conditions. Twenty-first century zoos are more sensitive to the needs of the animals, and many animals are housed in naturalistic habitats that simulate the animal's original ecosystem. Zookeepers also recognize the importance of mentally stimulating larger mammals through various enrichment activities in order to keep them alert and healthy. Large animal parks in which animals roam freely have also become increasingly popular.

Ethical Issues

Proponents of keeping animals in zoos claim that there is much to be gained in terms of science, conservation, and even the long term welfare of the animals themselves. In some zoos, extensive research is undertaken in the fields of zoology, biology, animal behavior, and veterinary medicine, providing valuable information that is useful in a variety of milieus (Bostock 1993). Many endangered species are bred through intricately designed captive breeding programs, in accordance with species survival plans to ensure the genetic diversity, and thus survivability, of the species. The successful captive breeding program of the highly endangered California condor by the San Diego Zoo produced enough animals that many were released back into the wild. Some zoos have also evolved from simple purveyors of facts about individual animals into educators, describing the ecosystems, environmental concerns, and policy issues surrounding the animal, thus attempting to provide a more complete learning experience to the public. Indeed up until the recent proliferation of cable TV programs dedicated to animals, visiting the zoo was often the public's first exposure to, and education about, exotic animals and related conservation issues. Educating the public, many supporters believe, is crucial for raising awareness of critical conservation and preservation issues. Finally proponents point to the fact that many zoo animals live longer in captivity than their wild counterparts, suggesting that zoos are actually beneficial to the animals themselves (Bostock 1993).

Opponents of zoos contest the claims that the animals are well-treated. Despite significant improvements in zoo-keeping practices, many zoos around the world still display animals in small cages and in sterile environments. Even in the United States, many animals are not provided the minimum standards required by the American Zoo and Aquarium Association (AZA). According to the Humane Society of the United States, only about 10 percent of more than 2,000 animal exhibitors licensed by the U.S. Department of Agriculture (USDA) are approved by the AZA, which has high standards for animal care. Opponents also doubt the legitimacy of scientific research, suggesting that such research is in fact not that common, and that most is geared solely toward the management of captive animals and cannot be extrapolated to wild populations (Hancocks 2001). Questions also arise concerning conservation efforts in zoos. For example, is the purpose of conservation to preserve genes, individual animals, entire populations, or ecosystems? And which species should be selected for captive breeding programs? Still others argue that there is much to be done in terms of educating the public, in that zoos tend to perpetuate an overly simplistic, dominionistic, if not positivistic, view of the natural world. The result is that zoos tend to ignore smaller yet more populous animals in favor of charismatic megafauna that most visitors find more interesting (Hancocks 2001).

Philosopher Dale Jamieson, in his now famous, and controversial, essay, "Against Zoos," argues that even if there are some benefits to zoos, there is an overwhelming ethical reason for not having them, namely the rights inherent in each animal to live freely and to develop its own potential. Furthermore he contends that capturing wild animals for the hungry zoo market often leads to the death of many other animals, often the mothers or adult males who protect the young. While zoo supporter Stephen Bostock agrees that capturing wild animals is one of zoo keeping's weaknesses, even calling for a ban on the trading and capturing of wild animals, he disagrees with the notion that only wild animals can live freely. Freedom, he suggests, describes an environment in which most of the animals' needs are cared for, and well-managed zoos can do just that.

As a result of the continued professionalization of zoos and zoo keeping, several international associations have developed codes of ethics by which member zoos must abide. Ethical standards focus on everything from minimum standards of animal care, responsibility to the animals, species survival plans, commitment to biodiversity and conservation efforts, and professional conduct. Member zoos found in violation of ethical standards face sanctions or dismissal from the association. Many ethical discussions regarding zoos will likely continue, but some claim that debating whether or not zoos should exist at all is one that should end. David Hancocks explains that zoos are here to stay, and that human energy should focus on how to improve them, and to develop a new relationship with animals and nature (2001).

ELIZABETH C. MCNIE

SEE ALSO *Animal Rights; Animal Welfare; Bioethics; Colonialism and Postcolonialism; Modernization.*

BIBLIOGRAPHY

Bostock, Stephen St. C. (1993). *Zoos and Animal Rights: The Ethics of Keeping Animals*. London: Routledge. Bostock makes a strong argument in favor of zoos and of their scientific, environmental, and societal benefits.

Hancocks, David. (2001). *A Different Nature: The Paradoxical World of Zoos and Their Uncertain Future*. Berkeley: University of California Press. Hancocks makes convincing suggestions regarding the need to change the goals of zoos and the way zoos are managed.

Jamieson, Dale. (1985). "Against Zoos." In *In Defense of Animals*, ed. Peter Singer. Oxford: Basic Blackwell.

Kisling Jr., Vernon N. (2001). "Ancient Collections and Menageries." In *Zoo and Aquarium History: Ancient Animal Collections to Zoological Gardens*, ed. Vernon N. Kisling Jr. Boca Raton, FL: CRC Press. This edition provides an excellent history of zoos and aquaria around the world.

Norton, Bryan G.; Michael Hutchins; Elizabeth F. Stevens; and Terry L. Maple, eds. (1995). *Ethics on the Ark: Zoos, Animal Welfare, and Wildlife Conservation*. Washington, DC: Smithsonian Institution Press. This book, published in cooperation with the AZA, includes chapters by many of the important authors on all sides of the ethics of zoos, animals, and conservation and is an excellent resource. Part of the Zoo and Aquarium Biology and Conservation Series.

INTERNET RESOURCES

American Zoo and Aquarian Association. "AZA Code of Professional Ethics." Available from http://www.aza.org/AboutAZA/CodeEthics/.

European Association of Zoos and Aquaria. "Code of Ethics." Available from http://www.eaza.net/info/2ethics.html.

Human Society of the United States. "Zoos." Available at http://www.hsus.org/wildlife/issues_facing_wildlife/zoos/.

South East Asian Zoos Association. "SEAZA Code of Ethics." Available from http://www.seaza.org/CommitteeWelfare.html.

APPENDIX CONTENTS

The *Encyclopedia of Science, Technology, and Ethics* includes five appendices designed to provide supporting materials for individual articles and further resources for readers.

APPENDIX I

SELECTIVE, ANNOTATED, GENERAL BIBLIOGRAPHY ON SCIENCE, TECHNOLOGY, AND ETHICS

This selected bibliography emphasizes works in English and in print during the early 2000s that are also accessible to the generally educated reader. The goal is to provide an introduction to some of the most useful efforts to lay out arguments relevant to science, technology, and ethics. Arguments may focus on science and technology as a whole or on some specific aspect of the science, technology, and ethics interaction.

The bibliography is divided into six sections:

1. Reference Works

2. Monographs and Edited Volumes: General Implications

3. Monographs and Edited Volumes: Specialized Approaches

4. Textbooks

5. Twentieth and Twenty-First Century Ethics

6. Journals

Reference works are alphabetized by title, and include a few items of marginal value, if only to steer readers away from some materials that might otherwise attract attention. Monographs and edited volumes are divided into those of a general orientation and those focused more on specific sciences or technologies. Some specific approaches that nevertheless have broader implications as well as textbooks that transcend the genre are included in the section on general monographs. This section thus constitutes the core resources in the bibliography. Supplementing these core sections is another of selected works on ethics that indicate the background traditions of reflection brought to bear on science and technology from the early twentieth century.

Like reference works, journals are alphabetized by title. All other works are alphabetized by author or editor. Multiple works by the same author are arranged chronologically by date of publication.

1. Reference Works

Applied Ethics: Critical Concepts in Philosophy. Ruth Chadwick and Doris Schroeder, eds. 6 vols. London: Routledge, 2002. Vol. 1 deals with the nature and scope of applied ethics. Vols. 2 and 3 focus on ethical issues in medicine, technology, and the life sciences. Vol. 4 is dedicated to environmental issues. Vol. 5 is on business and economics. Vol. 6 is on politics. Collects and reprints a large number of influential articles from the last half of the twentieth century. Each volume includes an introduction summarizing the historical context and trends.

The Blackwell Guide to the Philosophy of Computing and Information. Luciano Floridi, ed. Malden, MA: Blackwell, 2004. Twenty-six articles. Most directly relevant are those on "Computer Ethics," "Computer-mediated Communication and Human-Computer Interaction," "Internet Culture," "The Philosophy of AI and Its Critique," "Virtual Reality," and "Philosophy of Information Technology."

A Companion to Environmental Philosophy. Dale Jamieson, ed. Oxford: Blackwell Publishers, 2003. Pp. xvi, 531. Collects a preface and thirty-six essays arranged in four parts: cultural traditions, contemporary environmental ethics, environmental philosophy and its neighbors (e.g., literature, aesthetics, history, ecology, politics, and law), and problems in environmental philosophy.

A Companion to Ethics. Peter Singer, ed. Cambridge, MA: Blackwell, 1991. Pp. xxii, 565. Forty-seven chapters highlighting origins of ethics, major traditions, the Western philosophical tradition, basic theories of obligation, applied ethics (poverty, environmentalism, euthanasia, abortion, sex, personal relationships, equality, animals, business, crime and punishment, politics, and war), arguments concerning the nature of ethics (realism, intuitionism, naturalism, etc.), and challenges (feminism, evolution, Marxism, etc.). Although neither "science" nor "technology" appear in either the table of contents or the index, this provides useful background material for discussions in science, technology, and ethics.

A Companion to Genethics. Justine Burley and John Harris, eds. Oxford: Blackwell, 2002. A comprehensive look at the philosophical, ethical, social and political dimensions of developments in human genetics.

The Concise Encyclopedia of the Ethics of New Technologies. Ruth Chadwick, ed. San Diego: Academic Press, 2001. A selective examination of several contemporary technologies and the ethical dilemmas they present along with examples of frameworks for their assessment like environmental impact statements and different ethical theories. Arranged as 37 articles each with an outline, glossary, defining statement, and bibliography. A repackaging of selected articles from the 4-vol. *Encyclopedia of Applied Ethics.*

Encyclopedia of Applied Ethics. Ruth Chadwick, ed. 4 vols. San Diego: Academic Press, 1998. A major synthetic and informative reference work. Two hundred eighty one articles (averaging 5000 to 10,000 words) covering theories, concepts, and ethics related to medicine, science, the environment, law, education, politics, business, the media, social services, and social interactions. Two relevant spin-off collections are *The Concise Encyclopedia of the Ethics of New Technologies* (2001) and *The Concise Encyclopedia of Ethics in Politics and the Media* (2001).

Encyclopedia of Bioethics. Warren Reich, ed. First edition, 4 vols. New York: Macmillan Reference, 1978. Second edition, 5 vols. New York: Macmillan Reference, 1995. Third edition, 5 vols., Stephen G. Post, ed. New York: Macmillan, 2004. A model of scholarship and influence.

Encyclopedia of Ethical, Legal, and Policy Issues in Biotechnology. Thomas H. Murray and Maxwell J. Mehlman, eds. 2 vols. New York: John Wiley, 2000.

Encyclopedia of Ethics in Science and Technology. Nigel Barber. New York: Facts on File, 2002. A one-person product. Slightly better than Newton's *Social Issues* volume below, but similar.

Encyclopedia of Twentieth-Century Technology. Colin A. Hempstead, ed. 2 vols. New York: Routledge. Approximately 400 articles by about 175 authors. The focus is on technical descriptions, but there are a few articles on "Technology and Ethics" and related topics.

The Facts on File Encyclopedia of Science, Technology, and Society. Rudi Volti, ed. 3 vols. New York: Facts on File, 1999. Approximately 900 well crafted articles by 95 contributors, the majority of whom are historians of science and technology. Although the preface describes the focus as society as much as science or technology, the work is better or technical than social dimensions. There are no articles, for instance, on ethics, which is not even an indexed term.

Handbook of Science and Technology Studies. Jasanoff, Sheila, Gerald E. Markle, James C. Petersen, and Trevor Pinch, eds. Thousand Oaks, CA: Sage, 2004.

Science, Technology, and Society: An Encyclopedia. Sal Restivo, ed. New York: Oxford University Press, 2005.

Social Issues in Science and Technology: An Encyclopedia. David E. Newton. Santa Barbara, CA: ABC-CLIO, 1999. Approximately 350 entries, mostly 500–1000 words each. A one-person product of relatively high quality. Paperback version titled *From Global Warming to Dolly the Sheep: An Encyclopedia of Social Issues in Science and Technology.*

2. Monographs and Edited Volumes: General

Alcon, Paul A. *Practical Ethics for a Technological World.* Upper Saddle River, NJ: Prentice Hall, 2001. Pp. xiv, 239. Aims to be a guide to ethical decision making in the technological world; works back and forth to explore ethics and technology and their mutual interactions. Naive and weakly spiritual in orientation.

Allen, Anita L. *The New Ethics: A Guided Tour of the Twenty-First Century Moral Landscape.* New York: Miramax Books, 2004. Pp. xxxviii, 322. Overviews the contemporary ethical landscape focusing on widespread unethical behavior (e.g., lying, cheating, and corruption), new moral challenges presented by science, technology, and medicine, and complacency and apathy. Discusses ways to improve ethical behavior at work and in education. Concludes with sections on choosing well (e.g., consumption, family, and dying) and justice in multi-cultural societies.

Barbour, Ian. *Ethics in an Age of Technology*. San Francisco: Harper, 1993. Pp. xix, 312. An extended analysis of divergent ethical views of technology focusing on the values of justice, participation, and development. Considers case studies in agriculture, energy, genetic engineering, and computers. Argues in defense of environmental sustainability, appropriate technologies, and personal responsibility for promoting progressive change through education, political action, and the pursuit of alternative visions of the good life.

Barbour, Ian. *Religion and Science: Historical and Contemporary Issues*. San Francisco: Harper, 1997. Pp. xv, 368. This is a revised and expanded edition of *Religion in an Age of Science* (1990). Gives a broad overview of historical interactions between religion and science, and develops a typology of four ways of interacting: conflict, independence, dialogue, and integration. Defends dialogue and integration in both method and substantive forms of knowledge.

Bird, Stephanie, J., and Raymond Spier, eds. "The Role of Scientific Societies in Promoting Research Integrity." Theme issue, *Science and Engineering Ethics*, vol. 9, no. 2, April 2003. Pp. 158. Fourteen papers on the role professional scientific societies in promoting and implementing guidelines for research ethics. Includes examples, recommendations for further work, and strategies for evaluating existing programs.

Borgmann, Albert. *Holding On to Reality: The Nature of Information at the Turn of the Millennium*. Chicago, IL: The University of Chicago Press, 1999. Pp. 274. Explores, philosophically and historically, the relationship between things and signs, or reality and information, especially the rise of information as reality. Articulates and advocates a theoretical and ethical balance of signs and things that holds onto reality by averting a slide into hyperreality.

Borgmann, Albert. *Technology and the Character of Contemporary Life: A Philosophical Inquiry*. Chicago: University of Chicago Press, 1984. Borgmann's most general argument for a distinction between technological devices and focal things, each of which influences the development of different patterns of human behavior.

Borgmann, Albert. *Crossing the Postmodern Divide*. Chicago: University of Chicago Press, 1992. Pp. 173. A lucid and concise description of deep contemporary cultural challenges that traces them back to foundational thinkers in modern Western philosophy (e.g., Descartes) and presents a way forward that avoids the dehumanizing extremes of "hyperreality."

Buchanan, Richard, and Margolin, Victor eds. *Discovering Design: Explorations in Design Studies*. Chicago: University of Chicago Press, 1995. Pp. xxvi, 254. Includes the article "Prometheus of the Everyday: The Ecology of the Artificial and the Designer's Responsibility" by Ezio Manzini.

Callahan, Daniel. *The Tyranny of Survival: And Other Pathologies of Civilized Life*. New York, NY: Macmillan, 1973. Pp. xv, 284. Reprinted, University Press of America, 1985. Argues for a more realistic and sustainable aspiration than the quest for endless technological progress and unbounded individual freedom. Uses population growth and genetic technologies to illuminate technological change and argue for limiting technological excess and cultural hubris.

Callahan, Daniel. *What Kind of Life: The Limits of Medical Progress*. New York: Simon and Schuster, 1990. Pp. 318. Takes a synoptic view and argues that deep premises about health, illness, and life are fundamentally flawed and lead to insatiable expectations for healthier, longer lives that cannot be satisfied and that drive under-performing and increasingly expensive health care systems. Offers a new way to think of health that could help devise a more reasonable and just health care system that balances worthy aspirations with necessary limits.

Callahan, Daniel. *Setting Limits: Medical Goals in an Aging Society*. Washington, D.C.: Georgetown University Press, 1995. Pp. 272. Explores the shadows of medical progress and the attendant new ways of thinking about health, life, and aging (e.g., old age is to be overcome with the use of science and technology). Addresses such questions as the proper ends of medicine, what the young owe the old, the allocation of resources to the elderly, and care of the elderly dying. Seeks to stimulate a discussion on the future of health care for the aged and proposes a different way of understanding this issue. Concludes with responses to critics.

Callahan, Daniel. *The Troubled Dream of Life: In Search of a Peaceful Death*. Washington, D.C.: Georgetown University Press, 2000. Pp. 255. Integrates legal and policy issues of death and dying with deep philosophical questions about the meaning of death and its relation to self. Argues that many problems in the care of the dying, both in public attitudes and medical progress stem from mistaken views of death. Seeks to foster a common view of death by treating foundational issues rather than specific law or policy questions.

Callahan, Daniel. *What Price Better Health? Hazards of the Research Imperative*. Berkeley: University of California Press, 2003. Pp. xii, 329. Centered on the concept of the research imperative in medicine, which is a complex topic that refers to the inherent momentum of research and the view that the importance of research could trump moral values. Argues it is primarily a cultural (as opposed to a property inherent in the research community) problem that fuels most of the "shadows" or hazards of medical research. Chapters consider several issues including research as a moral obligation, enhancement, risks and benefits, human subjects research, and a distinction between doing good and doing well.

Collins, Harry, and Trevor Pinch. *The Golem: What You Should Know about Science*. 2nd ed. New York: Cambridge University Press, 1998. Pp. xix, 192. Directed to a general audience. Argues that science is akin to a clumsy and dangerous yet potentially helpful creature. Presents the actual workings of science to show that the authorization of knowledge claims is a political, complex, and messy process of persuasion that produces many controversies. Includes a description of the "experimenter's regress." Collects an introduction, conclusion, and seven case studies on the production and negotiation of new scientific knowledge, including experiments on relativity, the chemical transfer of memory, and solar neutrinos.

Collins, Harry, and Trevor Pinch. *The Golem At Large: What You Should Know About Technology*. New York: Cambridge University Press. 1998. Pp. xi, 163. Continues the social constructivist argument applied to technology.

Committee on Science, Engineering, and Public Policy (US). Panel on Scientific Responsibility and the Conduct of Research. *Responsible Science: Ensuring the Integrity of the Research Process*. 2 vols. Washington D.C.: National Academy Press, 1992 and 1993. Pp. xxiii, 199 (each vol.). Authorized by the National Research Council (whose members are drawn from the councils of the National Academy of Sciences, the National Academy of Engineering, and the Institute of Medicine). Reviews factors affecting the integrity of science and the research process in the US and institutional mechanisms for addressing allegations of misconduct. Recommends steps for reinforcing responsible research practices.

Ellul, Jacques. *The Technological Society*. Trans. John Wilkinson. New York: Knopf, 1964. Pp. xxxvi, 449, xiv. A classic examination of the social and moral consequences of the domination of "technique," or the totality of methods driven by the urge to absolute efficiency. Provides an historical overview and analyses of techni-

que and the economy and state. Features a chapter titled "The Characterology of Technique," which argues that modern technique is fundamentally new due to its pervasiveness, connection to modern science, large scale, "automatism," and "self-augmentation."

Federman, Daniel, Kathi E. Hanna, and Laura Lyman Rodriguez, eds. *Responsible Research: A Systems Approach to Protecting Research Participants*. Washington, DC: National Academies Press, 2002. Pp. xix, 290. An Institute of Medicine report commissioned by the Secretary of the Department of Health that offers a comprehensive review of the present system for protecting human participants and suggestions for strengthening it. Recommends gathering data and taking a diversity of approaches to maximize the protection of individuals participating in research. Emphasizes a systematic approach and the importance of institutional cultures, training, improved informed consent, and improved research review procedures.

Feenberg, Andrew. *Critical Theory of Technology*. New York: Oxford University Press, 1991. Pp. xi, 235. Updates critical social theory for a high-tech world.

Feenberg, Andrew. *Alternative Modernity: The Technical Turn in Philosophy and Social Theory*. Berkeley: University of California Press, 1995. Pp. xi, 251.

Feenberg, Andrew. *Questioning Technology*. New York, NY: Routledge, 1999. Pp. xvii, 243. A philosophy of technology that critiques essentialism and shows the centrality of technological design to modern society and democratic politics. Proceeds in three parts: the politicizing of technology, democratic rationalization, and technology and modernity.

Feenberg, Andrew, and Alastair Hannay, eds. *Technology and the Politics of Knowledge*. Bloomington: Indiana University Press, 1995. Collects 16 articles by Steven Vogel, Robert B. Pippin, Langdon Winner, Albert Borgmann, Hubert L. Dreyfus, Terry Winograd, Tom Rockmore, Don Ihde, Yaron Ezrahi, Donna Haraway, Helen Longino, Marcel Hénaff, Pieter Tijmes, Paul Dumouchel, and Bruno Latour.

Florman, Samuel C. *The Existential Pleasures of Engineering*. 2nd ed. New York, NY: St. Martin's Press, 1994 (1st edition, 1976). Pp. xviii, 205. Inquires into the nature of the contemporary engineering experience. Views it as vital and alive, something to be celebrated as a response to deep human impulses, and as a source of sensual and spiritual reward.

Fox, Warwick, ed. *Ethics and the Built Environment*. London: Routledge, 2000. Pp. xv, 240. Seeks a critical mass of ideas to initiate a field of study to rectify environmental ethics traditional disregard of the built environment. Collects seventeen essays arranged into three sections: the green imperative, building with greater sensitivity to people and places, and toward a theory of ethics of the built environment.

Fukuyama, Francis. *Our Posthuman Future: Consequences of the Biotechnology Revolution.* New York: Farrar, Straus, and Giroux, 2002. Pp. xiii, 256. Examines the some techniques and ethical issues, develops an understanding of natural human rights, and concludes with comments on the regulation of biotechnology and recommended policies for the future. Creates a taxonomy of concerns and argues that the greatest reasons to worry about biotechnology are not utilitarian but that the human essence will somehow be lost. Argues for greater political control over the uses of science and technology.

Fuller, Steve. *The Governance of Science: Ideology and the Future of the Open Society.* Buckingham, PA: Open University Press, 2000. Pp. xii, 167. Rejects communitarian and liberal ideologies of science in favor of a republican approach centered on the right to be wrong. Argues that the scaling up of science threatens this ideal and focuses on the challenges of multiculturalism and capitalism for the university as a republic of science. Proposes a new social contract for science.

Goujon, P., and Bertrand Heriard Dubreuil, eds. *Technology and Ethics: A European Quest for Responsible Engineering.* Leuven, Belgium: Peeters, 2001. Pp. xx, 616. Collects 37 essays in three sections as part of the core materials project for the development of courses in professional ethics, in order to serve as a European engineering ethics handbook. The three main sections consider the ethics of industrial engineers, institutional responsibility, and the social and political policy implications. Includes contributions by humanists, social scientists, and engineers.

Guston, David H. *Between Politics and Science: Assuring the Integrity and Productivity of Research.* New York, NY: Cambridge University Press, 2000. Pp. xvii, 213. Examines the deterioration of the post-World War II assumption in U.S. science policy that integrity and productivity were the automatic products of unfettered scientific inquiry. Shows how "boundary organizations" have developed since the 1980s to rebuild and maintain trust between politics and science. Shows the attention to detail necessary for designing such institutions to be effective.

Habermas, Jürgen. *Technik und Wissenschaft als "Ideologie."* [Technology and Science as "Ideology"]. Frankfurt am Main: Suhrkamp, 1968. Pp. 169.

Habermas, Jürgen. *The Future of Human Nature.* Oxford: Polity Press, 2003. Pp. 127, viii. Asks if there are post-metaphysical answers to the question: what is the good life? Expands this question beyond personal ethics to the questions of a species ethic posed by genetic technologies where a novel kind of self transformation poses the dilemma that the "ethical understanding of language-using agents is at stake *in its entirety.*" Concludes with a postscript and a reflection on faith and knowledge.

Harris, Charles E., Michael S. Pritchard, and Michael J. Rabins. *Engineering Ethics: Concepts and Cases.* 3rd ed. Belmont, CA: Wadsworth, 2005. Pp. xvii, 390. (1st ed., 1995). Analyzes the field of engineering ethics through ethical problem-solving strategies, generic topics of concern such as responsibility, honesty, and risk, and special topics such as professional societies, the environment, and international engineering contexts. Designed for classroom use, it includes case studies and an interactive CD-ROM.

Heidegger, Martin. *The Question Concerning Technology and Other Essays.* Trans. William Lovitt. New York: Harper and Row, 1977. Pp. xi, 182. A classic work in the philosophy of technology, argues that modern technology is more than merely instrumental means to ends, but rather it is a "challenging revealing" that hides Being and presents the world as a standing reserve of objects ready to hand. An ontological account of technology's fundamental impact on human experience.

Hickman, Larry A. *Philosophical Tools for Technological Culture: Putting Pragmatism to Work.* Bloomington: Indiana University Press, 2001. Pp. xi, 215. Argues that philosophy has a productive role to play as reformer and critic of technological culture between post-modern decontructionism and the ancient practice of grand system building. Draws inspiration from John Dewey to develop a kind of philosophy called "productive pragmatism." Takes up several issues including education, expertise, art, community, and responsibility.

Hughes, Thomas P. *Human-Built World: How to Think About Technology and Culture.* Chicago: University of Chicago Press, 2004. Pp. xii, 223. An extended bibliographic essay on the history of technology and its various interpretations across time. Draws primarily from literature, art, and architecture to trace the transformation in meanings of technology from the second creation to machine to systems, controls, and information, to culture. Concludes with comments on creating an ecotechnology.

Ihde, Don. *Technology and Lifeworld: From Garden to Earth.* A phenomenological analysis of human-technology relations that suggests the emergence of a new kind of ethical relationship between humans and the world.

Institute of Medicine National Research Council. *Integrity in Scientific Research: Creating an Environment that Promotes Responsible Conduct.* Washington, DC: The National Academies Press, 2002. Pp. xiv, 202. A report issued by the Institute of Medicine Committee on Assessing Integrity in Research Environments that defines the desired outcomes in research integrity and the teaching of research ethics and provides a set of initiatives to enhance integrity in research. Also considers methods for assessing those initiatives.

Johnson, Deborah G., ed. *Ethical Issues in Engineering.* Englewood Cliffs, NJ: Prentice Hall, 1991. Pp. viii, 392. Collects 32 articles providing historical and social context of engineering ethics, analyses of professional codes, and discussions of responsibilities to society, company loyalty, and obligations to clients.

Johnson, Deborah G. *Computer Ethics.* 3rd ed. Upper Saddle River, NJ: Prentice-Hall, 2001. Pp. xvi, 240. (1st ed. 1985; 2nd ed. 1994). Articulates the field of computer ethics with a focus on the core issues of professional ethics, privacy, property, accountability, and social implications and values. Includes two chapters on ethics and the internet. Each chapter includes short case studies, analysis, study questions, and suggested readings.

Jonas, Hans. *The Imperative of Responsibility: In Search of an Ethics for the Technological Age.* Trans. Hans Jonas and David Herr. Chicago: University of Chicago Press, 1984. Pp. xii, 255. (Originally published as *Das Prinzip Verantwortung: Versuch einer Ethik fuer die technologische Zivilisation.* Frankfurt am Main: Insel Verlag, 1979; and *Macht oder Ohnmacht der Subjektivitaet? Das Leib-Seele-Problem im Vorfeld des Prinzips Verantwortung.* Frankfurt am Main: Insel Verlag, 1981.) Rethinks the foundations of ethics in light of modern technology by developing a metaphysical theory of responsibility that takes account of the extended time and space horizons affected by technological action. Also introduces a philosophy of nature to bridge the chasm between "is" and "ought" and develops a "heuristics of fear" to counter the dangers of utopianism. Jonas' goal is to develop an ethics of responsibility capable of saving humanity from the excesses of its own Promethean power.

Jonas, Hans. *Mortality and Morality: A Search for the Good after Auschwitz.* Ed. Lawrence Vogel. Evanston, IL: Northwestern University Press, 1996. Pp. xi, 218. Considered the consummation of Jonas' quest to critique nihilism and develop an ethic capable of limiting the powers of modern technology. Jonas grounds an imperative of responsibility in the phenomenon of life and speculates on theology and faith after the Holocaust. Includes an introduction by Lawrence Vogel that provides philosophical and historical context.

Kass, Leon. *Life, Liberty, and the Defense of Dignity: The Challenge for Bioethics.* San Francisco: Encounter, 2002. Pp. 313. Argues that there is more to biotechnology than saving life and avoiding death, namely, the preservation of human dignity and human nature. Claims that this is a peculiar challenge for modern liberal democracies where the dangers lie close to cherished principles, especially individual freedom, equality, and social progress. Traces the root of the dangers to modern scientific, especially biological, thought.

Keulartz, Jozef, Michiel Korthals, Maartje Schermer, and Tsjalling Swierstra, eds. *Pragmatist Ethics for a Technological Culture.* Norwell, MA: Kluwer Academic Publishers, 2002. Pp. xxvi, 264. Argues that pragmatism can serve as a solid way to cope with questions of technology and human values. Includes twenty chapters arranged into prologue, epilogue, and sections on technology and ethics, the status of pragmatism, pragmatism and practices, and discourse ethics and deliberative democracy.

Kitcher, Philip. *Science, Truth, and Democracy.* Oxford: Oxford University Press, 2001. Pp. xiii, 219. Argues that epistemic values do not stand apart from or above other values and practical interests. This requires a new ideal of science beyond the neat separation of science from society. This ideal is labeled "well-ordered" science, which is set in a democratic framework that takes the proper notion of scientific significance to be that which would emerge from ideal deliberation among ideal agents. Then considers problems posed by lapses from the ideal and responsibilities of those who work on projects that conflict with the ideal.

Latour, Bruno. *Science in Action: How to Follow Scientists and Engineers through Society.* Cambridge, MA: Harvard University Press, 1987. Pp. 274. A classic anthropological study of the actual workings of science (rather than theoretical accounts of those workings or deference to a "black box" account) to understand how hypotheses become accepted facts. Emphasizes the importance of interpersonal interactions and rhetoric in both the literature and laboratory for the making of science. Furthered the social construction of science movement begun by Thomas Kuhn.

Latour, Bruno. *Politics of Nature: How to Bring the Sciences into Democracy*. Cambridge, MA: Harvard University Press, 2004. Pp. x, 307. Argues for an end to the dichotomy between nature and society and offers a new conceptual context for understanding political ecology and its promise to advance democracy that accounts for humans and non-humans as citizens. Claims that our conception of science is important both for our understanding of nature and politics.

Lightman, Alan, Daniel Sarewitz, and Christina Desser, eds. *Living with the Genie: Essays on Technology and the Quest for Human Mastery*. Washington, DC: Island Press, 2003. Pp. viii, 347. Examines the contrast between the rapid pace of technological change and the enduring core of humanness within the overarching argument that science and technology are the result of decisions and are thus fundamentally about voice and the allocation of power in democratic societies and the global economy. Collects a general introduction and sixteen essays that address topics at the interface of values, science, and technology such as artificial intelligence, HVAC systems, disability, death, happiness, and property rights.

Lowrance, William W. *Modern Science and Human Values*. New York, NY: Oxford University Press, 1985. Pp. xiv, 250. Examines how technical progress and expertise influence and are influenced by other parts of society. Argues that a more nuanced understanding of science, technology, and values is necessary for more effectively putting science and technology into the service of society. Themes include facts and values, expertise, decision-making, and science and technology in the polis.

McKibben, Bill. *Enough: Staying Human in an Engineered Age*. New York: Times Books, 2003. Pp. xiii, 271. Argues that aggressively pursuing certain new technologies (genetic engineering, robotics, and nanotechnology) will lead to a post-human era that impoverishes the meaning of being human. Explores how the technologies work and how to control them. Asks the central questions of whether people in the West lead sufficiently comfortable lives with sufficient technology now and whether controlling technologies is possible at all.

Melzer, Arthur M., Jerry Weinberger, and M. Richard Zinman, eds. *Technology in the Western Political Tradition*. Ithaca, NY: Cornell University Press, 1993. Pp. xv, 333. Presents a preface, introduction, and twelve essays that address the political character and implica-

tions of technology from classical antiquity through the nineteenth century and the meanings of technology for contemporary political life. An introduction by Leon Kass establishes "the problem of technology" as it provokes questions of human happiness at the same time that it undercuts the validity of answers to those questions. This leads to a need to rediscover the nontechnological conception of liberty and dignity in liberal democracies.

Mitcham, Carl, and Robert Mackey, eds. *Philosophy and Technology: Readings in the Philosophical Problems of Technology*. New York: Free Press, 1972. Paperback edition, 1983. Pp. xii, 403. A collection of 26 articles, some originally translated, that has remained in print for more than 30 years. The sections on "Ethical and Political Critiques," "Religious Critiques," and "Two Existentialist Critiques" are the most relevant.

National Academy of Engineering. *Emerging Technologies and Ethical Issues in Engineering*. Washington, D.C.: The National Academies Press, 2004. Pp. x, 155. Result of an NAE conference. Includes a keynote address by William A. Wulf and nine essays in three sections: emerging technologies, state of the art in engineering ethics, and ethics in engineering education.

National Academy of Sciences, National Academy of Engineering, and Institute of Medicine. *Responsible Science: Ensuring the Integrity of the Research Process*. Vol 1. Washington, DC: National Academy Press, 1992. Pp. xxiii, 1999. Result of a panel discussion to review factors affecting the integrity of research and recommend steps for reinforcing responsible research practices. Also reviews institutional mechanisms for addressing allegations of misconduct and considers the advantages and disadvantages of formal guidelines for the conduct of research.

National Academy of Sciences, National Academy of Engineering, and Institute of Medicine. *Responsible Science: Ensuring the Integrity of the Research Process*. Vol 2. Washington, DC: National Academy Press, 1993. Pp. xi, 275. See above for background. This volume includes background papers, samples of guidelines for the conduct of research, scientific research policies and practices, and policies and procedures for handling allegations of misconduct.

Postman, Neil. *Technopoly: The Surrender of Culture to Technology*. New York: Knopf, 1992. Pp. xii, 222. A broad-brush criticism of technological culture that updates arguments from the 1960s and 1970s.

President's Council on Bioethics. *Beyond Therapy: Biotechnology and the Pursuit of Happiness, A Report of the*

President's Council on Bioethics. New York: Regan Books, 2003. Pp. xiii, 328. A report by the U.S. bioethics commission with a foreword by its Chair, Leon Kass. Explores the ethical and social implications of using biotechnology for purposes of enhancement beyond therapy even as it problematizes this distinction. Includes chapters on "Better Children," "Superior Performance," "Ageless Bodies," "Happy Souls," and a conclusion.

Resnik, David B. *The Ethics of Science: An Introduction*. London: Routledge, 1998. Pp. x, 221. Develops a conceptual framework for understanding the ethics of scientific research and applies it to ethical questions in science. Seeks to clarify the nature of research ethics and the meaning of ethical behavior in science. Draws from several case studies and includes an appendix with 50 hypothetical case studies.

Sarewitz, Daniel. *Frontiers of Illusion: Science, Technology, and the Politics of Progress*. Philadelphia: Temple University Press, 1996. Pp. xi, 235. Deconstructs several "myths" instantiated in post-World War II U.S. science politics in order to gain clarity on the central questions of how science can best serve society, what science to pursue, and the relationship between scientific progress and human welfare. Concludes with a chapter titled "Toward a New Mythology," which includes policy recommendations for more explicitly integrating other values with epistemic pursuits in a democratic fashion.

Sassower, Raphael. *Technoscientific Angst: Ethics and Responsibility*. Minneapolis, MN: University of Minnesota Press, 1997. Pp. xv, 140. Relates the lessons of Auschwitz and Hiroshima to contemporary decision making about technoscience and the responsibility of intellectuals in a way that borrows from Hannah Arendt and Hans Jonas. Examines the anguish and angst felt by scientists but rarely exposed and the ambiguity concerning the responsibility of the technoscientific community in the face of mass destruction.

Sclove, Richard. *Democracy and Technology*. New York: Guilford Press, 1995. Pp. xiv, 338. Argues for democratic participation in technology, and proposes criteria for assessing engineering design in terms of the promotion of democracy.

Shrader-Frenchette, K. S. *Risk and Rationality: Philosophical Foundations for Populist Reforms*. Los Angeles: University of California Press, 1991. Pp. x, 312. Sketches a middle ground between the dominant sides of industrial charges of scientific illiteracy and populist charges of technological oppression. Proposes a new paradigm for making decisions about when the acceptance of public hazards is rational that includes more

trust in the judgments of non-experts. Proceeds through a general introduction to a discussion of problematic risk-evaluation strategies to proposed reform for risk evaluation.

Shrader-Frenchette, Kristin. *Ethics of Scientific Research*. Lanham, MD: Rowman & Littlefield, 1994. Pp. x, 243. Arranged in ten chapters: introduction to and history of research ethics, professional codes, objectivity, promoting the public good, handling conflicts, uncertainty, case study in conservation research, gender and racial biases, social responsibility of engineers, and public health research. Last three chapters are authored by Helen Longino, Carl Mitcham, and Carl Cranor respectively.

Shrader-Frechette, Kristin. *Environmental Justice: Creating Equality, Reclaiming Democracy*. Oxford: Oxford Press, 2002. Pp. xiii, 269. Argues not only for protecting nature but also for public-interest advocacy in the name of the people who are victimized by environmental injustices. Diagnoses, analyzes, and seeks to resolve environmental injustices. Chapters elucidate concepts of justice (e.g., distributive, participatory, and procedural) and focus on case studies such as future generations and nuclear waste disposal, poor peoples and land use decisions, and risky occupational environments. Concludes with steps to take action.

Spier, Raymond. *Ethics, Tools, and the Engineer*. New York, NY: CRC Press, 2001. Pp. xiv, 306. Employs an evolutionary biology framework to discuss ethics and engineering. Discusses the meaning of ethics, describes engineers as toolmakers and users, considers the control and proper use of tools, and speculates on the cloning of humans. Also discusses the hazard and operability (HAZOP) process as a gatekeeping operation.

Spier, Raymond, ed. *Science and Technology Ethics*. New York, NY: Routledge, 2002. Pp. viii, 247. Reexamines contemporary ethics, asking whether sufficient ethical guidelines exist to minimize the disruptions of science and technology and maximize their benefits. Proposes new approaches to science and engineering practices. Eleven essays examine science and engineering broadly, developments in biology and information technology, the military industry, and environmental responsibilities.

Stokes, Donald E. *Pasteur's Quadrant: Basic Science and Technological Innovation*. Washington, DC: Brookings Institution Press, 1997. Pp. xiv, 180. Examines and reconceptualizes the division between basic and applied research that is at the core of post-World War II U.S. science policy. Analyzes the ways in which understanding and use are often tightly intertwined in use-inspired basic research, presents this in a quadrant, and uses this to offer recommendations for a new contract between government and science.

Suzuki, David, and Peter Knudtson. *Genethics: The Clash between the New Genetics and Human Values.* Cambridge, MA: Harvard University Press, 1989. Proposes a set of genetic principles that emphasize individual rights and confidentiality with regard to genetic screening, caution in violating boundaries across species, and a ban on biological weapon development and the genetic manipulation of human germ cells.

Tenner, Edward. *Why Things Bite Back: Technology and the Revenge of Unintended Consequences.* New York: Alfred A. Knopf, 1996. Pp. xiii, 346. Explores the way in which technology, no matter how well designed, demands more human work despite promises to the contrary and introduces more chronic and insidious problems as the acute ones are never wholly resolved. These occurrences are explained as "revenge effects" that emerge from the interplay of technology, laws, customs, and habits. This stems largely from the inability to foresee future consequences of action.

Tiles, Mary, and Hans Oberdiek. *Living in a Technological Culture: Human Tools and Human Values.* London: Routledge, 1995. Pp. xi, 212. A philosophical reflection on technology, its many meanings, and its manifold relationships to culture. Examines conflicting visions of technology, facts and values, efficiency, science and the authority of experts, the transition from applied science to techno-science, and politics and responsibility.

Wenk, Edward, *Tradeoffs: Imperatives of Choice in a High-Tech World.* Baltimore, MD: Johns Hopkins University Press, 1989. Pp. xii, 238. Explores the neglect by economic and political institutions of the social impacts of new technologies and seeks to provide the "attentive public" with knowledge on how to direct and control technological applications. Argues that technology is more than hardware, but is an entire social system, that always entails side effects and tradeoffs that demand close attention to risk and uncertainties in a process of "looking before we leap." Explores public policy, private sector policies, their relationship, and the relationship of technology to science.

Whitbeck, Caroline. *Ethics in Engineering Practice and Research.* New York: Cambridge University Press, 1998. Pp. xx, 330. Uses a collection of case studies to address the professional and research responsibilities of engineers. Designed for classroom use, it includes a gen-

eral introduction to ethical concepts and offers interactive activities with Case Western Reserve University's Online Ethics Center for Engineering and Science at http://onlineethics.org/.

Wiener, Norbert. *The Human Use of Human Beings: Cybernetics and Society.* Boston, Houghton Mifflin, 1950. Gives an account of the purpose of a human life and four principles of justice. Offers a method of applied ethics and discusses questions and topics in computer ethics. Republished by Da Capo Press in 1988.

Winner, Langdon. *The Whale and the Reactor: A Search for Limits in an Age of High Technology.* Chicago: University of Chicago Press, 1986. Outlines a political philosophy of technology as a form of political action. Technologies are not just means but "forms of life." Includes the influential essay "Do Artifacts Have Politics?" (pp. 19-39), to which the answer is yes. Other chapters discuss failed attempts to introduce technological fixes into political life as well as the weakness of environmentalism, technology assessment, and appeals to values.

3. Monographs and Edited Volumes: Specialized

Abram, David. *The Spell of the Sensuous: Perception and Language in a More-than-Human World.* New York: Pantheon Books, 1996. A personal and phenomenological account of human being as fundamentally dependent on contact and conviviality with what is not human. Calls for a renewal of human relationships with the sensuous world in which technologies are rooted in order to reassess the human and technological relationship with natural places. Aimed at both environmental activists and scholars.

Adas, Michael. *Machines as the Measure of Men: Science, Technology and Ideologies of Western Dominance.* London: Cornell University Press, 1989.

Aman, Kenneth, ed. *Ethical Principles for Development: Needs, Capacities or Rights. Proceedings of the IDEA/Montclair Conference.* Upper Montclair, NJ: Institute for Critical Thinking, 1991.

Angell, Marcia. *Science on Trial: The Clash of Medical Evidence and the Law in the Breast Implant Case.* New York: W. W. Norton and Company, 1996. Pp. 268. Critical analysis of law's treatment of science in the case of breast implants by medical researcher and former journal editor. Inquires into the distinctions in the way science, the law, and the public regard evidence and weigh risk. Organized in ten chapters with a preface and afterword. Includes history, analysis of litigation and regulation, and the effects of corruption.

Arnhart, Larry. *Darwinian Natural Right: The Biological Ethics of Human Nature*. Albany: State University of New York Press, 1998. Defends a contemporary version of Aristotelian ethics using evolutionary biology.

Attfield, Judy. *Utility Reassessed: The Role of Ethics in the Practice of Design*. Manchester: Manchester University Press, 1999.

Barry, Robert L. and Gerard V. Bradley, eds. *Set No Limits: A Rebuttal to Daniel Callahan's Proposal to Limit Health Care for the Elderly*. Urbana: University of Illinois Press, 1991. Collects eight essays (including a preface and prologue) that criticize age-based rationing schemes for the allocation of health care resources. Argues that health care reforms are necessary but that it is not legally or morally justifiable to deprive people of life-sustaining care solely on the basis of their age. Considers moral and ethical, legal and jurisprudential, and public policy and economic aspects of age-based rationing.

Bavertz, Kurt, ed. *Sanctity of Life and Human Dignity*. Dordrecht: Kluwer, 1996. Pp. xix, 318. Engendered by a 1992 conference, compiles a general introduction and eighteen essays including an annotated bibliography and literature review. Sections include the concepts of human dignity, sanctity of life, and person, problems of critical care, and the role of the state.

Bayertz, Kurt. *GenEthics*. Cambridge, UK: Cambridge University Press, 1995. Clarifies the ethical dimensions generated by new human reproductive and genetic advancements. Most emphasis is on reproductive assisting technologies.

Beatley, Timothy. *Ethical Land Use*. Baltimore, MD: John Hopkins University Press, 1994.

Beauchamp, Tom, and Veatch, Robert eds. *Ethical Issues in Death and Dying*. 2nd ed. New York: Prentice Hall, 1996. Pp. xiv, 458. Gathers nine chapters of diverse resources pertaining to death and dying including essays, case studies, and government publications. Chapters are: definitions of death, truth-telling with dying patients, suicide, physician assisted suicide and euthanasia, forgoing treatment and causing death, decisions to forgo treatment involving once competent persons, decisions to forgo treatment involving never-competent patients, futile treatment and terminal care, and social reasons for limiting terminal care.

Bell, Robert. *Impure Science: Fraud, Compromise, and Political Influence in Scientific Research*. New York, NY: John Wiley & Sons, Inc., 1992. Pp. xvi, 301.

Explores how the pursuit of money and prestige have compromised and corrupted scientific research in the U.S. Uses many case studies (e.g., Breuning and Baltimore) to substantiate and illuminate argument. Concludes with recommendations.

Belsey, Andrew, and Ruth Chadwick, eds. *Ethical Issues in Journalism and the Media*. London: Routledge, 1992. Pp. xiii, 179. Eleven original essays on topics such as ethics and politics, owners, editors, and journalists, terrorism and reporting restrictions, objectivity, honesty, privacy, codes of conduct, and freedom of speech.

Benso, Silvia. *The Face of Things: A Different Side of Ethics*. Albany: State University of New York Press, 2000. Pp. xxxviii, 258. Tries to bridge Emmanuel Levinas' emphasis on "love without things" and Martin Heidegger's "things without love" by arguing for an ethics of festive things. Amazingly fails to reference the work of Albert Borgmann.

Berry, Wendell. *The Unsettling of America: Culture and Agriculture*. San Francisco: Sierra Club Books, 1977. Pp. ix, 228. A criticism of modern industrial agriculture and its ecological and cultural consequences. A third edition was published by University of California Press, 1996.

Bertrand, Claude-Jean, ed. *An Arsenal for Democracy: Media Accountability Systems*. Cresskill, NJ: Hampton Press, 2003. Pp. xi, 420. Provides information on a wide range of ways in which to democratize the news media and make it accountable to the public, primarily through media accountability systems. Posits these systems as intermediaries between total loss of social responsibility and strict legal regulation. Arranged in twenty-nine chapters including principles and rules, press councils, research, ombudsmen, and media accountability systems in seven countries.

Bertrand, Claude-Jean. *Media Ethics and Accountability Systems*. New Brunswick, NJ: Transaction, 2000.

Brody, Baruch. *The Ethics of Biomedical Research*. New York: Oxford, 1998. Pp. xiii, 386. Covers both animal and human subjects research including chapters on genetic research, research involving vulnerable subjects, drug/device approval process, and a concluding chapter with philosophical reflections. Features four appendices on international, European transnational, U.S., and other countries' research ethics policies.

Brunner, Ronald D., Christine H. Colburn, Christina M. Cromley, Roberta A. Klein, and Elizabeth A. Olson. *Finding Common Ground: Governance and Natural Resources in the American West*. New Haven: Yale University Press, 2002. Pp. xiii, 303. Designed to help broad audiences understand the potential of community-based initiatives for resolving public policy disputes in the name of the common interest. Organized into a general introduction, four case studies, and a conclusion that seeks to draw out the lessons learned from community-based initiatives of policy making.

Buchanan, Allen, Dan Brock, Norman Daniels, and Daniel Wikler. *From Choice to Chance: Genetics and Justice*. New York: Cambridge, 2000.

Bud, Robert. *The Uses of Life: A History of Biotechnology*. Cambridge: Cambridge University Press, 1993. Pp. xvii, 299. Explores the long history of biotechnology, emphasizing the past 100 years, from ancient conceptions to nineteenth century zymotechnology to human genome research. Also tracks the disparate meanings of the term over time and cultures.

Callicott, J. Baird. *In Defense of the Land Ethic: Essays in Environmental Philosophy*. Albany: State University of New York Press, 1989. Pp. x, 325. Takes the econcentrist standpoint, drawing from sociobiology and ecology, that modern values of Western civilization must be overhauled. Organized into five sections: animal liberation and environmental ethics, a holistic environmental ethic, a non-anthropocentric value theory for environmental ethics, American Indian environmental ethics, and environmental education, natural aesthetics, and E.T. The second section develops and defends Aldo Leopold's land ethic.

Casebeer, William D. *Natural Ethical Facts: Evolution, Connectionism, and Moral Cognition*. Cambridge, MA: MIT Press, 2003. Argues for a strong form of scientific ethics, recapitulating a neo-Aristotelian virtue theory using resources from evolutionary biology and cognitive neuroscience.

Chadwick, Ruth, Darren Shickle, Henk ten Have, and Urban Wiesing, eds. *The Ethics of Genetic Screening*. London: Kluwer, 1999. Pp. xvi, 255. Collects twenty-one essays resulting from a three-year multinational and multidisciplinary project known as Euroscreen. Opens with an overview of genetic screening and the ethical principles available for addressing developments in the field with special reference to the Wilson and Jungner principles on screening. Other topics include nation-specific perspectives on ethical debates, regulatory systems, and history.

Chiles, James R. *Inviting Disaster: An Inside Look at Catastrophes and Why They Happen*. New York, NY: HarperBusiness, 2002. Pp. xxx, 338. Compiles twelve chapters and an introduction that use major disasters to highlight how "smart," increasingly complex systems fail as the pace and scope of change overwhelms human capabilities of response and control.

Chubin, Daryl E and Ellen W. Chu, eds. *Science Off the Pedestal: Social Perspectives on Science and Technology*. Belmont, CA: Wadsworth Publishing Co., 1989. Pp. x, 196. Designed for classroom use. Presents a primarily U.S.-centered account of science as cultural force, way of knowing, and institutionalized activity to supplement more traditional science teaching. Collects fourteen chapters and a postscript in three parts: science, technology, and other social institutions, world views and politics of knowledge, and science and technology as public resources.

Cohen, Avner, and Steven P. Lee, eds. *Nuclear Weapons and the Future of Humanity: The Fundamental Questions*. Totowa, NJ: Rowman and Allanheld, 1986. Pp. xii, 496. The single best collection of articles on this topic. Collects twenty-five essays and an afterward by John Holdren. Topics include reflections on the present threat, the oddity of nuclear thinking, just war and morality, and reformations of social and political realities toward a non-nuclear future. NUC Ethics

Cook, Robert Lynn. *Code of Silence: Ethics of Disasters*. Jefferson City, MO: Trojan Publishing, 2003.

Council of Biology Editors, Inc. *Ethics and Policy in Scientific Publication*. Bethesda, MD: Council of Biology Editors, Inc., 1990. Pp. xiii, 290. Presents the results of a survey of Council members about nineteen scenarios to identify and define ethical issues in publishing research results. Also presents twenty-nine papers from a conference. Issues include misconduct, peer review, conflicts of interest, informed consent, and much more.

Crocker, David A., and Linden, Toby, eds. *Ethics of Consumption: The Good Life, Justice, and Global Stewardship*. Lanham, MD: Rowman and Littlefield, 1998. Pp. xviii, 585. Contains "The Road Not Taken: Friendship, Consumerism and Happiness" by Robert E. Lane, pp. 218-248.

Cutcliffe, Stephen H. *Ideas, Machines, and Values: An Introduction to Science, Technology, and Society Studies*. Lanham, MD: Rowman and Littlefield, 2000. Pp. xii, 179. A broad overview of STS as a field of study including its historical emergence, relationships to the philosophy, sociology, and history of science and technology, and programs, institutions, and journals in the field. Includes a chapter on interdisciplinarity and the current state of STS and comments on future directions for the field.

Cutcliffe, Stephen H., and Carl Mitcham, eds. *Visions of STS: Counterpoints in Science, Technology, and*

Society. Albany: State University of New York Press, 2001. Pp. vi, 170. Collects a general introduction on the historical background and challenges of STS and ten essays arranged in three sections: general perspectives, applications, and critiques. Aims to clarify the complexities and debates within STS that emerge from its interdisciplinary nature by presenting ten views of where STS is or where it should be heading.

Danielson, Peter. *Artificial Morality: Virtuous Robots for Virtual Games*. New York, NY: Routledge, 1992. Pp. xiv, 240. Engages in controversies about the adequacy of rational choice theories and builds moral robots to explore the role of artificial intelligence in the development of a claim that morality is person-made and rational. Shows that moral agents are rational in the sense that they successfully solve some social problems that amoral agents cannot solve.

Davis, Michael. *Ethics and the University*. New York, NY: Routledge, 1999. Pp. xii, 267. Organized in three parts: a broad introduction to ethics in the academy, research ethics, and teaching ethics.

Davis, Michael. *Profession, Code and Ethics*. Burlington, VT: Ashgate Publishing Co., 2002. Pp. ix, 256. Addressed at scholars, teachers, and students. Presents a definition of profession and argues that codes of ethics are inherent to the nature of professionalism. Collects fourteen chapters arranged in four parts: lawyers, engineers and scientists, police, and teaching ethics.

Davis, Michael. *Thinking Like an Engineer: Studies in the Ethics of a Profession*. New York, NY: Oxford University Press, 1998. Pp. xii, 240. Inquires into the nature of engineering and the ethical principles that guide it. Provides historical background, comments on codes of ethics and whistleblowing, and thoughts on protecting engineering judgment. Then supplies empirical work to support the philosophical account of engineering

Deane-Drummond, Celia, Bronislaw Szerszynski, and Robin Grove-White, eds. *Reordering Nature: Theology, Society, and the New Genetics*. London: T and T Clark, 2003.

De Waal, Frans. *Good Natured: The Origins of Right and Wrong in Humans and Other Animals*. Cambridge, MA: Harvard University Press, 1996.

Dreyfus, Hubert L. *On the Internet*. New York, NY: Routledge, 2001. Pp. ix, 127. Critiques certain aspects of the promise of the internet to extend and improve human interaction, especially distance learning. Grounds his critique in the history of Western philosophy and certain long-standing conceptions such as mind-body dualism. Looks to existentialism and its focus on embodiment as an important resource for theories of education. Argues distance education can work, but care must be made to implement it correctly.

Escobar, Arturo. *Encountering Development: The Making and Unmaking of the Third World*. New Jersey: Princeton University Press, 1995. Pp. ix, 290. A discursive poststructuralist critique of economics as the foundational structure of modernity. Argues that development and the "Third World" are being unmade due to repeated failures to achieve goals and aspires to imagine alternatives for a post-development era.

Evan, William M., and Mark Manion. *Minding the Machines: Preventing Technological Disasters*. Upper Saddle River, NJ: Prentice Hall, 2002. Pp. xxiv, 485. Offers explanations for why technological disasters occur and preventive measures to cover all areas of risk. Topics examined include: history and theories of disasters, strategic responses, design and organizational failures, sociocultural failures, responsibilities of institutions and individuals, and participatory technology and the role of the citizen. Also comments on legal system and private corporations and provides some case studies.

Farber, Paul Lawrence. *The Temptations of Evolutionary Ethics*. Berkeley: University of California Press, 1994.

Foster, Kenneth R., and Peter W. Huber. *Judging Science: Scientific Knowledge and the Federal Courts*. Cambridge, MA: MIT Press, 1997. Pp. 333. An extended commentary on scientific validity and the law's rules of evidence aimed at non-expert audiences. Explains the significance of the *Daubert* criteria and addresses the central question of when evidence presented as scientific should be considered reliable enough to be presented to a jury. Concludes with an attempt to reconcile the law's needs for workable rules of evidence with the views of scientific validity and reliability held in scientific disciplines.

Fukuyama, Francis. *Trust: The Social Virtues and the Creation of Prosperity*. New York: Free Press, 1995. A comparative historical study of high-trust and low-trust societies and their business and economic consequences.

Goldberg, Steven. *Culture Clash: Law and Science in America*. New York: New York University Press, 1994. Pp. xi, 255. Argues that law and culture are at the roots of the slippage between the promise of U.S. science and the reality of commercial technology. Organized into ten chapters including the constitutional status of and statutory framework for basic research, science and religion in the law, legal restrictions on new technology, the human genome, nuclear fusion, and artificial intelligence.

Goldschmidt, Walter. *As You Sow: Three Studies in the Social Consequences of Agribusiness*. New York: Universe Books, 1978. Pp. liv, 505. Examines the consequences of corporate agriculture for rural communities in the United States. Features an extended general introduction on "Agriculture and the Social Order" that traces the rise of agribusiness.

Gough, Michael, ed. *Politicizing Science: The Alchemy of Policymaking*. Stanford, CA: Hoover University Press, 2003. Pp. xxi, 313. Shows the ways in which the connections between politics and science can thwart the achievement of social goals. Collects a preface, introduction, and eleven essays written by scientists about specific cases of excessive politicization.

Gould, Stephen J. *Rocks of Ages: Science and Religion in the Fullness of Life*. New York: Ballantine Books, 2002. Pp. viii, 241. Discusses the relationship between religion and science, arguing that the two are non-overlapping magisteria (NOMA) that can work peacefully together but only if there is no attempt to synthesize them somehow or bring one under the domain of the other. Argues that science deals with facts and theories about nature, whereas religion deals with human values and ultimate meaning.

Graham, Gordon. *The Internet: A Philosophical Inquiry*. New York, NY: Routledge, 1999. Pp. ix, 179. Assesses the implications of the internet for concepts of identity, moral anarchy, censorship, community, democracy, virtual reality, and imagination. Opens by negotiating the extremes of luddism and technophilia.

Greenberg, Daniel S. *Science, Money, and Politics: Political Triumph and Ethical Erosion*. Chicago: University of Chicago Press, 2001. Pp. x, 530. Examines and seeks to explain the prosperity and autonomy of science in the United States from the end of World War II to the turn of the century. Argues that the scientific "metropolis" has successfully lobbied for political resources, especially money and independence, but in so doing it has eroded its ethical integrity through these strategies of acquiring support and in the conduct of research. Takes a thematic approach through twenty-eight chapters that take up beliefs, social characteristics, goals, and revealing episodes.

Greenberg, Daniel S. *The Politics of Pure Science*, 2nd edition. New York: New American Library, 1999 (1st edition, 1967). Pp. xxvii, 311. Draws from personal experience writing for the journal Science on the politics of science and focuses on basic research. Explains how this politics works without sliding into either reverence or cynicism. Divided into three sections that treat the scientific community, the shaping of science politics during and after World War II, and some more recent examples of science politics. Concludes with notes about the new politics of science that demands more accountability from the scientific community.

Hamelink, Cees J. *The Ethics of Cyberspace*. Thousand Oaks, CA: Sage, 2000.

Hargrove, Eugene. *Foundations of Environmental Ethics*. Englewood Cliffs, NJ: Prentice Hall, 1989. Pp. x, 229. Organized into three sections. "Traditional Positions" explores Greek and modern philosophy. "The Environmental Position" outlines aesthetic, scientific, and wildlife protection attitudes and treats the perennial issues of value such as instrumental versus intrinsic. "Philosophical and Ethical Implications" presents an ontological argument for environmental ethics and discusses "therapeutic nihilism" in the context of environmental management.

Harries, Karsten. *The Ethical Function of Architecture*. Cambridge, MA: MIT Press, 1997. Pp. xiii, 403. Argues that architecture faces a deep philosophical problem bound up with questions of interpretation, the good life, and genuine dwelling as technology transforms human experience away from a focus on place and community. Claims that architecture should help define a sense of place in a disorienting world by articulating a common ethos. Includes 123 illustrations.

Hayles, M. Katherine. *How We Became Posthuman*. Chicago: University of Chicago Press, 1999. Pp. xiv, 350. Drawing from the history of cybernetics and information theories, argues that the emergence of distributed cognition and the disembodiment of infromation both furthers and overturns the liberal humanist subject. "Posthuman" is used in multiple, sometimes ironic ways, but all of which connot some form of union of humans with intelligent machines. Argues that human identity is more than information, but relies also on its instantiation and seeks to foster a future that embraces information technology "without being seduced by fantasies of unlimited power and disembodied immortality, that recognizes and celebrates finitude [and material embeddedness] as a condition of human being" (p. 5).

Hefner, Philip. *Technology and Human Becoming*. Minneapolis: Fortress Press, 2003. Proposes a Christian theory of co-creation in the use of science and technology.

Heller, Agnes. *Beyond Justice*. Oxford, U.K.: Basil Blackwell, 1987. Pp. vi, 346. Critiques theoretical assumptions underlying traditional and modern notions of justice, argues that all claims to justice are rooted in other values such as freedom and life, and claims that, although justice may be a precondition of the good life, the good life is something beyond justice. Contains analytic, historical, and normative chapters.

Hendler, Sue, ed. *Planning Ethics: A Reader in Planning Theory, Practice and Education.* New Brunswick, NJ: Center for Urban Policy Research, 1995. Pp. xx, 374. Reflects and furthers the expansion of professional ethics to more public and global concerns. Collects fifteen essays in three parts, each of which is set in the context of ethical theory: planning theory, planning practice, and planning education. Intended for planners and philosophers.

Herkert, Joseph R., ed. *Social, Ethical, and Policy Implications of Engineering.* New York: Institute of Electrical and Electronics Engineers (IEEE) Press, 2000. Pp. xi, 339. Collects 35 articles arranged in three categories: the societal context of technology and engineering, ethical responsibilities of engineers, and engineering ethics and public policy. Emphasis is placed on the policy aspects of contemporary ethical issues. Aimed at engineering educators, students, and practitioners. All articles are reprinted from the *IEEE Technology and Society Magazine.*

Hess, David. *Science Studies: An Advanced Introduction.* New York: New York University Press, 1997. Pp. vii, 197. Focuses on U.S. topics and highlights cross-disciplinary misunderstandings in the field. Collects six chapters including a chapter that discuss the philosophy of science, sociology of science, social studies of knowledge, critical and cultural studies of science and technology, and a conclusion that primarily treats policy issues.

Heyd, David. *Genethics: Moral Issues in the Creation of People.* Berkeley: University of California Press, 1992. Attempts to resolve many ethical paradoxes in intergenerational justice raised by advances in medicine, genetic engineering, and demographic forecasting.

Higgs, Eric. *Nature by Design: People, Natural Process, and Ecological Restoration.* Cambridge, MA: MIT Press, 2003. Pp. xv, 341. Introduces concept and cases of ecological restoration. Focuses on the concern that restoration acts as an apology for technological excess and demonstrates a hubristic urge to manipulate nature to mirror cultural values. Proposes "focal restoration" as a preferred way of ensuring participation and engagement in restoration projects and highlighting the importance of responsible and intentional "wild" design.

Higgs, Eric, Andrew Light, and David Strong, eds. *Technology and the Good Life?* Chicago: University of Chicago Press, 2000. A collection of essays on the work of Albert Borgmann.

Hilgartner, Stephen. *Science on Stage: Expert Advice as Public Drama.* Stanford, CA: Stanford University Press, 2000. Pp. xvi, 214. Uses two National Academy of Science reports to examine the production and use of science advice in an age conflicted by a vision of expertise as both value-laden and objective. Employs the theoretical trope of the theater to investigate how advisory bodies produce credibility and authority by managing information and appearances in complex ways. Investigates the "boundary work" and rhetorical and narrative techniques at the borders of science and society and uses the idea of "stage management" to differentiate "back stage" from "front stage" elements of science advice.

Homan, Roger. *The Ethics of Social Research.* New York: Longman, 1991.

Howard, Ted, and Jeremy Rifkin. *Who Should Play God?: The Artificial Creation of Life and What It Means for the Future of the Human Race.* New York: Delacorte Press, 1977. Pp. 272. Introduces genetic engineering and its history, links it to the ideology of eugenics supporters, describes its likely forms of application, and concludes with recommendations. Staunchly opposes genetic engineering and reductionism, arguing that the choice is between preserving humans and other species as they are or launching a mass program of biological reengineering. Argues that genetic engineering is inherently anti-democratic and elitist and requires active public participation to prevent dehumanization.

Jasanoff, Sheila. *The Fifth Branch: Science Advisors as Policymakers.* Harvard University Press, 1990. Pp. xiii, 302. Draws from social studies of science, especially constructivist work, to present a conceptual framework and differentiated vocabulary for the dilemmas faced by science advisory committees. Argues for procedural reforms in the role of science advisors in public policy making. Addresses the question of the limits of participatory decision-making in an age of growing technological complexity and expert knowledge.

Jasanoff, Sheila. *Science at the Bar: Law, Science, and Technology in America.* Cambridge, MA: Harvard University Press, 1995. Pp. xvii, 285. A classic overview of law-science relationship from social studies of science perspective. Argues that the courts actively influence the production of science and technology and serve as democratizing agents, but are often constrained in this role by positivistic assumptions. Analyzes scientific and legal modes of reasoning and concludes with a prescriptive look ahead.

Kass, Leon. *Toward a More Natural Science: Biology and Human Affairs*. New York: Free Press, 1985. Pp. xiv, 370. An Aristotelian account that argues that science can go too far if it is not appropriately regulated by the wisdom contained in our emotional reactions to certain technological advances.

Kavka, Gregory S. *Moral Paradoxes of Nuclear Deterrence*. New York: Cambridge University Press, 1987. Pp. xii, 243. A tightly argued exploration of the major quandaries that defends nuclear deterrence, if subjected to proper restrictions, as morally justified. Highlights conflicts and dilemmas both within and between utilitarian and deontological ethics.

Kellert, Stephen R. *The Value of Life: Biological Diversity in Human Society*. Washington, DC: Island Press, 1996. A taxonomy of views of nature.

Kevles, Daniel J. *In the Name of Eugenics: Genetics and the Use of Human Heredity*. Berkeley: University of California Press, 1985. Seminal work on history of eugenics and the eugenic implications of new reproductive technologies. Reprinted by Harvard University Press, 1995.

Kimbell, Richard. *Assessing Technology: International Trends in Curriculum and Assessment*. Philadelphia, PA: Open University Press, 1997. Pp. xiv, 249. Explores the issues of assessment that have emerged with the technology curriculum in the U.K., especially the problems of process-centered assessment that involve evaluating students' capabilities in the process of design and development. Provides international comparisons to the U.S., Germany, Taiwan, and Australia. Concludes with general reflections.

Koehn, Daryl. *The Ground of Professional Ethics*. New York, NY: Routledge, 1994. Pp. x, 224. Confronts and rebuts the challenge to the authority and ethics of professionals by arguing that it rests on a secure and morally legitimating ground because and to the extent that these professions are structured to merit the trust of clients.

LaFollette, Marcel C. *Stealing into Print: Fraud, Plagiarism, and Misconduct in Scientific Publishing*. Berkeley: University of California Press, 1992. Pp. viii, 293. Focuses on how scientific misconduct affects communication practices and policies in the journals that disseminate the results of scientific research.

Layton, Edwin T. Jr. *The Revolt of the Engineers: Social Responsibility and the American Engineering Profession*. 2nd ed. Baltimore: Johns Hopkins University Press, 1986. Pp. xxii, 286. (First published 1971.) Classic examination of the professionalization of engineering in the U.S. from 1900-1945. Analyzes the tensions between business interests and technical expertise, and describes failed attempts during the first half of the 20th century to promote unity and autonomy (the "revolt") around an ideology of engineers as professional leaders of advanced civilization. A new preface comments briefly on post-World War II developments.

Levine, Robert. *Ethics and the Regulation of Clinical Research*. 2nd ed. New Haven: Yale, 1988.

Light, Andrew and Eric Katz, eds. *Environmental Pragmatism*. London: Routledge Press, 1998. Pp. xvi, 352. Presents environmental pragmatism as a way to direct the fruits of (open-ended, pluralistic, and context specific) philosophical inquiry toward practical resolution of environmental problems. Collects seventeen essays and a general introduction.

Marcus, Stephen J., ed. *Neuroethics: Mapping the Field: Conference Proceedings, May 13-14, 2002, San Francisco, California*. New York: Dana Press, 2002. Pp. vii, 367. Result of a conference composed of scientists, ethicists, humanists, and others on the personal and social implications of human brain research. Organized into five sections: notions of self, social policy, ethics, public discourse, and mapping the future. Also includes two speeches, one by Arthur Caplan that argues the main issue is equity rather than worries about enhancement, and an introduction mapping the new emerging field of neuroethics.

Margolin, Victor. *The Politics of the Artificial: Essays on Design and Design Studies*. Chicago: University of Chicago Press, 2002. Pp. 273.

Mason, Richard, Florence Mason, and Mary Culnan. *Ethics of Information Management*. Thousand Oaks, CA: Sage Publications, 1995.

McDonough, William, and Michael Braungart. *Cradle to Cradle: Remaking the Way We Make Things*. New York: North Point Press, 2002. Pp. 193.

McGee, Glenn. *The Perfect Baby: A Pragmatic Approach to Genetics*. New York: Rowman and Littlefield, 1997. Denies the necessity of a "genethics," arguing that the wisdom we need can be found in the everyday experience of parents.

Mehlman, Maxwell J., and Jeffrey R. Botkin. *Access to the Genome: The Challenge to Equality*. Washington, DC: Georgetown University Press, 1998. Summarizes the Human Genome Project and discuss its practical health applications and ethical and policy challenges such as bans, equal access, genetic handicapping, and genetic lotteries.

Mendelsohn, Everett, Merritt Roe Smith, and Peter Weingar, eds. *Science, Technology and the Military*, 2 vols. Boston: Kluwer, 1988. Pp. xxix, vii, 288; 274. Collects papers presented at 1987 conference with an introductory overview. Topics include war and the restructuring of physics, the military and technological development, industry, medicine, academy, and nuclear weapons and power.

Mepham, Ben., ed. *Food Ethics*. London: Routledge, 1996. Pp. xiv, 178. Collects ten essays and a select bibliography on such issues as food aid and trade, biotechnology, global hunger, consumer sovereignty, research ethics, and nutrition and health. Features an essay that presents an evaluative framework for ethical analysis of food biotechnologies.

Mirowski, Philip and Esther-Mirjam Sent, eds. *Science Bought and Sold: Essays in the Economics of Science*. Chicago: University of Chicago Press, 2002. Pp. ix, 573. Presents science as a deeply economic activity of investment and profit and shows the changing relations between science and economics. Collects a general introduction and nineteen original and reprinted essays arranged in six parts including science as a production process, science as a problem of information processing, contours of the globalized privatization regime, and the future of scientific credit.

Molotch, Harvey. *Where Stuff Comes From: How Toasters, Toilets, Cars, Computers and Many Other Things Come to Be As They Are*. New York: Routledge, 2003. Pp. 324.

Moulakis, Athanasios. *Beyond Utility: Liberal Education for a Technological Age*. Columbia: University of Missouri Press, 1994. Pp. viii, 171. Generated from experiences teaching a Humanities for Engineers course. Considers the larger purposes of liberal arts education and how they relate to the education of professionals. Addresses the controversy in education circles about tradeoffs between narrow, professional and broad, humanistic education.

National Academy of Engineering. *The Engineer of 2020: Visions of Engineering in the New Century*. Washington, D.C.: The National Academies Press, 2004. Pp. xv, 101. Result of a forward-looking conference about what engineering will and should be like in the future and to what extent engineers can shape that future. Includes an appendix with possible future scenarios.

Paradis, James, and George C. Williams, eds. *Evolution and Ethics*. Princeton, NJ: Princeton University Press, 1989. Contains the essay "A Sociobiological Expansion of Evolution and Ethics" by George Williams.

Pattyn, Bart, ed. *Media Ethics: Opening Social Dialogue*, Leuven, Belgium: Peeters, 2000. Contains the important article "An Intellectual History of Media Ethics" by Clifford Christians.

Pelletier, Louise, and Alberto Pérez-Gómez, eds. *Architecture, Ethics, and Technology*. Montreal: McGill-Queen's University Press, 1994.

Perrow, Charles. *Normal Accidents: Living with High-Risk Technologies*. New York: Basic Books, 1984. (Revised 1999, Princeton, NJ: Princeton University Press). Pp. x, 386. Traces six examples of modern industrial systems to argue that tight coupling and interactive complexity inevitably produce accidents, and that these are more important concerns than operator error or the failure of parts. Offers an assessment of these systems and recommendations for future action. Concludes with a discussion of high-risk decision making.

Peters, Ted. *Playing God?: Genetic Determinism and Human Freedom*. 2nd ed. New York: Routledge, 2003. Pp. xvii, 260. Rejects genetic determinism and argues that human nature is the product of genes, environment, and free will. Defends a Christian understanding of humans as future-oriented and cocreative as an ethic for guiding genetic research. Takes up questions of genetic manipulation beyond therapy, ethics and science in the "gay gene" controversy, and such issues as patenting genes, cloning, stem cell research, and germ-line intervention.

Postrel, Virginia. *The Future and Its Enemies: The Growing Conflict over Creativity, Enterprise, and Progress*. New York: Free Press, 1998. Pp. xviii, 265. A libertarian defense of technological innovation as basis for human freedom that portrays two alternative futures: one that is diverse, dynamic, decentralized and choice-driven and the other that is static, centralized, and controlled. Explores the clash between dynamism and stasis and defends the former over the latter. Has a companion website at www.dynamist.com.

Proctor, Robert N. *Value-Free Science? Purity and Power in Modern Knowledge*. Cambridge, MA: Harvard University Press, 1991. Pp. xi, 331. Traces the origin of value neutrality in the separation of theory and practice, the isolation of moral knowledge from natural philosophy, and the mechanical conception of the universe. Explores the exclusion of morals and politics in the social sciences, especially in Germany, and reviews more recent critiques of value-neutral science.

Reiss, Michael J., and Roger Straughan. *Improving Nature? The Science and Ethics of Genetic Engineering.* New York: Cambridge University Press, 1996. Covers a broad range of ethical and theological concerns in genetic engineering of microorganisms, plants, animals, and humans.

Resnik, David B. *Owning the Genome: A Moral Analysis of DNA Patenting.* Albany: State University of New York Press, 2004. Pp. xiii, 235. Examines the main arguments for and against different types and scopes of DNA patenting from both consequentialist and deontological perspectives. Argues that consequentialist arguments pertain to most issues, whereas deontological arguments have a more limited application. Claims that DNA patenting offers society many important benefits and poses a few important threats. Articulates and defends the precautionary principle in some areas and concludes with policy recommendations.

Rifkin, Jeremy. *Who Should Play God?: The Artificial Creation of Life and What It Means for the Future of the Human Race.* New York: Delacorte Press, 1977.

Rip, Arie, Thomas J. Misa, and Johan Schot, eds. *Managing Technology in Society: The approach of Constructive Technology Assessment.* London, England: Pinter, 1995. Pp. xii, 361. Explores the concept of critical technology assessment and the need for it in the goal of maximizing benefits and minimizing harms of technologies, uses case studies to argue that changing entrenched technologies and institutions is difficult but possible, discusses conditions for learning about experiences to try in other contexts, and argues that such policies will be context specific.

Roco, Mihail C., and William S. Bainbridge, eds. *Societal Implications of Nanoscience and Nanotechnology.* Dordrecht, The Netherlands: Kluwer, 2001. Pp. vii, 370. Collects articles from various contributors organized into five introductory chapters on nanotechnology goals and societal interactions, social science approaches to assessment, and recommendations. Chapter six provides topical considerations including education, medicine, environment, space, and security.

Rolston III, Holmes. *Conserving Natural Value.* New York: Columbia University Press, 1993. Pp. 259. A philosophical argument that seeks to balance natural and cultural values and considers the anthropocentric and intrinsic theories of value.

Sachs, Wolfgang ed. *The Development Dictionary: A Guide to Knowledge as Power.* London: Zed Books, 1992. Pp. 306. Compiles a general introduction and nineteen essays that deconstruct key terms in the modern development discourse such as needs, progress, science, technology, development, state, and environment. Argues that it is time to abandon the dominant development paradigm or "cast of mind."

Sarewitz, Daniel, Roger A. Pielke, Jr., and Radford Byerly, Jr, eds. *Prediction: Science, Decision Making, and the Future of Nature.* Washington, DC: Island Press, 2000. Pp. xv, 405. Addresses the application of scientific predictions to environmental problems, noting promises and limits, and pointing out that predictions are at once technical, political and social. Argues that the relationship of predictions to policy making is rocky due to the complexity of systems that generate uncertainty (and uncertainty about uncertainty) and the widely held and problematic assumption that predictions can simplify the decision-making process. Includes a general introduction and eighteen essays collected in six parts: prediction as a problem, natural hazards, politics, policy, prediction in perspective, and a conclusion.

Schlossberger, Eugene. *The Ethical Engineer.* Philadelphia, PA: Temple University Press, 1993. Pp. xii, 284. Addressed both to practicing professionals and engineering students. Uses illustrating cases to supplement the text. Includes an introduction to engineering ethics and ethical decision making, comments ethical theories and the sources of ethical decisions, issues such as honesty, good faith, employee-employer relations, and consulting.

Schmitz, David, and Elizabeth Willott. *Environmental Ethics: What Really Matters, What Really Works.* Oxford: Oxford University Press, 2002. Pp. xxi, 566. Collects classic essays in environmental ethics in fifteen topical areas, each introduced with questions for reflection and discussion.

Schumacher, E. F. *Small Is Beautiful: Economics as if People Mattered.* New York: Harper Perennial, 1989. (Originally published 1973, Harper & Row; reprint 1999, Hartley & Marks.) Pp. xxiii, 324. A critique of neo-classical economics, its conception of human nature and desires, natural resources, and its tendencies to globalize systems of production and distribution on massive scales. Defends small-scale, decentralized economies and includes the influential essay "Buddhist Economics," which challenges the goal displacement of growth-oriented economies that use technology to alienate human meaning by focusing on conceptions of "right livelihood" and celebrating the humanizing and liberating quality of work when scaled down and rooted in a community.

Schweber, S. S. *In the Shadow of the Bomb: Bethe, Oppenheimer, and the Moral Responsibility of the Scientist.* Princeton, NJ: Princeton University Press, 2000. Pp. xviii, 260. Examines the different reactions of two physicists to the moral dilemmas posed by the development and use of atomic weapons and questions of the professional responsibilities and public roles of scientists and engineers. Details the different roles played by Oppenheimer and Bethe, their foundations, and their consequences.

Sen, Amartya. *Development as Freedom.* New York: Anchor Books, 1999.

Sieber, Joan E., ed. *The Ethics of Social Research: Surveys and Experiments.* 2 vols. New York: Springer-Verlag, 1982. Pp. xii, 249 and x, 187. Designed to assist social scientists in preparing for and resolving ethical issues. Arranged as ten chapters in the first volume and seven in the second in four total sections: respect for the individual, protection of privacy and confidentiality, ethnographic fieldwork and beneficial reciprocity, and the roles of social scientists in research regulation and media relations.

Silver, Lee. *Remaking Eden: Cloning and Beyond in a Brave New World.* New York: Avon, 1998. Pp. viii, 317. Takes stock of the current state of reproduction and genetics (reprogenetics) technology to survey likely future scenarios. Argues that Huxley's dystopian vision of a "brave new world" is mistaken because individuals, not governments, will control reprogenetic technologies and that a society that values individual freedom above all else has difficulty justifying restrictions on the use of technologies by individuals. Surveys the changing meanings of parenthood, childhood, and the meaning of human life, dismisses many oppositions to reprogentic technologies, and concludes that such new technologies are inevitable as guaranteed by the global market.

Sismondo, Sergio. *An Introduction to Science and Technology Studies.* Malden, MA: Blackwell, 2004. Pp. vii, 202. Provides a clear overview of the field for readers unfamiliar with it. Intended for undergraduate or graduate classroom use. Organized into sixteen chapters that address historical and conceptual topics such as the Kuhnian revolution following the prehistory of STS, actor-network theory, social construction of knowledge, rhetoric and discourse, and expertise and the public understanding of science.

Sonnert, Gerhard. *Ivory Bridges: Connecting Science and Society.* Cambridge, MA: The MIT Press, 2002. Pp. x, 227. Scrutinizes the links between science and society beginning with a Jeffersonian concept of science policy, followed by a consideration of voluntary public interest associations of scientists, and concluding with questions of autonomy and responsibility.

Steinbock, Bonnie, ed. *Ethical and Legal Issues in Human Reproduction.* Hampshire, U.K.: Ashgate, 2002.

Stock, Greg. *Redesigning Humans.* New York: Houghton Mifflin, 2002. Strong defense of the genetic engineering of human beings.

Stone, Jeremy J. *"Every Man Should Try:" Adventures of a Public Interest Activist.* New York, NY: PublicAffairs, 1999. An autobiography that focuses on the development of the Federation of American Scientists. Provides an inside, personal look at some of the politics behind nuclear disarmament talks, reflections on why successes and failures occurred, and lessons about the complexities of public interest science.

Sutton, Victoria. *Law and Science: Cases and Materials.* Durham, NC: Carolina Academic Press, 2001. Pp. xxiv, 388. A legal casebook. Includes over sixty cases arranged into five chapters: an introduction, government, private sector, courts, and a future outlook.

Szerszynski, Bronislaw. *Nature, Technology, and the Sacred.* Malden, MA: Blackwell, 2005. Pp. xviii, 222. Uses the term "sacred" to understand the ways in which a range of religious framings are involved in ideas of and interactions with nature and technology. Argues that implicitly religious understandings of nature and technology are widespread in Western cultures. Begins with reflections on modernity and the disenchantment of the world, arguing against contemporary theorists who claim no such thing has occurred. Argues for a conscious reappropriation of sacral traditions and outlines the implications.

Thompson, Alison K., and Ruth F. Chadwick, eds. *Genetic Information: Acquisition, Access, and Control.* New York: Kluwer, 1999. Pp. xi, 335. Collects thirty essays arranged in five sections: eugenics, genetics and insurance, commercialization of genetic information, public awareness, and theoretical concerns.

Thompson, Paul B. *Agricultural Ethics: Research, Teaching, and Public Policy.* Ames: Iowa State University Press, 1998. Pp. xi, 239. Aims to provide an introduction to philosophical reflection on agriculture and food production by reflecting on food system issues with key concepts from ethics. Emphasizes the importance of technological change, ethical extensionism, and questions about the worth of the family farm. Organized in three sections: research, teaching, and public policy with a general introduction and conclusion.

Thompson, Paul B. *The Ethics of Aid and Trade: U.S. Food Policy, Foreign Competition, and the Social Contract*. New York: Cambridge University Press, 1992. Pp. x, 233. Explores the principles of U.S. agricultural policy and foreign aid, arguing that the traditional model of the nation-state should be replaced with the "trading state." Addresses protectionist challenges to foreign aid and development assistance in moral, economic, and political terms. Proposes a model of international relations with greater fluidity of material and intellectual exchange and creates a new interpretation of social contract theory that is geared to the goals of international trade and development policy.

Thomson, Norma, ed. *Instilling Ethics*. Lanham, MD: Rowman and Littlefield, 2000. Pp. xv, 239. Collects fourteen original articles arranged in three sections: sources of ethical reflection, modernity and the problems of ethical reflection, and instilling ethics today.

Valenstein, Elliot S. *Great and Desperate Cures: The Rise and Decline of Psychosurgery and Other Radical Treatments for Mental Illness*. New York: Basic Books, 1986. Pp. xiv, 338. Pursues the history of psychosurgery (e.g., lobotomy) as a cautionary tale, arguing that these procedures were very much a part of mainstream medicine and that the conditions that fostered their development are still active. Sets the tale in context with an opening chapter on the treatment of mental illness.

Verbeek, Peter-Paul. *What Things Do: Philosophical Reflections on Technology, Agency, and Design*. Robert P. Crease, trans. University Park: Pennsylvania State University Press, 2005. Pp. viii, 249. Develops an innovative approach to understanding the role of technological devices in lived experience and how they shape personality and society. Distinguishes analysis from classical philosophy of technology to develop an empirical, "postphenomenological" approach.

Wachs, Martin, ed. *Ethics in Planning*. New Brunswick, NJ: Center for Urban Policy Research, 1985. Pp., xxi, 372. The first compendium of works on ethics in planning. Collects a general introduction (with a fourfold taxonomy of ethical issues) and seventeen essays arranged in four sections: overview of ethical issues in urban planning and administration, corruption and whistle-blowing, ethical issues in policy making, and the emergence of an environmental ethics. Includes four appendices with relevant codes of ethics.

Walter, Jennifer K., and Eran P. Klein, eds. *The Story of Bioethics: From Seminal Works to Contemporary Explorations*. Washington, DC: Georgetown University Press, 2003. Pp. xv, 248.

Wilson, Edward O. *Consilience: The Unity of Knowledge*. New York: Alfred A. Knopf, 1998. Pp. 332. Seeks to develop a unification of knowledge according to the principles found in the natural sciences, especially sociobiology. Espouses a version of material reductionism and champions the Enlightenment ideals of objective knowledge, human progress, and the unity of truth.

Wilson, Edward O. *Sociobiology: The New Synthesis*. 25th Anniversary Edition. Cambridge, MA: Harvard University Press, 2000. Pp. xiii, 697. First published in 1975, established the field of sociobiology, also terms evolutionary psychology. For an overview of early controversies related to this topic, see Arthur L. Caplan, ed., *The Sociobiology Debate: Readings on the Ethical and Scientific Issues Concerning Sociobiology* (New York: Harper and Row, 1978).

4. Textbooks

Almond, Brenda, ed. *Introducing Applied Ethics*. Oxford: Blackwell, 1995. Includes more than twenty texts on family life, professional ethics, law, economics, and international relations. Little focus on science or technology.

Baum, Robert R., and Albert Flores, eds. *Ethical Problems in Engineering*. 2 vols. Troy, NY: Center for the Study of the Human Dimensions of Science and Technology, 1978. Although out of print, this remains a classic engineering ethics collection.

Beauchamp, Tom L., and James F. Childress. *Principles of Biomedical Ethics*. 5th ed. New York: Oxford University Press, 2001. Pp. xi, 454. One of the most influential textbooks in the bioethics field. The most developed use of principlism in bioethics, arranged in three parts that treat moral norms, character, and theories and outline the basic principles of respect for autonomy, nonmaleficence, beneficence, and justice as well as a chapter on professional-patient relationships.

Bowyer, Kevin W. *Ethics and Computing: Living Responsibly in a Computerized World*. Washington, DC: Institute of Electrical and Electronics Engineers (IEEE) Computer Society Press, 1996. Pp. xvi, 449. Examines issues central to computer ethics including hacking, privacy, computers in safety-critical systems, whistle blowing, intellectual property, environmental health, law, and equity. Includes case studies and exercises suitable for undergraduate courses.

Bulger, Ruth Ellen, Elizabeth Heitman, and Stanley Joel Reiser, eds. *The Ethical Dimensions of the Biological*

Sciences. New York: Cambridge University Press, 1993. Pp. xi, 294. Collects thirty-six articles including a general introduction addressed primarily to graduate students and faculty responsible for teaching ethics in science. Includes classic essays, seminal works, policy statements, and research guidelines that address such topics as the ethics of research and teaching, the qualifications for authorship, and the relationship of science, industry, and society. Each section includes questions for discussion.

Cheney, Darwin, ed. *Ethical Issues in Research.* Frederick, MD: University Publishing Group, 1993. Pp. xx, 237. Collects an overview and twenty-two chapters arranged in five parts: misrepresentation of data (U.S. and international perspectives), conflict of interest, research on human subjects, use of embryos and fetuses, and use of animals.

DesJardins, Joseph, eds. *Environmental Ethics: Concepts, Policy, and Theory.* London: Mayfield, 1999. Pp. xvi, 620. A broad overview with discussion and study questions following each of 18 chapters with classic essays arranged into four sections: context, basic concepts, policies and controversies, and philosophy and theory.

Edel, Abraham; Elizabeth Flower; and Finbarr W. O'Connor. *Critique of Applied Ethics: Reflections and Recommendations.* Philadelphia: Temple University Press, 1994. Pp. vi, 274. Surveys theories of applied ethics and argues that the stabilities of traditional morality must be combined with new knowledge to direct the rapid pace of techno-societal change. Divided into two sections: philosophical background and an analysis of practical problems. Conclusion emphasizes the importance of applying theories to complex and changing contexts.

Ermann, M. David, and Michele S. Shauf, eds. *Computers, Ethics, and Society.* 3rd ed. New York: Oxford University Press, 2003. Pp. vi, 249. Standard text covering ethical frameworks, personal decision making, politics, and professional responsibilities. First edition, 1990.

Erwin, Edward, Sidney Gendin, and Lowell Kleiman, eds. *Ethical Issues in Scientific Research: An Anthology.* New York, NY: Garland Publishing, 1994. Pp. xi, 413. Collects twenty-six essays in six sections: science and values, fraud and deception, human experimentation, animal research, genetics research, controversial research topics.

Elliott, Deni, and Judy E. Stern, eds. *Research Ethics: A Reader.* Hanover, NH: University Press of New England, 1997. Pp. xii, 319. A student reader with original and reprinted articles, essays, and case studies. Topics include teaching ethics, misconduct, conducting, reporting, and funding research, conflicts of interest, institutional responsibility, and animal and human experimentation.

Fleddermann, Charles B. *Engineering Ethics.* Upper Saddle River, NJ: Prentice Hall, 1999.

Gorman, Michael E., Matthew M. Mehallik, and Patricia Werhane. Ethical and Environmental Challenges to Engineering. Upper Saddle River, NJ: Prentice Hall, 2000.

Gunn, Alastair S., and P. Aarne Vesilind. *Hold Paramount: The Engineer's Responsibility to Society.* Pacific Grove, CA: Brooks/Cole, 2003. Pp. xiv, 160. Intended for use as a textbook. Includes cases studies, feature boxes, and discussion questions. Topics addressed include expertise and obligation, codes of ethics, terrorism, professional development, conflicts of interest, and much more.

Johnson, Deborah G., and Helen Nissenbaum, eds. *Computers, Ethics, and Social Values.* Upper Saddle River, NJ: Prentice Hall, 1995. Pp. vi, 714. Collects fifty-eight articles organized in seven chapters that seek to define and differentiate the field of computer ethics. Explores the significance of computers in terms of social values such as privacy, justice, democracy, and property. Examines computers in controversies involving traditional ethical notions such as crime, risk, and responsibility. Concludes with a look at the ethical issues of an increasingly networked information society.

Kaplan, David M., ed. *Readings in the Philosophy of Technology.* Lanham, MD: Rowman and Littlefield, 2004. Pp. xvi, 512Thirty-one readings, with sections on "Technology and Ethics," "Technology and Politics," and "Technology and Human Nature" most directly relevant.

Katz, Eric, Andrew Light, and William Thompson, eds. *Controlling Technology.* 2nd ed. Amherst, NY: Prometheus Books, 2003. Pp. 531. Thirty-four essays and a general introduction aimed at humanists, scientists, and engineers. Arranged to address fundamental issues at the intersection of technology and human values, especially democracy. Topics include human autonomy and freedom, the autonomy of technology, human equality, and respect for others. Arranged in eight parts including appropriate technology, technology, ethics, and politics, and computers, information, and virtual reality.

Light, Andrew, and Holmes Rolston III, eds. *Environmental Ethics: An Anthology.* Oxford: Blackwell Publishers, 2003. Pp. x, 554. Collects a general introduction and forty essays arranged in seven sections including definitions of environmental ethics, moral standing, the question of intrinsic value in nature, monism versus pluralism, and reframing environmental ethics. Includes a bibliographic essay by Clare Palmer that sketches the history and central issues of environmental ethics.

Loue, Sana. *Textbook of Research Ethics: Theory and Practice*. Dordrecht: Kluwer, 1999. Provides a brief history of human subjects research and reviews relevant ethical theories and principles. Refers to international documents and national policies and includes case studies and discussion exercises.

Macrina, Francis L. *Scientific Integrity: An Introductory Text with Cases*. Washington, DC: ASM Press, 1995. Pp. xxi, 283. Designed for students pursuing careers in biomedical research. Most chapters conclude with case studies and extended case studies are included in an appendix. Topics include use of animals, human experimentation, mentoring, authorship, ownership of data, and genetics.

Mappes, Thomas A., and David Degrazia. *Biomedical Ethics*. 5th ed. New York: McGraw Hill, 2000.

Martin, Mike W., and Roland Schinzinger. *Ethics in Engineering*. 4th ed. New York: McGraw-Hill, 2005. Pp. xi, 339. (1st ed., 1983.) This widely used text argues for conceiving of engineering as social experimentation and thus applies issues of informed consent to engineering practice. One of the earliest, most original, and widely used books in the field. See also Martin and Schinzinger's shorter version: *Introduction to Engineering Ethics* (Boston: McGraw-Hill, 2000).

Mitcham, Carl, and R. Shannon Duval. *Engineer's Toolkit: Engineering Ethics*. Upper Saddle River, NJ: Prentice Hall, 2000. Pp. x, 131. A short, elementary modular text.

Murphy, Timothy. *Case Studies in Biomedical Research Ethics*. Boston: MIT, 2004. Pp. xvii, 340. Intended as a text for instruction in biomedical research ethics. Collects over 100 case studies organized into nine topics including oversight and study design, informed consent, genetic research, and authorship and publication. Each topical area includes a general introduction and each case study includes study questions.

National Academy of Sciences, National Academy of Engineering, and Institute of Medicine. *On Being a Scientist: Responsible Conduct in Research*, 2nd ed. Washington, DC: National Academy Press, 1995. Pp. 27. Designed to stimulate group discussion, primarily in classrooms. Traces the history of thought about research

ethics through brief considerations of several topics including the social foundations of science, data, values in science, conflicts of interest, openness, misconduct, and authorship.

Penslar, Robin Levin, ed. *Research Ethics: Cases and Materials*. Bloomington: Indiana University Press, 1995. Pp. xvi, 278. A collection of case studies designed to aid faculty in raising and discussing ethically problematic aspects of conducting research. Arranged in three main sections that cover cases in biology, psychology, and history. Includes a general introduction to research ethics and ethical theory.

Scharff, Robert C., and Val Dusek, eds. *Philosophy of Technology: The Technological Condition: An Anthology*. Malden, MA: Blackwell, 2003. Pp. xi, 686. Fifty-five readings. Parts V, Technology and Human Ends," and VI, "Technology as Social Practice," constitute half the volume.

Schinzinger, Roland, and Mike W. Martin. *Introduction to Engineering Ethics*. 3rd ed. Boston: McGraw Hill, 2000. Pp. xi, 260. (1st ed. : 2nd ed.) Clarifies key concepts and provides case studies in the basic issues of engineering ethics, with an emphasis on the moral problems faced by engineers in the corporate setting. Includes an appendix with codes of engineering ethics from seven professional societies.

Seebauer, Edmund G., and Robert L. Barry. *Fundamentals of Ethics for Scientists and Engineers*. New York, NY: Oxford University Press, 2001. Pp. xvi, 269. An approach to education in technical ethics that develops a progressive "ethical serial" case study approach and highlights virtue theory. The first half focuses on ethical reasoning and the second half on applications. Organized in four units: foundational principles, resolving ethical conflicts, justice, and advanced topics (e.g., risk, resource allocation, and habit and intuition).

Sherlock, Richard, and John D. Morrey, eds. *Ethical Issues in Biotechnology*. Lanham, MD: Rowman and Littlefield, 2002. Pp. xiii, 643. Intended for use as a text book. Collects thirty-four essays arranged in six sections: fundamental issues, agricultural biotechnology, food biotechnology, animal biotechnology, human genetic testing and therapy, and human cloning and stem cell research. Includes overviews of basic ethics and science and concludes with study cases designed to spark classroom discussion.

Stern, Judy E. and Deni Elliot. *The Ethics of Scientific Research: A Guidebook for Course Development*. Hanover, NH: University Press of New England, 1997. Pp. x, 116. Result of a three-year project to produce a graduate level course in research ethics. Outlines course goals and plan and discusses how to train faculty to teach ethics and how to evaluate efforts. Concludes with a course reading list and extended case and topic bibliographies as well as a videography.

Tavani, Herman T. *Ethics and Technology: Ethical Issues in an Age of Information and Communication Technology*. New York: John Wiley and Sons, 2003. Pp. xxiv, 344. Introduces the relatively new field of Cyberethics. Discusses key concepts and terms, includes actual and hypothetical case studies, and provides review questions at the end of each chapter.

Unger, Stephen H. *Controlling Technology: Ethics and the Responsible Engineer*. 2nd ed. New York: John Wiley, 1994. Pp. xiv, 353. Argues that the democratic control of technology requires engineers to take responsibility for the consequences of their work. Includes case studies on successful and unsuccessful instances of engineering ethics, codes of ethics for engineers, the role of engineering societies in ethics, and engineering and law.

Veatch, Robert. *The Basics of Bioethics*. 2nd. ed. Upper Saddle River, NJ: Prentice Hall, 2003. Pp. xvii, 205. A brief survey that gives a broad introduction to the field. Covers the basics of ethics, Hippocratic oath, moral standing, patient rights, death and dying, social ethics (e.g., allocation of resources and human subjects research), human control of life and human nature, conflicts among principles, and a new chapter on the virtues (professional, secular, religious, and care) in bioethics.

Vesilind, P. Aarne, and Alastair S. Gunn. *Engineering, Ethics, and the Environment*. New York: Cambridge University Press, 1998.

Zimmerman, Michael E., J. Baird Callicot, George Sessions, Karen J. Warren, and John Clark, eds. *Environmental Philosophy: From Animal Rights to Radical Ecology*. 3rd edition. Upper Saddle River, NJ: Prentice-Hall, 2001. Pp. ix, 486. Collects thirty-two essays in four sections: environmental ethics, deep ecology, ecofeminism, and political ecology. Includes a brief general introduction that places environmental philosophy in historical and conceptual context. First edition, 1993.

5. Twentieth and Twenty-First Century Ethics

Baier, Kurt. *The Moral Point of View: A Rational Basis of Ethics*. Ithaca, NY: Cornell University Press, 1958. Pp. xii, 326. Argues that the distinctly moral perspective is the universalizability of rules and judgments.

Bauman, Zygmunt. *Postmodern Ethics*. Oxford: Blackwell, 1993. Pp. vi, 255. A sociologist's overview the postmodern rejection of the adequacy in ethics of rules, universality, and foundations, and the loss of the sense of self, with a brief statement of the positive possibilities opened by such a stance.

Broad, C.D. *Five Types of Ethical Theory*. London: Routledge and Kegan Paul, 1930. Pp. xxv, 288. An analytic assessment of the ethical theories of Spinoza, Joseph Butler, David Hume, Immanuel Kant, and Henry Sidgwick.

Dewey, John. *Human Nature and Conduct: An Introduction to Social Psychology*. New York: Henry Holt, 1922. Pp. vii, 336. Proposes a pragmatist ethics grounded in psychology. For two other statements of Dewey's pragmatist ethics, see *Ethics* (1908) with James Tufts and *Theory of Valuation* (1939).

Frankena, William K. *Ethics*. Second edition. Englewood Cliffs, NJ: Prentice Hall, 1973. Pp. xvi, 125. (First edition, 1963.) A widely used and influential textbook that defends a version of rule utilitarianism, that is, the moral theory that takes as foundational assessments of the consequences of rules for guiding human behavior. Gives fair consideration to both egoistic and deontological theories, but finds them wanting. No particular effort to consider science or technology, although rule utilitarianism is often the assumed justification for each.

Habermas, Jürgen. *The Theory of Communicative Action*. 2 vols. Trans. Thomas McCarthy. Boston: Beacon Press. 1984-1987.

Hare, R.M. *The Language of Morals*, 2nd ed. Oxford: Clarendon Press, 1961. Pp. viii, 2002. (First edition, 1952.) The single most influential book in meta-ethics. Concerned not with normative issues so much as the nature and function of moral discourse.

Jonsen, Albert R., and Stephen Toulmin. *The Abuse of Casuistry: A History of Moral Reasoning*. Berkeley: University of California Press, 1988. Pp. ix, 420. The title is misleading; this book is in fact a defense of casuistry against those who would too quickly abuse it in the name of principlist ethics. Grew out of the experience of the coauthors working with the National Commission for the Protection of Human Subjects of Biomedical and Behavioral Research, 1974-1978.

Kohlberg, Lawrence. *Essays on Moral Development*, vol. 1: *The Philosophy of Moral Development: Moral Stages and the Idea of Justice*. New York: Harper and Row, 1981. Pp. xxxv, 441. Collected papers providing the most complete statement of Kohlberg's influential theory (building on the work of Jean Piaget but based as well on his own empirical observations). Continued with vol. 2, *The Psychology of Moral Development: Moral*

Stages and the Life Cycle, and vol. 3, *Education and Moral Development: Moral Stages and Practice*.

Levinas, Emmanuel. *Totality and Infinity*. Trans. Alphonso Lingis. The Hague: Martinus Nijhoff, 1969. Pp. 307. (French original, 1961.) See also Adriaan T. Peperzak, ed., *Ethics as First Philosophy: The Significance of Emmanuel Levinas for Philosophy, Literature, and Religion* (New York: Routledge, 1995), which collects 21 original essays on Levinas' thought.

MacIntyre, Alasdair. *After Virtue: A Study in Moral Theory*. Second ed. Notre Dame, IN: University of Notre Dame Press, 1984. Pp. xi, 286. (First edition, 1981.) Three subsequent books in which MacIntyre extends his argument: *Whose Justice? Which Rationality?* (Notre Dame, IN: University of Notre Dame Press, 1988), *Three Rival Versions of Moral Enquiry: Encyclopedia, Genealogy, and Tradition* (Notre Dame, IN: University of Notre Dame Press, 1990), and *Dependent Rational Animals: Why Human Beings Need the Virtues* (1999).

Maritain, Jacques. *Integral Humanism: Temporal and Spiritual Problems of a New Christendom*. Trans. Joseph W. Evans. New York: Scribners, 1968. Pp. xii, 308. (French original, 1936.) An effort by one of the founders of Neothomism to develop a humanistic ethics that engages the modern world and responds to both liberalism and Marxism. No direct discussion of science and technology. Subsequent related efforts to restate the Thomistic perspective can be found in Yves R. Simon, *The Definition of Moral Virtue*, ed. Vukan Kuic (New York: Fordham University Press, 1986); and Ralph McInerny, *Ethica Thomistica: The Moral Philosophy of Thomas Aquinas*, revised edition (Washington, DC: Catholic University of America Press, 1997).

Moore, G.E. *Principia Ethica*. Cambridge, UK: Cambridge University Press, 1903. Pp. xxvii, 232. Although published during the first decade of the 20th century this book has exercised a strong influence over Anglo-American analytic ethics (comparable to the influence of Nietzsche's *On the Genealogy of Morals* on continental European phenomenological ethics). Argues that good is a unique, indefinable property that is directly intuited and for which nothing else can be substituted. It formulates in precise terms the so-called "naturalistic fallacy" (of identifying the good with the natural) and argues against naturalistic ethics, hedonism (meaning consequentialism), and metaphysical ethics (meaning the philosophy of Immanuel Kant). The long chapter five, "Ethics in Relation to Conduct," sets forth a program in practical ethics that anticipates applied ethics. The final chapter six, "The Ideal," distinguishes intrinsic goods in themselves, which Moore argues are exemplified in aesthetic enjoyments and personal affection, from extrinsic goods. Moore restates his argument in more textbook form in *Ethics* (London: Oxford University Press, 1912).

Münch, Richard. *The Ethics of Modernity: Formation and Transformation in Britain, France, Germany and the United States*. Lanham, MD: Rowman & Littlefield Publishers, Inc., 2001. Pp. xii, 281. A comparative interpretation of the common impulse of the transformation to modernism and its different expressions in four Western countries. Begins with an assessment of the West compared to the East and traces the formation of ethics through modern secularized and globalizing culture. Describes modern ethics as "instrumental activism," or the refusal to take the world as it is but rather to plan and intervene in it according to ideals. This creates a second world that is unpredictable and often brings unintended side effects that in turn call for more instrumental activism, or control.

Nietzsche, Friedrich. *Zur Genealogie der Moral* [On the genealogy of morals]. 1887. Although published in the last third of the nineteenth century, this book has exercised a strong influence over continental European phenomenological ethics (comparable to the influence of Moore's *Principia Ethica* on Anglo-American analytic ethics). Aiming to clarify his previous book, *Beyond Good and Evil* (1886), this volume, subtitled "A Polemic," is composed of three essays. The first distinguishes between moralities that has their origins in ruling classes (and distinguish between good and bad) and those formulated by the oppressed (who oppose good and evil). The second focuses on explicating the origins of guilt and bad conscience. The third criticizes ascetic ideals.

Ross, W.D. *The Right and the Good*. Oxford: Clarendon Press, 1930. Pp. vii, 176. Attempts to bridge deontological theories of the right and utilitarian theories of the good. Prima facie rights can on occasion be outweighed by anticipated bad consequences.

Scheler, Max. *Formalism in Ethics and Non-Formal Ethics of Values: A New Attempt toward the Foundation of an Ethical Personalism*. Trans. Manfred S. Frings and Roger L. Funk. Evanston, IL: Northwestern University Press, 1973. (Original German, 1913-1916.) Influential approach to ethics in the continental European phenomenological tradition. Criticizes Kantian formalism and defends the person as a source of substantive values, which range from sensible through vital and spiritual to the holy. For one subsequent statement of this approach emphasizing compassion as foundational for ethics see Werner Marx, *Towards a Phenomenological Ethics: Ethos and the Life-World* (Albany: State University of New York Press, 1992).

Scott, Charles E. *The Question of Ethics: Nietzsche, Foucault, Heidegger.* Bloomington: Indiana University Press, 1990. Pp. xii, 225. Argues that Nietzsche's questioning of ethics as a pathology is a fundamental part of ethics, an argument that he deepens with interpretations of Foucault and the problem of Heidegger's Nazism. Scott's thesis is that strong ethical commitments can create their own unethical behaviors, and that the questioning of ethics can (and must) be done on ethical not rejection of ethical grounds. Modest mentions of both science and technology. The argument is extended in Scott's *On the Advantages and Disadvantages of Ethics and Politics* (Bloomington: Indiana University Press, 1996), which includes more extended discussions of the ethical challenge of technology.

Toulmin, Stephen. *An Examination of the Place of Reason in Ethics.* Cambridge, UK: Cambridge University Press, 1950. Pp. xiv, 228. An attempt to develop a theory of moral reasoning in the analytic tradition that is perhaps the first instance to take explicit account of engineering and technology; see section 12.5, "Ethics and Engineering." Toulmin subsequently argues that attention to practical issues actually rescued ethics from abstraction in such articles as "The Recovery of Practical Philosophy," *American Scholar* 57, no. 3 (Summer 1971), pp. 337-352; and "How Medicine Saved the Life of Ethics," *Perspectives in Biology and Medicine* 25, no. 4 (Summer 1982), pp. 736-750. See also Toulmin's criticism of Enlightenment ethical rationalism in *Cosmopolis: The Hidden Agenda of Modernity* (New York: Free Press, 1990).

Williams, Bernard. *Ethics and the Limits of Philosophy.* Cambridge, MA: Harvard University Press, 1985. Pp. xiv, 230. An extended assessment of the limitations of modern ethics as "too much and too unknowingly caught up in ... administrative ideas of rationality" (p. 197). Argues that ethics needs to recover some of the resources of classical Greek philosophy while taking into account scientific knowledge in order to respond to the Socratic question of how one should live by making possible the pursuit of a meaningful life.

6. Journals

Bioethics, Publication of the International Association of Bioethics.

Bulletin of Science, Technology, and Society. Has been associated with the National Association for Science, Technology, and Society.

Environmental Ethics. Publication of the International Society for Environmental Ethics (ISEE).

Environmental Philosophy. Publication of the International Association for Environmental Philosophy (IAEP), University of North Texas.

Environmental Science and Policy. Published by Elsevier.

Ethics and Information Technology. Published by Kluwer.

Hastings Center Report. Publication of the Hastings Center.

IEEE Technology and Society Magazine. Publication of the Society on Social Implications of Technology of the Institute of Electrical and Electronics Engineers.

Philosophy and Public Affairs. Published by Blackwell-Synergy.

Philosophy and Public Policy Quarterly. Publication of The Institute for Philosophy and Public Policy, University of Maryland.

Science and Engineering Ethics. Published by Opragen.

Science and Public Policy. Published by Beechtree.

Science, Technology, and Human Values. Publication of the Society for Social Studies of Science (4S).

Techne: Research in Philosophy and Technology. An electronic journal published by the Society for Philosophy and Technology (SPT).

Technology and Culture. Publication of the Society for the History of Technology (SHOT).

Technology in Society. Published by Elsevier.

The American Journal of Bioethics. Publication of The American Journal of Bioethics at bioethics.net.

The New Atlantis. Publication of The Ethics and Public Policy Center.

COMPILED BY ADAM BRIGGLE AND
CARL MITCHAM

APPENDIX II

INTERNET RESOURCES ON SCIENCE, TECHNOLOGY, AND ETHICS

This listing of Internet Resources reflects the fact that science, technology, and ethics discussions tend to be divided according to scholarly communities, as summarized in the specialized introduction entries of the encyclopedia.

General

American Association for the Advancement of Science: http://www.aaas.org/. An international non-profit organization founded in 1848 to advance science and innovation, also publishes the journal Science. Site includes news, publications, career information, and statistics on indicators in research and development. It features several programs, including the "Dialogue on Science, Ethics and Religion" at http://www.aaas.org/spp/dser/.

Carnegie Council on Ethics and International Affairs: http://www.cceia.org/index.php. Contains publications and links. In-depth sections include environment, armed conflict, human rights, and global justice. Features an electronic forum for discussion.

Case Western Reserve Online Ethics Center for Engineering and Science: http://onlineethics.org/. Created with an NSF grant and geared to engineers, scientists, and students. Focuses on engineering and research ethics, diversity, and issues in computer and natural sciences. Features numerous case studies, original materials, links, and an extensive collection of codes of ethics.

European Group on Ethics in Science and New Technologies: http://europa.eu.int/comm/european_group_ethics/index_en.htm. Established in 1997 to advise the European Commission. Features its opinions on diverse subjects, publications, and links.

Institute for Global Ethics: http://www.globalethics.org/default.html. Promotes ethics at several levels through research, dialogue, and action. Provides educational program materials and organizational services. Features white papers and other publications.

Kurzweil AI: http://www.kurzweilai.net/index.html?flash=2. Non-flash version available at http://www.kurzweilai.net/index.html?flash=1. Investigates the accelerating growth of intelligence and knowledge and the growing intersection of various fields of research and technology and their impacts on society. Site includes news, publications, and editorials. Also features Ramona, a photorealistic avatar host, and an innovative networked presentation of information.

Loyola University Center for Ethics and Social Justice: http://www.luc.edu/ethics/. Founded in 1991, provides ethics education to individuals and organizations.

Sigma Xi, The Scientific Research Society: http://www.sigmaxi.org/. A chapter-based organization that promotes the health of the scientific enterprise, supports original research, honors scientific achievement, and publishes the journal American Scientist. Site features links to local chapters, information on meetings and events, publications, programs, and news, as well as the booklet "The Responsible Researcher," which supplements "Honor in Science."

UNESCO World Commission on the Ethics of Scientific Knowledge and Technology (COMEST): http://portal.unesco.org/shs/en/ev.php-URL_ID=6193&URL_DO=DO_TOPIC&URL_SECTION=201.html. Created in 1997 to mirror at the international level, national commissions on science, technology, and ethics. Site has information on its functions and full publications.

Agricultural Ethics

Food-Ethics.net: http://food-ethics.net/. A European Union project begun in 2003 that serves as a subject information gateway for professionals to facilitate access to high quality information on ethical principles of food and ethical traceability.

The Food Ethics Council: http://www.foodethics-council.org/index.html. Founded in 1998 to address a broad spectrum of issues from the use of antibiotics to intellectual property. Site includes publications, news, and project information.

Applied Ethics

Ethics Updates: http://ethics.acusd.edu/. Founded in 1994 and edited by Lawrence Hinman at University of San Diego. Site has diverse resources including videos, bibliographic essays, publications, and links arranged in three main groups; ethical theory, resources, and applied ethics.

EthicsWeb.ca: http://www.ethicsweb.ca/resources/. Developed as part of the W. Maurice Young Center. Includes information on topics, institutions, and publications in several areas including business, health care, research, and environmental ethics and resources on ethics in decision making.

Harvard Edmond J. Safra Foundation Center for Ethics: http://www.ethics.harvard.edu/. Features publications, information on ethics in the curriculum, and links to other institutions.

Santa Clara University Center for Applied Ethics: http://www.scu.edu/ethics/. Established in 1986 and has information on diverse subjects including biotechnology and healthcare ethics, business ethics, and government ethics. Also features perspectives on recent events, publications, and links.

University of British Columbia W. Maurice Young Center for Applied Ethics: http://www.ethics.ubc.ca/. Created in 1993 to study, train, and consult in a diverse range of applied ethics topics. Site includes working papers and other publications, information on trainings and courses, and news.

Bioethics

American Journal of Bioethics: http://www.bioethics.net/. Founded in 1993 and the most read source of information on bioethics. Site contains news, editorials, essays, and a discussion forum.

American Society for Bioethics and the Humanities: http://www.asbh.org/index.htm. A professional association founded in 1998 to provide research, teach-ing, and policy development in bioethics. Site features publications and links for members and non-members.

Bioethics: http://www.web-miner.com/bioethics. htm. A comprehensive site operated by Sharon Stoerger with annotated links to academic centers, government agencies, publications, and other resources.

Bioethics.com: http://bioethics.com/. Features news, commentaries, and links in nine categories including stem cell research, research ethics, and health care.

The Center for Bioethics and Human Dignity: http://www.cbhd.org/. A Christian organization founded in 1994. Site contains news, articles, and a topical listing of bioethics issues, each with a bibliography.

Council of Europe Bioethics Division: http://www.coe.int/T/E/Legal_affairs/Legal_co-operation/Bioethics/. Includes information on legal conventions and protocols as well as research projects.

The Hastings Center: http://www.thehastingscenter.-org/. Research institute founded in 1969 to study issues in biotechnology, health care, and the environment. Site includes news, research projects, publications, and a library.

Human Genome Project Ethical, Legal, and, Social Issues (ELSI): http://www.ornl.gov/sci/techresources/Human_Genome/elsi/elsi.shtml. Contains information on societal implications of genetic research, including links and articles on gene testing, gene therapy, privacy, patenting, forensics, courts, and behavior.

International Association of Bioethics: http://www.bioethics-international.org/. Focuses on networking and cross-cultural issues in bioethics and publishes two journals.

International Society of Bioethics: http://www.si-bi.org/ingles/home2.htm. Spanish organization founded in 1996. Site features links and a focus on Latin American bioethics.

President's Council on Bioethics: www.bioethics.-gov. Homepage for the U.S. federal bioethics panel created by George W. Bush in 2001. Site contains numerous publications, full texts of transcripts and meetings, and several other resources and links arranged topically.

UNESCO Bioethics Programme: http://portal.unesco.org/shs/en/ev.php-URL_ID=1372&URL_DO=DO_-TOPIC&URL_SECTION=201.html. Primarily responsible for the Secretariat of two advisory bodies: International Bioethics Committee and Intergovernmental Bioethics Committee. Site links to these bodies and contains general information.

Biotechethics

Biotechnology Watch: http://www.infoshop.org/biotechwatch.html. An activist organization skeptical of biotechnology applications that is part of the Alternative Media Project. Site contains news, links, and information on direct action campaigns.

Ethics for the Biotech Industry: http://www.biotechethics.ca/. A program of academic research that views biotechnology through business and professional ethics. Site contains resources and publications.

Ford Foundation Program on Biotechnology, Religion, and Ethics: http://cohesion.rice.edu/centersandinst/bioreliethics/fordgrant.cfm?doc_id=2378. An expired four-year project on religion and biotechnology. Site has contact information for researchers involved.

Transhuman.com: http://www.transhuman.com/. A pro-biotechnology group advocating for the use of biotechnology to overcome human limitations. Site contains a book store and resources on transhumanism.

Business Ethics

Better Business Bureau: http://www.bbb.org/. Founded in 1912 to solve marketplace problems through self-regulation and consumer education. Site contains news, resources, and connections to local BBB organizations.

Business and Human Rights Resource Centre: http://www.business-humanrights.org/Home. Promotes awareness and discussion on issues involving business and human rights, including resources, news, reports of corporate misconduct, and examples of best practice. Features an in-depth library arranged topically, including information on individual companies and laws.

Business Ethics: http://www.web-miner.com/busethics.htm. Site operated by Sharon Stoerger that contains annotated links to publications, professional societies, case studies, resources, centers, and more.

Business Ethics Magazine: http://www.business-ethics.com/. Homepage for the magazine. Site contains information on events and an extensive business ethics directory with contact information for various organizations.

Business for Social Responsibility: http://www.bsr.org/. Non-profit organization that provides information, tools, training and advisory services to make corporate social responsibility an integral part of business operations and strategies. Site contains information on advisory services, news, links, and reports.

European Business Ethics Network: http://www.eben.org/. An international collaboration dedicated to the promotion of business ethics. Site contains in-house information and external links.

Global Ethics: http://www.ethics.org/i_centers.html. Supports local groups in establishing ethics initiatives. Site features products, resources, and research on organizational ethics, character development, and ethics centers worldwide.

Institute of Business Ethics: http://www.ibe.org.uk/home.html. Founded in 1986 to promote ethical standards and share information. Site includes publications, events, training, news, information on how to create and implement codes of conduct, and resources on teaching business ethics.

International Business Ethics Institute: http://www.business-ethics.org/about.asp. Founded in 1994 to promote business ethics and corporate responsibility through public awareness, education, and fostering international business ethics organizations in companies. Site contains resources on education and professional services and publications.

Communication Ethics

Communication Ethics Limited: http://www.communication-ethics.com/. A consultancy-network and partner of the Institute of Communication Ethics. Site includes information on social justice, information integrity, and more.

Institute of Communication Ethics: http://www.communication-ethics.org.uk/. Offers education, research, and training in communication ethics. Site provides information for members, link to its journal *Ethical Space*, and information on events.

Computer Ethics

Computer Ethics: http://library.thinkquest.org/26658/. Provides basic understanding of ethical issues for internet users. Main portion has introduction to computer ethics, copyrights and licensing information, privacy issues, and censorship information. Users can submit content and create individual accounts. Has news, links, references, and information for teachers.

Ethics in Computing: http://ethics.csc.ncsu.edu/. Arranged topographically by speech issues, commerce, risks, privacy, computer abuse, social justice, intellectual property and basics. Each section has extensive information, links, references, and/or case studies.

The Research Center on Computing and Society: http://www.southernct.edu/organizations/rccs/index.html. Hosted by Southern Connecticut State University.

Features news, links, and resources for researchers, teachers, and students. Contains supplementary materials to be used with a computer ethics textbook.

Development Ethics

Development Studies Association: http://www.devstud.org.uk/studygroups/ethics.htm. Based in and Ireland and the U.K. to promote ethics and knowledge of international development. Site organized by working group topics including development ethics, women in development, sustainability, and information technology and development.

International Development Ethics Association: http://www.development-ethics.org/. Multi-disciplinary, cross-cultural group studying the ethics of global development. Site contains newsletter, links, and information on conferences and other events.

Engineering Ethics

Case Western Reserve Online Ethics Center for Engineering and Science: http://onlineethics.org/. Operating under an NSF grant and geared to engineers, scientists, and students. Focus on engineering and research ethics, diversity, and issues in computer and natural sciences. Features numerous case studies, original materials, links, and an extensive collection of codes of ethics.

National Institute for Engineering Ethics: http://www.niee.org/pd.cfm?pt=NIEE. Founded in 2001 as part of Texas Tech University. Site contains newsletter, links, educational resources, and products and services.

National Society of Professional Engineers: http://www.nspe.org/home.asp. Founded in 1934 and serves over 50,000 members. Site contains information on licensure, ethics, and law, products and services, educational materials, employment opportunities, a journal, links, information on events and conferences, and more.

Texas Tech University Engineering Ethics: http://www.niee.org/. Central hub that links three sites: Applied Ethics in Professional Practice (featuring the case of the month program), National Institute for Engineering Ethics, and the Murdough Center for Engineering Professionalism. Also features events, correspondence courses, videos and other resources, and ethics case studies.

University of Virginia Engineering Ethics: http://repo-nt.tcc.virginia.edu/ethics/. Disseminates engineering ethics cases studies and resources for students and faculties. Access to full case studies requires authorization from the University of Virginia.

Environmental Ethics

Environmental Protection Agency, Office of Environmental Justice: http://www.epa.gov/compliance/resources/ej.html. Features a frequently asked question section, newsletters and listservs, reports, publications, and information on policy and guiding documents.

Institute for Environment, Philosophy, and Public Policy: http://www.lancs.ac.uk/fss/ieppp/. Multi-disciplinary research group founded in 2000 at Lancaster University. Site contains news and events, information for current and prospective students, and research updates.

International Association for Environmental Philosophy: http://www.environmentalphilosophy.org/. Multi-disciplinary group studying broad range of topics in environmental philosophy. Site features news, newsletter, resources, links, and information on membership and events.

International Society for Environmental Ethics: http://www.cep.unt.edu/ISEE.html. Group founded in 1990 as the first major professional environmental ethics organization. Site features a listserv, newsletter, bibliography, selected books and articles, a syllabus project, and links.

University of North Texas Environmental Ethics: http://www.cep.unt.edu/. Features information on books, journals, educational and professional opportunities, links, news, and events.

Genethics

Genethics.ca: http://genethics.ca/index.html. A clearinghouse for social, ethical, and legal issues related to genomic knowledge and technology. Features topics (eugenics, patenting, DNA banking, gene therapy, GMOs, and many more), news, journals, conferences, and links to discussion forums.

Center for Economic and Social Aspects of Genomics: http://www.cesagen.lancs.ac.uk/. Based at the Universities of Lancaster and Cardiff to study the economic, social, and ethical implications of genomic research. Site features research projects, resources, newsletter, and events.

Information Ethics

Information Ethics, Inc.: http://www.info-ethics.com/. Contains resources, links, publications, and focuses on the ethics of software development. Links to a service branch that consults and trains clients.

International Center for Information Ethics: http://icie.zkm.de/. Platform for exchanging information on worldwide teaching and research. Features news, articles, links to institutions in the field, teaching resources, and publications.

Journalism Ethics

European Codes of Journalism Ethics: http://www.uta.fi/ethicnet/. A comprehensive databank offering resources for students, teachers, scholars, and practitioners. Arranged by links to thirty-five European countries (and the International Federation of Journalists) with contact information and codes of journalism ethics for each. Also features supplementary links.

Indiana University Journalism Ethics Cases Online: http://www.journalism.indiana.edu/Ethics/. Collects an extensive list of case studies in thirteen topical areas (including privacy, sensitive news topics, and workplace issues) to be used for students, teachers, practitioners, and media consumers.

Journalism Ethics: http://www.web-miner.com/journethics.htm. A comprehensive site operated by Sharon Stoerger with annotated links to articles, centers, and professional organizations. Many of the article links are broken.

Poynter Online Ethics: http://www.poynter.org/subject.asp?id=32. Includes columns, discussion, case studies and an extensive archive of ethics related stories. Also contains credibility and ethics bibliographies, codes of ethics, and ethics guidelines for publishing featuring seven core values.

Medical Ethics

American College of Physicians Center for Ethics and Professionalism: http://www.acponline.org/ethics/. Devoted to policy development and implementation. Features resources on end-of-life care, managed care ethics, and many other areas. Provides career related information, resources for students, advice for advocates, and services for various practitioners.

American Medical Association, Medical Ethics: http://www.ama-assn.org/ama/pub/category/2416.html. Arranged into eight areas that feature different aspects of AMA work in medical ethics. These include an interactive forum for analysis and discussion, an ethics working group, an effort to develop health care performance measures for ethics, and strategies for teaching and evaluating professionalism.

BMC Medical Ethics: http://www.biomedcentral.com/bmcmedethics/. An open access, peer-reviewed journal that considers articles on the ethics of medical research and practice.

Public Responsibility in Medicine and Research: http://www.primr.org/. Established in 1974 to implement ethical standards in research. Cite contains educational materials, resources, events, and information on certification of IRB professionals.

Stanford Center for Biomedical Ethics: http://scbe.stanford.edu/. Conducts interdisciplinary research and education in biomedical ethics and provides clinical and research ethics consultation. Site features news, events, job opportunities, newsletter, educational materials, and other resources.

Military Ethics

Joint Services Conference on Professional Ethics: http://www.usafa.af.mil/jscope/. An organization of military professionals, academics, and others formed to discuss ethical issues relevant to the military. Site made possible by the U.S. Air Force Academy and features general information, case studies, bibliography, core values of each military branch and links to past conferences.

Naval Academy Center for the Study of Professional Military Ethics: http://www.usna.edu/Ethics/. Formed in 1998 to promote ethical advancement of military leaders through research and education. Site contains events, publications, news, and links.

Nanoethics

Foresight Institute: http://www.foresight.org/. A member of the Foresight family of institutions formed to help society prepare for nanotechnology and other advanced technologies of the future. Site features news, events, quarterly newsletter, discussion, and information on research, public policy, and career opportunities.

Nanoscience and Technology Studies Societal and Ethical Implications: http://www.cla.sc.edu/cpecs/nirt/mission.html. Founded in 2001 at the University of South Carolina to research the ethical, legal, and social implications of nanotechnology. Site includes research, education, outreach, papers and other publications, links, and information on events and grants.

Nanotechnology Now: http://www.nanotech-now.com/. An up-to-the-minute news service on nanotechnology developments geared primarily for those in research and industry. Includes links to a consulting service and technology transfer and patenting service.

National Nanotechnology Initiative (NNI) Societal and Environmental Implications: http://www.nano.gov/html/facts/society.html. A multi-agency U.S. federal research and development project, part of which is devoted to the ethical, social, environmental, and legal implications of nanotechnology. Site contains links on societal and environmental implications for researchers and educational resources.

Neuroethics

Center for Cognitive Liberty and Ethics: http://www.cognitiveliberty.org/mission.html. A network of scholars promoting freedom of thought through research and advocacy based on core principles of privacy, autonomy, and choice. Site contains news, publications, and resources arranged topically.

Nuclear Ethics

Alsos Digital Library for Nuclear Issues: http://alsos.wlu.edu/. Named after the original Alsos Missions (1944-1945) that followed in the wake of Allied Armies in Europe to investigate the extent to which Nazi Germany was working on developing at atomic bomb. Includes a broad range of annotated references for the study of nuclear issues. This searchable collection includes books, articles, films, CD-ROMs, and websites.

Bulletin of the Atomic Scientists: http://www.the-bulletin.org/index.html. Founded in 1945 and educates citizens on national security issues, especially nuclear and other weapons of mass destruction. Site features extensive data on nuclear weapons capabilities around the globe, news, articles, links, current and past issues of the journal, and the doomsday clock.

Planning Ethics

American Planning Association: http://www.planning.org/. Includes information on ethics for professional planners including legislation and policy, careers, news, publications, research, conferences, consultant services, and information on creating local communities.

Professional Ethics

Illinois Institute of Technology Center for the Study of Ethics in the Professions: http://www.iit.edu/departments/csep/. Founded in 1976 to promote education and scholarship on professional ethics. Site features a library, codes of ethics, publications, and links.

Research Ethics

Central Office for Research Ethics Committees: http://www.corec.org.uk/. Organized for three main user groups: patients and the public, research ethics committee community, and applicants. Each section contains news, links, and information about and updates to relevant rules.

Office of Research Integrity: http://ori.dhhs.gov/. Oversees and directs Public Health Service (PHS) research integrity activities and promotes integrity in biomedical and behavioral research. Site contains information on policies, protocols for handling misconduct, links to related international organizations, educational materials, and conference and events announcements.

On Being a Scientist: Responsible Conduct in Research: http://www.nap.edu/readingroom/books/obas/. An on-line booklet published by the National Academy Press in 1994. Chapters span the spectrum from broad considerations such as "Values in Science" to narrower topics such as "The Allocation of Credit."

Plagiarism: http://www.web-miner.com/plagiarism.htm. A comprehensive site operated by Sharon Stoerger with annotated links to articles and resources for instructors and students.

Research Ethics: http://www.web-miner.com/researchethics.htm. A comprehensive site operated by Sharon Stoerger with annotated links to articles, case studies, policies and guidelines, and centers.

Rhetoric of Science and Technology

American Association for the Rhetoric of Science and Technology: http://aarst.jmccw.org/. Founded in 1992. Site features news, discussion forum, merchandise, pedagogical materials, links to similar organizations, and information on conferences and events.

Science and Technology Policy

American Association for the Advancement of Science: Science and Policy: http://www.aaas.org/programs/science_policy/. The Directorate of Science and Policy Programs operates eight programs at the interface of science, government, and society. Site links to these programs: ethics, and religion; fellowships; science, technology, and congress; research and development budget analysis; science, technology, and security policy; research competitiveness; scientific freedom and responsibility; and science and human rights.

Consortium for Science, Policy and Outcomes: http://www.cspo.org/. An intellectual network aimed at enhancing the contribution of science and technology to societal goals such as freedom, equality, and quality of life. Site features news, editorials, projects, education and outreach materials, and a library.

Ethics and Public Policy Center: http://www.eppc.org/about/. Established in 1976 to clarify and reinforce the bond between the Judeo-Christian moral tradition and the public debate over domestic and foreign policy issues. Site contains news, updates, publications, conferences, and events.

European Scientific and Technological Options Assessment: http://www.europarl.eu.int/stoa/default_en.htm. Provides independent assessments of the science and technology components of policy options faced by the European Parliament. Site contains newsletter, publication, work plans, workshops, and links to relevant network of experts.

Humanities/Policy: http://humanitiespolicy.unt.edu/. Network of scholars developing interdisciplinary approaches to integrating ethics and values with science to better meet societal goals. Site features information on policy, the humanities, projects, and resources for scientists and engineers.

New Directions in Science, Policy, and the Humanities: http://newdirections.unt.edu//. Fosters interdisciplinary networks including private sector and government to work toward solutions for environmental problems. Site features interdisciplinary resources, workshops, and project outcomes.

The National Academies Committee on Science, Engineering, and Public Policy (COSEPUP): http://www7.nationalacademies.org/cosepup/. Provides independent analyses of cross-cutting issues in science and technology policy, often for government agencies. Site includes publications, links, resources, and current and recent projects.

United States Office of Science and Technology Policy (OSTP): http://www.ostp.gov/. Established in 1976 to advise the President on science and technology aspects of public policy. Site contains news, outreach, projects, and information on science, technology, and government.

University of Colorado Center for Science and Technology Policy Research: http://sciencepolicy.colorado.edu/. Founded in 2001 to conduct research, education, and outreach to improve the relationship between societal needs and science and technology policies. Site features news, events, publications, educational materials, and several projects with various foci including water, climate, and carbon.

Science Fiction

Asimov's Science Fiction: http://www.asimovs.com/. Site features current and archived journals but also includes discussion forums, links, and other resources.

SciFi.com: http://www.scifi.com/. Site features listings on the television channel but also includes news, events, and pedagogical materials.

Science, Technology, and Art

Interdisciplinarity Resources: http://notes.utk.edu/bio/unistudy.nsf/0/5fd8d0b054118786852566fd008282be?OpenDocument. Maintained by the University of Tennessee, Knoxville, site links to several related projects and resources at the interface of science, technology, art, humanities, and culture.

Science, Technology, and Law

American Bar Association Section of Science and Technology Law: http://www.abanet.org/scitech/home.html. Provides updates, links, publications, and a search engine for documents related to science, technology, and law.

Cornell Law School Legal Ethics Library: http://www.law.cornell.edu/ethics/. A digital library that contains both the codes or rules setting standards for the professional conduct of lawyers and commentary on the law governing lawyers, organized by jurisdiction and topic. Also includes materials on multidisciplinary practice.

National Academies Science, Technology, and Law Program: http://www7.nationalacademies.org/stl/index.html. Established in 1992 to link the science and engineering communities with the law community. Site features links, events, contacts, and current studies.

Science, Technology, and Literature

Society for Literature and Science: http://sls.press.jhu.edu/. Site features a bulletin board, publication, links, educational materials, and a directory.

Science, Technology, and Society Studies

History of Science Society: http://www.hssonline.org/. HSS was founded in 1924 and is dedicated to understanding science, technology, medicine, and their interactions with society in historical context. The HSS site features publications, information on the profession, and educational and research materials.

Society for Philosophy and Technology: http://www.spt.org/ index.html. SPT was founded in 1980 to facilitate philosophically significant reflections on technology. The SPT site includes journal, newsletter, and links.

Society for Social Studies of Science: http://4sonline.org/. 4S grew out of a program on the public understanding of science at Harvard University in the 1960s. It is now an organization devoted to understanding science and technology. The 4S site features scholarly resources, information on the profession, conferences, and information for students.

COMPILED BY ADAM BRIGGLE AND
CARL MITCHAM

APPENDIX III

GLOSSARY OF TERMS

The following selection of terms and definitions is based on one originally authored by Caroline Whitbeck at the Online Ethics Center for Engineering and Science (OECES) at Case Western Reserve University (onlineethics.org) with the help of advisors to the OECES and discussed at greater length in her book, Ethics in Engineering Practice. Its aim is to introduce a number of concepts and terms that figure prominently in many discussions of science, technology, and ethics. Italicized terms within a glossary definition are defined in the glossary.

Academic Honesty and Academic Integrity: The maintenance of truthfulness and proper crediting of sources of ideas and expressions. Behaviors such as cheating on examinations and lab reports, or plagiarism of course papers and homework assignments violate academic integrity. Violations of academic integrity by students have the same character as violations of research integrity by scholars and research investigators (see *Research Ethics*). Other matters of academic integrity include honesty in writing letters of recommendation and in reporting institutional statistics.

Accountable: To be accountable is to be answerable or required to answer for one's actions. Sometimes the term *accountable* is used with a moral connotation (*normatively*), meaning morally required to answer for one's actions without specifying to whom one is accountable. More often accountable is used to describe the sociological fact that a person or organization in question is required to answer to a particular party by some rules or organizational structure. For example, "the principal is accountable to the school board" gives a description of the social facts without suggesting anything about the ethical legitimacy of the organizational structure.

Confusion arises when "responsible" is used as a synonym for accountable, especially in discussions of official responsibilities. When responsible is used as a synonym for accountable the preposition "to" is also involved, as in "each staff employee is responsible/accountable to a supervisor" (see *Responsibility*). Being a responsible person, that is, the sort of person who fulfills one's moral responsibilities, is an ideal of character, a virtue. Being accountable is not

a moral virtue but only a fact about one's social or organizational situation. Although it is often argued that people are more likely to behave responsibly if they are held accountable for their actions, there is no necessary link between being responsible and being accountable.

Administrative Law: *Administrative law* is constituted by that body of regulations, rules, orders, decisions, and policies that carry out the regulatory powers created of administrative agencies. In ordinary use, as contrasted with technical legal use, people often speak of administrative law as "regulation." For example, it is often pointed out that it is easier for regulatory agencies, such as U.S. Environmental Protection Agency or the U.S. Occupational Safety and Health Administration to update their regulations than it is to get Congress to pass new laws. In the technical legal sense, regulation is law has "the force of law." Administrative law, like all other forms of law, is subject to assessment and criticism in terms of ethics and justice. See also *Civil Law* and *Criminal Law*.

Affirmative Action: Positive steps to enhance the diversity of some group, often to remedy the cumulative effect of subtle as well as gross expressions of prejudice. In science and engineering affirmation actions often aim to enhance the participation of women and underrepresented minorities in these fields.

Applied versus Basic Research: Applied research is the investigation of phenomena to discover whether their properties are appropriate to a particular need or want, usually a human need or want. In contrast, basic research investigates phenomena without reference to particular needs and wants. Applied research is more closely associated with technology, engineering, invention, and development. Basic research is sometimes described as "pure research."

Assent: Assent is a variation of the concept of *Informed Consent* specifically used in reference to research subjects such as children or other persons without the full competence to provide informed consent. For instance, because children under 18 are below the legal age of consent, the U.S. Department of Health and Human Services (DHHS) requires additional protections when children are involved as subjects. Assent is defined as "a child's affirmative agreement to participate in research. Mere failure to object should not, absent affirmative agreement, be construed as assent" (45 CFR 46 Subpart D). In addition, the federal regulations require the permission of one or both parents or guardians of the child, depending on the nature of the research to be performed.

Authenticity: The character trait or virtue of being genuine and honest with oneself as well as others. Therefore, authenticity connotes not only candor, but an absence of hypocrisy or self-deception.

Autonomy: See *Right to Self-Determination*

Basic Research: See *Applied versus Basic Research*

Bias: An inclination that influences judgment is a bias. The term may be used in a merely descriptive way to mean an inclination, but more often it is used indicate an inclination that influences judgment but ought not to. *Prejudice* is a synonym for bias in this pejorative sense.

However, the bias that cannot be completely eliminated in the work of scientific investigators, in contrast to bias or prejudice that can and should be eliminated, is also an important topic in research ethics. For example, the way disciplinary training inclines people to interpret the results of an experiment in terms of the established categories of that discipline is a permanent feature of research, and one that must be taken into account in assessing responsible behavior in research. Of course, researchers may hold disciplinary biases and still be unbiased in other respects, that is, they may be impartial on the question of the truth or falsity of a particular research hypothesis.

Biotechnology: As defined by the U.S. government, biotechnology refers to any technique that uses living organisms (or parts of organisms) to make or modify products, to improve plants and animals, or to develop microorganisms for specific use. Biotechnology focuses on the practical applications of science (as opposed to doing basic research). Historically, biotechnology has had an impact in three main areas: health, food and agriculture, and environmental protection. Biotechnologists try to solve problems in these and other areas such as the need to cure or prevent illness, for clean water, and to preserve food.

Bribe: A bribe is something given or offered to a person or organization in a position of trust to induce such a person to behave in a way inconsistent with that trust. As C. E. Harris et al. (2000) point out, offering a bribe is not the same as capitulating to extortion (that is, capitulating to a demand under coercion or intimidation). It may be ethically justified to pay extortion in some circumstances, even though it would be wrong to offer a bribe. Bribes are paid to obtain something to which one does not have a right, such as a special advantage in awarding a contract. In contrast, extortion is paid to secure something to which one has a right, such as the return of expensive equipment one has legally brought into a country but which a corrupt customs official claims has been "lost."

Candor: Candor is the quality of being frank or open. The original, now obsolete sense of the term was of the virtue of purity or innocence. Although being open and unbiased is a positive quality, in some circumstances it is better to be discreet rather than candid with someone about a particular topic. Certainly, there are matters in which a person is morally required to keep something confidential, and therefore being candid with the wrong party about such a matter would be an ethical breach. See also *Authenticity*.

Challenge Study: A study in which researchers intentionally give subjects or patients pharmacological agents in order to induce and study psychiatric symptomology.

Civil Law: That body of law relating to contracts and suits, as contrasted with criminal law. Civil law covers suits of one party by another for such matters as breach of contract or negligence, and as such may have application in scientific and engineering contracts as well certain professional obligations. The standard of proof in civil cases is preponderance of evidence—a greater weight of evidence for than against. This is a weaker standard of proof than exists in criminal cases. Civil law, like all other forms of law, is subject to assessment and criticism in terms of ethics and justice. See also *Administrative Law* and *Criminal Law*.

Civil Rights: Rights associated with citizenship that one acquires simply by being a citizen. Not all of these are inalienable rights, however. See *Rights*. For example, according to the law in some states, a citizen may lose the right to vote if convicted of certain crimes.

Complainant: A person who raises concerns inside or outside her organizations about something she believes to be amiss. The term does not have the negative connotation of "complainer." The complainant is one who speaks up in some way about a problem. This speaking up may or may not include filing a formal charge. See also *Whistleblower*.

Confidential: That which is done or communicated in trust is confidential. Confidential information is information entrusted to another. The implication is that it is information that for some reason (from personal privacy to competitive advantage) the person entrusting the information does not wish some others to know. Thus confidential information is information to be shared only with a very limited group who are involved with furthering certain ends which the one entrusting the information wants served, such as treatment of a disease, or development and manufacture of a new product. Most professions recognize some duty to keep confidential a client's information, although such a duty has limits when the confidential information concerns a danger to others.

Conflict of Interest: Someone has a conflict of interest when that person is in a position of trust requiring the exercise judgment on behalf of others (people, institutions, etc.) and also has interests or obligations of the sort that might interfere with the exercise of such judgment, and which the person is morally required to either avoid or openly acknowledge.

The lesser requirement of open acknowledgment is usually adopted when it seems too burdensome to require that persons in positions of trust divest themselves of the interest that conflicts with a position of responsibility. For example, some journals require that authors disclose any substantial financial interests that might have biased their research assessment. Requiring investigators to divest themselves of investments that they may have made on the basis of their scientific judgment would be too burdensome, and might even suppress publication.

Dictionary definitions frequently apply the term only to conflicts between a person's private interests and those of a public office, and by extension with that person's professional obligations and responsibilities. However, there can also be conflicts of interest in which private interests do not enter. For example, the American Bar Association specifies as part of a general rule on conflict of interest that lawyers should not represent a client if such representation may be materially limited by the lawyer's responsibilities to another client or to a third party. There is no similar rule requiring engineers or engineering firms to avoid, say, building manufacturing facilities for, or supplying parts to, two companies that directly compete in the same market, although the engineering firm might need to be especially careful to avoid disclosing the proprietary information of one company to the other.

This example illustrates the point that one needs to look carefully at the nature of a professional's or public official's obligations and responsibilities in order to know when conflicting interests become a conflict of interest, that is, when a situation that requires discretion to handle the actual or potential conflict fairly is one that he is morally required to avoid altogether, or at least to disclose to all parties.

Policies requiring financial disclosure, that is disclosure of financial interests that might conflict with judgment as a researcher or as public official, are very commonly called a "conflict of interest policy," although such financial conflict of interest is only one specific type.

Contract: As used in ethics, the term contract means an explicit agreement that is freely entered into. Only a small number of these would qualify as legal contracts. A legal contract is a legally binding agreement among two or more parties. Breach of contract is the failure to fulfill a legal contract.

Copyright: A legal right (usually of the author or composer or publisher of a work) to exclusive publication production, sale, and/or distribution of some work for a specified period of time. What is protected by the copyright is the "expression," not the idea. Notice that taking another's idea without attribution may be plagiarism, so copyrights are not the equivalent of legal prohibition of plagiarism.

Cost-Benefit Analysis: To use cost-benefit analysis (also sometimes called risk-benefit or risk-cost-benefit analysis) requires that one consider only those consequences or the probability of consequences that can be quantified, such as number of deaths, days of illness, or monetary costs. Cost-benefit analysis is a technique taken from economics that weighs alternative actions in terms of such consequences. Its great strength is that it can introduce a measure of objectivity into sometimes complex or contentious decision-making. Its weakness is that it is not able to consider consequences such as the loss of moral integrity or human flourishing that do not lend themselves to quantification. See also *Risk*.

Criminal Law: That body of law relating to crimes (which can be classified as either felonies or misdemeanors). Crimes are offenses against the state; they are investigated by the police, prosecuted at public expense, and can be punished by incarceration as well as fines. The uses of science and technology to commit crimes are always subject to criminal prosecution and punishment. The standard of proof in criminal cases is stronger than the standard of proof in civil cases, which is preponderance of evidence and may extend to absence of reasonable doubt. Criminal law, like all other forms of law, is subject to ethical assessment and criticism. See also *Administrative Law* and *Civil Law*.

Data Selection: Involves emphasizing some data over other data or sometimes ignoring certain data. The term is primarily used when data selection is legitimate and clarifies research, as opposed to selection that falsifies the research. Selection of data for analysis or presentation is legitimate only when undertaken on the basis of clear criteria for thinking that in comparison with other related data it is less subject to confounding influences or "noise." Any selection must be disclosed in reporting the research.

Defendant: A party being sued in civil proceedings, or accused of a crime in criminal proceedings.

Dilemma: A dilemma involves a forced choice between courses of action (usually two) which are both unacceptable. Sometimes people will call any challenging moral problem a dilemma, but this is misleading. Only a few moral problems are dilemmas in the technical sense of the term. Calling moral problems "dilemmas" is confusing because it implies that the only possible responses are the two obvious (and unacceptable) ones; this tends to discourage real problem solving.

Discrimination: Discrimination in the common, morally relevant use of the term, is a failure to treat people fairly due to a bias against (or for), based on a characteristic such as race, religion, sex, national origin, sexual orientation, physical appearance, or disability that is irrelevant to the decision at hand (e.g., job skills or qualifications for public housing). Discrimination may be intentional or unintentional. Discrimination is a form of behavior that shows prejudice, but not the only form.

Due Process: That procedure or process required for a given judgment to be fair. Fairness here is specified in terms of the process rather than the outcome. For example, although it is desirable that those and only those who are guilty of a crime be punished for it, infallibility of judgment by the law courts cannot be guaranteed. The feasible goal is to try to ensure everyone a fair trial. Similarly, although it is hoped that important research does not go unrecognized, it is impossible to guarantee that the contributions of those who are "ahead of their time" will be recognized. The feasible goal is to ensure fair process in the reviewing of research proposals for funding or research results for publication.

Duty: See *Obligations*

Ethical Relativism: Ethical relativism or "relativism" may be used to indicate several different views. One view, which is also called "ethical subjectivism," is the view that the truth of some ethical judgment as applied to a person's behavior depends on whether the person believes the actions to be right or wrong. This view is commonly expressed as "there is no right or wrong, it's only a matter of opinion." Acceptance of this view undermines the claim of any ethical judgment to have validity. One who believes in ethical relativism in this sense would have to agree that there was nothing objectively wrong with a person torturing or killing another person, as long as the individual committing those actions sincerely believed it was not wrong to do so.

A second view, which is sometimes called "cultural relativism," is the view that ethical judgments and moral rules always reflect the cultural context from which they are derived and cannot be immediately applied to other cultural contexts. Some who hold this view are skeptical about even the possibility of saying that slavery is wrong in a slave-holding society and so are close to the ethical subjectivists. At the other extreme, some cultural relativists only put the burden of proof on those who think that they can generalize from one social context to another, but accept that the burden of proof is often met.

Many ethical philosophers, such as Alasdair MacIntyre (1988) and Annette Baier (1994), who do not consider themselves relativists, nevertheless argue that moralities are social products constructed by particular people in particular societal contexts and must be understood in relation to those societal contexts. For example, the Hippocratic oath specifies extensive duties toward those who have taught one medicine. In this oath physicians pledge to respect and care for their teachers as for their own parents. The societal context in which these duties of physicians were formulated was very different from what it is in industrialized nations today. It is implausible that the same duties should apply to physicians in all societies, but this does not mean that they did not have validity when the oath was first formulated. What makes the difference is not a person's or the group's opinion, but the social reality in which the person participated.

Ethics and Morals: The term *ethics* is used in several different ways. First, it may mean the study of morals, meaning individual or social forms of behavior. It is also the name for that branch of philosophy concerned with the nature of morals and moral evaluation: what is right and wrong, virtuous or vicious, good and bad, and beneficial or harmful (to oneself or others).

Second, ethics or morality may be used to mean the standards for ethical or moral behavior of a particular group, such as "Buddhist ethics" or "nursing ethics" or "Roman Catholic morality" or "the professional ethics of engineers in the United States." To give a description of such ethical codes and standards is "descriptive ethics." Descriptive ethics does not include a judgment as to whether the code or standards of behavior are ethically justified. The examination of the adequacy of moral or ethical values, standards, or judgments is *normative* ethics.

Third, some authors even use the terms *ethics* or *morality* more loosely to refer to any code of behavior, even one that no one regards as having any moral justification. For example, Robert Jackall (1988) describes what he calls the "ethics" or "morality" of a corporation and takes it to include such judgments as "What is right is what the guy above you wants from you." Such a judgment is about the most immediate way to survive in the organization, but does not pretend to be a statement about what is morally or ethically justified. It may be important to examine such codes of behavior and see how they affect the opportunities for moral action, but not every code of behavior has, or is even claimed to have, moral or ethical justification.

The term "moral" tends to be used for more practical elements, such as "moral problems" and "moral beliefs," and "ethical" tends to be used for more abstract and theoretical elements, such as "ethical principles," but the distinction is by no means hard and fast.

Evaluation: Evaluation can be either descriptive or normative. Descriptive evaluation may range from simple measurement to complex judgment about such things as the presence of mineral deposits. Normative evaluations involve judgments as to whether something is good or bad in some respects—a value judgment.

Explanation: Explanations of human actions typically make reference to the agent's reasons or motives for some action. For example, the student went to the bookstore to buy a text book. Causes are usually cited only for human actions that are not intentional, such as falling. A person's falling might be causally explained by the slipperiness of the road surface, the person having been pushed, or drugged, or having a heart attack. A person's falling with acceleration would be explained in terms of the gravitational field in which the person was falling. Notice that in certain contexts, notably ordinary life and law, it is often the unusual that is explained, whereas scientific explanation more commonly explains typical behavior.

Fabrication: In research ethics, fabrication means making up data, experiments, or other significant information in proposing, conducting, or reporting research. In engineering, fabrication has a benign connotation, meaning to make something. Sometimes the term is used to refer specifically to an intermediate stage between design and manufacture or construction.

Falsification: In research ethics falsification means changing or misrepresenting data or experiments, or misrepresenting other significant matters such as the credentials of an investigator in a research proposal. Unlike fabrication, the distinguishing of falsification data from legitimate data selection often requires judgment and an understanding of statistical methods.

Fraud: An intentional deception perpetrated to secure an unfair gain. Financial fraud, that is, a deception practiced on another party to cheat that party out of money, is the most commonly discussed type of fraud.

The terms "research fraud" or "scientific fraud" are used to mean an intentional deception about experiments or results, and is a type of research misconduct. In this case, the act may not include a financial transaction and there need not be an injured party or even anyone who was actually deceived. Therefore, so-called "research fraud" does not fit the legal criteria for a fraudulent act, which are discussed below.

In a civil law suit charging fraud there is a plaintiff who makes the charge against a defendant. To win a suit the plaintiff must prove five points, which are the five legal criteria for fraud: (1) the defendant made a false representation; (2) the defendant knew that the representation was false or at least recklessly disregarded whether it was true or false; (3) the defendant intended to induce belief in the misrepresentation; (4) the plaintiff had a reasonable belief in the misrepresentation; and (5) the plaintiff suffered damage as a result. As is illustrated in the case of "research fraud," the term "fraud" is often used informally to mean a misrepresentation but in which there may be no injured party who might become a plaintiff and so which satisfies only the first two criteria.

Glass Ceiling: The term *glass ceiling* indicates a barrier to advancement within an organization experienced by members of certain groups because of prejudice (including discomfort in their presence). This term is most often used when the organization recruits members of an affected group but then fails to promote them through the junior ranks on a comparable basis to other favored groups. If members of a group tend to leave the organization soon after entering, this is termed a "revolving door" rather than a "glass ceiling." The barrier in an organization may be different for different groups that are commonly victims of prejudice and usually is strongly influenced by so-called "corporate culture."

Good: The good is that which is rational to want or desire. A good knife is one that has characteristics it is rational to want in a device with one blade used for cutting. When considering what makes a good person, or good character, the matter becomes more complex, because it then becomes important to ask whether the traits under discussion are those that people would want in themselves, or those that they would want in others, and whether these are the same, or in what sort of society they might be the same.

Good Scientific Practice: See *Research Ethics or the Responsible Conduct of Research*

Human Rights: See *Rights*

Inherently Safe: The term *inherently safe* is applied to products, processes, and systems in which operational safety is independent of any user training or auxiliary devices. For instance, Elisha Otis's invention of the safety elevator in the 1850s was designed to function only when the lift cable disengaged a brake; if the cable failed, the brake was automatically engaged. Since the 1970s nuclear reactors that are designed to automatically shut down in the case of any malfunction are also described as inherently safe. With inherently safe technologies, any deviation from expected use or operation leads to a non-hazardous state.

Inherently safe manufacturing processes utilize machines that will not function unless workers have placed themselves in safe positions, as when for example both hands must be placed on two separate switches before a cutting operation can proceed. Another inherently safe process would minimize the use of hazardous materials or the time employed in their use, thus reducing the dependency of safety on worker training or protective equipment.

In contrast, the notion of inherently dangerous or unreasonably dangerous is a legal notion that applies to products, processes, and systems which, under normal operating conditions, entail some level of hazard. Inherently dangerous can entail strict liability in tort.

A related term "intrinsically safe" is applied to electrical or electronic equipment that is incapable of producing a dangerous spark or thermal effect under either normal or abnormal operating conditions.

Informed Consent: Describes the obligation of physicians or researchers to allow patients or subjects to be active participants in decision-making with regard their care or research in which they play a role. Informed consent is rooted in the concept of autonomous choice or the *Right of Self-Determination,* and requires five elements: (1) disclosure (of information to the patient/subject), (2) comprehension (by the patient/subject of the information being disclosed); (3) voluntariness (of the patient/subject in making his/her choice); (4) competence (of the patient/subject to make a decision); and (5) consent (by the patient/subject).

Legal Contract: See *Contract*

Legal Rights: See *Rights*

Liability: A person is liable when obligated by law to make satisfaction, compensation, or restitution for some act or injury. Liability is a legal notion indicating a legal debt or obligation. The liabilities most often at issues in discussions of science, technology, and ethics are those having to do with compensation for injury (to one's person, property, finances, or reputation) or to clean up toxic contamination. Legal liability to compensate for an injury or to clean up contamination does not necessarily require that one has caused the injury or contamination, or that one be guilty in a moral sense. Under the doctrine of strict liability a party may be liable without having been guilty of negligence.

Liberties: See *Rights*

Morals: See *Ethics and Morals*

Moral Agent: One whose actions are capable of moral evaluation. We may say that an avalanche killed three people, but the avalanche is not open to moral evaluation. The avalanche is an amoral force. A competent and reasonably mature human being is the most familiar example of a moral agent. In contrast, most so-called "lower" (that is, non-human) animals are generally understood to be amoral (although this is open to debate regarding species that have complex and flexible social relations, such as primates and dolphins).

Moral Integrity: *Moral integrity* is a complex and subtle ethical notion. As theologian Stanley Hauerwas (1981) has argued, it is central to all the other virtues but more fundamental than any single virtue. The root of the term integrity is wholeness. Moral wholeness rather than rigidity best captures the idea of moral integrity. For example, a person might discover that some long-held ethical belief, attitude, or rule of conduct was mistaken because the person came to see that it was incompatible with other, more fundamental ethical commitments. A person's moral integrity is central to a person's sense of meaning.

Philosophers such as John Ladd (1982) have argued that a person's moral integrity is a central aspect of that person's well-being. Therefore, leading another person to compromise moral integrity is a fundamental injury to that person. Concern for a person's moral integrity requires an understanding of the person's moral convictions and in this respect differs from merely respecting a person's *Right of Self-Determination,* which requires only that one refrain from restricting their actions.

Some professional codes of ethics uphold the right to refuse work that would compromise an individual professional's ethical commitments even when the act in question (say, performing an abortion or developing weapons systems) is something the profession as a whole has not ruled morally objectionable.

Moral Standing: A being's moral standing determines the extent to which its well-being must be ethically considered for its own sake. To say that some groups of beings have moral standing is to say that, as a moral matter, their well-being must be given some consideration. It does not decide the question of whether they have the same moral standing as people (and thus have "human" rights). The welfare of such beings as cattle, for example, might be consid-

ered for prudential reasons, but that would not require that they have moral standing. One might decide that it is important to feed one's cattle, just as one might decide it was stupid to throw one's stamp collection into the river, thinking of the cattle or stamps as investments, without believing that either deserves such or better treatment out of consideration for their own well-being.

Moral Values: See *Values*

Motive: That which moves a person to action. Typically these are emotions, desires, or concerns. Thus people say such things as, "The motive for the crime was revenge." However, it is often common to hear someone speak simply of the intended result as "the motive." For example, any of the following sentences might be used to convey the same thought: "Lee's motive in arising early was to avoid traffic." "Lee arose early to avoid traffic." "Lee arose early because he wanted to avoid traffic." In such cases we assume that a desire or concern to realize the intended state is the implied motive. The expression "mixed motives" is used most often to suggest, not just any combination off emotions, desires, and concerns, but more specifically a mixture of selfish and altruistic concerns.

Negligence: A failure to be sufficiently careful in relation to a matter about which one has a moral responsibility to exercise care. Some careless mistakes are negligent, as when a surgeon sews up a patient with surgical instruments inside. Others are not, as when one dribbles soup down the front of one's sweater. Negligence is a legal basis for the recovery of damages from a private or civil wrong or injury, what is called a tort. The failure to fulfill a recognized duty, or to act with less care than would a reasonably prudent person in the same circumstances is the mark of negligence.

Normative: Derived from the Latin *norma,* the name for a carpenter's square, and is a loose synonym for authoritative or required. It is sometimes used broadly to mean that which establishes or reflects any sort of standard, even a statistical one. But in ethics when a standard, judgment, or assessment is normative, it concerns respects in which something is right or wrong, good or bad. Value judgments are normative in the ethical sense, but the judgment that X is greener or heavier than Y is not. In the ethical sense normative is a close synonym for prescriptive, that which "makes or gives rules."

However, not all rules are ethically normative; they may simply establish order. For example, the statement "Put out your trash out for collection on Tuesdays" is a prescription about when to put out trash, but does not suggest or establish that there is anything in Tuesday trash collection that is superior to collection on some other day.

Obligations: Requirements arising from a person's situation or circumstances (e.g., relationships, knowledge, position) that specify what must or must not be done for some moral, legal, religious, or institutional reasons. For example, students may have an obligation see their advisor on or before registration day, simply because this is one of actions students in a particular institutional context are asked to perform. But it can also be argued that persons insofar as they are moral have an obligation to keep their promises, because this is one of things that being moral entails. Notice that usually statements of obligations specify what acts are required or forbidden without reference to the consequences of performing the act (except in so far as these consequences are a part of the characterization of the act itself).

Obligations can be more or less specific. That drivers are obligated to obey the traffic rules is much more specific than "Engineers have an obligation in their work to ensure the public safety." The second obligation names a responsibility that engineers have to achieve a certain end, namely safety of the public, but fails to specify what specific acts they should or should not perform in order to ensure safety.

A legal obligation is one that specifies what types of actions are permitted, forbidden, or required with certain state-enforced penalties attached for failures to comply.

Official Responsibility: See *Responsibility, Official*

Patents: A (special, alienable, prima facie) legal right granted by the government to use, or at least (in cases where other patents that such use would infringe) to bar others from using a device, design, or type of plant that one has created. In the United States restrictions last for 17 years for useful devices, and 14 years for designs. Specific provisions of U.S. patent law may soon change to bring it into conformity with the provisions of other techologically developed countries. To patent a device one must prove that it is useful, original, and not obvious. Patents are subject to challenge in court and may be upheld or overturned.

Paternalism: Derived from the Latin word for father (*pater*) and means acting as a parent toward someone who is not in fact one's child. Acting like a parent toward those who are not one's children may or may not be justified in particular circumstances. Parents need to make judgments about many areas of their young children's lives (with the particular areas depending on the age or maturity of the children), but adults assume the responsibility for making those decisions for themselves. Paternalism may be roughly defined (following Gert and Culver, 1976) as violating a moral rule of conduct toward someone or limiting that person's self-determination (and hence often infringing that person's rights) for what is perceived as being that person's own benefit.

Paternalism may be justified or unjustified, but the paternalistic treatment of adults always requires justification because of the infringement of the person's right that it entails. Paternalism in the treatment of clients most commonly arises in professional contexts where the professional has a face-to-face relationship with those whose well-being they seek to ensure, and in professions where practitioners are in positions of greater power than their clients.

The question of paternalism often arises in medicine and healthcare with respect to the treatment of patients. Because many engineers and scientists in industry must protect the safety and health of anonymous members of the public rather than identified clients, and usually do not occupy positions of greater power than their clients, paternalism is not a frequently discussed topic for engineers in industry. Nevertheless, issues of paternalism often do arise for engineers and scientists in relationships among co-workers and in educational contexts.

Plagiarism: Commonly defined as the unauthorized or unacknowledged appropriation of the words, graphics images, or ideas of another person. In some instances reference is also made to artistic creations such as music. Plagiarism is theft of credit and covers ideas as well as forms of expression and should be distinguished from copyright violation, which does not cover ideas and is a matter of intellectual property violation.

Plaintiff: An injured party suing someone (a defendant) in order to be compensated for an injury or loss.

Preferences: Statements about the person who has them. If statements of preference are false it is because they are not true about that person. Such statements of preferences should be distinguished from *Value Judgments*, which are statements about the thing being judged good or bad.

Economists often avoid getting into substantive value discussions by considering only what people want or prefer and how much they prefer some things over other things. What makes this confusing is that they sometimes speak about what people "value," but they mean only what people prefer, and not what people have reasons for thinking are good or bad in some respect. It is of course possible for someone to prefer something because they judge it to be good. However, many factors other than value judgments enter into people's preferences, such as their early conditioning, habits, and vivid personal experiences. (If this were not the case, there would not be such a large market for cigarettes, for example, since even most smokers do not judge smoking to be a good thing to do.)

Prejudice: Bias for or against someone or something that fails to take true account of their characteristics.

Principal: A principal in an engineering firm (or other company) is a co-owner, that is, a partner or stockholder.

Privacy: It is common to distinguish three species of privacy: physical, informational, and decisional. In addition, philosopher and legal theorist Anita Allen (2003) distinguishes dispositional privacy. Physical privacy is a restriction on the ability of others to experience a person through one or more of the five senses. Informational privacy is a restriction on facts about the person that are unknown or unknowable. Decisional privacy is the exclusion of others from decisions, such as healthcare decisions or marital decisions, made by the person and his group of intimates. Finally, dispositional privacy is a restriction on the ability of others to know a person's state of mind.

Claims to privacy find moral justification in a recognition that people need to have control over some matters that intimately relate to them in order to function as people and be responsible for their own actions. Foremost among these are rights to one's own body. If, for example, people were permitted to drug one another at will, that would effectively undercut the rest of moral life.

Just what a person is expected to do in order to respect another's privacy varies with culture. For example, expectations that people will knock on the door before entering certain areas assumes the existence of both doors and of expectations about the amount of so-called "private space" to which a person is entitled. In some contemporary cultures, parents oversee their children's affairs much more closely than in others. In traditional Chinese families, for example, it is expected that parents will do such things as read the mail addressed to their adolescent children as part of their responsible oversight of their children, whereas in Anglo-American culture such acts would be viewed as intrusions on the adolescent's privacy.

Questions of privacy have become particularly prominent as computers and other technological innovations have made it possible to collect, assemble, and transmit quantities of information in ways that previously were impossible. Once the questions of appropriate levels of privacy protection have been established, the question of how that level of privacy can be practically ensured is a matter of security.

Profession: An occupation, the practice of which directly influences human well-being, and requires mastery of a complex body of knowledge and specialized skills, requiring both formal education and practical experience.

Professional Engineering: In the United States a professional engineer (P.E.) is a person who is licensed to practice engineering in a particular state or U.S. territory after meeting all requirements of the law. To practice in multiple states or territories, the P.E. must be licensed in each state in which he or she wishes to practice.

Professional Responsibility: See *Responsibility, Professional*

Property: A property, from the Latin *proprius*, meaning "one's own" or "special," can refer to the key characteristic of a thing ("One property of water is to be a solvent") or that to which an individual has special rights. In this second sense, very different sorts of things may be regarded as property. Individual rights to property (other than clothing and other personal effects), especially the right to own land, is a major innovation in modern thought. Land was one important kind of property, physical objects that constitute "the fruit of one's labor" were another.

It was a short step from physical property to intellectual property, the fruit of one's intellectual labor, which was given some recognition in the U.S. Constitution (see *Copyright, Patent, Trademark, Trade Secret Patent*). (Notice that "ideas" cannot be owned by these means but only some "expression," design, or device.) The advent of electronic information has raised new issues and problems about intellectual property and rights to such property, because of the extreme ease with which electronic information can be copied and transmitted.

Proprietary/Property Rights: Proprietary rights, claims, etc. are the rights, claims, etc. of owners. Sorting out the rights that go with property ownership is complicated, both because of the variety of types of property, and because of the problem of sorting out conflicting claims regarding property and conflicts between property rights and other rights.

Prudence or Prudential Judgment: See *Values and Value Judgments*

Reparations: Benefits given to some person or group to make amends for damage done by previous injustice. For example, as a result of the "Civil Liberties Act of 1988," Japanese-Americans who were placed in internment camps during World War II were given a monetary payment (of about $20,000 each) as reparations. Because children may be damaged by injustice done to their forebears, e.g. because poverty undermines their health or limits their educational opportunities, arguments are made for reparations to descendants of those who were first injured, if the consequences of the injury are of the sort that pass from one generation to the next.

Research Ethics or the Responsible Conduct of Research: *Research ethics* or *responsible conduct of research* (RCR) are terms used broadly to refer to many ethically significant issues that arise in research, from fair apportionment of credit among members of a research team, to responsible behavior in submitting or reviewing grant applications and the responsible treatment of research subjects.

Since the U.S. government and institutional regulations regarding the treatment of human and animal research subjects predated the increased attention to and the regulation of matters of research integrity (including fair credit) that arose in the 1980s, some RCR resources address only issues of research integrity and not the treatment of research subjects. Similarly, laboratory safety has long been regulated by OSHA, and is not necessarily a matter of research integrity. Therefore, it too is often omitted from discussions of responsible behavior in research. For example, *On Being a Scientist: Responsible Conduct in Research* (1995), put out by the National Academy of Sciences, the National Academy of Engineering, and the Institute of Medicine, in both its first and second edition omitted any discussion of the treatment of research subjects or laboratory safety. The treatment of research sub-

jects and laboratory safety are nevertheless reasonably classed as matters of research ethics, even if they are not always included under this designation.

In Europe the term "good scientific practice" (GSP) covers much of the same territory as research ethics and RCR.

Research Misconduct: *Research misconduct* is a term used rather narrowly. It does not include all violations of standards of research ethics. In particular, it is not applied to violations of the norms for the use of human or animal subjects.

In the United States the three actions that have been the focus of misconduct definitions are fabrication, falsification, and plagiarism (FFP). In 1995 the Congressionally-mandated Commission on Research Integrity issued a report, "Integrity and Misconduct in Research," arguing that FFP did not cover all serious deviations from accepted practices, and proposed a broader definition of research misconduct as misappropriation, interference, and misrepresentation, but this definition was not adopted. After extensive public debate the U.S. Office of Science and Technology Policy in 2000 issued the following common definition: "Research misconduct is defined as fabrication, falsification, or plagiarism in proposing, performing, or reviewing research, or in reporting research results." See the U.S. Federal Policy on Research Misconduct http://onlineethics.org/fedresmis.html.

Responsible Conduct of Research: See *Research Ethics and Responsible Conduct of Research*

Responsibility: *Responsibility* is a complex concept with both non-moral and moral meanings, and at least forward- and back-looking forms. The moral and forward-looking sense of responsibility is the sense in which one is responsible for achieving (or maintaining) a good result in some matter. The idea is that one is entrusted with achieving or maintaining this outcome, and expected both to have relevant knowledge and skills, and to make a conscientious effort. However, despite one's best efforts, the result may not be achieved. For example, patients of responsible physicians may die. The work of a responsible engineer may result in an accident because the accident was not foreseeable, it was not possible to compensate for the factors causing the accident, or because others were unwilling to heed the engineer's warnings. The moral and backward-looking sense of responsibility is that in which a person or group deserves ethical evaluation for some act or outcome, that is deserves moral praise for a good outcome or blame for a bad one.

The moral sense of responsibility should not be confused with the causal sense of responsibility for some existing or past state of affairs. For example, when we say "the storm was responsible for three deaths and heavy property damage," meaning that it caused these outcomes, we do not mean to attribute moral responsibility to the storm. Storms do not have moral responsibilities, and are neither responsible nor irresponsible in the moral sense. However, when a moral agent is causally responsible for some outcome, that is some reason to think that the agent is morally responsible for it. Causal responsibility is not conclusive evidence of moral responsibility, however. If one's actions cause a terrible outcome only because of bad moral luck, in the form of a freak accident, then one is not morally responsible for the outcome.

Responsibility, Official: The responsibility one is assigned as a result of some job or office. Unfortunately, official responsibilities may require one to behave unethically. But although "It was my job" might be a reason, it is not a valid excuse for immoral behavior. However, even when the requirements of an official responsibility are ethically acceptable, the concept of official responsibility functions differently from moral responsibility. Official responsibility resembles moral responsibility in generating prescriptions for conduct—duties or at least statements about what someone "ought" to do. As philosopher John Ladd points out, moral and official responsibility differ in at least two respects: First, official responsibilities are "exclusionary"—if one person has a particular official responsibility, another person does not (unless, of course, it was part of the job description of both). Second, official responsibilities, together with whatever rights, duties, and requirements for accountability attend them, are all "alienable" (see *Rights*)—they can be given to or taken over by someone else. In contrast, if one has a moral responsibility to inform the public about some matter, then even if one is in the position to delegate that responsibility to someone else, one still must see that the responsibility is fulfilled, because one does not get rid of a moral responsibility by giving it to someone else.

Responsibility, Professional: Professional responsibility is a paradigm case of the moral responsibility that arises from the special knowledge that one possesses. It is mastery of a special body of advanced knowledge, particularly knowledge that bears

directly on the well-being of others, which demarcates a profession. As custodians of special knowledge that bears on human well-being, professionals are constrained by special moral responsibilities—that is, moral requirements to apply their knowledge in ways that benefit the rest of the society.

Right of Self-Determination: The right of self-determination equals the right to choose one's own actions or course of life, so long as doing so does not interfere unduly with the lives and actions of others.

Rights: Rights are claims that have some justification behind them. A moral right is a morally justified claim. A legal right is a legally justified claim. When we use the term "right" without specifying the nature of the justification, we usually mean a moral right.

Rights specify acts that are permitted, forbidden, or required. If they specify acts that the rights-holder may perform (such as vote, or drive a car), they are often called licenses. If they specify acts that others may not perform (as the right to life obliges others to refrain from killing the rights holder), they are called liberties or (in law) negative rights. If they specify what the rights-holder should receive, the law commonly calls them positive rights, although some philosophers call them claim rights.

Other major types of classifications of rights are:

- Alienable rights and inalienable rights. Alienable rights may be taken or given away. Inalienable rights cannot be taken or given away.

- Human rights and special rights. Human rights belong to all people, or all people who are competent to exercise them. (Another term that is a close synonym for human rights is "natural rights.") In contrast, a right that only belongs to some people is termed a "special" right.

- Absolute rights and prima facie rights. Absolute rights cannot be outweighed by other considerations; prima facie rights can be outweighed by other considerations. For example, many of those who oppose capital punishment say that the right to life is an absolute right, but those who believe that capital punishment is morally justified in some circumstances say it is only a prima facie right.

See also *Copyright* and *Propietary/Property Rights*.

Risk: *Risk* is used colloquially as a term for a danger that arises unpredictably, such as being struck by a car. Sometimes it is used for the likelihood of a particular danger or hazard, as when someone says, "You can reduce your risk of being hit by a car by crossing at the crosswalk."

When used in technical context, such as in the terms "risk assessment" or "risk management," the notion of risk is the probability or likelihood of some resulting harm (such as the likelihood of being killed by being struck by a car) multiplied by the magnitude of the harm. One can then compare, say, the average citizen's risk of death from crossing the street with such a person's risk of death from cancer in a given period. One could also compare the risk of harms of different magnitudes. For example, two monetary risks: the rather likely event of losing a quarter in a malfunctioning vending machine, and the comparatively unlikely loss of one's wallet due to robbery at gun point. It may turn out that there is a greater risk of monetary loss from malfunctioning vending machines than from robbery at gun point. Notice that use of the technical sense of risk requires that one be able to meaningfully quantify the resulting harms. For many harms this is difficult to do except in an arbitrary way.

Risk and Benefits: See *Costs and Benefits*

Safety: *Safety* involves freedom from danger. A property of a device or process is safe insofar as it limits the risk of accident below some specified acceptable level.

Scientific Misconduct: See *Falsification, Fraud, Plagiarism, Research Misconduct.*

Screening: Involves the testing of a large number of individuals in a way designed to identify those with a particular genetic trait, characteristic, or biological condition. Screening differs from other biological testing in that it is done without any indication that the condition tested for is one possessed by any particular individual who is screened.

Security: The security of a system is the extent of protection against some unwanted occurrence such as the invasion of privacy, theft, and the corruption of information or physical damage.

Self-Deception: A failure to make explicit, even to oneself, some truth about oneself (often one's behavior). It may take the form of making up some rationalization for a behavior that is inconsistent with one's sense of self, or it may take the form of failing to take notice of some of the features of the situation when it would be appropriate to do so. The latter phenomenon is one that psychologists call "denial." Self-deception is a barrier to *authenticity*.

Stakeholder: A person or group who can affect or be affected by an action. Responsible decision-making requires consideration of the effects on all stakeholders. Some stakeholders may not be morally entitled to consideration of the same aspects of their welfare, however. For example, a corporate decision may affect or be influenced by employees, stockholders, customers, suppliers, communities, some government agencies, and corporate competitors. Competitors are entitled to fairness in competition, but not to the same consideration as, say, employees.

Standard: An established basis of comparison in measuring or judging capacity, quantity, content, value, quality, etc. It may also be a specified set of safety or performance criteria that a device or process ought to possess. The meeting of safety or performance standards must generally be demonstrated by a series of tests conducted under predetermined conditions.

Standard of Care: The degree of care that a reasonably prudent person would exercise in some particular circumstances. In negligence law, if someone's conduct falls below such a standard, then the person may be liable in tort for injuries or damages resulting from his or her conduct. In professional malpractice cases, a standard of care is applied to measure the competence as well of the degree of care shown by a professional's actions.

Therapeutic Illusion: A condition under which research subjects falsely believe that taking part in a particular study will likely result in some direct therapeutic benefit for themselves.

Therapeutic Orphan: A term given to children to express the concern that a fear of harming individual children by exposing them to research results is harming children as a class by undermining efforts to gain knowledge about how to better treat them. The question of the best methods for determining safe and effective medications is a continuing problem for drug research.

Tort: A private or civil (as contrasted with criminal) wrong or injury. Sometimes "tort law" is used as a general designation to include provisions concerning breaches of contract as well as a failure in some duty. However, the term *tort* is commonly used more narrowly to refer only to specific failure in some recognized duty, or a failure to exercise reasonable prudence or care. In this narrower sense *tort* is contrasted with "breach of contract" (failure to fulfill a legal agreement).

Trademark: An officially registered and legally restricted name, symbol, or representation, the use of which is restricted to its owner.

Trade Secret: A device, method, or formula that gives one an advantage over the competition, and which must therefore be kept secret if it is to be of special value. It is legal to use reverse engineering to learn a competitor's trade secret. "Know how" concerning research procedures may function as something like a trade secret.

Trust: Confident reliance. We may have confidence in events, people, or circumstances, or at least in our beliefs and predictions about them, but if we do not in some way rely on them, our confidence alone does not amount to trust. Reliance is a source of risk, and risk differentiates trusting in something from merely being confident about it. When one is in full control of an outcome or otherwise immune from disappointment, trust is not necessary. Of course, it is possible to rely on other people or on circumstances simply because one lacks other options.

The bases for confidence in relying on some person may not be morally sound. Trust may be naive or otherwise ill-founded. In this case it is likely to be disappointed. Trust may also rest on a morally unsound foundation as when, for example, one party feigns *trustworthiness* or behaves reliably only because the other party dominates.

Trustworthiness: When *trust* is well-founded and if trust of another person or moral agent is morally sound, then it is based on trustworthiness. Put another way, that which deserves trust is trustworthy.

Values and Value Judgments: Value judgments judge things to be good or bad in some respect. Moral or ethical values are only one type of value, and moral evaluation is only one type of value judgment. Consider the following nine value judgments:

1. This is a good (important, significant) hypothesis.

2. That is a good (insightful or informative) article.

3. This is a good (beautiful, masterfully executed) symphony.

4. That is a good (prudent or effective) strategy.

5. This is a bad (stupid, short-sighted) idea.

6. That is a good (virtuous, of high moral character) person.

7. This is a bad (evil, vicious) motive.

8. That is a good (kind, generous or right-minded) act or deed.

9. That is the right thing to do.

The first two are judgments of epistemic or knowledge value. The third is an aesthetic judgment. The fourth and fifth are prudential judgments. The sixth, seventh, and eighth are moral judgments. The ninth is also a moral judgment that is similar in some respects to the eighth, although the presence of "the" rather than "a" in the ninth suggests that the act in question is uniquely acceptable.

Assertions such as the ninth are usually justified by an appeal to moral rules, often to the exclusion of any mention of consequences. There are other types of value and value judgments that also come into play in ethics, such as those related to religious value. Religious terms of evaluation include "sacred" and "holy," as contrasted with "profane" and "mundane." In addition to purely religious judgments, the practice of most religions also involves making moral or ethical judgments.

Virtues and Vices: Virtues are positive traits of moral character such as honesty, kindness, or being a courageous or responsible person. Vices are negative traits of moral character such as dishonesty or cowardice. Notice that these terms of moral evaluation are applied to people, rather than to their actions (which may be assessed in terms of rights, obligations, and moral rules) or to the outcomes they seek to achieve (which may be assessed in terms of responsibilities).

Washout Study: A study in which patients or subjects are removed from all psychiatric medication to study baseline states or pure effects of new drug treatment.

Whistleblower: A whistleblower is a person who takes a concern (such as one about safety, financial fraud, or mistreatment of research animals) outside of the organization in which the abuse or suspected abuse is occurring and with which the whistleblower is affiliated. Not all whistleblowing is equally adversarial to the affected organization, even though it is at least an embarrassment for an organization to be exposed as one that cannot correct its own problems.

There are many regulatory agencies, such as the U.S. Occupational Safety and Health Administration, that exist to perform oversight and to which whistleblowers can go anonymously. Going to those charged with oversight, such as regulatory agencies, is usually seen as much less adversarial than, say, going to the media. Some people have used the term *whistlerblower* for those who raise an issue within their organization, but the more general term for a person who raises an issue inside or outside an organization is *complainant*.

BIBLIOGRAPHY

Allen, Anita. (2003). "Privacy and Law." *Privacies: Philosophical Evaluations*, ed. Beate Roessler. Palo Alto, CA: Stanford University Press.

Baier, Annette. (1994). *Moral Prejudices: Essays on Ethics*. Cambridge, MA: Harvard University Press.

Committee on Science, Engineering, and Public Policy. (1995). *On Being a Scientist: Responsible Conduct in Research*, 2nd edition. Washington, DC: National Academy Press. The first edition appeared in 1988.

Gert, Bernard, and Charles M. Culver. (1976). "Paternalistic Behavior." *Philosophy and Public Affairs* 6: 45–57.

Harris, Charles E.; Michael S. Pritchard; and Michael J. Rabins. (2000). *Engineering Ethics: Concepts and Cases* 2nd edition. Belmont, CA: Wadsworth/Thomson Learning.

Jackall, Robert. (1988). *Moral Mazes: The World of Corporate Managers*. New York: Oxford University Press.

Hauerwas, Stanley. (1981). *A Community of Character: Toward a Constructive Christian Social Ethic*. Notre Dame: IN: University of Notre Dame Press.

Ladd, John. (1982). "The Distinction Between Rights and Responsibilities: A Defense." *Linacre Quarterly* 49: 121–142.

MacIntyre, Alasdair. (1988). *Whose Justice? Which Rationality?* Notre Dame, IN: University of Notre Dame Press.

Whitbeck, Caroline. (2006). *Ethics in Engineering Practice*, 2nd edition. New York: Cambridge University Press.

APPENDIX IV

CHRONOLOGY OF HISTORICAL EVENTS RELATED TO SCIENCE, TECHNOLOGY, AND ETHICS

The purpose of this chronology is not to provide an all-inclusive history of science, technology, and ethics. Rather the aim is to highlight enough of the most important developments to capture a sense of the timing and pace of both macro-level (e.g., shifts from pre-modern to modern forms of science) and micro-level (e.g., the interplay of contemporary thinkers) changes and interactions. By organizing information in a historical manner, the chronology provides a supplementary perspective for thinking about science, technology, and ethics. It enables users of the encyclopedia to orient specific article topics within the larger sweep of time that conditions and is in turn conditioned by various persons, events, organizations, and ideas. The compilers of this chronology are grateful to Dr. David Lee for allowing the use of material from his website at http://www.sciencetimeline.net/.

The Ancient World: From the First Tools to 550 B.C.E.

Paleolithic (Old Stone Age) ca. 2 million B.C.E.–13000 B.C.E.

ca. 2 million–10000	Hunting and gathering were the main forms of human sustenance.
ca. 1 million	Chipped or patterned stone tools were first used.
ca. 125000	The control of fire by humans is widespread.
ca. 40000	Specialized instruments, such as needles and harpoons came into use.
ca. 30000	Cro–Magnon Man inhabited the valleys of France.

Mesolithic (Middle Stone Age) 13000 B.C.E.–6000 B.C.E.

ca. 11000	Bands of hunters in Europe depicted animals in cave paintings.
ca. 10000	Humans first began practicing agriculture.
ca. 10000	Wolves were first domesticated.
ca. 9000	Sheep were domesticated in the Middle East.
ca. 7700	Farming people settled in the Fertile Crescent.
ca. 7000	Wheat was domesticated in Mesopotamia.
ca. 6500	The wheel was invented in the Tigris–Euphrates basin by Sumerians.

ca. 6300	The earliest dug-out canoes were being made.

Neolithic (New Stone Age) 6000 B.C.E.–3000 B.C.E.

ca. 5000–3500	Villagers in Mesopotamia began practicing irrigation.
ca. 4800	Astronomical calendar stones were being used on the Nabta plateau.
ca. 4400	The first loom was used in Egypt.
ca. 4000	Light wooden plows were used in Mesopotamia.
ca. 4000–3500	Copper smelting in minute quantities was introduced in Mesopotamia.
ca. 3600	Southwestern Asians began using bronze, which unlike smelted copper, can hold an edge.
ca. 3500	The Sumerian civilization was born, which featured animal drawn vehicles, bronze, and the cuneiform alphabet.
ca. 3400	The first dynasty began in Egypt.
ca. 3000	The Sahara desert changed from a fertile area to an arid desert due to over use.
ca 3200	Wheeled vehicles were used in Uruk.
ca. 3200	The Egyptians were using sailboats with masts and broad, square sails.

They also painted pictures of these boats.

ca. 3000   Cotton was being grown in India.

ca. 3000   The Egyptians used a writing material called papyrus, which was made from woven reeds.

Bronze and Iron Ages 3000 B.C.E. –550 B.C.E.

ca. 2900–2450   The Great Pyramid of Cheops was built at Giza, Egypt.

ca. 2850   The sphinx was built in Egypt.

ca. 2700   Cuneiform signs and numerals appeared on Sumerian tablets, with a slanted double wedge between number symbols to indicate the absence of a number, or zero.

ca. 2500   Bronze was used widely, enabling the dagger form to be stretched into swords.

ca. 2500   The Iron Age began in the Middle East.

ca. 2500–2000   The dome was first used in architecture.

ca. 2400   The short, composite bow was developed by mounted archers. Unstrung, it curved forward and could pierce armor at 100 yards.

2205   The Xia dynasty came to power in China.

ca. 2000   The Minoan civilization emerged on Crete.

ca. 2000   Chinese thinkers discovered magnetic attraction.

1792–1750   Hammurabi (d. 1750) was king of Mesopotamia and created his code of laws, including "eye for an eye."

ca. 1900–1600   Stonehenge was built in present day England.

ca. 1750   The Babylonians began to use advanced geometry to make astronomical studies.

ca. 1700   The Babylonians created the first windmills, which pumped water for irrigation.

ca. 1700   Judaism was founded by Abraham.

ca. 1650   The first use was made of phonetic signs, derived from Egyptian hieroglyphics, in the Serabit el Khadim inscriptions, in the Sinai peninsula.

ca. 1600   The Mycenaean civilization emerged on the Greek mainland. Rulers constructed hilltop fortresses and were buried with treasures acquired through trade.

ca. 1500   The Chinese began weaving with silk.

ca. 1450   The Egyptians invented the sundial.

ca. 1400   The Egyptians invented the water clock.

ca. 1380   The Egyptians built the first canal, merging the Nile and the Red Sea.

ca. 1200–1000   Iron smelting was introduced on an industrial scale in Armenia.

1193   Troy fell to the Greeks in the Trojan War.

ca. 1100   Modern alphabetic writing was prefigured in the Phoenician alphabet.

ca. 1000   The Olmec civilization flourished in Mesoamerica.

ca. 850   Homer wrote the *Iliad* and *Odyssey*.

ca. 630   Zarathustra (aka Zoroaster) (c. 630–c. 530), of present day Iran, founded the mystical religion of Zorastrianism, one of the first forms of monotheism.

ca. 600   Thales of Miletus (c. 624–c. 547) speculated that the basic stuff of nature is water. He also argued that logic should replace myth as the foundation of human understanding.

ca. 600   Anaximander of Miletus (611–547) discovered the ecliptic (the angle between the plane of the earth's rotation and the plane of the solar system).

597   Babylonian society, with the hanging gardens, reached its zenith.

The Classical World 550 B.C.E.–500 C.E.

World Events

509 B.C.E.– 476 C.E.   The Roman Empire.

492–400 B.C.E.   Classical Greece introduced the world to democracy and many of the great ideas of philosophy.

432–404 B.C.E.   Athens and Sparta engaged each other in the Peloponnesian war. Athens lost, signaling the decline of Greek power and the rise of the Hellenistic age.

334 B.C.E.   Alexander the Great (356–323) invaded Asia in the first of many victories that would eventually push his empire as far as India.

321–185 B.C.E.   The Maurya Dynasty ruled in India.

221 B.C.E.   China was unified under the "First Emperor," Qin Shi Huangdi (259–210). The Great Wall was built around this time.

49–44 B.C.E.   Julius Caesar (100–44) was declared dictator for life, marking the transfer of Rome from Republic to Empire.

ca. 7 B.C.E.– ca 33 C.E.   Jesus Christ lived in Palestine.

ca. 79 C.E.   Domitian (51–96) dedicated the Roman Colosseum.

79 C.E.   Mt. Vesuvius erupted, burying Hercaulenum and Pompeii.

313 C.E.   Constantine (272–337) became Christian and issued the Edict of Milan, which granted freedom of worship to all inhabitants of the Roman Empire.

415 C.E.   A mob of rioters burned down the Library of Alexandria, and much of the recorded knowledge of the western world was lost.

Technological Inventions

ca. 500 B.C.E.   The Persians constructed the first highways.

ca. 400 B.C.E.   An arrow–shooting catapult was developed at Syracuse. It deliberately and systematically utilized mechanical and physical principles to improve weaponry.

ca. 170 B.C.E.   Parchment, superior to papyrus because it can be printed on both sides and folded, was invented in Pergamon.

ca. 150 B.C.E.   Hipparchus of Nicaea (c. 190–c. 120) invented the astrolabe, which was widely used until the eighteenth century when the sextant was invented.

ca. 100 B.C.E.   Paper was first used in China.

ca. 260 B.C.E.   Archimedes of Syracuse (c. 287–212) invented many machines, including a pump, effective levers and compound pulleys, and a mechanical planetarium.

ca. 50 C.E.   Hero of Alexandria (c. 10–c. 70) invented the first steam engine (the aeolipile), many feedback control devices, and the first type of analogue computer programming.

105 C.E.   The Chinese court official Ts'ai Lun (d. 121) developed a method to make paper out of cotton rags.

271 C.E.   Chinese mathematicians invented the magnetic compass.

Philosophy and Ethics

528 B.C.E.   Buddhism was founded by Siddhartha (563–483), a former prince, in India.

ca. 500 B.C.E.   Lao Tzu, of China, founded the naturalistic philosophy of Taoism.

ca 500 B.C.E.   Heraclitus of Ephesus (540–475) maintained that permanence was an illusion and the only possible real state was the process of becoming.

ca. 450 B.C.E.   Anaxagoras (500–428) proposed the first clearly materialist philosophy that the universe is made entirely of matter in motion.

ca. 440 B.C.E.   Protagoras of Abdera (485–415) held that man is the measure of all things by which he meant that we only know what we perceive, not the thing perceived.

ca. 425 B.C.E.   Herodotus of Halicarnassus (c. 485–c. 420) wrote the first scientific history by asking questions rather than just telling what he thinks he knows.

399 B.C.E.   Socrates (469–399) drank hemlock as punishment for his subversive views. His fate demonstrated the conflict between the philosopher (knowledge) and the city and the paradox that a liberal education is at once radical (challenging conventions) and conservative (forming good citizens).

ca. 380 B.C.E. At his Academy in Athens, Plato (420–340) expounded his metaphysics based upon the doctrine of forms, or eternal ideas.

ca. 335 B.C.E. Aristotle (384–322) established the Lyceum in Athens and wrote on such varied topics as logic, ethics, physics, metaphysics, politics, epistemology, and biology.

ca. 300 B.C.E. Epicurus (341–270) adopted and expanded the philosophy of atomism, in which all happens purely by chance, raising questions of determinism and freedom.

ca. 50 B.C.E. Marcus Tullius Cicero (106–43) transformed Greek philosophy into a practical affair, suitable to Roman concerns about law, governance, and military strategy.

ca. 350 C.E. Christianity began to flourish and its doctrines became systematized during several ecumenical councils.

397 C.E. Augustine (354–430) wrote *The Confessions*.

Scientific Discoveries

ca. 530 B.C.E. Pythagoras (585–497) studied musical intervals and regarded mathematics as the study of ultimate, eternal reality, immanent in nature and the universe.

ca. 450 B.C.E. Empedocles of Agrigento (d. 433) explained physical changes as movements of the basic particles of which things consisted, Fire, Earth, Air, and Water.

ca. 430 B.C.E. Hippocrates of Chios (c. 460–c. 377) squared the lune, a major step toward squaring the circle.

ca. 420 B.C.E. Democritus of Abdera (c. 460–c. 370) developed atomic theory, which held that haphazard collisions of atoms accounted for the formation and dissolution of objects.

ca. 300 B.C.E. Epicurus (341–270) attempted to deal with the contradictions between constant atoms and the appearance of novel combinations.

ca. 300 B.C.E. Euclid (365–300) wrote "Elements," a treatise on geometry. He offered an axiomatic system based on a few "common notions" and five basic postulates.

300 B.C.E. The number of volumes in the Library of Alexandria reached 500,000.

ca. 260 B.C.E. Archimedes of Syracuse (c. 287–212) formulated the principle that a body immersed in fluid is buoyed up by a force equal to the weight of the displaced fluid.

45 B.C.E. Sosigenes of Alexandria designed a calendar of 365.25 days, which was introduced by Julius Caesar.

ca. 10 C.E. Strabo (c. 63 B.C.E. –c. 24 C.E.) published his *Geographia*, which served as an encyclopedia of geographical knowledge at that time.

ca. 50 C.E. Hero of Alexandria explained that the four elements consist of atoms. He also observed that heated air expanded and made contributions to optics and geometry.

127–141 C.E. Claudius Ptolemaeus (c. 85–c. 165), or Ptolemy, compiled a compendium of opinion and data on the stars. He rejected the Peripatetic physics of the heavens.

ca. 170 C.E. Claudius Galen (131–201) used pulse taking as a diagnostic, performed animal dissections, and wrote treatises on anatomy.

190 C.E. Chinese mathematicians calculated pi to five decimal places. Archimedes had previously done so in the third century B.C.E.

Age of Faith 500–1400

World Events

ca. 450–1200 Europe was in the Middle Ages.

ca. 500–900 The Mayan Civilization dominated much of Mesoamerica.

527–565 The Byzantium Empire spread under Justinian's rule.

541 The bubonic plague spread from Egypt throughout the Roman–Byzantine world.

581–907 The Sui and Tang Dynasties ruled in China.

610	Mohammed began to secretly preach at Mecca.
771	Charlemagne became the king of all Franks.
ca. 900–1000	The Vikings discovered Greenland.
960–1279	The Song Dynasty ruled in China.
1066	William of Normandy became the first king of England.
1095–1291	The Crusades.
1211–1223	Genghis Khan invaded China, Persia, and Russia.
1215	King John of England signed the Magna Carta, which limited the powers of the king and guaranteed certain political liberties.
1271	Marco Polo (1254-1324, Venetia) journeyed to China along the Silk Road.
1281–1919	The Ottoman Empire reached its zenith in the sixteenth century, but then declined until it was ultimately dissolved in the aftermath of World War I.
ca. 1300–1600	The Renaissance in Europe marked the end of the Middle Ages. It was a cultural movement that revived the works of ancient Greece and Rome.
1337–1453	France and England fought the Hundred Years' War.
1347–1351	The Black Death, bubonic plague, wiped out roughly a third of Europe's population.
1368–1644	The Ming Dynasty ruled in China.

Technological Inventions

700	Block printing was developed in Japan.
700	The Chinese invented porcelain.
ca. 770	Stirrups were introduced in Frankish lands, enabling the development of the armored knight and mounted shock combat, which vastly altered society.
ca. 770	Iron horseshoes were common.
ca. 850	The Moors in Spain prepared pure copper by reacting its salts with iron, a forerunner of electroplating.
867	Wang Jie printed the oldest book known, *The Diamond Sutra*, in China.
ca. 1045	The Chinese inventor Pi Sheng made moveable type of earthenware.
ca. 1100	The crossbow was developed in Europe and outlawed, in 1139, by the second Ecumenical Lateran Council, as one of the first formal attempts at arms control.
ca. 1250	Gunpowder became known in Europe, perhaps introduced from China through the Mongols.
ca. 1350	Cannons were used widely in European battles. Developments in gunpowder in helped speed the military adoption of cannons.

Philosophy and Ethics

ca. 1250	Albert of Bollstadt (c. 1200–1280, Bavaria), called Albertus Magnus, wrote commentaries on Aristotle and studied plant morphology and ecology.
1267–1268	Roger Bacon (1214–1294, England) championed empiricism and the modern scientific method, asserting that the only basis for certainty is experience, or verification.
1267–1273	Thomas Aquinas (1224–1274, born in Italy) composed a synthesis of Christianity and Aristotelian philosophy.

Scientific Discoveries

ca. 1000	Ibn Sina (980–1037, Persia), or Avicenna, studied medicine and geology and challenged Aristotelian conceptions of motion.
ca. 1000	Ibn al-Haitham (965–1038, Arabia), or al-Hazen, studied optics and challenged Ptolemy by insisting that the hypothetical spheres corresponded to real bodies.
ca. 1190	Moses ben Maimon (1135–1204, born in Spain), or Maimonides, studied astrological systems and maintained the separation of earthly and heavenly spheres.
ca. 1215	Robert Grosseteste (1168–1253, England) studied optics and analyzed the

inductive and experimental procedures of science.

ca. 1323　William Ockham (1285–1349, England) introduced the distinction between dynamic motion and kinematic motion.

1337　William Merle of Oxford made regular records of the weather.

Age of Discovery 1400–1750

World Events

1418　Prince Henry the Navigator (1394–1460, Portugal) began exploring Africa.

1431　Joan of Arc (1412–1431, France) was burned at the stake.

1486　Bartolomeu Dias (1450–1500, Portugal) sailed around the Cape of Good Hope.

1492–1504　Christopher Columbus (1451–1506, Italy) discovered the Caribbean islands.

1497　Vasco da Gama (1469–1524, Portugal) sailed around Africa, discovering a sea route to India.

1509　Michelangelo (1475–1564, Italy) painted the ceiling of the Sistine Chapel.

1517　Martin Luther (1483–1546, Germany) posted his ninety–five theses in Wittenberg, initiating the Reformation.

1519–1521　Ferdinand Magellan (c. 1470–1521, Portugal) circumnavigated the globe.

1547　Ivan IV (1530–1584), or Ivan the Terrible, became the first ruler of Russia to claim the title of tsar.

1607　Jamestown, Virginia was established as the first English colony in the New World.

1618–1648　The Thirty Years' War raged between Protestants and Catholics.

1619　The first slaves were transported to America.

1620　Pilgrims landed at Plymouth Rock, Massachusetts.

1644　The Ming Dynasty ends in China and the Manchus come to power.

1661　Louis XIV (1638–1715) became absolute monarch of France.

1660　The Royal Society of London was founded.

1682　Peter the Great (1672–1725) became tsar of Russia. His efforts at westernization led to the development of Russia as a major European power.

1704　Johann Sebastian Bach (1685–1750, Germany) began composing music.

Technological Inventions

1437　Johann Gutenberg (c. 1390–1468, Germany) became the first in Europe to print with movable type cast in molds.

1475　The first muzzle-loaded rifles were developed in Italy and Germany.

1502　Peter Henlein of Nuremberg (c. 1480–1542, Germany) constructed the first watch.

1568　Concrete, which had been used in ancient times, was resuscitated by the architect Philibert de l'Orme (c. 1510–1570, France), who publicized its composition.

ca 1595　Spectacle maker Zacharias Janssen (1580–1638, Netherlands) invented the compound microscope.

1592　Galileo (1564–1642, Tuscany) invented a thermometer.

1594　Alexander Cummings (England) invented the flush toilet under English patent number 814. The ancient Cretans, however, used flush toilets as early as ca 2000 B.C.E.

1605　Hans Lippershey (c. 1570–1619, Netherlands) developed the telescope.

1621　Dud Dudley (1599–1684, England) invented the first blast furnace.

1625　William Oughtred (1575–1660, England) invented the slide rule.

1654　Otto von Guericke (1602–1686, Germany) invented the vacuum pump and the Magdeburg hemispheres.

1707   Denis Papin (1647–c. 1712, France) invented the high–pressure boiler.

1718   James Puckle (1667–1724, England) invented the machine gun. The Puckle Gun was capable of firing nine rounds before being reloaded.

ca. 1730   Two different men, John Hadley (1682–1744, England) and Thomas Godfrey (1704–1749, American colonies) independently invent the Sextant.

Philosophy and Ethics

1503   Desiderius Erasmus (1466–1536, Netherlands) argued that the chief evil of the day was a blind respect for traditions without considering the true message of Christ.

1532   *The Prince* by Niccolò Machiavelli (1469–1527, Italy) was published. It presents an early form of utilitarianism and realpolitik, although Machiavelli was a Republican.

1583   Giordano Bruno (1548–1600, Italy) defended a decentralized, infinite universe, governed by the identity of fundamental laws, rather than two separate spheres.

1620   Francis Bacon (1561–1626, England) published *Novum Organum*, which modeled an early form of empiricism as superior to scholastic a priori methods.

1637   René Descartes (1596–1650, France) wrote *Discourse on Method*

1651   Thomas Hobbes (1588–1679, England), in *Leviathan*, argued that humans must surrender individual autonomy to the state in order to avoid constant war.

1690   John Locke (1632–1704, England) argued that the mind is a blank slate. He also defended a social contract theory of the state and individual property rights.

1710   George Berkeley (1685–1753, Ireland) developed idealism, which holds that qualities, not things, are perceived and that perception is relative to the perceiver.

1725   Giovanni Battista Vico (1668–1774, Italy) critiqued the methodology of the natural sciences and maintained that truth is an act made by humans.

1748   David Hume (1711–1776, Scotland) described the mind as a bundle of perceptions and argued that moral obligation is a function of human passion rather than reason.

Scientific Discoveries

ca. 1482   Leonardo da Vinci (1452–1519, Italy) studied the human body and improved and invented many instruments with a devotion to the Archimedean ideal of measurement.

1536   Philippus Aureolus Paracelsus (1493–1541, Switzerland) foreshadowed systematic, modern medicine by rejecting the bodily "humours" as explanatory terms in physiology.

1543   Nicolaus Copernicus (1473–1543, Poland) defended heliocentrism along NeoPlatonic lines.

1569   Gerard de Cremer (1512–1594, Flanders), or Gerardus Mercator, published the projection map of the world that bears his name.

1572   Tycho Brahe (1546–1601, Denmark) observed a supernova in the constellation *Cassiopeia*, now known as Tycho's star.

1583   Galileo Galilei (1564–1643, Tuscany) pioneered the scientific age due to his systematic, quantitative experiments and his mathematical analysis of their results.

1586   Simon Stevin (1548–1620, Denmark) maintained that perpetual motion was impossible and made contributions to physics and geometry.

1604   Johannes Kepler (1571–1630, Germany) held that the intensity of light varies inversely with the square of the distance from the source.

1619   Kepler stated the third law of planetary motion, argued that the planets' orbits were ellipses, and developed a universal law to explain both heavenly and earthly bodies.

1627   William Harvey (1578–1657, England) confirmed his observation that the blood circulates throughout the body, which he inferred from the structure of the venal valves.

ca. 1629   Pierre de Fermat (1601–1665, France) discovered the fundamental principle of analytic geometry and pioneered differential calculus.

1633   The Inquisition forced Galileo to recant his belief in Copernican theory.

1650   Archbishop Usher estimated by reading the Bible that the earth was created on October 23, 4004 B.C.E. at 9:00 am.

1654   Blaise Pascal (1623–1662, France) and Pierre de Fermat developed the foundation for the theory of probability.

1661   Robert Boyle (1627–1691, England) separated chemistry from alchemy, leading to the general abandonment of ancient concepts of matter.

1665   Robert Hooke (1635–1703, England) named and gave the first description of cells.

1665–1666   Newton (1643–1727, England) made discoveries in calculus, universal gravitation, and optics.

1674   Anton van Leeuwenhoek (1632–1723, Netherlands) reported his discovery of protozoa. He made contributions to the microscope and cell biology.

1675   Gottfried Wilhelm von Leibniz (1646–1716, Germany) developed differential calculus.

1684   Leibniz published his system of calculus, developed independently of Newton. It is Leibniz's notation that has been adopted.

1687   Newton argued that natural laws govern the behavior of earthly and heavenly bodies. These laws of motion the groundwork for classical mechanics.

1693   Edmund Halley (1656–1742, England) discovered the formula for the focus of a lens and suggested a measurement of the distance between the earth and the sun.

Age of Revolution 1750 - 1830

World Events

1756–1763   The Seven Years' War was the first "world war" involving most European countries and their colonies around the globe.

1762   Catherine the Great (1729–1796) ascended the Russian throne.

1775–1783   The American Revolution began with the battle of Lexington and Concord and ended with the Treaty of Paris.

1789–1799   The French Revolution began with the storming of the Bastille and culminated in a coup by Napoleon.

1796   Napoleon Bonaparte (1769–1821, France) defeated Austria in the first of a string of military victories in Europe.

1799   French troops under Bonaparte discovered the Rosetta Stone that permitted Thomas Young and Jean-François Champollian to decipher Egyptian hieroglyphs.

1800–1830   Several Latin American countries won their Independence. For example: Venezuela 1810, Argentina 1816, Peru 1821, Brazil 1822, and Bolivia 1825.

1803   The Louisiana Purchase ushered in an era of expansion in America.

1808   Johann Wolfgang von Goethe (1749–1832, Germany) wrote the first part of Faust, a cautionary tale about the corrupting force of the powers that knowledge can unlock.

Technological Inventions

1752   Benjamin Franklin (1706–1790, U.S.) invented a lightening conductor. He also invented the Franklin stove and bifocals.

1764   James Hargreaves (1720–1778, England) invented the spinning jenny.

1769	James Watt (1736–1819, Scotland) patented a new type of steam engine equipped with a simple centrifugal "governor" for safety.
1785	Edmund Cartwright (1743–1823, England) invented the power loom.
1794	Eli Whitney (1765–1825, U.S.) patented the cotton gin, which quickly and easily separated the fiber from the seeds and seedpods.
1796	Edward Jenner (1749–1823, England) developed the first system of vaccination, by infecting patients with cowpox in order to make them resistant to smallpox.
1800	Alessandro Volta (1745–1827, Italy) invented the electric battery, a device that stores energy and makes it available in an electric form.
1804	Richard Trevithick (1771–1833, England) built the first steam–powered locomotive.
1826	Samuel Morey (1762–1843, U.S.) invented the internal combustion engine.

Philosophy and Ethics

1762	Jean-Jacques Rousseau (1712–1778, Switzerland) argued that the only natural association for humans is the family and that society must form a social contract.
1776	Adam Smith (1723–1790, Scotland) advanced the idea that businesses survive through successful trading in pursuit of their self–interest.
1781 and 1787	Immanuel Kant (1724–1804, Prussia) wrote the *Critique of Pure Reason*, in which he distinguished sensory from a priori elements of reason.
1789	Jeremy Bentham (1748–1832, England) outlined an ethical system based on a hedonistic calculation of the utility of actions and the greatest happiness of all.
1807	Georg Wilhelm Friedrich Hegel (1770–1831, Germany) criticized the distinction of objective and subjective and developed a dialectical and comprehensive philosophy.

1819	Arthur Schopenhauer (1788–1860, Germany) developed a life philosophy centered on the concept of will.

Scientific Discoveries

1751	Benjamin Franklin published works on electricity and invented many terms still in use, including positive, negative, conductor, and battery.
1754	Jean Le Rond d'Alembert (1717–1783, France) formulated "D'Alembert's ratio."
1766	Henry Cavendish (1731–1810, England) isolated and described "inflammable air," later named hydrogen, and distinguished it from carbon dioxide.
1772	Daniel Rutherford (1749–1819, England) discovered nitrogen.
1774	Joseph Priestly (1733–1804, England) discovered sulphur dioxide, ammonia, and "dephlogisticated air," later named oxygen.
1780	Antoine Laurent Lavoisier (1743–1794, France) and Pierre-Simon Laplace (1749–1827, France) developed a theory of chemical and thermal phenomena based on the assumption that heat is a substance and held that respiration is a form of combustion.
1783	Lazare Nicolas Marguerite Carnot (1753–1823, France) specified the optimal and abstract conditions for the operation of machines.
1786	Kant suggested the doctrine of the unity and convertibility of forces.
1787	Lavoisier published a nomenclature of chemistry.
1791	Goethe began publishing works on optics and developed a holistic philosophy opposed to Newtonian reductionism's dependence on theoretical constructs.
1795	James Hutton (1726–1797, Scotland) wrote the earliest comprehensive treatise that is a geologic synthesis, featuring uniformitarianism as a guiding principle.

1801   John Dalton (1766–1844, England) formulated the law of gaseous expansion at constant pressure and the law of gaseous partial pressures.

1803   Dalton put forth his theory of the atom.

1808   Dalton published a periodic table based on atomic weights.

1809   Jean-Baptiste Monet de Lamarck (1744–1829, France) stated that acquired characteristics were heritable and was a proponent of evolution.

1811   Amedeo Avogadro (1776–1856, Italy) proposed that equal volumes of gases at the same temperature and pressure contain the same number of molecules.

1824   Sadi Carnot (1796–1832, France) established the fundamental theory of the internal combustion engine and initiated the modern theory of thermodynamics.

1829   Charles Lyell (1797–1875, England) expanded on the principle of uniformitarianism and constant change in geology, which was useful for developing theories of evolution.

The Age of Industry and Empire 1830–1910

World Events

1830   The first railroad came into operation, running between Liverpool and Manchester, England.

1839   China was defeated by Britain in the First Opium War.

1848   Europe was convulsed with political revolutions.

1848   Mexico ceded vast amounts of land to the U.S. at the end of the Mexican War.

1849   The California gold rush drew thousands of settlers out West.

1858   Britain imposed formal colonial rule on India.

1859   Edwin Drake (1819–1880, U.S.) discovers oil near Titusville, Pennsylvania, ushering in the massive exploitation of petroleum to fuel modern industrialization.

1869   The first transcontinental train route in U.S. was completed.

1869   A French company completed the Suez Canal, allowing water transport between Europe and Asia without circumnavigating Africa.

1871   Kaiser Wilhelm I was declared German Emperor and the North German Confederation was transformed into the German Empire (Deutsches Reich).

1861–1865   The American Civil War was fought between the Union and the Confederacy. Slavery was ended by the Thirteenth Amendment in 1865.

1880s   The French began using the first pesticide.

1884–1885   The Berlin Conference regulated and formalized the colonization of Africa by European countries.

1887   The U.S. Congress founded the National Institutes of Health (NIH).

1898   The U.S. gained control of Cuba and the Philippines in the Spanish American War.

1904   The New York City subway opened.

Technological Inventions

1831   Michael Faraday (1791–1867, England) invented the electrical generator and the Bunsen burner and performed pioneering experiments in electromagnetism.

1834   Thomas Davenport (1802–1851, U.S.) is generally credited with inventing the electric motor.

1835   Charles Babbage (1791–1871, England) started work on the first "analytical engine," a precursor to the modern computer that used punch cards.

1835   Samuel Colt (1814–1862, U.S.) invented the revolver pistol.

1837   Samuel F. B. Morse (1791–1872, U.S.) invented the electrical telegraph and Morse code.

1838	John Deere (1804–1886, U.S.) invented the first cast-steel plow, a significant improvement over iron plows.
1853	Henry Bessemer (1813–1898, England) and William Kelly (1811–1888, U.S.) invented the Bessemer steel process.
1860	J.J.E. Lenoir (1822–1900, France) developed the first practical internal combustion engine. It relied upon coal gas and was double–acting.
1866	Alfred Nobel (1833–1896, Sweden) patented dynamite. It consisted of a mixture of nitroglycerine with inert, absorbent clay such as kieselguhr.
1866	Wilhelm (1855–1919, Germany) and Carl Friedrich von Siemens (1872–1941, Germany) invented the open–hearth furnace.
1866	Christopher Sholes (1819–1890, U.S.) invented the first modern, practical typewriter.
1876	Alexander Graham Bell (1847–1922, Scotland–Canada–U.S.) invented the telephone.
1876	Nikolas August Otto (1832–1891, Germany) designed the first four-stroke piston engine.
1877	Thomas A. Edison (1847–1931, U.S.) developed the phonograph, or gramophone, the first device for recording and replaying sound.
1879	Edison achieved his goal of making the burning time of the electric light bulb long enough to be commercially viable.
1882	Nikola Tesla (1856–1943, Serbia–U.S.) built the first induction motor, invented the Tesla coil (a type of transformer), and performed work on rotating magnetic fields.
1883	Sir Joseph Swann (1828–1914, England) invented the first synthetic fiber.
1885	Carl Friedrich Benz (1844–1929, Germany) invented the gasoline-powered automobile. The work of

Gottlieb Daimler (1834–1900, Germany) was also important.

1885	George Eastman (1854–1932, U.S.) invented roll film, which brought photography into popular usage and was the basis for the later invention of motion picture film.
1898	Rudolf Diesel (1858–1913, France-Germany) received a patent for the diesel engine.
1903	Orville (1871–1948, U.S.) and Wilbur Wright (1867–1912, U.S.) achieved flight in a manned, gasoline power–driven, heavier–than–air flying machine.
1904	Building off the work of Heinrich Hertz (1857–1894, Germany) and James Clerk Maxwell (1831–1879, Scotland), Christian Huelsmeyer invented radar.

Philosophy and Ethics

1830	Auguste Comte (1798–1857, France) developed positivism, a belief that natural science comprises the whole of human knowledge.
1855	Herbert Spencer (1820–1903, England) attempted to generalize from Darwinian evolution a comprehensive account of human social and moral progress.
1861	John Stuart Mill (1806–1873, England) extended and refined Bentham's utilitarian moral theory.
1867	Karl Marx (1818–1883, Prussia) systematically critiqued capitalism and developed a philosophy of dialectical materialism to account for historical change.
1885	Friedrich Wilhelm Nietzsche (1844–1900, Germany) deconstructed all meta-narratives and advocated the transvaluation of values through the strength of will.
1890	William James (1842–1910, U.S.) developed psychological theory into a systematic science and advanced a philosophy of pragmatism.
1900	Sigmund Freud (1856–1939, Austria) developed a tripartite under-

standing of human being and emphasized the importance of unconscious forces.

1903   G.E. Moore (1873–1958, England) rejected the "naturalistic fallacy" and developed analytic philosophy.

Scientific Discoveries

1820   Hans Christian Orsted (1777–1851, Denmark) discovered the relationship between electricity and magnetism.

1831   Faraday discovered electromagnetic induction.

1839   Charles Goodyear (1800–1860, U.S.) discovered the vulcanization process that creates rubber.

1840   William Whewell (1794–1866, England) introduced the word "scientist" to distinguish science or natural philosophy from a priori reasoning and moral science.

1840   Louis Agassiz (1807–1873, Switzerland-U.S.) published a demonstration of the existence of a glacial epoch in the temperate zones.

1847   Hermann Ludwig Ferdinand von Helmholtz (1821–1894, Germany) formulated the law of the conservation of energy.

1854   George Boole (1815–1864, England) invented Boolean algebra, the foundation of all modern computer arithmetic.

1857   Louis Pasteur (1822–1895, France) demonstrated that lactic acid fermentation is carried out by living bacteria and performed work with chiral molecules.

1858   Rudolf Virchow (1821–1902, Germany) stated that "every cell originates from another cell." He made contributions to pathology, medicine, and anthropology.

1859   Charles Darwin (1809–1882, England) presented his theory of biological evolution by natural selection.

1866   Gregor Mendel (1822–1884, Austria) interpreted heredity in terms of a pairing of unit characters that could in practice be treated as indivisible and independent particles.

1888   Hertz discovered radio waves, verifying Maxwell's prediction of electromagnetic waves.

1891   Marie Eugene Dubois (1858-1940, Netherlands) discovered Javaman, now known as Homo erectus.

1893   Émile Durkheim (1858–1917, France) and Max Weber (1864–1920, Germany) founded sociology and explained religion in terms of its social functions.

1895   Wilhelm Conrad Röntgen (1845–1923, Germany) observed a new form of penetrating radiation, which he named X–rays.

1896   J.J. Thompson (1856–1940, England) discovered the electron, which had been posited earlier by G. Johnstone Stoney as a unit of charge in electrochemistry.

1900   Max Planck (1858–1947, Germany) developed Planck's Law of Black Body Radiation, a pioneering works in the development of quantum mechanics.

1905   Albert Einstein (1879–1955, Germany–U.S.) demonstrated that the presence of atoms could be confirmed by observing objects influenced by their fluctuations.

1905   Einstein developed the Special Theory of relativity.

The Modern World 1910–2004

World Events

1914–1919   World War I began with the assassination of Archduke Franz Ferdinand and ended when a vanquished Germany signed the Treaty of Versailles.

1917   The Russian Revolution gave rise to the USSR. The Bolsheviks became the Communist party and held power for most of the twentieth century.

1925   The "Monkey Trial" of John T. Scopes (1900–1970, U.S.) occurred in Tennessee after he taught evolution in his classroom.

1927 Charles Lindbergh (1902–1974, U.S.) flew solo across the Atlantic Ocean.

1929 Inflated by speculation with borrowed money, the U.S. stock market crashed in October, initiating the slide into the Great Depression that would last through the 1930s.

1930 The U.S. Food and Drug Administration (FDA) was created.

1931 The International Council for Science (ICSU) was founded.

1932 Aldous Huxley (1894–1963, England) published *Brave New World*, the classical formulation of a techno-scientific dystopia.

1939 Leo Szilard (1898–1964, Hungary–U.S.) and Eugene Paul Wigner (1902–1995, Hungary–U.S.) visited Einstein to discuss methods of averting a German atomic bomb.

1939 Britain and France declared war on Germany, signaling the beginning of World War II.

1941 Pearl Harbor, a U.S. naval base in Hawaii, was attacked by the Japanese.

1944 The liberation of mainland Europe from Nazi occupation commenced with the Battle of Normandy, D–Day, on June 6.

1945 On July 16, a plutonium atomic bomb was detonated at the Trinity Site in the New Mexico desert.

1945 On August 6 and August 9, the U.S. dropped atomic bombs on Hiroshima and Nagasaki, respectively. Hundreds of thousands were killed.

1945 World War II ended with the surrender of Germany and Japan.

1945 The United Nations was founded in San Francisco.

1947 The General Agreement on Tariffs and Trade, later renamed the World Trade Organization, signaled the beginning of institutionalized economic globalization.

1948 The state of Israel was proclaimed.

1949 The North Atlantic Treaty Organization was established to counter Soviet aggression.

1950–1953 The Korean War occurred between the communist North and anti-communist South and was a proxy war between the U.S. and the Soviet Union.

1950 The U.S. Congress established the National Science Foundation (NSF).

1955 Bertrand Russell (1872–1970, England) and Albert Einstein issued the Russell-Einstein Manifesto, which called for international arms control and peace.

1957 The USSR launched Sputnik I, the first artificial satellite to orbit earth. Sputnik II was launched shortly thereafter and carried the first living passenger, a dog named Laika.

1958 The U.S. Congress established the National Aeronautics and Space Administration (NASA).

1959 Richard Feynman (1918–1988, U.S.) delivered his now famous speech "There's Plenty of Room at the Bottom" that foreshadowed later developments in nanotechnology.

1960 The U.S. FDA approved the birth control pill.

1961–1975 The Vietnam War occurred between communist North Korea and its allies and South Korea and its allies, primarily the United States.

1962 Rachel Carson (1907–1964, U.S.) wrote *Silent Spring*, which detailed the negative impact of pesticides on the environment.

1962 Trofim Denisovich Lysenko (1898–1976, U.S.S.R.) was removed from his position as head of the Academy of Agricultural Sciences of the Soviet Union.

1963	The United States, Great Britain, and the Soviet Union signed the Limited Test Ban Treaty.
1966–1976	Mao Zedong (1893–1976) and his wife Jiang Qing (1914–1991) carried out the Cultural Revolution in China.
1966–1979	Workers at the U.S. Center for Disease Control and the World Health Organization eradicated smallpox worldwide with vaccinations and containment.
1968	The Nuclear Non–Proliferation Treaty took effect, prohibiting non-nuclear weapon States from possessing, manufacturing, or acquiring nuclear weapons.
1969	Neil A. Armstrong (b.1930, U.S.) became the first man to walk on Moon. He was accompanied by Edwin E. Aldrin, Jr. (b.1930, U.S.).
1970	The U.S. Environmental Protection Agency (EPA) was created.
1975	The Asilomar Conference established guidelines for the physical and biological containment of recombinant DNA (rDNA).
1976	The two U.S. Viking probes landed on Mars.
1979	On March 28, a reactor at the Three Mile Island Nuclear Generating Station (Pennsylvania, U.S.) suffered a partial core meltdown.
1979	The U.S. spacecraft Voyager 1 photographed Jupiter's rings.
1979	The first "test tube baby," Louise Brown (U.K.), was born using the technique of in vitro fertilization (IVF).
1984	A Union Carbide pesticide plant in Bhopal, India accidentally released forty tons of methyl isocyanate (MIC) into the surrounding environment.
1986	On January 28, the space shuttle Challenger exploded just seventy-three seconds after launch. The accident was caused by the failure of an O–ring seal in the right solid rocket booster.
1986	On April 26, the Chernobyl nuclear power plant in Ukraine (then part of the Soviet Union) suffered a catastrophic nuclear meltdown.
1988	The Intergovernmental Panel on Climate Change (IPCC) was created to assess climate science and the impacts of climate change.
1988	James Watson unilaterally sets aside three to five percent of the budget of the Human Genome Project to study Ethical, Legal, and Social Issues (ELSI) of genomic research.
1989	On March 24, the Exxon Valdez oil tanker spilled eleven million gallons of crude oil into Prince William Sound, Alaska. It was the worst oil spill in United States history.
1989	The Berlin wall was torn down, signaling the end of the cold war. Germany began the process of reunification.
1992	The United States and thirty–four other industrial nations met in Rio de Janeiro, Brazil to discuss world environmental concerns.
1993	The U.S. Supreme Court articulated its set of criteria for the admissibility of scientific expert testimony in the case of Daubert v. Merrell Dow.
1996	The Comprehensive Nuclear Test Ban Treaty was signed by seventy-one nations, banning all nuclear explosions in all environments for military or civilian purposes.
1997	The World Commission on the Ethics of Scientific Knowledge and Technology (COMEST) was created as a U.N. body.
2001	The U.S. President's Council on Bioethics was created as part of a decision by President George W. Bush (b. 1946) to fund limited stem cell research.
2001	On September 11, the World Trade Center and Pentagon were attacked by terrorists who had hijacked commercial airplanes.

2003   The Space Shuttle Columbia was lost when it exploded upon reentry.

Technological Inventions

1913   Henry Ford (1863–1947, U.S.) added the assembly line to his automobile plant at Highland Park, Michigan.

1916   Paul Langevin (1872–1946, France) achieved the first successful use of sonar.

1924   Robert Goddard (1882–1945, U.S.) built and launched the first liquid-fueled rocket.

1927   Vannevar Bush (1890–1974, U.S.) developed the Differential Analyzer, an analog computer, which sped the solution of problems related to the electric power network.

1927   Vladimir Zworykin (1889–1982, Russia–U.S.), Paul Nipkow (1860–1940, Germany), Philo T. Farnsworth (1906–1971, U.S.), and John Baird (1888–1946, Scotland) all contributed to the invention of television.

1935   IBM introduced a punch card machine with an arithmetic unit based on relays that could perform multiplication.

1936   Alan M. Turing (1912–1954, England) conceived the Turing machine, the abstract precursor of the computer that gave a mathematically precise definition to algorithm.

1936   Felix Wankel (1902–1988, Germany) designed a motor (the Wankel engine) that revolved around a central shaft, using a rotary piston instead of reciprocating pistons.

1938   Roy Plunkett (1910–1994, U.S.) accidentally invented Polytetrafluoroethylene, commonly known as Teflon, while working at DuPont.

1940   Igor Sikorsky (1889–1972, Russia–U.S.) invented the helicopter.

ca. 1942   John von Neumann (1903–1957, Hungary–U.S.) developed architecture for a computing machine that allows it to be reprogrammable.

1944   Howard W. Aiken (1900–1973, U.S.) and a team of engineers from IBM displayed the first widely known and influential large scale automatic digital computer.

1945   The atomic bomb was developed in Los Alamos as part of the top secret Manhattan Project by J. Robert Oppenheimer (1904–1967, U.S.), Hans Bethe (1906–2005, Germany–U.S.), Einstein, Enrico Fermi (1901–1954, Italy–U.S.), Richard Feynman (1918–1988, U.S.) and hundreds of other scientists.

1946   The Raytheon Corporation patented the microwave oven. It built the first commercial microwave in 1947, which measured six feet tall and weighed 750 pounds.

1947   Working at Bell Labs, John Bardeen (1908–1991, U.S.), Walter Brattain (1902–1987, U.S.), and William Shockley (1910–1989, England–U.S.) invented the transistor, a solid state semiconductor device used for amplification and switching.

1949   Francis Bacon (1904–1992, England) invented a fuel cell, an electrochemical device similar to a battery, employing only hydrogen and water.

1951   Carl Djerassi (b.1923, Austria), Gregory Pincus (1903–1967, U.S.), Min Church Chiang, and John Rock (1890–1984, U.S.) all contributed to the invention of the oral contraceptive pill. Margaret Sanger (1879–1976, U.S.) worked to educate women about different birth control methods.

1954   Joseph Murray (b.1919, U.S.) and J. Hartwell Harrison performed the first successful human organ transplant.

1955   The USS Nautilus (SSN–571), the first nuclear powered submarine, was launched.

1955	Enrico Fermi and Leo Szilard received a joint U.S. patent for the nuclear reactor. The first nuclear power plant began producing electricity in Obninsk, Russia, in 1954.	1981	NASA launched the first space shuttle, Columbia.
1958	Jack Kilby (b.1923, U.S.) (Texas Instruments) and Robert Noyce (1927–1990, U.S.) (Fairchild Semiconductor) developed the first integrated circuit, a microelectronic semiconductor device consisting of many interconnected transistors.	1981	Programmers at Microsoft Corporation developed a computer disk operating system, MS–DOS.
		1982	The FDA approved the first recombinant pharmaceutical, insulin. This allowed insulin to be used on a wide scale and reduced reactions to impurities.
1960	Theodore Maiman (b.1927, U.S.) invented the first operable laser, a device that uses generates a very collimated, monochromatic, and coherent beam of light.	1982	Sony and Philips Corporations introduced the compact disc (CD) player.
1969	Edward Hoff (b.1937, U.S.) and Intel Corp. developed the microprocessor, which is an electronic computer central processing unit (CPU) made from miniaturized transistors and other circuit elements on a single semiconductor intergrated circuit (chip).	1983	ARPANET changed its core networking protocols from NCP to TCP/IP, marking the start of the Internet.
		1985	Kary Banks Mullis (b. 1944, U.S.) and co-workers invented the polymerase chain reaction (PCR) which multiplies DNA sequences *in vitro*.
1969	The Advanced Research Projects Agency Network (ARPANET) of the U.S. Department of Defense was the world's first operational packet switching network and the progenitor of the global Internet.	1985	Alec Jeffreys (b. 1950, England) invented DNA fingerprinting, a technique to distinguish between two individuals using only samples of their DNA.
1970	The first useful optical fiber was invented by researchers at Corning Glass Works.	1988	Working at the Roussel Uclaf company, Etienne Baulieu (France) developed the RU–486 abortifacient or "abortion pill," Mifepristone.
1971	Bowmar released the first pocket-sized calculator, the 901B, with four functions and an eight–digit red LED display.	1989	Charles Bennett and Gilles Brassard built a quantum computer, a device that computes using superpostions and entaglement of quantum states.
1977	Steven Jobs (b.1955, U.S.) and Steven Wozniak (b.1950, U.S.) introduced the Apple II, initiating the widespread use of home computers.	1989	The World Wide Web was developed by Tim Berners-Lee (b. 1955, England). The current web can be traced back to a project at CERN (the European Organization for Particles Physics Research) called ENQUIRE. The primary underlying concept of hypertext came from earlier efforts such as Vannevar Bush's memex and Ted Nelson's (b. 1937, U.S.) Project Xanadu.
1979	The first commercial cellular phone service is started in Japan. Researchers at Bell Labs had been working on the technology since the late 1940s.		
1980	Heinrich Rohrer (b.1933, Switzerland) and Gerd Binnig (b. 1947, Germany) developed a "scanning tunneling microscope."	1990	W. French Anderson (U.S.) performed the first gene transplant on a human being, injecting engineered genes into a four–year–old to repair her faulty immune system.

1990	Scientists at NASA and the European Space Agency (ESA) launched the Hubble Space Telescope.
1993	The work of Ivan Getting and Bradford Parkinson led to the invention of the Global Positioning System (GPS).
1997	The digital versatile disk (DVD) was introduced.

Philosophy and Ethics

1910	Alfred North Whitehead (1861–1947, England) and Bertrand Russell put forth the theory that there is a discontinuity between a class and its members and attempted to overcome certain logical paradoxes by the formal device of branding them meaningless.
1918	Ludwig Wittgenstein (1889–1951, Austria) put forth his theory of language as "picturing" reality, which he later abandoned for language as a system or a game played amongst a community.
1929	John Dewey (1859–1952, U.S.) argued that an experimental approach to moral decision making promised to solve the fact/value gap that had been championed by several analytic philosophers.
1934	Karl R. Popper (1902–1994, Austria) advanced the theory that the test of an empirical system, the demarcation of the limit of scientific knowledge, is its "falsifiability" and not its "verifiability."
1949	Simon de Beauvoir (1908–1986, France) traced the oppression of women through literary and historic sources and argued that the male is objectified as a positive norm.
1951	Willard Van Orman Quine (1908–2000, U.S.) argued against reductionism and the distinction between analytic and synthetic.
1953	Martin Heidegger (1889–1976, Germany) argued that modern technology reveals the world as an undifferentiated standing reserve (Bestand) of energy and resources subordinated to the will of humans, thus, the culmination of modern nihilism. He contrasted the way in which technology conceals Being to the way in which Being is revealed by the language of poetry.
1958	Hannah Arendt (1906–1975, Germany-U.S.) analyzed the modern human condition marked by the hegemony of laboring and making over action and the revolt against natural limits.
1962	Thomas S. Kuhn (1922–1996, U.S.) argued that new scientific paradigms are formed and retained because they are useful and conform to the standards of a community of practitioners, not because they approximate reality.
1971	John Rawls (1921–2002, U.S.) outlined the social arrangement of the "veil of ignorance" that guarantees no interests will be sacrificed arbitrarily to the interests of others. His concept of "justice as fairness" presented a non-historical variation of the social contract theory.
1974	Robert Nozick (1938–2002, U.S.) claimed that direct action by the state is rarely warranted, and that justice should be evaluated by reference to the means by which social policies are implemented, rather than their consequences.
1974	Thomas Nagel (b. 1937, U.S.) attempted to reconcile the subjective elements of human life with the urge for objective, value free truth.
1975	Peter Singer (b.1946, Australia) argued that, since a difference of species entails no moral distinction between sentient beings, it is wrong to mistreat non–human animals.
1978	Mary Daly (b. 1928, U.S.) argued that women must create a separate culture in order to fully effect their power outside of a patriarichal society.
1979	Hans Jonas (1903–1993, born in Germany) formulated a new ethics

designed to save humanity from the excesses of its own technological powers.

1981 Jurgen Habermas (b. 1929, Germany) developed a theory of moral discourse and knowledge in society as part of a larger effort to develop a post–metaphysical normativity founded on interpersonal relationships.

1982 Richard Rorty (b. 1931, U.S.) distinguished between Platonic, Positivist, and Pragmatist notions of truth and the consequences for acting on each choice.

1986 Martha Nussbaum (b. 1947, U.S.) argued that the moral philosophy of Aristotle remains relevant in the examination of human emotions and decision making.

Scientific Discoveries

1910 Fritz Haber (1868–1934, Germany) and Carl Bosch (1874–1940, Germany) patented the Haber–Bosch process for producing ammonia from the nitrogen contained in air.

1911 Einstein made predictions about the influence of gravity on the propagation of light.

1913 Niels Bohr (1885–1962, Denmark) calculated closely the frequencies of the spectrum of atomic hydrogen, supporting his conception of electron orbitals and foreshadowing quantum mechanics.

1915 Einstein completed the mathematical generalization of the theory of relativity and attributed the magic of the theory to differential calculus. This theory replaced the Kepler–Newton theory of planetary motion.

1923 Sigmund Freud (1856–1939, Austria) argued that the functioning of the mental apparatus is best understood as being the result of the interaction among three agencies or structures, which he labeled id, ego, and superego.

1925 Werner Heisenberg (1901–1976, Germany) formulated matrix mechanics, the first formalization of quantum mechanics.

1926 Erwin Schrödinger (1887–1961, Austria) initiated the development of the final quantum theory by describing wave mechanics, which predicted the positions of the electrons.

1927 Heisenberg proposed the Uncertainty Principle, which states that one cannot simultaneously determine the position and momentum of a subatomic particle.

1929 Alexander Fleming (1881–1955, Scotland) discovered the antibiotic substance lysozyme and issued a publication about penicillin.

1937 Hans Adolf Krebs (1900–1981, Germany) discovered the citrus acid cycle, also known as the tricarboxylic acid cycle and the Krebs cycle.

1938 Otto Hahn (1879–1968, Germany), Lise Meitner (1878–1968, Austria-Germany), and co–workers discovered nuclear fission.

1942 Paul Herman Mueller discovered the insecticidal properties of DDT (Dichloro–diphenyl–trichloroethane). DDT was first synthesized in 1873.

1943 Selman Waksman (1888–1973, Russia-U.S.) discovered streptomycin, which was the first antibiotic remedy for tuberculosis. It was first isolated by Albert Schatz, Waksman's research student.

1944 Friedrich Hayek (1899–1992, Austria) argued that only the unorganized price system in a free market enables order to arise from the chaos of individual plans.

1945 Vannevar Bush presented his vision of the "memex," which foreshadowed personal computers and hypertext systems like the World Wide Web.

1947 Ilya Prigogine (1917–2003, Belgium) studied dissipative structures and the self-organization of open thermodynamic systems.

1947 Willard Libby (1908–1980, U.S.) and others develop radiocarbon dating.

1948 George Gamow (1904–1968, Russia–U.S.) and Ralph Alpher (b.1921, U.S.) published the big bang theory of how the universe began.

1950 Norbert Wiener (1894–1946, U.S.) popularized the social implications of the emerging field of cybernetics.

1953 Working with the x-ray research of Rosalind Franklin (1920–1958, England), James Watson (b.1928, U.S.-England) and Francis Crick (1916–2004, England) built a model of DNA showing that the structure was two paired, complementary strands, helical and anti-parallel, associated by secondary, noncovalent bonds.

1964 Louis Leaky (1903–1972, U.K.) identified and named *Homo habilis*.

1970 Stephen Hawking (b.1942, England) and Roger Penrose (b.1931, England) proved that the Universe must have had a beginning in time, on the basis of Einstein's theory of General Relativity.

1975 Edward O. Wilson (b.1929, U.S.) analyzed the social instincts of animals and humans, giving rise to sociobiology.

1976 Richard Dawkins (b.1941, England) argued that the gene (or the "meme" in cultural evolution) is the relevant unit of selection.

1984 Luc Montagnier (b.1932, France) and other scientists working at the Pasteur Institute isolated the human immunodeficiency virus, or HIV. Robert Gallo (b.1937, U.S.) published the discovery of the HIV virus in the same year.

1984 Joe Farman, Brian Gardiner, and Jonathan Shanklin (England) published their discovery of the ozone hole.

1996 At the Roslin Institute in Scotland, Ian Wilmut (b.1944, Scotland) and Keith Campbell (England) cloned a sheep, "Dolly" (1996–2003), from adult cells.

1997 The U.S. Pathfinder vehicle studied and photographed Mars.

1998 Robert Waterston, John E. Sulston, and numerous colleagues reported the mapping of the entire genome of *Caenorhabditis elegans*.

2000 Researchers at the Human Genome Project completed a rough draft of the nucleotide sequence of the human genome. The project was completed ahead of schedule due to advances in sequence analysis and computer technologies.

2000 Craig Venter (b.1946, U.S.) led a team which sequenced the genome of *Drosophila melanogaster*.

2004 Mars rovers Spirit and Opportunity sent back photos of the red planet and collected data that further supported the hypothesis that water was once prevalent on Mars.

2004 Researchers at Seoul National University in South Korea became the first to clone a human embryo and then cull master stem cells from it.

COMPILED BY ADAM BRIGGLE
AND CARL MITCHAM

APPENDIX V

ETHICS CODES AND RELATED DOCUMENTS

This selective collection of professional ethics codes related to technology, engineering, and science, in both the professional and corporate contexts, along with a few declarations and manifestos, indicates the wide range of responses that exist in the technical and intellectual communities to some of the issues covered in the Encyclopedia of Science, Technology, and Ethics. *Most well developed are codes of conduct in the medical professional (which are well documented in the* Encyclopedia of Bioethics *and thus not duplicated here) and the engineering profession, as is indicated by the number of official documents from engineering societies throughout the world. Two other major resources for professional codes of this and related types can be found at the Case Western Reserve Online Ethics Center for Engineering and Science (onlineethics.org) and the Illinois Institute of Technology Center for the Study of Ethics in the Professions (ethics.iit.edu).*

1. Architecture and Design

American Institute of Architects (AIA) Code of Ethics

American Institute of Graphic Arts (AIGA) Standards of Professional Practice

Industrial Designers Society of America (IDSA) Code of Ethics

2. Computers

Association for Computing Machinery Code of Ethics and Professional Conduct

Software Engineering Code of Ethics and Professional Practice

Ten Commandments of Computer Ethics of the Computer Ethics Institute

3. Engineering

Ethics Codes in Professional Engineering: Overview and Comparisons

U.S. Engineering Societies

Accreditation Board for Engineering and Technology (ABET) Code of Ethics

American Society of Civil Engineers (ASCE) Code of Ethics

American Society of Mechanical Engineers (ASME) Code of Ethics

Institute of Electrical and Electronic Engineers (IEEE) Code of Ethics

National Society of Professional Engineers (NSPE) Code of Ethics

Puerto Rico: Association of Engineers and Surveyors of Puerto Rico Code of Ethics

Non-U.S. Engineering Societies

AUSTRALIA

The Institution of Engineers Code of Ethics

BANGLADESH

The Institution of Engineers Code of Ethics

CANADA

Canadian Council of Professional Engineers Code of Ethics

Association of Professional Engineers of Ontario Code of Ethics

Canadian Information Processing Society Code of Ethics

CHILE

Association of Engineers of Chile Code of Ethics

CHINA

Chinese Mechanical Engineering Society Code of Ethics

Retired Engineers Association of the Nanjing Chemical-Industrial Corporation Code of Ethics

COLOMBIA

Columbia Society of Engineers Code of Ethics

COSTA RICA

Federal Association of Engineers and Architects of Costa Rica Code of Ethics

DOMINICAN REPUBLIC

Dominican Association of Engineers, Architects, and Surveyors Code of Ethics

FINLAND

Engineering Society of Finland Code of Ethics

FRANCE

National Council of Engineers and Scientists of France Charter of Ethics of the Engineer

GERMANY
  Association of German Engineers Code of Ethics
  Fundamentals of Engineering Ethics

HONDURAS
  Association of Civil Engineers of Honduras Code of
  Ethics

HONG KONG
  Hong Kong Institution of Engineers Code of Ethics

INDIA
  Indian Institute of Chemical Engineers Code of Ethics
  Indian National Academy of Engineering Code of
  Ethics
  India Society of Engineers Code of Ethics
  The Institution of Engineers Code of Ethics

IRELAND
  The Institution of Engineers of Ireland Code of
  Ethics

JAMAICA
  Jamaican Institution of Engineers Code of Ethics

JAPAN
  Science Council of Japan Code of Ethics

MEXICO
  Mexican Union of Associations of Engineers Code
  of Ethics

NEW ZEALAND
  The Institution of Engineers Code of Ethics

NORWAY
  Association of Norwegian Civil Engineers Code of
  Ethics

PAKISTAN
  The Institution of Engineers Code of Ethics

SINGAPORE
  The Institution of Engineers Code of Ethics

SRI LANKA
  The Institution of Engineers Code of Ethics

SWEDEN
  Swedish Federation of Civil Engineers Code of
  Ethics

SWITZERLAND
  Swiss Technical Association Code of Ethics

UNITED KINGDOM
  Institution of Civil Engineers Code of Ethics
  Institution of Mechanical Engineers Code of Ethics

VENEZUELA
  Association of Engineers of Venezuela Code of
  Ethics
  Transnational Engineering Societies
  Fédération Européenne d'Associations Nationales
  d'Ingénieurs (FEANI, European Federation of
  National Engineering Associations) Code of Ethics
  Founding Statement of the International Network of
  Engineers and Scientists for Global Responsibility
  (INES) Appeal to Engineers and Scientists
  Unión Panamericana de Asociaciones de Ingenieros
  (UPADI, Pan American Federation of
  Engineering Societies) Code of Ethics
  World Federation of Engineering Societies Model
  Code of Ethics

4. Corporations and NGOs

  Code of Conduct for NGOs
  Dow Corning Ethical Business Conduct
  Eaton Ethical Business Conduct
  Lockheed Martin Corporation Code of Ethics and
  Business Conduct
  Responsible Care Guiding Principles (Chemical
  Industry)

5. Declarations and Manifestos

  Einstein-Russell Manifesto (1955)
  Mount Carmel Declaration on Technology and
  Moral Responsibility (1974)
  Rio Declaration on Environment and Development
  (1992)
  Technorealism Manifesto (1998)
  Declaration on Science and the Use of Scientific
  Knowledge (1999)
  Declaration of Santo Domingo (1999)
  Rio de Janeiro Declaration on Ethics in Science and
  Technology (2003)
  Ahmedabad Declaration (2005)

6. Science

  Chemist's Code of Conduct of the American
  Chemical Society
  Code of Ethics of the American Anthropological
  Association
  Hippocratic Oath for Scientists (U.S. Student
  Pugwash Group)
  International Network of Engineers and Scientists
  for Global Responsibility

7. Government

  Definition of Research Misconduct from the U.S.
  Federal Register

THE AMERICAN INSTITUTE OF ARCHITECTS (AIA): 2004 CODE OF ETHICS AND PROFESSIONAL CONDUCT

• • •

Preamble

Members of The American Institute of Architects are dedicated to the highest standards of professionalism, integrity, and competence. This **Code of Ethics and Professional Conduct** states guidelines for the conduct of Members in fulfilling those obligations. The **Code** is arranged in three tiers of statements:

- Canons, Ethical Standards, and Rules of Conduct:

- Canons are broad principles of conduct.

- Ethical Standards (E.S.) are more specific goals toward which Members should aspire in professional performance and behavior.

- Rules of Conduct (Rule) are mandatory; violation of a Rule is grounds for disciplinary action by the Institute.

Rules of Conduct, in some instances, implement more than one Canon or Ethical Standard. The **Code** applies to the professional activities of all classes of Members, wherever they occur. It addresses responsibilities to the public, which the profession serves and enriches; to the clients and users of architecture and in the building industries, who help to shape the built environment; and to the art and science of architecture, that continuum of knowledge and creation which is the heritage and legacy of the profession. Commentary is provided for some of the Rules of Conduct. That commentary is meant to clarify or elaborate the intent of the rule. The commentary is not part of the *Code*. Enforcement will be determined by application of the Rules of Conduct alone; the commentary will assist those seeking to conform their conduct to the **Code** and those charged with its enforcement.

Statement in Compliance with Antitrust Law

The following practices are not, in themselves, unethical, unprofessional, or contrary to any policy of The American Institute of Architects or any of its components:

1. submitting, at any time, competitive bids or price quotations, including in circumstances where price is the sole or principal consideration in the selection of an architect;

2. providing discounts; or

3. providing free services.

Individual architects or architecture firms, acting alone and not on behalf of the Institute or any of its components, are free to decide for themselves whether or not to engage in any of these practices. Antitrust law permits the Institute, its components, or Members to advocate legislative or other government policies or actions relating to these practices. Finally, architects should continue to consult with state laws or regulations governing the practice of architecture.

CANON I

• • •

General Obligations

Members should maintain and advance their knowledge of the art and science of architecture, respect the body of architectural accomplishment, contribute to its growth, thoughtfully consider the social and environmental impact of their professional activities, and exercise learned and uncompromised professional judgment.

E.S. 1.1 Knowledge and Skill: Members should strive to improve their professional knowledge and skill.

Rule In practicing architecture, **1.101** Members shall demonstrate a consistent pattern of reasonable care and competence, and shall apply the technical knowledge and skill which is ordinarily applied by architects of good standing practicing in the same locality.

Commentary: By requiring a "consistent pattern" of adherence to the common law standard of competence, this rule allows for discipline of a Member who more than infrequently does not achieve that standard. Isolated instances of minor lapses would not provide the basis for discipline.

E.S. 1.2 Standards of Excellence: Members should continually seek to raise the standards of aesthetic excellence, architectural education, research, training, and practice.

E.S. 1.3 Natural and Cultural Heritage: Members should respect and help conserve their natural and cultural heritage while striving to improve the environment and the quality of life within it.

E.S. 1.4 Human Rights: Members should uphold human rights in all their professional endeavors.

Rule 1.401 Members shall not discriminate in their professional activities on the basis of race, religion, gender, national origin, age, disability, or sexual orientation.

E.S. 1.5 Allied Arts & Industries: Members should promote allied arts and contribute to the knowledge and capability of the building industries as a whole.

CANON II

• • •

Obligations to the Public

Members should embrace the spirit and letter of the law governing their professional affairs and should promote and serve the public interest in their personal and professional activities.

E.S. 2.1 Conduct: Members should uphold the law in the conduct of their professional activities.

Rule 2.101 Members shall not, in the conduct of their professional practice, knowingly violate the law.

Commentary: The violation of any law, local, state or federal, occurring in the conduct of a Member's professional practice, is made the basis for discipline by this rule. This includes the federal Copyright Act, which prohibits copying architectural works without the permission of the copyright owner. Allegations of violations of this rule must be based on an independent finding of a violation of the law by a court of competent jurisdiction or an administrative or regulatory body.

Rule 2.102 Members shall neither offer nor make any payment or gift to a public official with the intent of influencing the official's judgment in connection with an existing or prospective project in which the Members are interested.

Commentary: This rule does not prohibit campaign contributions made in conformity with applicable campaign financing laws.

Rule 2.103 Members serving in a public capacity shall not accept payments or gifts which are intended to influence their judgment.

Rule 2.104 Members shall not engage in conduct involving fraud or wanton disregard of the rights of others.

Commentary: This rule addresses serious misconduct whether or not related to a Member's professional practice. When an alleged violation of this rule is based on a violation of a law, or of fraud, then its proof must be based on an independent finding of a violation of the law or a finding of fraud by a court of competent jurisdiction or an administrative or regulatory body.

Rule 2.105 If, in the course of their work on a project, the Members become aware of a decision taken by their employer or client which violates any law or regulation and which will, in the Members' judgment, materially affect adversely the safety to the public of the finished project, the Members shall:

a. advise their employer or client against the decision,

b. refuse to consent to the decision, and

c. report the decision to the local building inspector or other public official charged with the enforcement of the applicable laws and regulations, unless the Members are able to cause the matter to be satisfactorily resolved by other means.

Commentary: This rule extends only to violations of the building laws that threaten the public safety. The obligation under this rule applies only to the safety of the finished project, an obligation coextensive with the usual undertaking of an architect.

Rule 2.106 Members shall not counsel or assist a client in conduct that the architect knows, or reasonably should know, is fraudulent or illegal.

E.S. 2.2 Public Interest Services: Members should render public interest professional services and encourage their employees to render such services.

E.S. 2.3 Civic Responsibility: Members should be involved in civic activities as citizens and professionals, and should strive to improve public appreciation and understanding of architecture and the functions and responsibilities of architects.

Rule 2.301 Members making public statements on architectural issues shall disclose when they are being compensated for making such statements or when they have an economic interest in the issue.

CANON III

• • •

Obligations to the Client

Members should serve their clients competently and in a professional manner, and should exercise unprejudiced and unbiased judgment when performing all professional services.

E.S. 3.1 Competence: Members should serve their clients in a timely and competent manner.

Rule 3.101 In performing professional services, Members shall take into account applicable laws and regulations. Members may rely on the advice of other qualified persons as to the intent and meaning of such regulations.

Rule 3.102 Members shall undertake to perform professional services only when they, together with those whom they may engage as consultants, are qualified by education, training, or experience in the specific technical areas involved.

Commentary: This rule is meant to ensure that Members not undertake projects that are beyond their professional capacity. Members venturing into areas that require expertise they do not possess may obtain that expertise by additional education, training, or through the retention of consultants with the necessary expertise.

Rule 3.103 Members shall not materially alter the scope or objectives of a project without the client's consent.

E.S. 3.2 Conflict of Interest: Members should avoid conflicts of interest in their professional practices and fully disclose all unavoidable conflicts as they arise.

Rule 3.201 A Member shall not render professional services if the Member's professional judgment could be affected by responsibilities to another project or person, or by the Member's own interests, unless all those who rely on the Member's judgment consent after full disclosure.

Commentary: This rule is intended to embrace the full range of situations that may present a Member with a conflict between his interests or responsibilities and the interest of others. Those who are entitled to disclosure may include a client, owner, employer, contractor, or others who rely on or are affected by the Member's professional decisions. A Member who cannot appropriately communicate about a conflict directly with an affected person must take steps to ensure that disclosure is made by other means.

Rule 3.202 When acting by agreement of the parties as the independent interpreter of building contract documents and the judge of contract performance, Members shall render decisions impartially.

Commentary: This rule applies when the Member, though paid by the owner and owing the owner loyalty, is nonetheless required to act with impartiality in fulfilling the architect's professional responsibilities.

E.S. 3.3 Candor and Truthfulness: Members should be candid and truthful in their professional communications and keep their clients reasonably informed about the clients' projects.

Rule 3.301 Members shall not intentionally or recklessly mislead existing or prospective clients about the results that can be achieved through the use of the Members' services, nor shall the Members state that they can achieve results by means that violate applicable law or this **Code.**

Commentary: This rule is meant to preclude dishonest, reckless, or illegal representations by a Member either in the course of soliciting a client or during performance.

E.S. 3.4 Confidentiality: Members should safeguard the trust placed in them by their clients.

Rule 3.401 Members shall not knowingly disclose information that would adversely affect their client or that they have been asked to maintain in confidence, except as other wise allowed or required by this **Code** or applicable law.

*Commentary: To encourage the full and open exchange of information necessary for a successful professional relationship, Members must recognize and respect the sensitive nature of confidential client communications. Because the law does not recognize an architect-client privilege, however, the rule permits a Member to reveal a confidence when a failure to do so would be unlawful or contrary to another ethical duty imposed by this **Code.***

CANON IV

• • •

Obligations to the Profession

Members should uphold the integrity and dignity of the profession.

E.S. 4.1 Honesty and Fairness: Members should pursue their professional activities with honesty and fairness.

Rule 4.101 Members having substantial information which leads to a reasonable belief that another Member has committed a violation of this **Code** which raises a serious question as to that Member's honesty, trustworthiness, or fitness as a Member, shall file a complaint with the National Ethics Council.

Commentary: Often, only an architect can recognize that the behavior of another architect poses a serious question as to that other's professional integrity. In those circumstances, the duty to the professional's calling requires that a complaint be filed. In most jurisdictions, a complaint that invokes professional standards is protected from a libel or slander action if the complaint was made in good faith. If in doubt, a Member should seek counsel before reporting on another under this rule.

Rule 4.102 Members shall not sign or seal drawings, specifications, reports, or other professional work for which they do not have responsible control.

Commentary: Responsible control means the degree of knowledge and supervision ordinarily required by the professional standard of care. With respect to the work of licensed consultants, Members may sign or seal such work if they have reviewed it, coordinated its preparation, or intend to be responsible for its adequacy.

Rule 4.103 Members speaking in their professional capacity shall not knowingly make false statements of material fact.

Commentary: This rule applies to statements in all professional contexts, including applications for licensure and AIA membership.

E.S. 4.2 Dignity and Integrity: Members should strive, through their actions, to promote the dignity and integrity of the profession, and to ensure that their representatives and employees conform their conduct to this **Code.**

Rule 4.201 Members shall not make misleading, deceptive, or false statements or claims about their professional qualifications, experience, or performance and shall accurately state the scope and nature of their responsibilities in connection with work for which they are claiming credit.

Commentary: This rule is meant to prevent Members from claiming or implying credit for work which they did not do, misleading others, and denying other participants in a project their proper share of credit.

Rule 4.202 Members shall make reasonable efforts to ensure that those over whom they have supervisory authority conform their conduct to this **Code.**

*Commentary: What constitutes "reasonable efforts" under this rule is a common sense matter. As it makes sense to ensure that those over whom the architect exercises supervision be made generally aware of the **Code**, it can also make sense to bring a particular provision to the attention of a particular employee when a situation is present which might give rise to violation.*

CANON V

• • •

Obligations to Colleagues

Members should respect the rights and acknowledge the professional aspirations and contributions of their colleagues.

E.S. 5.1 Professional Environment: Members should provide their associates and employees with a suitable working environment, compensate them fairly, and facilitate their professional development.

E.S. 5.2 Intern and Professional Development: Members should recognize and fulfill their obligation to nurture fellow professionals as they progress through all stages of their career, beginning with professional education in the academy, progressing through internship and continuing throughout their career.

E.S. 5.3 Professional Recognition: Members should build their professional reputation on the merits of their own service and performance and should recognize and give credit to others for the professional work they have performed.

Rule 5.301 Members shall recognize and respect the professional contributions of their employees, employers, professional colleagues, and business associates.

Rule 5.302 Members leaving a firm shall not, without the permission of their employer or partner, take designs, drawings, data, reports, notes, or other materials relating to the firm's work, whether or not performed by the Member.

Rule 5.303 A Member shall not unreasonably withhold permission from a departing employee or partner to take copies of designs, drawings, data, reports, notes, or other materials relating to work performed by the employee or partner that are not confidential.

Commentary: A Member may impose reasonable conditions, such as the payment of copying costs, on the right of departing persons to take copies of their work.

RULES OF APPLICATION, ENFORCEMENT, AND AMENDMENT

• • •

Application

The **Code of Ethics and Professional Conduct** applies to the professional activities of all members of the AIA.

Enforcement

The Bylaws of the Institute state procedures for the enforcement of the **Code of Ethics and Professional Conduct.** Such procedures provide that:

1. Enforcement of the **Code** is administered through a National Ethics Council, appointed by the AIA Board of Directors.

2. Formal charges are filed directly with the National Ethics Council by Members, components, or any-

one directly aggrieved by the conduct of the Members.

3. Penalties that may be imposed by the National Ethics Council are:

(a) Admonition
(b) Censure
(c) Suspension of membership for a period of time
(d) Termination of membership

4. Appeal procedures are available.

5. All proceedings are confidential, as is the imposition of an admonishment; however, all other penalties shall be made public.

Enforcement of Rules 4.101 and 4.202 refer to and support enforcement of other Rules. A violation of Rules 4.101 or 4.202 cannot be established without proof of a pertinent violation of at least one other Rule.

Amendment

The Code of Ethics and Professional Conduct may be amended by the convention of the Institute under the same procedures as are necessary to amend the Institute's Bylaws. The **Code** may also be amended by the AIA Board of Directors upon a two-thirds vote of the entire Board.

*2004 EDITION. *This copy of the* **Code of Ethics** *is current as of September 2004. Contact the General Counsel's Office for further information at (202) 626-7311.*

AMERICAN INSTITUTE OF GRAPHIC ARTS (AIGA) STANDARDS OF PROFESSIONAL PRACTICE

• • •

The purpose of the statement of policy on professional practice is to provide all AIGA members with a clear standard of professional conduct. AIGA encourages the highest level of professional conduct in design. The policy is not binding. Rather, it reflects the view AIGA on the kind of conduct that is in the best interest of the profession, clients, and the public.

For the purposes of this document the word "designer" means an individual, practicing design as a freelance or salaried graphic designer, or group of designers acting in partnership or other form of association.

The designer's professional responsibility

1.1 A designer shall at all times act in a way that supports the aims of the AIGA and its members, and encourages the highest standards of design and professionalism.

1.2 A designer shall not undertake, within the context of his or her professional practice, any activity that will compromise his or her status as a professional consultant.

The designer's responsibility to clients

2.1 A designer shall acquaint himself or herself with a client's business and design standards and shall act in the client's best interest within the limits of professional responsibility.

2.2 A designer shall not work simultaneously on assignments that create a conflict of interest without agreement of the clients or employers concerned, except in specific cases where it is the convention of a particular trade for a designer to work at the same time for various competitors.

2.3 A designer shall treat all work in progress prior to the completion of a project and all knowledge of a client's intentions, production methods, and business organization as confidential and shall not divulge such information in any manner whatsoever without the consent of the client. It is the designer's responsibility to ensure that all staff members act accordingly.

The designer's responsibility to other designers

3.1 Designers in pursuit of business opportunities should support fair and open competition based upon professional merit.

3.2 A designer shall not knowingly accept any professional assignment on which another designer has been or is working without notifying the other designer or until he or she is satisfied that any previous appointments have been properly terminated and that all materials relevant to the continuation of the project are the clear property of the client.

3.3 A designer must not attempt, directly or indirectly, to supplant another designer through unfair means; nor must he or she compete with another designer by means of unethical inducements.

3.4 A designer must be fair in criticism and shall not denigrate the work or reputation of a fellow designer.

3.5 A designer shall not accept instructions from a client that involve infringement of another person's property rights without permission, or consciously act in any manner involving any such infringement.

3.6 A designer working in a country other than his or her own shall observe the relevant Code of Conduct of the national society concerned.

Fees

4.1 A designer shall work only for a fee, a royalty, salary, or other agreed-upon form of compensation. A designer shall not retain any kickbacks, hidden discounts, commission, allowances, or payment in kind from contractors or suppliers.

4.2 A reasonable handling and administration charge may be added, with the knowledge and understanding of the client, as a percentage to all reimbursable items, billable to a client, that pass through the designer's account.

4.3 A designer who is financially concerned with any suppliers who may benefit from any recommendations made by the designer in the course of a project shall secure the approval of the client or employer of this fact in advance.

4.4 A designer who is asked to advise on the selection of designers or the consultants shall not base such advice in the receipt of payment from the designer or consultants recommended.

Publicity

5.1 Any self-promotion, advertising, or publicity must not contain deliberate misstatements of competence, experience, or professional capabilities. It must be fair both to clients and other designers.

5.2 A designer may allow a client to use his or her name for the promotion of work designed or services provided but only in a manner that is appropriate to the status of the profession.

Authorship

6.1 A designer shall not claim sole credit for a design on which other designers have collaborated.

6.2 When not the sole author of a design, it is incumbent upon a designer to clearly identify his or her specific responsibilities or involvement with the design. Examples of such work may not be used for publicity, display, or portfolio samples without clear identification of precise areas of authorship.

First published by AIGA, the professional association for design. www.aiga.org.

INDUSTRIAL DESIGNERS SOCIETY OF AMERICA (IDSA) CODE OF ETHICS

• • •

Recognizing that industrial designers affect the quality of life in our increasingly independent and complex society; that responsible ethi-cal decision making often requires conviction, courage and ingenuity in today's competitive business context: We, the members of the Industrial Designers Society of America, will endeavor to meet the standards set forth in this code, and strive to support and defend one another in doing so.

Fundamental Ethical Principles

We will uphold and advance the integrity of our profession by:

1. Supporting one another in achieving our goals of maintaining high professional standards and levels of competence, and honoring commitments we make to others;

2. Being honest and fair in serving the public, our clients, employers, peers, employees and students regardless of gender, race, creed, ethnic origin, age, disability or sexual orientation;

3. Striving to maintain sufficient knowledge of relevant current events and trends so as to be able to assess the economic and environmental effects of our decisions;

4. Using our knowledge and skill for the enrichment of human well-being, present and future; and

5. Supporting equality of rights under the law and opposing any denial or abridgement of equal rights by the United States or by any individual state on account of gender, race, creed, ethnic origin, age, disability or sexual orientation.

Articles of Ethical Practice

The following articles provide an outline of ethical guidelines designed to advance the quality of our profession. They provide general principles in which the "Ethics Advisory Council" can resolve more specific questions that may arise.

Article I: We are responsible to the public for their safety, and their economic and general well-being is our foremost professional concern. We will participate only in projects we judge to be ethically sound and in conformance with pertinent legal regulations; we will advise our clients and employers when we have serious reservations concerning projects we have been assigned.

Article II: We will provide our employers and clients with original and innovative design service of high quality; by serving their interests as faithful agents; by treating privileged information with discretion; by communicating effectively with their appropriate staff members; by avoiding conflicts of interest; and by establishing clear contractual understandings regarding obligations of both parties. Only with agreement of all concerned will we work on competing product lines simultaneously.

Article III: We will compete fairly with our colleagues by building our professional reputation primarily on the quality of our work; by issuing only truthful, objective and non-misleading public statements and promotional materials; by respecting competitors' contractual relationships with their clients; and by commenting only with candor and fairness regarding the character of work of other industrial designers.

Article IV: We will be responsible to our employees by facilitating their professional development insofar as possible; by establishing clear contractual understandings; by maintaining safe and appropriate work environments; by properly crediting work accomplished; and by providing fair and adequate compensation for salary and overtime hours.

Article V: We will be responsible to design education by holding as one of our fundamental concerns the education of design students; by advocating implementation of sufficiently inclusive curricula and requiring satisfactory proficiency to enable students to enter the profession with adequate knowledge and skills; by providing opportunities for internships (and collaboratives) with and observation of practicing designers; by respecting students' rights to ownership of their designs; and by fairly crediting them for work accomplished.

Article VI: We will advance the interests of our profession by abiding by this code; by providing a forum within the Society for the ongoing review of ethical concerns; and by publishing, as appropriate, interpretations of this Code.

COMPUTERS

• • •

ASSOCIATION FOR COMPUTING MACHINERY (ACM) CODE OF ETHICS AND PROFESSIONAL CONDUCT

• • •

Adopted by ACM Council 10/16/92.

Preamble

Commitment to ethical professional conduct is expected of every member (voting members, associate members, and student members) of the Association for Computing Machinery (ACM).

This Code, consisting of 24 imperatives formulated as statements of personal responsibility, identifies the elements of such a commitment. It contains many, but not all, issues professionals are likely to face. *Section 1* outlines fundamental ethical considerations, while *Section 2* addresses additional, more specific considerations of professional conduct. Statements in *Section 3* pertain more specifically to individuals who have a leadership role, whether in the workplace or in a volunteer capacity such as with organizations like ACM. Principles involving compliance with this Code are given in *Section 4*.

The Code shall be supplemented by a set of Guidelines, which provide explanation to assist members in dealing with the various issues contained in the Code. It is expected that the Guidelines will be changed more frequently than the Code.

The Code and its supplemented Guidelines are intended to serve as a basis for ethical decision making in the conduct of professional work. Secondarily, they may serve as a basis for judging the merit of a formal complaint pertaining to violation of professional ethical standards.

It should be noted that although computing is not mentioned in the imperatives of *Section 1*, the Code is concerned with how these fundamental imperatives apply to one's conduct as a computing professional. These imperatives are expressed in a general form to emphasize that ethical principles which apply to computer ethics are derived from more general ethical principles.

It is understood that some words and phrases in a code of ethics are subject to varying interpretations, and that any ethical principle may conflict with other ethical principles in specific situations. Questions related to ethical conflicts can best be answered by thoughtful consideration of fundamental principles, rather than reliance on detailed regulations.

Contents and Guidelines

1. GENERAL MORAL IMPERATIVES.

• • •

As an ACM member I will . . .

1.1 Contribute to society and human well-being.

This principle concerning the quality of life of all people affirms an obligation to protect fundamental human rights and to respect the diversity of all cultures. An essential aim of computing professionals is to minimize negative consequences of computing systems, including threats to health and safety. When designing or implementing systems, computing professionals must attempt to ensure that the products of their efforts will be used in socially responsible ways, will meet social needs, and will avoid harmful effects to health and welfare.

In addition to a safe social environment, human well-being includes a safe natural environment. Therefore, computing professionals who design and develop systems must be alert to, and make others aware of, any potential damage to the local or global environment.

1.2 Avoid harm to others.

"Harm" means injury or negative consequences, such as undesirable loss of information, loss of property, property damage, or unwanted environmental impacts. This principle prohibits use of computing technology in ways that result in harm to any of the following: users, the general public, employees, employers. Harmful actions include intentional destruction or modification of files and programs leading to serious loss of resources or unnecessary expenditure of human resources such as the time and effort required to purge systems of "computer viruses."

Well-intended actions, including those that accomplish assigned duties, may lead to harm unexpectedly. In such an event the responsible person or persons are obligated to undo or mitigate the negative consequences as much as possible. One way to avoid unintentional harm is to carefully consider potential impacts on all those affected by decisions made during design and implementation.

To minimize the possibility of indirectly harming others, computing professionals must minimize malfunctions by following generally accepted standards for system design and testing. Furthermore, it is often necessary to assess the social consequences of systems to project the likelihood of any serious harm to others. If system features are misrepresented to users, coworkers, or supervisors, the individual computing professional is responsible for any resulting injury.

In the work environment the computing professional has the additional obligation to report any signs of system dangers that might result in serious personal or social damage. If one's superiors do not act to curtail or mitigate such dangers, it may be necessary to "blow the whistle" to help correct the problem or reduce the risk. However, capricious or misguided reporting of violations can, itself, be harmful. Before reporting violations, all relevant aspects of the incident must be thoroughly assessed. In particular, the assessment of risk and responsibility must be credible. It is suggested that advice be sought from other computing professionals. See *principle 2.5* regarding thorough evaluations.

1.3 Be honest and trustworthy.

Honesty is an essential component of trust. Without trust an organization cannot function effectively. The honest computing professional will not make deliberately false or deceptive claims about a system or system design, but will instead provide full disclosure of all pertinent system limitations and problems.

A computer professional has a duty to be honest about his or her own qualifications, and about any circumstances that might lead to conflicts of interest.

Membership in volunteer organizations such as ACM may at times place individuals in situations where their statements or actions could be interpreted as carrying the "weight" of a larger group of professionals. An ACM member will exercise care to not misrepresent ACM or positions and policies of ACM or any ACM units.

1.4 Be fair and take action not to discriminate.

The values of equality, tolerance, respect for others, and the principles of equal justice govern this imperative.

Discrimination on the basis of race, sex, religion, age, disability, national origin, or other such factors is an explicit violation of ACM policy and will not be tolerated.

Inequities between different groups of people may result from the use or misuse of information and technology. In a fair society, all individuals would have equal opportunity to participate in, or benefit from, the use of computer resources regardless of race, sex, religion, age, disability, national origin or other such similar factors. However, these ideals do not justify unauthorized use of computer resources nor do they provide an adequate basis for violation of any other ethical imperatives of this code.

1.5 Honor property rights including copyrights and patent.

Violation of copyrights, patents, trade secrets and the terms of license agreements is prohibited by law in most circumstances. Even when software is not so protected, such violations are contrary to professional behavior. Copies of software should be made only with proper authorization. Unauthorized duplication of materials must not be condoned.

1.6 Give proper credit for intellectual property.

Computing professionals are obligated to protect the integrity of intellectual property. Specifically, one must not take credit for other's ideas or work, even in cases where the work has not been explicitly protected by copyright, patent, etc.

1.7 Respect the privacy of others.

Computing and communication technology enables the collection and exchange of personal information on a scale unprecedented in the history of civilization. Thus there is increased potential for violating the privacy of individuals and groups. It is the responsibility of professionals to maintain the privacy and integrity of data describing individuals. This includes taking precautions to ensure the accuracy of data, as well as protecting it from unauthorized access or accidental disclosure to inappropriate individuals. Furthermore, procedures must be established to allow individuals to review their records and correct inaccuracies.

This imperative implies that only the necessary amount of personal information be collected in a system, that retention and disposal periods for that information be clearly defined and enforced, and that personal information gathered for a specific purpose not be used for other purposes without consent of the individual(s).

These principles apply to electronic communications, including electronic mail, and prohibit procedures that capture or monitor electronic user data, including messages, without the permission of users or bona fide authorization related to system operation and maintenance. User data observed during the normal duties of system operation and maintenance must be treated with strictest confidentiality, except in cases where it is evidence for the violation of law, organizational regulations, or this Code. In these cases, the nature or contents of that information must be disclosed only to proper authorities.

1.8 Honor confidentiality.

The principle of honesty extends to issues of confidentiality of information whenever one has made an explicit promise to honor confidentiality or, implicitly, when private information not directly related to the performance of one's duties becomes available. The ethical concern is to respect all obligations of confidentiality to employers, clients, and users unless discharged from such obligations by requirements of the law or other principles of this Code.

2. MORE SPECIFIC PROFESSIONAL RESPONSIBILITIES.

• • •

As an ACM computing professional I will . . .

2.1 Strive to achieve the highest quality, effectiveness and dignity in both the process and products of professional work.

Excellence is perhaps the most important obligation of a professional. The computing professional must strive to achieve quality and to be cognizant of the serious negative consequences that may result from poor quality in a system.

2.2 Acquire and maintain professional competence.

Excellence depends on individuals who take responsibility for acquiring and maintaining professional competence. A professional must participate in setting standards for appropriate levels of competence, and strive to achieve those standards. Upgrading technical knowledge and competence can be achieved in several ways: doing independent study; attending seminars, conferences, or courses; and being involved in professional organizations.

2.3 Know and respect existing laws pertaining to professional work.

ACM members must obey existing local, state, province, national, and international laws unless there is a compelling ethical basis not to do so. Policies and procedures of the organizations in which one participates must also be obeyed. But compliance must be balanced with the recognition that sometimes existing laws and rules may be immoral or inappropriate and, therefore, must be challenged. Violation of a law or regulation may be ethical when that law or rule has inadequate moral basis or when it conflicts with another law judged to be more important. If one decides to violate a law or rule because it is viewed as unethical, or for any other reason, one must fully accept responsibility for one's actions and for the consequences.

2.4 Accept and provide appropriate professional review.

Quality professional work, especially in the computing profession, depends on professional reviewing and critiquing. Whenever appropriate, individual members should seek and utilize peer review as well as provide critical review of the work of others.

2.5 Give comprehensive and thorough evaluations of computer systems and their impacts, including analysis of possible risks.

Computer professionals must strive to be perceptive, thorough, and objective when evaluating, recommending, and presenting system descriptions and alternatives. Computer professionals are in a position of special trust, and therefore have a special responsibility to provide objective, credible evaluations to employers, clients, users, and the public. When providing evaluations the professional must also identify any relevant conflicts of interest, as stated in *imperative 1.3.*

As noted in the discussion of *principle 1.2* on avoiding harm, any signs of danger from systems must be reported to those who have opportunity and/or responsibility to resolve them. See the guidelines for *imperative 1.2* for more details concerning harm, including the reporting of professional violations.

2.6 Honor contracts, agreements, and assigned responsibilities.

Honoring one's commitments is a matter of integrity and honesty. For the computer professional this includes ensuring that system elements perform as intended. Also, when one contracts for work with another party, one has an obligation to keep that party properly informed about progress toward completing that work.

A computing professional has a responsibility to request a change in any assignment that he or she feels cannot be completed as defined. Only after serious consideration and with full disclosure of risks and concerns to the employer or client, should one accept the assignment. The major underlying principle here is the obligation to accept personal accountability for professional work. On some occasions other ethical principles may take greater priority.

A judgment that a specific assignment should not be performed may not be accepted. Having clearly identified one's concerns and reasons for that judgment, but failing to procure a change in that assignment, one may yet be obligated, by contract or by law, to proceed as directed. The computing professional's ethical judgment should be the final guide in deciding whether or not to proceed. Regardless of the decision, one must accept the responsibility for the consequences.

However, performing assignments "against one's own judgment" does not relieve the professional of responsibility for any negative consequences.

2.7 Improve public understanding of computing and its consequences.

Computing professionals have a responsibility to share technical knowledge with the public by encouraging understanding of computing, including the impacts of computer systems and their limitations. This imperative implies an obligation to counter any false views related to computing.

2.8 Access computing and communication resources only when authorized to do so.

Theft or destruction of tangible and electronic property is prohibited by *imperative 1.2*—"Avoid harm to others." Trespassing and unauthorized use of a computer or communication system is addressed by this imperative. Trespassing includes accessing communication networks and computer systems, or accounts and/or files associated with those systems, without explicit authorization to do so. Individuals and organizations have the right to restrict access to their systems so long as they do not violate the discrimination principle (see *1.4*). No one should enter or use another's computer system, software, or data files without permission. One must always have appropriate approval before using system resources, including communication ports, file space, other system peripherals, and computer time.

3. ORGANIZATIONAL LEADERSHIP IMPERATIVES.

• • •

As an ACM member and an organizational leader, I will . . .

BACKGROUND NOTE: This section draws extensively from the draft IFIP Code of Ethics, especially its sections on organizational ethics and international concerns. The ethical obligations of organizations tend to be neglected in most codes of professional conduct, perhaps because these codes are written from the perspective of the individual member. This dilemma is addressed by stating these imperatives from the perspective of the organizational leader. In this context "leader" is viewed as any organizational member who has leadership or educational responsibilities. These imperatives generally may apply to organizations as well as their leaders. In this context "organizations" are corporations, government agencies, and other "employers," as well as volunteer professional organizations.

3.1 Articulate social responsibilities of members of an organizational unit and encourage full acceptance of those responsibilities.

Because organizations of all kinds have impacts on the public, they must accept responsibilities to society. Organizational procedures and attitudes oriented toward quality and the welfare of society will reduce harm to members of the public, thereby serving public interest and fulfilling social responsibility. Therefore, organizational leaders must encourage full participation in meeting social responsibilities as well as quality performance.

3.2 Manage personnel and resources to design and build information systems that enhance the quality of working life.

Organizational leaders are responsible for ensuring that computer systems enhance, not degrade, the quality of working life. When implementing a computer system, organizations must consider the personal and professional development, physical safety, and human dignity of all workers. Appropriate human-computer ergonomic standards should be considered in system design and in the workplace.

3.3 Acknowledge and support proper and authorized uses of an organization's computing and communication resources.

Because computer systems can become tools to harm as well as to benefit an organization, the leadership has the responsibility to clearly define appropriate and inappropriate uses of organizational computing resources. While the number and scope of such rules should be minimal, they should be fully enforced when established.

3.4 Ensure that users and those who will be affected by a system have their needs clearly articulated during the assessment and design of requirements; later the system must be validated to meet requirements.

Current system users, potential users and other persons whose lives may be affected by a system must have their needs assessed and incorporated in the statement of requirements. System validation should ensure compliance with those requirements.

3.5 Articulate and support policies that protect the dignity of users and others affected by a computing system.

Designing or implementing systems that deliberately or inadvertently demean individuals or groups is ethically unacceptable. Computer professionals who are in decision-making positions should verify that systems are designed and implemented to protect personal privacy and enhance personal dignity.

3.6 Create opportunities for members of the organization to learn the principles and limitations of computer systems.

This complements the imperative on public understanding (2.7). Educational opportunities are essential to facilitate optimal participation of all organizational members. Opportunities must be available to all members to help them improve their knowledge and skills in computing, including courses that familiarize them with the consequences and limitations of particular types of systems. In particular, professionals must be made aware of the dangers of building systems around oversimplified models, the improbability of anticipating and designing for every possible operating condition, and other issues related to the complexity of this profession.

4. COMPLIANCE WITH THE CODE.

•••

As an ACM member I will . . .

4.1 Uphold and promote the principles of this Code.

The future of the computing profession depends on both technical and ethical excellence. Not only is it important for ACM computing professionals to adhere to the principles expressed in this Code, each member should encourage and support adherence by other members.

4.2 Treat violations of this code as inconsistent with membership in the ACM.

Adherence of professionals to a code of ethics is largely a voluntary matter. However, if a member does not follow this code by engaging in gross misconduct, membership in ACM may be terminated.

This Code and the supplemental Guidelines were developed by the Task Force for the Revision of the ACM Code of Ethics and

Professional Conduct: Ronald E. Anderson, Chair, Gerald Engel, Donald Gotterbarn, Grace C. Hertlein, Alex Hoffman, Bruce Jawer, Deborah G. Johnson, Doris K. Lidtke, Joyce Currie Little, Dianne Martin, Donn B. Parker, Judith A. Perrolle, and Richard S. Rosenberg. The Task Force was organized by ACM/SIGCAS and funding was provided by the ACM SIG Discretionary Fund. This Code and the supplemental Guidelines were adopted by the ACM Council on October 16, 1992.

SOFTWARE ENGINEERING CODE OF ETHICS AND PROFESSIONAL PRACTICE

•••

IEEE-CS/ACM Joint Task Force on Software Engineering Ethics and Professional Practices

PREAMBLE

•••

Computers have a central and growing role in commerce, industry, government, medicine, education, entertainment and society at large. Software engineers are those who contribute by direct participation or by teaching, to the analysis, specification, design, development, certification, maintenance and testing of software systems. Because of their roles in developing software systems, software engineers have significant opportunities to do good or cause harm, to enable others to do good or cause harm, or to influence others to do good or cause harm. To ensure, as much as possible, that their efforts will be used for good, software engineers must commit themselves to making software engineering a beneficial and respected profession. In accordance with that commitment, software engineers shall adhere to the following Code of Ethics and Professional Practice.

The Code contains eight Principles related to the behavior of and decisions made by professional software engineers, including practitioners, educators, managers, supervisors and policy makers, as well as trainees and students of the profession. The Principles identify the ethically responsible relationships in which individuals, groups, and organizations participate and the primary obligations within these relationships. The Clauses of each Principle are illustrations of some of the obligations included in these relationships. These obligations are founded in the software engineer's humanity, in special care owed to people affected by the work of software engineers, and in the unique elements of the practice of software engineering. The Code prescribes these as obligations of anyone claiming to be or aspiring to be a software engineer.

It is not intended that the individual parts of the Code be used in isolation to justify errors of omission or commission. The list of Principles and Clauses is not exhaustive. The Clauses should not be read as separating the acceptable from the unacceptable in professional conduct in all practical situations. The Code is not a simple ethical algorithm that generates ethical decisions. In some situations, standards may be in tension with each other or with standards from other sources. These situations require the software engineer to use ethical judgment to act in a manner which is most consistent with the spirit of the Code of Ethics and Professional Practice, given the circumstances.

Ethical tensions can best be addressed by thoughtful consideration of fundamental principles, rather than blind reliance on detailed regulations. These Principles should influence software engineers to consider broadly who is affected by their work; to examine if they and their colleagues are treating other human beings with due respect; to consider how the public, if reasonably well informed, would view their decisions; to analyze how the least empowered will be affected by their decisions; and to consider whether their acts would be judged worthy of the ideal professional working as a software engineer. In all these judgments concern for the health, safety and welfare of the public is primary; that is, the "Public Interest" is central to this Code.

The dynamic and demanding context of software engineering requires a code that is adaptable and relevant to new situations as they occur. However, even in this generality, the Code provides support for software engineers and managers of software engineers who need to take positive action in a specific case by documenting the ethical stance of the profession. The Code provides an ethical foundation to which individuals within teams and the team as a whole can appeal. The Code helps to define those actions that are ethically improper to request of a software engineer or teams of software engineers.

The Code is not simply for adjudicating the nature of questionable acts; it also has an important educational function. As this Code expresses the consensus of the profession on ethical issues, it is a means to educate both the public and aspiring professionals about the ethical obligations of all software engineers.

PRINCIPLES

• • •

Principle 1 PUBLIC: Software engineers shall act consistently with the public interest. In particular, software engineers shall, as appropriate:

1.01. Accept full responsibility for their own work.

1.02. Moderate the interests of the software engineer, the employer, the client and the users with the public good.

1.03. Approve software only if they have a well-founded belief that it is safe, meets specifications, passes appropriate tests, and does not diminish quality of life, diminish privacy or harm the environment. The ultimate effect of the work should be to the public good.

1.04. Disclose to appropriate persons or authorities any actual or potential danger to the user, the public, or the environment, that they reasonably believe to be associated with software or related documents.

1.05. Cooperate in efforts to address matters of grave public concern caused by software, its installation, maintenance, support or documentation.

1.06. Be fair and avoid deception in all statements, particularly public ones, concerning software or related documents, methods and tools.

1.07. Consider issues of physical disabilities, allocation of resources, economic disadvantage and other factors that can diminish access to the benefits of software.

1.08. Be encouraged to volunteer professional skills to good causes and to contribute to public education concerning the discipline.

Principle 2 CLIENT AND EMPLOYER: Software engineers shall act in a manner that is in the best interests of their client and employer, consistent with the public interest. In particular, software engineers shall, as appropriate:

2.01. Provide service in their areas of competence, being honest and forthright about any limitations of their experience and education.

2.02. Not knowingly use software that is obtained or retained either illegally or unethically.

2.03. Use the property of a client or employer only in ways properly authorized, and with the client's or employer's knowledge and consent.

2.04. Ensure that any document upon which they rely has been approved, when required, by someone authorized to approve it.

2.05. Keep private any confidential information gained in their professional work, where such confidentiality is consistent with the public interest and consistent with the law.

2.06. Identify, document, collect evidence and report to the client or the employer promptly if, in their opinion, a project is likely to fail, to prove too expensive, to violate intellectual property law, or otherwise to be problematic.

2.07. Identify, document, and report significant issues of social concern, of which they are aware, in software or related documents, to the employer or the client.

2.08. Accept no outside work detrimental to the work they perform for their primary employer.

2.09. Promote no interest adverse to their employer or client, unless a higher ethical concern is being compromised; in that case, inform the employer or another appropriate authority of the ethical concern.

Principle 3 PRODUCT: Software engineers shall ensure that their products and related modifications meet the highest professional standards possible. In particular, software engineers shall, as appropriate:

3.01. Strive for high quality, acceptable cost, and a reasonable schedule, ensuring significant trade-offs are clear to and accepted by the employer and the client, and are available for consideration by the user and the public.

3.02. Ensure proper and achievable goals and objectives for any project on which they work or propose.

3.03. Identify, define and address ethical, economic, cultural, legal and environmental issues related to work projects.

3.04. Ensure that they are qualified for any project on which they work or propose to work, by an appropriate combination of education, training, and experience.

3.05. Ensure that an appropriate method is used for any project on which they work or propose to work.

3.06. Work to follow professional standards, when available, that are most appropriate for the task at hand, departing from these only when ethically or technically justified.

3.07. Strive to fully understand the specifications for software on which they work.

3.08. Ensure that specifications for software on which they work have been well documented, satisfy the users' requirements and have the appropriate approvals.

3.09. Ensure realistic quantitative estimates of cost, scheduling, personnel, quality and outcomes on any project on which they work or propose to work and provide an uncertainty assessment of these estimates.

3.10. Ensure adequate testing, debugging, and review of software and related documents on which they work.

3.11. Ensure adequate documentation, including significant problems discovered and solutions adopted, for any project on which they work.

3.12. Work to develop software and related documents that respect the privacy of those who will be affected by that software.

3.13. Be careful to use only accurate data derived by ethical and lawful means, and use it only in ways properly authorized.

3.14. Maintain the integrity of data, being sensitive to outdated or flawed occurrences.

3.15. Treat all forms of software maintenance with the same professionalism as new development.

Principle 4 JUDGMENT: Software engineers shall maintain integrity and independence in their professional judgment. In particular, software engineers shall, as appropriate:

4.01. Temper all technical judgments by the need to support and maintain human values.

4.02. Only endorse documents either prepared under their supervision or within their areas of competence and with which they are in agreement.

4.03. Maintain professional objectivity with respect to any software or related documents they are asked to evaluate.

4.04. Not engage in deceptive financial practices such as bribery, double billing, or other improper financial practices.,/item>

4.05. Disclose to all concerned parties those conflicts of interest that cannot reasonably be avoided or escaped.

4.06. Refuse to participate, as members or advisors, in a private, governmental or professional body concerned with software related issues, in which they, their employers or their clients have undisclosed potential conflicts of interest.

Principle 5 MANAGEMENT: Software engineering managers and leaders shall subscribe to and promote an ethical approach to the management of software development and maintenance. In particular, those managing or leading software engineers shall, as appropriate:

5.01 Ensure good management for any project on which they work, including effective procedures for promotion of quality and reduction of risk.

5.02. Ensure that software engineers are informed of standards before being held to them.

5.03. Ensure that software engineers know the employer's policies and procedures for protecting passwords, files and information that is confidential to the employer or confidential to others.

5.04. Assign work only after taking into account appropriate contributions of education and experience tempered with a desire to further that education and experience.

5.05. Ensure realistic quantitative estimates of cost, scheduling, personnel, quality and outcomes on any project on which they work or propose to work, and provide an uncertainty assessment of these estimates.

5.06. Attract potential software engineers only by full and accurate description of the conditions of employment.

5.07. Offer fair and just remuneration.

5.08. Not unjustly prevent someone from taking a position for which that person is suitably qualified.

5.09. Ensure that there is a fair agreement concerning ownership of any software, processes, research, writing, or other intellectual property to which a software engineer has contributed.

5.10. Provide for due process in hearing charges of violation of an employer's policy or of this Code.

5.11. Not ask a software engineer to do anything inconsistent with this Code.

5.12. Not punish anyone for expressing ethical concerns about a project.

Principle 6 PROFESSION: Software engineers shall advance the integrity and reputation of the profession consistent with the public interest. In particular, software engineers shall, as appropriate:

6.01. Help develop an organizational environment favorable to acting ethically.

6.02. Promote public knowledge of software engineering.

6.03. Extend software engineering knowledge by appropriate participation in professional organizations, meetings and publications.

6.04. Support, as members of a profession, other software engineers striving to follow this Code.

6.05. Not promote their own interest at the expense of the profession, client or employer.

6.06. Obey all laws governing their work, unless, in exceptional circumstances, such compliance is inconsistent with the public interest.

6.07. Be accurate in stating the characteristics of software on which they work, avoiding not only false claims but also claims that might reasonably be supposed to be speculative, vacuous, deceptive, misleading, or doubtful.

6.08. Take responsibility for detecting, correcting, and reporting errors in software and associated documents on which they work.

6.09. Ensure that clients, employers, and supervisors know of the software engineer's commitment to this Code of ethics, and the subsequent ramifications of such commitment.

6.10. Avoid associations with businesses and organizations which are in conflict with this code.

6.11. Recognize that violations of this Code are inconsistent with being a professional software engineer.

6.12. Express concerns to the people involved when significant violations of this Code are detected unless this is impossible, counter-productive, or dangerous.

6.13. Report significant violations of this Code to appropriate authorities when it is clear that consultation with people involved in these significant violations is impossible, counter-productive or dangerous.

Principle 7 COLLEAGUES: Software engineers shall be fair to and supportive of their colleagues. In particular, software engineers shall, as appropriate:

7.01. Encourage colleagues to adhere to this Code.

7.02. Assist colleagues in professional development.

7.03. Credit fully the work of others and refrain from taking undue credit.

7.04. Review the work of others in an objective, candid, and properly-documented way.

7.05. Give a fair hearing to the opinions, concerns, or complaints of a colleague.

7.06. Assist colleagues in being fully aware of current standard work practices including policies and procedures for protecting passwords, files and other confidential information, and security measures in general.

7.07. Not unfairly intervene in the career of any colleague; however, concern for the employer, the client or public interest may compel software engineers, in good faith, to question the competence of a colleague.

7.08. In situations outside of their own areas of competence, call upon the opinions of other professionals who have competence in that area.

Principle 8 SELF: Software engineers shall participate in lifelong learning regarding the practice of their profession and shall promote an ethical approach to the practice of the profession. In particular, software engineers shall continually endeavor to:

8.01. Further their knowledge of developments in the analysis, specification, design, development, maintenance and testing of software and related documents, together with the management of the development process.

8.02. Improve their ability to create safe, reliable, and useful quality software at reasonable cost and within a reasonable time.

8.03. Improve their ability to produce accurate, informative, and well-written documentation.

8.04. Improve their understanding of the software and related documents on which they work and of the environment in which they will be used.

8.05. Improve their knowledge of relevant standards and the law governing the software and related documents on which they work.

8.06. Improve their knowledge of this Code, its interpretation, and its application to their work.

8.07. Not give unfair treatment to anyone because of any irrelevant prejudices.

8.08. Not influence others to undertake any action that involves a breach of this Code.

8.09. Recognize that personal violations of this Code are inconsistent with being a professional software engineer.

This Code was developed by the IEEE-CS/ACM joint task force on Software Engineering Ethics and Professional Practices (SEEPP):

Executive Committee: Donald Gotterbarn (Chair), Keith Miller and Simon Rogerson;

Members: Steve Barber, Peter Barnes, Ilene Burnstein, Michael Davis, Amr El-Kadi, N. Ben Fairweather, Milton Fulghum, N. Jayaram, Tom Jewett, Mark Kanko, Ernie Kallman, Duncan Langford, Joyce Currie Little, Ed Mechler, Manuel J. Norman, Douglas Phillips, Peter Ron Prinzivalli, Patrick Sullivan, John Weckert, Vivian Weil, S. Weisband and Laurie Honour Werth.

©1999 by the Institute of Electrical and Electronics Engineers, Inc. and the Association for Computing Machinery, Inc.

TEN COMMANDMENTS OF COMPUTER ETHICS OF THE COMPUTER ETHICS INSTITUTE

• • •

1. Thou shalt not use a computer to harm other people.

2. Thou shalt not interfere with other people's computer work.

3. Thou shalt not snoop around in other people's computer files.

4. Thou shalt not use a computer to steal.

5. Thou shalt not use a computer to bear false witness.

6. Thou shalt not copy or use proprietary software for which you have not paid.

7. Thou shalt not use other people's computer resources without authorization or proper compensation.

8. Thou shalt not appropriate other people's intellectual output.

9. Thou shalt think about the social consequences of the program you are writing or the system you are designing.

10. Thou shalt always use a computer in ways that ensure consideration and respect for your fellow humans.

Written by Ramon Barguin, pres., Computer Ethics Institute

ETHICS CODES IN PROFESSIONAL ENGINEERING: OVERVIEW AND COMPARISONS

•••

The development of ethics codes in professional engineering began in the late 1800s and continues into the present. It has been influenced by the development of ethics codes in other professions, especially medicine and law, but exhibits its own dynamics and characteristics. This historical dynamics is particularly apparent in the United States, in a movement toward responsibility for public safety, health, and welfare. Outside the United States the movement is not as well documented, but modest comparisons can be made between professional engineering codes in different countries.

Engineering Ethics Codes in General

A code of ethics—also known as a code of conduct—is the public expression of guidelines for behavior by a professional organization enforced in some manner by that organization. A professional code is, as it were, regionalized legislation. What law—as a set of rules for behavior articulated and enforced by the state—does for society as a whole, so codes of ethics do for what Alexis de Tocqueville referred to as public associations (*Democracy in America*, vol. II, book 2, chapter 5) and are now called non-governmental organizations (NGOs).

Thus in order for there to be engineering ethics codes there must first be organized associations of engineers. But as the comparison with law also suggests, this is a necessary but not sufficient condition for engineering ethics codes. There are states that are governed by custom or tradition rather than by law. Just as law is often preceded (and complemented) by more informal and even unconscious mores and social norms, so among engineers it might be that the general function served by a code of ethics could be met (as well as complemented) by more implicit social mores.

The comparison invites further consideration of the possibility of diverse forms of professional organization and diverse relationships between professional associations and ethics codes. Complementing comparative government is comparative professional ethics. One aspect of this comparison would have to include consideration of the relation between various engineering standards, "building codes" or "construction codes," and ethics. For instance, one can postulate an inverse relationship between construction and ethics codes. When construction codes are detailed and explicit, ethics codes can be correspondingly ambiguous, whereas when construction codes are loose or non-existent, the engineering would depend on a high degree of moral dedication not to cut corners.

ENGINEERING ASSOCIATIONS. Engineering associations arose during the eighteenth century in two distinct contexts. In the first they arose within the government as formal organizations of those military personnel especially trained to design and operate "engines of war" (hence the term "engineers") and fortifications. In 1716 in France state service took civilian form as the Corps des Ponts et Chaussées; three decades later, for the more effective training of manpower for this corps, there was established the famous Ecole des Ponts et Chaussées (1747). This was followed by the Ecole des Mines (1783) and the Ecole Polytechnique (1794), the latter founded to train officers for the French revolutionary army. (For a general assessment of the complexities of professional engineering in France, including reference to engineering ethics, see Didier 1999.)

In a second instance engineers came together in informal associations independent of government. In England in the late 1700s, John Smeaton, architect of the Eddystone Lighthouse, and colleagues "were accustomed to dine together every fortnight at the Crown and Anchor in the Strand, spending the evening in conversation on engineering subjects" (Smiles, 1861–1862, vol. 2, p. 474). This led to the informal formation of a club called The Society of Civil Engineers, the term "civil engineer" having been coined by Smeaton in 1768 to distinguish those engineers who were not soldiers. It was not until 1828 that this club was incorporated under Royal Charter as the Institute of Civil Engineers.

The implications of these two origins are quite different. In the French system the education or the school, established by the government, has been primary. One becomes an engineer by earning the special academic degree of engineer, and is then entitled to be called "engineer." The professional organization of such engineers is either the bureau or agency for which one works or some kind of alumni association. Under such circumstances a code of ethics stresses governmental or state service and can afford to be largely implicit.

In the British system the professional association is primary. One becomes an engineer not by earning a special academic degree but by meeting the standards for joining a professional organization. Indeed, academic courses of instruction in engineering are not set up in England until the early 1800s, and the engineering degree does not have its wholly unique curriculum but is simply specified as a kind of bachelor degree.

Under such circumstances it has been found more necessary to formulate an explicit code of ethics, which has tended to stress promotion of the profession over governmental service. For instance, the Royal Charter of the British Institution of Mechanical Engineers (founded 1847), gives the aims of the organization as "to encourage invention and research," "to hold meetings," "to print, publish and distribute the proceedings," and "to co-operate with universities" in "matters connected with mechanical engineering." According to the By-Laws of the Institution, members should conduct themselves "in order to facilitate the advancement of the science of mechanical engineering by preserving the respect in which the community holds persons who are engaged in the professional of mechanical engineering." In other words, the primary obligation of engineers is to the engineering profession rather than the state.

Engineering Ethics in the United States

Although the first institution of higher education in the to grant engineering degrees in the United States was the Military Academy at West Point (founded 1802), non-military engineering schools rapidly superseded it in influence, and the U.S. has largely followed the British model in its professional engineering organizations. The engineering degree is simply one type of the bachelor's degree, and to be a professional engineer is effectively constituted by membership in a professional engineering association such as the American Society for Civil Engineers (ASCE, founded 1852) or the American Society for Mechanical Engineers (ASME, founded 1880). (For background on the history of the engineering profession in the U.S., see Layton 1971.)

The professional codes of these associations initially highlighted professional loyalty and—no doubt reflecting a unique commitment to capitalistic enterprise—especially loyalty to a client or employer (and for most engi-

neers, it was an employer). For instance, the 1914 code of the ASME listed the first duty of the engineer to be a "faithful agent or trustee" of some employing client or corporation. Although Michael Davis (2002) has contested a too literal reading of this requirement, the ASME Committee on Code of Ethics (1915) in a contemporaneous commentary emphasized "protection of a client's or employer's interests" as an engineer's "first obligation." At the same time, engineers should also "endeavor to assist the public to a fair and correct general understanding of engineering matters." But across the twentieth century engineering educators returned repeatedly to the difficulties of communicating to engineers a broad conception of their professional responsibilities.

Following World War II, and especially during the 1970s, engineering ethics codes in the United States were subject to considerable discussion and revision to reflect a new awareness of and commitment not just to public education but to public welfare. The background of this new ferment regarding engineering ethics was concern over the enormous powers engineers now exercised, and public concern about a number of specifically technical catastrophes as well as environmental degradation associated with technical engineering developments. Well-known examples were the DC-10 crashes and Ford Pinto car accidents caused by designs that companies refused to correct because of economic constraints, even though engineers called them to attention.

Such experiences led to the development of a new category of technical hero, the "whistle blower" who transgresses company loyalty and goes public with allegations of wrong doing. Here one influential example involved the case of three engineers—Holger Hjortsvang, Max Blankenzee, and Robert Bruder—who, while working on the San Francisco Bay Area Rapid Transit (BART) in 1972, came to the conclusion that the system was unsafe. When the contractor refused to heed their warnings, they appealed to an oversight board and were fired. But the California Society of Professional Engineers supported them, and indeed a few months later a train had an accident of exactly the kind they had predicted.

Subsequent examples included Richard Parks blowing of the whistle (in 1983) on unsafe practices in the clean-up of the Three Mile Island nuclear disaster, and Roger Boisjolay's exposure of the warnings given to Morton-Thiokol and NASA before the launch of the space shuttle Challenger (of 1986). During the 1980s the Accreditation Board for Engineering and Technology also began to require that engineering programs include engineering ethics in their curricula (Stephan

2002), perhaps in part as a result of such cases and the problems they created for practicing engineers.

Engineering Ethics Outside the United States

Although engineering ethics developments in the United States have taken place independent of contact with developments in other countries, the problems with which United States engineers have been trying to deal transcend national boundaries. Moreover, engineering ethics outside the United States can provide welcome new perspectives for U.S. engineers while profiting as well from U.S. achievements.

For example, ethics codes in Canada and in Australia provide other important variations on the British model that have nevertheless historically placed stronger weight on public responsibility. In Europe there exists a multinational and transdiciplinary to develop an approach to engineering ethics that offers an alternative to the standard U.S. case study, individual responsibility emphasis (see Goujon and Hériard Dubreuil 2001). It is also the case that various transnational professional engineering associations such as the Unión Panaméricana de Asociaciones de Ingenieros (UPADI or Pan American Union of Associations of Engineers) and the Européenne d'Associations Nationales d'Ingénieurs (FEANI or European Federation of National Engineering Associations) are making important contributions to engineering ethics. For present purposes, however, and as a general introduction to the collection of professional ethics codes that follow, it is sufficient to consider six national cases of some particular interest: Germany, Japan, Hong Kong, Sweden, the Dominican Republic, and Chile. (For more on code developments in other countries, see Davis1990, 1991, and 1992.)

Philosophical Engineering Ethics in Germany

The development of engineering ethics in Germany has a much more developed theoretical base than in the United States. Nineteenth and early twentieth century attitudes toward engineering were influenced by philosophers such as Immanuel Kant and G. W. F. Hegel, and by the German notion of education as *Bildung*, formation or growth, understood as the perfection of human nature through culture. Ernst Kapp and Friedrich Dessauer, for instance, argued that like the classics and the humanities, the experience of technological creativity could contribute to the development of a higher moral consciousness.

Immediately after World War II, however, because some of its members had been compromised by involvement with National Socialism, the Verein Deutscher Ingenieure (VDI or Association of German Engineers) developed its first explicit ethics code, the "Engineer's Confessions," which exhibited a distinctly religious character. It also undertook to promote a new philosophical reflection among engineers by establishing a *Mensch und Technik* [Humans and technology] committee, an initiative that has led to a more sustained dialogue between engineers and philosophers than in any other country.

During the 1960s and 1980s the discussion of technology and philosophy became a publicly debated issue. The role of the engineer and the impact on society was discussed during an international conference organized by the German Commission for UNESCO. The public became involved and concerns for the environment were brought up and discussed by committees and groups throughout the country. In 1980 the VDI wrote "Future Tasks" which discussed societal, political, and ethical goals such as improving the possibilities for life, as well as what was technically possible.

This work in turn led to replacement of what had become the dated "Engineer's Confessions" and to further interdisciplinary engineering-philosophy research, especially on the theoretical basis of technology assessment. With regard to professional ethics, one *Mensch und Technik* working committee report in 1980 proposed simply that "The aim of all engineers is the improvement of the possibilities of life for all humanity by the development and appropriate application of technical means." With regard to the foundations of technology assessment, a second working committee in 1986 identified eight fields of value (environmental quality, health, safety, functionality, economics, living standards, personal development, and social quality), mapped out their interrelations, and developed a draft set of recommendations for their implementation in the design of technical products and projects. (For a more extended discussion of these developments, see Huning and Mitcham 1993.)

In 2002, no doubt with influence from movements toward globalization, a new generation of philosophers and engineers simplified the VDI ethics code, and stressed raising ethics awareness and conflict resolution at the levels of both individual practice and oppositions between principles. "In the case of conflicting values," engineers are encouraged to give priority "to the values of humanity over the dynamics of nature, to issues of human rights over technology implementation and exploitation, to public welfare over private interests, and to safety and security over functionality and profitability."

Science and Engineering Ethics Combined in Japan

Engineering ethics codes in Japan, the second major World War II defeated power, exhibited a quite different genesis. To begin with, engineering became professionally organized only after World War II and did so in much closer association with science. In Japan science and engineering have not been treated as much as separate enterprises as they have been in Europe or the United States.

Moreover, the first and most influential code-like document is the "Statement on Atomic Research in Japan" issued by the Japanese Science Council (which includes both scientists and engineers) in 1954. This statement sets forth what have become known as "The Three Principles for the Peaceful Use of Atomic Energy": All research shall be conducted with full openness to the public, shall be democratically administered, and shall be carried out under the autonomous control of the Japanese themselves.

As is readily apparent, these principles reflect the desire of Japanese during the 1950s to distance themselves from United States interests (recall that the Allied occupation ended in 1952) and policy. Immediately after World War II, the U.S. prohibited all Japanese research in aviation, atomic energy, and any other war-related area. But by 1951, following the Communist victory in China and the outbreak of the Korean War, U.S. policy began to shift toward encouraging certain kinds of military-related science and engineering and the incorporation of Japan into the Western alliance.

Indeed, Japanese scientists and engineers recognized that the Three Principles were in opposition to, for example, the U.S. policy of secrecy in atomic research, and in order to avoid publicity and the possible development of opposition, the JSC statement was not initially translated into English. It is also a policy which, although formulated by scientists and engineers themselves, was readily adopted by the government, thus perhaps reflecting the greater social prestige and political influence of the Japanese technical community in comparison with that in the United States.

Beginning in the 1980s scientists and engineers developed a new interest in ethics reflective of but with continuing distinctions from interests in the United States. This is illustrated, for instance, by the JSC declarations on "The Basic Principles of International Scientific Exchange" (1988) and "The New Science Scheme: Science for Society and the Fusion of Humanity and Natural Sciences" (2003), both of which have emphasized a responsibility on the part of scientists and engineers to promote sound scientific development and

to help educate the public about important issues related to scientific and technological development. In 1999 there was also established the Japan Accreditation Board for Engineering Education (JABEE), an agency that has given special attention to engineering ethics education.

Engineering Ethics as Institutional Protection in Hong Kong

Another special case in Asia that can be briefly mentioned is that of the Hong Kong Institution of Engineers (HKIE, founded 1947). As a British Crown Colony, the professional organization of engineers in Hong Kong originally developed not just on the British model but as a branch of British institutions. With the realization that Hong Kong would in the near future (in 1997) be returned to Chinese sovereignty, however, local engineers in the 1970s began to provide Hong Kong with a truly independent engineering association. Part of this activity involved some intensive discussion of professional ethics, with a special conference being organized in 1980 on "Professional Ethics in the Modern World."

In 1994 the HKIE formally adopted a set of "Rules of Conduct" that differed in a few key respects from the parent organization. Although the primary obligation remained the responsibility to the profession, this was modified by the following statement: "When working in a country other than Hong Kong [the Hong Kong engineer should] order his [or her] conduct according to the existing recognized standards of conduct in that country, except that he should abide by these rules as applicable in the absence of local standards."

The basis of this modification had been clearly spelled out in previous discussions. At an inter-professional symposium in December 1985, F.Y. Kan of the Hong Kong Institute of Surveyors identified the role of his professional association as the promotion of the status of surveyors and the usefulness of the profession. "So far," he is reported to have said,

> the role [has] not changed but, with the Sino-British agreement in operation [to return Hong Kong to Chinese sovereignty in 1992], there might be a tendency to a far-reaching effect on the professions. There was, therefore, a need to break away from U.K. qualifications. However, professional competence must be maintained and this could bring institutions into the political field. (Luscher, 1986, p. 39)

In a world in which engineering easily comes into contact with the political field—something that is increasingly likely to be the case not only in Hong

Kong—it is increasingly important for engineers to think about ethical issues, and to do with awareness of what is happening in their profession throughout the world.

Engineering Ethics as Social Reform in Sweden

In Europe there has also been some desire to establish professional engineering independence of various pressures from other nations. In this regard Sweden provides an instructive case study of a neutral country that used its engineering prowess to provide itself with a strong military by relying on a well-developed domestic weapons industry. One of the leading weapons producing corporations has been Bofors, a primary supplier of advanced field artillery, anti-aircraft artillery, and ship artillery to the Swedish armed forces. Known not only domestically but internationally for such technologies, in the 1960s Bofors increased its exports. In principle, exports of military weapons were prohibited. But the government can legally waive this restriction for special cases, which nevertheless became increasingly questionable.

An engineer named Ingvar Bratt began working for Bofors in 1969 and participated in projects including a missile and anti-aircraft gun which were delivered to Malaysia in 1977. During the 1970s and 1980s, however, Bratt became politically active, and by 1982 was publicly opposed all weapons exports, even approved ones.

Rumors arose that unapproved countries had acquired Bofors missile technology presumably through an approved third-party country. Bratt discovered evidence in a Bofors' office near his own that missiles had in fact been exported to Singapore. He shared this information with a journalist who that Singapore was an arms dealer. This suggested Singapore as a possible approved country through which the unapproved countries such as Dubai and Bahrain were receiving arms. In 1984 Bratt left Bofors and helped to pursue further evidence of these illegal activities.

This exposé contributed to development of a new code of engineering ethics, one that downplayed company loyalty, a focus of the previous code, and emphasized responsibility to "humanity, the environment, and society." In response to the view that engineers were often those who contributed to social or environmental problems, the new ethics codes stressed the social and ecological responsibility of engineers, promoting the idea that engineers might play a more positive role in society. (This section draws heavily on Welin 1991.)

Engineering Ethics to Resist Corruption in the Dominican Republic

Engineers ethics codes in developing countries provide still another point of comparison. Concern for engineering and ethics has emerged in the Dominican Republic in response to numerous engineering failures and catastrophes that have occurred there as a result of professional negligence.

Engineering, architecture, and surveying were first introduced to the Domincan Republic by Spanish conquistadores in the early 1500s. But engineering was not a formal course of study until the 1900s, and there was not much difference between engineers and architects until 1945 when the first engineering organization was formed.

During the 1980s many engineered structures failed, which led to increased calls for governmental regulation. But a civil engineer, Orlando Franco Batlle, also argued that part of the problem rested with a weak tradition in professional engineering ethics, and promoted new guidelines for the ethical and responsible exercise of the civil engineering profession. In this effort he was inspired by the code of the American Society of Civil Engineers.

The Colegio Dominicano de Ingenieros, Arquitectos y Agrimensores (CODIA or Dominican Association of Engineers, Architects, and Surveyors) had in the 1960s created a code of ethics to promote national interest and relationships within the profession and with clients. But there was no mention of public safety, health, or well-being. There was also no reference to responsibility or concern for the negative effects of engineering on society or the environment. Yet Franco Batlle's argument was unable to bring about a change in this code.

However, whether a reform of the professional ethics code would have any substantial impact on the problem of substandard work remained questionable. A survey in 1990 among CODIA members revealed that most had not even read the existing code, and if they had did not take it seriously. Engineering was thought to be simply the best paying job in the country, with medicine is the most prestigious. This implied that engineers had chosen their profession for economic benefit—and, in fact, two thirds of the engineering professors thought that societal interest was secondary to self-interest. (This section adapts research by César Cuello Nieto, 1992.)

Engineering Ethics as Alternative Development in Chile

A second comparison from the perspective of engineering ethics in a developing country is provided by Chile. In Chile, as in many countries other than the United States, professional codes such as the "Code of

Professional Ethics of the Engineers of the Colegio de Ingenieros de Chile [Association of Engineers of Chile]," actually have the force of law, as a result of having been formulated, in this case, in response to general legislation calling for such codes in all professional organizations. Although the Colegio was founded in 1958, its code was not formulated until it was required by the authoritarian regime of Augusto Pinochette (1973-1990). At the same time, as Marcos García de la Huerta (1991) has argued, Pinochette's two-decade dictatorship severely compromised almost all professional practices. This is a degradation that García de la Huerta has himself worked to overcome by publishing what is probably the first textbook on engineering ethics in Latin America (García de la Huerta and Mitcham 2001).

The Chilean code, like many others, includes little by way of positive guidance. There is, for instance, no mention of any responsibility to public safety, health, and welfare. Instead, the code consists primarily of an extended list of actions that are contrary to sound professional conduct, and that are thus punishable by professional censure. Among many unremarkable canons against conflict of interest, graft, and more, however, is one rejecting "actions or failures to act that favor or permit the unnecessary use of foreign engineering for objectives and work for which Chilean engineering is sufficient and adequate." Such a canon, emphasizing national interests, can also be found in other codes throughout Asia and Latin America, from India to Venezuela.

It is important to note that such a canon need not have simply nationalistic implications. Judith Sutz, for example, a computer scientist in Uruguay, in an essay raising important questions about the directions of information technology research in Latin America, argues that

> The basic question is, What do Latin American engineers want? Do they want to seek original solutions to indigenous problems? Or do they only want to identify with that which is more modern, more sophisticated, more powerful—disregarding real usefulness—in order to feel like they "live" in the developed world? (Sutz 1993, p. 304)

Many countries experience a serious difficulty in addressing their own real problems. Driven by what René Girard (1965) calls mimetic desire, engineers and scientists often devote themselves to high-tech research that brings international prestige rather than to less glamorous but more useful tasks. One serious challenge to professional engineering in the age of globalization will be the extent to which various national and cultural differences can be maintained in the face of such pressures.

Acknowledgments

This article has drawn heavily on a research grant from the National Science Foundation Grant (NSF #DIR 8721989), June 1988 to November 1990, with supplemental support from the Philosophy and Technology Studies Center at Polytechnic University, the Science, Technology, and Society Program of Pennsylvania State University. This research has been further extended by participation in the development of a Humanitarian Engineering Program at the Colorado School of Mines funded by a grant from the William and Flora Hewlett Foundation. Michelle DeBacker, a research assistant, has also made numerous contributions to the preparation of this introduction and the collection of codes.

CARL MITCHAM

BIBLIOGRAPHY

Committee on Code of Ethics, American Society of Mechanical Engineers. (1915). "Report of Committee on Code of Ethics," *American Society of Mechanical Engineers Transactions*, vol. 36, pp. 23–27.

Cuello Nieto, César. (1992). "Engineering and Ethics in the Dominican Republic." In Carl Mitcham, *Engineering Ethics thoughout the World: Introduction, Documentation, Commentary, and Bibliography*. University Park: Science, Technology, and Society Program, Pennsylvania State University, pp. II-2–II-18.

Davis, Michael. (1990). "Ethics Around the World," special theme issue, *Perspectives on the Professions*, vol. 10, no. 1 (August), pp. 1–10. Articles on professional ethics in Australia, Canada, the USSR, and Great Britain, with occasional mentions of engineering.

Davis, Michael. (1991). "Ethics Around the World, Part 2," special theme issue, *Perspectives on the Professions*, vol. 11, no. 1 (August), pp. 1–10. Articles on professional ethics in Argentina, Chile, Hong Kong, Sweden, New Zealand, Costa Rica, and Egypt, with occasional mentions of engineering.

Davis, Michael. (1992). "Ethics Around the World, Part 2," special theme issue, *Perspectives on the Professions*, vol. 12, no. 1 (August), pp. 1–10. Articles on professional ethics in Asia (China, India, and Japan), Japan, Albania, Eastern Europe, Peru, and Argentina, with occasional mentions of engineering.

Davis, Michael. (2002). "Three Myths about Codes of Engineers Ethics," in *Profession, Code, and Ethics*. Burlington, VT: Ashgate, pp. 121–131.

Didier, Christelle. (1999). "Engineering Ethics in France: A Historical Perspective," *Technology in Society*, vol. 21, no. 4, pp. 471–486.

García de la Huerta, Marcos. (1991). "The Ethical Codes of Dictatorship: Ethics in Chile," *Perspectives on the Professions*, vol. 11, no. 1 (August).

García de la Huerta, Marcos, and Carl Mitcham. (2001). *Laética en la profesión de inteniero: Ingeniería y ciudadanía*

[Ethics in the engineering profession: Engineering and citizenship]. Santiago, Chile: Departamento de Estudios Humanísticos, Facultad de Ciencias Físicas y Matemáticas, Universidad de Chile.

Girard, René. (1965). *Deceit, Desire, and the Novel: Self and Other in Literary Structure*. Trans. Yvonne Freccero. Baltimore: Johns Hopkins University Press.

Goujon, Philippe, and Bertrand Hérierd Dubreuil, eds. (2001). *Technology and Ethics: A European Quest for Responsible Engineering*. Leuven, Belgium: Peeters.

Layton, Edwin T. Jr. (1971). *The Revolt of the Engineers*. Cleveland: Press of Case Western Reserve University. Revised edition, Baltimore: Johns Hopkins University Press, 1986.

Luscher, D.S. (1986). "The Changing Role of the Professional Institutions in Hong Kong," *Journal of the Hong Kong Institution of Engineers*, vol. 14, no. 2 (February), pp. 39–40.

Smiles, Samuel. (1861–1862). *The Lives of the Engineers, With an Account of Their Principal Works, Comprising also a History of Inland Communication in Britain*. 3 vols. London: J. Murray, 1861–1862.

Stephan, Karl D. (2002). "All This and Engineering Too: A History of Accreditation Requirements," *IEEE Technology and Society Magazine*, vol. 21, no. 3 (Fall), pp. 8–15.

Sutz, Judith Sutz. (1993). "The Social Implications of Information Technologies: A Latin American Perspective." In Carl Mitcham, ed., *Philosophy and Technology, vol. 10: Spanish Language Contributions to the Philosophy of Technology* Boston: Kluwer.

Welin, Stellan. (1991). "Ethics in Sweden," *Perspectives on the Professions*, vol. 11, no. 1 (August).

ACCREDITATION BOARD FOR ENGINEERING AND TECHNOLOGY (ABET) CODE OF ETHICS

...

As approved by the Board of Directors on October 30, 1999

Preamble

The Accreditation Board for Engineering and Technology, Inc. (ABET) requires ethical conduct by each volunteer and staff member engaged in fulfilling the mission of ABET. The organization requires that every volunteer and staff member exhibit the highest standards of professionalism, honesty, and integrity. The services provided by ABET require impartiality, fairness, and equity. All persons involved with ABET activities must perform their duties under the highest standards of ethical behavior. It is the purpose of this document to detail the ethical standards under which we agree to operate.

The ABET Guidelines for Interpretation of the Canons

The ABET guidelines for interpretation of the Canons represent the objectives toward which its volunteers and staff members should strive. They are principles which those involved in accreditation activities can reference in specific situations. In addition, they provide interpretive guidance to the ABET Professional Development Committee.

1. ABET volunteers and staff members agree to accept responsibility in making accreditation decisions and credential evaluations consistent with approved criteria and the safety, health, and welfare of the public and to disclose promptly factors that may directly or indirectly conflict with these duties and/or may endanger the public. a). All those involved in ABET activities shall recognize that the lives, safety, health and welfare of the general public are dependent upon a pool of qualified graduate professionals to continue the work of their profession. b). Programs shall not receive accreditation that do not meet the criteria as set forth by the profession through ABET in the areas of engineering, technology, computing, and applied science. c). If ABET volunteers or staff members have knowledge of or reason to believe that an accredited program may be non-compliant with the appropriate criteria, they shall present such information to ABET in writing and shall cooperate with ABET in furnishing such further information or assistance as may be required. d). If credential evaluation staff members have reason to believe that the credentials submitted for evaluation are not authentic or information submitted in support of an evaluation is misleading, they shall cooperate with ABET or any other entities affected by this process to verify the validity of facts and proof the authenticity of the academic documents in question.

2. ABET volunteers and staff members agree to perform services only in areas of our competence. All those involved in ABET activities shall undertake accreditation assignments only when qualified by education and/or experience in the specific technical field involved.

3. ABET volunteers and staff members agree to act as faithful agents or trustees of ABET, avoiding conflicts of interest and disclosing them to affected parties when they exist. a). All those involved in ABET activities shall avoid all known conflicts of interest when representing ABET in any situation. b). They shall disclose all known or potential conflicts of interest that could influence or appear to influence their judgment or the quality of their services. c). They shall not serve as a consultant in accreditation matters to a program or institution while serving as a member or alternate of a commission or the Board of Directors. Program evaluators who have or will serve as consultants must disclose this to ABET per the ABET Conflict of Interest Policy and may not participate in any deliberations regarding ABET matters for that institution. d). They shall not undertake any assignments or take part in any discussions that would knowingly create a conflict of interest between them and ABET or between them and the institutions seeking programmatic accreditation. e). They shall not solicit

or accept gratuities, directly or indirectly, from programs under review for accreditation or from individuals/entities when credentials are under evaluation. f). They shall not solicit or accept any contribution, directly or indirectly, to influence the accreditation decision of programs or the outcome of credential evaluations.

4. ABET volunteers and staff members agree to keep confidential all matters relating to accreditation decisions and credential evaluations unless by doing so we endanger the public or are required by law to disclose information. a). All those involved in ABET activities shall treat information coming to them in the course of their assignments as confidential, and shall not use such information as a means of making personal profit under any circumstances. b). They shall not reveal confidential information or findings except as authorized or required by law or court order. c). They shall only reveal confidential information or findings in their entirety where required to do so and then only with the prior consent of ABET and the institution/programs involved.

5. ABET volunteers and staff members agree to issue either public or internal statements only in an objective and truthful manner. a). All those involved in ABET activities shall be objective and truthful in reports, statements or testimony. They shall include all relevant and pertinent information in such reports, statements, or testimony and shall avoid any act tending to promote their own interest at the expense of the integrity of the process. b). They shall issue no statements, criticisms, or arguments on accreditation matters which are inspired or paid for by an interested party, or parties, unless they preface their comments by identifying themselves, by disclosing the identities of the party or parties on whose behalf they are speaking, and by revealing the existence of any financial interest they may have in matters under discussion. c). They shall not use statements containing a misrepresentation of fact or omitting a material fact. d. They shall admit their own errors when proven wrong and refrain from distorting or altering the facts to justify their mistakes or decisions.

6. ABET volunteers and staff members agree to conduct ourselves honorably, responsibly, ethically, and lawfully so as to enhance the reputation, and usefulness of ABET. a). All those involved in accreditation activities and credential evaluations shall refrain from any conduct that deceives the public. b). They shall not falsify or permit misrepresenta-

tion of their, or their associates', academic or professional qualifications. c). They shall not maliciously or falsely, directly or indirectly, injure the professional reputation, prospects, practice or employment of another. If they believe others are guilty of unethical or illegal behavior, they shall present such information to the proper authority for action.

7. ABET volunteers and staff members agree to treat fairly all persons regardless of race, religion, gender, disability, age, national origin, marital status or political affiliation. All those involved in accreditation activities and credential evaluations shall act with fairness and justice to all parties.

8. ABET volunteers and staff members agree to assist colleagues and co-workers in their professional development and to support them in following this code of conduct. a). ABET will provide broad dissemination of these canons of conduct to its volunteers, staff, representative organizations, and other stakeholders impacted by accreditation and credential evaluations. b). ABET will provide training in the use and understanding of the Code of Conduct for all new volunteers and staff members. c). All those involved in accreditation matters and credential evaluations shall continue their professional development throughout their service with ABET and shall provide/participate in opportunities for the professional and ethical development of all stakeholders.

9. Through its Committee on Professional Development, ABET will provide a mechanism for the prompt and fair adjudication of alleged violations of the Code of Conduct. Persons found to be in violation of the ABET Code of Conduct may be subject to any of a number of sanctions including being declared ineligible for service in further activities on behalf of ABET.

Fundamental Canons

Now, therefore, as a volunteers and/or staff member of the Accreditation Board for Engineering and Technology, Inc., and/or its member societies and having read and understood the above stated Guidelines, I _____ do hereby commit myself to the highest ethical and professional conduct and agree:

1. to accept responsibility in making accreditation decisions and credential evaluations consistent with approved criteria and the safety, health, and welfare of the public and to disclose promptly, factors that may directly or indirectly conflict with these duites and/or may endanger the public;

2. to perform services only in areas of my competence;

3. to act as a faithful agent or trustee of ABET avoiding conflicts of interest and disclosing them to affected parties including but not limited to, ABET when they exist;

4. to keep confidential all matters relating to accreditation decisions and credential evaluations unless by doing so we harm the public or are required by law to disclose information;

5. to issue either public or internal statements only in an objective and truthful manner;

6. to conduct myself honorably, responsibly, ethically, and lawfully so as to enhance the reputation and effectiveness of ABET;

7. to treat fairly all persons regardless of race, religion, gender, disability, age, national origin, marital status, or political affiliation;

8. to assist colleagues and co-workers in their professional development and to support them in following this code of conduct;

9. to support a mechanism for the prompt and fair adjudication of alleged violations of these canons.

© ABET. Used with permission.

AMERICAN SOCIETY OF CIVIL ENGINEERS (ASCE) CODE OF ETHICS

• • •

Fundamental Principles

Engineers uphold and advance the integrity, honor and dignity of the engineering profession by:

(1) using their knowledge and skill for the enhancement of human welfare and the environment;

(2) being honest and impartial and serving with fidelity the public, their employers and clients;

(3) striving to increase the competence and prestige of the engineering profession; and

(4) supporting the professional and technical societies of their disciplines.

Fundamental Canons

(1) *Engineers shall* hold paramount the safety, health and welfare of the public and shall strive to comply with the principles of sustainable development3 in the performance of their professional duties.

(2) *Engineers shall* perform services only in areas of their competence.

(3) *Engineers shall* issue public statements only in an objective and truthful manner.

(4) *Engineers shall* act in professional matters for each employer or client as faithful agents or trustees, and shall avoid conflicts of interest.

(5) *Engineers shall* build their professional reputation on the merit of their services and shall not compete unfairly with others.

(6) *Engineers shall* act in such a manner as to uphold and enhance the honor, integrity, and dignity of the engineering profession.

(7) *Engineers shall* continue their professional development throughout their careers, and shall provide opportunities for the professional development of those engineers under their supervision.

Guidelines to Practice Under the Fundamental Canons of Ethics

CANON 1.

• • •

Engineers shall hold paramount the safety, health and welfare of the public and shall strive to comply with the principles of sustainable development in the performance of their professional duties.

(a) Engineers shall recognize that the lives, safety, health and welfare of the general public are dependent upon engineering judgments, decisions and practices incorporated into structures, machines, products, processes and devices.

(b) Engineers shall approve or seal only those design documents, reviewed or prepared by them, which are determined to be safe for public health and welfare in conformity with accepted engineering standards.

(c) Engineers whose professional judgment is overruled under circumstances where the safety, health and welfare of the public are endangered, or the principles of sustainable development ignored, shall inform their clients or employers of the possible consequences.

(d) Engineers who have knowledge or reason to believe that another person or firm may be in violation of any of the provisions of Canon 1 shall present such information to the proper authority in writing and shall cooperate with the proper authority in furnishing such further information or assistance as may be required.

(e) Engineers should seek opportunities to be of constructive service in civic affairs and work for the advancement of the safety, health and well-being of their communities, and the protection of the environment through the practice of sustainable development.

(f) Engineers should be committed to improving the environment by adherence to the principles of sustainable development so as to enhance the quality of life of the general public.

CANON 2.

• • •

Engineers shall perform services only in areas of their competence.

(a) Engineers shall undertake to perform engineering assignments only when qualified by education or experience in the technical field of engineering involved.

(b) Engineers may accept an assignment requiring education or experience outside of their own fields of competence, provided their services are restricted to those phases of the project in which they are qualified. All other phases of such project shall be performed by qualified associates, consultants, or employees.

(c) Engineers shall not affix their signatures or seals to any engineering plan or document dealing with subject matter in which they lack competence by virtue of education or experience or to any such plan or document not reviewed or prepared under their supervisory control.

CANON 3.

• • •

Engineers shall issue public statements only in an objective and truthful manner.

(a) Engineers should endeavor to extend the public knowledge of engineering and sustainable development, and shall not participate in the dissemination of untrue, unfair or exaggerated statements regarding engineering.

(b) Engineers shall be objective and truthful in professional reports, statements, or testimony. They shall include all relevant and pertinent information in such reports, statements, or testimony.

(c) Engineers, when serving as expert witnesses, shall express an engineering opinion only when it is founded upon adequate knowledge of the facts, upon a background of technical competence, and upon honest conviction.

(d) Engineers shall issue no statements, criticisms, or arguments on engineering matters which are inspired or paid for by interested parties, unless they indicate on whose behalf the statements are made.

(e) Engineers shall be dignified and modest in explaining their work and merit, and will avoid any act tending to promote their own interests at the expense of the integrity, honor and dignity of the profession.

CANON 4.

• • •

Engineers shall act in professional matters for each employer or client as faithful agents or trustees, and shall avoid conflicts of interest.

(a) Engineers shall avoid all known or potential conflicts of interest with their employers or clients and shall promptly inform their employers or clients of any business association, interests, or circumstances which could influence their judgment or the quality of their services.

(b) Engineers shall not accept compensation from more than one party for services on the same project, or for services pertaining to the same project, unless the circumstances are fully disclosed to and agreed to, by all interested parties.

(c) Engineers shall not solicit or accept gratuities, directly or indirectly, from contractors, their agents, or other parties dealing with their clients or employers in connection with work for which they are responsible.

(d) Engineers in public service as members, advisors, or employees of a governmental body or department shall not participate in considerations or actions with respect to services solicited or provided by them or their organization in private or public engineering practice.

(e) Engineers shall advise their employers or clients when, as a result of their studies, they believe a project will not be successful.

(f) Engineers shall not use confidential information coming to them in the course of their assignments as a means of making personal profit if such action is adverse to the interests of their clients, employers or the public.

(g) Engineers shall not accept professional employment outside of their regular work or interest without the knowledge of their employers.

CANON 5.

• • •

Engineers shall build their professional reputation on the merit of their services and shall not compete unfairly with others.

(a) Engineers shall not give, solicit or receive either directly or indirectly, any political contribution, gratuity, or unlawful consideration in order to secure work, exclusive of securing salaried positions through employment agencies.

(b) Engineers should negotiate contracts for professional services fairly and on the basis of demonstrated competence and qualifications for the type of professional service required.

(c) Engineers may request, propose or accept professional commissions on a contingent basis only under circumstances in which their professional judgments would not be compromised.

(d) Engineers shall not falsify or permit misrepresentation of their academic or professional qualifications or experience.

(e) Engineers shall give proper credit for engineering work to those to whom credit is due, and shall recognize the proprietary interests of others. Whenever possible, they shall name the person or persons who may be responsible for designs, inventions, writings or other accomplishments.

(f) Engineers may advertise professional services in a way that does not contain misleading language or is in any other manner derogatory to the dignity of the profession. Examples of permissible advertising are as follows:

Professional cards in recognized, dignified publications, and listings in rosters or directories published by responsible organizations, provided that the cards or listings are consistent in size and content and are in a section of the publication regularly devoted to such professional cards.

Brochures which factually describe experience, facilities, personnel and capacity to render service, providing they are not misleading with respect to the engineer's participation in projects described.

Display advertising in recognized dignified business and professional publications, providing it is factual and is not misleading with respect to the engineer's extent of participation in projects described.

A statement of the engineers' names or the name of the firm and statement of the type of service posted on projects for which they render services.

Preparation or authorization of descriptive articles for the lay or technical press, which are factual and dignified. Such articles shall not imply anything more than direct participation in the project described.

Permission by engineers for their names to be used in commercial advertisements, such as may be published by contractors, material suppliers, etc., only by means of a modest, dignified notation acknowledging the engineers' participation in the project described. Such permission shall not include public endorsement of proprietary products.

(g) Engineers shall not maliciously or falsely, directly or indirectly, injure the professional reputation, prospects, practice or employment of another engineer or indiscriminately criticize another's work.

(h) Engineers shall not use equipment, supplies, laboratory or office facilities of their employers to carry on outside private practice without the consent of their employers.

CANON 6.

• • •

Engineers shall act in such a manner as to uphold and enhance the honor, integrity, and dignity of the engineering profession.

Engineers shall not knowingly act in a manner which will be derogatory to the honor, integrity, or dignity of the engineering profession or knowingly engage in business or professional practices of a fraudulent, dishonest or unethical nature.

CANON 7.

• • •

Engineers shall continue their professional development throughout their careers, and shall provide opportunities for the professional development of those engineers under their supervision.

(a) Engineers should keep current in their specialty fields by engaging in professional practice, participating in continuing education courses, reading in the technical literature, and attending professional meetings and seminars.

(b) Engineers should encourage their engineering employees to become registered at the earliest possible date.

(c) Engineers should encourage engineering employees to attend and present papers at professional and technical society meetings.

(d) Engineers shall uphold the principle of mutually satisfying relationships between employers and employees with respect to terms of employment including professional grade descriptions, salary ranges, and fringe benefits.

As adopted September 2, 1914, and most recently amended November 10, 1996.

(1) The American Society of Civil Engineers adopted THE FUNDAMENTAL PRINCIPLES of the ABET Code of Ethics of Engineers as accepted by the Accreditation Board for Engineering and Technology, Inc. (ABET). (By ASCE Board of Direction action April 12-14, 1975)

(2) In November 1996, the ASCE Board of Direction adopted the following definition of Sustainable Development: "ustainable Development is the challenge of meeting human needs for natural resources, industrial products, energy, food, transportation, shelter, and effective waste management while conserving and protecting environmental quality and the natural resource base essential for future development."

© 2005 ASCE. Reprinted with permission.

AMERICAN SOCIETY OF MECHANICAL ENGINEERS (ASME) CODE OF ETHICS

• • •

ASME requires ethical practice by each of its members and has adopted the following Code of Ethics of Engineers as referenced in the ASME Constitution, Article C2.1.1.

Code of ethics of engineers

THE FUNDAMENTAL PRINCIPLES   Engineers uphold and advance the integrity, honor and dignity of the engineering profession by:

(I) using their knowledge and skill for the enhancement of human welfare;

(II) being honest and impartial, and serving with fidelity the public, their employers and clients; and

(III) striving to increase the competence and prestige of the engineering profession.

THE FUNDAMENTAL CANONS

(1) Engineers shall hold paramount the safety, health and welfare of the public in the performance of their professional duties.

(2) Engineers shall perform services only in the areas of their competence.

(3) Engineers shall continue their professional development throughout their careers and shall provide opportunities for the professional and ethical development of those engineers under their supervision.

(4) Engineers shall act in professional matters for each employer or client as faithful agents or trustees, and shall avoid conflicts of interest or the appearance of conflicts of interest.

(5) Engineers shall build their professional reputation on the merit of their services and shall not compete unfairly with others.

(6) Engineers shall associate only with reputable persons or organizations.

(7) Engineers shall issue public statements only in an objective and truthful manner.

(8) Engineers shall consider environmental impact in the performance of their professional duties.

(9) Engineers shall consider sustainable development in the performance of their professional duties.

The Board on Professional Practice and Ethics maintains an archive of interpretations to the ASME Code of Ethics (P-15.7). These interpretations shall serve as guidance to the user of the ASME Code of Ethics and are available on the Board's website or upon request.
Responsibility: Council on Member Affairs/Board on Professional Practice and Ethics
Adopted: March 7, 1976
Revised several times

INSTITUTE OF ELECTRICAL AND ELECTRONIC ENGINEERS (IEEE) CODE OF ETHICS

• • •

We, the members of the IEEE, in recognition of the importance of our technologies in affecting the quality of life throughout the world, and in accepting a personal obligation to our profession, its members and the communities we serve, do hereby commit ourselves to the highest ethical and professional conduct and agree:

1. to accept responsibility in making engineering decisions consistent with the safety, health and welfare of the public, and to disclose promptly factors that might endanger the public or the environment;

2. to avoid real or perceived conflicts of interest whenever possible, and to disclose them to affected parties when they do exist;

3. to be honest and realistic in stating claims or estimates based on available data;

4. to reject bribery in all its forms;

5. to improve the understanding of technology, its appropriate application, and potential consequences;

6. to maintain and improve our technical competence and to undertake technological tasks for others only if qualified by training or experience, or after full disclosure of pertinent limitations;

7. to seek, accept, and offer honest criticism of technical work, to acknowledge and correct errors, and to credit properly the contributions of others;

8. to treat fairly all persons regardless of such factors as race, religion, gender, disability, age, or national origin;

9. to avoid injuring others, their property, reputation, or employment by false or malicious action;

10. to assist colleagues and co-workers in their professional development and to support them in following this code of ethics.

Approved by the IEEE Board of Directors
August 1990
© *1990; reprinted with permission of IEEE.*

NATIONAL SOCIETY OF PROFESSIONAL ENGINEERS (NSPE) CODE OF ETHICS

• • •

Preamble

Engineering is an important and learned profession. As members of this profession, engineers are expected to exhibit the highest standards of honesty and integrity. Engineering has a direct and vital impact on the quality of life for all people. Accordingly, the services provided by engineers require honesty, impartiality, fairness, and equity, and must be dedicated to the protection of the public health, safety, and welfare. Engineers must perform under a standard of professional behavior that requires adherence to the highest principles of ethical conduct.

I. Fundamental Canons

Engineers, in the fulfillment of their professional duties, shall:

1. Hold paramount the safety, health and welfare of the public.

2. Perform services only in areas of their competence.

3. Issue public statements only in an objective and truthful manner.

4. Act for each employer or client as faithful agents or trustees.

5. Avoid deceptive acts.

6. Conduct themselves honorably, responsibly, ethically, and lawfully so as to enhance the honor, reputation, and usefulness of the profession.

II. Rules of Practice

1. Engineers shall hold paramount the safety, health, and welfare of the public.

(a) If engineers' judgment is overruled under circumstances that endanger life or property, they shall notify their employer or client and such other authority as may be appropriate.

(b) Engineers shall approve only those engineering documents that are in conformity with applicable standards.

(c) Engineers shall not reveal facts, data, or information without the prior consent of the client or employer except as authorized or required by law or this Code.

(d) Engineers shall not permit the use of their name or associate in business ventures with any person or firm that they believe are engaged in fraudulent or dishonest enterprise.

(e) Engineers shall not aid or abet the unlawful practice of engineering by a person or firm.

(f) Engineers having knowledge of any alleged violation of this Code shall report thereon to appropriate professional bodies and, when relevant, also to public authorities, and cooperate with the proper authorities in furnishing such information or assistance as may be required.

(2) Engineers shall perform services only in the areas of their competence.

(a) Engineers shall undertake assignments only when qualified by education or experience in the specific technical fields involved.

(b) Engineers shall not affix their signatures to any plans or documents dealing with subject matter in which they lack competence, nor to any plan or document not prepared under their direction and control.

(c) Engineers may accept assignments and assume responsibility for coordination of an entire project and sign and seal the engineering documents for the entire project, provided that each technical segment is signed and sealed only by the qualified engineers who prepared the segment.

(3) Engineers shall issue public statements only in an objective and truthful manner.

(a) Engineers shall be objective and truthful in professional reports, statements, or testimony. They shall include all relevant and pertinent information in such reports, statements, or testimony, which should bear the date indicating when it was current.

(b) Engineers may express publicly technical opinions that are founded upon knowledge of the facts and competence in the subject matter.

(c) Engineers shall issue no statements, criticisms, or arguments on technical matters that are inspired or paid for by interested parties, unless they have prefaced their comments by explicitly identifying the interested parties on whose behalf they are speaking, and by revealing the existence of any interest the engineers may have in the matters.

(4) Engineers shall act for each employer or client as faithful agents or trustees.

(a) Engineers shall disclose all known or potential conflicts of interest that could influence or appear to influence their judgment or the quality of their services.

(b) Engineers shall not accept compensation, financial or otherwise, from more than one party for services on the same project, or for services pertaining to the same project, unless the circumstances are fully disclosed and agreed to by all interested parties.

(c) Engineers shall not solicit or accept financial or other valuable consideration, directly or indirectly, from outside agents in connection with the work for which they are responsible.

(d) Engineers in public service as members, advisors, or employees of a governmental or quasi-governmental body or department shall not participate in decisions with respect to services solicited or provided by them or their organizations in private or public engineering practice.

(e) Engineers shall not solicit or accept a contract from a governmental body on which a principal or officer of their organization serves as a member.

(5) Engineers shall avoid deceptive acts.

(a) Engineers shall not falsify their qualifications or permit misrepresentation of their or their associates' qualifications. They shall not misrepresent or exaggerate their responsibility in or for the subject matter of prior assignments. Brochures or other presentations incident to the solicitation of employment shall not misrepresent pertinent facts concerning employers, employees, associates, joint venturers, or past accomplishments.

(b) Engineers shall not offer, give, solicit or receive, either directly or indirectly, any contribution to influence the award of a contract by public authority, or which may be reasonably construed by the public as having the effect of intent to influencing the awarding of a contract. They shall not offer any gift or other valuable consideration in order to secure work. They shall not pay a commission, percentage, or brokerage fee in order to secure work, except to a bona fide employee or bona fide established commercial or marketing agencies retained by them.

III. Professional Obligations

(1) Engineers shall be guided in all their relations by the highest standards of honesty and integrity.

(a) Engineers shall acknowledge their errors and shall not distort or alter the facts.

(b) Engineers shall advise their clients or employers when they believe a project will not be successful.

(c) Engineers shall not accept outside employment to the detriment of their regular work or interest. Before accepting any outside engineering employment they will notify their employers.

(d) Engineers shall not attempt to attract an engineer from another employer by false or misleading pretenses.

(e) Engineers shall not promote their own interest at the expense of the dignity and integrity of the profession.

(2) Engineers shall at all times strive to serve the public interest.

(a) Engineers shall seek opportunities to participate in civic affairs; career guidance for youths; and work for the advancement of the safety, health, and well-being of their community.

(b) Engineers shall not complete, sign, or seal plans and/or specifications that are not in conformity with applicable engineering standards. If the client or employer insists on such unprofessional conduct, they shall notify the proper authorities and withdraw from further service on the project.

(c) Engineers shall endeavor to extend public knowledge and appreciation of engineering and its achievements.

(3) Engineers shall avoid all conduct or practice that deceives the public.

(a) Engineers shall avoid the use of statements containing a material misrepresentation of fact or omitting a material fact.

(b) Consistent with the foregoing, engineers may advertise for recruitment of personnel.

(c) Consistent with the foregoing, engineers may prepare articles for the lay or technical press, but such articles shall not imply credit to the author for work performed by others.

(4) Engineers shall not disclose, without consent, confidential information concerning the business affairs or technical processes of any present or former client or employer, or public body on which they serve.

(a) Engineers shall not, without the consent of all interested parties, promote or arrange for new employment or practice in connection with a specific project for which the engineer has gained particular and specialized knowledge.

(b) Engineers shall not, without the consent of all interested parties, participate in or represent an adversary interest in connection with a specific project or proceeding in which the engineer has gained particular specialized knowledge on behalf of a former client or employer.

(5) Engineers shall not be influenced in their professional duties by conflicting interests.

(a) Engineers shall not accept financial or other considerations, including free engineering designs, from material or equipment suppliers for specifying their product.

(b) Engineers shall not accept commissions or allowances, directly or indirectly, from contractors or other parties dealing with clients or employers of the engineer in connection with work for which the engineer is responsible.

(6) Engineers shall not attempt to obtain employment or advancement or professional engagements by untruthfully criticizing other engineers, or by other improper or questionable methods.

(a) Engineers shall not request, propose, or accept a commission on a contingent basis under circumstances in which their judgment may be compromised.

(b) Engineers in salaried positions shall accept part-time engineering work only to the extent consistent with policies of the employer and in accordance with ethical considerations.

(c) Engineers shall not, without consent, use equipment, supplies, laboratory, or office facilities of an employer to carry on outside private practice.

(7) Engineers shall not attempt to injure, maliciously or falsely, directly or indirectly, the professional reputation, prospects, practice, or employment of other engineers. Engineers who believe others are guilty of unethical or illegal practice shall present such information to the proper authority for action.

(a) Engineers in private practice shall not review the work of another engineer for the same client, except with the knowledge of such engineer, or unless the connection of such engineer with the work has been terminated.

(b) Engineers in governmental, industrial, or educational employ are entitled to review and evaluate the work of other engineers when so required by their employment duties.

(c) Engineers in sales or industrial employ are entitled to make engineering comparisons of represented products with products of other suppliers.

(8) Engineers shall accept personal responsibility for their professional activities, provided, however, that engineers may seek indemnification for services arising out of their practice for other than gross negligence, where the engineer's interests cannot otherwise be protected.

(a) Engineers shall conform with state registration laws in the practice of engineering.

(b) Engineers shall not use association with a none-ngineer, a corporation, or partnership as a "cloak" for unethical acts.

(9) Engineers shall give credit for engineering work to those to whom credit is due, and will recognize the proprietary interests of others.

(a) Engineers shall, whenever possible, name the person or persons who may be individually responsible for designs, inventions, writings, or other accomplishments.

(b) Engineers using designs supplied by a client recognize that the designs remain the property of the client and may not be duplicated by the engineer for others without express permission.

(c) Engineers, before undertaking work for others in connection with which the engineer may make improvements, plans, designs, inventions, or other records that may justify copyrights or patents, should enter into a positive agreement regarding ownership.

(d) Engineers' designs, data, records, and notes referring exclusively to an employer's work are the employer's property. The employer should indemnify the engineer for use of the information for any purpose other than the original purpose.

(e) Engineers shall continue their professional development throughout their careers and should keep current in their specialty fields by engaging in professional practice, participating in continuing education courses, reading in the technical literature, and attending professional meetings and seminars.

—As Revised January 2003

"By order of the United States District Court for the District of Columbia, former Section 11(c) of the NSPE Code of Ethics prohibiting competitive bidding, and all policy statements, opinions, rulings or other guidelines interpreting its scope, have been rescinded as unlawfully interfering with the legal right of engineers, protected under the antitrust laws, to provide price information to prospective clients; accordingly, nothing contained in the NSPE Code of Ethics, policy statements, opinions, rulings or other guidelines prohibits the submission of price quotations or competitive bids for engineering services at any time or in any amount."

Statement by NSPE Executive Committee

In order to correct misunderstandings which have been indicated in some instances since the issuance of the Supreme Court decision and the entry of the Final Judg-ment, it is noted that in its decision of April 25, 1978, the Supreme Court of the United States declared: "The Sherman Act does not require competitive bidding."

It is further noted that as made clear in the Supreme Court decision:

(1) Engineers and firms may individually refuse to bid for engineering services.

(2) Clients are not required to seek bids for engineering services.

(3) Federal, state, and local laws governing procedures to procure engineering services are not affected, and remain in full force and effect.

(4) State societies and local chapters are free to actively and aggressively seek legislation for professional selection and negotiation procedures by public agencies.

(5) State registration board rules of professional conduct, including rules prohibiting competitive bidding for engineering services, are not affected and remain in full force and effect. State registration boards with authority to adopt rules of professional conduct may adopt rules governing procedures to obtain engineering services.

(6) As noted by the Supreme Court, "nothing in the judgment prevents NSPE and its members from attempting to influence governmental action ..."

NOTE: In regard to the question of application of the Code to corporations vis-à-vis real persons, business form or type should not negate nor influence conformance of individuals to the Code. The Code deals with professional services, which services must be performed by real persons. Real persons in turn establish and implement policies within business structures. The Code is clearly written to apply to the Engineer and items incumbent on members of NSPE to endeavor to live up to its provisions. This applies to all pertinent sections of the Code.

PUERTO RICO: ASSOCIATION OF ENGINEERS AND SURVEYORS OF PUERTO RICO CODE OF ETHICS

• • •

College of Engineers and Surveyors of Puerto Rico

RULES OF ETHICS

In order to maintain and extol the integrity, honor, and dignity of their professions, in accordance with the highest moral and ethical professional norms of conduct, the **Engineer** and the **Surveyor**:

1. Should consider their principal function as professionals as that of serving humanity. Their relation as professional and client, and as professional and patron, should be subject to their fundamental function of promoting the wellbeing of humanity and that of protecting the public interest.

2. They will be honest and impartial and will serve faithfully in the development of their professional functions, always maintaining their independence of criteria which constitutes the base of professionalism.

3. They will strive to improve the competence and the prestige of engineering and surveying.

RULES OF PROFESSIONAL ETHICS

The **Engineer** and the **Surveyor**, in fulfilling their professional duties, must:

RULE I: Protect, above all other consideration, the security, environment, health, and wellbeing of the community in the execution of their professional responsibilities.

RULE II: Provide services only in their areas of competence.

RULE III: Make public declarations only in a true and objective form.

RULE IV: Act in professional matters for each patron or client as faithful agents or fiduciaries, and avoid conflicts of interests or the mere appearance of these, always maintaining independence of criteria as the base of professionalism.

RULE V: Build their professional reputation on the merit of their services and not compete disloyally with others.

RULE VI: Not participate in deceitful acts in the pursuit of employment and in offering professional services.

RULE VII: Act with the decorum that sustains and enhances the honor, integrity, and dignity of their professions.

RULE VIII: Associate only with persons and organizations of good reputation.

RULE IX: Continue their professional development throughout their careers and promote opportunities for the professional and ethical development of the engineers and surveyors under their supervision.

RULE X: Strive to and accept to take professional actions only in conformity with the applicable laws and with these Rules.

NORMS OF PRACTICE RULE I: Protect, above all other consideration, the security, environment, health, and wellbeing of the community in the execution of their professional responsibilities.

The **Engineer** and the **Surveyor**:

a. Will recognize that the life, the security, the environment, the health, and the wellbeing of the community depend on the judgments, decisions, and professional practices incorporated in systems, structures, machines, processes, products, and artifacts.

b. Will approve, seal, stamp, or certify, when appropriate, only those documents revised or prepared by those who understand that they are safe for the environment, health, and wellbeing of the community in conformity with the accepted standards.

c. When their professional judgment might have been repealed in circumstances where the security, environment, health, or wellbeing of the community are put in danger, they will inform their clients or patrons of the possible consequences. If the threat to the security, environment, health, or wellbeing of the community continues, they will inform the concerned authorities about the matter.

d. When they have knowledge or sufficient reason to believe that another engineer or surveyor is violating the dispositions of this Code, or that a person or firm is putting in danger the security, environment, health, or wellbeing of the community, they will present such information in writing to the concerned authorities and will cooperate with said authorities by providing what information or assistance that might be required by them.

e. They will serve constructively in civic matters and will work for the advancement of the security, environment, health, and wellbeing of their communities.

f. They will promise to better the environment and do all that which might be within their reach to enhance the quality of life.

RULE II: **Provide services only in their areas of competence.**

The **Engineer** and the **Surveyor:**

a. Will only undertake those jobs for which they are qualified by education or experience in the specific technical fields which are being dealt with.

b. Will be able to accept a charge that requires education and experience outside of their fields of competence always and whenever their services are restricted to those phases of the project for which they are qualified. All the other phases of such a project will be executed by qualified associates, consultants, or employees who will approve, seal, stamp, or certify, where necessary, the concerned documents.

c. They will not approve, seal, stamp, or certify, where necessary, any plan or document that deals with some material en which they do not have competence by virtue of their education or experience.

RULE III: **Make public declarations only in a true and objective form.**

The **Engineer** and the Surveyor:

a. Will be objective and true in professional reports, declarations, or testimonies. They will include all relevant or pertinent information in said reports, declarations, or testimonies.

b. Will undertake to make public knowledge the reach and the practice of their professions and will not participate in the dissemination of false, unjust, or exaggerated declarations.

c. When they serve as technical witnesses, experts, or technicians in any forum, they will express a professional opinion only when it is founded in an adequate knowledge of the facts of the controversy, in a technical competence about the material in question, and in an honest conviction of the exactitude and propriety of their testimonies.

d. They will not make declarations, critiques, or arguments about materials of their respective professions that are motivated or paid by an interested party or parties, unless in these commentaries their author is identified, and the identity of the party or parties whose interest is being spoken about is revealed, as well as the existence of any pecuniary interest that they might have in the matters under discussion.

e. They will be serious and restrained in explaining their work and merits, and will avoid any act tending to promote their own interest at the expense of the integrity, honor, and dignity of their profession or of another individual.

f. They will express publicly a professional opinion about technical matters only when that opinion is founded upon an adequate knowledge of the facts, and competence in these matters.

RULE IV: **Act in professional matters for each patron or client as faithful agents or fiduciaries, and avoid conflicts of interests or the mere appearance of these, always maintaining independence of criteria as the base of professionalism.**

The **Engineer** and the **Surveyor:**

a. Will avoid all known or potential conflicts of interest with their patrons or clients and will inform in a prompt manner said patrons or clients about any business relation, interests, or circumstances that might influence their judgment or the quality of their services.

b. Will not undertake any charge that might, knowingly, create a potential conflict of interest among them and their clients or patrons.

c. Will not accept compensation from third parties for services rendered in a project, or for services pertaining to the same project, unless the circumstances are completely revealed, and agreed upon by all interested parties.

d. Will not solicit or accept significant gratuities, directly or indirectly, from contractors or their agents or other parties in relation to work that is realized for patrons or clients for which they are responsible.

e. Will not solicit or accept considerations or compensations of any kind for specifying products or materials or suppliers of equipment, without divulging it to their clients or patrons.

f. Those who are in public service as members, advisors, or employers of a governmental body or department will not participate in decisions related to professional services solicited or provided by them or by their organizations in professional practice, be it private or public.

g. Will not solicit or accept contracts for professional services from a governmental body en which an individual or official of their organizations serves as a member.

h. When, as a result of their studies, they understand that a project will not be successful, they will make such an opinion part of the report to their patron or client.

i. Will treat all information that arrives to them in the course of their professional duties as confidential and will not use such information as a means to achieve personal benefit if such an action is adverse to the interests of their clients, of their patrons, of the commissions or committees to which they belong, or of the public.

j. Will not reveal confidential information concerning business matters or technical processes of any patron or bidder, current or previous, under evaluation, without their consent, except when it might be required by law.

k. Will not duplicate designs that are supplied to them by their clients for others, without the express authorization of their client and of the designer, considering the relevant contracts and laws.

l. Before undertaking work for others, in which they can make renovations, plans, designs, inventions, or other registers, that can justify obtaining author's rights or patents, will arrive at an agreement in relation to the rights of the respective parts.

m. Will not participate in or represent an adversary interest without the consent of the interested parts, in relation to a specific project or matter in which they have gained a particular specialized knowledge in the name of a former patron or client.

RULE V: **Build their professional reputation on the merit of their services and not compete disloyally with others.**

The **Engineer** and the **Surveyor:**

a. Will not offer, give, solicit, or receive, directly or indirectly, any monetary contribution or contribution of any other type directed at influencing the granting of a contract by a public authority. They will not offer any gift or any other type of consideration of worth with the aim of obtaining work. They will not pay a commission, percent, or rights of brokerage with the aim of obtaining work except to a bonafide employee or to commercial agencies or to established marketing agencies, bonafide and contracted by them for this reason.

b. Will negotiate contracts for professional services on the base of professional competence and demonstrated qualifications for the type of professional service required and then for just and reasonable honorariums.

c. Will not solicit, propose, or accept professional commissions on a base contingent upon circumstances

in which their professional judgment may be seen as compromised.

d. Will not attempt to recruit an employee from other patron by means of false or deceitful representations.

e. Will not maliciously or falsely damage, directly or indirectly, the professional reputation, the prospects, the practice, or the employment of another engineer or surveyor, nor will criticize indiscriminately the work of these people.

f. Will not use the equipment, supplies, laboratory, or office of their patrons in order to execute exterior private practice without their consent.

g. Will not take advantage of the advantages of a salaried position in order to disloyally compete with colleagues who exercise the profession privately.

h. Will not attempt to supplant, nor will supplant another engineer or surveyor, after a professional position has been offered to him or her, nor will compete unjustly with said person.

i. The professionals who act as subcontractors on a project or who in some capacity utilize the services of another professional will not be able to retain for themselves the professional honorariums charges without having attended to the payment of the honorariums of their collaborators at least in a form equitable or proportional to their own; or in any manner deprive or further that their professional companions do not receive just or equitable pay for their services.

j. Will not approve, seal, stamp, or certify, according to the case, nor authorize the presentation of plans, specifications, calculations, opinions, briefs, or reports that have not been elaborated by them or by others under their direct responsibility. Furthermore, they will give credit for the engineering, surveying, or architectural work to those who have done it.

RULE VI: **Not participate in deceitful acts in the pursuit of employment and in offering professional services.**

The **Engineer** and the **Surveyor:**

a. Will not falsify or permit the misrepresentation of their academic or professional qualifications, nor that of their associates or employees. They will not misrepresent or exaggerate the degree of their responsibility in previous positions or concerning the matters that these positions entailed. The folders or types of presentations created for the purpose of soliciting employment will not represent the pertinent facts concerning previous patrons, employees, associates, employers, or achievements.

b. Will announce their professional services without self-praise and without deceitful language, and in a manner in which the dignity of their professions is not diminished. Some examples of permissible announcements are as follows:

1. Professional announcements in recognized publications, and listings in registries or directories published by responsible organizations, as long as the announcements and registries are consequent in size and content and are in a section of the publication dedicated regularly to such professional announcements.

2. Brochures that in fact describe the experience, installations, personnel, and capacity to render services, as long as they are not deceitful with respect to the participation of the professionals in the projects described.

3. Announcements in recognized professional and business magazines, as long as they refer to facts, do not contain self-praising expressions or implications, and are not deceitful with respect to the degree of participation of the professionals in the projects described.

4. A declaration of the names of the professionals or the name of the firm and the type of service, announced in projects for which the professionals render service.

5. The preparation or authorization of descriptive articles for the press that refer to facts, are serious, and are free of implicated praise. Such articles will imply nothing more that the direct participation of the professionals in the project described.

6. The authorization of professionals so that their names may be used in commercial announcements, such as those that can be published by contractors, suppliers of materials, etc., only through a serious and restrained annotation, recognizing the participation of the professionals in the project described. Such authorization will not include the public endorsement of brand-name products.

RULE VII: **Act with the decorum that sustains and enhances the honor, integrity, and dignity of their professions.**

The **Engineer** and the **Surveyor:**

a. Will not act, knowingly, in such a manner that might be harmful to the honor, integrity, and dignity of their professions.

b. Will not associate with, employ, or in any other way utilize in practice any person to render professional services as an engineer, surveyor, or architect, unless that person is an engineer, a surveyor, or an architect recognized by valid authorities as being able to render such services.

c. Will not associate their name with the practice of their profession with non-professionals or with persons or entities that are not professionals legally authorized to exercise the professions of engineering, surveying, or architecture.

d. Will not share honorariums except with engineers, surveyors, or architects who have been their collaborators in works of engineering, surveying, and architecture.

e. Will admit and accept their own errors when they are demonstrated to them and will abstain from distorting or altering the facts with the purpose of justifying their decisions.

f. Will cooperate en extending the efficacy of their professions through the exchange of information and experience with other engineers, architects, and surveyors, and with students of these professions.

g. Will not compromise their professional criteria for any other particular interest.

RULE VIII: **Associate only with persons and organizations of good reputation.**

The **Engineer** and the **Surveyor:**

a. Will not associate with or permit the use of their names or that of their firms, knowingly, with businesses run by any other person or firm that they know or have sufficient reason to believe might be involved in professional or business practices of a fraudulent or dishonest nature.

b. Will not use the association with natural or juridical persons to hide unethical acts.

RULE IX: **Continue their professional development throughout their careers and promote opportunities for the professional and ethical development of the engineers and surveyors under their supervision.**

The **Engineer** and the **Surveyor:**

a. Will keep themselves up to date in their fields of specialty by exercising professional practice, participating in continuing education courses, reading technical literature, and attending professional meeting and seminars.

b. Will encourage the engineers and surveyors in their employ to further their education.

c. Will encourage their graduate employees in training in engineering and surveying to obtain their professional licenses as quickly as possible.

d. Will encourage the engineers and surveyors in their employ to attend and present papers in meetings or professional and technical societies.

e. Will support the principle of mutually satisfactory relations between patrons and employees with respect to the conditions of employment, including a description of professional degree, and scales of salary and benefits.

RULE X: **Strive to and accept to take professional actions only in conformity with the applicable laws and with these Rules.**

The **Engineer** and the Surveyor:

a. Will carry out what is laid out in the laws that govern the practice and direction of engineering and surveying, according to reforms, with the rule of the College of Engineers and Surveyors of Puerto Rico (CIAPR) and of the Examination Board of Engineers, Architects, and Surveyors, and with the agreements and directives legitimately adopted by the General Assembly and Governing Board of CIAPR.

b. Will appear at any interview, administrative investigation, viewing, or procedure, before the Tribunal of Discipline and Professional Ethics or the Commission of Defense of the Profession of CIAPR to which they have been duly cited by the College, be it as a witness, plaintiff, or defendant.

Approved at the Annual Ordinary Assembly celebrated Saturday, August 20, 1994, in the El Conquistador Hotel, Fajardo, Puerto Rico.

Engineer José R. Rodríguez Perazza, President

Engineer Benigno Despiau, Secretary

TRANSLATED BY JAMES A. LYNCH

THE INSTITUTION OF ENGINEERS CODE OF ETHICS

• • •

National Headquarters
11 National Circuit
Barton, Australian Capital Territory
2600 Australia

Founded 1919
Members: 45,000

Code of Ethics

Preamble

The further development of civilization, the conservation and management of natural resources, and the improvement of the standards of living of mankind are greatly affected by the work of the Engineer. For that work to be fully effective it is necessary not only that Engineers strive constantly to widen their knowledge and improve their skill but also that the community be willing to recognize the integrity and trust the judgment of members of the Profession of Engineering.

For this to happen, the Profession must be recognized in the community for

its skill in using technical expertise for the enhancement of human welfare

its loyalty to the community, to employers and clients

its honesty and impartiality in professional practice

Engineers shall so order their lives and work as to merit this trust.

To this end all members of the Institution are required to comply with the Code of Ethics set out hereunder; to give active support to the proper regulation of the qualifications, employment and practice of the Profession; and to promote the development and application of technology in the public interest.

Members acting in accordance with this Code will have the support of the Institution.

CODE

• • •

(1) The responsibility of Engineers for the welfare, health and safety of the community shall at all times come before their responsibility to the Profession, to sectional or private interests, or to other Engineers.

(2) Engineers shall act so as to uphold and enhance the honor, integrity and dignity of the Profession.

(3) Engineers shall perform work only in their areas of competence.

(4) Engineers shall build their professional reputation on merit and shall not compete unfairly.

(5) Engineers shall apply their skill and knowledge in the interests of their employer or client for whom they shall act, in professional matters, as faithful agents or trustees.

(6) Engineers shall give evidence, express opinions or make statements in an objective and truthful manner and on the basis of adequate knowledge.

(7) Engineers shall continue their professional development throughout their careers and shall actively assist and encourage Engineers under their direction to advance their knowledge and experience.

INTERPRETATIONS

It has been found in the past that inquiries are often received by the Institution from Engineers seeking guidance on the way in which the Code of Ethics applies in particular situations. The following interpretations are for the guidance and information of individual members as to the Institution's attitudes toward the implementation of this Code.

Clause 1:

The responsibility of Engineers for the welfare, health and safety of the community shall at all times come before their responsibility to the Profession, to sectional or private interests, or to other Engineers.

The principle here is that the interests of the community have priority over the interests of others. It follows that a member:

(a) shall avoid assignments that may create a conflict between the interests of his client or employer and the public interest;

(b) shall work in conformity with acceptable engineering standards and not in such a manner as to jeopardize the public welfare, health or safety;

(c) shall endeavor at all times to maintain engineering services essential to public welfare;

(d) shall in the course of his professional life endeavor to promote the wellbeing of the community. If his judgment is over-ruled in this matter he should inform his client or employer of the possible consequences (and, if appropriate, notify the proper authority of the situation);

(e) shall, if he considers that by so doing he can constructively advance the wellbeing of the community, contribute to public discussion on engineering matters in his area of competence.

Clause 2:

Engineers shall act so as to uphold and enhance the honor, integrity and dignity of the Profession.

The principle here is that the Profession should endeavor by its behavior to merit the highest esteem of the community. It follows therefore that a member:

(a) Shall not involve himself with any business or professional practice which he knows to be of a fraudulent or dishonest nature;

(b) shall not use association with other persons, corporations or partnerships to conceal unethical acts;

(c) shall not continue in partnership with, nor act in professional matters with, any Engineer who has been removed from membership of the Institution because of unprofessional conduct.

Clause 3:

Engineers shall perform work only in their areas of competence.

To this end the Institution has determined that:

(a) a member shall inform his employer or client, and make appropriate recommendations on obtaining further advice, if an assignment requires qualifications and experience outside his field of competence; and

(b) in the practice of Consulting Engineering a member shall not describe himself, nor permit himself to be described, nor act as a Consulting Engineer unless he is a Corporate Member, occupies a position of professional independence, is prepared to design and supervise engineering work or act as an unbiased and independent adviser on engineering matters, and conduct his practice in strict compli-

ance with the conditions approved by the Council of the Institution.

Clause 4:

Engineers shall build their professional reputation on merit and shall not compete unfairly.

The principle here is that Engineers shall not act improperly in a professional sense to gain a benefit. It follows that a member:

(a) shall only approach prospective clients or employers with due regard to his professional independence and to this Code of Ethics;

(b) shall neither pay nor offer directly or indirectly inducements to secure work;

(c) shall promote the principle of selection of consulting engineers by clients upon the basis of merit, and shall not compete with other consulting engineers on the basis of fees alone. It shall not be a breach of the Code of Ethics for a member, upon an inquiry made in that behalf by a client or prospective client, to provide information as to the basis upon which he usually charges fees for particular types of work. Also it shall not be a breach of the Code of Ethics for a member to submit a proposal for the carrying out of work which proposal includes, in addition to a technical proposal and an indication of the resources which the member can provide, information as to the basis upon which fees will be charged or as to the amount of the fees for the work which is proposed to be done. In this respect it is immaterial whether or not the member is aware that other engineers may have been requested to submit proposals, including fee proposals, for the same work;

(d) shall promote the principle of engagement of engineers upon the basis of merit. He shall uphold the principle of adequate and appropriate remuneration for professional engineering staff and shall give due consideration to terms of employment which have the approval of the profession's appropriate association;

(e) shall not attempt to supplant another Engineer, employed or consulting, who has been appointed;

(f) in the practice of Consulting Engineering, shall not undertake professional work on a basis which involves a speculative fee or remuneration which is conditional on implementation of the work. This does not preclude competitions conducted within Australia provided that such competitions are conducted in accordance with conditions approved by the Institution;

(g) shall neither falsify nor misrepresent his or his associate's qualifications, experience and prior responsibility;

(h) shall neither maliciously nor carelessly do anything to injure, directly or indirectly, the reputation, prospects or business of others;

(i) shall not use the advantages of a privileged position to compete unfairly with other Engineers;

(j) shall exercise due restraint in explaining his own work and shall refrain from unfair criticism of the work of another Engineer;

(k) shall give proper credit for professional work to those to whom credit is due and acknowledge the contribution of subordinates and others;

(l) may properly use circumspect advertising (which includes direct approaches to prospective clients by any means) to announce his practice and availability. The medium or other form of communication used and the content of the announcement shall be dignified, becoming to a professional engineer and free from any matter that could bring disrepute to the profession. Information given must be truthful, factual and free from ostentatious or laudatory expressions or implications.

Clause 5:

Engineers shall apply their skill and knowledge in the interests of their employer or client for whom they shall act, in professional matters, as faithful agents or trustees.

It follows that a member:

(a) shall at all times avoid all known or potential conflicts of interest. He should keep his employer or client fully informed on all matters, including financial interests, which could lead to such a conflict, in no circumstance should he participate in any decision which could involve him in conflict of interest;

(b) shall, when acting as administrator of a contract, be impartial as between the parties in the interpretation of the contract. This requirement of impartiality shall not diminish the duty of engineers to apply their skill and knowledge in the interests of the employer or client;

(c) shall not accept compensation, financial or otherwise, from more than one party for services on the same project, unless the circumstances are fully disclosed to, and agreed to, by all interested parties;

(d) shall neither solicit nor accept financial or other valuable considerations, including free engineering designs, from material or equipment suppliers for specifying their products;

(e) shall neither solicit nor accept gratuities, directly or indirectly, from contractors, their agents, or other parties dealing with his client or employer in connection with work for which he is responsible;

(f) shall advise his client or employer when as a result of his studies he believes that a project will not be viable;

(g) shall neither disclose nor use confidential information gained in the course of his employment without express permission.

Clause 6:

Engineers shall give evidence, express opinions or make statements in an objective and truthful manner and on the basis of adequate knowledge.

It follows that:

(a) a member's professional reports, statements or testimony before any tribunal shall be objective and accurate. He shall express an opinion only on the basis of adequate knowledge and technical competence in the area, but this shall not preclude a considered speculation based intuitively on experience and wide relevant knowledge;

(b) a member shall reveal the existence of any interest, pecuniary or otherwise, that could be taken to affect his judgment in a technical matter about which he is making a statement or giving evidence.

Clause 7:

Engineers shall continue their professional development throughout their careers and shall actively assist and encourage those under their direction to advance their knowledge and experience.

The principle here is that Engineers shall strive to widen their knowledge and improve their skill in order to achieve a continuing improvement of the Profession. It follows therefore that a member:

(a) shall encourage his professional employees and subordinates to further their education; and

(b) shall take a positive interest in, and encourage his fellow Engineers actively to support the Institution and other Professional Engineering organizations which further the general interests of the Profession. In this regard the Councils of The Institution of Engineers, Australia. The Association of Professional Engineers, Australia, and The Association of Consulting Engineers, Australia, have jointly advised and recommend to all Professional Engineers in Australia that the interests of the community and of their profession will be best served by full individual membership and active support for each of these respective organizations for which the member is eligible.

NOTES

This code is promulgated in a small blue four-page pamphlet. On the cover it states that the code was "Approved by the Council of The Institution of Engineers, Australia to be effective from 1 August 1981. Adopted by the Association of Consulting Engineers, Australia. Adopted by the Federal Council of the Association of Professional Engineers, Australia."

BANGLADESH

• • •

THE INSTITUTION OF ENGINEERS CODE OF ETHICS

• • •

Ramna, Dhaka-1000
Bangladesh

Founded: 1948
Members: 10,000

Professional Conduct and Code of Ethics

A. Professional Conducts:All Corporate Members as well as Associate Members, Students and Affiliates are required to order their conduct so as to uphold the reputation of the Institution and the dignity of the profession of Engineers and shall observe and be found by the Code of Ethics. Any alleged breach of this Code by a Corporate Member or an Associate Member or a Student or an Affiliate may be brought before the Council, which shall be investigated with the knowledge of the member. If the Council considers the charge proved, action will be taken by suspension from office, expulsion or admonition by a letter or posting his/her name with description of his/her offence.

B. Code of Ethics:

(1) A member's responsibility to his employer and to the profession shall have full regards to the public interest.

(2) A member shall order his conduct so as to uphold the dignity, standing and reputation of the profession.

(3) A member shall discharge his duties to his employer with complete fidelity. He shall not accept remuneration for services rendered other than from his employer or with his employer's permission.

(4) A member shall not maliciously or recklessly injure or attempt to injure, whether directly, or indirectly, the professional reputation, prospects, or business of another member.

(5) A member shall not improperly canvass or solicit professional employment nor offer to make by way of commission or otherwise payment for the introduction of such employment.

(6) A member shall not, in a self-laudatory language or in any manner derogatory to the dignity of the profession, or professional bodies, advertise or write articles for publication, nor shall he authorize such advertisements to be written or published by any other person.

(7) A member, without disclosing the fact to his employer in writing, shall not be a director of nor have a substantial financial interest in, nor be an agent for any company, firm or person carrying on any contracting, consulting or manufacturing business which is or may be involved in the work to which his employment relates, nor shall he receive directly or indirectly any royalty, gratuity or commission on any article or process used in or for the purposes of the work in respect of which he is employed unless or until such royalty, gratuity or commission has been authorized in writing by his employer.

(8) A member shall not use the advantages of a salaried position to compete unfairly with other engineers.

(9) A member in connection with work in a country other than his own shall order his conduct according to these Rules, so far as they are applicable, but where there are recognized standards of professional conduct, he shall adhere to them.

(10) A member who shall be convicted by a competent tribunal of a criminal offence which in the opinion of the Disciplinary Body renders him unfit to be a member shall be deemed to have been guilty of improper conduct.

(11) A member shall not, directly or indirectly, attempt to supplant another member, nor shall he intervene or attempt to intervene in or in connection with engineering work of any kind to which his knowledge has already been entrusted to another member.

(12) A member shall not be the medium of payments made on his employer's behalf unless so requested by his employer, nor shall he in connection with work on which he is employed place contracts or orders except with the authority of and on behalf of his employer.

(13) A member shall not knowingly compete on the basis of Professional charges with another member.

NOTES

Although the Institution of Engineers, Bangladesh, dates its founding from 1948, the country of Bangladesh did not come into existence until 1971 when East Pakistan declared its independence of West Pakistan. See also the notes for Institution of Engineers, Pakistan.

CANADA

• • •

CANADIAN COUNCIL OF PROFESSIONAL ENGINEERS CODE OF ETHICS

• • •

Suite 401, 116 Albert Street
Ottawa, Ontario
Canada K1P 5G3

Founded: 1936
Members: 12 consultant associations representing over 137,000 professional engineers

CODE OF ETHICS

• • •

Preamble

Provincial and territorial associations of Professional Engineers are responsible for the regulation of the practice of engineering in Canada. Each association has been established under a Professional Engineering Act of its provincial or territorial legislature and serves as the licensing authority for engineers practicing within its jurisdiction. The Canadian Council of Professional Engineers (CCPE) is the national federation of these associations. CCPE provides a coordinating function among the provincial and territorial associations, fostering mutual recognition among them and encouraging the greatest possible commonality of operation in their licensing functions.

CCPE issues national guidelines on various subjects as a means to achieve coordination among its constituent member associations. Such guidelines are an expression of general guiding principles which have a broad basis of consensus, while recognizing and supporting the autonomy of each constituent association to administer the Professional Engineering Act within its jurisdiction. CCPE guidelines enunciate the principles of an issue but leave the detailed applications, policies, practices and exceptions to the judgment of the constituent associations.

CODE OF ETHICS

• • •

Professional engineers shall conduct themselves in an honorable and ethical manner. Professional engineers shall uphold the values of truth, honesty and trustworthiness and safeguard human life and welfare and the environment. In keeping with these basic tenets, professional engineers shall:

(1) hold paramount the safety, health and welfare of the public and the protection of the environment and promote health and safety within the workplace;

(2) offer services, advise on or undertake engineering assignments only in areas of their competence and practice in a careful and diligent manner;

(3) act as faithful agents of their clients or employers, maintain confidentiality and avoid conflicts of interest;

(4) keep themselves informed in order to maintain their competence, strive to advance the body of knowledge within which they practice and provide opportunities for the professional development of their subordinates;

(5) conduct themselves with fairness, courtesy and good faith towards clients, colleagues and others, give credit where it is due and accept, as well as give, honest and fair professional criticism;

(6) present clearly to employers and clients the possible consequences if engineering decisions or judgments are overruled or disregarded;

(7) report to their association or other appropriate agencies any illegal or unethical engineering decisions or practices by engineers or others; and

(8) be aware of and ensure that clients and employers are made aware of societal and environmental consequences of actions or projects and endeavor to interpret engineering issues to the public in an objective and truthful manner.

NOTES

This code, adopted November 1991, is the outcome of a workshop on professional issues held in November 1989.

ASSOCIATION OF PROFESSIONAL ENGINEERS OF ONTARIO CODE OF ETHICS

• • •

1155 Yonge Street
Toronto, Ontario
Canada M4T 2Y5

Founded: 1922
Members: 61,000

(1) In this section, "negligence" means an act or an omission in the carrying out of the work of a practitioner that constitutes a failure to maintain the standards that a reasonable and prudent practitioner would maintain in the circumstances.

(2) For the purposes of the Act and this Regulation, "professional misconduct" means

(a) negligence;

(b) failure to make reasonable provision for the safeguarding of life, health or property of a person who may be affected by the work for which the practitioner is responsible;

(c) failure to act to correct or report a situation that the practitioner believes may endanger the safety or the welfare of the public;

(d) failure to make responsible provision for complying with applicable statutes, regulations, standards, codes, by-laws and rules in connection with work being undertaken by or under the responsibility of the practitioner;

(e) signing or sealing a final drawing, specification, plan, report or other document not actually prepared or checked by the practitioner;

(f) failure of a practitioner to present clearly to his employer the consequences to be expected from a deviation proposed in work, if the professional engineering judgment of the practitioner is overruled by non-technical authority in cases where the practitioner is responsible for the technical adequacy of professional engineering work;

(g) breach of the act or regulations, other than an action that is solely a breach of the code of ethics;

(h) undertaking work the practitioner is not competent to perform by virtue of his training and experience;

(i) failure to make prompt, voluntary and complete disclosure of an interest, direct or indirect that might in any way be, or be construed as, prejudicial to the professional judgment of the practitioner in rendering service to the public, to an employer or to a client, and in particular without limiting the generality of the foregoing, carrying out any of the following acts without making such a prior disclosure:

1. Accepting compensation in any form for a particular service from more than one party.

2. Submitting a tender or acting as a contractor in respect of work upon which the practitioner may be performing as a professional engineer.

3. Participating in the supply of material or equipment to be used by the employer or client of the practitioner.

4. Contracting in the practitioner's own right to perform professional engineering services for other than the practitioner's employer.

5. Expressing opinions or making statements concerning matters within the practice of professional engineering of public interest where the opinions or statements are inspired or paid for by other interests;

(j) conduct or an act relevant to the practice of professional engineering that, having regard to all the circumstances would reasonably be regarded by the engineering profession as disgraceful, dishonorable or unprofessional;

(k) failure by a practitioner to abide by the terms, conditions or limitations of the practitioner's license, limited license, temporary license or certificate;

(l) failure to supply documents or information requested by an investigator acting under section 34 of the Act;

(m) permitting, counseling or assisting a person who is not a practitioner to engage in the practice or professional engineering except as provided for in the Act or the regulations.

CODE OF ETHICS

• • •

The following is the Code of Ethics of the Association:

1. It is the duty of a practitioner to the public, to his employer, to his clients, to other members of his profession, and to himself to act at all times with,

---

i. fairness and loyalty to his associates, employers, clients, subordinates and employees.

ii. fidelity to public needs, and

iii. devotion to high ideals of personal honor and professional integrity.

2. A practitioner shall,

i. regard his duty to public welfare as paramount,

ii. endeavor at all times to enhance the public regard for his profession by extending the public knowledge thereof and discouraging untrue, unfair or exaggerated statements with respect to professional engineering,

iii. not express publicly, or while he is serving as a witness before a court, commission or other tribunal, opinions on professional engineering matters that are not founded on adequate knowledge and honest conviction,

iv. endeavor to keep his license, temporary license, limited license or certificate of authorization, as the case may be, permanently displayed in his place of business.

3. A practitioner shall act in professional engineering matters for each employer as a faithful agent or trustee and shall regard as confidential information obtained by him as to the business affairs, technical methods or processes of an employer and avoid or disclose a conflict of interest that might influence his actions or judgment.

4. A practitioner must disclose immediately to his client any interest, direct or indirect, that might be construed as prejudicial in any way to the professional judgment of the practitioner in rendering service to the client.

5. A practitioner who is an employee-engineer and is contracting in his own name to perform professional engineering work for other than his employer, must provide his client with a written statement of the nature of his status as an employee and the attendant limitations on his services to the client, must satisfy himself that the work will not conflict with his duty to his employer, and must inform his employer of the work.

6. A practitioner must cooperate in working with other professionals engaged on a project.

7. A practitioner shall,

i. conduct himself towards other practitioners with courtesy and good faith,

ii. not accept an engagement to review the work of another practitioner for the same employer except with the knowledge of the other practitioner or except where the connection of the other practitioner with the work has been terminated,

iii. not maliciously injure the reputation or business of another practitioner,

iv. not attempt to gain an advantage over the other practitioners by paying or accepting a commission in securing professional engineering work, and

v. give proper credit for engineering work, uphold the principle of adequate compensation for engineering work, provide opportunity for professional development and advancement of his associates and subordinates, and extend the effectiveness of the profession through the interchange of engineering information and experience.

8. A practitioner shall maintain the honor and integrity of his profession and without fear or favor expose before the proper tribunals unprofessional, dishonest or unethical conduct by any other practitioner.

NOTES

These two sections 86 and 91 are from Ontario Regulation 538/84 made under the Professional Engineers Act, 1984.

CANADIAN INFORMATION PROCESSING SOCIETY (CIPS) CODE OF ETHICS

• • •

430 King Street West, Suite 205
Toronto, Ontario
Canada M5V 1L5

Founded: 1958
Members: 6,000

CODE OF ETHICS AND STANDARDS OF CONDUCT

• • •

Foreword

The field of information processing has a large impact on society. In turn society has the right to demand that practitioners in this field act in a manner which recognizes their responsibilities toward society, to demand that the practitioners are of the highest caliber, and to demand that a mechanism exist to protect society from those practitioners who do not, or can not, live up to these responsibilities. The standards contained in this document, and our agreement to adhere to these standards, is the response of the Canadian Information Processing Society to these rightful demands.

Introduction

This document describes the code of Ethics and Standards of Conduct of the members of the Canadian Information Processing Society, with respect to their professional activities. It should not be construed to deny the existence of other ethical or legal obligations equally imperative, although not specifically mentioned.

First, the general standards and high ideals of the members of CIPS are described in the form of a Code of Ethics. Second, specific rules, the Standards of Conduct, elaborate each element of the Code in a manner which assists determination of whether or not specific activities of an individual violate the Code. They are intended to establish a minimum acceptable level of conduct, below which an individual may be said to be unethical. Third, there is a procedure which details the steps the society will follow in determining whether or not a violation of the rules has occurred, what disciplinary action is possible, and under what circumstances information will be released.

In total, this document describes the professional behavior that members of CIPS demand of themselves and their peers. All members agree to live up to these standards when the join the Society, and reaffirm this commitment each time they renew their membership.

The Code of Ethics and Standards of Conduct deal with matters that are subject to judgment and are difficult to state absolutely. They contain words such as "authority," "competence," and "faithful" which must be judged in light of the professional and moral standards in effect at a given time and place. The enforcement procedures require peers to interpret the areas requiring judgment at the specific time of the complaint using the guidelines contained in this document.

Code of Ethics

The following statements are agreed to by all members of CIPS as a condition of membership.

I acknowledge that my position as an information processing professional carries with it certain important obligations, and I will take diligent personal responsibility for their discharge.

P) To the public: I will endeavor to protect the public interest and strive to promote understanding of information processing and its application, but will not represent myself as an authority on topics in which I lack competence.

M) To myself and my profession: I will guard my competence and effectiveness as a valuable possession, and work at maintaining them despite changing circumstances and requirements. Furthermore, I will maintain high personal standards of moral responsibility, character, and integrity when acting in my professional capacity.

F) To my colleagues: I will treat my colleagues with integrity and respect, and hold their right to success to be as important as my own. I will contribute to the professional knowledge of information processing to the best of my ability.

E) To my employer and management: I will give faithful service to further my employer's legitimate best interests through management's direction.

C) To my clients: I will give frank and careful counsel on matters within my competence, and guard my client's confidential information and private matters absolutely. In my capacity of provider of product or serve, I will provide good value for my compensation, and will endeavor to protect the user of my product or service against consequential loss or harm.

S) To my students: I will provide scholarly education to my students in a sympathetic and helpful manner.

STANDARDS OF CONDUCT
• • •

The Code of Ethics is a set of ideals to which CIPS members aspire. The Standards of Conduct is intended to be more practicably enforceable.

The following statements are agreed to by all members of CIPS as a condition of membership.

Due to my obligation to the public:

P1) I will not unreasonably withhold information pertinent to a public issue relating to computing.

P2) I will not disseminate, nor allow to go unchallenged, false or misleading information that I believe may have significant consequence.

P3) I will not offer information or advice that I know to be false or misleading, of whose accuracy is beyond my competence to judge.

P4) I will not seek to acquire, through my position or special knowledge, for my own or other's use, information that is not rightly mine to possess.

P5) I will obey the laws of the country, and will not counsel, aid, or assist any person to act in any way contrary to these laws.

P6) I will endeavor to enhance public understanding of information processing, particularly its current capabilities and limitations, and the role of the computer as tool, not an authority.

Due to my obligation to myself and my profession:

M1) I will not knowingly allow my competence to fall short of that necessary for reasonable execution of my duties.

M2) I will conduct my professional affairs in such a manner as to cause no harm to the stature of the profession.

M3) I will take appropriate action on reasonably certain knowledge of unethical conduct on the part of a colleague.

Due to my obligation to my colleagues:

F1) I will not unreasonably withhold information pertinent to my work or profession.

F2) I will give full acknowledgement to the work of others.

Due to my obligation to my employer and to my management:

E1) I will accept responsibility for my work, and for informing others with a right and need to know of pertinent parts of my work.

E2) I will not accept work that I do not feel competent to perform to a reasonable level of management satisfaction.

E3) I will guard the legitimate confidentiality of my employer's private information.

E4) I will respect and guard my employer's (and his supplier's) proprietary interest, particularly with regards to data and software.

E5) I will respect the commercial aspect of my obligation to my employer.

Due to my obligation to my clients:

C1) I will be careful to ensure that proper expertise and current professional knowledge is made available.

C2) I will avoid conflicts of interest and give notice of potential conflicts of interest.

C3) I acknowledge that statements E1 to E5, cast in the employee/employer context, are also applicable in the consultant/client context.

Due to my obligation to my students:

S1) I will maintain my knowledge of information processing in those areas that I teach to a level exceeding curriculum requirements.

S2) I will treat my students respectfully as junior scholars, worthy of significant effort on my part.

Enforcement Procedures

It is essential that the Code of Ethics and Standards of Conduct be supported with clear, orderly, and reasonable enforcement procedures if the Society is to be able to discipline members who violate the Standards of Conduct. The enforcement procedures must be equitable to all parties, and must ensure that no actions are taken in an arbitrary or malicious manner. The following Enforcement Procedures have been designed with these points in mind.

The Complaint

The complaint must:

— be against a single individual, and

— be in writing, and

— cite the specific clause of the Standards of Conduct that is alleged to have been violated, and

— describe the specific action in question, and

— describe, in general terms, the substantial negative effect of that action upon the profession, the Society, a business, or an individual, and

— contain a statement that the specific action of the accused in question is or is not already or imminently [to the best knowledge of the complainant(s)] the subject of legal proceedings, and

— contain a signed statement that the facts are true to the best knowledge of the complainant(s).

This complaint must be sent to the National President of CIPS. The National President, or his delegate, will review the complaint to determine if it meets the above criteria. If it doesn't, it will be returned to the complainant(s) for possible change and re-submission. If the specific action of the accused is (imminently) the subject of legal proceedings, no further action will be taken unto those proceedings are concluded. If the complaint is not rejected then, subject to legal advice, the accused member will be notified (by Registered Mail to last known address), provided with a copy of the complaint, and allowed 30 days to prepare a written rebuttal of the complaint if so desired. The President of the Section the accused belongs to will be notified. The rebuttal should address the same points as the complaint, and must also include a statement that the facts contained in the rebuttal are true to the best knowledge of the accused.

The National President of CIPS or his delegate shall review the complaint and, if available, the rebuttal, to determine if there is sufficient evidence to hold a full hearing. If it is determined that a full hearing is warranted, the full information will be forwarded to a three member Hearing Committee appointed within 30 days of the receipt of the rebuttal or of the last date allowed for receipt of the rebuttal.

The Hearing Process

The Hearing Committee shall adhere to the following procedure:

— The Hearing Committee will attempt to interview, at the expense of CIPS, the complainant(s), and the accused, plus any other parties with relevant information. The number of people interviewed, and the extent of the effort to secure interviews, is a matter of judgment by the Hearing Committee. The Hearing Committee will decide if the accused may be present during the interviews. If the accused is not allowed to be present during the interviews, the accused shall be provided with notes documenting the substance of the interviews.

— The accused will be afforded the opportunity for a full hearing, with the complainant present if desired by the accused.

— The Hearing Committee shall have the services of legal counsel available as required. The accused, and the complainant, may obtain counsel, at their own expense, if either or both so desire.

— The Hearing Committee, after full and complete deliberation, will rule in writing as to the individual case.

Additional rules and procedures shall be established by the Hearing Committee as required in their judgment.

The Hearing Committee ruling may be:

1) a clearing of charges, or

2) a warning statement to the accused, or

3) suspension of national and local membership for a specified period of time, or

4) revocation of the current membership of the accused in the Society, and a statement of the accused's eligibility for other grades of membership.

5) Such other ruling as the Hearing Committee in its discretion sees fit (e.g.: change letterhead, business cards to delete reference to being a member of CIPS).

The Hearing Committee will prepare an opinion on the particular case that will cover the facts of the case, the action taken, and the reason for that action. This will be reviewed by the Executive Committee of the National Board of CIPS and by legal counsel at the discretion of the Executive Committee. When approved, this opinion will be sent to the accused, who may consider exercising the Appeal Process.

Due diligence should be used to provide this opinion to the accused within 120 days of the receipt of the complaint by the Hearing Committee. If this is not possible, a letter should be sent to the National President of CIPS, with copies to the accused and complainant(s), requesting an extension of this limit, and stating the reason for this request.

The Appeal Process

If not satisfied with the ruling of the Hearing Committee, the accused may appeal to the Executive Committee of the National Board of CIPS within 30 days of issuance of the Hearing Committee opinion. If appealed, the following procedure will be used.

— The Executive Committee, at its next scheduled meeting, or at a special session, shall review the opinion, and any other information available, and shall determine if:

1) a substantive procedural error has been committed by the Hearing Committee, or

2) substantial new evidence has been produced.

— The accused and the complainant are permitted legal counsel at the Executive Committee appeal session.

— The Executive Committee shall determine if, in its sole judgment, one of the two above noted criteria have been established, in which case the council shall refer the matter back to the previous or a new Hearing Committee for further proceedings.

— The decision of the Executive Committee shall be final: there shall be no further appeal.

Publication and Record Retention

After the Appeal Process and any further proceedings have been exhausted, or after completion of the time allowed to initiate an Appeal Process, the Opinion will be published in the appropriate CIPS publication if the ruling was a suspension or revocation of membership, and will be published at the request of the accused, if the ruling was a clearing of charges or issuing of warning statement.

The record of the Hearing Committee and all appropriate supporting documentation will be retained by National for five years. Response to queries may include statistical information that does not reveal detail about a specific complaint, such as the number of complaints processed, provided the approval of the Executive Committee is obtained, or responses may include copies of information previously published.

Any other information may be released only with the written permission of the Executive Committee, the accused, and the accuser(s).

NOTES

Dated January 1985. Published and promulgated on a two-sided letter-sized sheet.

CHILE

• • •

ASSOCIATION OF ENGINEERS OF CHILE CODE OF ETHICS

• • •

Avenida Santa María 1508
Casilla 13745
Santiago, Chile

Founded: 1958
Members: 18,000

Code of Professional Ethics of the Engineers of the Association of Engineers of Chile

Title I. On General Norms

1ST ARTICLE. The Code of Professional Ethics establishes the responsibilities and regulates the rights and obligations as well as the conduct of engineers.

2ND ARTICLE. It is the imperative obligation of the engineer to maintain a level of professional conduct raised to the highest moral level in defense of the prestige and prerogatives of his profession.

The norms of this Code apply to all engineering activities and professional specialization does not liberate from them.

The engineer enrolled in the Association of Engineers ought to accept, to know, and faithfully to fulfill this Code of Ethics.

Title II. On the Exercise of the Profession

3RD ARTICLE. Engineers are obligated to respect in their professional action, the dispositions of Law 12.851, the Professional Fee Schedule, and the dispositions of the present Code, and also, the agreements of the General Counsel and the appropriate Provincial Counsel.

4TH ARTICLE. Acts contrary to Professional Ethics are the following:

a) To act contrary to the decorum and prestige of the profession, contrary to the discipline of the Institution or contrary to the respect and solidarity that ought to be preserved among the members themselves.

b) To promote or to collaborate in the promulgation of laws or other norms of a legal character, resolutions, judgments or measures that infringe the rights of the engineering profession, of the Association of Engineers, or of one or more colleagues.

c) To concur with deliberate omissions that produce some of the effects indicated in the preceding letter.

d) To permit actions or omissions that favor or permit the unnecessary use of foreign engineering for objectives and work for which Chilean engineering is sufficient and adequate.

e) Engineers are obligated to denounce to the Association all persons who exercise engineering functions without the legal capacity for it, as well as to denounce all acts that indicate transgression of the norms of the Code.

f) To sign off on studies, projects, plans, specifications, reports, judgments or authorizations that have not been personally executed, studied or reviewed and to falsify consultations, the performance of jobs or the work of an organization, society or institution of any nature, in that which by law requires engagement of an association engineer.

g) To give or to receive commissions or other non-contractual benefits through managing, keeping, or granting appointments of any kind.

h) To participate directly or indirectly in the granting of professional titles that infringe or harm the prestige and professional quality of the engineer, of conformity with the principles of technology, of Engineering, laws or regulations in force.

i) To undertake some professional work, be it individually, associated with other colleagues or third parties, or as a member of a legal or def facto association, in return for the payment of a fee less than the minimum established by the Professional Fee Schedule, and to agree or to pay other colleagues, fees less than the minimum established in the Schedule of the Association.

j) To make use of or to utilize studies, projects, plans, reports or other documents related to engineering without the authorization of their authors or owners.

Title III. Relations with Colleagues and Other Professionals

5TH ARTICLE. Acts between engineers and other professionals considered contrary to professional ethics:

a) To publicize opinions that harm the prestige of a colleague.

b) To replace or try to replace a colleague, without his prior consent, in the rendering of previously engaged professional services.

c) To take undue advantage of performing a job to obtain particular clients.

d) To promote one's own appointment to a public or particular job that a colleague exercises, when this person has not manifested an intention to give it up.

e) In the formulation of proposals, public as well as private, the engineer is prohibited: to give or to solicit any information prior to the request for proposal, which would seem to leave the proposer in a favored situation with respect to others; to try to obtain a favorable decision for oneself, or for a third party, by discrediting other bidders on the proposal; or to find out about or to decide a proposal, outside established procedures on the principles or regulations that regulate such decision making.

Title IV. Relations with Directors and Clients

6TH ARTICLE. Acts considered contrary to professional ethics between engineers, directors or employers, are the following:

a) As an employee, functionary or executive of a business or organization, to accept for personal gain commissions, rebates, discounts or other benefits provided, from contractors or from persons interested in the sale of materials, equipment or services, or in the performance of work that has been entrusted to you.

b) To reveal proprietary data of a technical, financial, or personal character concerning interests confidential to your study or case.

c) To act with partiality in discharging the function of specialist, or arbiter, or to one who interpreters or awards contracts, grants, or jobs.

d) To divulge without proper authorization procedures, processes, or characteristics of equipment, that are protected by patents or contracts that establish the obligation to protect professional secrets.

TRANSLATED BY CARL MITCHAM

NOTES

This code is published and promulgated in a small pamphlet entitled *Estatutos y Códigos de Etica Profesional del Colegio de Ingenieros de Chile A.G.* [Statutes and codes of professional ethics of the Association of Engineers of Chile, Inc.] (Santiago, Chile: Colegio de Ingenieros de Chile A.G., n.d.). The pamphlet contains twenty unnumbered pages.

The first section of the pamphlet contains the statutes or by-laws of the Association (10 pages) followed by an official letter of recognition (dated 16 July 1981) from the Assistant Secretary of Economics, Development, and Reconstruction.

The second section contains the code of professional ethics of the Association of Engineers (Law 12.851—2 pages, translated here) along with a printing of the Code of Professional Ethics of the Pan American Union of Associations of Engineers (2 pages).

CHINA, PEOPLE'S REPUBLIC OF

• • •

CHINESE MECHANICAL ENGINEERING SOCIETY CODE OF ETHICS

• • •

Chapter 1. General Rules and Information

1. The Chinese Mechanical Engineering Society is a national mechanical scientists and technicians organization. It is part of the Chinese Science and Technology Society.

2. The Society is located in Peiking.

3. Our Society encourages dialectic materialism. Our goal is to unite the majority of mechanical technicians to promote mechanical industry, advance technological service, accelerate new technological research, produce more scientists, and speed up national modernization.

4. The duty of our Society is:

4.1 To open technology exchange, organize research and technical investigations and encourage the exploration and application of mechanical technology.

4.2 To offer scientific research such as proofing (theory, design), criticism and comments on equipment (machine and tools), information, etc., and to accept corporations, companies, and agencies' entrust, and to offer technology information service.

4.3 To expand technical training: offer higher education for professional technicians in order to raise the majority of technicians' knowledge levels and practice abilities.

4.4 To spread science and advance technology and science management.

4.5 To open a worldwide technology exchange and develop a good relationship with foreign technology organizations.

4.6 To control technology information: edit and publish scientific magazines and collect reports and technical documents.

4.7 To honor the scientists and technical reporters who contribute to society.

4.8 To deal with the activities and services for economical construction, and increase the majority of scientists', technical benefits and activities.

4.9 To protect the technician's right to express suggestions, ideas, and criticisms.

Chapter 2. Membership

5. Individual membership: anyone who recognizes our regulations, meets the following standards, and obtains our society's permission will become a member of our society. The individual also must:

5.1 Have been educated at a level equal to or above that of engineer, technician, professor, assistant professor, or other technical position.

5.2 Be a scientist or technician with an education level above a master's degree.

5.3 Be a college graduate with a mechanical major who has worked with related material for at least 3 years, has a certain technical knowledge level, and has the ability to work individually. However, if one does not have a college degree, an exception may be made if the individual has had many years worth of work experience which equals or surpasses our standard knowledge level.

5.4 Be a technician with extraordinary distribution.

5.5 Earnestly support the society, the chairman, the director, the manager who works with the mechanic technical organization, and the management.

6. The process for an individual to join the society is as follows: Send in the application, be introduced by other current members, and have recommendations from the company or from another technical society. After being approved, the individual will transport to our society and become a member. The individual will then be classified into whichever expert organization fits his/her work level.

7. Organizations as members: any organization, corporation or research center which earnestly support our society, and has employees who are experts in our field or related fields, can be accepted as a member.

8. Preparatory members and student members: Preparatory members: any mechanical science, technical, or managerial officers who are under 35 years of age and are college graduates or technology school graduates may commence work for a period. They may send in the application to our society, and after approval will become a member immediately.

Student members are required to have: had a mechanical major in college, received good grades junior and senior years, graduated from college, the ability to transform from student membership to preparing membership.

9. Foreign membership:

Any foreign mechanics and science technicians who are friendly to our country and want to communicate, exchange information, and participate, must send in our application, go through two members introductions or have a recommendation letter from a division of our society. The individual may also have membership in his/her own country's Mechanical Engineering Society which has participated with our society. After our approval, the individual may become a member immediately.

10. Our society may accept any well-known and respectable mechanical science technician, specialist, or scientist with great scientific accomplishment into our society.

11. The member must adhere to the rights and duties of the individual in his local technical organization.

   11.1 Members have the right to vote, and to be voted on.

   11.2 Members have the right to criticize and suggest new ideas to our society.

   11.3 Members have the right to join related technical activities.

   11.4 Members have the priority to obtain any related technical information date.

   11.5 Members must obey our society's regulation.

   11.6 Members must perform, follow and support our society's decisions and entrusted work.

   11.7 Members may join the society's different types of activities.

   11.8 Members must pay the membership fee according to regulations.

12. The Foreign Member's right and duty: The foreign member:

   12.1 May be invited to join our society and attend a science technology conference meeting, or other international technical activity, and have the meeting's registration fee reduced.

   12.2 Has the right to obtain our society's related technical information.

   12.3 Has the private right to publish and submit reports/articles in our society's magazine.

   12.4 May obtain the help of the society with the arrangement of technical visits.

   12.5 Must support our society's goals and accept the duties entrusted to him/her by our society.

   12.6 Must pay the membership fee according to the regulation.

13. Membership Card: You must get permission from the state engineering society to have a membership card. From this society, you can get Chinese Technical Engineer Prepared Membership card. The student membership card is issued by an organization member.

14. Individual and organization members have to pay annual membership fees. The payment methods and fee amounts are determined by negotiations between the society and local branches. If a member (including foreign members) does not pay the membership fee in the current year, he will be revoked of his membership rights. After failing to pay for two years the membership is automatically cancelled. Once the fee is paid we will reissue the membership card back to you.

15. Members have the right to withdraw from the society if leaving the university will cancel the student membership.

   Prepared members over the age of 35 years will also have their membership cancelled.

16. Any one who loses his/her political rights will naturally lose membership.

17. If a member's work address changes, s/he should connect with the local branch of the society.

Chapter 3. National Congress

18. The society's highest leading organization is national congress. Its jobs are:

   18.1 Checking and grading national council's work report.

   18.2 Deciding the next goal and plan of the society.

   18.3 Vote and select next direction of the council.

   18.4 Comment, check and discuss the society's regulations honoring the Scientists and societal members who have contributed to technological development.

19. National Congress is called by the national council.

20. National Representation Conference representatives are selected by National Council members and the experts in the society (people who work in the specific field).

Chapter 4. National Council

21. The National council is the leading organization after the National Congress. Its duties are:

   21.1 Execute nation congress's decision.

   21.2 Document a working report and record long term plans and work goals.

   21.3 Correct and review the society's regulations.

   21.4 Arrange the next date for the National Representative Conference.

22. The current national council members have been elected by previous council members and experts democratically; "absorb" new elected "members" and several national (or foreign) famous scientist,

expert. Then, through national representative conference's voting, produce new council members.

The total number of national council is around one hundred. They should have experience with technological research and science management. Have good moral standard, anxiously working, have good health, which can join the society's real practice wok. Any member can not been council member for over two terms.During the term, if council member can work due to accident or any other reason, after the board of national council's credential, the council member can be replaced.

23. Nation council select (vote) a director of the council, vice director, and a secretary. Board of national council. The director of the council can only work one term, then he will be one of the next term's board of national council.

24. The duty of the board of national council:

    24.1 Execute all jobs, work which given by national council.

    24.2 Make working plan and goal.

    24.3 Comment committees' working report.

    24.4 Hire people who are going work for the committee.

    24.5 Agreed, forbidden contract and negation.

    24.6 The board national council have conference every year.

25. According to the request of national council, setting several committee. Committee member works under national council's leading.

26. The national managing directors have the secretary department, the senior secretary will response for all regular works. All these senior secretary are given by the mechanical industry department.

Chapter 6. Society of Special Fields

27. We will set few major departments. These departments are responses in science study activity, engineer study or technical study activity. The national managing directors will decide how to set plan, how to regulate it and how to cancel it if it is needed.

28. The managing directors is a leader department its duty is:

    28.1 Perform the duty which is given by the national managing directors.

    28.2 Set the rule or major study activities and economic budget.

    28.3 Response in organization of different study activities.

    28.4 Give people some career advice thought state study society and city studysociety.

    28.5 Support the worker in this study society.

    28.6 The meeting of the board of directors of major study will has once a year.

29. Members in the board of directors of major study have to have the good health, and the honor technical degree. These members are introduced or elected by local departments. The board of directors should have no more than fifty people, and are elected every 4 years.

30. The board of directors has one president, three vise president, few secretaries and others. They will start their duty after the meeting of the meeting of the board of directors.

The president can not perform over one term.

31. The board of directors has two main parts: regular department (includes the secretary department, the accenting department and others beside research department) and the research department.

32. If it is necessary, the board of directors could be changed to few small boards. The small board of directors will be easy to manage and regulate.

33. The representative conference is the highest organization in state, city and local area. Any meeting, production and activity must follow the local rule.

34. According to research activity, we can set some direct and indirect relate departments to help our major study activity.

35. The duty of technical engineer study society is given by state, city or local department.

36. State, city or local department also advice the technical engineer study society to perform the job well.

Chapter 8. Relationship with Leader

37. This study society is lead by Chinese national science and mechanical systems.

Chapter 9. Fee

38. Our income comes from:

    38.1 The contribution of other co-level science research department (or companies).

    38.2 The mechanical industrial system and relate or dependent department.

    38.3 The income comes from the case research and activities.

38.4 The membership fee.

38.5 The national system, foreign system or personal contribution.

RETIRED ENGINEERS ASSOCIATION OF THE NANKING CHEMICAL-INDUSTRIAL CORPORATION CODE OF ETHICS

• • •

First Chapter: General Whole Principle

1. This Association is named "The Retired Engineers Association of The Nanking Chemical-Industrial Corporation." We simple call it "The Retired Association of Nanking Chemical."

2. "The Retired Association of Nanking Chemical" is a system which is consisted by retired engineers (include high technical workers) and under the leading of the communism party. This system will help the Corporation's development. It is a part of The Engineers Association of The Nanking Chemical-Industrial Corporation. Their action will open and develop the technology the technology. Making greater progress; take one more step forward.

3. The principle of this association is to combine and organize all retired engineers. Just as "The Older have some thing to feet; to learn and to practice". According to company's need, to do some Technical help and service.

Chapter 2: Duty

4. Must follow the policy that "Economic growth will dependent on the develop of technology. Technical work must face to economic growth." Manifest the point of "blooming in profusion; using all resources;" execute the democratize in this association. To have a good quality service.

5.1 Face to economical construction, explicate these retired engineers technical knowledge. Supply some suggestions to decisions of different departments improvement. To become a good helper.

5.2 For the company's business, They need to help this company to develop their own technology and learn some new knowledge from the advance countries.

5.3 For science developing, and helping those young engineers, We should offer some classes which can help younger to learn more experience.

5.5 Combine all strains; collect and exchange the sciences information; At some time, should learn English and translate them to Chinese (for us to learn to use).

5.6 Friendly to neighbor companies and related companies. This can help us to learn technology or exchange technology with them.

5.7 Tells the company what ideas do they suggest and what do they want. Study policy, technology, visit and help new members are very necessary.

5.8 Respect the older engineers emotion; respect their life, their health. Set up friendship.

CHAPTER 3: MEMBERSHIP

• • •

6. If you are a junior engineer or above, with a good health and must under 70 year old. And if you agree our associate principle you can fill a application form. Then we will give you a membership card after we discuss your case. For reach a good quality service, we will invite some special technical retired engineers to join with us.

7. The power of member:

7.1 Right for election; Right to be elected; Right to be cancelled.

7.2 Have the right to hear or get the new information and resources.

7.3 Have right to give the suggestion, to criticize the incorrect decision which made by association.

7.4 Have right to join the science, technical research; right to get pay.

7.5 Have right to though this association to tell self-request or other members request.

8. Responsibility of members.

8.1 Respect the principle of association, and execute the decision of association.

8.2 Join the active of association; hand the job that other associative ask for.

8.3 Keep professional morality. Maintain and protect the prefect and reputation. Never be allowed to damage the reputation of our association.

8.4 For some secret science information with a mark "Secret", no one be allowed to divulge a secret.

8.5 Must pay the membership fee on time.

9. Member has right to drop-out the membership. Member can fill out a application for drop-out. He (she) should return his (her) membership card after the association's agreement.

10. If member with no reason, and never perform any member's obligation in one year. He (she) will be cancelled from the membership and be requested to return membership card.

11. Any member who damage the principle of association, violate the benefit and reputation of our association, and also doesn't listen to advise, membership will be cancelled, or be punished.

12. Any member who performed illegal activity and be punished or get in jail, will be cancelled from membership.

13. Any decision of cancelling membership will notice to all members. This is the reference for some department in the future.

CHAPTER 4

• • •

14. Membership meeting is the most powerful in the association. This meeting has one in two years. Date of meeting can be changed if it is necessary.

15. Duty and responsibility of membership meeting:

15.1 Decide the main working principle and duty.

15.2 Listen and exam the working report and economic report of a aboard directors.

15.3 Fix and declare the principle of association.

15.4 Select the new director of board.

16. The board of directors will selected by members. The chairman of the board of directors will selected by the boards of director. The board of directors includes one chairman, one secretary and few wise-chairmen.

17. The chairman of board of directors has right to control the board and has right to use one wise-chairman work with him.

18. We will invest some consul for performing advises.

19. Set two people work in secretary apartment everyday. We will add more departments if we need.

CHAPTER 5

• • •

20. Our active fee from:

20.1 National or some related departments' help

20.2 Income of science resources and technology services

20.3 Membership fee

20.4 Receive subscribe money from corporation or personal.

20.5 Other current income.

21. Active fee will use for:

21.1 Perform the duty and develop activity

21.2 For engineers' additional perform payment.

21.3 Some request office supply expense.

21.4 Expense of some professional (senior) engineers training younger and performing technical service.

21.5 Other expenses.

22. Must set up a strong business rule and oversee the rule. The money will be controlled by the board of directors. Any one who want to use money should go to the board. Though wise-chairman, filled out a application. He (she) can use the money only if the application be agreed.

COLOMBIA

• • •

COLOMBIAN SOCIETY OF ENGINEERS

• • •

Carrera 4 N. 10-41
Bogotá, Colombia

Founded: 1887
Members: 1800

Code of Professional Ethics

The honor and dignity of the profession ought to be for the Engineer his or her major pride; as a result, in order to extol the profession, he should conform his conduct to the following norms that constitute his Code of Professional Ethics:

1. To exercise the profession as well as the activities derived from it with decorum, dignity, and integrity.

2. To always work under the assumption that the exercise of the profession constitutes not only a technical activity but also a social function.

3. To always act honorably and loyally with persons or entities to which services are offered.

4. To abstain from receiving gratuities and rewards other than the agreed upon salary or honorarium.

5. To not use with colleagues unfair methods of competition such as under bidding or offering professional services at a lower than standard price.

6. To try neither to supplant another engineer when a contract has already been awarded or a position determined nor to replace an honorable and competent employee.

7. To abstain from an intervention that would unfairly affect the professional reputation of a colleague.

8. To limit advertised services exclusively to those for which one is qualified by academic education or professional experience.

9. To not propose competitive bidding in which the value of the professional honorarium will be one of the factors that determines the selection of engineering consulting services, nor to participate in such competitive bidding.

10. Finally, to have due respect and consideration for all colleagues.

TRANSLATED BY JUAN LUCENA
AND CARL MITCHAM

COSTA RICA

• • •

FEDERAL ASSOCIATION OF ENGINEERS AND ARCHITECTS OF COSTA RICA CODE OF ETHICS

• • •

Apartado 2346
1000 San José, Costa Rica

Founded: 1971
Members: 6,000

Code of Professional Ethics

The following acts are unethical:

A. In relation to the Profession:

a) To perform in bad faith acts that have been established as contrary to good techniques or to incur voluntary omissions even if it be in compliance with the orders of authorities or mandates.

b) To accept a job knowing that it may lend itself to malice or fraud or may be against the general interest;

c) To sign plans, specifications, recommendations, records or reports which have not been executed, studied, or seen personally, except those documents which, in themselves are objects of public faith and must be exercised personally. (As reformed in session 3-82 A.E.R.)

d) To associate one's name with propaganda or activities with persons who appear unqualified as professionals, to honor disproportionately persons or things to commercial or political ends.

e) To receive or give commissions or other benefits for promoting, obtaining or determining plans of any class or in the assignment of professional jobs.

f) To violate or comply with others in violating the laws of the Federated Association or the Codes, Norms, and Rules which are indicated here, in relation to the exercise of the profession.

B. In relation to Colleagues:

a) To utilize ideas, plans or technical documents without consent of the authors.

b) To participate in competitions of price or with a price that is less than that which is established as the minimum by the Federated Association to contract a professional job.

c) To attempt to injure, falsely or maliciously, directly or indirectly the professional reputation, situation or business of another member of the Federated Association.

d) To attempt to supplant fraudulently another engineer or architect after he has made definitive steps in his occupation.

e) To use favors or offer commissions in order to obtain professional work, directly or indirectly.

f) To nominate or intervene so that another should be nominated to be in charge of technical jobs that must be undertaken by a professional, when nominee does not have needed qualifications.

g) To compete unloyally with one's colleagues who work on contract by using the advantages of a position in a company.

h) To promote propaganda in language that is boastful or in any way that affects the dignity of the profession.

i) To establish or influence the establishment of honorariums or remunerations for engineering or architecture, when such honorariums or remunerations obviously present a compensation that is inadequate for the importance and responsibility of the services to be rendered.

j) To act in any manner or compromise oneself in any manner or practice which serves to discredit the honor and dignity of the profession of engineering and architecture.

C. In Relation to the Constituents or Employers:

a) To accept for one's own benefit commissions, discounts, or bonuses from materials providers, contractors or persons concerned in the execution of a job.

b) To reveal reserved technical, financial or personal data about the confidential interests in his study or his contract which is under his care for constituents or employers.

c) To act on behalf of his constituents or employers in a professional capacity or other manner which is not the manner of a loyal and non-prejudiced agent, as trustee, expert or arbiter in any contract or engineering or architectural job.

TRANSLATED BY ANNA H. LYNCH

NOTES

A note at the top of this statement reads as follows:

The assembly of representatives of the Federated College of Engineers and Architects of Costa Rica, based on the mandates of the "Ley Organica del Colegio" number 4925 dated 17 December 1971, reformed by (the) number 5361 dated 16 October 1973, article 23, incise d), in session number 7-74 A.E.R. on the 24th of May, 1974, agreed to approve the following Code of Professional Ethics of the Federated College of Engineers and Architects of Costa Rica, which says the following:

At the bottom it is noted "Approved in the assembly of representatives in meeting on the 21st of May, 1974."

Following the code are two notes, as follows:

This code is in force as of its publication in the Official Diary. San Jose, June, 1974, -Carlos Alejandro Garcia Bonilla, Executive Director. Reformed by the Assembly of Representatives of the Federated College in session number 4-76 A.E.R. 4th Article, Thursday, the 4th of March, 1976 with the addition of incision f (to Article A).

When formed in 1971, the Colegio Federado unified five professional associations.

A.E.R. stands for Asemblea Extraordinaria de Representantes.

DOMINICAN REPUBLIC

• • •

DOMINICAN ASSOCIATION OF ENGINEERS, ARCHITECTS, AND SURVEYORS CODE OF ETHICS

• • •

Calle Padre Billini No. 58
Zona Colonial, Apartado 1514
Santo Domingo, Dominican Republic

Founded: 1945
Members: 10,300

Code of Professional Ethics

It is considered contrary to ethics and incompatible with the dignified exercise of the profession for a member of the Dominican Association of Engineers, Architects, and Surveyors:

1st To act in any way that tends to diminish the honor, dignity, respect, honesty, ability and other attributes that support the full exercise of the profession.

2nd To violate, to permit the violation of, or to influence the violation of the laws and regulations related to the exercise of the profession.

3rd To utilize positions in official, semi-official, autonomous or private organizations or institutions to act with disloyalty contrary to the genuine national interests or that would have consequences contrary to the good involvement of professionals.

4th To receive, offer, or confer improper commissions, or to utilize influences in conflict with legitimate competence in order to secure the conference of contracts, works, or the execution of projects as a special favor, or as a favor to ones associates or partners.

5th To offer oneself for the performance of functions or specialties for which one does not have reasonable capacity and experience.

6th To present or talk about oneself in laudatory terms or in any form that acts against the dignity and seriousness of the profession.

7th To exempt oneself by convenience, collusion, or ties of friendship or family from fulfilling the duties that his position or job requires him to do or to respect.

8th To offer, solicit, or render professional services for remunerations below those established as a minimum in Professional Fee Schedule of the Dominican Association of Engineers, Architects, and Surveyors.

9th To sign without permission surveys, calculations, designs or any other intellectual work that is the fruit of the labor of other professionals.

10th To make oneself responsible for works or projects which are not under one's immediate direction, revision, or supervision.

11th To take charge of a work without having completed all technical studies necessary for its correct execution, or when for the realization of such a work there have been appointed terms, prices, or any other conditions in conflict with the good practice of the profession.

12th To use the inherent advantages of a remunerated position in order to compete with the practicing professional independently of other professionals.

13th To act against the reputation and/or legitimate rights and interests of other professionals.

14th To acquire interests that directly or indirectly collide with those of the interests of the company or clients that employ one's services, or to take charge without the knowledge of interested parties of works in which there exist antagonistic interests.

15th To contravene deliberately the principles of justice and loyalty in one's relations with clients, personnel subordinates, and workers; in relation to the last, in a special manner in that relevant to maintaining equitable work conditions and to their just participation in profits.

16th To supplant or intend to supplant a colleague in a particular contract after a definitive decision has been made to employ him for this contract, and to substitute through political or ideological arrangements of a discriminatory or arbitrary character a professional colleague who has been terminated or suspended from his functions.

17th To propitiate, serve as instrument for, or support with one's name the unjust replacement of Dominican professionals by foreign companies or persons settled in the country, or to do the same if living abroad.

18th To intend by any means to undermine and/or slight the prestige of the Dominican Association of Engineers, Architects, and Surveyors, and in any form to contribute, support, or encourage that there be abolished or eliminated the laws, rules, principles, ends, and purposes of the Association without the consent of its competent organs, or to provoke in any way the disintegration or weakening of the instituted organs of the Association.

19th To intend to pervert the principles, ends, and purposes of the Dominican Association of Engineers, Architects, and Surveyors, and in any form to contribute, support, or encourage the abolition of the laws and rules of the Association without the consent of its competent organs or in any way to support the disintegration or weakening of the instituted organs of the Association.

TRANSLATED BY CÉSAR CUELLO NIETO AND
CARL MITCHAM

NOTES

According to a parenthetical note following the code, "This code of ethics was approved by the Assembly of Representatives of the Dominican Association of Engineers, Architects, and Surveyors in session 11 October 1969."

FINLAND

• • •

ENGINEERING SOCIETY OF FINLAND CODE OF ETHICS

• • •

Banvaktsg. 2
00520 Helsinki, Finland

Founded 1880
Members: 2,440

Code of Honor

In full knowledge of my rights and duties as a graduate engineer or architect, I will, in all my acts and deeds, obey the rules of life contained in this code of honor.

In my profession, I will not accept bribes. I will be tolerant. I know my duty to be the service of both my country and mankind as a whole.

In the recognition that my own knowledge and skills are inherited from the efforts of individuals over millennia, it is my desire to develop technology and engineering further, and especially to strive to teach the younger generation of engineers and architects the skills and traditions of my profession.

In addition to the development of technology, I will also be responsible for its right application and use, so that its consequences cause damage neither to society nor the individual.

I will participate only in honorable enterprises and deeds and will not take part in activities detrimental to the reputation or honor of engineers and architects.

I will respect the right of another to his ideas, publications and other results of his creativity.

I will strive to protect the interests and good name of every honorable engineer and architect, but if duty demands, I will not shrink from declaring the truth about anyone who has forfeited his right to this profession.

In my activities and strivings for position, I will use only loyal measures and will not attempt to damage my colleagues by unjustified criticism, and if I observe such an attempt, I will do my best to defeat it.

The employer or client for whom I am working can be assured that I will faithfully serve his best interests.

I will do my work well in order to justify honorable payment and will promote the development of my subordinates, as well as the quality of their working conditions and their remuneration.

I regard the participation of engineers and architects in public life, at local and national levels, to be an important factor in the development of our society.

I will continually cultivate my professional knowledge and competence and develop my personality by all means available; and I will remember that in my life and work I also represent the whole professional body of architects and graduate engineers.

NOTES

This code is promulgated in English as a one-page document with decorative border suitable for framing.

At the bottom it states that the code was "adopted in the meeting of the Council of The Engineering Society of Finland—STS, 16th December, 1966."

FRANCE

• • •

CONSEIL NATIONAL DES INGÉNIEURS ET DES SCIENTIFIQUES DE FRANCE (CNISF)

• • •

NATIONAL COUNCIL OF ENGINEERS AND SCIENTISTS OF FRANCE

• • •

Charter of Ethics of the Engineer

Preamble

As they become more and more powerful, technologies promote major changes in everyday life, in the transformation of our society and its environment, while they also bring with them risks of serious harms. Additionally, while their complexity makes them difficult to comprehend, and the force of information increases, misinformation can introduce in public opinion exaggerated worries about security, with baseless psychoses and irrational fears.

Consequently engineers must assume an essential double role in society, first as those who control these technologies in service of the human community, and second as those who diffuse information about the real possibilities and limitations and assessments of the benefits and the risks they generate.

Because of the special characteristics of the exercise of their profession, engineers must conduct themselves with a certain rigor; it becomes more and more imperative that they explicitly clarify the reference points used and reasons for their conduct. This is why the National Council of the Engineers and the Scientists of France has produced a Charter of Ethics. This Charter must be considered as the profession of faith of all those who are listed in the Registry of French Engineers created by the CNISF.

As a reference for engineers, the Charter will help engineering students prepare for the exercise of their profession. It will enable the values that guide engineers to be better comprehended by everyone.

The Charter annuls and replaces the old CNISF "code of ethics."

The term "code of ethics" will henceforth be reserved for documents that define the correct professional conduct in each of the fields of engineering and whose non-observance could entail the application of sanctions.

The CNISF thanks in advance all those who, through their contributions, help the Charter become known, appreciated, enduring, and improved.

Engineers in Society

- Engineers are responsible citizens establish the link between science, technology, and the human community; they are involved in civic action for the common good.

- Engineers spread their knowledge and pass on their experience to serve society.

- Engineers are conscious and make society aware of the impact of technological achievements on the environment.

- Engineers act to ensure the "sustainable development" of resources.

Engineers and Their Abilities

- Engineers are a source of innovation and the engine of progress.

- Engineers are objective and methodical in their procedures and judgments. They attempt to explain the foundations of their decisions.

- Engineers regularly update their knowledge and their abilities according to the evolution of science and technology.

- Engineers listen to their peers; they are open to all other disciplines.

- Engineers know how to admit their mistakes, take them into account, and learn lessons for the future.

Engineers and Their Profession

- Engineers fully use their abilities, while being conscious of their limitations.

- Engineers loyally respect the culture and values of their companies and those of their peers and clients. They would not act contrary to their professional conscience. If need be, they accept the consequences of any contradictions that may arise.

- Engineers respect the opinions of their professional peers. They listen and are open in discussions.

- Engineers behave toward their collaborators with loyalty and equality without any discrimination. They encourage them to develop their abilities and help them to fully realize the potential in their professions.

Engineers and Their Assignments

- Engineers try to attain the best result in utilizing the best means available and in the integration of human, economic, financial, social, and environmental dimensions.

- Engineers take into account all the constraints that their assignments impose, especially with respect to health, safety, and the environment.

- Engineers integrate in their analyses and decisions the ensemble of legitimate interests of their assignments, as well as consequences of any kind on other persons and their welfare. They anticipate risks and the probabilities; they work hard to take advantage of them and to eliminate negative effects.

- Engineers are rigorous in analysis, methods, and in making decision and solution choices.

- Faced with unexpected situations, engineers immediately take permitted initiatives to create better conditions, and directly inform the appropriate persons.

TRANSLATED BY CARL MITCHAM

GERMANY

• • •

ASSOCIATION OF GERMAN ENGINEERS CODE OF ETHICS

• • •

Graf-Recke-Strasse 84
Postfach 1139
W-4000 Düsseldorf 1, Germany

Founded: 1856
Membership: 95,000

ENGINEER'S CONFESSIONS

• • •

The *ENGINEER* should pursue his profession with respect for values beyond science and knowledge and with humbleness toward the Almighty who governs his earthly existence.

The *ENGINEER* should place professional work at the service of humanity and maintain the profession in those same principles of honesty, justice, and impartiality that are the law for all people.

The *ENGINEER* should work with respect for the dignity of human life and so as to fulfill his service to his fellowmen without regard for distinctions of origin, social rank, and worldview.

The *ENGINEER* should not bow down to those who disregard human rights and misuse the essence of technology; he should be a loyal co-worker for human morality and culture.

The *ENGINEER* should always work together with his professional colleagues for a sensible development of technology; he should respect their activity just as he expects them to rightly value his own creativity.

The *ENGINEER* should place the honor of his whole profession above economic advantage; he should behave so that his profession is accorded in all public arenas with as much respect and recognition as it deserves.

Düsseldorf, May 12th 1950

TRANSLATED BY CARL MITCHAM

FUNDAMENTALS OF ENGINEERING ETHICS

• • •

Preface

Natural sciences and engineering are important forces shaping our future. They exert both positive and negative influences upon our world. We all contribute to these changes. The engineering professions, however, have a particular responsibility in structuring these processes. Hence in 1950, the Association of Engineers VDI in Germany presented a document on the specific professional responsibilities of engineers.

Recently the VDI Executive Board passed the new document "Fundamentals of Engineering Ethics." They are intended to offer to all engineers, as creators of technology, orientation and support as they face conflicting professional responsibilities.

These fundamentals have been proposed by the "VDI philosophers" together with representatives of other disciplines within the VDI Committee on People and Technology.

I hope that this document may strengthen awareness and commitment in dealing with ethical issues of the engineering professions.

Dusseldorf, March 2002

Prof. Dr.-Ing. Hubertus Christ, President of the VDI

O. PREAMBLE

• • •

Engineers recognize natural sciences and engineering as important powers shaping society and human life today and tomorrow. Therefore engineers are aware of their specific responsibility. They orient their professional actions towards fundamentals and criteria of ethics and implement them into practice. The fundamentals suggested here offer such orientation and support for engineers as they are confronted with conflicting professional responsibilities.

The Association of Engineers in Germany (VDI)

• contributes to raising awareness about engineering ethics,

• offers consultancy and conflict resolution, and

• assists in all controversies related to issues of responsibility in engineering.

1. *Responsibilities*

1.1 Engineers are responsible for their professional actions and the resulting outcomes. According to professional standards, they fulfill their tasks as they correspond to their competencies and qualifications. Engineers perform these tasks and actions carrying both individual and shared responsibilities.

1.2 Engineers are responsible for their actions to the engineering community, to political and societal institutions as well as to their employers, customers, and technology users.

1.3 Engineers know the relevant laws and regulations of their countries. They honor them insofar as they do not contradict universal ethical principles. They are committed to applying them in their professional environment. Beyond such application they invest their professional and critical competencies into improving and developing further these laws and regulations.

1.4 Engineers are committed to developing sensible technology and technical solutions. They accept responsibility for quality, reliability, and safety of new technical products and processes. Their responsibilities include technical documentation as well as informing customers about both appropriate use and possible dangers of misuse of new technical solutions.

They furthermore include:

• defining the technical characteristics of such products and processes,

• suggesting alternative technical solutions and approaches, and

- taking into consideration the possibilities of unwanted technological developments and deliberate misuse of products and processes.

2. Orientation

2.1 Engineers are aware of the embeddedness of technical systems into their societal, economic and ecological context. Therefore they design technology corresponding to the criteria and values implied: the societal, economic and ecological feasibility of technical systems; their usability and safety; their contribution to health, personal development and welfare of the citizens; their impact on the lives of future generations (as previously outlined in the VDI Document 3780).

2.2 The fundamental orientation in designing new technological solutions is to maintain today and for future generations, the options of acting in freedom and responsibility.

Engineers thus avoid actions which may compel them to accept given constraints (e.g. the arbitrary pressures of crises or the forces of short-term profitability). On the contrary, engineers consider the values of individual freedom and their corresponding societal, economic, and ecological conditions the main prerequisites to the welfare of all citizens within modern society—excluding extrinsic or dogmatic control.

2.3 Engineers orient their professional responsibility on the same fundamentals of ethics as everybody else within society. Therefore engineers should not create products which are obviously to be used in unethical ways (e.g., products banned by international agreement). Furthermore they may not accept far-reaching dangers or uncontrollable risks caused by their technical solutions.

2.4 In cases of conflicting values, engineers give priority:

- to the values of humanity over the dynamics of nature,
- to issues of human rights over technology implementation and exploitation,
- to public welfare over private interests, and
- to safety and security over functionality and profitability of their technical solutions.

Engineers, however, are careful not to adopt such criteria or indicators in any dogmatic manner. They seek public dialogue in order to find acceptable balance and consensus concerning these conflicting values.

3. Implementation

3.1 Engineers are committed to keeping up and continually developing further their professional skills and competencies.

3.2 In cases of conflicting values, they are expected to analyze and weigh controversial views through discussions that cross borders of disciplines and cultures. In this way they acquire and strengthen their ability to play an active part in such technology assessment.

3.3 In all countries, national laws and regulations exist which concern technology use, working conditions, and the natural environment. Engineers are aware of the relevance of engineering ethics for these laws and regulations.

Many of these laws today take up controversial issues related to open questions in engineering sciences and ethics. Engineers are challenged to invest their professional judgment into substantiating such questions.

Concerning national laws, the sequence of priorities is as follows: national laws have priority over professional regulations, such professional regulations have priority over individual contracts.

3.4 There may be cases when engineers are involved into professional conflicts which they cannot resolve co-operatively with their employers or customers. These engineers may apply to the appropriate professional institutions which are prepared to follow up such ethical conflicts. As a last resort, engineers may consider to directly inform the public about such conflicts or to refuse co-operation altogether. To prevent such escalating developments from taking place, engineers support the founding of these supporting professional institutions, in particular within the VDI.

3.5 Engineers are committed to educational activities in schools, universities, enterprises and professional institutions with the aims of promoting and structuring technology education, and enhancing ethical reflection on technology.

3.6 Engineers contribute to developing further and continually adapting these fundamentals of engineering ethics, and they participate in the discussions corresponding.

Fundamentals of Engineering Ethics Summary

- Engineers are responsible for their professional actions and tasks corresponding to their competencies and qualifications while carrying both individual and shared responsibilities.
- Engineers are committed to developing sensible and sustainable technological systems.
- Engineers are aware of the embeddedness of technical systems into their societal, economic and ecological context, and their impact on the lives of future generations.

- Engineers avoid actions which may compel them to accept given constraints and thus lead to reducing their individual responsibility.

- Engineers base their actions on the same ethical principles as everybody else within society. They honor national laws and regulations concerning technology use, working conditions, and the natural environment.

- Engineers discuss controversial views and values across the borders of disciplines and cultures.

- Engineers apply to their professional institutions in cases of conflicts concerning engineering ethics.

- Engineers contribute to defining and developing further relevant laws and regulations as well as political concepts in their countries.

- Engineers are committed to keeping up and continually developing further their professional skills and competencies.

- Engineers are committed to enhancing critical reflection on technology within schools, universities, enterprises, and professional institutions.

TRANSLATED BY CARL MITCHAM

HONDURAS

• • •

ASSOCIATION OF CIVIL ENGINEERS OF HONDURAS CODE OF ETHICS

• • •

CONSIDERING:

That it is urgent that the Code of Professional Ethics be put into practice to guard and sanction the professional conduct of the members of the association;

CONSIDERING:

That the standards that regulate the subject as established by the Organic Law contain guidelines that are general and not concrete ones dealing with particulars;

CONSIDERING:

That it is the obligation of the Directing Council to propose to the General Assembly Regulations of the Association that conform to the Organic Law and to promulgate resolutions that will insure compliance with these Regulations;

CONSIDERING:

That it is necessary to have a Code of Professional Ethics that meets the needs of the growing Association of Civil Engineers of Honduras (CICH);

THEREFORE:

The 38th Regular General Assembly of the Association of Civil Engineers of Honduras (CICH), using the power conferred by Article 16, section (c) of the Organic Law,

AGREES

To the following:

CODE OF PROFESSIONAL ETHICS

• • •

Chapter I

FUNDAMENTAL PRINCIPLES

Engineers ought to maintain and respect the integrity, honor, and dignity of the engineering profession:

I. Utilizing their knowledge and ability to improve human welfare.

II. Being honest and impartial and faithfully serving the public, their employees, and clients.

III. Striving to improve the capability and the prestige of the profession.

IV. Supporting technical and professional societies within their disciplines.

Chapter II

STANDARDS OF ETHICS

Article 1.

—Any colleague who transgresses from one or more of the duties or obligations stipulated by the present code in either his personal character or his engineering firm is considered in contempt of the ethics.

Article 2.

—The ethical misdeeds may be considered "slight," "serious," "grave," or "very grave."

Article 3.

It is the responsibility of the Honor Tribunal of the Association of Civil Engineers of Honduras to determine the qualification that corresponds to a transgression or a group of transgressions incurred by a colleague.

If more than one transgression is committed by the same student it cannot be qualified as "slight" even though each error considered individually may merit such qualification.

Article 4.

—Ethical transgressions are:

A) Toward the Profession:

a) To act in any way that serves to diminish the honor, respectability, and the virtues of honesty, integrity, and truthfulness that should serve as the basis for a full and complete exercise of the profession;

b) To exercise bad faith, engage in acts contrary to good technique, or to be involved in culpable omissions, even if it is done in order to comply with orders from superiors or to comply with commands;

c) To accept a job knowing that it is may lend itself to an evil deceit or be against the general good;

d) To sign as author any title for free or purchased plans, specifications, judgments, accounts, or any other professional information laid out by others;

e) To take charge of projects or works which are not under his immediate direction, review or supervision;

f) To associate with or to have his name linked with propaganda or activities involving people or entities who exercise or practice the engineering profession illegally;

g) To put himself forward for employment in specializations and operations for which he has no capacity, preparation, and reasonable experience;

TRANSLATED BY CARL MITCHAM AND
ANNA H. LYNCH

HONG KONG

• • •

THE HONG KONG INSTITUTION OF ENGINEERS CODE OF ETHICS

• • •

9/F Island Centre
No. 1 Great George Street
Causeway Bay, Hong Kong

Founded: 1975
Membership: 7,376

Rules of Conduct

Introduction

The Ordinance and Constitution make it clear that members are required to conduct themselves in a manner which is becoming to professional engineers, as may be seen from the following general statement from clause (1) of Article 12 of the Constitution:

> "Every member shall at all times so order his conduct as to uphold the dignity and reputation of the Institution and act with fairness and integrity towards all persons with whom his work is connected and towards other members."

The Council, in clause (3) of Article 12 of the Constitution, is required to make specific rules which are to be observed by members, and such rules have been drawn up and approved by the Council. These rules, given below, set the standard for the conduct of all Institution members, though they are not wholly relevant to Students.

If members have any comments to make on the application of these rules to the real life situation it would be appreciated if they would send their contributions to the Secretary, preferably before the end of August, for the consideration of the Rules of Conduct Working Party.

Rules of Conduct

Rule 1: Responsibility to the Profession. A member of the Institution shall order his conduct so as to uphold the dignity, standing and reputation of the profession. In pursuance of which a member shall, inter alia:

1.1 discharge his professional responsibilities with integrity, dignity, fairness and courtesy;

1.2 not allow himself to be advertised in self-laudatory language nor in any manner derogatory to the dignity of his profession, nor improperly solicit professional work for himself or others;

1.3 give opinions in his professional capacity that are, to the best of his ability, objective, reliable and honest;

1.4 take reasonable steps to avoid damage to the environment and the waste of natural resources or the products of human skill and industry;

1.5 ensure adequate development of his professional competence;

1.6 accept responsibility for his actions and ensure that persons to whom he delegates authority are sufficiently competent to carry the associated responsibility;

1.7 not undertake responsibility which he himself is not qualified and competent to discharge;

1.8 treat colleagues and co-workers fairly and not misuse the advantage of position;

1.9 when working in a country other than Hong Kong order his conduct according to the existing recognized standards of conduct in that country, except that he should abide by these rules as applicable in the absence of local standards.

1.10 when working within the field of another profession pay due attention to the ethics of that profession.

Rule 2. Responsibility to Colleagues. A member of the Institution shall not maliciously or recklessly injure nor attempt to injure whether directly or indirectly the professional reputation of another engineer, and shall foster the mutual advancement of the profession. In pursuance of which a member shall, inter alia:

2.1 where appropriate seek, accept and offer honest criticism of work and properly credit the contributions of others;

2.2 seek to further the interchange of information and experience with other engineers;

2.3 assist and support colleagues and engineering trainees in their professional development;

2.4 not abuse his connection with the Institution to further his business interest;

2.5 not maliciously or falsely injure the professional reputation, prospects or practice of another member provided however that he shall bring to the notice of the Institution any evidence of unethical, illegal or unfair professional practice;

2.6 support the aims and activities of the Institution.

Rule 3. Responsibility to Employers or Clients. A member of the Institution shall discharge his duties to his employer or client with integrity. In pursuance of which a member shall, inter alia:

3.1 offer complete loyalty to his employer or client, past and present, in all matters concerning remuneration and in all business affairs and at the same time act with fairness between his employer or client and any other part concerned;

3.2 inform his employer or client in writing of any conflict between his personal or financial interest and faithful service to his employer or client;

3.3 not accept any financial or contractual obligation on behalf of his employer or client without their authority;

3.4 where possible advise those concerned of the consequences to be expected if his engineering judgment, in areas of his responsibility, is overruled by non-technical authority;

3.5 advise his employer or client in anticipating the possible consequences of relevant developments that come to his knowledge;

3.6 neither give nor accept any gift, payment or service of more than nominal value to or from those having business relationships with his employer or client without consent of the latter;

3.7 where necessary co-operate with, or arrange for the services of, other experts wherever an employer's or client's interest might best be served thereby.

Rule 4. Responsibility to the Public. A member of the Institution in discharging his responsibilities to his employer and the profession shall at all times be governed by the overriding interest of the general public, in particular their welfare, health and safety. In pursuance of which a member shall, inter alia:

4.1 seek to protect the safety, health and welfare of the public;

4.2 when making a public statement professionally, try to ensure that both his qualification to make the statement and his association with a benefiting party are made known to the recipients of the statement;

4.3 seek to extend public understanding of the engineering profession.

NOTES

Published in *Hong Kong Engineer* 12(7) (July 1984): 7–8.

INDIA

• • •

INDIAN INSTITUTE OF CHEMICAL ENGINEERS CODE OF ETHICS

• • •

Dr. H.L. Roy Building
Raja Subodh Mullick Road

Post Box No. 17001
Calcutta 700032, India
Founded: 1947

CODE OF ETHICS FOR MEMBERS

• • •

INDIAN INSTITUTE OF CHEMICAL ENGINEERS EXPECTS ALL ITS INDIVIDUAL MEMBERS TO BE GUIDED IN THEIR PROFESSIONAL LIFE AND CONDUCT BY THE FOLLOWING CODE OF ETHICS

1. Members shall be guided by the highest standards of integrity in all their professional dealings.

2. The members shall uphold the dignity of the profession and the reputation of the Institute.

3. The members shall avoid sensationalism and misleading claims and statements. In making first publication concerning inventions, discoveries or improvements in their fields, the members shall use the channels of recognized scientific societies or standard technical publications or periodicals.

4. The members shall endeavor at all times to give credit for work to those who, as far as their knowledge goes, are the real authors of such work.

5. The members shall provide sufficient opportunity and take responsibility for the training and development of other engineers under their change.

6. If a member considers another member guilty of unethical practice, he shall present the information to the Council of the Institute. He shall endeavor to avoid, under all circumstances, injuring the reputation of any member directly or indirectly.

7. The members shall not misrepresent their qualifications to clients, employers or others with whom they come in contact in their profession.

8. The members shall not divulge or make use of any confidential information or findings of clients, employers, or professional committees/commissions to which they are appointed as members for their personal gain without prior consent of the concerned authority.

9. The members shall uphold the principle that unreasonably low professional charges encourage inferior and unreliable work. This does not, however, preclude them from honorary work for professional/national advancement.

10. The members should inform their clients or employers of any interest in a business which may compete with or affect the interest of their clients or employers.

11. The members shall refuse to undertake for compensation work which they believe will be unprofitable to clients, without first advising the clients as to the improbability of successful results.

12. When called upon to undertake the use of inventions, equipment, processes and products in which a member has a financial interest, he shall make his status clear before engagement.

13. The members shall always give complete and accurate reports for promotion of business/enterprises and avoid unnecessary claims.

14. The members shall not indulge in any occupation which is contrary to law or public welfare.

INDIAN NATIONAL ACADEMY OF ENGINEERING CODE OF ETHICS

• • •

c/o Institution of Engineers (India) Bldg.
Bahadur Shah Zafar Marg
New Delhi 110002, India

Founded: 1987
Members: 128

Obligation

As a Fellow of the Indian National Academy of Engineering, I shall follow the code of ethics, maintain integrity in research and publications, uphold the cause of Engineering and the dignity of the Academy, endeavor to be objective in judgment, and strive for the enrichment of human values and thoughts.

Signature

Name in full

NOTES

This code is in the form of an obligation which has to be signed by every Fellow upon admission to the Academy. S.N. Mitra, Honorary Secretary of the Academy, explains the undefined reference to "the code of ethics" by simply noting (in a letter dated October 19, 1990) that "We do not have any elaborate Code of

Ethics for the Fellows of our Academy. We have only the Obligation Form, which, in a sense, is a summarized version of the Code of Ethics."

INDIA SOCIETY OF ENGINEERS CODE OF ETHICS

• • •

12-B Netaji Subhas Road
Calcutta 700001, India

Founded: 1934
Members: 8,000

Code of Ethics for Members of Indian Society of Engineers

The most important rules for a Corporate Member in a Professional sphere to follow, in India or abroad, is the code of practice for the society of which he is a member. This is the following:

i) A Corporate Member should observe the principles of honesty, justice, and courtesy in his profession. His personal conduct should uphold his Professional reputation, he should avoid adverse Questions affecting brother associations/Professionals, and he should uphold the dignity and honor of the Society.

ii) A Corporate Member will co-operate with others in his profession by fair interchange of information and experience and endeavor to protect the profession from misrepresentation and misunderstanding, and will not divulge any confidential finding or actions of an engineering commission or committee, as a Member without obtaining permission from the Authority.

iii) A Corporate Member will not directly or indirectly make damage to the reputation or practice of another Corporate Member or criticize technically without proper forum of Engineering Society or Engineering Press.

iv) A Corporate Member will neither misrepresent his Qualification and misguide his employer or client or to the profession, nor disclose trade secrets or technical affairs of his client or employer without proper Authority.

v) A Corporate Member will not review works of another Corporate Member at the same time for the same client, except with the consent of the other Member.

vi) A Corporate Member will, if he considers another Corporate Member is guilty of unethical, illegal or unfair practices, inform the Council of the Society in writing with necessary documents for action.

vii) A Corporate Member shall always confirm the National Interest in his own Professional Engineering areas.

THE INSTITUTION OF ENGINEERS (INDIA) CODE OF ETHICS

• • •

8 Gokhale Road
Calcutta 700020, India

Founded: 1920
Members: 300,000

CODE OF ETHICS FOR CORPORATE MEMBERS

• • •

Foreword

"The task of ethics," said Jacques Maritain, "is a humble one but it is also magnanimous in carrying the mutable application of immutable moral principles even in the midst of agonies of an unhappy world as far as there is in it a gleam of humanity." To uphold the concept of professional conduct amongst Corporate Members, the Institution introduced the professional Conduct Rules for Corporate Members on August 30 th, 1944. They were replaced by the Code of Ethics for Corporate Members on October 15th, 1954. The Code was revised consistent with the changing needs of the profession on August 12th, 1962.

A Corporate Member should allow the principles of honesty, justice and courtesy to guide him in the practice of his profession and in his personal conduct. He should not merely observe them passively, but should apply them dynamically in the discharge of his duties to the public and the profession.

He should scrupulously guard his professional reputation and avoid association with any enterprise of questionable character. He should uphold the dignity and honor of the Institution.

The Code

1. A Corporate Member will cooperate with others in his profession by the free interchange of informa-

tion and experience and will contribute to the work of engineering institutions to the maximum effectiveness he is capable of.

2. A Corporate Member will endeavor to protect the engineering profession from misrepresentation and misunderstanding.

3. A Corporate Member will refrain from expressing publicly an opinion on an engineering subject unless he is informed of the facts relating to that subject.

4. A Corporate Member will express an opinion only when it is founded on adequate knowledge and honest conviction if he is serving as a witness before a court or commission.

5. A Corporate Member will not divulge any confidential findings or actions of an engineering commission or committee, of which he is a member, without obtaining official consent.

6. A Corporate Member will take care that credit for engineering work is given to those to whom credit is properly due.

7. A Corporate Member will not offer his professional services by advertisement or through any commercial advertising media, or solicit professional work either directly, or through an agent or in any other manner derogatory to the dignity of the profession.

8. A Corporate Member will not directly or indirectly injure the professional reputation or practice of another Corporate Member.

9. A Corporate Member will exercise due restraint in criticizing the work of another Corporate Member and remember that the proper forum for technical criticism is an engineering society or the engineering press.

10. A Corporate Member will not try to supplant another Corporate Member in a particular employment.

11. A Corporate Member will not compete unfairly with another Corporate Member by charging fees below those customary for others in his profession practicing in the same field and in the same area.

12. A Corporate Member will not associate in work with an engineer who does not conform to ethical practices.

13. A Corporate Member will act in professional matters for his client or employer as faithful agent or trustee.

14. A Corporate Member will not misrepresent his qualifications to a client or employer or to the profession.

15. A Corporate Member will not disclose information concerning the business or technical affairs of his client or employer without his consent.

16. A Corporate Member will present clearly the consequences to be expected if his professional judgment is overruled by the non-professional authority where he is responsible for the professional adequacy of work.

17. A Corporate Member will act with fairness and justice between his client or employer and the contractor when dealing with contracts.

18. A Corporate Member will not be financially interested in the bids of a contractor on competitive work for which he is employed as an engineer unless he has the written consent of his client or employer.

19. A Corporate Member will not resolve any commission, discount, or other indirect profit in connection with any work with which he is entrusted.

20. A Corporate Member will make his status clear to his client or employer before undertaking an engagement if he may be called upon to decide on the use of inventions or equipment or any other thing in which he may have a financial interest.

21. A Corporate Member will immediately inform his client or employer of any interest in a business which may compete with or affect the business of his client or employer.

22. A Corporate Member will not allow an interest in any business to affect the engineering work for which he is employed or may be called upon to perform.

23. A Corporate Member will engage, or advise engaging, engineering experts and specialists when in his judgment such services are in the interests of his client or employer.

24. A Corporate Member will not review the work of another Corporate Member for the same client except with the knowledge of the second Corporate Member, unless such engineering engagement or the work which is subject to review is terminated.

25. A Corporate Member will not accept financial or other compensation from more than one interested party for the same service, or for services pertaining to the same work, without the consent of all interested parties.

26. A Corporate Member will subscribe to the principles of appropriate and adequate compensation for those engaged in engineering work, including those in subordinate positions.

28. A Corporate Member will endeavor to provide opportunity for the professional development and advancement of engineers in his employ.

29. A Corporate Member will, if he considers that another Corporate Member is guilty of unethical,

illegal or unfair practice, present the information to the Council of the Institution for action.

30. A Corporate Member who is engaged in engineering work in a country abroad will order his conduct according to the professional standards and customs of that country, adhering as closely as is practicable to the principles of this Code.

NOTES

This code is published and promulgated in a pocket-sized pamphlet.

IRELAND

• • •

THE INSTITUTION OF ENGINEERS OF IRELAND CODE OF ETHICS

• • •

22 Clyde Road, Ballsbridge
Dublin 4, Ireland

Founded: 1835
Members: 5,900

STANDARDS OF PROFESSIONAL CONDUCT

• • •

Part I: FUNDAMENTAL PRINCIPLES

1. Every corporate member of the Institution shall order his conduct so as to serve the public interest and uphold the honor and standing of The Institution and of the Engineering Profession.

2. *In his relations with his employers, clients, professional colleagues, subordinates and others with whom he works, and with the public*, he shall maintain high standards of conduct and integrity.

3. *In his relations with an employer or client* he shall act at all times as a faithful agent or trustee, using all his professional skill and experience and making freely available his sincere opinion and advice in the proper interest of his employer or client. He shall do nothing directly or indirectly which might conflict or appear to conflict with those interests or might influence or appear to influence his opinion or advice.

4. *In his relations with another engineer* he shall respect his dignity and professional standing and shall do nothing directly or indirectly to injure maliciously

his reputation, practice, employment or livelihood or to lessen the satisfaction that he obtains from his work. He shall never compete unfairly for any engagement or appointment. He shall ensure, so far as he is able, that an engineer receives credit for his professional achievements and the financial and other rewards to which he is entitled, and that a subordinate is provided with opportunities to develop his talents and exercise his skill.

5. *In his relations with all others with whom he works* he shall act with justice and impartiality and with respect for their rights and dignity as citizens and human beings.

6. *In his relations with the public* he shall apply his skill and experience to the common good and the advancement of human welfare and shall perform his professional duties and express his professional opinions with proper regard for true economy and for the safety, health and welfare of the public. Should he come to the conclusion after full consultation with his employer or client that any work required of him by them is likely to be seriously injurious to the public welfare or to create a hazard to the health or safety of the community he has a duty to put his opinion on record and to inform The Institution of this action.

7. *As an independent expert or arbitrator* he shall act with complete impartiality, uninfluenced by any personal consideration.

8. He has a duty to maintain his knowledge up-to-date in relation to that branch of engineering in which he practices.

Part II: GUIDE TO PROFESSIONAL CONDUCT

1. He shall not divulge any confidential information regarding the business affairs, technical processes or financial standing of his clients or employers without their consent. He shall not use information obtained in the course of his assignment for the purpose of making personal profit if such action is contrary to the best interest of his client, his employer or the public. He shall not divulge without authoritative permission any unpublished information obtained by him as a member of an investigating commission or advisory board.

2. His remuneration shall be restricted to his fee, commission or salary (including bonuses, etc.). Where his remuneration is by fee it shall be in accordance with the Conditions of Engagement and Scale of Fees published jointly with the Association of Consulting Engineers of Ireland as in force from time to time. He

shall not knowingly compete with another Chartered Engineer on the basis of professional charges.

3. He shall not receive any royalty or commission on any article or process used on his recommendation on work for which he is responsible unless such payment has the full consent of his client or employer.

4. He shall not while acting in a professional capacity be at the same time a director or substantial shareholder in any contracting, manufacturing or distributing business with which he may have dealings on behalf of his client or employer without divulging the full facts in writing to his client or employer, and obtaining his written agreement thereto.

5. He shall not advertise his practice or his availability except in accordance with such Code of Practice as may be in force from time to time. Under no circumstances shall he pay an agent to introduce clients to him.

6. He shall not practice as a consultant in the following circumstances:

   (a) in partnership with one who is not professionally qualified in engineering or an allied profession;

   (b) as Principal or one of the major shareholders of a limited liability Company unless the Company has the prior approval of the Council of The Institution.

7. A member shall not use the advantage of a salaried position to compete unfairly with other engineers. His outside activities in the engineering field should normally be confined to branches of engineering for which he has special qualifications. He shall not undertake as a part-time consultant any work which he might subsequently have to review in the course of his salaried employment.

8. When acting as a Consultant a member shall not attempt to supplant another Chartered Engineer nor shall he take over or review the work of another Chartered Engineer acting as a Consultant, without either having the written consent of such Engineer or having fully satisfied himself that such Engineer's association with the work has been terminated and his account fully discharged.

NOTES

Approved by the Council of The Institution of Engineers of Ireland at its Meeting of 15th October, 1971.

Published and promulgated as a four-page pamphlet.

Under revision as of November 1991.

JAMAICA

• • •

JAMAICAN INSTITUTION OF ENGINEERS CODE OF ETHICS

• • •

P.O. Box 122, Kingston
10 Jamaica

Founded: 1960
Members: 500

CODE OF ETHICS

• • •

A Professional Engineer

1. owes certain duties to the public, to his employers, to other members of his profession and to himself and shall act at all times with:

   (a) fidelity to public needs;

   (b) fairness and loyalty to his associates, employers, clients, subordinates and employees: and

   (c) devotion to high ideals of personal honor and professional integrity.

2. shall express opinions on engineering matters only on the basis of adequate knowledge and honest conviction.

3. shall have proper regard for the safety health and welfare of the public in the performance of his professional duties.

4. shall endeavor to extend public understanding of engineering and its place in society.

5. shall not be associated with enterprises contrary to the public interest or sponsored by persons of questionable integrity, or which does not conform to the basic principles of the code.

6. shall sign and/or seal only those plans, specifications and reports actually prepared by him or under his direct professional supervision.

7. shall act for his client or employer as a faithful agent or trustee.

8. shall not disclose confidential information pertaining to the interests of his clients or employers without their consent.

9. shall present clearly to his clients or employers the consequences to be expected if his professional judgment is over-ruled by non-technical authority in matters pertaining to work for which he is professionally responsible.

10. shall not undertake any assignment which may create a conflict of interest with his clients or employers without the full knowledge of his clients or employers.

11. shall not accept remuneration for services rendered other than from his client or employer.

12. shall conduct himself towards other professional engineers with courtesy, fairness and good faith.

13. shall not compete unfairly with another engineer by attempting to obtain employment, advancement or professional engagements by competitive bidding, by taking advantage of a salaried position or by criticizing other engineers.

14. shall undertake only such work as he is competent to perform by virtue of his training and experience.

15. shall not advertise his work or merit in a self-laudatory manner and shall avoid all conduct or practice likely to discredit or unfavorably reflect upon the dignity or honor of the profession.

16. shall advise his Association or Institution or the Council of any practice by another Professional Engineer which he believes to be contrary to the Code of Ethics.

GUIDE TO PRACTICE UNDER THE CODE OF ETHICS

• • •

GENERAL:

ARTICLE 1. A Professional Engineer owes certain duties to thepublic, to his employers, to other members of his profession and to himself and shall act at all times with:

(a) fidelity to public needs;

(b) fairness and loyalty to his associates, employers, clients, subordinates and employees; and

(c) devotion to high ideals of personal honor and professional integrity.

DUTIES OF THE PROFESSIONAL ENGINEER TO THE PUBLIC

• • •

A Professional Engineer

ARTICLE 2. shall express opinions on engineering matters onlyon the basis of adequate knowledge and honest conviction.

(a) He shall ensure, to the best of his ability, the statements on engineering matters attributed to him are not misleading and properly reflect his professional opinion;

(b) He shall not express publicly or while he is serving as a witness before a court, commission or other tribunal opinions on professional engineering matters that are not founded on adequate knowledge and honest conviction.

ARTICLE 3. shall have proper regard for the safety health andwelfare of the public in the performance of his professional duties.

(a) He shall notify the proper authorities of any situation which he considers, on the basis of his professional knowledge, to be a danger to public safety or health.

(b) He shall complete, sign, or seal only those plans and/or specifications which reflect proper regard for the safety and health of the public.

ARTICLE 4. shall endeavor to extend public understanding of engineering and its place in society.

(a) He shall endeavor at all times to enhance the public regard for, and its understanding of, his profession by extending the public knowledge thereof and discouraging untrue, unfair or exaggerated statements with respect to professional engineering.

(b) He shall not give opinions or make statements on professional engineering projects connected with public policy where such statements are inspired or paid for by private interests unless he clearly discloses on whose behalf he is giving the opinions or making the statements.

ARTICLE 5. shall not be associated with enterprises contrary to the public interest or sponsored by persons of questionable integrity, or persons who do not conform to the basic principles of the code.

(a) He shall conform with registration laws in his practice of engineering.

(b) He shall not sanction the publication of his reports in part or in whole in a manner calculated to mislead and if it comes to his knowledge that they are so published, he shall take immediate steps to correct any false impressions given by them.

ARTICLE 6. shall sign and/or seal only those plans, specification and reports actually prepared by him or under his direct professional supervision.

DUTIES OF THE PROFESSIONAL ENGINEER TO HIS CLIENT OR EMPLOYER:
A Professional Engineer

ARTICLE 7. shall act for his client or employer as a faithful agent or trustee.

(a) He shall be realistic and honest in all estimates, reports, statements, and testimony.

(b) He shall admit and accept his own errors when proven obviously wrong and refrain from distorting or altering the facts in an attempt to justify his decision.

(c) He shall advise his client or employer when he believes a project will not be successful.

(d) He shall not accept outside employment to the detriment of his regular work or interest, or without the consent of his employer.

(e) He shall not attempt to attract an engineer from another employer by unfair methods.

(f) He shall engage, or advise engaging, experts and specialists when such services are in his clients or employer's best interests.

ARTICLE 8. shall not disclose confidential information pertaining to the interests of his clients or employers without their consent.

(a) He shall not use information coming to him confidentially in the course of his assignment as a means of making personal gain except with the knowledge and consent of his client or employer.

(b) He shall not divulge, without official consent, any confidential findings resulting from studies or actions of any commission or board of which he is a member or for which he is acting.

ARTICLE 9. shall present clearly to his clients or employers the consequences to be expected if his professional judgment is over-ruled by non-technical authority in matters pertaining to work for which he is professionally responsible.

ARTICLE 10. shall not undertake any assignment which may create a conflict of interest with his clients or employers without the full knowledge of his clients or employers.

ARTICLE 11. shall not undertake any assignment which may create a conflict of interest with his clients or employers without the full knowledge of his clients or employers.

(a) He shall inform his client or employer of any business connections, interests, or circumstances which may be deemed as influencing his judgment or the quality of his services to his client or employer.

(b) When in public service as a member, advisor or employee of a governmental body or department, he shall not participate in considerations or actions with respect to services provided by him or his organization in private engineering practice.

(c) He shall not solicit or accept an engineering contract from a governmental body on which a principal or officer of his organization serves as a member.

ARTICLE 11. shall not accept remuneration for services rendered other than from his client or employer.

(a) He shall not accept compensation from more than one interest party for the same service or for services pertaining to the same work, under circumstances that may involve a conflict of interest, without the consent of all interested parties.

(b) He shall not accept any royalty or commission on any article or process used on the work for which he is responsible without the consent of his client or employer.

(c) He shall not undertake work at a fee or salary below the accepted standards of the profession in the area.

(d) He shall not tender on competitive work upon which he may be acting as a consulting engineer.

(e) He shall not act as consulting engineer in respect of any work upon which he may be the contractor.

DUTIES OF THE PROFESSIONAL ENGINEER
TO THE PROFESSION
• • •

A Professional Engineer

ARTICLE 12. shall conduct himself towards other professional engineers with courtesy, fairness and good faith.

(a) He shall not accept any engagement to review the work of another professional engineer for the same employer or client except with the knowledge of such engineer, unless such engineer's engagement on the work has been terminated.

(b) He shall not maliciously injure the reputation or business of another professional engineer.

ARTICLE 13. shall not compete unfairly with another engineer by attempting to obtain employment, advancement or professional engagements by competitive bidding, by taking advantage of a salaried position, or by criticizing other engineers.

(a) He shall not attempt to supplant another engineer in a particular employment after becoming aware that definite steps have been taken toward the other's employment.

(b) He shall not offer to pay, either directly or indirectly, any commission, political contribution, or a gift or other consideration in order to secure professional engineering work.

(c) He shall not solicit or submit engineering proposals on the basis of competitive bidding.

(d) He shall not use equipment, supplies, laboratory, or office facilities of his employer to carry on outside private practice without consent.

ARTICLE 14. shall undertake only such work as he is competent to perform by virtue of his training and experience.

(a) He shall not misrepresent his qualifications.

ARTICLE 15. shall not advertise his work or merit in a self-laudatory manner, and shall avoid all conduct or practice likely to discredit or unfavorably reflect upon the dignity or honor of the profession.

(a) Circumspect advertising may be properly employed by the Engineer to announce his practice and availability. Only those media shall be used as are necessary to reach directly an interested and potential client or employer, and such media shall in themselves be dignified, reputable and characteristically free of any factor or circumstance that would bring disrepute to the profession or to the professional using them. The substance of such advertising shall be limited to fact and shall contain no statement or offer intended to discredit or displace another engineer, either specifically or by implication.

ARTICLE 16. shall advise his Association or Institution or the Council of any practice by another Professional Engineer which he believes to be contrary to the Code of Ethics.

NOTES

Adopted by the Jamaica Institution of Engineers September 1986.

The JIE is a non-profit professional organization, comprised of members who are Engineers from all the various disciplines of Engineering, including Civil, Electrical, Mechanical, Chemical, Industrial and Agricultural.

The Institution is currently involved in a six-year program of technical co-operation (concluding in 1993) with the Canadian Society for Civil Engineering (CSCE), and funded by the Canadian International Development Agency (CIDA). The main objective being the improvement of technical expertise within the JIE community as regards to Civil Engineering aspects of transportation infrastructure and other topics.

JAPAN

• • •

SCIENCE COUNCIL OF JAPAN CODE OF ETHICS

• • •

7-22-34 Roppongi
Minatoku, Tokyo 106

Founded: 1949
Members: 210

Statement on "Charter for Scientific Researchers"

PREAMBLE

• • •

In order to promote the sound development of scientific research in Japan, the Science Council of Japan (JSC) recommended twice, in 1962 and in 1976, that the government prepare for the enactment of a Basic Act on Scientific Research to define its responsibility and urged the government to enact such a law. The Council has prepared and hereby issues a "Charter for Scientific Researchers" to complement the proposed Basic Act on Scientific Research, and itself resolves to abide by this "Charter." The Council thus makes public the responsibility of scientific researchers themselves, and expects the researchers of Japan to accomplish their tasks in accordance with the spirit of the "Charter."

CHARTER FOR SCIENTIFIC
RESEARCHERS

• • •

Science enriches human life by the rational search for truth with actual evidence and also by applying the results in practical use. The search for truth in scientific research and the application of its results belong to the highest intellectual activities of human beings. Scientific researchers who are engaged in these activities are required to be sincere toward reality, exclude arbitrary decisions and keep their minds pure and strict toward truth.

It is not only the demand of human society but also the duty of scientific researchers to promote the sound development of science and the beneficial application of its results. To fulfill their duty, scientific researchers are required to act upon the following five points:

1. To be conscious of the significance and aim of his or her own research and to contribute to the welfare of mankind and world peace.

2. To defend the freedom of scientific research and to respect originality in research and development.

3. To attach importance to the harmonious development between various fields of science and to propagate the scientific attitude and knowledge among the general public.

4. To guard against disregard and abuse of scientific research and to strive to eliminate such dangers.

5. To place great value on the international nature of scientific research and to endeavor to promote interchanges with the scientific community of the world.

Purport and Process Leading to Adoption of "Charter for Scientific Researchers"

Explanatory note by Special Committee for Promotion of Science

In January 1975, at the opening of the tenth term of the Science Council of Japan, it was decided to take up for examination a proposal for formulation of a Charter for Scientists. Deliberations on this question have continued until now.

From the time of its establishment in 1949, JSC has constantly kept the rights and responsibilities of scientists under consideration, and pledged that it will strive to contribute to world peace and the welfare of mankind, based on the conviction that science provides the foundation for a cultured and peaceful nation.

The Council has continued to deliberate important questions relating to the sciences, and has made many recommendations and issued a wide range of statements. In 1962 and again in 1976 it recommended that preparations be made for legislation of a Basic Act for Scientific Research. The purpose of such an Act would be to define the responsibility of the State for the development of scientific research in Japan, and as complementary to this, the Science Council of Japan declared by resolution, as a representative body of scientific researchers, that it would adopt a "Charter for Scientific Researchers" (provisional name) setting out the responsibility of scientists toward the general public. The

"Charter" would declare that scientific researchers must be conscious of the purposes of scientific research and their own social responsibilities, and devote themselves to the sound development of scientific research such as will meet the expectations of the people; that they accept it as their responsibility to protest against any oppression of freedom of scientific research, and make clear the damage which disregard and/or abuse of science and technology would cause to human society, thus to protect the welfare of the nation and the people.

The 18th session of the UNESCO General Conference in October, 1974 adopted a Recommendation on the Status of Scientific Researchers concerned mainly with the rights and status of scientific researchers, and the 70th session of the JSC General Meeting followed this up with its renewed recommendation to the Japanese government for a Basic Act for Scientific Research, in the desire to carry into effect in this country the spirit and contents of the UNESCO Recommendation as soon as possible.

In the hope that the proposed Charter could be drafted during the Council's 10th term, discussions were taken up among the Members, and a subcommittee on a "Charter for Scientific Researchers" was established in the Special Committee for Man and Science, which also had the responsibilities of scientific researchers under consideration. First, second and third drafts of the "Charter" were submitted to scientific researchers all over Japan through the members of JCS and through various academic societies and associations, seeking their comments.

The draft of an "Appeal to Examine the Responsibilities of Scientific Researchers" was presented to the 73rd session of the General Meeting in October, 1977 during the last session of the 10th term. The need for further examination was acknowledged, and it was agreed that the drafting of the "Charter for Scientific Researchers" should be completed as soon as possible in the 11th term.

Basic deliberations during the 11th term (1978-1981) highlighted the following three targets:

(1) high evaluation of creativity, originality and foresightedness in scientific researchers

(2) respect for human dignity and awareness of social responsibility among scientific researchers

(3) emphasis on global concept and on scientific cooperation with developing countries.

On points (1) and (2), it was decided that the Special Committee for Promotion of Science should bear the main responsibility for examining basic policy, and that the draft of the "Charter for Scientific Researchers" should be prepared by the newly appointed Subcommittee within the Special Committee.

Accordingly, the Subcommittee took up the results from considerations in the Council's previous term, and examined also documents from overseas relating to charters for scientists, and literature on the status and responsibility of scientists. Further comments from Members of JSC were received through questionnaires on the requirements, character and content of the "Charter." Based on these, the first draft was completed in February, 1979 and a consensus sought among scientific researchers. The first draft was deliberated at each Division of JSC meeting in that month. Based on these investigations, the Subcommittee presented the second draft of the "Charter" to the 77th session of the general meeting held in May. After receiving opinions on the second draft and making several amendments, the Special Committee for Promotion of Science submitted a draft of the Charter for Scientific Researchers on the second day of meeting of the 79th session of the General Meeting on 24 April, 1980. Seven Members spoke in approval of the draft, which, with minor verbal modification, was then adopted unanimously.

The Science Council of Japan hereby presents the "Charter for Scientific Researchers," with its resolution to abide by it, setting out the responsibilities of scientists toward the general public, and expresses the hope that scientific researchers will carry on their tasks in the spirit of this "Charter."

NOTES

Was founded as the governmental organization representative of all Japanese scientists to promote and reflect scientific development throughout national life, industry and administration, to co-ordinate scientific research and to link scientific organizations abroad.

MEXICO

• • •

MEXICAN UNION OF ASSOCIATIONS OF ENGINEERS CODE OF ETHICS

• • •

Code of Ethics of the Mexican Engineer (UMAI)

Contributed by Araceli Solano

The Code of Mexican Professional Engineering Ethics was published July 1, 1983, and signed by the witness, the Certified Licensed Miguel de la Madrid Hurtado, Constitutional President of the United Mexican States, which is transcribed below.

CONSIDERING THAT:

1. Mexican engineers sustain their conduct with the respect and love for the fatherland.

2. Engineers in our country have achieved the practice of their profession thanks to the opportunity that the Mexican nation affords them.

3. For their preparation they have a great obligation to contribute to the satisfaction of the needs and improvement of the quality of life of the Mexican people, with the moral conviction and responsibility of sustaining a development in accordance with social justice.

4. It is a duty to foster a favorable atmosphere for the development of activity in accordance with the Code of Ethics that specifies social obligations that make possible the respect of each professional for the rest, in search of a just and harmonious human conviviality within each nation and among nations.

5. Universal principles and our greatest traditions consider as a solemn duty both international solidarity and respect for the moral values of other peoples, in particular in those places where engineers forward their education or eventually exercises their profession.

6. The diverse codes of professional ethics of colleges and associations of engineers come together on one and the same conception.

7. The Union of Mexican Engineers has acknowledged principles and norms of conduct.

The Ordinary General Assembly of UMAI adopts the following Code of Professional Ethics of the Mexican Engineer:

Engineers recognize that the greatest merit is work, for which reason they will exercise their profession committed to service to Mexican society, caring for the wellbeing and progress of the majority. When transforming nature for the benefit of humanity, engineers should augment their awareness that the world is the living space of man and that their interest in the universe is a guarantee of the triumph of the spirit and of the knowledge of reality in order to make it more just and happy. Engineers should refuse work that has as its goal a crime against the general interest; in this way they will avoid situations which implicate dangers or constitute a threat to the environment, to life, health, or other rights of the human being. It is an inescapable duty of the engineer to sustain the prestige of the profession and strive for its proper exercise; likewise, to maintain a professional conduct cemented in capability, honor, strength, moderation, magnanimity, modesty, forthrightness, and justice, with consciousness of subordinating the wellbeing of the individual to the wellbeing of society. Engineers should procure the constant perfection of their knowledge, in particular that of their profession, divulge their wisdom, share their experience, provide opportunities for the education and enablement of workers, bestow recognition, moral and material support to the educational institution where they realized their studies; in this way they will return to society the opportunities that they have received. It is the responsibility of engineers that their work be realized with efficiency and aid to legal dispositions. In particular, they will ensure the fulfillment of the norms of protection of workers established in Mexican labor legislation. In the exercise of their profession, engineers must fulfill with diligence the commitments that they have assumed and will develop with dedication and loyalty the jobs assigned to them, avoiding putting personal interests first in the attention to the matters that are entrusted to them, or colluding in order to exercise disloyal competition to the detriment of those who received their services. They will observe decorous conduct, treating with respect, diligence, impartiality, and rectitude the persons with whom they have a relation, particularly their collaborators, abstaining from deviance and abuses of authority and from disposing or authorizing a subordinate to illicit conduct, such as unduly favoring third parties. Engineers must safeguard the interests of the institution or person for whom they are working and make good use of the resources that have been assigned to them for the undertaking of their work. They will fulfill the orders that in the exercise of their powers their superiors dictate to them, will respect and make respected their position and work; if they disagree with their superiors they will have the obligation to manifest before them the reasons for their disagreement. Engineers will have as a norm the creation and promotion of national technology; they will take special care to ensure that the transfer of technology adapted to our conditions conforms to the established legal framework. It is obligatory to keep as a professional secret the confidential data that they learn in the exercise of their profession, except when they might be required by a competent authority.

TRANSLATED BY JAMES A. LYNCH

NEW ZEALAND
THE INSTITUTION OF ENGINEERS
CODE OF ETHICS

• • •

P.O. Box 12241101
Molesworth Street

Wellington,
New Zealand

Founded: 1914
Members: 6,047

Code of Ethics

Protection of Life and Safeguarding People: Members have a duty of care to protect life and to safeguard people.

Guidelines

To satisfy this clause you need to:

1.1 Give priority to the safety and well-being of the community and have regard to this principle in assessing duty to clients and colleagues.

1.2 Be responsible for ensuring that reasonable steps are taken to minimize the risk of loss of life, injury or suffering which may result from the work or the effects of your work.

1.3 Draw the attention of those affected to the level and significance of risk associated with the work.

1.4 Assess and minimize potential dangers involved in the construction, manufacture and use of your products or projects.

Professionalism and Integrity

Members shall undertake their duties with professionalism and integrity and shall work within their levels of competence.

Guidelines

To satisfy this clause you need to:

2.1 Exercise initiative, skill and judgment to the best of your ability for the benefit of your employer or client.

2.2 Give engineering decisions, recommendations or opinions that are honest, objective and factual. If these are ignored or rejected you should ensure that those affected are made aware of the possible consequences.

In particular, where vested with the power to make decisions binding on both parties under a contract between principal and contractor, act fairly and impartially as between the parties and (after any appropriate consultation with the parties) make such decisions independently of either party in accordance with your own professional judgment.

2.3 Accept personal responsibility for work done by you or under your supervision or direction and take reasonable steps to ensure that anyone working under your authority is both competent to carry out the assigned tasks and accepts a like personal responsibility.

2.4 Ensure you do not misrepresent your areas or levels of experience or competence.

2.5 Take care not to disclose confidential information relating to your work or knowledge of your employer or client without the agreement of those parties.

2.6 Disclose any financial or other interest that may, or may be seen to, impair your professional judgment.

2.7 Ensure that you do not promise to, give to, or accept from any third party anything of substantial value by way of inducement.

2.8 First inform another member before reviewing their work and refrain from criticizing the work of other professionals without due cause.

2.9 Uphold the reputation of the Institution and its members, and support other members as they seek to comply with the Code of Ethics.

2.10 Follow a recognized professional practice (Model Conditions of Engagement are available) in communicating with your client on commercial matters.

Society and Community Well-Being

Members shall actively contribute to the well-being of society and, when involved in any engineering project or application of technology, shall, where appropriate, recognize the need to identify, inform and consult affected parties.

Guidelines

To satisfy this clause you need to:

3.1 Apply skill, judgment and initiative to contribute positively to the well-being of society.

3.2 Recognize in all your work your obligation to anticipate possible conflicts and endeavor to resolve them responsibly, and where necessary utilize the experience of the Institution and colleagues for guidance.

3.3 Treat people with dignity and have consideration for the values and cultural sensitivities of all groups within the community affected by your work.

3.4 Endeavour to be fully informed about relevant public policies, community needs, and perceptions, which affect your work.

3.5 As a citizen, use your knowledge and experience to contribute helpfully to public debate and to community affairs except where constrained by contractual or employment obligations.

Sustainable Management and Care of the Environment

Members shall be committed to the need for sustainable management of the planet's resources and seek to minimize adverse environmental impacts of their engineering works or applications of technology for both present and future generations.

Guidelines

To satisfy this clause you need to:

4.1 Be committed to the efficient use of resources.

4.2 Minimize the generation of waste and encourage environmentally sound reuse, recycling and disposal.

4.3 Recognize adverse impacts of your work on the environment and seek to avoid or mitigate them.

4.4 Recognize the long-term imperative of sustainable management throughout your work. (Sustainable Management is often defined as meeting the needs of the present without compromising the ability of future generations to meet their own needs).

Promotion of Engineering Knowledge

Members shall continue the development of their own and the profession's knowledge, skill and expertise in the art and science of engineering and technology, and shall share and exchange advances for the benefit of society.

Guidelines

To satisfy this clause you need to:

5.1 Seek and encourage excellence in your own and others' practice of the art and science of engineering and technology.

5.2 Contribute to the collective wisdom of the profession and art of engineering and technology in which you practice.

5.3 Improve and update your understanding of the science and art of engineering and technology and encourage the exchange of knowledge with your professional colleagues.

5.4 Wherever possible share information about your experiences and in particular about successes and failures.

NOTES

Approved by Council 5 July 1996.

NORWAY

• • •

ASSOCIATION OF NORWEGIAN CIVIL ENGINEERS CODE OF ETHICS

• • •

RELATION OF THE CIVIL ENGINEER TO SOCIETY

• • •

1. In their professional work, the members shall promote a community-oriented and harmonious technical and industrial development.

2. The members shall execute their work according to sound technical principles. Proper consideration must be given to economic and human factors, to the influence of the work on the environment and the community, and to other demands dictated by circumstances.

3. Professional (technical) questions must be dealt with in a factual and objective manner. The members must attempt to give the public a correct understanding of technical matters and to counteract erroneous conceptions.

RELATION OF CIVIL ENGINEER TO EMPLOYER AND CLIENT

• • •

1. The members shall protect the interests of their employers and clients in matters which have been entrusted to them, as long as this does not contradict general ethical fundamental principles.

2. The members are not allowed to receive compensation from anypartner in a group-deal unless all other partners are also aware of this. The members must not use their professional position to obtain personal advantages.

RELATION OF CIVIL ENGINEER TO COLLEAGUES AND CO-WORKERS

• • •

1. The members shall protect the professional reputation of their colleagues and co-workers against unfair criticism, slander, or false accusations. They should contribute to the fact that whosoever has executed a technical assignment should also receive the acknowledgement and compensation for this.

2. The members should not engage in disloyal competition. The rightful ownership of others with regard to plans, drawings, ideas, inventions, etc., should be respected.

3. A member is not allowed to take over a position after a colleague if there is reason to believe that the latter was unfairly dismissed or in some other manner deprived of his work for reasons which contradict the general ethical fundamental principles.

4. A member is not allowed to take over an assignment which has been entrusted to a colleague without first informing the latter and without ascertaining that there are reasonable grounds for the client's solicitation.

5. Members are not allowed to advertise their activities or to offer their services in an unworthy or misleading manner or to attempt to obtain assignments with improper methods.

TRANSLATED BY BIRGITTA D. KNUTTGEN

NOTES

This code is promulgated by means of a one-page type-written and photocopied document.

An introductory note states that the code was "passed by the Board of Governors of the NIF [Norwegian Civil Engineers Association] on June 26, 1970, as a supplement to paragraph 8, point 1, of the statutes."

PAKISTAN

• • •

THE INSTITUTION OF ENGINEERS CODE OF ETHICS

• • •

Engineering Centre
Gulberg - III
Lahore, Pakistan
Founded: 1948

Professional Ethics and Code of Conduct

ARTICLE 1

To maintain, uphold and advance the honor and dignity of the engineering profession in accordance with this Code, a member shall:

(a) uphold the Ideology of Pakistan;

(b) be honest, impartial and serve the country, his employer, clients and the public at large with devotion;

(c) strive to increase the competence and prestige of the engineering profession;

(d) use his knowledge and skill for the advancement and welfare of mankind;

(e) promote and ensure the maximum utilization of human and material resources of Pakistan for achieving self-reliance; and

(f) not sacrifice the national interest for any personal gain.

ARTICLE 2

(1) A member shall be guided in all professional matters by the highest standards of integrity and act as a faithful agent or a trustee for each of his client and employer.

(2) A member shall:

(a) be realistic and honest in all estimates, reports, statements, and testimony and shall carry out his professional duties without fear or favor;

(b) admit and accept his own errors when proved and shall refrain from distorting or altering the facts justifying his decision or action;

(c) advise his client or employer honestly about the viability of the project entrusted to him;

(d) not accept any other employment to the detriment of his regular work or interest without the consent of his employer;

(e) not attempt to attract an engineer from another employer by false or misleading pretenses;

(f) not restrain an employee from obtaining a better position with another employer; and

(g) not endeavor to promote his personal interest at the expense of the dignity and integrity of the profession.

ARTICLE 3

A member shall have utmost regard for the safety, health, and welfare of the public in the performance of his professional duties and for that purpose he shall:

(a) regard his duty to the public welfare as paramount;

(b) seek opportunities to be of service in civic affairs and work for the advancement of the safety, health, and well-being of the community;

(c) not undertake, prepare, sign, approve, or authenticate any plan, design or specifications which are not safe for the safety, health, and welfare of a person or persons, or are not in conformity with the accepted engineering standards and if any client or an employer insists on such

unprofessional conduct, he shall notify the authorities concerned and withdraw from further service on the project; and

(d) point out the consequences to his client or the employer if his engineering judgment is overruled by any non-technical person.

ARTICLE 4

(1) A member shall avoid all acts or practices likely to discredit the dignity or honor of the profession and for that purpose he shall not advertise his professional services in a manner derogatory to the dignity of the profession. He may, however, utilize the following means of identification:

(i) professional cards and listing in recognized and dignified publications and classified section of the telephone directories;

(ii) sign boards at the site of his office or projects for which he renders services; and

(iii) brochures, business cards, letterheads, and other factual representations of experience, facilities, personnel and capacity to render services.

(2) A member shall write articles for recognized publications but such articles should be dignified, free from ostentations or laudatory implications, based on factual conclusions and should not imply other than his direct participation in the work described unless credit is given to others for their share of the work.

(3) A member shall not allow himself to be listed for employment using exaggerated statements of his qualifications.

ARTICLE 5

(1) A member shall endeavor to extend public knowledge and appreciation of the engineering profession, propagate the achievements of the profession and protect it from misrepresentation and misunderstanding.

ARTICLE 6

(1) A member shall express an opinion of an engineering subject only when founded on adequate knowledge, experience, and honest conviction.

ARTICLE 7

(1) A member shall undertake engineering assignments only when he possesses adequate qualifications,

training, and experience. He shall engage or advise for engaging of the experts and specialists whenever the client's or employer's interests are best served by the service.

(2) A member shall not discourage the necessity of other appropriate engineering services, designs, plans, or specifications or limit-free competition by specifying materials of particular make or model.

ARTICLE 8

(1) A member shall not disclose confidential information concerning the business affairs or technical processes of any present or former client or employer without his consent.

ARTICLE 9

(1) A member shall uphold the principles of appropriate and adequate compensation for those engaged in engineering work and for that purpose he shall not:

   (a) undertake or agree to perform any engineering service free except for civic, charitable, religious, or non-profit organizations or institutions;

   (b) undertake professional engineering work at a remuneration below the accepted standards of the profession in the discipline; and

   (c) accept remuneration from either an employee or employment agency for giving employment.

(2) A member shall offer remuneration in accordance with the qualifications and experience of an engineer employed by him.

(3) A member working in any sales section or department shall not offer or give engineering consultation, designs, or advice, other than specifically applying to the equipment being sold in that section or department.

ARTICLE 10

(1) A member shall not accept compensation, financial, or otherwise, from more than one party for the same service, or for services pertaining to the same work unless all interested parties give their consent to such compensation.

(2) A member shall not accept:

   (a) financial or other considerations, including free engineering design, from material or equipment suppliers for specifying their products; and

   (b) commissions or allowances, directly or indirectly from contractors or other parties dealing with his clients or employer in connection with work for which he is professionally responsible.

ARTICLE 11

(1) A member shall not compete unfairly with another member or engineer by attempting to obtain employment, professional engagements or personal gains by taking advantage of his superior position or by criticizing other engineers or by any other improper means or methods.

(2) An engineer shall not attempt to supplant another engineer in a particular employment after becoming aware that definite steps have been taken towards other's employment.

(3) A member shall not accept part-time engineering work at a fee or remuneration less than that of the recognized standard for a similar work and without the consent of his employer if he is already in another employment.

(4) A member shall not utilize equipment, supplies, laboratory, or office facilities of his employer or client for the purpose of private practice without his consent.

ARTICLE 12

(1) A member shall not attempt to injure, maliciously or falsely, directly or indirectly, the professional reputation, prospects, practices, or employment of another engineer or member.

(2) A member engaged in private practice shall not review the work of another engineer for the same client, except with knowledge of such engineer or, unless the connection of such engineer with the work has been terminated; provided that a member shall be entitled to review and evaluate the work of other engineers when so required by his employment duties.

(3) A member employed in any sales or industrial concern shall be entitled to make engineering comparisons of his products with products of other suppliers.

ARTICLE 13

(1) A member shall not associate with or allow the use of his name by an enterprise of questionable character; nor will he become professionally associated with engineers who do not conform to ethical practices or with persons not legally qualified to render the professional services for which the association is intended.

(2) A member shall strictly comply with the bye-laws, orders, and instructions issued by the Institution of Engineers (Pakistan) from time to time in professional practice and shall not use the association with a non-engineering corporation, or partnership as a cloak for any unethical act or acts.

ARTICLE 14

(1) A member shall give credit for engineering work to those to whom credit is due, recognize the proprietary interests of others and disclose the name of a person or persons who may be responsible for his designs, inventions, specifications, writings, or other accomplishments.

(2) When a member uses designs, plans, specifications, data, and notes supplied to him by a client or an employer or are prepared by him in reference to such client or the employer's work such designs, plans, specifications, data, and notes shall remain the property of the client and shall not be duplicated by a member for any use without the express permission of the client.

(3) Before undertaking any work on behalf of a person or persons for making improvements, plans, designs, inventions, or specifications which may justify copyright or patent, a member shall get ownership of such improvements, plans, designs, inventions, or specifications determined for the purpose of registration under the relevant copyright and patent laws.

ARTICLE 15

(1) A member shall disseminate professional knowledge by interchanging information and experience with other members or engineers and students to provide them opportunity for the professional development and advancement of engineers under his supervision.

(2) A member shall encourage his engineering employees to improve their knowledge, attend and present papers at professional meetings, and provide a prospective engineering employee with complete information on working conditions and his proposed status of employment and after employment keep him informed of any change in such conditions.

ARTICLE 16

A member employed abroad shall order his conduct according to this Code, so far as this is applicable, and the laws and regulations of the country of his employment.

ARTICLE 17

A member shall report unethical professional practices of an engineer or a member with substantiating data to the Institution of Engineers (Pakistan) as a witness, if required.

NOTES

This code is published in a booklet entitled The Institution of Engineers, Pakistan: Revised Constitution and By-Laws (Lahore, Pakistan: The Institution of Engineers, Pakistan, 1981). The booklet contains 88 numbered pages.

Part I, "Constitution," covers pp. 1-24. Part II, "By-Laws," as amended by the 174th Central Council Meeting held at Karachi 28-29 August 1980, covers pp. 28-81. This second part includes, as chapter II, "Membership," section 17 (last section), the "Professional Ethics and Code of Conduct" (pp. 35-42).

The code itself is prefaced with the statement that "The following Code of Conduct has been approved by the Central Council which shall apply to all members of the Institution of Engineers (Pakistan). This Code of Conduct is identical to the Code of Conduct approved by the Pakistan Engineering Council for its members."

The Institution of Engineers, Pakistan, is the successor to The Institution of Engineers, India, as a result of the independence and partition of these two countries. In the words of the "Preamble" of the Constitution: "whereas the Institution of Engineers (India) registered under the Indian Companies Act 1913 and incorporated by the Royal Charter 1935 existing immediately before the 14th of August, 1947 had its jurisdiction throughout India, has now its jurisdiction limited within the territory under the sovereignty of the Government of the Republic of India and had/has no successor other than 'The Institute of Engineers, Pakistan' anywhere within the territory forming Pakistan ... 'The Institute of Engineers, Pakistan' is and shall be entitled to all rights or interests as might have accrued to or as might have deemed to accrue to the same as duly and legally constituted successor of 'The Institution of Engineers, India' in Pakistan" (pp. 1-2).

As the "Preface" notes, there was a further reorganization of the Institution of Engineers, Pakistan, in 1973 as a result of "the separation of East Pakistan" (p. vii).

SINGAPORE

* * *

THE INSTITUTION OF ENGINEERS, SINGAPORE CODE OF ETHICS

* * *

Rules for Professional Conduct

These rules shall apply to all forms of engineering employment, and for the purpose of these Rules the term "Employer" shall include the term "Client".

All members of the Institution are enjoined to conform with the letter and the spirit of the Rules set out hereunder.

(1) A member, in his responsibility to his Employer and to the profession, shall have full regard to the public interest.

(2) A member shall order his conduct so as to uphold the dignity, standing, and reputation of the profession.

(3) A member shall discharge his duties to his Employer with complete fidelity.

In whatever capacity he is engaged, he shall assiduously apply this skill and knowledge in the interests of his Employer. If he is confronted by a problem which calls for knowledge and experience which he does not possess, he shall not hesitate to inform his Employer of the fact, and shall make an appropriate recommendation as to the desirability of obtaining further advice. He shall not accept remuneration for services rendered other than from his Employer or with his Employer's permission.

(4) If called upon to give evidence or otherwise to speak on a matter of fact, he shall speak what he believes to be the truth, irrespective of its effect on his own interest, the interests of other Engineers, or other sectional interest.

(5) A member shall not maliciously or recklessly injure or attempt to injure, whether directly or indirectly, the professional reputation, prospects, or business of another Engineer.

Unless he is convinced that his duty to the public or his employer compels him to do so, he shall not express opinions which reflect on the ability or integrity of another Engineer.

(6) A member shall not improperly canvass or solicit professional employment nor offer to make by way of commission or otherwise payment for the introduction of such employment.

(7) A member shall not, in self-laudatory language in any manner derogatory to the dignity of the profession, advertise or write articles for publication, nor shall he authorize such advertisements to be written or published by any other person.

(8) A member, without disclosing the fact to his Employer in writing, shall not be a director of nor have substantial financial interest in, nor be agent for any company, firm or person carrying on any contracting, consulting or manufacturing business which is or may be involved in the work to which his employment relates; nor shall he receive directly or indirectly any royalty, gratuity or commission on any article or process used in or for the purpose of the work in respect of which he is employed unless or until such royalty, gratuity, or commission has been authorized in writing by his Employer.

He shall not report upon or make recommendation on any tender from a company or firm in which he has any substantial interest or on tenders which include such a tender unless specifically requested to do so in writing by his Employer. In this case, he shall maintain an attitude of complete impartiality.

(9) A member shall not use the advantages of a salaried position to compete unfairly with Engineers in private practice to the detriment of salaried engineers.

(10) A member who shall be convicted by a competent tribunal of a criminal offence which in the opinion of the disciplinary body renders him unfit to be a member shall be guilty of improper conduct.

(11) A member shall not, directly or indirectly, attempt to supplant another Engineer; nor shall he intervene or attempt to intervene in or in connection with engineering work of any kind which to his knowledge has been entrusted to another Engineer.

(12) A member shall not be the medium of payments made on his Employer's behalf unless so requested by his Employer; nor shall he in connection with work on which he is employed place contracts or orders except with the authority of and on behalf of his Employer.

(13) When in a position of authority over other Engineers, he shall take every care to afford to those under his direction every reasonable opportunity to advance their knowledge and experience.

He shall ensure that proper credit is given to any subordinate who has contributed in any material way to work for which he is responsible.

(14) A member shall not use for his personal gain or advantage, nor shall he disclose, any confidential information which he may acquire as a result of special opportunities arising out of work for his employer.

(15) In the preparation of plans, specification and contract documents, and on the supervision of construction work, a member shall assiduously watch and conserve the interests of his employer. However, in the interpretation of contract documents, he shall maintain an attitude of scrupulous impartiality as between his employer on the one hand, and the contractor on the other, and shall, as far as he can, ensure that each party in the contract shall discharge his respective duties and enjoy his respective rights as set down in the contract agreement.

SRI LANKA
THE INSTITUTION OF ENGINEERS
CODE OF ETHICS

● ● ●

120/15 Wijerama Mawatha
Colombo 7, Sri Lanka

BY-LAWS–APPENDIX I

● ● ●

1989

● ● ●

FORWARD

The need for professional ethics is recognized in most professions and the by-laws of the Institution of Engineers, Sri Lanka, require its members to observe certain rules of conduct.

This Code was approved by the General Membership at the Annual General Meeting held on 31st October, 1989.

For society to recognize the integrity and to trust the judgment of engineers they are required to comply with the Code of Ethics set out in this booklet.

Members acting in accordance with this Code would create an image that would stand out as a beacon of competence as well as of uprightness and integrity.

D.G. SENADHIPATHY, PRESIDENT 1990/91
1st March, 1991

CODE OF ETHICS

● ● ●

Clause 1. Engineers shall hold paramount the safety, health and welfare of the public and proper utilization of funds in the performance of their professional duties. It shall precedence over their responsibility to the profession, to sectional or private interests, to employers or to other Engineers.

Clause 2. Engineers shall always act in such a manner as to uphold and enhance the honor, integrity and dignity of the profession while safeguarding public interest at all times.

Clause 3. Engineers shall build their reputation on merit and shall not compete unfairly.

Clause 4. Engineers shall perform professional services only in the areas of their competence.

Clause 5. Engineers shall apply their skills and knowledge in the interest of their employer or client for whom they shall act, in professional matters, as faithful agents or trustees, so far as they do not conflict with the other requirements listed here and the general public interest.

Clause 6. Engineers shall give evidence, express opinions or make statements in an objective and truthful manner.

Clause 7. Engineers shall continue their professional development throughout their careers and shall actively assist and encourage engineers under their direction to advance their knowledge and experience.

RULES

● ● ●

Clause 1

Engineers shall hold paramount the safety, health and welfare of the public and proper utilization of the funds in the performance of their professional duties. It shall take precedence over their responsibility to the profession, sectional or private interests, to employers or to other engineers.

As the first requirement places the interests of the community above all other, Engineers—

Rule 1.1 shall be objective and truthful in professional reports, statements or testimony. They shall include all relevant and pertinent information in such reports, statements or testimony.

Rule 1.2 shall endeavor at all times to maintain engineering services essential to public welfare.

Rule 1.3 shall work in conformity with recognized engineering standards so as not to jeopardize the public welfare, health or safety.

Rule 1.4 shall not participate in assignments that would create conflict of interest between their clients or employers, and the public.

Rule 1.5 shall, in the event of their judgment being overruled in matters pertaining to welfare, health or safety of the community, inform their clients or employers of the possible consequences and bring to their notice their (Engineers') obligations as professionals to inform the relevant authority.

Rule 1.6 Shall contribute to public discussion on engineering matters in their areas of competence if they consider that by so doing they can constructively advance the well-being of the community.

Rule 1.7 having knowledge of any alleged violation of this Code shall co-operate with the proper authorities in furnishing such information or assistance as may be required.

Rule 1.8 shall not knowingly participate in any act which will result in waste or misappropriation of public funds.

Clause 2

Engineers shall always act in such a manner as to uphold and enhance the honor, integrity and dignity of the profession while safeguarding public interest at all times.

This requires that the profession should endeavor by its behavior to merit the highest esteem of the community. It follows therefore that engineers—

Rule 2.1 shall not involve themselves with any business or professional practice which they know to be fraudulent or dishonest in nature.

Rule 2.2 shall not use association with other persons, corporations or partnerships to conceal unethical acts.

Rule 2.3 shall not continue in partnership with, or act in professional matters with any engineer who has been removed from membership of this Institution because of improper conduct.

Clause 3

Engineers shall build their reputation on merit and shall not compete unfairly.

This requirement is to ensure that engineers shall not seek to gain a benefit by improper means. It follows that engineers—

Rule 3.1 shall neither pay nor offer, directly or indirectly, inducements including political contribution.

Rule 3.2 shall promote the principle of engagement of engineers upon the basis of merit. They shall uphold the principle of adequate and appropriate remuneration for professional engineering staff and shall give due consideration to terms of engagement which have the approval of the Professional's appropriate association.

Rule 3.3 shall not attempt to supplant another engineer, employed or consulting, who has been appointed.

Rule 3.4 shall neither falsify nor misrepresent their own or their associate's qualifications, experience and prior responsibilities.

Rule 3.5 shall not maliciously do anything to injure, directly, or indirectly, the reputation, prospects or business of other.

Rule 3.6 shall not use the advantage of a privileged position to compete unfairly with other engineers.

Rule 3.7 shall exercise due restraint in explaining their own work and shall refrain from unfair criticism of the work of other engineers.

Rule 3.8 shall give proper credit for professional work to those to whom credit is due and acknowledge the contribution of subordinates and others.

Clause 4

Engineers shall perform professional services only in the areas of their competence.

To this end engineers—

Rule 4.1 shall undertake assignments only when qualified by education and experience in the specific technical fields involved. If an assignment requires qualification and experience outside their fields of competence they shall engage competent professionals with necessary qualifications and experience and keep the employers and clients informed of such arrangements.

Rule 4.2 shall not affix their signature to any plans or documents dealing with subject matter in which they lack competence, or to any plan or document not prepared under their direction or control.

Clause 5

Engineers shall apply their skills and knowledge in the interest of their employer or client for whom they shall act, in professional matters, as faithful agents or trustees, so far as they do not conflict with other requirements listed here and the general public interest.

It follows that engineers—

Rule 5.1 shall at all times avoid all known or potential conflicts of interest. They should keep their employees or clients dully informed on all matters, including financial interests, which could lead to such a conflict, and in no circumstances should they participate in any decision which could involve them in conflict of interest.

Rule 5.2 shall when acting as administrators of a contract be impartial as between the parties in the interpretation of the contract.

Rule 5.3 shall not accept compensation, financial or otherwise from more than one party for services on the same project, unless the circumstances are fully disclosed and agreed to, by all interested parties.

Rule 5.4 shall neither solicit nor accept financial or other valuable consideration, including free engineering designs, from material or equipment suppliers for specifying their products (except such designs obtained with the knowledge and consent of the employer or client).

Rule 5.5 shall neither solicit nor accept gratuities, directly or indirectly from contractors or their agents, or other parties dealing with their clients or employers in connection with work for which they are responsible.

Rule 5.6 Shall advise their clients or employers when as a result of their studies they believe that a project will not be viable.

Rule 5.7 Shall neither disclose nor use confidential information gained in the course of their employment without express permission (except where public interest and safety are involved).

Rule 5.8 shall not complete, sign, or seal plans and/or specifications that are not of a design safe to the public health and welfare and in conformity with accepted engineering standards. If the client or employer insists on such unprofessional conduct, they shall notify the proper authorities and withdraw from further service on the project.

Clause 6

Engineers shall give evidence, express opinion or make statements in an objective and truthful manner.

It follows that—

Rule 6.1 engineers' professional reports, statements or testimony before any tribunal shall be objective and such opinions shall be expressed only on the basis of adequate knowledge and technical competence in the area, but this does not preclude a considered speculation based intuitively on experience and wide relevant knowledge.

Rule 6.2 engineers shall reveal the existence of any interest, pecuniary or otherwise that could be taken to effect their judgment in a technical matter about which they are making a statement or giving evidence.

Clause 7

Engineers shall continue their professional development throughout their careers and shall actively assist and encourage engineers under their direction to advance their knowledge and experience.

The requirement here is that engineers shall strive to widen their knowledge and improve their skill in order to achieve a continuing improvement of the profession. It follows therefore that engineers—

Rule 7.1 shall encourage their professional employees and subordinates to further their education, and

Rule 7.2 shall take a positive interest in and encourage their fellow engineers actively to support the Institution and other professional engineering bodies which further the general interest of the profession.

GUIDELINES FOR PROFESSIONAL CONDUCT

● ● ●

1. Engineers shall be guided in all their professional relations by the highest standards of integrity.

   a. Engineers shall admit and accept their own errors when proven wrong and refrain from distorting or altering the facts in an attempt to justify their decision.

   b. Engineers shall advise their clients or employers when they believe a project will not be successful

   c. Engineers shall not accept assignments outside their employment to the detriment of their regular work or interest. Before accepting any assignments outside their employment they will notify their employers and obtain their prior permission.

2. Engineers shall at all times strive to serve the public interest.

   a. Engineers shall seek opportunities to be of constructive service in civil affairs and work for the advancement of the safety, health and well being of their community.

   b. In public or private sector employment engineers shall refrain from participating knowingly in any act that will result in waste or misappropriation of employers funds.

3. Engineers shall refrain from all conduct or practice which is likely to discredit the profession or deceive the public.

   a. Engineers shall refrain from using statements containing material misrepresentation of fact, or omitting material fact.

   b. Engineers shall refrain from showmanship, or self-laudation or from attempting to attract clients thereby and making derogatory statements about others. Consistent with the foregoing Engineers may advertise for recruitment of personnel.

   c. Consistent with the foregoing: Engineers may publish articles in the press or in technical journals but such articles shall not imply credit to the author for work performed by other.

4. Engineers shall not disclose confidential information concerning the business affairs or technical processes of employers without their consent.

5. Engineers shall not be influenced in their professional duties by conflicting interests.

   a. Engineers shall not accept financial or other consideration, from material or equipment suppliers for specifying their product.

   b. Engineers shall not accept commissions or allowances, directly or indirectly from contractors or other parties in connection with work for which the Engineer is responsible.

   c. Consistent with the foregoing Engineers may publish articles in the press on in technical journals but such articles shall not imply credit to the author for work performed by other.

6. Engineers shall uphold the principle of appropriate and adequate compensation for those engaged in engineering work.

   a. Engineers shall not accept remuneration from either an employee or employment agency for giving employment.

   b. Engineers, when employing other engineers, shall offer a salary according to professional qualifications, experience and recognized standards.

7. Engineers shall not compete unfairly with other engineers to obtain employment or advancement in employment or in seeking professional engagements by taking advantage of their position, by criticizing other engineers, or by other improper or questionable means.

   a. Engineers shall not request, propose, or accept a professional commission under circumstances in which their professional judgment may be compromised.

   b. Engineers in salaried position shall accept part-time engineering work only with the expressed permission of the employer and at recognized rates for such work.

   c. Engineers shall not use equipment, supplies, laboratory, or office facilities of an employer to carry out outside private work without the consent of the employer.

8. Engineers shall not attempt to injure, maliciously or falsely, (directly or indirectly) the professional reputation, prospects practice or employment of other engineers, nor indiscriminately criticize other engineers' work. Engineers who believe others are guilty of unethical or illegal practice shall present such information to the proper authority for action.

   a. Engineers in private practice shall not review the work of another engineer for the same client, except with the knowledge of such engineer, or unless the connection of such engineer with the work has been terminated for un-ethical practices.

   b. Engineers in governmental, industrial or educational employ are entitled to review and evaluate the work of other engineers when so required by their employers.

   c. Engineers in sales or industrial employ shall not criticize products of other manufactures which are similar to their own.

9. Engineers shall accept personal responsibility for their professional activities.

   a. Engineers shall conform with state registration laws in the practice of engineering.

   b. Engineers shall not use association with a non-engineer, a corporation, or partnership, as a "cloak" for unethical acts, but must accept personal responsibility for their own professional acts.

10. Engineers shall give credit for engineering work of other engineers to whom credit is due, and will recognize the proprietary interests of others.

    a. Engineers shall, when possible, name the person or persons who may be individually responsible for designs, inventions, writings, or other accomplishments.

    b. Engineers using designs supplied by client shall recognize that the designs remain the property of the client and shall not be duplicated by the Engineer for others without expressed permission.

    c. Engineers, before undertaking work for others which may result in the engineers producing inventions, plans, designs, improvements or other such, which may justify copyrights or patents, should enter into a position agreement regarding ownership.

    d. Engineers' designs, data, records, and notes referring exclusively to an employer's work shall not be sued for another client unless with the expressed permission of the employer for whom such work was carried out.

11. Engineers shall cooperate in extending the effectiveness of the profession by interchanging information and experience with other Engineers and Students, and will endeavor to provide opportunity for the professional development and advancement of engineers under their supervision.

    a. Engineers shall encourage Engineer employees' efforts to improve their education.

    b. Engineers shall encourage Engineer employees to attend and present papers at professional and technical society meetings.

    c. Engineers shall urge Engineer employees to become registered engineers at the earliest possible date.

    d. Engineers shall assign a professional engineer duties of a nature to utilize his full training and experience, in so far as is possible, and delegate lesser functions to sub-professionals or to technicians.

    e. Engineers shall provide a prospective employee with complete information on working conditions and proposed status of employment, and after engaging will keep such employees informed of any proposed changes.

SWEDEN

• • •

SWEDISH FEDERATION OF CIVIL ENGINEERS CODE OF ETHICS

• • •

Code of Honor for Civil Engineers

1. The civil engineer should feel, while practicing his profession, a personal responsibility that technology will be utilized in a fashion which benefits humanity, the environment, and society.

2. The civil engineer should strive to improve technology and technical expertise in the direction of a more efficient utilization of resources without detrimental side effects.

3. The civil engineer should be prepared to share his knowledge in public and private contexts in order to reach the best possible basis for a decision and to illustrate the capacities and the risks of technology.

4. The civil engineer should not work within or collaborate with corporations or organizations of a questionable character or ones whose goals are in conflict with the civil engineer's personal convictions.

5. The civil engineer should show complete loyalty to employers and co-workers. Any difficulties in this respect should be dealt with in an open discussion and, first of all, at the place of work.

6. The civil engineer is not permitted to use improper methods in the competition for employments, assignments, or commissions and, furthermore, must not attempt to damage the reputation of colleagues through unjustified accusations.

7. The civil engineer should respect the confidential nature of especially entrusted information, as well as the rights of others with regard to ideas, inventions, research, plans, and designs.

8. The civil engineer is not allowed to favor unauthorized interests and should openly account for financial and other interests that could affect the trust in his impartiality and judgment.

9. The civil engineer should, privately and in public, in speech and in writing, strive for an objective mode of presentation and avoid incorrect, misleading, or exaggerated statements.

10. The civil engineer should actively support colleagues who find themselves in trouble because of actions of the kind described in these rules and should prevent any violations against the rules, according to his best judgment.

TRANSLATED BY BIRGITTA D. KNUTTGEN

NOTES

This code is promulgated by means of a one-page document with a simple double-line border suitable for framing. As a kind of preface to the code it includes the following statement:

Technology and the natural sciences are powerful tools in the service of humankind, for better and for worse. They have thoroughly transformed society and will continue to have a profound effect on humankind also in the future.

The civil engineers are the bearers and managers of technical knowledge. Therefore, they are also given the special responsibility of ensuring that technology will be used in the best interests of society and humankind and that it will be transferred to future generations in an improved state.

In 1929, the Swedish Technological Association established a Code of Honor. The developments in society and technology have warranted a revision of the code. To provide support for the personal decision-making of a civil engineer with regard to ethical considerations, The Swedish Federation of Civil Engineers, on the 15th of November, 1988, established the following.

1. With his specialized knowledge and competence as his guide, the STV Expert safeguards the legitimate concern of his employers. He does not overestimate his own abilities.

2. The STV Expert, in fulfilling his assignments, bears in mind the dignity of his profession. He does not participate in any procedure that could be injurious to this dignity.

3. The STV Expert is committed to maintaining professional secrecy in all aspects of his assignments.

4. The STV Expert, in his capacity as expert or arbitrator, is committed to being strictly objective. Should the danger of a conflict of interest arise, he is obliged to refuse or give up his position.

5. The STV Expert accepts no remunerations or personal privileges from any third party. As the representative or advisor to an employer, he acts with complete independence.

6. The STV Expert observes the appropriate technical standards. He is obliged to constantly further his studies in order to remain at the level of expertise required by his profession.

7. The STV Expert charges the customary fee for his area of expertise.14 March 1990.

SWITZERLAND

• • •

SWISS TECHNICAL ASSOCIATION CODE OF ETHICS

• • •

Swiss Technical Association Honor Code

With reference to your inquiry of 10 April 1990, we regret to inform you that our association has not adopted an actual honor code. On the other hand, we have a so-called Chamber of Experts (architects, engineers), whose members can be consulted for expert opinions of all kinds. The members of this chapter are subject to an honor code. We have enclosed the version currently in force.

HONOR CODE FOR STV EXPERTS

• • •

The STV Expert is committed to uphold and apply the following principles:

UNITED KINGDOM

• • •

INSTITUTION OF CIVIL ENGINEERS CODE OF ETHICS

• • •

1-7 Great George Street, Westminster
London SW1P 3AA, England

Founded: 1818
Royal Charter: 1828
Members: 4,500

Rules for Professional Conduct

Made by the Council on 19 March 1963, and modified in 1971, 1973 and 1982 in accordance with By-law 32.

Expressions used in these Rules shall have the meaning if any assigned to them by the By-laws, Regulations, and Rules of the Institution. These Rules apply to all forms of engineering employment, and for the purpose of these Rules the term "Employer" shall include the term "Client."

1. A member, in his responsibility to his Employer and to the profession, shall have full regard to the public interest, particularly in matters of health and safety.

2. A member shall discharge his professional responsibilities with integrity.

3. A member shall discharge his duties to his Employer with complete fidelity. He shall not accept remuneration for services rendered other than from his Employer or with his Employer's permission.

4. A member shall not maliciously or recklessly injure or attempt to injure, whether directly or indirectly, the professional reputation, prospects or business of another Engineer.

5. A member shall not improperly canvass or solicit professional employment nor offer to make by way of commission or otherwise payment for the introduction of such employment.

6. A member shall not, in self-laudatory language or in any manner derogatory to the dignity of the profession, advertise or write articles for publication, nor shall he authorize such advertisements to be written or published by any person.

7. A member, without disclosing the fact to his Employer in writing, shall not be a director of nor have a substantial financial interest in, nor be agent for any company, firm or person carrying on any contracting, consulting or manufacturing business which is or may be involved in the work to which his employment relates; nor shall he receive directly or indirectly any royalty, gratuity or commission on any article or process used in or for the purposes of the work in respect of which he is employed unless or until such royalty, gratuity or commission has been authorized in writing by his Employer.

8. A member shall not use the advantages of a salaried position to compete unfairly with other engineers.

9. A member in connection with work in a country other than his own shall order his conduct according to these Rules, so far as they are applicable; but where there are recognized standards of professional conduct, he shall adhere to them.

10. A member who shall be convicted by a competent tribunal of a criminal offence which in the opinion of the disciplinary body renders him unfit to be a member shall be deemed to have been guilty of improper conduct.

11. A member shall not, directly or indirectly, attempt to supplant another Engineer, nor shall he intervene or attempt to intervene in or in connection with engineering work of any kind which to his knowledge has already been entrusted to another Engineer.

12. A member shall not be the medium of payments made on his Employer's behalf unless so requested by his Employer, nor shall he in connection with work on which he is employed place contracts or orders except with the authority of and on behalf of his Employer.

13. A member shall afford such assistance as he may reasonably be able to give to further the Education and Training of candidates for the Profession.

NOTES

This code is promulgated in the last page of a 28-page yellow pamphlet (1985) that includes its Royal Charter, By-laws, Regulations, and Rules.

By-law 32, to which reference is made in the preliminary indication of adoption dates, reads as follows: "Without prejudice to the generality of the last preceding By-law the Council may, for the purpose of ensuring the fulfillment of this requirement, make, amend, and rescind Rules to be observed by Corporate and Non-Corporate Members, with regard to their conduct in any respect which may be relevant to their position or intended position as members of the Institution and may publish directions or pronouncements as to specific conduct which is to be regarded as proper or as improper (as the case may be)."

The current SCET (Institution of Civil Engineers and Technicians) code is a result of past mergers with other professional organizations, i.e. in 1984 with The Institution of Municipal Engineers and again in 1989 with The Society of Civil Engineering Technicians and the Board of Incorporated Engineers and Technicians.

INSTITUTION OF MECHANICAL ENGINEERS CODE OF ETHICS

• • •

1 Birdcage Walk, Westminster
London SW1H 9JJ, England

Founded: 1847
Members: 78,000

PROFESSIONAL CODE OF CONDUCT

• • •

1. INTRODUCTION

The Institution of Mechanical Engineering as a learned body has three main functions: to promote the development of mechanical engineer, to govern the qualifications of its members, and to control their professional conduct. This leaflet, issued by the Professional Affairs Board of the Institution is concerned with the third of these functions and its surrounding circumstances. In this connection members are also referred to other relevant guides issued by the Institution, viz: Health & Safety at Work, Professional Engineers and Trade Unions (PAB 2/83) Guide for Consultancy and Product Liability.

2. CLASSES OF MEMBERSHIP

Under By-Law 2, the membership is divided into Corporate Members, (those entitled to be heard and vote at annual, ordinary, and special meetings) and Non-Corporate Members (with no such privileges except at ordinary meetings where they may be heard on mechanical engineering or allied subjects). The former group consists of three classes of persons viz:

Fellows

Members

The latter group (Non-Corporate Members) consists of six classes of persons viz:

Honorary Fellows who when elected Honorary Fellows were not already Corporate Members.

Companions

Associates

Associate Members

Graduates

Students

3. ABBREVIATED TITLES AND DESCRIPTION OF MEMBERSHIP (BY-LAW 6)

Corporate members may abbreviate their titles to Hon-FIMechE, FIMechE, or MIMechE as applicable, while Non-Corporate members may not use abbreviated titles except Honorary Fellows (Hon FIMechE) and Associate Members (AMIMechE) and, in certain cases of long-standing membership, Companions (CIMechE). By-LAW 6 (iii) states "a member shall not use or permit to be used any of the said titles or abbreviations in letters larger or bolder than those used in the name of the member which they follow."

4. CONDUCT OF MEMBERS (BY-LAW 32)

4.1 The Professional Conduct of all members is governed by By-Law 32 and its associated Rules of Conduct. Extracts are given below.

32. (i) In order to facilitate the advancement of the science of mechanical engineering by preserving the respect in which the community holds persons who are engaged in the profession of mechanical engineering, every member of any class shall at all times so order his conduct as to uphold the dignity and reputation of the Institution and act with fairness and integrity towards all persons with whom his work is connected and towards other members.

(ii) Every Corporate Member shall at all times so order his conduct as to uphold the dignity and reputation of his profession, and to safeguard the public interest in matters of safety and health and otherwise. He shall exercise his professional skill and judgment to the best of his ability and discharge his professional responsibilities with integrity.

4.2 By-Law 32 (iii) allows the Council of the Institution to make, vary, or rescind Rules of Conduct for any class of member provided approval is received at a Special Meeting of Corporate members.

The only Rules of Conduct so approved are repeated below:

Pursuant to By-Law 32.

In these Rules, 'member' means a member of any class referred to in By-Law 2, and 'employer' includes 'client'.

1. A member whose professional advice is not accepted shall take all reasonable steps to ensure that the person overruling or neglecting his advice is aware of any danger which the member believes may result from such overruling or neglect.

2. A member shall not recklessly or maliciously injure or attempt to injure whether directly or indirectly the professional reputation, prospects, or business of another.

3. A member shall inform his employer in writing of any conflict between his personal interest and faithful service to his employer.

4. A member shall not improperly disclose any information concerning the business of his employer or any past employer.

5. A member shall not solicit or accept remuneration in connection with professional services rendered to his employer other than from his employer or with his employer's consent; nor shall he receive directly or indirectly any royalty, gratuity, or commission on any article or process used in or for the purpose of the work in respect of which he is employed unless or until such royalty, gratuity, or commission has been authorized in writing by his employer.

6. Where a member of any class has been (a) adjudicated bankrupt or (b) convicted of an offense, he shall be deemed to have been guilty of improper conduct if the circumstances of the offense are such as to constitute a breach of the By-Laws or of these Rules.

4.3 Members frequently seek guidance from the Institution over Rule 1. In considering this question of engineer should have a clear understanding of what he is accountable for. This is best achieved by reference to agreed written terms of reference. If an engineer, in the course of his duties makes a decision which is overruled by his employer and this, in his view, would be detrimental to public health and safety, then his obligation to his Institution will be discharged by issuing a written statement to his employer setting out the reasons why he believes public health and safety will be affected. As an employee, an engineer has no authority to direct his employer, therefore he cannot be held responsible for his employer's conduct. If the employer's action should prove to be detrimental to public health and safety, then this would be a matter for adjudication by the Courts.

Employed engineers finding themselves in such a situation are advised, in the first instance, to seek the view of fellow members of the Institution with whom they work. If further guidance is required, then the Professional Services Manager should be approached at Institution H.Q. Self-employed consulting engineers are able to resolve their own conflicts with professional obligations by being able to choose assignments and methods of working.

4.4 The obligations arising from the Institution's Codes and Rules of Conduct may be interpreted as requiring each member to behave so as:

to maintain and develop his Professional competence in the engineering field in which he practices;

to accept personal responsibility for his work and for those for whom he is accountable;

to give objective and reliable advice on matters within his field of practice when called upon to do so;

to avoid giving professional advice in engineering matters outside his competence;

to avoid malicious injury to the reputation, prospects, or business of others;

to avoid self-laudatory language in advertising his services or in published articles.

5. HEALTH AND SAFETY AT WORK ACT 1974

The 1974 Act imposes statutory duties on all persons at work and failure to comply with these may lead to criminal proceedings against them. All members are therefore expected to be familiar with the provisions of the Act and to read the guidance Booklet published by the Institution. Membership of a Professional body imposes on members the additional obligation of bringing the attention of their colleagues to the requirements of the 1974 Act.

6. PROFESSIONAL NEGLIGENCE

Professional negligence is discussed in the IMechE Booklet "Product Liability". A court judgment going against a member accused of professional negligence may under By-Law 33 (Disciplinary action), be taken as prima facie evidence of improper conduct. However, this is not necessarily so and will always depend upon all the circumstances of the case.

7. THE ENGINEER AS AN EXPERT WITNESS

A member called upon to testify as an Expert Witness should remember that he has a professional obligation to assist the court in reaching an equitable verdict and not to act as an advocate for whoever pays his fee. Guidance on this subject is provided in the IMechE Booklet "Guide for Consultancy".

8. TRADE UNION AND INDUSTRIAL ACTION

The act of joining a Trade Union is not contrary to the Institution's Rules of Professional Conduct. Any member of the Institution is free to join or not to join a Trade Union and if he so wishes to join, then the choice of a Union lies with him, but he is advised, where possible, to join one which supports his professional obligations and status.

Members are not forbidden to engage in industrial action provided such action does not conflict with their professional obligations as set out in the By-Laws. It is also important to exhaust the negotiating procedures before considering action. The Employment Act 1982 makes special provision to protect professional employees from dismissal arising from a conflict between professional obligation and obligations to a Trade Union. Guidance on all aspects of Union membership is given in IMechE leaflet reference PAB 2/83.

9. ADVERTISING AND USE OF SITE BOARDS

Advice on advertising and use of site boards is provided in the IMechE Booklet "Guide for Consultancy".

10. EXPULSION AND OTHER DISCIPLINARY ACTION (BY-LAW 33)

By-Law 33 provides the Council of the Institution with powers to investigate allegations of improper conduct lodged against any member and allows disciplinary action to be taken where a member is found guilty.

By-Law 33(i) is reproduced below:

33. (i) For the purposes of this By-Law improper conduct shall mean:

(a) the making of any false representation in applying for election or transfer to any class of membership of the Institution, or

(b) any breach of these By-Laws or of any Regulation or Rule or direction made or given thereunder, or

(c) any conduct injurious to the Institution.

Under the Disciplinary Regulations pursuant to By-Law 33, two Committees are appointed to investigate and adjudicate upon allegations of improper conduct: they are the Investigating Panel and the Disciplinary Board. Where the Investigating Panel finds that there is a prima facie case to answer, the accused member will be invited to put forward his observations in writing to the Panel for further consideration. If a prima facie case is still evident and the matter is not trivial, then the case goes to the Disciplinary Board for a full hearing. The accused member will be given a full and fair opportunity of being heard, of calling witnesses, and of cross-examining any witnesses testifying before the Board. He will be given the opportunity of being represented by a lawyer or by any other member of the Institution of his own choice. The full procedure covering disciplinary action is set out in the Institution's Disciplinary Rules.

For convenience the use of the words "he" or "his" in the text of this leaflet is to be read as being applicable to both sexes.

November, 1983

NOTES

Promulgation is by means of a two-sided yellow leaflet. At the top in a bold box is the statement: "Members should keep this leaflet for future reference."

VENEZUELA

• • •

ASSOCIATION OF ENGINEERS OF VENEZUELA CODE OF ETHICS

• • •

Apartado 2006, Bosque Los Caobos
Caracas 101, Venezuela

Founded: 1861
Members: 7,000

Code of Professional Ethics

It is considered contrary to ethics and incompatible with the dignified exercise of the profession for a member of the Academy of Engineers of Venezuela:

1. To act in any way that serves to diminish the honor, respectability, and the virtues of honesty, integrity, and truthfulness that should serve as the base for a full and complete exercise of the profession.

2. To violate or to permit the violation of laws, ordinances, and regulations related to professional activity.

3. To neglect the maintenance and improvement of his technical knowledge thus becoming unworthy of the trust society places in the professional activity.

4. To put himself forward for employment in specializations and operations for which the applicant has no capacity, preparation, and reasonable experience, so as to describe or advertise himself in laudatory terms or in any manner which goes against the dignity and seriousness of the profession.

5. To neglect because of friendship, convenience, or coercion the fulfillment of contractual obligations when it is his job to respect and to fulfill them.

6. To offer, solicit, or borrow professional services by means of payments below those established as the minimum by the Academy of Engineers of Venezuela.

7. To lay out projects or prepare reports with negligence or inattention or with overly optimistic criteria.

8. To sign plans laid out by others and to take responsibility for projects or works that are not under his immediate direction, review, or supervision.

9. To take charge of works without having undertaken all of the necessary technical studies for their correct execution, when for their realization schedules incompatible with good professional practice have been set up.

10. To concur deliberately or to invite competitive bidding.

11. To offer, to give, or to receive commissions or loans and to solicit influences or to use them for obtaining or securing professional work or for creating privileged positions in their performance.

12. To use the advantages inherent in a job to compete with the independent practice of other professionals.

13. To act against the reputation or the legitimate interests of other professionals.

14. To acquire interests which directly or indirectly clash with the interests of the company or client that employs his services, or to take charge without knowledge about those parties interested in works in which conflicting interests exist.

15. To act deliberately against the principles of justice and loyalty in his relations with clients, subordinate personnel, and workers, especially in relation to the last, in reference to the maintenance of fair working conditions and their just participation in profits.

16. To intervene directly or indirectly to the destruction of natural resources, or to neglect the corresponding action to avoid the production of products that contribute to environmental deterioration.

17. To act in any way that would permit or facilitate contracting with foreign companies for studies or projects, construction or inspection of works, when in the judgment of the Academy of Engineers there exists in Venezuela the capacity to perform these tasks.

TRANSLATED BY ANNA H. LYNCH AND CARL MITCHAM

NOTES

This code is promulgated in two forms:

(1) It is included in a booklet entitled Reglamento interno [Internal regulation] (Caracas: Colegio de Ingenieros de Venezuela, 1988). This is a booklet of 137 numbered pages.

Following a prefatory note and table of contents, the first major part of this booklet prints decree 444 (24 November 1958), "Ley de Ejercicio de la Ingenieria, la Arquitectura y Profesiones Afines" [Law on the practice of engineering, architecture and related professions], pp. 3-12. This is followed by a commentary which includes both general considerations on the history and development of the Association and remarks on each article in the decree (pp. 15-27), with a one page summary of "Conclusions from the First Interamerican Workshop of University Professionals," Montevideo, November 1957.

Then comes the code of ethics (pp. 29-30). This printing of the code notes that point 15 was adopted on June 27, 1957; point 16 on October 4, 1976; and point 17 on June 27, 1980.

The second major of the booklet contains, in accord with article 21 of the law of 1958, the by-laws of the Association (pp. 31-132) as of August 13, 1984.

(2) The code is also printed as a separate, one-page document suitable for framing.

FÉDÉRATION EUROPÉENNE D'ASSOCIATIONS NATIONALES D'INGÉNIEURS (FEANI, EUROPEAN FEDERATION OF NATIONAL ENGINEERING ASSOCIATIONS) CODE OF ETHICS

• • •

The FEANI Code of Conduct is additional to and does not take the place of any Code of Ethics to which the registrant might be subject in his own country.

All persons listed in the FEANI Register have the obligation to be conscious of the importance of science and technology for mankind and of their own social responsibilities when engaged in their professional activities.

They exercise their profession in accordance with the normal rules of good conduct of European societies, respecting particularly the professional rights and the dignity of all those with whom they work.

They thereby undertake to comply with and maintain the following code of ethics.

1. Personal Ethics

The Engineer shall maintain his competence at the highest level, with a view to providing excellence of services in accordance with what is regarded as good practice in his profession and having regard to the laws of the country in which he is working.

His professional integrity and intellectual honesty shall be the guarantees of his impartiality of analysis, judgment and consequent decision. He shall consider himself bound in conscience by any business confidentiality agreement into which he has freely entered.

He shall not accept any payment except those agreed with his relevant employer.

He shall display his commitment to the engineering profession by taking part in the activities of its Associations, notably those which promote the profession and contribute to the continuing training of their members.

He shall use only titles to which he has a right.

2. Professional Ethics

The Engineer shall accept assignments only within the area of his competence.

Beyond this limit, he shall seek the collaboration of appropriate experts.

He is responsible for organizing and executing his assignments.

He must obtain a clear definition of the services required of him. Executing his assignments, he shall take all necessary steps to overcome any difficulties encountered whilst ensuring the safety of persons and property.

He shall take remuneration corresponding to the service rendered and the responsibilities assumed.

He shall try to ensure that the remuneration of each be consonant with the service rendered and the responsibilities assumed.

He strives for a high level of technical achievement which will also contribute to and promote a healthy and agreeable environment for his fellowmen.

3. Social Responsibility

The Engineer shall

- respect the personal rights of his superiors, colleagues and subordinates by taking due account of their requirements and aspirations, provided they conform to the laws and ethics of their professions,

- be conscious of nature, environment, safety and health and work to the benefit and welfare of mankind,

- provide the general public with clear information, only in his field of competence, to enable a proper understanding of technical matters of public interest,

- treat with the utmost respect the traditional and cultural values of the countries in which he exercises his profession.

N.B. : In this text, "he" and "his" are taken respectively for "he/she" and "his/her."

FOUNDING STATEMENT OF THE INTERNATIONAL NETWORK OF ENGINEERS AND SCIENTISTS FOR GLOBAL RESPONSIBILITY (INES)

• • •

November 29, 1991

Rapid changes in our environment and our societies are forcing us to become more conscious of our role in the world. Science and technology are employed in a worldwide competition for military and economic power. The impacts of this competition have global implications. We have entered a phase in which global developments are in conflict with basic requirements for human survival. Large stocks of weapons for mass destruction, the over-exploitation of common limited resources, and a heavily unbalanced world economy provide fundamental challenges to human civilization and may even threaten its further existence. The end of the cold war and the progress towards democracy and national self-determination in many regions provide important opportunities to resolve long-standing threats to international security. Dismantling the vast nuclear and conventional arsenals and demilitarizing international relations remains a high priority. However, after the decline of international bipolar divisions, many major problems remain. Regional and inter-communal conflicts, together with the proliferation of weapons technologies, threaten local and global security. Newly recognized problems such as climate change, ozone depletion and loss of species diversity raise new challenges regarding energy use, population growth and other aspects of development. Gross inequalities and injustice between and within industrialized and developing countries undermine military, economic, social and environmental security.

Developments in science and technology have helped to create global interdependence and to make us more profoundly aware of the planet's condition. Many engineers and scientists play a key role in both the processes that threaten international security and those that provide hope for the future. International organizations and norms are being developed to tackle common problems, and many structures for regional cooperation are emerging to overcome national divisions.

The engineering and scientific community is intrinsically international, with informal networks and channels of communication. However most existing professional organizations are highly specialized. It is now time to establish a multidisciplinary international network of engineers and scientists for global responsibility to promote the following aims:

- to encourage and facilitate international communication among engineers and scientists seeking to promote international peace and security, justice and sustainable development, and working for a responsible use of science and technology. This includes:
- to work for the reduction of military spending and for the transfer of resources thus liberated to the satisfaction of basic needs,
- to promote environmentally sound technologies, taking into account long-term effects,
- to enhance the awareness of ethical principles among engineers and scientists, and to support those who have been victimized for acting upon such principles.

In order to accomplish these aims, members and bodies of the network will

Promote collaborative and interdisciplinary research relating to such issues,

Publicize relevant research, contribute to education and scientific training and inform the public and professional colleagues,

Facilitate and undertake expert and responsible contributions to relevant policy debates, and advocate changes in national and international policies pertinent to the above aims.

We are convinced that it is our continuous task to reflect on values and standards of behavior which take into account basic human needs and our interrelationship with the biosphere. Membership of the network is open to non-governmental organizations and individual engineers and scientists. It will be a network seeking to provide a central resource for, and to promote coordination among, its members. We hope that the synergy of different approaches will facilitate steps from vision toward action.

UNIÓN PANAMERICANA DE ASOCIACIONES DE INGENIEROS

• • •

(UPADI, PANAMERICAN UNION OF ASSOCIATIONS OF ENGINEERS)

• • •

CODE OF PROFESSIONAL ETHICAL CONDUCT

• • •

I. The Fundamentals

1. The Code of Professional Ethics adopted by UPADI member organizations is intended to establish the responsibilities, to regulate rights, and to fix norms of conduct that should be observed by all engineers, both within their professional circles as well as within the larger society, nationally as well as internationally.

2. It is the imperative duty of the Pan-American engineers to maintain their professional and moral conduct at the highest level, in defense of the prestige and rights of the profession, and to be vigilant regarding the correct and proper practice and observation at all times of the dignity, integrity, and respect and loyal adherence to this code.

3. The UPADI engineers shall constantly seek to improve their knowledge and their profession, communicating and sharing their knowledge and experience, in an attempt to provide opportunities for the professional development of their colleagues.

II. Professional Practices

1. The practice of the engineering profession shall be understood exclusively in terms of engineers who hold university titles qualifying them in diverse specialties, in accordance with the current legislation in each country.

2. The practice of engineering shall be considered first and foremost a social function. Projects which might be used against the public interest should be refused, thus avoiding situations which involve danger and constitute a threat to life, health and the environment, or affect property and other human rights.

3. Professional practice implies obligatory service in whatever form the professional assumes: Individual, in society or in a dependent relationship.

4. The formation of professional prestige of engineers shall be based on ability and honesty.

III. Acts Contrary to Ethics

To be considered unethical and incompatible with the dignity of the profession:

1. To act against the honor decorum and prestige of the profession and against the dignity and solidarity which the engineers should guard within their professional circles.

2. To intervene directly or indirectly in the destruction of natural resources or to fail to engage in activity corresponding to the avoidance of the production of anything that contributes to the deterioration of the environment.

3. To permit or contribute to the committing of injustices against engineers.

4. To falsely attribute errors to other engineers.

5. To attempt to substitute of replace other engineers in the offering of professional services.

6. To authorize with one's firm, studies, projects, plans, specifications, reports, or professional opinions that have not been personally developed, executed, controlled or authenticated.

7. To offer or lend professional services for remuneration below the standards already established in the respective tariffs.

8. To utilize studies, projects, plans, reports or other documents that are not subject to public domain, without authorization from its authors or owners.

9. To reveal information reserved of a technical, financial or professional nature, as well as divulge, without proper authorization, procedures, processes or group characteristics that are protected by patents or contracts which establish obligations of professional secrecy.

10. To commit deliberate omissions or negligence in professional activities.

11. To fail to respect the norms established by the authorities and institutions of engineering of the country in which one is executing work.

IV. Organization and Control

1. The offering of professional service involves security and the well-being of the community and is of the character of public service. Thus said, it is necessary that the engineers of each country are matriculated in colleges, counsels or associations is obligatory.

2. The integration and government of these organizations shall be exercised by those same who are matriculated in these organizations and who should fulfill and follow this Code of Professional Ethics.

THE WORLD FEDERATION OF ENGINEERING SOCIETIES MODEL CODE OF ETHICS

• • •

Final version adopted in 2001

I. BROAD PRINCIPLES

Ethics is generally understood as the discipline or field of study dealing with moral duty or obligation. This typically gives rise to a set of governing principles or values which in turn are used to judge the appropriateness of particular conducts or behaviors. These principles are usually presented either as broad guiding principles of an idealistic or inspirational nature, or, alternatively, as a detailed and specific set of rules couched in legalistic or imperative terms to make them more enforceable. Professions that have been given the privilege and responsibility of self regulation, including the engineering profession, have tended to opt for the first alternative, espousing sets of underlying principles as codes of professional ethics which form the basis and framework for responsible professional practice. Arising from this context, professional codes of ethics have sometimes been incorrectly interpreted as a set of "rules" of conduct intended for passive observance. A more appropriate use by practicing professionals is to interpret the essence of the underlying principles within their daily decision-making situations in a dynamic manner, responsive to the need of the situation. As a consequence, a code of professional ethics is more than a minimum standard of conduct; rather, it is a set of principles which should guide professionals in their daily work.

In summary, the model Code presented herein expresses the expectations of engineers and society in discriminating engineers' professional responsibilities. The Code is based on broad principles of truth, honesty and trustworthiness, respect for human life and welfare, fairness, openness, competence and accountability. Some of these broader ethical principles or issues deemed more universally applicable are not specifically defined in the Code although they are understood to be applicable as well. Only those tenets deemed to be particularly applicable to the practice of professional engineering are specified. Nevertheless, certain ethical principles or issues not commonly considered to be part of professional ethics should be implicitly accepted to judge the engineer's professional performance.

Issues regarding the environment and sustainable development know no geographical boundaries. The engineers and citizens of all nations should know and respect the environmental ethic. It is desirable, therefore, that engineers in each nation continue to observe the philosophy of the Principles of Environmental Ethics delineated in Section III of this code.

II. PRACTICE PROVISION ETHICS.

Professional engineers shall:

- hold paramount the safety, health and welfare of the public and the protection of both the natural and the built environment in accordance with the Principles of Sustainable Development;

- promote health and safety within the workplace;

- offer services, advise on or undertake engineering assignments only in areas of their competence and practice in a careful and diligent manner;

- act as faithful agents of their clients or employers, maintain confidentially and disclose conflicts of interest;

- keep themselves informed in order to maintain their competence, strive to advance the body of knowledge within which they practice and provide opportunities for the professional development of their subordinates and fellow practitioners;

- conduct themselves with fairness, and good faith towards clients, colleagues and others, give credit where it is due and accept, as well as give, honest and fair professional criticism;

- be aware of and ensure that clients and employers are made aware of societal and environmental consequences of actions or projects and endeavor to interpret engineering issues to the public in an objective and truthful manner;

- present clearly to employers and clients the possible consequences of overruling or disregarding of engineering decisions or judgment;

- report to their association and/or appropriate agencies any illegal or unethical engineering decisions or practices of engineers or others.

III. ENVIRONMENTAL ENGINEERING ETHICS

Engineers, as they develop any professional activity, shall:

- try with the best of their ability, courage, enthusiasm and dedication, to obtain a superior technical achievement, which will contribute to and promote a healthy and agreeable surrounding for all people, in open spaces as well as indoors;

- strive to accomplish the beneficial objectives of their work with the lowest possible consumption of raw materials and energy and the lowest production of wastes and any kind of pollution;

- discuss in particular the consequences of their proposals and actions, direct or indirect, immediate or long term, upon the health of people, social equity and the local system of values;

- study thoroughly the environment that will be affected, assess all the impacts that might arise in the structure, dynamics and aesthetics of the ecosystems involved, urbanized or natural, as well as in the pertinent socioeconomic systems, and select the best alternative for development that is both environmentally sound and sustainable;

- promote a clear understanding of the actions required to restore and, if possible, to improve the environment that may be disturbed, and include them in their proposals;

- reject any kind of commitment that involves unfair damages for human surroundings and nature, and aim for the best possible technical, social, and political solution;

- be aware that the principles of eco-systemic interdependence, diversity maintenance, resource recovery and inter-relational harmony form the basis of humankind's continued existence and that each of these bases poses a threshold of sustainability that should not be exceeded.

IV. CONCLUSION

Always remember that war, greed, misery and ignorance, plus natural disasters and human induced pollution and destruction of resources, are the main causes of the progressive impairment of the environment and that engineers, as an active member of society, deeply involved in the promotion of development, must use our talent, knowledge and imagination to assist society in removing those evils and improving the quality of life for all people.

INTERPRETATION OF THE CODE OF ETHICS

• • •

The interpretive articles which follow expand on and discuss some of the more difficult and interrelated components of the Code especially related to the Practice Provisions. No attempt is made to expand on all clauses of the Code, nor is the elaboration presented on a clause-by-clause basis. The objective of this approach is to broaden the interpretation, rather than narrow its focus. The ethics of professional engineering is an integrated whole and cannot be reduced to fixed "rules". Therefore, the issues and questions arising from the Code are discussed in a general framework, drawing on any and all portions of the Code to demonstrate their interrelationship and to expand on the basic intent of the Code.

Sustainable Development and Environment

Engineers shall strive to enhance the quality of the biophysical and socioeconomic urban environment and the one of buildings and spaces and to promote the principles of sustainable development.

Engineers shall seek opportunities to work for the enhancement of safety, health, and the social welfare of both their local community and the global community through the practice of sustainable development.

Engineers whose recommendations are overruled or ignored on issues of safety, health, welfare, or sustainable development shall inform their contractor or employer of the possible consequences.

Protection of the Public and the Environment

Professional Engineers shall hold paramount the safety, health and welfare of the public and the protection of the environment. This obligation to the safety, health and welfare of the general public, which includes one's own work environment, is often dependent upon engineering judgments, risk assessments, decisions and practices incorporated into structures, machines, products, processes and devices. Therefore, engineers must control and ensure that what they are involved with is in conformity with accepted engineering practice, standards and applicable codes, and would be considered safe based on peer adjudication. This responsibility extends to include all and any situations which an engineer encounters and includes an obligation to advise the appropriate authority if there is reason to believe that any engineering activity, or its products, processes, etc. do not confirm with the above stated conditions.

The meaning of paramount in this basic tenet is that all other requirements of the Code are subordinate if protection of public safety, the environment or other substantive public interests are involved.

Faithful Agent of Clients and Employers

Engineers shall act as faithful agents or trustees of their clients and employers with objectivity, fairness and justice to all parties. With respect to the handling of confidential or proprietary information, the concept of ownership of the information and protecting that party's rights is appropriate. Engineers shall not reveal facts, data or information obtained in a professional capacity without the prior consent of its owner. The only exception to respecting confidentially and maintaining a trustee's position is in instances where the public interest or the environment is at risk as discussed in the preceding section; but even in these circumstances, the engineer should endeavor to have the client and/or employer appropriately redress the situation, or at least, in the absence of a compelling reason to the contrary, should make every reasonable effort to contact them and explain clearly the potential risks, prior to informing the appropriate authority.

Professional Engineers shall avoid conflict of interest situations with employers and clients but, should such conflict arise, it is the engineer's responsibility to fully disclose, without delay, the nature of the conflict to the party(ies) with whom the conflict exists. In these circumstances where full disclosure is insufficient, or seen to be insufficient, to protect all parties' interests, as well as the public, the engineer shall withdraw totally from the issue or use extraordinary means, involving independent parties if possible, to monitor the situation. For example, it is inappropriate to act simultaneously as agent for both the provider and the recipient of professional services. If client's and employer's interests are at odds, the engineer shall attempt to deal fairly with both. If the conflict of interest is between the intent of a corporate employer and a regulatory standard, the engineer must attempt to reconcile the difference, and if that is unsuccessful, it may become necessary to inform.

Being a faithful agent or trustee includes the obligation of engaging, or advising to engage, experts or specialists when such services are deemed to be in the client's or employer's best interests. It also means being accurate, objective and truthful in making public statements on behalf of the client or employer when required to do so, while respecting the client's and employer's rights of confidentiality and proprietary information.

Being a faithful agent includes not using a previous employer's or client's specific privileged or proprietary information and trade practices or process information, without the owner's knowledge and consent. However, general technical knowledge, experience and expertise gained by the engineer through involvement with the previous work may be freely used without consent or subsequent undertakings.

Competence and Knowledge

Professional Engineers shall offer services, advise on or undertake engineering assignments only in areas of their competence by virtue of their training and experi-ence. This includes exercising care and communicating clearly in accepting or interpreting assignments, and in setting expected outcomes. It also includes the responsibility to obtain the services of an expert if required or, if the knowledge is unknown, to proceed only with full disclosure of the circumstances and, if necessary, of the experimental nature of the activity to all parties involved. Hence, this requirement is more than simply duty to a standard of care, it also involves acting with honesty and integrity with one's client or employer and one's self. Professional Engineers have the responsibility to remain abreast of developments and knowledge in their area of expertise, that is, to maintain their own competence. Should there be a technologically driven or individually motivated shift in the area of technical activity, it is the engineer's duty to attain and maintain competence in all areas of involvement including being knowledgeable with the, technical and legal framework and regulations governing their work. In effect, it requires a personal commitment to ongoing professional development, continuing education and self-testing.

In addition to maintaining their own competence, Professional Engineers have an obligation to strive to contribute to the advancement of the body of knowledge within which they practice, and to the profession in general. Moreover, within the framework of the practice of their profession, they are expected to participate in providing opportunities to further the professional development of their colleagues.

This competence requirement of the Code extends to include an obligation to the public, the profession and one's peers, that opinions on engineering issues are expressed honestly and only in areas of one's competence. It applies equally to reporting or advising on professional matters and to issuing public statements. This requires honesty with one's self to present issues fairly, accurately and with appropriate qualifiers and disclaimers, and to avoid personal, political and other non-technical biases. The latter is particularly important for public statements or when involved in a technical forum.

Fairness and Integrity in the Workplace

Honesty, integrity, continuously updated competence, devotion to service and dedication to enhancing the life quality of society are cornerstones of professional responsibility. Within this framework, engineers shall be objective and truthful and include all known and pertinent information on professional reports, statements and testimony. They shall accurately and objectively represent their clients, employers, associates and themselves consistent with their academic, experience and professional qualifications. This tenet is more than 'not misrepresenting'; it also implies disclosure of all relevant information and issues, especially when serving in an advisory capacity or as an expert witness. Similarly, fairness, honesty and accuracy in advertising are expected.

If called upon to verify another engineer's work, there is an obligation to inform (or make every effort to inform) the other engineer, whether the other engineer is still actively involved or not. In this situation, and in any circumstance, engineers shall give proper recognition and credit where credit is due and accept, as well as give, honest and fair criticism on professional matters, all the while maintaining dignity and respect for everyone involved.

Engineers shall not accept nor offer covert payment or other considerations for the purpose of securing, or as remuneration for engineering assignments. Engineers should prevent their personal or political involvement from influencing or compromising their professional role or responsibility.

Consistent with the Code, and having attempted to remedy any situation within their organization, engineers are obligated to report to their association or other appropriate agency any illegal or unethical engineering decisions by engineers or others. Care must be taken not to enter into legal arrangements which compromise this obligation.

Professional Accountability and Leadership

Engineers have a duty to practice in a careful and diligent manner and accept responsibility, and be accountable for their actions. This duty is not limited to design, or its supervision and management, but applies to all areas of practice. For example, it includes construction supervision and management, preparation of shop drawings, engineering reports, feasibility studies, environmental impact assessments, engineering developmental work, etc.

The signing and sealing of engineering documents indicates the taking of responsibility for the work. This practice is required for all types of engineering endeavor, regardless where or for whom the work is done, including but not limited to, privately and publicly owned firms, crown corporations, and government agencies/departments. There are no exceptions; signing and sealing documents is appropriate whenever engineering principles have been used and public welfare may be at risk.

Taking responsibility for engineering activity includes being accountable for one's own work and, in the case of a senior engineer, accepting responsibility for the work of a team. The latter implies responsible supervision where the engineer is actually in a position to review, modify and direct the entirety of the engineering work. This concept requires setting reasonable limits on the extent of activities, and the number of engineers and others, whose work can be supervised by the responsible engineer. The practice of a "symbolic" responsibility or supervision is the situation where an engineer, say with the title of "chief engineer", takes full responsibility for all engineering on behalf of a large corporation, utility or government agency/department, even though the engineer may not be aware of many of the engineering activities or decisions being made daily throughout the firm or department. The essence of this approach is that the firm is taking the responsibility of default, whether engineering supervision or direction is applied or not.

Engineers have a duty to advise their employer and, if necessary, their clients and even their professional association, in that order, in situations when the overturning of an engineering decision may result in breaching their duty to safeguard the public. The initial action is to discuss the problem with the supervisor/employer. If the employer does not adequately respond to the engineer's concern, then the client must be advised in the case of a consultancy situation, or the most senior officer should be informed in the case of a manufacturing process plant or government agency. Failing this attempt to rectify the situation the engineer must advise in confidence his professional association of his concerns.

In the same order as mentioned above, the engineer must report unethical engineering activity undertaken by other engineers or by non-engineers. This extends to include for example, situations in which senior officials of a firm make "executive" decisions which clearly and substantially alter the engineering aspects of the work, or protection of the public welfare or the environment arising from the work.

Because of the rapid advancements in technology and the increasing ability of engineering activities to impact on the environment, engineers have an obligation to be mindful of the effect that their decisions will have on the environment and the well-being of society, and to report any concerns of this nature in the same manner as previously mentioned. Further to the above, with the rapid advancement of technology in today's world and the possible social impacts on large populations of people, engineers must endeavor to foster the public's understanding of technical issues and the role of Engineering more than ever before.

Sustainable development is the challenge of meeting current human needs for natural resources, industrial products, energy, food, transportation, shelter, and effective waste management while conserving and, if possible, enhancing the Earth's environmental quality, natural resources, ethical, intellectual, working and affectionate capabilities of people and socioeconomic bases, essential for the human needs of future generations. The proper observance to these principles will considerably help to the eradication of the world poverty.

CODE OF CONDUCT FOR NGOs

• • •

Preamble

(1) The following represents the work of several non-governmental organizations (NGOs) working from late 1991 through the NGO Conference in Paris, the outcomes of the Agenda Ya Wananchi, from meeting during the New York PrepCom and in the intervening months up to and including the Global Forum in Rio de Janeiro in June, 1992.

(2) The goal of this NGO Code of Conduct process is to eventually have a Code that NGOs can sign on to.

(3) We pledge to continue to engage in the process to analyze and deepen this activity and make recommendations that groups may adopt.

(4) There has been a dramatic growth of community groups and NGOs during the past 10 years. The work of community and citizen groups and organisations and NGOs now constitutes the best option for citizen action to change the forces against a sustainable future.

(5) In order to build up our constituency base, to truly serve the people within our community/organization, certain ethical and accountable agreements need to be acknowledged.

Principles

(6) An NGO Code of Conduct could contain the following principles:

(7) National and local NGOs (in North and South) should:

(a) be rooted in issues at home

(b) have some definable constituency or membership

(c) have open democratic working systems, gender parity, consultative problem-solving, non-discriminatory practices

(d) have clear conflict of interest guidelines

(e) have a code of ethics for staff

(f) publish an annual report and audited financial statements

(g) be non-profit, non-party political

(h) foster justice and equity, alleviate poverty and preserve cultural integrity

(i) endeavor to enhance the total environment - physical, biological and human

(j) have a fair wage structure, with a credible scale between highest and lowest paid worker

(k) be truly with people and not impose their agendas on them

(l) base all their work on the resources available to the people, their expertise, existing institutions, culture and religions; be self-sufficient while remaining open to the assistance offered by their various partners

(m) avoid being corrupted both materially and spiritually

(n) facilitate people's efforts

(o) share information with all members; set up necessary mechanisms to gather and exchange experiences; and get actively involved in environmental education (awareness-building) and training

(p) articulate a broad political framework and code of ethics to guide their internal operations and their work with community groups and people's organisations, as well as their relations with the South, NGOs and the North

(q) ensure the highest levels of accountability, starting with their own constituencies - the people. This includes uncompromising evaluations involving the participation of the local populations. Campaigns

(8) Northern and Southern NGOs often have non-project or non-funding based relationships. Generally, these relationships are the basis for campaigns to protest certain social or environmental problems in a Northern or Southern country; or the campaigns may be on international issues, like the World Bank's Global Environmental Facility (GEF).

(9) This treaty should be designed to make clear the process of consultation and decision-making

among all the participants to facilitate a process of dialogue between Northern and Southern NGOs on campaigns. At this point, we have only questions, not answers:

(a) The overriding principle this treaty seeks to ensure is consultation among NGOs before anyone takes a position that might affect another. But that is not as easy as it seems.

(b) If a group in one country sends out an international action alert about a problem in its country, what obligation does it have to first assure that there is a consensus among the NGOs in that country about that problem? Conversely, what obligation has a group that receives an action alert to first assure that the alert is the result of a consensus position in the country of origin before responding to the action alert?

(c) Who has the obligation to compile a reasonable list of NGOs in each country (without a list it is not possible for groups elsewhere to consult with NGOs in one country before taking positions on issues that might affect that country)?

(d) What constitutes reasonable consultation? How many groups is "enough"?

(e) How long should the consultation process be allowed to take? Can deadlines be set for responses if there is a hearing or legislative action coming up? What if there is no response - is that consultation?

(f) Can a contact person be chosen in each region or country to facilitate communications and consultation? How would that person be chosen? In a crisis, may that person speak for their constituency without consultation?

(g) What if groups within a region disagree? Who gets listened to? What if regions disagree? Declaration of Solidarity

(10) Before making public expression of solidarity for NGOs and individuals a proper consultation process should be undertaken to ensure the safety of the affected parties.

Regarding NGOs working outside their country

(11) Northern and Southern NGOs should collaborate on the basis of:

(a) equitable and genuine partnership

(b) two-way flow of all information, ideas and experiences

(c) financial transparency.

(12) Southern NGOs not Northern NGOs have the major responsibility for activities within their own countries.

(13) Northern NGOs when working in the South must have transparent advisory systems within the country of operation; there must be transparent criteria for selection of working partners.

(14) Northern NGOs should monitor Northern government/corporate activity in their host country.

(15) Northern NGOs in their host country should live in an appropriate comparative level as counterpart NGOs, not in expatriate style.

(16) Northern NGOs should develop effective policy on international issues.

(17) Because development groups get most of their funding from their national governments, most Northern NGOs hardly question the policies and activities of their governments in the South. On the contrary, they have become accessories to the hidden agendas pursued by their governments and transnational corporations in gaining control over the resources of the South. In order for Northern NGOs to be able to forge genuine people-to-people solidarity, they should:

(a) build a relationship that is based on mutual respect and collaboration as equal partners, and that fosters self-determination and self-reliance

(b) use their comparative advantage of easy access to information and pass it on to their partners in the South

(c) challenge their governments and educate the public in order to change the prevailing inequitable international economic order and development paradigms which have been largely responsible for the deteriorating global environment

(d) campaign for genuine grassroots democracy in their own countries

(e) campaign for sustainable life-styles based on their own local resources as much as possible, and paying fair (ecological) prices for imported products. Action Plan for Follow-Up

(18) Regional focal points to publicize and maximize NGO input

(19) Broad correspondence

(20) 1993 meeting to prepare final copy for widespread adoption.

DOW CORNING ETHICAL BUSINESS CONDUCT

• • •

Dow Corning's Responsibilities to Employees:

All relations with employees will be guided by our belief that the dignity of the individual is primary.

Opportunity without bias will be afforded each employee in relation to demonstrated ability, initiative and potential.

Management practices will be consistent with our intent to provide continuing employment for all productive employees.

Qualified citizens of countries where we do business will be hired and trained for available positions consistent with their capabilities.

We will strive to create and maintain a work environment that fosters honesty, personal growth, teamwork, open communications and dedication to our vision and values.

We will provide a safe, clean and pleasant work environment that at minimum meets all applicable laws and regulations.

The privacy of an individual's records will be respected. Employees may review their own records upon request.

Management will provide, communicate and implement a Problem Resolution Process for use by all employees to identify and resolve business ethics and employee conduct problems and other disagreements between employees.

Our Responsibilities as Dow Corning Employees:

Employees will treat Dow Corning proprietary information as a valued asset and diligently protect it from loss or negligent disclosure.

Employees will respect our commitment to protect the confidentiality of information entrusted to us by customers, suppliers and others in our business dealings.

The proprietary information of others will be obtained only through the use of legal and ethical methods.

Employees will not engage in activities that either jeopardize or conflict with the company's interests. Recognizing and avoiding conflicts of interest is the responsibility of each employee.

When a potential conflict of interest exists, the employee is obligated to bring the situation to the attention of Dow Corning management for resolution.

Employees will use or authorize company resources only for legitimate business purposes.

The cost of goods or services purchased for Dow Corning must be reasonable and in line with competitive standards.

Employees will not engage in bribery, price fixing, kickbacks, collusion or any related practice that might be, or give the appearance of being, illegal or unethical.

Employees will avoid contacts with competitors, suppliers, government agencies and other parties that are, or appear to be, engaging in unfair competition or the restriction of free trade.

Business interactions with our competitors will be limited to those necessary for buyer-seller agreements, licensing agreements or matters of general interest to industry or society. All such interactions will be documented.

Relations with Customers, Distributors, Suppliers

We are committed to providing products and services that meet the requirements of our customers. We will provide information and support necessary to effectively use our products.

Business integrity is a criterion for selecting and retaining those who represent Dow Corning.

Dow Corning will regularly encourage its distributors, agents and other representatives to conduct their business on our behalf in a legal and ethical manner.

The purchase of goods and services will be based on quality, price, service, ability to supply and the supplier's adherence to legal and ethical business practices.

Environmental, Product Stewardship and Social Responsibility

- We are committed to the responsible management of chemicals through our support and practice of the principles of Responsible Care.

- Environmental consideration will be integrated into all appropriate business decisions and will be guided by Dow Corning's Principles of Environmental Management.

- We will continually strive to assure that our products and services are safe, efficacious and accurately represented for their intended uses. We will fully represent the nature and characteristics of our raw materials, intermediates and products—including toxicity and other potential hazards—to our employees, suppliers, transporters and customers.

- We will build and maintain positive relationships with communities where we have a presence. Our efforts will focus on education, civic, cultural, environmental, and health and safety programs.

*A registered trademark of the Chemical Manufacturers Association.

International Business Guidelines

Dow Corning will be a responsible corporate citizen wherever we do business. We recognize that laws, business practices and customs differ from country to country. If legal conflicts arise in or between locations where we do business, or if conflicts with this Code present themselves, we will seek reasonable ways to resolve the differences. Failing timely resolution, we will remove ourselves from the particular business situation. Dow Corning employees will not authorize or give payments or gifts to government employees or their beneficiaries or anyone else in order to obtain or retain business. Facilitating payments to expedite the performance of routine services are strongly discouraged. In countries where local business practice dictates such payments and there is no alternative, facilitating payments are to be for the minimum amount necessary and must be accurately documented and recorded. No contributions to political parties or candidates will be given by Dow Corning, even in countries where such contributions are legal. Dow Corning considers its technology and know-how to be valuable assets and encourages their inter-company and transborder transfer to achieve its overall business objectives. Dow Corning, its subsidiaries and its majority-owned joint ventures expect to pay or receive fair compensation for the value provided or received for the technology or know-how transferred.

Financial Responsibilities

Dow Corning funds will be used only for purposes that arc legal and ethical and all transactions will be properly and accurately recorded.

We will maintain a system of internal accounting controls for Dow Corning and assure that all involved employees are fully apprised of that system.

Dow Corning encourages the free flow of funds for investment, borrowing, dividending and the return of capital throughout the world.

Dow Corning Corporation, its subsidiaries and its majority-owned joint venture companies will strive to establish and maintain inter-company prices and fees for goods and services comparable to those which would prevail in open-market transactions between unrelated parties. Within this context, the goal is to have inter-company prices and fees for goods and services that meet all applicable laws and are mutually agreed upon by the Dow Corning entities involved.

We will not participate in any financial arrangement where the perceived intent of the transaction would be a violation of this Code of Conduct.

Dow Corning Values

Integrity: Our integrity is demonstrated in our ethical conduct and in our respect for the values cherished by the society of which we are a part.

Employees: Our employees are the source from which our ideas, actions and performance flow. The full potential of our people is best realized in an environment that breeds fairness, self-fulfillment, teamwork and dedication to excellence.

Customers: Our relationship with each customer is entered in the spirit of a long-term partnership and is predicated on making the customer's interests our interests.

Quality: Our never-ending quest for quality performance is based on our understanding of our customers' needs and our willingness and capability to fulfill those needs.

Technology: Our advancement of chemistry and related sciences in our chosen fields is the Value that most differentiates Dow Corning.

Environment: Our commitment to the safekeeping of the natural environment is founded on our appreciation of it as the basis for the existence of life.

Safety: Our attention to safety is based on our full-time commitment to injury-free work, individual self-worth and a consideration for the well being of others.

Profit. Our long-term profit growth is essential to our long-term existence. How our profits are derived, and the purposes for which they are used, are influenced by our Values and our shareholders.

Used by permission. © Dow Corning.

EATON ETHICAL BUSINESS CONDUCT

• • •

Eaton Corporation's commitment to the highest degree of integrity and honesty in the conduct of its business affairs is stated in the following letter. This letters, and prior versions of it, have been distributed periodically to Eaton employees since 1976.

Eaton Corporation
Eaton Center
Cleveland, OH 44114-2584
September 1, 1996

Dear Fellow Employee:

Eaton has always had a well-deserved reputation for honesty and integrity—a reputation which we have all helped build and maintain. My purpose in writing is to reaffirm Eaton's commitment to the highest standards of ethical behavior. I particularly want to emphasize that our standards remain constant even as Eaton experiences new international growth and evolution into a truly global company.

If you're concerned about any particular situation involving ethics, please don't hesitate to contact your supervisor or another member of management.

Here are the broad concepts that we regard as fundamental principles of ethical business behavior.

Obeying the Law—We respect and obey the laws of the cities, states and countries where we operate.

Competition—We respect the rights of competitors, customers and suppliers. The only competitive advantages we seek are those gained through superior research, engineering, manufacturing and marketing. We do not engage in unfair or illegal trade practices.

Conflicts of Interest—We expect Eaton employees to avoid any association which might conflict with their loyalty to the company or compromise their judgment. Under this guideline, it would be a conflict of interest for an Eaton employee to work simultaneously for a competitor, supplier or a customer.

Government Contracts—Eaton's customers include national, state and local governments. We take care to comply with the special laws, rules and regulations which govern these contracts.

Payments to Government Personnel—We do not make illegal payments to government officials of any country. In the case of U.S. federal government employees, we must comply with the stringent rules on business gratuities that they are permitted to accept.

Kickbacks and Gratuities—We do not offer or accept kickbacks or bribes, or gifts of substantial value.

Political Contributions—Our policy prohibits company contributions to political candidates or parties even where such contributions are lawful. We encourage individual employees to be involved in the political process and make personal contributions as they see fit.

It is important that the policies and principles set forth in this letter be understood and followed on a consistent basis by each of us. Our reputation for integrity is an important corporate asset. The principles as outlined are designed to help us protect that asset. Anyone violating these principles will face appropriate disciplinary action. Your commitment to ethical behavior is essential if Eaton is to maintain the highest degree of honesty and integrity in its business activities.

Used by permission. © Eaton Corp.

LOCKHEED MARTIN CORP. CODE OF ETHICS AND BUSINESS CONDUCT

• • •

Introduction
Dear Colleague:

This booklet, Setting the Standard, has been adopted by the Lockheed Martin Board of Directors as our Company's Code of Ethics and Business Conduct. It summarizes the virtues and principles that are to guide our actions in business. We expect our agents, consultants, contractors, representatives, and suppliers to be guided by them as well.

There are numerous resources available to assist you in meeting the challenge of performing your duties and responsibilities. There can be no better course of action for you than to apply common sense and sound judgment to the manner in which you conduct yourself. However, do not hesitate to use the resources that are available whenever it is necessary to seek clarification.

Lockheed Martin aims to "set the standard" for ethical business conduct. We will achieve this through six virtues: Honesty, Integrity, Respect, Trust, Responsibility, and Citizenship.

Honesty: to be truthful in all our endeavors; to be honest and forthright with one another and with our customers, communities, suppliers, and shareholders.

Integrity: to say what we mean, to deliver what we promise, and to stand for what is right.

Respect: to treat one another with dignity and fairness, appreciating the diversity of our workforce and the uniqueness of each employee.

Trust: to build confidence through teamwork and open, candid communication.

Responsibility: to speak up - without fear of retribution - and report concerns in the work place, including violations of laws, regulations and company policies, and seek clarification and guidance whenever there is doubt.

Citizenship: : to obey all the laws of the United States and the other countries in which we do business and to do our part to make the communities in which we live better.

You can count on us to do everything in our power to meet Lockheed Martin's standards. We are counting on you to do the same. We are confident that our trust in you is well placed and we are determined to be worthy of your trust.

Daniel M. Tellep
Norman R. Augustine
Bernard L. Schwartz

June 1996

Treat in an Ethical Manner Those to Whom Lockheed Martin Has an Obligation

We are committed to the ethical treatment of those to whom we have an obligation.

For our employees we are committed to honesty, just management, and fairness, providing a safe and healthy environment, and respecting the dignity due everyone.

For our customers we are committed to produce reliable products and services, delivered on time, at a fair price.

For the communities in which we live and work we are committed to acting as concerned and responsible neighbors, reflecting all aspects of good citizenship.

For our shareholders we are committed to pursuing sound growth and earnings objectives and to exercising prudence in the use of our assets and resources.

For our suppliers we are committed to fair competition and the sense of responsibility required of a good customer.

Obey the Law

We will conduct our business in accordance with all applicable laws and regulations. The laws and regulations related to contracting with the United States government are far reaching and complex, thus placing burdens on Lockheed Martin that are in addition to those faced by companies without extensive government contracts. Compliance with the law does not comprise our entire ethical responsibility. Rather, it is a minimum, absolutely essential condition for performance of our duties.

Promote a Positive Work Environment

All employees want and deserve a work place where they feel respected, satisfied, and appreciated. Harassment or discrimination of any kind and especially involving race, color, religion, gender, age, national origin, disability, and veteran or marital status is unacceptable in our work place environment.

Providing an environment that supports the honesty, integrity, respect, trust, responsibility, and citizenship of every employee permits us the opportunity to achieve excellence in our work place. While everyone who works for the Company must contribute to the creation and maintenance of such an environment, our executives and management personnel assume special responsibility for fostering a context for work that will bring out the best in all of us.

Work Safely: Protect Yourself and Your Fellow Employees

We are committed to providing a drug-free, safe, and healthy work environment. Each of us is responsible for compliance with environmental, health, and safety laws and regulations. Observe posted warnings and regulations. Report immediately to the appropriate management any accident or injury sustained on the job, or any environmental or safety concern you may have.

Keep Accurate and Complete Records

We must maintain accurate and complete Company records. Transactions between the Company and outside individuals and organizations must be promptly and accurately entered in our books in accordance with generally accepted accounting practices and principles. No one should rationalize or even consider misrepresenting facts or falsifying records. It is illegal, will not be tolerated, and will result in disciplinary action.

Record Costs Properly

Employees and their supervisors are responsible for ensuring that labor and material costs are accurately recorded and charged on the Company's records. These costs include, but are not limited to, normal contract work, work related to independent research and development, and bid and proposal activities.

Strictly Adhere to All Antitrust Laws

Antitrust is a blanket term for strict federal and state laws that protect the free enterprise system. The laws deal with agreements and practices "in restraint of trade" such as price fixing and boycotting suppliers or customers, for example. They also bar pricing intended to run a competitor out of business; disparaging, misrepresenting, or harassing a competitor; stealing trade secrets; bribery, and kickbacks.

Antitrust laws are vigorously enforced. Violations may result in severe penalties such as forced sales of parts of businesses and significant fines for the Company. There may also be sanctions against individual employees including substantial fines and prison sentences. These laws also apply to international operations and transactions related to imports into and exports from the United States. Employees involved in any dealings with competitors are expected to know that U.S. and foreign antitrust laws may apply to their activities and to consult with the Legal Department prior to negotiating with or entering into any arrangement with a competitor.

Know and Follow the Law When Involved in International Business

The Foreign Corrupt Practices Act (FCPA), a federal statute, prohibits offering anything of value to foreign officials for the purpose of improperly influencing an official decision. It also prohibits unlawful political contributions to obtain or retain business. Finally, it prohibits the use of false records or accounts in the conduct of foreign business. Employees involved in international operations must be familiar with the FCPA. You must also be familiar with the terms and conditions of 1976 Securities and Exchange Commission and Federal Trade Commission consent decrees resulting from past issues. The FCPA and the consent decrees govern the conduct of all Lockheed Martin employees throughout the world.

If you are not familiar with documents or laws, consult with the Legal Department prior to negotiating any foreign transaction.

International transfers of equipment or technology are subject to other U.S. Government regulations like the International Traffic and Arms Regulations (ITAR), which may contain prior approval and reporting requirements. If you participate in this business activity, you should know, understand, and strictly comply with these regulations.

It may be illegal to enter into an agreement to refuse to deal with potential or actual customers or suppliers, or otherwise to engage in or support restrictive international trade practices or boycotts.

It is also important that employees doing business in foreign countries know and abide by the laws of those countries.

Follow the Rules in Using or Working with Former Government Personnel

U.S. government laws and regulations governing the employment or services from former military and civilian government personnel prohibit conflicts of interest ("working both sides of the street"). These laws and rules must be faithfully and fully observed.

Follow the Law and Use Common Sense in Political Contributions and Activities

Federal law prohibits corporations from donating corporate funds, goods, or services—directly or indirectly—to candidates for federal offices. This includes employees' work time. As a matter of policy we will not make political contributions in foreign countries.

Carefully Bid, Negotiate, and Perform Contracts

We must comply with the laws and regulations that govern the acquisition of goods and services by our customers. We will compete fairly and ethically for all business opportunities. In circumstances where there is reason to believe that the release or receipt of non- public information is unauthorized, do not attempt to obtain and do not accept such information from any source.

Appropriate steps should be taken to recognize and avoid organizational conflicts in which one business unit's activities may preclude the pursuit of a related activity by another Company business unit.

If you are involved in proposals, bid preparations, or contract negotiations, you must be certain that all statements, communications, and representations to prospective customers are accurate and truthful. Once awarded, all contracts must be performed in compliance with specifications, requirements, and clauses.

Avoid Illegal and Questionable Gifts or Favors

To Government Personnel:

Federal, state and local government departments and agencies are governed by laws and regulations concerning acceptance by their employees of entertainment, meals, gifts, gratuities, and other things of value from firms and persons with whom those departments and agencies do business or over whom they have regulatory authority. It is the general policy of Lockheed Martin to strictly comply with those laws and regulations. With regard to all federal Executive Branch employees and any other government employees who work for customers or potential customers of the Corporation, it is the policy of Lockheed Martin to prohibit its employees from giving them things of value. Permissible exceptions are offering Lockheed Martin advertising or promotional items of *nominal value* such as a coffee mug, calendar, or similar item displaying the *Company logo*, and providing modest refreshments such as soft drinks, coffee, and donuts on an occasional basis in connection with business activities. "Nominal value" is $10.00 or less. (Note: Even though this policy may be more restrictive than the U.S. Government's own policy with regard to federal Executive Branch employees, this policy shall govern the conduct of all Lockheed Martin employees.) Legislative, judicial, and state and local government personnel are subject to different restrictions; both the regulations and Corporate Policies pertaining to them must be consulted before courtesies are offered.

To Non-Government Personnel:

As long as it doesn't violate the standards of conduct of the recipient's organization, it's an acceptable practice to provide meals, refreshments, and entertainment of reasonable value in conjunction with business discussions with non-government personnel. Gifts, other than those of nominal value ($50.00 or less), to private individuals or companies are prohibited unless specifically approved by the appropriate Ethics Officer or Corporate Office of Ethics and Business Conduct.

To Foreign Government Personnel and Public Officials:

The Company may be restricted from giving meals, gifts, gratuities, entertainment, or other things of value to personnel of foreign governments and foreign public officials by the Foreign Corrupt Practices Act and by laws of foreign countries. Employees must discuss such situations with the Legal Counsel and consult the Hospitality Guidelines (maintained by the Legal Department) prior to making any gifts or providing any gratuities other than advertising items.

To Lockheed Martin Personnel:

Lockheed Martin employees may accept meals, refreshments, or entertainment of nominal value in connection with business discussions. While it is difficult to define "nominal" by means of a specific dollar amount, a common sense determination should dictate what would be considered lavish, extravagant, or frequent. It is the personal responsibility of each employee to ensure that his or her acceptance of such meals, refreshments, or entertainment is proper and could not reasonably be construed in any way as an attempt by the offering party to secure favorable treatment.

Lockheed Martin employees are not permitted to accept funds in any form or amount, or any gift that has a retail or exchange value of $20 or more from individuals, companies, or representatives of companies having or seeking business relationships with Lockheed Martin. If you have any questions about the propriety of a gift, gratuity, or item of value, contact your Ethics Officer or the Corporate Office of Ethics and Business Conduct for guidance.

If you buy goods or services for Lockheed Martin, or are involved in the procurement process, you must treat all suppliers uniformly and fairly. In deciding among competing suppliers, you must objectively and impartially weigh all facts and avoid even the appearance of favoritism. Established routines and procedures should be followed in the procurement of all goods and services.

Steer Clear of Conflicts of Interest

Playing favorites or having conflicts of interest—in practice or in appearance—runs counter to the fair treatment to which we are all entitled. Avoid any relationship, influence, or activity that might impair, or even appear to impair, your ability to make objective and fair decisions when performing your job. When in doubt, share the facts of the situation with your supervisor, Legal Department, or Ethics Officer.

Here are some ways a conflict of interest could arise:

- Employment by a competitor or potential competitor, regardless of the nature of the employment, while employed by Lockheed Martin.

- Acceptance of gifts, payment, or services from those seeking to do business with Lockheed Martin.

- Placement of business with a firm owned or controlled by an employee or his/her family.

- Ownership of, or substantial interest in, a company which is a competitor or a supplier.

- Acting as a consultant to a Lockheed Martin customer or supplier.

Maintain the Integrity of Consultants, Agents, and Representatives

Business integrity is a key standard for the selection and retention of those who represent Lockheed Martin. Agents, representatives, or consultants must certify their willingness to comply with the Company's policies and procedures and must never be retained to circumvent our values and principles. Paying bribes or kickbacks, engaging in industrial espionage, obtaining the proprietary data of a third party, or gaining inside information or influence are just a few examples of what could give us an unfair competitive advantage in a government procurement and could result in violations of law.

Protect Proprietary Information

Proprietary company information may not be disclosed to anyone without proper authorization. Keep proprietary documents protected and secure. In the course of normal business activities, suppliers, customers, and competitors may sometimes divulge to you information that is proprietary to their business. Respect these confidences.

Obtain and Use Company and Customer Assets Wisely

Proper use of company and customer property, facilities, and equipment is your responsibility. Use and maintain these assets with the utmost care and respect, guarding against waste and abuse. Be cost-conscious and alert to opportunities for improving performance while reducing costs. The use of company time, material, or facilities for purposes not directly related to company business, or the removal or borrowing of company property without permission, is prohibited.

All employees are responsible for complying with requirements of software copyright licenses related to software packages used in fulfilling job requirements.

Do Not Engage in Speculative or Insider Trading

In our role as a U.S. corporation and a major government contractor, we must always be alert to and comply with the security laws and regulations of the United States.

It is against the law for employees to buy or sell Lockheed Martin stock based on "insider" information about or involving the Company. Play it safe: don't speculate in the securities of Lockheed Martin when you are aware of information affecting the company's business that has not been publicly released or in situations where trading would call your judgment into question. This includes all varieties of stock trading such as options, puts and calls, straddles, selling short, etc. Two simple rules can help protect you in this area: (1) Don't use non-public information for personal gain. (2) Don't pass along such information to someone else who has no need to know.

This guidance also applies to the securities of other companies (suppliers, vendors, subcontractors, etc.) for which you receive information in the course of your employment at Lockheed Martin.

For More Information:

In order to support a comprehensive Ethics and Business Conduct Program, Lockheed Martin has developed education and communication programs in many subject areas.

These programs have been developed to provide employees with job-specific information to raise their level of awareness and sensitivity to key issues.

Interactive video training modules and related brochures are planned to be available on the following topics:

Antitrust ComplianceLabor Charging

Domestic Consultants Leveraging Differences

Drug-Free WorkplaceMaterial Costs

Environment, Health, and SafetyOrganizational Conflicts of Interest

Ethics Procurement

Ex-Government Employees Procurement Integrity

Export Control Product Substitution

Foreign Corrupt Practices Act

Government PropertySecurity

International ConsultantsSexual Harassment

International Military Sales Software License Compliance

Kickbacks re On Thin Ethical Ice When You Hear . . .

"Well, maybe just this once . . ."

"No one will ever know . . ."

"It doesn't matter how it gets done as long as it gets done."

You can probably think of many more phrases that raise warning flags. If you find yourself using any of these expressions, take the Quick Quiz on the following page and make sure you are on solid ethical ground.

Quick Quiz—When In Doubt, Ask Yourself . . .

Are my actions legal?

Am I being fair and honest?

Will my action stand the test of time?

How will I feel about myself afterwards?

How will it look in the newspaper?

Will I sleep soundly tonight?

What would I tell my child to do?

If you are still not sure what to do, ask . . . and keep asking until you are certain you are doing the right thing.

Our Goal: An Ethical Work Environment

We have established the Office of Vice President - Ethics and Business Conduct to underscore our commitment to ethical conduct throughout our Company.

This office reports directly to the Office of the Chairman and the Audit and Ethics Committee of the Board of Directors, and oversees a vigorous corporate-wide effort to promote a positive, ethical work environment for all employees.

Our Ethics Officers operate confidential ethics helplines at each operating company, as well as at the corporate level. You are urged to use these resources whenever you have a question or concern that cannot be readily addressed within your work group or through your supervisor.

In addition, if you need information on how to contact your local Ethics Officer - or wish to discuss a matter of concern with the corporate Office of Ethics and Business Conduct - you are encouraged to use one of the following confidential means of communication:

Call: 1-800-LM ETHIC (1-800-563-8442)

For the Hearing or Speech Impaired: (1-800-441- 7457)

Write: Office of Ethics and Business Conduct

Office of Ethics and Business Conduct

Lockheed Martin Corporation

P.O. Box 34143 Bethesda, MD 20827-0143

Fax: 818-876-2082

Internet E-Mail:Corporate.Ethics@den.mmc.com

When you contact your Company Ethics Officer or the Corporate Office of Ethics and Business Conduct:

- You will be treated with dignity and respect.
- Your communication will be protected to the greatest extent possible.
- Your concerns will be seriously addressed and, if not resolved at the time you call, you will be informed of the outcome.
- You need not identify yourself.
- Remember, there's never a penalty for using the HelpLine. People in a position of authority can't stop you; if they try, they're subject to disciplinary action up to and including dismissal.

Used by permission. © Lockheed Martin.

RESPONSIBLE CARE GUIDING PRINCIPLES (CHEMICAL INDUSTRY)

• • •

Our industry creates products and services that make life better for people around the world - both today and in the future. The benefits of our industry are accompanied by enduring commitments to Responsible Care in the management of chemicals worldwide. We will make continuous progress toward the vision of no accidents, injuries, or harm to the environment and will publicly report our global health, safety, and environmental performance. We will lead our companies in ethical ways that increasingly benefit society, the economy and the environment while adhering to the following principles:

1. To seek and incorporate public input regarding our products and operations.
2. To provide chemicals that can be manufactured, transported, used and disposed of safely.
3. To make health, safety, the environment and resource conservation critical considerations for all new and existing products and processes.
4. To provide information on health or environmental risks and pursue protective measures for employees, the public and other key stakeholders.

5. To work with customers, carriers, suppliers, distributors and contractors to foster the safe use, transport and disposal of chemicals.

6. To operate our facilities in a manner that protects the environment and the health and safety of our employees and the public.

7. To support education and research on the health, safety and environmental effects of our products and processes.

8. To work with others to resolve problems associated with past handling and disposal practices.

9. To lead in the development of responsible laws, regulations, and standards that safeguard the community, workplace and environment.

10. To practice Responsible Care by encouraging and assisting others to adhere to these principles and practices.

EINSTEIN-RUSSELL MANIFESTO
(1955)

•••

In the tragic situation which confronts humanity, we feel that scientists should assemble in conference to appraise the perils that have arisen as a result of the development of weapons of mass destruction, and to discuss a resolution in the spirit of the appended draft.

We are speaking on this occasion, not as members of this or that nation, continent, or creed, but as human beings, members of the species Man, whose continued existence is in doubt. The world is full of conflicts; and, overshadowing all minor conflicts, the titanic struggle between Communism and anti-Communism.

Almost everybody who is politically conscious has strong feelings about one or more of these issues; but we want you, if you can, to set aside such feelings and consider yourselves only as members of a biological species which has had a remarkable history, and whose disappearance none of us can desire.

We shall try to say no single word which should appeal to one group rather than to another. All, equally, are in peril, and, if the peril is understood, there is hope that they may collectively avert it.

We have to learn to think in a new way. We have to learn to ask ourselves, not what steps can be taken to give military victory to whatever group we prefer, for there no longer are such steps; the question we have to ask ourselves is: what steps can be taken to prevent a military contest of which the issue must be disastrous to all parties?

The general public, and even many men in positions of authority, have not realized what would be involved in a war with nuclear bombs. The general public still thinks in terms of the obliteration of cities. It is understood that the new bombs are more powerful than the old, and that, while one A-bomb could obliterate Hiroshima, one H-bomb could obliterate the largest cities, such as London, New York, and Moscow.

No doubt in an H-bomb war great cities would be obliterated. But this is one of the minor disasters that would have to be faced. If everybody in London, New York, and Moscow were exterminated, the world might,

in the course of a few centuries, recover from the blow. But we now know, especially since the Bikini test, that nuclear bombs can gradually spread destruction over a very much wider area than had been supposed.

It is stated on very good authority that a bomb can now be manufactured which will be 2,500 times as powerful as that which destroyed Hiroshima. Such a bomb, if exploded near the ground or under water, sends radio-active particles into the upper air. They sink gradually and reach the surface of the earth in the form of a deadly dust or rain. It was this dust which infected the Japanese fishermen and their catch of fish.

No one knows how widely such lethal radioactive particles might be diffused, but the best authorities are unanimous in saying that a war with H-bombs might possibly put an end to the human race. It is feared that if many H-bombs are used there will be universal death, sudden only for a minority, but for the majority a slow torture of disease and disintegration.

Many warnings have been uttered by eminent men of science and by authorities in military strategy. None of them will say that the worst results are certain. What they do say is that these results are possible, and no one can be sure that they will not be realized. We have not yet found that the views of experts on this question depend in any degree upon their politics or prejudices. They depend only, so far as our researches have revealed, upon the extent of the particular expert's knowledge. We have found that the men who know most are the most gloomy.

Here, then, is the problem which we present to you, stark and dreadful and inescapable: Shall we put an end to the human race; or shall mankind renounce war? People will not face this alternative because it is so difficult to abolish war.

The abolition of war will demand distasteful limitations of national sovereignty. But what perhaps impedes understanding of the situation more than anything else is that the term "mankind" feels vague and abstract. People scarcely realize in imagination that the danger is to themselves and their children and their grandchildren, and not only to a dimly apprehended humanity. They can scarcely bring themselves to grasp that they, individually, and those whom they love are in imminent danger of perishing agonizingly. And so they hope that perhaps war may be allowed to continue provided modern weapons are prohibited.

This hope is illusory. Whatever agreements not to use H-bombs had been reached in time of peace, they would no longer be considered binding in time of war, and both sides would set to work to manufacture H-bombs as soon as war broke out, for, if one side manufactured the bombs and the other did not, the side that manufactured them would inevitably be victorious.

Although an agreement to renounce nuclear weapons as part of a general reduction of armaments would not afford an ultimate solution, it would serve certain important purposes. First: any agreement between East and West is to the good in so far as it tends to diminish tension. Second: the abolition of thermo-nuclear weapons, if each side believed that the other had carried it out sincerely, would lessen the fear of a sudden attack in the style of Pearl Harbor, which at present keeps both sides in a state of nervous apprehension. We should, therefore, welcome such an agreement though only as a first step. Most of us are not neutral in feeling, but, as human beings, we have to remember that, if the issues between East and West are to be decided in any manner that can give any possible satisfaction to anybody, whether Communist or anti-Communist, whether Asian or European or American, whether White or Black, then these issues must not be decided by war. We should wish this to be understood, both in the East and in the West. There lies before us, if we choose, continual progress in happiness, knowledge, and wisdom. Shall we, instead, choose death, because we cannot forget our quarrels? We appeal, as human beings, to human beings: Remember your humanity, and forget the rest. If you can do so, the way lies open to a new Paradise; if you cannot, there lies before you the risk of universal death.

Resolution

We invite this Congress, and through it the scientists of the world and the general public, to subscribe to the following resolution:

> "In view of the fact that in any future world war nuclear weapons will certainly be employed, and that such weapons threaten the continued existence of mankind, we urge the Governments of the world to realize, and to acknowledge publicly, that their purpose cannot be furthered by a world war, and we urge them, consequently, to find peaceful means for the settlement of all matters of dispute between them."

Max Born

Perry W. Bridgman

Albert Einstein

Leopold Infeld

Frederic Joliot-Curie

Herman J. Muller

Linus Pauling

Cecil F. Powell

Joseph Rotblat

Bertrand Russell

Hideki Yukawa

MOUNT CARMEL DECLARATION ON TECHNOLOGY AND MORAL RESPONSIBILITY (1974)

• • •

We, meeting at Haifa to celebrate the fiftieth anniversary of the Technion-Israel Institute of Technology, deeply troubled by the threats to the welfare and survival of the human species that are increasingly posed by improvident uses of applied science and technology, offer the following Declaration for consideration and adoption. It is addressed, most urgently, to all whom it concerns, to governments and other political agencies, to administrators and managers, experts and laymen, educators and students, to all who have the power to influence decisions or the right to be consulted about them.

1. We recognize the great contributions of technology to the improvement of the human condition. Yet continued intensification and extension of technology has unprecedented potentialities for evil as well as good. Technological consequences are now so ramified and interconnected, so sweeping in unforeseen results, so grave in magnitude of the irreversible changes they induce, as to constitute a threat to the very survival of the species.

2. While actions at the level of community and state are urgently needed, legitimate local interests must not take precedence over the common interest of *all* human beings in justice, happiness, and peace. Responsible control of technology by social systems and institutions is an urgent *global* concern, overriding all conflicts of interest and all divergencies in religion, race or political allegiance. Ultimately all must benefit from the promise of technology, or all must suffer—even perish—together.

3. Technological applications and innovations result from human actions. As such, they demand politi-

cal, social, economic, ecological and above all *moral* evaluation. No technology is morally "neutral."

4. Human beings, both as individuals and as members or agents of social institutions, bear the sole responsibility for abuses of technology. Invocation of supposedly inflexible laws of technological inertia and technological transformation is an evasion of moral and political responsibility.

5. Creeds and moral philosophies that teach respect for human dignity can, in spite of all differences, unite in actions to cope with the problems posed by new technologies. It is an urgent task to work toward new codes for guidance in an age of pervasive technology.

6. Every technological undertaking must respect basic human rights and cherish human dignity. We must not gamble with human survival. We must not degrade people into *things* used by machines: every technological innovation must be judged by its contributions to the development of genuinely free and creative *persons*.

7. The "developed" and the "developing" nations have different priorities but an ultimate convergence of shared interests:

For the developed nations: rejection of expansion at all costs and the selfish satisfaction of ever-multiplying desires–and adoption of policies of *principled restraint*—with unstinting assistance to the unfortunate and the underprivileged.

For the developing nations: complementary but appropriately modified policies of principled restraint, especially in population growth, and a determination to avoid repeating the excesses and follies of the more "developed" economies.

Absolute priority should be given to the relief of human misery, the eradication of hunger and disease, the abolition of social injustice and the achievement of lasting peace.

8. These problems and their implications need to be discussed and investigated by all educational institutions and all media of communication. They call for intense and imaginative research enlisting the cooperation of humanists and social scientists, as well as natural scientists and technologists. Better technology is needed, but will not suffice to solve the problems caused by intensive uses of technology. We need guardian *disciplines* to monitor and assess technological innovations, with especial attention to their moral implications.

9. Implementation of these purposes will demand improved social institutions through the active participation of statesmen and their expert advisers, and the informed understanding and consent of those most directly affected—especially the young, who have the greatest stake in the future.

10. This agenda calls for sustained work on three distinct but connected tasks: the development of "guardian disciplines" for watching, modifying, improving, and restraining the human consequences of technology (a special but not exclusive responsibility of the scientists and technologists who originate technological innovations); the confluence of varying moral codes in common action; and the creation of improved educational and social institutions.

Without minimizing the prevalence of human irrationality and the potency of envy and hate, we have sufficient faith in ourselves and our fellows to hope for a future in which all can have a chance to close the gap between aspiration and reality—a chance to become at last truly human.

No agenda is more urgent for human welfare and survival. This declaration, henceforth to be called the Mount Carmel Declaration on Technology and Moral Responsibility, is proclaimed in Jerusalem on this day, Wednesday, the twenty-fifth of December, 1974, in the Residence of the President of the State of Israel.

RIO DECLARATION ON ENVIRONMENT AND DEVELOPMENT (1992)

• • •

The United Nations Conference on Environment and Development,

Having met at Rio de Janeiro from 3 to 14 June 1992,

Reaffirming the Declaration of the United Nations Conference on the Human Environment, adopted at Stockholm on 16 June 1972, and seeking to build upon it,

With the goal of establishing a new and equitable global partnership through the creation of new levels of cooperation among States, key sectors of societies and people,

Working towards international agreements which respect the interests of all and protect the integrity of the global environmental and developmental system,

Recognizing the integral and interdependent nature of the Earth, our home,

Proclaims that:

Principle 1

Human beings are at the centre of concerns for sustainable development. They are entitled to a healthy and productive life in harmony with nature.

Principle 2

States have, in accordance with the Charter of the United Nations and the principles of international law, the sovereign right to exploit their own resources pursuant to their own environmental and developmental policies, and the responsibility to ensure that activities within their jurisdiction or control do not cause damage to the environment of other States or of areas beyond the limits of national jurisdiction.

Principle 3

The right to development must be fulfilled so as to equitably meet developmental and environmental needs of present and future generations.

Principle 4

In order to achieve sustainable development, environmental protection shall constitute an integral part of the development process and cannot be considered in isolation from it.

Principle 5

All States and all people shall cooperate in the essential task of eradicating poverty as an indispensable requirement for sustainable development, in order to decrease the disparities in standards of living and better meet the needs of the majority of the people of the world.

Principle 6

The special situation and needs of developing countries, particularly the least developed and those most environmentally vulnerable, shall be given special priority. International actions in the field of environment and development should also address the interests and needs of all countries.

Principle 7

States shall cooperate in a spirit of global partnership to conserve, protect and restore the health and integrity of the Earth's ecosystem. In view of the different contributions to global environmental degradation, States have common but differentiated responsibilities.

The developed countries acknowledge the responsibility that they bear in the international pursuit of sustainable development in view of the pressures their societies place on the global environment and of the technologies and financial resources they command.

Principle 8

To achieve sustainable development and a higher quality of life for all people, States should reduce and eliminate unsustainable patterns of production and consumption and promote appropriate demographic policies.

Principle 9

States should cooperate to strengthen endogenous capacity-building for sustainable development by improving scientific understanding through exchanges of scientific and technological knowledge, and by enhancing the development, adaptation, diffusion and transfer of technologies, including new and innovative technologies.

Principle 10

Environmental issues are best handled with the participation of all concerned citizens, at the relevant level. At the national level, each individual shall have appropriate access to information concerning the environment that is held by public authorities, including information on hazardous materials and activities in their communities, and the opportunity to participate in decision-making processes. States shall facilitate and encourage public awareness and participation by making information widely available. Effective access to judicial and administrative proceedings, including redress and remedy, shall be provided.

Principle 11

States shall enact effective environmental legislation. Environmental standards, management objectives and priorities should reflect the environmental and developmental context to which they apply. Standards applied by some countries may be inappropriate and of unwarranted economic and social cost to other countries, in particular developing countries.

Principle 12

States should cooperate to promote a supportive and open international economic system that would lead to economic growth and sustainable development in all countries, to better address the problems of environmental degradation. Trade policy measures for environmental purposes should not constitute a means of arbitrary or unjustifiable discrimination or a disguised restriction on international trade. Unilateral actions to deal with environmental challenges outside the jurisdiction of the importing country should be avoided. Environmental measures addressing transboundary or global environmental problems should, as far as possible, be based on an international consensus.

Principle 13

States shall develop national law regarding liability and compensation for the victims of pollution and other environmental damage. States shall also cooperate in an expeditious and more determined manner to develop further international law regarding liability and compensation for adverse effects of environmental damage caused by activities within their jurisdiction or control to areas beyond their jurisdiction.

Principle 14

States should effectively cooperate to discourage or prevent the relocation and transfer to other States of any activities and substances that cause severe environmental degradation or are found to be harmful to human health.

Principle 15

In order to protect the environment, the precautionary approach shall be widely applied by States according to their capabilities. Where there are threats of serious or irreversible damage, lack of full scientific certainty shall not be used as a reason for postponing cost-effective measures to prevent environmental degradation.

Principle 16

National authorities should endeavour to promote the internalization of environmental costs and the use of economic instruments, taking into account the approach that the polluter should, in principle, bear the cost of pollution, with due regard to the public interest and without distorting international trade and investment.

Principle 17

Environmental impact assessment, as a national instrument, shall be undertaken for proposed activities that are likely to have a significant adverse impact on the environment and are subject to a decision of a competent national authority.

Principle 18

States shall immediately notify other States of any natural disasters or other emergencies that are likely to produce sudden harmful effects on the environment of those States. Every effort shall be made by the international community to help States so afflicted.

Principle 19

States shall provide prior and timely notification and relevant information to potentially affected States on activities that may have a significant adverse transboundary environmental effect and shall consult with those States at an early stage and in good faith.

Principle 20

Women have a vital role in environmental management and development. Their full participation is therefore essential to achieve sustainable development.

Principle 21

The creativity, ideals and courage of the youth of the world should be mobilized to forge a global partnership in order to achieve sustainable development and ensure a better future for all.

Principle 22

Indigenous people and their communities and other local communities have a vital role in environmental management and development because of their knowledge and traditional practices. States should recognize and duly support their identity, culture and interests and enable their effective participation in the achievement of sustainable development.

Principle 23

The environment and natural resources of people under oppression, domination and occupation shall be protected.

Principle 24

Warfare is inherently destructive of sustainable development. States shall therefore respect international law providing protection for the environment in times of armed conflict and cooperate in its further development, as necessary.

Principle 25

Peace, development and environmental protection are interdependent and indivisible.

Principle 26

States shall resolve all their environmental disputes peacefully and by appropriate means in accordance with the Charter of the United Nations.

Principle 27

States and people shall cooperate in good faith and in a spirit of partnership in the fulfillment of the principles embodied in this Declaration and in the further development of international law in the field of sustainable development.

TECHNOREALISM MANIFESTO (1998)

• • •

1. Technologies are not neutral.

A great misconception of our time is the idea that technologies are completely free of bias—that because they are inanimate artifacts, they don't promote certain kinds of behaviors over others. In truth, technologies come loaded with both intended and unintended social, political, and economic leanings. Every tool provides its users with a particular manner of seeing the world and specific ways of interacting with others. It is important for each of us to consider the biases of various technologies and to seek out those that reflect our values and aspirations.

2. The Internet is revolutionary, but not Utopian.

The Net is an extraordinary communications tool that provides a range of new opportunities for people, communities, businesses, and government. Yet as cyberspace becomes more populated, it increasingly resembles society at large, in all its complexity. For every empowering or enlightening aspect of the wired life, there will also be dimensions that are malicious, perverse, or rather ordinary.

3. Government has an important role to play on the electronic frontier.

Contrary to some claims, cyberspace is not formally a place or jurisdiction separate from Earth. While governments should respect the rules and customs that have arisen in cyberspace, and should not stifle this new world with inefficient regulation or censorship, it is foolish to say that the public has no sovereignty over what an errant citizen or fraudulent corporation does online. As the representative of the people and the guardian of democratic values, the state has the right and responsibility to help integrate cyberspace and conventional society.

Technology standards and privacy issues, for example, are too important to be entrusted to the marketplace alone. Competing software firms have little interest in preserving the open standards that are essential to a fully functioning interactive network. Markets encourage innovation, but they do not necessarily insure the public interest.

4. Information is not knowledge.

All around us, information is moving faster and becoming cheaper to acquire, and the benefits are manifest. That said, the proliferation of data is also a serious challenge, requiring new measures of human discipline and skepticism. We must not confuse the thrill of acquiring or distributing information quickly with the more daunting task of converting it into knowledge and wisdom. Regardless of how advanced our computers become, we should never use them as a substitute for our own basic cognitive skills of awareness, perception, reasoning, and judgment.

5. Wiring the schools will not save them.

The problems with America's public schools—disparate funding, social promotion, bloated class size, crumbling infrastructure, lack of standards—have almost nothing to do with technology. Consequently, no amount of technology will lead to the educational revolution prophesied by President Clinton and others. The art of teaching cannot be replicated by computers, the Net, or by "distance learning." These tools can, of course, augment an already high-quality educational experience. But to rely on them as any sort of panacea would be a costly mistake.

6. Information wants to be protected.

It's true that cyberspace and other recent developments are challenging our copyright laws and frameworks for protecting intellectual property. The answer, though, is not to scrap existing statutes and principles. Instead, we must update old laws and interpretations so that information receives roughly the same protection it did in the context of old media. The goal is the same: to give authors sufficient control over their work so that they have an incentive to create, while maintaining the right of the public to make fair use of that information. In neither context does information want "to be free." Rather, it needs to be protected.

7. The public owns the airwaves; the public should benefit from their use.

The recent digital spectrum giveaway to broadcasters underscores the corrupt and inefficient misuse of public resources in the arena of technology. The citizenry should benefit and profit from the use of public frequencies, and should retain a portion of the spectrum for educational, cultural, and public access uses. We should demand more for private use of public property.

8. Understanding technology should be an essential component of global citizenship.

In a world driven by the flow of information, the interfaces—and the underlying code—that make information visible are becoming enormously powerful social forces. Understanding their strengths and limitations, and even participating in the creation of better tools, should be an important part of being an involved citizen. These tools affect our lives as much as laws do, and we should subject them to a similar democratic scrutiny.

DECLARATION ON SCIENCE AND THE USE OF SCIENTIFIC KNOWLEDGE (1999)

• • •

Preamble

1. We all live on the same planet and are part of the biosphere. We have come to recognize that we are in a situation of increasing interdependence, and that our future is intrinsically linked to the preservation of the global life-support systems and to the survival of all forms of life. The nations and the scientists of the world are called upon to acknowledge the urgency of using knowledge from all fields of science in a responsible manner to address human needs and aspirations without misusing this knowledge. We seek active collaboration across all the fields of scientific endeavor, that is the natural sciences such as the physical, earth and biological sciences, the biomedical and engineering sciences, and the social and human sciences. While the *Framework for Action* emphasizes the promise and the

dynamism of the natural sciences but also their potential adverse effects, and the need to understand their impact on and relations with society, the commitment to science, as well as the challenges and the responsibilities set out in this Declaration, pertain to all fields of the sciences. All cultures can contribute scientific knowledge of universal value. The sciences should be at the service of humanity as a whole, and should contribute to providing everyone with a deeper understanding of nature and society, a better quality of life and a sustainable and healthy environment for present and future generations.

2. Scientific knowledge has led to remarkable innovations that have been of great benefit to humankind. Life expectancy has increased strikingly, and cures have been discovered for many diseases. Agricultural output has risen significantly in many parts of the world to meet growing population needs. Technological developments and the use of new energy sources have created the opportunity to free humankind from arduous labour. They have also enabled the generation of an expanding and complex range of industrial products and processes. Technologies based on new methods of communication, information handling and computation have brought unprecedented opportunities and challenges for the scientific endeavor as well as for society at large. Steadily improving scientific knowledge on the origin, functions and evolution of the universe and of life provides humankind with conceptual and practical approaches that profoundly influence its conduct and prospects.

3. In addition to their demonstrable benefits the applications of scientific advances and the development and expansion of human activity have also led to environmental degradation and technological disasters, and have contributed to social imbalance or exclusion. As one example, scientific progress has made it possible to manufacture sophisticated weapons, including conventional weapons and weapons of mass destruction. There is now an opportunity to call for a reduction in the resources allocated to the development and manufacture of new weapons and to encourage the conversion, at least partially, of military production and research facilities to civilian use. The United Nations General Assembly has proclaimed the year 2000 as International Year for the Culture of Peace and the year 2001 as United Nations Year of Dialogue among Civilizations as steps towards a lasting peace; the scientific community, together with

other sectors of society, can and should play an essential role in this process.

4. Today, whilst unprecedented advances in the sciences are foreseen, there is a need for a vigorous and informed democratic debate on the production and use of scientific knowledge. The scientific community and decision-makers should seek the strengthening of public trust and support for science through such a debate. Greater interdisciplinary efforts, involving both natural and social sciences, are a prerequisite for dealing with ethical, social, cultural, environmental, gender, economic and health issues. Enhancing the role of science for a more equitable, prosperous and sustainable world requires the long-term commitment of all stakeholders, public and private, through greater investment, the appropriate review of investment priorities, and the sharing of scientific knowledge.

5. Most of the benefits of science are unevenly distributed, as a result of structural asymmetries among countries, regions and social groups, and between the sexes. As scientific knowledge has become a crucial factor in the production of wealth, so its distribution has become more inequitable. What distinguishes the poor (be it people or countries) from the rich is not only that they have fewer assets, but also that they are largely excluded from the creation and the benefits of scientific knowledge.

6. We, participants in the *World Conference on Science for the Twenty-first Century: A New Commitment*, assembled in Budapest, Hungary, from 26 June to 1 July 1999 under the aegis of the United Nations Educational, Scientific and Cultural Organization (UNESCO) and the International Council for Science (ICSU):

Considering:

7. where the natural sciences stand today and where they are heading, what their social impact has been and what society expects from them,

8. that in the twenty-first century science must become a shared asset benefiting all peoples on a basis of solidarity, that science is a powerful resource for understanding natural and social phenomena, and that its role promises to be even greater in the future as the growing complexity of the relationship between society and the environment is better understood,

9. the ever-increasing need for scientific knowledge in public and private decision-making, including notably the influential role to be played by science in the formulation of policy and regulatory decisions,

10. that access to scientific knowledge for peaceful purposes from a very early age is part of the right to education belonging to all men and women, and that science education is essential for human development, for creating endogenous scientific capacity and for having active and informed citizens,

11. that scientific research and its applications may yield significant returns towards economic growth and sustainable human development, including poverty alleviation, and that the future of humankind will become more dependent on the equitable production, distribution and use of knowledge than ever before,

12. that scientific research is a major driving force in the field of health and social care and that greater use of scientific knowledge would considerably improve human health,

13. the current process of globalization and the strategic role of scientific and technological knowledge within it,

14. the urgent need to reduce the gap between the developing and developed countries by improving scientific capacity and infrastructure in developing countries,

15. that the information and communication revolution offers new and more effective means of exchanging scientific knowledge and advancing education and research,

16. the importance for scientific research and education of full and open access to information and data belonging to the public domain,

17. the role played by the social sciences in the analysis of social transformations related to scientific and technological developments and the search for solutions to the problems generated in the process,

18. the recommendations of major conferences convened by the organizations of the United Nations system and others, and of the meetings associated with the World Conference on Science,

19. that scientific research and the use of scientific knowledge should respect human rights and the dignity of human beings, in accordance with the Universal Declaration of Human Rights and in the light of the Universal Declaration on the Human Genome and Human Rights,

20. that some applications of science can be detrimental to individuals and society, the environment and human health, possibly even threatening the continuing existence of the human species, and that the contribution of science is indispensable to the cause of peace and development, and to global safety and security,

21. that scientists with other major actors have a special responsibility for seeking to avert applications of science which are ethically wrong or have an adverse impact,

22. the need to practice and apply the sciences in line with appropriate ethical requirements developed on the basis of an enhanced public debate,

23. that the pursuit of science and the use of scientific knowledge should respect and maintain life in all its diversity, as well as the life-support systems of our planet,

24. that there is a historical imbalance in the participation of men and women in all science-related activities,

25. that there are barriers which have precluded the full participation of other groups, of both sexes, including disabled people, indigenous peoples and ethnic minorities, hereafter referred to as disadvantaged groups,

26. that traditional and local knowledge systems, as dynamic expressions of perceiving and understanding the world, can make, and historically have made, a valuable contribution to science and technology, and that there is a need to preserve, protect, research and promote this cultural heritage and empirical knowledge,

27. that a new relationship between science and society is necessary to cope with such pressing global problems as poverty, environmental degradation, inadequate public health, and food and water security, in particular those associated with population growth,

28. the need for a strong commitment to science on the part of governments, civil society and the productive sector, as well as an equally strong commitment of scientists to the well-being of society,

   Proclaim the following:

1. Science for knowledge; knowledge for progress

29. The inherent function of the scientific endeavor is to carry out a comprehensive and thorough inquiry into nature and society, leading to new knowledge. This new knowledge provides educational, cultural and intellectual enrichment and leads to technological advances and economic benefits. Promoting fundamental and problem-oriented research is essential for achieving endogenous development and progress.

30. Governments, through national science policies and in acting as catalysts to facilitate interaction and communication between stakeholders, should give recognition to the key role of scientific research in the acquisition of knowledge, in the training of scientists and in the education of the public. Scientific research funded by the private sector has become a crucial factor for socio-economic development, but this cannot exclude the need for publicly-funded research. Both sectors should work in close collaboration and in a complementary manner in the financing of scientific research for long-term goals.

2. Science for peace

31. The essence of scientific thinking is the ability to examine problems from different perspectives and seek explanations of natural and social phenomena, constantly submitted to critical analysis. Science thus relies on critical and free thinking, which is essential in a democratic world. The scientific community, sharing a long-standing tradition that transcends nations, religions and ethnicity, should promote, as stated in the Constitution of UNESCO, the "intellectual and moral solidarity of mankind", which is the basis of a culture of peace. Worldwide cooperation among scientists makes a valuable and constructive contribution to global security and to the development of peaceful interactions between different nations, societies and cultures, and could give encouragement to further steps in disarmament, including nuclear disarmament.

32. Governments and society at large should be aware of the need to use natural and social sciences and technology as tools to address the root causes and impacts of conflict. Investment in scientific research which addresses them should be increased.

3. Science for development

33. Today, more than ever, science and its applications are indispensable for development. All levels of government and the private sector should provide

enhanced support for building up an adequate and evenly distributed scientific and technological capacity through appropriate education and research programs as an indispensable foundation for economic, social, cultural and environmentally sound development. This is particularly urgent for developing countries. Technological development requires a solid scientific basis and needs to be resolutely directed towards safe and clean production processes, greater efficiency in resource use and more environmentally friendly products. Science and technology should also be resolutely directed towards prospects for better employment, improving competitiveness and social justice. Investment in science and technology aimed both at these objectives and at a better understanding and safeguarding of the planet's natural resource base, biodiversity and life-support systems must be increased. The objective should be a move towards sustainable development strategies through the integration of economic, social, cultural and environmental dimensions.

34. Science education, in the broad sense, without discrimination and encompassing all levels and modalities, is a fundamental prerequisite for democracy and for ensuring sustainable development. In recent years, worldwide measures have been undertaken to promote basic education for all. It is essential that the fundamental role played by women in the application of scientific development to food production and health care be fully recognized, and efforts made to strengthen their understanding of scientific advances in these areas. It is on this platform that science education, communication and popularization need to be built. Special attention still needs to be given to marginalized groups. It is more than ever necessary to develop and expand science literacy in all cultures and all sectors of society as well as reasoning ability and skills and an appreciation of ethical values, so as to improve public participation in decision-making related to the application of new knowledge. Progress in science makes the role of universities particularly important in the promotion and modernization of science teaching and its coordination at all levels of education. In all countries, and in particular the developing countries, there is a need to strengthen scientific research in higher education, including postgraduate programs, taking into account national priorities.

35. The building of scientific capacity should be supported by regional and international cooperation, to ensure both equitable development and the spread and utilization of human creativity without discrimination of any kind against countries, groups or individuals. Cooperation between developed and developing countries should be carried out in conformity with the principles of full and open access to information, equity and mutual benefit. In all efforts of cooperation, diversity of traditions and cultures should be given due consideration. The developed world has a responsibility to enhance partnership activities in science with developing countries and countries in transition. Helping to create a critical mass of national research in the sciences through regional and international cooperation is especially important for small States and least developed countries. Scientific structures, such as universities, are essential for personnel to be trained in their own country with a view to a subsequent career in that country. Through these and other efforts conditions conducive to reducing or reversing the brain drain should be created. However, no measures adopted should restrict the free circulation of scientists.

36. Progress in science requires various types of cooperation at and between the intergovernmental, governmental and non-governmental levels, such as: multilateral projects; research networks, including South-South networking; partnerships involving scientific communities of developed and developing countries to meet the needs of all countries and facilitate their progress; fellowships and grants and promotion of joint research; programs to facilitate the exchange of knowledge; the development of internationally recognized scientific research centers, particularly in developing countries; international agreements for the joint promotion, evaluation and funding of mega-projects and broad access to them; international panels for the scientific assessment of complex issues; and international arrangements for the promotion of postgraduate training. New initiatives are required for interdisciplinary collaboration. The international character of fundamental research should be strengthened by significantly increasing support for long-term research projects and for international collaborative projects, especially those of global interest. In this respect particular attention should be given to the need for continuity of support for research. Access to these facilities for scientists from developing countries should be actively supported and open to all on the basis of scientific merit. The use of information

and communication technology, particularly through networking, should be expanded as a means of promoting the free flow of knowledge. At the same time, care must be taken to ensure that the use of these technologies does not lead to a denial or restriction of the richness of the various cultures and means of expression.

37. For all countries to respond to the objectives set out in this Declaration, in parallel with international approaches, in the first place national strategies and institutional arrangements and financing systems need to be set up or revised to enhance the role of sciences in sustainable development within the new context. In particular they should include: a long-term national policy on science to be developed together with the major public and private actors; support to science education and scientific research; the development of cooperation between R&D institutions, universities and industry as part of national innovation systems; the creation and maintenance of national institutions for risk assessment and management, vulnerability reduction, safety and health; and incentives for investment, research and innovation. Parliaments and governments should be invited to provide a legal, institutional and economic basis for enhancing scientific and technological capacity in the public and private sectors and facilitate their interaction. Science decision-making and priority-setting should be made an integral part of overall development planning and the formulation of sustainable development strategies. In this context, the recent initiative by the major G-8 creditor countries to embark on the process of reducing the debt of certain developing countries will be conducive to a joint effort by the developing and developed countries towards establishing appropriate mechanisms for the funding of science in order to strengthen national and regional scientific and technological research systems.

38. Intellectual property rights need to be appropriately protected on a global basis, and access to data and information is essential for undertaking scientific work and for translating the results of scientific research into tangible benefits for society. Measures should be taken to enhance those relationships between the protection of intellectual property rights and the dissemination of scientific knowledge that are mutually supportive. There is a need to consider the scope, extent and application of intellectual property rights in relation to the equi-

table production, distribution and use of knowledge. There is also a need to further develop appropriate national legal frameworks to accommodate the specific requirements of developing countries and traditional knowledge and its sources and products, to ensure their recognition and adequate protection on the basis of the informed consent of the customary or traditional owners of this knowledge.

4. Science in society and science for society

39. The practice of scientific research and the use of knowledge from that research should always aim at the welfare of humankind, including the reduction of poverty, be respectful of the dignity and rights of human beings, and of the global environment, and take fully into account our responsibility towards present and future generations. There should be a new commitment to these important principles by all parties concerned.

40. A free flow of information on all possible uses and consequences of new discoveries and newly developed technologies should be secured, so that ethical issues can be debated in an appropriate way. Each country should establish suitable measures to address the ethics of the practice of science and of the use of scientific knowledge and its applications. These should include due process procedures for dealing with dissent and dissenters in a fair and responsive manner. The World Commission on the Ethics of Scientific Knowledge and Technology of UNESCO could provide a means of interaction in this respect.

41. All scientists should commit themselves to high ethical standards, and a code of ethics based on relevant norms enshrined in international human rights instruments should be established for scientific professions. The social responsibility of scientists requires that they maintain high standards of scientific integrity and quality control, share their knowledge, communicate with the public and educate the younger generation. Political authorities should respect such action by scientists. Science curricula should include science ethics, as well as training in the history and philosophy of science and its cultural impact.

42. Equal access to science is not only a social and ethical requirement for human development, but also essential for realizing the full potential of scientific communities worldwide and for orienting scientific progress towards meeting the needs of humankind.

The difficulties encountered by women, constituting over half of the world's population, in entering, pursuing and advancing in a career in the sciences and in participating in decision-making in science and technology should be addressed urgently. There is an equally urgent need to address the difficulties faced by disadvantaged groups which preclude their full and effective participation.

43. Governments and scientists of the world should address the complex problems of poor health and increasing inequalities in health between different countries and between different communities within the same country with the objective of achieving an enhanced, equitable standard of health and improved provision of quality health care for all. This should be undertaken through education, by using scientific and technological advances, by developing robust long-term partnerships between all stakeholders and by harnessing programs to the task.

•••

44. We, participants in the *World Conference on Science for the Twenty-first Century: A New Commitment*, commit ourselves to making every effort to promote dialogue between the scientific community and society, to remove all discrimination with respect to education for and the benefits of science, to act ethically and cooperatively within our own spheres of responsibility, to strengthen scientific culture and its peaceful application throughout the world, and to promote the use of scientific knowledge for the well-being of populations and for sustainable peace and development, taking into account the social and ethical principles illustrated above.

45. We consider that the Conference document *Science Agenda—Framework for Action* gives practical expression to a new commitment to science, and can serve as a strategic guide for partnership within the United Nations system and between all stakeholders in the scientific endeavor in the years to come.

46. We therefore adopt this *Declaration on Science and the Use of Scientific Knowledge* and agree upon the *Science Agenda—Framework for Action* as a means of achieving the goals set forth in the Declaration, and call upon UNESCO and ICSU to submit both documents to the General Conference of UNESCO and to the General Assembly of ICSU. The United Nations General Assembly will also be seized of these documents. The purpose is to enable both UNESCO and ICSU to identify and implement follow-up action in their respective programs, and to mobilize the support of all partners, particularly those in the United Nations system, in order to reinforce international coordination and cooperation in science.

DECLARATION OF SANTO DOMINGO (1999)

•••

We, the Heads of State and/or Government of the States, Countries and Territories of the Association of Caribbean States (ACS), meeting in the City of Santo Domingo de Guzmán, Dominican Republic, on 16 and 17 April 1999;

Committed to the principles and objectives enshrined in the Convention Establishing the ACS, and recognizing the validity of the Declaration of Principles and Plan of Action on Tourism, Trade and Transport resulting from the historic First Summit held in Port of Spain, Trinidad and Tobago, in August 1995 and the priorities identified for promoting regional integration, functional co-operation and co-ordination among the Member States and Associate Members of the ACS;

Have decided to analyze the progress made by the ACS from Port-of-Spain 1995 to Santo Domingo 1999 and determine the projection of the Caribbean Region into the 21st Century; and therefore:

1. We identify tourism as the activity where the Association has achieved the most significant progress. We recognize that sustainable tourism constitutes an adequate response to the challenges of increasing rates of growth in employment and foreign exchange earnings, protecting and preserving the environment and natural resources, protecting cultural patrimony and values. We support community participation, as well as the involvement of local interests in aspects of the tourism development process, such as policy making, planning, management, ownership and the sharing of benefits generated by this activity. In this respect, we adopt the Declaration on the Sustainable Tourism Zone of the Caribbean (STZC).

2. We reiterate our commitment to work jointly for the consolidation of an enhanced economic space

for trade and investment, based on the principles of the World Trade Organization (WTO), for which we shall continue to encourage integration and co-operation measures that permit the strengthening of intra-regional trade and investment.

3. We note with satisfaction the progress yielded in the area of trade liberalization and economic integration in the sub-regional and bilateral spheres among the Member States and Associate Members of the ACS. Within the framework of Article XX of the Convention Establishing the ACS, the interested countries will continue to encourage according to their priorities, trade agreements and tariff preferences, as identified in the initiative to establish the Caribbean Preferential Tariff (CPT).

4. We reiterate that the rationalization and definition of regional transport policies are among the highest priorities of the ACS Plan of Action. We consider that transport must be the fundamental instrument for the development of tourism and trade in the region. In this respect, we emphasize our commitment to the objectives of the program "Uniting the Caribbean by Air and Sea".

5. Based on the fulfillment of commitments made in Agenda 21, we support the activities for the protection and conservation of the environment and natural resources. In addition, we support the effort of CARICOM to have the Caribbean Sea declared a Special Area in the context of Sustainable Development, and instruct that this subject be included in the Caribbean Environmental Strategy. For this purpose, a high level meeting of experts will be convened to study this topic. Participation in this meeting will be open to all members of the Association.

6. We consider the Caribbean Sea an invaluable asset and agree to give special priority to its preservation. We therefore deplore its ecological degradation and reject its continuous use for the transport of nuclear and toxic waste that may in any way cause a greater degradation of the Caribbean Sea.

7. We express our deepest solidarity with the countries and territories of the ACS affected by natural disasters in recent years, as well as by the extensive losses of lives and material resources, caused by these phenomena, which have increased their difficulties in implementing their programs of economic and social development.

8. We instruct the national authorities responsible for the prevention, mitigation and preparation for disasters, to put into practice, as soon as possible, the implementation mechanisms of the Regional Co-operation Agreement in the Area of Natural Disasters, signed by the Ministers of Foreign Affairs. In this respect, special focus will be placed on strengthening co-operation with the Caribbean Disaster and Emergency Response Agency (CDERA) and the Central American Co-ordination Centre for the Prevention of Natural Disasters (CEPREDENAC).

9. We emphasize the importance of co-operation in science and technology as the basis for the promotion of sustainable development of the region and in this respect, we observe with satisfaction the progress made in the development of the Co-operation Mechanism in the area of Science and Technology.

10. We recognize the efforts to widen regional collaboration and co-operation with respect to the linguistic integration program, the promotion of the teaching of the official languages of the ACS and the development of programs of integration, co-operation and exchanges in the areas of education and culture. Similarly, we express our support for the activities being developed in the region with regard to the preservation of the cultural patrimony, and the promotion and defense of our cultural values.

11. We appreciate the importance of international co-operation for the development of the peoples and economies of the region, and we take note of the renewed effort by the ACS Special Fund to work in this direction.

12. We are aware that globalization constitutes for the region an enormous challenge, that entails risks and opportunities. We therefore reiterate our interest in strengthening consultation and co-ordination of our positions in all those issues of mutual interest in the international agenda.

13. We agree that, faced with the rapid globalization process, multilateralism is the indispensable response for dealing with its challenges and utilizing its advantages, and in particular, for ensuring the effective exercise of the juridical equality of the States. We are aware moreover that the transparent and democratic functioning of multilateral bodies should be based on international law.

14. We reiterate our categorical rejection of all unilateral coercive measures, as well as the extraterritorial application of national laws by any State, since this is contrary to International Law, and more-

over threatens the sovereignty of States and international co-existence. In this context, we reiterate our exhortation to the Government of the United States of America to put an end to the application of the Helms-Burton Law, in accordance with the Resolutions approved by the United Nations General Assembly.

15. We reaffirm our commitment to the preservation, consolidation and strengthening of democracy, political pluralism and the Rule of Law, as an ideal framework that allows respect for the defense and promotion of all human rights, including the right to development and basic liberties. In this respect, we reiterate that civic participation is an indispensable element in the creation of a new political culture. We also reiterate respect for the principles of sovereignty and non-intervention, in addition to the right of all peoples to build their own political system in peace, stability and justice.

16. We reiterate moreover the need to implement social and economic measures aimed at achieving integrated and harmonious development, based on equity, social justice, the raising of the standards of living of the population, and the eradication of poverty, with the human being as the fundamental focus of development plans.

17. We renew our commitment to work for the sustainable development of the Caribbean through co-operation and integration.

18. We recognize the differences in the size and levels of development of the economies of the countries of the ACS and attach special significance to the vulnerability of the small economies of our region. We will take into consideration these differences in the treatment of the countries in the activities being developed within the framework of the ACS. We will search for means, complementary with suitable internal policies that would afford opportunities to encourage participation and further the level of development of the small and less developed economies.

19. We urge the international community to strengthen programs of technical and financial assistance, human resource training, and the transfer of technology, in order to improve the opportunities for the small and less developed economies to prosper in the international system.

20. In this context, we agree that there is a need to promote co-operation and concerted action among the Member States and Associate Members of the ACS, so as to increase the negotiating capability of our region in international fora.

21. We reaffirm the principles adopted at the First ACS Summit, with regard to the international problem of the illicit traffic of drugs and related crimes, which represents a serious threat to tourism, trade and transport, and indeed, endangers the sovereignty and security of each State.

22. We reiterate the principles governing international co-operation for dealing with the international problem of the illicit traffic of drugs and related crimes, including shared responsibility, the global, integrated and balanced approach, unrestricted respect for the principles of International Law, in particular those of sovereignty and territorial integrity. We therefore strongly reject every type of intervention in the internal matters of States and the extraterritorial application of domestic laws and unilateral measures. In this respect, we agree that programs, actions and results must be considered within an agreed intergovernmental framework.

23. We are aware of the great wealth of the cultural diversity in the Caribbean region and as a result, we agree to increase efforts in defense of our cultural identity, to protect and promote its expressions, given that culture is one of the fundamental bases for the integration of the Caribbean peoples.

24. We reiterate the commitment of our governments to work in close collaboration in order to contribute to the success of the European Union/Latin America and the Caribbean Summit, which constitutes an exceptional opportunity for promoting concerted action among ACS Members, increasing co-operation and enhancing existing dialogues and agreements between the two regions. To this end, we will promote the Latin American and Caribbean proposal, adopted in Mexico City, in December 1998, aimed at identifying inter-regional co-operation activities that contribute to enhancing relations with the countries of the European Union.

25. This Summit will also be a special occasion to establish a direct and frank dialogue with the leaders of the European Union, in order to advance in a decisive manner economic relations between both regions, especially in the areas of trade and investment, as well as to promote the convergence of efforts to restore international financial stability

and to redress the continued imbalances that might provoke a global recession.

26. We call for the optimization of the potential and opportunities provided by the sectoral links among the programs of the ACS and collaboration with relevant regional and national organizations, in order to ensure increasing complementarity among the activities of Member States and Associate Members.

27. We express our deep gratitude to the President of the Dominican Republic, His Excellency Leonel Fernández, and to the Government and people of the Dominican Republic, for the warmth, friendliness and lavish hospitality accorded to us throughout the Second Summit.

To give impetus to the goals and objectives outlined in this Declaration, we agree to adopt and execute the attached Plan of Action.

RIO DE JANEIRO DECLARATION ON ETHICS IN SCIENCE AND TECHNOLOGY (2003)

• • •

We, the Ministers and Higher Authorities of Science and Technology of South America, gathered in Rio de Janeiro on this 4th day of December, 2003, to reflect upon the limits that ethics impose on the production and use of scientific knowledge,

Considering:

the Declaration on the Use of Scientific Knowledge, signed in Budapest in 1999, that placed science in its social and international context as an instrument for the well being of all peoples, and called upon all countries to work for the good of humanity;

the overwhelming process of economic globalization and the growing impact of scientific development and technological innovations on our societies;

that the South American countries represented at this meeting recognize the need in the elaboration of their management policies for scientific and technological development to pay special attention to the ethical implications, so that principles founded upon such policies may serve

as guidance for efforts to achieve the well-being of their peoples and their autonomy as nations;

that a more democratic and far-reaching application of this knowledge requires national and regional development projects that include society as a whole;

that such projects must be viewed from the harmonic perspective of our peoples' common international interests, in order to confront the current trends of globalization in the realm of science, technology, economics, politics, and culture;

that the ethical and human conscience that grows at the heart of our societies impels us to prioritize, in the distribution of the benefits of knowledge to all, especially to women and children as well as all facets of excluded and marginalized segments of society, and the production of knowledge by women;

that the principles of democracy and social justice should govern international relations, serving as a reference for fraternity among countries, nations, and peoples;

that democracy, independence, and respect not only for individual and regional differences but also for the right and the struggle for peace, must reflect, within our countries, the same struggle for liberty, respect for human rights and, fundamentally, access for all to the intangible and practical benefits of human knowledge in culture, the arts, science and technology, through education and democratization of the results of economic development;

that we must defend an international system that elects to combat hunger and exclusion, especially exclusion from all forms of knowledge, as the highest priority, promoting universal quality education and that assures the right of all to healthcare, education, and housing while at the same time hinders abuses of power, condemns discrimination, and denounces intolerance and all other conditions or interests that may lead to war and the breakdown of democratic structures;

that free access to scientific knowledge and to effective participation in its creation, as well as the technological development and innovation, allowing the integration of our efforts, especially with respect to the establishment of an effective network of scientific and technological cooperation;

recognizing that the scientific and technological component forms the basis of the so-called "knowledge economy" - the economy of the third millennium - and that improved scientific and technological capacity will allow the participation in this economy and therefore in development; and

Facing limits imposed by international trade rules which, most of the time, do not consider the interests of the developing countries and their populations, and that our countries will also face competition from those countries possessing technology, as well as their transnational companies, the main beneficiaries of so-called "globalization".

Do recommend:

that the foundational activities for science and technology, such as education, scientific research, culture and technological development, be recognized and treated as public goods, and that an effort be made to diffuse knowledge, placing it at the disposal of humanity, especially the communities of the Third World;

that the governments of the Region support UNESCO in its efforts to allow the sectors and activities which constitute the "knowledge economy" (education, science, and culture) to contribute to socio-economic development in order to ensure the effective democratization of the components of knowledge generated by the digital industry and to render more flexible trade practices in the international regime of intellectual property, particularly in public health;

that the governments devote greater attention to the treatment given to science and technology in the context of the international trade rules and negotiations, adopting new critical approaches to the rules in effect and generating innovative proposals that increase access for the countries in the Region to knowledge and its benefits;

that our governments promote and stimulate the dissemination of information and knowledge through significant investments in R&D, information technology, robotics and computer science, software and hardware, popularizing the sources and the means of information as well as promoting universal access for all citizens;

that our governments support the increase in the use and production of software, seeking autonomy and cost reductions for the countries of the region;

that national and regional research groups be established with the objective of studying alternatives for the production of low-cost personal computers, aimed at universalizing usage of such computers, as well as implementing projects for regional cooperation in this field;

Do further recommend:

that attention be given to non-proprietary treatment of software, transmissions, and other digital technologies essential to ensuring the linguistic-cultural diversity of countries with relatively low representation on the Internet as well as in the use of electronic databases;

that an international network of scientific and technological knowledge be created, public in nature and freely accessible, also linked to databases on patents and inventions;

that a fund be established for the promotion of education, science, and culture in cyberspace, in support of networks of public schools, universities and research institutes in the countries of the Region, whose objective would be to promote science in the classroom and its popularization;

that the protection of individual rights and freedoms be promoted in measures relating to the fight against terrorism and to the promotion of a culture of cybersecurity;

that nations work together for the creation of an international consensus for the conversion of a portion of the payment of the external debt of developing countries into national investments in science and technology;

that our governments consider, the development of capacities which allow people to have access to new knowledge that make possible their productive participation in new sectors, if technological change so demands;

that the commitment to create spaces of cooperation in science and technology among our countries be reiterated, in both the public and private sectors, taking into account the ethical, political, social, and economic challenges they face;

that the essential role of the United Nations System's specialized agencies, particularly UNESCO, be recognized in supporting the elaboration of effective policies and guidelines in the field of ethics of Science and Technology and in technical cooperation through the exchange of international specialists, resource mobilization programs for the promotion of integrated interdisciplinary approaches to cooperation for development in science and technology and for the transfer of technological knowledge;

that UNESCO's work in the field of Ethics of Science and Technology and its role as focal point and legitimate participant in the worldwide debate over this issue be recognized and supported;

that the establishment, by UNESCO, of a mechanism that integrates and proposes dialogue on issues related to the Ethics of Science and Technology among our Governments be supported in order to promote the creation of programs for the teaching of ethics in basic, secondary and higher education and teacher training programs in this area; and the establishment of a network of governmental and non-governmental institutions in this area be supported;

that the work of COMEST as an independent advisory body of UNESCO regarding issues of Ethics in Science and Technology be recognized and that participation in this Commission be improved by the continued inclusion of representatives from all continents;

that the recommendations set forth by COMEST in such areas as the teaching of ethics, outer space, energy, and water be examined, in order to reinforce and to incorporate where necessary this ethical reflection in national and regional policies, in strategies, and in projects;

that States, organizations and other institutions interested in promoting and deepening reflection on the ethics of science be encouraged to create national and institutional commissions on scientific ethics;

that States be urged to implement, within the shortest time possible, the Universal Declaration on the Human Genome and Human Rights, approved in 1997 at the United Nations General Assembly;

and that the International Declaration on Human Genetic Data, approved at the 32nd UNESCO General Conference, be supported.

Thus, the Ministers and Higher Authorities of Science and Technology of South America, gathered in Rio de Janeiro, request the Heads of State and Government to confirm the growing importance of the ethical dimension of Science and Technology for the promotion of sustainable and equitable development, supporting the strengthening of cooperation in Science and Technology, above all with respect to their ethical implications, among the countries of South America, under the terms of the present Declaration.

The signatories hereby agree to transmit this Declaration to the Secretary General of the United Nations, as well as to the Director-General of UNESCO.

Rio de Janeiro, December 4, 2003

Signatories:

ROBERTO AMARAL—Minister of Science and Technology of Brazil

TULIO DEL BONO—Secretary of Science and Technology of Argentina

LUIS ALBERTO LIMA—President of the National Council of Science and Technology (CONCYT) of Paraguay

MARIA DEL ROSÁRIO GUERRA—Director of the Colombian Institute for Development of Science and Technology (CONCIENCIAS)

BENJAMIN MARTICORENA—President of the National Council of Science and Technology (CONCYTEC) of Peru

CPLP Authorities:

JOÃO BATISTA NGANDAJINA—Minister of Science and Technology of Angola

MARIA DE FÁTIMA SILVA BARBOSA—Minister of National Education of Guinea-Bissau

LÍDIA MARIA ARTHUR BRITO—Minister of Higher Education, Science and Technology of Mozambique

MARIA DA GRAÇA CARVALHO—Minister of Science and Higher Education of Portugal

MARIA DE FÁTIMA SILVA BARBOSA—Minister of National Education of Guinea-Bissau

AHMEDABAD DECLARATION
(2005)

• • •

This Declaration was made on January 20th, 2005, by more than 800 learners, thinkers and practitioners from over 40 countries, engaged in education for sustainable development, at the *Education for a Sustainable Future* conference held at Centre for Environment Education, Ahmedabad, India.

As the first international gathering of the United Nations Decade of Education for Sustainable Development (DESD), we warmly welcome this Decade that highlights the potential of action education to move people towards sustainable lifestyles and policies.

If the world's peoples are to enjoy a high quality of life, we must move quickly toward a sustainable future. Although most indicators point away from sustainability, growing grassroots efforts worldwide are taking on the enormous task of changing this trend.

We accept our responsibility and we urge all people to join us in doing all we can to pursue the principles of the Decade with humility, inclusivity, and a strong sense of humanity. We invite wide participation through networks, partnerships, and institutions.

As we gather in the city where Mahatma Gandhi lived and worked, we remember his words: "Education for life; education through life; education throughout life". These words underscore our commitment to the ideal of education that is participatory and lifelong.

We firmly believe that a key to sustainable development is the empowerment of all people, according to the principles of equity and social justice, and that a key to such empowerment is action-oriented education.

ESD implies a shift from viewing education as a delivery mechanism, to the recognition that we are all learners as well as teachers. ESD must happen in villages and cities, schools and universities, corporate offices and assembly lines, and in the offices of ministers and civil servants. All must struggle with how to live and work in a way that protects the environment, advances social justice, and promotes economic fairness for present and future generations. We must learn how to resolve conflicts, create a caring society, and live in peace.

ESD must start with examining our own lifestyles and our willingness to model and advance sustainability in our communities. We pledge to share our diverse experiences and collective knowledge to refine the vision of sustainability while continually expanding its practice. Through our actions we will add substance and vigor to the UNDESD processes.

We are optimistic that the objectives of the Decade will be realized and move forward from Ahmedabad in a spirit of urgency, commitment, hope, and enthusiasm.

CHEMIST'S CODE OF CONDUCT OF THE AMERICAN CHEMICAL SOCIETY

• • •

The American Chemical Society expects its members to adhere to the highest ethical standards. Indeed, the federal Charter of the Society (1937) explicitly lists among its objectives **"the improvement of the qualifications and usefulness of chemists through high standards of professional ethics, education and attainments"**

Chemists have professional obligations to the public, to colleagues, and to science. One expression of these obligations is embodied in "The Chemist's Creed," approved by the ACS Council in 1965. The principles of conduct enumerated below are intended to replace "The Chemist's Creed". They were prepared by the Council Committee on Professional Relations, approved by the Council (March 16, 1994), and adopted by the Board of Directors (June 3, 1994) for the guidance of society members in various professional dealings, especially those involving conflicts of interest.

Chemists Acknowledge Responsibilities To:

- The Public: Chemists have a professional responsibly to serve the public interest and welfare and to further knowledge of science. Chemists should actively be concerned with the health and welfare of co-workers, consumer and the community. Public comments on scientific matters should be made with care and precision, without unsubstantiated, exaggerated, or premature statements.

- The Science of Chemistry: Chemists should seek to advance chemical science, understand the limitations of their knowledge, and respect the truth. Chemists should ensure that their scientific contributions, and those of the collaborators, are thorough, accurate, and a unbiased in design, implementation, and presentation.

- The Profession: Chemists should remain current with developments in their field, share ideas and information, keep accurate and complete laboratory records, maintain integrity in all conduct and publications, and give due credit to the contributions of others. Conflicts of interest and scientific misconduct, such as fabrication, falsification, and plagiarism, are incompatible with this Code.

- The Employer: Chemists should promote and protect the legitimate interests of their employers, perform work honestly and competently, fulfill obligations, and safeguard proprietary information.

- Employees: Chemists, as employers, should treat subordinates with respect for their professionalism and concern for their well-being, and provide them with a safe, congenial working environment, fair compensation, and proper acknowledgment of their scientific contributions.

- Students: Chemists should regard the tutelage of students as a trust conferred by society for the promotion of the student's learning and professional development. Each student should be treated respectfully and without exploitation.

- Associates: Chemists should treat associates with respect, regardless of the level of their formal education, encourage them, learn with them, share ideas honestly, and give credit for their contributions.

- Clients: Chemists should serve clients faithfully and incorruptibly, respect confidentiality, advise honestly, and charge fairly.

- The Environment: Chemists should understand and anticipate the environmental consequences of their work. Chemists have responsibility to avoid pollution and to protect the environment.

CODE OF ETHICS OF THE AMERICAN ANTHROPOLOGICAL ASSOCIATION

• • •

I. Preamble

Anthropological researchers, teachers and practitioners are members of many different communities, each with its own moral rules or codes of ethics. Anthropologists have moral obligations as members of other groups, such as the family, religion, and community, as well as the profession. They also have obligations to the scholarly discipline, to the wider society and culture, and to the human species, other species, and the environment. Furthermore, fieldworkers may develop close relationships with persons or animals with whom they work, generating an additional level of ethical considerations.

In a field of such complex involvements and obligations, it is inevitable that misunderstandings, conflicts, and the need to make choices among apparently incompatible values will arise. Anthropologists are responsible for grappling with such difficulties and struggling to resolve them in ways compatible with the principles stated here. The purpose of this Code is to foster discussion and education. The American Anthropological Association (AAA) does not adjudicate claims for unethical behavior.

The principles and guidelines in this Code provide the anthropologist with tools to engage in developing and maintaining an ethical framework for all anthropological work.

II. Introduction

Anthropology is a multidisciplinary field of science and scholarship, which includes the study of all aspects of humankind—archaeological, biological, linguistic and sociocultural. Anthropology has roots in the natural and social sciences and in the humanities, ranging in approach from basic to applied research and to scholarly interpretation.

As the principal organization representing the breadth of anthropology, the American Anthropological Association (AAA) starts from the position that generating and appropriately utilizing knowledge (i.e., publishing, teaching, developing programs, and informing policy) of the peoples of the world, past and present, is a worthy goal; that the generation of anthropological knowledge is a dynamic process using many different and ever-evolving approaches; and that for moral and practical reasons, the generation and utilization of knowledge should be achieved in an ethical manner.

The mission of American Anthropological Association is to advance all aspects of anthropological research and to foster dissemination of anthropological knowledge through publications, teaching, public education, and application. An important part of that mission is to help educate AAA members about ethical obligations and challenges involved in the generation, dissemination, and utilization of anthropological knowledge.

The purpose of this Code is to provide AAA members and other interested persons with guidelines for making ethical choices in the conduct of their anthropological work. Because anthropologists can find themselves in complex situations and subject to more than one code of ethics, the AAA Code of Ethics provides a framework, not an ironclad formula, for making decisions.

Persons using the Code as a guideline for making ethical choices or for teaching are encouraged to seek out illustrative examples and appropriate case studies to enrich their knowledge base.

Anthropologists have a duty to be informed about ethical codes relating to their work, and ought periodically to receive training on current research activities and ethical issues. In addition, departments offering anthropology degrees should include and require ethical training in their curriculums.

No code or set of guidelines can anticipate unique circumstances or direct actions in specific situations. The individual anthropologist must be willing to make carefully considered ethical choices and be prepared to make clear the assumptions, facts and issues on which those choices are based. These guidelines therefore address *general* contexts, priorities and relationships which should be considered in ethical decision making in anthropological work.

III. Research

In both proposing and carrying out research, anthropological researchers must be open about the purpose(s), potential impacts, and source(s) of support for research projects with funders, colleagues, persons studied or providing information, and with relevant parties affected by the research. Researchers must expect to utilize the results of their work in an appropriate fashion and disseminate the results through appropriate and timely activities. Research fulfilling these expectations is ethical, regardless of the source of funding (public or private) or purpose (i.e., "applied," "basic," "pure," or "proprietary").

Anthropological researchers should be alert to the danger of compromising anthropological ethics as a condition to engage in research, yet also be alert to proper demands of good citizenship or host-guest relations. Active contribution and leadership in seeking to shape public or private sector actions and policies may be as ethically justifiable as inaction, detachment, or noncooperation, depending on circumstances. Similar principles hold for anthropological researchers employed or otherwise affiliated with nonanthropological institutions, public institutions, or private enterprises.

A. Responsibility to people and animals with whom anthropological researchers work and whose lives and cultures they study.

1. Anthropological researchers have primary ethical obligations to the people, species, and materials they study and to the people with whom they work. These obligations can supersede the goal of seeking new knowledge, and can lead to decisions not to undertake or to discontinue a research project when the primary obligation conflicts with other responsibilities, such as those owed to sponsors or clients. These ethical obligations include:

- To avoid harm or wrong, understanding that the development of knowledge can lead to change which may be positive or negative for the people or animals worked with or studied

- To respect the well-being of humans and nonhuman primates

- To work for the long-term conservation of the archaeological, fossil, and historical records

- To consult actively with the affected individuals or group(s), with the goal of establishing a working relationship that can be beneficial to all parties involved

2. Anthropological researchers must do everything in their power to ensure that their research does not harm the safety, dignity, or privacy of the people with whom they work, conduct research, or perform other professional activities. Anthropological researchers working with animals must do everything in their power to ensure that the research does not harm the safety, psychological well-being or survival of the animals or species with which they work.

3. Anthropological researchers must determine in advance whether their hosts/providers of information wish to remain anonymous or receive recognition, and make every effort to comply with those wishes. Researchers must present to their research participants the possible impacts of the choices, and make clear that despite their best efforts, anonymity may be compromised or recognition fail to materialize.

4. Anthropological researchers should obtain in advance the informed consent of persons being studied, providing information, owning or controlling access to material being studied, or otherwise identified as having interests which might be impacted by the research. It is

understood that the degree and breadth of informed consent required will depend on the nature of the project and may be affected by requirements of other codes, laws, and ethics of the country or community in which the research is pursued. Further, it is understood that the informed consent process is dynamic and continuous; the process should be initiated in the project design and continue through implementation by way of dialogue and negotiation with those studied. Researchers are responsible for identifying and complying with the various informed consent codes, laws and regulations affecting their projects. Informed consent, for the purposes of this code, does not necessarily imply or require a particular written or signed form. It is the quality of the consent, not the format, that is relevant.

5. Anthropological researchers who have developed close and enduring relationships (i.e., covenantal relationships) with either individual persons providing information or with hosts must adhere to the obligations of openness and informed consent, while carefully and respectfully negotiating the limits of the relationship.

6. While anthropologists may gain personally from their work, they must not exploit individuals, groups, animals, or cultural or biological materials. They should recognize their debt to the societies in which they work and their obligation to reciprocate with people studied in appropriate ways.

B. Responsibility to scholarship and science

1. Anthropological researchers must expect to encounter ethical dilemmas at every stage of their work, and must make good-faith efforts to identify potential ethical claims and conflicts in advance when preparing proposals and as projects proceed. A section raising and responding to potential ethical issues should be part of every research proposal.

2. Anthropological researchers bear responsibility for the integrity and reputation of their discipline, of scholarship, and of science. Thus, anthropological researchers are subject to the general moral rules of scientific and scholarly conduct: they should not deceive or knowingly misrepresent (i.e., fabricate evidence, falsify, plagiarize), or attempt to prevent reporting of misconduct, or obstruct the scientific/scholarly research of others.

3. Anthropological researchers should do all they can to preserve opportunities for future fieldworkers to follow them to the field.

4. Anthropological researchers should utilize the results of their work in an appropriate fashion, and

whenever possible disseminate their findings to the scientific and scholarly community.

5. Anthropological researchers should seriously consider all reasonable requests for access to their data and other research materials for purposes of research. They should also make every effort to insure preservation of their fieldwork data for use by posterity.

C. Responsibility to the public

1. Anthropological researchers should make the results of their research appropriately available to sponsors, students, decision makers, and other nonanthropologists. In so doing, they must be truthful; they are not only responsible for the factual content of their statements but also must consider carefully the social and political implications of the information they disseminate. They must do everything in their power to insure that such information is well understood, properly contextualized, and responsibly utilized. They should make clear the empirical bases upon which their reports stand, be candid about their qualifications and philosophical or political biases, and recognize and make clear the limits of anthropological expertise. At the same time, they must be alert to possible harm their information may cause people with whom they work or colleagues.

2. Anthropologists may choose to move beyond disseminating research results to a position of advocacy. This is an individual decision, but not an ethical responsibility.

IV. Teaching

Responsibility to students and trainees

While adhering to ethical and legal codes governing relations between teachers/mentors and students/trainees at their educational institutions or as members of wider organizations, anthropological teachers should be particularly sensitive to the ways such codes apply in their discipline (for example, when teaching involves close contact with students/trainees in field situations). Among the widely recognized precepts which anthropological teachers, like other teachers/mentors, should follow are:

1. Teachers/mentors should conduct their programs in ways that preclude discrimination on the basis of sex, marital status, "race," social class, political convictions, disability, religion, ethnic background, national origin, sexual orientation, age, or other criteria irrelevant to academic performance.

2. Teachers'/mentors' duties include continually striving to improve their teaching/training techniques; being available and responsive to student/trainee interests; counseling students/ trainees realistically regarding career opportunities; conscientiously supervising, encouraging, and supporting students'/trainees' studies; being fair, prompt, and reliable in communicating evaluations; assisting students/trainees in securing research support; and helping students/trainees when they seek professional placement.

3. Teachers/mentors should impress upon students/trainees the ethical challenges involved in every phase of anthropological work; encourage them to reflect upon this and other codes; encourage dialogue with colleagues on ethical issues; and discourage participation in ethically questionable projects.

4. Teachers/mentors should publicly acknowledge student/trainee assistance in research and preparation of their work; give appropriate credit for coauthorship to students/trainees; encourage publication of worthy student/trainee papers; and compensate students/trainees justly for their participation in all professional activities.

5. Teachers/mentors should beware of the exploitation and serious conflicts of interest which may result if they engage in sexual relations with students/trainees. They must avoid sexual liaisons with students/trainees for whose education and professional training they are in any way responsible.

V. Application

1. The same ethical guidelines apply to all anthropological work. That is, in both proposing and carrying out research, anthropologists must be open with funders, colleagues, persons studied or providing information, and relevant parties affected by the work about the purpose(s), potential impacts, and source(s) of support for the work. Applied anthropologists must intend and expect to utilize the results of their work appropriately (i.e., publication, teaching, program and policy development) within a reasonable time. In situations in which anthropological knowledge is applied, anthropologists bear the same responsibility to be open and candid about their skills and intentions, and monitor the effects of their work on all persons affected. Anthropologists may be involved in many types of work, frequently affecting individuals and groups with diverse and sometimes conflicting interests. The individual anthropologist must make carefully considered ethical choices and be prepared to make clear the assumptions, facts and issues on which those choices are based.

2. In all dealings with employers, persons hired to pursue anthropological research or apply anthropological knowledge should be honest about their qualifications, capabilities, and aims. Prior to making any professional commitments, they must review the purposes of prospective employers, taking into consideration the employer's past activities and future goals. In working for governmental agencies or private businesses, they should be especially careful not to promise or imply acceptance of conditions contrary to professional ethics or competing commitments.

3. Applied anthropologists, as any anthropologist, should be alert to the danger of compromising anthropological ethics as a condition for engaging in research or practice. They should also be alert to proper demands of hospitality, good citizenship and guest status. Proactive contribution and leadership in shaping public or private sector actions and policies may be as ethically justifiable as inaction, detachment, or noncooperation, depending on circumstances.

VI. Epilogue

Anthropological research, teaching, and application, like any human actions, pose choices for which anthropologists individually and collectively bear ethical responsibility. Since anthropologists are members of a variety of groups and subject to a variety of ethical codes, choices must sometimes be made not only between the varied obligations presented in this code but also between those of this code and those incurred in other statuses or roles. This statement does not dictate choice or propose sanctions. Rather, it is designed to promote discussion and provide general guidelines for ethically responsible decisions.

HIPPOCRATIC OATH FOR SCIENTISTS (U.S. STUDENT PUGWASH GROUP VERSION)

• • •

"I promise to work for a better world, where science and technology are used in socially responsible ways.

I will not use my education for any purpose intended to harm human beings or the environment. Throughout my career, I will consider the ethical implications of my work before I take action. While the demands placed upon me may be great, I sign this declaration because I recognize that individual responsibility is the first step on the path to peace."

INTERNATIONAL NETWORK OF ENGINEERS AND SCIENTISTS FOR GLOBAL RESPONSIBILITY (INES)

• • •

Appeal to Engineers and Scientists (1995)

APPEAL   Science and technology influence the social, economic and political development of civilization throughout the world. In many ways science and technology have made our life easier, richer and safer. However, science and technology can be used for destructive purposes and are key factors in the current growth economy that is threatening the viability of the biosphere and of human societies.

In its origins, science is a search for truth about our world. Its results can be used for good and misused for evil. Technological consequences are now so powerful and interconnected, so sweeping in unforeseen results, that they endanger basic requirements for sustaining life on earth. Without adherence to generally accepted ethical standards, science and technology can damage the future of society and life itself.

The greatest challenge of our time is to enable to all members of the world population to live in dignity in a manner that is sustainable for humankind and nature. In meeting this challenge science and technology—if used in the right way—play a decisive role by providing the necessary means or by analyzing the various consequences of human activities.

The web of humanity and life as a whole must not be endangered by vested interests. Knowledge gives power, and power may corrupt and be used for destructive purposes. Therefore, social structures and institutions on local, national, regional and global levels are urgently needed to promote responsible uses of science and technology. We appeal to engineers and scientists to respect human rights and human dignity unconditionally.

Secrecy of scientific and technological research allows its misuse. Our vision is a science which seeks truth in open discourse.

In the last decades several initiatives promoting ethical pledges of scientists have been launched. The values underlying these pledges can form the foundation of a worldwide community of responsibility among scientists and engineers. In adherence to the UNESCO Declaration for Scientific Professionals of November 1974, we have attempted to harmonize existing pledges into the following code of ethics:

PLEDGE   1. I acknowledge as a scientist or engineer that I have a special responsibility for the future of humankind. I share a duty to sustain life as a whole. I therefore pledge to reflect upon my scientific work and its possible consequences in advance and to judge it according to ethical standards. I will do this even though it is not possible to foresee all possible consequences and even if I have no direct influence on them.

2. I pledge to use my knowledge and abilities for the protection and enrichment of life. I will respect human rights, and the dignity and importance of all forms of life in their interconnectedness. I am aware that curiosity and pressure to succeed may lead me into conflict with that objective. If there are indications that my work could pose severe threats to human life or to the environment, I will abstain until appropriate assessment and precautionary actions have been taken. If necessary and appropriate, I will inform the public.

3. I pledge not to take part in the development and production of weapons of mass destruction and of weapons that are banned by international conventions. Aware that even conventional arms can contribute to mass destruction, I will support political efforts to bring arms production, arms trade, and the transfer of military technology under strict international control.

4. I pledge to be truthful and to subject the assumptions, methods, findings and goals of my work, including possible impacts on humanity and on the environment, to open and critical discussion. To the best of my ability I shall contribute to public understanding of science. I shall support public participation in a critical discussion of the funding priorities and uses of science and technology. I will carefully consider the arguments from such discussions which question my work or its impact.

5. I pledge to support the open publication and discussion of scientific research. Since the results of science ultimately belong to humankind, I will conscientiously consider my participation in secret research projects that serve military or economic interests. I will not participate in secret research projects if I conclude that society will be injured thereby. Should I decide to participate in any secret research, I will continuously reflect upon its implications for society and the environment.

6. I pledge to enhance the awareness of ethical principles and the resulting obligations among scientists and engineers. I will join fellow scientists and others willing to take responsibility. I will support those who might experience professional disadvantages in attempting to live up to the principles of this pledge. I will support the establishment and the work of institutions that enable scientists to exercise their responsibilities more effectively according to this pledge.

7. I pledge to support research projects, whether in basic or applied science, that contribute to the solution of vital problems of humankind, including poverty, violations of human rights, armed conflicts and environmental degradation.

8. I acknowledge my duty to present and future generations, and pledge that the fulfillment of this duty will not be influenced by material advantages or political, national or economic loyalties.

The above text incorporates material and ideas from the following declarations:

- The Mount Carmel Declaration on Technology and Moral Responsibility (Haifa, 1974)
- The Biologists Pledge (MIT, 1987)
- Hippocratic Oath for Scientists (Nuclear Age Peace Foundation, (1987)
- The Buenos Aires Oath (Buenos Aires, 1988)
- The Uppsala Code of Ethics for Scientists (Uppsala, 1984)
- Hippocratic Oath for Scientists, Engineers and Executives (Inst. for Social Inventions, 1987)
- Scientists Pledge Not to Take Part in Military-Directed Research (SANA, London, 1991)
- Appeal to Scientists (Wittenberg, 1989)
- A Pledge for Scientists (Berlin, 1984)
- The Toronto Resolution (Toronto, 1991)

We see these declarations as a part of a wider movement which has expressed itself in particular in the Declaration of a Global Ethic of the Parliament of the World's Religions (Chicago, 1993) and in the Trieste Declaration of Human Duties (Trieste, 1994).

7. GOVERNMENT

•••

DEFINITION OF RESEARCH MISCONDUCT FROM THE U.S. FEDERAL REGISTER

•••

Research misconduct is defined as fabrication, falsification, or plagiarism in proposing, performing, or reviewing research, or in reporting research results.

1. No rights, privileges, benefits or obligations are created or abridged by issuance of this policy alone. The creation or abridgment of rights, privileges, benefits or obligations, if any, shall occur only upon implementation of this policy by the Federal agencies.

2. Research, as used herein, includes all basic, applied, and demonstration research in all fields of science, engineering, and mathematics. This includes, but is not limited to, research in economics, education, linguistics, medicine, psychology, social sciences, statistics, and research involving human subjects or animals.

Fabrication is making up data or results and recording or reporting them.

Falsification is manipulating research materials, equipment, or processes, or changing or omitting data or results such that the research is not accurately represented in the research record.

3. The research record is the record of data or results that embody the facts resulting from scientific inquiry, and includes, but is not limited to, research proposals, laboratory records, both physical and electronic, progress reports, abstracts, theses, oral presentations, internal reports, and journal articles.

Plagiarism is the appropriation of another person's ideas, processes, results, or words without giving appropriate credit.

Research misconduct does not include honest error or differences of opinion.

INDEX

*Page references to entire articles are in **boldface**. References to tables, figures, and illustrations are denoted by italics, sometimes followed by t (table) or f (figure).*

2001: A Space Odyssey (movie), 1694

••• A

AAA (American Anthropological Association), 1213–1214
AAAS. *See* American Association for the Advancement of Science (AAAS)
Aarhus Convention on Access to Information, Public Participation in Decision-making, and Access to Justice in Environmental Matters, 677
Abduction
　Peirce, Charles Sanders, 1395
　scientific ethics, 1728
ABET, Inc. (Accreditation Board for Engineering and Technology), 626
Abortion, **1–6**
　Anscombe, G. E. M, 83
　ethics of evidence, lviii
　fetal research, 767–768
　genetic counseling, 840–841
　Hindu perspectives, 920
　human cloning, 939
　Japanese perspectives, 1073
　Jewish perspectives, 1081
　medical ethics, 1185–1186
　morning-after pills, 235
　pre-implantation genetic diagnosis, 1055–1056
　right to life, 1637
Absolute/relative distinction, 1321
Absolute safety, 1674
Abstraction
　Christian perspectives, 333
　level of, 1000
Abstract networks, 1307
Abundance, 1771–1773
Abuse of science
　Bacon, Francis, 167
　computer hacking, 396

corruption, 433
Daoism, 467
development ethics, 514–515
double effect and dual use, 544
informed consent, 1016
social classes, 346–347
social indicators, 1809
statistics, 1868
Wells, H. G., 2062
See also Misconduct in science
Academy of Sciences, 1147
Acastos (Murdoch), 1250
Acceleration. *See* Speed
Acceptable daily intakes, 1676
Access
　communications technology, 367, 374
　digital divide, 525–526
　digital libraries, 528
　information ethics, 1007–1008
　Internet, 1051
　medical treatment, 194
　networks, 1309
　terrorism, 1931–1935
Accidents
　Apollo program, 87
　aviation, 159–160
　Bhopal case, 182–183, 668, 1913
　building collapse, 266–268
　Challenger and *Columbia*, 1332–1333, 1618, 1833, 1835–1838
　Chernobyl accident, 312–317, 1336
　DC-10 case, 472–474
　responsibility, 1618
　ships, 1766
　Three-Mile Island, 1952–1957
　West Gate Bridge, 255
　See also Normal accidents
Accountability
　publicly funded science, 1697
　research, **6–7**

science policy, 1703
　See also Responsibility
Accounting, **7–11,** 400
Accreditation, engineering, 626, 1073–1074
Accreditation Board for Engineering and Technology (ABET, Inc.), 626
Accuracy, 1313
AC/DC competition, 592
ACGIH (American Conference of Governmental Industrial Hygienists), 740–741
Acid mine drainage, **11–12**
Ackoff, Russell L., 1881
ACM (Association of Computing Machines), 2030
Acoustics, 1256
ACS (American Chemical Society), 309
ACTG 076, 349
Action
　knowledge, 1522
　philosophy of, 103–105
　science and technology as, 163–164
Action Group on Erosion, Technology and Concentration (ETC), 189, 1260
Action responsibility, 1620, 1622(*f*1)
Active *vs.* passive euthanasia, 712
Active *vs.* passive rights, 808
Activism
　animals testing, 436
　bioengineering ethics, 192
　Blackett, Patrick, 238
　disabled persons rights, 533
　Gandhi, Mohandes Karamchand, 986–987
　Greenpeace, 252
　Rifkin, Jeremy, 231
　Rotblat, Joseph, 1043
　Russell, Bertrand, 1666–1667
　Sakharov, Andrei, 1677–1678
　Sanger, Margaret, 233, 1678–1680

Canadian Environmental Protection Act (CEPA), 283
Canadian Networks of Centers of Excellence (NCE), 284
Canadian Psychological Association, 1538
Cancer, **285–288**
  alternative medicine, 386
  biostatistics, 218, 221, 222*t*
  breast cancer treatment, 221*f*
  Chernobyl accident, 315
  death rates, 220*f*
  DES (diethylstilbestrol), 503
  genetic screening, 196
  National Cancer Institute, 1272
  radiation, 1565
  radiation exposure, 1566–1567
  radiation treatment, 1335–1336
  randomized clinical trials, 219
Canning, food, 777
Cannon, Walter B., 1873
Canon law, 1097
Capacity
  engineering, 1597, 1597*f*
  neuroscience research, 1312
  responsibility, 1617
Capacity and demand, 1597, 1597*f*
Çapek, Karel, 1691
Capital
  contracts, 587
  private property, 1526
Capitalism, **288–291**
  alienation, 53
  Bernal, J. D., 179
  colonialism, 354–355
  critical social theory, 447, 449
  decentralization, 1436
  liberalism, 1122
  market theory, 1162
  Marxism, 377–378
  Marx's criticism of, 1165–1169, 1438
  political economy, 1439
  Rand, Ayn, 1582
  secularization, 1737
  socialism and, 1814, 1815
  speed, 1846
  work, 2075–2076
  See also Market theory
Capital punishment. *See* Death penalty
Caplan, Arthur, 904
Capra, Fritjof, 1883
Captain America, 1895, 1896–1897
Captive animals, 79, 80–82
Carbonates, water, 2051
Carbon-14 dating, 95
Carbon dioxide, 150–151, 1444
Carbon emissions, 1467–1468
Carbon sequestration technologies, 1467
Carcinogens, 285, 286, 937

Cardiopulmonary resuscitation, 1397–1398
Cardiovascular disease, 1755, 1756
Care, ethics of. *See* Ethics of care
Caring responsibility, 1622
Carnap, Rudolf, 1138, 1139
Carnegie Endowment for International Peace, 1545
Carnes, Mark, 1242
Carrey, Jim, *1841*
Carrying capacity, 895
Carson, Rachel, *291,* **291–293**
  DDT, 475
  environmental regulation, 671
  plastics, 1419
  waste, 2047–2048
Carter, Jimmy
  missile defense systems, 1216
  nuclear waste reprocessing, 1348
Cartesian definition of profession, 1516–1517
Cartographers, 856
Cartwright, Edmund, 996
Caruso, David, 618
Cascade of uncertainty, 1993
Case-based casuistry, 91
Case-control studies
  biostatistics, 219, 222*t*
  epidemiology, 680(*t2*), 681
  probability, 1495
  social constructivism, 1790
  See also Clinical trials
Cash crops, 1161
Casualties. *See* Fatalities
Casuistry
  applied ethics, 91
  business ethics, 275
  engineering ethics, 629
Catastrophes, 1997
Categorical imperative, 1489–1490
Categorical reasoning, 1104–1105
Catholic Church
  Anscombe, G. E. M, 83
  birth control, 233, 235–236
  Christianity and science, 339
  colonialism, 356–357
  dignity, 529
  education, 595–596
  eugenics, opposition to, 709
  Galilei, Galileo, 814–815, 1068
  movie codes, 1240
  natural law, 1291–1292
  responsibility, 1610–1611
  Thomism, 1948
Cattell, James McKeen, 1563
Caudwell, Christopher, 1170–1171
Causal responsibility, 1616
Causation
  Aristotle, 106

  epidemiology, 681–682, 682*t*, 683*t*
  ethology, 705
  free will and determinism, 797
Cause and effect
  determinism, 511, 513
  Hume, David, 964
  unintended consequences, 1996–1997
CBAC (Canadian Biotechnology Advisory Committee), 283
CBD (Convention on Biological Diversity), 188
Cellular technology, 1571
Censorship
  advertising, 21
  challenged books, *1006*
  communications ethics, 369
  information ethics, 1007–1008
  Internet regulation, 374, 1051
  movies, 1239–1240
  scientific information, 2057
Centers for Disease Control and Prevention (CDC)
  biosecurity, 217
  emergent infectious diseases, 611–613
Centipede Game, *821,* 823–824
Central Europe, **293–297**
Centralization, 1436
Central limit theorem, 1499, 1499(*f8*), 1501, 1858
Central Pacific railroad, 1576
Centre for Technology, Innovation and Culture, 1683–1684
Centro Intercultural de Documentación (CIDOC), 981
CEPA (Canadian Environmental Protection Act), 283
CEQ (Council on Environmental Quality), 661
Certainty, 500–501
  See also Uncertainty
Certification of nuclear power plants, 1344
Cesium-137, *313*
CFC's, 1232–1235, 1973
CGIAR (Consultative Group on International Agricultural Research), 886
Chagnon, Napoleon A., 1213
Chakrabarty, Ananda, 1032
Challenger accident, xxxviii, 1332–1333, 1618, 1833, **1835–1838**
Chance
  decision theory, 486
  mathematical theory of communication, 1002
  quantum theory, 1504
Chandler, Alfred D., Jr., 1151–1152
Change and development, **297–301**
  biodiversity, 186–187

Classical pragmatism, 1468–1470
Classical probability, 1494
Classical theory of statistical inference, 1503
Classics, 594–595
Classification and taxonomy
  biodiversity, 186
  bioengineering, 190
  business ethics cases, 274
  Diagnostic and Statistical Manual of Mental Disorders, 521, 522
  early humans, 214
  fingerprinting, 784
  information, *1001*
  Islam and knowledge, 1064
  nongovernmental organizations, 1326
  periodic table of the elements, 307–308
  pollution, 1442–1443
  race, 1561–1564
  systems, 1880
  toxicology, 1594
Class issues, **345–347**
  Communism, 377–378
  environmental justice, 665–666
  leisure class, 2023–2024
  material culture, 1175
  Mill, John Stuart, 1202
  popular culture, 1451–1452
  work, 2074
  See also Socioeconomic issues
Clausius, Rudolf, 620
Clean Air Act, 1329
Cleanup. *See* Ecological restoration
Cleanup, environmental. *See* Remediation
Clean Vehicles program, 2000
Clean Water Act, 574–575
Clergy, 595
Client-centered therapy, 1537
Climate, 489
Climate change. *See* Global climate change
Cline, Martin, 830
Clinical Center, NIH, 1273
Clinical psychology, 1536, 1537
Clinical trials, **347–350**
  bioethics, 194
  cancer treatments, 287
  complementary and alternative medicine, 386
  DES (diethylstilbestrol), 502–503
  drugs, 547
  gene therapy, 830, 831
  meta-analysis, 1191–1192
  randomized, 218–219, 221*f*, 225
  statistics, 1867–1868
  uncertainty, 1994
  women, exclusion of, 1755, 1756

  See also Case-control studies; Human subjects research
Clinton, Bill
  environmental justice, 666
  environmental regulation, 671
  nanotechnology, 1259
Clock of the Long Now, 1847
Cloning. *See* Human cloning
Closed societies, 1450
Closed systems
  robot toys, 1657
  uncertainty, 1992
The Closing of the American Mind (Bloom), 1360–1361
CMB (cosmic microwave background), 438, 439–440
Coal, 995
Coastal issues, 2003
Coates, Gerry, 142
Coauthorship, 1412, 1600
Cochlear implants, 1531
Cochrane, Willard, 43–44
Cochrane Collaboration, 1191–1192, 1866
Code of Chivalry. *See* Chivalric code
Code of Hammurabi, 263, 1590
Codes, 1748–1749
Codes of ethics, **350–353**
  accountants, 10
  American Nuclear Society, 1673–1674
  American Psychological Association, 1537–1538
  American Society for Quality, 1598
  American Sociological Association, 1824
  archaeologists, 96–97
  ASME International, 1590
  Association for Computing Machinery, 126
  bioengineering ethics, 193
  biological weapons research, 210–211
  chemistry, 310
  computer industry, 394–395, 400
  deontology, 498
  design, 505
  discourse ethics, 536
  engineering, 430, 625–626, 1020, 1022
  engineering ethics, 627
  European engineers, 633–634
  genetic counselors, 840
  human subjects research, 284
  information ethics, 1009, 1009*t*
  Institute of Electrical and Electronics Engineers, 1869
  Institution of Professional Engineers New Zealand, 142

  International Council for Science, 1043
  Japan, 1072, 1073
  magazine publishing, 19
  military, 1197
  modeling, 1219
  National Society of Professional Engineers, 1674
  nutrition professionals, 1350
  operations research, 1363
  police, 1431
  professional engineering organizations, 1513–1514
  radio industry, 1571
  research ethics, 1024, 1025
  responsible conduct of research, 1624
  Shintoism, 1764
  Society of German Engineers, 865
  Society of Toxicology, 1592–1593
  sociologists, 1826
  tourism, 1971
  trust, 1981
  UK Association of Humanistic Psychology Practitioners, 1541
  zoos, 2089
Cognitive activity, science as, 1696
Cognitive development, 1729–1730
Cognitive enhancers, 1543–1545
Cognitive prostheses, 1531
Cognitive science
  altruism, 62–63
  computer modeling, 1731
  emotions, 1638–1639
  information overload, 1011
  IQ tests, 1057–1061
  Simon, Herbert A., 1770
  speed, 1845
Cognitive values, 744–745
Cohen, Ben, 649
Cohen, Stanley, 228
Cohort studies, 219, 222*t*, 680, 680(*t*3)
Coin toss, 1002
Cold launches, 1837
Cold War
  biological weapons, 208
  foreign aid, 1370
  information security, 1932
  nuclear weapons, 169
  restricted travel of scientists, 1041
  security, 1740
Cole, Luke W., 665
Cole, Simon, 737
Collaborative assurance, 1796
Collaborative science, 1602
Collateral damage, 1996
Collected Commentaries on Medicinal Herbs (Tao Hongjing), 469
Collective choices, 1436
Collective dose, 1568

geological surveys, 1270
hazardous technologies, 182
hazards, 901–902
HIV/AIDS prevention, 926–927
human research subjects, 962
Ibero-American perspectives, 973–976
India, 987–988, 989
information and computing technology, 525
intellectual property, 1033, 1046–1047
investment in science research, 358
Malaysia, 1067
market theory, 1161
modernization, 1222–1223
Montreal Protocol, 1234
Moon Treaty, 362, 363
museums of science and technology, 1252
neoliberalism, 1304–1306
nutrition, 1350–1351
operations research, 1364
Organization for Economic Co-operation and Development, 1368–1371
political risk assessment, 1440–1442
pollution, 1446
product safety and liability, 1511
smoking, 683
space resources, 362, 363
technology and power, 2024
technology transfer, 1913–1914
unintended consequences, 1997
urbanization, 2008–2009
vaccines and vaccination, 2018–2019
waste, 2049
Western science, 765
World Bank, 2079–2080
World Health Organization projects, 2083–2084
Development, behavioral, 852–853
Developmental psychobiology, 852
The Development Dictionary: A Guide to Knowledge as Power, 513
Development ethics, **513–519**
developmental state theories, 1437
feminist perspectives, 765
Japanese perspectives, 1072–1073
limits, 1133–1134
Man and the Biosphere Programme, 2002
neoliberalism, 1305–1306
Singapore, 1776–1778
See also Change and development; Progress
Deviance, normalization of, 1837
Device paradigm, 1912
Dewey, John, *519*, **519–521**
axiology, 1729

certainty, 1991
experimentation, 729
liberalism, 1123
pragmatism, 1469, 1470
praxiology, 1473
social engineering, 1804
values, 2021
Diabetes, 228
Diagnosis
biostatistics, 219–223, 222t
body, 241
genetic, 833, 847
persistent vegetative state, 1397
reliability, 1596
Diagnostic and Statistical Manual of Mental Disorders (DSM), **521–525,** 1596
Dialectical freedom, 791–792
Dialectical materialism. *See* Materialism
Dialogics, 1745
Dichloro-diphenyl-trichloroethane. *See* DDT
Dick, Philip K., 1694
Dickens, Charles, 1720–1721
Diebold, John, 146, 148
Diet. *See* Nutrition and science
Dietary supplements, 1351, 1352
Die Totale Mobilmachung (Jünger), 1092
Dietrich, John H., 945
Difference principle, 44, 1588
Différend, 1145–1146
Differentiated responsibility/capability, 1234
Diffusion flame, 769
Digestive system, 303
Digital divide, 367, **525–527**, 1029
Digital effects, 1842
Digital libraries, **527–528**
Digital literacy, 1718
Digital materials
information ethics, 1007
reproducibility, 397, 1050
Digital revolution, 112
Digital Rights Management (DRM), 1126
Dignity, **528–530**
death with, 1398
German perspectives, 866–867
Kass, Leon, 109
privacy, 1491
Dilation and curettage, 1
Dilemma of control, 984, 1997–1998
Diminishing marginal utility, 569
Direct action environmentalism, 1329
Direct current, 592
Direct democracy, **530–531**
Direct Democracy Campaign (DDC), 530
Directorate for Biological, Behavioral, and Social Sciences, 1285

Directorate for Computer and Information Science and Engineering, 1285
Directorate for Engineering, 1285
Directorate for Scientific, Technological and International Affairs (STIA), 1284
Disabilities, **531–534**
genetic counseling, 839–841
Nazi medicine, 1301
Disappearances, 958
Disarmament
biological weapons, 208
Limited Nuclear Test Ban Treaty, 1131–1132, 1340
Pugwash Conferences, 1551–1552
Russell, Bertrand, 1667
Sakharov, Andrei, 1677–1678
Disaster relief, 948–949
Disasters. *See* Hazards
Disch, Thomas M., 1693
Discipline and Punish (Foucault), 786
Disclosure
conflicts of interest, 404
genetic information, 833
informed consent, 1017
Discourse, postmodernism, 1145–1146
Discourse ethics, **534–539**
See also Rhetoric of science and technology
Discourse Ethics: Notes on Philosophical Justification (Habermas), 535
Discourse on Method (Descartes), 500
The Discourse on the Arts and Sciences (Rousseau), 1660
The Discourse on the Origin and Foundation of Inequality Among Men (Rousseau), 1660
Discovery (Space Shuttle), *1833*
Discretionary power, 734–736
Discretion in planning, 1415
Discrimination and bias
beauty standards, 436
genetic information, 847
genetic screening, 833
HIV/AIDS, persons with, 28, 926
IQ tests, 1058–1059
land ethics, 1115
neuroscience research, 1312–1313
organ transplant allocation, 1372
privacy, 1491
sex and gender, 1753–1757
sex selection, 1758–1760
See also Feminism; Race
Discriminatory environmentalism, 667–669
Disease. *See* Health and disease
Disengagement, 1406
Disinformation, 1000
Disintegrity, ecological, 575–576
Displays, 1252

European Convention on Human Rights, 677
European Federation of National Engineering Associations (FEANI), 1514
European Science and Technology Observatory, 1907–1908
European technology assessment, **1906–1908**
European Union
    nuclear waste repositories, 1348
    precautionary principle, 1474–1475, 1476
    public understanding of science, 1549–1550
Euthanasia, **710–713**
    medical ethics, 1186–1187
    natural law, 1294
    right to die, 1635
    right to life, 1637
Euthanasia in the Netherlands, **713–715**
Euthyphro (Plato), 1421–1422
Evaluative indicators, 1809
Evaluative judgments, 1887–1888
Even-Odd game, 822
Events in probability, 1494–1495
Evidence
    expertise, 733–734
    forensic science, 784–785
    police, 1431
    scientific, 1710, 1712
Evil
    Augustine, 140
    banality of, 105
    Jung, Carl Gustav, 1091
EVIST (Ethics and Values in Science and Technology), 1284–1285
Evolution, **715–720**
    aggression, 32–33
    biophilia, 213–215
    Butler, Samuel, 281–282
    Christian perspectives, 329, 335, 339
    Darwin, Charles, 470
    Dewey, John, 519–520
    dominance, 541
    ethics, 702
    ethology, 705
    free will, 798
    Galton, Francis, 816–817, 816–818
    game theory, 820, 824–825
    gender bias, 762–763
    genetic code, 1425
    globalization of disease, 198
    Haldane, J. B. S., 893–894
    Holocaust, 931
    human nature, 955–956
    Jewish perspectives, 1082
    Lewis, C. S., 1120
    life, diversity of, 1128
    linear dynamics, 388–389

    management as, 1153–1154
    Marxism, 1169
    morality, 116–117, 1729, 1730
    prediction, 1481
    probability, 1505
    rational choice theory, 1586–1587
    secularization, 1737
    selfish genes, 1741–1743
    social, 1814
    social Darwinism, 1800–1803
    sociobiology, 1820–1821
    Spencer, Herbert, 1848–1849
    statistical inference, 1864
    technological, 802
    theodicy, 1937–1938
    Thomism, 1947–1948
    tradeoffs, 1974
    Treat, Mary, 1979
    See also Natural selection
Evolutionary psychology
    genetics and behavior, 852
    selfish genes, 1742
    sex and gender, 1757
    sociobiology, 1822
Evolutionary Stable Strategy (ESS), 1821
Evolution-creation debate, **720–723**
EVS (Ethics and Values Studies), 1286, 1287
Ex ante regulation, 1708
Excavations, archaeological, 94–97
Exchange value, 2020
Exclusiveness of property, 1525
Exhibitions, 1251
Existence values, 650
Existentialism, **723–727**
    alienation, 53–54
    Brun, Jean, 801–802
    Buddhism, 258
    feminist, 1753
    Heidegger, Martin, 912–914
    Jaspers, Karl, 1074–1075
    life, concept of, 1130
    Ortega y Gasset, José, 1373–1374
    Pascal's Wager, 1388
Existential Thomism, 1946–1947
Expansionist view of science and values, 818
Expected utility theory, 487–488, 1643
Experience, 964
Experimental psychology, 1536, 1537
Experimentation, **727–731**
    Hobbes, Thomas, 929–930
    Newton, Isaac, 1321
    Peirce, Charles Sanders, 1395–1396
    reliability, 1595–1599
    social science, 638–639
Expertise, **731–739**
    applied ethics, 91–92

    Aristotle, 107
    choice behavior, 326
    demonopolization, 1648
    Dreyfus, Hubert L., 1405
    enquete commissions, 642–644
    environmental economics, 652
    environmental regulation, 673
    euthanasia, 714
    expert witnesses, 785
    interactive model of expert/lay relations, 1548
    knowledge, lxiii
    knowledge as power, 1910
    law, 1711
    management, 1153
    participation of laypeople, 1380–1384
    pragmatism, 1470
    risk perception, 1645
    Singapore, 1777–1778
    See also Technocracy
Expertocracy, 1899–1900
Exploratorium, 1251, 1365
Exploratory data analysis (EDA), 1866
Explosions, ship, 1766
Exponential distribution, 1504
Ex poste regulation, 1708–1709
Exposure limits, **739–742**, 740*f*
    radiation, 1335, 1565–1569, 1567*t*
    toxicology, 1593–1595
Expressed sequence tags (ESTs), 943
Extensionism, 76–77, 656
Extensivity, 62–63
Extermination, 1301–1302
External auditing, 7
Externalists, 1730
Externalities
    environmental economics, 652
    innovation, 1905
    market theory, 1163
    pollution, 1444–1445
    unintended consequences, 1996
External rhetorics of science and technology, 1627–1628
External values, 1316–1317
Extinction
    biodiversity, 186–187
    climate change, 871
    indigenous peoples, 991
Extraction, energy, 621
Extrapolation of data, 1594
Extraterrestrial life, 1533–1534, 1834
Extrinsic values, 2022

• • • F

FAA (Federal Aviation Administration), 472–473
"Fable of the Bees" (Mandeville), 639

● ● ● H

and disease, 903
double effect, 543, 544
Durkheim, Émile, 551
embryonic research, 1054–1055
emotions, 1637–1638
environmental ethics, 656
evolutionary, 116–117, 1729, 1730
free will, 388
gender essentialism, 759
Hobbes, Thomas, 928
human nature, 955–956
Hume, David, 965
information ethics, 1007
legalism, 1710
master-slave dialectic, 1323–1324
natural law legality *vs.* moral theories, 1290
religious *vs.* moral obligations, 1106–1107
rights-based, 1633–1634
Singapore, 1777–1778
social functions, 1144
socialism, 1815
sociology of, 1823–1824
Spencer, Herbert, 1848–1849
universal moral responsibility, 1622–1623
as unnatural, 1297–1298
video games, 2030
virtual reality, 2035–2036
virtue ethics, 2041–2042
Moral philosophy, 598, 599, 600
Moral psychology, 1729–1730
Moral self-ownership, 1491
Moral sense, 1077–1078
Moral status
embryos, 608
human clones, 942
knowledge, lx
natural law, 1294
robots, 1656
robot toys, 1657–1658
Moral value *vs.* instrumental value, 1888
Moratoria on technological innovation, 1905
More, Thomas, **1236–1237,** *1237,* 2010
Morgagni, Giovanni, 904
Morgan, Thomas Hunt, 844
Morgenstern, Oskar, 486–487, 1506
Morning-after pills, 235
Morphologies, engineering, 636
Morrill Land Grant Act, 599
Morris, Charles, 1745, 2020
Morris, Henry, 721
Morris, William, **1237–1239,** *1238*
Mortality
developed countries, 34
epidemiology, 681
relative frequency, 1501

Moses, Robert, xl, 254
Mosquitoes
DDT, 475–476
malaria, 2083
Motherhood, 749, 761
Motion Picture Association of America, 1241
Motion Picture Producers and Distributors Association of America, 1240
Motion studies
management, 1152
rational goal management model, 1156
Motivation, 278
Mounier, Emmanuel, 799–800
Mou Zongsan, 409
Movies, **1239–1244**
Benjamin, Walter, 175
Brecht, Bertolt, 250–251
critical social theory, 448
information ethics in, 1008*t*
postmodernism, 1463
science fiction, 1693
special effects, 1840–1843
violence in, 368–369
Muhammad, 1061, 1064
Muir, John, 418, 670, 1327, 2067
Muller, Hermann J., 1147
Müller, Johannes, 864
Müller, Paul, 475
Multiagent artificial intelligence systems, 117
Multiculturalism
communication ethics, 367–368
immigration reduction, 895
open society, 1360–1361
See also Pluralism
Multidisciplinarity
Earth Systems Engineering and Management, 566–567
interdisciplinarity as distinct from, 1034
planning, 1414
praxiology, 1473
Multifactor Emotional Intelligence Scale, 618
Multilateral Nuclear Force (MNF), 1341
Multinationals. *See* Corporations
Multiple identities, 1275
Multiple intelligence, 617
Multiple-use management, **1245–1246,** 1245*f*
Multiple uses, 1317–1319
Multi-user virtual reality, 2036
Multivariate techniques, 1862
Multiversities, 599–600
Mumford, Lewis, 1199, **1246–1249,** *1247*
Murder

Communist regimes, 377
death penalty, 481
euthanasia, 711
Murdoch, Iris, **1249–1251,** *1250*
Murrah Federal Building bombing, *267,* 268–269
Murray, Janet, 1718
Museums of science and technology, **1251–1254,** 1365
Music, **1254–1258**
Augustine, 139
Internet file sharing, 1050, 1051
phenomenology, 1402
"Music and Technique" (Adorno), 1257
Mussolini, Benito, 750–751, *752*
Mustard gas, 303
Mutagenesis, 836
Mutation, 1894
Mutual assured destruction, 1216
Mutual Film Corp. v. Ohio Industrial Commission, 1239–1240, 1241
Myers, Greg, 1627
Myth of the Machine (Mumford), 1248–1249
Mythology
Popol Vuh, 357
popular culture, 1452
Prometheus, *1523,* 1523–1524
wilderness, 2067–2068
Yekuana, 992–993
Myths, 1746

● ● ● N

Nacos, Brigitte L., 1928
Nader, Ralph, 151, 1508
NAE (National Academy of Engineering), 1263–1264
Naess, Arne
eco-philosophy, 1683
environmental ethics, influence on, 654–655
self-realization, 657
Nagasaki. *See* Hiroshima and Nagasaki
Nagel, Thomas, 717
Nair, Indira, 697–698
Nanoethics, **1259–1262**
Nanotechnology, **xli–xlviii**
defensive technologies, xlvii
doomsday scenarios, 1164
Dutch perspectives, 555
law of accelerating returns, xliii
McKibben, Bill, xlvi
nanobots, xliv
nonbiological intelligence, xlv
privacy, 1491
technology assessment, 1907
technology trends, xlii

NAPA (National Academy of Public Administration), 1288–1289

Narain, Sunita, 987

Narcotics
  Huxley, Aldous, 969
  regulation, 547
  stroke and drug abuse, 222t

NARMS (National Antimicrobial Resistance Monitoring System), 613

NAS. See National Academy of Sciences (NAS)

NASA. See National Aeronautics and Space Administration (NASA)

Nash, John F., Jr., 819–820, 1506

Nash equilibrium, 819–825

National academies, **1263–1265**

National Academy of Engineering (NAE), 1263–1264

National Academy of Public Administration (NAPA), 1288–1289

National Academy of Sciences (NAS)
  atomic bombs, 135
  overview, 1263, 1264
  polygraph study, 1447
  scientific information censorship, 2057
  technical advice to government, 493

National Aeronautics and Space Administration (NASA), **1265–1267**
  Apollo program, 86–88
  Challenger and *Columbia* accidents, 1835–1838
  Federal Aviation Administration's partnership with, 159
  space telescopes, 1838–1840
  uncertainty, 1994

National Antimicrobial Resistance Monitoring System (NARMS), 613

National Association of Science Writers, 1086

National Bioethics Advisory Commission (NBAC), 203

National Board for Certified Counselors (NBCC), 1920

National Cancer Institute, 1272

National Center for Infectious Diseases, 613

National Commission for the Protection of Human Subjects of Biomedical and Behavioral Research, 202, 961, 1025

National Crime Information Center (NCIC), 1431–1432

National Endowment for the Humanities, 1283–1284

National Environmental Policy Act, 661, 671, 1329

National geological surveys, **1268–1271**

National Highway Traffic Safety Administration (NHTSA), 783

National Historical Preservation Act, 95

National Human Genome Research Institute (NHGRI), 846

National Institute for Occupational Safety and Health (NIOSH), 740

National Institutes of Health (NIH), **1271–1274**
  bioethics committees, 203–204
  Darsee case, 1208–1209
  misconduct in science, 1206
  peer review, 1391
  recombinant DNA research, 119–120, 1022
  research ethics education, 1604–1605
  research integrity, 1608
  research priorities, 906
  responsible conduct of research, 1624
  tradeoffs, 1972

Nationalism, **1275–1280**
  fascism, 750–754
  printing press, 1717

Nationalization of railroads, 1577

National Nanotechnology Initiative, U.S., 1259

National parks, 671, **1280–1283**, *1281*

National Park Service Act, 1280

National Public Radio (NPR), 1573

National Research Act, 767

National Research Council (NRC)
  Committee on Science and Technology for Countering Terrorism, 1933, 1934–1935
  overview, 1263–1265
  science policy, 1704
  technical advice to government, 493

National Science Advisory Board for Biosecurity (NSABB), 1932

National Science Board (NSB), 1688

National Science Foundation (NSF), **1283–1288**
  engineering ethics workshops, 626
  nanotechnology, 1261
  pseudoscience forms, 1532
  science and engineering indicators, 1688–1689
  Second Merit Review Criterion, 1288–1289
  technological innovation, 1904

National security
  airport security technology, 212
  basic issues, 1740
  biosecurity, 217
  chemical weapons, 306
  double effect and dual use, 545
  information ethics, 1008
  ordinary security *vs.*, 1739
  post-September 11 legislation, 1007t
  terrorism and science, 1931–1935

Union of Concerned Scientists, 2000–2001

National Security Advisory Board for Biosecurity(NSABB), 545, 1932

National Socialists. See Nazi medicine; Nazis

National Society of Professional Engineers (NSPE), 1514, 1674

National Telecommunications and Information Administration (NTIA), 1572–1573

National Traffic and Motor Vehicle Safety Act, 1508

National Transportation Safety Board (NTSB), 159

Nations, law of, 1097–1098

Nation states, 1762

Native Americans
  anthropology, 991
  colonialism, 356
  Earth, 560
  Kinzua Dam, 464
  Native American Graves Protection and Repatriation Act, 95–96
  Yekuana, 992–993

NATO, 1341–1342

Natural background radiation, 1566

Natural capital, 569, 570

Natural disasters. See Hazards

Natural history
  eighteenth century science and philosophy, 638–639
  museums, 1251

Naturalism
  pragmatism, 1470
  probability, 1505

Naturalistic empirical science, 1403

Naturalistic fallacy, 1728, 1742, 2049

Naturalistic moral tradition, 716

Natural law, **1289–1295**
  anti-posthumanism, 1461
  Lewis, C. S., 1121
  Locke, John, 1136–1137
  reproduction, 1629
  Spencer, Herbert, 1848–1849
  Thomas Aquinas, 1296, 1943, 1944

Natural liberty, 1786–1787

Natural philosophy
  Bacon, Francis, 165–167
  Boyle, Robert, 243–244
  education, 595–596
  Galenic medicine, 812–813
  Galilei, Galileo, 813–815
  Newton, Isaac, 1320–1321

Natural resources
  biodiversity, 189
  Chipko movement, 1878, 1879
  Christian perspectives, 334–335
  colonialism, 354–355, 356, 358

Neuroscience, *continued*
    neuroethics, 1310–1315
Neuse River, 465
Neutrality in science and technology, **1316–1319**
    axiology, 162–164
    fact/value dichotomy, 743–746
    progress, 1522
    technology, 883
    See also Objectivity
Nevada
    Hoover Dam, *464*
    Yucca Mountain repository, 1346–1347
New Atlantis (Bacon), **132–134,** 2011–2012
New eugenics, 709
Newfoundland, 1875
New Humanism, 421
Newman, John Henry, 70
New School Calvinism, 598–599
Newspapers. *See* Journalism
New technologies, **xlviii–lii**
    distributed responsibility, lii
    ethical assessment, xlix, xxx
    legal issues, 1712–1713
    meta-autonomy, l
    risk regulation, 1708–1710
    sociotechno-systems, li
New Testament, 338
Newton, Isaac, **1319–1323,** *1320*
    probability, 1502
    Scientific Revolution, 1733–1734
New Wave science fiction, 1693
New Worlds (magazine), 1693
New York, New Haven and Hartford railroad, *1575*
New York City
    bridges, 254
    Citicorp building, 268
New York Herald-Tribune, 254
New Zealand. *See* Australia and New Zealand
Neyman-Pearson theory, 1862, 1865
NGOs. *See* Nongovernmental organizations (NGOs)
NHGRI (National Human Genome Research Institute), 846
NHTSA (National Highway Traffic Safety Administration), 783
Nicomachean Ethics (Aristotle), 1095–1096
Niebuhr, H. Richard, 327, 1611
Nietzsche, Friedrich W., **1323–1324**
    conformism, 53
    Earth, 559
    existentialism, 723, 726
    life philosophy, 1129
    modernism, reaction against, 1462

Will to Power, 752–753
Nightingale, Florence, **1324–1326,** *1325,* 1864, 1865
NIH. *See* National Institutes of Health (NIH)
Nihilism, 1361
Nineteen Eighty-Four (Orwell), 1230
Ninov, Victor, 1211–1212
NIOSH (National Institute for Occupational Safety and Health), 740
Nipah virus, 610–611
Nitroglycerin, 308
Nobel, Alfred, 308
Nobel prizes
    Becquerel, Antoine-Henri, 1334
    Bethe, Hans, 180
    Blackett, Patrick, 237
    creation of, 308
    Curie, Marie and Pierre, 1334
    Müller, Paul, 475
    NIH scientists, 1274
    Pauling, Linus, 1389, 1390
    Penzias, Arnio, 438
    Proigogine, Ilya, 389
    Pugwash Conferences, 1550–1551
    Rotblat, Joseph, 1659
    Sakharov, Andrei, 1677
    Watson, James, 2054
    Wilson, Robert, 438
Noddings, Nel, 696
Nodes, 970–971
No-fault liability, 1610
Noise pollution, 161
Non-collision course collisions, 1331–1332
Noncooperative game theory, 1506
Non-discrimination, 1477
Non-Euclidean space, 1830–1831
Nongovernmental organizations (NGOs), **1326–1330,** *1328t*
    bioethics, 204–205
    civil society, 344
    Ibero-American science research, 978
    information ethics, 1005–1006
    operations research, 1363
    public policy centers, 1545–1547
    waste, 2047
    work issues, 2076–2077
    See also Specific organizations
Nonionizing radiation, 1565–1566
Nonlethal weapons technology, 1100
Nonnormativism, 904
Nonparametric methods, 1863
Nonpatterned theories of justice, 1096
Non-power, ethics of, 608, 792, 801
Nonprofit status, 1546–1547
Nontreatment, 1986–1988
Non-use values, 650
Nonviolent resistance

Chipko movement, 1878, 1879
    environmental justice, 666
    Gandhi, Mohandas, 826–827
Non-voluntary euthanasia, 710–711
Noonan, John T., Jr., 2
Nordic perspectives, **1680–1686**
No-return decisions, 1837
Normal, William, 728
Normal accidents, xix, **1331–1334**
Normal distribution
    economics, 585–586
    Galilei, anticipation of by, 1864
    probability, 1498–1499, *1499(f7),* 1501
    statistics, 1858, *1859(f2),* 1862
Normalization of deviance, 1837
Normative ethics, 1727
Normative structure of science, 1817
Normativism
    equality, 685–686
    Foucault, Michel, 787
    health and disease, 904
    Peirce, Charles Sanders, 1395–1396
    technical functions judgments, 1887–1889
    teleology, 1290–1291
Norms of science, 1607–1608
Norplant®, 236
North American technological history, 1850–1851
North Atlantic Treaty Organization (NATO), 1341–1342
North Carolina, 465
Northridge earthquake, 563
Norway. *See* Scandinavia
Norwegian University of Science and Technology, 1684
Nosologies, 904
Noss, Reed, 575
Notation, musical, 1255
Notice and take-down provision, 1709
Not-in-my-backyard (NIMBY)
    bridges, 254
    public participation, 412
Nozick, Robert
    justice, 1094, 1096
    Rawls, criticism of, 1589
    rights-based theories of risk, 1643
    rights theory, 1632
    social justice theory, 1246
NPR (National Public Radio), 1573
NRC. *See* National Research Council (NRC); Nuclear Regulatory Commission (NRC)
NSABB (National Science Advisory Board for Biosecurity), 545, 1932
NSB (National Science Board), 1688
NSF. *See* National Science Foundation (NSF)

V1: 1–462; V2: 463–1109; V3: 1111–1672; V4: 1673–2090

Organization for Economic Co-operation and Development, 1370
Protocol of San Salvador, 677
Protogenetics, 842–843
Prout, Henry Goslee, 1614
Proximate causation, 705
Prudent avoidance, 1566
Prussia, 2057
Pseudoevents, 1452
Pseudonyms, 1825
Pseudorandom, 1504
Pseudoscience, **1532–1535,** 1728
Psyche, 1091
Psychedelic drugs. *See* Narcotics
Psychiatry, 708
Psychoanalysis
    Adler and Jung, 1537
    Freud, Sigmund, 804–805, 1536, 1539–1540
Psychobiology, developmental, 852–853
Psychology, **1535–1542**
    advertising, 20
    aggression, 32–33
    altruism, 61–63
    assisted reproduction technology, 124–125
    biophilia, 214
    children and families affected by HIV/AIDS, *30*
    consciousness research, 411
    consumerism, 427
    criminal dysfunction, 482
    criminology, 444–445
    Darwin, Charles, 471
    death and dying, 478
    depression medication, 230
    Diagnostic and Statistical Manual of Mental Disorders, 521–524
    emotion, 614–616
    homosexuality, 935
    human research subjects, 962
    information overload, 1012
    invention, 1053
    IQ test factors, 1059
    Jung, Carl Gustav, 1090–1091
    Kübler-Ross, Elizabeth, 478
    misconduct cases, 1214
    moral cognitive development, 1729–1730
    overview, 1535–1538
    prosthetics, 1531
    robot toys, 1657–1658
    Skinner, B. F., *1781,* 1781–1782
    speed, 1844–1845
    stress, 1873–1874
    violence, 2031–2033
Psychometric research, 1645
Psychopharmacology, 547, 615, **1542–1545**

Psychosurgery, 1314
Psychotherapeutics, 1314–1315
Ptolemaic system, *558*
Publication practices
    authorship, 1718–1719
    contracts and ethics, 431
    government funding, 1932
    open access, 1361
    research ethics, 1600, 1602
Public Broadcasting Act, 1573
"Public Engagement with Science and Technology" (PEST), 1549
Public goods game, 821, 1488–1489
Public health
    Chernobyl accident, 314–315
    emergent infectious diseases, 610–614
    epidemiology, 679–684
    exposure limits, 739–742, 740f
    funding, 1934
    Germany, 1302
    hormesis, 935–938
    industrialization and, 1350
    National Institutes of Health, 1271–1274
    terrorism, 1932
    World Health Organization, 2082–2085
Public Health and Marine Hospital Service, 1272
Public Health Service, U.S., 1986–1988
Public Health Service Act, 1273
Public interest
    Cicero's Creed, 341
    codes of ethics, 352, 1020
    genethics, 835
    Georgia Basin Futures Project, 857–861
    governance of science, 879
    responsibility, 1613
    telecommunications, 372
    television, 373
    waste, 2049–2050
    whistleblowing, 2063
Public lands, 1268
Public opinion
    brain death, 246
    consumerism, 428
    Hiroshima and Nagasaki bombings, 921
    origins of life, 720
    Silent Spring (Carson), 292–293
    speed, 1847
    statistics, 1868
Public policy
    Association for Computing Machinery, 125–126
    Baruch Plan, 168–169
    bioethics commissions, 202–204, 206

complementary and alternative medicine, 387
constructive technology assessment, 423–424
Earth Systems Engineering and Management, 566–567
ecological integrity, 575
ecological management, 581–582
embryonic research, 1055
emergent infectious disease management, 612–614
engineering ethics, 630
enquete commissions, 642–644
environmental ethics, 658
environmental impact assessment, 661
expertise, 732
exposure limits, 741
future generations, 809
Georgia Basin Futures Project, 860–861
global climate change, 872–873
intellectual property, 1031–1032
Iran trade, 1021
Kyoto Protocol, 1046
Lasswell, Harold D., 1111–1112
models, 1219
nanotechnology, 1261–1262
participation, 1380–1384
planning ethics, 1415
pollution, 1445
post-September 11, 1007t
prediction, 1480
privacy and the Internet, 1050
professional engineering organizations, 1514
psychology, 1538
public policy centers, 1545–1547
Pugwash Conferences, 1551–1552
research ethics, 1604
royal commissions, 1662–1663
scientific advice to lawmakers, 493–494
scientific research investment, 271
scientism, 1735–1736
sensitivity analysis, 1751–1752
social theory of science and technology, 1819–1820
speed, 1847
technical knowledge requirement, 342–343
tradeoffs, 1975
trust, 1981–1982
unemployment compensation, 2075
Union of Concerned Scientists, 2000–2001
in vitro fertilization, 1056
water, 2052, 2082
wildlife management, 2070

••• T